Digital Signal Processing

Principles, Algorithms, and Applications

Third Edition

John G. Proakis
Northeastern University

Dimitris G. Manolakis
Boston College

PRENTICE HALL
Upper Saddle River, New Jersey 07458

Library of Congress Cataloging-in-Publication Data

Proakis, John G.
    Digital signal processing: principles, algorithms, and
applications / John G. Proakis, Dimitris G. Manolakis.
        p. cm.
    Includes bibliographical references and index.
    ISBN 0-13-373762-4
    1. Signal processing—Digital techniques.   I. Manolakis, Dimitris
G. II. Title.
    TK5102.9.P757 1996   95-9117
        621.382′2—dc20   CIP

Acquisitions editor: Tom Robbins
Editorial/production supervision: ETP/Harrison
Full service coordinator: Irwin Zucker
Cover design: Bruce Kensalaar
Manufacturing buyer: Donna Sullivan

© 1996 by Prentice-Hall, Inc.
Simon & Schuster/A Viacom Company
Upper Saddle River, New Jersey 07458

Printed in the United States of America

10 9 8 7 6 5

ISBN 0-13-373762-4

Prentice-Hall International (UK) Limited, *London*
Prentice-Hall of Australia Pty. Limited, *Sydney*
Prentice-Hall Canada Inc, *Toronto*
Prentice-Hall Hispanoamericana, S.A., *Mexico*
Prentice-Hall of India Private Limited, *New Delhi*
Prentice-Hall of Japan, Inc., *Tokyo*
Simon & Schuster Asia Pte. Ltd., *Singapore*
Editora Prentice-Hall do Brasil, Ltda., *Rio de Janeiro*

Contents

2 DISCRETE-TIME SIGNALS AND SYSTEMS   43

Preface

This book was developed based on our teaching of undergraduate and graduate level courses in digital signal processing over the past several years. In this book we present the fundamentals of discrete-time signals, systems, and modern digital processing algorithms and applications for students in electrical engineering, computer engineering, and computer science. The book is suitable for either a one-semester or a two-semester undergraduate level course in discrete systems and digital signal processing. It is also intended for use in a one-semester first-year graduate-level course in digital signal processing.

It is assumed that the student in electrical and computer engineering has had undergraduate courses in advanced calculus (including ordinary differential equations), and linear systems for continuous-time signals, including an introduction to the Laplace transform. Although the Fourier series and Fourier transforms of periodic and aperiodic signals are described in Chapter 4, we expect that many students may have had this material in a prior course.

A balanced coverage is provided of both theory and practical applications. A large number of well designed problems are provided to help the student in mastering the subject matter. A solutions manual is available for the benefit of the instructor and can be obtained from the publisher.

The third edition of the book covers basically the same material as the second edition, but is organized differently. The major difference is in the order in which the DFT and FFT algorithms are covered. Based on suggestions made by several reviewers, we now introduce the DFT and describe its efficient computation immediately following our treatment of Fourier analysis. This reorganization has also allowed us to eliminate repetition of some topics concerning the DFT and its applications.

In Chapter 1 we describe the operations involved in the analog-to-digital conversion of analog signals. The process of sampling a sinusoid is described in some detail and the problem of aliasing is explained. Signal quantization and digital-to-analog conversion are also described in general terms, but the analysis is presented in subsequent chapters.

Chapter 2 is devoted entirely to the characterization and analysis of linear time-invariant (shift-invariant) discrete-time systems and discrete-time signals in the time domain. The convolution sum is derived and systems are categorized according to the duration of their impulse response as a finite-duration impulse response (FIR) and as an infinite-duration impulse response (IIR). Linear time-invariant systems characterized by difference equations are presented and the solution of difference equations with initial conditions is obtained. The chapter concludes with a treatment of discrete-time correlation.

The z-transform is introduced in Chapter 3. Both the bilateral and the unilateral z-transforms are presented, and methods for determining the inverse z-transform are described. Use of the z-transform in the analysis of linear time-invariant systems is illustrated, and important properties of systems, such as causality and stability, are related to z-domain characteristics.

Chapter 4 treats the analysis of signals and systems in the frequency domain. Fourier series and the Fourier transform are presented for both continuous-time and discrete-time signals. Linear time-invariant (LTI) discrete systems are characterized in the frequency domain by their frequency response function and their response to periodic and aperiodic signals is determined. A number of important types of discrete-time systems are described, including resonators, notch filters, comb filters, all-pass filters, and oscillators. The design of a number of simple FIR and IIR filters is also considered. In addition, the student is introduced to the concepts of minimum-phase, mixed-phase, and maximum-phase systems and to the problem of deconvolution.

The DFT, its properties and its applications, are the topics covered in Chapter 5. Two methods are described for using the DFT to perform linear filtering. The use of the DFT to perform frequency analysis of signals is also described.

Chapter 6 covers the efficient computation of the DFT. Included in this chapter are descriptions of radix-2, radix-4, and split-radix fast Fourier transform (FFT) algorithms, and applications of the FFT algorithms to the computation of convolution and correlation. The Goertzel algorithm and the chirp-z transform are introduced as two methods for computing the DFT using linear filtering.

Chapter 7 treats the realization of IIR and FIR systems. This treatment includes direct-form, cascade, parallel, lattice, and lattice-ladder realizations. The chapter includes a treatment of state-space analysis and structures for discrete-time systems, and examines quantization effects in a digital implementation of FIR and IIR systems.

Techniques for design of digital FIR and IIR filters are presented in Chapter 8. The design techniques include both direct design methods in discrete time and methods involving the conversion of analog filters into digital filters by various transformations. Also treated in this chapter is the design of FIR and IIR filters by least-squares methods.

Chapter 9 focuses on the sampling of continuous-time signals and the reconstruction of such signals from their samples. In this chapter, we derive the sampling theorem for bandpass continuous-time-signals and then cover the A/D and D/A conversion techniques, including oversampling A/D and D/A converters.

Chapter 10 provides an indepth treatment of sampling-rate conversion and its applications to multirate digital signal processing. In addition to describing decimation and interpolation by integer factors, we present a method of sampling-rate conversion by an arbitrary factor. Several applications to multirate signal processing are presented, including the implementation of digital filters, subband coding of speech signals, transmultiplexing, and oversampling A/D and D/A converters.

Linear prediction and optimum linear (Wiener) filters are treated in Chapter 11. Also included in this chapter are descriptions of the Levinson–Durbin algorithm and Schür algorithm for solving the normal equations, as well as the AR lattice and ARMA lattice-ladder filters.

Power spectrum estimation is the main topic of Chapter 12. Our coverage includes a description of nonparametric and model-based (parametric) methods. Also described are eigen-decomposition-based methods, including MUSIC and ESPRIT.

At Northeastern University, we have used the first six chapters of this book for a one-semester (junior level) course in discrete systems and digital signal processing.

A one-semester senior level course for students who have had prior exposure to discrete systems can use the material in Chapters 1 through 4 for a quick review and then proceed to cover Chapter 5 through 8.

In a first-year graduate level course in digital signal processing, the first five chapters provide the student with a good review of discrete-time systems. The instructor can move quickly through most of this material and then cover Chapters 6 through 9, followed by either Chapters 10 and 11 or by Chapters 11 and 12.

We have included many examples throughout the book and approximately 500 homework problems. Many of the homework problems can be solved numerically on a computer, using a software package such as MATLAB©. These problems are identified by an asterisk. Appendix D contains a list of MATLAB functions that the student can use in solving these problems. The instructor may also wish to consider the use of a supplementary book that contains computer based exercises, such as the books *Digital Signal Processing Using MATLAB* (P.W.S. Kent, 1996) by V. K. Ingle and J. G. Proakis and *Computer-Based Exercises for Signal Processing Using MATLAB* (Prentice Hall, 1994) by C. S. Burrus et al.

The authors are indebted to their many faculty colleagues who have provided valuable suggestions through reviews of the first and second editions of this book. These include Drs. W. E. Alexander, Y. Bresler, J. Deller, V. Ingle, C. Keller, H. Lev-Ari, L. Merakos, W. Mikhael, P. Monticciolo, C. Nikias, M. Schetzen, H. Trussell, S. Wilson, and M. Zoltowski. We are also indebted to Dr. R. Price for recommending the inclusion of split-radix FFT algorithms and related suggestions. Finally, we wish to acknowledge the suggestions and comments of many former graduate students, and especially those by A. L. Kok, J. Lin and S. Srinidhi who assisted in the preparation of several illustrations and the solutions manual.

John G. Proakis
Dimitris G. Manolakis

 1

Introduction

Digital signal processing is an area of science and engineering that has developed rapidly over the past 30 years. This rapid development is a result of the significant advances in digital computer technology and integrated-circuit fabrication. The digital computers and associated digital hardware of three decades ago were relatively large and expensive and, as a consequence, their use was limited to general-purpose non-real-time (off-line) scientific computations and business applications. The rapid developments in integrated-circuit technology, starting with medium-scale integration (MSI) and progressing to large-scale integration (LSI), and now, very-large-scale integration (VLSI) of electronic circuits has spurred the development of powerful, smaller, faster, and cheaper digital computers and special-purpose digital hardware. These inexpensive and relatively fast digital circuits have made it possible to construct highly sophisticated digital systems capable of performing complex digital signal processing functions and tasks, which are usually too difficult and/or too expensive to be performed by analog circuitry or analog signal processing systems. Hence many of the signal processing tasks that were conventionally performed by analog means are realized today by less expensive and often more reliable digital hardware.

We do not wish to imply that digital signal processing is the proper solution for all signal processing problems. Indeed, for many signals with extremely wide bandwidths, real-time processing is a requirement. For such signals, analog or, perhaps, optical signal processing is the only possible solution. However, where digital circuits are available and have sufficient speed to perform the signal processing, they are usually preferable.

Not only do digital circuits yield cheaper and more reliable systems for signal processing, they have other advantages as well. In particular, digital processing hardware allows programmable operations. Through software, one can more easily modify the signal processing functions to be performed by the hardware. Thus digital hardware and associated software provide a greater degree of flexibility in system design. Also, there is often a higher order of precision achievable with digital hardware and software compared with analog circuits and analog signal processing systems. For all these reasons, there has been an explosive growth in digital signal processing theory and applications over the past three decades.

In this book our objective is to present an introduction of the basic analysis tools and techniques for digital processing of signals. We begin by introducing some of the necessary terminology and by describing the important operations associated with the process of converting an analog signal to digital form suitable for digital processing. As we shall see, digital processing of analog signals has some drawbacks. First, and foremost, conversion of an analog signal to digital form, accomplished by sampling the signal and quantizing the samples, results in a distortion that prevents us from reconstructing the original analog signal from the quantized samples. Control of the amount of this distortion is achieved by proper choice of the sampling rate and the precision in the quantization process. Second, there are finite precision effects that must be considered in the digital processing of the quantized samples. While these important issues are considered in some detail in this book, the emphasis is on the analysis and design of digital signal processing systems and computational techniques.

1.1 SIGNALS, SYSTEMS, AND SIGNAL PROCESSING

A *signal* is defined as any physical quantity that varies with time, space, or any other independent variable or variables. Mathematically, we describe a signal as a function of one or more independent variables. For example, the functions

$$s_1(t) = 5t$$
$$s_2(t) = 20t^2 \tag{1.1.1}$$

describe two signals, one that varies linearly with the independent variable t (time) and a second that varies quadratically with t. As another example, consider the function

$$s(x, y) = 3x + 2xy + 10y^2 \tag{1.1.2}$$

This function describes a signal of two independent variables x and y that could represent the two spatial coordinates in a plane.

The signals described by (1.1.1) and (1.1.2) belong to a class of signals that are precisely defined by specifying the functional dependence on the independent variable. However, there are cases where such a functional relationship is unknown or too highly complicated to be of any practical use.

For example, a speech signal (see Fig. 1.1) cannot be described functionally by expressions such as (1.1.1). In general, a segment of speech may be represented to a high degree of accuracy as a sum of several sinusoids of different amplitudes and frequencies, that is, as

$$\sum_{i=1}^{N} A_i(t) \sin[2\pi F_i(t)t + \theta_i(t)] \tag{1.1.3}$$

where $\{A_i(t)\}$, $\{F_i(t)\}$, and $\{\theta_i(t)\}$ are the sets of (possibly time-varying) amplitudes, frequencies, and phases, respectively, of the sinusoids. In fact, one way to interpret the information content or message conveyed by any short time segment of the

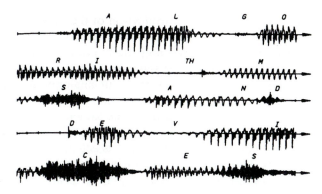

Figure 1.1 Example of a speech signal.

speech signal is to measure the amplitudes, frequencies, and phases contained in the short time segment of the signal.

Another example of a natural signal is an electrocardiogram (ECG). Such a signal provides a doctor with information about the condition of the patient's heart. Similarly, an electroencephalogram (EEG) signal provides information about the activity of the brain.

Speech, electrocardiogram, and electroencephalogram signals are examples of information-bearing signals that evolve as functions of a single independent variable, namely, time. An example of a signal that is a function of two independent variables is an image signal. The independent variables in this case are the spatial coordinates. These are but a few examples of the countless number of natural signals encountered in practice.

Associated with natural signals are the means by which such signals are generated. For example, speech signals are generated by forcing air through the vocal cords. Images are obtained by exposing a photographic film to a scene or an object. Thus signal generation is usually associated with a *system* that responds to a stimulus or force. In a speech signal, the system consists of the vocal cords and the vocal tract, also called the vocal cavity. The stimulus in combination with the system is called a *signal source*. Thus we have speech sources, images sources, and various other types of signal sources.

A *system* may also be defined as a physical device that performs an operation on a signal. For example, a filter used to reduce the noise and interference corrupting a desired information-bearing signal is called a system. In this case the filter performs some operation(s) on the signal, which has the effect of reducing (filtering) the noise and interference from the desired information-bearing signal.

When we pass a signal through a system, as in filtering, we say that we have processed the signal. In this case the processing of the signal involves filtering the noise and interference from the desired signal. In general, the system is characterized by the type of operation that it performs on the signal. For example, if the operation is linear, the system is called linear. If the operation on the signal is nonlinear, the system is said to be nonlinear, and so forth. Such operations are usually referred to as *signal processing*.

For our purposes, it is convenient to broaden the definition of a system to include not only physical devices, but also software realizations of operations on a signal. In digital processing of signals on a digital computer, the operations performed on a signal consist of a number of mathematical operations as specified by a software program. In this case, the program represents an implementation of the system in *software*. Thus we have a system that is realized on a digital computer by means of a sequence of mathematical operations; that is, we have a digital signal processing system realized in software. For example, a digital computer can be programmed to perform digital filtering. Alternatively, the digital processing on the signal may be performed by digital *hardware* (logic circuits) configured to perform the desired specified operations. In such a realization, we have a physical device that performs the specified operations. In a broader sense, a digital system can be implemented as a combination of digital hardware and software, each of which performs its own set of specified operations.

This book deals with the processing of signals by digital means, either in software or in hardware. Since many of the signals encountered in practice are analog, we will also consider the problem of converting an analog signal into a digital signal for processing. Thus we will be dealing primarily with digital systems. The operations performed by such a system can usually be specified mathematically. The method or set of rules for implementing the system by a program that performs the corresponding mathematical operations is called an *algorithm*. Usually, there are many ways or algorithms by which a system can be implemented, either in software or in hardware, to perform the desired operations and computations. In practice, we have an interest in devising algorithms that are computationally efficient, fast, and easily implemented. Thus a major topic in our study of digital signal processing is the discussion of efficient algorithms for performing such operations as filtering, correlation, and spectral analysis.

1.1.1 Basic Elements of a Digital Signal Processing System

Most of the signals encountered in science and engineering are analog in nature. That is, the signals are functions of a continuous variable, such as time or space, and usually take on values in a continuous range. Such signals may be processed directly by appropriate analog systems (such as filters or frequency analyzers) or frequency multipliers for the purpose of changing their characteristics or extracting some desired information. In such a case we say that the signal has been processed directly in its analog form, as illustrated in Fig. 1.2. Both the input signal and the output signal are in analog form.

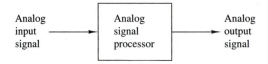

Figure 1.2   Analog signal processing.

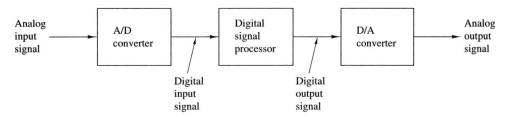

Figure 1.3   Block diagram of a digital signal processing system.

Digital signal processing provides an alternative method for processing the analog signal, as illustrated in Fig. 1.3. To perform the processing digitally, there is a need for an interface between the analog signal and the digital processor. This interface is called an *analog-to-digital (A/D) converter.* The output of the A/D converter is a digital signal that is appropriate as an input to the digital processor.

The digital signal processor may be a large programmable digital computer or a small microprocessor programmed to perform the desired operations on the input signal. It may also be a hardwired digital processor configured to perform a specified set of operations on the input signal. Programmable machines provide the flexibility to change the signal processing operations through a change in the software, whereas hardwired machines are difficult to reconfigure. Consequently, programmable signal processors are in very common use. On the other hand, when signal processing operations are well defined, a hardwired implementation of the operations can be optimized, resulting in a cheaper signal processor and, usually, one that runs faster than its programmable counterpart. In applications where the digital output from the digital signal processor is to be given to the user in analog form, such as in speech communications, we must provide another interface from the digital domain to the analog domain. Such an interface is called a *digital-to-analog (D/A) converter.* Thus the signal is provided to the user in analog form, as illustrated in the block diagram of Fig. 1.3. However, there are other practical applications involving signal analysis, where the desired information is conveyed in digital form and no D/A converter is required. For example, in the digital processing of radar signals, the information extracted from the radar signal, such as the position of the aircraft and its speed, may simply be printed on paper. There is no need for a D/A converter in this case.

1.1.2 Advantages of Digital over Analog Signal Processing

There are many reasons why digital signal processing of an analog signal may be preferable to processing the signal directly in the analog domain, as mentioned briefly earlier. First, a digital programmable system allows flexibility in reconfiguring the digital signal processing operations simply by changing the program.

Reconfiguration of an analog system usually implies a redesign of the hardware followed by testing and verification to see that it operates properly.

Accuracy considerations also play an important role in determining the form of the signal processor. Tolerances in analog circuit components make it extremely difficult for the system designer to control the accuracy of an analog signal processing system. On the other hand, a digital system provides much better control of accuracy requirements. Such requirements, in turn, result in specifying the accuracy requirements in the A/D converter and the digital signal processor, in terms of word length, floating-point versus fixed-point arithmetic, and similar factors.

Digital signals are easily stored on magnetic media (tape or disk) without deterioration or loss of signal fidelity beyond that introduced in the A/D conversion. As a consequence, the signals become transportable and can be processed off-line in a remote laboratory. The digital signal processing method also allows for the implementation of more sophisticated signal processing algorithms. It is usually very difficult to perform precise mathematical operations on signals in analog form but these same operations can be routinely implemented on a digital computer using software.

In some cases a digital implementation of the signal processing system is cheaper than its analog counterpart. The lower cost may be due to the fact that the digital hardware is cheaper, or perhaps it is a result of the flexibility for modifications provided by the digital implementation.

As a consequence of these advantages, digital signal processing has been applied in practical systems covering a broad range of disciplines. We cite, for example, the application of digital signal processing techniques in speech processing and signal transmission on telephone channels, in image processing and transmission, in seismology and geophysics, in oil exploration, in the detection of nuclear explosions, in the processing of signals received from outer space, and in a vast variety of other applications. Some of these applications are cited in subsequent chapters.

As already indicated, however, digital implementation has its limitations. One practical limitation is the speed of operation of A/D converters and digital signal processors. We shall see that signals having extremely wide bandwidths require fast-sampling-rate A/D converters and fast digital signal processors. Hence there are analog signals with large bandwidths for which a digital processing approach is beyond the state of the art of digital hardware.

1.2 CLASSIFICATION OF SIGNALS

The methods we use in processing a signal or in analyzing the response of a system to a signal depend heavily on the characteristic attributes of the specific signal. There are techniques that apply only to specific families of signals. Consequently, any investigation in signal processing should start with a classification of the signals involved in the specific application.

1.2.1 Multichannel and Multidimensional Signals

As explained in Section 1.1, a signal is described by a function of one or more independent variables. The value of the function (i.e., the dependent variable) can be a real-valued scalar quantity, a complex-valued quantity, or perhaps a vector. For example, the signal

$$s_1(t) = A \sin 3\pi t$$

is a real-valued signal. However, the signal

$$s_2(t) = A e^{j3\pi t} = A \cos 3\pi t + j A \sin 3\pi t$$

is complex valued.

In some applications, signals are generated by multiple sources or multiple sensors. Such signals, in turn, can be represented in vector form. Figure 1.4 shows the three components of a vector signal that represents the ground acceleration due to an earthquake. This acceleration is the result of three basic types of elastic waves. The primary (P) waves and the secondary (S) waves propagate within the body of rock and are longitudinal and transversal, respectively. The third type of elastic wave is called the surface wave, because it propagates near the ground surface. If $s_k(t)$, $k = 1, 2, 3$, denotes the electrical signal from the kth sensor as a function of time, the set of $p = 3$ signals can be represented by a vector $\mathbf{S}_3(t)$, where

$$\mathbf{S}_3(t) = \begin{bmatrix} s_1(t) \\ s_2(t) \\ s_3(t) \end{bmatrix}$$

We refer to such a vector of signals as a *multichannel signal*. In electrocardiography, for example, 3-lead and 12-lead electrocardiograms (ECG) are often used in practice, which result in 3-channel and 12-channel signals.

Let us now turn our attention to the independent variable(s). If the signal is a function of a single independent variable, the signal is called a *one-dimensional* signal. On the other hand, a signal is called *M-dimensional* if its value is a function of M independent variables.

The picture shown in Fig. 1.5 is an example of a two-dimensional signal, since the intensity or brightness $I(x, y)$ at each point is a function of two independent variables. On the other hand, a black-and-white television picture may be represented as $I(x, y, t)$ since the brightness is a function of time. Hence the TV picture may be treated as a three-dimensional signal. In contrast, a color TV picture may be described by three intensity functions of the form $I_r(x, y, t)$, $I_g(x, y, t)$, and $I_b(x, y, t)$, corresponding to the brightness of the three principal colors (red, green, blue) as functions of time. Hence the color TV picture is a three-channel, three-dimensional signal, which can be represented by the vector

$$\mathbf{I}(x, y, t) = \begin{bmatrix} I_r(x, y, t) \\ I_g(x, y, t) \\ I_b(x, y, t) \end{bmatrix}$$

In this book we deal mainly with single-channel, one-dimensional real- or complex-valued signals and we refer to them simply as signals. In mathematical

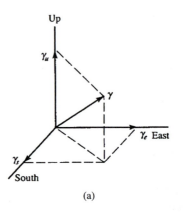

(a)

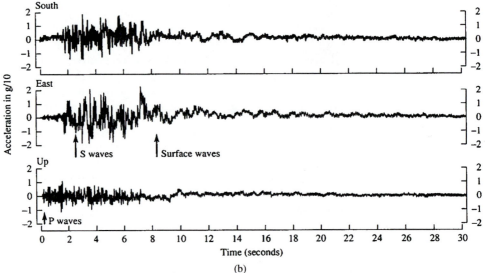

(b)

Figure 1.4 Three components of ground acceleration measured a few kilometers from the epicenter of an earthquake. (From *Earthquakes*, by B. A. Bold, ©1988 by W. H. Freeman and Company. Reprinted with permission of the publisher.)

terms these signals are described by a function of a single independent variable. Although the independent variable need not be time, it is common practice to use *t* as the independent variable. In many cases the signal processing operations and algorithms developed in this text for one-dimensional, single-channel signals can be extended to multichannel and multidimensional signals.

1.2.2 Continuous-Time Versus Discrete-Time Signals

Signals can be further classified into four different categories depending on the characteristics of the time (independent) variable and the values they take. *Continuous-time signals* or *analog signals* are defined for every value of time and

Figure 1.5   Example of a two-dimensional signal.

they take on values in the continuous interval (a, b), where a can be $-\infty$ and b can be ∞. Mathematically, these signals can be described by functions of a continuous variable. The speech waveform in Fig. 1.1 and the signals $x_1(t) = \cos \pi t$, $x_2(t) = e^{-|t|}$, $-\infty < t < \infty$ are examples of analog signals. *Discrete-time signals* are defined only at certain specific values of time. These time instants need not be equidistant, but in practice they are usually taken at equally spaced intervals for computational convenience and mathematical tractability. The signal $x(t_n) = e^{-|t_n|}$, $n = 0, \pm 1, \pm 2, \ldots$ provides an example of a discrete-time signal. If we use the index n of the discrete-time instants as the independent variable, the signal value becomes a function of an integer variable (i.e., a sequence of numbers). Thus a discrete-time signal can be represented mathematically by a sequence of real or complex numbers. To emphasize the discrete-time nature of a signal, we shall denote such a signal as $x(n)$ instead of $x(t)$. If the time instants t_n are equally spaced (i.e., $t_n = nT$), the notation $x(nT)$ is also used. For example, the sequence

$$x(n) = \begin{cases} 0.8^n, & \text{if } n \geq 0 \\ 0, & \text{otherwise} \end{cases} \tag{1.2.1}$$

is a discrete-time signal, which is represented graphically as in Fig. 1.6.

In applications, discrete-time signals may arise in two ways:

1. By selecting values of an analog signal at discrete-time instants. This process is called *sampling* and is discussed in more detail in Section 1.4. All measuring instruments that take measurements at a regular interval of time provide discrete-time signals. For example, the signal $x(n)$ in Fig. 1.6 can be obtained

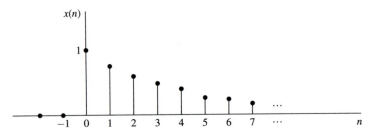

Figure 1.6 Graphical representation of the discrete time signal $x(n) = 0.8^n$ for $n > 0$ and $x(n) = 0$ for $n < 0$.

by sampling the analog signal $x(t) = 0.8^t$, $t \geq 0$ and $x(t) = 0$, $t < 0$ once every second.

2. By accumulating a variable over a period of time. For example, counting the number of cars using a given street every hour, or recording the value of gold every day, results in discrete-time signals. Figure 1.7 shows a graph of the Wölfer sunspot numbers. Each sample of this discrete-time signal provides the number of sunspots observed during an interval of 1 year.

1.2.3 Continuous-Valued Versus Discrete-Valued Signals

The values of a continuous-time or discrete-time signal can be continuous or discrete. If a signal takes on all possible values on a finite or an infinite range, it

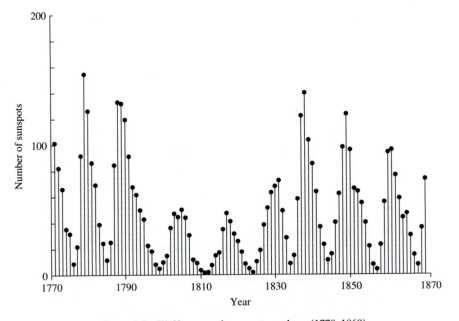

Figure 1.7 Wölfer annual sunspot numbers (1770–1869).

is said to be continuous-valued signal. Alternatively, if the signal takes on values from a finite set of possible values, it is said to be a discrete-valued signal. Usually, these values are equidistant and hence can be expressed as an integer multiple of the distance between two successive values. A discrete-time signal having a set of discrete values is called a *digital signal*. Figure 1.8 shows a digital signal that takes on one of four possible values.

In order for a signal to be processed digitally, it must be discrete in time and its values must be discrete (i.e., it must be a digital signal). If the signal to be processed is in analog form, it is converted to a digital signal by sampling the analog signal at discrete instants in time, obtaining a discrete-time signal, and then by *quantizing* its values to a set of discrete values, as described later in the chapter. The process of converting a continuous-valued signal into a discrete-valued signal, called *quantization*, is basically an approximation process. It may be accomplished simply by rounding or truncation. For example, if the allowable signal values in the digital signal are integers, say 0 through 15, the continuous-value signal is quantized into these integer values. Thus the signal value 8.58 will be approximated by the value 8 if the quantization process is performed by truncation or by 9 if the quantization process is performed by rounding to the nearest integer. An explanation of the analog-to-digital conversion process is given later in the chapter.

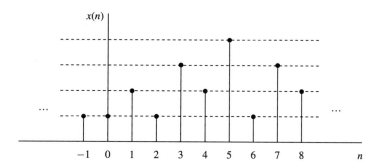

Figure 1.8   Digital signal with four different amplitude values.

1.2.4 Deterministic Versus Random Signals

The mathematical analysis and processing of signals requires the availability of a mathematical description for the signal itself. This mathematical description, often referred to as the *signal model*, leads to another important classification of signals. Any signal that can be uniquely described by an explicit mathematical expression, a table of data, or a well-defined rule is called *deterministic*. This term is used to emphasize the fact that all past, present, and future values of the signal are known precisely, without any uncertainty.

In many practical applications, however, there are signals that either cannot be described to any reasonable degree of accuracy by explicit mathematical formulas, or such a description is too complicated to be of any practical use. The lack of such a relationship implies that such signals evolve in time in an unpredictable manner. We refer to these signals as *random*. The output of a noise generator, the seismic signal of Fig. 1.4, and the speech signal in Fig. 1.1 are examples of random signals.

Figure 1.9 shows two signals obtained from the same noise generator and their associated histograms. Although the two signals do not resemble each other visually, their histograms reveal some similarities. This provides motivation for

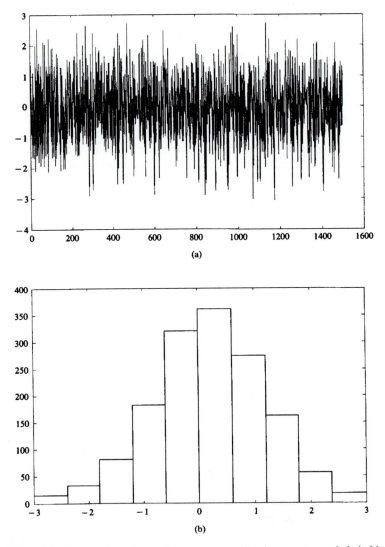

Figure 1.9 Two random signals from the same signal generator and their histograms.

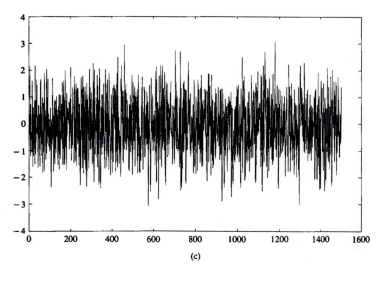

(c)

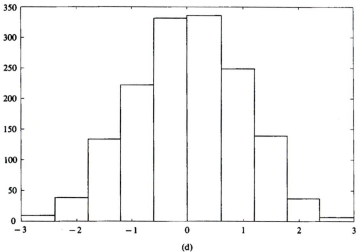

(d)

Figure 1.9 *Continued*

the analysis and description of random signals using *statistical* techniques instead of explicit formulas. The mathematical framework for the theoretical analysis of random signals is provided by the theory of probability and stochastic processes. Some basic elements of this approach, adapted to the needs of this book, are presented in Appendix A.

It should be emphasized at this point that the classification of a *real-world* signal as deterministic or random is not always clear. Sometimes, both approaches lead to meaningful results that provide more insight into signal behavior. At other times, the wrong classification may lead to erroneous results, since some mathematical tools may apply only to deterministic signals while others may apply only to random signals. This will become clearer as we examine specific mathematical tools.

1.3 THE CONCEPT OF FREQUENCY IN CONTINUOUS-TIME AND DISCRETE-TIME SIGNALS

The concept of frequency is familiar to students in engineering and the sciences. This concept is basic in, for example, the design of a radio receiver, a high-fidelity system, or a spectral filter for color photography. From physics we know that frequency is closely related to a specific type of periodic motion called harmonic oscillation, which is described by sinusoidal functions. The concept of frequency is directly related to the concept of time. Actually, it has the dimension of inverse time. Thus we should expect that the nature of time (continuous or discrete) would affect the nature of the frequency accordingly.

1.3.1 Continuous-Time Sinusoidal Signals

A simple harmonic oscillation is mathematically described by the following continuous-time sinusoidal signal:

$$x_a(t) = A\cos(\Omega t + \theta), \quad -\infty < t < \infty \tag{1.3.1}$$

shown in Fig. 1.10. The subscript a used with $x(t)$ denotes an analog signal. This signal is completely characterized by three parameters: A is the *amplitude* of the sinusoid, Ω is the *frequency* in radians per second (rad/s), and θ is the *phase* in radians. Instead of Ω, we often use the frequency F in cycles per second or hertz (Hz), where

$$\Omega = 2\pi F \tag{1.3.2}$$

In terms of F, (1.3.1) can be written as

$$x_a(t) = A\cos(2\pi F t + \theta), \quad -\infty < t < \infty \tag{1.3.3}$$

We will use both forms, (1.3.1) and (1.3.3), in representing sinusoidal signals.

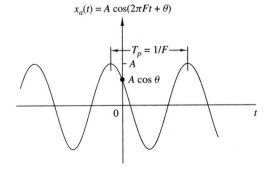

Figure 1.10   Example of an analog sinusoidal signal.

The analog sinusoidal signal in (1.3.3) is characterized by the following properties:

A1. For every fixed value of the frequency F, $x_a(t)$ is periodic. Indeed, it can easily be shown, using elementary trigonometry, that

$$x_a(t + T_p) = x_a(t)$$

where $T_p = 1/F$ is the fundamental period of the sinusoidal signal.

A2. Continuous-time sinusoidal signals with distinct (different) frequencies are themselves distinct.

A3. Increasing the frequency F results in an increase in the rate of oscillation of the signal, in the sense that more periods are included in a given time interval.

We observe that for $F = 0$, the value $T_p = \infty$ is consistent with the fundamental relation $F = 1/T_p$. Due to continuity of the time variable t, we can increase the frequency F, without limit, with a corresponding increase in the rate of oscillation.

The relationships we have described for sinusoidal signals carry over to the class of complex exponential signals

$$x_a(t) = Ae^{j(\Omega t + \theta)} \tag{1.3.4}$$

This can easily be seen by expressing these signals in terms of sinusoids using the Euler identity

$$e^{\pm j\phi} = \cos\phi \pm j\sin\phi \tag{1.3.5}$$

By definition, frequency is an inherently positive physical quantity. This is obvious if we interpret frequency as the number of cycles per unit time in a periodic signal. However, in many cases, only for mathematical convenience, we need to introduce negative frequencies. To see this we recall that the sinusoidal signal (1.3.1) may be expressed as

$$x_a(t) = A\cos(\Omega t + \theta) = \frac{A}{2}e^{j(\Omega t + \theta)} + \frac{A}{2}e^{-j(\Omega t + \theta)} \tag{1.3.6}$$

which follows from (1.3.5). Note that a sinusoidal signal can be obtained by adding two equal-amplitude complex-conjugate exponential signals, sometimes called phasors, illustrated in Fig. 1.11. As time progresses the phasors rotate in opposite directions with angular frequencies $\pm\Omega$ radians per second. Since a *positive frequency* corresponds to counterclockwise uniform angular motion, a *negative frequency* simply corresponds to clockwise angular motion.

For mathematical convenience, we use both negative and positive frequencies throughout this book. Hence the frequency range for analog sinusoids is $-\infty < F < \infty$.

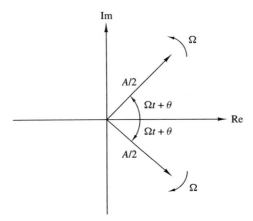

Figure 1.11 Representation of a cosine function by a pair of complex-conjugate exponentials (phasors).

1.3.2 Discrete-Time Sinusoidal Signals

A discrete-time sinusoidal signal may be expressed as

$$x(n) = A\cos(\omega n + \theta), \quad -\infty < n < \infty \tag{1.3.7}$$

where n is an integer variable, called the sample number, A is the *amplitude* of the sinusoid, ω is the *frequency* in radians per sample, and θ is the *phase* in radians.

If instead of ω we use the frequency variable f defined by

$$\omega \equiv 2\pi f \tag{1.3.8}$$

the relation (1.3.7) becomes

$$x(n) = A\cos(2\pi f n + \theta), \quad -\infty < n < \infty \tag{1.3.9}$$

The frequency f has dimensions of cycles per sample. In Section 1.4, where we consider the sampling of analog sinusoids, we relate the frequency variable f of a discrete-time sinusoid to the frequency F in cycles per second for the analog sinusoid. For the moment we consider the discrete-time sinusoid in (1.3.7) independently of the continuous-time sinusoid given in (1.3.1). Figure 1.12 shows a sinusoid with frequency $\omega = \pi/6$ radians per sample ($f = \frac{1}{12}$ cycles per sample) and phase $\theta = \pi/3$.

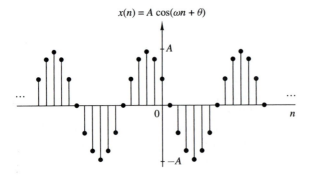

Figure 1.12 Example of a discrete-time sinusoidal signal ($\omega = \pi/6$ and $\theta = \pi/3$).

In contrast to continuous-time sinusoids, the discrete-time sinusoids are characterized by the following properties:

B1. *A discrete-time sinusoid is periodic only if its frequency f is a rational number.*

By definition, a discrete-time signal $x(n)$ is periodic with period $N(N > 0)$ if and only if

$$x(n + N) = x(n) \qquad \text{for all } n \tag{1.3.10}$$

The smallest value of N for which (1.3.10) is true is called the *fundamental period*.

The proof of the periodicity property is simple. For a sinusoid with frequency f_0 to be periodic, we should have

$$\cos[2\pi f_0(N + n) + \theta] = \cos(2\pi f_0 n + \theta)$$

This relation is true if and only if there exists an integer k such that

$$2\pi f_0 N = 2k\pi$$

or, equivalently,

$$f_0 = \frac{k}{N} \tag{1.3.11}$$

According to (1.3.11), a discrete-time sinusoidal signal is periodic only if its frequency f_0 can be expressed as the ratio of two integers (i.e., f_0 is rational).

To determine the fundamental period N of a periodic sinusoid, we express its frequency f_0 as in (1.3.11) and cancel common factors so that k and N are relatively prime. Then the fundamental period of the sinusoid is equal to N. Observe that a small change in frequency can result in a large change in the period. For example, note that $f_1 = 31/60$ implies that $N_1 = 60$, whereas $f_2 = 30/60$ results in $N_2 = 2$.

B2. *Discrete-time sinusoids whose frequencies are separated by an integer multiple of 2π are identical.*

To prove this assertion, let us consider the sinusoid $\cos(\omega_0 n + \theta)$. It easily follows that

$$\cos[(\omega_0 + 2\pi)n + \theta] = \cos(\omega_0 n + 2\pi n + \theta) = \cos(\omega_0 n + \theta) \tag{1.3.12}$$

As a result, all sinusoidal sequences

$$x_k(n) = A\cos(\omega_k n + \theta), \qquad k = 0, 1, 2, \ldots \tag{1.3.13}$$

where

$$\omega_k = \omega_0 + 2k\pi, \qquad -\pi \leq \omega_0 \leq \pi$$

are *indistinguishable* (i.e., *identical*). On the other hand, the sequences of any two sinusoids with frequencies in the range $-\pi \leq \omega \leq \pi$ or $-\frac{1}{2} \leq f \leq \frac{1}{2}$ are distinct. Consequently, discrete-time sinusoidal signals with frequencies $|\omega| \leq \pi$ or $|f| \leq \frac{1}{2}$

are unique. Any sequence resulting from a sinusoid with a frequency $|\omega| > \pi$, or $|f| > \frac{1}{2}$, is identical to a sequence obtained from a sinusoidal signal with frequency $|\omega| < \pi$. Because of this similarity, we call the sinusoid having the frequency $|\omega| > \pi$ an *alias* of a corresponding sinusoid with frequency $|\omega| < \pi$. Thus we regard frequencies in the range $-\pi \le \omega \le \pi$, or $-\frac{1}{2} \le f \le \frac{1}{2}$ as unique and all frequencies $|\omega| > \pi$, or $|f| > \frac{1}{2}$, as aliases. The reader should notice the difference between discrete-time sinusoids and continuous-time sinusoids, where the latter result in distinct signals for Ω or F in the entire range $-\infty < \Omega < \infty$ or $-\infty < F < \infty$.

B3. *The highest rate of oscillation in a discrete-time sinusoid is attained when $\omega = \pi$ (or $\omega = -\pi$) or, equivalently, $f = \frac{1}{2}$ (or $f = -\frac{1}{2}$).*

To illustrate this property, let us investigate the characteristics of the sinusoidal signal sequence

$$x(n) = \cos \omega_0 n$$

when the frequency varies from 0 to π. To simplify the argument, we take values of $\omega_0 = 0$, $\pi/8$, $\pi/4$, $\pi/2$, π corresponding to $f = 0$, $\frac{1}{16}$, $\frac{1}{8}$, $\frac{1}{4}$, $\frac{1}{2}$, which result in periodic sequences having periods $N = \infty$, 16, 8, 4, 2, as depicted in Fig. 1.13. We note that the period of the sinusoid decreases as the frequency increases. In fact, we can see that the rate of oscillation increases as the frequency increases.

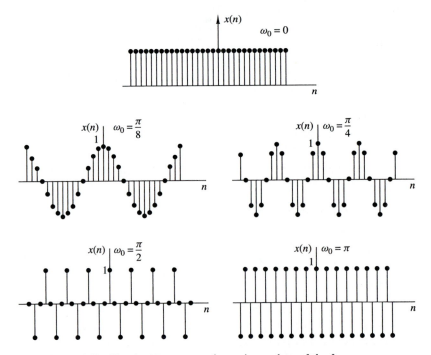

Figure 1.13   Signal $x(n) = \cos \omega_0 n$ for various values of the frequency ω_0.

To see what happens for $\pi \leq \omega_0 \leq 2\pi$, we consider the sinusoids with frequencies $\omega_1 = \omega_0$ and $\omega_2 = 2\pi - \omega_0$. Note that as ω_1 varies from π to 2π, ω_2 varies from π to 0. It can be easily seen that

$$x_1(n) = A \cos \omega_1 n = A \cos \omega_0 n$$

$$x_2(n) = A \cos \omega_2 n = A \cos(2\pi - \omega_0)n \tag{1.3.14}$$

$$= A \cos(-\omega_0 n) = x_1(n)$$

Hence ω_2 is an alias of ω_1. If we had used a sine function instead of a cosine function, the result would basically be the same, except for a 180° phase difference between the sinusoids $x_1(n)$ and $x_2(n)$. In any case, as we increase the relative frequency ω_0 of a discrete-time sinusoid from π to 2π, its rate of oscillation decreases. For $\omega_0 = 2\pi$ the result is a constant signal, as in the case for $\omega_0 = 0$. Obviously, for $\omega_0 = \pi$ (or $f = \frac{1}{2}$) we have the highest rate of oscillation.

As for the case of continuous-time signals, negative frequencies can be introduced as well for discrete-time signals. For this purpose we use the identity

$$x(n) = A \cos(\omega n + \theta) = \frac{A}{2} e^{j(\omega n + \theta)} + \frac{A}{2} e^{-j(\omega n + \theta)} \tag{1.3.15}$$

Since discrete-time sinusoidal signals with frequencies that are separated by an integer multiple of 2π are identical, it follows that the frequencies in any interval $\omega_1 \leq \omega \leq \omega_1 + 2\pi$ constitute *all* the existing discrete-time sinusoids or complex exponentials. Hence the frequency range for discrete-time sinusoids is finite with duration 2π. Usually, we choose the range $0 \leq \omega \leq 2\pi$ or $-\pi \leq \omega \leq \pi (0 \leq f \leq 1, -\frac{1}{2} \leq f \leq \frac{1}{2})$, which we call the *fundamental range*.

1.3.3 Harmonically Related Complex Exponentials

Sinusoidal signals and complex exponentials play a major role in the analysis of signals and systems. In some cases we deal with sets of *harmonically related* complex exponentials (or sinusoids). These are sets of periodic complex exponentials with fundamental frequencies that are multiples of a single positive frequency. Although we confine our discussion to complex exponentials, the same properties clearly hold for sinusoidal signals. We consider harmonically related complex exponentials in both continuous time and discrete time.

Continuous-time exponentials.   The basic signals for continuous-time, harmonically related exponentials are

$$s_k(t) = e^{jk\Omega_0 t} = e^{j2\pi k F_0 t} \qquad k = 0, \pm 1, \pm 2, \ldots \tag{1.3.16}$$

We note that for each value of k, $s_k(t)$ is periodic with fundamental period $1/(kF_0) = T_p/k$ or fundamental frequency kF_0. Since a signal that is periodic with period T_p/k is also periodic with period $k(T_p/k) = T_p$ for any positive integer k, we see that all of the $s_k(t)$ have a common period of T_p. Furthermore, according to Section 1.3.1, F_0 is allowed to take any value and all members of the set are distinct, in the sense that if $k_1 \neq k_2$, then $s_{k1}(t) \neq s_{k2}(t)$.

From the basic signals in (1.3.16) we can construct a linear combination of harmonically related complex exponentials of the form

$$x_a(t) = \sum_{k=-\infty}^{\infty} c_k s_k(t) = \sum_{k=-\infty}^{\infty} c_k e^{jk\Omega_0 t} \tag{1.3.17}$$

where c_k, $k = 0, \pm 1, \pm 2, \ldots$ are arbitrary complex constants. The signal $x_a(t)$ is periodic with fundamental period $T_p = 1/F_0$, and its representation in terms of (1.3.17) is called the *Fourier series* expansion for $x_a(t)$. The complex-valued constants are the Fourier series coefficients and the signal $s_k(t)$ is called the kth harmonic of $x_a(t)$.

Discrete-time exponentials. Since a discrete-time complex exponential is periodic if its relative frequency is a rational number, we choose $f_0 = 1/N$ and we define the sets of harmonically related complex exponentials by

$$s_k(n) = e^{j2\pi k f_0 n}, \qquad k = 0, \pm 1, \pm 2, \ldots \tag{1.3.18}$$

In contrast to the continuous-time case, we note that

$$s_{k+N}(n) = e^{j2\pi n(k+N)/N} = e^{j2\pi n} s_k(n) = s_k(n)$$

This means that, consistent with (1.3.10), there are only N distinct periodic complex exponentials in the set described by (1.3.18). Furthermore, all members of the set have a common period of N samples. Clearly, we can choose any consecutive N complex exponentials, say from $k = n_0$ to $k = n_0 + N - 1$ to form a harmonically related set with fundamental frequency $f_0 = 1/N$. Most often, for convenience, we choose the set that corresponds to $n_0 = 0$, that is, the set

$$s_k(n) = e^{j2\pi kn/N}, \qquad k = 0, 1, 2, \ldots, N - 1 \tag{1.3.19}$$

As in the case of continuous-time signals, it is obvious that the linear combination

$$x(n) = \sum_{k=0}^{N-1} c_k s_k(n) = \sum_{k=0}^{N-1} c_k e^{j2\pi kn/N} \tag{1.3.20}$$

results in a periodic signal with fundamental period N. As we shall see later, this is the Fourier series representation for a periodic discrete-time sequence with Fourier coefficients $\{c_k\}$. The sequence $s_k(n)$ is called the kth harmonic of $x(n)$.

Example 1.3.1

Stored in the memory of a digital signal processor is one cycle of the sinusoidal signal

$$x(n) = \sin\left(\frac{2\pi n}{N} + \theta\right)$$

where $\theta = 2\pi q/N$, where q and N are integers.

(a) Determine how this table of values can be used to obtain values of harmonically related sinusoids having the same phase.

(b) Determine how this table can be used to obtain sinusoids of the same frequency but different phase.

Solution

(a) Let $x_k(n)$ denote the sinusoidal signal sequence

$$x_k(n) = \sin\left(\frac{2\pi nk}{N} + \theta\right)$$

This is a sinusoid with frequency $f_k = k/N$, which is harmonically related to $x(n)$. But $x_k(n)$ may be expressed as

$$x_k(n) = \sin\left[\frac{2\pi(kn)}{N} + \theta\right]$$

$$= x(kn)$$

Thus we observe that $x_k(0) = x(0)$, $x_k(1) = x(k)$, $x_k(2) = x(2k)$, and so on. Hence the sinusoidal sequence $x_k(n)$ can be obtained from the table of values of $x(n)$ by taking every kth value of $x(n)$, beginning with $x(0)$. In this manner we can generate the values of all harmonically related sinusoids with frequencies $f_k = k/N$ for $k = 0, 1, \ldots, N-1$.

(b) We can control the phase θ of the sinusoid with frequency $f_k = k/N$ by taking the first value of the sequence from memory location $q = \theta N/2\pi$, where q is an integer. Thus the initial phase θ controls the starting location in the table and we wrap around the table each time the index (kn) exceeds N.

1.4 ANALOG-TO-DIGITAL AND DIGITAL-TO-ANALOG CONVERSION

Most signals of practical interest, such as speech, biological signals, seismic signals, radar signals, sonar signals, and various communications signals such as audio and video signals, are analog. To process analog signals by digital means, it is first necessary to convert them into digital form, that is, to convert them to a sequence of numbers having finite precision. This procedure is called *analog-to-digital (A/D) conversion*, and the corresponding devices are called *A/D converters (ADCs)*.

Conceptually, we view A/D conversion as a three-step process. This process is illustrated in Fig. 1.14.

1. *Sampling.* This is the conversion of a continuous-time signal into a discrete-time signal obtained by taking "samples" of the continuous-time signal at discrete-time instants. Thus, if $x_a(t)$ is the input to the sampler, the output is $x_a(nT) \equiv x(n)$, where T is called the *sampling interval*.

2. *Quantization.* This is the conversion of a discrete-time continuous-valued signal into a discrete-time, discrete-valued (digital) signal. The value of each

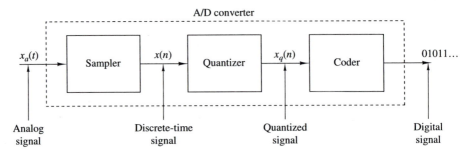

Figure 1.14   Basic parts of an analog-to-digital (A/D) converter.

signal sample is represented by a value selected from a finite set of possible values. The difference between the unquantized sample $x(n)$ and the quantized output $x_q(n)$ is called the quantization error.

3. *Coding.* In the coding process, each discrete value $x_q(n)$ is represented by a b-bit binary sequence.

Although we model the A/D converter as a sampler followed by a quantizer and coder, in practice the A/D conversion is performed by a single device that takes $x_a(t)$ and produces a binary-coded number. The operations of sampling and quantization can be performed in either order but, in practice, sampling is always performed before quantization.

In many cases of practical interest (e.g., speech processing) it is desirable to convert the processed digital signals into analog form. (Obviously, we cannot listen to the sequence of samples representing a speech signal or see the numbers corresponding to a TV signal.) The process of converting a digital signal into an analog signal is known as *digital-to-analog (D/A) conversion.* All D/A converters "connect the dots" in a digital signal by performing some kind of interpolation, whose accuracy depends on the quality of the D/A conversion process. Figure 1.15 illustrates a simple form of D/A conversion, called a zero-order hold or a staircase approximation. Other approximations are possible, such as linearly connecting a pair of successive samples (linear interpolation), fitting a quadratic through three successive samples (quadratic interpolation), and so on. Is there an optimum (ideal) interpolator? For signals having a *limited frequency content* (finite bandwidth), the sampling theorem introduced in the following section specifies the optimum form of interpolation.

Sampling and quantization are treated in this section. In particular, we demonstrate that sampling does not result in a loss of information, nor does it introduce distortion in the signal if the signal bandwidth is finite. In principle, the analog signal can be reconstructed from the samples, provided that the sampling rate is sufficiently high to avoid the problem commonly called *aliasing.* On the other hand, quantization is a noninvertible or irreversible process that results in signal distortion. We shall show that the amount of distortion is dependent on

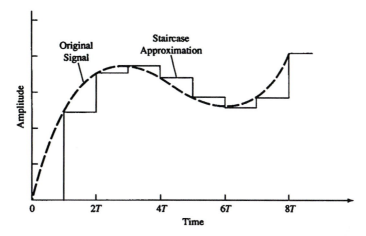

Figure 1.15 Zero-order hold digital-to-analog (D/A) conversion.

the accuracy, as measured by the number of bits, in the A/D conversion process. The factors affecting the choice of the desired accuracy of the A/D converter are cost and sampling rate. In general, the cost increases with an increase in accuracy and/or sampling rate.

1.4.1 Sampling of Analog Signals

There are many ways to sample an analog signal. We limit our discussion to *periodic* or *uniform sampling*, which is the type of sampling used most often in practice. This is described by the relation

$$x(n) = x_a(nT), \qquad -\infty < n < \infty \tag{1.4.1}$$

where $x(n)$ is the discrete-time signal obtained by "taking samples" of the analog signal $x_a(t)$ every T seconds. This procedure is illustrated in Fig. 1.16. The time interval T between successive samples is called the *sampling period* or *sample interval* and its reciprocal $1/T = F_s$ is called the *sampling rate* (samples per second) or the *sampling frequency* (hertz).

Periodic sampling establishes a relationship between the time variables t and n of continuous-time and discrete-time signals, respectively. Indeed, these variables are linearly related through the sampling period T or, equivalently, through the sampling rate $F_s = 1/T$, as

$$t = nT = \frac{n}{F_s} \tag{1.4.2}$$

As a consequence of (1.4.2), there exists a relationship between the frequency variable F (or Ω) for analog signals and the frequency variable f (or ω) for discrete-time signals. To establish this relationship, consider an analog sinusoidal signal of the form

$$x_a(t) = A\cos(2\pi F t + \theta) \tag{1.4.3}$$

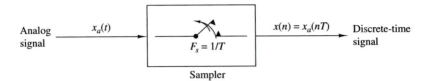

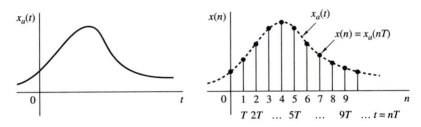

Figure 1.16   Periodic sampling of an analog signal.

which, when sampled periodically at a rate $F_s = 1/T$ samples per second, yields

$$x_a(nT) \equiv x(n) = A\cos(2\pi FnT + \theta)$$
$$= A\cos\left(\frac{2\pi nF}{F_s} + \theta\right) \tag{1.4.4}$$

If we compare (1.4.4) with (1.3.9), we note that the frequency variables F and f are linearly related as

$$f = \frac{F}{F_s} \tag{1.4.5}$$

or, equivalently, as

$$\omega = \Omega T \tag{1.4.6}$$

The relation in (1.4.5) justifies the name *relative* or *normalized frequency*, which is sometimes used to describe the frequency variable f. As (1.4.5) implies, we can use f to determine the frequency F in hertz only if the sampling frequency F_s is known.

We recall from Section 1.3.1 that the range of the frequency variable F or Ω for continuous-time sinusoids are

$$-\infty < F < \infty$$
$$-\infty < \Omega < \infty \tag{1.4.7}$$

However, the situation is different for discrete-time sinusoids. From Section 1.3.2 we recall that

$$-\tfrac{1}{2} < f < \tfrac{1}{2}$$
$$-\pi < \omega < \pi \tag{1.4.8}$$

By substituting from (1.4.5) and (1.4.6) into (1.4.8), we find that the frequency of the continuous-time sinusoid when sampled at a rate $F_s = 1/T$ must fall in the range

$$-\frac{1}{2T} = -\frac{F_s}{2} \le F \le \frac{F_s}{2} = \frac{1}{2T} \tag{1.4.9}$$

or, equivalently,

$$-\frac{\pi}{T} = -\pi F_s \le \Omega \le \pi F_s = \frac{\pi}{T} \tag{1.4.10}$$

These relations are summarized in Table 1.1.

TABLE 1.1   RELATIONS AMONG FREQUENCY VARIABLES

Continuous-time signals	Discrete-time signals
$\Omega = 2\pi F$ $\dfrac{\text{radians}}{\text{sec}}$ Hz	$\omega = 2\pi f$ $\dfrac{\text{radians}}{\text{sample}}$ $\dfrac{\text{cycles}}{\text{sample}}$
$\omega = \Omega T, f = F/F_s$ $\Omega = \omega/T, F = f \cdot F_s$	$-\pi \le \omega \le \pi$ $-\frac{1}{2} \le f \le \frac{1}{2}$
$-\infty < \Omega < \infty$ $-\infty < F < \infty$	$-\pi/T \le \Omega \le \pi/T$ $-F_2/2 \le F \le F_s/2$

From these relations we observe that the fundamental difference between continuous-time and discrete-time signals is in their range of values of the frequency variables F and f, or Ω and ω. Periodic sampling of a continuous-time signal implies a mapping of the infinite frequency range for the variable F (or Ω) into a finite frequency range for the variable f (or ω). Since the highest frequency in a discrete-time signal is $\omega = \pi$ or $f = \frac{1}{2}$, it follows that, with a sampling rate F_s, the corresponding highest values of F and Ω are

$$F_{\max} = \frac{F_s}{2} = \frac{1}{2T}$$
$$\Omega_{\max} = \pi F_s = \frac{\pi}{T} \tag{1.4.11}$$

Therefore, sampling introduces an ambiguity, since the highest frequency in a continuous-time signal that can be uniquely distinguished when such a signal is sampled at a rate $F_s = 1/T$ is $F_{\max} = F_s/2$, or $\Omega_{\max} = \pi F_s$. To see what happens to frequencies above $F_s/2$, let us consider the following example.

Example 1.4.1

The implications of these frequency relations can be fully appreciated by considering the two analog sinusoidal signals

$$x_1(t) = \cos 2\pi (10)t$$
$$x_2(t) = \cos 2\pi (50)t \tag{1.4.12}$$

which are sampled at a rate $F_s = 40$ Hz. The corresponding discrete-time signals or sequences are

$$x_1(n) = \cos 2\pi \left(\frac{10}{40}\right) n = \cos \frac{\pi}{2} n$$

$$x_2(n) = \cos 2\pi \left(\frac{50}{40}\right) n = \cos \frac{5\pi}{2} n$$

(1.4.13)

However, $\cos 5\pi n/2 = \cos(2\pi n + \pi n/2) = \cos \pi n/2$. Hence $x_2(n) = x_1(n)$. Thus the sinusoidal signals are identical and, consequently, indistinguishable. If we are given the sampled values generated by $\cos(\pi/2)n$, there is some ambiguity as to whether these sampled values correspond to $x_1(t)$ or $x_2(t)$. Since $x_2(t)$ yields exactly the same values as $x_1(t)$ when the two are sampled at $F_s = 40$ samples per second, we say that the frequency $F_2 = 50$ Hz is an *alias* of the frequency $F_1 = 10$ Hz at the sampling rate of 40 samples per second.

It is important to note that F_2 is not the only alias of F_1. In fact at the sampling rate of 40 samples per second, the frequency $F_3 = 90$ Hz is also an alias of F_1, as is the frequency $F_4 = 130$ Hz, and so on. All of the sinusoids $\cos 2\pi(F_1 + 40k)t$, $k = 1$, 2, 3, 4, ... sampled at 40 samples per second, yield identical values. Consequently, they are all aliases of $F_1 = 10$ Hz.

In general, the sampling of a continuous-time sinusoidal signal

$$x_a(t) = A \cos(2\pi F_0 t + \theta)$$

(1.4.14)

with a sampling rate $F_s = 1/T$ results in a discrete-time signal

$$x(n) = A \cos(2\pi f_0 n + \theta)$$

(1.4.15)

where $f_0 = F_0/F_s$ is the relative frequency of the sinusoid. If we assume that $-F_s/2 \le F_0 \le F_s/2$, the frequency f_0 of $x(n)$ is in the range $-\frac{1}{2} \le f_0 \le \frac{1}{2}$, which is the frequency range for discrete-time signals. In this case, the relationship between F_0 and f_0 is one-to-one, and hence it is possible to identify (or reconstruct) the analog signal $x_a(t)$ from the samples $x(n)$.

On the other hand, if the sinusoids

$$x_a(t) = A \cos(2\pi F_k t + \theta)$$

(1.4.16)

where

$$F_k = F_0 + kF_s, \qquad k = \pm1, \pm2, \ldots$$

(1.4.17)

are sampled at a rate F_s, it is clear that the frequency F_k is outside the fundamental frequency range $-F_s/2 \le F \le F_s/2$. Consequently, the sampled signal is

$$x(n) \equiv x_a(nT) = A \cos \left(2\pi \frac{F_0 + kF_s}{F_s} n + \theta\right)$$

$$= A \cos(2\pi n F_0/F_s + \theta + 2\pi kn)$$

$$= A \cos(2\pi f_0 n + \theta)$$

which is identical to the discrete-time signal in (1.4.15) obtained by sampling (1.4.14). Thus an infinite number of continuous-time sinusoids is represented by sampling the *same* discrete-time signal (i.e., by the same set of samples). Consequently, if we are given the sequence $x(n)$, an ambiguity exists as to which continuous-time signal $x_a(t)$ these values represent. Equivalently, we can say that the frequencies $F_k = F_0 + kF_s$, $-\infty < k < \infty$ (k integer) are indistinguishable from the frequency F_0 after sampling and hence they are aliases of F_0. The relationship between the frequency variables of the continuous-time and discrete-time signals is illustrated in Fig. 1.17.

An example of aliasing is illustrated in Fig. 1.18, where two sinusoids with frequencies $F_0 = \frac{1}{8}$ Hz and $F_1 = -\frac{7}{8}$ Hz yield identical samples when a sampling rate of $F_s = 1$ Hz is used. From (1.4.17) it easily follows that for $k = -1$, $F_0 = F_1 + F_s = (-\frac{7}{8} + 1)$ Hz $= \frac{1}{8}$ Hz.

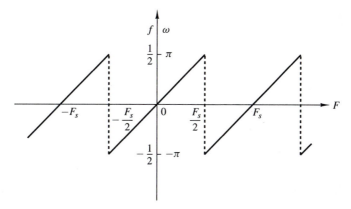

Figure 1.17 Relationship between the continuous-time and discrete-time frequency variables in the case of periodic sampling.

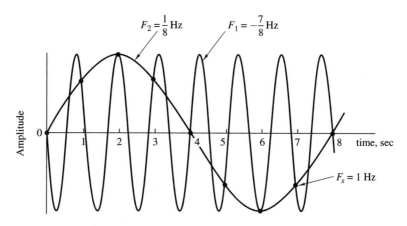

Figure 1.18 Illustration of aliasing.

Since $F_s/2$, which corresponds to $\omega = \pi$, is the highest frequency that can be represented uniquely with a sampling rate F_s, it is a simple matter to determine the mapping of any (alias) frequency above $F_s/2$ ($\omega = \pi$) into the equivalent frequency below $F_s/2$. We can use $F_s/2$ or $\omega = \pi$ as the pivotal point and reflect or "fold" the alias frequency to the range $0 \leq \omega \leq \pi$. Since the point of reflection is $F_s/2$ ($\omega = \pi$), the frequency $F_s/2$ ($\omega = \pi$) is called the *folding frequency*.

Example 1.4.2

Consider the analog signal

$$x_a(t) = 3 \cos 100\pi t$$

(a) Determine the minimum sampling rate required to avoid aliasing.

(b) Suppose that the signal is sampled at the rate $F_s = 200$ Hz. What is the discrete-time signal obtained after sampling?

(c) Suppose that the signal is sampled at the rate $F_s = 75$ Hz. What is the discrete-time signal obtained after sampling?

(d) What is the frequency $0 < F < F_s/2$ of a sinusoid that yields samples identical to those obtained in part (c)?

Solution

(a) The frequency of the analog signal is $F = 50$ Hz. Hence the minimum sampling rate required to avoid aliasing is $F_s = 100$ Hz.

(b) If the signal is sampled at $F_s = 200$ Hz, the discrete-time signal is

$$x(n) = 3 \cos \frac{100\pi}{200} n = 3 \cos \frac{\pi}{2} n$$

(c) If the signal is sampled at $F_s = 75$ Hz, the discrete-time signal is

$$x(n) = 3 \cos \frac{100\pi}{75} n = 3 \cos \frac{4\pi}{3} n$$

$$= 3 \cos \left(2\pi - \frac{2\pi}{3} \right) n$$

$$= 3 \cos \frac{2\pi}{3} n$$

(d) For the sampling rate of $F_s = 75$ Hz, we have

$$F = f F_s = 75 f$$

The frequency of the sinusoid in part (c) is $f = \frac{1}{3}$. Hence

$$F = 25 \text{ Hz}$$

Clearly, the sinusoidal signal

$$y_a(t) = 3 \cos 2\pi F t$$

$$= 3 \cos 50\pi t$$

sampled at $F_s = 75$ samples/s yields identical samples. Hence $F = 50$ Hz is an alias of $F = 25$ Hz for the sampling rate $F_s = 75$ Hz.

1.4.2 The Sampling Theorem

Given any analog signal, how should we select the sampling period T or, equivalently, the sampling rate F_s? To answer this question, we must have some information about the characteristics of the signal to be sampled. In particular, we must have some general information concerning the *frequency content* of the signal. Such information is generally available to us. For example, we know generally that the major frequency components of a speech signal fall below 3000 Hz. On the other hand, television signals, in general, contain important frequency components up to 5 MHz. The information content of such signals is contained in the amplitudes, frequencies, and phases of the various frequency components, but detailed knowledge of the characteristics of such signals is not available to us prior to obtaining the signals. In fact, the purpose of processing the signals is usually to extract this detailed information. However, if we know the maximum frequency content of the general class of signals (e.g., the class of speech signals, the class of video signals, etc.), we can specify the sampling rate necessary to convert the analog signals to digital signals.

Let us suppose that any analog signal can be represented as a sum of sinusoids of different amplitudes, frequencies, and phases, that is,

$$x_a(t) = \sum_{i=1}^{N} A_i \cos(2\pi F_i t + \theta_i) \tag{1.4.18}$$

where N denotes the number of frequency components. All signals, such as speech and video, lend themselves to such a representation over any short time segment. The amplitudes, frequencies, and phases usually change slowly with time from one time segment to another. However, suppose that the frequencies do not exceed some known frequency, say F_{max}. For example, $F_{max} = 3000$ Hz for the class of speech signals and $F_{max} = 5$ MHz for television signals. Since the maximum frequency may vary slightly from different realizations among signals of any given class (e.g., it may vary slightly from speaker to speaker), we may wish to ensure that F_{max} does not exceed some predetermined value by passing the analog signal through a filter that severely attenuates frequency components above F_{max}. Thus we are certain that no signal in the class contains frequency components (having significant amplitude or power) above F_{max}. In practice, such filtering is commonly used prior to sampling.

From our knowledge of F_{max}, we can select the appropriate sampling rate. We know that the highest frequency in an analog signal that can be unambiguously reconstructed when the signal is sampled at a rate $F_s = 1/T$ is $F_s/2$. Any frequency above $F_s/2$ or below $-F_s/2$ results in samples that are identical with a corresponding frequency in the range $-F_s/2 \le F \le F_s/2$. To avoid the ambiguities resulting from aliasing, we must select the sampling rate to be sufficiently high. That is, we must select $F_s/2$ to be greater than F_{max}. Thus to avoid the problem of aliasing, F_s is selected so that

$$F_s > 2F_{max} \tag{1.4.19}$$

where F_{max} is the largest frequency component in the analog signal. With the sampling rate selected in this manner, any frequency component, say $|F_i| < F_{max}$, in the analog signal is mapped into a discrete-time sinusoid with a frequency

$$-\frac{1}{2} \le f_i = \frac{F_i}{F_s} \le \frac{1}{2} \tag{1.4.20}$$

or, equivalently,

$$-\pi \le \omega_i = 2\pi f_i \le \pi \tag{1.4.21}$$

Since, $|f| = \frac{1}{2}$ or $|\omega| = \pi$ is the highest (unique) frequency in a discrete-time signal, the choice of sampling rate according to (1.4.19) avoids the problem of aliasing. In other words, the condition $F_s > 2F_{max}$ ensures that all the sinusoidal components in the analog signal are mapped into corresponding discrete-time frequency components with frequencies in the fundamental interval. Thus all the frequency components of the analog signal are represented in sampled form without ambiguity, and hence the analog signal can be reconstructed without distortion from the sample values using an "appropriate" interpolation (digital-to-analog conversion) method. The "appropriate" or ideal interpolation formula is specified by the *sampling theorem*.

Sampling Theorem. If the highest frequency contained in an analog signal $x_a(t)$ is $F_{max} = B$ and the signal is sampled at a rate $F_s > 2F_{max} \equiv 2B$, then $x_a(t)$ can be exactly recovered from its sample values using the interpolation function

$$g(t) = \frac{\sin 2\pi Bt}{2\pi Bt} \tag{1.4.22}$$

Thus $x_a(t)$ may be expressed as

$$x_a(t) = \sum_{n=-\infty}^{\infty} x_a\left(\frac{n}{F_s}\right) g\left(t - \frac{n}{F_s}\right) \tag{1.4.23}$$

where $x_a(n/F_s) = x_a(nT) \equiv x(n)$ are the samples of $x_a(t)$.

When the sampling of $x_a(t)$ is performed at the minimum sampling rate $F_s = 2B$, the reconstruction formula in (1.4.23) becomes

$$x_a(t) = \sum_{n=-\infty}^{\infty} x_a\left(\frac{n}{2B}\right) \frac{\sin 2\pi B(t - n/2B)}{2\pi B(t - n/2B)} \tag{1.4.24}$$

The sampling rate $F_N = 2B = 2F_{max}$ is called the *Nyquist rate*. Figure 1.19 illustrates the ideal D/A conversion process using the interpolation function in (1.4.22).

As can be observed from either (1.4.23) or (1.4.24), the reconstruction of $x_a(t)$ from the sequence $x(n)$ is a complicated process, involving a weighted sum of the interpolation function $g(t)$ and its time-shifted versions $g(t-nT)$ for $-\infty < n < \infty$, where the weighting factors are the samples $x(n)$. Because of the complexity and the infinite number of samples required in (1.4.23) or (1.4.24), these reconstruction

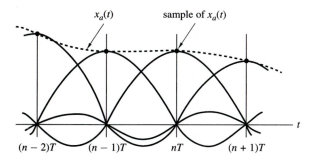

Figure 1.19 Ideal D/A conversion (interpolation).

formulas are primarily of theoretical interest. Practical interpolation methods are given in Chapter 9.

Example 1.4.3

Consider the analog signal

$$x_a(t) = 3\cos 50\pi t + 10\sin 300\pi t - \cos 100\pi t$$

What is the Nyquist rate for this signal?

Solution   The frequencies present in the signal above are

$$F_1 = 25 \text{ Hz}, \qquad F_2 = 150 \text{ Hz}, \qquad F_3 = 50 \text{ Hz}$$

Thus $F_{max} = 150$ Hz and according to (1.4.19),

$$F_s > 2F_{max} = 300 \text{ Hz}$$

The Nyquist rate is $F_N = 2F_{max}$. Hence

$$F_N = 300 \text{ Hz}$$

Discussion   It should be observed that the signal component $10\sin 300\pi t$, sampled at the Nyquist rate $F_N = 300$, results in the samples $10\sin \pi n$, which are identically zero. In other words, we are sampling the analog sinusoid at its zero-crossing points, and hence we miss this signal component completely. This situation would not occur if the sinusoid is offset in phase by some amount θ. In such a case we have $10\sin(300\pi t + \theta)$ sampled at the Nyquist rate $F_N = 300$ samples per second, which yields the samples

$$10\sin(\pi n + \theta) = 10(\sin \pi n \cos \theta + \cos \pi n \sin \theta)$$

$$= 10\sin \theta \cos \pi n$$

$$= (-1)^n 10\sin \theta$$

Thus if $\theta \neq 0$ or π, the samples of the sinusoid taken at the Nyquist rate are not all zero. However, we still cannot obtain the correct amplitude from the samples when the phase θ is unknown. A simple remedy that avoids this potentially troublesome situation is to sample the analog signal at a rate higher than the Nyquist rate.

Example 1.4.4

Consider the analog signal

$$x_a(t) = 3\cos 2000\pi t + 5\sin 6000\pi t + 10\cos 12{,}000\pi t$$

(a) What is the Nyquist rate for this signal?

(b) Assume now that we sample this signal using a sampling rate $F_s = 5000$ samples/s. What is the discrete-time signal obtained after sampling?

(c) What is the analog signal $y_a(t)$ we can reconstruct from the samples if we use ideal interpolation?

Solution

(a) The frequencies existing in the analog signal are

$$F_1 = 1 \text{ kHz}, \qquad F_2 = 3 \text{ kHz}, \qquad F_3 = 6 \text{ kHz}$$

Thus $F_{\max} = 6$ kHz, and according to the sampling theorem,

$$F_s > 2F_{\max} = 12 \text{ kHz}$$

The Nyquist rate is

$$F_N = 12 \text{ kHz}$$

(b) Since we have chosen $F_s = 5$ kHz, the folding frequency is

$$\frac{F_s}{2} = 2.5 \text{ kHz}$$

and this is the maximum frequency that can be represented uniquely by the sampled signal. By making use of (1.4.2) we obtain

$$x(n) = x_a(nT) = x_a\left(\frac{n}{F_s}\right)$$

$$= 3\cos 2\pi(\tfrac{1}{5})n + 5\sin 2\pi(\tfrac{3}{5})n + 10\cos 2\pi(\tfrac{6}{5})n$$

$$= 3\cos 2\pi(\tfrac{1}{5})n + 5\sin 2\pi(1 - \tfrac{2}{5})n + 10\cos 2\pi(1 + \tfrac{1}{5})n$$

$$= 3\cos 2\pi(\tfrac{1}{5})n + 5\sin 2\pi(-\tfrac{2}{5})n + 10\cos 2\pi(\tfrac{1}{5})n$$

Finally, we obtain

$$x(n) = 13\cos 2\pi(\tfrac{1}{5})n - 5\sin 2\pi(\tfrac{2}{5})n$$

The same result can be obtained using Fig. 1.17. Indeed, since $F_s = 5$ kHz, the folding frequency is $F_s/2 = 2.5$ kHz. This is the maximum frequency that can be represented uniquely by the sampled signal. From (1.4.17) we have $F_0 = F_k - kF_s$. Thus F_0 can be obtained by subtracting from F_k an integer multiple of F_s such that $-F_s/2 \le F_0 \le F_s/2$. The frequency F_1 is less than $F_s/2$ and thus it is not affected by aliasing. However, the other two frequencies are above the folding frequency and they will be changed by the aliasing effect. Indeed,

$$F_2' = F_2 - F_s = -2 \text{ kHz}$$

$$F_3' = F_3 - F_s = 1 \text{ kHz}$$

From (1.4.5) it follows that $f_1 = \tfrac{1}{5}$, $f_2 = -\tfrac{2}{5}$, and $f_3 = \tfrac{1}{5}$, which are in agreement with the result above.

(c) Since only the frequency components at 1 kHz and 2 kHz are present in the sampled signal, the analog signal we can recover is

$$y_a(t) = 13 \cos 2000\pi t - 5 \sin 4000\pi t$$

which is obviously different from the original signal $x_a(t)$. This distortion of the original analog signal was caused by the aliasing effect, due to the low sampling rate used.

Although aliasing is a pitfall to be avoided, there are two useful practical applications based on the exploitation of the aliasing effect. These applications are the stroboscope and the sampling oscilloscope. Both instruments are designed to operate as aliasing devices in order to represent high frequencies as low frequencies.

To elaborate, consider a signal with high-frequency components confined to a given frequency band $B_1 < F < B_2$, where $B_2 - B_1 \equiv B$ is defined as the bandwidth of the signal. We assume that $B << B_1 < B_2$. This condition means that the frequency components in the signal are much larger than the bandwidth B of the signal. Such signals are usually called passband or narrowband signals. Now, if this signal is sampled at a rate $F_s \geq 2B$, but $F_s << B_1$, then all the frequency components contained in the signal will be aliases of frequencies in the range $0 < F < F_s/2$. Consequently, if we observe the frequency content of the signal in the fundamental range $0 < F < F_s/2$, we know precisely the frequency content of the analog signal since we know the frequency band $B_1 < F < B_2$ under consideration. Consequently, if the signal is a narrowband (passband) signal, we can reconstruct the original signal from the samples, provided that the signal is sampled at a rate $F_s > 2B$, where B is the bandwidth. This statement constitutes another form of the sampling theorem, which we call the *passband form* in order to distinguish it from the previous form of the sampling theorem, which applies in general to all types of signals. The latter is sometimes called the *baseband form*. The *passband form* of the sampling theorem is described in detail in Section 9.1.2.

1.4.3 Quantization of Continuous-Amplitude Signals

As we have seen, a digital signal is a sequence of numbers (samples) in which each number is represented by a finite number of digits (finite precision).

The process of converting a discrete-time continuous-amplitude signal into a digital signal by expressing each sample value as a finite (instead of an infinite) number of digits, is called *quantization*. The error introduced in representing the continuous-valued signal by a finite set of discrete value levels is called *quantization error* or *quantization noise*.

We denote the quantizer operation on the samples $x(n)$ as $Q[x(n)]$ and let $x_q(n)$ denote the sequence of quantized samples at the output of the quantizer. Hence

$$x_q(n) = Q[x(n)]$$

Then the quantization error is a sequence $e_q(n)$ defined as the difference between the quantized value and the actual sample value. Thus

$$e_q(n) = x_q(n) - x(n) \qquad (1.4.25)$$

We illustrate the quantization process with an example. Let us consider the discrete-time signal

$$x(n) = \begin{cases} 0.9^n, & n \geq 0 \\ 0, & n < 0 \end{cases}$$

obtained by sampling the analog exponential signal $x_a(t) = 0.9^t$, $t \geq 0$ with a sampling frequency $F_s = 1$ Hz (see Fig. 1.20(a)). Observation of Table 1.2, which shows the values of the first 10 samples of $x(n)$, reveals that the description of the sample value $x(n)$ requires n significant digits. It is obvious that this signal cannot

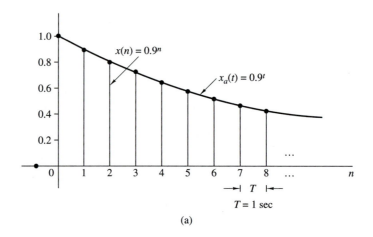

(a)

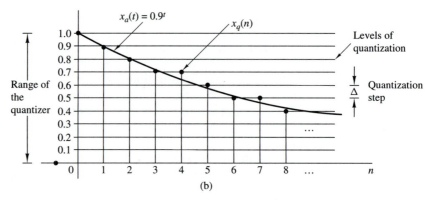

(b)

Figure 1.20   Illustration of quantization.

TABLE 1.2 NUMERICAL ILLUSTRATION OF QUANTIZATION WITH ONE
SIGNIFICANT DIGIT USING TRUNCATION OR ROUNDING

n	$x(n)$ Discrete-time signal	$x_q(n)$ (Truncation)	$x_q(n)$ (Rounding)	$e_q(n) = x_q(n) - x(n)$ (Rounding)
0	1	1.0	1.0	0.0
1	0.9	0.9	0.9	0.0
2	0.81	0.8	0.8	−0.01
3	0.729	0.7	0.7	−0.029
4	0.6561	0.6	0.7	0.0439
5	0.59049	0.5	0.6	0.00951
6	0.531441	0.5	0.5	−0.031441
7	0.4782969	0.4	0.5	0.0217031
8	0.43046721	0.4	0.4	−0.03046721
9	0.387420489	0.3	0.4	0.012579511

be processed by using a calculator or a digital computer since only the first few samples can be stored and manipulated. For example, most calculators process numbers with only eight significant digits.

However, let us assume that we want to use only one significant digit. To eliminate the excess digits, we can either simply discard them (*truncation*) or discard them by rounding the resulting number (*rounding*). The resulting quantized signals $x_q(n)$ are shown in Table 1.2. We discuss only quantization by rounding, although it is just as easy to treat truncation. The rounding process is graphically illustrated in Fig. 1.20b. The values allowed in the digital signal are called the *quantization levels*, whereas the distance Δ between two successive quantization levels is called the *quantization step size* or *resolution*. The rounding quantizer assigns each sample of $x(n)$ to the nearest quantization level. In contrast, a quantizer that performs truncation would have assigned each sample of $x(n)$ to the quantization level below it. The quantization error $e_q(n)$ in rounding is limited to the range of $-\Delta/2$ to $\Delta/2$, that is,

$$-\frac{\Delta}{2} \leq e_q(n) \leq \frac{\Delta}{2} \tag{1.4.26}$$

In other words, the instantaneous quantization error cannot exceed half of the quantization step (see Table 1.2).

If $x_{\min}$ and $x_{\max}$ represent the minimum and maximum value of $x(n)$ and L is the number of quantization levels, then

$$\Delta = \frac{x_{\max} - x_{\min}}{L - 1} \tag{1.4.27}$$

We define the *dynamic range* of the signal as $x_{\max} - x_{\min}$. In our example we have $x_{\max} = 1$, $x_{\min} = 0$, and $L = 11$, which leads to $\Delta = 0.1$. Note that if the dynamic range is fixed, increasing the number of quantization levels, L results in a decrease of the quantization step size. Thus the quantization error decreases and the accuracy of the quantizer increases. In practice we can reduce the quantization error to an insignificant amount by choosing a sufficient number of quantization levels.

Theoretically, quantization of analog signals always results in a loss of information. This is a result of the ambiguity introduced by quantization. Indeed, quantization is an irreversible or noninvertible process (i.e., a many-to-one mapping) since all samples in a distance $\Delta/2$ about a certain quantization level are assigned the same value. This ambiguity makes the exact quantitative analysis of quantization extremely difficult. This subject is discussed further in Chapter 9, where we use statistical analysis.

1.4.4 Quantization of Sinusoidal Signals

Figure 1.21 illustrates the sampling and quantization of an analog sinusoidal signal $x_a(t) = A \cos \Omega_0 t$ using a rectangular grid. Horizontal lines within the range of the quantizer indicate the allowed levels of quantization. Vertical lines indicate the sampling times. Thus, from the original analog signal $x_a(t)$ we obtain a discrete-time signal $x(n) = x_a(nT)$ by sampling and a discrete-time, discrete-amplitude signal $x_q(nT)$ after quantization. In practice, the staircase signal $x_q(t)$ can be obtained by using a zero-order hold. This analysis is useful because sinusoids are used as test signals in A/D converters.

If the sampling rate F_s satisfies the sampling theorem, quantization is the only error in the A/D conversion process. Thus we can evaluate the quantization error

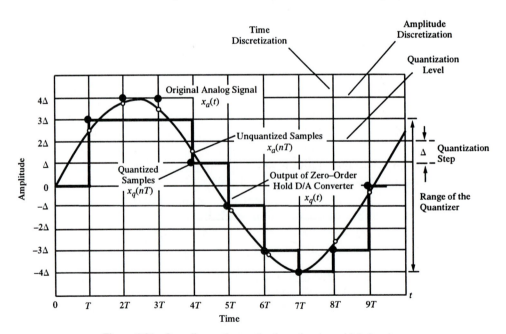

Figure 1.21 Sampling and quantization of a sinusoidal signal.

by quantizing the analog signal $x_a(t)$ instead of the discrete-time signal $x(n) = x_a(nT)$. Inspection of Fig. 1.21 indicates that the signal $x_a(t)$ is almost linear between quantization levels (see Fig. 1.22). The corresponding quantization error $e_q(t) = x_a(t) - x_q(t)$ is shown in Fig. 1.22. In Fig. 1.22, τ denotes the time that $x_a(t)$ stays within the quantization levels. The mean-square error power P_q is

$$P_q = \frac{1}{2\tau} \int_{-\tau}^{\tau} e_q^2(t)dt = \frac{1}{\tau} \int_0^{\tau} e_q^2(t)dt \tag{1.4.28}$$

Since $e_q(t) = (\Delta/2\tau)t, \; -\tau \le t \le \tau$, we have

$$P_q = \frac{1}{\tau} \int_0^{\tau} \left(\frac{\Delta}{2\tau}\right)^2 t^2 dt = \frac{\Delta^2}{12} \tag{1.4.29}$$

If the quantizer has b bits of accuracy and the quantizer covers the entire range $2A$, the quantization step is $\Delta = 2A/2^b$. Hence

$$P_q = \frac{A^2/3}{2^{2b}} \tag{1.4.30}$$

The average power of the signal $x_a(t)$ is

$$P_x = \frac{1}{T_p} \int_0^{T_p} (A \cos \Omega_0 t)^2 dt = \frac{A^2}{2} \tag{1.4.31}$$

The quality of the output of the A/D converter is usually measured by the *signal-to-quantization noise ratio (SQNR)*, which provides the ratio of the signal power to the noise power:

$$\text{SQNR} = \frac{P_x}{P_q} = \frac{3}{2} \cdot 2^{2b}$$

Expressed in decibels (dB), the SQNR is

$$\text{SQNR(dB)} = 10 \log_{10} \text{SQNR} = 1.76 + 6.02b \tag{1.4.32}$$

This implies that the SQNR increases approximately 6 dB for every bit added to the word length, that is, for each doubling of the quantization levels.

Although formula (1.4.32) was derived for sinusoidal signals, we shall see in Chapter 9 that a similar result holds for every signal whose dynamic range spans the range of the quantizer. This relationship is extremely important because it dictates

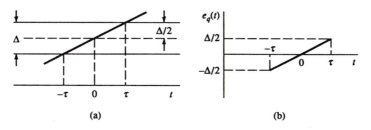

Figure 1.22   The quantization error $e_q(t) = x_a(t) - x_q(t)$.

the number of bits required by a specific application to assure a given signal-to-noise ratio. For example, most compact disc players use a sampling frequency of 44.1 kHz and 16-bit sample resolution, which implies a SQNR of more than 96 dB.

1.4.5 Coding of Quantized Samples

The coding process in an A/D converter assigns a unique binary number to each quantization level. If we have L levels we need at least L different binary numbers. With a word length of b bits we can create 2^b different binary numbers. Hence we have $2^b \geq L$, or equivalently, $b \geq \log_2 L$. Thus the number of bits required in the coder is the smallest integer greater than or equal to $\log_2 L$. In our example it can easily be seen that we need a coder with $b = 4$ bits. Commercially available A/D converters may be obtained with finite precision of $b = 16$ or less. Generally, the higher the sampling speed and the finer the quantization, the more expensive the device becomes.

1.4.6 Digital-to-Analog Conversion

To convert a digital signal into an analog signal we can use a digital-to-analog (D/A) converter. As stated previously, the task of a D/A converter is to interpolate between samples.

The sampling theorem specifies the optimum interpolation for a bandlimited signal. However, this type of interpolation is too complicated and, hence impractical, as indicated previously. From a practical viewpoint, the simplest D/A converter is the zero-order hold shown in Fig. 1.15, which simply holds constant the value of one sample until the next one is received. Additional improvement can be obtained by using linear interpolation as shown in Fig. 1.23 to connect successive samples with straight-line segments. The zero-order hold and linear interpolator are analyzed in Section 9.3. Better interpolation can be achieved by using more sophisticated higher-order interpolation techniques.

In general, suboptimum interpolation techniques result in passing frequencies above the folding frequency. Such frequency components are undesirable and are usually removed by passing the output of the interpolator through a proper analog

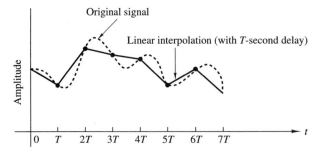

Figure 1.23   Linear point connector (with T-second delay).

filter, which is called a *postfilter* or *smoothing filter*. Thus D/A conversion usually involves a suboptimum interpolator followed by a postfilter. D/A converters are treated in more detail in Section 9.3.

1.4.7 Analysis of Digital Signals and Systems Versus Discrete-Time Signals and Systems

We have seen that a digital signal is defined as a function of an integer independent variable and its values are taken from a finite set of possible values. The usefulness of such signals is a consequence of the possibilities offered by digital computers. Computers operate on numbers, which are represented by a string of 0's and 1's. The length of this string (*word length*) is fixed and finite and usually is 8, 12, 16, or 32 bits. The effects of finite word length in computations cause complications in the analysis of digital signal processing systems. To avoid these complications, we neglect the quantized nature of digital signals and systems in much of our analysis and consider them as discrete-time signals and systems.

In Chapters 6, 7, and 9 we investigate the consequences of using a finite word length. This is an important topic, since many digital signal processing problems are solved with small computers or microprocessors that employ fixed-point arithmetic. Consequently, one must look carefully at the problem of finite-precision arithmetic and account for it in the design of software and hardware that performs the desired signal processing tasks.

1.5 SUMMARY AND REFERENCES

In this introductory chapter we have attempted to provide the motivation for digital signal processing as an alternative to analog signal processing. We presented the basic elements of a digital signal processing system and defined the operations needed to convert an analog signal into a digital signal ready for processing. Of particular importance is the sampling theorem, which was introduced by Nyquist (1928) and later popularized in the classic paper by Shannon (1949). The sampling theorem as described in Section 1.4.2 is derived in Chapter 4. Sinusoidal signals were introduced primarily for the purpose of illustrating the aliasing phenomenon and for the subsequent development of the sampling theorem.

Quantization effects that are inherent in the A/D conversion of a signal were also introduced in this chapter. Signal quantization is best treated in statistical terms, as described in Chapters 6, 7, and 9.

Finally, the topic of signal reconstruction, or D/A conversion, was described briefly. Signal reconstruction based on staircase or linear interpolation methods is treated in Section 9.3.

There are numerous practical applications of digital signal processing. The book edited by Oppenheim (1978) treats applications to speech processing, image processing, radar signal processing, sonar signal processing, and geophysical signal processing.

PROBLEMS

1.1 Classify the following signals according to whether they are (1) one- or multi-dimensional; (2) single or multichannel, (3) continuous time or discrete time, and (4) analog or digital (in amplitude). Give a brief explanation.

    (a) Closing prices of utility stocks on the New York Stock Exchange.

    (b) A color movie.

    (c) Position of the steering wheel of a car in motion relative to car's reference frame.

    (d) Position of the steering wheel of a car in motion relative to ground reference frame.

    (e) Weight and height measurements of a child taken every month.

1.2 Determine which of the following sinusoids are periodic and compute their fundamental period.

    (a) $\cos 0.01\pi n$ **(b)** $\cos\left(\pi\dfrac{30n}{105}\right)$ **(c)** $\cos 3\pi n$ **(d)** $\sin 3n$ **(e)** $\sin\left(\pi\dfrac{62n}{10}\right)$

1.3 Determine whether or not each of the following signals is periodic. In case a signal is periodic, specify its fundamental period.

    (a) $x_a(t) = 3\cos(5t + \pi/6)$

    (b) $x(n) = 3\cos(5n + \pi/6)$

    (c) $x(n) = 2\exp[j(n/6 - \pi)]$

    (d) $x(n) = \cos(n/8)\cos(\pi n/8)$

    (e) $x(n) = \cos(\pi n/2) - \sin(\pi n/8) + 3\cos(\pi n/4 + \pi/3)$

1.4 (a) Show that the fundamental period N_p of the signals

$$s_k(n) = e^{j2\pi kn/N}, \qquad k = 0, 1, 2, \ldots$$

    is given by $N_p = N/GCD(k, N)$, where GCD is the greatest common divisor of k and N.

    (b) What is the fundamental period of this set for $N = 7$?

    (c) What is it for $N = 16$?

1.5 Consider the following analog sinusoidal signal:

$$x_a(t) = 3\sin(100\pi t)$$

    (a) Sketch the signal $x_a(t)$ for $0 \le t \le 30$ ms.

    (b) The signal $x_a(t)$ is sampled with a sampling rate $F_s = 300$ samples/s. Determine the frequency of the discrete-time signal $x(n) = x_a(nT)$, $T = 1/F_s$, and show that it is periodic.

    (c) Compute the sample values in one period of $x(n)$. Sketch $x(n)$ on the same diagram with $x_a(t)$. What is the period of the discrete-time signal in milliseconds?

    (d) Can you find a sampling rate F_s such that the signal $x(n)$ reaches its peak value of 3? What is the minimum F_s suitable for this task?

1.6 A continuous-time sinusoid $x_a(t)$ with fundamental period $T_p = 1/F_0$ is sampled at a rate $F_s = 1/T$ to produce a discrete-time sinusoid $x(n) = x_a(nT)$.

    (a) Show that $x(n)$ is periodic if $T/T_p = k/N$ (i.e., T/T_p is a rational number).

    (b) If $x(n)$ is periodic, what is its fundamental period T_p in seconds?

(c) Explain the statement: $x(n)$ is periodic if its fundamental period T_p, in seconds, is equal to an integer number of periods of $x_a(t)$.

1.7 An analog signal contains frequencies up to 10 kHz.
  (a) What range of sampling frequencies allows exact reconstruction of this signal from its samples?
  (b) Suppose that we sample this signal with a sampling frequency $F_s = 8$ kHz. Examine what happens to the frequency $F_1 = 5$ kHz.
  (c) Repeat part (b) for a frequency $F_2 = 9$ kHz.

1.8 An analog electrocardiogram (ECG) signal contains useful frequencies up to 100 Hz.
  (a) What is the Nyquist rate for this signal?
  (b) Suppose that we sample this signal at a rate of 250 samples/s. What is the highest frequency that can be represented uniquely at this sampling rate?

1.9 An analog signal $x_a(t) = \sin(480\pi t) + 3\sin(720\pi t)$ is sampled 600 times per second.
  (a) Determine the Nyquist sampling rate for $x_a(t)$.
  (b) Determine the folding frequency.
  (c) What are the frequencies, in radians, in the resulting discrete time signal $x(n)$?
  (d) If $x(n)$ is passed through an ideal D/A converter, what is the reconstructed signal $y_a(t)$?

1.10 A digital communication link carries binary-coded words representing samples of an input signal

$$x_a(t) = 3\cos 600\pi t + 2\cos 1800\pi t$$

The link is operated at 10,000 bits/s and each input sample is quantized into 1024 different voltage levels.
  (a) What is the sampling frequency and the folding frequency?
  (b) What is the Nyquist rate for the signal $x_a(t)$?
  (c) What are the frequencies in the resulting discrete-time signal $x(n)$?
  (d) What is the resolution Δ?

1.11 Consider the simple signal processing system shown in Fig. P1.11. The sampling periods of the A/D and D/A converters are $T = 5$ ms and $T' = 1$ ms, respectively. Determine the output $y_a(t)$ of the system, if the input is

$$x_a(t) = 3\cos 100\pi t + 2\sin 250\pi t \qquad (t \text{ in seconds})$$

The postfilter removes any frequency component above $F_s/2$.

Figure P1.11

1.12 (a) Derive the expression for the discrete-time signal $x(n)$ in Example 1.4.2 using the periodicity properties of sinusoidal functions.
  (b) What is the analog signal we can obtain from $x(n)$ if in the reconstruction process we assume that $F_s = 10$ kHz?

1.13 The discrete-time signal $x(n) = 6.35 \cos(\pi/10)n$ is quantized with a resolution (a) $\Delta = 0.1$ or (b) $\Delta = 0.02$. How many bits are required in the A/D converter in each case?

1.14 Determine the bit rate and the resolution in the sampling of a seismic signal with dynamic range of 1 volt if the sampling rate is $F_s = 20$ samples/s and we use an 8-bit A/D converter? What is the maximum frequency that can be present in the resulting digital seismic signal?

1.15* *Sampling of sinusoidal signals: aliasing* Consider the following continuous-time sinusoidal signal

$$x_a(t) = \sin 2\pi F_0 t, \qquad -\infty < t < \infty$$

Since $x_a(t)$ is described mathematically, its sampled version can be described by values every T seconds. The sampled signal is described by the formula

$$x(n) = x_a(nT) = \sin 2\pi \frac{F_0}{F_s} n, \qquad -\infty < n < \infty$$

where $F_s = 1/T$ is the sampling frequency.
(a) Plot the signal $x(n)$, $0 \le n \le 99$ for $F_s = 5$ kHz and $F_0 = 0.5, 2, 3,$ and 4.5 kHz. Explain the similarities and differences among the various plots.
(b) Suppose that $F_0 = 2$ kHz and $F_s = 50$ kHz.
(1) Plot the signal $x(n)$. What is the frequency f_0 of the signal $x(n)$?
(2) Plot the signal $y(n)$ created by taking the even-numbered samples of $x(n)$. Is this a sinusoidal signal? Why? If so, what is its frequency?

1.16* *Quantization error in A/D conversion of a sinuoidal signal* Let $x_q(n)$ be the signal obtained by quantizing the signal $x(n) = \sin 2\pi f_0 n$. The quantization error power P_q is defined by

$$P_q = \frac{1}{N} \sum_{n=0}^{N-1} e^2(n) = \frac{1}{N} \sum_{n=0}^{N-1} [x_q(n) - x(n)]^2$$

The "quality" of the quantized signal can be measured by the signal-to-quantization noise ratio (SQNR) defined by

$$\text{SQNR} = 10 \log_{10} \frac{P_x}{P_q}$$

where P_x is the power of the unquantized signal $x(n)$.
(a) For $f_0 = 1/50$ and $N = 200$, write a program to quantize the signal $x(n)$, using truncation, to 64, 128, and 256 quantization levels. In each case plot the signals $x(n)$, $x_q(n)$, and $e(n)$ and compute the corresponding SQNR.
(b) Repeat part (a) by using rounding instead of truncation.
(c) Comment on the results obtained in parts (a) and (b).
(d) Compare the experimentally measured SQNR with the theoretical SQNR predicted by formula (1.4.32) and comment on the differences and similarities.

2

Discrete-Time Signals and Systems

In Chapter 1 we introduced the reader to a number of important types of signals and described the sampling process by which an analog signal is converted to a discrete-time signal. In addition, we presented in some detail the characteristics of discrete-time sinusoidal signals. The sinusoid is an important elementary signal that serves as a basic building block in more complex signals. However, there are other elementary signals that are important in our treatment of signal processing. These discrete-time signals are introduced in this chapter and are used as basis functions or building blocks to describe more complex signals.

The major emphasis in this chapter is the characterization of discrete-time systems in general and the class of linear time-invariant (LTI) systems in particular. A number of important time-domain properties of LTI systems are defined and developed, and an important formula, called the convolution formula, is derived which allows us to determine the output of an LTI system to any given arbitrary input signal. In addition to the convolution formula, difference equations are introduced as an alternative method for describing the input–output relationship of an LTI system, and in addition, recursive and nonrecursive realizations of LTI systems are treated.

Our motivation for the emphasis on the study of LTI systems is twofold. First, there is a large collection of mathematical techniques that can be applied to the analysis of LTI systems. Second, many practical systems are either LTI systems or can be approximated by LTI systems. Because of its importance in digital signal processing applications and its close resemblance to the convolution formula, we also introduce the correlation between two signals. The autocorrelation and crosscorrelation of signals are defined and their properties are presented.

2.1 DISCRETE-TIME SIGNALS

As we discussed in Chapter 1, a discrete-time signal $x(n)$ is a function of an independent variable that is an integer. It is graphically represented as in Fig. 2.1. It is important to note that a discrete-time signal is *not defined* at instants between

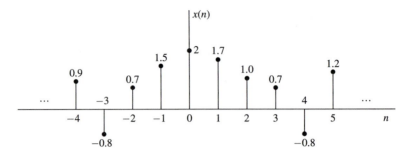

Figure 2.1   Graphical representation of a discrete-time signal.

two successive samples. Also, it is incorrect to think that $x(n)$ is equal to zero if n is not an integer. Simply, the signal $x(n)$ is not defined for noninteger values of n.

In the sequel we will assume that a discrete-time signal is defined for every integer value n for $-\infty < n < \infty$. By tradition, we refer to $x(n)$ as the "nth sample" of the signal even if the signal $x(n)$ is inherently discrete time (i.e., not obtained by sampling an analog signal). If, indeed, $x(n)$ was obtained from sampling an analog signal $x_a(t)$, then $x(n) \equiv x_a(nT)$, where T is the sampling period (i.e., the time between successive samples).

Besides the graphical representation of a discrete-time signal or sequence as illustrated in Fig. 2.1, there are some alternative representations that are often more convenient to use. These are:

1. Functional representation, such as

$$x(n) = \begin{cases} 1, & \text{for } n = 1, 3 \\ 4, & \text{for } n = 2 \\ 0, & \text{elsewhere} \end{cases} \tag{2.1.1}$$

2. Tabular representation, such as

n	$\cdots$	-2	-1	0	1	2	3	4	5	$\cdots$
$x(n)$	$\cdots$	0	0	0	1	4	1	0	0	$\cdots$

3. Sequence representation

An infinite-duration signal or sequence with the time origin ($n = 0$) indicated by the symbol $\uparrow$ is represented as

$$x(n) = \{\ldots 0, 0, 1, 4, 1, 0, 0, \ldots\} \tag{2.1.2}$$
$$\phantom{x(n) = \{\ldots 0, 0, }\uparrow$$

A sequence $x(n)$, which is zero for $n < 0$, can be represented as

$$x(n) = \{0, 1, 4, 1, 0, 0, \ldots\} \tag{2.1.3}$$
$$\phantom{x(n) = \{}\uparrow$$

The time origin for a sequence $x(n)$, which is zero for $n < 0$, is understood to be the first (leftmost) point in the sequence.

A finite-duration sequence can be represented as

$$x(n) = \{3, -1, -2, 5, 0, 4, -1\} \qquad (2.1.4)$$
$$\uparrow$$

whereas a finite-duration sequence that satisfies the condition $x(n) = 0$ for $n < 0$ can be represented as

$$x(n) = \{0, 1, 4, 1\} \qquad (2.1.5)$$
$$\uparrow$$

The signal in (2.1.4) consists of seven samples or points (in time), so it is called or identified as a seven-point sequence. Similarly, the sequence given by (2.1.5) is a four-point sequence.

2.1.1 Some Elementary Discrete-Time Signals

In our study of discrete-time signals and systems there are a number of basic signals that appear often and play an important role. These signals are defined below.

1. The *unit sample sequence* is denoted as $\delta(n)$ and is defined as

$$\delta(n) \equiv \begin{cases} 1, & \text{for } n = 0 \\ 0, & \text{for } n \neq 0 \end{cases} \qquad (2.1.6)$$

In words, the unit sample sequence is a signal that is zero everywhere, except at $n = 0$ where its value is unity. This signal is sometimes referred to as a *unit impulse*. In contrast to the analog signal $\delta(t)$, which is also called a unit impulse and is defined to be zero everywhere except $t = 0$, and has unit area, the unit sample sequence is much less mathematically complicated. The graphical representation of $\delta(n)$ is shown in Fig. 2.2.

2. The *unit step signal* is denoted as $u(n)$ and is defined as

$$u(n) \equiv \begin{cases} 1, & \text{for } n \geq 0 \\ 0, & \text{for } n < 0 \end{cases} \qquad (2.1.7)$$

Figure 2.3 illustrates the unit step signal.

3. The *unit ramp signal* is denoted as $u_r(n)$ and is defined as

$$u_r(n) \equiv \begin{cases} n, & \text{for } n \geq 0 \\ 0, & \text{for } n < 0 \end{cases} \qquad (2.1.8)$$

This signal is illustrated in Fig. 2.4.

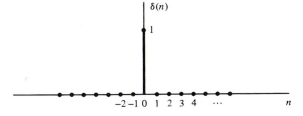

Figure 2.2 Graphical representation of the unit sample signal.

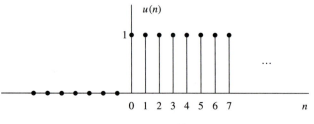

Figure 2.3 Graphical representation of the unit step signal.

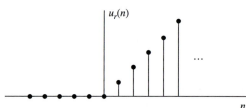

Figure 2.4 Graphical representation of the unit ramp signal.

4. The *exponential signal* is a sequence of the form

$$x(n) = a^n \qquad \text{for all } n \qquad (2.1.9)$$

If the parameter a is real, then $x(n)$ is a real signal. Figure 2.5 illustrates $x(n)$ for various values of the parameter a.

When the parameter a is complex valued, it can be expressed as

$$a \equiv re^{j\theta}$$

where r and θ are now the parameters. Hence we can express $x(n)$ as

$$\begin{aligned} x(n) &= r^n e^{j\theta n} \\ &= r^n(\cos\theta n + j\sin\theta n) \end{aligned} \qquad (2.1.10)$$

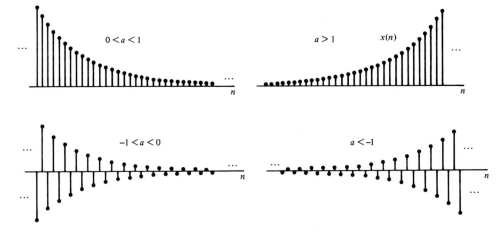

Figure 2.5 Graphical representation of exponential signals.

Since $x(n)$ is now complex valued, it can be represented graphically by plotting the real part

$$x_R(n) \equiv r^n \cos \theta n \qquad (2.1.11)$$

as a function of n, and separately plotting the imaginary part

$$x_I(n) \equiv r^n \sin \theta n \qquad (2.1.12)$$

as a function of n. Figure 2.6 illustrates the graphs of $x_R(n)$ and $x_I(n)$ for $r = 0.9$ and $\theta = \pi/10$. We observe that the signals $x_R(n)$ and $x_I(n)$ are a damped (decaying exponential) cosine function and a damped sine function. The angle variable θ is simply the frequency of the sinusoid, previously denoted by the (normalized) frequency variable ω. Clearly, if $r = 1$, the damping disappears and $x_R(n)$, $x_I(n)$, and $x(n)$ have a fixed amplitude, which is unity.

Alternatively, the signal $x(n)$ given by (2.1.10) can be represented graphically by the amplitude function

$$|x(n)| = A(n) \equiv r^n \qquad (2.1.13)$$

and the phase function

$$\angle x(n) = \phi(n) \equiv \theta n \qquad (2.1.14)$$

Figure 2.7 illustrates $A(n)$ and $\phi(n)$ for $r = 0.9$ and $\theta = \pi/10$. We observe that the phase function is linear with n. However, the phase is defined only over the interval $-\pi < \theta \leq \pi$ or, equivalently, over the interval $0 \leq \theta < 2\pi$. Consequently, by convention $\phi(n)$ is plotted over the finite interval $-\pi < \theta \leq \pi$ or $0 \leq \theta < 2\pi$. In other words, we subtract multiplies of 2π from $\phi(n)$ before plotting. In one case, $\phi(n)$ is constrained to the range $-\pi < \theta \leq \pi$ and in the other case $\phi(n)$ is constrained to the range $0 \leq \theta < 2\pi$. The subtraction of multiples of 2π from $\phi(n)$ is equivalent to interpreting the function $\phi(n)$ as $\phi(n)$, modulo 2π. The graph for $\phi(n)$, modulo 2π, is shown in Fig. 2.7b.

2.1.2 Classification of Discrete-Time Signals

The mathematical methods employed in the analysis of discrete-time signals and systems depend on the characteristics of the signals. In this section we classify discrete-time signals according to a number of different characteristics.

Energy signals and power signals. The energy E of a signal $x(n)$ is defined as

$$E \equiv \sum_{n=-\infty}^{\infty} |x(n)|^2 \qquad (2.1.15)$$

We have used the magnitude-squared values of $x(n)$, so that our definition applies to complex-valued signals as well as real-valued signals. The energy of a signal can be finite or infinite. If E is finite (i.e., $0 < E < \infty$), then $x(n)$ is called an *energy*

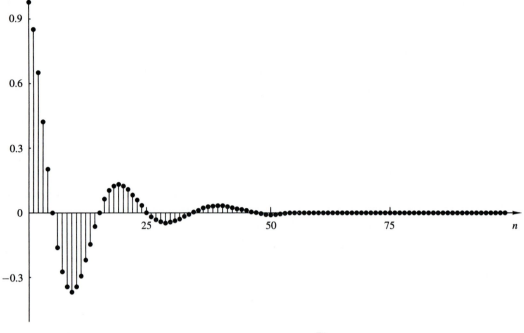

(a) Graph of $x_R(n) = (0.9)^n \cos \frac{\pi n}{10}$

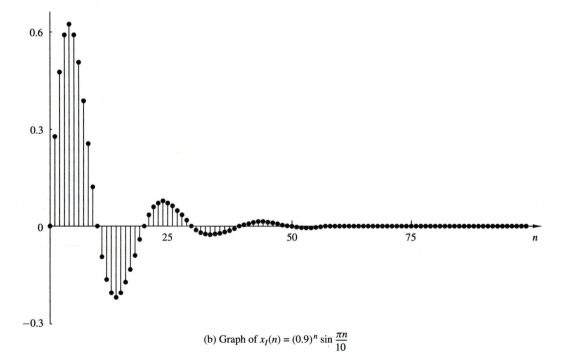

(b) Graph of $x_I(n) = (0.9)^n \sin \frac{\pi n}{10}$

Figure 2.6 Graph of the real and imaginary components of a complex-valued exponential signal.

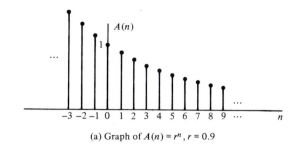

(a) Graph of $A(n) = r^n$, $r = 0.9$

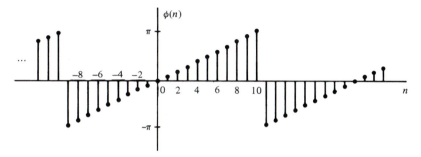

(b) Graph of $\phi(n) = \frac{\pi}{10}n$, modulo 2π plotted in the range $(-\pi, \pi)$

Figure 2.7   Graph of amplitude and phase function of a complex-valued exponential signal: (a) graph of $A(n) = r^n$, $4 = 0.9$; (b) graph of $\phi(n) = (\pi/10)n$, modulo 2π plotted in the range $(-\pi, \pi]$.

signal. Sometimes we add a subscript x to E and write E_x to emphasize that E_x is the energy of the signal $x(n)$.

Many signals that possess infinite energy, have a finite average power. The average power of a discrete-time signal $x(n)$ is defined as

$$P = \lim_{N \to \infty} \frac{1}{2N+1} \sum_{n=-N}^{N} |x(n)|^2 \tag{2.1.16}$$

If we define the signal energy of $x(n)$ over the finite interval $-N \le n \le N$ as

$$E_N \equiv \sum_{n=-N}^{N} |x(n)|^2 \tag{2.1.17}$$

then we can express the signal energy E as

$$E \equiv \lim_{N \to \infty} E_N \tag{2.1.18}$$

and the average power of the signal $x(n)$ as

$$P \equiv \lim_{N \to \infty} \frac{1}{2N+1} E_N \tag{2.1.19}$$

Clearly, if E is finite, $P = 0$. On the other hand, if E is infinite, the average power P may be either finite or infinite. If P is finite (and nonzero), the signal is called a *power signal*. The following example illustrates such a signal.

Example 2.1.1

Determine the power and energy of the unit step sequence. The average power of the unit step signal is

$$P = \lim_{N \to \infty} \frac{1}{2N+1} \sum_{n=0}^{N} u^2(n)$$

$$= \lim_{N \to \infty} \frac{N+1}{2N+1} = \lim_{N \to \infty} \frac{1 + 1/N}{2 + 1/N} = \frac{1}{2}$$

Consequently, the unit step sequence is a power signal. Its energy is infinite.

Similarly, it can be shown that the complex exponential sequence $x(n) = Ae^{j\omega_0 n}$ has average power A^2, so it is a power signal. On the other hand, the unit ramp sequence is neither a power signal nor an energy signal.

Periodic signals and aperiodic signals. As defined on Section 1.3, a signal $x(n)$ is periodic with period $N(N > 0)$ if and only if

$$x(n + N) = x(n) \text{ for all } n \tag{2.1.20}$$

The smallest value of N for which (2.1.20) holds is called the (fundamental) period. If there is no value of N that satisfies (2.1.20), the signal is called *nonperiodic* or *aperiodic*.

We have already observed that the sinusoidal signal of the form

$$x(n) = A \sin 2\pi f_0 n \tag{2.1.21}$$

is periodic when f_0 is a rational number, that is, if f_0 can be expressed as

$$f_0 = \frac{k}{N} \tag{2.1.22}$$

where k and N are integers.

The energy of a periodic signal $x(n)$ over a single period, say, over the interval $0 \le n \le N - 1$, is finite if $x(n)$ takes on finite values over the period. However, the energy of the periodic signal for $-\infty \le n \le \infty$ is infinite. On the other hand, the average power of the periodic signal is finite and it is equal to the average power over a single period. Thus if $x(n)$ is a periodic signal with fundamental period N and takes on finite values, its power is given by

$$P = \frac{1}{N} \sum_{n=0}^{N-1} |x(n)|^2 \tag{2.1.23}$$

Consequently, periodic signals are power signals.

Symmetric (even) and antisymmetric (odd) signals.   A real-valued signal $x(n)$ is called symmetric (even) if

$$x(-n) = x(n) \tag{2.1.24}$$

On the other hand, a signal $x(n)$ is called antisymmetric (odd) if

$$x(-n) = -x(n) \tag{2.1.25}$$

We note that if $x(n)$ is odd, then $x(0) = 0$. Examples of signals with even and odd symmetry are illustrated in Fig. 2.8.

We wish to illustrate that any arbitrary signal can be expressed as the sum of two signal components, one of which is even and the other odd. The even signal component is formed by adding $x(n)$ to $x(-n)$ and dividing by 2, that is,

$$x_e(n) = \tfrac{1}{2}[x(n) + x(-n)] \tag{2.1.26}$$

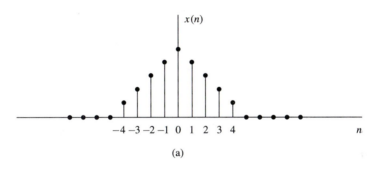

(a)

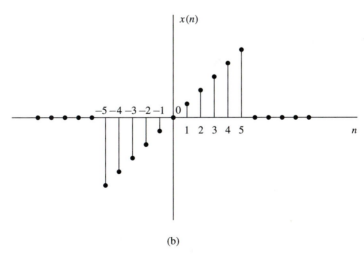

(b)

Figure 2.8 Example of even (a) and odd (b) signals.

Clearly, $x_e(n)$ satisfies the symmetry condition (2.1.24). Similarly, we form an odd signal component $x_o(n)$ according to the relation

$$x_o(n) = \tfrac{1}{2}[x(n) - x(-n)] \qquad (2.1.27)$$

Again, it is clear that $x_o(n)$ satisfies (2.1.25); hence it is indeed odd. Now, if we add the two signal components, defined by (2.1.26) and (2.1.27), we obtain $x(n)$, that is,

$$x(n) = x_e(n) + x_o(n) \qquad (2.1.28)$$

Thus any arbitrary signal can be expressed as in (2.1.28).

2.1.3 Simple Manipulations of Discrete-Time Signals

In this section we consider some simple modifications or manipulations involving the independent variable and the signal amplitude (dependent variable).

Transformation of the independent variable (time).   A signal $x(n)$ may be shifted in time by replacing the independent variable n by $n - k$, where k is an integer. If k is a positive integer, the time shift results in a delay of the signal by k units of time. If k is a negative integer, the time shift results in an advance of the signal by $|k|$ units in time.

Example 2.1.2

A signal $x(n)$ is graphically illustrated in Fig. 2.9a. Show a graphical representation of the signals $x(n-3)$ and $x(n+2)$.

Solution   The signal $x(n-3)$ is obtained by delaying $x(n)$ by three units in time. The result is illustrated in Fig. 2.9b. On the other hand, the signal $x(n+2)$ is obtained by advancing $x(n)$ by two units in time. The result is illustrated in Fig. 2.9c. Note that delay corresponds to shifting a signal to the right, whereas advance implies shifting the signal to the left on the time axis.

If the signal $x(n)$ is stored on magnetic tape or on a disk or, perhaps, in the memory of a computer, it is a relatively simple operation to modify the base by introducing a delay or an advance. On the other hand, if the signal is not stored but is being generated by some physical phenomenon in real time, it is not possible to advance the signal in time, since such an operation involves signal samples that have not yet been generated. Whereas it is always possible to insert a delay into signal samples that have already been generated, it is physically impossible to view the future signal samples. Consequently, in real-time signal processing applications, the operation of advancing the time base of the signal is physically unrealizable.

Another useful modification of the time base is to replace the independent variable n by $-n$. The result of this operation is a *folding* or a *reflection* of the signal about the time origin $n = 0$.

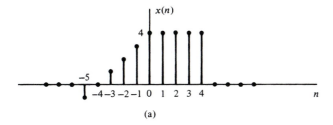

(a)

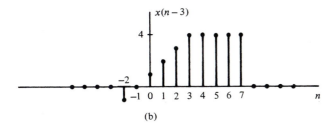

(b)

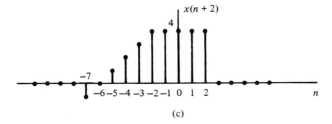

(c)

Figure 2.9 Graphical representation of a signal, and its delayed and advanced versions.

Example 2.1.3

Show the graphical representation of the signal $x(-n)$ and $x(-n+2)$, where $x(n)$ is the signal illustrated in Fig. 2.10a.

Solution   The new signal $y(n) = x(-n)$ is shown in Fig. 2.10b. Note that $y(0) = x(0)$, $y(1) = x(-1)$, $y(2) = x(-2)$, and so on. Also, $y(-1) = x(1)$, $y(-2) = x(2)$, and so on. Therefore, $y(n)$ is simply $x(n)$ reflected or folded about the time origin $n = 0$. The signal $y(n) = x(-n+2)$ is simply $x(-n)$ delayed by two units in time. The resulting signal is illustrated in Fig. 2.10c. A simple way to verify that the result in Fig. 2.10c is correct is to compute samples, such as $y(0) = x(2)$, $y(1) = x(1)$, $y(2) = x(0)$, $y(-1) = x(3)$, and so on.

It is important to note that the operations of folding and time delaying (or advancing) a signal are not commutative. If we denote the time-delay operation by TD and the folding operation by FD, we can write

$$\text{TD}_k[x(n)] = x(n-k) \qquad k > 0$$
$$\text{FD}[x(n)] = x(-n) \tag{2.1.29}$$

Now

$$\text{TD}_k\{\text{FD}[x(n)]\} = \text{TD}_k[x(-n)] = x(-n+k) \tag{2.1.30}$$

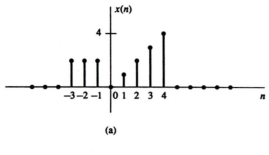

(a)

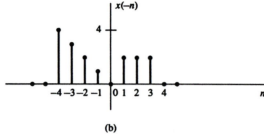

(b)

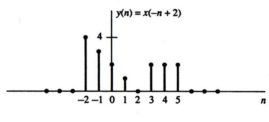

(c)

Figure 2.10 Graphical illustration of the folding and shifting operations.

whereas

$$FD\{TD_k[x(n)]\} = FD[x(n - k)] = x(-n - k) \qquad (2.1.31)$$

Note that because the signs of n and k in $x(n-k)$ and $x(-n+k)$ are different, the result is a shift of the signals $x(n)$ and $x(-n)$ to the right by k samples, corresponding to a time delay.

A third modification of the independent variable involves replacing n by μn, where μ is an integer. We refer to this time-base modification as *time scaling* or *down-sampling*.

Example 2.1.4

Show the graphical representation of the signal $y(n) = x(2n)$, where $x(n)$ is the signal illustrated in Fig. 2.11a.

Solution We note that the signal $y(n)$ is obtained from $x(n)$ by taking every other sample from $x(n)$, starting with $x(0)$. Thus $y(0) = x(0)$, $y(1) = x(2)$, $y(2) = x(4), \ldots$ and $y(-1) = x(-2)$, $y(-2) = x(-4)$, and so on. In other words, we have skipped

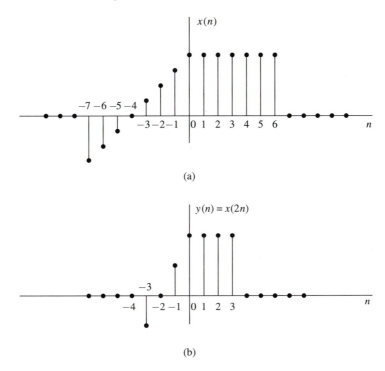

Figure 2.11   Graphical illustration of down-sampling operation.

the odd-numbered samples in $x(n)$ and retained the even-numbered samples. The resulting signal is illustrated in Fig. 2.11b.

If the signal $x(n)$ was originally obtained by sampling an analog signal $x_a(t)$, then $x(n) = x_a(nT)$, where T is the sampling interval. Now, $y(n) = x(2n) = x_a(2Tn)$. Hence the time-scaling operation described in Example 2.1.4 is equivalent to changing the sampling rate from $1/T$ to $1/2T$, that is, to decreasing the rate by a factor of 2. This is a *downsampling* operation.

Addition, multiplication, and scaling of sequences.   Amplitude modifications include *addition, multiplication,* and *scaling* of discrete-time signals.

Amplitude scaling of a signal by a constant A is accomplished by multiplying the value of every signal sample by A. Consequently, we obtain

$$y(n) = Ax(n) \qquad -\infty < n < \infty$$

The *sum* of two signals $x_1(n)$ and $x_2(n)$ is a signal $y(n)$, whose value at any instant is equal to the sum of the values of these two signals at that instant, that is,

$$y(n) = x_1(n) + x_2(n) \qquad -\infty < n < \infty$$

The *product* of two signals is similarly defined on a sample-to-sample basis as

$$y(n) = x_1(n)x_2(n) \qquad -\infty < n < \infty$$

2.2 DISCRETE-TIME SYSTEMS

In many applications of digital signal processing we wish to design a device or an algorithm that performs some prescribed operation on a discrete-time signal. Such a device or algorithm is called a discrete-time system. More specifically, a *discrete-time system* is a device or algorithm that operates on a discrete-time signal, called the *input* or *excitation*, according to some well-defined rule, to produce another discrete-time signal called the *output* or *response* of the system. In general, we view a system as an operation or a set of operations performed on the input signal $x(n)$ to produce the output signal $y(n)$. We say that the input signal $x(n)$ is *transformed* by the system into a signal $y(n)$, and express the general relationship between $x(n)$ and $y(n)$ as

$$y(n) \equiv T[x(n)] \tag{2.2.1}$$

where the symbol T denotes the transformation (also called an operator), or processing performed by the system on $x(n)$ to produce $y(n)$. The mathematical relationship in (2.2.1) is depicted graphically in Fig. 2.12.

There are various ways to describe the characteristics of the system and the operation it performs on $x(n)$ to produce $y(n)$. In this chapter we shall be concerned with the time-domain characterization of systems. We shall begin with an input–output description of the system. The input–output description focuses on the behavior at the terminals of the system and ignores the detailed internal construction or realization of the system. Later, in Section 7.5, we introduce the state-space description of a system. In this description we develop mathematical equations that not only describe the input–output behavior of the system but specify its internal behavior and structure.

2.2.1 Input–Output Description of Systems

The input–output description of a discrete-time system consists of a mathematical expression or a rule, which explicitly defines the relation between the input and output signals (*input–output relationship*). The exact internal structure of the system is either unknown or ignored. Thus the only way to interact with the system is by using its input and output terminals (i.e., the system is assumed to be a "black box" to the user). To reflect this philosophy, we use the graphical representa-

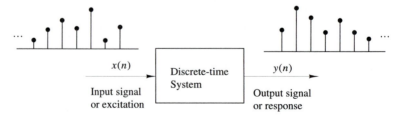

Figure 2.12   Block diagram representation of a discrete-time system.

tion depicted in Fig. 2.12, and the general input–output relationship in (2.2.1) or, alternatively, the notation

$$x(n) \xrightarrow{\ T\ } y(n) \tag{2.2.2}$$

which simply means that $y(n)$ is the response of the system T to the excitation $x(n)$. The following examples illustrate several different systems.

Example 2.2.1

Determine the response of the following sytems to the input signal

$$x(n) = \begin{cases} |n|, & -3 \le n \le 3 \\ 0, & \text{otherwise} \end{cases}$$

(a) $y(n) = x(n)$

(b) $y(n) = x(n-1)$

(c) $y(n) = x(n+1)$

(d) $y(n) = \frac{1}{3}[x(n+1) + x(n) + x(n-1)]$

(e) $y(n) = max\{x(n+1), x(n), x(n-1)\}$

(f) $y(n) = \sum_{k=-\infty}^{n} x(k) = x(n) + x(n-1) + x(n-2) + \cdots$ (2.2.3)

Solution   First, we determine explicitly the sample values of the input signal

$$x(n) = \{\ldots, 0, 3, 2, 1, 0, 1, 2, 3, 0, \ldots\}$$
$$\uparrow$$

Next, we determine the output of each system using its input–output relationship.

(a) In this case the output is exactly the same as the input signal. Such a system is known as the *identity* system.

(b) This system simply delays the input by one sample. Thus its output is given by

$$x(n) = \{\ldots, 0, 3, 2, 1, 0, 1, 2, 3, 0, \ldots\}$$
$$\uparrow$$

(c) In this case the system "advances" the input one sample into the future. For example, the value of the output at time $n = 0$ is $y(0) = x(1)$. The response of this system to the given input is

$$x(n) = \{\ldots, 0, 3, 2, 1, 0, 1, 2, 3, 0, \ldots\}$$
$$\uparrow$$

(d) The output of this system at any time is the mean value of the present, the immediate past, and the immediate future samples. For example, the output at time $n = 0$ is

$$y(0) = \frac{1}{3}[x(-1) + x(0) + x(1)] = \frac{1}{3}[1 + 0 + 1] = \frac{2}{3}$$

Repeating this computation for every value of n, we obtain the output signal

$$y(n) = \{\ldots, 0, 1, \tfrac{5}{3}, 2, 1, \tfrac{2}{3}, 1, 2, \tfrac{5}{3}, 1, 0, \ldots\}$$
$$\uparrow$$

(e) This system selects as its output at time n the maximum value of the three input samples $x(n-1)$, $x(n)$, and $x(n+1)$. Thus the response of this system to the input signal $x(n)$ is

$$y(n) = \{0, 3, 3, 3, 2, 1, 2, 3, 3, 3, 0, \ldots\}$$
$$\uparrow$$

(f) This system is basically an *accumulator* that computes the running sum of all the past input values up to present time. The response of this system to the given input is

$$y(n) = \{\ldots, 0, 3, 5, 6, 6, 7, 9, 12, 0, \ldots\}$$
$$\uparrow$$

We observe that for several of the systems considered in Example 2.2.1 the output at time $n = n_0$ depends not only on the value of the input at $n = n_0$ [i.e., $x(n_0)$], but also on the values of the input applied to the system before and after $n = n_0$. Consider, for instance, the accumulator in the example. We see that the output at time $n = n_0$ depends not only on the input at time $n = n_0$, but also on $x(n)$ at times $n = n_0 - 1$, $n_0 - 2$, and so on. By a simple algebraic manipulation the input–output relation of the accumulator can be written as

$$y(n) = \sum_{k=-\infty}^{n} x(k) = \sum_{k=-\infty}^{n-1} x(k) + x(n) \qquad (2.2.4)$$
$$= y(n-1) + x(n)$$

which justifies the term *accumulator*. Indeed, the system computes the current value of the output by adding (accumulating) the current value of the input to the previous output value.

There are some interesting conclusions that can be drawn by taking a close look into this apparently simple system. Suppose that we are given the input signal $x(n)$ for $n \geq n_0$, and we wish to determine the output $y(n)$ of this system for $n \geq n_0$. For $n = n_0, n_0 + 1, \ldots$, (2.2.4) gives

$$y(n_0) = y(n_0 - 1) + x(n_0)$$

$$y(n_0 + 1) = y(n_0) + x(n_0 + 1)$$

and so on. Note that we have a problem in computing $y(n_0)$, since it depends on $y(n_0 - 1)$. However,

$$y(n_0 - 1) = \sum_{k=-\infty}^{n_0-1} x(k)$$

that is, $y(n_0 - 1)$ "summarizes" the effect on the system from all the inputs which had been applied to the system before time n_0. Thus the response of the system for $n \geq n_0$ to the input $x(n)$ that is applied at time n_0 is the combined result of this input and all inputs that had been applied previously to the system. Consequently, $y(n)$, $n \geq n_0$ is not uniquely determined by the input $x(n)$ for $n \geq n_0$.

The additional information required to determine $y(n)$ for $n \geq n_0$ is the *initial condition* $y(n_0 - 1)$. This value summarizes the effect of all previous inputs to the system. Thus the initial condition $y(n_0 - 1)$ together with the input sequence $x(n)$ for $n \geq n_0$ uniquely determine the output sequence $y(n)$ for $n \geq n_0$.

If the accumulator had no excitation prior to n_0, the initial condition is $y(n_0 - 1) = 0$. In such a case we say that the system is *initially relaxed*. Since $y(n_0-1) = 0$, the output sequence $y(n)$ depends only on the input sequence $x(n)$ for $n \geq n_0$.

It is customary to assume that every system is relaxed at $n = -\infty$. In this case, if an input $x(n)$ is applied at $n = -\infty$, the corresponding output $y(n)$ is *solely* and *uniquely* determined by the given input.

Example 2.2.2

The accumulator described by (2.2.3) is excited by the sequence $x(n) = nu(n)$. Determine its output under the condition that:

(a) It is initially relaxed [i.e., $y(-1) = 0$].
(b) Initially, $y(-1) = 1$.

Solution   The output of the system is defined as

$$y(n) = \sum_{k=-\infty}^{n} x(k) = \sum_{k=-\infty}^{-1} x(k) + \sum_{k=0}^{n} x(k)$$

$$= y(-1) + \sum_{k=0}^{n} x(k)$$

But

$$\sum_{k=0}^{n} x(k) = \frac{n(n+1)}{2}$$

(a) If the system is initially relaxed, $y(-1) = 0$ and hence

$$y(n) = \frac{n(n+1)}{2} \qquad n \geq 0$$

(b) On the other hand, if the initial condition is $y(-1) = 1$, then

$$y(n) = 1 + \frac{n(n+1)}{2} = \frac{n^2+n+2}{2} \qquad n \geq 0$$

2.2.2 Block Diagram Representation of Discrete-Time Systems

It is useful at this point to introduce a block diagram representation of discrete-time systems. For this purpose we need to define some basic building blocks that can be interconnected to form complex systems.

An adder.   Figure 2.13 illustrates a system (adder) that performs the addition of two signal sequences to form another (the sum) sequence, which we denote

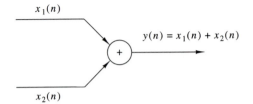

Figure 2.13 Graphical representation of an adder.

as $y(n)$. Note that it is not necessary to store either one of the sequences in order to perform the addition. In other words, the addition operation is *memoryless*.

A constant multiplier. This operation is depicted by Fig. 2.14, and simply represents applying a scale factor on the input $x(n)$. Note that this operation is also memoryless.

Figure 2.14 Graphical representation of a constant multiplier.

A signal multiplier. Figure 2.15 illustrates the multiplication of two signal sequences to form another (the product) sequence, denoted in the figure as $y(n)$. As in the preceding two cases, we can view the multiplication operation as memoryless.

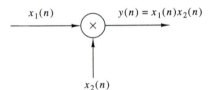

Figure 2.15 Graphical representation of a signal multiplier.

A unit delay element. The unit delay is a special system that simply delays the signal passing through it by one sample. Figure 2.16 illustrates such a system. If the input signal is $x(n)$, the output is $x(n-1)$. In fact, the sample $x(n-1)$ is stored in memory at time $n-1$ and it is recalled from memory at time n to form

$$y(n) = x(n-1)$$

Thus this basic building block requires memory. The use of the symbol z^{-1} to denote the unit of delay will become apparent when we discuss the z-transform in Chapter 3.

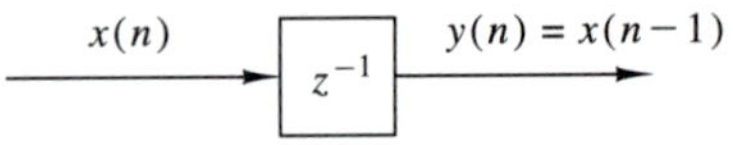

$$x(n) \quad \boxed{z^{-1}} \quad y(n) = x(n-1)$$

Figure 2.16 Graphical representation of the unit delay element.

A unit advance element. In contrast to the unit delay, a unit advance moves the input $x(n)$ ahead by one sample in time to yield $x(n+1)$. Figure 2.17 illustrates this operation, with the operator z being used to denote the unit advance.

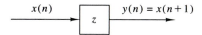

$$x(n) \qquad\qquad y(n) = x(n+1)$$

Figure 2.17   Graphical representation of the unit advance element.

We observe that any such advance is physically impossible in real time, since, in fact, it involves looking into the future of the signal. On the other hand, if we store the signal in the memory of the computer, we can recall any sample at any time. In such a nonreal-time application, it is possible to advance the signal $x(n)$ in time.

Example 2.2.3

Using basic building blocks introduced above, sketch the block diagram representation of the discrete-time system described by the input–output relation.

$$y(n) = \tfrac{1}{4}y(n-1) + \tfrac{1}{2}x(n) + \tfrac{1}{2}x(n-1) \qquad (2.2.5)$$

where $x(n)$ is the input and $y(n)$ is the output of the system.

Solution   According to (2.2.5), the output $y(n)$ is obtained by multiplying the input $x(n)$ by 0.5, multiplying the previous input $x(n-1)$ by 0.5, adding the two products, and then adding the previous output $y(n-1)$ multiplied by $\tfrac{1}{4}$. Figure 2.18a illustrates this block diagram realization of the system. A simple rearrangement of (2.2.5), namely,

$$y(n) = \tfrac{1}{4}y(n-1) + \tfrac{1}{2}[x(n) + x(n-1)] \qquad (2.2.6)$$

leads to the block diagram realization shown in Fig. 2.18b. Note that if we treat "the system" from the "viewpoint" of an input–output or an external description, we are not concerned about how the system is realized. On the other hand, if we adopt an

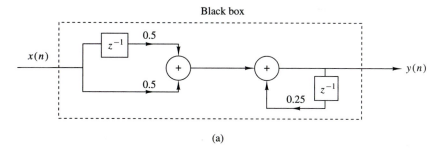

(a)

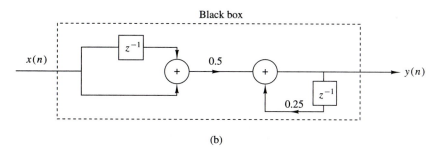

(b)

Figure 2.18   Block diagram realizations of the system $y(n) = 0.25y(n-1) + 0.5x(n) + 0.5x(n-1)$.

internal description of the system, we know exactly how the system building blocks are configured. In terms of such a realization, we can see that a system is *relaxed* at time $n = n_0$ if the outputs of all the *delays* existing in the system are zero at $n = n_0$ (i.e., all memory is *filled* with zeros).

2.2.3 Classification of Discrete-Time Systems

In the analysis as well as in the design of systems, it is desirable to classify the systems according to the general properties that they satisfy. In fact, the mathematical techniques that we develop in this and in subsequent chapters for analyzing and designing discrete-time systems depend heavily on the general characteristics of the systems that are being considered. For this reason it is necessary for us to develop a number of properties or categories that can be used to describe the general characteristics of systems.

We stress the point that for a system to possess a given property, the property must hold for every possible input signal to the system. If a property holds for some input signals but not for others, the system does not possess that property. Thus a counterexample is sufficient to prove that a system does not possess a property. However, to prove that the system has some property, we must prove that this property holds for every possible input signal.

Static versus dynamic systems. A discrete-time system is called *static* or memoryless if its output at any instant n depends at most on the input sample at the same time, but not on past or future samples of the input. In any other case, the system is said to be *dynamic* or to have memory. If the output of a system at time n is completely determined by the input samples in the interval from $n - N$ to $n(N \geq 0)$, the system is said to have *memory* of duration N. If $N = 0$, the system is static. If $0 < N < \infty$, the system is said to have *finite memory*, whereas if $N = \infty$, the system is said to have *infinite memory*.

The systems described by the following input–output equations

$$y(n) = ax(n) \tag{2.2.7}$$

$$y(n) = nx(n) + bx^3(n) \tag{2.2.8}$$

are both static or memoryless. Note that there is no need to store any of the past inputs or outputs in order to compute the present output. On the other hand, the systems described by the following input–output relations

$$y(n) = x(n) + 3x(n - 1) \tag{2.2.9}$$

$$y(n) = \sum_{k=0}^{n} x(n - k) \tag{2.2.10}$$

$$y(n) = \sum_{k=0}^{\infty} x(n - k) \tag{2.2.11}$$

are dynamic systems or systems with memory. The systems described by (2.2.9) and (2.2.10) have finite memory, whereas the system described by (2.2.11) has infinite memory.

We observe that static or memoryless systems are described in general by input–output equations of the form

$$y(n) = T[x(n), n] \tag{2.2.12}$$

and they do not include delay elements (memory).

Time-invariant versus time-variant systems.   We can subdivide the general class of systems into the two broad categories, time-invariant systems and time-variant systems. A system is called time-invariant if its input–output characteristics do not change with time. To elaborate, suppose that we have a system T in a relaxed state which, when excited by an input signal $x(n)$, produces an output signal $y(n)$. Thus we write

$$y(n) = T[x(n)] \tag{2.2.13}$$

Now suppose that the same input signal is delayed by k units of time to yield $x(n-k)$, and again applied to the same system. If the characteristics of the system do not change with time, the output of the relaxed system will be $y(n-k)$. That is, the output will be the same as the response to $x(n)$, except that it will be delayed by the same k units in time that the input was delayed. This leads us to define a time-invariant or shift-invariant system as follows.

Definition.   A relaxed system T is *time invariant* or *shift invariant* if and only if

$$x(n) \xrightarrow{\ T\ } y(n)$$

implies that

$$x(n-k) \xrightarrow{\ T\ } y(n-k) \tag{2.2.14}$$

for every input signal $x(n)$ and every time shift k.

To determine if any given system is time invariant, we need to perform the test specified by the preceding definition. Basically, we excite the system with an arbitrary input sequence $x(n)$, which produces an output denoted as $y(n)$. Next we delay the input sequence by same amount k and recompute the output. In general, we can write the output as

$$y(n, k) = T[x(n-k)]$$

Now if this output $y(n, k) = y(n-k)$, for all possible values of k, the system is time invariant. On the other hand, if the output $y(n, k) \neq y(n-k)$, even for one value of k, the system is time variant.

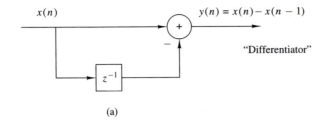

(a)

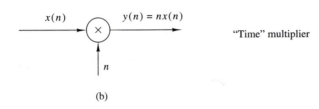

(b)

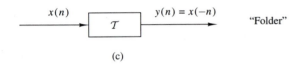

(c)

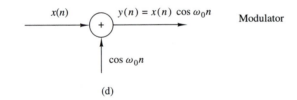

(d)

Figure 2.19 Examples of a time-invariant (a) and some time-variant systems (b)–(d).

Example 2.2.4

Determine if the systems shown in Fig. 2.19 are time invariant or time variant.

Solution

(a) This system is described by the input–output equations

$$y(n) = T[x(n)] = x(n) - x(n-1) \qquad (2.2.15)$$

Now if the input is delayed by k units in time and applied to the system, it is clear from the block diagram that the output will be

$$y(n, k) = x(n - k) - x(n - k - 1) \qquad (2.2.16)$$

On the other hand, from (2.2.14) we note that if we delay $y(n)$ by k units in time, we obtain

$$y(n - k) = x(n - k) - x(n - k - 1) \qquad (2.2.17)$$

Since the right-hand sides of (2.2.16) and (2.2.17) are identical, it follows that $y(n, k) = y(n - k)$. Therefore, the system is time invariant.

(b) The input–output equation for this system is

$$y(n) = T[x(n)] = nx(n) \qquad (2.2.18)$$

The response of this system to $x(n - k)$ is

$$y(n, k) = nx(n - k) \qquad (2.2.19)$$

Now if we delay $y(n)$ in (2.2.18) by k units in time, we obtain

$$y(n - k) = (n - k)x(n - k)$$
$$= nx(n - k) - kx(n - k) \qquad (2.2.20)$$

This system is time variant, since $y(n, k) \neq y(n - k)$.

(c) This system is described by the input–output relation

$$y(n) = T[x(n)] = x(-n) \qquad (2.2.21)$$

The response of this system to $x(n - k)$ is

$$y(n, k) = T[x(n - k)] = x(-n - k) \qquad (2.2.22)$$

Now, if we delay the output $y(n)$, as given by (2.2.21), by k units in time, the result will be

$$y(n - k) = x(-n + k) \qquad (2.2.23)$$

Since $y(n, k) \neq y(n - k)$, the system is time variant.

(d) The input–output equation for this system is

$$y(n) = x(n) \cos \omega_0 n \qquad (2.2.24)$$

The response of this system to $x(n - k)$ is

$$y(n, k) = x(n - k) \cos \omega_0 n \qquad (2.2.25)$$

If the expression in (2.2.24) is delayed by k units and the result is compared to (2.2.25), it is evident that the system is time variant.

Linear versus nonlinear systems.   The general class of systems can also be subdivided into linear systems and nonlinear systems. A linear system is one that satisfies the *superposition principle*. Simply stated, the principle of superposition requires that the response of the system to a weighted sum of signals be equal to the corresponding weighted sum of the responses (outputs) of the system to each of the individual input signals. Hence we have the following definition of linearity.

Definition.   A relaxed T system is linear if and only if

$$T[a_1 x_1(n) + a_2 x_2(n)] = a_1 T[x_1(n)] + a_2 T[x_2(n)] \qquad (2.2.26)$$

for any arbitrary input sequences $x_1(n)$ and $x_2(n)$, and any arbitrary constants a_1 and a_2.

Figure 2.20 gives a pictorial illustration of the superposition principle.

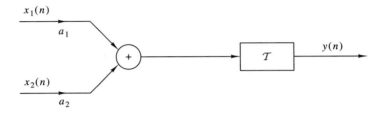

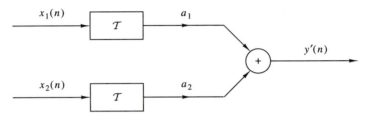

Figure 2.20 Graphical representation of the superposition principle. T is linear if and only if $y(n) = y'(n)$.

The superposition principle embodied in the relation (2.2.26) can be separated into two parts. First, suppose that $a_2 = 0$. Then (2.2.26) reduces to

$$T[a_1 x_1(n)] = a_1 T[x_1(n)] = a_1 y_1(n) \qquad (2.2.27)$$

where

$$y_1(n) = T[x_1(n)]$$

The relation (2.2.27) demonstrates the *multiplicative* or *scaling property* of a linear system. That is, if the response of the system to the input $x_1(n)$ is $y_1(n)$, the response to $a_1 x_1(n)$ is simply $a_1 y_1(n)$. Thus any scaling of the input results in an identical scaling of the corresponding output.

Second, suppose that $a_1 = a_2 = 1$ in (2.2.26). Then

$$\begin{aligned} T[x_1(n) + x_2(n)] &= T[x_1(n)] + T[x_1(n)] \\ &= y_1(n) + y_2(n) \end{aligned} \qquad (2.2.28)$$

This relation demonstrates the *additivity property* of a linear system. The additivity and multiplicative properties constitute the superposition principle as it applies to linear systems.

The linearity condition embodied in (2.2.26) can be extended arbitrarily to any weighted linear combination of signals by induction. In general, we have

$$x(n) = \sum_{k=1}^{M-1} a_k x_k(n) \xrightarrow{T} y(n) = \sum_{k=1}^{M-1} a_k y_k(n) \qquad (2.2.29)$$

where

$$y_k(n) = T[x_k(n)] \qquad k = 1, 2, \ldots, M-1 \qquad (2.2.30)$$

We observe from (2.2.27) that if $a_1 = 0$, then $y(n) = 0$. In other words, a re-laxed, linear system with zero input produces a zero output. If a system produces a nonzero output with a zero input, the system may be either nonrelaxed or non-linear. If a relaxed system does not satisfy the superposition principle as given by the definition above, it is called *nonlinear*.

Example 2.2.5

Determine if the systems described by the following input–output equations are linear or nonlinear.

(a) $y(n) = nx(n)$ (b) $y(n) = x(n^2)$ (c) $y(n) = x^2(n)$

(d) $y(n) = Ax(n) + B$ (e) $y(n) = e^{x(n)}$

Solution

(a) For two input sequences $x_1(n)$ and $x_2(n)$, the corresponding outputs are

$$y_1(n) = nx_1(n)$$
$$y_2(n) = nx_2(n)$$
(2.2.31)

A linear combination of the two input sequences results in the output

$$y_3(n) = T[a_1x_1(n) + a_2x_2(n)] = n[a_1x_1(n) + a_2x_2(n)]$$
$$= a_1nx_1(n) + a_2nx_2(n)$$
(2.2.32)

On the other hand, a linear combination of the two outputs in (2.2.31) results in the output

$$a_1y_1(n) + a_2y_2(n) = a_1nx_1(n) + a_2nx_2(n)$$
(2.2.33)

Since the right-hand sides of (2.2.32) and (2.2.33) are identical, the system is linear.

(b) As in part (a), we find the response of the system to two separate input signals $x_1(n)$ and $x_2(n)$. The result is

$$y_1(n) = x_1(n^2)$$
$$y_2(n) = x_2(n^2)$$
(2.2.34)

The output of the system to a linear combination of $x_1(n)$ and $x_2(n)$ is

$$y_3(n) = T[a_1x_1(n) + a_2x_2(n)] = a_1x_1(n^2) + a_2x_2(n^2)$$
(2.2.35)

Finally, a linear combination of the two outputs in (2.2.36) yields

$$a_1y_1(n) + a_2y_2(n) = a_1x_1(n^2) + a_2x_2(n^2)$$
(2.2.36)

By comparing (2.2.35) with (2.2.36), we conclude that the system is linear.

(c) The output of the system is the square of the input. (Electronic devices that have such an input–output characteristic and are called square-law devices.) From our previous discussion it is clear that such a system is memoryless. We now illustrate that this system is nonlinear.

The responses of the system to two separate input signals are

$$y_1(n) = x_1^2(n)$$
$$y_2(n) = x_2^2(n)$$
(2.2.37)

The response of the system to a linear combination of these two input signals is

$$
\begin{aligned}
y_3(n) &= T[a_1 x_1(n) + a_2 x_2(n)] \\
&= [a_1 x_1(n) + a_2 x_2(n)]^2 \\
&= a_1^2 x_1^2(n) + 2a_1 a_2 x_1(n) x_2(n) + a_2^2 x_2^2(n)
\end{aligned}
$$
(2.2.38)

On the other hand, if the system is linear, it would produce a linear combination of the two outputs in (2.2.37), namely,

$$a_1 y_1(n) + a_2 y_2(n) = a_1 x_1^2(n) + a_2 x_2^2(n)$$
(2.2.39)

Since the actual output of the system, as given by (2.2.38), is not equal to (2.2.39), the system is nonlinear.

(d) Assuming that the system is excited by $x_1(n)$ and $x_2(n)$ separately, we obtain the corresponding outputs

$$y_1(n) = Ax_1(n) + B$$
$$y_2(n) = Ax_2(n) + B$$
(2.2.40)

A linear combination of $x_1(n)$ and $x_2(n)$ produces the output

$$
\begin{aligned}
y_3(n) &= T[a_1 x_1(n) + a_2 x_2(n)] \\
&= A[a_1 x_1(n) + a_2 x_2(n)] + B \\
&= Aa_1 x_1(n) + a_2 Ax_2(n) + B
\end{aligned}
$$
(2.2.41)

On the other hand, if the system were linear, its output to the linear combination of $x_1(n)$ and $x_2(n)$ would be a linear combination of $y_1(n)$ and $y_2(n)$, that is,

$$a_1 y_1(n) + a_2 y_2(n) = a_1 Ax_1(n) + a_1 B + a_2 Ax_2(n) + a_2 B$$
(2.2.42)

Clearly, (2.2.41) and (2.2.42) are different and hence the system fails to satisfy the linearity test.

The reason that this system fails to satisfy the linearity test is not that the system is nonlinear (in fact, the system is described by a linear equation) but the presence of the constant B. Consequently, the output depends on both the input excitation and on the parameter $B \neq 0$. Hence, for $B \neq 0$, the system is not relaxed. If we set $B = 0$, the system is now relaxed and the linearity test is satisfied.

(e) Note that the system described by the input–output equation

$$y(n) = e^{x(n)}$$
(2.2.43)

is relaxed. If $x(n) = 0$, we find that $y(n) = 1$. This is an indication that the system is nonlinear. This, in fact, is the conclusion reached when the linearity test, is applied.

Causal versus noncausal systems.   We begin with the definition of causal discrete-time systems.

Definition.   A system is said to be *causal* if the output of the system at any time n [i.e., $y(n)$] depends only on present and past inputs [i.e., $x(n)$, $x(n-1)$, $x(n-2), \ldots$], but does not depend on future inputs [i.e., $x(n+1)$, $x(n+2), \ldots$]. In mathematical terms, the output of a causal system satisfies an equation of the form

$$y(n) = F[x(n), x(n-1), x(n-2), \ldots] \tag{2.2.44}$$

where $F[\cdot]$ is some arbitrary function.

If a system does not satisfy this definition, it is called *noncausal*. Such a system has an output that depends not only on present and past inputs but also on future inputs.

It is apparent that in real-time signal processing applications we cannot observe future values of the signal, and hence a noncausal system is physically unrealizable (i.e., it cannot be implemented). On the other hand, if the signal is recorded so that the processing is done off-line (nonreal time), it is possible to implement a noncausal system, since all values of the signal are available at the time of processing. This is often the case in the processing of geophysical signals and images.

Example 2.2.6

Determine if the systems described by the following input–output equations are causal or noncausal.

(a) $y(n) = x(n) - x(n-1)$ **(b)** $y(n) = \sum_{k=-\infty}^{n} x(k)$ **(c)** $y(n) = ax(n)$

(d) $y(n) = x(n) + 3x(n+4)$ **(e)** $y(n) = x(n^2)$ **(f)** $y(n) = x(2n)$

(g) $y(n) = x(-n)$

Solution   The systems described in parts (a), (b), and (c) are clearly causal, since the output depends only on the present and past inputs. On the other hand, the systems in parts (d), (e), and (f) are clearly noncausal, since the output depends on future values of the input. The system in (g) is also noncausal, as we note by selecting, for example, $n = -1$, which yields $y(-1) = x(1)$. Thus the output at $n = -1$ depends on the input at $n = 1$, which is two units of time into the future.

Stable versus unstable systems.   Stability is an important property that must be considered in any practical application of a system. Unstable systems usually exhibit erratic and extreme behavior and cause overflow in any practical implementation. Here, we define mathematically what we mean by a stable system, and later, in Section 2.3.6, we explore the implications of this definition for linear, time-invariant systems.

Definition.   An arbitrary relaxed system is said to be bounded input–bounded output (BIBO) stable if and only if every bounded input produces a bounded output.

The conditions that the input sequence $x(n)$ and the output sequence $y(n)$ are bounded is translated mathematically to mean that there exist some finite numbers, say M_x and M_y, such that

$$|x(n)| \leq M_x < \infty \qquad |y(n)| \leq M_y < \infty \qquad (2.2.45)$$

for all n. If, for some bounded input sequence $x(n)$, the output is unbounded (infinite), the system is classified as unstable.

Example 2.2.7

Consider the nonlinear system described by the input–output equation

$$y(n) = y^2(n-1) + x(n)$$

As an input sequence we select the bounded signal

$$x(n) = C\delta(n)$$

where C is a constant. We also assume that $y(-1) = 0$. Then the output sequence is

$$y(0) = C, \quad y(1) = C^2, \quad y(2) = C^4, \quad \ldots, \quad y(n) = C^{2^n}$$

Clearly, the output is unbounded when $1 < |C| < \infty$. Therefore, the system is BIBO unstable, since a bounded input sequence has resulted in an unbounded output.

2.2.4 Interconnection of Discrete-Time Systems

Discrete-time systems can be interconnected to form larger systems. There are two basic ways in which systems can be interconnected: in cascade (series) or in parallel. These interconnections are illustrated in Fig. 2.21. Note that the two interconnected systems are different.

In the cascade interconnection the output of the first system is

$$y_1(n) = T_1[x(n)] \qquad (2.2.46)$$

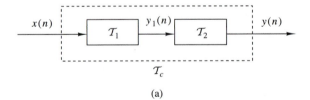

(a)

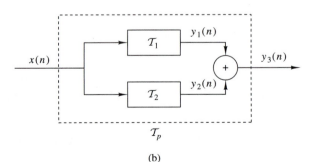

(b)

Figure 2.21 Cascade (a) and parallel (b) interconnections of systems.

and the output of the second system is

$$y(n) = T_2[y_1(n)]$$
$$= T_2\{T_1[x(n)]\}$$

(2.2.47)

We observe that systems T_1 and T_2 can be combined or consolidated into a single overall system

$$T_c \equiv T_2 T_1$$

(2.2.48)

Consequently, we can express the output of the combined system as

$$y(n) = T_c[x(n)]$$

In general, the order in which the operations T_1 and T_2 are performed is important. That is,

$$T_2 T_1 \neq T_1 T_2$$

for arbitrary systems. However, if the systems T_1 and T_2 are linear and time invariant, then (a) T_c is time invariant and (b) $T_2 T_1 = T_1 T_2$, that is, the order in which the systems process the signal is not important. $T_2 T_1$ and $T_1 T_2$ yield identical output sequences.

The proof of (a) follows. The proof of (b) is given in Section 2.3.4. To prove time invariance, suppose that T_1 and T_2 are time invariant; then

$$x(n-k) \xrightarrow{T_1} y_1(n-k)$$

and

$$y_1(n-k) \xrightarrow{T_2} y(n-k)$$

Thus

$$x(n-k) \xrightarrow{T_c = T_2 T_1} y(n-k)$$

and therefore, T_c is time invariant.

In the parallel interconnection, the output of the system T_1 is $y_1(n)$ and the output of the system T_2 is $y_2(n)$. Hence the output of the parallel interconnection is

$$y_3(n) = y_1(n) + y_2(n)$$
$$= T_1[x(n)] + T_2[x(n)]$$
$$= (T_1 + T_2)[x(n)]$$
$$= T_p[x(n)]$$

where $T_p = T_1 + T_2$.

In general, we can use parallel and cascade interconnection of systems to construct larger, more complex systems. Conversely, we can take a larger system and break it down into smaller subsystems for purposes of analysis and implementation. We shall use these notions later, in the design and implementation of digital filters.

2.3 ANALYSIS OF DISCRETE-TIME LINEAR TIME-INVARIANT SYSTEMS

In Section 2.2 we classified systems in accordance with a number of characteristic properties or categories, namely: linearity, causality, stability, and time invariance. Having done so, we now turn our attention to the analysis of the important class of linear, time-invariant (LTI) systems. In particular, we shall demonstrate that such systems are characterized in the time domain simply by their response to a unit sample sequence. We shall also demonstrate that any arbitrary input signal can be decomposed and represented as a weighted sum of unit sample sequences. As a consequence of the linearity and time-invariance properties of the system, the response of the system to any arbitrary input signal can be expressed in terms of the unit sample response of the system. The general form of the expression that relates the unit sample response of the system and the arbitrary input signal to the output signal, called the convolution sum or the convolution formula, is also derived. Thus we are able to determine the output of any linear, time-invariant system to any arbitrary input signal.

2.3.1 Techniques for the Analysis of Linear Systems

There are two basic methods for analyzing the behavior or response of a linear system to a given input signal. One method is based on the direct solution of the input–output equation for the system, which, in general, has the form

$$y(n) = F[y(n-1), y(n-2), \ldots, y(n-N), x(n), x(n-1), \ldots, x(n-M)]$$

where $F[\cdot]$ denotes some function of the quantities in brackets. Specifically, for an LTI system, we shall see later that the general form of the input–output relationship is

$$y(n) = -\sum_{k=1}^{N} a_k y(n-k) + \sum_{k=0}^{M} b_k x(n-k) \tag{2.3.1}$$

where $\{a_k\}$ and $\{b_k\}$ are constant parameters that specify the system and are independent of $x(n)$ and $y(n)$. The input–output relationship in (2.3.1) is called a difference equation and represents one way to characterize the behavior of a discrete-time LTI system. The solution of (2.3.1) is the subject of Section 2.4.

The second method for analyzing the behavior of a linear system to a given input signal is first to decompose or resolve the input signal into a sum of elementary signals. The elementary signals are selected so that the response of the system to each signal component is easily determined. Then, using the linearity property of the system, the responses of the system to the elementary signals are added to obtain the total response of the system to the given input signal. This second method is the one described in this section.

To elaborate, suppose that the input signal $x(n)$ is resolved into a weighted sum of elementary signal components $\{x_k(n)\}$ so that

$$x(n) = \sum_k c_k x_k(n) \tag{2.3.2}$$

where the $\{c_k\}$ are the set of amplitudes (weighting coefficients) in the decomposition of the signal $x(n)$. Now suppose that the response of the system to the elementary signal component $x_k(n)$ is $y_k(n)$. Thus,

$$y_k(n) \equiv T[x_k(n)] \tag{2.3.3}$$

assuming that the system is relaxed and that the response to $c_k x_k(n)$ is $c_k y_k(n)$, as a consequence of the scaling property of the linear system.

Finally, the total response to the input $x(n)$ is

$$
\begin{aligned}
y(n) = T[x(n)] &= T\left[\sum_k c_k x_k(n)\right] \\
&= \sum_k c_k T[x_k(n)] \\
&= \sum_k c_k y_k(n)
\end{aligned}
\tag{2.3.4}
$$

In (2.3.4) we used the additivity property of the linear system.

Although to a large extent, the choice of the elementary signals appears to be arbitrary, our selection is heavily dependent on the class of input signals that we wish to consider. If we place no restriction on the characteristics of the input signals, its resolution into a weighted sum of unit sample (impulse) sequences proves to be mathematically convenient and completely general. On the other hand, if we restrict our attention to a subclass of input signals, there may be another set of elementary signals that is more convenient mathematically in the determination of the output. For example, if the input signal $x(n)$ is periodic with period N, we have already observed in Section 1.3.5 that a mathematically convenient set of elementary signals is the set of exponentials

$$x_k(n) = e^{j\omega_k n} \qquad k = 0, 1, \ldots, N-1 \tag{2.3.5}$$

where the frequencies $\{\omega_k\}$ are harmonically related, that is,

$$\omega_k = \left(\frac{2\pi}{N}\right)k \qquad k = 0, 1, \ldots, N-1 \tag{2.3.6}$$

The frequency $2\pi/N$ is called the fundamental frequency, and all higher-frequency components are multiples of the fundamental frequency component. This subclass of input signals is considered in more detail later.

For the resolution of the input signal into a weighted sum of unit sample sequences, we must first determine the response of the system to a unit sample sequence and then use the scaling and multiplicative properties of the linear system to determine the formula for the output given any arbitrary input. This development is described in detail as follows.

2.3.2 Resolution of a Discrete-Time Signal into Impulses

Suppose we have an arbitrary signal $x(n)$ that we wish to resolve into a sum of unit sample sequences. To utilize the notation established in the preceding section, we select the elementary signals $x_k(n)$ to be

$$x_k(n) = \delta(n - k) \tag{2.3.7}$$

where k represents the delay of the unit sample sequence. To handle an arbitrary signal $x(n)$ that may have nonzero values over an infinite duration, the set of unit impulses must also be infinite, to encompass the infinite number of delays.

Now suppose that we multiply the two sequences $x(n)$ and $\delta(n - k)$. Since $\delta(n - k)$ is zero everywhere except at $n = k$, where its value is unity, the result of this multiplication is another sequence that is zero everywhere except at $n = k$, where its value is $x(k)$, as illustrated in Fig. 2.22. Thus

$$x(n)\delta(n - k) = x(k)\delta(n - k) \tag{2.3.8}$$

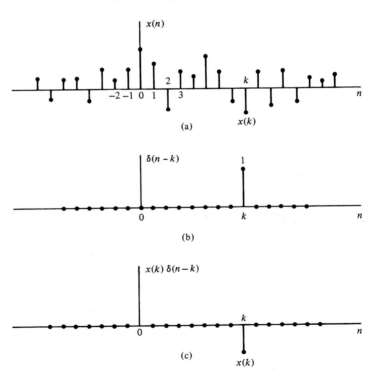

Figure 2.22 Multiplication of a signal $x(n)$ with a shifted unit sample sequence.

is a sequence that is zero everywhere except at $n = k$, where its value is $x(k)$. If we were to repeat the multiplication of $x(n)$ with $\delta(n - m)$, where m is another delay $(m \neq k)$, the result will be a sequence that is zero everywhere except at $n = m$, where its value is $x(m)$. Hence

$$x(n)\delta(n - m) = x(m)\delta(n - m) \tag{2.3.9}$$

In other words, each multiplication of the signal $x(n)$ by a unit impulse at some delay k, [i.e., $\delta(n - k)$], in essence picks out the single value $x(k)$ of the signal $x(n)$ at the delay where the unit impulse is nonzero. Consequently, if we repeat this multiplication over all possible delays, $-\infty < k < \infty$, and sum all the product sequences, the result will be a sequence equal to the sequence $x(n)$, that is,

$$x(n) = \sum_{k=-\infty}^{\infty} x(k)\delta(n - k) \tag{2.3.10}$$

We emphasize that the right-hand side of (2.3.10) is the summation of an infinite number of unit sample sequences where the unit sample sequence $\delta(n - k)$ has an amplitude value of $x(k)$. Thus the right-hand side of (2.3.10) gives the resolution of or decomposition of any arbitrary signal $x(n)$ into a weighted (scaled) sum of shifted unit sample sequences.

Example 2.3.1

Consider the special case of a finite-duration sequence given as

$$x(n) = \{2, 4, 0, 3\}$$
$$\uparrow$$

Resolve the sequence $x(n)$ into a sum of weighted impulse sequences.

Solution   Since the sequence $x(n)$ is nonzero for the time instants $n = -1, 0, 2$, we need three impulses at delays $k = -1, 0, 2$. Following (2.3.10) we find that

$$x(n) = 2\delta(n + 1) + 4\delta(n) + 3\delta(n - 2)$$

2.3.3 Response of LTI Systems to Arbitrary Inputs: The Convolution Sum

Having resolved an arbitrary input signal $x(n)$ into a weighted sum of impulses, we are now ready to determine the response of any relaxed linear system to any input signal. First, we denote the response $y(n, k)$ of the system to the input unit sample sequence at $n = k$ by the special symbol $h(n, k)$, $-\infty < k < \infty$. That is,

$$y(n, k) \equiv h(n, k) = T[\delta(n - k)] \tag{2.3.11}$$

In (2.3.11) we note that n is the time index and k is a parameter showing the location of the input impulse. If the impulse at the input is scaled by an amount $c_k \equiv x(k)$, the response of the system is the correspondingly scaled output, that is,

$$c_k h(n, k) = x(k)h(n, k) \tag{2.3.12}$$

Finally, if the input is the arbitrary signal $x(n)$ that is expressed as a sum of weighted impulses, that is,

$$x(n) = \sum_{k=-\infty}^{\infty} x(k)\delta(n-k) \tag{2.3.13}$$

then the response of the system to $x(n)$ is the corresponding sum of weighted outputs, that is,

$$y(n) = T[x(n)] = T\left[\sum_{k=-\infty}^{\infty} x(k)\delta(n-k)\right]$$

$$= \sum_{k=-\infty}^{\infty} x(k)T[\delta(n-k)] \tag{2.3.14}$$

$$= \sum_{k=-\infty}^{\infty} x(k)h(n,k)$$

Clearly, (2.3.14) follows from the superposition property of linear systems, and is known as the *superposition summation*.

We note that (2.3.14) is an expression for the response of a linear system to any arbitrary input sequence $x(n)$. This expression is a function of both $x(n)$ and the responses $h(n,k)$ of the system to the unit impulses $\delta(n-k)$ for $-\infty < k < \infty$. In deriving (2.3.14) we used the linearity property of the system but not its time-invariance property. Thus the expression in (2.3.14) applies to any relaxed linear (time-variant) system.

If, in addition, the system is time invariant, the formula in (2.3.14) simplifies considerably. In fact, if the response of the LTI system to the unit sample sequence $\delta(n)$ is denoted as $h(n)$, that is,

$$h(n) \equiv T[\delta(n)] \tag{2.3.15}$$

then by the time-invariance property, the response of the system to the delayed unit sample sequence $\delta(n-k)$ is

$$h(n-k) = T[\delta(n-k)] \tag{2.3.16}$$

Consequently, the formula in (2.3.14) reduces to

$$y(n) = \sum_{k=-\infty}^{\infty} x(k)h(n-k) \tag{2.3.17}$$

Now we observe that the relaxed LTI system is completely characterized by a single function $h(n)$, namely, its response to the unit sample sequence $\delta(n)$. In contrast, the general characterization of the output of a time-variant, linear system requires an infinite number of unit sample response functions, $h(n,k)$, one for each possible delay.

The formula in (2.3.17) that gives the response $y(n)$ of the LTI system as a function of the input signal $x(n)$ and the unit sample (impulse) response $h(n)$ is called a *convolution sum*. We say that the input $x(n)$ is convolved with the impulse response $h(n)$ to yield the output $y(n)$. We shall now explain the procedure for computing the response $y(n)$, both mathematically and graphically, given the input $x(n)$ and the impulse response $h(n)$ of the system.

Suppose that we wish to compute the output of the system at some time instant, say $n = n_0$. According to (2.3.17), the response at $n = n_0$ is given as

$$y(n_0) = \sum_{k=-\infty}^{\infty} x(k)h(n_0 - k) \tag{2.3.18}$$

Our first observation is that the index in the summation is k, and hence both the input signal $x(k)$ and the impulse response $h(n_0 - k)$ are functions of k. Second, we observe that the sequences $x(k)$ and $h(n_0 - k)$ are multiplied together to form a product sequence. The output $y(n_0)$ is simply the sum over all values of the product sequence. The sequence $h(n_0 - k)$ is obtained from $h(k)$ by, first, folding $h(k)$ about $k = 0$ (the time origin), which results in the sequence $h(-k)$. The folded sequence is then shifted by n_0 to yield $h(n_0 - k)$. To summarize, the process of computing the convolution between $x(k)$ and $h(k)$ involves the following four steps.

1. *Folding.* Fold $h(k)$ about $k = 0$ to obtain $h(-k)$.
2. *Shifting.* Shift $h(-k)$ by n_0 to the right (left) if n_0 is positive (negative), to obtain $h(n_0 - k)$.
3. *Multiplication.* Multiply $x(k)$ by $h(n_0 - k)$ to obtain the product sequence $v_{n_0}(k) \equiv x(k)h(n_0 - k)$.
4. *Summation.* Sum all the values of the product sequence $v_{n_0}(k)$ to obtain the value of the output at time $n = n_0$.

We note that this procedure results in the response of the system at a single time instant, say $n = n_0$. In general, we are interested in evaluating the response of the system over all time instants $-\infty < n < \infty$. Consequently, steps 2 through 4 in the summary must be repeated, for all possible time shifts $-\infty < n < \infty$.

In order to gain a better understanding of the procedure for evaluating the convolution sum, we shall demonstrate the process graphically. The graphs will aid us in explaining the four steps involved in the computation of the convolution sum.

Example 2.3.2

The impulse response of a linear time-invariant system is

$$h(n) = \{1, 2, 1, -1\} \tag{2.3.19}$$
$$\uparrow$$

Determine the response of the system to the input signal

$$x(n) = \{1, 2, 3, 1\} \tag{2.3.20}$$
$$\uparrow$$

Solution   We shall compute the convolution according to the formula (2.3.17), but we shall use graphs of the sequences to aid us in the computation. In Fig. 2.23a we illustrate the input signal sequence $x(k)$ and the impulse response $h(k)$ of the system, using k as the time index in order to be consistent with (2.3.17).

The first step in the computation of the convolution sum is to fold $h(k)$. The folded sequence $h(-k)$ is illustrated in Fig. 2.23b. Now we can compute the output at $n = 0$, according to (2.3.17), which is

$$y(0) = \sum_{k=-\infty}^{\infty} x(k)h(-k) \tag{2.3.21}$$

Since the shift $n = 0$, we use $h(-k)$ directly without shifting it. The product sequence

$$v_0(k) \equiv x(k)h(-k) \tag{2.3.22}$$

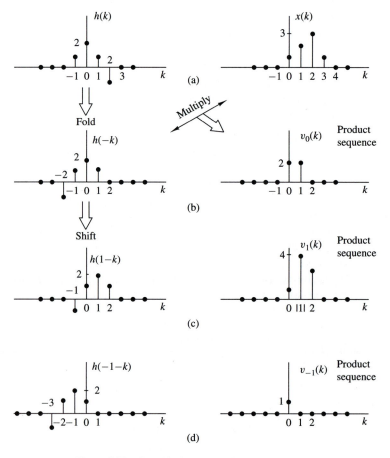

Figure 2.23   Graphical computation of convolution.

is also shown in Fig. 2.23b. Finally, the sum of all the terms in the product sequence yields

$$y(0) = \sum_{h=-\infty}^{\infty} v_0(k) = 4$$

We continue the computation by evaluating the response of the system at $n = 1$. According to (2.3.17),

$$y(1) = \sum_{h=-\infty}^{\infty} x(k)h(1 - k) \tag{2.3.23}$$

The sequence $h(1 - k)$ is simply the folded sequence $h(-k)$ shifted to the right by one unit in time. This sequence is illustrated in Fig. 2.23c. The product sequence

$$v_1(k) = x(k)h(1 - k) \tag{2.3.24}$$

is also illustrated in Fig. 2.23c. Finally, the sum of all the values in the product sequence yields

$$y(1) = \sum_{k=-\infty}^{\infty} v_1(k) = 8$$

In a similar manner, we obtain $y(2)$ by shifting $h(-k)$ two units to the right, forming the product sequence $v_2(k) = x(k)h(2 - k)$ and then summing all the terms in the product sequence obtaining $y(2) = 8$. By shifting $h(-k)$ farther to the right, multiplying the corresponding sequence, and summing over all the values of the resulting product sequences, we obtain $y(3) = 3$, $y(4) = -2$, $y(5) = -1$. For $n > 5$, we find that $y(n) = 0$ because the product sequences contain all zeros. Thus we have obtained the response $y(n)$ for $n > 0$.

Next we wish to evaluate $y(n)$ for $n < 0$. We begin with $n = -1$. Then

$$y(-1) = \sum_{k=-\infty}^{\infty} x(k)h(-1 - k) \tag{2.3.25}$$

Now the sequence $h(-1 - k)$ is simply the folded sequence $h(-k)$ shifted one time unit to the left. The resulting sequence is illustrated in Fig. 2.23d. The corresponding product sequence is also shown in Fig. 2.23d. Finally, summing over the values of the product sequence, we obtain

$$y(-1) = 1$$

From observation of the graphs of Fig. 2.23, it is clear that any further shifts of $h(-1 - k)$ to the left always results in an all-zero product sequence, and hence

$$y(n) = 0 \qquad \text{for } n \leq -2$$

Now we have the entire response of the system for $-\infty < n < \infty$, which we summarize below as

$$y(n) = \{\ldots, 0, 0, 1, 4, 8, 8, 3, -2, -1, 0, 0, \ldots\} \tag{2.3.26}$$
$$\uparrow$$

In Example 2.3.2 we illustrated the computation of the convolution sum, using graphs of the sequences to aid us in visualizing the steps involved in the computation procedure.

Before working out another example, we wish to show that the convolution operation is commutative in the sense that it is irrelevant which of the two sequences is folded and shifted. Indeed, if we begin with (2.3.17) and make a change in the variable of the summation, from k to m, by defining a new index $m = n - k$, then $k = n - m$ and (2.3.17) becomes

$$y(n) = \sum_{m=-\infty}^{\infty} x(n - m)h(m) \tag{2.3.27}$$

Since m is a dummy index, we may simply replace m by k so that

$$y(n) = \sum_{k=-\infty}^{\infty} x(n - k)h(k) \tag{2.3.28}$$

The expression in (2.3.28) involves leaving the impulse response $h(k)$ unaltered, while the input sequence is folded and shifted. Although the output $y(n)$ in (2.3.28) is identical to (2.3.17), the product sequences in the two forms of the convolution formula are not identical. In fact, if we define the two product sequences as

$$v_n(k) = x(k)h(n - k)$$

$$w_n(k) = x(n - k)h(k)$$

it can be easily shown that

$$v_n(k) = w_n(n - k)$$

and therefore,

$$y(n) = \sum_{k=-\infty}^{\infty} v_n(k) = \sum_{k=-\infty}^{\infty} w_n(n - k)$$

since both sequences contain the same sample values in a different arrangement.

Example 2.3.3

Determine the output $y(n)$ of a relaxed linear time-invariant system with impulse response

$$h(n) = a^n u(n), |a| < 1$$

when the input is a unit step sequence, that is,

$$x(n) = u(n)$$

Solution In this case both $h(n)$ and $x(n)$ are infinite-duration sequences. We use the form of the convolution formula given by (2.3.28) in which $x(k)$ is folded. The

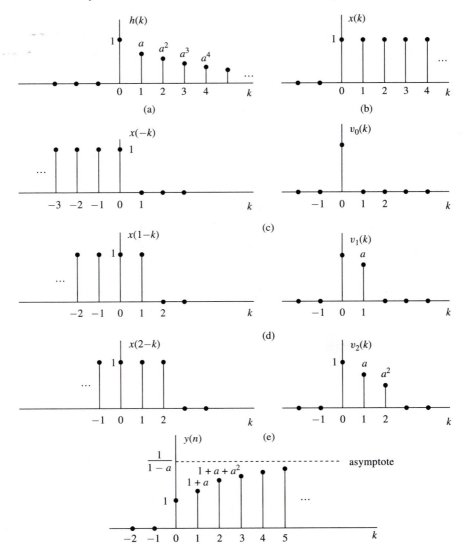

Figure 2.24   Graphical computation of convolution in Example 2.3.3.

sequences $h(k)$, $x(k)$, and $x(-k)$ are shown in Fig. 2.24. The product sequences $v_0(k)$, $v_1(k)$, and $v_2(k)$ corresponding to $x(-k)h(k)$, $x(1-k)h(k)$, and $x(2-k)h(k)$ are illustrated in Fig. 2.24c, d, and e, respectively. Thus we obtain the outputs

$$y(0) = 1$$
$$y(1) = 1 + a$$
$$y(2) = 1 + a + a^2$$

Clearly, for $n > 0$, the output is

$$y(n) = 1 + a + a^2 + \cdots + a^n$$

$$= \frac{1 - a^{n+1}}{1 - a} \tag{2.3.29}$$

On the other hand, for $n < 0$, the product sequences consist of all zeros. Hence

$$y(n) = 0 \qquad n < 0$$

A graph of the output $y(n)$ is illustrated in Fig. 2.24f for the case $0 < a < 1$. Note the exponential rise in the output as a function of n. Since $|a| < 1$, the final value of the output as n approaches infinity is

$$y(\infty) = \lim_{n \to \infty} y(n) = \frac{1}{1 - a} \tag{2.3.30}$$

To summarize, the convolution formula provides us with a means for computing the response of a relaxed, linear time-invariant system to any arbitrary input signal $x(n)$. It takes one of two equivalent forms, either (2.3.17) or (2.3.28), where $x(n)$ is the input signal to the system, h(n) is the impulse response of the system, and $y(n)$ is the *output* of the system in response to the input signal $x(n)$. The evaluation of the convolution formula involves four operations, namely: *folding* either the impulse response as specified by (2.3.17) or the input sequence as specified by (2.3.28) to yield either $h(-k)$ or $x(-k)$, respectively, *shifting* the folded sequence by n units in time to yield either $h(n - k)$ or $x(n - k)$, *multiplying* the two sequences to yield the product sequence, either $x(k)h(n - k)$ or $x(n - k)h(k)$, and finally *summing* all the values in the product sequence to yield the output $y(n)$ of the system at time n. The folding operation is done only once. However, the other three operations are repeated for all possible shifts $-\infty < n < \infty$ in order to obtain $y(n)$ for $-\infty < n < \infty$.

2.3.4 Properties of Convolution and the Interconnection of LTI Systems

In this section we investigate some important properties of convolution and interpret these properties in terms of interconnecting linear time-invariant systems. We should stress that these properties hold for every input signal.

It is convenient to simplify the notation by using an asterisk to denote the convolution operation. Thus

$$y(n) = x(n) * h(n) \equiv \sum_{k=-\infty}^{\infty} x(k)h(n - k) \tag{2.3.31}$$

In this notation the sequence following the asterisk [i.e., the impulse response $h(n)$] is folded and shifted. The input to the system is $x(n)$. On the other hand, we also showed that

$$y(n) = h(n) * x(n) \equiv \sum_{k=-\infty}^{\infty} h(k)x(n - k) \tag{2.3.32}$$

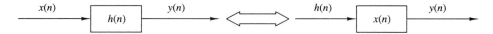

Figure 2.25   Interpretation of the commutative property of convolution.

In this form of the convolution formula, it is the input signal that is folded. Alternatively, we may interpret this form of the convolution formula as resulting from an interchange of the roles of $x(n)$ and $h(n)$. In other words, we may regard $x(n)$ as the impulse response of the system and $h(n)$ as the excitation or input signal. Figure 2.25 illustrates this interpretation.

We can view convolution more abstractly as a mathematical operation between two signal sequences, say $x(n)$ and $h(n)$, that satisfies a number of properties. The property embodied in (2.3.31) and (2.3.32) is called the commutative law.

Commutative law

$$x(n) * h(n) = h(n) * x(n) \qquad (2.3.33)$$

Viewed mathematically, the convolution operation also satisfies the associative law, which can be stated as follows.

Associative law

$$[x(n) * h_1(n)] * h_2(n) = x(n) * [h_1(n) * h_2(n)] \qquad (2.3.34)$$

From a physical point of view, we can interpret $x(n)$ as the input signal to a linear time-invariant system with impulse response $h_1(n)$. The output of this system, denoted as $y_1(n)$, becomes the input to a second linear time-invariant system with impulse response $h_2(n)$. Then the output is

$$y(n) = y_1(n) * h_2(n)$$
$$= [x(n) * h_1(n)] * h_2(n)$$

which is precisely the left-hand side of (2.3.34). Thus the left-hand side of (2.3.34) corresponds to having two linear time-invariant systems in cascade. Now the right-hand side of (2.3.34) indicates that the input $x(n)$ is applied to an equivalent system having an impulse response, say $h(n)$, which is equal to the convolution of the two impulse responses. That is,

$$h(n) = h_1(n) * h_2(n)$$

and

$$y(n) = x(n) * h(n)$$

Furthermore, since the convolution operation satisfies the commutative property, one can interchange the order of the two systems with responses $h_1(n)$ and $h_2(n)$ without altering the overall input–output relationship. Figure 2.26 graphically illustrates the associative property.

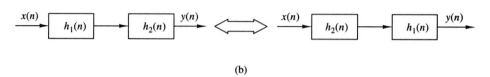

(a)

(b)

Figure 2.26 Implications of the associative (a) and the associative and commutative (b) properties of convolution.

Example 2.3.4

Determine the impulse response for the cascade of two linear time-invariant systems having impulse responses

$$h_1(n) = (\tfrac{1}{2})^n u(n)$$

and

$$h_2(n) = (\tfrac{1}{4})^n u(n)$$

Solution   To determine the overall impulse response of the two systems in cascade, we simply convolve $h_1(n)$ with $h_2(n)$. Hence

$$h(n) = \sum_{k=-\infty}^{\infty} h_1(k)h_2(n-k)$$

where $h_2(n)$ is folded and shifted. We define the product sequence

$$v_n(k) = h_1(k)h_2(n-k)$$
$$= (\tfrac{1}{2})^k (\tfrac{1}{4})^{n-k}$$

which is nonzero for $k \geq 0$ and $n - k \geq 0$ or $n \geq k \geq 0$. On the other hand, for $n < 0$, we have $v_n(k) = 0$ for all k, and hence

$$h(n) = 0, n < 0$$

For $n \geq k \geq 0$, the sum of the values of the product sequence $v_n(k)$ over all k yields

$$h(n) = \sum_{k=0}^{n} (\tfrac{1}{2})^k (\tfrac{1}{4})^{n-k}$$
$$= (\tfrac{1}{4})^n \sum_{k=0}^{n} 2^k$$
$$= (\tfrac{1}{4})^n (2^{n+1} - 1)$$
$$= (\tfrac{1}{2})^n [2 - (\tfrac{1}{2})^n], n \geq 0$$

The generalization of the associative law to more than two systems in cascade follows easily from the discussion given above. Thus if we have L linear time-invariant systems in cascade with impulse responses $h_1(n), h_2(n), \ldots, h_L(n)$, there is an equivalent linear time-invariant system having an impulse response that is equal to the $(L-1)$-fold convolution of the impulse responses. That is,

$$h(n) = h_1(n) * h_2(n) * \cdots * h_L(n) \qquad (2.3.35)$$

The commutative law implies that the order in which the convolutions are performed is immaterial. Conversely, any linear time-invariant system can be decomposed into a cascade interconnection of subsystems. A method for accomplishing the decomposition will be described later.

A third property that is satisfied by the convolution operation is the distributive law, which may be stated as follows.

Distributive law

$$x(n) * [h_1(n) + h_2(n)] = x(n) * h_1(n) + x(n) * h_2(n) \qquad (2.3.36)$$

Interpreted physically, this law implies that if we have two linear time-invariant systems with impulse responses $h_1(n)$ and $h_2(n)$ excited by the same input signal $x(n)$, the sum of the two responses is identical to the response of an overall system with impulse response

$$h(n) = h_1(n) + h_2(n)$$

Thus the overall system is viewed as a parallel combination of the two linear time-invariant systems as illustrated in Fig. 2.27.

The generalization of (2.3.36) to more than two linear time-invariant systems in parallel follows easily by mathematical induction. Thus the interconnection of L linear time-invariant systems in parallel with impulse responses $h_1(n)$, $h_2(n), \ldots, h_L(n)$ and excited by the same input $x(n)$ is equivalent to one overall system with impulse response

$$h(n) = \sum_{j=1}^{L} h_j(n) \qquad (2.3.37)$$

Conversely, any linear time-invariant system can be decomposed into a parallel interconnection of subsystems.

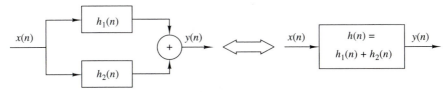

Figure 2.27 Interpretation of the distributive property of convolution: two LTI systems connected in parallel can be replaced by a single system with $h(n) = h_1(n) + h_2(n)$.

2.3.5 Causal Linear Time-Invariant Systems

In Section 2.2.3 we defined a causal system as one whose output at time n depends only on present and past inputs but does not depend on future inputs. In other words, the output of the system at some time instant n, say $n = n_0$, depends only on values of $x(n)$ for $n \leq n_0$.

In the case of a linear time-invariant system, causality can be translated to a condition on the impulse response. To determine this relationship, let us consider a linear time-invariant system having an output at time $n = n_0$ given by the convolution formula

$$y(n_0) = \sum_{k=-\infty}^{\infty} h(k)x(n_0 - k)$$

Suppose that we subdivide the sum into two sets of terms, one set involving present and past values of the input [i.e., $x(n)$ for $n \leq n_0$] and one set involving future values of the input [i.e., $x(n)$, $n > n_0$]. Thus we obtain

$$y(n_0) = \sum_{k=0}^{\infty} h(k)x(n_0 - k) + \sum_{k=-\infty}^{-1} h(k)x(n_0 - k)$$

$$= [h(0)x(n_0) + h(1)x(n_0 - 1) + h(2)x(n_0 - 2) + \cdots]$$

$$+ [h(-1)x(n_0 + 1) + h(-2)x(n_0 + 2) + \cdots]$$

We observe that the terms in the first sum involve $x(n_0)$, $x(n_0 - 1), \ldots$, which are the present and past values of the input signal. On the other hand, the terms in the second sum involve the input signal components $x(n_0 + 1)$, $x(n_0 + 2), \ldots$. Now, if the output at time $n = n_0$ is to depend only on the present and past inputs, then, clearly, the impulse response of the system must satisfy the condition

$$h(n) = 0 \qquad n < 0 \qquad\qquad (2.3.38)$$

Since $h(n)$ is the response of the relaxed linear time-invariant system to a unit impulse applied at $n = 0$, it follows that $h(n) = 0$ for $n < 0$ is both a necessary and a sufficient condition for causality. Hence an *LTI system is causal if and only if its impulse response is zero for negative values of n.*

Since for a causal system, $h(n) = 0$ for $n < 0$, the limits on the summation of the convolution formula may be modified to reflect this restriction. Thus we have the two equivalent forms

$$y(n) = \sum_{k=0}^{\infty} h(k)x(n - k) \qquad\qquad (2.3.39)$$

$$= \sum_{k=-\infty}^{n} x(k)h(n - k) \qquad\qquad (2.3.40)$$

As indicated previously, causality is required in any real-time signal processing application, since at any given time n we have no access to future values of the input signal. Only the present and past values of the input signal are available in computing the present output.

It is sometimes convenient to call a sequence that is zero for $n < 0$, a *causal sequence*, and one that is nonzero for $n < 0$ and $n > 0$, a *noncausal sequence*. This terminology means that such a sequence could be the unit sample response of a causal or a noncausal system, respectively.

If the input to a causal linear time-invariant system is a causal sequence [i.e., if $x(n) = 0$ for $n < 0$], the limits on the convolution formula are further restricted. In this case the two equivalent forms of the convolution formula become

$$y(n) = \sum_{k=0}^{n} h(k)x(n-k) \tag{2.3.41}$$

$$= \sum_{k=0}^{n} x(k)h(n-k) \tag{2.3.42}$$

We observe that in this case, the limits on the summations for the two alternative forms are identical, and the upper limit is growing with time. Clearly, the response of a causal system to a causal input sequence is causal, since $y(n) = 0$ for $n < 0$.

Example 2.3.5

Determine the unit step response of the linear time-invariant system with impulse response

$$h(n) = a^n u(n) \qquad |a| < 1$$

Solution   Since the input signal is a unit step, which is a causal signal, and the system is also causal, we can use one of the special forms of the convolution formula, either (2.3.41) or (2.3.42). Since $x(n) = 1$ for $n \geq 0$, (2.3.41) is simpler to use. Because of the simplicity of this problem, one can skip the steps involved with sketching the folded and shifted sequences. Instead, we use direct substitution of the signals sequences in (2.3.41) and obtain

$$y(n) = \sum_{k=0}^{n} a^k$$

$$= \frac{1 - a^{n+1}}{1 - a}$$

and $y(n) = 0$ for $n < 0$. We note that this result is identical to that obtained in Example 2.3.3. In this simple case, however, we computed the convolution algebraically without resorting to the detailed procedure outlined previously.

2.3.6 Stability of Linear Time-Invariant Systems

As indicated previously, stability is an important property that must be considered in any practical implementation of a system. We defined an arbitrary relaxed system as BIBO stable if and only if its output sequence $y(n)$ is bounded for every bounded input $x(n)$.

If $x(n)$ is bounded, there exists a constant M_x such that

$$|x(n)| \le M_x < \infty$$

Similarly, if the output is bounded, there exists a constant M_y such that

$$|y(n)| < M_y < \infty$$

for all n.

Now, given such a bounded input sequence $x(n)$ to a linear time-invariant system, let us investigate the implications of the definition of stability on the characteristics of the system. Toward this end, we work again with the convolution formula

$$y(n) = \sum_{k=-\infty}^{\infty} h(k)x(n-k)$$

If we take the absolute value of both sides of this equation, we obtain

$$|y(n)| = \left| \sum_{k=-\infty}^{\infty} h(k)x(n-k) \right|$$

Now, the absolute value of the sum of terms is always less than or equal to the sum of the absolute values of the terms. Hence

$$|y(n)| \le \sum_{k=-\infty}^{\infty} |h(k)||x(n-k)|$$

If the input is bounded, there exists a finite number M_x such that $|x(n)| \le M_x$. By substituting this upper bound for $x(n)$ in the equation above, we obtain

$$|y(n)| \le M_x \sum_{k=-\infty}^{\infty} |h(k)|$$

From this expression we observe that the output is bounded if the impulse response of the system satisfies the condition

$$S_h \equiv \sum_{k=-\infty}^{\infty} |h(k)| < \infty \qquad (2.3.43)$$

That is, *a linear time-invariant system is stable if its impulse response is absolutely summable.* This condition is not only sufficient but it is also necessary to ensure the stability of the system. Indeed, we shall show that if $S_h = \infty$, there is a bounded input for which the output is not bounded. We choose the bounded input

$$x(n) = \begin{cases} \dfrac{h^*(-n)}{|h^*(-n)|} & h(n) \ne 0 \\ 0, & h(n) = 0 \end{cases}$$

where $h^*(n)$ is the complex conjugate of $h(n)$. It is sufficient to show that there is one value of n for which $y(n)$ is unbounded. For $n = 0$ we have

$$y(0) = \sum_{k=-\infty}^{\infty} x(-k)h(k) = \sum_{k=-\infty}^{\infty} \frac{|h(k)|^2}{|h(k)|} = S_h$$

Thus, if $S_h = \infty$, a bounded input produces an unbounded output since $y(0) = \infty$.

The condition in (2.3.43) implies that the impulse response $h(n)$ goes to zero as n approaches infinity. As a consequence, the output of the system goes to zero as n approaches infinity if the input is set to zero beyond $n > n_0$. To prove this, suppose that $|x(n)| < M_x$ for $n < n_0$ and $x(n) = 0$ for $n \geq n_0$. Then, at $n = n_0 + N$, the system output is

$$y(n_0 + N) = \sum_{k=-\infty}^{N-1} h(k)x(n_0 + N - k) + \sum_{k=N}^{\infty} h(k)x(n_0 + N - k)$$

But the first sum is zero since $x(n) = 0$ for $n \geq n_0$. For the remaining part, we take the absolute value of the output, which is

$$|y(n_0 + N)| = \left| \sum_{k=N}^{\infty} h(k)x(n_0 + N - k) \right| \leq \sum_{k=N}^{\infty} |h(k)||x(n_0 + N - k)|$$

$$\leq M_x \sum_{k=N}^{\infty} |h(k)|$$

Now, as N approaches infinity,

$$\lim_{N \to \infty} \sum_{k=N}^{\infty} |h(n)| = 0$$

and hence

$$\lim_{N \to \infty} |y(n_0 + N)| = 0$$

This result implies that any excitation at the input to the system, which is of a finite duration, produces an output that is "transient" in nature; that is, its amplitude decays with time and dies out eventually, when the system is stable.

Example 2.3.6

Determine the range of values of the parameter a for which the linear time-invariant system with impulse response

$$h(n) = a^n u(n)$$

is stable.

Solution   First, we note that the system is causal. Consequently, the lower index on the summation in (2.3.43) begins with $k = 0$. Hence

$$\sum_{k=0}^{\infty} |a^k| = \sum_{k=0}^{\infty} |a|^k = 1 + |a| + |a|^2 + \cdots$$

Clearly, this geometric series converges to

$$\sum_{k=0}^{\infty} |a|^k = \frac{1}{1 - |a|}$$

provided that $|a| < 1$. Otherwise, it diverges. Therefore, the system is stable if $|a| < 1$. Otherwise, it is unstable. In effect, $h(n)$ must decay exponentially toward zero as n approaches infinity for the system to be stable.

Example 2.3.7

Determine the range of values of a and b for which the linear time-invariant system with impulse response

$$h(n) = \begin{cases} a^n, & n \geq 0 \\ b^n, & n < 0 \end{cases}$$

is stable.

Solution   This system is noncasual. The condition on stability given by (2.3.43) yields

$$\sum_{n=-\infty}^{\infty} |h(n)| = \sum_{n=0}^{\infty} |a|^n + \sum_{n=-\infty}^{-1} |b|^n$$

From Example 2.3.6 we have already determined that the first sum converges for $|a| < 1$. The second sum can be manipulated as follows:

$$\sum_{n=-\infty}^{-1} |b|^n = \sum_{n=1}^{\infty} \frac{1}{|b|^n} = \frac{1}{|b|} \left(1 + \frac{1}{|b|} + \frac{1}{|b|^2} + \cdots \right)$$

$$= \beta(1 + \beta + \beta^2 + \cdots) = \frac{\beta}{1 - \beta}$$

where $\beta = 1/|b|$ must be less than unity for the geometric series to converge. Consequently, the system is stable if both $|a| < 1$ and $|b| > 1$ are satisfied.

2.3.7 Systems with Finite-Duration and Infinite-Duration Impulse Response

Up to this point we have characterized a linear time-invariant system in terms of its impulse response $h(n)$. It is also convenient, however, to subdivide the class of linear time-invariant systems into two types, those that have a finite-duration impulse response (FIR) and those that have an infinite-duration impulse response (IIR). Thus an FIR system has an impulse response that is zero outside of some finite time interval. Without loss of generality, we focus our attention on causal FIR systems, so that

$$h(n) = 0 \qquad n < 0 \text{ and } n \geq M$$

The convolution formula for such a system reduces to

$$y(n) = \sum_{k=0}^{M-1} h(k)x(n-k)$$

A useful interpretation of this expression is obtained by observing that the output at any time n is simply a weighted linear combination of the input signal samples $x(n), x(n-1), \ldots, x(n-M+1)$. In other words, the system simply weights, by the values of the impulse response $h(k)$, $k = 0, 1, \ldots, M-1$, the most recent M signal samples and sums the resulting M products. In effect, the system acts as a *window* that views only the most recent M input signal samples in forming the output. It neglects or simply "forgets" all prior input samples [i.e., $x(n-M)$,

$x(n-M-1), \ldots]$. Thus we say that an FIR system has a finite memory of length-M samples.

In contrast, an IIR linear time-invariant system has an infinite-duration impulse response. Its output, based on the convolution formula, is

$$y(n) = \sum_{k=0}^{\infty} h(k)x(n-k)$$

where causality has been assumed, although this assumption is not necessary. Now, the system output is a weighted [by the impulse response $h(k)$] linear combination of the input signal samples $x(n)$, $x(n-1)$, $x(n-2), \ldots$. Since this weighted sum involves the present and all the past input samples, we say that the system has an infinite memory.

We investigate the characteristics of FIR and IIR systems in more detail in subsequent chapters.

2.4 DISCRETE-TIME SYSTEMS DESCRIBED BY DIFFERENCE EQUATIONS

Up to this point we have treated linear and time-invariant systems that are characterized by their unit sample response $h(n)$. In turn, $h(n)$ allows us to determine the output $y(n)$ of the system for any given input sequence $x(n)$ by means of the convolution summation,

$$y(n) = \sum_{k=-\infty}^{\infty} h(k)x(n-k) \tag{2.4.1}$$

In general, then, we have shown that any linear time-invariant system is characterized by the input–output relationship in (2.4.1). Moreover, the convolution summation formula in (2.4.1) suggests a means for the realization of the system. In the case of FIR systems, such a realization involves additions, multiplications, and a finite number of memory locations. Consequently, an FIR system is readily implemented directly, as implied by the convolution summation.

If the system is IIR, however, its practical implementation as implied by convolution is clearly impossible, since it requires an infinite number of memory locations, multiplications, and additions. A question that naturally arises, then, is whether or not it is possible to realize IIR systems other than in the form suggested by the convolution summation. Fortunately, the answer is yes, there is a practical and computationally efficient means for implementing a family of IIR systems, as will be demonstrated in this section. Within the general class of IIR systems, this family of discrete-time systems is more conveniently described by difference equations. This family or subclass of IIR systems is very useful in a variety of practical applications, including the implementation of digital filters, and the modeling of physical phenomena and physical systems.

2.4.1 Recursive and Nonrecursive Discrete-Time Systems

As indicated above, the convolution summation formula expresses the output of the linear time-invariant system explicitly and only in terms of the input signal. However, this need not be the case, as is shown here. There are many systems where it is either necessary or desirable to express the output of the system not only in terms of the present and past values of the input, but also in terms of the already available past output values. The following problem illustrates this point.

Suppose that we wish to compute the *cumulative average* of a signal $x(n)$ in the interval $0 \le k \le n$, defined as

$$y(n) = \frac{1}{n+1} \sum_{k=0}^{n} x(k) \qquad n = 0, 1, \ldots \tag{2.4.2}$$

As implied by (2.4.2), the computation of $y(n)$ requires the storage of all the input samples $x(k)$ for $0 \le k \le n$. Since n is increasing, our memory requirements grow linearly with time.

Our intuition suggests, however, that $y(n)$ can be computed more efficiently by utilizing the previous output value $y(n-1)$. Indeed, by a simple algebraic rearrangement of (2.4.2), we obtain

$$(n+1)y(n) = \sum_{k=0}^{n-1} x(k) + x(n)$$

$$= ny(n-1) + x(n)$$

and hence

$$y(n) = \frac{n}{n+1} y(n-1) + \frac{1}{n+1} x(n) \tag{2.4.3}$$

Now, the cumulative average $y(n)$ can be computed recursively by multiplying the previous output value $y(n-1)$ by $n/(n+1)$, multiplying the present input $x(n)$ by $1/(n+1)$, and adding the two products. Thus the computation of $y(n)$ by means of (2.4.3) requires two multiplications, one addition, and one memory location, as illustrated in Fig. 2.28. This is an example of a *recursive system*. In general, a system whose output $y(n)$ at time n depends on any number of past output values $y(n-1)$, $y(n-2)$, $\ldots$ is called a recursive system.

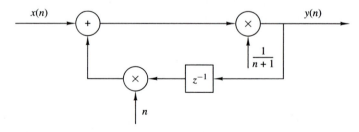

Figure 2.28 Realization of a recursive cumulative averaging system.

To determine the computation of the recursive system in (2.4.3) in more detail, suppose that we begin the process with $n = 0$ and proceed forward in time. Thus, according to (2.4.3), we obtain

$$y(0) = x(0)$$

$$y(1) = \tfrac{1}{2}y(0) + \tfrac{1}{2}x(1)$$

$$y(2) = \tfrac{2}{3}y(1) + \tfrac{1}{3}x(2)$$

and so on. If one grows fatigued with this computation and wishes to pass the problem to someone else at some time, say $n = n_0$, the only information that one needs to provide his or her successor is the past value $y(n_0 - 1)$ and the new input samples $x(n), x(n + 1), \ldots$. Thus the successor begins with

$$y(n_0) = \frac{n_0}{n_0 + 1}y(n_0 - 1) + \frac{1}{n_0 + 1}x(n_0)$$

and proceeds forward in time until some time, say $n = n_1$, when he or she becomes fatigued and passes the computational burden to someone else with the information on the value $y(n_1 - 1)$, and so on.

The point we wish to make in this discussion is that if one wishes to compute the response (in this case, the cumulative average) of the system (2.4.3) to an input signal $x(n)$ applied at $n = n_0$, we need the value $y(n_0 - 1)$ and the input samples $x(n)$ for $n \geq n_0$. The term $y(n_0 - 1)$ is called the *initial condition* for the system in (2.4.3) and contains all the information needed to determine the response of the system for $n \geq n_0$ to the input signal $x(n)$, independent of what has occurred in the past.

The following example illustrates the use of a (nonlinear) recursive system to compute the square root of a number.

Example 2.4.1   Square-Root Algorithm

Many computers and calculators compute the square root of a positive number A, using the iterative algorithm

$$s_n = \frac{1}{2}\left(s_{n-1} + \frac{A}{s_{n-1}}\right) \qquad n = 0, 1, \ldots$$

where s_{-1} is an initial guess (estimate) of $\sqrt{A}$. As the iteration converges we have $s_n \approx s_{n-1}$. Then it easily follows that $s_n \approx \sqrt{A}$.

Consider now the recursive system

$$y(n) = \frac{1}{2}\left[y(n-1) + \frac{x(n)}{y(n-1)}\right] \tag{2.4.4}$$

which is realized as in Fig. 2.29. If we excite this system with a step of amplitude A [i.e., $x(n) = Au(n)$] and use as an initial condition $y(-1)$ an estimate of $\sqrt{A}$, the response $y(n)$ of the system will tend toward $\sqrt{A}$ as n increases. Note that in contrast to the system (2.4.3), we do not need to specify exactly the initial condition. A rough estimate is sufficient for the proper performance of the system. For example, if we

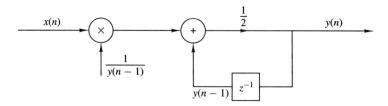

Figure 2.29 Realization of the square-root system.

let $A = 2$ and $y(-1) = 1$, we obtain $y(0) = \frac{3}{2}$, $y(1) = 1.4166667$, $y(2) = 1.4142157$. Similarly, for $y(-1) = 1.5$, we have $y(0) = 1.416667$, $y(1) = 1.4142157$. Compare these values with the $\sqrt{2}$, which is approximately 1.4142136.

We have now introduced two simple recursive systems, where the output $y(n)$ depends on the previous output value $y(n - 1)$ and the current input $x(n)$. Both systems are causal. In general, we can formulate more complex causal recursive systems, in which the output $y(n)$ is a function of several past output values and present and past inputs. The system should have a finite number of delays or, equivalently, should require a finite number of storage locations to be practically implemented. Thus the output of a causal and practically realizable recursive system can be expressed in general as

$$y(n) = F[y(n-1), y(n-2), \ldots, y(n-N), x(n), x(n-1), \ldots, x(n-M)] \quad (2.4.5)$$

where $F[\cdot]$ denotes some function of its arguments. This is a recursive equation specifying a procedure for computing the system output in terms of previous values of the output and present and past inputs.

In contrast, if $y(n)$ depends only on the present and past inputs, then

$$y(n) = F[x(n), x(n-1), \ldots, x(n-M)] \quad (2.4.6)$$

Such a system is called *nonrecursive*. We hasten to add that the causal FIR systems described in Section 2.3.7 in terms of the convolution sum formula have the form of (2.4.6). Indeed, the convolution summation for a causal FIR system is

$$y(n) = \sum_{k=0}^{M} h(k)x(n - k)$$
$$= h(0)x(n) + h(1)x(n-1) + \cdots + h(M)x(n-M)$$
$$= F[x(n), x(n-1), \ldots, x(n-M)]$$

where the function $F[\cdot]$ is simply a linear weighted sum of present and past inputs and the impulse response values $h(n)$, $0 \le n \le M$, constitute the weighting coefficients. Consequently, the causal linear time-invariant FIR systems described by the convolution formula in Section 2.3.7, are nonrecursive. The basic differences between nonrecursive and recursive systems are illustrated in Fig. 2.30. A simple inspection of this figure reveals that the fundamental difference between these two

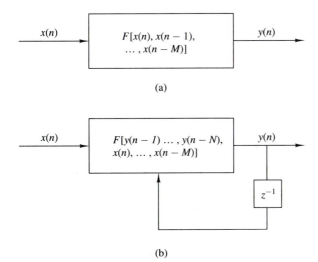

(a)

(b)

Figure 2.30 Basic form for a causal and realizable (a) nonrecursive and (b) recursive system.

systems is the feedback loop in the recursive system, which feeds back the output of the system into the input. This feedback loop contains a delay element. The presence of this delay is crucial for the realizability of the system, since the absence of this delay would force the system to compute $y(n)$ in terms of $y(n)$, which is not possible for discrete-time systems.

The presence of the feedback loop or, equivalently, the recursive nature of (2.4.5) creates another important difference between recursive and nonrecursive systems. For example, suppose that we wish to compute the output $y(n_0)$ of a system when it is excited by an input applied at time $n = 0$. If the system is recursive, to compute $y(n_0)$, we first need to compute all the previous values $y(0)$, $y(1), \ldots, y(n_0 - 1)$. In contrast, if the system is nonrecursive, we can compute the output $y(n_0)$ immediately without having $y(n_0 - 1)$, $y(n_0 - 2), \ldots$. In conclusion, the output of a recursive system should be computed in order [i.e., $y(0)$, $y(1)$, $y(2), \ldots$], whereas for a nonrecursive system, the output can be computed in any order [i.e., $y(200)$, $y(15)$, $y(3)$, $y(300)$, etc.]. This feature is desirable in some practical applications.

2.4.2 Linear Time-Invariant Systems Characterized by Constant-Coefficient Difference Equations

In Section 2.3 we treated linear time-invariant systems and characterized them in terms of their impulse responses. In this subsection we focus our attention on a family of linear time-invariant systems described by an input–output relation called a difference equation with constant coeffficients. Systems described by constant-coefficient linear difference equations are a subclass of the recursive and nonrecursive systems introduced in the preceding subsection. To bring out the important ideas, we begin by treating a simple recursive system described by a first-order difference equation.

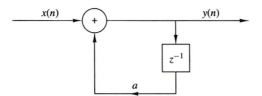

Figure 2.31 Block diagram realization of a simple recursive system.

Suppose that we have a recursive system with an input–output equation

$$y(n) = ay(n-1) + x(n) \qquad (2.4.7)$$

where a is a constant. Figure 2.31 shows a block diagram realization of the system. In comparing this system with the cumulative averaging system described by the input–output equation (2.4.3), we observe that the system in (2.4.7) has a constant coefficient (independent of time), whereas the system described in (2.4.3) has time-variant coefficients. As we will show, (2.4.7) is an input–output equation for a linear time-invariant system, whereas (2.4.3) describes a linear time-variant system.

Now, suppose that we apply an input signal $x(n)$ to the system for $n \geq 0$. We make no assumptions about the input signal for $n < 0$, but we do assume the existence of the initial condition $y(-1)$. Since (2.4.7) describes the system output implicitly, we must solve this equation to obtain an explicit expression for the system output. Suppose that we compute successive values of $y(n)$ for $n \geq 0$, beginning with $y(0)$. Thus

$$y(0) = ay(-1) + x(0)$$

$$y(1) = ay(0) + x(1) = a^2 y(-1) + ax(0) + x(1)$$

$$y(2) = ay(1) + x(2) = a^3 y(-1) + a^2 x(0) + ax(1) + x(2)$$

$$\vdots \qquad \vdots$$

$$y(n) = ay(n-1) + x(n)$$

$$= a^{n+1} y(-1) + a^n x(0) + a^{n-1} x(1) + \cdots + ax(n-1) + x(n)$$

or, more compactly,

$$y(n) = a^{n+1} y(-1) + \sum_{k=0}^{n} a^k x(n-k) \qquad n \geq 0 \qquad (2.4.8)$$

The response $y(n)$ of the system as given by the right-hand side of (2.4.8) consists of two parts. The first part, which contains the term $y(-1)$, is a result of the initial condition $y(-1)$ of the system. The second part is the response of the system to the input signal $x(n)$.

If the system is initially relaxed at time $n = 0$, then its memory (i.e., the output of the delay) should be zero. Hence $y(-1) = 0$. Thus a recursive system is relaxed if it starts with zero initial conditions. Because the memory of the system describes, in some sense, its "state," we say that the system is at zero state and its corresponding output is called the *zero-state response* or *forced response*, and is denoted by $y_{zs}(n)$. Obviously, the zero-state response or forced response of the system (2.4.7) is given by

$$y_{zs}(n) = \sum_{k=0}^{n} a^k x(n-k) \qquad n \ge 0 \tag{2.4.9}$$

It is interesting to note that (2.4.9) is a convolution summation involving the input signal convolved with the impulse response

$$h(n) = a^n u(n) \tag{2.4.10}$$

We also observe that the system described by the first-order difference equation in (2.4.7) is causal. As a result, the lower limit on the convolution summation in (2.4.9) is $k = 0$. Furthermore, the condition $y(-1) = 0$ implies that the input signal can be assumed causal and hence the upper limit on the convolution summation in (2.4.9) is n, since $x(n-k) = 0$ for $k > n$. In effect, we have obtained the result that the relaxed recursive system described by the first-order difference equation in (2.4.7), is a linear time-invariant IIR system with impulse response given by (2.4.10).

Now, suppose that the system described by (2.4.7) is initially nonrelaxed [i.e., $y(-1) \ne 0$] and the input $x(n) = 0$ for all n. Then the output of the system with zero input is called the *zero-input response* or *natural response* and is denoted by $y_{zi}(n)$. From (2.4.7), with $x(n) = 0$ for $-\infty < n < \infty$, we obtain

$$y_{zi}(n) = a^{n+1} y(-1) \qquad n \ge 0 \tag{2.4.11}$$

We observe that a recursive system with nonzero initial condition is nonrelaxed in the sense that it can produce an output without being excited. Note that the zero-input response is due to the memory of the system.

To summarize, the zero-input response is obtained by setting the input signal to zero, making it independent of the input. It depends only on the nature of the system and the initial condition. Thus the zero-input response is a characteristic of the system itself, and it is also known as the *natural* or *free response* of the system. On the other hand, the zero-state response depends on the nature of the system and the input signal. Since this output is a response forced upon it by the input signal, it is usually called the *forced response* of the system. In general, the total response of the system can be expressed as $y(n) = y_{zi}(n) + y_{zs}(n)$.

The system described by the first-order difference equation in (2.4.7) is the simplest possible recursive system in the general class of recursive systems described by linear constant-coefficient difference equations. The general form for such an equation is

$$y(n) = -\sum_{k=1}^{N} a_k y(n-k) + \sum_{k=0}^{M} b_k x(n-k) \tag{2.4.12}$$

or, equivalently,

$$\sum_{k=0}^{N} a_k y(n-k) = \sum_{k=0}^{M} b_k x(n-k) \qquad a_0 \equiv 1 \tag{2.4.13}$$

The integer N is called the *order* of the difference equation or the order of the system. The negative sign on the right-hand side of (2.4.12) is introduced as a matter of convenience to allow us to express the difference equation in (2.4.13) without any negative signs.

Equation (2.4.12) expresses the output of the system at time n directly as a weighted sum of past outputs $y(n-1)$, $y(n-2), \ldots, y(n-N)$ as well as past and present input signals samples. We observe that in order to determine $y(n)$ for $n \geq 0$, we need the input $x(n)$ for all $n \geq 0$, and the initial conditions $y(-1)$, $y(-2), \ldots, y(-N)$. In other words, the initial conditions summarize all that we need to know about the past history of the response of the system to compute the present and future outputs. The general solution of the N-order constant-coefficient difference equation is considered in the following subsection.

At this point we restate the properties of linearity, time invariance, and stability in the context of recursive systems described by linear constant-coefficient difference equations. As we have observed, a recursive system may be relaxed or nonrelaxed, depending on the initial conditions. Hence the definitions of these properties must take into account the presence of the initial conditions.

We begin with the definition of linearity. A system is linear if it satisfies the following three requirements:

1. The total response is equal to the sum of the zero-input and zero-state responses [i.e., $y(n) = y_{zi}(n) + y_{zs}(n)$].
2. The principle of superposition applies to the zero-state response (*zero-state linear*).
3. The principle of superposition applies to the zero-input response (*zero-input linear*).

A system that does not satisfy *all three* separate requirements is by definition nonlinear. Obviously, for a relaxed system, $y_{zi}(n) = 0$, and thus requirement 2, which is the definition of linearity given in Section 2.2.4, is sufficient.

We illustrate the application of these requirements by a simple example.

Example 2.4.2

Determine if the recursive system defined by the difference equation

$$y(n) = ay(n-1) + x(n)$$

is linear.

Solution   By combining (2.4.9) and (2.4.11), we obtain (2.4.8), which can be expressed as

$$y(n) = y_{zi}(n) + y_{zs}(n)$$

Thus the first requirement for linearity is satisfied.

To check for the second requirement, let us assume that $x(n) = c_1 x_1(n) + c_2 x_2(n)$. Then (2.4.9) gives

$$y_{zs}(n) = \sum_{k=0}^{n} a^k [c_1 x_1(n-k) + c_2 x_2(n-k)]$$

$$= c_1 \sum_{k=0}^{n} a^k x_1(n-k) + c_2 \sum_{k=0}^{n} a^k x_2(n-k)$$

$$= c_1 y_{zs}^{(1)}(n) + c_2 y_{zs}^{(2)}(n)$$

Hence $y_{zs}(n)$ satisfies the principle of superposition, and thus the system is zero-state linear.

Now let us assume that $y(-1) = c_1 y_1(-1) + c_2 y_2(-1)$. From (2.4.11) we obtain

$$y_{zi}(n) = a^{n+1} [c_1 y_1(-1) + c_2 y_2(-1)]$$

$$= c_1 a^{n+1} y_1(-1) + c_2 a^{n+1} y_2(-1)$$

$$= c_1 y_{zi}^{(1)}(n) + c_2 y_{zi}^{(2)}(n)$$

Hence the system is zero-input linear.

Since the system satisfies all three conditions for linearity, it is linear.

Although it is somewhat tedious, the procedure used in Example 2.4.2 to demonstrate linearity for the system described by the first-order difference equation, carries over directly to the general recursive systems described by the constant-coefficient difference equation given in (2.4.13). Hence, a recursive system described by the linear difference equation in (2.4.13) also satisfies all three conditions in the definition of linearity, and therefore it is linear.

The next question that arises is whether or not the causal linear system described by the linear constant-coefficient difference equation in (2.4.13) is time invariant. This is fairly easy, when dealing with systems described by explicit input–output mathematical relationships. Clearly, the system described by (2.4.13) is time invariant because the coefficients a_k and b_k are constants. On the other hand, if one or more of these coefficients depends on time, the system is time variant, since its properties change as a function of time. Thus we conclude that *the recursive system described by a linear constant-coefficient difference equation is linear and time invariant.*

The final issue is the stability of the recursive system described by the linear, constant-coefficient difference equation in (2.4.13). In Section 2.3.6 we introduced the concept of bounded input–bounded output (BIBO) stability for relaxed systems. For nonrelaxed systems that may be nonlinear, BIBO stability should be viewed with some care. However, in the case of a linear time-invariant recursive system described by the linear constant-coefficient difference equation in (2.4.13), it suffices to state that such a system is BIBO stable if and only if for every bounded input and every bounded initial condition, the total system response is bounded.

Example 2.4.3

Determine if the linear time-invariant recursive system described by the difference equation given in (2.4.7) is stable.

Solution Let us assume that the input signal $x(n)$ is bounded in amplitude, that is, $|x(n)| \le M_x < \infty$ for all $n \ge 0$. From (2.4.8) we have

$$|y(n)| \le |a^{n+1}y(-1)| + \left| \sum_{k=0}^{n} a^k x(n-k) \right|, \qquad n \ge 0$$

$$\le |a|^{n+1}|y(-1)| + M_x \sum_{k=0}^{n} |a|^k, \qquad n \ge 0$$

$$\le |a|^{n+1}|y(-1)| + M_x \frac{1 - |a|^{n+1}}{1 - |a|} = M_y, \ n \ge 0$$

If n is finite, the bound M_y is finite and the output is bounded independently of the value of a. However, as $n \to \infty$, the bound M_y remains finite only if $|a| < 1$ because $|a|^n \to 0$ as $n \to \infty$. Then $M_y = M_x/(1 - |a|)$.

Thus the system is stable only if $|a| < 1$.

For the simple first-order system in Example 2.4.3, we were able to express the condition for BIBO stability in terms of the system parameter a, namely $|a| < 1$. We should stress, however, that this task becomes more difficult for higher-order systems. Fortunately, as we shall see in subsequent chapters, other simple and more efficient techniques exist for investigating the stability of recursive systems.

2.4.3 Solution of Linear Constant-Coefficient Difference Equations

Given a linear constant-coefficient difference equation as the input–output relationship describing a linear time-invariant system, our objective in this subsection is to determine an explicit expression for the output $y(n)$. The method that is developed is termed the *direct method*. An alternative method based on the z-transform is described in Chapter 3. For reasons that will become apparent later, the z-transform approach is called the *indirect method*.

Basically, the goal is to determine the output $y(n)$, $n \ge 0$, of the system given a specific input $x(n)$, $n \ge 0$, and a set of initial conditions. The direct solution method assumes that the total solution is the sum of two parts:

$$y(n) = y_h(n) + y_p(n)$$

The part $y_h(n)$ is known as the *homogeneous* or *complementary* solution, whereas $y_p(n)$ is called the *particular* solution.

The homogeneous solution of a difference equation. We begin the problem of solving the linear constant-coefficient difference equation given by

(2.4.13) by assuming that the input $x(n) = 0$. Thus we will first obtain the solution to the *homogeneous difference equation*

$$\sum_{k=0}^{N} a_k y(n-k) = 0 \qquad (2.4.14)$$

The procedure for solving a linear constant-coefficient difference equation directly is very similar to the procedure for solving a linear constant-coefficient differential equation. Basically, we assume that the solution is in the form of an exponential, that is,

$$y_h(n) = \lambda^n \qquad (2.4.15)$$

where the subscript h on $y(n)$ is used to denote the solution to the homogeneous difference equation. If we substitute this assumed solution in (2.4.14), we obtain the polynomial equation

$$\sum_{k=0}^{N} a_k \lambda^{n-k} = 0$$

or

$$\lambda^{n-N}(\lambda^N + a_1 \lambda^{N-1} + a_2 \lambda^{N-2} + \cdots + a_{N-1}\lambda + a_N) = 0 \qquad (2.4.16)$$

The polynomial in parentheses is called the *characteristic polynomial* of the system. In general, it has N roots, which we denote as $\lambda_1, \lambda_2, \ldots, \lambda_N$. The roots can be real or complex valued. In practice the coefficients $a_1, a_2, \ldots, a_N$ are usually real. Complex-valued roots occur as complex-conjugate pairs. Some of the N roots may be identical, in which case we have multiple-order roots.

For the moment, let us assume that the roots are distinct, that is, there are no multiple-order roots. Then the most general solution to the homogeneous difference equation in (2.4.14) is

$$y_h(n) = C_1 \lambda_1^n + C_2 \lambda_2^n + \cdots + C_N \lambda_N^n \qquad (2.4.17)$$

where $C_1, C_2, \ldots, C_N$ are weighting coefficients.

These coefficients are determined from the initial conditions specified for the system. Since the input $x(n) = 0$, (2.4.17) can be used to obtain the *zero-input response* of the system. The following examples illustrate the procedure.

Example 2.4.4

Determine the homogeneous solution of the system described by the first-order difference equation

$$y(n) + a_1 y(n-1) = x(n) \qquad (2.4.18)$$

Solution   The assumed solution obtained by setting $x(n) = 0$ is

$$y_h(n) = \lambda^n$$

When we substitute this solution in (2.4.18), we obtain [with $x(n) = 0$]

$$\lambda^n + a_1\lambda^{n-1} = 0$$

$$\lambda^{n-1}(\lambda + a_1) = 0$$

$$\lambda = -a_1$$

Therefore, the solution to the homogeneous difference equation is

$$y_h(n) = C\lambda^n = C(-a_1)^n \qquad (2.4.19)$$

The zero-input response of the system can be determined from (2.4.18) and (2.4.19). With $x(n) = 0$, (2.4.18) yields

$$y(0) = -a_1 y(-1)$$

On the other hand, from (2.4.19) we have

$$y_h(0) = C$$

and hence the zero-input response of the system is

$$y_{zi}(n) = (-a_1)^{n+1}y(-1) \qquad n \geq 0 \qquad (2.4.20)$$

With $a = -a_1$, this result is consistent with (2.4.11) for the first-order system, which was obtained earlier by iteration of the difference equation.

Example 2.4.5

Determine the zero-input response of the system described by the homogeneous second-order difference equation

$$y(n) - 3y(n-1) - 4y(n-2) = 0 \qquad (2.4.21)$$

Solution First we determine the solution to the homogeneous equation. We assume the solution to be the exponential

$$y_h(n) = \lambda^n$$

Upon substitution of this solution into (2.4.21), we obtain the characteristic equation

$$\lambda^n - 3\lambda^{n-1} - 4\lambda^{n-2} = 0$$

$$\lambda^{n-2}(\lambda^2 - 3\lambda - 4) = 0$$

Therefore, the roots are $\lambda = -1, 4$, and the general form of the solution to the homogeneous equation is

$$\begin{aligned} y_h(n) &= C_1\lambda_1^n + C_2\lambda_2^n \\ &= C_1(-1)^n + C_2(4)^n \end{aligned} \qquad (2.4.22)$$

The zero-input response of the system can be obtained from the homogenous solution by evaluating the constants in (2.4.22), given the initial conditions $y(-1)$ and $y(-2)$. From the difference equation in (2.4.21) we have

$$y(0) = 3y(-1) + 4y(-2)$$

$$\begin{aligned} y(1) &= 3y(0) + 4y(-1) \\ &= 3[3y(-1) + 4y(-2)] + 4y(-1) \\ &= 13y(-1) + 12y(-2) \end{aligned}$$

On the other hand, from (2.4.22) we obtain

$$y(0) = C_1 + C_2$$

$$y(1) = -C_1 + 4C_2$$

By equating these two sets of relations, we have

$$C_1 + C_2 = 3y(-1) + 4y(-2)$$

$$-C_1 + 4C_2 = 13y(-1) + 12y(-2)$$

The solution of these two equations is

$$C_1 = -\tfrac{1}{5}y(-1) + \tfrac{4}{5}y(-2)$$

$$C_2 = \tfrac{16}{5}y(-1) + \tfrac{16}{5}y(-2)$$

Therefore, the zero-input response of the system is

$$y_{zi}(n) = [-\tfrac{1}{5}y(-1) + \tfrac{4}{5}y(-2)](-1)^n$$

$$+[\tfrac{16}{5}y(-1) + \tfrac{16}{5}y(-2)](4)^n \qquad n \geq 0 \tag{2.4.23}$$

For example, if $y(-2) = 0$ and $y(-1) = 5$, then $C_1 = -1$, $C_2 = 16$, and hence

$$y_{zi}(n) = (-1)^{n+1} + (4)^{n+2} \qquad n \geq 0$$

These examples illustrate the method for obtaining the homogeneous solution and the zero-input response of the system when the characteristic equation contains distinct roots. On the other hand, if the characteristic equation contains multiple roots, the form of the solution given in (2.4.17) must be modified. For example, if λ_1 is a root of multiplicity m, then (2.4.17) becomes

$$y_h(n) = C_1\lambda_1^n + C_2 n\lambda_1^n + C_3 n^2\lambda_1^n + \cdots + C_m n^{m-1}\lambda_1^n$$

$$+ C_{m+1}\lambda_{m+1}^n + \cdots + C_N\lambda_n \tag{2.4.24}$$

The particular solution of the difference equation.   The particular solution $y_p(n)$ is required to satisfy the difference equation (2.4.13) for the specific input signal $x(n)$, $n \geq 0$. In other words, $y_p(n)$ is any solution satisfying

$$\sum_{k=0}^{N} a_k y_p(n - k) = \sum_{k=0}^{M} b_k x(n - k) \qquad a_0 = 1 \tag{2.4.25}$$

To solve (2.4.25), we assume for $y_p(n)$, a form that depends on the form of the input $x(n)$. The following example illustrates the procedure.

Example 2.4.6

Determine the particular solution of the first-order difference equation

$$y(n) + a_1 y(n - 1) = x(n), \qquad |a_1| < 1 \tag{2.4.26}$$

when the input $x(n)$ is a unit step sequence, that is,

$$x(n) = u(n)$$

Solution   Since the input sequence $x(n)$ is a constant for $n \geq 0$, the form of the solution that we assume is also a constant. Hence the assumed solution of the difference equation to the forcing function $x(n)$, called the *particular solution* of the difference equation, is

$$y_p(n) = Ku(n)$$

where K is a scale factor determined so that (2.4.26) is satisfied. Upon substitution of this assumed solution into (2.4.26), we obtain

$$Ku(n) + a_1 Ku(n-1) = u(n)$$

To determine K, we must evaluate this equation for any $n \geq 1$, where none of the terms vanish. Thus

$$K + a_1 K = 1$$

$$K = \frac{1}{1 + a_1}$$

Therefore, the particular solution to the difference equation is

$$y_p(n) = \frac{1}{1 + a_1} u(n) \tag{2.4.27}$$

In this example, the input $x(n)$, $n \geq 0$, is a constant and the form assumed for the particular solution is also a constant. If $x(n)$ is an exponential, we would assume that the particular solution is also an exponential. If $x(n)$ were a sinusoid, then $y_p(n)$ would also be a sinusoid. Thus our assumed form for the particular solution takes the basic form of the signal $x(n)$. Table 2.1 provides the general form of the particular solution for several types of excitation.

Example 2.4.7

Determine the particular solution of the difference equation

$$y(n) = \tfrac{5}{6}y(n-1) - \tfrac{1}{6}y(n-2) + x(n)$$

when the forcing function $x(n) = 2^n$, $n \geq 0$ and zero elsewhere.

TABLE 2.1   GENERAL FORM OF THE PARTICULAR SOLUTION FOR SEVERAL TYPES OF INPUT SIGNALS

Input Signal, $x(n)$	Particular Solution, $y_p(n)$
A (constant)	K
AM^n	KM^n
An^M	$K_0 n^M + K_1 n^{M-1} + \cdots + K_M$
$A^n n^M$	$A^n(K_0 n^M + K_1 n^{M-1} + \cdots + K_M)$
$\left\{ \begin{array}{l} A\cos\omega_0 n \\ A\sin\omega_0 n \end{array} \right\}$	$K_1 \cos\omega_0 n + K_2 \sin\omega_0 n$

Solution   The form of the particular solution is

$$y_p(n) = K2^n \qquad n \geq 0$$

Upon substitution of $y_p(n)$ into the difference equation, we obtain

$$K2^n u(n) = \tfrac{5}{6}K2^{n-1}u(n-1) - \tfrac{1}{6}K2^{n-2}u(n-2) + 2^n u(n)$$

To determine the value of K, we can evaluate this equation for any $n \geq 2$, where none of the terms vanish. Thus we obtain

$$4K = \tfrac{5}{6}(2K) - \tfrac{1}{6}K + 4$$

and hence $K = \tfrac{8}{5}$. Therefore, the particular solution is

$$y_p(n) = \tfrac{8}{5}2^n \qquad n \geq 0$$

We have now demonstrated how to determine the two components of the solution to a difference equation with constant coefficients. These two components are the homogeneous solution and the particular solution. From these two components, we construct the total solution from which we can obtain the zero-state response.

The total solution of the difference equation.   The linearity property of the linear constant-coefficient difference equation allows us to add the homogeneous solution and the particular solution in order to obtain the *total solution*. Thus

$$y(n) = y_h(n) + y_p(n)$$

The resultant sum $y(n)$ contains the constant parameters $\{C_i\}$ embodied in the homogeneous solution component $y_h(n)$. These constants can be determined to satisfy the initial conditions. The following example illustrates the procedure.

Example 2.4.8

Determine the total solution $y(n)$, $n \geq 0$, to the difference equation

$$y(n) + a_1 y(n-1) = x(n) \qquad\qquad (2.4.28)$$

when $x(n)$ is a unit step sequence [i.e., $x(n) = u(n)$] and $y(-1)$ is the initial condition.

Solution   From (2.4.19) of Example 2.4.4, the homogeneous solution is

$$y_h(n) = C(-a_1)^n$$

and from (2.4.26) of Example 2.4.6, the particular solution is

$$y_p(n) = \frac{1}{1 + a_1}$$

Consequently, the total solution is

$$y(n) = C(-a_1)^n + \frac{1}{1 + a_1} \qquad n \geq 0 \qquad\qquad (2.4.29)$$

where the constant C is determined to satisfy the initial condition $y(-1)$.

In particular, suppose that we wish to obtain the zero-state response of the system described by the first-order difference equation in (2.4.28). Then we set $y(-1) = 0$. To evaluate C, we evaluate (2.4.28) at $n = 0$ obtaining

$$y(0) + a_1 y(-1) = 1$$

$$y(0) = 1$$

On the other hand, (2.4.29) evaluated at $n = 0$ yields

$$y(0) = C + \frac{1}{1 + a_1}$$

Consequently,

$$C + \frac{1}{1 + a_1} = 1$$

$$C = \frac{a_1}{1 + a_1}$$

Substitution for C into (2.4.29) yields the zero-state response of the system

$$y_{zs}(n) = \frac{1 - (-a_1)^{n+1}}{1 + a_1} \qquad n \geq 0$$

If we evaluate the parameter C in (2.4.29) under the condition that $y(-1) \neq 0$, the total solution will include the zero-input response as well as the zero-state response of the system. In this case (2.4.28) yields

$$y(0) + a_1 y(-1) = 1$$

$$y(0) = -a_1 y(-1) + 1$$

On the other hand, (2.4.29) yields

$$y(0) = C + \frac{1}{1 + a_1}$$

By equating these two relations, we obtain

$$C + \frac{1}{1 + a_1} = -a_1 y(-1) + 1$$

$$C = -a_1 y(-1) + \frac{a_1}{1 + a_1}$$

Finally, if we substitute this value of C into (2.4.29), we obtain

$$y(n) = (-a_1)^{n+1} y(-1) + \frac{1 - (-a_1)^{n+1}}{1 + a_1} \qquad n \geq 0 \tag{2.4.30}$$

$$= y_{zi}(n) + y_{zs}(n)$$

We observe that the system response as given by (2.4.30) is consistent with the response $y(n)$ given in (2.4.8) for the first-order system (with $a = -a_1$), which was obtained by solving the difference equation iteratively. Furthermore, we note that the value of the constant C depends both on the initial condition $y(-1)$ and on the excitation function. Consequently, the value of C influences both the zero-input response and the zero-state response. On the other hand, if we wish to obtain the zero-state response only, we simply solve for C under the condition that $y(-1) = 0$, as demonstrated in Example 2.4.8.

We further observe that the particular solution to the difference equation can be obtained from the zero-state response of the system. Indeed, if $|a_1| < 1$, which is the condition for stability of the system, as will be shown in Section 2.4.4, the limiting value of $y_{zs}(n)$ as n approaches infinity, is the particular solution, that is,

$$y_p(n) = \lim_{n \to \infty} y_{zs}(n) = \frac{1}{1 + a_1}$$

Since this component of the system response does not go to zero as n approaches infinity, it is usually called the *steady-state response* of the system. This response persists as long as the input persists. The component that dies out as n approaches infinity is called the *transient response* of the system.

Example 2.4.9

Determine the response $y(n)$, $n \geq 0$, of the system described by the second-order difference equation

$$y(n) - 3y(n-1) - 4y(n-2) = x(n) + 2x(n-1) \tag{2.4.31}$$

when the input sequence is

$$x(n) = 4^n u(n)$$

Solution   We have already determined the solution to the homogeneous difference equation for this system in Example 2.4.5. From (2.4.22) we have

$$y_h(n) = C_1(-1)^n + C_2(4)^n \tag{2.4.32}$$

The particular solution to (2.4.31) is assumed to be an exponential sequence of the same form as $x(n)$. Normally, we could assume a solution of the form

$$y_p(n) = K(4)^n u(n)$$

However, we observe that $y_p(n)$ is already contained in the homogeneous solution, so that this particular solution is redundant. Instead, we select the particular solution to be linearly independent of the terms contained in the homogeneous solution. In fact, we treat this situation in the same manner as we have already treated multiple roots in the characteristic equation. Thus we assume that

$$y_p(n) = Kn(4)^n u(n) \tag{2.4.33}$$

Upon substitution of (2.4.33) into (2.4.31), we obtain

$$Kn(4)^n u(n) - 3K(n-1)(4)^{n-1} u(n-1) - 4K(n-2)(4)^{n-2} u(n-2)$$

$$= (4)^n u(n) + 2(4)^{n-1} u(n-1)$$

To determine K, we evaluate this equation for any $n \geq 2$, where none of the unit step terms vanish. To simplify the arithmetic, we select $n = 2$, from which we obtain $K = \frac{6}{5}$. Therefore,

$$y_p(n) = \frac{6}{5}n(4)^n u(n) \tag{2.4.34}$$

The total solution to the difference equation is obtained by adding (2.4.32) to (2.4.34). Thus

$$y(n) = C_1(-1)^n + C_2(4)^n + \tfrac{6}{5}n(4)^n \qquad n \geq 0 \qquad (2.4.35)$$

where the constants C_1 and C_2 are determined such that the initial conditions are satisfied. To accomplish this, we return to (2.4.31), from which we obtain

$$y(0) = 3y(-1) + 4y(-2) + 1$$

$$y(1) = 3y(0) + 4y(-1) + 6$$

$$= 13y(-1) + 12y(-2) + 9$$

On the other hand, (2.4.35) evaluated at $n = 0$ and $n = 1$ yields

$$y(0) = C_1 + C_2$$

$$y(1) = -C_1 + 4C_2 + \tfrac{24}{5}$$

We can now equate these two sets of relations to obtain C_1 and C_2. In so doing, we have the response due to initial conditions $y(-1)$ and $y(-2)$ (the zero-input response), and the zero-state or forced response.

Since we have already solved for the zero-input response in Example 2.4.5, we can simplify the computations above by setting $y(-1) = y(-2) = 0$. Then we have

$$C_1 + C_2 = 1$$

$$-C_1 + 4C_2 + \tfrac{24}{5} = 9$$

Hence $C_1 = -\tfrac{1}{25}$ and $C_2 = \tfrac{26}{25}$. Finally, we have the zero-state response to the forcing function $x(n) = (4)^n u(n)$ in the form

$$y_{zs}(n) = -\tfrac{1}{25}(-1)^n + \tfrac{26}{25}(4)^n + \tfrac{6}{5}n(4)^n \qquad n \geq 0 \qquad (2.4.36)$$

The total response of the system, which includes the response to arbitrary initial conditions, is the sum of (2.4.23) and (2.4.36).

2.4.4 The Impulse Response of a Linear Time-Invariant Recursive System

The impulse response of a linear time-invariant system was previously defined as the response of the system to a unit sample excitation [i.e., $x(n) = \delta(n)$]. In the case of a recursive system, $h(n)$ is simply equal to the zero-state response of the system when the input $x(n) = \delta(n)$ and the system is initially relaxed.

For example, in the simple first-order recursive system given in (2.4.7), the zero-state response given in (2.4.8), is

$$y_{zs}(n) = \sum_{k=0}^{n} a^k x(n - k) \qquad (2.4.37)$$

With $x(n) = \delta(n)$ is substituted into (2.4.37), we obtain

$$y_{zs}(n) = \sum_{k=0}^{n} a^k \delta(n - k)$$

$$= a^n \qquad n \geq 0$$

Hence the impulse response of the first-order recursive system described by (2.4.7) is

$$h(n) = a^n u(n) \tag{2.4.38}$$

as indicated in Section 2.4.2.

In the general case of an arbitrary, linear time-invariant recursive system, the zero-state response expressed in terms of the convolution summation is

$$y_{zs}(n) = \sum_{k=0}^{n} h(k)x(n-k) \qquad n \geq 0 \tag{2.4.39}$$

When the input is an impulse [i.e., $x(n) = \delta(n)$], (2.4.39) reduces to

$$y_{zs}(n) = h(n) \tag{2.4.40}$$

Now, let us consider the problem of determining the impulse response $h(n)$ given a linear constant-coefficient difference equation description of the system. In terms of our discussion in the preceding subsection, we have established the fact that the total response of the system to any excitation function consists of the sum of two solutions of the difference equation: the solution to the homogeneous equation plus the particular solution to the excitation function. In the case where the excitation is an impulse, the particular solution is zero, since $x(n) = 0$ for $n > 0$, that is,

$$y_p(n) = 0$$

Consequently, the response of the system to an impulse consists only of the solution to the homogeneous equation, with the $\{C_k\}$ parameters evaluated to satisfy the initial conditions dictated by the impulse. The following example illustrates the procedure for obtaining $h(n)$ given the difference equation for the system.

Example 2.4.10

Determine the impulse response $h(n)$ for the system described by the second-order difference equation

$$y(n) - 3y(n-1) - 4y(n-2) = x(n) + 2x(n-1) \tag{2.4.41}$$

Solution We have already determined in Example 2.4.5 that the solution to the homogeneous difference equation for this system is

$$y_h(n) = C_1(-1)^n + C_2(4)^n \qquad n \geq 0 \tag{2.4.42}$$

Since the particular solution is zero when $x(n) = \delta(n)$, the impulse response of the system is simply given by (2.4.42), where C_1 and C_2 must be evaluated to satisfy (2.4.41). For $n = 0$ and $n = 1$, (2.4.41) yields

$$y(0) = 1$$

$$y(1) = 3y(0) + 2 = 5$$

where we have imposed the conditions $y(-1) = y(-2) = 0$, since the system must be relaxed. On the other hand, (2.4.42) evaluated at $n = 0$ and $n = 1$ yields

$$y(0) = C_1 + C_2$$

$$y(1) = -C_1 + 4C_2$$

By solving these two sets of equations for C_1 and C_2, we obtain

$$C_1 = -\tfrac{1}{5}, \qquad C_2 = \tfrac{6}{5}$$

Therefore, the impulse response of the system is

$$h(n) = [-\tfrac{1}{5}(-1)^n + \tfrac{6}{5}(4)^n]u(n)$$

We make the observation that both the simple first-order recursive system and the second-order recursive system have impulse responses that are infinite in duration. In other words, both of these recursive systems are IIR systems. In fact, due to the recursive nature of the system, any recursive system described by a linear constant-coefficient difference equation is an IIR system. The converse is not true, however. That is, not every linear time-invariant IIR system can be described by a linear constant-coefficient difference equation. In other words, recursive systems described by linear constant-coefficient difference equations are a subclass of linear time-invariant IIR systems.

The extension of the approach that we have demonstrated for determining the impulse response of the first- and second-order systems, generalizes in a straightforward manner. When the system is described by an Nth-order linear difference equation of the type given in (2.4.13), the solution of the homogeneous equation is

$$y_h(n) = \sum_{k=1}^{N} C_k \lambda_k^n$$

when the roots $\{\lambda_k\}$ of the characteristic polynomial are distinct. Hence the impulse response of the system is identical in form, that is,

$$h(n) = \sum_{k=1}^{N} C_k \lambda_k^n \qquad (2.4.43)$$

where the parameters $\{C_k\}$ are determined by setting the initial conditions $y(-1) = \cdots = y(-N) = 0$.

This form of $h(n)$ allows us to easily relate the stability of a system, described by an Nth-order difference equation, to the values of the roots of the characteristic polynomial. Indeed, since BIBO stability requires that the impulse response be absolutely summable, then, for a causal system, we have

$$\sum_{n=0}^{\infty} |h(n)| = \sum_{n=0}^{\infty} \left| \sum_{k=1}^{N} C_k \lambda_k^n \right| \leq \sum_{k=1}^{N} |C_k| \sum_{n=0}^{\infty} |\lambda_k|^n$$

Now if $|\lambda_k| < 1$ for all k, then

$$\sum_{n=0}^{\infty} |\lambda_k|^n < \infty$$

and hence

$$\sum_{n=0}^{\infty} |h(n)| < \infty$$

On the other hand, if one or more of the $|\lambda_k| \geq 1$, $h(n)$ is no longer absolutely summable, and consequently, the system is unstable. Therefore, a necessary and sufficient condition for the stability of a causal IIR system described by a linear constant-coefficient difference equation, is that all roots of the characteristic polynomial be less than unity in magnitude. The reader may verify that this condition carries over to the case where the system has roots of multiplicity m.

2.5 IMPLEMENTATION OF DISCRETE-TIME SYSTEMS

Our treatment of discrete-time systems has been focused on the time-domain characterization and analysis of linear time-invariant systems described by constant-coefficient linear difference equations. Additional analytical methods are developed in the next two chapters, where we characterize and analyze LTI systems in the frequency domain. Two other important topics that will be treated later are the design and implementation of these systems.

In practice, system design and implementation are usually treated jointly rather than separately. Often, the system design is driven by the method of implementation and by implementation constraints, such as cost, hardware limitations, size limitations, and power requirements. At this point, we have not as yet developed the necessary analysis and design tools to treat such complex issues. However, we have developed sufficient background to consider some basic implementation methods for realizations of LTI systems described by linear constant-coefficient difference equations.

2.5.1 Structures for the Realization of Linear Time-Invariant Systems

In this subsection we describe structures for the realization of systems described by linear constant-coefficient difference equations. Additional structures for these systems are introduced in Chapter 7.

As a beginning, let us consider the first-order system

$$y(n) = -a_1 y(n-1) + b_0 x(n) + b_1 x(n-1) \tag{2.5.1}$$

which is realized as in Fig. 2.32a. This realization uses separate delays (memory) for both the input and output signal samples and it is called a *direct form I structure*. Note that this system can be viewed as two linear time-invariant systems in cascade. The first is a nonrecursive, system described by the equation

$$v(n) = b_0 x(n) + b_1 x(n-1) \tag{2.5.2}$$

whereas the second is a recursive system described by the equation

$$y(n) = -a_1 y(n-1) + v(n) \tag{2.5.3}$$

However, as we have seen in Section 2.3.4, if we interchange the order of the cascaded linear time-invariant systems, the overall system response remains the same. Thus if we interchange the order of the recursive and nonrecursive systems, we obtain an alternative structure for the realization of the system described by (2.5.1). The resulting system is shown in Fig. 2.32b. From this figure we obtain the two difference equations

$$w(n) = -a_1 w(n-1) + x(n) \qquad (2.5.4)$$

$$y(n) = b_0 w(n) + b_1 w(n-1) \qquad (2.5.5)$$

which provide an alternative algorithm for computing the output of the system described by the single difference equation given in (2.5.1). In other words, the two difference equations (2.5.4) and (2.5.5) are equivalent to the single difference equation (2.5.1).

A close observation of Fig. 2.32 reveals that the two delay elements contain the same input $w(n)$ and hence the same output $w(n-1)$. Consequently, these two elements can be merged into one delay, as shown in Fig. 2.32c. In contrast

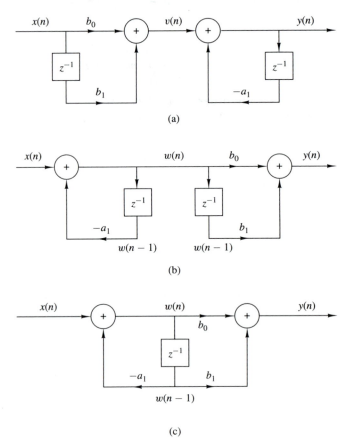

Figure 2.32 Steps in converting from the direct form I realization in (a) to the direct form II realization in (c).

to the direct form I structure, this new realization requires only one delay for the auxiliary quantity $w(n)$, and hence it is more efficient in terms of memory requirements. It is called the *direct form II structure* and it is used extensively in practical applications.

These structures can readily be generalized for the general linear time-invariant recursive system described by the difference equation

$$y(n) = -\sum_{k=1}^{N} a_k y(n-k) + \sum_{k=0}^{M} b_k x(n-k) \tag{2.5.6}$$

Figure 2.33 illustrates the direct form I structure for this system. This structure requires $M+N$ delays and $N+M+1$ multiplications. It can be viewed as the cascade of a nonrecursive system

$$v(n) = \sum_{k=0}^{M} b_k x(n-k) \tag{2.5.7}$$

and a recursive system

$$y(n) = -\sum_{k=1}^{N} a_k y(n-k) + v(n) \tag{2.5.8}$$

By reversing the order of these two systems as was previously done for the first-order system, we obtain the direct form II structure shown in Fig. 2.34 for

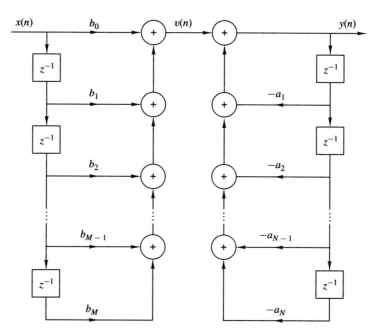

Figure 2.33   Direct form I structure of the system described by (2.5.6).

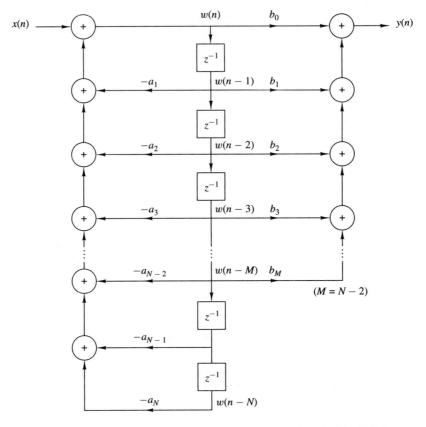

Figure 2.34   Direct form II structure for the system described by (2.5.6).

$N > M$. This structure is the cascade of a recursive system

$$w(n) = -\sum_{k=1}^{N} a_k w(n-k) + x(n) \tag{2.5.9}$$

followed by a nonrecursive system

$$y(n) = \sum_{k=0}^{M} b_k w(n-k) \tag{2.5.10}$$

We observe that if $N \geq M$, this structure requires a number of delays equal to the order N of the system. However, if $M > N$, the required memory is specified by M. Figure 2.34 can easily by modified to handle this case. Thus the direct form II structure requires $M + N + 1$ multiplications and $\max\{M, N\}$ delays. Because it requires the minimum number of delays for the realization of the system described by (2.5.6), it is sometimes called a *canonic form*.

A special case of (2.5.6) occurs if we set the system parameters $a_k = 0$, $k = 1, \ldots, N$. Then the input–output relationship for the system reduces to

$$y(n) = \sum_{k=0}^{M} b_k x(n - k) \tag{2.5.11}$$

which is a nonrecursive linear time-invariant system. This system views only the most recent $M + 1$ input signal samples and, prior to addition, weights each sample by the appropriate coefficient b_k from the set $\{b_k\}$. In other words, the system output is basically a *weighted moving average* of the input signal. For this reason it is sometimes called a *moving average (MA) system.* Such a system is an FIR system with an impulse response h(k) equal to the coefficients b_k, that is,

$$h(k) = \begin{cases} b_k, & 0 \le k \le M \\ 0, & \text{otherwise} \end{cases} \tag{2.5.12}$$

If we return to (2.5.6) and set $M = 0$, the general linear time-invariant system reduces to a "purely recursive" system described by the difference equation

$$y(n) = -\sum_{k=1}^{N} a_k y(n - k) + b_0 x(n) \tag{2.5.13}$$

In this case the system output is a weighted linear combination of N past outputs and the present input.

Linear time-invariant systems described by a second-order difference equation are an important subclass of the more general systems described by (2.5.6) or (2.5.10) or (2.5.13). The reason for their importance will be explained later when we discuss quantization effects. Suffice to say at this point that second-order systems are usually used as basic building blocks for realizing higher-order systems.

The most general second-order system is described by the difference equation

$$\begin{aligned} y(n) = {} & -a_1 y(n - 1) - a_2 y(n - 2) + b_0 x(n) \\ & + b_1 x(n - 1) + b_2 x(n - 2) \end{aligned} \tag{2.5.14}$$

which is obtained from (2.5.6) by setting $N = 2$ and $M = 2$. The direct form II structure for realizing this system is shown in Fig. 2.35a. If we set $a_1 = a_2 = 0$, then (2.5.14) reduces to

$$y(n) = b_0 x(n) + b_1 x(n - 1) + b_2 x(n - 2) \tag{2.5.15}$$

which is a special case of the FIR system described by (2.5.11). The structure for realizing this system is shown in Fig. 2.35b. Finally, if we set $b_1 = b_2 = 0$ in (2.5.14), we obtain the purely recursive second-order system described by the difference equation

$$y(n) = -a_1 y(n - 1) - a_2 y(n - 2) + b_0 x(n) \tag{2.5.16}$$

which is a special case of (2.5.13). The structure for realizing this system is shown in Fig. 2.35c.

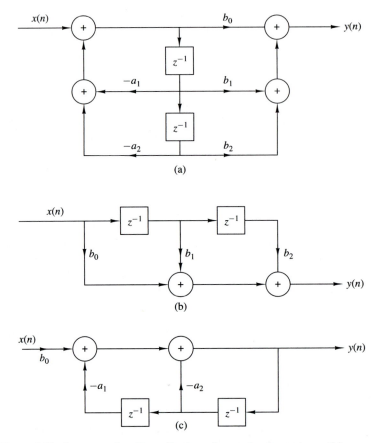

Figure 2.35 Structures for the realization of second-order systems: (a) general second-order system; (b) FIR system; (c) "purely recursive system"

2.5.2 Recursive and Nonrecursive Realizations of FIR Systems

We have already made the distinction between FIR and IIR systems, based on whether the impulse response $h(n)$ of the system has a finite duration, or an infinite duration. We have also made the distinction between recursive and nonrecursive systems. Basically, a causal recursive system is described by an input–output equation of the form

$$y(n) = F[y(n-1), \ldots, y(n-N), x(n), \ldots, x(n-M)] \qquad (2.5.17)$$

and for a linear time-invariant system specifically, by the difference equation

$$y(n) = -\sum_{k=1}^{N} a_k y(n-k) + \sum_{k=0}^{M} b_k x(n-k) \qquad (2.5.18)$$

On the other hand, causal nonrecursive systems do not depend on past values of the output and hence are described by an input–output equation of the form

$$y(n) = F[x(n), x(n-1), \ldots, x(n-M)] \tag{2.5.19}$$

and for linear time-invariant systems specifically, by the difference equation in (2.5.18) with $a_k = 0$ for $k = 1, 2, \ldots, N$.

In the case of FIR systems, we have already observed that it is always possible to realize such systems nonrecursively. In fact, with $a_k = 0$, $k = 1, 2, \ldots, N$, in (2.5.18), we have a system with an input–output equation

$$y(n) = \sum_{k=0}^{M} b_k x(n-k) \tag{2.5.20}$$

This is a nonrecursive and FIR system. As indicated in (2.5.12), the impulse response of the system is simply equal to the coefficients $\{b_k\}$. Hence every FIR system can be realized nonrecursively. On the other hand, any FIR system can also be realized recursively. Although the general proof of this statement is given later, we shall give a simple example to illustrate the point.

Suppose that we have an FIR system of the form

$$y(n) = \frac{1}{M+1} \sum_{k=0}^{M} x(n-k) \tag{2.5.21}$$

for computing the *moving average* of a signal $x(n)$. Clearly, this system is FIR with impulse response

$$h(n) = \frac{1}{M+1} \qquad 0 \le n \le M$$

Figure 2.36 illustrates the structure of the nonrecursive realization of the system. Now, suppose that we express (2.5.21) as

$$y(n) = \frac{1}{M+1} \sum_{k=0}^{M} x(n-1-k)$$

$$+ \frac{1}{M+1}[x(n) - x(n-1-M)]$$

$$= y(n-1) + \frac{1}{M+1}[x(n) - x(n-1-M)] \tag{2.5.22}$$

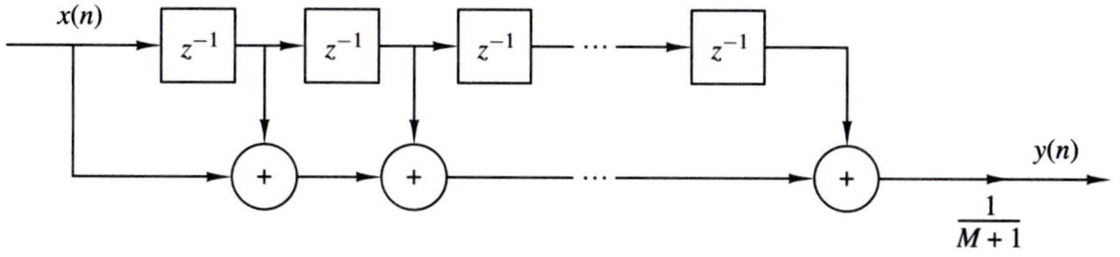

Figure 2.36   Nonrecursive realization of an FIR moving average system.

Now, (2.5.22) represents a recursive realization of the FIR system. The structure of this recursive realization of the moving average system is illustrated in Fig. 2.37.

In summary, we can think of the terms FIR and IIR as general characteristics that distinguish a type of linear time-invariant system, and of the terms *recursive* and *nonrecursive* as descriptions of the structures for realizing or implementing the system.

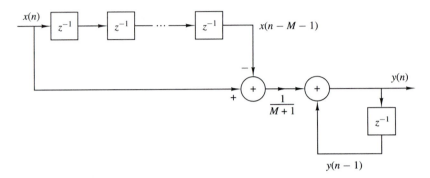

Figure 2.37   Recursive realization of an FIR moving average system.

2.6 CORRELATION OF DISCRETE-TIME SIGNALS

A mathematical operation that closely resembles convolution is correlation. Just as in the case of convolution, two signal sequences are involved in correlation. In contrast to convolution, however, our objective in computing the correlation between the two signals is to measure the degree to which the two signals are similar and thus to extract some information that depends to a large extent on the application. Correlation of signals is often encountered in radar, sonar, digital communications, geology, and other areas in science and engineering.

To be specific, let us suppose that we have two signal sequences $x(n)$ and $y(n)$ that we wish to compare. In radar and active sonar applications, $x(n)$ can represent the sampled version of the transmitted signal and $y(n)$ can represent the sampled version of the received signal at the output of the analog-to-digital (A/D) converter. If a target is present in the space being searched by the radar or sonar, the received signal $y(n)$ consists of a delayed version of the transmitted signal, reflected from the target, and corrupted by additive noise. Figure 2.38 depicts the radar signal reception problem.

We can represent the received signal sequence as

$$y(n) = \alpha x(n - D) + w(n) \tag{2.6.1}$$

where α is some attenuation factor representing the signal loss involved in the round-trip transmission of the signal $x(n)$, D is the round-trip delay, which is

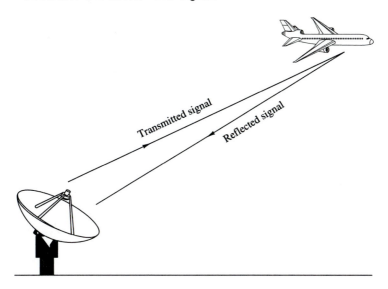

Figure 2.38   Radar target detection.

assumed to be an integer multiple of the sampling interval, and $w(n)$ represents the additive noise that is picked up by the antenna and any noise generated by the electronic components and amplifiers contained in the front end of the receiver. On the other hand, if there is no target in the space searched by the radar and sonar, the received signal $y(n)$ consists of noise alone.

Having the two signal sequences, $x(n)$, which is called the reference signal or transmitted signal, and $y(n)$, the received signal, the problem in radar and sonar detection is to compare $y(n)$ and $x(n)$ to determine if a target is present and, if so, to determine the time delay D and compute the distance to the target. In practice, the signal $x(n - D)$ is heavily corrupted by the additive noise to the point where a visual inspection of $y(n)$ does not reveal the presence or absence of the desired signal reflected from the target. Correlation provides us with a means for extracting this important information from $y(n)$.

Digital communications is another area where correlation is often used. In digital communications the information to be transmitted from one point to another is usually converted to binary from, that is, a sequence of zeros and ones, which are then transmitted to the intended receiver. To transmit a 0 we can transmit the signal sequence $x_0(n)$ for $0 \leq n \leq L-1$, and to transmit a 1 we can transmit the signal sequence $x_1(n)$ for $0 \leq n \leq L - 1$, where L is some integer that denotes the number of samples in each of the two sequences. Very often, $x_1(n)$ is selected to be the negative of $x_0(n)$. The signal received by the intended receiver may be represented as

$$y(n) = x_i(n) + w(n) \qquad i = 0, 1 \qquad 0 \leq n \leq L - 1 \qquad (2.6.2)$$

where now the uncertainty is whether $x_0(n)$ or $x_1(n)$ is the signal component in $y(n)$, and $w(n)$ represents the additive noise and other interference inherent in any communication system. Again, such noise has its origin in the electronic components contained in the front end of the receiver. In any case, the receiver knows the possible transmitted sequences $x_0(n)$ and $x_1(n)$ and is faced with the task of comparing the received signal $y(n)$ with both $x_0(n)$ and $x_1(n)$ to determine which of the two signals better matches $y(n)$. This comparison process is performed by means of the correlation operation described in the following subsection.

2.6.1 Crosscorrelation and Autocorrelation Sequences

Suppose that we have two real signal sequences $x(n)$ and $y(n)$ each of which has finite energy. The *crosscorrelation* of $x(n)$ and $y(n)$ is a sequence $r_{xy}(l)$, which is defined as

$$r_{xy}(l) = \sum_{n=-\infty}^{\infty} x(n)y(n-l) \qquad l = 0, \pm 1, \pm 2, \ldots \tag{2.6.3}$$

or, equivalently, as

$$r_{xy}(l) = \sum_{n=-\infty}^{\infty} x(n+l)y(n) \qquad l = 0, \pm 1, \pm 2, \ldots \tag{2.6.4}$$

The index l is the (time) shift (or *lag*) parameter and the subscripts xy on the crosscorrelation sequence $r_{xy}(l)$ indicate the sequences being correlated. The order of the subscripts, with x preceding y, indicates the direction in which one sequence is shifted, relative to the other. To elaborate, in (2.6.3), the sequence $x(n)$ is left unshifted and $y(n)$ is shifted by l units in time, to the right for l positive and to the left for l negative. Equivalently, in (2.6.4), the sequence $y(n)$ is left unshifted and $x(n)$ is shifted by l units in time, to the left for l positive and to the right for l negative. But shifting $x(n)$ to the left by l units relative to $y(n)$ is equivalent to shifting $y(n)$ to the right by l units relative to $x(n)$. Hence the computations (2.6.3) and (2.6.4) yield identical crosscorrelation sequences.

    If we reverse the roles of $x(n)$ and $y(n)$ in (2.6.3) and (2.6.4) and therefore reverse the order of the indices xy, we obtain the crosscorrelation sequence

$$r_{yx}(l) = \sum_{n=-\infty}^{\infty} y(n)x(n-l) \tag{2.6.5}$$

or, equivalently,

$$r_{yx}(l) = \sum_{n=-\infty}^{\infty} y(n+l)x(n) \tag{2.6.6}$$

By comparing (2.6.3) with (2.6.6) or (2.6.4) with (2.6.5), we conclude that

$$r_{xy}(l) = r_{yx}(-l) \tag{2.6.7}$$

Therefore, $r_{yx}(l)$ is simply the folded version of $r_{xy}(l)$, where the folding is done with respect to $l = 0$. Hence, $r_{yx}(l)$ provides exactly the same information as $r_{xy}(l)$, with respect to the similarity of $x(n)$ to $y(n)$.

Example 2.6.1

Determine the crosscorrelation sequence $r_{xy}(l)$ of the sequences

$$x(n) = \{\ldots, 0, 0, 2, -1, 3, 7, 1, 2, -3, 0, 0, \ldots\}$$
$$\uparrow$$

$$y(n) = \{\ldots, 0, 0, 1, -1, 2, -2, 4, 1, -2, 5, 0, 0, \ldots\}$$
$$\uparrow$$

Solution   Let us use the definition in (2.6.3) to compute $r_{xy}(l)$. For $l = 0$ we have

$$r_{xy}(0) = \sum_{n=-\infty}^{\infty} x(n)y(n)$$

The product sequence $v_0(n) = x(n)y(n)$ is

$$v_0(n) = \{\ldots, 0, 0, 2, 1, 6, -14, 4, 2, 6, 0, 0, \ldots\}$$
$$\uparrow$$

and hence the sum over all values of n is

$$r_{xy}(0) = 7$$

For $l > 0$, we simply shift $y(n)$ to the right relative to $x(n)$ by l units, compute the product sequence $v_l(n) = x(n)y(n - l)$, and finally, sum over all values of the product sequence. Thus we obtain

$$r_{xy}(1) = 13, \qquad r_{xy}(2) = -18, \qquad r_{xy}(3) = 16, \qquad r_{xy}(4) = -7$$

$$r_{xy}(5) = 5, \qquad r_{xy}(6) = -3, \qquad r_{xy}(l) = 0, \qquad l \geq 7$$

For $l < 0$, we shift $y(n)$ to the left relative to $x(n)$ by l units, compute the product sequence $v_l(n) = x(n)y(n - l)$, and sum over all values of the product sequence. Thus we obtain the values of the crosscorrelation sequence

$$r_{xy}(-1) = 0, \qquad r_{xy}(-2) = 33, \qquad r_{xy}(-3) = -14, \qquad r_{xy}(-4) = 36$$

$$r_{xy}(-5) = 19, \qquad r_{xy}(-6) = -9, \qquad r_{xy}(-7) = 10, \qquad r_{xy}(l) = 0, \ l \leq -8$$

Therefore, the crosscorrelation sequence of $x(n)$ and $y(n)$ is

$$r_{xy}(l) = \{10, -9, 19, 36, -14, 33, 0, 7, 13, -18, 16, -7, 5, -3\}$$
$$\uparrow$$

The similarities between the computation of the crosscorrelation of two sequences and the convolution of two sequences is apparent. In the computation of convolution, one of the sequences is folded, then shifted, then multiplied by the other sequence to form the product sequence for that shift, and finally, the values of the product sequence are summed. Except for the folding operation, the computation of the crosscorrelation sequence involves the same operations: shifting one of the sequences, multiplication of the two sequences, and summing over all values of the product sequence. Consequently, if we have a computer program that performs convolution, we can use it to perform crosscorrelation by providing as inputs to the program, the sequence $x(n)$ and the folded sequence $y(-n)$. Then the convolution of $x(n)$ with $y(-n)$ yields the crosscorrelation $r_{xy}(l)$, that is,

$$r_{xy}(l) = x(l) * y(-l) \tag{2.6.8}$$

In the special case where $y(n) = x(n)$, we have the *autocorrelation* of $x(n)$, which is defined as the sequence

$$r_{xx}(l) = \sum_{n=-\infty}^{\infty} x(n)x(n-l) \tag{2.6.9}$$

or, equivalently, as

$$r_{xx}(l) = \sum_{n=-\infty}^{\infty} x(n+l)x(n) \tag{2.6.10}$$

In dealing with finite-duration sequences, it is customary to express the autocorrelation and crosscorrelation in terms of the finite limits on the summation. In particular, if $x(n)$ and $y(n)$ are causal sequences of length N [i.e., $x(n) = y(n) = 0$ for $n < 0$ and $n \geq N$], the crosscorrelation and autocorrelation sequences may be expressed as

$$r_{xy}(l) = \sum_{n=i}^{N-|k|-1} x(n)y(n-l) \tag{2.6.11}$$

and

$$r_{xx}(l) = \sum_{n=i}^{N-|k|-1} x(n)x(n-l) \tag{2.6.12}$$

where $i = l$, $k = 0$ for $l \geq 0$, and $i = 0$, $k = l$ for $l < 0$.

2.6.2 Properties of the Autocorrelation and Crosscorrelation Sequences

The autocorrelation and crosscorrelation sequences have a number of important properties that we now present. To develop these properties, let us assume that we have two sequences $x(n)$ and $y(n)$ with finite energy from which we form the linear combination,

$$ax(n) + by(n-l)$$

where a and b are arbitrary constants and l is some time shift. The energy in this signal is

$$\sum_{n=-\infty}^{\infty} [ax(n) + by(n-l)]^2 = a^2 \sum_{n=-\infty}^{\infty} x^2(n) + b^2 \sum_{n=-\infty}^{\infty} y^2(n-l)$$

$$+ 2ab \sum_{n=-\infty}^{\infty} x(n)y(n-l) \tag{2.6.13}$$

$$= a^2 r_{xx}(0) + b^2 r_{yy}(0) + 2ab r_{xy}(l)$$

First, we note that $r_{xx}(0) = E_x$ and $r_{yy}(0) = E_y$, which are the energies of $x(n)$ and $y(n)$, respectively. It is obvious that

$$a^2 r_{xx}(0) + b^2 r_{yy}(0) + 2ab r_{xy}(l) \geq 0 \qquad (2.6.14)$$

Now, assuming that $b \neq 0$, we can divide (2.6.14) by b^2 to obtain

$$r_{xx}(0) \left(\frac{a}{b}\right)^2 + 2r_{xy}(l) \left(\frac{a}{b}\right) + r_{yy}(0) \geq 0$$

We view this equation as a quadratic with coefficients $r_{xx}(0)$, $2r_{xy}(l)$, and $r_{yy}(0)$. Since the quadratic is nonnegative, it follows that the discriminant of this quadratic must be nonpositive, that is,

$$4[r_{xy}^2(l) - r_{xx}(0)r_{yy}(0)] \leq 0$$

Therefore, the crosscorrelation sequence satisfies the condition that

$$|r_{xy}(l)| \leq \sqrt{r_{xx}(0)r_{yy}(0)} = \sqrt{E_x E_y} \qquad (2.6.15)$$

In the special case where $y(n) = x(n)$, (2.6.15) reduces to

$$|r_{xx}(l)| \leq r_{xx}(0) = E_x \qquad (2.6.16)$$

This means that the autocorrelation sequence of a signal attains its maximum value at zero lag. This result is consistent with the notion that a signal matches perfectly with itself at zero shift. In the case of the crosscorrelation sequence, the upper bound on its values is given in (2.6.15).

Note that if any one or both of the signals involved in the crosscorrelation are scaled, the shape of the crosscorrelation sequence does not change, only the amplitudes of the crosscorrelation sequence are scaled accordingly. Since scaling is unimportant, it is often desirable, in practice, to normalize the autocorrelation and crosscorrelation sequences to the range from -1 to 1. In the case of the autocorrelation sequence, we can simply divide by $r_{xx}(0)$. Thus the normalized autocorrelation sequence is defined as

$$\rho_{xx}(l) = \frac{r_{xx}(l)}{r_{xx}(0)} \qquad (2.6.17)$$

Similarly, we define the normalized crosscorrelation sequence

$$\rho_{xy}(l) = \frac{r_{xy}(l)}{\sqrt{r_{xx}(0)r_{yy}(0)}} \qquad (2.6.18)$$

Now $|\rho_{xx}(l)| \leq 1$ and $|\rho_{xy}(l)| \leq 1$, and hence these sequences are independent of signal scaling.

Finally, as we have already demonstrated, the crosscorrelation sequence satisfies the property

$$r_{xy}(l) = r_{yx}(-l)$$

With $y(n) = x(n)$, this relation results in the following important property for the autocorrelation sequence

$$r_{xx}(l) = r_{xx}(-l) \tag{2.6.19}$$

Hence the autocorrelation function is an even function. Consequently, it suffices to compute $r_{xx}(l)$ for $l \geq 0$.

Example 2.6.2

Compute the autocorrelation of the signal

$$x(n) = a^n u(n), 0 < a < 1$$

Solution   Since $x(n)$ is an infinite-duration signal, its autocorrelation also has infinite duration. We distinguish two cases.

If $l \geq 0$, from Fig. 2.39 we observe that

$$r_{xx}(l) = \sum_{n=l}^{\infty} x(n)x(n-l) = \sum_{n=l}^{\infty} a^n a^{n-l} = a^{-l} \sum_{n=l}^{\infty} (a^2)^n$$

Since $a < 1$, the infinite series *converges* and we obtain

$$r_{xx}(l) = \frac{1}{1-a^2} a^l \qquad l \geq 0$$

For $l < 0$ we have

$$r_{xx}(l) = \sum_{n=0}^{\infty} x(n)x(n-l) = a^{-l} \sum_{n=0}^{\infty} (a^2)^n = \frac{1}{1-a^2} a^{-l} \qquad l < 0$$

But when l is negative, $a^{-l} = a^{|l|}$. Thus the two relations for $r_{xx}(l)$ can be combined into the following expression:

$$r_{xx}(l) = \frac{1}{1-a^2} a^{|l|} \qquad -\infty < l < \infty \tag{2.6.20}$$

The sequence $r_{xx}(l)$ is shown in Fig. 2.42(d). We observe that

$$r_{xx}(-l) = r_{xx}(l)$$

and

$$r_{xx}(0) = \frac{1}{1-a^2}$$

Therefore, the normalized autocorrelation sequence is

$$\rho_{xx}(l) = \frac{r_{xx}(l)}{r_{xx}(0)} = a^{|l|} \qquad -\infty < l < \infty \tag{2.6.21}$$

2.6.3 Correlation of Periodic Sequences

In Section 2.6.1 we defined the crosscorrelation and autocorrelation sequences of energy signals. In this section we consider the correlation sequences of power signals and, in particular, periodic signals.

Let $x(n)$ and $y(n)$ be two power signals. Their crosscorrelation sequence is defined as

$$r_{xy}(l) = \lim_{M \to \infty} \frac{1}{2M+1} \sum_{n=-M}^{M} x(n)y(n-l) \tag{2.6.22}$$

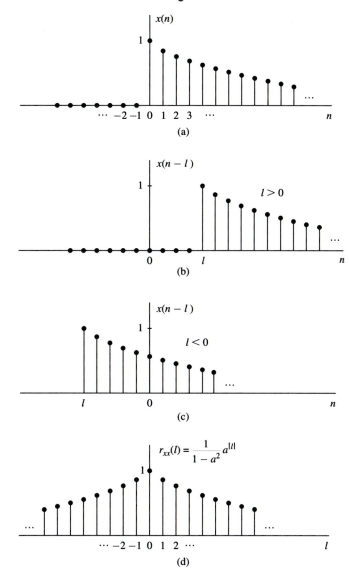

Figure 2.39 Computation of the autocorrelation of the signal $x(n) = a^n$, $0 < a < 1$.

If $x(n) = y(n)$, we have the definition of the autocorrelation sequence of a power signal as

$$r_{xx}(l) = \lim_{M \to \infty} \frac{1}{2M + 1} \sum_{n=-M}^{M} x(n)x(n - l) \tag{2.6.23}$$

In particular, if $x(n)$ and $y(n)$ are two periodic sequences, each with period N, the averages indicated in (2.6.22) and (2.6.23) over the infinite interval, are identical to the averages over a single period, so that (2.6.22) and (2.6.23) reduce to

$$r_{xy}(l) = \frac{1}{N} \sum_{n=0}^{N-1} x(n)y(n-l) \tag{2.6.24}$$

and

$$r_{xx}(l) = \frac{1}{N} \sum_{n=0}^{N-1} x(n)x(n-l) \tag{2.6.25}$$

It is clear that $r_{xy}(l)$ and $r_{xx}(l)$ are periodic correlation sequences with period N. The factor $1/N$ can be viewed as a normalization scale factor.

In some practical applications, correlation is used to identify periodicities in an observed physical signal which may be corrupted by random interference. For example, consider a signal sequence $y(n)$ of the form

$$y(n) = x(n) + w(n) \tag{2.6.26}$$

where $x(n)$ is a periodic sequence of some unknown period N and $w(n)$ represents an additive random interference. Suppose that we observe M samples of $y(n)$, say $0 \le n \le M - 1$, where $M >> N$. For all practical purposes, we can assume that $y(n) = 0$ for $n < 0$ and $n \ge M$. Now the autocorrelation sequence of $y(n)$, using the normalization factor of $1/M$, is

$$r_{yy}(l) = \frac{1}{M} \sum_{n=0}^{M-1} y(n)y(n-l) \tag{2.6.27}$$

If we substitute for $y(n)$ from (2.6.26) into (2.6.27) we obtain

$$r_{yy}(l) = \frac{1}{M} \sum_{n=0}^{M-1} [x(n) + w(n)][x(n-l) + w(n-l)]$$

$$= \frac{1}{M} \sum_{n=0}^{M-1} x(n)x(n-l)$$

$$+ \frac{1}{M} \sum_{n=0}^{M-1} [x(n)w(n-l) + w(n)x(n-l)] \tag{2.6.28}$$

$$+ \frac{1}{M} \sum_{n=0}^{M-1} w(n)w(n-l)$$

$$= r_{xx}(l) + r_{xw}(l) + r_{wx}(l) + r_{ww}(l)$$

The first factor on the right-hand side of (2.6.28) is the autocorrelation sequence of $x(n)$. Since $x(n)$ is periodic, its autocorrelation sequence exhibits the same periodicity, thus containing relatively large peaks at $l = 0$, N, $2N$, and so on. However, as the shift l approaches M, the peaks are reduced in amplitude due to the fact that we have a finite data record of M samples so that many of the products $x(n)x(n-l)$ are zero. Consequently, we should avoid computing $r_{yy}(l)$ for large lags, say, $l > M/2$.

The crosscorrelations $r_{xw}(l)$ and $r_{wx}(l)$ between the signal $x(n)$ and the additive random interference are expected to be relatively small as a result of the expectation that $x(n)$ and $w(n)$ will be totally unrelated. Finally, the last term on the right-hand side of (2.6.28) is the autocorrelation sequence of the random sequence $w(n)$. This correlation sequence will certainly contain a peak at $l = 0$, but because of its random characteristics, $r_{ww}(l)$ is expected to decay rapidly toward zero. Consequently, only $r_{xx}(l)$ is expected to have large peaks for $l > 0$. This behavior allows us to detect the presence of the periodic signal $x(n)$ buried in the interference $w(n)$ and to identify its period.

An example that illustrates the use of autocorrelation to identify a hidden periodicity in an observed physical signal is shown in Fig. 2.40. This figure illustrates the autocorrelation (normalized) sequence for the Wölfer sunspot numbers for $0 \le l \le 20$, where any value of l corresponds to one year. These numbers are given in Table 2.2 for the 100-year period 1770–1869. There is clear evidence in this figure that a periodic trend exists, with a period of 10 to 11 years.

Example 2.6.3

Suppose that a signal sequence $x(n) = \sin(\pi/5)n$, for $0 \le n \le 99$ is corrupted by an additive noise sequence $w(n)$, where the values of the additive noise are selected independently from sample to sample, from a uniform distribution over the range

TABLE 2.2   YEARLY WÖLFER SUNSPOT NUMBERS

1770	101	1795	21	1820	16	1845	40
1771	82	1796	16	1821	7	1846	62
1772	66	1797	6	1822	4	1847	98
1773	35	1798	4	1823	2	1848	124
1774	31	1799	7	1824	8	1849	96
1775	7	1800	14	1825	17	1850	66
1776	20	1801	34	1826	36	1851	64
1777	92	1802	45	1827	50	1852	54
1778	154	1803	43	1828	62	1853	39
1779	125	1804	48	1829	67	1854	21
1780	85	1805	42	1830	71	1855	7
1781	68	1806	28	1831	48	1856	4
1782	38	1807	10	1832	28	1857	23
1783	23	1808	8	1833	8	1858	55
1784	10	1809	2	1834	13	1859	94
1785	24	1810	0	1835	57	1860	96
1786	83	1811	1	1836	122	1861	77
1787	132	1812	5	1837	138	1862	59
1788	131	1813	12	1838	103	1863	44
1789	118	1814	14	1839	86	1864	47
1790	90	1815	35	1840	63	1865	30
1791	67	1816	46	1841	37	1866	16
1792	60	1817	41	1842	24	1867	7
1793	47	1818	30	1843	11	1868	37
1794	41	1819	24	1844	15	1869	74

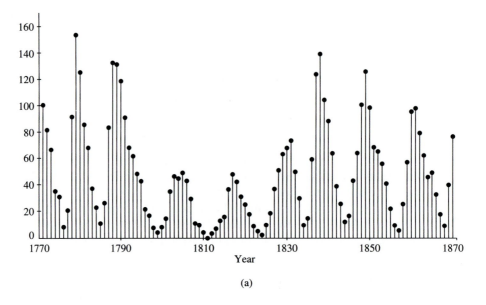

(a)

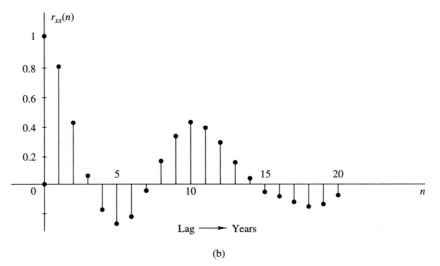

(b)

Figure 2.40 Identification of periodicity in the Wölfer sunspot numbers: (a) annual Wölfer sunspot numbers; (b) autocorrelation sequence.

$(-\Delta/2, \Delta/2)$, where Δ is a parameter of the distribution. The observed sequence is $y(n) = x(n) + w(n)$. Determine the autocorrelation sequence $r_{yy}(n)$ and thus determine the period of the signal $x(n)$.

Solution The assumption is that the signal sequence $x(n)$ has some unknown period that we are attempting to determine from the noise-corrupted observations $\{y(n)\}$. Although $x(n)$ is periodic with period 10, we have only a finite-duration sequence of length $M = 100$ [i.e., 10 periods of $x(n)$]. The noise power level P_w in the sequence $w(n)$ is determined by the parameter Δ. We simply state that $P_w = \Delta^2/12$. The signal power level is $P_x = \frac{1}{2}$. Therefore, the signal-to-noise ratio (SNR) is defined as

$$\frac{P_x}{P_w} = \frac{\frac{1}{2}}{\Delta^2/12} = \frac{6}{\Delta^2}$$

Usually, the SNR is expressed on a logarithmic scale in decibels (dB) as $10\log_{10}$ (P_x/P_w).

Figure 2.41 illustrates a sample of a noise sequence $w(n)$, and the observed sequence $y(n) = x(n) + w(n)$ when the SNR $= 1$ dB. The autocorrelation sequence

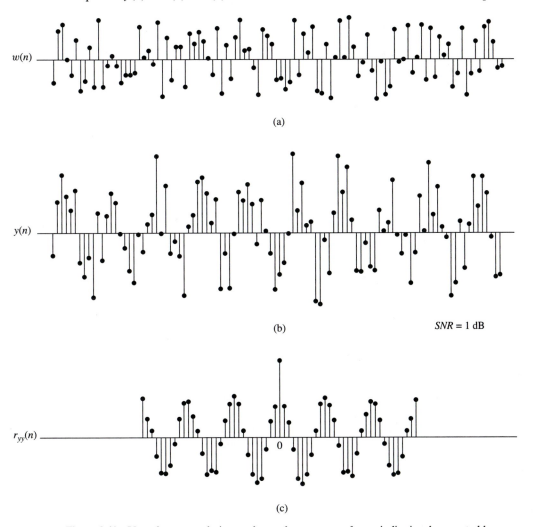

Figure 2.41 Use of autocorrelation to detect the presence of a periodic signal corrupted by noise.

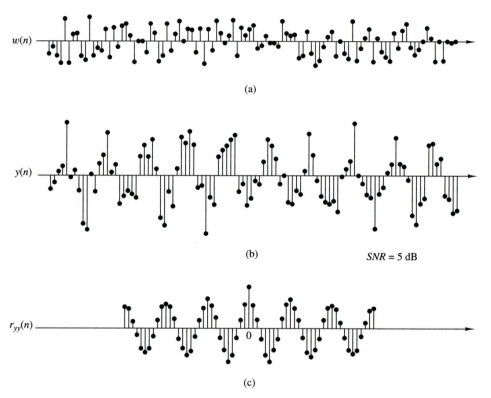

(a)

(b) $SNR = 5$ dB

(c)

Figure 2.42 Use of autocorrelation to detect the presence of a periodic signal corrupted by noise.

$r_{yy}(l)$ is illustrated in Fig. 2.41c. We observe that the periodic signal $x(n)$, embedded in $y(n)$, results in a periodic autocorrelation function $r_{xx}(l)$ with period $N = 10$. The effect of the additive noise is to add to the peak value at $l = 0$, but for $l \neq 0$, the correlation sequence $r_{ww}(l) \approx 0$ as a result of the fact that values of $w(n)$ were generated independently. Such noise is usually called *white noise*. The presence of this noise explains the reason for the large peak at $l = 0$. The smaller, nearly equal peaks at $l = \pm 10, \pm 20, \ldots$ are due the periodic characteristics of $x(n)$.

Figure 2.42 illustrates the noise sequence $w(n)$, the noise-corrupted signal $y(n)$, and the autocorrelation sequence $r_{yy}(l)$ for the same signal, within which is embedded a signal at a smaller noise level. In this case, the SNR $= 5$ dB. Even with this relatively small noise level, the periodicity of the signal is not easily determined from observation of $y(n)$. However, it is clearly evident from observation of the autocorrelation sequence $r_{yy}(n)$.

2.6.4 Computation of Correlation Sequences

As indicated on Section 2.6.1, the procedure for computing the crosscorrelation sequence between $x(n)$ and $y(n)$ involves shifting one of the sequences, say $x(n)$, to obtain $x(n - l)$, multiplying the shifted sequence by $y(n)$ to obtain the product sequence $y(n)x(n - l)$, and then summing all the values of the product sequence to obtain $r_{yx}(l)$. This procedure is repeated for different values of the lag l. Except for the folding operation that is involved in convolution, these basic operations for computing the correlation sequence are identical to those in convolution.

The procedure for computing the convolution is directly applicable to computing the correlation of two sequences. Specifically, if we fold $y(n)$ to obtain $y(-n)$, then the convolution of $x(n)$ with $y(-n)$ is identical to the crosscorrelation of $x(n)$ with $y(n)$. That is,

$$r_{xy}(l) = x(n) * y(-n)|_{n=l} \tag{2.6.29}$$

As a consequence, the computational procedure described for convolution can be applied directly to the computation of the correlation sequence.

We now describe an algorithm that can be easily programmed to compute the crosscorrelation sequence of two finite-duration signals $x(n)$, $0 \le n \le N - 1$, and $y(n)$, $0 \le n \le M - 1$.

The algorithm computes $r_{xy}(l)$ for positive lags. According to the relation $r_{xy}(-l) = r_{yx}(l)$, the values of $r_{xy}(l)$ for negative lags can be obtained by using the same algorithm for positive lags, and interchanging the roles of $x(n)$ and $y(n)$. We observe that if $M \le N$, $r_{xy}(l)$ can be computed by the relations

$$r_{xy}(l) = \begin{cases} \displaystyle\sum_{n=l}^{M-1+l} x(n)y(n - l), & 0 \le l \le N - M \\ \displaystyle\sum_{n=l}^{N-1} x(n)y(n - l), & N - M < l \le N - 1 \end{cases} \tag{2.6.30}$$

On the other hand, if $M > N$, the formula for the crosscorrelation becomes

$$r_{xy}(l) = \sum_{n=l}^{N-1} x(n)y(n - l) \qquad 0 \le l \le N - 1 \tag{2.6.31}$$

The formulas in (2.6.30) and (2.6.31) can be combined and computed by means of the following simple algorithm illustrated in the flowchart in Fig. 2.43. By interchanging the roles of $x(n)$ and $y(n)$ and recomputing the crosscorrelation sequence, we obtain the values of $r_{xy}(l)$ corresponding to negative shifts l.

If we wish to compute the autocorrelation sequence $r_{xx}(l)$, we set $y(n) = x(n)$ and $M = N$ in (2.6.31). The computation of $r_{xx}(l)$ can be done by means of the same algorithm for positive shifts only.

2.6.5 Input–Output Correlation Sequences

In this section we derive two input–output relationships for LTI systems in the "correlation domain." Let us assume that a signal $x(n)$ with known autocorrelation $r_{xx}(l)$ is applied to an LTI system with impulse response $h(n)$, producing the

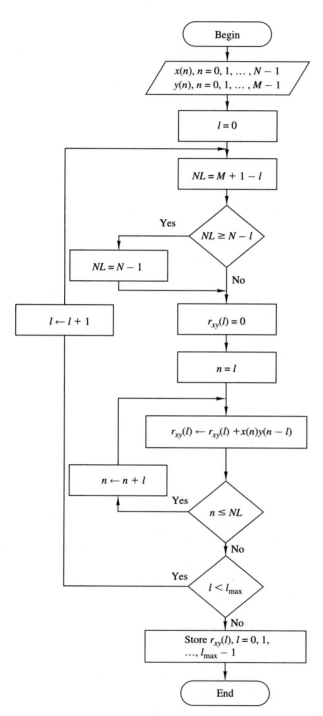

Figure 2.43 Flowchart for software implementation of crosscorrelation.

output signal

$$y(n) = h(n) * x(n) = \sum_{k=-\infty}^{\infty} h(k)x(n-k)$$

The crosscorrelation between the output and the input signal is

$$r_{yx}(l) = y(l) * x(-l) = h(l) * [x(l) * x(-l)]$$

or

$$r_{yx}(l) = h(l) * r_{xx}(l) \qquad (2.6.32)$$

where we have used (2.6.8) and the properties of convolution. Hence the crosscorrelation between the input and the output of the system is the convolution of the impulse response with the autocorrelation of the input sequence. Alternatively, $r_{yx}(l)$ may be viewed as the output of the LTI system when the input sequence is $r_{xx}(l)$. This is illustrated in Fig. 2.44. If we replace l by $-l$ in (2.6.32), we obtain

$$r_{xy}(l) = h(-l) * r_{xx}(l)$$

The autocorrelation of the output signal can be obtained by using (2.6.8) with $x(n) = y(n)$ and the properties of convolution. Thus we have

$$
\begin{aligned}
r_{yy}(l) &= y(l) * y(-l) \\
&= [h(l) * x(l)] * [h(-l) * x(-l)] \\
&= [h(l) * h(-l)] * [x(l) * x(-l)] \\
&= r_{hh}(l) * r_{xx}(l)
\end{aligned}
\qquad (2.6.33)
$$

The autocorrelation $r_{hh}(l)$ of the impulse response $h(n)$ exists if the system is stable. Furthermore, the stability insures that the system does not change the type (energy or power) of the input signal. By evaluating (2.6.33) for $l = 0$ we obtain

$$r_{yy}(0) = \sum_{k=-\infty}^{\infty} r_{hh}(k)r_{xx}(k) \qquad (2.6.34)$$

which provides the energy (or power) of the output signal in terms of autocorrelations. These relationships hold for both energy and power signals. The direct derivation of these relationships for energy and power signals, and their extensions to complex signals, are left as exercises for the student.

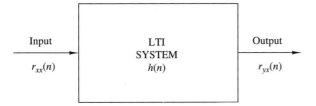

Figure 2.44   Input–output relation for crosscorrelation $r_{yx}(n)$.

2.7 SUMMARY AND REFERENCES

The major theme of this chapter is the characterization of discrete-time signals and systems in the time domain. Of particular importance is the class of linear time-invariant (LTI) systems which are widely used in the design and implementation of digital signal processing systems. We characterized LTI systems by their unit sample response $h(n)$ and derived the convolution summation, which is a formula for determining the response $y(n)$ of the system characterized by $h(n)$ to any given input sequence $x(n)$.

The class of LTI systems characterized by linear difference equations with constant coefficients is by far the most important of the LTI systems in the theory and application of digital signal processing. The general solution of a linear difference equation with constant coefficients was derived in this chapter and shown to consist of two components: the solution of the homogeneous equation which represents the natural response of the system when the input is zero, and the particular solution, which represents the response of the system to the input signal. From the difference equation, we also demonstrated how to derive the unit sample response of the LTI system.

Linear time-invariant systems were generally subdivided into FIR (finite-duration impulse response) and IIR (infinite-duration impulse response) depending on whether $h(n)$ has finite duration or infinite duration, respectively. The realizations of such systems were briefly described. Furthermore, in the realization of FIR systems, we made the distinction between recursive and nonrecursive realizations. On the other hand, we observed that IIR systems can be implemented recursively, only.

There are a number of texts on discrete-time signals and systems. We mention as examples the books by McGillem and Cooper (1984), Oppenheim and Willsky (1983), and Siebert (1986). Linear constant-coefficient difference equations are treated in depth in the books by Hildebrand (1952) and Levy and Lessman (1961).

The last topic in this chapter, on correlation of discrete-time signals, plays an important role in digital signal processing, especially in applications dealing with digital communications, radar detection and estimation, sonar, and geophysics. In our treatment of correlation sequences, we avoided the use of statistical concepts. Correlation is simply defined as a mathematical operation between two sequences, which produces another sequence, called either the *crosscorrelation sequence* when the two sequences are different, or the *autocorrelation sequence* when the two sequences are identical.

In practical applications in which correlation is used, one (or both) of the sequences is (are) contaminated by noise and, perhaps, by other forms of interference. In such a case, the noisy sequence is called a *random sequence* and is characterized in statistical terms. The corresponding correlation sequence becomes a function of the statistical characteristics of the noise and any other interference.

The statistical characterization of sequences and their correlation is treated in Appendix A. Supplementary reading on probabilistic and statistical concepts dealing with correlation can be found in the books by Davenport (1970), Helstrom (1990), Papoulis (1984), and Peebles (1987).

PROBLEMS

2.1 A discrete-time signal $x(n)$ is defined as

$$x(n) = \begin{cases} 1 + \dfrac{n}{3}, & -3 \le n \le -1 \\ 1, & 0 \le n \le 3 \\ 0, & \text{elsewhere} \end{cases}$$

(a) Determine its values and sketch the signal $x(n)$.
(b) Sketch the signals that result if we:
    (1) First fold $x(n)$ and then delay the resulting signal by four samples.
    (2) First delay $x(n)$ by four samples and then fold the resulting signal.
(c) Sketch the signal $x(-n + 4)$.
(d) Compare the results in parts (b) and (c) and derive a rule for obtaining the signal $x(-n + k)$ from $x(n)$.
(e) Can you express the signal $x(n)$ in terms of signals $\delta(n)$ and $u(n)$?

2.2 A discrete-time signal $x(n)$ is shown in Fig. P2.2. Sketch and label carefully each of the following signals.

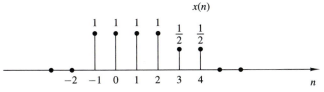

Figure P2.2

(a) $x(n - 2)$ **(b)** $x(4 - n)$ **(c)** $x(n + 2)$ **(d)** $x(n)u(2 - n)$
(e) $x(n - 1)\delta(n - 3)$ **(f)** $x(n^2)$ **(g)** even part of $x(n)$
(h) odd part of $x(n)$

2.3 Show that
(a) $\delta(n) = u(n) - u(n - 1)$
(b) $u(n) = \sum_{k=-\infty}^{n} \delta(k) = \sum_{k=0}^{\infty} \delta(n - k)$

2.4 Show that any signal can be decomposed into an even and an odd component. Is the decomposition unique? Illustrate your arguments using the signal

$$x(n) = \{2, 3, \underset{\uparrow}{4}, 5, 6\}$$

2.5 Show that the energy (power) of a real-valued energy (power) signal is equal to the sum of the energies (powers) of its even and odd components.

2.6 Consider the system

$$y(n) = T[x(n)] = x(n^2)$$

(a) Determine if the system is time invariant.

(b) To clarify the result in part (a) assume that the signal

$$x(n) = \begin{cases} 1, & 0 \le n \le 3 \\ 0, & \text{elsewhere} \end{cases}$$

is applied into the system.
  (1) Sketch the signal $x(n)$.
  (2) Determine and sketch the signal $y(n) = T[x(n)]$.
  (3) Sketch the signal $y_2'(n) = y(n - 2)$.
  (4) Determine and sketch the signal $x_2(n) = x(n - 2)$.
  (5) Determine and sketch the signal $y_2(n) = T[x_2(n)]$.
  (6) Compare the signals $y_2(n)$ and $y(n - 2)$. What is your conclusion?
(c) Repeat part (b) for the system

$$y(n) = x(n) - x(n - 1)$$

Can you use this result to make any statement about the time invariance of this system? Why?
(d) Repeat parts (b) and (c) for the system

$$y(n) = T[x(n)] = nx(n)$$

2.7 A discrete-time system can be
  (1) Static or dynamic
  (2) Linear or nonlinear
  (3) Time invariant or time varying
  (4) Causal or noncausal
  (5) Stable or unstable
Examine the following systems with respect to the properties above.
  (a) $y(n) = \cos[x(n)]$
  (b) $y(n) = \sum_{k=-\infty}^{n+1} x(k)$
  (c) $y(n) = x(n) \cos(\omega_0 n)$
  (d) $y(n) = x(-n + 2)$
  (e) $y(n) = \text{Trun}[x(n)]$, where $\text{Trun}[x(n)]$ denotes the integer part of $x(n)$, obtained by truncation
  (f) $y(n) = \text{Round}[x(n)]$, where $\text{Round}[x(n)]$ denotes the integer part of $x(n)$ obtained by rounding
  Remark: The systems in parts (e) and (f) are quantizers that perform truncation and rounding, respectively.
  (g) $y(n) = |x(n)|$
  (h) $y(n) = x(n)u(n)$
  (i) $y(n) = x(n) + nx(n + 1)$
  (j) $y(n) = x(2n)$
  (k) $y(n) = \begin{cases} x(n), & \text{if } x(n) \ge 0 \\ 0, & \text{if } x(n) < 0 \end{cases}$

  (l) $y(n) = x(-n)$
  (m) $y(n) = \text{sign}[x(n)]$
  (n) The ideal sampling system with input $x_a(t)$ and output $x(n) = x_a(nT)$, $-\infty < n < \infty$

2.8 Two discrete-time systems T_1 and T_2 are connected in cascade to form a new system T as shown in Fig. P2.8. Prove or disprove the following statements.

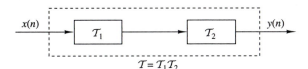

$$T = T_1 T_2$$ **Figure P2.8**

(a) If T_1 and T_2 are linear, then T is linear (i.e., the cascade connection of two linear systems is linear).

(b) If T_1 and T_2 are time invariant, then T is time invariant.

(c) If T_1 and T_2 are causal, then T is causal.

(d) If T_1 and T_2 are linear and time invariant, the same holds for T.

(e) If T_1 and T_2 are linear and time invariant, then interchanging their order does not change the system T.

(f) As in part (e) except that T_1, T_2 are now time varying. (*Hint:* Use an example.)

(g) If T_1 and T_2 are nonlinear, then T is nonlinear.

(h) If T_1 and T_2 are stable, then T is stable.

(i) Show by an example that the inverse of parts (c) and (h) do not hold in general.

2.9 Let T be an LTI, relaxed, and BIBO stable system with input $x(n)$ and output $y(n)$. Show that:

(a) If $x(n)$ is periodic with period N [i.e., $x(n) = x(n + N)$ for all $n \geq 0$], the output $y(n)$ tends to a periodic signal with the same period.

(b) If $x(n)$ is bounded and tends to a constant, the output will also tend to a constant.

(c) If $x(n)$ is an energy signal, the output $y(n)$ will also be an energy signal.

2.10 The following input–output pairs have been observed during the operation of a time-invariant system:

$$x_1(n) = \{1, 0, 2\} \overset{T}{\longleftrightarrow} y_1(n) = \{0, 1, 2\}$$
$$\uparrow \qquad\qquad\qquad \uparrow$$

$$x_2(n) = \{0, 0, 3\} \overset{T}{\longleftrightarrow} y_2(n) = \{0, 1, 0, 2\}$$
$$\uparrow \qquad\qquad\qquad \uparrow$$

$$x_3(n) = \{0, 0, 0, 1\} \overset{T}{\longleftrightarrow} y_3(n) = \{1, 2, 1\}$$
$$\uparrow \qquad\qquad\qquad\quad \uparrow$$

Can you draw any conclusions regarding the linearity of the system. What is the impulse response of the system?

2.11 The following input–output pairs have been observed during the operation of a linear system:

$$x_1(n) = \{-1, 2, 1\} \overset{T}{\longleftrightarrow} y_1(n) = \{1, 2, -1, 0, 1\}$$
$$\uparrow \qquad\qquad\qquad\quad \uparrow$$

$$x_2(n) = \{1, -1, -1\} \overset{T}{\longleftrightarrow} y_2(n) = \{-1, 1, 0, 2\}$$
$$\uparrow \qquad\qquad\qquad\quad \uparrow$$

$$x_3(n) = \{0, 1, 1\} \overset{T}{\longleftrightarrow} y_3(n) = \{1, 2, 1\}$$
$$\uparrow \qquad\qquad\qquad \uparrow$$

Can you draw any conclusions about the time invariance of this system?

2.12 The only available information about a system consists of N input–output pairs, of signals $y_i(n) = T[x_i(n)]$, $i = 1, 2, \ldots, N$.

(a) What is the class of input signals for which we can determine the output, using the information above, if the system is known to be linear?

(b) The same as above, if the system is known to be time invariant.

2.13 Show that the necessary and sufficient condition for a relaxed LTI system to be BIBO stable is

$$\sum_{n=-\infty}^{\infty} |h(n)| \leq M_h < \infty$$

for some constant M_n.

2.14 Show that:

(a) A relaxed linear system is causal if and only if for any input $x(n)$ such that

$$x(n) = 0 \text{ for } n < n_0 \Rightarrow y(n) = 0 \qquad \text{for } n < n_0$$

(b) A relaxed LTI system is causal if and only if

$$h(n) = 0 \qquad \text{for } n < 0$$

2.15 (a) Show that for any real or complex constant a, and any finite integer numbers M and N, we have

$$\sum_{n=M}^{N} a^n = \begin{cases} \dfrac{a^M - a^{N+1}}{1-a}, & \text{if } a \neq 1 \\ N - M + 1, & \text{if } a = 1 \end{cases}$$

(b) Show that if $|a| < 1$, then

$$\sum_{n=0}^{\infty} a^n = \frac{1}{1-a}$$

2.16 (a) If $y(n) = x(n) * h(n)$, show that $\sum_y = \sum_x \sum_h$, where $\sum_x = \sum_{n=-\infty}^{\infty} x(n)$.

(b) Compute the convolution $y(n) = x(n) * h(n)$ of the following signals and check the correctness of the results by using the test in (a).

    (1) $x(n) = \{1, 2, 4\}, h(n) = \{1, 1, 1, 1, 1\}$

    (2) $x(n) = \{1, 2, -1\}, h(n) = x(n)$

    (3) $x(n) = \{0, 1, -2, 3, -4\}, h(n) = \{\frac{1}{2}, \frac{1}{2}, 1, \frac{1}{2}\}$

    (4) $x(n) = \{1, 2, 3, 4, 5\}, h(n) = \{1\}$

    (5) $x(n) = \{1, -2, 3\}, h(n) = \{0, 0, 1, 1, 1, 1\}$
           ↑   ↑

    (6) $x(n) = \{0, 0, 1, 1, 1, 1\}, h(n) = \{1, -2, 3\}$
           ↑   ↑

    (7) $x(n) = \{0, 1, 4, -3\}, h(n) = \{1, 0, -1, -1\}$
           ↑   ↑

    (8) $x(n) = \{1, 1, 2\}, h(n) = u(n)$
           ↑

    (9) $x(n) = \{1, 1, 0, 1, 1\}, h(n) = \{1, -2, -3, 4\}$
              ↑   ↑

    (10) $x(n) = \{1, 2, 0, 2, 1\} h(n) = x(n)$
             ↑

    (11) $x(n) = (\frac{1}{2})^n u(n), h(n) = (\frac{1}{4})^n u(n)$

2.17 Compute and plot the convolutions $x(n) * h(n)$ and $h(n) * x(n)$ for the pairs of signals shown in Fig. P2.17.

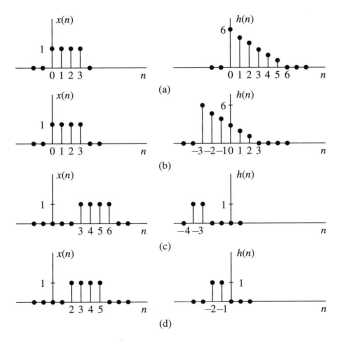

Figure P2.17

2.18 Determine and sketch the convolution $y(n)$ of the signals

$$x(n) = \begin{cases} \frac{1}{3}n, & 0 \le n \le 6 \\ 0, & \text{elsewhere} \end{cases}$$

$$h(n) = \begin{cases} 1, & -2 \le n \le 2 \\ 0, & \text{elsewhere} \end{cases}$$

(a) Graphically
(b) Analytically

2.19 Compute the convolution $y(n)$ of the signals

$$x(n) = \begin{cases} \alpha^n, & -3 \le n \le 5 \\ 0, & \text{elsewhere} \end{cases}$$

$$h(n) = \begin{cases} 1, & 0 \le n \le 4 \\ 0, & \text{elsewhere} \end{cases}$$

2.20 Consider the following three operations.
   (a) Multiply the integer numbers: 131 and 122.
   (b) Compute the convolution of signals: $\{1, 3, 1\} * \{1, 2, 2\}$.
   (c) Multiply the polynomials: $1 + 3z + z^2$ and $1 + 2z + 2z^2$.
   (d) Repeat part (a) for the numbers 1.31 and 12.2.
   (e) Comment on your results.

2.21 Compute the convolution $y(n) = x(n) * h(n)$ of the following pairs of signals.
   (a) $x(n) = a^n u(n)$, $h(n) = b^n u(n)$ when $a \ne b$ and when $a = b$
   (b) $x(n) = \begin{cases} 1, & n = -2, 0, 1 \\ 2, & n = -1 \\ 0, & \text{elsewhere} \end{cases}$
   $h(n) = \delta(n) - \delta(n - 1) + \delta(n - 4) + \delta(n - 5)$

    (c) $x(n) = u(n+1) - u(n-4) - \delta(n-5)$
         $h(n) = [u(n+2) - u(n-3)] \cdot (3 - |n|)$
    (d) $x(n) = u(n) - u(n-5)$
         $h(n) = u(n-2) - u(n-8) + u(n-11) - u(n-17)$

2.22 Let $x(n)$ be the input signal to a discrete-time filter with impulse response $h_i(n)$ and let $y_i(n)$ be the corresponding output.

    (a) Compute and sketch $x(n)$ and $y_i(n)$ in the following cases, using the same scale in all figures.

$$x(n) = \{1, 4, 2, 3, 5, 3, 3, 4, 5, 7, 6, 9\}$$

$$h_1(n) = \{1, 1\}$$

$$h_2(n) = \{1, 2, 1\}$$

$$h_3(n) = \{\tfrac{1}{2}, \tfrac{1}{2}\}$$

$$h_4(n) = \{\tfrac{1}{4}, \tfrac{1}{2}, \tfrac{1}{4}\}$$

$$h_5(n) = \{\tfrac{1}{4}, -\tfrac{1}{2}, \tfrac{1}{4}\}$$

    Sketch $x(n)$, $y_1(n)$, $y_2(n)$ on one graph and $x(n)$, $y_3(n)$, $y_4(n)$, $y_5(n)$ on another graph

    (b) What is the difference between $y_1(n)$ and $y_2(n)$, and between $y_3(n)$ and $y_4(n)$?

    (c) Comment on the smoothness of $y_2(n)$ and $y_4(n)$. Which factors affect the smoothness?

    (d) Compare $y_4(n)$ with $y_5(n)$. What is the difference? Can you explain it?

    (e) Let $h_6(n) = \{\tfrac{1}{2}, -\tfrac{1}{2}\}$. Compute $y_6(n)$. Sketch $x(n)$, $y_2(n)$, and $y_6(n)$ on the same figure and comment on the results.

2.23 The discrete-time system

$$y(n) = n y(n-1) + x(n) \quad n \geq 0$$

    is at rest [i.e., $y(-1) = 0$]. Check if the system is linear time invariant and BIBO stable.

2.24 Consider the signal $y(n) = a^n u(n)$, $0 < a < 1$.

    (a) Show that any sequence $x(n)$ can be decomposed as

$$x(n) = \sum_{n=-\infty}^{\infty} c_k y(n-k)$$

    and express c_k in terms of $x(n)$.

    (b) Use the properties of linearity and time invariance to express the output $y(n) = T[x(n)]$ in terms of the input $x(n)$ and the signal $g(n) = T[y(n)]$, where $T[\cdot]$ is an LTI system.

    (c) Express the impulse response $h(n) = T[\delta(n)]$ in terms of $g(n)$.

2.25 Determine the zero-input response of the system described by the second-order difference equation

$$x(n) - 3y(n-1) - 4y(n-2) = 0$$

2.26 Determine the particular solution of the difference equation

$$y(n) = \tfrac{5}{6} y(n-1) - \tfrac{1}{6} y(n-2) + x(n)$$

when the forcing function is $x(n) = 2^n u(n)$.

2.27 Determine the response $y(n)$, $n \geq 0$, of the system described by the second-order difference equation

$$y(n) - 3y(n-1) - 4y(n-2) = x(n) + 2x(n-1)$$

to the input $x(n) = 4^n u(n)$.

2.28 Determine the impulse response of the following causal system:

$$y(n) - 3y(n-1) - 4y(n-2) = x(n) + 2x(n-1)$$

2.29 Let $x(n)$, $N_1 \leq n \leq N_2$ and $h(n)$, $M_1 \leq n \leq M_2$ be two finite-duration signals.
  (a) Determine the range $L_1 \leq n \leq L_2$ of their convolution, in terms of N_1, N_2, M_1 and M_2.
  (b) Determine the limits of the cases of partial overlap from the left, full overlap, and partial overlap from the right. For convenience, assume that $h(n)$ has shorter duration than $x(n)$.
  (c) Illustrate the validity of your results by computing the convolution of the signals

$$x(n) = \begin{cases} 1, & -2 \leq n \leq 4 \\ 0, & \text{elsewhere} \end{cases}$$

$$h(n) = \begin{cases} 2, & -1 \leq n \leq 2 \\ 0, & \text{elsewhere} \end{cases}$$

2.30 Determine the impulse response and the unit step response of the systems described by the difference equation
  (a) $y(n) = 0.6y(n-1) - 0.08y(n-2) + x(n)$
  (b) $y(n) = 0.7y(n-1) - 0.1y(n-2) + 2x(n) - x(n-2)$

2.31 Consider a system with impulse response

$$h(n) = \begin{cases} (\tfrac{1}{2})^n, & 0 \leq n \leq 4 \\ 0, & \text{elsewhere} \end{cases}$$

Determine the input $x(n)$ for $0 \leq n \leq 8$ that will generate the output sequence

$$y(n) = \{1, 2, 2.5, 3, 3, 3, 2, 1, 0, \ldots\}$$
$$\uparrow$$

2.32 Consider the interconnection of LTI systems as shown in Fig. P2.32.
  (a) Express the overall impulse response in terms of $h_1(n)$, $h_2(n)$, $h_3(n)$, and $h_4(n)$.
  (b) Determine $h(n)$ when

$$h_1(n) = \{\tfrac{1}{2}, \tfrac{1}{4}, \tfrac{1}{2}\}$$

$$h_2(n) = h_3(n) = (n+1)u(n)$$

$$h_4(n) = \delta(n-2)$$

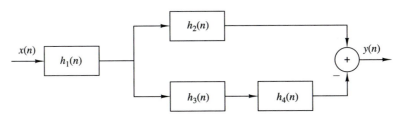

Figure P2.32

(c) Determine the response of the system in part (b) if

$$x(n) = \delta(n+2) + 3\delta(n-1) - 4\delta(n-3)$$

2.33 Consider the system in Fig. P2.33 with $h(n) = a^n u(n)$, $-1 < a < 1$. Determine the response $y(n)$ of the system to the excitation

$$x(n) = u(n+5) - u(n-10)$$

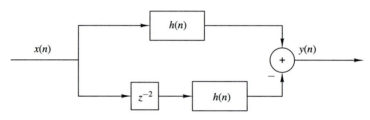

Figure P2.33

2.34 Compute and sketch the step response of the system

$$y(n) = \frac{1}{M} \sum_{k=0}^{M-1} x(n-k)$$

2.35 Determine the range of values of the parameter a for which the linear time-invariant system with impulse response

$$h(n) = \begin{cases} a^n, & n \geq 0, n \text{ even} \\ 0, & \text{otherwise} \end{cases}$$

is stable.

2.36 Determine the response of the system with impulse response

$$h(n) = a^n u(n)$$

to the input signal

$$x(n) = u(n) - u(n-10)$$

(*Hint:* The solution can be obtained easily and quickly by applying the linearity and time-invariance properties to the result in Example 2.3.5.)

2.37 Determine the response of the (relaxed) system characterized by the impulse response

$$h(n) = (\tfrac{1}{2})^n u(n)$$

to the input signal

$$x(n) = \begin{cases} 1, & 0 \leq n < 10 \\ 0, & \text{otherwise} \end{cases}$$

2.38 Determine the response of the (relaxed) system characterized by the impulse response

$$h(n) = (\tfrac{1}{2})^n u(n)$$

to the input signals

(a) $x(n) = 2^n u(n)$

(b) $x(n) = u(-n)$

2.39 Three systems with impulse responses $h_1(n) = \delta(n) - \delta(n-1)$, $h_2(n) = h(n)$, and $h_3(n) = u(n)$, are connected in cascade.
  (a) What is the impulse response, $h_c(n)$, of the overall system?
  (b) Does the order of the interconnection affect the overall system?

2.40 (a) Prove and explain graphically the difference between the relations

$$x(n)\delta(n - n_0) = x(n_0)\delta(n - n_0) \qquad \text{and} \qquad x(n) * \delta(n - n_0) = x(n - n_0)$$

  (b) Show that a discrete-time system, which is described by a convolution summation, is LTI and relaxed,
  (c) What is the impulse response of the system described by $y(n) = x(n - n_0)$?

2.41 Two signals $s(n)$ and $v(n)$ are related through the following difference equations

$$s(n) + a_1 s(n - 1) + \cdots + a_N s(n - N) = b_0 v(n)$$

Design the block diagram realization of:
  (a) The system that generates $s(n)$ when excited by $v(n)$.
  (b) The system that generates $v(n)$ when excited by $s(n)$.
  (c) What is the impulse response of the cascade interconnection of systems in parts (a) and (b)?

2.42 Compute the zero-state response of the system described by the difference equation

$$y(n) + \tfrac{1}{2}y(n - 1) = x(n) + 2x(n - 2)$$

to the input

$$x(n) = \{1, 2, 3, 4, 2, 1\}$$
$$\uparrow$$

by solving the difference equation recursively.

2.43 Determine the direct form II realization for each of the following LTI systems.
  (a) $2y(n) + y(n - 1) - 4y(n - 3) = x(n) + 3x(n - 5)$
  (b) $y(n) = x(n) - x(n - 1) + 2x(n - 2) - 3x(n - 4)$

2.44 Consider the discrete-time system shown in Fig. P2.44.

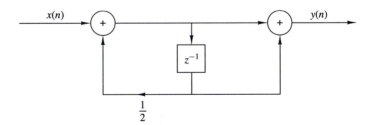

Figure P2.44

  (a) Compute the 10 first samples of its impulse response.
  (b) Find the input–output relation.
  (c) Apply the input $x(n) = \{1, 1, 1, \ldots\}$ and compute the first 10 samples of the output.
$$\uparrow$$

(d) Compute the first 10 samples of the output for the input given in part (c) by using convolution.

(e) Is the system causal? Is it stable?

2.45 Consider the system described by the difference equation

$$y(n) = ay(n-1) + bx(n)$$

(a) Determine b in terms of a so that

$$\sum_{n=-\infty}^{\infty} h(n) = 1$$

(b) Compute the zero-state step response $s(n)$ of the system and choose b so that $s(\infty) = 1$.

(c) Compare the values of b obtained in parts (a) and (b). What did you notice?

2.46 A discrete-time system is realized by the structure shown in Fig. P2.46.

(a) Determine the impulse response.

(b) Determine a realization for its inverse system, that is, the system which produces $x(n)$ as an output when $y(n)$ is used as an input.

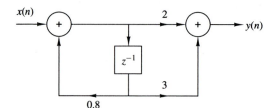

Figure P2.46

2.47 Consider the discrete-time system shown in Fig. P2.47.

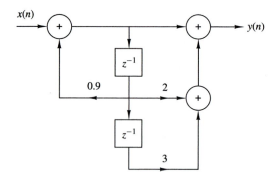

Figure P2.47

(a) Compute the first six values of the impulse response of the system.

(b) Compute the first six values of the zero-state step response of the system.

(c) Determine an analytical expression for the impulse response of the system.

2.48 Determine and sketch the impulse response of the following systems for $n = 0, 1, \ldots, 9$.

(a) Fig. P2.48(a).

(b) Fig. P2.48(b).

(c) Fig. P2.48(c).

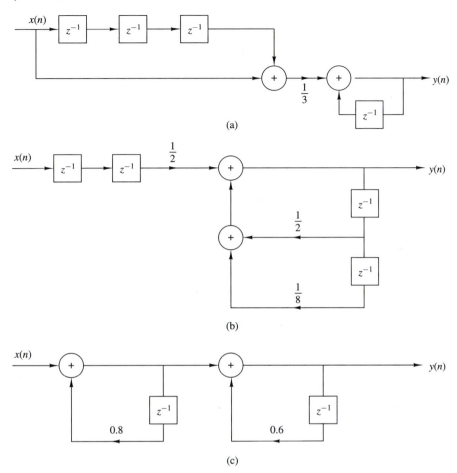

Figure P2.48

(d) Classify the systems above as FIR or IIR.

(e) Find an explicit expression for the impulse response of the system in part (c).

2.49 Consider the systems shown in Fig. P2.49.

(a) Determine and sketch their impulse responses $h_1(n)$, $h_2(n)$, and $h_3(n)$.

(b) Is it possible to choose the coefficients of these systems in such a way that

$$h_1(n) = h_2(n) = h_3(n)$$

2.50 Consider the system shown in Fig. P2.50.

(a) Determine its impulse response $h(n)$.

(b) Show that $h(n)$ is equal to the convolution of the following signals.

$$h_1(n) = \delta(n) + \delta(n-1)$$

$$h_2(n) = (\tfrac{1}{2})^n u(n)$$

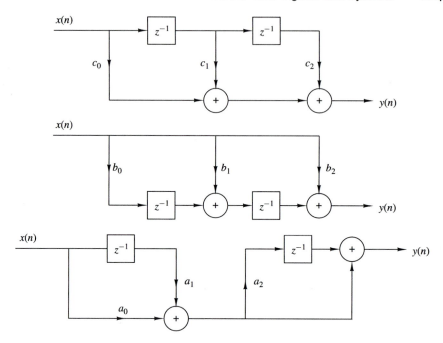

Figure P2.49

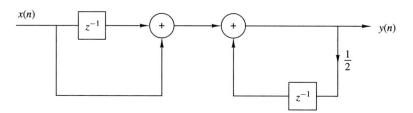

Figure P2.50

2.51 Compute the sketch the convolution $y_i(n)$ and correlation $r_i(n)$ sequences for the following pair of signals and comment on the results obtained.

(a) $x_1(n) = \{1, 2, 4\}$ $h_1(n) = \{1, 1, 1, 1, 1\}$
             ↑   ↑

(b) $x_2(n) = \{0, 1, -2, 3, -4\}$ $h_2(n) = \{\frac{1}{2}, 1, 2, 1, \frac{1}{2}\}$
             ↑   ↑

(c) $x_3(n) = \{1, 2, 3, 4\}$ $h_3(n) = \{4, 3, 2, 1\}$
             ↑   ↑

(d) $x_4(n) = \{1, 2, 3, 4\}$ $h_4(n) = \{1, 2, 3, 4\}$
             ↑   ↑

2.52 The zero-state response of a causal LTI system to the input $x(n) = \{1, 3, 3, 1\}$ is $y(n) = \{1, 4, 6, 4, 1\}$. Determine its impulse response.
             ↑

2.53 Prove by direct substitution the equivalence of equations (2.5.9) and (2.5.10), which describe the direct form II structure, to the relation (2.5.6), which describes the direct form I structure.

2.54 Determine the response $y(n)$, $n \geq 0$ of the system described by the second-order difference equation

$$y(n) - 4y(n-1) + 4y(n-2) = x(n) - x(n-1)$$

when the input is

$$x(n) = (-1)^n u(n)$$

and the initial conditions are $y(-1) = y(-2) = 0$.

2.55 Determine the impulse response $h(n)$ for the system described by the second-order difference equation

$$y(n) - 4y(n-1) + 4y(n-2) = x(n) - x(n-1)$$

2.56 Show that any discrete-time signal $x(n)$ can be expressed as

$$x(n) = \sum_{k=-\infty}^{\infty} [x(k) - x(k-1)]u(n-k)$$

where $u(n-k)$ is a unit step delayed by k units in time, that is,

$$u(n-k) = \begin{cases} 1, & n \geq k \\ 0, & \text{otherwise} \end{cases}$$

2.57 Show that the output of an LTI system can be expressed in terms of its unit step response $s(n)$ as follows.

$$y(n) = \sum_{k=-\infty}^{\infty} [s(k) - s(k-1)]x(n-k)$$

$$= \sum_{k=-\infty}^{\infty} [x(k) - x(k-1)]s(n-k)$$

2.58 Compute the correlation sequences $r_{xx}(l)$ and $r_{xy}(l)$ for the following signal sequences.

$$x(n) = \begin{cases} 1, & n_0 - N \leq n \leq n_0 + N \\ 0, & \text{otherwise} \end{cases}$$

$$y(n) = \begin{cases} 1, & -N \leq n \leq N \\ 0, & \text{otherwise} \end{cases}$$

2.59 Determine the autocorrelation sequences of the following signals.
   (a) $x(n) = \{1, 2, 1, 1\}$
$\uparrow$

   (b) $y(n) = \{1, 1, 2, 1\}$
$\uparrow$

What is your conclusion?

2.60 What is the normalized autocorrelation sequence of the signal $x(n)$ given by

$$x(n) = \begin{cases} 1, & -N \leq n \leq N \\ 0, & \text{otherwise} \end{cases}$$

2.61 An audio signal $s(t)$ generated by a loudspeaker is reflected at two different walls with reflection coefficients r_1 and r_2. The signal $x(t)$ recorded by a microphone close to the loudspeaker, after sampling, is

$$x(n) = s(n) + r_1 s(n - k_1) + r_2 s(n - k_2)$$

where k_1 and k_2 are the delays of the two echoes.
(a) Determine the autocorrelation $r_{xx}(l)$ of the signal $x(n)$.
(b) Can we obtain r_1, r_2, k_1, and k_2 by observing $r_{xx}(l)$?
(c) What happens if $r_2 = 0$?

2.62* *Time-delay estimation in radar* Let $x_a(t)$ be the transmitted signal and $y_a(t)$ be the received signal in a radar system, where

$$y_a(t) = ax_a(t - t_d) + v_a(t)$$

and $v_a(t)$ is additive random noise. The signals $x_a(t)$ and $y_a(t)$ are sampled in the receiver, according to the sampling theorem, and are processed digitally to determine the time delay and hence the distance of the object. The resulting discrete-time signals are

$$x(n) = x_a(nT)$$
$$y(n) = y_a(nT) = ax_a(nT - DT) + v_a(nT)$$
$$\overset{\Delta}{=} ax(n - D) + v(n)$$

(a) Explain how we can measure the delay D by computing the crosscorrelation $r_{xy}(l)$.
(b) Let $x(n)$ be the 13-point *Barker sequence*

$$x(n) = \{+1, +1, +1, +1, +1, -1, -1, +1, +1, -1, +1, -1, +1\}$$

and $v(n)$ be a Gaussian random sequence with zero mean and variance $\sigma^2 = 0.01$. Write a program that generates the sequence $y(n)$, $0 \le n \le 199$ for $a = 0.9$ and $D = 20$. Plot the signals $x(n)$, $y(n)$, $0 \le n \le 199$.
(c) Compute and plot the crosscorrelation $r_{xy}(l)$, $0 \le l \le 59$. Use the plot to estimate the value of the delay D.
(d) Repeat parts (b) and (c) for $\sigma^2 = 0.1$ and $\sigma^2 = 1$.
(e) Repeat parts (b) and (c) for the signal sequence

$$x(n) = \{-1, -1, -1, +1, +1, +1, +1, -1,$$
$$+ 1, -1, +1, +1, -1, -1, +1\}$$

which is obtained from the four-stage feedback shift register shown in Fig. P2.62.

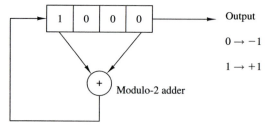

Figure P2.61   Linear feedback shift register.

Note that $x(n)$ is just one period of the periodic sequence obtained from the feedback shift register.

(f) Repeat parts (b) and (c) for a sequence of period $N = 2^7 - 1$, which is obtained from a seven-stage feedback shift register. Table 2.3 gives the stages connected to the modulo-2 adder for (maximal-length) shift-register sequences of length $N = 2^m - 1$.

TABLE 2.3 SHIFT-REGISTER
CONNECTIONS FOR GENERATING
MAXIMAL-LENGTH SEQUENCES

m	Stages Connected to Modulo-2 Adder
1	1
2	1, 2
3	1, 3
4	1, 4
5	1, 4
6	1, 6
7	1, 7
8	1, 5, 6, 7
9	1, 6
10	1, 8
11	1, 10
12	1, 7, 9, 12
13	1, 10, 11, 13
14	1, 5, 9, 14
15	1, 15
16	1, 5, 14, 16
17	1, 15

2.63* *Implementation of LTI systems*   Consider the recursive discrete-time system described by the difference equation

$$y(n) = -a_1 y(n-1) - a_2 y(n-2) + b_0 x(n)$$

where $a_1 = -0.8$, $a_2 = 0.64$, and $b_0 = 0.866$.

(a) Write a program to compute and plot the impulse response $h(n)$ of the system for $0 \le n \le 49$.

(b) Write a program to compute and plot the zero-state step response $s(n)$ of the system for $0 \le n \le 100$.

(c) Define an FIR system with impulse response $h_{FIR}(n)$ given by

$$h_{FIR}(n) = \begin{cases} h(n), & 0 \le n \le 19 \\ 0, & \text{elsewhere} \end{cases}$$

where $h(n)$ is the impulse response computed in part (a). Write a program to compute and plot its step response.

(d) Compare the results obtained in parts (b) and (c) and explain their similarities and differences.

2.64* Write a computer program that computes the overall impulse response $h(n)$ of the system shown in Fig. P2.64 for $0 \leq n \leq 99$. The systems T_1, T_2, T_3, and T_4 are specified by

$$T_1 : h_1(n) = \{1, \tfrac{1}{2}, \tfrac{1}{4}, \tfrac{1}{8}, \tfrac{1}{16}, \tfrac{1}{32}\}$$
$$\uparrow$$

$$T_2 : h_2(n) = \{1, 1, 1, 1, 1\}$$
$$\uparrow$$

$$T_3 : y_3(n) = \tfrac{1}{4}x(n) + \tfrac{1}{2}x(n-1) + \tfrac{1}{4}x(n-2)$$

$$T_4 : y(n) = 0.9y(n-1) - 0.81y(n-2) + v(n) + v(n-1)$$

Plot $h(n)$ for $0 \leq n \leq 99$.

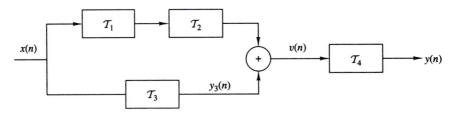

Figure P2.64

3

The Z-Transform and Its Application to the Analysis of LTI Systems

Transform techniques are an important tool in the analysis of signals and linear time-invariant (LTI) systems. In this chapter we introduce the z-transform, develop its properties, and demonstrate its importance in the analysis and characterization of linear time-invariant systems.

The z-transform plays the same role in the analysis of discrete-time signals and LTI systems as the Laplace transform does in the analysis of continuous-time signals and LTI systems. For example, we shall see that in the z-domain (complex z-plane) the convolution of two time-domain signals is equivalent to multiplication of their corresponding z-transforms. This property greatly simplifies the analysis of the response of an LTI system to various signals. In addition, the z-transform provides us with a means of characterizing an LTI system, and its response to various signals, by its pole–zero locations.

We begin this chapter by defining the z-transform. Its important properties are presented in Section 3.2. In Section 3.3 the transform is used to characterize signals in terms of their pole–zero patterns. Section 3.4 describes methods for inverting the z-transform of a signal so as to obtain the time-domain representation of the signal. The one-sided z-transform is treated in Section 3.5 and used to solve linear difference equations with nonzero initial conditions. The chapter concludes with a discussion on the use of the z-transform in the analysis of LTI systems.

3.1 THE Z-TRANSFORM

In this section we introduce the z-transform of a discrete-time signal, investigate its convergence properties, and briefly discuss the inverse z-transform.

3.1.1 The Direct z-Transform

The z-transform of a discrete-time signal $x(n)$ is defined as the power series

$$X(z) \equiv \sum_{n=-\infty}^{\infty} x(n)z^{-n} \tag{3.1.1}$$

where z is a complex variable. The relation (3.1.1) is sometimes called the *direct z-transform* because it transforms the time-domain signal $x(n)$ into its complex-plane representation $X(z)$. The inverse procedure [i.e., obtaining $x(n)$ from $X(z)$] is called the *inverse z-transform* and is examined briefly in Section 3.1.2 and in more detail in Section 3.4.

For convenience, the z-transform of a signal $x(n)$ is denoted by

$$X(z) \equiv Z\{x(n)\} \tag{3.1.2}$$

whereas the relationship between $x(n)$ and $X(z)$ is indicated by

$$x(n) \xleftrightarrow{z} X(z) \tag{3.1.3}$$

Since the z-transform is an infinite power series, it exists only for those values of z for which this series converges. The *region of convergence* (ROC) of $X(z)$ is the set of all values of z for which $X(z)$ attains a finite value. Thus any time we cite a z-transform we should also indicate its ROC.

We illustrate these concepts by some simple examples.

Example 3.1.1

Determine the z-transforms of the following *finite-duration* signals.

    (a) $x_1(n) = \{1, 2, 5, 7, 0, 1\}$

    (b) $x_2(n) = \{1, 2, 5, 7, 0, 1\}$
               $\uparrow$

    (c) $x_3(n) = \{0, 0, 1, 2, 5, 7, 0, 1\}$

    (d) $x_4(n) = \{2, 4, 5, 7, 0, 1\}$
               $\uparrow$

    (e) $x_5(n) = \delta(n)$

    (f) $x_6(n) = \delta(n - k), k > 0$

    (g) $x_7(n) = \delta(n + k), k > 0$

Solution From definition (3.1.1), we have

    (a) $X_1(z) = 1 + 2z^{-1} + 5z^{-2} + 7z^{-3} + z^{-5}$, ROC: entire z-plane except $z = 0$

    (b) $X_2(z) = z^2 + 2z + 5 + 7z^{-1} + z^{-3}$, ROC: entire z-plane except $z = 0$ and $z = \infty$

    (c) $X_3(z) = z^{-2} + 2z^{-3} + 5z^{-4} + 7z^{-5} + z^{-7}$, ROC: entire z-plane except $z = 0$

    (d) $X_4(z) = 2z^2 + 4z + 5 + 7z^{-1} + z^{-3}$, ROC: entire z-plane except $z = 0$ and $z = \infty$

    (e) $X_5(z) = 1$[i.e., $\delta(n) \xleftrightarrow{z} 1$], ROC: entire z-plane

    (f) $X_6(z) = z^{-k}$[i.e., $\delta(n - k) \xleftrightarrow{z} z^{-k}$], $k > 0$, ROC: entire z-plane except $z = 0$

    (g) $X_7(z) = z^k$[i.e., $\delta(n + k) \xleftrightarrow{z} z^k$], $k > 0$, ROC: entire z-plane except $z = \infty$

From this example it is easily seen that the ROC of a *finite-duration signal* is the entire z-plane, except possibly the points $z = 0$ and/or $z = \infty$. These points are excluded, because $z^k (k > 0)$ becomes unbounded for $z = \infty$ and $z^{-k} (k > 0)$ becomes unbounded for $z = 0$.

From a mathematical point of view the z-transform is simply an alternative representation of a signal. This is nicely illustrated in Example 3.1.1, where we see that the coefficient of z^{-n}, in a given transform, is the value of the signal at time n. In other words, the exponent of z contains the time information we need to identify the samples of the signal.

In many cases we can express the sum of the finite or infinite series for the z-transform in a closed-form expression. In such cases the z-transform offers a compact alternative representation of the signal.

Example 3.1.2

Determine the z-transform of the signal

$$x(n) = (\tfrac{1}{2})^n u(n)$$

Solution   The signal $x(n)$ consists of an infinite number of nonzero values

$$x(n) = \{1, (\tfrac{1}{2}), (\tfrac{1}{2})^2, (\tfrac{1}{2})^3, \ldots, (\tfrac{1}{2})^n, \ldots\}$$

The z-transform of $x(n)$ is the infinite power series

$$X(z) = 1 + \tfrac{1}{2} z^{-1} + (\tfrac{1}{2})^2 z^{-2} + (\tfrac{1}{2})^n z^{-n} + \cdots$$

$$= \sum_{n=0}^{\infty} (\tfrac{1}{2})^n z^{-n} = \sum_{n=0}^{\infty} (\tfrac{1}{2} z^{-1})^n$$

This is an infinite geometric series. We recall that

$$1 + A + A^2 + A^3 + \cdots = \frac{1}{1 - A} \qquad \text{if } |A| < 1$$

Consequently, for $|\tfrac{1}{2} z^{-1}| < 1$, or equivalently, for $|z| > \tfrac{1}{2}$, $X(z)$ converges to

$$X(z) = \frac{1}{1 - \tfrac{1}{2} z^{-1}} \qquad \text{ROC: } |z| > \tfrac{1}{2}$$

We see that in this case, the z-transform provides a compact alternative representation of the signal $x(n)$.

Let us express the complex variable z in polar form as

$$z = re^{j\theta} \tag{3.1.4}$$

where $r = |z|$ and $\theta = \measuredangle z$. Then $X(z)$ can be expressed as

$$X(z)|_{z=re^{j\theta}} = \sum_{n=-\infty}^{\infty} x(n) r^{-n} e^{-j\theta n}$$

In the ROC of $X(z)$, $|X(z)| < \infty$. But

$$|X(z)| = \left| \sum_{n=-\infty}^{\infty} x(n)r^{-n}e^{-j\theta n} \right| \qquad (3.1.5)$$

$$\leq \sum_{n=-\infty}^{\infty} |x(n)r^{-n}e^{-j\theta n}| = \sum_{n=-\infty}^{\infty} |x(n)r^{-n}|$$

Hence $|X(z)|$ is finite if the sequence $x(n)r^{-n}$ is absolutely summable.

The problem of finding the ROC for $X(z)$ is equivalent to determining the range of values of r for which the sequence $x(n)r^{-n}$ is absolutely summable. To elaborate, let us express (3.1.5) as

$$|X(z)| \leq \sum_{n=-\infty}^{-1} |x(n)r^{-n}| + \sum_{n=0}^{\infty} \left| \frac{x(n)}{r^n} \right| \qquad (3.1.6)$$

$$\leq \sum_{n=1}^{\infty} |x(-n)r^n| + \sum_{n=0}^{\infty} \left| \frac{x(n)}{r^n} \right|$$

If $X(z)$ converges in some region of the complex plane, both summations in (3.1.6) must be finite in that region. If the first sum in (3.1.6) converges, there must exist values of r small enough such that the product sequence $x(-n)r^n$, $1 \leq n < \infty$, is absolutely summable. Therefore, the ROC for the first sum consists of all points in a circle of some radius r_1, where $r_1 < \infty$, as illustrated in Fig. 3.1a. On the other hand, if the second sum in (3.1.6) converges, there must exist values of r large enough such that the product sequence $x(n)/r^n$, $0 \leq n < \infty$, is absolutely summable. Hence the ROC for the second sum in (3.1.6) consists of all points outside a circle of radius $r > r_2$, as illustrated in Fig. 3.1b.

Since the convergence of $X(z)$ requires that both sums in (3.1.6) be finite, it follows that the ROC of $X(z)$ is generally specified as the annular region in the z-plane, $r_2 < r < r_1$, which is the common region where both sums are finite. This region is illustrated in Fig. 3.1c. On the other hand, if $r_2 > r_1$, there is no common region of convergence for the two sums and hence $X(z)$ does not exist.

The following examples illustrate these important concepts.

Example 3.1.3

Determine the z-transform of the signal

$$x(n) = \alpha^n u(n) = \begin{cases} \alpha^n, & n \geq 0 \\ 0, & n < 0 \end{cases}$$

Solution From the definition (3.1.1) we have

$$X(z) = \sum_{n=0}^{\infty} \alpha^n z^{-n} = \sum_{n=0}^{\infty} (\alpha z^{-1})^n$$

If $|\alpha z^{-1}| < 1$ or equivalently, $|z| > |\alpha|$, this power series converges to $1/(1 - \alpha z^{-1})$.

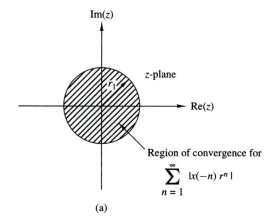

(a)

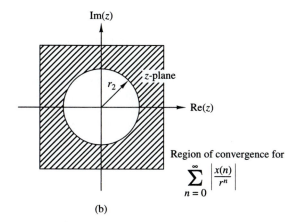

(b)

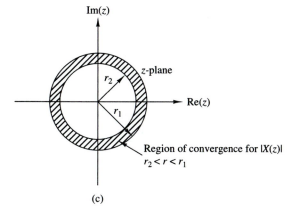

(c)

Figure 3.1   Region of convergence for $X(z)$ and its corresponding causal and anticausal components.

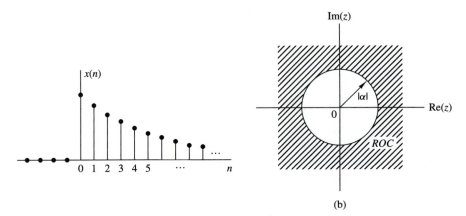

(b)

Figure 3.2 The exponential signal $x(n) = \alpha^n u(n)$ (a), and the ROC of its z-transform (b).

Thus we have the z-transform pair

$$x(n) = \alpha^n u(n) \stackrel{z}{\longleftrightarrow} X(z) = \frac{1}{1 - \alpha z^{-1}} \qquad \text{ROC: } |z| > |\alpha| \qquad (3.1.7)$$

The ROC is the exterior of a circle having radius $|\alpha|$. Figure 3.2 shows a graph of the signal $x(n)$ and its corresponding ROC. Note that, in general, α need not be real.

If we set $\alpha = 1$ in (3.1.7), we obtain the z-transform of the unit step signal

$$x(n) = u(n) \stackrel{z}{\longleftrightarrow} X(z) = \frac{1}{1 - z^{-1}} \qquad \text{ROC: } |z| > 1 \qquad (3.1.8)$$

Example 3.1.4

Determine the z-transform of the signal

$$x(n) = -\alpha^n u(-n-1) = \begin{cases} 0, & n \geq 0 \\ -\alpha^n, & n \leq -1 \end{cases}$$

Solution   From the definition (3.1.1) we have

$$X(z) = \sum_{n=-\infty}^{-1} (-\alpha^n) z^{-n} = -\sum_{l=1}^{\infty} (\alpha^{-1} z)^l$$

where $l = -n$. Using the formula

$$A + A^2 + A^3 + \cdots = A(1 + A + A^2 + \cdots) = \frac{A}{1 - A}$$

when $|A| < 1$ gives

$$X(z) = -\frac{\alpha^{-1} z}{1 - \alpha^{-1} z} = \frac{1}{1 - \alpha z^{-1}}$$

provided that $|\alpha^{-1} z| < 1$ or, equivalently, $|z| < |\alpha|$. Thus

$$x(n) = -\alpha^n u(-n-1) \stackrel{z}{\longleftrightarrow} X(z) = -\frac{1}{1 - \alpha z^{-1}} \qquad \text{ROC: } |z| < |\alpha| \qquad (3.1.9)$$

The ROC is now the interior of a circle having radius $|\alpha|$. This is shown in Fig. 3.3.

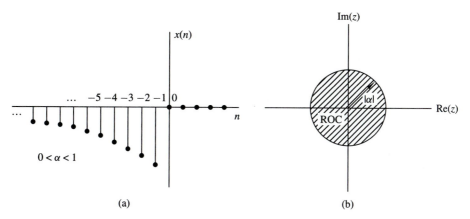

Figure 3.3 Anticausal signal $x(n) = -\alpha^n u(-n - 1)$ (a), and the ROC of its z-transform (b).

Examples 3.1.3 and 3.1.4 illustrate two very important issues. The first concerns the uniqueness of the z-transform. From (3.1.7) and (3.1.9) we see that the causal signal $\alpha^n u(n)$ and the anticausal signal $-\alpha^n u(-n - 1)$ have identical closed-form expressions for the z-transform, that is,

$$Z\{\alpha^n u(n)\} = Z\{-\alpha^n u(-n - 1)\} = \frac{1}{1 - \alpha z^{-1}}$$

This implies that a closed-form expression for the z-transform does not uniquely specify the signal in the time domain. The ambiguity can be resolved only if in addition to the closed-form expression, the ROC is specified. In summary, *a discrete-time signal $x(n)$ is uniquely determined by its z-transform $X(z)$ and the region of convergence of $X(z)$.* In this text the term "z-transform" is used to refer to both the closed-form expression and the corresponding ROC. Example 3.1.3 also illustrates the point that *the ROC of a causal signal is the exterior of a circle of some radius r_2 while the ROC of an anticausal signal is the interior of a circle of some radius r_1.* The following example considers a sequence that is nonzero for $-\infty < n < \infty$.

Example 3.1.5

Determine the z-transform of the signal

$$x(n) = \alpha^n u(n) + b^n u(-n - 1)$$

Solution From definition (3.1.1) we have

$$X(z) = \sum_{n=0}^{\infty} \alpha^n z^{-n} + \sum_{n=-\infty}^{-1} b^n z^{-n} = \sum_{n=0}^{\infty} (\alpha z^{-1})^n + \sum_{l=1}^{\infty} (b^{-1} z)^l$$

The first power series converges if $|\alpha z^{-1}| < 1$ or $|z| > |\alpha|$. The second power series converges if $|b^{-1} z| < 1$ or $|z| < |b|$.

In determining the convergence of $X(z)$, we consider two different cases.

Case 1 $|b| < |\alpha|$:   In this case the two ROC above do not overlap, as shown in Fig. 3.4(a). Consequently, we cannot find values of z for which both power series converge simultaneously. Clearly, in this case, $X(z)$ does not exist.

Case 2 $|b| > |\alpha|$:   In this case there is a ring in the z-plane where both power series converge simultaneously, as shown in Fig. 3.4(b). Then we obtain

$$X(z) = \frac{1}{1 - \alpha z^{-1}} - \frac{1}{1 - b z^{-1}}$$

$$= \frac{b - \alpha}{\alpha + b - z - \alpha b z^{-1}} \tag{3.1.10}$$

The ROC of $X(z)$ is $|\alpha| < |z| < |b|$.

This example shows that *if there is a ROC for an infinite duration two-sided signal, it is a ring (annular region) in the z-plane.* From Examples 3.1.1, 3.1.3, 3.1.4, and 3.1.5, we see that the ROC of a signal depends on both its duration (finite or infinite) and on whether it is causal, anticausal, or two-sided. These facts are summarized in Table 3.1.

One special case of a two-sided signal is a signal that has infinite duration on the right side but not on the left [i.e., $x(n) = 0$ for $n < n_0 < 0$]. A second case is a signal that has infinite duration on the left side but not on the

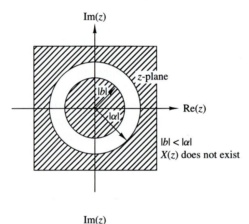

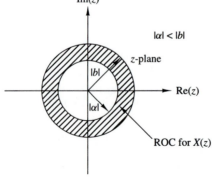

Figure 3.4 ROC for z-transform in Example 3.1.5.

TABLE 3.1 CHARACTERISTIC FAMILIES OF SIGNALS WITH THEIR CORRESPONDING ROC

Signal	ROC

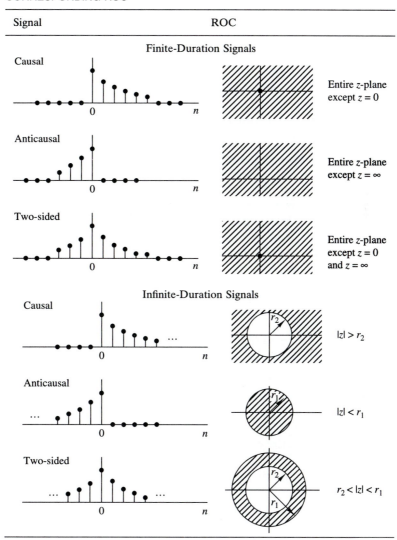

right [i.e., $x(n) = 0$ for $n > n_1 > 0$]. A third special case is a signal that has finite duration on both the left and right sides [i.e., $x(n) = 0$ for $n < n_0 < 0$ and $n > n_1 > 0$]. These types of signals are sometimes called *right-sided, left-sided,* and *finite-duration two-sided*, signals, respectively. The determination of the ROC for these three types of signals is left as an exercise for the reader (Problem 3.5).

Finally, we note that the *z*-transform defined by (3.1.1) is sometimes referred to as the *two-sided* or *bilateral z-transform*, to distinguish it from the *one-sided* or *unilateral z-transform* given by

$$X^+(z) = \sum_{n=0}^{\infty} x(n)z^{-n} \qquad (3.1.11)$$

The one-sided z-transform is examined in Section 3.5. In this text we use the expression z-transform exclusively to mean the two-sided z-transform defined by (3.1.1). The term "two-sided" will be used only in cases where we want to resolve any ambiguities. Clearly, if $x(n)$ is causal [i.e., $x(n) = 0$ for $n < 0$], the one-sided and two-sided z-transforms are equivalent. In any other case, they are different.

3.1.2 The Inverse z-Transform

Often, we have the z-transform $X(z)$ of a signal and we must determine the signal sequence. The procedure for transforming from the z-domain to the time domain is called the *inverse z-transform*. An inversion formula for obtaining $x(n)$ from $X(z)$ can be derived by using the *Cauchy integral theorem*, which is an important theorem in the theory of complex variables.

To begin, we have the z-transform defined by (3.1.1) as

$$X(z) = \sum_{k=-\infty}^{\infty} x(k)z^{-k} \qquad (3.1.12)$$

Suppose that we multiply both sides of (3.1.12) by z^{n-1} and integrate both sides over a closed contour within the ROC of $X(z)$ which encloses the origin. Such a contour is illustrated in Fig. 3.5. Thus we have

$$\oint_C X(z)z^{n-1}dz = \oint_C \sum_{k=-\infty}^{\infty} x(k)z^{n-1-k}dz \qquad (3.1.13)$$

where C denotes the closed contour in the ROC of $X(z)$, taken in a counterclockwise direction. Since the series converges on this contour, we can interchange the order of integration and summation on the right-hand side of (3.1.13). Thus

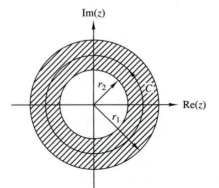

Figure 3.5 Contour C for integral in (3.1.13).

(3.1.13) becomes

$$\oint_C X(z) z^{n-1} dz = \sum_{k=-\infty}^{\infty} x(k) \oint_C z^{n-1-k} dz \tag{3.1.14}$$

Now we can invoke the Cauchy integral theorem, which states that

$$\frac{1}{2\pi j} \oint_C z^{n-1-k} \, dz = \begin{cases} 1, & k = n \\ 0, & k \neq n \end{cases} \tag{3.1.15}$$

where C is any contour that encloses the origin. By applying (3.1.15), the right-hand side of (3.1.14) reduces to $2\pi j x(n)$ and hence the desired inversion formula

$$x(n) = \frac{1}{2\pi j} \oint_C X(z) z^{n-1} \, dz \tag{3.1.16}$$

Although the contour integral in (3.1.16) provides the desired inversion formula for determining the sequence $x(n)$ from the z-transform, we shall not use (3.1.16) directly in our evaluation of inverse z-transforms. In our treatment we deal with signals and systems in the z-domain which have rational z-transforms (i.e., z-transforms that are a ratio of two polynomials). For such z-transforms we develop a simpler method for inversion that stems from (3.1.16) and employs a table lookup.

3.2 PROPERTIES OF THE Z-TRANSFORM

The z-transform is a very powerful tool for the study of discrete-time signals and systems. The power of this transform is a consequence of some very important properties that the transform possesses. In this section we examine some of these properties.

In the treatment that follows, it should be remembered that when we combine several z-transforms, the ROC of the overall transform is, at least, the intersection of the ROC of the individual transforms. This will become more apparent later, when we discuss specific examples.

Linearity.   If

$$x_1(n) \xleftrightarrow{\ z\ } X_1(z)$$

and

$$x_2(n) \xleftrightarrow{\ z\ } X_2(z)$$

then

$$x(n) = a_1 x_1(n) + a_2 x_2(n) \xleftrightarrow{\ z\ } X(z) = a_1 X_1(z) + a_2 X_2(z) \tag{3.2.1}$$

for any constants a_1 and a_2. The proof of this property follows immediately from the definition of linearity and is left as an exercise for the reader.

The linearity property can easily be generalized for an arbitrary number of signals. Basically, it implies that the z-transform of a linear combination of signals is the same linear combination of their z-transforms. Thus the linearity property helps us to find the z-transform of a signal by expressing the signal as a sum of elementary signals, for each of which, the z-transform is already known.

Example 3.2.1

Determine the *z*-transform and the ROC of the signal

$$x(n) = [3(2^n) - 4(3^n)]u(n)$$

Solution If we define the signals

$$x_1(n) = 2^n u(n)$$

and

$$x_2(n) = 3^n u(n)$$

then $x(n)$ can be written as

$$x(n) = 3x_1(n) - 4x_2(n)$$

According to (3.2.1), its *z*-transform is

$$X(z) = 3X_1(z) - 4X_2(z)$$

From (3.1.7) we recall that

$$a^n u(n) \overset{z}{\longleftrightarrow} \frac{1}{1 - az^{-1}} \qquad \text{ROC: } |z| > |\alpha| \qquad (3.2.2)$$

By setting $\alpha = 2$ and $\alpha = 3$ in (3.2.2), we obtain

$$x_1(n) = 2^n u(n) \overset{z}{\longleftrightarrow} X_1(z) = \frac{1}{1 - 2z^{-1}} \qquad \text{ROC: } |z| > 2$$

$$x_2(n) = 3^n u(n) \overset{z}{\longleftrightarrow} X_2(z) = \frac{1}{1 - 3z^{-1}} \qquad \text{ROC: } |z| > 3$$

The intersection of the ROC of $X_1(z)$ and $X_2(z)$ is $|z| > 3$. Thus the overall transform $X(z)$ is

$$X(z) = \frac{3}{1 - 2z^{-1}} - \frac{4}{1 - 3z^{-1}} \qquad \text{ROC: } |z| > 3$$

Example 3.2.2

Determine the *z*-transform of the signals

(a) $x(n) = (\cos \omega_0 n)u(n)$
(b) $x(n) = (\sin \omega_0 n)u(n)$

Solution

(a) By using Euler's identity, the signal $x(n)$ can be expressed as

$$x(n) = (\cos \omega_0 n)u(n) = \tfrac{1}{2}e^{j\omega_0 n}u(n) + \tfrac{1}{2}e^{-j\omega_0 n}u(n)$$

Thus (3.2.1) implies that

$$X(z) = \tfrac{1}{2}Z\{e^{j\omega_0 n}u(n)\} + \tfrac{1}{2}Z\{e^{-j\omega_0 n}u(n)\}$$

If we set $\alpha = e^{\pm j\omega_0} (|\alpha| = |e^{\pm j\omega_0}| = 1)$ in (3.2.2), we obtain

$$e^{j\omega_0 n} u(n) \overset{z}{\longleftrightarrow} \frac{1}{1 - e^{j\omega_0} z^{-1}} \qquad \text{ROC: } |z| > 1$$

and

$$e^{-j\omega_0 n} u(n) \overset{z}{\longleftrightarrow} \frac{1}{1 - e^{-j\omega_0} z^{-1}} \qquad \text{ROC: } |z| > 1$$

Thus

$$X(z) = \frac{1}{2} \frac{1}{1 - e^{j\omega_0} z^{-1}} + \frac{1}{2} \frac{1}{1 - e^{-j\omega_0} z^{-1}} \qquad \text{ROC: } |z| > 1$$

After some simple algebraic manipulations we obtain the desired result, namely,

$$(\cos \omega_0 n) u(n) \overset{z}{\longleftrightarrow} \frac{1 - z^{-1} \cos \omega_0}{1 - 2z^{-1} \cos \omega_0 + z^{-2}} \qquad \text{ROC: } |z| > 1 \qquad (3.2.3)$$

(b) From Euler's identity,

$$x(n) = (\sin \omega_0 n) u(n) = \frac{1}{2j} [e^{j\omega_0 n} u(n) - e^{-j\omega_0 n} u(n)]$$

Thus

$$X(z) = \frac{1}{2j} \left(\frac{1}{1 - e^{j\omega_0} z^{-1}} - \frac{1}{1 - e^{-j\omega_0} z^{-1}} \right) \qquad \text{ROC: } |z| > 1$$

and finally,

$$(\sin \omega_0 n) u(n) \overset{z}{\longleftrightarrow} \frac{z^{-1} \sin \omega_0}{1 - 2z^{-1} \cos \omega_0 + z^{-2}} \qquad \text{ROC: } |z| > 1 \qquad (3.2.4)$$

Time shifting.   If

$$x(n) \overset{z}{\longleftrightarrow} X(z)$$

then

$$x(n - k) \overset{z}{\longleftrightarrow} z^{-k} X(z) \qquad (3.2.5)$$

The ROC of $z^{-k} X(z)$ is the same as that of $X(z)$ except for $z = 0$ if $k > 0$ and $z = \infty$ if $k < 0$. The proof of this property follows immediately from the definition of the z-transform given in (3.1.1)

The properties of linearity and time shifting are the key features that make the z-transform extremely useful for the analysis of discrete-time LTI systems.

Example 3.2.3

By applying the time-shifting property, determine the z-transform of the signals $x_2(n)$ and $x_3(n)$ in Example 3.1.1 from the z-transform of $x_1(n)$.

Solution   It can easily be seen that

$$x_2(n) = x_1(n + 2)$$

and

$$x_3(n) = x_1(n-2)$$

Thus from (3.2.5) we obtain

$$X_2(z) = z^2 X_1(z) = z^2 + 2z + 5 + 7z^{-1} + z^{-3}$$

and

$$X_3(z) = z^{-2} X_1(z) = z^{-2} + 2z^{-3} + 5z^{-4} + 7z^{-5} + z^{-7}$$

Note that because of the multiplication by z^2, the ROC of $X_2(z)$ does not include the point $z = \infty$, even if it is contained in the ROC of $X_1(z)$.

Example 3.2.3 provides additional insight in understanding the meaning of the shifting property. Indeed, if we recall that the coefficient of z^{-n} is the sample value at time n, it is immediately seen that delaying a signal by $k(k > 0)$ samples [i.e., $x(n) \to x(n-k)$] corresponds to multiplying all terms of the z-transform by z^{-k}. The coefficient of z^{-n} becomes the coefficient of $z^{-(n+k)}$.

Example 3.2.4

Determine the transform of the signal

$$x(n) = \begin{cases} 1, & 0 \le n \le N-1 \\ 0, & \text{elsewhere} \end{cases} \tag{3.2.6}$$

Solution   We can determine the z-transform of this signal by using the definition (3.1.1). Indeed,

$$X(z) = \sum_{n=0}^{N-1} 1 \cdot z^{-n} = 1 + z^{-1} + \cdots + z^{-(N-1)} = \begin{cases} N, & \text{if } z = 1 \\ \dfrac{1 - z^{-N}}{1 - z^{-1}}, & \text{if } z \ne 1 \end{cases} \tag{3.2.7}$$

Since $x(n)$ has finite duration, its ROC is the entire z-plane, except $z = 0$.

Let us also derive this transform by using the linearity and time shifting properties. Note that $x(n)$ can be expressed in terms of two unit step signals

$$x(n) = u(n) - u(n-N)$$

By using (3.2.1) and (3.2.5) we have

$$X(z) = Z\{u(n)\} - Z\{u(n-N)\} = (1 - z^{-N})Z\{u(n)\} \tag{3.2.8}$$

However, from (3.1.8) we have

$$Z\{u(n)\} = \frac{1}{1 - z^{-1}} \qquad \text{ROC: } |z| > 1$$

which, when combined with (3.2.8), leads to (3.2.7).

Example 3.2.4 helps to clarify a very important issue regarding the ROC of the combination of several z-transforms. If the linear combination of several signals has finite duration, the ROC of its z-transform is exclusively dictated by the finite-duration nature of this signal, not by the ROC of the individual transforms.

Scaling in the z-domain.   If

$$x(n) \xleftrightarrow{\ z\ } X(z) \qquad \text{ROC: } r_1 < |z| < r_2$$

then

$$a^n x(n) \overset{z}{\longleftrightarrow} X(a^{-1}z) \qquad \text{ROC: } |a|r_1 < |z| < |a|r_2 \qquad (3.2.9)$$

for any constant a, real or complex.

Proof. From the definition (3.1.1)

$$Z\{a^n x(n)\} = \sum_{n=-\infty}^{\infty} a^n x(n) z^{-n} = \sum_{n=-\infty}^{\infty} x(n)(a^{-1}z)^{-n}$$

$$= X(a^{-1}z)$$

Since the ROC of $X(z)$ is $r_1 < |z| < r_2$, the ROC of $X(a^{-1}z)$ is

$$r_1 < |a^{-1}z| < r_2$$

or

$$|a|r_1 < |z| < |a|r_2$$

To better understand the meaning and implications of the scaling property, we express a and z in polar form as $a = r_0 e^{j\omega_0}$, $z = re^{j\omega}$, and we introduce a new complex variable $w = a^{-1}z$. Thus $Z\{x(n)\} = X(z)$ and $Z\{a^n x(n)\} = X(w)$. It can easily be seen that

$$w = a^{-1}z = \left(\frac{1}{r_0}r\right) e^{j(\omega - \omega_0)}$$

This change of variables results in either shrinking (if $r_0 > 1$) or expanding (if $r_0 < 1$) the z-plane in combination with a rotation (if $\omega_0 \neq 2k\pi$) of the z-plane (see Fig. 3.6). This explains why we have a change in the ROC of the new transform where $|a| < 1$. The case $|a| = 1$, that is, $a = e^{j\omega_0}$ is of special interest because it corresponds only to rotation of the z-plane.

Example 3.2.5

Determine the z-transforms of the signals

(a) $x(n) = a^n (\cos \omega_0 n) u(n)$

(b) $x(n) = a^n (\sin \omega_0 n) u(n)$

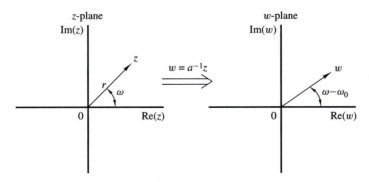

Figure 3.6 Mapping of the z-plane to the w-plane via the transformation $w = a^{-1}z$, $a = r_0 e^{j\omega_0}$.

Solution

(a) From (3.2.3) and (3.2.9) we easily obtain

$$a^n(\cos \omega_0 n)u(n) \xleftrightarrow{z} \frac{1 - az^{-1}\cos \omega_0}{1 - 2az^{-1}\cos \omega_0 + a^2 z^{-2}} \qquad |z| > |a| \qquad (3.2.10)$$

(b) Similarly, (3.2.4) and (3.2.9) yield

$$a^n(\sin \omega_0 n)u(n) \xleftrightarrow{z} \frac{az^{-1}\sin \omega_0}{1 - 2az^{-1}\cos \omega_0 + a^2 z^{-2}} \qquad |z| > |a| \qquad (3.2.11)$$

Time reversal. If

$$x(n) \xleftrightarrow{z} X(z) \qquad \text{ROC: } r_1 < |z| < r_2$$

then

$$x(-n) \xleftrightarrow{z} X(z^{-1}) \qquad \text{ROC: } \frac{1}{r_2} < |z| < \frac{1}{r_1} \qquad (3.2.12)$$

Proof. From the definition (3.1.1), we have

$$Z\{x(-n)\} = \sum_{n=-\infty}^{\infty} x(-n)z^{-n} = \sum_{l=-\infty}^{\infty} x(l)(z^{-1})^{-l} = X(z^{-1})$$

where the change of variable $l = -n$ is made. The ROC of $X(z^{-1})$ is

$$r_1 < |z^{-1}| < r_2 \quad \text{or equivalently} \quad \frac{1}{r_2} < |z| < \frac{1}{r_1}$$

Note that the ROC for $x(n)$ is the inverse of that for $x(-n)$. This means that if z_0 belongs to the ROC of $x(n)$, then $1/z_0$ is in the ROC for $x(-n)$.

An intuitive proof of (3.2.12) is the following. When we fold a signal, the coefficient of z^{-n} becomes the coefficient of z^n. Thus, folding a signal is equivalent to replacing z by z^{-1} in the z-transform formula. In other words, reflection in the time domain corresponds to inversion in the z-domain.

Example 3.2.6

Determine the z-transform of the signal

$$x(n) = u(-n)$$

Solution   It is known from (3.1.8) that

$$u(n) \xleftrightarrow{z} \frac{1}{1 - z^{-1}} \qquad \text{ROC: } |z| > 1$$

By using (3.2.12), we easily obtain

$$u(-n) \xleftrightarrow{z} \frac{1}{1 - z} \qquad \text{ROC: } |z| < 1 \qquad (3.2.13)$$

Differentiation in the z-domain. If

$$x(n) \xleftrightarrow{z} X(z)$$

then

$$nx(n) \xleftrightarrow{z} -z\frac{dX(z)}{dz} \tag{3.2.14}$$

Proof. By differentiating both sides of (3.1.1), we have

$$\frac{dX(z)}{dz} = \sum_{n=-\infty}^{\infty} x(n)(-n)z^{-n-1} = -z^{-1}\sum_{n=-\infty}^{\infty}[nx(n)]z^{-n}$$

$$= -z^{-1}Z\{nx(n)\}$$

Note that both transforms have the same ROC.

Example 3.2.7

Determine the z-transform of the signal

$$x(n) = na^n u(n)$$

Solution   The signal $x(n)$ can be expressed as $nx_1(n)$, where $x_1(n) = a^n u(n)$. From (3.2.2) we have that

$$x_1(n) = a^n u(n) \xleftrightarrow{z} X_1(z) = \frac{1}{1-az^{-1}} \qquad \text{ROC: } |z| > |a|$$

Thus, by using (3.2.14), we obtain

$$na^n u(n) \xleftrightarrow{z} X(z) = -z\frac{dX_1(z)}{dz} = \frac{az^{-1}}{(1-az^{-1})^2} \qquad \text{ROC: } |z| > |a| \tag{3.2.15}$$

If we set $a = 1$ in (3.2.15), we find the z-transform of the unit ramp signal

$$nu(n) \xleftrightarrow{z} \frac{z^{-1}}{(1-z^{-1})^2} \qquad \text{ROC: } |z| > 1 \tag{3.2.16}$$

Example 3.2.8

Determine the signal $x(n)$ whose z-transform is given by

$$X(z) = \log(1 + az^{-1}) \qquad |z| > |a|$$

Solution   By taking the first derivative of $X(z)$, we obtain

$$\frac{dX(z)}{dz} = \frac{-az^{-2}}{1+az^{-1}}$$

Thus

$$-z\frac{dX(z)}{dz} = az^{-1}\left[\frac{1}{1-(-a)z^{-1}}\right] \qquad |z| > |a|$$

The inverse z-transform of the term in brackets is $(-a)^n$. The multiplication by z^{-1} implies a time delay by one sample (time shifting property), which results in $(-a)^{n-1}u(n-1)$. Finally, from the differentiation property we have

$$nx(n) = a(-a)^{n-1}u(n-1)$$

or

$$x(n) = (-1)^{n+1} \frac{a^n}{n} u(n-1)$$

Convolution of two sequences.   If

$$x_1(n) \overset{z}{\longleftrightarrow} X_1(z)$$

$$x_2(n) \overset{z}{\longleftrightarrow} X_2(z)$$

then

$$x(n) = x_1(n) * x_2(n) \overset{z}{\longleftrightarrow} X(z) = X_1(z)X_2(z) \tag{3.2.17}$$

The ROC of $X(z)$ is, at least, the intersection of that for $X_1(z)$ and $X_2(z)$.

Proof. The convolution of $x_1(n)$ and $x_2(n)$ is defined as

$$x(n) = \sum_{k=-\infty}^{\infty} x_1(k)x_2(n-k)$$

The z-transform of $x(n)$ is

$$X(z) = \sum_{n=-\infty}^{\infty} x(n)z^{-n} = \sum_{n=-\infty}^{\infty} \left[\sum_{k=-\infty}^{\infty} x_1(k)x_2(n-k) \right] z^{-n}$$

Upon interchanging the order of the summations and applying the time-shifting property in (3.2.5), we obtain

$$X(z) = \sum_{k=-\infty}^{\infty} x_1(k) \left[\sum_{n=-\infty}^{\infty} x_2(n-k)z^{-n} \right]$$

$$= X_2(z) \sum_{k=-\infty}^{\infty} x_1(k)z^{-k} = X_2(z)X_1(z)$$

Example 3.2.9

Compute the convolution $x(n)$ of the signals

$$x_1(n) = \{1, -2, 1\}$$

$$x_2(n) = \begin{cases} 1, & 0 \le n \le 5 \\ 0, & \text{elsewhere} \end{cases}$$

Solution   From (3.1.1), we have

$$X_1(z) = 1 - 2z^{-1} + z^{-2}$$

$$X_2(z) = 1 + z^{-1} + z^{-2} + z^{-3} + z^{-4} + z^{-5}$$

According to (3.2.17), we carry out the multiplication of $X_1(z)$ and $X_2(z)$. Thus

$$X(z) = X_1(z)X_2(z) = 1 - z^{-1} - z^{-6} + z^{-7}$$

Hence

$$x(n) = \{1, -1, 0, 0, 0, 0, -1, 1\}$$
$$\uparrow$$

The same result can also be obtained by noting that

$$X_1(z) = (1 - z^{-1})^2$$

$$X_2(z) = \frac{1 - z^{-6}}{1 - z^{-1}}$$

Then

$$X(z) = (1 - z^{-1})(1 - z^{-6}) = 1 - z^{-1} - z^{-6} + z^{-7}$$

The reader is encouraged to obtain the same result explicitly by using the convolution summation formula (time-domain approach).

The convolution property is one of the most powerful properties of the z-transform because it converts the convolution of two signals (time domain) to multiplication of their transforms. Computation of the convolution of two signals, using the z-transform, requires the following steps:

1. Compute the z-transforms of the signals to be convolved.

$$X_1(z) = Z\{x_1(n)\}$$

$$\text{(time domain} \longrightarrow z\text{-domain)}$$

$$X_2(z) = Z\{x_2(n)\}$$

2. Multiply the two z-transforms.

$$X(z) = X_1(z)X_2(z) \qquad (z\text{-domain})$$

3. Find the inverse z-transform of $X(z)$.

$$x(n) = Z^{-1}\{X(z)\} \qquad (z\text{-domain} \longrightarrow \text{time domain})$$

This procedure is, in many cases, computationally easier than the direct evaluation of the convolution summation.

Correlation of two sequences.   If

$$x_1(n) \overset{z}{\longleftrightarrow} X_1(z)$$

$$x_2(n) \overset{z}{\longleftrightarrow} X_2(z)$$

then

$$r_{x_1 x_2}(l) = \sum_{n=-\infty}^{\infty} x_1(n)x_2(n - l) \overset{z}{\longleftrightarrow} R_{x_1 x_2}(z) = X_1(z)X_2(z^{-1}) \qquad (3.2.18)$$

Proof. We recall that

$$r_{x_1 x_2}(l) = x_1(l) * x_2(-l)$$

Using the convolution and time-reversal properties, we easily obtain

$$R_{x_1 x_2}(z) = Z\{x_1(l)\}Z\{x_2(-l)\} = X_1(z)X_2(z^{-1})$$

The ROC of $R_{x_1 x_2}(z)$ is at least the intersection of that for $X_1(z)$ and $X_2(z^{-1})$.

As in the case of convolution, the crosscorrelation of two signals is more easily done via polynomial multiplication according to (3.2.18) and then inverse transforming the result.

Example 3.2.10

Determine the autocorrelation sequence of the signal

$$x(n) = a^n u(n), \quad -1 < a < 1$$

Solution   Since the autocorrelation sequence of a signal is its correlation with itself, (3.2.18) gives

$$R_{xx}(z) = Z\{r_{xx}(l)\} = X(z)X(z^{-1})$$

From (3.2.2) we have

$$X(z) = \frac{1}{1 - az^{-1}} \qquad \text{ROC: } |z| > |a| \qquad \text{(causal signal)}$$

and by using (3.2.15), we obtain

$$X(z^{-1}) = \frac{1}{1 - az} \qquad \text{ROC: } |z| < \frac{1}{|a|} \qquad \text{(anticausal signal)}$$

Thus

$$R_{xx}(z) = \frac{1}{1 - az^{-1}} \frac{1}{1 - az} = \frac{1}{1 - a(z + z^{-1}) + a^2} \qquad \text{ROC: } |a| < |z| < \frac{1}{|a|}$$

Since the ROC of $R_{xx}(z)$ is a ring, $r_{xx}(l)$ is a two-sided signal, even if $x(n)$ is causal.

To obtain $r_{xx}(l)$, we observe that the *z*-transform of the sequence in Example 3.1.5 with $b = 1/a$ is simply $(1 - a^2)R_{xx}(z)$. Hence it follows that

$$r_{xx}(l) = \frac{1}{1 - a^2} a^{|l|} \qquad -\infty < l < \infty$$

The reader is encouraged to compare this approach with the time-domain solution of the same problem given in Section 2.6.

Multiplication of two sequences.   If

$$x_1(n) \xleftrightarrow{z} X_1(z)$$

$$x_2(n) \xleftrightarrow{z} X_2(z)$$

then

$$x(n) = x_1(n)x_2(n) \xleftrightarrow{z} X(z) = \frac{1}{2\pi j}\oint_C X_1(v)X_2\left(\frac{z}{v}\right)v^{-1}dv \qquad (3.2.19)$$

where C is a closed contour that encloses the origin and lies within the region of convergence common to both $X_1(v)$ and $X_2(1/v)$.

Proof. The z-transform of $x_3(n)$ is

$$X(z) = \sum_{n=-\infty}^{\infty} x(n)z^{-n} = \sum_{n=-\infty}^{\infty} x_1(n)x_2(n)z^{-n}$$

Let us substitute the inverse transform

$$x_1(n) = \frac{1}{2\pi j} \oint_C X_1(v)v^{n-1}dv$$

for $x_1(n)$ in the z-transform $X(z)$ and interchange the order of summation and integration. Thus we obtain

$$X(z) = \frac{1}{2\pi j} \oint_C X_1(v) \left[\sum_{n=-\infty}^{\infty} x_2(n) \left(\frac{z}{v}\right)^{-n} \right] v^{-1}dv$$

The sum in the brackets is simply the transform $X_2(z)$ evaluated at z/v. Therefore,

$$X(z) = \frac{1}{2\pi j} \oint_C X_1(v)X_2\left(\frac{z}{v}\right) v^{-1}dv$$

which is the desired result.

To obtain the ROC of $X(z)$ we note that if $X_1(v)$ converges for $r_{1l} < |v| < r_{1u}$ and $X_2(z)$ converges for $r_{2l} < |z| < r_{2u}$, then the ROC of $X_2(z/v)$ is

$$r_{2l} < \left|\frac{z}{v}\right| < r_{2u}$$

Hence the ROC for $X(z)$ is at least

$$r_{1l}r_{2l} < |z| < r_{1u}r_{2u} \qquad (3.2.20)$$

Although this property will not be used immediately, it will prove useful later, especially in our treatment of filter design based on the window technique, where we multiply the impulse response of an IIR system by a finite-duration "window" which serves to truncate the impulse response of the IIR system.

For complex-valued sequences $x_1(n)$ and $x_2(n)$ we can define the product sequence as $x(n) = x_1(n)x_2^*(n)$. Then the corresponding complex convolution integral becomes

$$x(n) = x_1(n)x_2^*(n) \overset{z}{\longleftrightarrow} X(z) = \frac{1}{2\pi j} \oint_C X_1(v)X_2^*\left(\frac{z^*}{v^*}\right) v^{-1}dv \qquad (3.2.21)$$

The proof of (3.2.21) is left as an exercise for the reader.

Parseval's relation.   If $x_1(n)$ and $x_2(n)$ are complex-valued sequences, then

$$\sum_{n=-\infty}^{\infty} x_1(n)x_2^*(n) = \frac{1}{2\pi j} \oint_C X_1(v)X_2^*\left(\frac{1}{v^*}\right) v^{-1}dv \qquad (3.2.22)$$

provided that $r_{1l}r_{2l} < 1 < r_{1u}r_{2u}$, where $r_{1l} < |z| < r_{1u}$ and $r_{2l} < |z| < r_{2u}$ are the ROC of $X_1(z)$ and $X_2(z)$. The proof of (3.2.22) follows immediately by evaluating $X(z)$ in (3.2.21) at $z = 1$.

The Initial Value Theorem.   If $x(n)$ is *causal* [i.e., $x(n) = 0$ for $n < 0$], then

$$x(0) = \lim_{z \to \infty} X(z) \tag{3.2.23}$$

Proof. Since $x(n)$ is causal, (3.1.1) gives

$$X(z) = \sum_{n=0}^{\infty} x(n)z^{-n} = x(0) + x(1)z^{-1} + x(2)z^{-2} + \cdots$$

Obviously, as $z \to \infty$, $z^{-n} \to 0$ since $n > 0$ and (3.2.23) follows.

All the properties of the *z*-transform presented in this section are summarized in Table 3.2 for easy reference. They are listed in the same order as they have been introduced in the text. The conjugation properties and Parseval's relation are left as exercises for the reader.

We have now derived most of the *z*-transforms that are encountered in many practical applications. These *z*-transform pairs are summarized in Table 3.3 for easy reference. A simple inspection of this table shows that these *z*-transforms are all *rational functions* (i.e., ratios of polynomials in z^{-1}). As will soon become apparent, rational *z*-transforms are encountered not only as the *z*-transforms of various important signals but also in the characterization of discrete-time linear time-invariant systems described by constant-coefficient difference equations.

3.3 RATIONAL *Z*-TRANSFORMS

As indicated in Section 3.2, an important family of *z*-transforms are those for which $X(z)$ is a rational function, that is, a ratio of two polynomials in z^{-1} (or z). In this section we discuss some very important issues regarding the class of rational *z*-transforms.

3.3.1 Poles and Zeros

The *zeros* of a *z*-transform $X(z)$ are the values of z for which $X(z) = 0$. The *poles* of a *z*-transform are the values of z for which $X(z) = \infty$. If $X(z)$ is a rational function, then

$$X(z) = \frac{N(z)}{D(z)} = \frac{b_0 + b_1 z^{-1} + \cdots + b_M z^{-M}}{a_0 + a_1 z^{-1} + \cdots + a_N z^{-N}} = \frac{\displaystyle\sum_{k=0}^{M} b_k z^{-k}}{\displaystyle\sum_{k=0}^{N} a_k z^{-k}} \tag{3.3.1}$$

If $a_0 \neq 0$ and $b_0 \neq 0$, we can avoid the negative powers of z by factoring out the terms $b_0 z^{-M}$ and $a_0 z^{-N}$ as follows:

$$X(z) = \frac{N(z)}{D(z)} = \frac{b_0 z^{-M}}{a_0 z^{-N}} \frac{z^M + (b_1/b_0)z^{M-1} + \cdots + b_M/b_0}{z^N + (a_1/a_0)z^{N-1} + \cdots + a_N/a_0}$$

TABLE 3.2 PROPERTIES OF THE Z-TRANSFORM

Property	Time Domain	z-Domain	ROC
Notation	$x(n)$ $x_1(n)$ $x_2(n)$	$X(z)$ $X_1(z)$ $X_2(z)$	ROC: $r_2 < \|z\| < r_1$ ROC_1 ROC_2
Linearity	$a_1 x_1(n) + a_2 x_2(n)$	$a_1 X_1(z) + a_2 X_2(z)$	At least the intersection of ROC_1 and ROC_2
Time shifting	$x(n-k)$	$z^{-k} X(z)$	That of $X(z)$, except $z = 0$ if $k > 0$ and $z = \infty$ if $k < 0$
Scaling in the z-domain	$a^n x(n)$	$X(a^{-1}z)$	$\|a\|r_2 < \|z\| < \|a\|r_1$
Time reversal	$x(-n)$	$X(z^{-1})$	$\dfrac{1}{r_1} < \|z\| < \dfrac{1}{r_2}$
Conjugation	$x^*(n)$	$X^*(z^*)$	ROC
Real part	$\text{Re}\{x(n)\}$	$\frac{1}{2}[X(z) + X^*(z^*)]$	Includes ROC
Imaginary part	$\text{Im}\{x(n)\}$	$\frac{1}{2}[X(z) - X^*(z^*)]$	Includes ROC
Differentiation in the z-domain	$nx(n)$	$-z\dfrac{dX(z)}{dz}$	$r_2 < \|z\| < r_1$
Convolution	$x_1(n) * x_2(n)$	$X_1(z)X_2(z)$	At least, the intersection of ROC_1 and ROC_2
Correlation	$r_{x_1 x_2}(l) = x_1(l) * x_2(-l)$	$R_{x_1 x_2}(z) = X_1(z)X_2(z^{-1})$	At least, the intersection of ROC of $X_1(z)$ and $X_2(z^{-1})$
Initial value theorem	If $x(n)$ causal	$x(0) = \lim\limits_{z \to \infty} X(z)$	
Multiplication	$x_1(n)x_2(n)$	$\dfrac{1}{2\pi j}\oint_C X_1(v)X_2\left(\dfrac{z}{v}\right) v^{-1}\,dv$	At least $r_{1l}r_{2l} < \|z\| < r_{1u}r_{2u}$
Parseval's relation	$\displaystyle\sum_{n=-\infty}^{\infty} x_1(n)x_2^*(n) = \dfrac{1}{2\pi j}\oint_C X_1(v)X_2^*(1/v^*)v^{-1}\,dv$		

TABLE 3.3 SOME COMMON Z-TRANSFORM PAIRS

	Signal, $x(n)$	z-Transform, $X(z)$	ROC
1	$\delta(n)$	1	All z
2	$u(n)$	$\dfrac{1}{1-z^{-1}}$	$\|z\| > 1$
3	$a^n u(n)$	$\dfrac{1}{1-az^{-1}}$	$\|z\| > \|a\|$
4	$na^n u(n)$	$\dfrac{az^{-1}}{(1-az^{-1})^2}$	$\|z\| > \|a\|$
5	$-a^n u(-n-1)$	$\dfrac{1}{1-az^{-1}}$	$\|z\| < \|a\|$
6	$-na^n u(-n-1)$	$\dfrac{az^{-1}}{(1-az^{-1})^2}$	$\|z\| < \|a\|$
7	$(\cos \omega_0 n)u(n)$	$\dfrac{1-z^{-1}\cos\omega_0}{1-2z^{-1}\cos\omega_0+z^{-2}}$	$\|z\| > 1$
8	$(\sin \omega_0 n)u(n)$	$\dfrac{z^{-1}\sin\omega_0}{1-2z^{-1}\cos\omega_0+z^{-2}}$	$\|z\| > 1$
9	$(a^n \cos \omega_0 n)u(n)$	$\dfrac{1-az^{-1}\cos\omega_0}{1-2az^{-1}\cos\omega_0+a^2z^{-2}}$	$\|z\| > \|a\|$
10	$(a^n \sin \omega_0 n)u(n)$	$\dfrac{az^{-1}\sin\omega_0}{1-2az^{-1}\cos\omega_0+a^2z^{-2}}$	$\|z\| > \|a\|$

Since $N(z)$ and $D(z)$ are polynomials in z, they can be expressed in factored form as

$$X(z) = \frac{N(z)}{D(z)} = \frac{b_0}{a_0} z^{-M+N} \frac{(z-z_1)(z-z_2)\cdots(z-z_M)}{(z-p_1)(z-p_2)\cdots(z-p_N)}$$

$$X(z) = Gz^{N-M} \frac{\displaystyle\prod_{k=1}^{M}(z-z_k)}{\displaystyle\prod_{k=1}^{N}(z-p_k)} \tag{3.3.2}$$

where $G \equiv b_0/a_0$. Thus $X(z)$ has M finite zeros at $z = z_1, z_2, \ldots, z_M$ (the roots of the numerator polynomial), N finite poles at $z = p_1, p_2, \ldots, p_N$ (the roots of the denominator polynomial), and $|N - M|$ zeros (if $N > M$) or poles (if $N < M$) at the origin $z = 0$. Poles or zeros may also occur at $z = \infty$. A zero exists at $z = \infty$ if $X(\infty) = 0$ and a pole exists at $z = \infty$ if $X(\infty) = \infty$. If we count the poles and zeros at zero and infinity, we find that $X(z)$ has exactly the same number of poles as zeros.

We can represent $X(z)$ graphically by a *pole–zero plot* (or *pattern*) in the complex plane, which shows the location of poles by crosses ($\times$) and the location of zeros by circles ($\circ$). The multiplicity of multiple-order poles or zeros is indicated by a number close to the corresponding cross or circle. Obviously, by definition, the ROC of a z-transform should not contain any poles.

Example 3.3.1

Determine the pole–zero plot for the signal

$$x(n) = a^n u(n) \qquad a > 0$$

Solution   From Table 3.3 we find that

$$X(z) = \frac{1}{1 - az^{-1}} = \frac{z}{z - a} \qquad \text{ROC: } |z| > a$$

Thus $X(z)$ has one zero at $z_1 = 0$ and one pole at $p_1 = a$. The pole–zero plot is shown in Fig. 3.7. Note that the pole $p_1 = a$ is not included in the ROC since the z-transform does not converge at a pole.

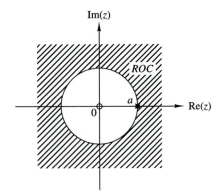

Figure 3.7   Pole–zero plot for the causal exponential signal $x(n) = a^n u(n)$.

Example 3.3.2

Determine the pole–zero plot for the signal

$$x(n) = \begin{cases} a^n, & 0 \le n \le M - 1 \\ 0, & \text{elsewhere} \end{cases}$$

where $a > 0$.

Solution   From the definition (3.1.1) we obtain

$$X(z) = \sum_{n=0}^{M-1} (az^{-1})^n = \frac{1 - (az^{-1})^M}{1 - az^{-1}} = \frac{z^M - a^M}{z^{M-1}(z - a)}$$

Since $a > 0$, the equation $z^M = a^M$ has M roots at

$$z_k = ae^{j2\pi k/M} \qquad k = 0, 1, \ldots, M - 1$$

The zero $z_0 = a$ cancels the pole at $z = a$. Thus

$$X(z) = \frac{(z - z_1)(z - z_2) \cdots (z - z_{M-1})}{z^{M-1}}$$

which has $M - 1$ zeros and $M - 1$ poles, located as shown in Fig. 3.8 for $M = 8$. Note that the ROC is the entire z-plane except $z = 0$ because of the $M - 1$ poles located at the origin.

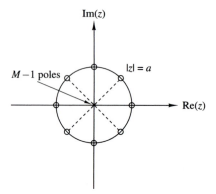

Figure 3.8   Pole–zero pattern for the finite-duration signal $x(n) = a^n$, $0 \le n \le M - 1 (a > 0)$, for $M = 8$.

Clearly, if we are given a pole–zero plot, we can determine $X(z)$, by using (3.3.2), to within a scaling factor G. This is illustrated in the following example.

Example 3.3.3

Determine the z-transform and the signal that corresponds to the pole–zero plot of Fig. 3.9.

Solution   There are two zeros $(M = 2)$ at $z_1 = 0$, $z_2 = r \cos \omega_0$ and two poles $(N = 2)$ at $p_1 = re^{j\omega_0}$, $p_2 = re^{-j\omega_0}$. By substitution of these relations into (3.3.2), we obtain

$$X(z) = G\frac{(z - z_1)(z - z_2)}{(z - p_1)(z - p_2)} = G\frac{z(z - r \cos \omega_0)}{(z - re^{j\omega_0})(z - re^{-j\omega_0})} \qquad \text{ROC: } |z| > r$$

After some simple algebraic manipulations, we obtain

$$X(z) = G\frac{1 - rz^{-1} \cos \omega_0}{1 - 2rz^{-1} \cos \omega_0 + r^2z^{-2}} \qquad \text{ROC: } |z| > r$$

From Table 3.3 we find that

$$x(n) = G(r^n \cos \omega_0 n)u(n)$$

From Example 3.3.3, we see that the product $(z - p_1)(z - p_2)$ results in a polynomial with real coefficients, when p_1 and p_2 are complex conjugates. In

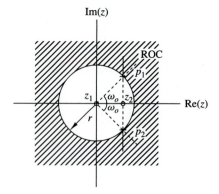

Figure 3.9   Pole-zero pattern for Example 3.3.3.

general, if a polynomial has real coefficients, its roots are either real or occur in complex-conjugate pairs.

As we have seen, the z-transform $X(z)$ is a complex function of the complex variable $z = \mathrm{Re}(z) + j \, \mathrm{Im}(z)$. Obviously, $|X(z)|$, the magnitude of $X(z)$, is a real and positive function of z. Since z represents a point in the complex plane, $|X(z)|$ is a two-dimensional function and describes a "surface." This is illustrated in Fig. 3.10(a) for the z-transform

$$X(z) = \frac{z^{-1} - z^{-2}}{1 + 1.2732z^{-1} + 0.81z^{-2}} \tag{3.3.3}$$

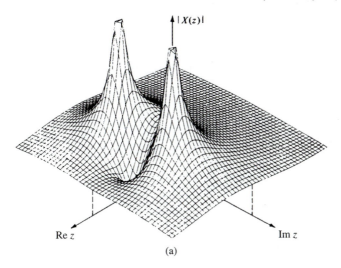

(a)

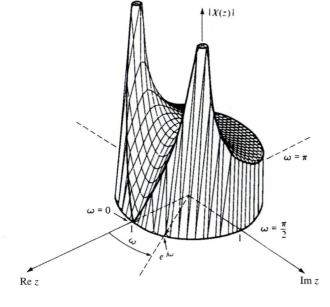

Figure 3.10 Graph of $|X(z)|$ for the z-transform in (3.3.3). [Reproduced with permission from *Introduction to Systems Analysis*, by T. H. Glisson, © 1985 by McGraw-Hill Book Company.]

which has one zero at $z_1 = 1$ and two poles at p_1, $p_2 = 0.9e^{\pm j\pi/4}$. Note the high peaks near the singularities (poles) and the deep valley close to the zero. Figure 3.10(b) illustrates the graph of $|X(z)|$ for $z = e^{j\omega}$.

3.3.2 Pole Location and Time-Domain Behavior for Causal Signals

In this subsection we consider the relation between the z-plane location of a pole pair and the form (shape) of the corresponding signal in the time domain. The discussion is based generally on the collection of z-transform pairs given in Table 3.3 and the results in the preceding subsection. We deal exclusively with real, causal signals. In particular, we see that the characteristic behavior of causal signals depends on whether the poles of the transform are contained in the region $|z| < 1$, or in the region $|z| > 1$, or on the circle $|z| = 1$. Since the circle $|z| = 1$ has a radius of 1, it is called the *unit circle*.

If a real signal has a z-transform with one pole, this pole has to be real. The only such signal is the real exponential

$$x(n) = a^n u(n) \xleftrightarrow{z} X(z) = \frac{1}{1 - az^{-1}} \qquad \text{ROC: } |z| > |a|$$

having one zero at $z_1 = 0$ and one pole at $p_1 = a$ on the real axis. Figure 3.11

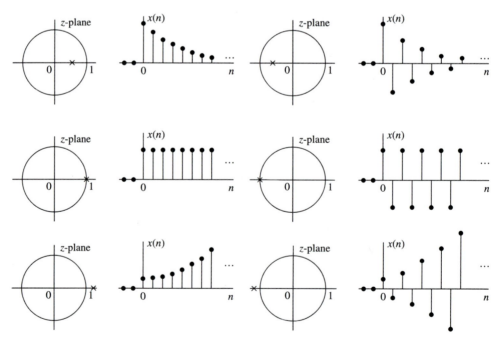

Figure 3.11 Time-domain behavior of a single-real pole causal signal as a function of the location of the pole with respect to the unit circle.

illustrates the behavior of the signal with respect to the location of the pole relative to the unit circle. The signal is decaying if the pole is inside the unit circle, fixed if the pole is on the unit circle, and growing if the pole is outside the unit circle. In addition, a negative pole results in a signal that alternates in sign. Obviously, causal signals with poles outside the unit circle become unbounded, cause overflow in digital systems, and in general, should be avoided.

A causal real signal with a double real pole has the form

$$x(n) = na^n u(n)$$

(see Table 3.3) and its behavior is illustrated in Fig. 3.12. Note that in contrast to the single-pole signal, a double real pole on the unit circle results in an unbounded signal.

Figure 3.13 illustrates the case of a pair of complex-conjugate poles. According to Table 3.3, this configuration of poles results in an exponentially weighted sinusoidal signal. The distance r of the poles from the origin determines the envelope of the sinusoidal signal and their angle with the real positive axis, its relative frequency. Note that the amplitude of the signal is growing if $r > 1$, constant if $r = 1$ (sinusoidal signals), and decaying if $r < 1$.

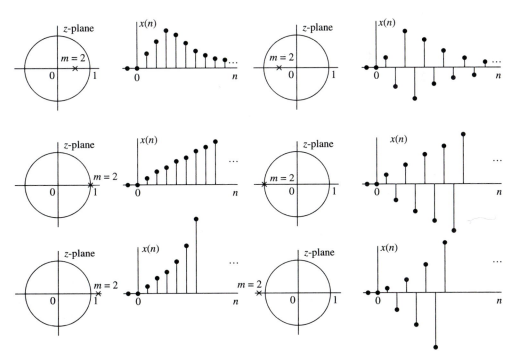

Figure 3.12 Time-domain behavior of causal signals corresponding to a double ($m = 2$) real pole, as a function of the pole location.

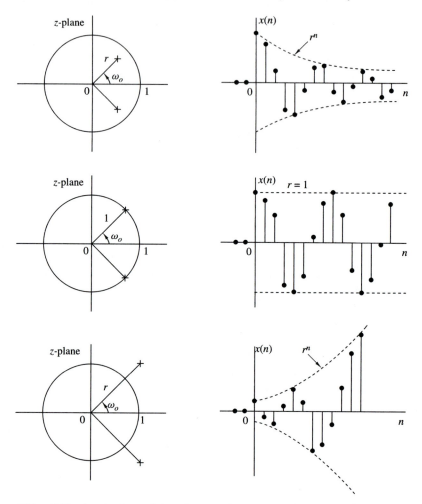

Figure 3.13   A pair of complex-conjugate poles corresponds to causal signals with oscillatory behavior.

Finally, Fig. 3.14 shows the behavior of a causal signal with a double pair of poles on the unit circle. This reinforces the corresponding results in Fig. 3.12 and illustrates that multiple poles on the unit circle should be treated with great care.

To summarize, causal real signals with simple real poles or simple complex-conjugate pairs of poles, which are inside or on the unit circle are always bounded in amplitude. Furthermore, a signal with a pole (or a complex-conjugate pair of poles) near the origin decays more rapidly than one associated with a pole near (but inside) the unit circle. Thus the time behavior of a signal depends strongly on the location of its poles relative to the unit circle. Zeros also affect the behavior of a signal but not as strongly as poles. For example, in the

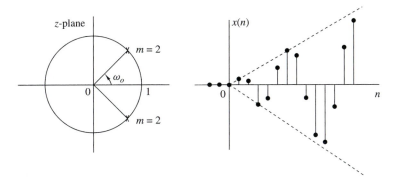

Figure 3.14 Causal signal corresponding to a double pair of complex-conjugate poles on the unit circle.

case of sinusoidal signals, the presence and location of zeros affects only their phase.

At this point, it should be stressed that everything we have said about causal signals applies as well to causal LTI systems, since their impulse response is a causal signal. Hence if a pole of a system is outside the unit circle, the impulse response of the system becomes unbounded and, consequently, the system is unstable.

3.3.3 The System Function of a Linear Time-Invariant System

In Chapter 2 we demonstrated that the output of a (relaxed) linear time-invariant system to an input sequence $x(n)$ can be obtained by computing the convolution of $x(n)$ with the unit sample response of the system. The convolution property, derived in Section 3.2, allows us to express this relationship in the z-domain as

$$Y(z) = H(z)X(z) \tag{3.3.4}$$

where $Y(z)$ is the z-transform of the output sequence $y(n)$, $X(z)$ is the z-transform of the input sequence $x(n)$ and $H(z)$ is the z-transform of the unit sample response $h(n)$.

If we know $h(n)$ and $x(n)$, we can determine their corresponding z-transforms $H(z)$ and $X(z)$, multiply them to obtain $Y(z)$, and therefore determine $y(n)$ by evaluating the inverse z-transform of $Y(z)$. Alternatively, if we know $x(n)$ and we observe the output $y(n)$ of the system, we can determine the unit sample response by first solving for $H(z)$ from the relation

$$H(z) = \frac{Y(z)}{X(z)} \tag{3.3.5}$$

and then evaluating the inverse z-transform of $H(z)$.

Since

$$H(z) = \sum_{n=-\infty}^{\infty} h(n)z^{-n} \tag{3.3.6}$$

it is clear that $H(z)$ represents the z-domain characterization of a system, whereas $h(n)$ is the corresponding time-domain characterization of the system. In other words, $H(z)$ and $h(n)$ are equivalent descriptions of a system in the two domains. The transform $H(z)$ is called the *system function*.

The relation in (3.3.5) is particularly useful in obtaining $H(z)$ when the system is described by a linear constant-coefficient difference equation of the form

$$y(n) = - \sum_{k=1}^{N} a_k y(n-k) + \sum_{k=0}^{M} b_k x(n-k) \tag{3.3.7}$$

In this case the system function can be determined directly from (3.3.7) by computing the z-transform of both sides of (3.3.7). Thus, by applying the time-shifting property, we obtain

$$Y(z) = - \sum_{k=1}^{N} a_k Y(z) z^{-k} + \sum_{k=0}^{M} b_k X(z) z^{-k}$$

$$Y(z) \left(1 + \sum_{k=1}^{N} a_k z^{-k} \right) = X(z) \left(\sum_{k=0}^{M} b_k z^{-k} \right)$$

$$\frac{Y(z)}{X(z)} = \frac{\displaystyle\sum_{k=0}^{M} b_k z^{-k}}{1 + \displaystyle\sum_{k=1}^{N} a_k z^{-k}}$$

or, equivalently,

$$H(z) = \frac{\displaystyle\sum_{k=0}^{M} b_k z^{-k}}{1 + \displaystyle\sum_{k=1}^{N} a_k z^{-k}} \tag{3.3.8}$$

Therefore, a linear time-invariant system described by a constant-coefficient difference equation has a rational system function.

This is the general form for the system function of a system described by a linear constant-coefficient difference equation. From this general form we obtain two important special forms. First, if $a_k = 0$ for $1 \le k \le N$, (3.3.8) reduces to

$$\begin{aligned} H(z) &= \sum_{k=0}^{M} b_k z^{-k} \\ &= \frac{1}{z^M} \sum_{k=0}^{M} b_k z^{M-k} \end{aligned} \tag{3.3.9}$$

In this case, $H(z)$ contains M zeros, whose values are determined by the system parameters $\{b_k\}$, and an Mth-order pole at the origin $z = 0$. Since the system contains only trivial poles (at $z = 0$) and M nontrivial zeros, it is called an *all-zero system*. Clearly, such a system has a finite-duration impulse response (FIR), and it is called an FIR system or a moving average (MA) system.

On the other hand, if $b_k = 0$ for $1 \leq k \leq M$, the system function reduces to

$$H(z) = \frac{b_0}{1 + \displaystyle\sum_{k=1}^{N} a_k z^{-k}}$$

$$= \frac{b_0 z^N}{\displaystyle\sum_{k=0}^{N} a_k z^{N-k}} \qquad a_0 \equiv 1 \tag{3.3.10}$$

In this case $H(z)$ consists of N poles, whose values are determined by the system parameters $\{a_k\}$ and an Nth-order zero at the origin $z = 0$. We usually do not make reference to these trivial zeros. Consequently, the system function in (3.3.10) contains only nontrivial poles and the corresponding system is called an *all-pole system*. Due to the presence of poles, the impulse response of such a system is infinite in duration, and hence it is an IIR system.

The general form of the system function given by (3.3.8) contains both poles and zeros, and hence the corresponding system is called a *pole–zero system*, with N poles and M zeros. Poles and/or zeros at $z = 0$ and $z = \infty$ are implied but are not counted explicitly. Due to the presence of poles, a pole–zero system is an IIR system.

The following example illustrates the procedure for determining the system function and the unit sample response from the difference equation.

Example 3.3.4

Determine the system function and the unit sample response of the system described by the difference equation

$$y(n) = \tfrac{1}{2}y(n-1) + 2x(n)$$

Solution   By computing the z-transform of the difference equation, we obtain

$$Y(z) = \tfrac{1}{2}z^{-1}Y(z) + 2X(z)$$

Hence the system function is

$$\frac{Y(z)}{X(z)} \equiv H(z) = \frac{2}{1 - \tfrac{1}{2}z^{-1}}$$

This system has a pole at $z = \tfrac{1}{2}$ and a zero at the origin. Using Table 3.3 we obtain the inverse transform

$$h(n) = 2(\tfrac{1}{2})^n u(n)$$

This is the unit sample response of the system.

We have now demonstrated that rational z-transforms are encountered in commonly used systems and in the characterization of linear time-invariant systems. In Section 3.4 we describe several methods for determining the inverse z-transform of rational functions.

3.4 INVERSION OF THE *Z*-TRANSFORM

As we saw in Section 3.1.2, the inverse z-transform is formally given by

$$x(n) = \frac{1}{2\pi j} \oint_C X(z) z^{n-1} dz \tag{3.4.1}$$

where the integral is a contour integral over a closed path C that encloses the origin and lies within the region of convergence of $X(z)$. For simplicity, C can be taken as a circle in the ROC of $X(z)$ in the z-plane.

There are three methods that are often used for the evaluation of the inverse z-transform in practice:

1. Direct evaluation of (3.4.1), by contour integration.
2. Expansion into a series of terms, in the variables z, and z^{-1}.
3. Partial-fraction expansion and table lookup.

3.4.1 The Inverse *z*-Transform by Contour Integration

In this section we demonstrate the use of the Cauchy residue theorem to determine the inverse z-transform directly from the contour integral.

Cauchy residue theorem.   Let $f(z)$ be a function of the complex variable z and C be a closed path in the z-plane. If the derivative $df(z)/dz$ exists on and inside the contour C and if $f(z)$ has no poles at $z = z_0$, then

$$\frac{1}{2\pi j} \oint_C \frac{f(z)}{z - z_0} dz = \begin{cases} f(z_0), & \text{if } z_0 \text{ is inside } C \\ 0, & \text{if } z_0 \text{ is outside } C \end{cases} \tag{3.4.2}$$

More generally, if the $(k+1)$-order derivative of $f(z)$ exists and $f(z)$ has no poles at $z = z_0$, then

$$\frac{1}{2\pi j} \oint_C \frac{f(z)}{(z - z_0)^k} dz = \begin{cases} \dfrac{1}{(k-1)!} \dfrac{d^{k-1} f(z)}{dz^{k-1}} \bigg|_{z=z_0}, & \text{if } z_0 \text{ is inside } C \\ 0, & \text{if } z_0 \text{ is outside } C \end{cases} \tag{3.4.3}$$

The values on the right-hand side of (3.4.2) and (3.4.3) are called the residues of the pole at $z = z_0$. The results in (3.4.2) and (3.4.3) are two forms of the *Cauchy residue theorem*.

We can apply (3.4.2) and (3.4.3) to obtain the values of more general contour integrals. To be specific, suppose that the integrand of the contour integral is

$P(z) = f(z)/g(z)$, where $f(z)$ has no poles inside the contour C and $g(z)$ is a polynomial with distinct (simple) roots $z_1, z_2, \ldots, z_n$ inside C. Then

$$\frac{1}{2\pi j} \oint_C \frac{f(z)}{g(z)} dz = \frac{1}{2\pi j} \oint_C \left[\sum_{i=1}^{n} \frac{A_i(z)}{z - z_i} \right] dz$$

$$= \sum_{i=1}^{n} \frac{1}{2\pi j} \oint_C \frac{A_i(z)}{z - z_i} dz \qquad (3.4.4)$$

$$= \sum_{i=1}^{n} A_i(z_i)$$

where

$$A_i(z) = (z - z_i) P(z) = (z - z_i) \frac{f(z)}{g(z)} \qquad (3.4.5)$$

The values $\{A_i(z_i)\}$ are residues of the corresponding poles at $z = z_i$, $i = 1, 2, \ldots, n$. Hence the value of the contour integral is equal to the sum of the residues of all the poles inside the contour C.

We observe that (3.4.4) was obtained by performing a partial-fraction expansion of the integrand and applying (3.4.2). When $g(z)$ has multiple-order roots as well as simple roots inside the contour, the partial-fraction expansion, with appropriate modifications, and (3.4.3) can be used to evaluate the residues at the corresponding poles.

In the case of the inverse z-transform, we have

$$x(n) = \frac{1}{2\pi j} \oint_C X(z) z^{n-1} dz$$

$$= \sum_{\text{all poles } \{z_i\} \text{ inside } C} [\text{residue of } X(z)z^{n-1} \text{ at } z = z_i] \qquad (3.4.6)$$

$$= \sum_{i} (z - z_i) X(z) z^{n-1} |_{z=z_i}$$

provided that the poles $\{z_i\}$ are simple. If $X(z)z^{n-1}$ has no poles inside the contour C for one or more values of n, then $x(n) = 0$ for these values.

The following example illustrates the evaluation of the inverse z-transform by use of the Cauchy residue theorem.

Example 3.4.1

Evaluate the inverse z-transform of

$$X(z) = \frac{1}{1 - az^{-1}} \qquad |z| > |a|$$

using the complex inversion integral.

Solution   We have

$$x(n) = \frac{1}{2\pi j} \oint_C \frac{z^{n-1}}{1 - az^{-1}} dz = \frac{1}{2\pi j} \oint_C \frac{z^n dz}{z - a}$$

where C is a circle at radius greater than $|a|$. We shall evaluate this integral using (3.4.2) with $f(z) = z^n$. We distinguish two cases.

1. If $n \geq 0$, $f(z)$ has only zeros and hence no poles inside C. The only pole inside C is $z = a$. Hence

$$x(n) = f(z_0) = a^n \qquad n \geq 0$$

2. If $n < 0$, $f(z) = z^n$ has an nth-order pole at $z = 0$, which is also inside C. Thus there are contributions from both poles. For $n = -1$ we have

$$x(-1) = \frac{1}{2\pi j} \oint_C \frac{1}{z(z-a)} dz = \frac{1}{z-a}\bigg|_{z=0} + \frac{1}{z}\bigg|_{z=a} = 0$$

If $n = -2$, we have

$$x(-2) = \frac{1}{2\pi j} \oint_C \frac{1}{z^2(z-a)} dz = \frac{d}{dz}\left(\frac{1}{z-a}\right)\bigg|_{z=0} + \frac{1}{z^2}\bigg|_{z=a} = 0$$

By continuing in the same way we can show that $x(n) = 0$ for $n < 0$. Thus

$$x(n) = a^n u(n)$$

3.4.2 The Inverse z-Transform by Power Series Expansion

The basic idea in this method is the following: Given a z-transform $X(z)$ with its corresponding ROC, we can expand $X(z)$ into a power series of the form

$$X(z) = \sum_{n=-\infty}^{\infty} c_n z^{-n} \tag{3.4.7}$$

which converges in the given ROC. Then, by the uniqueness of the z-transform, $x(n) = c_n$ for all n. When $X(z)$ is rational, the expansion can be performed by long division.

To illustrate this technique, we will invert some z-transforms involving the same expression for $X(z)$, but different ROC. This will also serve to emphasize again the importance of the ROC in dealing with z-transforms.

Example 3.4.2

Determine the inverse z-transform of

$$X(z) = \frac{1}{1 - 1.5z^{-1} + 0.5z^{-2}}$$

when

(a) ROC: $|z| > 1$

(b) ROC: $|z| < 0.5$

Solution

(a) Since the ROC is the exterior of a circle, we expect $x(n)$ to be a causal signal. Thus we seek a power series expansion in negative powers of z. By dividing the numerator of $X(z)$ by its denominator, we obtain the power series

$$X(z) = \frac{1}{1 - \frac{3}{2}z^{-1} + \frac{1}{2}z^{-2}} = 1 + \frac{3}{2}z^{-1} + \frac{7}{4}z^{-2} + \frac{15}{8}z^{-3} + \frac{31}{16}z^{-4} + \cdots$$

By comparing this relation with (3.1.1), we conclude that

$$x(n) = \{1, \tfrac{3}{2}, \tfrac{7}{4}, \tfrac{15}{8}, \tfrac{31}{16}, \ldots\}$$
$$\uparrow$$

Note that in each step of the long-division process, we eliminate the lowest-power term of z^{-1}.

(b) In this case the ROC is the interior of a circle. Consequently, the signal $x(n)$ is anticausal. To obtain a power series expansion in positive powers of z, we perform the long division in the following way:

$$
\begin{array}{r}
2z^2 + 6z^3 + 14z^4 + 30z^5 + 62z^6 + \cdots \\
\tfrac{1}{2}z^{-2} - \tfrac{3}{2}z^{-1} + 1 \overline{)\, 1 } \\
\underline{1 - 3z + 2z^2} \\
3z - 2z^2 \\
\underline{3z - 9z^2 + 6z^3} \\
7z^2 - 6z^3 \\
\underline{7z^2 - 21z^3 + 14z^4} \\
15z^3 - 14z^4 \\
\underline{15z^3 - 45z^4 + 30z^5} \\
31z^4 - 30z^5
\end{array}
$$

Thus

$$X(z) = \frac{1}{1 - \frac{3}{2}z^{-1} + \frac{1}{2}z^{-2}} = 2z^2 + 6z^3 + 14z^4 + 30z^5 + 62z^6 + \cdots$$

In this case $x(n) = 0$ for $n \geq 0$. By comparing this result to (3.1.1), we conclude that

$$x(n) = \{\cdots 62, 30, 14, 6, 2, 0, 0\}$$
$$\uparrow$$

We observe that in each step of the long-division process, the lowest-power term of z is eliminated. We emphasize that in the case of anticausal signals we simply carry out the long division by writing down the two polynomials in "reverse" order (i.e., starting with the most negative term on the left).

From this example we note that, in general, the method of long division will not provide answers for $x(n)$ when n is large because the long division becomes tedious. Although, the method provides a direct evaluation of $x(n)$, a closed-form solution is not possible, except if the resulting pattern is simple enough to infer the general term $x(n)$. Hence this method is used only if one wished to determine the values of the first few samples of the signal.

Example 3.4.3

Determine the inverse z-transform of

$$X(z) = \log(1 + az^{-1}) \qquad |z| > |a|$$

Solution Using the power series expansion for $\log(1 + x)$, with $|x| < 1$, we have

$$X(z) = \sum_{n=1}^{\infty} \frac{(-1)^{n+1} a^n z^{-n}}{n}$$

Thus

$$x(n) = \begin{cases} (-1)^{n+1} \dfrac{a^n}{n}, & n \geq 1 \\ 0, & n \leq 0 \end{cases}$$

Expansion of irrational functions into power series can be obtained from tables.

3.4.3 The Inverse z-Transform by Partial-Fraction Expansion

In the table lookup method, we attempt to express the function $X(z)$ as a linear combination

$$X(z) = \alpha_1 X_1(z) + \alpha_2 X_2(z) + \cdots + \alpha_K X_K(z) \tag{3.4.8}$$

where $X_1(z), \ldots, X_K(z)$ are expressions with inverse transforms $x_1(n), \ldots, x_K(n)$ available in a table of z-transform pairs. If such a decomposition is possible, then $x(n)$, the inverse z-transform of $X(z)$, can easily be found using the linearity property as

$$x(n) = \alpha_1 x_1(n) + \alpha_2 x_2(n) + \cdots + \alpha_K x_K(n) \tag{3.4.9}$$

This approach is particularly useful if $X(z)$ is a rational function, as in (3.3.1). Without loss of generality, we assume that $a_0 = 1$, so that (3.3.1) can be expressed as

$$X(z) = \frac{N(z)}{D(z)} = \frac{b_0 + b_1 z^{-1} + \cdots + b_M z^{-M}}{1 + a_1 z^{-1} + \cdots + a_N z^{-N}} \tag{3.4.10}$$

Note that if $a_0 \neq 1$, we can obtain (3.4.10) from (3.3.1) by dividing both numerator and denominator by a_0.

A rational function of the form (3.4.10) is called *proper* if $a_N \neq 0$ and $M < N$. From (3.3.2) it follows that this is equivalent to saying that the number of finite zeros is less than the number of finite poles.

An improper rational function ($M \geq N$) can always be written as the sum of a polynomial and a proper rational function. This procedure is illustrated by the following example.

Example 3.4.4

Express the improper rational transform

$$X(z) = \frac{1 + 3z^{-1} + \frac{11}{6} z^{-2} + \frac{1}{3} z^{-3}}{1 + \frac{5}{6} z^{-1} + \frac{1}{6} z^{-2}}$$

in terms of a polynomial and a proper function.

Solution   First, we note that we should reduce the numerator so that the terms z^{-2} and z^{-3} are eliminated. Thus we should carry out the long division with these two polynomials written in *reverse* order. We stop the division when the order of the remainder becomes z^{-1}. Then we obtain

$$X(z) = 1 + 2z^{-1} + \frac{\frac{1}{6}z^{-1}}{1 + \frac{5}{6}z^{-1} + \frac{1}{6}z^{-2}}$$

In general, any improper rational function ($M \geq N$) can be expressed as

$$X(z) = \frac{N(z)}{D(z)} = c_0 + c_1 z^{-1} + \cdots + c_{M-N} z^{-(M-N)} + \frac{N_1(z)}{D(z)} \tag{3.4.11}$$

The inverse z-transform of the polynomial can easily be found by inspection. We focus our attention on the inversion of proper rational transforms, since any improper function can be transformed into a proper function by using (3.4.11). We carry out the development in two steps. First, we perform a partial fraction expansion of the proper rational function and then we invert each of the terms.

Let $X(z)$ be a proper rational function, that is,

$$X(z) = \frac{N(z)}{D(z)} = \frac{b_0 + b_1 z^{-1} + \cdots + b_M z^{-M}}{1 + a_1 z^{-1} + \cdots + a_N z^{-N}} \tag{3.4.12}$$

where

$$a_N \neq 0 \qquad \text{and} \qquad M < N$$

To simplify our discussion we eliminate negative powers of z by multiplying both the numerator and denominator of (3.4.12) by z^N. This results in

$$X(z) = \frac{b_0 z^N + b_1 z^{N-1} + \cdots + b_M z^{N-M}}{z^N + a_1 z^{N-1} + \cdots + a_N} \tag{3.4.13}$$

which contains only positive powers of z. Since $N > M$, the function

$$\frac{X(z)}{z} = \frac{b_0 z^{N-1} + b_1 z^{N-2} + \cdots + b_M z^{N-M-1}}{z^N + a_1 z^{N-1} + \cdots + a_N} \tag{3.4.14}$$

is also always proper.

Our task in performing a partial-fraction expansion is to express (3.4.14) or, equivalently, (3.4.12) as a sum of simple fractions. For this purpose we first factor the denominator polynomial in (3.4.14) into factors that contain the poles $p_1, p_2, \ldots, p_N$ of $X(z)$. We distinguish two cases.

Distinct poles.   Suppose that the poles $p_1, p_2, \ldots, p_N$ are all different (distinct). Then we seek an expansion of the form

$$\frac{X(z)}{z} = \frac{A_1}{z - p_1} + \frac{A_2}{z - p_2} + \cdots + \frac{A_N}{z - p_N} \tag{3.4.15}$$

The problem is to determine the coefficients $A_1, A_2, \ldots, A_N$. There are two ways to solve this problem, as illustrated in the following example.

Example 3.4.5

Determine the partial-fraction expansion of the proper function

$$X(z) = \frac{1}{1 - 1.5z^{-1} + 0.5z^{-2}} \tag{3.4.16}$$

Solution   First we eliminate the negative powers, by multiplying both numerator and denominator by z^2. Thus

$$X(z) = \frac{z^2}{z^2 - 1.5z + 0.5}$$

The poles of $X(z)$ are $p_1 = 1$ and $p_2 = 0.5$. Consequently, the expansion of the form (3.4.15) is

$$\frac{X(z)}{z} = \frac{z}{(z-1)(z-0.5)} = \frac{A_1}{z-1} + \frac{A_2}{z-0.5} \tag{3.4.17}$$

A very simple method to determine A_1 and A_2 is to multiply the equation by the denominator term $(z-1)(z-0.5)$. Thus we obtain

$$z = (z - 0.5)A_1 + (z - 1)A_2 \tag{3.4.18}$$

Now if we set $z = p_1 = 1$ in (3.4.18), we eliminate the term involving A_2. Hence

$$1 = (1 - 0.5)A_1$$

Thus we obtain the result $A_1 = 2$. Next we return to (3.4.18) and set $z = p_2 = 0.5$, thus eliminating the term involving A_1, so we have

$$0.5 = (0.5 - 1)A_2$$

and hence $A_2 = -1$. Therefore, the result of the partial-fraction expansion is

$$\frac{X(z)}{z} = \frac{2}{z-1} - \frac{1}{z-0.5} \tag{3.4.19}$$

The example given above suggests that we can determine the coefficients A_1, $A_2, \ldots, A_N$, by multiplying both sides of (3.4.15) by each of the terms $(z - p_k)$, $k = 1, 2, \ldots, N$, and evaluating the resulting expressions at the corresponding pole positions, $p_1, p_2, \ldots, p_N$. Thus we have, in general,

$$\frac{(z - p_k)X(z)}{z} = \frac{(z - p_k)A_1}{z - p_1} + \cdots + A_k + \cdots + \frac{(z - p_k)A_N}{z - p_N} \tag{3.4.20}$$

Consequently, with $z = p_k$, (3.4.20) yields the kth coefficient as

$$A_k = \frac{(z - p_k)X(z)}{z}\bigg|_{z=p_k} \qquad k = 1, 2, \ldots, N \tag{3.4.21}$$

Example 3.4.6

Determine the partial-fraction expansion of

$$X(z) = \frac{1 + z^{-1}}{1 - z^{-1} + 0.5z^{-2}} \tag{3.4.22}$$

Solution   To eliminate negative powers of z in (3.4.22), we multiply both numerator and denominator by z^2. Thus

$$\frac{X(z)}{z} = \frac{z+1}{z^2 - z + 0.5}$$

The poles of $X(z)$ are complex conjugates

$$p_1 = \tfrac{1}{2} + j\tfrac{1}{2}$$

and

$$p_2 = \tfrac{1}{2} - j\tfrac{1}{2}$$

Since $p_1 \neq p_2$, we seek an expansion of the form (3.4.15). Thus

$$\frac{X(z)}{z} = \frac{z+1}{(z-p_1)(z-p_2)} = \frac{A_1}{z-p_1} + \frac{A_2}{z-p_2}$$

To obtain A_1 and A_2, we use the formula (3.4.21). Thus we obtain

$$A_1 = \frac{(z-p_1)X(z)}{z}\bigg|_{z=p_1} = \frac{z+1}{z-p_2}\bigg|_{z=p_1} = \frac{\tfrac{1}{2} + j\tfrac{1}{2} + 1}{\tfrac{1}{2} + j\tfrac{1}{2} - \tfrac{1}{2} + j\tfrac{1}{2}} = \tfrac{1}{2} - j\tfrac{3}{2}$$

$$A_2 = \frac{(z-p_2)X(z)}{z}\bigg|_{z=p_2} = \frac{z+1}{z-p_1}\bigg|_{z=p_2} = \frac{\tfrac{1}{2} - j\tfrac{1}{2} + 1}{\tfrac{1}{2} - j\tfrac{1}{2} - \tfrac{1}{2} - j\tfrac{1}{2}} = \tfrac{1}{2} + j\tfrac{3}{2}$$

The expansion (3.4.15) and the formula (3.4.21) hold for both real and complex poles. The only constraint is that all poles be distinct. We also note that $A_2 = A_1^*$. It can be easily seen that this is a consequence of the fact that $p_2 = p_1^*$. In other words, *complex-conjugate poles result in complex-conjugate coefficients in the partial-fraction expansion.* This simple result will prove very useful later in our discussion.

Multiple-order poles.   If $X(z)$ has a pole of multiplicity l, that is, it contains in its denominator the factor $(z - p_k)^l$, then the expansion (3.4.15) is no longer true. In this case a different expansion is needed. First, we investigate the case of a double pole (i.e., $l = 2$).

Example 3.4.7

Determine the partial-fraction expansion of

$$X(z) = \frac{1}{(1 + z^{-1})(1 - z^{-1})^2} \tag{3.4.23}$$

Solution   First, we express (3.4.23) in terms of positive powers of z, in the form

$$\frac{X(z)}{z} = \frac{z^2}{(z+1)(z-1)^2}$$

$X(z)$ has a simple pole at $p_1 = -1$ and a double pole $p_2 = p_3 = 1$. In such a case the appropriate partial-fraction expansion is

$$\frac{X(z)}{z} = \frac{z^2}{(z+1)(z-1)^2} = \frac{A_1}{z+1} + \frac{A_2}{z-1} + \frac{A_3}{(z-1)^2} \tag{3.4.24}$$

The problem is to determine the coefficients A_1, A_2, and A_3.

We proceed as in the case of distinct poles. To determine A_1, we multiply both sides of (3.4.24) by $(z + 1)$ and evaluate the result at $z = -1$. Thus (3.4.24) becomes

$$\frac{(z + 1)X(z)}{z} = A_1 + \frac{z + 1}{z - 1}A_2 + \frac{z + 1}{(z - 1)^2}A_3$$

which, when evaluated at $z = -1$, yields

$$A_1 = \left.\frac{(z + 1)X(z)}{z}\right|_{z=-1} = \frac{1}{4}$$

Next, if we multiply both sides of (3.4.24) by $(z - 1)^2$, we obtain

$$\frac{(z - 1)^2 X(z)}{z} = \frac{(z - 1)^2}{z + 1}A_1 + (z - 1)A_2 + A_3 \tag{3.4.25}$$

Now, if we evaluate (3.4.25) at $z = 1$, we obtain A_3. Thus

$$A_3 = \left.\frac{(z - 1)^2 X(z)}{z}\right|_{z=1} = \frac{1}{2}$$

The remaining coefficient A_2 can be obtained by differentiating both sides of (3.4.25) with respect to z and evaluating the result at $z = 1$. Note that it is not necessary formally to carry out the differentiation of the right-hand side of (3.4.25), since all terms except A_2 vanish when we set $z = 1$. Thus

$$A_2 = \frac{d}{dz}\left[\frac{(z - 1)^2 X(z)}{z}\right]_{z=1} = \frac{3}{4} \tag{3.4.26}$$

The generalization of the procedure in the example above to the case of an lth-order pole $(z - p_k)^l$ is straightforward. The partial-fraction expansion must contain the terms

$$\frac{A_{1k}}{z - p_k} + \frac{A_{2k}}{(z - p_k)^2} + \cdots + \frac{A_{lk}}{(z - p_k)^l}$$

The coefficients $\{A_{ik}\}$ can be evaluated through differentiation as illustrated in Example 3.4.7 for $l = 2$.

Now that we have performed the partial-fraction expansion, we are ready to take the final step in the inversion of $X(z)$. First, let us consider the case in which $X(z)$ contains distinct poles. From the partial-fraction expansion (3.4.15), it easily follows that

$$X(z) = A_1\frac{1}{1 - p_1 z^{-1}} + A_2\frac{1}{1 - p_2 z^{-1}} + \cdots + A_N\frac{1}{1 - p_N z^{-1}} \tag{3.4.27}$$

The inverse z-transform, $x(n) = Z^{-1}\{X(z)\}$, can be obtained by inverting each term in (3.4.27) and taking the corresponding linear combination. From Table 3.3 it follows that these terms can be inverted using the formula

$$Z^{-1}\left\{\frac{1}{1 - p_k z^{-1}}\right\} = \begin{cases} (p_k)^n u(n), & \text{if ROC: } |z| > |p_k| \\ & \text{(causal signals)} \\ -(p_k)^n u(-n - 1), & \text{if ROC: } |z| < |p_k| \\ & \text{(anticausal signals)} \end{cases} \tag{3.4.28}$$

If the signal $x(n)$ is causal, the ROC is $|z| > p_{max}$, where $p_{max} = \max\{|p_1|,$ $|p_2|, \ldots, |p_N|\}$. In this case all terms in (3.4.27) result in causal signal components and the signal $x(n)$ is given by

$$x(n) = (A_1 p_1^n + A_2 p_2^n + \cdots + A_N p_N^n)u(n) \tag{3.4.29}$$

If all poles are real, (3.4.29) is the desired expression for the signal $x(n)$. Thus a causal signal, having a z-transform that contains real and distinct poles, is a linear combination of real exponential signals.

Suppose now that all poles are distinct but some of them are complex. In this case some of the terms in (3.4.27) result in complex exponential components. However, if the signal $x(n)$ is real, we should be able to reduce these terms into real components. If $x(n)$ is real, the polynomials appearing in $X(z)$ have real coefficients. In this case, as we have seen in Section 3.3, if p_j is a pole, its complex conjugate p_j^* is also a pole. As was demonstrated in Example 3.4.6, the corresponding coefficients in the partial-fraction expansion are also complex conjugates. Thus the contribution of two complex-conjugate poles is of the form

$$x_k(n) = [A_k(p_k)^n + A_k^*(p_k^*)^n]u(n) \tag{3.4.30}$$

These two terms can be combined to form a real signal component. First, we express A_j and p_j in polar form (i.e., amplitude and phase) as

$$A_k = |A_k|e^{j\alpha_k} \tag{3.4.31}$$

$$p_k = r_k e^{j\beta_k} \tag{3.4.32}$$

where α_k and β_k are the phase components of A_k and p_k. Substitution of these relations into (3.4.30) gives

$$x_k(n) = |A_k|r_k^n[e^{j(\beta_k n + \alpha_k)} + e^{-j(\beta_k n + \alpha_k)}]u(n)$$

or, equivalently,

$$x_k(n) = 2|A_k|r_k^n \cos(\beta_k n + \alpha_k)u(n) \tag{3.4.33}$$

Thus we conclude that

$$Z^{-1}\left(\frac{A_k}{1 - p_k z^{-1}} + \frac{A_k^*}{1 - p_k^* z^{-1}}\right) = 2|A_k|r_k^n \cos(\beta_k n + \alpha_k)u(n) \tag{3.4.34}$$

if the ROC is $|z| > |p_k| = r_k$.

From (3.4.34) we observe that each pair of complex-conjugate poles in the z-domain results in a causal sinusoidal signal component with an exponential envelope. The distance r_k of the pole from the origin determines the exponential weighting (growing if $r_k > 1$, decaying if $r_k < 1$, constant if $r_k = 1$). The angle of the poles with respect to the positive real axis provides the frequency of the sinusoidal signal. The zeros, or equivalently the numerator of the rational transform, affect only indirectly the amplitude and the phase of $x_k(n)$ through A_k.

In the case of *multiple* poles, either real or complex, the inverse transform of terms of the form $A/(z - p_k)^n$ is required. In the case of a double pole the following transform pair (see Table 3.3) is quite useful:

$$Z^{-1}\left\{\frac{pz^{-1}}{(1-pz^{-1})^2}\right\} = np^n u(n) \tag{3.4.35}$$

provided that the ROC is $|z| > |p|$. The generalization to the case of poles with higher multiplicity is left as an exercise for the reader.

Example 3.4.8

Determine the inverse z-transform of

$$X(z) = \frac{1}{1 - 1.5z^{-1} + 0.5z^{-2}}$$

if

(a) ROC: $|z| > 1$
(b) ROC: $|z| < 0.5$
(c) ROC: $0.5 < |z| < 1$

Solution   This is the same problem that we treated in Example 3.4.2. The partial-fraction expansion for $X(z)$ was determined in Example 3.4.5. The partial-fraction expansion of $X(z)$ yields

$$X(z) = \frac{2}{1-z^{-1}} - \frac{1}{1-0.5z^{-1}} \tag{3.4.36}$$

To invert $X(z)$ we should apply (3.4.28) for $p_1 = 1$ and $p_2 = 0.5$. However, this requires the specification of the corresponding ROC.

(a) In case when the ROC is $|z| > 1$, the signal $x(n)$ is causal and both terms in (3.4.36) are causal terms. According to (3.4.28), we obtain

$$x(n) = 2(1)^n u(n) - (0.5)^n u(n) = (2 - 0.5^n)u(n) \tag{3.4.37}$$

which agrees with the result in Example 3.4.2(a).

(b) When the ROC is $|z| < 0.5$, the signal $x(n)$ is anticausal. Thus both terms in (3.4.36) result in anticausal components. From (3.4.28) we obtain

$$x(n) = [-2 + (0.5)^n]u(-n-1) \tag{3.4.38}$$

(c) In this case the ROC $0.5 < |z| < 1$ is a ring, which implies that the signal $x(n)$ is two-sided. Thus one of the terms corresponds to a causal signal and the other to an anticausal signal. Obviously, the given ROC is the overlapping of the regions $|z| > 0.5$ and $|z| < 1$. Hence the pole $p_2 = 0.5$ provides the causal part and the pole $p_1 = 1$ the anticausal. Thus

$$x(n) = -2(1)^n u(-n-1) - (0.5)^n u(n) \tag{3.4.39}$$

Example 3.4.9

Determine the causal signal $x(n)$ whose z-transform is given by

$$X(z) = \frac{1 + z^{-1}}{1 - z^{-1} + 0.5z^{-2}}$$

Solution   In Example 3.4.6 we have obtained the partial-fraction expansion as

$$X(z) = \frac{A_1}{1 - p_1 z^{-1}} + \frac{A_2}{1 - p_2 z^{-1}}$$

where

$$A_1 = A_2^* = \tfrac{1}{2} - j\tfrac{3}{2}$$

and

$$p_1 = p_2^* = \tfrac{1}{2} + j\tfrac{1}{2}$$

Since we have a pair of complex-conjugate poles, we should use (3.4.34). The polar forms of A_1 and p_1 are

$$A_1 = \frac{\sqrt{10}}{2} e^{-j71.565}$$

$$p_1 = \frac{1}{\sqrt{2}} e^{j\pi/4}$$

Hence

$$x(n) = \sqrt{10} \left(\frac{1}{\sqrt{2}}\right)^n \cos\left(\frac{\pi n}{4} - 71.565°\right) u(n)$$

Example 3.4.10

Determine the causal signal $x(n)$ having the z-transform

$$X(z) = \frac{1}{(1 + z^{-1})(1 - z^{-1})^2}$$

Solution   From Example 3.4.7 we have

$$X(z) = \frac{1}{4} \frac{1}{1 + z^{-1}} + \frac{3}{4} \frac{1}{1 - z^{-1}} + \frac{1}{2} \frac{z^{-1}}{(1 - z^{-1})^2}$$

By applying the inverse transform relations in (3.4.28) and (3.4.35), we obtain

$$x(n) = \frac{1}{4}(-1)^n u(n) + \frac{3}{4} u(n) + \frac{1}{2} n u(n) = \left[\frac{1}{4}(-1)^n + \frac{3}{4} + \frac{n}{2}\right] u(n)$$

3.4.4 Decomposition of Rational z-Transforms

At this point it is appropriate to discuss some additional issues concerning the decomposition of rational z-transforms, which will prove very useful in the implementation of discrete-time systems.

Suppose that we have a rational z-transform $X(z)$ expressed as

$$X(z) = \frac{\displaystyle\sum_{k=0}^{M} b_k z^{-k}}{1 + \displaystyle\sum_{k=1}^{N} a_k z^{-k}} = b_0 \frac{\displaystyle\prod_{k=1}^{M}(1 - z_k z^{-1})}{\displaystyle\prod_{k=1}^{N}(1 - p_k z^{-1})} \tag{3.4.40}$$

where, for simplicity, we have assumed that $a_0 \equiv 1$. If $M \geq N$ [i.e., $X(z)$ is improper], we convert $X(z)$ to a sum of a polynomial and a proper function

$$X(z) = \sum_{k=0}^{M-N} c_k z^{-k} + X_{pr}(z) \qquad (3.4.41)$$

If the poles of $X_{pr}(z)$ are distinct, it can be expanded in partial fractions as

$$X_{pr}(z) = A_1 \frac{1}{1 - p_1 z^{-1}} + A_2 \frac{1}{1 - p_2 z^{-1}} + \cdots + A_N \frac{1}{1 - p_N z^{-1}} \qquad (3.4.42)$$

As we have already observed, there may be some complex-conjugate pairs of poles in (3.4.42). Since we usually deal with real signals, we should avoid complex coefficients in our decomposition. This can be achieved by grouping and combining terms containing complex-conjugate poles, in the following way:

$$\frac{A}{1 - p z^{-1}} + \frac{A^*}{1 - p^* z^{-1}} = \frac{A - A p^* z^{-1} + A^* - A^* p z^{-1}}{1 - p z^{-1} - p^* z^{-1} + p p^* z^{-2}}$$
$$= \frac{b_0 + b_1 z^{-1}}{1 + a_1 z^{-1} + a_2 z^{-2}} \qquad (3.4.43)$$

where

$$b_0 = 2 \operatorname{Re}(A), \qquad a_1 = -2 \operatorname{Re}(p)$$
$$b_1 = 2 \operatorname{Re}(A p*), \qquad a_2 = |p|^2 \qquad (3.4.44)$$

are the desired coefficients. Obviously, any rational transform of the form (3.4.43) with coefficients given by (3.4.44), which is the case when $a_1^2 - 4a_2 < 0$, can be inverted using (3.4.34). By combining (3.4.41), (3.4.42), and (3.4.43) we obtain a partial-fraction expansion for the z-transform with *distinct* poles that contains real coefficients. The general result is

$$X(z) = \sum_{k=0}^{M-N} c_k z^{-k} + \sum_{k=1}^{K_1} \frac{b_k}{1 + a_k z^{-1}} + \sum_{k=1}^{K_2} \frac{b_{0k} + b_{1k} z^{-1}}{1 + a_{1k} z^{-1} + a_{2k} z^{-2}} \qquad (3.4.45)$$

where $K_1 + 2K_2 = N$. Obviously, if $M = N$, the first term is just a constant, and when $M < N$, this term vanishes. When there are also multiple poles, some additional higher-order terms should be included in (3.4.45).

An alternative form is obtained by expressing $X(z)$ as a product of simple terms as in (3.4.40). However, the complex-conjugate poles and zeros should be combined to avoid complex coefficients in the decomposition. Such combinations result in second-order rational terms of the following form:

$$\frac{(1 - z_k z^{-1})(1 - z_k^* z^{-1})}{(1 - p_k z^{-1})(1 - p_k^* z^{-1})} = \frac{1 + b_{1k} z^{-1} + b_{2k} z^{-2}}{1 + a_{1k} z^{-1} + a_{2k} z^{-2}} \qquad (3.4.46)$$

where

$$b_{1k} = -2 \operatorname{Re}(z_k), \qquad a_{1k} = -2 \operatorname{Re}(p_k)$$
$$b_{2k} = |z_k|^2, \qquad a_{2k} = |p_k|^2 \qquad (3.4.47)$$

Assuming for simplicity that $M = N$, we see that $X(z)$ can be decomposed in the following way:

$$X(z) = b_0 \prod_{k=1}^{K_1} \frac{1 + b_k z^{-1}}{1 + a_k z^{-1}} \prod_{k=1}^{K_2} \frac{1 + b_{1k} z^{-1} + b_{2k} z^{-2}}{1 + a_{1k} z^{-1} + a_{2k} z^{-2}} \qquad (3.4.48)$$

where $N = K_1 + 2K_2$. We will return to these important forms in Chapters 7 and 8.

3.5 THE ONE-SIDED Z-TRANSFORM

The two-sided z-transform requires that the corresponding signals be specified for the entire time range $-\infty < n < \infty$. This requirement prevents its use for a very useful family of practical problems, namely the evaluation of the output of nonrelaxed systems. As we recall, these systems are described by difference equations with nonzero initial conditions. Since the input is applied at a finite time, say n_0, both input and output signals are specified for $n \geq n_0$, but by no means are zero for $n < n_0$. Thus the two-sided z-transform cannot be used. In this section we develop the one-sided z-transform which can be used to solve difference equations with initial conditions.

3.5.1 Definition and Properties

The *one-sided* or *unilateral* z-transform of a signal $x(n)$ is defined by

$$X^+(z) \equiv \sum_{n=0}^{\infty} x(n) z^{-n} \qquad (3.5.1)$$

We also use the notations $Z^+\{x(n)\}$ and

$$x(n) \xrightarrow{z^+} X^+(z)$$

The one-sided z-transform differs from the two-sided transform in the lower limit of the summation, which is always zero, whether or not the signal $x(n)$ is zero for $n < 0$ (i.e., causal). Due to this choice of lower limit, the one-sided z-transform has the following characteristics:

1. It does not contain information about the signal $x(n)$ for negative values of time (i.e., for $n < 0$).
2. It is *unique* only for causal signals, because only these signals are zero for $n < 0$.
3. The one-sided z-transform $X^+(z)$ of $x(n)$ is identical to the two-sided z-transform of the signal $x(n)u(n)$. Since $x(n)u(n)$ is causal, the ROC of its transform, and hence the ROC of $X^+(z)$, is always the exterior of a circle. Thus when we deal with one-sided z-transforms, it is not necessary to refer to their ROC.

Example 3.5.1

Determine the one-sided z-transform of the signals in Example 3.1.1.

Solution   From the definition (3.5.1), we obtain

$$x_1(n) = \{1, 2, 5, 7, 0, 1\} \overset{z^+}{\longleftrightarrow} X_1^+(z) = 1 + 2z^{-1} + 5z^{-2} + 7z^{-3} + z^{-5}$$
$$\uparrow$$

$$x_2(n) = \{1, 2, 5, 7, 0, 1\} \overset{z^+}{\longleftrightarrow} X_2^+(z) = 5 + 7z^{-1} + z^{-3}$$
$$\uparrow$$

$$x_3(n) = \{0, 0, 1, 2, 5, 7, 0, 1\} \overset{z^+}{\longleftrightarrow} X_3^+(z) = z^{-2} + 2z^{-3} + 5z^{-4} + 7z^{-5} + z^{-7}$$
$$\uparrow$$

$$x_4(n) = \{2, 4, 5, 7, 0, 1\} \overset{z^+}{\longleftrightarrow} X_4^+(z) = 5 + 7z^{-1} + z^{-3}$$
$$\uparrow$$

$$x_5(n) = \delta(n) \overset{z^+}{\longleftrightarrow} X_5^+(z) = 1$$

$$x_6(n) = \delta(n - k), \quad k > 0 \overset{z^+}{\longleftrightarrow} X_6^+(z) = z^{-k}$$

$$x_7(n) = \delta(n + k), \quad k > 0 \overset{z^+}{\longleftrightarrow} X_7^+(z) = 0$$

Note that for a noncausal signal, the one-sided z-transform is not unique. Indeed, $X_2^+(z) = X_4^+(z)$ but $x_2(n) \neq x_4(n)$. Also for anticausal signals, $X^+(z)$ is always zero.

Almost all properties we have studied for the two-sided z-transform carry over to the one-sided z-transform with the exception of the *shifting* property.

Shifting Property

Case 1: Time Delay   If

$$x(n) \overset{z^+}{\longleftrightarrow} X^+(z)$$

then

$$x(n - k) \overset{z^+}{\longleftrightarrow} z^{-k}[X^+(z) + \sum_{n=1}^{k} x(-n)z^n] \quad k > 0 \tag{3.5.2}$$

In case $x(n)$ is causal, then

$$x(n - k) \overset{z^+}{\longleftrightarrow} z^{-k}X^+(z) \tag{3.5.3}$$

Proof. From the definition (3.5.1) we have

$$Z^+\{x(n - k)\} = z^{-k} \left[\sum_{l=-k}^{-1} x(l)z^{-l} + \sum_{l=0}^{\infty} x(l)z^{-l} \right]$$

$$= z^{-k} \left[\sum_{l=-1}^{-k} x(l)z^{-l} + X^+(z) \right]$$

By changing the index from l to $n = -l$, the result in (3.5.2) is easily obtained.

Example 3.5.2

Determine the one-sided z-transform of the signals

(a) $x(n) = a^n u(n)$

(b) $x_1(n) = x(n-2)$ where $x(n) = a^n$

Solution

(a) From (3.5.1) we easily obtain

$$X^+(z) = \frac{1}{1 - az^{-1}}$$

(b) We will apply the shifting property for $k = 2$. Indeed, we have

$$Z^+\{x(n-2)\} = z^{-2}[X^+(z) + x(-1)z + x(-2)z^2]$$
$$= z^{-2}X^+(z) + x(-1)z^{-1} + x(-2)$$

Since $x(-1) = a^{-1}$, $x(-2) = a^{-2}$, we obtain

$$X_1^+(z) = \frac{z^{-2}}{1 - az^{-1}} + a^{-1}z^{-1} + a^{-2}$$

The meaning of the shifting property can be intuitively explained if we write (3.5.2) as follows:

$$Z^+\{x(n-k)\} = [x(-k) + x(-k+1)z^{-1} + \cdots + x(-1)z^{-k+1}]$$
$$+ z^{-k}X^+(z) \qquad k > 0 \tag{3.5.4}$$

To obtain $x(n-k)(k > 0)$ from $x(n)$, we should shift $x(n)$ by k samples to the right. Then k "new" samples, $x(-k), x(-k+1), \ldots, x(-1)$, enter the positive time axis with $x(-k)$ located at time zero. The first term in (3.5.4) stands for the z-transform of these samples. The "old" samples of $x(n-k)$ are the same as those of $x(n)$ simply shifted by k samples to the right. Their z-transform is obviously $z^{-k}X^+(z)$, which is the second term in (3.5.4).

Case 2: Time advance   If

$$x(n) \xleftrightarrow{z^+} X^+(z)$$

then

$$x(n+k) \xleftrightarrow{z^+} z^k\left[X^+(z) - \sum_{n=0}^{k-1} x(n)z^{-n}\right] k > 0 \tag{3.5.5}$$

Proof. From (3.5.1) we have

$$Z^+\{x(n+k)\} = \sum_{n=0}^{\infty} x(n+k)z^{-n} = z^k \sum_{l=k}^{\infty} x(l)z^{-l}$$

where we have changed the index of summation from n to $l = n + k$. Now, from

(3.5.1) we obtain

$$X^+(z) = \sum_{l=0}^{\infty} x(l)z^{-l} = \sum_{l=0}^{k-1} x(l)z^{-l} + \sum_{l=k}^{\infty} x(l)z^{-l}$$

By combining the last two relations, we easily obtain (3.5.5).

Example 3.5.3

With $x(n)$, as given in Example 3.5.2, determine the one-sided z-transform of the signal

$$x_2(n) = x(n+2)$$

Solution   We will apply the shifting theorem for $k = 2$. From (3.5.5), with $k = 2$, we obtain

$$Z^+\{x(n+2)\} = z^2 X^+(z) - x(0)z^2 - x(1)z$$

But $x(0) = 1$, $x(1) = a$, and $X^+(z) = 1/(1 - az^{-1})$. Thus

$$Z^+\{x(n+2)\} = \frac{z^2}{1 - az^{-1}} - z^2 - az$$

The case of a time advance can be intuitively explained as follows. To obtain $x(n+k)$, $k > 0$, we should shift $x(n)$ by k samples to the left. As a result, the samples $x(0)$, $x(1)$, ..., $x(k-1)$ "leave" the positive time axis. Thus we first remove their contribution to the $X^+(z)$, and then multiply what remains by z^k to compensate for the shifting of the signal by k samples.

The importance of the shifting property lies in its application to the solution of difference equations with constant coefficients and nonzero initial conditions. This makes the one-sided z-transform a very useful tool for the analysis of recursive linear time-invariant discrete-time systems.

An important theorem useful in the analysis of signals and systems is the final value theorem.

Final Value Theorem.   If

$$x(n) \xleftrightarrow{z^+} X^+(z)$$

then

$$\lim_{n \to \infty} x(n) = \lim_{z \to 1}(z - 1)X^+(z) \tag{3.5.6}$$

The limit in (3.5.6) exists if the ROC of $(z - 1)X^+(z)$ includes the unit circle.

The proof of this theorem is left as an exercise for the reader.

This theorem is useful when we are interested in the asymptotic behavior of a signal $x(n)$ and we know its z-transform, but not the signal itself. In such cases, especially if it is complicated to invert $X^+(z)$, we can use the final value theorem to determine the limit of $x(n)$ as n goes to infinity.

Example 3.5.4

The impulse response of a relaxed linear time-invariant system is $h(n) = \alpha^n u(n)$, $|\alpha| < 1$. Determine the value of the step response of the system as $n \to \infty$.

Solution   The step response of the system is

$$y(n) = h(n) * x(n)$$

where

$$x(n) = u(n)$$

Obviously, if we excite a causal system with a causal input the output will be causal. Since $h(n)$, $x(n)$, $y(n)$ are causal signals, the one-sided and two-sided z-transforms are identical. From the convolution property (3.2.17) we know that the z-transforms of $h(n)$ and $x(n)$ must be multiplied to yield the z-transform of the output. Thus

$$Y(z) = \frac{1}{1 - \alpha z^{-1}} \frac{1}{1 - z^{-1}} = \frac{z^2}{(z-1)(z-\alpha)} \qquad \text{ROC: } |z| > |\alpha|$$

Now

$$(z-1)Y(z) = \frac{z^2}{z-\alpha} \qquad \text{ROC: } |z| > |\alpha|$$

Since $|\alpha| < 1$ the ROC of $(z-1)Y(z)$ includes the unit circle. Consequently, we can apply (3.5.6) and obtain

$$\lim_{n \to \infty} y(n) = \lim_{z \to 1} \frac{z^2}{z-\alpha} = \frac{1}{1-\alpha}$$

3.5.2 Solution of Difference Equations

The one-sided z-transform is a very efficient tool for the solution of difference equations with nonzero initial conditions. It achieves that by reducing the difference equation relating the two time-domain signals to an equivalent algebraic equation relating their one-sided z-transforms. This equation can be easily solved to obtain the transform of the desired signal. The signal in the time domain is obtained by inverting the resulting z-transform. We will illustrate this approach with two examples.

Example 3.5.5

The well-known Fibonacci sequence of integer numbers is obtained by computing each term as the sum of the two previous ones. The first few terms of the sequence are

$$1, 1, 2, 3, 5, 8, \ldots$$

Determine a closed-form expression for the nth term of the Fibonacci sequence.

Solution   Let $y(n)$ be the nth term of the Fibonacci sequence. Clearly, $y(n)$ satisfies the difference equation

$$y(n) = y(n-1) + y(n-2) \tag{3.5.7}$$

with initial conditions

$$y(0) = y(-1) + y(-2) = 1 \tag{3.5.8a}$$

$$y(1) = y(0) + y(-1) = 1 \tag{3.5.8b}$$

From (3.5.8b) we have $y(-1) = 0$. Then (3.5.8a) gives $y(-2) = 1$. Thus we have to determine $y(n)$, $n \geq 0$, which satisfies (3.5.7), with initial conditions $y(-1) = 0$ and $y(-2) = 1$.

By taking the one-sided *z*-transform of (3.5.7) and using the shifting property (3.5.2), we obtain

$$Y^+(z) = [z^{-1}Y^+(z) + y(-1)] + [z^{-2}Y^+(z) + y(-2) + y(-1)z^{-1}]$$

or

$$Y^+(z) = \frac{1}{1 - z^{-1} - z^2} = \frac{z^2}{z^2 - z - 1} \tag{3.5.9}$$

where we have used the fact that $y(-1) = 0$ and $y(-2) = 1$.

We can invert $Y^+(z)$ by the partial-fraction expansion method. The poles of $Y^+(z)$ are

$$p_1 = \frac{1 + \sqrt{5}}{2} \qquad p_2 = \frac{1 - \sqrt{5}}{2}$$

and the corresponding coefficients are $A_1 = p_1/\sqrt{5}$ and $A_2 = -p_2/\sqrt{5}$. Therefore,

$$y(n) = \left[\frac{1 + \sqrt{5}}{2\sqrt{5}} \left(\frac{1 + \sqrt{5}}{2} \right)^n - \frac{1 - \sqrt{5}}{2\sqrt{5}} \left(\frac{1 - \sqrt{5}}{2} \right)^n \right] u(n)$$

or, equivalently,

$$y(n) = \frac{1}{\sqrt{5}} \left(\frac{1}{2} \right)^{n+1} \left[\left(1 + \sqrt{5} \right)^{n+1} - \left(1 - \sqrt{5} \right)^{n+1} \right] u(n) \tag{3.5.10}$$

Example 3.5.6

Determine the step response of the system

$$y(n) = \alpha y(n - 1) + x(n) \qquad -1 < \alpha < 1 \tag{3.5.11}$$

when the initial condition is $y(-1) = 1$.

Solution By taking the one-sided *z*-transform of both sides of (3.5.11), we obtain

$$Y^+(z) = \alpha[z^{-1}Y^+(z) + y(-1)] + X^+(z)$$

Upon substitution for $y(-1)$ and $X^+(z)$ and solving for $Y^+(z)$, we obtain the result

$$Y^+(z) = \frac{\alpha}{1 - \alpha z^{-1}} + \frac{1}{(1 - \alpha z^{-1})(1 - z^{-1})} \tag{3.5.12}$$

By performing a partial-fraction expansion and inverse transforming the result, we have

$$y(n) = \alpha^{n+1} u(n) + \frac{1 - \alpha^{n+1}}{1 - \alpha} u(n)$$

$$= \frac{1}{1 - \alpha} (1 - \alpha^{n+2}) u(n) \tag{3.5.13}$$

3.6 ANALYSIS OF LINEAR TIME-INVARIANT SYSTEMS IN THE Z-DOMAIN

In Section 3.4.3 we introduced the system function of a linear time-invariant system and related it to the unit sample response and to the difference equation description of systems. In this section we describe the use of the system function in the determination of the response of the system to some excitation signal. Furthermore, we extend this method of analysis to nonrelaxed systems. Our attention is focused on the important class of pole–zero systems represented by linear constant-coefficient difference equations with arbitrary initial conditions.

We also consider the topic of stability of linear time-invariant systems and describe a test for determining the stability of a system based on the coefficients of the denominator polynomial in the system function. Finally, we provide a detailed analysis of second-order systems, which form the basic building blocks in the realization of higher-order systems.

3.6.1 Response of Systems with Rational System Functions

Let us consider a pole–zero system described by the general linear constant-coefficient difference equation in (3.3.7) and the corresponding system function in (3.3.8). We represent $H(z)$ as a ratio of two polynomials $B(z)/A(z)$, where $B(z)$ is the numerator polynomial that contains the zeros of $H(z)$, and $A(z)$ is the denominator polynomial that determines the poles of $H(z)$. Furthermore, let us assume that the input signal $x(n)$ has a rational z-transform $X(z)$ of the form

$$X(z) = \frac{N(z)}{Q(z)} \tag{3.6.1}$$

This assumption is not overly restrictive, since, as indicated previously, most signals of practical interest have rational z-transforms.

If the system is initially relaxed, that is, the initial conditions for the difference equation are zero, $y(-1) = y(-2) = \cdots = y(-N) = 0$, the z-transform of the output of the system has the form

$$Y(z) = H(z)X(z) = \frac{B(z)N(z)}{A(z)Q(z)} \tag{3.6.2}$$

Now suppose that the system contains simple poles $p_1, p_2, \ldots, p_N$ and the z-transform of the input signal contains poles $q_1, q_2, \ldots, q_L$, where $p_k \neq q_m$ for all $k = 1, 2, \ldots, N$ and $m = 1, 2, \ldots, L$. In addition, we assume that the zeros of the numerator polynomials $B(z)$ and $N(z)$ do not coincide with the poles $\{p_k\}$ and $\{q_k\}$, so that there is no pole–zero cancellation. Then a partial-fraction expansion of $Y(z)$ yields

$$Y(z) = \sum_{k=1}^{N} \frac{A_k}{1 - p_k z^{-1}} + \sum_{k=1}^{L} \frac{Q_k}{1 - q_k z^{-1}} \tag{3.6.3}$$

The inverse transform of $Y(z)$ yields the output signal from the system in the form

$$y(n) = \sum_{k=1}^{N} A_k(p_k)^n u(n) + \sum_{k=1}^{L} Q_k(q_k)^n u(n) \tag{3.6.4}$$

We observe that the output sequence $y(n)$ can be subdivided into two parts. The first part is a function of the poles $\{p_k\}$ of the system and is called the *natural response* of the system. The influence of the input signal on this part of the response is through the scale factors $\{A_k\}$. The second part of the response is a function of the poles $\{q_k\}$ of the input signal and is called the *forced response* of the system. The influence of the system on this response is exerted through the scale factors $\{Q_k\}$.

We should emphasize that the scale factors $\{A_k\}$ and $\{Q_k\}$ are functions of both sets of poles $\{p_k\}$ and $\{q_k\}$. For example, if $X(z) = 0$ so that the input is zero, then $Y(z) = 0$, and consequently, the output is zero. Clearly, then, the natural response of the system is zero. This implies that the natural response of the system is different from the zero-input response.

When $X(z)$ and $H(z)$ have one or more poles in common or when $X(z)$ and/or $H(z)$ contain multiple-order poles, then $Y(z)$ will have multiple-order poles. Consequently, the partial-fraction expansion of $Y(z)$ will contain factors of the form $1/(1 - p_l z^{-1})^k$, $k = 1, 2, \ldots, m$, where m is the pole order. The inversion of these factors will produce terms of the form $n^{k-1} p_l^n$ in the output $y(n)$ of the system, as indicated in Section 3.4.2.

3.6.2 Response of Pole–Zero Systems with Nonzero Initial Conditions

Suppose that the signal $x(n)$ is applied to the pole–zero system at $n = 0$. Thus the signal $x(n)$ is assumed to be causal. The effects of all previous input signals to the system are reflected in the initial conditions $y(-1), y(-2), \ldots, y(-N)$. Since the input $x(n)$ is causal and since we are interested in determining the output $y(n)$ for $n \geq 0$, we can use the one-sided z-transform, which allows us to deal with the initial conditions. Thus the one-sided z-transform of (3.4.7) becomes

$$Y^+(z) = -\sum_{k=1}^{N} a_k z^{-k} \left[Y^+(z) + \sum_{n=1}^{k} y(-n) z^n \right] + \sum_{k=0}^{M} b_k z^{-k} X^+(z) \tag{3.6.5}$$

Since $x(n)$ is causal, we can set $X^+(z) = X(z)$. In any case (3.6.5) may be expressed as

$$Y^+(z) = \frac{\sum_{k=0}^{M} b_k z^{-k}}{1 + \sum_{k=1}^{N} a_k z^{-k}} X(z) - \frac{\sum_{k=1}^{N} a_k z^{-k} \sum_{n=1}^{k} y(-n) z^n}{1 + \sum_{k=1}^{N} a_k z^{-k}} \tag{3.6.6}$$

$$= H(z) X(z) + \frac{N_0(z)}{A(z)}$$

where

$$N_0(z) = -\sum_{k=1}^{N} a_k z^{-k} \sum_{n=1}^{k} y(-n)z^n \tag{3.6.7}$$

From (3.6.6) it is apparent that the output of the system with nonzero initial conditions can be subdivided into two parts. The first is the zero-state response of the system, defined in the z-domain as

$$Y_{zs}(z) = H(z)X(z) \tag{3.6.8}$$

The second component corresponds to the output resulting from the nonzero initial conditions. This output is the zero-input response of the system, which is defined in the z-domain as

$$Y_{zi}^{+}(z) = \frac{N_0(z)}{A(z)} \tag{3.6.9}$$

Hence the total response is the sum of these two output components, which can be expressed in the time domain by determining the inverse z-transforms of $Y_{zs}(z)$ and $Y_{zi}(z)$ separately, and then adding the results. Thus

$$y(n) = y_{zs}(n) + y_{zi}(n) \tag{3.6.10}$$

Since the denominator of $Y_{zi}^{+}(z)$, is $A(z)$, its poles are $p_1, p_2, \ldots, p_N$. Consequently, the zero-input response has the form

$$y_{zi}(n) = \sum_{k=1}^{N} D_k (p_k)^n u(n) \tag{3.6.11}$$

This can be added to (3.6.4) and the terms involving the poles $\{p_k\}$ can be combined to yield the total response in the form

$$y(n) = \sum_{k=1}^{N} A_k'(p_k)^n u(n) + \sum_{k=1}^{L} Q_k(q_k)^n u(n) \tag{3.6.12}$$

where, by definition,

$$A_k' = A_k + D_k \tag{3.6.13}$$

This development indicates clearly that the effect of the initial conditions is to alter the natural response of the system through modification of the scale factors $\{A_k\}$. There are no new poles introduced by the nonzero initial conditions. Furthermore, there is no effect on the forced response of the system. These important points are reinforced in the following example.

Example 3.6.1

Determine the unit step response of the system described by the difference equation

$$y(n) = 0.9y(n-1) - 0.81y(n-2) + x(n)$$

under the following initial conditions:

(a) $y(-1) = y(-2) = 0$
(b) $y(-1) = y(-2) = 1$

Solution   The system function is

$$H(z) = \frac{1}{1 - 0.9z^{-1} + 0.81z^{-2}}$$

This system has two complex-conjugate poles at

$$p_1 = 0.9e^{j\pi/3} \qquad p_2 = 0.9e^{-j\pi/3}$$

The z-transform of the unit step sequence is

$$X(z) = \frac{1}{1 - z^{-1}}$$

Therefore,

$$Y_{zs}(z) = \frac{1}{(1 - 0.9e^{j\pi/3}z^{-1})(1 - 0.9e^{-j\pi/3}z^{-1})(1 - z^{-1})}$$

$$= \frac{0.542 - j0.049}{1 - 0.9e^{j\pi/3}z^{-1}} + \frac{0.542 + j0.049}{1 - 0.9e^{-j\pi/3}z^{-1}} + \frac{1.099}{1 - z^{-1}}$$

and hence the zero-state response is

$$y_{zs}(n) = \left[1.099 + 1.088(0.9)^n \cos\left(\frac{\pi}{3}n - 5.2°\right)\right]u(n)$$

(a) Since the initial conditions are zero in this case, we conclude that $y(n) = y_{zs}(n)$.

(b) For the initial conditions $y(-1) = y(-2) = 1$, the additional component in the z-transform is

$$Y_{zi}(z) = \frac{N_0(z)}{A(z)} = \frac{0.09 - 0.81z^{-1}}{1 - 0.9z^{-1} + 0.81z^{-2}}$$

$$= \frac{0.026 + j0.4936}{1 - 0.9e^{j\pi/3}z^{-1}} + \frac{0.026 - j0.4936}{1 - 0.9e^{-j\pi/3}z^{-1}}$$

Consequently, the zero-input response is

$$y_{zi}(n) = 0.988(0.9)^n \cos\left(\frac{\pi}{3}n + 87°\right)u(n)$$

In this case the total response has the z-transform

$$Y(z) = Y_{zs}(z) + Y_{zi}(z)$$

$$= \frac{1.099}{1 - z^{-1}} + \frac{0.568 + j0.445}{1 - 0.9e^{j\pi/3}z^{-1}} + \frac{0.568 - j0.445}{1 - 0.9e^{-j\pi/3}z^{-1}}$$

The inverse transform yields the total response in the form

$$y(n) = 1.099u(n) + 1.44(0.9)^n \cos\left(\frac{\pi}{3}n + 38°\right)u(n)$$

3.6.3 Transient and Steady-State Responses

As we have seen from our previous discussion, the response of a system to a given input can be separated into two components, the natural response and the forced response. The natural response of a causal system has the form

$$y_{nr}(n) = \sum_{k=1}^{N} A_k (p_k)^n u(n) \tag{3.6.14}$$

where $\{p_k\}$, $k = 1, 2, \ldots, N$ are the poles of the system and $\{A_k\}$ are scale factors that depend on the initial conditions and on the characteristics of the input sequence.

If $|p_k| < 1$ for all k, then, $y_{nr}(n)$ decays to zero as n approaches infinity. In such a case we refer to the natural response of the system as the *transient response*. The rate at which $y_{nr}(n)$ decays toward zero depends on the magnitude of the pole positions. If all the poles have small magnitudes, the decay is very rapid. On the other hand, if one or more poles are located near the unit circle, the corresponding terms in $y_{nr}(n)$ will decay slowly toward zero and the transient will persist for a relatively long time.

The forced response of the system has the form

$$y_{fr}(n) = \sum_{k=1}^{L} Q_k (q_k)^n u(n) \tag{3.6.15}$$

where $\{q_k\}$, $k = 1, 2, \ldots, L$ are the poles in the forcing function and $\{Q_k\}$ are scale factors that depend on the input sequence and on the characteristics of the system. If all the poles of the input signal fall inside the unit circle, $y_{fr}(n)$ will decay toward zero as n approaches infinity, just as in the case of the natural response. This should not be surprising since the input signal is also a transient signal. On the other hand, when the causal input signal is a sinusoid, the poles fall on the unit circle and consequently, the forced response is also a sinusoid that persists for all $n \geq 0$. In this case, the forced response is called the *steady-state response* of the system. Thus, for the system to sustain a steady-state output for $n \geq 0$, the input signal must persist for all $n \geq 0$.

The following example illustrates the presence of the steady-state response.

Example 3.6.2

Determine the transient and steady-state responses of the system characterized by the difference equation

$$y(n) = 0.5y(n-1) + x(n)$$

when the input signal is $x(n) = 10\cos(\pi n/4)u(n)$. The system is initially at rest (i.e., it is relaxed).

Solution   The system function for this system is

$$H(z) = \frac{1}{1 - 0.5z^{-1}}$$

and therefore the system has a pole at $z = 0.5$. The z-transform of the input signal is (from Table 3.3)

$$X(z) = \frac{10(1 - (1/\sqrt{2})z^{-1})}{1 - \sqrt{2}z^{-1} + z^{-2}}$$

Consequently,

$$Y(z) = H(z)X(z)$$

$$= \frac{10(1 - (1/\sqrt{2})z^{-1})}{(1 - 0.5z^{-1})(1 - e^{j\pi/4}z^{-1})(1 - e^{-j\pi/4}z^{-1})}$$

$$= \frac{6.3}{1 - 0.5z^{-1}} + \frac{6.78e^{-j28.7°}}{1 - e^{j\pi/4}z^{-1}} + \frac{6.78e^{j28.7°}}{1 - e^{-j\pi/4}z^{-1}}$$

The natural or transient response is

$$y_{nr}(n) = 6.3(0.5)^n u(n)$$

and the forced or steady-state response is

$$y_{fr}(n) = [6.78e^{-j28.7}(e^{j\pi n/4}) + 6.78e^{j28.7}e^{-j\pi n/4}]u(n)$$

$$= 13.56 \cos\left(\frac{\pi}{4}n - 28.7°\right)u(n)$$

Thus we see that the steady-state response persists for all $n \geq 0$, just as the input signal persists for all $n \geq 0$.

3.6.4 Causality and Stability

As defined previously, a causal linear time-invariant system is one whose unit sample response $h(n)$ satisfies the condition

$$h(n) = 0 \qquad n < 0$$

We have also shown that the ROC of the z-transform of a causal sequence is the exterior of a circle. Consequently, *a linear time-invariant system is causal if and only if the ROC of the system function is the exterior of a circle of radius $r < \infty$, including the point $z = \infty$.*

The stability of a linear time-invariant system can also be expressed in terms of the characteristics of the system function. As we recall from our previous discussion, a necessary and sufficient condition for a linear time-invariant system to be BIBO stable is

$$\sum_{n=-\infty}^{\infty} |h(n)| < \infty$$

In turn, this condition implies that $H(z)$ must contain the unit circle within its ROC. Indeed, since

$$H(z) = \sum_{n=-\infty}^{\infty} h(n)z^{-n}$$

it follows that

$$|H(z)| \leq \sum_{n=-\infty}^{\infty} |h(n)z^{-n}| = \sum_{n=-\infty}^{\infty} |h(n)||z^{-n}|$$

When evaluated on the unit circle (i.e., $|z| = 1$),

$$|H(z)| \leq \sum_{n=-\infty}^{\infty} |h(n)|$$

Hence, if the system is BIBO stable, the unit circle is contained in the ROC of $H(z)$. The converse is also true. Therefore, *a linear time-invariant system is BIBO stable if and only if the ROC of the system function includes the unit circle.*

We should stress, however, that the conditions for causality and stability are different and that one does not imply the other. For example, a causal system may be stable or unstable, just as a noncausal system may be stable or unstable. Similarly, an unstable system may be either causal or noncausal, just as a stable system may be causal or noncausal.

For a causal system, however, the condition on stability can be narrowed to some extent. Indeed, a causal system is characterized by a system function $H(z)$ having as a ROC the exterior of some circle of radius r. For a stable system, the ROC must include the unit circle. Consequently, a causal and stable system must have a system function that converges for $|z| > r < 1$. Since the ROC cannot contain any poles of $H(z)$, it follows that *a causal linear time-invariant system is BIBO stable if and only if all the poles of $H(z)$ are inside the unit circle.*

Example 3.6.3

A linear time-invariant system is characterized by the system function

$$H(z) = \frac{3 - 4z^{-1}}{1 - 3.5z^{-1} + 1.5z^{-2}}$$

$$= \frac{1}{1 - \frac{1}{2}z^{-1}} + \frac{2}{1 - 3z^{-1}}$$

Specify the ROC of $H(z)$ and determine $h(n)$ for the following conditions:

(a) The system is stable.

(b) The system is causal.

(c) The system is anticausal.

Solution   The system has poles at $z = \frac{1}{2}$ and $z = 3$.

(a) Since the system is stable, its ROC must include the unit circle and hence it is $\frac{1}{2} < |z| < 3$. Consequently, $h(n)$ is noncausal and is given as

$$h(n) = (\tfrac{1}{2})^n u(n) - 2(3)^n u(-n - 1)$$

(b) Since the system is causal, its ROC is $|z| > 3$. In this case

$$h(n) = (\tfrac{1}{2})^n u(n) + 2(3)^n u(n)$$

This system is unstable.

(c) If the system is anticausal, its ROC is $|z| < 0.5$. Hence

$$h(n) = -[(\tfrac{1}{2})^n + 2(3)^n] u(-n - 1)$$

In this case the system is unstable.

3.6.5 Pole–Zero Cancellations

When a *z*-transform has a pole that is at the same location as a zero, the pole is canceled by the zero and, consequently, the term containing that pole in the inverse *z*-transform vanishes. Such pole–zero cancellations are very important in the analysis of pole–zero systems.

Pole–zero cancellations can occur either in the system function itself or in the product of the system function with the *z*-transform of the input signal. In the first case we say that the order of the system is reduced by one. In the latter case we say that the pole of the system is suppressed by the zero in the input signal, or vice versa. Thus, by properly selecting the position of the zeros of the input signal, it is possible to suppress one or more system modes (pole factors) in the response of the system. Similarly, by proper selection of the zeros of the system function, it is possible to suppress one or more modes of the input signal from the response of the system.

When the zero is located very near the pole but not exactly at the same location, the term in the response has a very small amplitude. For example, nonexact pole–zero cancellations can occur in practice as a result of insufficient numerical precision used in representing the coefficients of the system. Consequently, one should not attempt to stabilize an inherently unstable system by placing a zero in the input signal at the location of the pole.

Example 3.6.4

Determine the unit sample response of the system characterized by the difference equation

$$y(n) = 2.5y(n-1) - y(n-2) + x(n) - 5x(n-1) + 6x(n-2)$$

Solution   The system function is

$$H(z) = \frac{1 - 5z^{-1} + 6z^{-2}}{1 - 2.5z^{-1} + z^{-2}}$$

$$= \frac{1 - 5z^{-1} + 6z^{-2}}{(1 - \frac{1}{2}z^{-1})(1 - 2z^{-1})}$$

This system has poles at $p_1 = 2$ and $p_1 = \frac{1}{2}$. Consequently, at first glance it appears that the unit sample response is

$$Y(z) = H(z)X(z) = \frac{1 - 5z^{-1} + 6z^{-2}}{(1 - \frac{1}{2}z^{-1})(1 - 2z^{-1})}$$

$$= z\left(\frac{A}{z - \frac{1}{2}} + \frac{B}{z - 2}\right)$$

By evaluating the constants at $z = \frac{1}{2}$ and $z = 2$, we find that

$$A = \frac{5}{2} \qquad B = 0$$

The fact that $B = 0$ indicates that there exists a zero at $z = 2$ which cancels the pole at $z = 2$. In fact, the zeros occur at $z = 2$ and $z = 3$. Consequently, $H(z)$

reduces to

$$H(z) = \frac{1 - 3z^{-1}}{1 - \frac{1}{2}z^{-1}} = \frac{z - 3}{z - \frac{1}{2}}$$

$$= 1 - \frac{2.5z^{-1}}{1 - \frac{1}{2}z^{-1}}$$

and therefore

$$h(n) = \delta(n) - 2.5(\tfrac{1}{2})^{n-1}u(n - 1)$$

The reduced-order system obtained by canceling the common pole and zero is characterized by the difference equation

$$y(n) = \tfrac{1}{2}y(n - 1) + x(n) - 3x(n - 1)$$

Although the original system is also BIBO stable due to the pole–zero cancellation, in a practical implementation of this second-order system, we may encounter an instability due to imperfect cancellation of the pole and the zero.

Example 3.6.5

Determine the response of the system

$$y(n) = \tfrac{5}{6}y(n - 1) - \tfrac{1}{6}y(n - 2) + x(n)$$

to the input signal $x(n) = \delta(n) - \tfrac{1}{3}\delta(n - 1)$.

Solution   The system function is

$$H(z) = \frac{1}{1 - \frac{5}{6}z^{-1} + \frac{1}{6}z^{-2}}$$

$$= \frac{1}{\left(1 - \frac{1}{2}z^{-1}\right)\left(1 - \frac{1}{3}z^{-1}\right)}$$

This system has two poles, one at $z = \frac{1}{2}$ and the other at $z = \frac{1}{3}$. The z-transform of the input signal is

$$X(z) = 1 - \tfrac{1}{3}z^{-1}$$

In this case the input signal contains a zero at $z = \frac{1}{3}$ which cancels the pole at $z = \frac{1}{3}$. Consequently,

$$Y(z) = H(z)X(z)$$

$$Y(z) = \frac{1}{1 - \frac{1}{2}z^{-1}}$$

and hence the response of the system is

$$y(n) = (\tfrac{1}{2})^n u(n)$$

Clearly, the mode $(\frac{1}{3})^n$ is suppressed from the output as a result of the pole–zero cancellation.

3.6.6 Multiple-Order Poles and Stability

As we have observed, a necessary and sufficient condition for a causal linear time-invariant system to be BIBO stable is that all its poles lie inside the unit circle. The input signal is bounded if its z-transform contains poles $\{q_k\}$, $k = 1, 2, \ldots, L$, which satisfy the condition $|q_k| \leq 1$ for all k. We note that the forced response of the system, given in (3.6.15), is also bounded, even when the input signal contains one or more distinct poles on the unit circle.

In view of the fact that a bounded input signal may have poles on the unit circle, it might appear that a stable system may also have poles on the unit circle. This is not the case, however, since such a system produces an unbounded response when excited by an input signal that also has a pole at the same position on the unit circle. The following example illustrates this point.

Example 3.6.6

Determine the step response of the causal system described by the difference equation

$$y(n) = y(n-1) + x(n)$$

Solution   The system function for the system is

$$H(z) = \frac{1}{1 - z^{-1}}$$

We note that the system contains a pole on the unit circle at $z = 1$. The z-transform of the input signal $x(n) = u(n)$ is

$$X(z) = \frac{1}{1 - z^{-1}}$$

which also contains a pole at $z = 1$. Hence the output signal has the transform

$$Y(z) = H(z)X(z)$$

$$= \frac{1}{(1 - z^{-1})^2}$$

which contains a double pole at $z = 1$.

The inverse z-transform of $Y(z)$ is

$$y(n) = (n+1)u(n)$$

which is a ramp sequence. Thus $y(n)$ is unbounded, even when the input is bounded. Consequently, the system is unstable.

Example 3.6.6 demonstrates clearly that BIBO stability requires that the system poles be strictly inside the unit circle. If the system poles are all inside the unit circle and the excitation sequence $x(n)$ contains one or more poles that coincide with the poles of the system, the output $Y(z)$ will contain multiple-order poles. As indicated previously, such multiple-order poles result in an output sequence that contains terms of the form

$$A_k n^b (p_k)^n u(n)$$

where $0 \leq b \leq m - 1$ and m is the order of the pole. If $|p_k| < 1$, these terms decay to zero as n approaches infinity because the exponential factor $(p_k)^n$ dominates the term n^b. Consequently, no bounded input signal can produce an unbounded output signal if the system poles are all inside the unit circle.

Finally, we should state that the only useful systems which contain poles on the unit circle are the digital oscillators discussed in Chapter 4. We call such systems *marginally stable*.

3.6.7 The Schür–Cohn Stability Test

We have stated previously that the stability of a system is determined by the position of the poles. The poles of the system are the roots of the denominator polynomial of $H(z)$, namely,

$$A(z) = 1 + a_1 z^{-1} + a_2 z^{-2} + \cdots + a_N z^{-N} \tag{3.6.16}$$

When the system is causal all the roots of $A(z)$ must lie inside the unit circle for the system to be stable.

There are several computational procedures that aid us in determining if any of the roots of $A(z)$ lie outside the unit circle. These procedures are called *stability criteria*. Below we describe the Schür–Cohn test procedure for the stability of a system characterized by the system function $H(z) = B(z)/A(z)$.

Before we describe the Schür–Cohn test we need to establish some useful notation. We denote a polynomial of degree m by

$$A_m(z) = \sum_{k=0}^{m} a_m(k) z^{-k} \qquad a_m(0) = 1 \tag{3.6.17}$$

The *reciprocal* or *reverse polynomial* $B_m(z)$ of degree m is defined as

$$\begin{aligned} B_m(z) &= z^{-m} A_m(z^{-1}) \\ &= \sum_{k=0}^{m} a_m(m-k) z^{-k} \end{aligned} \tag{3.6.18}$$

We observe that the coefficients of $B_m(z)$ are the same as those of $A_m(z)$, but in reverse order.

In the Schür–Cohn stability test, to determine if the polynomial $A(z)$ has all its roots inside the unit circle, we compute a set of coefficients, called *reflection coefficients*, $K_1, K_2, \ldots, K_N$ from the polynomials $A_m(z)$. First, we set

$$A_N(z) = A(z)$$

and $\tag{3.6.19}$

$$K_N = a_N(N)$$

Then we compute the lower-degree polynomials $A_m(z)$, $m = N, N-1, N-2, \ldots, 1$, according to the recursive equation

$$A_{m-1}(z) = \frac{A_m(z) - K_m B_m(z)}{1 - K_m^2} \tag{3.6.20}$$

where the coefficients K_m are defined as

$$K_m = a_m(m) \tag{3.6.21}$$

The Schür–Cohn stability test states that *the polynomial $A(z)$ given by (3.6.16) has all its roots inside the unit circle if and only if the coefficients K_m satisfy the condition $|K_m| < 1$ for all $m = 1, 2, \ldots, N$.*

We shall not provide a proof of the Schür–Cohn test at this point. The theoretical justification for this test is given in Chapter 11. We illustrate the computational procedure with the following example.

Example 3.6.7

Determine if the system having the system function

$$H(z) = \frac{1}{1 - \frac{7}{4}z^{-1} - \frac{1}{2}z^{-2}}$$

is stable.

Solution We begin with $A_2(z)$, which is defined as

$$A_2(z) = 1 - \frac{7}{4}z^{-1} - \frac{1}{2}z^{-2}$$

Hence

$$K_2 = -\frac{1}{2}$$

Now

$$B_2(z) = -\frac{1}{2} - \frac{7}{4}z^{-1} + z^{-2}$$

and

$$A_1(z) = \frac{A_2(z) - K_2 B_2(z)}{1 - K_2^2}$$

$$= 1 - \frac{7}{2}z^{-1}$$

Therefore,

$$K_1 = -\frac{7}{2}$$

Since $|K_1| > 1$ it follows that the system is unstable. This fact is easily established in this example, since the denominator is easily factored to yield the two poles at $p_1 = -2$ and $p_2 = \frac{1}{4}$. However, for higher-degree polynomials, the Schür–Cohn test provides a simpler test for stability than direct factoring of $H(z)$.

The Schür–Cohn stability test can be easily programmed in a digital computer and it is very efficient in terms of arithmetic operations. Specifically, it requires only N^2 multiplications to determine the coefficients $\{K_m\}$, $m = 1, 2, \ldots, N$. The recursive equation in (3.6.20) can be expressed in terms of the polynomial coefficients by expanding the polynomials in both sides of (3.6.20) and equating the coefficients corresponding to equal powers. Indeed, it is easily established that (3.6.20) is equivalent to the following algorithm: Set

$$a_N(k) = a_k \qquad k = 1, 2, \ldots, N \tag{3.6.22}$$

$$K_N = a_N(N) \tag{3.6.23}$$

Then, for $m = N, N - 1, \ldots, 1$, compute

$$K_m = a_m(m) \qquad a_{m-1}(0) = 1$$

and

$$a_{m-1}(k) = \frac{a_m(k) - K_m b_m(k)}{1 - K_m^2} \qquad k = 1, 2, \ldots, m-1 \tag{3.6.24}$$

where

$$b_m(k) = a_m(m-k) \qquad k = 0, 1, \ldots, m \tag{3.6.25}$$

This recursive algorithm for the computation of the coefficients $\{K_m\}$ finds application in various signal processing problems, especially in speech signal processing.

3.6.8 Stability of Second-Order Systems

In this section we provide a detailed analysis of a system having two poles. As we shall see in Chapter 7, two-pole systems form the basic building blocks for the realization of higher-order systems.

Let us consider a causal two-pole system described by the second-order difference equation

$$y(n) = -a_1 y(n-1) - a_2 y(n-2) + b_0 x(n) \tag{3.6.26}$$

The system function is

$$H(z) = \frac{Y(z)}{X(z)} = \frac{b_0}{1 + a_1 z^{-1} + a_2 z^{-1}}$$

$$= \frac{b_0 z^2}{z^2 + a_1 z + a_2} \tag{3.6.27}$$

This system has two zeros at the origin and poles at

$$p_1, p_2 = -\frac{a_1}{2} \pm \sqrt{\frac{a_1^2 - 4a_2}{4}} \tag{3.6.28}$$

The system is BIBO stable if the poles lie inside the unit circle, that is, if $|p_1| < 1$ and $|p_2| < 1$. These conditions can be related to the values of the coefficients a_1 and a_2. In particular, the roots of a quadratic equation satisfy the relations

$$a_1 = -(p_1 + p_2) \tag{3.6.29}$$

$$a_2 = p_1 p_2 \tag{3.6.30}$$

From (3.6.29) and (3.6.30) we easily obtain the conditions that a_1 and a_2 must satisfy for stability. First, a_2 must satisfy the condition

$$|a_2| = |p_1 p_2| = |p_1||p_2| < 1 \tag{3.6.31}$$

The condition for a_1 can be expressed as

$$|a_1| < 1 + a_2 \tag{3.6.32}$$

The conditions in (3.6.31) and (3.6.32) can also be derived from the Schür–Cohn stability test. From the recursive equations in (3.6.22) through (3.6.25), we find that

$$K_1 = \frac{a_1}{1 + a_2} \tag{3.6.33}$$

and

$$K_2 = a_2 \tag{3.6.34}$$

The system is stable if and only if $|K_1| < 1$ and $|K_2| < 1$. Consequently,

$$-1 < a_2 < 1$$

or equivalently $|a_2| < 1$, which agrees with (3.6.31). Also,

$$-1 < \frac{a_1}{1 + a_2} < 1$$

or, equivalently,

$$a_1 < 1 + a_2$$
$$a_1 > -1 - a_2$$

which are in agreement with (3.6.32). Therefore, a two-pole system is stable if and only if the coefficients a_1 and a_2 satisfy the conditions in (3.6.31) and (3.6.32).

The stability conditions given in (3.6.31) and (3.6.32), define a region in the coefficient plane (a_1, a_2), which is in the form of a triangle, as shown in Fig. 3.15. The system is stable if and only if the point (a_1, a_2) lies inside the triangle, which we call the *stability triangle*.

The characteristics of the two-pole system depend on the location of the poles or, equivalently, on the location of the point (a_1, a_2) in the stability triangle. The poles of the system may be real or complex conjugate, depending on the value of the discriminant $\Delta = a_1^2 - 4a_2$. The parabola $a_2 = a_1^2/4$ splits the stability

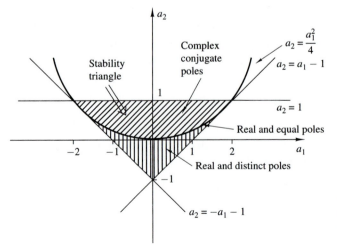

Figure 3.15   Region of stability (stability triangle) in the (a_1, a_2) coefficient plane for a second-order system.

triangle into two regions, as illustrated in Fig. 3.15. The region below the parabola $(a_1^2 > 4a_2)$ corresponds to real and distinct poles. The points on the parabola $(a_1^2 = 4a_2)$ result in real and equal (double) poles. Finally, the points above the parabola correspond to complex-conjugate poles.

Additional insight into the behavior of the system can be obtained from the unit sample responses for these three cases.

Real and distinct poles ($a_1^2 = 4a_2$).   Since p_1, p_2 are real and $p_1 \neq p_2$, the system function can be expressed in the form

$$H(z) = \frac{A_1}{1 - p_1 z^{-1}} + \frac{A_2}{1 - p_2 z^{-1}} \tag{3.6.35}$$

where

$$A_1 = \frac{b_0 p_1}{p_1 - p_2} \qquad A_2 = \frac{-b_0 p_2}{p_1 - p_2} \tag{3.6.36}$$

Consequently, the unit sample response is

$$h(n) = \frac{b_0}{p_1 - p_2}(p_1^{n+1} - p_2^{n+1})u(n) \tag{3.6.37}$$

Therefore, the unit sample response is the difference of two decaying exponential sequences. Figure 3.16 illustrates a typical graph for $h(n)$ when the poles are distinct.

Real and equal poles ($a_1^2 = 4a_2$).   In this case $p_1 = p_2 = p = -a_1/2$. The system function is

$$H(z) = \frac{b_0}{(1 - p z^{-1})^2} \tag{3.6.38}$$

and hence the unit sample response of the system is

$$h(n) = b_0(n + 1)p^n u(n) \tag{3.6.39}$$

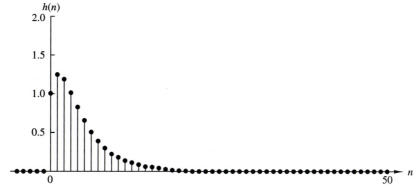

Figure 3.16   Plot of $h(n)$ given by (3.6.37) with $p_1 = 0.5$, $p_2 = 0.75$; $h(n) = [1/(p_1 - p_2)](p_1^{n+1} - p_2^{n+1})u(n)$.

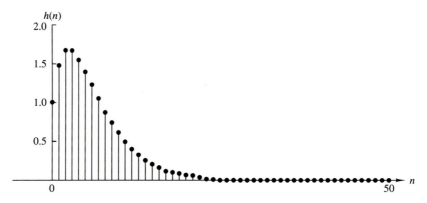

Figure 3.17   Plot of $h(n)$ given by (3.6.39) with $p = \frac{3}{4}$; $h(n) = (n+1)p^n u(n)$.

We observe that $h(n)$ is the product of a ramp sequence and a real decaying exponential sequence. The graph of $h(n)$ is shown in Fig. 3.17.

Complex-conjugate poles ($a_1^2 < 4a_2$).   Since the poles are complex conjugate, the system function can be factored and expressed as

$$H(z) = \frac{A}{1 - pz^{-1}} + \frac{A^*}{1 - p^*z^{-1}}$$

$$= \frac{A}{1 - re^{j\omega_0}z^{-1}} + \frac{A^*}{1 - re^{-j\omega_0}z^{-1}} \tag{3.6.40}$$

where $p = re^{j\omega}$ and $0 < \omega_0 < \pi$. Note that when the poles are complex conjugates, the parameters a_1 and a_2 are related to r and ω_0 according to

$$a_1 = -2r\cos\omega_0$$

$$a_2 = r^2 \tag{3.6.41}$$

The constant A in the partial-fraction expansion of $H(z)$ is easily shown to be

$$A = \frac{b_0 p}{p - p^*} = \frac{b_0 re^{j\omega_0}}{r(e^{j\omega_0} - e^{-j\omega_0})}$$

$$= \frac{b_0 e^{j\omega_0}}{j2\sin\omega_0} \tag{3.6.42}$$

Consequently, the unit sample response of a system with complex-conjugate poles is

$$h(n) = \frac{b_0 r^n}{\sin\omega_0} \frac{e^{j(n+1)\omega_0} - e^{-j(n+1)\omega_0}}{2j} u(n)$$

$$= \frac{b_0 r^n}{\sin\omega_0} \sin(n+1)\omega_0 u(n) \tag{3.6.43}$$

In this case $h(n)$ has an oscillatory behavior with an exponentially decaying envelope when $r < 1$. The angle ω_0 of the poles determines the frequency of oscillation and the distance r of the poles from the origin determines the rate of

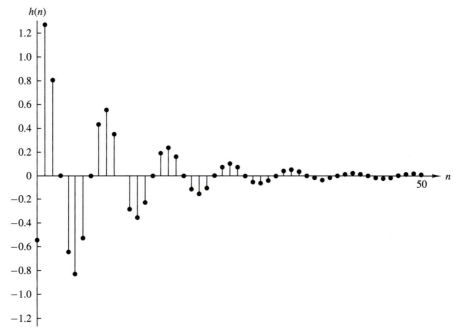

Figure 3.18 Plot of $h(n)$ given by (3.6.43) with $b_0 = 1$, $\omega_0 = \pi/4$, $r = 0.9$; $h(n) = [b_0 r^n /(\sin \omega_0)] \sin[(n + 1)\omega_0] u(n)$.

decay. When r is close to unity, the decay is slow. When r is close to the origin, the decay is fast. A typical graph of $h(n)$ is illustrated in Fig. 3.18.

3.7 SUMMARY AND REFERENCES

The z-transform plays the same role in discrete-time signals and systems as the Laplace transform does in continuous-time signals and systems. In this chapter we derived the important properties of the z-transform, which are extremely useful in the analysis of discrete-time systems. Of particular importance is the convolution property, which transforms the convolution of two sequences into a product of their z-transforms.

In the context of LTI systems, the convolution property results in the product of the z-transform $X(z)$ of the input signal with the system function $H(z)$, where the latter is the z-transform of the unit sample response of the system. This relationship allows us to determine the output of an LTI system in response to an input with transform $X(z)$ by computing the product $Y(z) = H(z)X(z)$ and then determining the inverse z-transform of $Y(z)$ to obtain the output sequence $y(n)$.

We observed that many signals of practical interest have rational z-transforms. Moreover, LTI systems characterized by constant-coefficient linear difference equations, also possess rational system functions. Consequently, in determining the inverse *z*-transform, we naturally emphasized the inversion of rational transforms. For such transforms, the partial-fraction expansion method is relatively easy to apply, in conjunction with the ROC, to determine the corresponding sequence in the time domain. The one-sided *z*-transform was introduced to solve for the response of causal systems excited by causal input signals with nonzero initial conditions.

Finally, we considered the characterization of LTI systems in the *z*-transform domain. In particular, we related the pole–zero locations of a system to its time-domain characteristics and restated the requirements for stability and causality of LTI systems in terms of the pole locations. We demonstrated that a causal system has a system function $H(z)$ with a ROC $|z| > r_1$, where $0 < r_1 \leq \infty$. In a stable and causal system, the poles of $H(z)$ lie inside the unit circle. On the other hand, if the system is noncausal, the condition for stability requires that the unit circle be contained in the ROC of $H(z)$. Hence a noncausal stable LTI system has a system function with poles both inside and outside the unit circle with an annular ROC that includes the unit circle. The Schür–Cohn test for the stability of a causal LTI system was described and the stability of second-order system was considered in some detail.

An excellent comprehensive treatment of the *z*-transform and its application to the analysis of LTI systems is given in the text by Jury (1964). The Schür–Cohn test for stability is treated in several texts. Our presentation was given in the context of reflection coefficients which are used in linear predictive coding of speech signals. The text by Markel and Gray (1976) is a good reference for the Schür–Cohn test and its application to speech signal processing.

PROBLEMS

3.1 Determine the *z*-transform of the following signals.

   (a) $x(n) = \{3, 0, 0, 0, 0, 6, 1, -4\}$
                 ↑

   (b) $x(n) = \begin{cases} (\frac{1}{2})^n, & n \geq 5 \\ 0, & n \leq 4 \end{cases}$

3.2 Determine the *z*-transforms of the following signals and sketch the corresponding pole–zero patterns.

   (a) $x(n) = (1 + n)u(n)$
   (b) $x(n) = (a^n + a^{-n})u(n)$, a real
   (c) $x(n) = (-1)^n 2^{-n} u(n)$
   (d) $x(n) = (na^n \sin \omega_0 n)u(n)$
   (e) $x(n) = (na^n \cos \omega_0 n)u(n)$
   (f) $x(n) = Ar^n \cos(\omega_0 n + \phi)u(n), 0 < r < 1$
   (g) $x(n) = \frac{1}{2}(n^2 + n)(\frac{1}{3})^{n-1}u(n-1)$
   (h) $x(n) = (\frac{1}{2})^n[u(n) - u(n-10)]$

3.3 Determine the z-transforms and sketch the ROC of the following signals.

(a) $x_1(n) = \begin{cases} (\frac{1}{3})^n, & n \geq 0 \\ (\frac{1}{2})^{-n}, & n < 0 \end{cases}$

(b) $x_2(n) = \begin{cases} (\frac{1}{3})^n - 2^n, & n \geq 0 \\ 0, & n < 0 \end{cases}$

(c) $x_3(n) = x_1(n + 4)$

(d) $x_4(n) = x_1(-n)$

3.4 Determine the z-transform of the following signals.

(a) $x(n) = n(-1)^n u(n)$

(b) $x(n) = n^2 u(n)$

(c) $x(n) = -na^n u(-n - 1)$

(d) $x(n) = (-1)^n \left(\cos \frac{\pi}{3} n \right) u(n)$

(e) $x(n) = (-1)^n u(n)$

(f) $x(n) = \{1, 0, -1, 0, 1, -1, \ldots\}$
       $\uparrow$

3.5 Determine the regions of convergence of right-sided, left-sided, and finite-duration two-sided sequences.

3.6 Express the z-transform of

$$y(n) = \sum_{k=-\infty}^{n} x(k)$$

in terms of $X(z)$. [*Hint*: Find the difference $y(n) - y(n - 1)$.]

3.7 Compute the convolution of the following signals by means of the z-transform.

$$x_1(n) = \begin{cases} (\frac{1}{3})^n, & n \geq 0 \\ (\frac{1}{2})^{-n}, & n < 0 \end{cases}$$

$$x_2(n) = (\frac{1}{2})^n u(n)$$

3.8 Use the convolution property to:

(a) Express the z-transform of

$$y(n) = \sum_{k=-\infty}^{n} x(k)$$

in terms of $X(z)$.

(b) Determine the z-transform of $x(n) = (n + 1)u(n)$. [*Hint*: Show first that $x(n) = u(n) * u(n)$.]

3.9 The z-transform $X(z)$ of a real signal $x(n)$ includes a pair of complex-conjugate zeros and a pair of complex-conjugate poles. What happens to these pairs if we multiply $x(n)$ by $e^{j\omega_0 n}$? (*Hint*: Use the scaling theorem in the z-domain.)

3.10 Apply the final value theorem to determine $x(\infty)$ for the signal

$$x(n) = \begin{cases} 1, & \text{if } n \text{ is even} \\ 0, & \text{otherwise} \end{cases}$$

3.11 Using long division, determine the inverse z-transform of

$$X(z) = \frac{1 + 2z^{-1}}{1 - 2z^{-1} + z^{-2}}$$

if (a) $x(n)$ is causal and (b) $x(n)$ is anticausal.

3.12 Determine the causal signal $x(n)$ having the z-transform

$$X(z) = \frac{1}{(1 - 2z^{-1})(1 - z^{-1})^2}$$

3.13 Let $x(n)$ be a sequence with z-transform $X(z)$. Determine, in terms of $X(z)$, the z-transforms of the following signals.

(a) $x_1(n) = \begin{cases} x\left(\dfrac{n}{2}\right), & \text{if } n \text{ even} \\ 0, & \text{if } n \text{ odd} \end{cases}$

(b) $x_2(n) = x(2n)$

3.14 Determine the causal signal $x(n)$ if its z-transform $X(z)$ is given by:

(a) $X(z) = \dfrac{1 + 3z^{-1}}{1 + 3z^{-1} + 2z^{-2}}$

(b) $X(z) = \dfrac{1}{1 - z^{-1} + \frac{1}{2}z^{-2}}$

(c) $X(z) = \dfrac{z^{-6} + z^{-7}}{1 - z^{-1}}$

(d) $X(z) = \dfrac{1 + 2z^{-2}}{1 + z^{-2}}$

(e) $X(z) = \dfrac{1}{4} \dfrac{1 + 6z^{-1} + z^{-2}}{(1 - 2z^{-1} + 2z^{-2})(1 - 0.5z^{-1})}$

(f) $X(z) = \dfrac{2 - 1.5z^{-1}}{1 - 1.5z^{-1} + 0.5z^{-2}}$

(g) $X(z) = \dfrac{1 + 2z^{-1} + z^{-2}}{1 + 4z^{-1} + 4z^{-2}}$

(h) $X(z)$ is specified by a pole–zero pattern in Fig. P3.14. The constant $G = \frac{1}{4}$.

(i) $X(z) = \dfrac{1 - \frac{1}{2}z^{-1}}{1 + \frac{1}{2}z^{-1}}$

(j) $X(z) = \dfrac{1 - az^{-1}}{z^{-1} - a}$

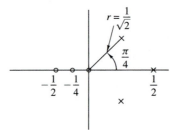

Figure P3.14

3.15 Determine all possible signals $x(n)$ associated with the z-transform

$$X(z) = \frac{5z^{-1}}{(1 - 2z^{-1})(3 - z^{-1})}$$

3.16 Determine the convolution of the following pairs of signals by means of the z-transform.

(a) $x_1(n) = (\frac{1}{4})^n u(n-1), \quad x_2(n) = [1 + (\frac{1}{2})^n]u(n)$

(b) $x_1(n) = u(n), \quad x_2(n) = \delta(n) + (\frac{1}{2})^n u(n)$

(c) $x_1(n) = (\frac{1}{2})^n u(n), \quad x_2(n) = \cos \pi n u(n)$

(d) $x_1(n) = nu(n), \quad x_2(n) = 2^n u(n-1)$

3.17 Prove the final value theorem for the one-sided z-transform.

3.18 If $X(z)$ is the z-transform of $x(n)$, show that:

(a) $Z\{x^*(n)\} = X^*(z^*)$

(b) $Z\{\text{Re}[x(n)]\} = \frac{1}{2}[X(z) + X^*(z^*)]$

(c) $Z\{\text{Im}[x(n)]\} = \frac{1}{2}[X(z) - X^*(z^*)]$

(d) If

$$x_k(n) = \begin{cases} x\left(\frac{n}{k}\right), & \text{if } n/k \text{ integer} \\ 0, & \text{otherwise} \end{cases}$$

then

$$X_k(z) = X(z^k)$$

(e) $Z\{e^{j\omega_0 n} x(n)\} = X(ze^{-j\omega_0})$

3.19 By first differentiating $X(z)$ and then using appropriate properties of the z-transform, determine $x(n)$ for the following transforms.

(a) $X(z) = \log(1 - 2z), \quad |z| < \frac{1}{2}$

(b) $X(z) = \log(1 - z^{-1}), \quad |z| > \frac{1}{2}$

3.20 (a) Draw the pole–zero pattern for the signal

$$x_1(n) = (r^n \sin \omega_0 n) u(n) \qquad 0 < r < 1$$

(b) Compute the z-transform $X_2(z)$, which corresponds to the pole–zero pattern in part (a).

(c) Compare $X_1(z)$ with $X_2(z)$. Are they identical? If not, indicate a method to derive $X_1(z)$ from the pole–zero pattern.

3.21 Show that the roots of a polynomial with real coefficients are real or form complex-conjugate pairs. The inverse is not true, in general.

3.22 Prove the convolution and correlation properties of the z-transform using only its definition.

3.23 Determine the signal $x(n)$ with z-transform

$$X(z) = e^z + e^{1/z} \qquad |z| \neq 0$$

3.24 Determine, in closed form, the causal signals $x(n)$ whose z-transforms are given by:

(a) $X(z) = \dfrac{1}{1 + 1.5z^{-1} - 0.5z^{-2}}$

(b) $X(z) = \dfrac{1}{1 - 0.5z^{-1} + 0.6z^{-2}}$

Partially check your results by computing $x(0)$, $x(1)$, $x(2)$, and $x(\infty)$ by an alternative method.

3.25 Determine all possible signals that can have the following z-transforms.

(a) $X(z) = \dfrac{1}{1 - 1.5z^{-1} + 0.5z^{-2}}$

(b) $X(z) = \dfrac{1}{1 - \frac{1}{2}z^{-1} + \frac{1}{4}z^{-2}}$

3.26 Determine the signal $x(n)$ with z-transform

$$X(z) = \frac{3}{1 - \frac{10}{3}z^{-1} + z^{-2}}$$

if $X(z)$ converges on the unit circle.

3.27 Prove the complex convolution relation given by (3.2.22).

3.28 Prove the conjugation properties and Parseval's relation for the z-transform given in Table 3.2.

3.29 In Example 3.4.1 we solved for $x(n)$, $n < 0$, by performing contour integrations for each value of n. In general, this procedure proves to be tedious. It can be avoided by making a transformation in the contour integral from z-plane to the $w = 1/z$ plane. Thus a circle of radius R in the z-plane is mapped into a circle of radius $1/R$ in the w-plane. As a consequence, a pole inside the unit circle in the z-plane is mapped into a pole outside the unit circle in the w-plane. By making the change of variable $w = 1/z$ in the contour integral, determine the sequence $x(n)$ for $n < 0$ in Example 3.4.1.

3.30 Let $x(n)$, $0 \le n \le N - 1$ be a finite-duration sequence, which is also real-valued and even. Show that the zeros of the polynomial $X(z)$ occur in mirror-image pairs about the unit circle. That is, if $z = re^{j\theta}$ is a zero of $X(z)$, then $z = (1/r)e^{j\theta}$ is also a zero.

3.31 Compute the convolution of the following pair of signals in the time domain and by using the one-sided z-transform.
(a) $x_1(n) = \{1, 1, 1, 1, 1, 1\}$, $x_2(n) = \{1, 1, 1\}$
$\qquad\quad\uparrow \qquad\qquad\qquad\qquad\quad \uparrow$

(b) $x_1(n) = (\frac{1}{2})^n u(n)$, $x_2(n) = (\frac{1}{3})^n u(n)$
(c) $x_1(n) = \{1, 2, 3, 4\}$, $x_2(n) = \{4, 3, 2, 1\}$
$\qquad\quad\uparrow \qquad\qquad\qquad\qquad\quad \uparrow$

(d) $x_1(n) = \{1, 1, 1, 1, 1, 1\}$, $x_2(n) = \{1, 1, 1\}$
$\qquad\quad\uparrow \qquad\qquad\qquad\qquad\quad \uparrow$

Did you obtain the same results by both methods? Explain.

3.32 Determine the one-sided z-transform of the constant signal $x(n) = 1$, $-\infty < n < \infty$.

3.33 Prove that the Fibonacci sequence can be thought of as the impulse response of the system described by the difference equation $y(n) = y(n - 1) + y(n - 2) + x(n)$. Then determine $h(n)$ using z-transform techniques.

3.34 Use the one-sided z-transform to determine $y(n)$, $n \ge 0$ in the following cases.
(a) $y(n) + \frac{1}{2}y(n - 1) - \frac{1}{4}y(n - 2) = 0$; $y(-1) = y(-2) = 1$
(b) $y(n) - 1.5y(n - 1) + 0.5y(n - 2) = 0$; $y(-1) = 1, y(-2) = 0$
(c) $y(n) = \frac{1}{2}y(n - 1) + x(n)$
$\qquad x(n) = (\frac{1}{3})^n u(n)$, $y(-1) = 1$
(d) $y(n) = \frac{1}{4}y(n - 2) + x(n)$
$\qquad x(n) = u(n)$
$\qquad y(-1) = 0$; $y(-2) = 1$

3.35 Show that the following systems are equivalent.
(a) $y(n) = 0.2y(n - 1) + x(n) - 0.3x(n - 1) + 0.02x(n - 2)$
(b) $y(n) = x(n) - 0.1x(n - 1)$

3.36 Consider the sequence $x(n) = a^n u(n)$, $-1 < a < 1$. Determine at least two sequences that are not equal to $x(n)$ but have the same autocorrelation.

3.37 Compute the unit step response of the system with impulse response

$$h(n) = \begin{cases} 3^n, & n < 0 \\ (\frac{2}{5})^n, & n \geq 0 \end{cases}$$

3.38 Compute the zero-state response for the following pairs of systems and input signals.

(a) $h(n) = (\frac{1}{3})^n u(n)$, $x(n) = (\frac{1}{2})^n \left(\cos \frac{\pi}{3} n \right) u(n)$

(b) $h(n) = (\frac{1}{2})^n u(n)$, $x(n) = (\frac{1}{3})^n u(n) + (\frac{1}{2})^{-n} u(-n-1)$

(c) $y(n) = -0.1 y(n-1) + 0.2 y(n-2) + x(n) + x(n-1)$
    $x(n) = (\frac{1}{3})^n u(n)$

(d) $y(n) = \frac{1}{2} x(n) - \frac{1}{2} x(n-1)$
    $x(n) = 10 \left(\cos \frac{\pi}{2} n \right) u(n)$

(e) $y(n) = -y(n-2) + 10 x(n)$
    $x(n) = 10 \left(\cos \frac{\pi}{2} n \right) u(n)$

(f) $h(n) = (\frac{2}{5})^n u(n)$, $x(n) = u(n) - u(n-7)$

(g) $h(n) = (\frac{1}{2})^n u(n)$, $x(n) = (-1)^n$, $-\infty < n < \infty$

(h) $h(n) = (\frac{1}{2})^n u(n)$, $x(n) = (n+1)(\frac{1}{4})^n u(n)$

3.39 Consider the system

$$H(z) = \frac{1 - 2z^{-1} + 2z^{-2} - z^{-3}}{(1 - z^{-1})(1 - 0.5z^{-1})(1 - 0.2z^{-1})} \qquad \text{ROC: } 0.5 < |z| < 1$$

(a) Sketch the pole–zero pattern. Is the system stable?

(b) Determine the impulse response of the system.

3.40 Compute the response of the system

$$y(n) = 0.7 y(n-1) - 0.12 y(n-2) + x(n-1) + x(n-2)$$

to the input $x(n) = n u(n)$. Is the system stable?

3.41 Determine the impulse response and the step response of the following causal systems. Plot the pole–zero patterns and determine which of the systems are stable.

(a) $y(n) = \frac{3}{4} y(n-1) - \frac{1}{8} y(n-2) + x(n)$

(b) $y(n) = y(n-1) - 0.5 y(n-2) + x(n) + x(n-1)$

(c) $H(z) = \dfrac{z^{-1}(1 + z^{-1})}{(1 - z^{-1})^3}$

(d) $y(n) = 0.6 y(n-1) - 0.08 y(n-2) + x(n)$

(e) $y(n) = 0.7 y(n-1) - 0.1 y(n-2) + 2x(n) - x(n-2)$

3.42 Let $x(n)$ be a causal sequence with z-transform $X(z)$ whose pole–zero plot is shown in Fig. P3.42. Sketch the pole–zero plots and the ROC of the following sequences:

(a) $x_1(n) = x(-n+2)$

(b) $x_2(n) = e^{j(\pi/3)n} x(n)$

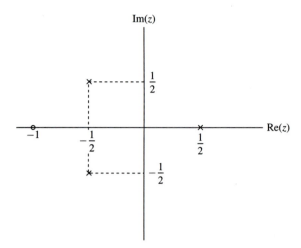

Figure P3.42

3.43 We want to design a causal discrete-time LTI system with the property that if the input is

$$x(n) = (\tfrac{1}{2})^n u(n) - \tfrac{1}{4}(\tfrac{1}{2})^{n-1} u(n-1)$$

then the output is

$$y(n) = (\tfrac{1}{3})^n u(n)$$

(a) Determine the impulse response $h(n)$ and the system function $H(z)$ of a system that satisfies the foregoing conditions.

(b) Find the difference equation that characterizes this system.

(c) Determine a realization of the system that requires the minimum possible amount of memory.

(d) Determine if the system is stable.

3.44 Determine the stability region for the causal system

$$H(z) = \frac{1}{1 + a_1 z^{-1} + a_2 z^{-2}}$$

by computing its poles and restricting them to be inside the unit circle.

3.45 Consider the system

$$H(z) = \frac{z^{-1} + \tfrac{1}{2} z^{-2}}{1 - \tfrac{3}{5} z^{-1} + \tfrac{2}{25} z^{-2}}$$

Determine:

(a) The impulse response

(b) The zero-state step response

(c) The step response if $y(-1) = 1$ and $y(-2) = 2$

3.46 Determine the system function, impulse response, and zero-state step response of the system shown in Fig P3.46.

3.47 Consider the causal system

$$y(n) = -a_1 y(n-1) + b_0 x(n) + b_1 x(n-1)$$

Determine:

(a) The impulse response

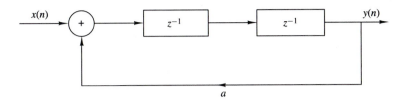

Figure P3.46

(b) The zero-state step response
(c) The step response if $y(-1) = A \neq 0$
(d) The response to the input

$$x(n) = \cos \omega_0 n \qquad 0 \leq n < \infty$$

3.48 Determine the zero-state response of the system

$$y(n) = \tfrac{1}{2}y(n-1) + 4x(n) + 3x(n-1)$$

to the input

$$x(n) = e^{j\omega_0 n} u(n)$$

What is the steady-state response of the system?

3.49 Consider the causal system defined by the pole–zero pattern shown in Fig. P3.49.
   (a) Determine the system function and the impulse response of the system given that $H(z)|_{z=1} = 1$.
   (b) Is the system stable?
   (c) Sketch a possible implementation of the system and determine the corresponding difference equations.

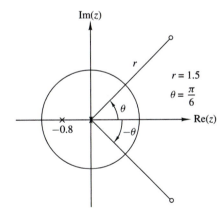

Figure P3.49

3.50 An FIR LTI system has an impulse response $h(n)$, which is real valued, even, and has finite duration of $2N + 1$. Show that if $z_1 = re^{j\omega_0}$ is a zero of the system, then $z_1 = (1/r)e^{j\omega_0}$ is also a zero.

3.51 Consider an LTI discrete-time system whose pole–zero pattern is shown in Fig. P3.51.
   (a) Determine the ROC of the system function $H(z)$ if the system is known to be stable.

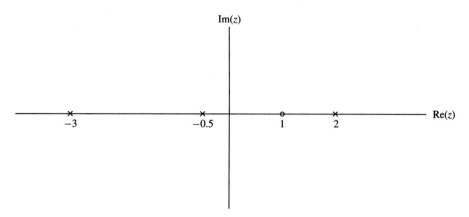

Figure P3.51

(b) It is possible for the given pole–zero plot to correspond to a causal and stable system? If so, what is the appropriate ROC?

(c) How many possible systems can be associated with this pole–zero pattern?

3.52 Let $x(n)$ be a causal sequence.

(a) What conclusion can you draw about the value of its z-transform $X(z)$ at $z = \infty$?

(b) Use the result in part (a) to check which of the following transforms cannot be associated with a causal sequence.

$$\text{(i)} \ X(z) = \frac{(z - \frac{1}{2})^4}{(z = \frac{1}{3})^3} \qquad \text{(ii)} \ X(z) = \frac{(1 - \frac{1}{2}z^{-1})^2}{(1 - \frac{1}{3}z^{-1})} \qquad \text{(iii)} \ X(z) = \frac{(z - \frac{1}{3})^2}{(z - \frac{1}{2})^3}$$

3.53 A causal pole–zero system is BIBO stable if its poles are inside the unit circle. Consider now a pole–zero system that is BIBO stable and has its poles inside the unit circle. Is the system always causal? [*Hint:* Consider the systems $h_1(n) = a^n u(n)$ and $h_2(n) = a^n u(n + 3)$, $|a| < 1$.]

3.54 Let $x(n)$ be an anticausal signal [i.e., $x(n) = 0$ for $n > 0$]. Formulate and prove an initial value theorem for anticausal signals.

3.55 The step response of an LTI system is

$$s(n) = (\tfrac{1}{3})^{n-2} u(n + 2)$$

(a) Find the system function $H(z)$ and sketch the pole–zero plot.

(b) Determine the impulse response $h(n)$.

(c) Check if the system is causal and stable.

3.56 Use contour integration to determine the sequence $x(n)$ whose z-transform is given by

(a) $X(z) = \dfrac{1}{1 - \frac{1}{2}z^{-1}} \qquad |z| > \frac{1}{2}$

(b) $X(z) = \dfrac{1}{1 - \frac{1}{2}z^{-1}} \qquad |z| < \frac{1}{2}$

(c) $X(z) = \dfrac{z - a}{1 - az}$ $|z| > |1/a|$

(d) $X(z) = \dfrac{1 - \frac{1}{4}z^{-1}}{1 - \frac{1}{6}z^{-1} - \frac{1}{6}z^{-2}}$ $|z| > \frac{1}{2}$

3.57 Let $x(n)$ be a sequence with z-transform

$$X(z) = \frac{1 - a^2}{(1 - az)(1 - az^{-1})} \qquad \text{ROC: } a < |z| < 1/a$$

with $0 < a < 1$. Determine $x(n)$ by using contour integration

3.58 The z-transform of a sequence $x(n)$ is given by

$$X(z) = \frac{z^{20}}{(z - \frac{1}{2})(z - 2)^5(z + \frac{5}{2})^2(z + 3)}$$

Furthermore it is known that $X(z)$ converges for $|z| = 1$.

(a) Determine the ROC of $X(z)$.

(b) Determine $x(n)$ at $n = -18$. (*Hint:* Use contour integration.)

4

Frequency Analysis of Signals and Systems

The Fourier transform is one of several mathematical tools that is useful in the analysis and design of LTI systems. Another is the Fourier series. These signal representations basically involve the decomposition of the signals in terms of sinusoidal (or complex exponential) components. With such a decomposition, a signal is said to be represented in the *frequency domain*.

As we shall demonstrate, most signals of practical interest can be decomposed into a sum of sinusoidal signal components. For the class of periodic signals, such a decomposition is called a *Fourier series*. For the class of finite energy signals, the decomposition is called the *Fourier transform*. These decompositions are extremely important in the analysis of LTI systems because the response of an LTI system to a sinusoidal input signal is a sinusoid of the same frequency but of different amplitude and phase. Furthermore, the linearity property of the LTI system implies that a linear sum of sinusoidal components at the input produces a similar linear sum of sinusoidal components at the output, which differ only in the amplitudes and phases from the input sinusoids. This characteristic behavior of LTI systems renders the sinusoidal decomposition of signals very important. Although many other decompositions of signals are possible, only the class of sinusoidal (or complex exponential) signals possess this desirable property in passing through an LTI system.

We begin our study of frequency analysis of signals with the representation of continuous-time periodic and aperiodic signals by means of the Fourier series and the Fourier transform, respectively. This is followed by a parallel treatment of discrete-time periodic and aperiodic signals. The properties of the Fourier transform are described in detail and a number of time-frequency dualities are presented.

4.1 FREQUENCY ANALYSIS OF CONTINUOUS-TIME SIGNALS

It is well known that a prism can be used to break up white light (sunlight) into the colors of the rainbow (see Fig. 4.1a). In a paper submitted in 1672 to the Royal Society, Isaac Newton used the term *spectrum* to describe the *continuous* bands

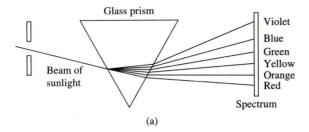

(a)

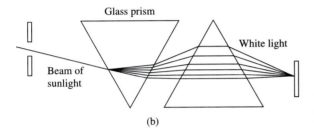

(b)

Figure 4.1 (a) Analysis and (b) synthesis of the white light (sunlight) using glass prisms.

of colors produced by this apparatus. To understand this phenomenon, Newton placed another prism upside-down with respect to the first, and showed that the colors blended back into white light, as in Fig. 4.1b. By inserting a slit between the two prisms and blocking one or more colors from hitting the second prism, he showed that the remixed light is no longer white. Hence the light passing through the first prism is simply analyzed into its component colors without any other change. However, only if we mix again all of these colors do we obtain the original white light.

Later, Joseph Fraunhofer (1787–1826), in making measurements of light emitted by the sun and stars, discovered that the spectrum of the observed light consists of distinct color lines. A few years later (mid-1800s) Gustav Kirchhoff and Robert Bunsen found that each chemical element, when heated to incandescence, radiated its own distinct color of light. As a consequence, each chemical element can be identified by its own *line spectrum*.

From physics we know that each color corresponds to a specific frequency of the visible spectrum. Hence the analysis of light into colors is actually a form of *frequency analysis*.

Frequency analysis of a signal involves the resolution of the signal into its frequency (sinusoidal) components. Instead of light, our signal waveforms are basically functions of time. The role of the prism is played by the Fourier analysis tools that we will develop: the Fourier series and the Fourier transform. The recombination of the sinusoidal components to reconstruct the original signal is basically a Fourier synthesis problem. The problem of signal analysis is basically the same for the case of a signal waveform and for the case of the light from heated chemical compositions. Just as in the case of chemical compositions, different signal waveforms have different spectra. Thus the spectrum provides an "identity" or a signature for the signal in the sense that no other signal has the same spectrum. As we will see, this attribute is related to the mathematical treatment of frequency-domain techniques.

If we decompose a waveform into sinusoidal components, in much the same way that a prism separates white light into different colors, the sum of these sinusoidal components results in the original waveform. On the other hand, if any of these components is missing, the result is a different signal.

In our treatment of frequency analysis, we will develop the proper mathematical tools ("prisms") for the decomposition of signals ("light") into sinusoidal frequency components (colors). Furthermore, the tools ("inverse prisms") for synthesis of a given signal from its frequency components will also be developed.

The basic motivation for developing the frequency analysis tools is to provide a mathematical and pictorial representation for the frequency components that are contained in any given signal. As in physics, the term *spectrum* is used when referring to the frequency content of a signal. The process of obtaining the spectrum of a given signal using the basic mathematical tools described in this chapter is known as *frequency* or *spectral analysis*. In contrast, the process of determining the spectrum of a signal in practice, based on actual measurements of the signal, is called *spectrum estimation*. This distinction is very important. In a practical problem the signal to be analyzed does not lend itself to an exact mathematical description. The signal is usually some information-bearing signal from which we are attempting to extract the relevant information. If the information that we wish to extract can be obtained either directly or indirectly from the spectral content of the signal, we can perform *spectrum estimation* on the information-bearing signal, and thus obtain an estimate of the signal spectrum. In fact, we can view spectral estimation as a type of spectral analysis performed on signals obtained from physical sources (e.g., speech, EEG, ECG, etc.). The instruments or software programs used to obtain spectral estimates of such signals are known as *spectrum analyzers*.

Here, we will deal with spectral analysis. However, in Chapter 12 we shall treat the subject of power spectrum estimation.

4.1.1 The Fourier Series for Continuous-Time Periodic Signals

In this section we present the frequency analysis tools for continuous-time periodic signals. Examples of periodic signals encountered in practice are square waves, rectangular waves, triangular waves, and of course, sinusoids and complex exponentials.

The basic mathematical representation of periodic signals is the Fourier series, which is a linear weighted sum of harmonically related sinusoids or complex exponentials. Jean Baptiste Joseph Fourier (1768–1830), a French mathematician, used such trigonometric series expansions in describing the phenomenon of heat conduction and temperature distribution through bodies. Although his work was motivated by the problem of heat conduction, the mathematical techniques that he developed during the early part of the nineteenth century now find application in a variety of problems encompassing many different fields, including optics, vibrations in mechanical systems, system theory, and electromagnetics.

From Chapter 1 we recall that a linear combination of harmonically related complex exponentials of the form

$$x(t) = \sum_{k=-\infty}^{\infty} c_k e^{j2\pi k F_0 t} \tag{4.1.1}$$

is a periodic signal with fundamental period $T_p = 1/F_0$. Hence we can think of the exponential signals

$$\{e^{j2\pi k F_0 t} \qquad k = 0, \pm 1, \pm 2, \ldots\}$$

as the basic "building blocks" from which we can construct periodic signals of various types by proper choice of the fundamental frequency and the coefficients $\{c_k\}$. F_0 determines the fundamental period of x(t) and the coefficients $\{c_k\}$ specify the shape of the waveform.

Suppose that we are given a periodic signal $x(t)$ with period T_p. We can represent the periodic signal by the series (4.1.1), called a *Fourier series,* where the fundamental frequency F_0 is selected to be the reciprocal of the given period T_p. To determine the expression for the coefficients $\{c_k\}$, we first multiply both sides of (4.1.1) by the complex exponential

$$e^{-j2\pi F_0 l t}$$

where l is an integer and then integrate both sides of the resulting equation over a single period, say from 0 to T_p, or more generally, from t_0 to $t_0 + T_p$, where t_0 is an arbitrary but mathematically convenient starting value. Thus we obtain

$$\int_{t_0}^{t_0+T_p} x(t)e^{-j2\pi l F_0 t}\, dt = \int_{t_0}^{t_0+T_p} e^{-j2\pi l F_0 t} \left(\sum_{k=-\infty}^{\infty} c_k e^{+j2\pi k F_0 t} \right) dt \tag{4.1.2}$$

To evaluate the integral on the right-hand side of (4.1.2), we interchange the order of the summation and integration and combine the two exponentials. Hence

$$\sum_{k=-\infty}^{\infty} c_k \int_{t_0}^{t_0+T_p} e^{j2\pi F_0 (k-l)t}\, dt = \sum_{k=-\infty}^{\infty} c_k \left[\frac{e^{j2\pi F_0(k-l)t}}{j2\pi F_0(k-l)} \right]_{t_0}^{t_0+T_p} \tag{4.1.3}$$

For $k \neq l$, the right-hand side of (4.1.3) evaluated at the lower and upper limits, t_0 and $t_0 + T_p$, respectively, yields zero. On the other hand, if $k = l$, we have

$$\int_{t_0}^{t_0+T_p} dt = t \Big]_{t_0}^{t_0+T_p} = T_p$$

Consequently, (4.1.2) reduces to

$$\int_{t_0}^{t_0+T_p} x(t)e^{-j2\pi l F_0 t}\, dt = c_l T_p$$

and therefore the expression for the Fourier coefficients in terms of the given periodic signal becomes

$$c_l = \frac{1}{T_p} \int_{t_0}^{t_0+T_p} x(t)e^{-j2\pi l F_0 t}\,dt$$

Since t_0 is arbitrary, this integral can be evaluated over any interval of length T_p, that is, over any interval equal to the period of the signal $x(t)$. Consequently, the integral for the Fourier series coefficients will be written as

$$c_l = \frac{1}{T_p} \int_{T_p} x(t)e^{-j2\pi l F_0 t}\,dt \tag{4.1.4}$$

An important issue that arises in the representation of the periodic signal $x(t)$ by the Fourier series is whether or not the series converges to $x(t)$ for every value of t, that is, if the signal $x(t)$ and its Fourier series representation

$$\sum_{k=-\infty}^{\infty} c_k e^{j2\pi k F_0 t} \tag{4.1.5}$$

are equal at every value of t. The so-called *Dirichlet conditions* guarantee that the series (4.1.5) will be equal to $x(t)$, except at the values of t for which $x(t)$ is discontinuous. At these values of t, (4.1.5) converges to the midpoint (average value) of the discontinuity. The Dirichlet conditions are:

1. The signal $x(t)$ has a finite number of discontinuities in any period.
2. The signal $x(t)$ contains a finite number of maxima and minima during any period.
3. The signal $x(t)$ is absolutely integrable in any period, that is,

$$\int_{T_p} |x(t)|\,dt < \infty \tag{4.1.6}$$

All periodic signals of practical interest satisfy these conditions.

The weaker condition, that the signal has finite energy in one period,

$$\int_{T_p} |x(t)|^2\,dt < \infty \tag{4.1.7}$$

guarantees that the energy in the difference signal

$$e(t) = x(t) - \sum_{k=-\infty}^{\infty} c_k e^{j2\pi k F_0 t}$$

is zero, although $x(t)$ and its Fourier series may not be equal for all values of t. Note that (4.1.6) implies (4.1.7), but not vice versa. Also, both (4.1.7) and the Dirichlet conditions are sufficient but not necessary conditions (i.e., there are signals that have a Fourier series representation but do not satisfy these conditions).

In summary, if $x(t)$ is periodic and satisfies the Dirichlet conditions, it can be represented in a Fourier series as in (4.1.1), where the coefficients are specified by (4.1.4). These relations are summarized below.

FREQUENCY ANALYSIS OF CONTINUOUS-TIME PERIODIC SIGNALS

Synthesis equation	$x(t) = \sum\limits_{k=-\infty}^{\infty} c_k e^{j2\pi k F_0 t}$	(4.1.8)
Analysis equation	$c_k = \dfrac{1}{T_p} \int_{T_p} x(t) e^{-j2\pi k F_0 t} dt$	(4.1.9)

In general, the Fourier coefficients c_k are complex valued. Moreover, it is easily shown that if the periodic signal is real, c_k and c_{-k} are complex conjugates. As a result, if

$$c_k = |c_k| e^{j\theta_k}$$

then

$$c_{-k} = |c_k|^{-j\theta_k}$$

Consequently, the Fourier series may also be represented in the form

$$x(t) = c_0 + 2\sum_{k=1}^{\infty} |c_k| \cos(2\pi k F_0 t + \theta_k) \tag{4.1.10}$$

where c_0 is real valued when $x(t)$ is real.

Finally, we should indicate that yet another form for the Fourier series can be obtained by expanding the cosine function in (4.1.10) as

$$\cos(2\pi k F_0 t + \theta_k) = \cos 2\pi k F_0 t \cos \theta_k - \sin 2\pi k F_0 t \sin \theta_k$$

Consequently, we can rewrite (4.1.10) in the form

$$x(t) = a_0 + \sum_{k=1}^{\infty} (a_k \cos 2\pi k F_0 t - b_k \sin 2\pi k F_0 t) \tag{4.1.11}$$

where

$$a_0 = c_0$$

$$a_k = 2|c_k| \cos \theta_k$$

$$b_k = 2|c_k| \sin \theta_k$$

The expressions in (4.1.8), (4.1.10), and (4.1.11) constitute three equivalent forms for the Fourier series representation of a real periodic signal.

4.1.2 Power Density Spectrum of Periodic Signals

A periodic signal has infinite energy and a finite average power, which is given as

$$P_x = \frac{1}{T_p} \int_{T_p} |x(t)|^2 dt \tag{4.1.12}$$

If we take the complex conjugate of (4.1.8) and substitute for $x^*(t)$ in (4.1.12), we obtain

$$P_x = \frac{1}{T_p} \int_{T_p} x(t) \sum_{k=-\infty}^{\infty} c_k^* e^{-j2\pi k F_0 t} \, dt$$

$$= \sum_{k=-\infty}^{\infty} c_k^* \left[\frac{1}{T_p} \int_{T_p} x(t) e^{-j2\pi k F_0 t} \, dt \right] \qquad (4.1.13)$$

$$= \sum_{k=-\infty}^{\infty} |c_k|^2$$

Therefore, we have established the relation

$$P_x = \frac{1}{T_p} \int_{T_p} |x(t)|^2 dt = \sum_{k=-\infty}^{\infty} |c_k|^2 \qquad (4.1.14)$$

which is called *Parseval's relation* for power signals.

To illustrate the physical meaning of (4.1.14), suppose that $x(t)$ consists of a single complex exponential

$$x(t) = c_k e^{j2\pi k F_0 t}$$

In this case, all the Fourier series coefficients except c_k are zero. Consequently, the average power in the signal is

$$P_x = |c_k|^2$$

It is obvious that $|c_k|^2$ represents the power in the kth harmonic component of the signal. Hence the total average power in the periodic signal is simply the sum of the average powers in all the harmonics.

If we plot the $|c_k|^2$ as a function of the frequencies kF_0, $k = 0, \pm1, \pm2, \ldots$, the diagram that we obtain shows how the power of the periodic signal is distributed among the various frequency components. This diagram, which is illustrated in Fig. 4.2, is called the *power density spectrum* of the periodic signal $x(t)$. Since the

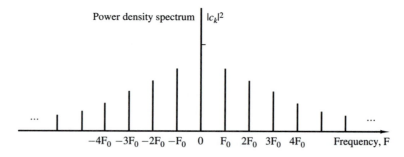

Figure 4.2 Power density spectrum of a continuous-time periodic signal.

---

*This function is also called the *power spectral density* or, simply, the *power spectrum*.

power in a periodic signal exists only at discrete values of frequencies (i.e., $F = 0$, $\pm F_0$, $\pm 2F_0, \ldots$), the signal is said to have a *line spectrum.* The spacing between two consecutive spectral lines is equal to the reciprocal of the fundamental period T_p, whereas the shape of the spectrum (i.e., the power distribution of the signal), depends on the time-domain characteristics of the signal.

As indicated in the preceding section, the Fourier series coefficients $\{c_k\}$ are complex valued, that is, they can be represented as

$$c_k = |c_k| e^{j\theta_k}$$

where

$$\theta_k = \measuredangle c_k$$

Instead of plotting the power density spectrum, we can plot the magnitude voltage spectrum $\{|c_k|\}$ and the phase spectrum $\{\theta_k\}$ as a function of frequency. Clearly, the power spectral density in the periodic signal is simply the square of the magnitude spectrum. The phase information is totally destroyed (or does not appear) in the power spectral density.

If the periodic signal is real valued, the Fourier series coefficients $\{c_k\}$ satisfy the condition

$$c_{-k} = c_k^*$$

Consequently, $|c_k|^2 = |c_k^*|^2$. Hence the power spectrum is a symmetric function of frequency. This condition also means that the magnitude spectrum is symmetric (even function) about the origin and the phase spectrum is an odd function. As a consequence of the symmetry, it is sufficient to specify the spectrum of a real periodic signal for positive frequencies only. Furthermore, the total average power can be expressed as

$$P_x = c_0^2 + 2 \sum_{k=1}^{\infty} |c_k|^2 \tag{4.1.15}$$

$$= a_0^2 + \frac{1}{2} \sum_{k=1}^{\infty} (a_k^2 + b_k^2) \tag{4.1.16}$$

which follows directly from the relationships given in Section 4.1.1 among $\{a_k\}$, $\{b_k\}$, and $\{c_k\}$ coefficients in the Fourier series expressions.

Example 4.1.1

Determine the Fourier series and the power density spectrum of the rectangular pulse train signal illustrated in Fig. 4.3.

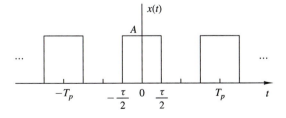

Figure 4.3   Continuous-time periodic train of rectangular pulses.

Solution The signal is periodic with fundamental period T_p and, clearly, satisfies the Dirichlet conditions. Consequently, we can represent the signal in the Fourier series given by (4.1.8) with the Fourier coefficients specified by (4.1.9).

Since $x(t)$ is an even signal [i.e., $x(t) = x(-t)$], it is convenient to select the integration interval from $-T_p/2$ to $T_p/2$. Thus (4.1.9) evaluated for $k = 0$ yields

$$c_0 = \frac{1}{T_p} \int_{-T_p/2}^{T_p/2} x(t)dt = \frac{1}{T_p} \int_{-\tau/2}^{\tau/2} A\,dt = \frac{A\tau}{T_p} \tag{4.1.17}$$

The term c_0 represents the average value (dc component) of the signal $x(t)$. For $k \neq 0$ we have

$$
\begin{aligned}
c_k &= \frac{1}{T_p} \int_{-\tau/2}^{\tau/2} A e^{-j2\pi k F_0 t} dt = \frac{A}{T_p} \left[\frac{e^{-j2\pi k F_0 t}}{-j2\pi k F_0} \right]_{-\tau/2}^{\tau/2} \\
&= \frac{A}{\pi F_0 k T_p} \frac{e^{j\pi k F_0 \tau} - e^{-j\pi k F_0 \tau}}{j2} \\
&= \frac{A\tau}{T_p} \frac{\sin \pi k F_0 \tau}{\pi k F_0 \tau} \qquad k = \pm 1, \pm 2, \ldots
\end{aligned}
\tag{4.1.18}
$$

It is interesting to note that the right-hand side of (4.1.18) has the form $(\sin\phi)/\phi$, where $\phi = \pi k F_0 \tau$. In this case ϕ takes on discrete values since F_0 and τ are fixed and the index k varies. However, if we plot $(\sin\phi)/\phi$ with ϕ as a continuous parameter over the range $-\infty < \phi < \infty$, we obtain the graph shown in Fig. 4.4. We observe that this function decays to zero as $\phi \to \pm\infty$, has a maximum value of unity at $\phi = 0$, and is zero at multiples of π (i.e., at $\phi = m\pi$, $m = \pm 1, \pm 2, \ldots$). It is clear that the Fourier coefficients given by (4.1.18) are the sample values of the $(\sin\phi)/\phi$ function for $\phi = \pi k F_0 \tau$ and scaled in amplitude by $A\tau/T_p$.

Since the periodic function $x(t)$ is even, the Fourier coefficients c_k are real. Consequently, the phase spectrum is either zero, when c_k is positive, or π when c_k is negative. Instead of plotting the magnitude and phase spectra separately, we may simply plot $\{c_k\}$ on a single graph, showing both the positive and negative values c_k on the graph. This is commonly done in practice when the Fourier coefficients $\{c_k\}$ are real.

Figure 4.5 illustrates the Fourier coefficients of the rectangular pulse train when T_p is fixed and the pulse width τ is allowed to vary. In this case $T_p = 0.25$ second, so that $F_0 = 1/T_p = 4$ Hz and $\tau = 0.05T_p$, $\tau = 0.1T_p$, and $\tau = 0.2T_p$. We observe that the effect of decreasing τ while keeping T_p fixed is to spread out the signal power over the frequency range. The spacing between adjacent spectral lines is $F_0 = 4$ Hz, independent of the value of the pulse width τ.

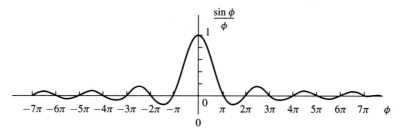

Figure 4.4 The function $(\sin\phi)/\phi$.

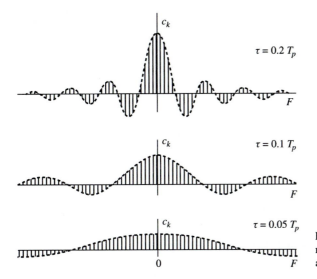

Figure 4.5   Fourier coefficients of the rectangular pulse train with T_p is fixed and the pulse width τ varies.

On the other hand, it is also instructive to fix τ and vary the period T_p when $T_p > \tau$. Figure 4.6 illustrates this condition when $T_p = 5\tau$, $T_p = 10\tau$, and $T_p = 20\tau$. In this case, the spacing between adjacent spectral lines decreases as T_p increases. In the limit as $T_p \to \infty$, the Fourier coefficients c_k approach zero due to the factor of T_p in the denominator of (4.1.18). This behavior is consistent with the fact that as $T_p \to \infty$ and τ remains fixed, the resulting signal is no longer a power signal. Instead,

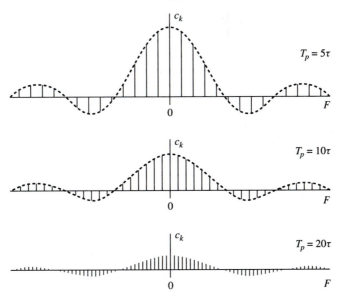

Figure 4.6   Fourier coefficient of a rectangular pulse train with fixed pulse width τ and varying period T_p.

it becomes an energy signal and its average power is zero. The spectra of finite energy signals are described in the next section.

We also note that if $k \neq 0$ and $\sin(\pi k F_0 \tau) = 0$, then $c_k = 0$. The harmonics with zero power occur at frequencies kF_0 such that $\pi(kF_0)\tau = m\pi$, $m = \pm1, \pm2, \dots$, or at $kF_0 = m/\tau$. For example, if $F_0 = 4$ Hz and $\tau = 0.2T_p$, it follows that the spectral components at ±20 Hz, ±40 $Hz, \dots$ have zero power. These frequencies correspond to the Fourier coefficients c_k, $k = \pm5, \pm10, \pm15, \dots$. On the other hand, if $\tau = 0.1T_p$, the spectral components with zero power are $k = \pm10, \pm20, \pm30, \dots$.

The power density spectrum for the rectangular pulse train is

$$|c_k|^2 = \begin{cases} \left(\dfrac{A\tau}{T_p}\right)^2, & k = 0 \\[2ex] \left(\dfrac{A\tau}{T_p}\right)^2 \left(\dfrac{\sin \pi k F_0 \tau}{\pi k F_0 \tau}\right)^2, & k = \pm1, \pm2, \dots \end{cases} \tag{4.1.19}$$

4.1.3 The Fourier Transform for Continuous-Time Aperiodic Signals

In Section 4.1.1 we developed the Fourier series to represent a periodic signal as a linear combination of harmonically related complex exponentials. As a consequence of the periodicity, we saw that these signals possess line spectra with equidistant lines. The line spacing is equal to the fundamental frequency, which in turn is the inverse of the fundamental period of the signal. We can view the fundamental period as providing the number of lines per unit of frequency (line density), as illustrated in Fig. 4.6.

With this interpretation in mind, it is apparent that if we allow the period to increase without limit, the line spacing tends toward zero. In the limit, when the period becomes infinite, the signal becomes aperiodic and its spectrum becomes continuous. This argument suggests that the spectrum of an aperiodic signal will be the envelope of the line spectrum in the corresponding periodic signal obtained by repeating the aperiodic signal with some period T_p.

Let us consider an aperiodic signal $x(t)$ with finite duration as shown in Fig. 4.7a. From this aperiodic signal, we can create a periodic signal $x_p(t)$ with period T_p, as shown in Fig. 4.7b. Clearly, $x_p(t) = x(t)$ in the limit as $T_p \to \infty$, that is,

$$x(t) = \lim_{T_p \to \infty} x_p(t)$$

This interpretation implies that we should be able to obtain the spectrum of $x(t)$ from the spectrum of $x_p(t)$ simply by taking the limit as $T_p \to \infty$.

We begin with the Fourier series representation of $x_p(t)$,

$$x_p(t) = \sum_{k=-\infty}^{\infty} c_k e^{j2\pi k F_0 t} \qquad F_0 = \frac{1}{T_p} \tag{4.1.20}$$

where

$$c_k = \frac{1}{T_p} \int_{-T_p/2}^{T_p/2} x_p(t) e^{-j2\pi k F_0 t} dt \tag{4.1.21}$$

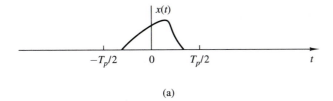

(a)

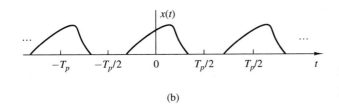

(b)

Figure 4.7 (a) Aperiodic signal $x(t)$ and (b) periodic signal $x_p(t)$ constructed by repeating $x(t)$ with a period T_p.

Since $x_p(t) = x(t)$ for $-T_p/2 \le t \le T_p/2$, (4.1.21) can be expressed as

$$c_k = \frac{1}{T_p} \int_{-T_p/2}^{T_p/2} x(t) e^{-j2\pi k F_0 t} dt \tag{4.1.22}$$

It is also true that $x(t) = 0$ for $|t| > T_p/2$. Consequently, the limits on the integral in (4.1.22) can be replaced by $-\infty$ and ∞. Hence

$$c_k = \frac{1}{T_p} \int_{-\infty}^{\infty} x(t) e^{-j2\pi k F_0 t} dt \tag{4.1.23}$$

Let us now define a function $X(F)$, called the *Fourier transform* of $x(t)$, as

$$X(F) = \int_{-\infty}^{\infty} x(t) e^{-j2\pi F t} dt \tag{4.1.24}$$

$X(F)$ is a function of the continuous variable F. It does not depend on T_p or F_0. However, if we compare (4.1.23) and (4.1.24), it is clear that the Fourier coefficients c_k can be expressed in terms of $X(F)$ as

$$c_k = \frac{1}{T_p} X(k F_0)$$

or equivalently,

$$T_p c_k = X(k F_0) = X\left(\frac{k}{T_p}\right) \tag{4.1.25}$$

Thus the Fourier coefficients are samples of $X(F)$ taken at multiples of F_0 and scaled by F_0 (multiplied by $1/T_p$). Substitution for c_k from (4.1.25) into (4.1.20) yields

$$x_p(t) = \frac{1}{T_p} \sum_{k=-\infty}^{\infty} X\left(\frac{k}{T_p}\right) e^{j2\pi k F_0 t} \tag{4.1.26}$$

We wish to take the limit of (4.1.26) as T_p approaches infinity. First, we define $\Delta F = 1/T_p$. With this substitution, (4.1.26) becomes

$$x_p(t) = \sum_{k=-\infty}^{\infty} X(k\Delta F)e^{j2\pi k\Delta F t}\Delta F \tag{4.1.27}$$

It is clear that in the limit as T_p approaches infinity, $x_p(t)$ reduces to $x(t)$. Also, ΔF becomes the differential dF and $k\Delta F$ becomes the continuous frequency variable F. In turn, the summation in (4.1.27) becomes an integral over the frequency variable F. Thus

$$\lim_{T_p \to \infty} x_p(t) = x(t) = \lim_{\Delta F \to 0} \sum_{k=-\infty}^{\infty} X(k\Delta F)e^{-j2\pi k\Delta F t}\Delta F$$
$$x(t) = \int_{-\infty}^{\infty} X(F)e^{j2\pi F t}dF \tag{4.1.28}$$

This integral relationship yields $x(t)$ when $X(F)$ is known, and it is called the *inverse Fourier transform*.

This concludes our heuristic derivation of the Fourier transform pair given by (4.1.24) and (4.1.28) for an aperiodic signal $x(t)$. Although the derivation is not mathematically rigorous, it led to the desired Fourier transform relationships with relatively simple intuitive arguments. In summary, the frequency analysis of continuous-time aperiodic signals involves the following Fourier transform pair.

FREQUENCY ANALYSIS OF CONTINUOUS-TIME APERIODIC SIGNALS

Synthesis equation inverse transform	$x(t) = \displaystyle\int_{-\infty}^{\infty} X(F)e^{j2\pi F t}dF$	(4.1.29)
Analysis equation direct transform	$X(F) = \displaystyle\int_{-\infty}^{\infty} x(t)e^{-j2\pi F t}dt$	(4.1.30)

It is apparent that the essential difference between the Fourier series and the Fourier transform is that the spectrum in the latter case is continuous and hence the synthesis of an aperiodic signal from its spectrum is accomplished by means of integration instead of summation.

Finally, we wish to indicate that the Fourier transform pair in (4.1.29) and (4.1.30) can be expressed in terms of the radian frequency variable $\Omega = 2\pi F$. Since $dF = d\Omega/2\pi$, (4.1.29) and (4.1.30) become

$$x(t) = \frac{1}{2\pi}\int_{-\infty}^{\infty} X(\Omega)e^{j\Omega t}d\Omega \tag{4.1.31}$$

$$X(\Omega) = \int_{-\infty}^{\infty} x(t)e^{-j\Omega t}dt \tag{4.1.32}$$

The set of conditions that guarantee the existence of the Fourier transform is the

Dirichlet conditions, which may be expressed as:

1. The signal $x(t)$ has a finite number of finite discontinuities.
2. The signal $x(t)$ has a finite number of maxima and minima.
3. The signal $x(t)$ is absolutely integrable, that is,

$$\int_{-\infty}^{\infty} |x(t)| dt < \infty \tag{4.1.33}$$

The third condition follows easily from the definition of the Fourier transform, given in (4.1.30). Indeed,

$$|X(F)| = \left| \int_{-\infty}^{\infty} x(t) e^{-j2\pi Ft} dt \right| \le \int_{-\infty}^{\infty} |x(t)| dt$$

Hence $|X(F)| < \infty$ if (4.1.33) is satisfied.

A weaker condition for the existence of the Fourier transform is that $x(t)$ has finite energy; that is,

$$\int_{-\infty}^{\infty} |x(t)|^2 dt < \infty \tag{4.1.34}$$

Note that if a signal $x(t)$ is absolutely integrable, it will also have finite energy. That is, if

$$\int_{-\infty}^{\infty} |x(t)| dt < \infty$$

then

$$E_x = \int_{-\infty}^{\infty} |x(t)|^2 dt < \infty \tag{4.1.35}$$

However, the converse is not true. That is, a signal may have finite energy but may not be absolutely integrable. For example, the signal

$$x(t) = \frac{\sin 2\pi t}{\pi t} \tag{4.1.36}$$

is square integrable but is not absolutely integrable. This signal has the Fourier transform

$$X(F) = \begin{cases} 1, & |F| \le 1 \\ 0, & |F| > 1 \end{cases} \tag{4.1.37}$$

Since this signal violates (4.1.33), it is apparent that the Dirichlet conditions are sufficient but not necessary for the existence of the Fourier transform. In any case, nearly all finite energy signals have a Fourier transform, so that we need not worry about the pathological signals, which are seldom encountered in practice.

4.1.4 Energy Density Spectrum of Aperiodic Signals

Let $x(t)$ be any finite energy signal with Fourier transform $X(F)$. Its energy is

$$E_x = \int_{-\infty}^{\infty} |x(t)|^2 dt$$

which, in turn, may be expressed in terms of $X(F)$ as follows:

$$E_x = \int_{-\infty}^{\infty} x(t)x^*(t)dt$$

$$= \int_{-\infty}^{\infty} x(t)dt \left[\int_{-\infty}^{\infty} X^*(F)e^{-j2\pi Ft}dF \right]$$

$$= \int_{-\infty}^{\infty} X^*(F)dF \left[\int_{-\infty}^{\infty} x(t)e^{-j2\pi Ft}dt \right]$$

$$= \int_{-\infty}^{\infty} |X(F)|^2 dF$$

Therefore, we conclude that

$$E_x = \int_{-\infty}^{\infty} |x(t)|^2 dt = \int_{-\infty}^{\infty} |X(F)|^2 dF \qquad (4.1.38)$$

This is *Parseval's relation* for aperiodic, finite energy signals and expresses the principle of conservation of energy in the time and frequency domains.

The spectrum $X(F)$ of a signal is in general, complex valued. Consequently, it is usually expressed in polar forms as

$$X(F) = |X(F)|e^{j\Theta(F)}$$

where $|X(F)|$ is the magnitude spectrum and $\Theta(F)$ is the phase spectrum,

$$\Theta(F) = \measuredangle X(F)$$

On the other hand, the quantity

$$S_{xx}(F) = |X(F)|^2 \qquad (4.1.39)$$

which is the integrand in (4.1.38), represents the distribution of energy in the signal as a function of frequency. Hence $S_{xx}(F)$ is called the *energy density spectrum* of $x(t)$. The integral of $S_{xx}(F)$ over all frequencies gives the total energy in the signal. Viewed in another way, the energy in the signal $x(t)$ over a band of frequencies $F_1 \leq F \leq F_1 + \Delta F$ is

$$\int_{F_1}^{F_1 + \Delta F} S_{xx}(F)dF$$

From (4.1.39) we observe that $S_{xx}(F)$ does not contain any phase information [i.e., $S_{xx}(F)$ is purely real and nonnegative]. Since the phase spectrum of $x(t)$ is not contained in $S_{xx}(F)$, it is impossible to reconstruct the signal given $S_{xx}(F)$.

Finally, as in the case of Fourier series, it is easily shown that if the signal $x(t)$ is real, then

$$|X(-F)| = |X(F)| \qquad (4.1.40)$$

$$\measuredangle X(-F) = -\measuredangle X(F) \qquad (4.1.41)$$

By combining (4.1.40) and (4.1.39), we obtain

$$S_{xx}(-F) = S_{xx}(F) \tag{4.1.42}$$

In other words, the energy density spectrum of a real signal has even symmetry.

Example 4.1.2

Determine the Fourier transform and the energy density spectrum of a rectangular pulse signal defined as

$$x(t) = \begin{cases} A, & |t| \le \tau/2 \\ 0, & |t| > \tau/2 \end{cases} \tag{4.1.43}$$

and illustrated in Fig. 4.8(a).

Solution   Clearly, this signal is aperiodic and satisfies the Dirichlet conditions. Hence its Fourier transform exists. By applying (4.1.30), we find that

$$X(F) = \int_{-\tau/2}^{\tau/2} A e^{-j2\pi Ft} dt = A\tau \frac{\sin \pi F\tau}{\pi F\tau} \tag{4.1.44}$$

We observe that $X(F)$ is real and hence it can be depicted graphically using only one diagram, as shown in Fig. 4.8(b). Obviously, $X(F)$ has the shape of the $(\sin\phi)/\phi$ function shown in Fig. 4.4. Hence the spectrum of the rectangular pulse is the envelope of the line spectrum (Fourier coefficients) of the periodic signal obtained by

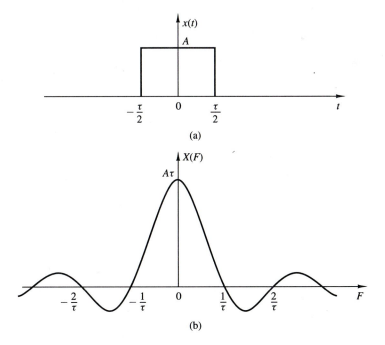

Figure 4.8   (a) Rectangular pulse and (b) its Fourier transform.

periodically repeating the pulse with period T_p as in Fig. 4.3. In other words, the Fourier coefficients c_k in the corresponding periodic signal $x_p(t)$ are simply samples of $X(F)$ at frequencies $kF_0 = k/T_p$. Specifically,

$$c_k = \frac{1}{T_p} X(kF_0) = \frac{1}{T_p} X\left(\frac{k}{T_p}\right) \tag{4.1.45}$$

From (4.1.44) we note that the zero crossings of $X(F)$ occur at multiples of $1/\tau$. Furthermore, the width of the main lobe, which contains most of the signal energy, is equal to $2/\tau$. As the pulse duration τ decreases (increases), the main lobe becomes broader (narrower) and more energy is moved to the higher (lower) frequencies, as illustrated in Fig. 4.9. Thus as the signal pulse is expanded (compressed) in time, its transform is compressed (expanded) in frequency. This behavior, between the time function and its spectrum, is a type of uncertainty principle that appears in different forms in various branches of science and engineering.

Finally, the energy density spectrum of the rectangular pulse is

$$S_{xx}(F) = (A\tau)^2 \left(\frac{\sin \pi F \tau}{\pi F \tau}\right)^2 \tag{4.1.46}$$

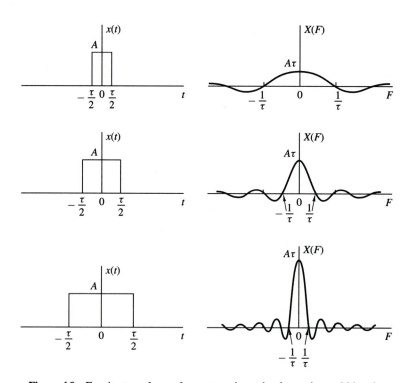

Figure 4.9   Fourier transform of a rectangular pulse for various width values.

4.2 FREQUENCY ANALYSIS OF DISCRETE-TIME SIGNALS

In Section 4.1 we developed the Fourier series representation for continuous-time periodic (power) signals and the Fourier transform for finite energy aperiodic signals. In this section we repeat the development for the class of discrete-time signals.

As we have observed from the discussion of Section 4.1, the Fourier series representation of a continuous-time periodic signal can consist of an infinite number of frequency components, where the frequency spacing between two successive harmonically related frequencies is $1/T_p$, and where T_p is the fundamental period. Since the frequency range for continuous-time signals extends from $-\infty$ to ∞, it is possible to have signals that contain an infinite number of frequency components. In contrast, the frequency range for discrete-time signals is unique over the interval $(-\pi, \pi)$ or $(0, 2\pi)$. A discrete-time signal of fundamental period N can consist of frequency components separated by $2\pi/N$ radians or $f = 1/N$ cycles. Consequently, the Fourier series representation of the discrete-time periodic signal will contain at most N frequency components. This is the basic difference between the Fourier series representations for continuous-time and discrete-time periodic signals.

4.2.1 The Fourier Series for Discrete-Time Periodic Signals

Suppose that we are given a periodic sequence $x(n)$ with period N, that is, $x(n) = x(n + N)$ for all n. The Fourier series representation for $x(n)$ consists of N harmonically related exponential functions

$$e^{j2\pi kn/N} \qquad k = 0, 1, \ldots, N - 1$$

and is expressed as

$$x(n) = \sum_{k=0}^{N-1} c_k e^{j2\pi kn/N} \tag{4.2.1}$$

where the $\{c_k\}$ are the coefficients in the series representation.

To derive the expression for the Fourier coefficients, we use the following formula:

$$\sum_{n=0}^{N-1} e^{j2\pi kn/N} = \begin{cases} N, & k = 0, \pm N, \pm 2N, \ldots \\ 0, & \text{otherwise} \end{cases} \tag{4.2.2}$$

Note the similarity of (4.2.2) with the continuous-time counterpart in (4.1.3). The proof of (4.2.2) follows immediately from the application of the geometric summation formula

$$\sum_{n=0}^{N-1} a^n = \begin{cases} N, & a = 1 \\ \dfrac{1 - a^N}{1 - a}, & a \neq 1 \end{cases} \tag{4.2.3}$$

The expression for the Fourier coefficients c_k can be obtained by multiplying both sides of (4.2.1) by the exponential $e^{-j2\pi ln/N}$ and summing the product from $n = 0$ to $n = N - 1$. Thus

$$\sum_{n=0}^{N-1} x(n)e^{-j2\pi ln/N} = \sum_{n=0}^{N-1}\sum_{k=0}^{N-1} c_k e^{j2\pi(k-l)n/N} \qquad (4.2.4)$$

If we perform the summation over n first, in the right-hand side of (4.2.4), we obtain

$$\sum_{n=0}^{N-1} e^{j2\pi(k-l)n/N} = \begin{cases} N, & k - l = 0, \pm N, \pm 2N, \ldots \\ 0, & \text{otherwise} \end{cases} \qquad (4.2.5)$$

where we have made use of (4.2.2). Therefore, the right-hand side of (4.2.4) reduces to Nc_l and hence

$$c_l = \frac{1}{N}\sum_{n=0}^{N-1} x(n)e^{-j2\pi ln/N} \qquad l = 0, 1, \ldots, N - 1 \qquad (4.2.6)$$

Thus we have the desired expression for the Fourier coefficients in terms of the signal $x(n)$.

The relationships (4.2.1) and (4.2.6) for the frequency analysis of discrete-time signals are summarized below.

FREQUENCY ANALYSIS OF DISCRETE-TIME PERIODIC SIGNALS

Synthesis equation	$x(n) = \displaystyle\sum_{k=0}^{N-1} c_k e^{j2\pi kn/N}$	(4.2.7)
Analysis equation	$c_k = \dfrac{1}{N}\displaystyle\sum_{n=0}^{N-1} x(n)e^{-j2\pi kn/N}$	(4.2.8)

Equation (4.2.7) is often called the *discrete-time Fourier series* (DTFS). The Fourier coefficients $\{c_k\}$, $k = 0, 1, \ldots, N - 1$ provide the description of $x(n)$ in the frequency domain, in the sense that c_k represents the amplitude and phase associated with the frequency component

$$s_k(n) = e^{j2\pi kn/N} = e^{j\omega_k n}$$

where $\omega_k = 2\pi k/N$.

We recall from Section 1.3.3 that the functions $s_k(n)$ are periodic with period N. Hence $s_k(n) = s_k(n + N)$. In view of this periodicity, it follows that the Fourier coefficients c_k, when viewed beyond the range $k = 0, 1, \ldots, N - 1$, also satisfy a periodicity condition. Indeed, from (4.2.8), which holds for every value of k, we have

$$c_{k+N} = \frac{1}{N}\sum_{n=0}^{N-1} x(n)e^{-j2\pi(k+N)n/N} = \frac{1}{N}\sum_{n=0}^{N-1} x(n)e^{-j2\pi kn/N} = c_k \qquad (4.2.9)$$

Therefore, the Fourier series coefficients $\{c_k\}$ form a periodic sequence when extended outside of the range $k = 0, 1, \ldots, N - 1$. Hence

$$c_{k+N} = c_k$$

that is, $\{c_k\}$ is a periodic sequence with fundamental period N. *Thus the spectrum of a signal $x(n)$, which is periodic with period N, is a periodic sequence with period N.* Consequently, any N consecutive samples of the signal or its spectrum provide a complete description of the signal in the time or frequency domains.

Although the Fourier coefficients form a periodic sequence, we will focus our attention on the single period with range $k = 0, 1, \ldots, N - 1$. This is convenient, since in the frequency domain, this amounts to covering the fundamental range $0 \leq \omega_k = 2\pi k/N < 2\pi$, for $0 \leq k \leq N - 1$. In contrast, the frequency range $-\pi < \omega_k = 2\pi k/N \leq \pi$, corresponds to $-N/2 < k \leq N/2$, which creates an inconvenience when N is odd. Clearly, if we use a sampling frequency F_s, the range $0 \leq k \leq N - 1$ corresponds to the frequency range $0 \leq F < F_s$.

Example 4.2.1

Determine the spectra of the signals

(a) $x(n) = \cos \sqrt{2}\pi n$
(b) $x(n) = \cos \pi n/3$
(c) $x(n)$ is periodic with period $N = 4$ and

$$x(n) = \{1, 1, 0, 0\}$$
$$\uparrow$$

Solution

(a) For $\omega_0 = \sqrt{2}\pi$, we have $f_0 = 1/\sqrt{2}$. Since f_0 is not a rational number, the signal is not periodic. Consequently, this signal cannot be expanded in a Fourier series. Nevertheless, the signal does possess a spectrum. Its spectral content consists of the single frequency component at $\omega = \omega_0 = \sqrt{2}\pi$.

(b) In this case $f_0 = \frac{1}{6}$ and hence $x(n)$ is periodic with fundamental period $N = 6$. From (4.2.8) we have

$$c_k = \frac{1}{6}\sum_{n=0}^{5} x(n)e^{-j2\pi kn/6} \qquad k = 0, 1, \ldots, 5$$

However, $x(n)$ can be expressed as

$$x(n) = \cos \frac{2\pi n}{6} = \frac{1}{2}e^{j2\pi n/6} + \frac{1}{2}e^{-j2\pi n/6}$$

which is already in the form of the exponential Fourier series in (4.2.7). In comparing the two exponential terms in $x(n)$ with (4.2.7), it is apparent that $c_1 = \frac{1}{2}$. The second exponential in $x(n)$ corresponds to the term $k = -1$ in (4.2.7). However, this term can also be written as

$$e^{-j2\pi n/6} = e^{j2\pi(5-6)n/6} = e^{j2\pi(5n)/6}$$

which means that $c_{-1} = c_5$. But this is consistent with (4.2.9), and our previous observation that the Fourier series coefficients form a periodic sequence of period N. Consequently, we conclude that

$$c_0 = c_2 = c_3 = c_4 = 0$$

$$c_1 = \tfrac{1}{2} \qquad c_5 = \tfrac{1}{2}$$

(c) From (4.2.8), we have

$$c_k = \frac{1}{4} \sum_{n=0}^{3} x(n) e^{-j2\pi kn/4} \qquad k = 0, 1, 2, 3$$

or

$$c_k = \tfrac{1}{4}(1 + e^{-j\pi k/2}) \qquad k = 0, 1, 2, 3$$

For $k = 0, 1, 2, 3$ we obtain

$$c_0 = \tfrac{1}{2} \qquad c_1 = \tfrac{1}{4}(1 - j) \qquad c_2 = 0 \qquad c_3 = \tfrac{1}{4}(1 + j)$$

The magnitude and phase spectra are

$$|c_0| = \frac{1}{2} \qquad |c_1| = \frac{\sqrt{2}}{4} \qquad |c_2| = 0 \qquad |c_3| = \frac{\sqrt{2}}{4}$$

$$\angle c_0 = 0 \qquad \angle c_1 = -\frac{\pi}{4} \qquad \angle c_2 = \text{undefined} \qquad \angle c_3 = \frac{\pi}{4}$$

Figure 4.10 illustrates the spectral content of the signals in (b) and (c).

4.2.2 Power Density Spectrum of Periodic Signals

The average power of a discrete-time periodic signal with period N was defined in (2.1.23) as

$$P_x = \frac{1}{N} \sum_{n=0}^{N-1} |x(n)|^2 \qquad (4.2.10)$$

We shall now derive an expression for P_x in terms of the Fourier coefficient $\{c_k\}$.
If we use the relation (4.2.7) in (4.2.10), we have

$$P_x = \frac{1}{N} \sum_{n=0}^{N-1} x(n) x^*(n)$$

$$= \frac{1}{N} \sum_{n=0}^{N-1} x(n) \left(\sum_{k=0}^{N-1} c_k^* e^{-j2\pi kn/N} \right)$$

Now, we can interchange the order of the two summations and make use of (4.2.8), obtaining

$$P_x = \sum_{k=0}^{N-1} c_k^* \left[\frac{1}{N} \sum_{n=0}^{N-1} x(n) e^{-j2\pi kn/N} \right]$$

$$= \sum_{k=0}^{N-1} |c_k|^2 = \frac{1}{N} \sum_{n=0}^{N-1} |x(n)|^2$$

$$(4.2.11)$$

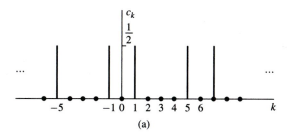

(a)

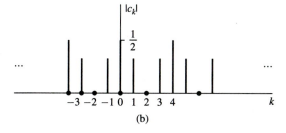

(b)

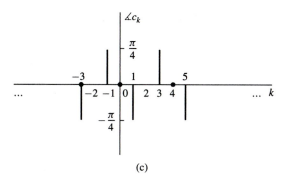

(c)

Figure 4.10 Spectra of the periodic signals discussed in Example 4.2.1 (b) and (c).

which is the desired expression for the average power in the periodic signal. In other words, the average power in the signal is the sum of the powers of the individual frequency components. We view (4.2.11) as a Parseval's relation for discrete-time periodic signals. The sequence $|c_k|^2$ for $k = 0, 1, \ldots, N - 1$ is the distribution of power as a function of frequency and is called the *power density spectrum* of the periodic signal.

If we are interested in the energy of the sequence $x(n)$ over a single period, (4.2.11) implies that

$$E_N = \sum_{n=0}^{N-1} |x(n)|^2 = N \sum_{k=0}^{N-1} |c_k|^2 \tag{4.2.12}$$

which is consistent with our previous results for continuous-time periodic signals.

If the signal $x(n)$ is real [i.e., $x^*(n) = x(n)$], then, proceeding as in Section 4.2.1, we can easily show that

$$c_k^* = c_{-k} \tag{4.2.13}$$

or equivalently,

$$|c_{-k}| = |c_k| \qquad \text{(even symmetry)} \qquad (4.2.14)$$

$$-\measuredangle c_{-k} = \measuredangle c_k \qquad \text{(odd symmetry)} \qquad (4.2.15)$$

These symmetry properties for the magnitude and phase spectra of a periodic signal, in conjunction with the periodicity property, have very important implications on the frequency range of discrete-time signals.

Indeed, by combining (4.2.9) with (4.2.14) and (4.2.15), we obtain

$$|c_k| = |c_{N-k}| \qquad (4.2.16)$$

and

$$\measuredangle c_k = -\measuredangle c_{N-k} \qquad (4.2.17)$$

More specifically, we have

$$
\begin{aligned}
|c_0| &= |c_N|, & \measuredangle c_0 &= -\measuredangle c_N = 0 \\
|c_1| &= |c_{N-1}|, & \measuredangle c_1 &= -\measuredangle c_{N-1} \\
|c_{N/2}| &= |c_{N/2}|, & \measuredangle c_{N/2} &= 0 & \text{if } N \text{ is even} \\
|c_{(N-1)/2}| &= |c_{(N+1)/2}|, & \measuredangle c_{(N-1)/2} &= -\measuredangle c_{(N+1)/2} & \text{if } N \text{ is odd}
\end{aligned}
\qquad (4.2.18)
$$

Thus, for a real signal, the spectrum c_k, $k = 0, 1, \ldots, N/2$ for N even, or $k = 0, 1, \ldots, (N-1)/2$ for N odd, completely specifies the signal in the frequency domain. Clearly, this is consistent with the fact that the highest relative frequency that can be represented by a discrete-time signal is equal to π. Indeed, if $0 \le \omega_k = 2\pi k/N \le \pi$, then $0 \le k \le N/2$.

By making use of these symmetry properties of the Fourier series coefficients of a real signal, the Fourier series in (4.2.7) can also be expressed in the alternative forms

$$x(n) = c_0 + 2 \sum_{k=1}^{L} |c_k| \cos\left(\frac{2\pi}{N} kn + \theta_k\right) \qquad (4.2.19)$$

$$= a_0 + \sum_{k=1}^{L} \left(a_k \cos\frac{2\pi}{N} kn - b_k \sin\frac{2\pi}{N} kn\right) \qquad (4.2.20)$$

where $a_0 = c_0$, $a_k = 2|c_k| \cos\theta_k$, $b_k = 2|c_k| \sin\theta_k$, and $L = N/2$ if N is even and $L = (N-1)/2$ if N is odd.

Finally, we note that as in the case of continuous-time signals, the power density spectrum $|c_k|^2$ does not contain any phase information. Furthermore, the spectrum is discrete and periodic with a fundamental period equal to that of the signal itself.

Example 4.2.2   Periodic "Square-Wave" Signal

Determine the Fourier series coefficients and the power density spectrum of the periodic signal shown in Fig. 4.11.

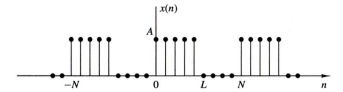

Figure 4.11 Discrete-time periodic square-wave signal.

Solution By applying the analysis equation (4.2.8) to the signal shown in Fig. 4.11, we obtain

$$c_k = \frac{1}{N} \sum_{n=0}^{N-1} x(n) e^{-j2\pi kn/N} = \frac{1}{N} \sum_{n=0}^{L-1} A e^{-j2\pi kn/N} \qquad k = 0, 1, \ldots, N-1$$

which is a geometric summation. Now we can use (4.2.3) to simplify the summation above. Thus we obtain

$$c_k = \frac{A}{N} \sum_{n=0}^{L-1} (e^{-j2\pi k/N})^n = \begin{cases} \dfrac{AL}{N}, & k = 0 \\[2ex] \dfrac{A}{N} \dfrac{1 - e^{-j2\pi kL/N}}{1 - e^{-j2\pi k/N}}, & k = 1, 2, \ldots, N-1 \end{cases}$$

The last expression can be simplified further if we note that

$$\frac{1 - e^{-j2\pi kL/N}}{1 - e^{-j2\pi k/N}} = \frac{e^{-j\pi kL/N}}{e^{-j\pi k/N}} \frac{e^{j\pi kL/N} - e^{-j\pi kL/N}}{e^{j\pi k/N} - e^{-j\pi k/N}}$$

$$= e^{-j\pi k(L-1)/N} \frac{\sin(\pi kL/N)}{\sin(\pi k/N)}$$

Therefore,

$$c_k = \begin{cases} \dfrac{AL}{N}, & k = 0, +N, \pm 2N, \ldots \\[2ex] \dfrac{A}{N} e^{-j\pi k(L-1)/N} \dfrac{\sin(\pi kL/N)}{\sin(\pi k/N)}, & \text{otherwise} \end{cases} \tag{4.2.21}$$

The power density spectrum of this periodic signal is

$$|c_k|^2 = \begin{cases} \left(\dfrac{AL}{N}\right)^2, & k = 0, +N, \pm 2N, \ldots \\[2ex] \left(\dfrac{A}{N}\right)^2 \left(\dfrac{\sin \pi kL/N}{\sin \pi k/N}\right)^2, & \text{otherwise} \end{cases} \tag{4.2.22}$$

Figure 4.12 illustrates the plots of $|c_k|^2$ for $L = 5$ and 7, $N = 40$ and 60, and $A = 1$.

4.2.3 The Fourier Transform of Discrete-Time Aperiodic Signals

Just as in the case of continuous-time aperiodic energy signals, the frequency analysis of discrete-time aperiodic finite-energy signals involves a Fourier transform of the time-domain signal. Consequently, the development in this section parallels to a large extent, that given in Section 4.1.3.

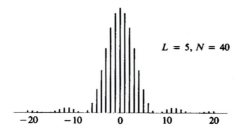

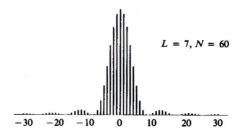

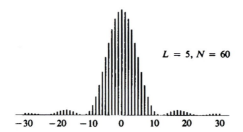

Figure 4.12 Plot of the power density spectrum given by (4.2.22).

The Fourier transform of a finite-energy discrete-time signal $x(n)$ is defined as

$$X(\omega) = \sum_{n=-\infty}^{\infty} x(n)e^{-j\omega n} \tag{4.2.23}$$

Physically, $X(\omega)$ represents the frequency content of the signal $x(n)$. In other words, $X(\omega)$ is a decomposition of $x(n)$ into its frequency components.

We observe two basic differences between the Fourier transform of a discrete-time finite-energy signal and the Fourier transform of a finite-energy analog signal. First, for continuous-time signals, the Fourier transform, and hence the spectrum of the signal, have a frequency range of $(-\infty, \infty)$. In contrast, the frequency range for a discrete-time signal is unique over the frequency interval of $(-\pi, \pi)$ or, equivalently, $(0, 2\pi)$. This property is reflected in the Fourier transform of the signal. Indeed, $X(\omega)$ is periodic with period 2π, that is,

$$
\begin{aligned}
X(\omega + 2\pi k) &= \sum_{n=-\infty}^{\infty} x(n)e^{-j(\omega + 2\pi k)n} \\
&= \sum_{n=-\infty}^{\infty} x(n)e^{-j\omega n}e^{-j2\pi kn} \qquad (4.2.24) \\
&= \sum_{n=-\infty}^{\infty} x(n)e^{-j\omega n} = X(\omega)
\end{aligned}
$$

Hence $X(\omega)$ is periodic with period 2π. But this property is just a consequence of the fact that the frequency range for any discrete-time signal is limited to $(-\pi, \pi)$ or $(0, 2\pi)$, and any frequency outside this interval is equivalent to a frequency within the interval.

The second basic difference is also a consequence of the discrete-time nature of the signal. Since the signal is discrete in time, the Fourier transform of the signal involves a summation of terms instead of an integral, as in the case of continuous-time signals.

Since $X(\omega)$ is a periodic function of the frequency variable ω, it has a Fourier series expansion, provided that the conditions for the existence of the Fourier series, described previously, are satisfied. In fact, from the definition of the Fourier transform $X(\omega)$ of the sequence $x(n)$, given by (4.2.23), we observe that $X(\omega)$ has the form of a Fourier series. The Fourier coefficients in this series expansion are the values of the sequence $x(n)$.

To demonstrate this point, let us evaluate the sequence $x(n)$ from $X(\omega)$. First, we multiply both sides (4.2.23) by $e^{j\omega m}$ and integrate over the interval $(-\pi, \pi)$. Thus we have

$$
\int_{-\pi}^{\pi} X(\omega)e^{j\omega m}\,d\omega = \int_{-\pi}^{\pi} \left[\sum_{n=-\infty}^{\infty} x(n)e^{-j\omega n} \right] e^{j\omega m}\,d\omega \qquad (4.2.25)
$$

The integral on the right-hand side of (4.2.25) can be evaluated if we can interchange the order of summation and integration. This interchange can be made if the series

$$
X_N(\omega) = \sum_{n=-N}^{N} x(n)e^{-j\omega n}
$$

converges uniformly to $X(\omega)$ as $N \to \infty$. Uniform convergence means that, for every ω, $X_N(\omega) \to X(\omega)$, as $N \to \infty$. The convergence of the Fourier transform is discussed in more detail in the following section. For the moment, let us assume that the series converges uniformly, so that we can interchange the order of summation and integration in (4.2.25). Then

$$
\int_{-\pi}^{\pi} e^{j\omega(m-n)}\,d\omega = \begin{cases} 2\pi, & m = n \\ 0, & m \neq n \end{cases}
$$

Consequently,

$$\sum_{n=-\infty}^{\infty} x(n) \int_{-\pi}^{\pi} e^{j\omega(m-n)} d\omega = \begin{cases} 2\pi x(m), & m = n \\ 0, & m \neq n \end{cases} \tag{4.2.26}$$

By combining (4.2.25) and (4.2.26), we obtain the desired result that

$$x(n) = \frac{1}{2\pi} \int_{-\pi}^{\pi} X(\omega) e^{j\omega n} d\omega \tag{4.2.27}$$

If we compare the integral in (4.2.27) with (4.1.9), we note that this is just the expression for the Fourier series coefficient for a function that is periodic with period 2π. The only difference between (4.1.9) and (4.2.27) is the sign on the exponent in the integrand, which is a consequence of our definition of the Fourier transform as given by (4.2.23). Therefore, the Fourier transform of the sequence $x(n)$, defined by (4.2.23), has the form of a Fourier series expansion.

In summary, the *Fourier transform pair for discrete-time signals* is as follows.

FREQUENCY ANALYSIS OF DISCRETE-TIME APERIODIC SIGNALS

Synthesis equation inverse transform	$x(n) = \dfrac{1}{2\pi} \displaystyle\int_{2\pi} X(\omega) e^{j\omega n} d\omega$ (4.2.28)
Analysis equation direct transform	$X(\omega) = \displaystyle\sum_{n=-\infty}^{\infty} x(n) e^{-j\omega n}$ (4.2.29)

4.2.4 Convergence of the Fourier Transform

In the derivation of the inverse transform given by (4.2.28), we assumed that the series

$$X_N(\omega) = \sum_{n=-N}^{N} x(n) e^{-j\omega n} \tag{4.2.30}$$

converges uniformly to $X(\omega)$, given in the integral of (4.2.28), as $N \to \infty$. By uniform convergence we mean that for each ω,

$$\lim_{N \to \infty} \left\{ \sup_{\omega} |X(\omega) - X_N(\omega)| \right\} = 0 \tag{4.2.31}$$

Uniform convergence is guaranteed if $x(n)$ is absolutely summable. Indeed, if

$$\sum_{n=-\infty}^{\infty} |x(n)| < \infty \tag{4.2.32}$$

then

$$|X(\omega)| = \left| \sum_{n=-\infty}^{\infty} x(n) e^{-j\omega n} \right| \leq \sum_{n=-\infty}^{\infty} |x(n)| < \infty$$

Hence (4.2.32) is a sufficient condition for the existence of the discrete-time Fourier transform. We note that this is the discrete-time counterpart of the third Dirichlet condition for the Fourier transform of continuous-time signals. The first two conditions do not apply due to the discrete-time nature of $\{x(n)\}$.

Some sequences are not absolutely summable, but they are square summable. That is, they have finite energy

$$E_x = \sum_{n=-\infty}^{\infty} |x(n)|^2 < \infty \tag{4.2.33}$$

which is a weaker condition than (4.2.32). We would like to define the Fourier transform of finite-energy sequences, but we must relax the condition of uniform convergence. For such sequences we can impose a mean-square convergence condition:

$$\lim_{N \to \infty} \int_{-\pi}^{\pi} |X(\omega) - X_N(\omega)|^2 d\omega = 0 \tag{4.2.34}$$

Thus the energy in the error $X(\omega) - X_N(\omega)$ tends toward zero, but the error $|X(\omega) - X_N(\omega)|$ does not necessarily tend to zero. In this way we can include finite-energy signals in the class of signals for which the Fourier transform exists.

Let us consider an example from the class of finite-energy signals. Suppose that

$$X(\omega) = \begin{cases} 1, & |\omega| \le \omega_c \\ 0, & \omega_c < |\omega| \le \pi \end{cases} \tag{4.2.35}$$

The reader should remember that $X(\omega)$ is periodic with period 2π. Hence (4.2.35) represents only one period of $X(\omega)$. The inverse transform of $X(\omega)$ results in the sequence

$$x(n) = \frac{1}{2\pi} \int_{-\pi}^{\pi} X(\omega)e^{j\omega n} d\omega$$

$$= \frac{1}{2\pi} \int_{-\omega_c}^{\omega_c} e^{j\omega n} d\omega = \frac{\sin \omega_c n}{\pi n} \qquad n \ne 0$$

For $n = 0$, we have

$$x(0) = \frac{1}{2\pi} \int_{-\omega_c}^{\omega_c} d\omega = \frac{\omega_c}{\pi}$$

Hence

$$x(n) = \begin{cases} \dfrac{\omega_c}{\pi}, & n = 0 \\ \dfrac{\omega_c}{\pi} \dfrac{\sin \omega_c n}{\omega_c n}, & n \ne 0 \end{cases} \tag{4.2.36}$$

This transform pair is illustrated in Fig. 4.13.

Sometimes, the sequence $\{x(n)\}$ in (4.2.36) is expressed as

$$x(n) = \frac{\sin \omega_c n}{\pi n} \qquad -\infty < n < \infty \tag{4.2.37}$$

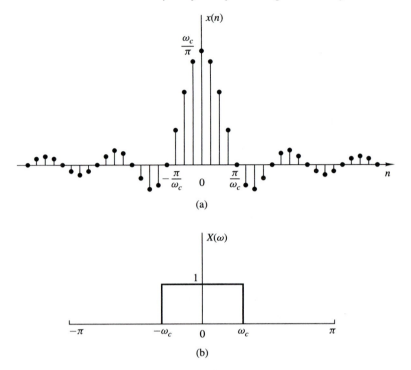

Figure 4.13   Fourier transform pair in (4.2.35) and (4.2.36).

with the understanding that at $n = 0$, $x(n) = \omega_c/\pi$. We should emphasize, however, that $(\sin \omega_c n)/\pi n$ is not a continuous function, and hence L'Hospital's rule cannot be used to determine $x(0)$.

Now let us consider the determination of the Fourier transform of the sequence given by (4.2.37). The sequence $\{x(n)\}$ is not absolutely summable. Hence the infinite series

$$\sum_{n=-\infty}^{\infty} x(n)e^{-j\omega n} = \sum_{n=-\infty}^{\infty} \frac{\sin \omega_c n}{\pi n} e^{-j\omega n} \qquad (4.2.38)$$

does not converge uniformly for all ω. However, the sequence $\{x(n)\}$ has a finite energy $E_x = \omega_c/\pi$ as will be shown in Section 4.3. Hence the sum in (4.2.38) is guaranteed to converge to the $X(\omega)$ given by (4.2.35) in the mean-square sense.

To elaborate on this point, let us consider the finite sum

$$X_N(\omega) = \sum_{n=-N}^{N} \frac{\sin \omega_c n}{\pi n} e^{-j\omega n} \qquad (4.2.39)$$

Figure 4.14 shows the function $X_N(\omega)$ for several values of N. We note that there is a significant oscillatory overshoot at $\omega = \omega_c$, independent of the value of N. As

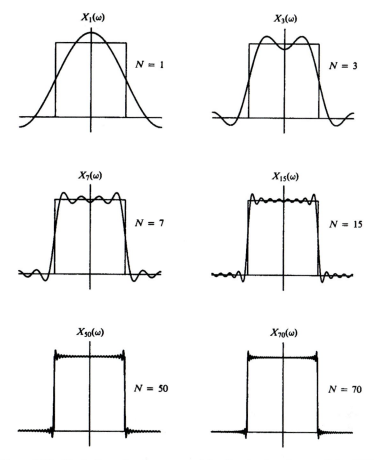

Figure 4.14 Illustration of convergence of the Fourier transform and the Gibbs phenomenon at the point of discontinuity.

N increases, the oscillations become more rapid, but the size of the ripple remains the same. One can show that as $N \rightarrow \infty$, the oscillations converge to the point of the discontinuity at $\omega = \omega_c$, but their amplitude does not go to zero. However, (4.2.34) is satisfied, and therefore $X_N(\omega)$ converges to $X(\omega)$ in the mean-square sense.

The oscillatory behavior of the approximation $X_N(\omega)$ to the function $X(\omega)$ at a point of discontinuity of $X(\omega)$ is called the *Gibbs phenomenon*. A similar effect is observed in the truncation of the Fourier series of a continuous-time periodic signal, given by the synthesis equation (4.1.8). For example, the truncation of the Fourier series for the periodic square-wave signal in Example 4.1.1, gives rise to the same oscillatory behavior in the finite-sum approximation of $x(t)$. The Gibbs phenomenon will be encountered again in the design of practical, discrete-time FIR systems considered in Chapter 8.

4.2.5 Energy Density Spectrum of Aperiodic Signals

Recall that the energy of a discrete-time signal $x(n)$ is defined as

$$E_x = \sum_{n=-\infty}^{\infty} |x(n)|^2 \tag{4.2.40}$$

Let us now express the energy E_x in terms of the spectral characteristic $X(\omega)$. First we have

$$E_x = \sum_{n=-\infty}^{\infty} x(n)x^*(n) = \sum_{n=-\infty}^{\infty} x(n) \left[\frac{1}{2\pi} \int_{-\pi}^{\pi} X^*(\omega)e^{-j\omega n}d\omega \right]$$

If we interchange the order of integration and summation in the equation above, we obtain

$$E_x = \frac{1}{2\pi} \int_{-\pi}^{\pi} X^*(\omega) \left[\sum_{n=-\infty}^{\infty} x(n)e^{-j\omega n} \right] d\omega$$

$$= \frac{1}{2\pi} \int_{-\pi}^{\pi} |X(\omega)|^2 d\omega$$

Therefore, the energy relation between $x(n)$ and $X(\omega)$ is

$$E_x = \sum_{n=-\infty}^{\infty} |x(n)|^2 = \frac{1}{2\pi} \int_{-\pi}^{\pi} |X(\omega)|^2 d\omega \tag{4.2.41}$$

This is Parseval's relation for discrete-time aperiodic signals with finite energy.

The spectrum $X(\omega)$ is, in general, a complex-valued function of frequency. It may be expressed as

$$X(\omega) = |X(\omega)|e^{j\Theta(\omega)} \tag{4.2.42}$$

where

$$\Theta(\omega) = \measuredangle X(\omega)$$

is the phase spectrum and $|X(\omega)|$ is the magnitude spectrum.

As in the case of continuous-time signals, the quantity

$$S_{xx}(\omega) = |X(\omega)|^2 \tag{4.2.43}$$

represents the distribution of energy as a function of frequency, and it is called the *energy density spectrum* of $x(n)$. Clearly, $S_{xx}(\omega)$ does not contain any phase information.

Suppose now that the signal $x(n)$ is real. Then it easily follows that

$$X^*(\omega) = X(-\omega) \tag{4.2.44}$$

or equivalently,

$$|X(-\omega)| = |X(\omega)| \qquad \text{(even symmetry)} \tag{4.2.45}$$

and

$$\angle X(-\omega) = -\angle X(\omega) \qquad \text{(odd symmetry)} \qquad (4.2.46)$$

From (4.2.43) it also follows that

$$S_{xx}(-\omega) = S_{xx}(\omega) \qquad \text{(even symmetry)} \qquad (4.2.47)$$

From these symmetry properties we conclude that the frequency range of real discrete-time signals can be limited further to the range $0 \le \omega \le \pi$ (i.e., one-half of the period). Indeed, if we know $X(\omega)$ in the range $0 \le \omega \le \pi$, we can determine it for the range $-\pi \le \omega < 0$ using the symmetry properties given above. As we have already observed, similar results hold for discrete-time periodic signals. Therefore, the frequency-domain description of a real discrete-time signal is completely specified by its spectrum in the frequency range $0 \le \omega \le \pi$.

Usually, we work with the fundamental interval $0 \le \omega \le \pi$ or $0 \le F \le F_s/2$, expressed in Hertz. We sketch more than half a period only when required by the specific application.

Example 4.2.3

Determine and sketch the energy density spectrum $S_{xx}(\omega)$ of the signal

$$x(n) = a^n u(n) \qquad -1 < a < 1$$

Solution   Since $|a| < 1$, the sequence $x(n)$ is absolutely summable, as can be verified by applying the geometric summation formula,

$$\sum_{n=-\infty}^{\infty} |x(n)| = \sum_{n=0}^{\infty} |a|^n = \frac{1}{1 - |a|} < \infty$$

Hence the Fourier transform of $x(n)$ exists and is obtained by applying (4.2.29). Thus

$$X(\omega) = \sum_{n=0}^{\infty} a^n e^{-j\omega n} = \sum_{n=0}^{\infty} (ae^{-j\omega})^n$$

Since $|ae^{-j\omega}| = |a| < 1$, use of the geometric summation formula again yields

$$X(\omega) = \frac{1}{1 - ae^{-j\omega}}$$

The energy density spectrum is given by

$$S_{xx}(\omega) = |X(\omega)|^2 = X(\omega)X^*(\omega) = \frac{1}{(1 - ae^{-j\omega})(1 - ae^{j\omega})}$$

or, equivalently, as

$$S_{xx}(\omega) = \frac{1}{1 - 2a\cos\omega + a^2}$$

Note that $S_{xx}(-\omega) = S_{xx}(\omega)$ in accordance with (4.2.47).

Figure 4.15 shows the signal $x(n)$ and its corresponding spectrum for $a = 0.5$ and $a = -0.5$. Note that for $a = -0.5$ the signal has more rapid variations and as a result its spectrum has stronger high frequencies.

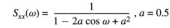

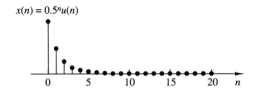

$x(n) = 0.5^n u(n)$

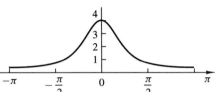

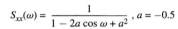

$x(n) = (-0.5)^n u(n)$

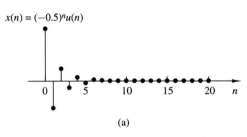

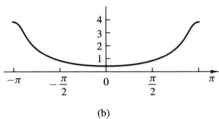

(a)   (b)

Figure 4.15 (a) Sequence $x(n) = (\frac{1}{2})^n u(n)$ and $x(n) = (-\frac{1}{2})^n u(n)$; (b) their energy density spectra.

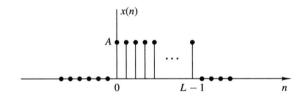

Figure 4.16 Discrete-time rectangular pulse.

Example 4.2.4

Determine the Fourier transform and the energy density spectrum of the sequence

$$x(n) = \begin{cases} A, & 0 \le n \le L - 1 \\ 0, & \text{otherwise} \end{cases} \tag{4.2.48}$$

which is illustrated in Fig. 4.16.

Solution Before computing the Fourier transform, we observe that

$$\sum_{n=-\infty}^{\infty} |x(n)| = \sum_{n=0}^{L-1} |A| = L|A| < \infty$$

Hence $x(n)$ is absolutely summable and its Fourier transform exists. Furthermore, we note that $x(n)$ is a finite-energy signal with $E_x = |A|^2 L$.

The Fourier transform of this signal is

$$X(\omega) = \sum_{n=0}^{L-1} A e^{-j\omega n}$$

$$= A \frac{1 - e^{-j\omega L}}{1 - e^{-j\omega}}$$

$$= Ae^{-j(\omega/2)(L-1)} \frac{\sin(\omega L/2)}{\sin(\omega/2)} \qquad (4.2.49)$$

For $\omega = 0$ the transform in (4.2.49) yields $X(0) = AL$, which is easily established by setting $\omega = 0$ in the defining equation for $X(\omega)$, or by using L'Hospital's rule in (4.2.49) to resolve the indeterminate form when $\omega = 0$.

The magnitude and phase spectra of $x(n)$ are

$$|X(\omega)| = \begin{cases} |A|L, & \omega = 0 \\ |A| \left| \dfrac{\sin(\omega L/2)}{\sin(\omega/2)} \right|, & \text{otherwise} \end{cases} \qquad (4.2.50)$$

and

$$\angle X(\omega) = \angle A - \frac{\omega}{2}(L-1) + \angle \frac{\sin(\omega L/2)}{\sin(\omega/2)} \qquad (4.2.51)$$

where we should remember that the phase of a real quantity is zero if the quantity is positive and π if it is negative.

The spectra $|X(\omega)|$ and $\angle X(\omega)$ are shown in Fig. 4.17 for the case $A = 1$ and $L = 5$. The energy density spectrum is simply the square of the expression given in (4.2.50).

There is an interesting relationship that exists between the Fourier transform of the constant amplitude pulse in Example 4.2.4 and the periodic rectangular

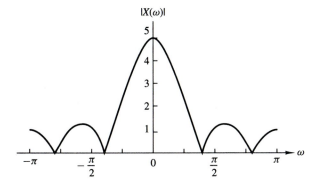

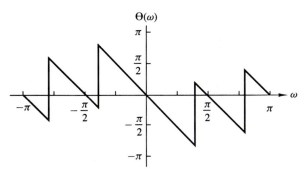

Figure 4.17 Magnitude and phase of Fourier transform of the discrete-time rectangular pulse in Fig. 4.16.

wave considered in Example 4.2.2. If we evaluate the Fourier transform as given in (4.2.49) at a set of equally spaced (harmonically related) frequencies

$$\omega_k = \frac{2\pi}{N}k \qquad k = 0, 1, \ldots, N - 1$$

we obtain

$$X\left(\frac{2\pi}{N}k\right) = Ae^{-j(\pi/N)k(L-1)}\frac{\sin[(\pi/N)kL]}{\sin[(\pi/N)k]} \tag{4.2.52}$$

If we compare this result with the expression for the Fourier series coefficients given in (4.2.21) for the periodic rectangular wave, we find that

$$X\left(\frac{2\pi}{N}k\right) = Nc_k \qquad k = 0, 1, \ldots, N - 1 \tag{4.2.53}$$

To elaborate, we have established that the Fourier transform of the rectangular pulse, which is identical with a single period of the periodic rectangular pulse train, evaluated at the frequencies $\omega = 2\pi k/N$, $k = 0, 1, \ldots, N - 1$, which are identical to the harmonically related frequency components used in the Fourier series representation of the periodic signal, is simply a multiple of the Fourier coefficients $\{c_k\}$ at the corresponding frequencies.

The relationship given in (4.2.53) for the Fourier transform of the rectangular pulse evaluated at $\omega = 2\pi k/N$, $k = 0, 1, \ldots, N - 1$, and the Fourier coefficients of the corresponding periodic signal, is not only true for these two signals but, in fact, holds in general. This relationship is developed further in Chapter 5.

4.2.6 Relationship of the Fourier Transform to the z-Transform

The z-transform of a sequence $x(n)$ is defined as

$$X(z) = \sum_{n=-\infty}^{\infty} x(n)z^{-n} \qquad \text{ROC: } r_2 < |z| < r_1 \tag{4.2.54}$$

where $r_2 < |z| < r_1$ is the region of convergence of $X(z)$. Let us express the complex variable z in polar form as

$$z = re^{j\omega} \tag{4.2.55}$$

where $r = |z|$ and $\omega = \angle z$. Then, within the region of convergence of $X(z)$, we can substitute $z = re^{j\omega}$ into (4.2.54). This yields

$$X(z)|_{z=re^{j\omega}} = \sum_{n=-\infty}^{\infty} [x(n)r^{-n}]e^{-j\omega n} \tag{4.2.56}$$

From the relationship in (4.2.56) we note that $X(z)$ can be interpreted as the Fourier transform of the signal sequence $x(n)r^{-n}$. The weighting factor r^{-n} is growing with n if $r < 1$ and decaying if $r > 1$. Alternatively, if $X(z)$ converges for

$|z| = 1$, then

$$X(z)|_{z=e^{j\omega}} \equiv X(\omega) = \sum_{n=-\infty}^{\infty} x(n)e^{-j\omega n} \tag{4.2.57}$$

Therefore, the Fourier transform can be viewed as the z-transform of the sequence evaluated on the unit circle. If $X(z)$ does not converge in the region $|z| = 1$ [i.e., if the unit circle is not contained in the region of convergence of $X(z)$], the Fourier transform $X(\omega)$ does not exist.

We should note that the existence of the z-transform requires that the sequence $\{x(n)r^{-n}\}$ be absolutely summable for some value of r, that is,

$$\sum_{n=-\infty}^{\infty} |x(n)r^{-n}| < \infty \tag{4.2.58}$$

Hence if (4.2.58) converges only for values of $r > r_0 > 1$, the z-transform exists, but the Fourier transform does not exist. This is the case, for example, for causal sequences of the form $x(n) = a^n u(n)$, where $|a| > 1$.

There are sequences, however, that do not satisfy the requirement in (4.2.58), for example, the sequence

$$x(n) = \frac{\sin \omega_c n}{\pi n} \qquad -\infty < n < \infty \tag{4.2.59}$$

This sequence does not have a z-transform. Since it has a finite energy, its Fourier transform converges in the mean-square sense to the discontinuous function $X(\omega)$, defined as

$$X(\omega) = \begin{cases} 1, & |\omega| < \omega_c \\ 0, & \omega_c < |\omega| \leq \pi \end{cases} \tag{4.2.60}$$

In conclusion, the existence of the z-transform requires that (4.2.58) be satisfied for some region in the z-plane. If this region contains the unit circle, the Fourier transform $X(\omega)$ exists. However, the existence of the Fourier transform, which is defined for finite energy signals, does not necessarily ensure the existence of the z-transform.

4.2.7 The Cepstrum

Let us consider a sequence $\{x(n)\}$ having a z-transform $X(z)$. We assume that $\{x(n)\}$ is a stable sequence so that $X(z)$ converges on the unit circle. The *complex cepstrum* of the sequence $\{x(n)\}$ is defined as the sequence $\{c_x(n)\}$, which is the inverse z-transform of $C_x(z)$, where

$$C_x(z) = \ln X(z) \tag{4.2.61}$$

The complex cepstrum exists if $C_x(z)$ converges in the annular region $r_1 < |z| < r_2$, where $0 < r_1 < 1$ and $r_2 > 1$. Within this region of convergence, $C_x(z)$ can be represented by the Laurent series

$$C_x(z) = \ln X(z) = \sum_{n=-\infty}^{\infty} c_x(n)z^{-n} \tag{4.2.62}$$

where

$$c_x(n) = \frac{1}{2\pi j} \int_C \ln X(z) z^{n-1} dz \tag{4.2.63}$$

C is a closed contour about the origin and lies within the region of convergence. Clearly, if $C_x(z)$ can be represented as in (4.2.62), the complex cepstrum sequence $\{c_x(n)\}$ is stable. Furthermore, if the complex cepstrum exists, $C_x(z)$ converges on the unit circle and hence we have

$$C_x(\omega) = \ln X(\omega) = \sum_{n=-\infty}^{\infty} c_x(n) e^{-j\omega n} \tag{4.2.64}$$

where $\{c_x(n)\}$ is the sequence obtained from the inverse Fourier transform of $\ln X(\omega)$, that is,

$$c_x(n) = \frac{1}{2\pi} \int_{-\pi}^{\pi} \ln X(\omega) e^{j\omega n} d\omega \tag{4.2.65}$$

If we express $X(\omega)$ in terms of its magnitude and phase, say

$$X(\omega) = |X(\omega)| e^{j\theta(\omega)} \tag{4.2.66}$$

then

$$\ln X(\omega) = \ln |X(\omega)| + j\theta(\omega) \tag{4.2.67}$$

By substituting (4.2.67) into (4.2.65), we obtain the complex cepstrum in the form

$$c_x(n) = \frac{1}{2\pi} \int_{-\pi}^{\pi} [\ln |X(\omega)| + j\theta(\omega)] e^{j\omega n} d\omega \tag{4.2.68}$$

We can separate the inverse Fourier transform in (4.2.68) into the inverse Fourier transforms of $\ln |X(\omega)|$ and $\theta(\omega)$:

$$c_m(n) = \frac{1}{2\pi} \int_{-\pi}^{\pi} \ln |X(\omega)| e^{j\omega n} d\omega \tag{4.2.69}$$

$$c_\theta(n) = \frac{1}{2\pi} \int_{-\pi}^{\pi} \theta(\omega) e^{j\omega n} d\omega \tag{4.2.70}$$

In some applications, such as speech signal processing, only the component $c_m(n)$ is computed. In such a case the phase of $X(\omega)$ is ignored. Therefore, the sequence $\{x(n)\}$ cannot be recovered from $\{c_m(n)\}$. That is, the transformation from $\{x(n)\}$ to $\{c_m(n)\}$ is not invertible.

In speech signal processing, the (real) cepstrum has been used to separate and thus to estimate the spectral content of the speech from the pitch frequency of the speech. The complex cepstrum is used in practice to separate signals that are convolved. The process of separating two convolved signals is called *deconvolution* and the use of the complex cepstrum to perform the separation is called *homomorphic deconvolution*. This topic is discussed in Section 4.6.

4.2.8 The Fourier Transform of Signals with Poles on the Unit Circle

As was shown in Section 4.2.6, the Fourier transform of a sequence $x(n)$ can be determined by evaluating its z-transform $X(z)$ on the unit circle, provided that the unit circle lies within the region of convergence of $X(z)$. Otherwise, the Fourier transform does not exist.

There are some aperiodic sequences that are neither absolutely summable nor square summable. Hence their Fourier transforms do not exist. One such sequence is the unit step sequence, which has the z-transform

$$X(z) = \frac{1}{1 - z^{-1}}$$

Another such sequence is the causal sinusoidal signal sequence $x(n) = (\cos \omega_0 n) u(n)$. This sequence has the z-transform

$$X(z) = \frac{1 - z^{-1} \cos \omega_0}{1 - 2z^{-1} \cos \omega_0 + z^{-2}}$$

Note that both of these sequences have poles on the unit circle.

For sequences such as these two examples, it is sometimes useful to extend the Fourier transform representation. This can be accomplished, in a mathematically rigorous way, by allowing the Fourier transform to contain impulses at certain frequencies corresponding to the location of the poles of $X(z)$ that lie on the unit circle. The impulses are functions of the continuous frequency variable ω and have infinite amplitude, zero width, and unit area. An impulse can be viewed as the limiting form of a rectangular pulse of height $1/a$ and width a, in the limit as $a \to 0$. Thus, by allowing impulses in the spectrum of a signal, it is possible to extend the Fourier transform representation to some signal sequences that are neither absolutely summable nor square summable.

The following example illustrates the extension of the Fourier transform representation for three sequences.

Example 4.2.5

Determine the Fourier transform of the following signals.

  (a) $x_1(n) = u(n)$
  (b) $x_2(n) = (-1)^n u(n)$
  (c) $x_3(n) = (\cos \omega_0 n) u(n)$

by evaluating their z-transforms on the unit circle.

Solution

  (a) From Table 4.3 we find that

$$X_1(z) = \frac{1}{1 - z^{-1}} = \frac{z}{z - 1} \qquad \text{ROC: } |z| > 1$$

$X_1(z)$ has a pole, $p_1 = 1$, on the unit circle, but converges for $|z| > 1$.

If we evaluate $X_1(z)$ on the unit circle, except at $z = 1$, we obtain

$$X_1(\omega) = \frac{e^{j\omega/2}}{2j\sin(\omega/2)} = \frac{1}{2\sin(\omega/2)} e^{j(\omega - \pi/2)} \qquad \omega \neq 2\pi k \quad k = 0, 1, \ldots$$

At $\omega = 0$ and multiples of 2π, $X_1(\omega)$ contains impulses of area π.

Hence the presence of a pole at $z = 1$ (i.e., at $\omega = 0$) creates a problem only when we want to compute $|X_1(\omega)|$ at $\omega = 0$, because $|X_1(\omega)| \to \infty$ as $\omega \to 0$. For any other value of ω, $X_1(\omega)$ is finite (i.e., well behaved). Although, at first glance one might expect the signal to have zero-frequency components at all frequencies except at $\omega = 0$, this is not the case. This happens because the signal $x_1(n)$ is not a constant for all $-\infty < n < \infty$. Instead, it is *turned on* at $n = 0$. This abrupt jump creates all frequency components existing in the range $0 < \omega \leq \pi$. Generally, all signals which start at a finite time have nonzero-frequency components everywhere in the frequency axis from zero up to the folding frequency.

(b) From Table 3.3 we find that the z-transform of $a^n u(n)$ with $a = -1$ reduces to

$$X_2(z) = \frac{1}{1 + z^{-1}} = \frac{z}{z + 1} \qquad \text{ROC: } |z| > 1$$

which has a pole at $z = -1 = e^{j\pi}$. The Fourier transform evaluated at frequencies other than $\omega = \pi$ and multiples of 2π is

$$X_2(\omega) = \frac{e^{j\omega/2}}{2\cos(\omega/2)} \qquad \omega \neq 2\pi(k + \tfrac{1}{2}) \quad k = 0, 1, \ldots$$

In this case the impulses occurs at $\omega = \pi + 2\pi k$.

Hence the magnitude is

$$|X_2(\omega)| = \frac{1}{2|\cos(\omega/2)|} \qquad \omega \neq 2\pi k + \pi \quad k = 0, 1, \ldots$$

and the phase is

$$\angle X_2(\omega) = \begin{cases} \dfrac{\omega}{2}, & \text{if } \cos \dfrac{\omega}{2} \geq 0 \\[2ex] \dfrac{\omega}{2} + \pi, & \text{if } \cos \dfrac{\omega}{2} < 0 \end{cases}$$

Note that due to the presence of the pole at $a = -1$ (i.e., at frequency $\omega = \pi$), the magnitude of the Fourier transform becomes infinite. Now $|X(\omega)| \to \infty$ as $\omega \to \pi$. We observe that $(-1)^n u(n) = (\cos \pi n) u(n)$, which is the fastest possible oscillating signal in discrete time.

(c) From the discussion above, it follows that $X_3(\omega)$ is infinite at the frequency component $\omega = \omega_0$. Indeed, from Table 3.3, we find that

$$x_3(n) = (\cos \omega_0 n) u(n) \overset{z}{\longleftrightarrow} X_3(z) = \frac{1 - z^{-1} \cos \omega_0}{1 - 2z^{-1} \cos \omega_0 + z^{-2}} \qquad \text{ROC: } |z| > 1$$

The Fourier transform is

$$X_3(\omega) = \frac{1 - e^{-j\omega} \cos \omega_0}{(1 - e^{-j(\omega - \omega_0)})(1 - e^{j(\omega + \omega_0)})} \qquad \omega \neq \pm \omega_0 + 2\pi k \quad k = 0, 1, \ldots$$

The magnitude of $X_3(\omega)$ is given by

$$|X_3(\omega)| = \frac{|1 - e^{-j\omega} \cos \omega_0|}{|1 - e^{-j(\omega-\omega_0)}||1 - e^{-j(\omega+\omega_0)}|} \qquad \omega \neq \pm\omega_0 + 2\pi k \quad k = 0, 1, \ldots$$

Now if $\omega = -\omega_0$ or $\omega = \omega_0$, $|X_3(\omega)|$ becomes infinite. For all other frequencies, the Fourier transform is well behaved.

4.2.9 The Sampling Theorem Revisited

To process a continuous-time signal using digital signal processing techniques, it is necessary to convert the signal into a sequence of numbers. As was discussed in Section 1.4, this is usually done by sampling the analog signal, say $x_a(t)$, periodically every T seconds to produce a discrete-time signal $x(n)$ given by

$$x(n) = x_a(nT) \qquad -\infty < n < \infty \tag{4.2.71}$$

The relationship (4.2.71) describes the sampling process in the time domain. As discussed in Chapter 1, the sampling frequency $F_s = 1/T$ must be selected large enough such that the sampling does not cause any loss of spectral information (no aliasing). Indeed, if the spectrum of the analog signal can be recovered from the spectrum of the discrete- time signal, there is no loss of information. Consequently, we investigate the sampling process by finding the relationship between the spectra of signals $x_a(t)$ and $x(n)$.

If $x_a(t)$ is an aperiodic signal with finite energy, its (voltage) spectrum is given by the Fourier transform relation

$$X_a(F) = \int_{-\infty}^{\infty} x_a(t) e^{-j2\pi F t} dt \tag{4.2.72}$$

whereas the signal $x_a(t)$ can be recovered from its spectrum by the inverse Fourier transform

$$x_a(t) = \int_{-\infty}^{\infty} X_a(F) e^{j2\pi F t} dF \tag{4.2.73}$$

Note that utilization of all frequency components in the infinite frequency range $-\infty < F < \infty$ is necessary to recover the signal $x_a(t)$ if the signal $x_a(t)$ is not bandlimited.

The spectrum of a discrete-time signal $x(n)$, obtained by sampling $x_a(t)$, is given by the Fourier transform relation

$$X(\omega) = \sum_{n=-\infty}^{\infty} x(n) e^{-j\omega n} \tag{4.2.74}$$

or, equivalently,

$$X(f) = \sum_{n=-\infty}^{\infty} x(n) e^{-j2\pi f n} \tag{4.2.75}$$

The sequence $x(n)$ can be recovered from its spectrum $X(\omega)$ or $X(f)$ by the inverse transform

$$x(n) = \frac{1}{2\pi} \int_{-\pi}^{\pi} X(\omega)e^{j\omega n}\,d\omega$$

$$= \int_{-1/2}^{1/2} X(f)e^{j2\pi fn}\,df$$

(4.2.76)

In order to determine the relationship between the spectra of the discrete-time signal and the analog signal, we note that periodic sampling imposes a relationship between the independent variables t and n in the signals $x_a(t)$ and $x(n)$, respectively. That is,

$$t = nT = \frac{n}{F_s}$$

(4.2.77)

This relationship in the time domain implies a corresponding relationship between the frequency variables F and f in $X_a(F)$ and $X(f)$, respectively.

Indeed, substitution of (4.2.77) into (4.2.73) yields

$$x(n) \equiv x_a(nT) = \int_{-\infty}^{\infty} X_a(F)e^{j2\pi nF/F_s}\,dF$$

(4.2.78)

If we compare (4.2.76) with (4.2.78), we conclude that

$$\int_{-1/2}^{1/2} X(f)e^{j2\pi fn}\,df = \int_{-\infty}^{\infty} X_a(F)e^{j2\pi nF/F_s}\,dF$$

(4.2.79)

From the development in Chapter 1 we know that periodic sampling imposes a relationship between the frequency variables F and f of the corresponding analog and discrete-time signals, respectively. That is,

$$f = \frac{F}{F_s}$$

(4.2.80)

With the aid of (4.2.80), we can make a simple change in variable in (4.2.79), and obtain the result

$$\frac{1}{F_s} \int_{-F_s/2}^{F_s/2} X\left(\frac{F}{F_s}\right) e^{j2\pi nF/F_s}\,dF = \int_{-\infty}^{\infty} X_a(F)e^{j2\pi nF/F_s}\,dF$$

(4.2.81)

We now turn our attention to the integral on the right-hand side of (4.2.81). The integration range of this integral can be divided into an infinite number of intervals of width F_s. Thus the integral over the infinite range can be expressed as a sum of integrals, that is,

$$\int_{-\infty}^{\infty} X_a(F)e^{j2\pi nF/F_s}\,dF = \sum_{k=-\infty}^{\infty} \int_{(k-1/2)F_s}^{(k+1/2)F_s} X_a(F)e^{j2\pi nF/F_s}\,dF$$

(4.2.82)

We observe that $X_a(F)$ in the frequency interval $(k - \frac{1}{2})F_s$ to $(k + \frac{1}{2})F_s$ is identical to $X_a(F - kF_s)$ in the interval $-F_s/2$ to $F_s/2$. Consequently,

$$\sum_{k=-\infty}^{\infty} \int_{(k-1/2)F_s}^{(k+1/2)F_s} X_a(F)e^{j2\pi n F/F_s}\, dF = \sum_{k=-\infty}^{\infty} \int_{-F_s/2}^{F_s/2} X_a(F - kF_s)e^{j2\pi n F/F_s}\, dF$$

$$= \int_{-F_s/2}^{F_s/2} \left[\sum_{k=-\infty}^{\infty} X_a(F - kF_s) \right] e^{j2\pi n F/F_s}\, dF$$

$$(4.2.83)$$

where we have used the periodicity of the exponential, namely,

$$e^{j2\pi n (F + kF_s)/F_s} = e^{j2\pi n F/F_s}$$

Comparing (4.2.83), (4.2.82), and (4.2.81), we conclude that

$$X\left(\frac{F}{F_s}\right) = F_s \sum_{k=-\infty}^{\infty} X_a(F - kF_s) \qquad (4.2.84)$$

or, equivalently,

$$X(f) = F_s \sum_{k=-\infty}^{\infty} X_a[(f - k)F_s] \qquad (4.2.85)$$

This is the desired relationship between the spectrum $X(F/F_s)$ or $X(f)$ of the discrete-time signal and the spectrum $X_a(F)$ of the analog signal. The right-hand side of (4.2.84) or (4.2.85) consists of a periodic repetition of the scaled spectrum $F_s X_a(F)$ with period F_s. This periodicity is necessary because the spectrum $X(f)$ or $X(F/F_s)$ of the discrete-time signal is periodic with period $f_p = 1$ or $F_p = F_s$.

For example, suppose that the spectrum of a band-limited analog signal is as shown in Fig. 4.18(a). The spectrum is zero for $|F| \geq B$. Now, if the sampling frequency F_s is selected to be greater than $2B$, the spectrum $X(F/F_s)$ of the discrete-time signal will appear as shown in Fig. 4.18(b). Thus, if the sampling frequency F_s is selected such that $F_s \geq 2B$, where $2B$ is the Nyquist rate, then

$$X\left(\frac{F}{F_s}\right) = F_s X_a(F) \qquad |F| \leq F_s/2 \qquad (4.2.86)$$

In this case there is no aliasing and therefore, the spectrum of the discrete-time signal is identical (within the scale factor F_s) to the spectrum of the analog signal, within the fundamental frequency range $|F| \leq F_s/2$ or $|f| \leq \frac{1}{2}$.

On the other hand, if the sampling frequency F_s is selected such that $F_s < 2B$, the periodic continuation of $X_a(F)$ results in spectral overlap, as illustrated in Fig. 4.18(c) and (d). Thus the spectrum $X(F/F_s)$ of the discrete-time signal contains aliased frequency components of the analog signal spectrum $X_a(F)$. The end result is that the aliasing which occurs prevents us from recovering the original signal $x_a(t)$ from the samples.

Given the discrete-time signal $x(n)$ with the spectrum $X(F/F_s)$, as illustrated in Fig. 4.18(b), with no aliasing, it is now possible to reconstruct the original analog

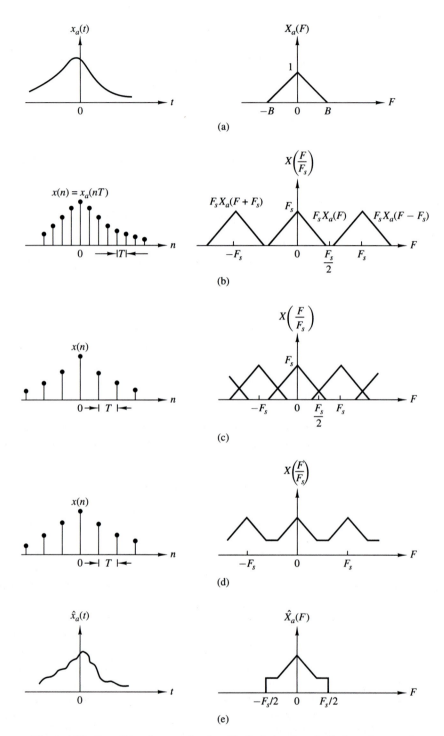

Figure 4.18 Sampling of an analog bandlimited signal and aliasing of spectral components.

signal from the samples $x(n)$. Since in the absence of aliasing

$$X_a(F) = \begin{cases} \dfrac{1}{F_s} X\left(\dfrac{F}{F_s}\right), & |F| \le F_s/2 \\ 0, & |F| > F_s/2 \end{cases} \tag{4.2.87}$$

and by the Fourier transform relationship (4.2.75),

$$X\left(\frac{F}{F_s}\right) = \sum_{n=-\infty}^{\infty} x(n)e^{-j2\pi Fn/F_s} \tag{4.2.88}$$

the inverse Fourier transform of $X_a(F)$ is

$$x_a(t) = \int_{-F_s/2}^{F_s/2} X_a(F)e^{j2\pi Ft}dF \tag{4.2.89}$$

Let us assume that $F_s = 2B$. With the substitution of (4.2.87) into (4.2.89), we have

$$\begin{aligned} x_a(t) &= \frac{1}{F_s} \int_{-F_s/2}^{F_s/2} \left[\sum_{n=-\infty}^{\infty} x(n)e^{-j2\pi Fn/F_s}\right] e^{j2\pi Ft}dF \\ &= \frac{1}{F_s} \sum_{n=-\infty}^{\infty} x(n) \int_{-F_s/2}^{F_s/2} e^{j2\pi F(t-n/F_s)}dF \tag{4.2.90} \\ &= \sum_{n=-\infty}^{\infty} x_a(nT)\frac{\sin(\pi/T)(t-nT)}{(\pi/T)(t-nT)} \end{aligned}$$

where $x(n) = x_a(nT)$ and where $T = 1/F_s = 1/2B$ is the sampling interval. This is the reconstruction formula given by (1.4.24) in our discussion of the sampling theorem.

The reconstruction formula in (4.2.90) involves the function

$$g(t) = \frac{\sin(\pi/T)t}{(\pi/T)t} = \frac{\sin 2\pi Bt}{2\pi Bt} \tag{4.2.91}$$

appropriately shifted by nT, $n = 0, \pm1, \pm2, \ldots$, and multiplied or weighted by the corresponding samples $x_a(nT)$ of the signal. We call (4.2.90) an interpolation formula for reconstructing $x_a(t)$ from its samples, and $g(t)$, given in (4.2.91), is the interpolation function. We note that at $t = kT$, the interpolation function $g(t-nT)$ is zero except at $k = n$. Consequently, $x_a(t)$ evaluated at $t = kT$ is simply the sample $x_a(kT)$. At all other times the weighted sum of the time shifted versions of the interpolation function combine to yield exactly $x_a(t)$. This combination is illustrated in Fig. 4.19.

The formula in (4.2.90) for reconstructing the analog signal $x_a(t)$ from its samples is called the *ideal interpolation formula.* It forms the basis for the *sampling theorem,* which can be stated as follows.

Sampling Theorem.   A bandlimited continuous-time signal, with highest frequency (bandwidth) B Hertz, can be uniquely recovered from its samples provided that the sampling rate $F_s \ge 2B$ samples per second.

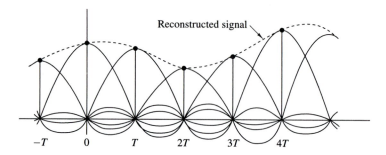

Figure 4.19   Reconstruction of a continuous-time signal using ideal interpolation.

According to the sampling theorem and the reconstruction formula in (4.2.90), the recovery of $x_a(t)$ from its samples $x(n)$, requires an infinite number of samples. However, in practice we use a finite number of samples of the signal and deal with finite-duration signals. As a consequence, we are concerned only with reconstructing a finite-duration signal from a finite number of samples.

When aliasing occurs due to too low a sampling rate, the effect can be described by a multiple folding of the frequency axis of the frequency variable F for the analog signal. Figure 4.20(a) shows the spectrum $X_a(F)$ of an analog signal. According to (4.2.84), sampling of the signal with a sampling frequency F_s results in a periodic repetition of $X_a(F)$ with period F_s. If $F_s < 2B$, the shifted replicas of $X_a(F)$ overlap. The overlap that occurs within the fundamental frequency range $-F_s/2 \leq F \leq F_s/2$, is illustrated in Fig. 4.20(b). The corresponding spectrum of the discrete-time signal within the fundamental frequency range, is obtained by adding all the shifted portions within the range $|f| \leq \frac{1}{2}$, to yield the spectrum shown in Fig. 4.20(c).

A careful inspection of Fig. 4.20(a) and (b) reveals that the aliased spectrum in Fig. 4.20(c) can be obtained by folding the original spectrum like an accordian with pleats at every odd multiple of $F_s/2$. Consequently, the frequency $F_s/2$ is called the *folding frequency*, as indicated in Chapter 1. Clearly, then, periodic sampling automatically forces a folding of the frequency axis of an analog signal at odd multiples of $F_s/2$, and this results in the relationship $F = fF_s$ between the frequencies for continuous-time signals and discrete-time signals. Due to the folding of the frequency axis, the relationship $F = fF_s$ is not truly linear, but piecewise linear, to accommodate for the aliasing effect. This relationship is illustrated in Fig. 4.21.

If the analog signal is bandlimited to $B \leq F_s/2$, the relationship between f and F is linear and one-to-one. In other words, there is no aliasing. In practice, prefiltering with an antialiasing filter is usually employed prior to sampling. This ensures that frequency components of the signal above $F \geq B$ are sufficiently attenuated so that, if aliased, they cause negligible distortion on the desired signal.

The relationships among the time-domain and frequency-domain functions $x_a(t)$, $x(n)$, $X_a(F)$, and $X(f)$ are summarized in Fig. 4.22. The relationships for

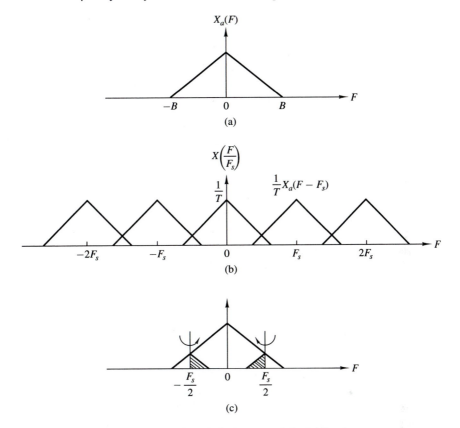

Figure 4.20   Illustration of aliasing around the folding frequency.

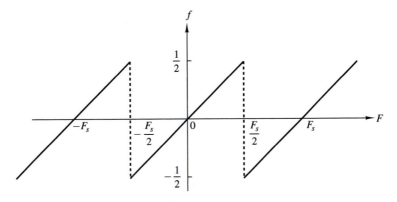

Figure 4.21   Relationship between frequency variables F and f.

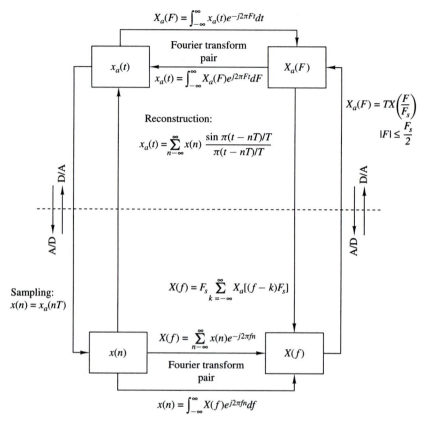

Figure 4.22 Time-domain and frequency-domain relationships for sampled signals.

recovering the continuous-time functions, $x_a(t)$ and $X_a(F)$, from the discrete-time quantities $x(n)$ and $X(f)$, assume that the analog signal is bandlimited and that it is sampled at the Nyquist rate (or faster).

The following examples serve to illustrate the problem of the aliasing of frequency components.

Example 4.2.6 Aliasing in Sinusoidal Signals

The continuous-time signal

$$x_a(t) = \cos 2\pi F_0 t = \tfrac{1}{2} e^{j2\pi F_0 t} + \tfrac{1}{2} e^{-j2\pi F_0 t}$$

has a discrete spectrum with spectral lines at $F = \pm F_0$, as shown in Fig. 4.23(a). The process of sampling this signal with a sampling frequency F_s introduces replicas of the spectrum about multiples of F_s. This is illustrated in Fig. 4.23(b) for $F_s/2 < F_0 < F_s$.

To reconstruct the continuous-time signal, we should select the frequency components inside the fundamental frequency range $|F| \leq F_s/2$. The resulting spectrum

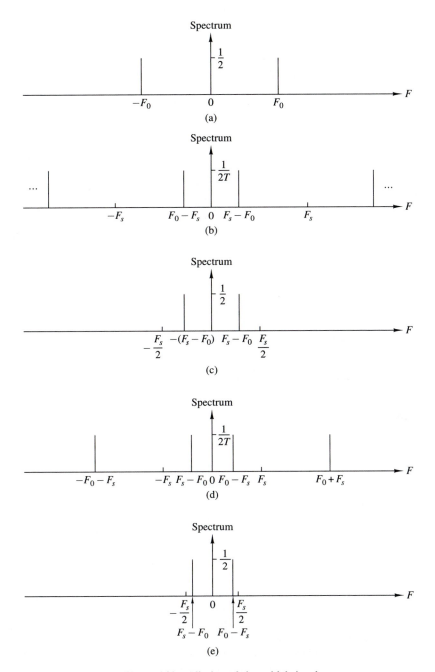

Figure 4.23   Aliasing of sinusoidal signals.

is shown in Fig. 4.23(c). The reconstructed signal is

$$x_a(t) = \cos 2\pi (F_s - F_0)t$$

Now, if F_s is selected such that $F_s < F_0 < 3F_s/2$, the spectrum of the sampled signal is shown in Fig. 4.23(d). The reconstructed signal, shown in Fig. 4.23(e), is

$$x_a(t) = \cos 2\pi (F_0 - F_s)t$$

In both cases, aliasing has occurred, so that the frequency of the reconstructed signal is an aliased version of the frequency of the original signal.

Example 4.2.7   Sampling a Nonbandlimited Signal

Consider the continuous-time signal

$$x_a(t) = e^{-A|t|}   \qquad A > 0$$

whose spectrum is given by

$$X_a(F) = \frac{2A}{A^2 + (2\pi F)^2}$$

Determine the spectrum of the sampled signal $x(n) \equiv x_a(nT)$.

Solution   If we sample $x_a(t)$ with a sampling frequency $F_s = 1/T$, we have

$$x(n) = x_a(nT) = e^{-AT|n|} = (e^{-AT})^{|n|}   \qquad -\infty < n < \infty$$

The spectrum of $x(n)$ can be found easily if we use a direct computation of the Fourier transform. We find that

$$X\left(\frac{F}{F_s}\right) = \frac{1 - e^{-2AT}}{1 - 2e^{-AT}\cos 2\pi FT + e^{-2AT}}   \qquad T = \frac{1}{F_s}$$

Clearly, since $\cos 2\pi FT = \cos 2\pi (F/F_s)$ is periodic with period F_s, so is $X(F/F_s)$.

Since $X_a(F)$ is not bandlimited, aliasing cannot be avoided. The spectrum of the reconstructed signal $\hat{x}_a(t)$ is

$$\hat{X}_a(F) = \begin{cases} TX\left(\dfrac{F}{F_s}\right), & |F| \le \dfrac{F_s}{2} \\[2mm] 0, & |F| > \dfrac{F_s}{2} \end{cases}$$

Figure 4.24(a) shows the original signal $x_a(t)$ and its spectrum $X_a(F)$ for $A = 1$. The sampled signal $x(n)$ and its spectrum $X(F/F_s)$ are shown in Fig. 4.24(b) for $F_s = 1$ Hz. The aliasing distortion is clearly noticeable in the frequency domain. The reconstructed signal $\hat{x}_a(t)$ is shown in Fig. 4.24(c). The distortion due to aliasing can be reduced significantly by increasing the sampling rate. For example, Fig. 4.24(d) illustrates the reconstructed signal corresponding to a sampling rate $F_s = 20$ Hz. It is interesting to note that in every case $x_a(nT) = \hat{x}_a(nT)$, but $x_a(t) \ne \hat{x}_a(t)$ at other values of time.

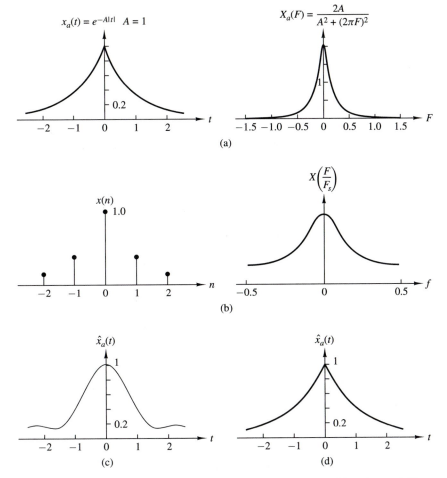

Figure 4.24 (a) Analog signal $x_a(t)$ and its spectrum $X_a(F)$; (b) $x(n) = x_a(nT)$ and the spectrum of $x(n)$ for $A = 1$ and $F_s = 1$ Hz; (c) reconstructed signal $\hat{x}_a(t)$ for $F_s = 1$ Hz; (d) reconstructed signal $\hat{x}_a(t)$ for $F_s = 20$ Hz.

4.2.10 Frequency-Domain Classification of Signals: The Concept of Bandwidth

Just as we have classified signals according to their time-domain characteristics, it is also desirable to classify signals according to their frequency-domain characteristics. It is common practice to classify signals in rather broad terms according to their frequency content.

In particular, if a power signal (or energy signal) has its power density spectrum (or its energy density spectrum) concentrated about zero frequency, such a signal is called a *low-frequency signal*. Figure 4.25(a) illustrates the spectral characteristics of such a signal. On the other hand, if the signal power density

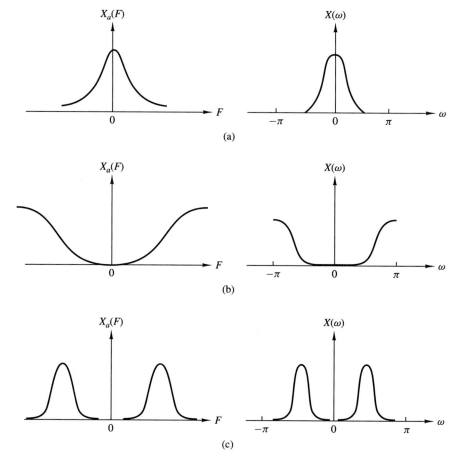

Figure 4.25 (a) Low-frequency, (b) high-frequency, and (c) medium-frequency signals.

spectrum (or the energy density spectrum) is concentrated at high frequencies, the signal is called a *high-frequency signal.* Such a signal spectrum is illustrated in Fig. 4.25(b). A signal having a power density spectrum (or an energy density spectrum) concentrated somewhere in the broad frequency range between low frequencies and high frequencies is called a *medium-frequency signal* or a *bandpass signal.* Figure 4.25(c) illustrates such a signal spectrum.

In addition to this relatively broad frequency-domain classification of signals, it is often desirable to express quantitatively the range of frequencies over which the power or energy density spectrum is concentrated. This quantitative measure is called the *bandwidth* of a signal. For example, suppose that a continuous-time signal has 95% of its power (or energy) density spectrum concentrated in the frequency range $F_1 \leq F \leq F_2$. Then the 95% bandwidth of the signal is $F_2 - F_1$. In a similar manner, we may define the 75% or 90% or 99% bandwidth of the signal.

In the case of a bandpass signal, the term *narrowband* is used to describe the signal if its bandwidth $F_2 - F_1$ is much smaller (say, by a factor of 10 or more) than the median frequency $(F_2 + F_1)/2$. Otherwise, the signal is called *wideband*.

We shall say that a signal is *bandlimited* if its spectrum is zero outside the frequency range $|F| \geq B$. For example, a continuous-time finite-energy signal $x(t)$ is bandlimited if its Fourier transform $X(F) = 0$ for $|F| > B$. A discrete-time finite-energy signal $x(n)$ is said to be *(periodically) bandlimited* if

$$|X(\omega)| = 0 \qquad \text{for } \omega_0 < |\omega| < \pi$$

Similarly, a periodic continuous-time signal $x_p(t)$ is periodically bandlimited if its Fourier coefficients $c_k = 0$ for $|k| > M$, where M is some positive integer. A periodic discrete-time signal with fundamental period N is periodically bandlimited if the Fourier coefficients $c_k = 0$ for $k_0 < |k| < N$. Figure 4.26 illustrates the four types of bandlimited signals.

By exploiting the duality between the frequency domain and the time domain, we can provide similar means for characterizing signals in the time domain. In particular, a signal $x(t)$ will be called *time-limited* if

$$x(t) = 0 \qquad |t| > \tau$$

If the signal is periodic with period T_p, it will be called *periodically time-limited* if

$$x_p(t) = 0 \qquad \tau < |t| < T_p/2$$

If we have a discrete-time signal $x(n)$ of finite duration, that is,

$$x(n) = 0 \qquad |n| > N$$

it is also called time-limited. When the signal is periodic with fundamental period N, it is said to be periodically time-limited if

$$x(n) = 0 \qquad n_0 < |n| < N$$

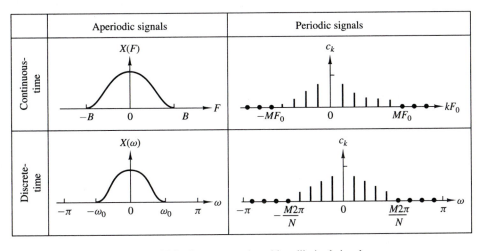

Figure 4.26   Some examples of bandlimited signals.

We state, without proof, that no *signal can be time-limited and bandlimited simultaneously*. Furthermore, a reciprocal relationship exists between the time duration and the frequency duration of a signal. To elaborate, if we have a short-duration rectangular pulse in the time domain, its spectrum has a width that is inversely proportional to the duration of the time- domain pulse. The narrower the pulse becomes in the time domain, the larger the bandwidth of the signal becomes. Consequently, the product of the time duration and the bandwidth of a signal cannot be made arbitrarily small. A short-duration signal has a large bandwidth and a small bandwidth signal has a long duration. Thus, for any signal, the time–bandwidth product is fixed and cannot be made arbitrarily small.

Finally, we note that we have discussed frequency analysis methods for periodic and aperiodic signals with finite energy. However, there is a family of deterministic aperiodic signals with finite power. These signals consist of a linear superposition of complex exponentials with nonharmonically related frequencies, that is,

$$x(n) = \sum_{k=1}^{M} A_k e^{j\omega_k n}$$

where $\omega_1, \omega_2, \ldots, \omega_M$ are nonharmonically related. These signals have discrete spectra but the distances among the lines are nonharmonically related. Signals with discrete nonharmonic spectra are sometimes called quasi-periodic.

4.2.11 The Frequency Ranges of Some Natural Signals

The frequency analysis tools that we have developed in this chapter are usually applied to a variety of signals that are encountered in practice (e.g., seismic, biological, and electromagnetic signals). In general, the frequency analysis is performed for the purpose of extracting information from the observed signal. For example, in the case of biological signals, such as an ECG signal, the analytical tools are used to extract information relevant for diagnostic purposes. In the case of seismic signals, we may be interested in detecting the presence of a nuclear explosion or in determining the characteristics and location of an earthquake. An electromagnetic signal, such as a radar signal reflected from an airplane, contains information on the position of the plane and its radial velocity. These parameters can be estimated from observation of the received radar signal.

In processing any signal for the purpose of measuring parameters or extracting other types of information, one must know approximately the range of frequencies contained by the signal. For reference, Tables 4.1, 4.2, and 4.3 give approximate limits in the frequency domain for biological, seismic, and electromagnetic signals.

4.2.12 Physical and Mathematical Dualities

In the previous sections of the chapter we have introduced several methods for the frequency analysis of signals. Several methods were necessary to accommodate the

TABLE 4.1   FREQUENCY RANGES OF SOME BIOLOGICAL SIGNALS

Type of Signal	Frequency Range (Hz)
Electroretinogram[a]	0–20
Electronystagmogram[b]	0–20
Pneumogram[c]	0–40
Electrocardiogram (ECG)	0–100
Electroencephalogram (EEG)	0–100
Electromyogram[d]	10–200
Sphygmomanogram[e]	0–200
Speech	100–4000

[a]A graphic recording of retina characteristics.
[b]A graphic recording of involuntary movement of the eyes.
[c]A graphic recording of respiratory activity.
[d]A graphic recording of muscular action, such as muscular contraction.
[e]A recording of blood pressure.

TABLE 4.2   FREQUENCY RANGES OF SOME SEISMIC SIGNALS

Type of Signal	Frequency Range (Hz)
Wind noise	100–1000
Seismic exploration signals	10–100
Earthquake and nuclear explosion signals	0.01–10
Seismic noise	0.1–1

TABLE 4.3   FREQUENCY RANGES OF ELECTROMAGNETIC SIGNALS

Type of Signal	Wavelength (m)	Frequency Range (Hz)
Radio broadcast	10^4–10^2	3×10^4–3×10^6
Shortwave radio signals	10^2–10^{-2}	3×10^6–3×10^{10}
Radar, satellite communications, space communications, common-carrier microwave	1–10^{-2}	3×10^8–3×10^{10}
Infrared	10^{-3}–10^{-6}	3×10^{11}–3×10^{14}
Visible light	3.9×10^{-7}–8.1×10^{-7}	3.7×10^{14}–7.7×10^{14}
Ultraviolet	10^{-7}–10^{-8}	3×10^{15}–3×10^{16}
Gamma rays and x-rays	10^{-9}–10^{-10}	3×10^{17}–3×10^{18}

different types of signals. To summarize, the following frequency analysis tools have been introduced:

1. The Fourier series for continuous-time periodic signals.
2. The Fourier transform for continuous-time aperiodic signals.
3. The Fourier series for discrete-time periodic signals.
4. The Fourier transform for discrete-time aperiodic signals.

Figure 4.27 summarizes the analysis and synthesis formulas for these types of signals.

As we have already indicated several times, there are two time-domain characteristics that determine the type of signal spectrum we obtain. These are whether the time variable is continuous or discrete, and whether the signal is periodic or aperiodic. Let us briefly summarize the results of the previous sections.

Continuous-time signals have aperiodic spectra. A close inspection of the Fourier series and Fourier transform analysis formulas for continuous-time signals does not reveal any kind of periodicity in the spectral domain. This lack of periodicity is a consequence of the fact that the complex exponential $\exp(j2\pi Ft)$ is a function of the continuous variable t, and hence it is not periodic in F. Thus the frequency range of continuous-time signals extends from $F = 0$ to $F = \infty$.

Discrete-time signals have periodic spectra. Indeed, both the Fourier series and the Fourier transform for discrete-time signals are periodic with period $\omega = 2\pi$. As a result of this periodicity, the frequency range of discrete-time signals is finite and extends from $\omega = -\pi$ to $\omega = \pi$ radians, where $\omega = \pi$ corresponds to the highest possible rate of oscillation.

Periodic signals have discrete spectra. As we have observed, periodic signals are described by means of Fourier series. The Fourier series coefficients provide the "lines" that constitute the discrete spectrum. The line spacing ΔF or Δf is equal to the inverse of the period T_p or N, respectively, in the time domain. That is, $\Delta F = 1/T_p$ for continuous-time periodic signals and $\Delta f = 1/N$ for discrete-time signals.

Aperiodic finite energy signals have continuous spectra. This property is a direct consequence of the fact that both $X(F)$ and $X(\omega)$ are functions of $\exp(j2\pi Ft)$ and $\exp(j\omega n)$, respectively, which are continuous functions of the variables F and ω. The continuity in frequency is necessary to break the harmony and thus create aperiodic signals.

In summary, we can conclude that *periodicity with "period" α in one domain automatically implies discretization with "spacing" of $1/\alpha$ in the other domain, and vice versa.*

If we keep in mind that "period" in the frequency domain means the frequency range, "spacing" in the time domain is the sampling period T, line spacing in the frequency domain is ΔF, then $\alpha = T_p$ implies that $1/\alpha = 1/T_p = \Delta F$, $\alpha = N$ implies that $\Delta f = 1/N$, and $\alpha = F_s$ implies that $T = 1/F_s$.

These time-frequency dualities are apparent from observation of Fig. 4.27. We stress, however, that the illustrations used in this figure do not correspond to any actual transform pairs. Thus any comparison among them should be avoided.

A careful inspection of Fig. 4.27 also reveals some mathematical symmetries and dualities among the several frequency analysis relationships. In particular,

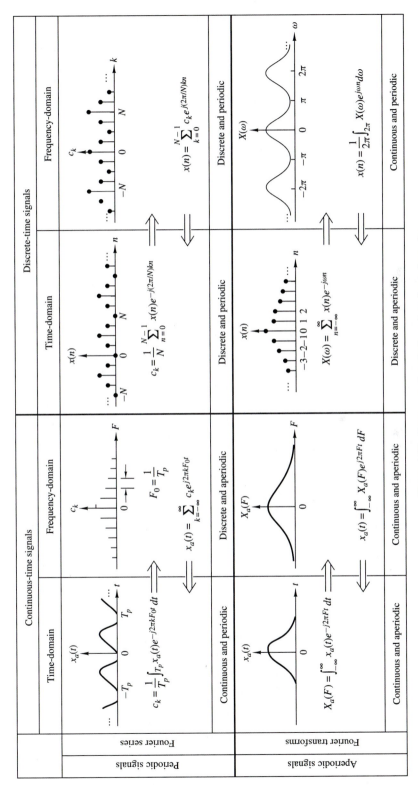

Figure 4.27 Summary of analysis and synthesis formulas.

we observe that there are dualities between the following analysis and synthesis equations:

1. The analysis and synthesis equations of the continuous-time Fourier transform.
2. The analysis and synthesis equations of the discrete-time Fourier series.
3. The analysis equation of the continuous-time Fourier series and the synthesis equation of the discrete-time Fourier transform.
4. The analysis equation of the discrete-time Fourier transform and the synthesis equation of the continuous-time Fourier series.

Note that all dual relations differ only in the sign of the exponent of the corresponding complex exponential. It is interesting to note that this change in sign can be thought of either as a folding of the signal or a folding of the spectrum, since

$$e^{-j2\pi Ft} = e^{j2\pi(-F)t} = e^{j2\pi F(-t)}$$

If we turn our attention now to the spectral density of signals, we recall that we have used the term *energy density spectrum* for characterizing finite-energy aperiodic signals and the term *power density spectrum* for periodic signals. This terminology is consistent with the fact that periodic signals are power signals and aperiodic signals with finite energy are energy signals.

4.3 PROPERTIES OF THE FOURIER TRANSFORM FOR DISCRETE-TIME SIGNALS

The Fourier transform for aperiodic finite-energy discrete-time signals described in the preceding section possesses a number of properties that are very useful in reducing the complexity of frequency analysis problems in many practical applications. In this section we develop the important properties of the Fourier transform. Similar properties hold for the Fourier transform of aperiodic finite-energy continuous-time signals.

For convenience, we adopt the notation

$$X(\omega) \equiv F\{x(n)\} = \sum_{n=-\infty}^{\infty} x(n)e^{-j\omega n} \qquad (4.3.1)$$

for the direct transform (analysis equation) and

$$x(n) \equiv F^{-1}\{X(\omega)\} = \frac{1}{2\pi} \int_{2\pi} X(\omega)e^{j\omega n} d\omega \qquad (4.3.2)$$

for the inverse transform (synthesis equation). We also refer to $x(n)$ and $X(\omega)$ as a *Fourier transform pair* and denote this relationship with the notation

$$x(n) \overset{F}{\longleftrightarrow} X(\omega) \qquad (4.3.3)$$

Recall that $X(\omega)$ is periodic with period 2π. Consequently, any interval of length 2π is sufficient for the specification of the spectrum. Usually, we plot the spectrum in the fundamental interval $[-\pi, \pi]$. We emphasize that all the spectral information contained in the fundamental interval is necessary for the complete description or characterization of the signal. For this reason, the range of integration in (4.3.2) is always 2π, independent of the specific characteristics of the signal within the fundamental interval.

4.3.1 Symmetry Properties of the Fourier Transform

When a signal satisfies some symmetry properties in the time domain, these properties impose some symmetry conditions on its Fourier transform. Exploitation of any symmetry characteristics leads to simpler formulas for both the direct and inverse Fourier transform. A discussion of various symmetry properties and the implications of these properties in the frequency domain is given here.

Suppose that both the signal $x(n)$ and its transform $X(\omega)$ are complex-valued functions. Then they can be expressed in rectangular form as

$$x(n) = x_R(n) + jx_I(n) \tag{4.3.4}$$

$$X(\omega) = X_R(\omega) + jX_I(\omega) \tag{4.3.5}$$

By substituting (4.3.4) and $e^{-j\omega} = \cos\omega - j\sin\omega$ into (4.3.1) and separating the real and imaginary parts, we obtain

$$X_R(\omega) = \sum_{n=-\infty}^{\infty} [x_R(n)\cos\omega n + x_I(n)\sin\omega n] \tag{4.3.6}$$

$$X_I(\omega) = -\sum_{n=-\infty}^{\infty} [x_R(n)\sin\omega n - x_I(n)\cos\omega n] \tag{4.3.7}$$

In a similar manner, by substituting (4.3.5) and $e^{j\omega} = \cos\omega + j\sin\omega$ into (4.3.2), we obtain

$$x_R(n) = \frac{1}{2\pi}\int_{2\pi} [X_R(\omega)\cos\omega n - X_I(\omega)\sin\omega n]d\omega \tag{4.3.8}$$

$$x_I(n) = \frac{1}{2\pi}\int_{2\pi} [X_R(\omega)\sin\omega n + X_I(\omega)\cos\omega n]d\omega \tag{4.3.9}$$

Now, let us investigate some special cases.

Real signals. If $x(n)$ is real, then $x_R(n) = x(n)$ and $x_I(n) = 0$. Hence (4.3.6) and (4.3.7) reduce to

$$X_R(\omega) = \sum_{n=-\infty}^{\infty} x(n)\cos\omega n \tag{4.3.10}$$

and

$$X_I(\omega) = - \sum_{n=-\infty}^{\infty} x(n) \sin \omega n \qquad (4.3.11)$$

Since $\cos(-\omega n) = \cos \omega n$ and $\sin(-\omega n) = -\sin \omega n$, it follows from (4.3.10) and (4.3.11) that

$$X_R(-\omega) = X_R(\omega) \qquad \text{(even)} \qquad (4.3.12)$$

$$X_I(-\omega) = -X_I(\omega) \qquad \text{(odd)} \qquad (4.3.13)$$

If we combine (4.3.12) and (4.3.13) into a single equation, we have

$$X^*(\omega) = X(-\omega) \qquad (4.3.14)$$

In this case we say that the spectrum of a real signal has *Hermitian symmetry*.

With the aid of Fig. 4.28, we observe that the magnitude and phase spectra for real signals are

$$|X(\omega)| = \sqrt{X_R^2(\omega) + X_I^2(\omega)} \qquad (4.3.15)$$

$$\angle X|\omega| = \tan^{-1} \frac{X_I(\omega)}{X_R(\omega)} \qquad (4.3.16)$$

As a consequence of (4.3.12) and (4.3.13), the magnitude and phase spectra also possess the symmetry properties

$$|X(\omega)| = |X(-\omega)| \qquad \text{(even)} \qquad (4.3.17)$$

$$\angle X(-\omega) = -\angle X(\omega) \qquad \text{(odd)} \qquad (4.3.18)$$

In the case of the inverse transform of a real-valued signal [i.e., $x(n) = x_R(n)$], (4.3.8) implies that

$$x(n) = \frac{1}{2\pi} \int_{2\pi} [X_R(\omega) \cos \omega n - X_I(\omega) \sin \omega n] d\omega \qquad (4.3.19)$$

Since both products $X_R(\omega) \cos \omega n$ and $X_I(\omega) \sin \omega n$ are even functions of ω, we have

$$x(n) = \frac{1}{\pi} \int_0^{\pi} [X_R(\omega) \cos \omega n - X_I(\omega) \sin \omega n] d\omega \qquad (4.3.20)$$

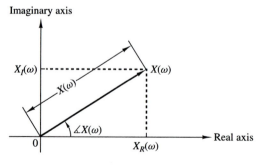

Figure 4.28 Magnitude and phase functions.

Real and even signals. If $x(n)$ is real and even [i.e., $x(-n) = x(n)$], then $x(n) \cos \omega n$ is even and $x(n) \sin \omega n$ is odd. Hence, from (4.3.10), (4.3.11), and (4.3.20) we obtain

$$X_R(\omega) = x(0) + 2 \sum_{n=1}^{\infty} x(n) \cos \omega n \qquad \text{(even)} \qquad (4.3.21)$$

$$X_I(\omega) = 0 \qquad (4.3.22)$$

$$x(n) = \frac{1}{\pi} \int_0^{\pi} X_R(\omega) \cos \omega n \, d\omega \qquad (4.3.23)$$

Thus real and even signals possess real-valued spectra, which, in addition, are even functions of the frequency variable ω.

Real and odd signals. If $x(n)$ is real and odd [i.e., $x(-n) = -x(n)$], then $x(n) \cos \omega n$ is odd and $x(n) \sin \omega n$ is even. Consequently, (4.3.10), (4.3.11) and (4.3.20) imply that

$$X_R(\omega) = 0 \qquad (4.3.24)$$

$$X_I(\omega) = -2 \sum_{n=1}^{\infty} x(n) \sin \omega n \qquad \text{(odd)} \qquad (4.3.25)$$

$$x(n) = -\frac{1}{\pi} \int_0^{\pi} X_I(\omega) \sin \omega n \, d\omega \qquad (4.3.26)$$

Thus real-valued odd signals possess purely imaginary-valued spectral characteristics, which, in addition, are odd functions of the frequency variable ω.

Purely imaginary signals. In this case $x_R(n) = 0$ and $x(n) = j x_I(n)$. Thus (4.3.6), (4.3.7), and (4.3.9) reduce to

$$X_R(\omega) = \sum_{n=-\infty}^{\infty} x_I(n) \sin \omega n \qquad \text{(odd)} \qquad (4.3.27)$$

$$X_I(\omega) = \sum_{n=-\infty}^{\infty} x_I(n) \cos \omega n \qquad \text{(even)} \qquad (4.3.28)$$

$$x_I(n) = \frac{1}{\pi} \int_0^{\pi} [X_R(\omega) \sin \omega n + X_I(\omega) \cos \omega n] \, d\omega \qquad (4.3.29)$$

If $x_I(n)$ is odd [i.e., $x_I(-n) = -x_I(n)$], then

$$X_R(\omega) = 2 \sum_{n=1}^{\infty} x_I(n) \sin \omega n \qquad \text{(odd)} \qquad (4.3.30)$$

$$X_I(\omega) = 0 \qquad (4.3.31)$$

$$x_I(n) = \frac{1}{\pi} \int_0^{\pi} X_R(\omega) \sin \omega n \, d\omega \qquad (4.3.32)$$

Similarly, if $x_I(n)$ is even [i.e., $x_I(-n) = x_I(n)$], we have

$$X_R(\omega) = 0 \tag{4.3.33}$$

$$X_I(\omega) = x_I(0) + 2 \sum_{n=1}^{\infty} x_I(n) \cos \omega n \qquad \text{(even)} \tag{4.3.34}$$

$$x_I(n) = \frac{1}{\pi} \int_0^{\pi} X_I(\omega) \cos \omega n \ d\omega \tag{4.3.35}$$

An arbitrary, possibly complex-valued signal $x(n)$ can be decomposed as

$$x(n) = x_R(n) + jx_I(n) = x_R^e(n) + x_R^o(n) + j[x_I^e(n) + x_I^o(n)]$$
$$= x_e(n) + x_o(n) \tag{4.3.36}$$

where, by definition,

$$x_e(n) = x_R^e(n) + jx_I^e(n) = \tfrac{1}{2}[x(n) + x^*(-n)]$$

$$x_o(n) = x_R^o(n) + jx_I^o(n) = \tfrac{1}{2}[x(n) - x^*(-n)]$$

The superscripts e and o denote the even and odd signal components, respectively. We note that $x_e(n) = x_e(-n)$ and $x_o(-n) = -x_o(n)$. From (4.3.36) and the Fourier transform properties established above, we obtain the following relationships:

$$x(n) = [x_R^e(n) + jx_I^e(n)] + [x_R^o(n) + jx_I^o(n)]$$

$$X(\omega) = [X_R^e(\omega) + jX_I^e(\omega)] + [X_R^o(\omega) + jX_I^o(\omega)] \tag{4.3.37}$$

These symmetry properties of the Fourier transform are summarized in Table 4.4 and in Fig. 4.29. They are often used to simplify Fourier transform calculations in practice.

Example 4.3.1

Determine and sketch $X_R(\omega)$, $X_I(\omega)$, $|X(\omega)|$, and $\angle X(\omega)$ for the Fourier transform

$$X(\omega) = \frac{1}{1 - ae^{-j\omega}} \qquad -1 < a < 1 \tag{4.3.38}$$

Solution By multiplying both the numerator and denominator of (4.3.38) by the complex conjugate of the denominator, we obtain

$$X(\omega) = \frac{1 - ae^{j\omega}}{(1 - ae^{-j\omega})(1 - ae^{j\omega})} = \frac{1 - a\cos\omega - ja\sin\omega}{1 - 2a\cos\omega + a^2}$$

This expression can be subdivided into real and imaginary parts. Thus we obtain

$$X_R(\omega) = \frac{1 - a\cos\omega}{1 - 2a\cos\omega + a^2}$$

$$X_I(\omega) = -\frac{a\sin\omega}{1 - 2a\cos\omega + a^2}$$

Substitution of the last two equations into (4.3.15) and (4.3.16) yields the magnitude and phase spectra as

$$|X(\omega)| = \frac{1}{\sqrt{1 - 2a\cos\omega + a^2}} \tag{4.3.39}$$

TABLE 4.4 SYMMETRY PROPERTIES OF THE DISCRETE-TIME
FOURIER TRANSFORM

Sequence	DTFT
$x(n)$	$X(\omega)$
$x^*(n)$	$X^*(-\omega)$
$x^*(-n)$	$X^*(\omega)$
$x_R(n)$	$X_e(\omega) = \frac{1}{2}[X(\omega) + X^*(-\omega)]$
$jx_I(n)$	$X_o(\omega) = \frac{1}{2}[X(\omega) - X^*(-\omega)]$
$x_e(n) = \frac{1}{2}[x(n) + x^*(-n)]$	$X_R(\omega)$
$x_o(n) = \frac{1}{2}[x(n) - x^*(-n)]$	$jX_I(\omega)$
Real Signals	
	$X(\omega) = X^*(-\omega)$
Any real signal	$X_R(\omega) = X_R(-\omega)$
$x(n)$	$X_I(\omega) = -X_I(-\omega)$
	$|X(\omega)| = |X(-\omega)|$
	$\angle X(\omega) = -\angle X(-\omega)$
$x_e(n) = \frac{1}{2}[x(n) + x(-n)]$	$X_R(\omega)$
(real and even)	(real and even)
$x_o(n) = \frac{1}{2}[x(n) - x(-n)]$	$jX_I(\omega)$
(real and odd)	(imaginary and odd)

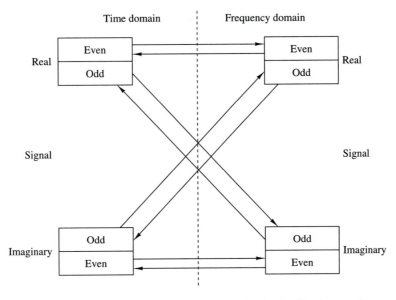

Figure 4.29 Summary of symmetry properties for the Fourier transform.

and

$$\angle X(\omega) = -\tan^{-1} \frac{a \sin \omega}{1 - a \cos \omega} \tag{4.3.40}$$

Figures 4.30 and 4.31 show the graphical representation of these spectra for $a = 0.8$. The reader can easily verify that as expected, all symmetry properties for the spectra of real signals apply to this case.

Example 4.3.2

Determine the Fourier transform of the signal

$$x(n) = \begin{cases} A, & -M \leq n \leq M \\ 0, & \text{elsewhere} \end{cases} \tag{4.3.41}$$

Solution Clearly, $x(-n) = x(n)$. Thus $x(n)$ is a real and even signal. From (4.3.21) we obtain

$$X(\omega) = X_R(\omega) = A \left(1 + 2 \sum_{n=1}^{M} \cos \omega n \right)$$

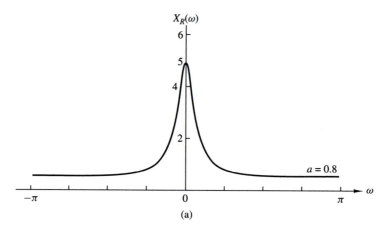

(a)

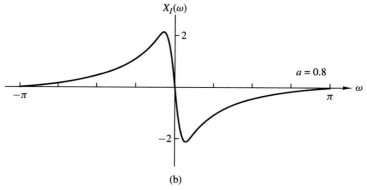

(b)

Figure 4.30 Graph of $X_R(\omega)$ and $X_I(\omega)$ for the transform in Example 4.3.1.

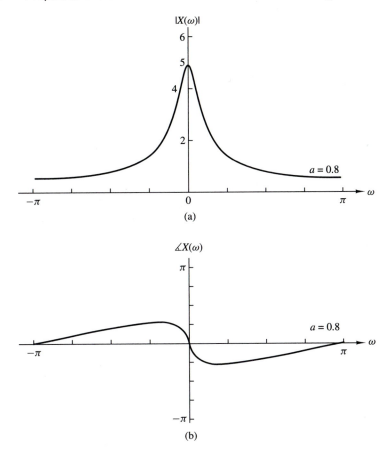

Figure 4.31   Magnitude and phase spectra of the transform in Example 4.3.1.

If we use the identity given in Problem 4.13, we obtain the simpler form

$$X(\omega) = A \frac{\sin(M + \frac{1}{2})\omega}{\sin(\omega/2)}$$

Since $X(\omega)$ is real, the magnitude and phase spectra are given by

$$|X(\omega)| = \left| A \frac{\sin(M + \frac{1}{2})\omega}{\sin(\omega/2)} \right| \qquad (4.3.42)$$

and

$$\angle X(\omega) = \begin{cases} 0, & \text{if } X(\omega) > 0 \\ \pi, & \text{if } X(\omega) < 0 \end{cases} \qquad (4.3.43)$$

Figure 4.32 shows the graphs for $X(\omega)$.

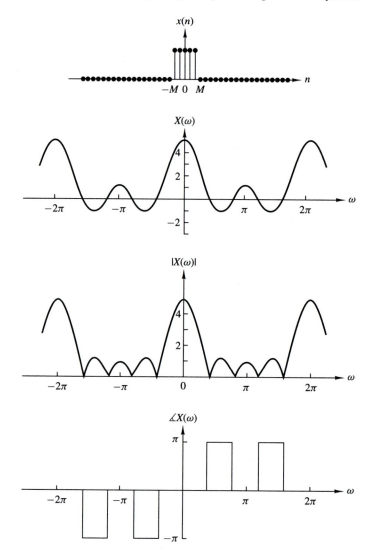

Figure 4.32 Spectral characteristics of rectangular pulse in Example 4.3.2.

4.3.2 Fourier Transform Theorems and Properties

In this section we introduce several Fourier transform theorems and illustrate their use in practice by examples.

Linearity.   If

$$x_1(n) \overset{F}{\longleftrightarrow} X_1(\omega)$$

and

$$x_2(n) \xleftrightarrow{F} X_2(\omega)$$

then

$$a_1 x_1(n) + a_2 x_2(n) \xleftrightarrow{F} a_1 X_1(\omega) + a_2 X_2(\omega) \qquad (4.3.44)$$

Simply stated, the Fourier transformation, viewed as an operation on a signal $x(n)$, is a linear transformation. Thus the Fourier transform of a linear combination of two or more signals is equal to the same linear combination of the Fourier transforms of the individual signals. This property is easily proved by using (4.3.1). The linearity property makes the Fourier transform suitable for the study of linear systems.

Example 4.3.3

Determine the Fourier transform of the signal

$$x(n) = a^{|n|} \qquad -1 < a < 1 \qquad (4.3.45)$$

Solution   First, we observe that $x(n)$ can be expressed as

$$x(n) = x_1(n) + x_2(n)$$

where

$$x_1(n) = \begin{cases} a^n, & n \geq 0 \\ 0, & n < 0 \end{cases}$$

and

$$x_2(n) = \begin{cases} a^{-n}, & n < 0 \\ 0, & n \geq 0 \end{cases}$$

Beginning with the definition of the Fourier transform in (4.3.1), we have

$$X_1(\omega) = \sum_{n=-\infty}^{\infty} x_1(n) e^{-j\omega n} = \sum_{n=0}^{\infty} a^n e^{-j\omega n} = \sum_{n=0}^{\infty} (ae^{-j\omega})^n$$

The summation is a geometric series that converges to

$$X_1(\omega) = \frac{1}{1 - ae^{-j\omega}}$$

provided that

$$|ae^{-j\omega}| = |a| \cdot |e^{-j\omega}| = |a| < 1$$

which is a condition that is satisfied in this problem. Similarly, the Fourier transform of $x_2(n)$ is

$$X_2(\omega) = \sum_{n=-\infty}^{\infty} x_2(n) e^{-j\omega n} = \sum_{n=-\infty}^{-1} a^{-n} e^{-j\omega n}$$

$$= \sum_{n=-\infty}^{-1} (ae^{j\omega})^{-n} = \sum_{k=1}^{\infty} (ae^{j\omega})^k$$

$$= \frac{ae^{j\omega}}{1 - ae^{j\omega}}$$

By combining these two transforms, we obtain the Fourier transform of $x(n)$ in the form

$$X(\omega) = X_1(\omega) + X_2(\omega)$$

$$= \frac{1 - a^2}{1 - 2a \cos \omega + a^2} \qquad (4.3.46)$$

Figure 4.33 illustrates $x(n)$ and $X(\omega)$ for the case in which $a = 0.8$.

Time shifting.   If

$$x(n) \xleftrightarrow{F} X(\omega)$$

then

$$x(n - k) \xleftrightarrow{F} e^{-j\omega k} X(\omega) \qquad (4.3.47)$$

The proof of this property follows immediately from the Fourier transform of $x(n - k)$ by making a change in the summation index. Thus

$$F\{x(n - k)\} = X(\omega)e^{-j\omega k}$$

$$= |X(\omega)|e^{j[\angle X(\omega) - \omega k]}$$

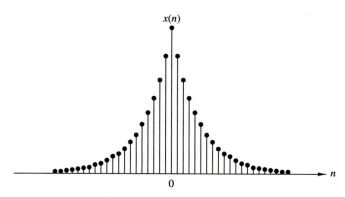

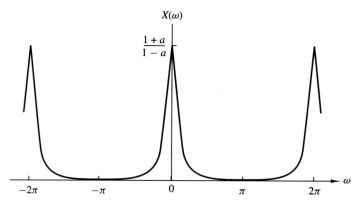

Figure 4.33   Sequence $x(n)$ and its Fourier transform in Example 4.3.3 with $a = 0.8$.

This relation means that if a signal is shifted in the time domain by k samples, its magnitude spectrum remains unchanged. However, the phase spectrum is changed by an amount $-\omega k$. This result can easily be explained if we recall that the frequency content of a signal depends only on its shape. From a mathematical point of view, we can say that shifting by k in the time domain, is equivalent to multiplying the spectrum by $e^{-j\omega k}$ in the frequency domain.

Time reversal.   If

$$x(n) \xleftrightarrow{F} X(\omega)$$

then

$$x(-n) \xleftrightarrow{F} X(-\omega) \tag{4.3.48}$$

This property can be established by performing the Fourier transformation of $x(-n)$ and making the summation index. Thus

$(l)e^{j\omega l} = X(-\omega)$

18) we obtain

$|X(-\omega)|e^{j\angle X(-\omega)}$

$\angle X(\omega)$

origin in time, its magnitude spectrum n undergoes a change in sign (phase

(ω)

$X_2(\omega)$

$$x(n) = x_1(n) * x_2(n) \xleftrightarrow{F} X(\omega) = X_1(\omega)X_2(\omega) \tag{4.3.49}$$

To prove (4.3.49), we recall the convolution formula

$$x(n) = x_1(n) * x_2(n) = \sum_{k=-\infty}^{\infty} x_1(k)x_2(n-k)$$

By multiplying both sides of this equation by the exponential $\exp(-j\omega n)$ and summing over all n, we obtain

$$X(\omega) = \sum_{n=-\infty}^{\infty} x(n)e^{-j\omega n} = \sum_{n=-\infty}^{\infty} \left[\sum_{k=-\infty}^{\infty} x_1(k)x_2(n-k) \right] e^{-j\omega n}$$

① Does TA happen to have the answers for all the Hw ?

② where can I get more examples like the one below !?

Q & A.

After interchanging the order of the summations and making a simple change in the summation index, the right-hand side of this equation reduces to the product $X_1(\omega)X_2(\omega)$. Thus (4.3.49) is established.

The convolution theorem is one of the most powerful tools in linear systems analysis. That is, if we convolve two signals in the time domain, then this is equivalent to multiplying their spectra in the frequency domain. In later chapters we will see that the convolution theorem provides an important computational tool for many digital signal processing applications.

Example 4.3.4

By use of (4.3.49), determine the convolution of the sequences

$$x_1(n) = x_2(n) = \{1, 1, 1\}$$
$$\uparrow$$

Solution   By using (4.3.21), we obtain

$$X_1(\omega) = X_2(\omega) = 1 + 2\cos\omega$$

Then

$$
\begin{aligned}
X(\omega) &= X_1(\omega)X_2(\omega) = (1 + 2\cos\omega)^2 \\
&= 3 + 4\cos\omega + 2\cos 2\omega \\
&= 3 + 2(e^{j\omega} + e^{-j\omega}) + (e^{j2\omega} + e^{-j2\omega})
\end{aligned}
$$

Hence the convolution of $x_1(n)$ with $x_2(n)$ is

$$x(n) = \{1\ 2\ 3\ 2\ 1\}$$
$$\uparrow$$

Figure 4.34 illustrates the foregoing relationships.

The correlation theorem.   If

$$x_1(n) \xleftrightarrow{F} X_1(\omega)$$

and

$$x_2(n) \xleftrightarrow{F} X_2(\omega)$$

then

$$r_{x_1 x_2}(m) \xleftrightarrow{F} S_{x_1 x_2}(\omega) = X_1(\omega)X_2(-\omega) \qquad (4.3.50)$$

The proof of (4.3.50) is similar to the proof of (4.3.49). In this case, we have

$$r_{x_1 x_2}(n) = \sum_{k=-\infty}^{\infty} x_1(k)x_2(k-n)$$

By multiplying both sides of this equation by the exponential $\exp(-j\omega n)$ and summing over all n, we obtain

$$S_{x_1 x_2}(\omega) = \sum_{n=-\infty}^{\infty} r_{x_1 x_2}(n)e^{-j\omega n} = \sum_{n=-\infty}^{\infty}\left[\sum_{k=-\infty}^{\infty} x_1(k)x_2(k-n)\right]e^{-j\omega n}$$

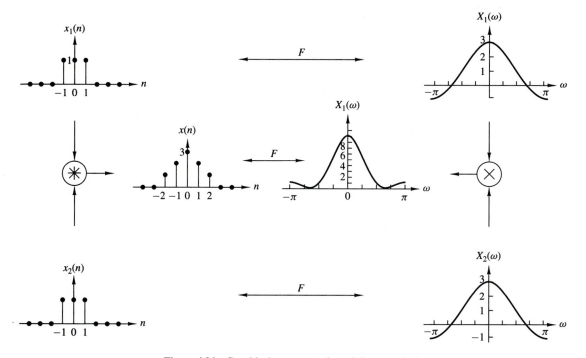

Figure 4.34 Graphical representation of the convolution property.

Finally, we interchange the order of the summations and make a change in the summation index. Thus we find that the right-hand side of the equation above reduces to $X_1(\omega)X_2(-\omega)$. The function $S_{x_1 x_2}(\omega)$ is called the *cross-energy density spectrum* of the signals $x_1(n)$ and $x_2(n)$.

The Wiener–Khintchine theorem.   Let $x(n)$ be a real signal. Then

$$r_{xx}(l) \xleftrightarrow{\ F\ } S_{xx}(\omega) \tag{4.3.51}$$

That is, the energy spectral density of an energy signal is the Fourier transform of its autocorrelation sequence. This is a special case of (4.3.50).

This is a very important result. It means that the autocorrelation sequence of a signal and its energy spectral density contain the same information about the signal. Since, neither of these contains any phase information, it is impossible to uniquely reconstruct the signal from the autocorrelation function or the energy density spectrum.

Example 4.3.5

Determine the energy density spectrum of the signal

$$x(n) = a^n u(n) \qquad -1 < a < 1$$

Solution From Example 2.6.2 we found that the autocorrelation function for this signal is

$$r_{xx}(l) = \frac{1}{1 - a^2} a^{|l|} \qquad \infty < l < \infty$$

By using the result in (4.3.46) for the Fourier transform of $a^{|l|}$, derived in Example 4.3.3, we have

$$F\{r_{xx}(l)\} = \frac{1}{1 - a^2} F\{a^{|l|}\} = \frac{1}{1 - 2a \cos \omega + a^2}$$

Thus, according to the Wiener–Khintchine theorem,

$$S_{xx}(\omega) = \frac{1}{1 - 2a \cos \omega + a^2}$$

Frequency shifting.   If

$$x(n) \xleftrightarrow{F} X(\omega)$$

then

$$e^{j\omega_0 n} x(n) \xleftrightarrow{F} X(\omega - \omega_0) \qquad (4.3.52)$$

This property is easily proved by direct substitution into the analysis equation (4.3.1). According to this property, multiplication of a sequence $x(n)$ by $e^{j\omega_0 n}$ is equivalent to a frequency translation of the spectrum $X(\omega)$ by ω_0. This frequency translation is illustrated in Fig. 4.35. Since the spectrum $X(\omega)$ is periodic, the shift ω_0 applies to the spectrum of the signal in every period.

The modulation theorem.   If

$$x(n) \xleftrightarrow{F} X(\omega)$$

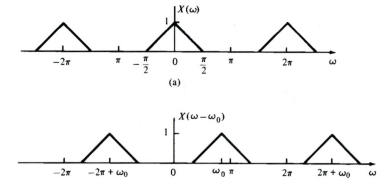

Figure 4.35 Illustration of the frequency-shifting property of the Fourier transform.

then

$$x(n) \cos \omega_0 n \xleftrightarrow{F} \tfrac{1}{2}[X(\omega + \omega_0) + X(\omega - \omega_0)] \qquad (4.3.53)$$

To prove the modulation theorem, we first express the signal $\cos \omega_0 n$ as

$$\cos \omega_0 n = \tfrac{1}{2}(e^{j\omega_0 n} + e^{-j\omega_0 n})$$

Upon multiplying $x(n)$ by these two exponentials and using the frequency-shifting property described in the preceding section, we obtain the desired result in (4.3.53).

Although the property given in (4.3.52) can also be viewed as (complex) modulation, in practice we prefer to use (4.3.53) because the signal $x(n) \cos \omega_0 n$ is real. Clearly, in this case the symmetry properties (4.3.12) and (4.3.13) are preserved.

The modulation theorem is illustrated in Fig. 4.36, which contains a plot of the spectra of the signals $x(n)$, $y_1(n) = x(n) \cos 0.5\pi n$ and $y_2(n) = x(n) \cos \pi n$.

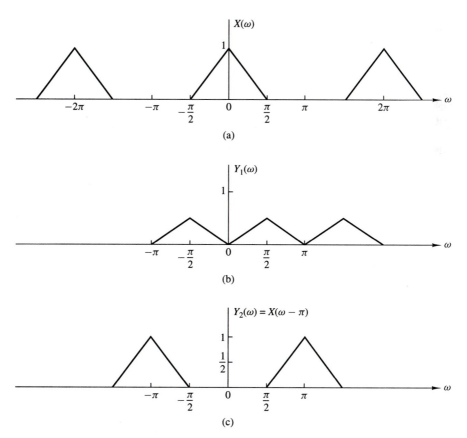

Figure 4.36 Graphical representation of the modulation theorem.

Parseval's theorem. If

$$x_1(n) \xleftrightarrow{F} X_1(\omega)$$

and

$$x_2(n) \xleftrightarrow{F} X_2(\omega)$$

then

$$\sum_{n=-\infty}^{\infty} x_1(n)x_2^*(n) = \frac{1}{2\pi} \int_{-\pi}^{\pi} X_1(\omega)X_2^*(\omega)d\omega \qquad (4.3.54)$$

To prove this theorem, we use (4.3.1) to eliminate $X_1(\omega)$ on the right-hand side of (4.3.54). Thus we have

$$\frac{1}{2\pi} \int_{2\pi} \left[\sum_{n=-\infty}^{\infty} x_1(n)e^{-j\omega n} \right] X_2^*(\omega)d\omega$$

$$= \sum_{n=-\infty}^{\infty} x_1(n) \frac{1}{2\pi} \int_{2\pi} X_2^*(\omega)e^{-j\omega n}d\omega = \sum_{n=-\infty}^{\infty} x_1(n)x_2^*(n)$$

In the special case where $x_2(n) = x_1(n) = x(n)$, Parseval's relation (4.3.54) reduces to

$$\sum_{n=-\infty}^{\infty} |x(n)|^2 = \frac{1}{2\pi} \int_{2\pi} |X(\omega)|^2 d\omega \qquad (4.3.55)$$

We observe that the left-hand side of (4.3.55) is simply the energy E_x of the signal $x(n)$. It is also equal to the autocorrelation of $x(n)$, $r_{xx}(l)$, evaluated at $l = 0$. The integrand in the right-hand side of (4.3.55) is equal to the energy density spectrum, so the integral over the interval $-\pi \le \omega \le \pi$ yields the total signal energy. Therefore, we conclude that

$$E_x = r_{xx}(0) = \sum_{n=-\infty}^{\infty} |x(n)|^2 = \frac{1}{2\pi} \int_{2\pi} |X(\omega)|^2 d\omega = \frac{1}{2\pi} \int_{-\pi}^{\pi} S_{xx}(\omega)d\omega \qquad (4.3.56)$$

Multiplication of two sequences (Windowing theorem). If

$$x_1(n) \xleftrightarrow{F} X_1(\omega)$$

and

$$x_2(n) \xleftrightarrow{F} X_2(\omega)$$

then

$$x_3(n) \equiv x_1(n)x_2(n) \xleftrightarrow{F} X_3(\omega) = \frac{1}{2\pi} \int_{-\pi}^{\pi} X_1(\lambda) X_2(\omega - \lambda)d\lambda \qquad (4.3.57)$$

The integral on the right-hand side of (4.3.57) represents the convolution of the Fourier transforms $X_1(\omega)$ and $X_2(\omega)$. This relation is the dual of the time-domain convolution. In other words, the multiplication of two time-domain sequences is equivalent to the convolution of their Fourier transforms. On the other hand, the convolution of two time-domain sequences is equivalent to the multiplication of their Fourier transforms.

To prove (4.3.57) we begin with the Fourier transform of $x_3(n) = x_1(n)x_2(n)$ and use the formula for the inverse transform, namely,

$$x_1(n) = \frac{1}{2\pi} \int_{-\pi}^{\pi} X_1(\lambda)e^{j\lambda n}d\lambda$$

Thus, we have

$$
\begin{aligned}
X_3(\omega) &= \sum_{n=-\infty}^{\infty} x_3(n)e^{-j\omega n} = \sum_{n=-\infty}^{\infty} x_1(n)x_2(n)e^{-j\omega n} \\
&= \sum_{n=-\infty}^{\infty} \left[\frac{1}{2\pi} \int_{-\pi}^{\pi} X_1(\lambda)e^{j\lambda n}d\lambda \right] x_2(n)e^{-j\omega n} \\
&= \frac{1}{2\pi} \int_{-\pi}^{\pi} X_1(\lambda)d\lambda \left[\sum_{n=-\infty}^{\infty} x_2(n)e^{-j(\omega-\lambda)n} \right] \\
&= \frac{1}{2\pi} \int_{-\pi}^{\pi} X_1(\lambda) X_2(\omega - \lambda)d\lambda
\end{aligned}
$$

The convolution integral in (4.3.57) is known as the *periodic convolution* of $X_1(\omega)$ and $X_2(\omega)$ because it is the convolution of two periodic functions having the same period. We note that the limits of integration extend over a single period. Furthermore, we note that due to the periodicity of the Fourier transform for discrete-time signals, there is no "perfect" duality between the time and frequency domains with respect to the convolution operation, as in the case of continuous-time signals. Indeed, convolution in the time domain (aperiodic summation) is equivalent to multiplication of continuous periodic Fourier transforms. However, multiplication of aperiodic sequences is equivalent to periodic convolution of their Fourier transforms.

The Fourier transform pair in (4.3.57) will prove useful in our treatment of FIR filter design based on the window technique.

Differentiation in the frequency domain.   If

$$x(n) \overset{F}{\longleftrightarrow} X(\omega)$$

then

$$nx(n) \overset{F}{\longleftrightarrow} j\frac{dX(\omega)}{d\omega} \tag{4.3.58}$$

To prove this property, we use the definition of the Fourier transform in (4.3.1) and differentiate the series term by term with respect to ω. Thus we obtain

$$
\begin{aligned}
\frac{dX(\omega)}{d\omega} &= \frac{d}{d\omega}\left[\sum_{n=-\infty}^{\infty} x(n)e^{-j\omega n}\right] \\
&= \sum_{n=-\infty}^{\infty} x(n)\frac{d}{d\omega}e^{-j\omega n} \\
&= -j\sum_{n=-\infty}^{\infty} nx(n)e^{-j\omega n}
\end{aligned}
$$

Now we multiply both sides of the equation by j to obtain the desired result in (4.3.58).

The properties derived in this section are summarized in Table 4.5, which serves as a convenient reference. Table 4.6 illustrates some useful Fourier transform pairs that will be encountered in later chapters.

TABLE 4.5   PROPERTIES OF THE FOURIER TRANSFORM FOR DISCRETE-TIME SIGNALS

Property	Time Domain	Frequency Domain
Notation	$x(n)$	$X(\omega)$
	$x_1(n)$	$X_1(\omega)$
	$x_2(n)$	$X_2(\omega)$
Linearity	$a_1 x_1(n) + a_2 x_2(n)$	$a_1 X_1(\omega) + a_2 X_2(\omega)$
Time shifting	$x(n-k)$	$e^{-j\omega k}X(\omega)$
Time reversal	$x(-n)$	$X(-\omega)$
Convolution	$x_1(n) * x_2(n)$	$X_1(\omega)X_2(\omega)$
Correlation	$r_{x_1 x_2}(l) = x_1(l) * x_2(-l)$	$S_{x_1 x_2}(\omega) = X_1(\omega)X_2(-\omega)$
		$= X_1(\omega)X_2^*(\omega)$
		[if $x_2(n)$ is real]
Wiener–Khintchine theorem	$r_{xx}(l)$	$S_{xx}(\omega)$
Frequency shifting	$e^{j\omega_0 n}x(n)$	$X(\omega - \omega_0)$
Modulation	$x(n)\cos\omega_0 n$	$\frac{1}{2}X(\omega + \omega_0) + \frac{1}{2}X(\omega - \omega_0)$
Multiplication	$x_1(n)x_2(n)$	$\dfrac{1}{2\pi}\displaystyle\int_{-\pi}^{\pi} X_1(\lambda)X_2(\omega - \lambda)d\lambda$
Differentiation in the frequency domain	$nx(n)$	$j\dfrac{dX(\omega)}{d\omega}$
Conjugation	$x^*(n)$	$X^*(-\omega)$
Parseval's theorem	$\displaystyle\sum_{n=-\infty}^{\infty} x_1(n)x_2^*(n) = \dfrac{1}{2\pi}\displaystyle\int_{-\pi}^{\pi} X_1(\omega)X_2^*(\omega)d\omega$	

TABLE 4.6 SOME USEFUL FOURIER TRANSFORM PAIRS FOR DISCRETE-TIME APERIODIC SIGNALS

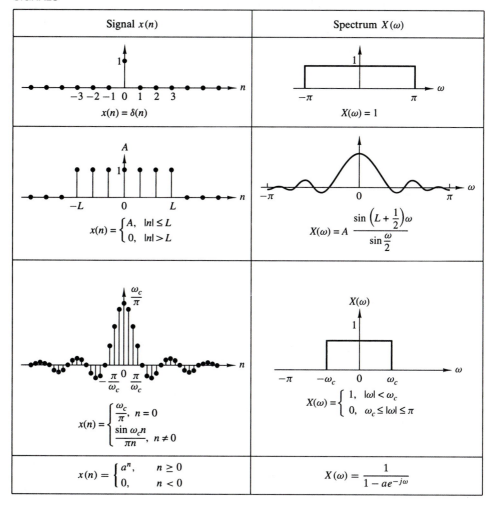

Signal $x(n)$	Spectrum $X(\omega)$
$x(n) = \delta(n)$	$X(\omega) = 1$
$x(n) = \begin{cases} A, & |n| \leq L \\ 0, & |n| > L \end{cases}$	$X(\omega) = A\,\dfrac{\sin\left(L+\frac{1}{2}\right)\omega}{\sin\frac{\omega}{2}}$
$x(n) = \begin{cases} \dfrac{\omega_c}{\pi}, & n = 0 \\ \dfrac{\sin\omega_c n}{\pi n}, & n \neq 0 \end{cases}$	$X(\omega) = \begin{cases} 1, & |\omega| < \omega_c \\ 0, & \omega_c \leq |\omega| \leq \pi \end{cases}$
$x(n) = \begin{cases} a^n, & n \geq 0 \\ 0, & n < 0 \end{cases}$	$X(\omega) = \dfrac{1}{1 - ae^{-j\omega}}$

4.4 FREQUENCY-DOMAIN CHARACTERISTICS OF LINEAR TIME-INVARIANT SYSTEMS

In this section we develop the characterization of linear time-invariant systems in the frequency domain. The basic excitation signals in this development are the complex exponentials and sinusoidal functions. The characteristics of the system are described by a function of the frequency variable ω called the frequency response, which is the Fourier transform of the impulse response $h(n)$ of the system.

The frequency response function completely characterizes a linear time-invariant system in the frequency domain. This allows us to determine the steady-state response of the system to any arbitrary weighted linear combination of sinusoids or complex exponentials. Since periodic sequences, in particular, lend themselves to a Fourier series decomposition as a weighted sum of harmonically related complex exponentials, it becomes a simple matter to determine the response of a linear time-invariant system to this class of signals. This methodology is also applied to aperiodic signals since such signals can be viewed as a superposition of infinitesimal size complex exponentials.

4.4.1 Response to Complex Exponential and Sinusoidal Signals: The Frequency Response Function

In Chapter 2, it was demonstrated that the response of any relaxed linear time-invariant system to an arbitrary input signal $x(n)$, is given by the convolution sum formula

$$y(n) = \sum_{k=-\infty}^{\infty} h(k)x(n-k) \tag{4.4.1}$$

In this input–output relationship, the system is characterized in the time domain by its unit sample response $\{h(n), -\infty < n < \infty\}$.

To develop a frequency-domain characterization of the system, let us excite the system with the complex exponential

$$x(n) = Ae^{j\omega n} \qquad -\infty < n < \infty \tag{4.4.2}$$

where A is the amplitude and ω is any arbitrary frequency confined to the frequency interval $[-\pi, \pi]$. By substituting (4.4.2) into (4.4.1), we obtain the response

$$\begin{aligned} y(n) &= \sum_{k=-\infty}^{\infty} h(k)[Ae^{j\omega(n-k)}] \\ &= A\left[\sum_{k=-\infty}^{\infty} h(k)e^{-j\omega k}\right] e^{j\omega n} \end{aligned} \tag{4.4.3}$$

We observe that the term in brackets in (4.4.3) is a function of the frequency variable ω. In fact, this term is the Fourier transform of the unit sample response $h(k)$ of the system. Hence we denote this function as

$$H(\omega) = \sum_{k=-\infty}^{\infty} h(k)e^{-j\omega k} \tag{4.4.4}$$

Clearly, the function $H(\omega)$ exists if the system is BIBO stable, that is, if

$$\sum_{n=-\infty}^{\infty} |h(n)| < \infty$$

With the definition in (4.4.4), the response of the system to the complex exponential given in (4.4.2) is

$$y(n) = AH(\omega)e^{j\omega n} \tag{4.4.5}$$

We note that the response is also in the form of a complex exponential with the same frequency as the input, but altered by the multiplicative factor $H(\omega)$.

As a result of this characteristic behavior, the exponential signal in (4.4.2) is called an *eigenfunction* of the system. In other words, an eigenfunction of a system is an input signal that produces an output that differs from the input by a constant multiplicative factor. The multiplicative factor is called an *eigenvalue* of the system. In this case, a complex exponential signal of the form (4.4.2) is an eigenfunction of a linear time-invariant system, and $H(\omega)$ evaluated at the frequency of the input signal is the corresponding eigenvalue.

Example 4.4.1

Determine the output sequence of the system with impulse response

$$h(n) = (\tfrac{1}{2})^n u(n) \tag{4.4.6}$$

when the input is the complex exponential sequence

$$x(n) = Ae^{j\pi n/2} \qquad -\infty < n < \infty$$

Solution First we evaluate the Fourier transform of the impulse response $h(n)$, and then we use (4.4.5) to determine $y(n)$. From Example 4.2.3 we recall that

$$H(\omega) = \sum_{n=-\infty}^{\infty} h(n)e^{-j\omega n} = \frac{1}{1 - \tfrac{1}{2}e^{-j\omega}} \tag{4.4.7}$$

At $\omega = \pi/2$, (4.4.7) yields

$$H\left(\frac{\pi}{2}\right) = \frac{1}{1 + j\tfrac{1}{2}} = \frac{2}{\sqrt{5}} e^{-j26.6°}$$

and therefore the output is

$$y(n) = A\left(\frac{2}{\sqrt{5}}e^{-j26.6°}\right)e^{j\pi n/2}$$

$$y(n) = \frac{2}{\sqrt{5}}Ae^{j(\pi n/2 - 26.6°)} \qquad -\infty < n < \infty \tag{4.4.8}$$

This example clearly illustrates that the only effect of the system on the input signal is to scale the amplitude by $2/\sqrt{5}$ and shift the phase by $-26.6°$. Thus the output is also a complex exponential of frequency $\pi/2$, amplitude $2A/\sqrt{5}$, and phase $-26.6°$.

If we alter the frequency of the input signal, the effect of the system on the input also changes and hence the output changes. In particular, if the input sequence is a complex exponential of frequency π, that is,

$$x(n) = Ae^{j\pi n} \qquad -\infty < n < \infty \tag{4.4.9}$$

then, at $\omega = \pi$,

$$H(\pi) = \frac{1}{1 - \tfrac{1}{2}e^{-j\pi}} = \frac{1}{\tfrac{3}{2}} = \frac{2}{3}$$

and the output of the system is

$$y(n) = \tfrac{2}{3} A e^{j\pi n} \qquad -\infty < n < \infty \tag{4.4.10}$$

We note that $H(\pi)$ is purely real [i.e., the phase associated with $H(\omega)$ is zero at $\omega = \pi$]. Hence, the input is scaled in amplitude by the factor $H(\pi) = \tfrac{2}{3}$, but the phase shift is zero.

In general, $H(\omega)$ is a complex-valued function of the frequency variable ω. Hence it can be expressed in polar form as

$$H(\omega) = |H(\omega)| e^{j\Theta(\omega)} \tag{4.4.11}$$

where $|H(\omega)|$ is the magnitude of $H(\omega)$ and

$$\Theta(\omega) = \angle H(\omega)$$

which is the phase shift imparted on the input signal by the system at the frequency ω.

Since $H(\omega)$ is the Fourier transform of $\{h(k)\}$, it follows that $H(\omega)$ is a periodic function with period 2π. Furthermore, we can view (4.4.4) as the exponential Fourier series expansion for $H(\omega)$, with $h(k)$ as the Fourier series coefficients. Consequently, the unit impulse $h(k)$ is related to $H(\omega)$ through the integral expression

$$h(k) = \frac{1}{2\pi} \int_{-\pi}^{\pi} H(\omega) e^{j\omega k} d\omega \tag{4.4.12}$$

For a linear time-invariant system with a real-valued impulse response, the magnitude and phase functions possess symmetry properties which are developed as follows. From the definition of $H(\omega)$, we have

$$\begin{aligned}
H(\omega) &= \sum_{k=-\infty}^{\infty} h(k) e^{-j\omega k} \\
&= \sum_{k=-\infty}^{\infty} h(k) \cos \omega k - j \sum_{k=-\infty}^{\infty} h(k) \sin \omega k \\
&= H_R(\omega) + j H_I(\omega) \\
&= \sqrt{H_R^2(\omega) + H_I^2(\omega)}\, e^{j \tan^{-1}[H_I(\omega)/H_R(\omega)]}
\end{aligned} \tag{4.4.13}$$

where $H_R(\omega)$ and $H_I(\omega)$ denote the real and imaginary components of $H(\omega)$, defined as

$$\begin{aligned}
H_R(\omega) &= \sum_{k=-\infty}^{\infty} h(k) \cos \omega k \\
H_I(\omega) &= -\sum_{k=-\infty}^{\infty} h(k) \sin \omega k
\end{aligned} \tag{4.4.14}$$

It is clear from (4.4.12) that the magnitude and phase of $H(\omega)$, expressed in terms of $H_R(\omega)$ and $H_I(\omega)$, are

$$\begin{aligned}
|H(\omega)| &= \sqrt{H_R^2(\omega) + H_I^2(\omega)} \\
\Theta(\omega) &= \tan^{-1} \frac{H_I(\omega)}{H_R(\omega)}
\end{aligned} \tag{4.4.15}$$

We note that $H_R(\omega) = H_R(-\omega)$ and $H_I(\omega) = -H_I(-\omega)$, so that $H_R(\omega)$ is an even function of ω and $H_I(\omega)$ is an odd function of ω. As a consequence, it follows that $|H(\omega)|$ is an even function of ω and $\Theta(\omega)$ is an odd function of ω. Hence, if we know $|H(\omega)|$ and $\Theta(\omega)$ for $0 \leq \omega \leq \pi$, we also know these functions for $-\pi \leq \omega \leq 0$.

Example 4.4.2   Moving Average Filter

Determine the magnitude and phase of $H(\omega)$ for the three-point moving average (MA) system

$$y(n) = \tfrac{1}{3}[x(n+1) + x(n) + x(n-1)]$$

and plot these two functions for $0 \leq \omega \leq \pi$.

Solution   Since

$$h(n) = \{\tfrac{1}{3}, \tfrac{1}{3}, \tfrac{1}{3}\}$$
$$\uparrow$$

it follows that

$$H(\omega) = \tfrac{1}{3}(e^{j\omega} + 1 + e^{-j\omega}) = \tfrac{1}{3}(1 + 2\cos\omega)$$

Hence

$$|H(\omega)| = \tfrac{1}{3}|1 + 2\cos\omega| \qquad (4.4.16)$$

$$\Theta(\omega) = \begin{cases} 0, & 0 \leq \omega \leq 2\pi/3 \\ \pi, & 2\pi/3 \leq \omega < \pi \end{cases}$$

Figure 4.37 illustrates the graphs of the magnitude and phase of $H(\omega)$. As indicated previously, $|H(\omega)|$ is an even function of frequency and $\Theta(\omega)$ is an odd function of

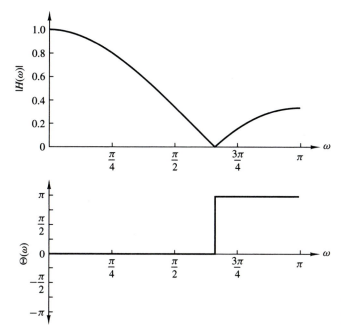

Figure 4.37 Magnitude and phase responses for the MA system in Example 4.4.2.

frequency. It is apparent from the frequency response characteristic $H(\omega)$ that this moving average filter smooths the input data, as we would expect from the input–output equation.

The symmetry properties satisfied by the magnitude and phase functions of $H(\omega)$, and the fact that a sinusoid can be expressed as a sum or difference of two complex-conjugate exponential functions, imply that the response of a linear time-invariant system to a sinusoid is similar in form to the response when the input is a complex exponential. Indeed, if the input is

$$x_1(n) = Ae^{j\omega n}$$

the output is

$$y_1(n) = A|H(\omega)|e^{j\Theta(\omega)}e^{j\omega n}$$

On the other hand, if the input is

$$x_2(n) = Ae^{-j\omega n}$$

the response of the system is

$$y_2(n) = A|H(-\omega)|e^{j\Theta(-\omega)}e^{-j\omega n}$$
$$= A|H(\omega)|e^{-j\Theta(\omega)}e^{-j\omega n}$$

where, in the last expression, we have made use of the symmetry properties $|H(\omega)| = |H(-\omega)|$ and $\Theta(\omega) = -\Theta(-\omega)$. Now, by applying the superposition property of the linear time-invariant system, we find that the response of the system to the input

$$x(n) = \tfrac{1}{2}[x_1(n) + x_2(n)] = A \cos \omega n$$

is

$$y(n) = \tfrac{1}{2}[y_1(n) + y_2(n)]$$
$$y(n) = A|H(\omega)| \cos[\omega n + \Theta(\omega)]$$

(4.4.17)

Similarly, if the input is

$$x(n) = \frac{1}{j2}[x_1(n) - x_2(n)] = A \sin \omega n$$

the response of the system is

$$y(n) = \frac{1}{j2}[y_1(n) - y_2(n)]$$
$$y(n) = A|H(\omega)| \sin[\omega n + \Theta(\omega)]$$

(4.4.18)

It is apparent from this discussion that $H(\omega)$, or equivalently, $|H(\omega)|$ and $\Theta(\omega)$, completely characterize the effect of the system on a sinusoidal input signal of any arbitrary frequency. Indeed, we note that $|H(\omega)|$ determines the amplification ($|H(\omega)| > 1$) or attenuation ($|H(\omega)| < 1$) imparted by the system on the input sinusoid. The phase $\Theta(\omega)$ determines the amount of phase shift imparted by the system on the input sinusoid. Consequently, by knowing $H(\omega)$, we are able to determine the response of the system to any sinusoidal input signal. Since $H(\omega)$ specifies the response of the system in the frequency domain, it is called the *frequency response* of the system. Correspondingly, $|H(\omega)|$ is called the *magnitude response* and $\Theta(\omega)$ is called the *phase response* of the system.

If the input to the system consists of more than one sinusoid, the superposition property of the linear system can be used to determine the response. The following examples illustrate the use of the superposition property.

Example 4.4.3

Determine the response of the system in Example 4.4.1 to the input signal

$$x(n) = 10 - 5\sin\frac{\pi}{2}n + 20\cos\pi n \qquad -\infty < n < \infty$$

Solution The frequency response of the system is given in (4.4.7) as

$$H(\omega) = \frac{1}{1 - \frac{1}{2}e^{-j\omega}}$$

The first term in the input signal is a fixed signal component corresponding to $\omega = 0$. Thus

$$H(0) = \frac{1}{1 - \frac{1}{2}} = 2$$

The second term in $x(n)$ has a frequency $\pi/2$. At this frequency the frequency response of the system is

$$H\left(\frac{\pi}{2}\right) = \frac{2}{\sqrt{5}}e^{-j26.6^\circ}$$

Finally, the third term in $x(n)$ has a frequency $\omega = \pi$. At this frequency

$$H(\pi) = \frac{2}{3}$$

Hence the response of the system to $x(n)$ is

$$y(n) = 20 - \frac{10}{\sqrt{5}}\sin\left(\frac{\pi}{2}n - 26.6^\circ\right) + \frac{40}{3}\cos\pi n \qquad -\infty < n < \infty$$

Example 4.4.4

A linear time-invariant system is described by the following difference equation:

$$y(n) = ay(n-1) + bx(n) \qquad 0 < a < 1$$

(a) Determine the magnitude and phase of the frequency response $H(\omega)$ of the system.

(b) Choose the parameter b so that the maximum value of $|H(\omega)|$ is unity, and sketch $|H(\omega)|$ and $\angle H(\omega)$ for $a = 0.9$.

(c) Determine the output of the system to the input signal

$$x(n) = 5 + 12 \sin \frac{\pi}{2} n - 20 \cos \left(\pi n + \frac{\pi}{4} \right)$$

Solution The impulse response of the system is

$$h(n) = ba^n u(n)$$

Since $|a| < 1$, the system is BIBO stable and hence $H(\omega)$ exists.

(a) The frequency response is

$$H(\omega) = \sum_{n=-\infty}^{\infty} h(n) e^{-j\omega n}$$

$$= \frac{b}{1 - ae^{-j\omega}}$$

Since

$$1 - ae^{-j\omega} = (1 - a \cos \omega) + ja \sin \omega$$

it follows that

$$|1 - ae^{-j\omega}| = \sqrt{(1 - a \cos \omega)^2 + (a \sin \omega)^2}$$

$$= \sqrt{1 + a^2 - 2a \cos \omega}$$

and

$$\angle (1 - ae^{-j\omega}) = \tan^{-1} \frac{a \sin \omega}{1 - a \cos \omega}$$

Therefore,

$$|H(\omega)| = \frac{|b|}{\sqrt{1 + a^2 - 2a \cos \omega}}$$

$$\angle H(\omega) = \Theta(\omega) = \angle b - \tan^{-1} \frac{a \sin \omega}{1 - a \cos \omega}$$

(b) Since the parameter a is positive, the denominator of $|H(\omega)|$ attains a minimum at $\omega = 0$. Therefore, $|H(\omega)|$ attains its maximum value at $\omega = 0$. At this frequency we have

$$|H(0)| = \frac{|b|}{1 - a} = 1$$

which implies that $b = \pm (1 - a)$. We choose $b = 1 - a$, so that

$$|H(\omega)| = \frac{1 - a}{\sqrt{1 + a^2 - 2a \cos \omega}}$$

and

$$\Theta(\omega) = - \tan^{-1} \frac{a \sin \omega}{1 - a \cos \omega}$$

The frequency response plots for $|H(\omega)|$ and $\Theta(\omega)$ are illustrated in Fig. 4.38. We observe that this system attenuates high frequency signals.

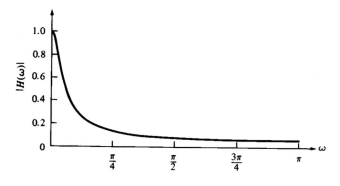

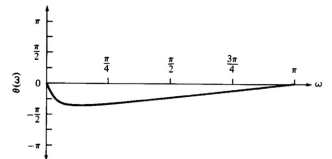

Figure 4.38 Magnitude and phase responses for the system in Example 4.4.4 with $a = 0.9$.

(c) The input signal consists of components of frequencies $\omega = 0$, $\pi/2$, and π. For $\omega = 0$, $|H(0)| = 1$ and $\Theta(0) = 0$. For $\omega = \pi/2$,

$$\left| H\left(\frac{\pi}{2}\right) \right| = \frac{1-a}{\sqrt{1+a^2}} = \frac{0.1}{\sqrt{1.81}} = 0.074$$

$$\Theta\left(\frac{\pi}{2}\right) = -\tan^{-1} a = -42°$$

For $\omega = \pi$,

$$|H(\pi)| = \frac{1-a}{1+a} = \frac{0.1}{1.9} = 0.053$$

$$\Theta(\pi) = 0$$

Therefore, the output of the system is

$$y(n) = 5|H(0)| + 12\left| H\left(\frac{\pi}{2}\right) \right| \sin\left[\frac{\pi}{2}n + \Theta\left(\frac{\pi}{2}\right)\right]$$

$$- 20|H(\pi)| \cos\left[\pi n + \frac{\pi}{4} + \Theta(\pi)\right]$$

$$= 5 + 0.888 \sin\left(\frac{\pi}{2}n - 42°\right) - 1.06 \cos\left(\pi n + \frac{\pi}{4}\right) \qquad -\infty < n < \infty$$

In the most general case, if the input to the system consists of an arbitrary linear combination of sinusoids of the form

$$x(n) = \sum_{i=1}^{L} A_i \cos(\omega_i n + \phi_i) \qquad -\infty < n < \infty$$

where $\{A_i\}$ and $\{\phi_i\}$ are the amplitudes and phases of the corresponding sinusoidal components, then the response of the system is simply

$$y(n) = \sum_{i=1}^{L} A_i |H(\omega_i)| \cos[\omega_i n + \phi_i + \Theta(\omega_i)] \qquad (4.4.19)$$

where $|H(\omega_i)|$ and $\Theta(\omega_i)$ are the magnitude and phase, respectively, imparted by the system to the individual frequency components of the input signal.

It is clear that depending on the frequency response $H(\omega)$ of the system, input sinusoids of different frequencies will be affected differently by the system. For example, some sinusoids may be completely suppressed by the system if $H(\omega) = 0$ at the frequencies of these sinusoids. Other sinusoids may receive no attenuation (or perhaps, some amplification) by the system. In effect, we can view the linear time-invariant system functioning as a *filter* to sinusoids of different frequencies, passing some of the frequency components to the output and suppressing or preventing other frequency components from reaching the output. In fact, as discussed in Chapter 8, the basic digital filter design problem involves determining the parameters of a linear time-invariant system to achieve a desired frequency response $H(\omega)$.

4.4.2 Steady-State and Transient Response to Sinusoidal Input Signals

In the discussion in the preceding section, we determined the response of a linear time-invariant system to exponential and sinusoidal input signals applied to the system at $n = -\infty$. We usually call such signals eternal exponentials or eternal sinusoids, because they were applied at $n = -\infty$. In such a case, the response that we observe at the output of the system is the steady-state response. There is no transient response in this case.

On the other hand, if the exponential or sinusoidal signal is applied at some finite time instant, say at $n = 0$, the response of the system consists of two terms, the transient response and the steady-state response. To demonstrate this behavior, let us consider, as an example, the system described by the first-order difference equation

$$y(n) = ay(n-1) + x(n) \qquad (4.4.20)$$

This system was considered in Section 2.4.2. Its response to any input $x(n)$ applied at $n = 0$ is given by (2.4.8) as

$$y(n) = a^{n+1} y(-1) + \sum_{k=0}^{n} a^k x(n-k) \qquad n \geq 0 \qquad (4.4.21)$$

where $y(-1)$ is the initial condition.

Now, let us assume that the input to the system is the complex exponential

$$x(n) = Ae^{j\omega n} \qquad n \geq 0 \qquad (4.4.22)$$

applied at $n = 0$. When we substitute (4.4.22) into (4.4.21), we obtain

$$
\begin{aligned}
y(n) &= a^{n+1}y(-1) + A\sum_{k=0}^{n} a^{k}e^{j\omega(n-k)} \\
&= a^{n+1}y(-1) + A\left[\sum_{k=0}^{n}(ae^{-j\omega})^{k}\right]e^{j\omega n} \\
&= a^{n+1}y(-1) + A\frac{1 - a^{n+1}e^{-j\omega(n+1)}}{1 - ae^{-j\omega}}e^{j\omega n} \qquad n \geq 0 \\
&= a^{n+1}y(-1) - \frac{Aa^{n+1}e^{-j\omega(n+1)}}{1 - ae^{-j\omega}}e^{j\omega n} + \frac{A}{1 - ae^{-j\omega}}e^{j\omega n} \qquad n \geq 0
\end{aligned}
\qquad (4.4.23)
$$

We recall that the system in (4.4.20) is BIBO stable if $|a| < 1$. In this case the two terms involving a^{n+1} in (4.4.23) decay toward zero as n approaches infinity. Consequently, we are left with the steady-state response

$$
\begin{aligned}
y_{ss}(n) &= \lim_{n\to\infty} y(n) = \frac{A}{1 - ae^{-j\omega}}e^{j\omega n} \\
&= AH(\omega)e^{j\omega n}
\end{aligned}
\qquad (4.4.24)
$$

The first two terms in (4.4.23) constitute the transient response of the system, that is,

$$y_{tr}(n) = a^{n+1}y(-1) - \frac{Aa^{n+1}e^{-j\omega(n+1)}}{1 - ae^{-j\omega}}e^{j\omega n} \qquad n \geq 0 \qquad (4.4.25)$$

which decay toward zero as n approaches infinity. The first term in the transient response is the zero-input response of the system and the second term is the transient produced by the exponential input signal.

In general, all linear time-invariant BIBO systems behave in a similar fashion when excited by a complex exponential, or by a sinusoid at $n = 0$ or at some other finite time instant. That is, the transient response decays toward zero as $n \to \infty$, leaving only the steady-state response that we determined in the preceding section. In many practical applications, the transient response of the system is unimportant, and therefore it is usually ignored in dealing with the response of the system to sinusoidal inputs.

4.4.3 Steady-State Response to Periodic Input Signals

Suppose that the input to a stable linear time-invariant system is a periodic signal $x(n)$ with fundamental period N. Since such a signal exists from $-\infty < n < \infty$, the total response of the system at any time instant n, is simply equal to the steady-state response.

To determine the response $y(n)$ of the system, we make use of the Fourier series representation of the periodic signal, which is

$$x(n) = \sum_{k=0}^{N-1} c_k e^{j2\pi kn/N} \qquad k = 0, 1, \ldots, N-1 \qquad (4.4.26)$$

where the $\{c_k\}$ are the Fourier series coefficients. Now the response of the system to the complex exponential signal

$$x_k(n) = c_k e^{j2\pi kn/N} \qquad k = 0, 1, \ldots, N-1$$

is

$$y_k(n) = c_k H\left(\frac{2\pi}{N}k\right) e^{j2\pi kn/N} \qquad k = 0, 1, \ldots, N-1 \qquad (4.4.27)$$

where

$$H\left(\frac{2\pi k}{N}\right) = H(\omega)|_{\omega = 2\pi k/N} \qquad k = 0, 1, \ldots, N-1$$

By using the superposition principle for linear systems, we obtain the response of the system to the periodic signal $x(n)$ in (4.4.26) as

$$y(n) = \sum_{k=0}^{N-1} c_k H\left(\frac{2\pi k}{N}\right) e^{j2\pi kn/N} \qquad -\infty < n < \infty \qquad (4.4.28)$$

This result implies that the response of the system to the periodic input signal $x(n)$ is also periodic with the same period N. The Fourier series coefficients for $y(n)$ are

$$d_k \equiv c_k H\left(\frac{2\pi k}{N}\right) \qquad k = 0, 1, \ldots, N-1 \qquad (4.4.29)$$

Hence, the linear system can change the shape of the periodic input signal by scaling the amplitude and shifting the phase of the Fourier series components, but it does not affect the period of the periodic input signal.

4.4.4 Response to Aperiodic Input Signals

The convolution theorem, given in (4.3.49), provides the desired frequency-domain relationship for determining the output of an LTI system to an aperiodic finite-energy signal. If $\{x(n)\}$ denotes the input sequence, $\{y(n)\}$ denotes the output sequence, and $\{h(n)\}$ denotes the unit sample response of the system, then from the convolution theorem, we have

$$Y(\omega) = H(\omega)X(\omega) \qquad (4.4.30)$$

where $Y(\omega)$, $X(\omega)$, and $H(\omega)$ are the corresponding Fourier transforms of $\{y(n)\}$, $\{x(n)\}$, and $\{h(n)\}$, respectively. From this relationship we observe that the spectrum of the output signal is equal to the spectrum of the input signal multiplied by the frequency response of the system.

If we express $Y(\omega)$, $H(\omega)$, and $X(\omega)$ in polar form, the magnitude and phase of the output signal can be expressed as

$$|Y(\omega)| = |H(\omega)||X(\omega)| \tag{4.4.31}$$

$$\angle Y(\omega) = \angle X(\omega) + \angle H(\omega) \tag{4.4.32}$$

where $|H(\omega)|$ and $\angle H(\omega)$ are the magnitude and phase responses of the system.

By its very nature, a finite-energy aperiodic signal contains a continuum of frequency components. The linear time-invariant system, through its frequency response function, attenuates some frequency components of the input signal and amplifies other frequency components. Thus the system acts as a *filter* to the input signal. Observation of the graph of $|H(\omega)|$ shows which frequency components are amplified and which are attenuated. On the other hand, the angle of $H(\omega)$ determines the phase shift imparted in the continuum of frequency components of the input signal as a function of frequency. If the input signal spectrum is changed by the system in an undesirable way, we say that the system has caused *magnitude and phase distortion.*

We also observe that *the output of a linear time-invariant system cannot contain frequency components that are not contained in the input signal.* It takes either a linear time-variant system or a nonlinear system to create frequency components that are not necessarily contained in the input signal.

Figure 4.39 illustrates the time-domain and frequency-domain relationships that can be used in the analysis of BIBO-stable LTI systems. We observe that in time-domain analysis, we deal with the convolution of the input signal with the impulse response of the system to obtain the output sequence of the system. On the other hand, in frequency-domain analysis, we deal with the input signal spectrum $X(\omega)$ and the frequency response $H(\omega)$ of the system, which are related through multiplication, to yield the spectrum of the signal at the output of the system.

We can use the relation in (4.4.30) to determine the spectrum $Y(\omega)$ of the output signal. Then the output sequence $\{y(n)\}$ can be determined from the inverse Fourier transform

$$y(n) = \frac{1}{2\pi} \int_{-\pi}^{\pi} Y(\omega)e^{j\omega n} d\omega \tag{4.4.33}$$

However, this method is seldom used. Instead, the z-transform introduced in Chapter 3 is a simpler method for solving the problem of determining the output sequence $\{y(n)\}$.

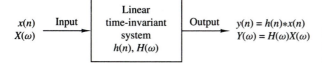

Figure 4.39   Time- and frequency-domain input–output relationships in LTI systems.

Let us return to the basic input–output relation in (4.4.30) and compute the squared magnitude of both sides. Thus we obtain

$$|Y(\omega)|^2 = |H(\omega)|^2 |X(\omega)|^2$$

$$S_{yy}(\omega) = |H(\omega)|^2 S_{xx}(\omega)$$

(4.4.34)

where $S_{xx}(\omega)$ and $S_{yy}(\omega)$ are the energy density spectra of the input and output signals, respectively. By integrating (4.4.34) over the frequency range $(-\pi, \pi)$, we obtain the energy of the output signal as

$$E_y = \frac{1}{2\pi} \int_{-\pi}^{\pi} S_{yy}(\omega)d\omega$$

$$= \frac{1}{2\pi} \int_{-\pi}^{\pi} |H(\omega)|^2 S_{xx}(\omega)d\omega$$

(4.4.35)

Example 4.4.5

A linear time-invariant system is characterized by its impulse response

$$h(n) = (\tfrac{1}{2})^n u(n)$$

Determine the spectrum and the energy density spectrum of the output signal when the system is excited by the signal

$$x(n) = (\tfrac{1}{4})^n u(n)$$

Solution   The frequency response function of the system

$$H(\omega) = \sum_{n=0}^{\infty} (\tfrac{1}{2})^n e^{-j\omega n}$$

$$= \frac{1}{1 - \tfrac{1}{2}e^{-j\omega}}$$

Similarly, the input sequence $\{x(n)\}$ has a Fourier transform

$$X(\omega) = \frac{1}{1 - \tfrac{1}{4}e^{-j\omega}}$$

Hence the spectrum of the signal at the output of the system is

$$Y(\omega) = H(\omega)X(\omega)$$

$$= \frac{1}{(1 - \tfrac{1}{2}e^{-j\omega})(1 - \tfrac{1}{4}e^{-j\omega})}$$

The corresponding energy density spectrum is

$$S_{yy}(\omega) = |Y(\omega)|^2 = |H(\omega)|^2 |X(\omega)|^2$$

$$= \frac{1}{(\tfrac{5}{4} - \cos\omega)(\tfrac{17}{16} - \tfrac{1}{2}\cos\omega)}$$

4.4.5 Relationships Between the System Function and the Frequency Response Function

From the discussion in Section 4.2.6 we know that if the system function $H(z)$ converges on the unit circle, we can obtain the frequency response of the system by evaluating $H(z)$ on the unit circle. Thus

$$H(\omega) = H(z)|_{z=e^{j\omega}} = \sum_{n=-\infty}^{\infty} h(n)e^{-j\omega n} \tag{4.4.36}$$

In the case where $H(z)$ is a rational function of the form $H(z) = B(z)/A(z)$, we have

$$H(\omega) = \frac{B(\omega)}{A(\omega)} = \frac{\displaystyle\sum_{k=0}^{M} b_k e^{-j\omega k}}{1 + \displaystyle\sum_{k=1}^{N} a_k e^{-j\omega k}} \tag{4.4.37}$$

$$= b_0 \frac{\displaystyle\prod_{k=1}^{M}(1 - z_k e^{-j\omega})}{\displaystyle\prod_{k=1}^{N}(1 - p_k e^{-j\omega})} \tag{4.4.38}$$

where the $\{a_k\}$ and $\{b_k\}$ are real, but $\{z_k\}$ and $\{p_k\}$ may be complex-valued.

It is sometimes desirable to express the magnitude squared of $H(\omega)$ in terms of $H(z)$. First, we note that

$$|H(\omega)|^2 = H(\omega)H^*(\omega)$$

For the rational system function given by (4.4.38), we have

$$H^*(\omega) = b_0 \frac{\displaystyle\prod_{k=1}^{M}(1 - z_k^* e^{j\omega})}{\displaystyle\prod_{k=1}^{N}(1 - p_k^* e^{j\omega})} \tag{4.4.39}$$

It follows that $H^*(\omega)$ is obtained by evaluating $H^*(1/z^*)$ on the unit circle, where for a rational system function,

$$H^*(1/z^*) = b_0 \frac{\displaystyle\prod_{k=1}^{M}(1 - z_k^* z)}{\displaystyle\prod_{k=1}^{N}(1 - p_k^* z)} \tag{4.4.40}$$

However, when $\{h(n)\}$ is real or, equivalently, the coefficients $\{a_k\}$ and $\{b_k\}$ are real, complex-valued poles and zeros occur in complex-conjugate pairs. In this case, $H^*(1/z^*) = H(z^{-1})$. Consequently, $H^*(\omega) = H(-\omega)$, and

$$|H(\omega)|^2 = H(\omega)H^*(\omega) = H(\omega)H(-\omega) = H(z)H(z^{-1})|_{z=e^{j\omega}} \qquad (4.4.41)$$

According to the correlation theorem for the z-transform (see Table 3.2), the function $H(z)H(z^{-1})$ is the z-transform of the autocorrelation sequence $\{r_{hh}(m)\}$ of the unit sample response $\{h(n)\}$. Then it follows from the Wiener–Khintchine theorem that $|H(\omega)|^2$ is the Fourier transform of $\{r_{hh}(m)\}$.

Similarly, if $H(z) = B(z)/A(z)$, the transforms $D(z) = B(z)B(z^{-1})$ and $C(z) = A(z)A(z^{-1})$ are the z-transforms of the autocorrelation sequences $\{c_l\}$ and $\{d_l\}$, where

$$c_l = \sum_{k=0}^{N-|l|} a_k a_{k+l} \qquad -N \le l \le N \qquad (4.4.42)$$

$$d_l = \sum_{k=0}^{M-|l|} b_k b_{k+l} \qquad -M \le l \le M \qquad (4.4.43)$$

Since the system parameters $\{a_k\}$ and $\{b_k\}$ are real valued, it follows that $c_l = c_{-l}$ and $d_l = d_{-l}$. By using this symmetry property, $|H(\omega)|^2$ may be expressed as

$$|H(\omega)|^2 = \frac{d_0 + 2\sum_{k=1}^{M} d_k \cos k\omega}{c_0 + 2\sum_{k=1}^{N} c_k \cos k\omega} \qquad (4.4.44)$$

Finally, we note that $\cos k\omega$ can be expressed as a polynomial function of $\cos\omega$. That is,

$$\cos k\omega = \sum_{m=0}^{k} \beta_m (\cos\omega)^m \qquad (4.4.45)$$

where $\{\beta_m\}$ are the coefficients in the expansion. Consequently, the numerator and denominator of $|H(\omega)|^2$ can be viewed as polynomial functions of $\cos\omega$. The following example illustrates the foregoing relationships.

Example 4.4.6

Determine $|H(\omega)|^2$ for the system

$$y(n) = -0.1y(n-1) + 0.2y(n-2) + x(n) + x(n-1)$$

Solution   The system function is

$$H(z) = \frac{1+z^{-1}}{1+0.1z^{-1}-0.2z^{-2}}$$

and its ROC is $|z| > 0.5$. Hence $H(\omega)$ exists. Now

$$H(z)H(z^{-1}) = \frac{1+z^{-1}}{1+0.1z^{-1}-0.2z^{-2}} \cdot \frac{1+z}{1+0.1z-0.2z^2}$$

$$= \frac{2+z+z^{-1}}{1.05+0.08(z+z^{-1})-0.2(z^{-2}+z^{-2})}$$

By evaluating $H(z)H(z^{-1})$ on the unit circle, we obtain

$$|H(\omega)|^2 = \frac{2 + 2\cos\omega}{1.05 + 0.16\cos\omega - 0.4\cos 2\omega}$$

However, $\cos 2\omega = 2\cos^2\omega - 1$. Consequently, $|H(\omega)|^2$ may be expressed as

$$|H(\omega)|^2 = \frac{2(1 + \cos\omega)}{1.45 + 0.16\cos\omega - 0.8\cos^2\omega}$$

We note that given $H(z)$, it is straightforward to determine $H(z^{-1})$ and then $|H(\omega)|^2$. However, the inverse problem of determining $H(z)$ given $|H(\omega)|^2$ or the corresponding impulse response $\{h(n)\}$, is not straightforward. Since $|H(\omega)|^2$ does not contain the phase information in $H(\omega)$, it is not possible to uniquely determine $H(z)$.

To elaborate on the point, let us assume that the N poles and M zeros of $H(z)$ are $\{p_k\}$ and $\{z_k\}$, respectively. The corresponding poles and zeros of $H(z^{-1})$ are $\{1/p_k\}$ and $\{1/z_k\}$, respectively. Given $|H(\omega)|^2$ or, equivalently, $H(z)H(z^{-1})$, we can determine different system functions $H(z)$ by assigning to $H(z)$, a pole p_k or its reciprocal $1/p_k$, and a zero z_k or its reciprocal $1/z_k$. For example, if $N = 2$ and $M = 1$, the poles and zeros of $H(z)H(z^{-1})$ are $\{p_1, p_2, 1/p_1, 1/p_2\}$ and $\{z_1, 1/z_1\}$. If p_1 and p_2 are real, the possible poles for $H(z)$ are $\{p_1, p_2\}$, $\{1/p_1, 1/p_2\}$, $\{p_1, 1/p_2\}$, and $\{p_2, 1/p_1\}$ and the possible zeros are $\{z_1\}$ or $\{1/z_1\}$. Therefore, there are eight possible choices of system functions, all of which result in the same $|H(\omega)|^2$. Even if we restrict the poles of $H(z)$ to be inside the unit circle, there are still two different choices for $H(z)$, depending on whether we pick the zero $\{z_1\}$ or $\{1/z_1\}$. Therefore, we cannot determine $H(z)$ uniquely given only the magnitude response $|H(\omega)|$.

4.4.6 Computation of the Frequency Response Function

In evaluating the magnitude response and the phase response as functions of frequency, it is convenient to express $H(\omega)$ in terms of its poles and zeros. Hence we write $H(\omega)$ in factored form as

$$H(\omega) = b_0 \frac{\displaystyle\prod_{k=1}^{M}(1 - z_k e^{-j\omega k})}{\displaystyle\prod_{k=1}^{N}(1 - p_k e^{-j\omega k})} \tag{4.4.46}$$

or, equivalently, as

$$H(\omega) = b_0 e^{j\omega(N-M)} \frac{\displaystyle\prod_{k=1}^{M}(e^{j\omega} - z_k)}{\displaystyle\prod_{k=1}^{N}(e^{j\omega} - p_k)} \tag{4.4.47}$$

Let us express the complex-valued factors in (4.4.47) in polar form as

$$e^{j\omega} - z_k = V_k(\omega)e^{j\Theta_k(\omega)} \qquad (4.4.48)$$

and

$$e^{j\omega} - p_k = U_k(\omega)e^{j\Phi_k(\omega)} \qquad (4.4.49)$$

where

$$V_k(\omega) \equiv |e^{j\omega} - z_k|, \quad \Theta_k(\omega) \equiv \measuredangle(e^{j\omega} - z_k) \qquad (4.4.50)$$

and

$$U_k(\omega) \equiv |e^{j\omega} - p_k|, \quad \Phi_k(\omega) = \measuredangle(e^{j\omega} - p_k) \qquad (4.4.51)$$

The magnitude of $H(\omega)$ is equal to the product of magnitudes of all terms in (4.4.47). Thus, using (4.4.48) through (4.4.51), we obtain

$$|H(\omega)| = |b_0|\frac{V_1(\omega)\cdots V_M(\omega)}{U_1(\omega)U_2(\omega)\cdots U_N(\omega)} \qquad (4.4.52)$$

since the magnitude of $e^{j\omega(N-M)}$ is 1.

The phase of $H(\omega)$ is the sum of the phases of the numerator factors, minus the phases of the denominator factors. Thus, by combining (4.4.48) through (4.4.51), we have

$$\measuredangle H(\omega) = \measuredangle b_0 + \omega(N - M) + \Theta_1(\omega) + \Theta_2(\omega) + \cdots + \Theta_M(\omega)$$
$$- [\Phi_1(\omega) + \Phi_2(\omega) + \cdots + \Phi_N(\omega)] \qquad (4.4.53)$$

The phase of the gain term b_0 is zero or π, depending on whether b_0 is positive or negative. Clearly, if we know the zeros and the poles of the system function $H(z)$, we can evaluate the frequency response from (4.4.52) and (4.4.53).

There is a geometric interpretation of the quantities appearing in (4.4.52) and (4.4.53). Let us consider a pole p_k and a zero z_k located at points A and B of the z-plane, as shown in Fig. 4.40(a). Assume that we wish to compute $H(\omega)$ at a specific value of frequency ω. The given value of ω determines the angle of $e^{j\omega}$ with the positive real axis. The tip of the vector $e^{j\omega}$ specifies a point L on the unit circle. The evaluation of the Fourier transform for the given value of ω is equivalent to evaluating the z-transform at the point L of the complex plane. Let us draw the vectors **AL** and **BL** from the pole and zero locations to the point L, at which we wish to compute the Fourier transform. From Fig. 4.40(a) it follows that

$$\mathbf{CL} = \mathbf{CA} + \mathbf{AL}$$

and

$$\mathbf{CL} = \mathbf{CB} + \mathbf{BL}$$

However, $\mathbf{CL} = e^{j\omega}$, $\mathbf{CA} = p_k$ and $\mathbf{CB} = z_k$. Thus

$$\mathbf{AL} = e^{j\omega} - p_k \qquad (4.4.54)$$

and

$$\mathbf{BL} = e^{j\omega} - z_k \qquad (4.4.55)$$

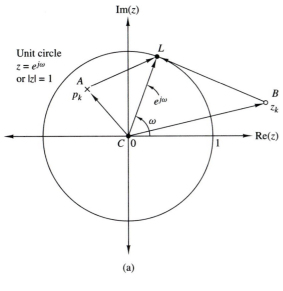

(a)

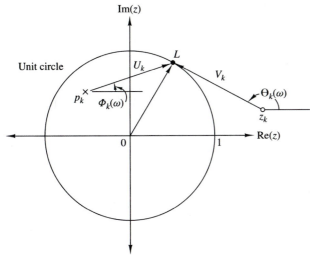

(b)

Figure 4.40 Geometric interpretation of the contribution of a pole and a zero to the Fourier transform (1) magnitude: the factor V_k/U_k, (2) phase: the factor $\Theta_k - \Phi_k$.

By combining these relations with (4.4.48) and (4.4.49), we obtain

$$\mathbf{AL} = e^{j\omega} - p_k = U_k(\omega)e^{j\Phi_k(\omega)} \tag{4.4.56}$$

$$\mathbf{BL} = e^{j\omega} - z_k = V_k(\omega)e^{j\Theta_k(\omega)} \tag{4.4.57}$$

Thus $U_k(\omega)$ is the length of $\mathbf{AL}$, that is, the distance of the pole p_k from the point L corresponding to $e^{j\omega}$, whereas $V_k(\omega)$ is the distance of the zero z_k from the same point L. The phases $\Phi_k(\omega)$ and $\Theta_k(\omega)$ are the angles of the vectors $\mathbf{AL}$ and $\mathbf{BL}$

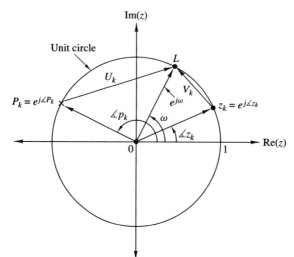

Figure 4.41 A zero on the unit circle causes $|H(\omega)| = 0$ and $\omega = \measuredangle z_k$. In contrast, a pole on the unit circle results in $|H(\omega)| = \infty$ at $\omega = \measuredangle p_k$.

with the positive real axis, respectively. These geometric interpretations are shown in Fig. 4.40(b).

Geometric interpretations are very useful in understanding how the location of poles and zeros affects the magnitude and phase of the Fourier transform. Suppose that a zero, say z_k, and a pole, say p_k, are on the unit circle as shown in Fig. 4.41. We note that at $\omega = \measuredangle z_k$, $V_k(\omega)$ and consequently $|H(\omega)|$ become zero. Similarly, at $\omega = \measuredangle p_k$ the length $U_k(\omega)$ becomes zero and hence $|H(\omega)|$ becomes infinite. Clearly, the evaluation of phase in these cases has no meaning.

From this discussion we can easily see that the presence of a zero close to the unit circle causes the magnitude of the frequency response, at frequencies that correspond to points of the unit circle close to that point, to be small. In contrast, the presence of a pole close to the unit circle causes the magnitude of the frequency response to be large at frequencies close to that point. Thus poles have the opposite effect of zeros. Also, placing a zero close to a pole cancels the effect of the pole, and vice versa. This can be also seen from (4.4.47), since if $z_k = p_k$, the terms $e^{j\omega} - z_k$ and $e^{j\omega} - p_k$ cancel. Obviously, the presence of both poles and zeros in a transform results in a greater variety of shapes for $|H(\omega)|$ and $\measuredangle H(\omega)$. This observation is very important in the design of digital filters. We conclude our discussion with the following example illustrating these concepts.

Example 4.4.7

Evaluate the frequency response of the system described by the system function

$$H(z) = \frac{1}{1 - 0.8z^{-1}} = \frac{z}{z - 0.8}$$

Solution   Clearly, $H(z)$ has a zero at $z = 0$ and a pole at $p = 0.8$. Hence the frequency response of the system is

$$H(\omega) = \frac{e^{j\omega}}{e^{j\omega} - 0.8}$$

The magnitude response is

$$|H(\omega)| = \frac{|e^{j\omega}|}{|e^{j\omega} - 0.8|} = \frac{1}{\sqrt{1.64 - 1.6 \cos \omega}}$$

and the phase response is

$$\theta(\omega) = \omega - \tan^{-1} \frac{\sin \omega}{\cos \omega - 0.8}$$

The magnitude and phase responses are illustrated in Fig. 4.42. Note that the peak of the magnitude response occurs at $\omega = 0$, the point on the unit circle closest to the pole located at 0.8.

If the magnitude response in (4.4.52) is expressed in decibels,

$$|H(\omega)|_{dB} = 20 \log_{10} |b_0| + 20 \sum_{k=1}^{M} \log_{10} V_k(\omega) - 20 \sum_{k=1}^{N} \log_{10} U_k(\omega) \qquad (4.4.58)$$

Thus the magnitude response is expressed as a sum of the magnitude factors in $|H(\omega)|$.

4.4.7 Input–Output Correlation Functions and Spectra

In Section 2.6.5 we developed several correlation relationships between the input and the output sequences of an LTI system. Specifically, we derived the following equations:

$$r_{yy}(m) = r_{hh}(m) * r_{xx}(m) \qquad (4.4.59)$$

$$r_{yx}(m) = h(m) * r_{xx}(m) \qquad (4.4.60)$$

where $r_{xx}(m)$ is the autocorrelation sequence of the input signal $\{x(n)\}$, $r_{yy}(m)$ is the autocorrelation sequence of the output $\{y(n)\}$, $r_{hh}(m)$ is the autocorrelation sequence of the impulse response $\{h(n)\}$, and $r_{yx}(m)$ is the crosscorrelation sequence between the output and the input signals. Since (4.4.59) and (4.4.60) involve the convolution operation, the z-transform of these equations yields

$$\begin{aligned} S_{yy}(z) &= S_{hh}(z) S_{xx}(z) \\ &= H(z) H(z^{-1}) S_{xx}(z) \end{aligned} \qquad (4.4.61)$$

$$S_{yx}(z) = H(z) S_{xx}(z) \qquad (4.4.62)$$

If we substitute $z = e^{j\omega}$ in (4.4.62), we obtain

$$\begin{aligned} S_{yx}(\omega) &= H(\omega) S_{xx}(\omega) \\ &= H(\omega) |X(\omega)|^2 \end{aligned} \qquad (4.4.63)$$

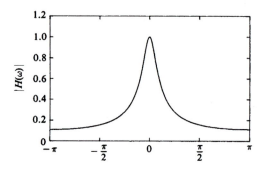

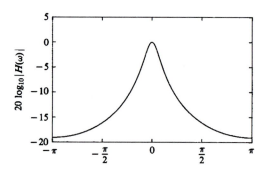

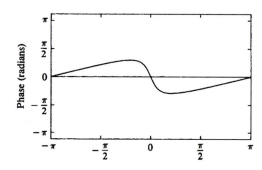

Figure 4.42 Magnitude and phase of system with $H(z) = 1/(1 - 0.8z^{-1})$.

where $S_{yx}(\omega)$ is the cross-energy density spectrum of $\{y(n)\}$ and $\{x(n)\}$. Similarly, evaluating $S_{yy}(z)$ on the unit circle yields the energy density spectrum of the output signal as

$$S_{yy}(\omega) = |H(\omega)|^2 S_{xx}(\omega) \qquad (4.4.64)$$

where $S_{xx}(\omega)$ is the energy density spectrum of the input signal.

Since $r_{yy}(m)$ and $S_{yy}(\omega)$ are a Fourier transform pair, it follows that

$$r_{yy}(m) = \frac{1}{2\pi} \int_{-\pi}^{\pi} S_{yy}(\omega) e^{j\omega m} d\omega \qquad (4.4.65)$$

The total energy in the output signal is simply

$$E_y = \frac{1}{2\pi} \int_{-\pi}^{\pi} S_{yy}(\omega)d\omega = r_{yy}(0)$$

$$= \frac{1}{2\pi} \int_{-\pi}^{\pi} |H(\omega)|^2 \, S_{xx}(\omega)d\omega$$

(4.4.66)

The result in (4.4.66) may be used to easily prove that $E_y \geq 0$.

Finally, we note that if the input signal has a flat spectrum [i.e., $S_{xx}(\omega) = E_x = $ constant for $\pi \leq \omega \leq -\pi$], (4.4.63) reduces to

$$S_{yx}(\omega) = H(\omega)E_x$$

(4.4.67)

where E_x is the constant value of the spectrum. Hence

$$H(\omega) = \frac{1}{E_x} S_{yx}(\omega)$$

(4.4.68)

or, equivalently,

$$h(n) = \frac{1}{E_x} r_{yx}(m)$$

(4.4.69)

The relation in (4.4.69) implies that $h(n)$ can be determined by exciting the input to the system by a spectrally flat signal $\{x(n)\}$, and crosscorrelating the input with the output of the system. This method is useful in measuring the impulse response of an unknown system.

4.4.8 Correlation Functions and Power Spectra for Random Input Signals

This development parallels the derivations in Section 4.4.7, with the exception that we now deal with statistical moments of the input and output signals of an LTI system. The various statistical parameters are introduced in Appendix A.

Let us consider a discrete-time linear time-invariant system with unit sample response $\{h(n)\}$ and frequency response $H(f)$. For this development we assume that $\{h(n)\}$ is real. Let $x(n)$ be a sample function of a stationary random process $X(n)$ that excites the system and let $y(n)$ denote the response of the system to $x(n)$.

From the convolution summation that relates the output to the input we have

$$y(n) = \sum_{k=-\infty}^{\infty} h(k)x(n-k)$$

(4.4.70)

Since $x(n)$ is a random input signal, the output is also a random sequence. In other words, for each sample sequence $x(n)$ of the process $X(n)$, there is a corresponding sample sequence $y(n)$ of the output random process $Y(n)$. We wish to relate the statistical characteristics of the output random process $Y(n)$ to the statistical characterization of the input process and the characteristics of the system.

The expected value of the output $y(n)$ is

$$m_y \equiv E[y(n)] = E\left[\sum_{k=-\infty}^{\infty} h(k)x(n-k)\right]$$

$$= \sum_{k=-\infty}^{\infty} h(k)E[x(n-k)] \qquad (4.4.71)$$

$$m_y = m_x \sum_{k=-\infty}^{\infty} h(k)$$

From the Fourier transform relationship

$$H(\omega) = \sum_{k=-\infty}^{\infty} h(k)e^{-j\omega k} \qquad (4.4.72)$$

we have

$$H(0) = \sum_{k=-\infty}^{\infty} h(k) \qquad (4.4.73)$$

which is the dc gain of the system. The relationship in (4.4.73) allows us to express the mean value in (4.4.71) as

$$m_y = m_x H(0) \qquad (4.4.74)$$

The autocorrelation sequence for the output random process is

$$\gamma_{yy}(m) = E[y^*(n)y(n+m)]$$

$$= E\left[\sum_{k=-\infty}^{\infty} h(k)x^*(n-k) \sum_{j=-\infty}^{\infty} h(j)x(n+m-j)\right]$$

$$= \sum_{k=-\infty}^{\infty} \sum_{j=-\infty}^{\infty} h(k)h(j)E[x^*(n-k)x(n+m-j)] \qquad (4.4.75)$$

$$= \sum_{k=-\infty}^{\infty} \sum_{j=-\infty}^{\infty} h(k)h(j)\gamma_{xx}(k-j+m)$$

This is the general form for the autocorrelation of the output in terms of the autocorrelation of the input and the impulse response of the system.

A special form of (4.4.75) is obtained when the input random process is white, that is, when $m_x = 0$ and

$$\gamma_{xx}(m) = \sigma_x^2 \delta(m) \qquad (4.4.76)$$

where $\sigma_x^2 \equiv \gamma_{xx}(0)$ is the input signal power. Then (4.4.75) reduces to

$$\gamma_{yy}(m) = \sigma_x^2 \sum_{k=-\infty}^{\infty} h(k)h(k+m) \qquad (4.4.77)$$

Under this condition the output process has the average power

$$\gamma_{yy}(0) = \sigma_x^2 \sum_{n=-\infty}^{\infty} h^2(n) = \sigma_x^2 \int_{-1/2}^{1/2} |H(f)|^2 df \qquad (4.4.78)$$

where we have applied Parseval's theorem.

The relationship in (4.4.75) can be transformed into the frequency domain by determining the power density spectrum of $\gamma_{yy}(m)$. We have

$$
\begin{aligned}
\Gamma_{yy}(\omega) &= \sum_{m=-\infty}^{\infty} \gamma_{yy}(m)e^{-j\omega m} \\
&= \sum_{m=-\infty}^{\infty} \left[\sum_{k=-\infty}^{\infty} \sum_{l=-\infty}^{\infty} h(k)h(l)\gamma_{xx}(k-l+m) \right] e^{-j\omega m} \\
&= \sum_{k=-\infty}^{\infty} \sum_{l=-\infty}^{\infty} h(k)h(l) \left[\sum_{m=-\infty}^{\infty} \gamma_{xx}(k-l+m)e^{-j\omega m} \right] \\
&= \Gamma_{xx}(f) \left[\sum_{k=-\infty}^{\infty} h(k)e^{j\omega k} \right] \left[\sum_{l=-\infty}^{\infty} h(l)e^{-j\omega l} \right] \\
&= |H(\omega)|^2 \Gamma_{xx}(\omega)
\end{aligned}
\tag{4.4.79}
$$

This is the desired relationship for the power density spectrum of the output process, in terms of the power density spectrum of the input process and the frequency response of the system.

The equivalent expression for continuous-time systems with random inputs is

$$
\Gamma_{yy}(F) = |H(F)|^2 \Gamma_{xx}(F)
\tag{4.4.80}
$$

where the power density spectra $\Gamma_{yy}(F)$ and $\Gamma_{xx}(F)$ are the Fourier transforms of the autocorrelation functions $\gamma_{yy}(\tau)$ and $\gamma_{xx}(\tau)$, respectively, and where $H(F)$ is the frequency response of the system, which is related to the impulse response by the Fourier transform, that is,

$$
H(F) = \int_{-\infty}^{\infty} h(t)e^{-j2\pi Ft} dt
\tag{4.4.81}
$$

As a final exercise, we determine the crosscorrelation of the output $y(n)$ with the input signal $x(n)$. If we multiply both sides of (4.4.70) by $x^*(n-m)$ and take the expected value, we obtain

$$
\begin{aligned}
E[y(n)x^*(n-m)] &= E\left[\sum_{k=-\infty}^{\infty} h(k)x^*(n-m)x(n-k) \right] \\
\gamma_{yx}(m) &= \sum_{k=-\infty}^{\infty} h(k)E[x^*(n-m)x(n-k)] \\
&= \sum_{k=-\infty}^{\infty} h(k)\gamma_{xx}(m-k)
\end{aligned}
\tag{4.4.82}
$$

Since (4.4.82) has the form of a convolution, the frequency-domain equivalent expression is

$$
\Gamma_{yx}(\omega) = H(\omega)\Gamma_{xx}(\omega)
\tag{4.4.83}
$$

In the special case where $x(n)$ is white noise, (4.4.83) reduces to

$$
\Gamma_{yx}(\omega) = \sigma_x^2 H(\omega)
\tag{4.4.84}
$$

where σ_x^2 is the input noise power. This result means that an unknown system with frequency response $H(\omega)$ can be identified by exciting the input with white noise, crosscorrelating the input sequence with the output sequence to obtain $\gamma_{yx}(m)$, and finally, computing the Fourier transform of $\gamma_{yx}(m)$. The result of these computations is proportional to $H(\omega)$.

4.5 LINEAR TIME-INVARIANT SYSTEMS AS FREQUENCY-SELECTIVE FILTERS

The term *filter* is commonly used to describe a device that discriminates, according to some attribute of the objects applied at its input, what passes through it. For example, an air filter allows air to pass through it but prevents dust particles that are present in the air from passing through. An oil filter performs a similar function, with the exception that oil is the substance allowed to pass through the filter, while particles of dirt are collected at the input to the filter and prevented from passing through. In photography, an ultraviolet filter is often used to prevent ultraviolet light, which is present in sunlight and which is not a part of visible light, from passing through and affecting the chemicals on the film.

As we have observed in the preceding section, a linear time-invariant system also performs a type of discrimination or filtering among the various frequency components at its input. The nature of this filtering action is determined by the frequency response characteristics $H(\omega)$, which in turn depends on the choice of the system parameters (e.g., the coefficients $\{a_k\}$ and $\{b_k\}$ in the difference equation characterization of the system). Thus, by proper selection of the coefficients, we can design frequency-selective filters that pass signals with frequency components in some bands while they attenuate signals containing frequency components in other frequency bands.

In general, a linear time-invariant system modifies the input signal spectrum $X(\omega)$ according to its frequency response $H(\omega)$ to yield an output signal with spectrum $Y(\omega) = H(\omega)X(\omega)$. In a sense, $H(\omega)$ acts as a *weighting function* or a *spectral shaping function* to the different frequency components in the input signal. When viewed in this context, any linear time-invariant system can be considered to be a frequency-shaping filter, even though it may not necessarily completely block any or all frequency components. Consequently, the terms "linear time-invariant system" and "filter" are synonymous and are often used interchangeably.

We use the term *filter* to describe a linear time-invariant system used to perform spectral shaping or frequency-selective filtering. Filtering is used in digital signal processing in a variety of ways. For example, removal of undesirable noise from desired signals, spectral shaping such as equalization of communication channels, signal detection in radar, sonar, and communications, and for performing spectral analysis of signals, and so on.

4.5.1 Ideal Filter Characteristics

Filters are usually classified according to their frequency-domain characteristics as lowpass, highpass, bandpass, and bandstop or band-elimination filters. The ideal magnitude response characteristics of these types of filters are illustrated in Fig. 4.43. As shown, these ideal filters have a constant-gain (usually taken as unity-gain) passband characteristic and zero gain in their stopband.

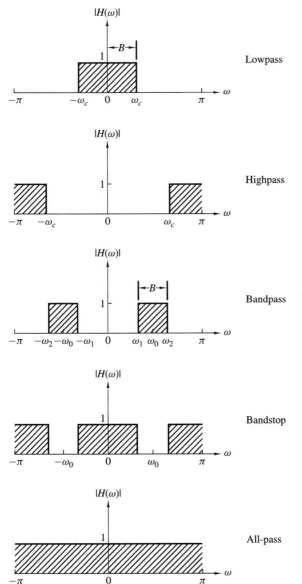

Figure 4.43 Magnitude responses for some ideal frequency-selective discrete-time filters.

Another characteristic of an ideal filter is a linear phase response. To demonstrate this point, let us assume that a signal sequence $\{x(n)\}$ with frequency components confined to the frequency range $\omega_1 < \omega < \omega_2$ is passed through a filter with frequency response

$$H(\omega) = \begin{cases} Ce^{-j\omega n_0}, & \omega_1 < \omega < \omega_2 \\ 0, & \text{otherwise} \end{cases} \tag{4.5.1}$$

where C and n_0 are constants. The signal at the output of the filter has a spectrum

$$\begin{aligned} Y(\omega) &= X(\omega)H(\omega) \\ &= CX(\omega)e^{-j\omega n_0} \qquad \omega_1 < \omega < \omega_2 \end{aligned} \tag{4.5.2}$$

By applying the scaling and time-shifting properties of the Fourier transform, we obtain the time-domain output

$$y(n) = Cx(n - n_0) \tag{4.5.3}$$

Consequently, the filter output is simply a delayed and amplitude-scaled version of the input signal. A pure delay is usually tolerable and is not considered a distortion of the signal. Neither is amplitude scaling. Therefore, ideal filters have a linear phase characteristic within their passband, that is,

$$\Theta(\omega) = -\omega n_0 \tag{4.5.4}$$

The derivative of the phase with respect to frequency has the units of delay. Hence we can define the signal delay as a function of frequency as

$$\tau_g(\omega) = -\frac{d\Theta(\omega)}{d\omega} \tag{4.5.5}$$

$\tau_g(\omega)$ is usually called the *envelope delay* or the *group delay* of the filter. We interpret $\tau_g(\omega)$ as the time delay that a signal component of frequency ω undergoes as it passes from the input to the output of the system. Note that when $\Theta(\omega)$ is linear as in (4.5.4), $\tau_g(\omega) = n_0 = $ constant. In this case all frequency components of the input signal undergo the same time delay.

In conclusion, ideal filters have a constant magnitude characteristic and a linear phase characteristic within their passband. In all cases, such filters are not physically realizable but serve as a mathematical idealization of practical filters. For example, the ideal lowpass filter has an impulse response

$$h_{lp}(n) = \frac{\sin \omega_c \pi n}{\pi n} \qquad -\infty < n < \infty \tag{4.5.6}$$

We note that this filter is not causal and it is not absolutely summable and therefore it is also unstable. Consequently, this ideal filter is physically unrealizable. Nevertheless, its frequency response characteristics can be approximated very closely by practical, physically realizable filters, as will be demonstrated in Chapter 8.

In the following discussion, we treat the design of some simple digital filters by the placement of poles and zeros in the z-plane. We have already described how the location of poles and zeros affects the frequency response characteristics of the system. In particular, in Section 4.4.6 we presented a graphical method for computing the frequency response characteristics from the pole–zero plot. This same approach can be used to design a number of simple but important digital filters with desirable frequency response characteristics.

The basic principle underlying the pole–zero placement method is to locate poles near points of the unit circle corresponding to frequencies to be emphasized, and to place zeros near the frequencies to be deemphasized. Furthermore, the following constraints must be imposed:

1. All poles should be placed inside the unit circle in order for the filter to be stable. However, zeros can be placed anywhere in the z-plane.
2. All complex zeros and poles must occur in complex-conjugate pairs in order for the filter coefficients to be real.

From our previous discussion we recall that for a given pole–zero pattern, the system function $H(z)$ can be expressed as

$$H(z) = \frac{\sum_{k=0}^{M} b_k z^{-k}}{1 + \sum_{k=1}^{N} a_k z^{-k}} = b_0 \frac{\prod_{k=1}^{M}(1 - z_k z^{-1})}{\prod_{k=1}^{N}(1 - p_k z^{-1})} \qquad (4.5.7)$$

where b_0 is a gain constant selected to normalize the frequency response at some specified frequency. That is, b_0 is selected such that

$$|H(\omega_0)| = 1 \qquad (4.5.8)$$

where ω_0 is a frequency in the passband of the filter. Usually, N is selected to equal or exceed M, so that the filter has more nontrivial poles than zeros.

In the next section, we illustrate the method of pole–zero placement in the design of some simple lowpass, highpass, and bandpass filters, digital resonators, and comb filters. The design procedure is facilitated when carried out interactively on a digital computer with a graphics terminal.

4.5.2 Lowpass, Highpass, and Bandpass Filters

In the design of lowpass digital filters, the poles should be placed near the unit circle at points corresponding to low frequencies (near $\omega = 0$) and zeros should be placed near or on the unit circle at points corresponding to high frequencies (near $\omega = \pi$). The opposite holds true for highpass filters.

Figure 4.44 illustrates the pole–zero placement of three lowpass and three highpass filters. The magnitude and phase responses for the single-pole filter with system function

$$H_1(z) = \frac{1 - a}{1 - az^{-1}} \qquad (4.5.9)$$

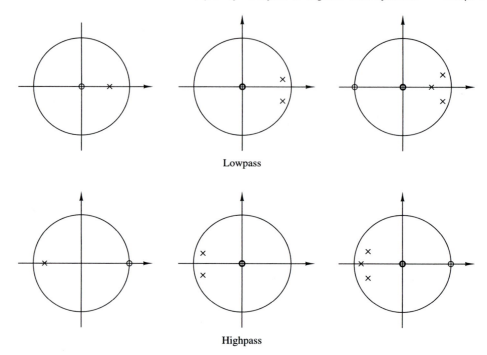

Figure 4.44   Pole–zero patterns for several lowpass and highpass filters.

are illustrated in Fig. 4.45 for $a = 0.9$. The gain G was selected as $1 - a$, so that the filter has unity gain at $\omega = 0$. The gain of this filter at high frequencies is relatively small.

The addition of a zero at $z = -1$ further attenuates the response of the filter at high frequencies. This leads to a filter with a system function

$$H_2(z) = \frac{1 - a}{2} \frac{1 + z^{-1}}{1 - az^{-1}} \tag{4.5.10}$$

and a frequency response characterstic that is also illustrated in Fig. 4.45. In this case the magnitude of $H_2(\omega)$ goes to zero at $\omega = \pi$.

Similarly, we can obtain simple highpass filters by reflecting (folding) the pole–zero locations of the lowpass filters about the imaginary axis in the z-plane. Thus we obtain the system function

$$H_3(z) = \frac{1 - a}{2} \frac{1 - z^{-1}}{1 + az^{-1}} \tag{4.5.11}$$

which has the frequency response characteristics illustrated in Fig. 4.46 for $a = 0.9$.

Example 4.5.1

A two-pole lowpass filter has the system function

$$H(z) = \frac{b_0}{(1 - pz^{-1})^2}$$

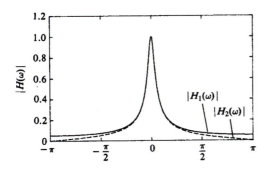

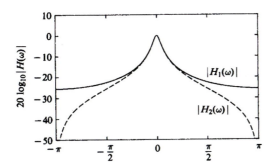

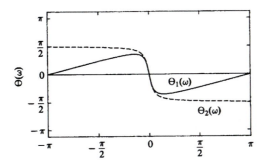

Figure 4.45 Magnitude and phase response of (1) a single-pole filter and (2) a one-pole, one-zero filter; $H_1(z) = (1 - a)/(1 - az^{-1})$, $H_2(z) = [(1 - a)/2][(1 + z^{-1})/(1 - az^{-1})]$ and $a = 0.9$.

Determine the values of b_0 and p such that the frequency response $H(\omega)$ satisfies the conditions

$$H(0) = 1$$

and

$$\left| H\left(\frac{\pi}{4}\right) \right|^2 = \frac{1}{2}$$

Solution   At $\omega = 0$ we have

$$H(0) = \frac{b_0}{(1 - p)^2} = 1$$

Hence

$$b_0 = (1 - p)^2$$

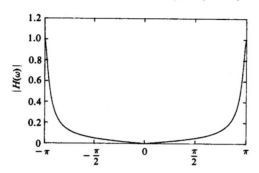

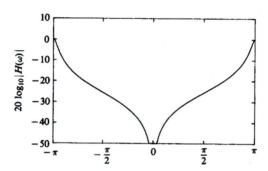

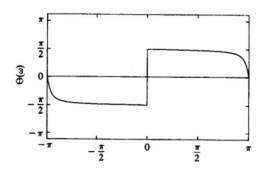

Figure 4.46 Magnitude and phase response of a simple highpass filter; $H(z) = [(1-a)/2][(1-z^{-1})/(1+az^{-1})]$ with $a = 0.9$.

At $\omega = \pi/4$,

$$H\left(\frac{\pi}{4}\right) = \frac{(1-p)^2}{(1-pe^{-j\pi/4})^2}$$

$$= \frac{(1-p)^2}{(1-p\cos(\pi/4) + jp\sin(\pi/4))^2}$$

$$= \frac{(1-p)^2}{(1-p/\sqrt{2} + jp/\sqrt{2})^2}$$

Hence

$$\frac{(1-p)^4}{[(1-p/\sqrt{2})^2 + p^2/2]^2} = \frac{1}{2}$$

or, equivalently,

$$\sqrt{2}(1 - p)^2 = 1 + p^2 - \sqrt{2}p$$

The value of $p = 0.32$ satisfies this equation. Consequently, the system function for the desired filter is

$$H(z) = \frac{0.46}{(1 - 0.32z^{-1})^2}$$

The same principles can be applied for the design of bandpass filters. Basically, the bandpass filter should contain one or more pairs of complex-conjugate poles near the unit circle, in the vicinity of the frequency band that constitutes the passband of the filter. The following example serves to illustrate the basic ideas.

Example 4.5.2

Design a two-pole bandpass filter that has the center of its passband at $\omega = \pi/2$, zero in its frequency response characteristic at $\omega = 0$ and $\omega = \pi$, and its magnitude response is $1/\sqrt{2}$ at $\omega = 4\pi/9$.

Solution   Clearly, the filter must have poles at

$$p_{1,2} = re^{\pm j\pi/2}$$

and zeros at $z = 1$ and $z = -1$. Consequently, the system function is

$$H(z) = G\frac{(z - 1)(z + 1)}{(z - jr)(z + jr)}$$

$$= G\frac{z^2 - 1}{z^2 + r^2}$$

The gain factor is determined by evaluating the frequency response $H(\omega)$ of the filter at $\omega = \pi/2$. Thus we have

$$H\left(\frac{\pi}{2}\right) = G\frac{2}{1 - r^2} = 1$$

$$G = \frac{1 - r^2}{2}$$

The value of r is determined by evaluating $H(\omega)$ at $\omega = 4\pi/9$. Thus we have

$$\left|H\left(\frac{4\pi}{9}\right)\right|^2 = \frac{(1 - r^2)^2}{4}\frac{2 - 2\cos(8\pi/9)}{1 + r^4 + 2r^2\cos(8\pi/9)} = \frac{1}{2}$$

or, equivalently,

$$1.94(1 - r^2)^2 = 1 - 1.88r^2 + r^4$$

The value of $r^2 = 0.7$ satisfies this equation. Therefore, the system function for the desired filter is

$$H(z) = 0.15\frac{1 - z^{-2}}{1 + 0.7z^{-2}}$$

Its frequency response is illustrated in Fig. 4.47.

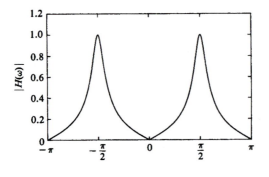

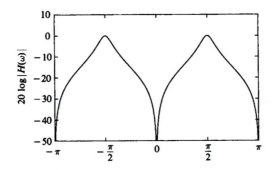

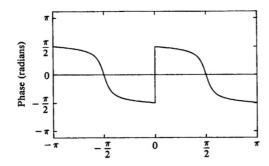

Figure 4.47 Magnitude and phase response of a simple bandpass filter in Example 4.5.2; $H(z) = 0.15[(1 - z^{-2})/(1 + 0.7z^{-2})]$.

It should be emphasized that the main purpose of the foregoing methodology for designing simple digital filters by pole–zero placement is to provide insight into the effect that poles and zeros have on the frequency response characteristic of systems. The methodology is not intended as a good method for designing digital filters with well-specified passband and stopband characteristics. Systematic methods for the design of sophisticated digital filters for practical applications are discussed in Chapter 8.

A simple lowpass-to-highpass filter transformation. Suppose that we have designed a prototype lowpass filter with impulse response $h_{lp}(n)$. By using the frequency translation property of the Fourier transform, it is possible to convert the prototype filter to either a bandpass or a highpass filter. Frequency transformations for converting a prototype lowpass filter into a filter of another type are described in detail in Section 8.3. In this section we present a simple-frequency transformation for converting a lowpass filter into a highpass filter, and vice versa.

If $h_{lp}(n)$ denotes the impulse response of a lowpass filter with frequency response $H_{lp}(\omega)$, a highpass filter can be obtained by translating $H_{lp}(\omega)$ by π radians (i.e., replacing ω by $\omega - \pi$). Thus

$$H_{hp}(\omega) = H_{lp}(\omega - \pi) \qquad (4.5.12)$$

where $H_{hp}(\omega)$ is the frequency response of the highpass filter. Since a frequency translation of π radians is equivalent to multiplication of the impulse response $h_{lp}(n)$ by $e^{j\pi n}$, the impulse response of the highpass filter is

$$h_{hp}(n) = (e^{j\pi})^n h_{lp}(n) = (-1)^n h_{lp}(n) \qquad (4.5.13)$$

Therefore, the impulse response of the highpass filter is simply obtained from the impulse response of the lowpass filter by changing the signs of the odd-numbered samples in $h_{lp}(n)$. Conversely,

$$h_{lp}(n) = (-1)^n h_{hp}(n) \qquad (4.5.14)$$

If the lowpass filter is described by the difference equation

$$y(n) = -\sum_{k=1}^{N} a_k y(n - k) + \sum_{k=0}^{M} b_k x(n - k) \qquad (4.5.15)$$

its frequency response is

$$H_{lp}(\omega) = \frac{\displaystyle\sum_{k=0}^{M} b_k e^{-j\omega k}}{1 + \displaystyle\sum_{k=1}^{N} a_k e^{-j\omega k}} \qquad (4.5.16)$$

Now, if we replace ω by $\omega - \pi$, in (4.5.16), then

$$H_{hp}(\omega) = \frac{\displaystyle\sum_{k=0}^{M} (-1)^k b_k e^{-j\omega k}}{1 + \displaystyle\sum_{k=1}^{N} (-1)^k a_k e^{-j\omega k}} \qquad (4.5.17)$$

which corresponds to the difference equation

$$y(n) = -\sum_{k=1}^{N} (-1)^k a_k y(n - k) + \sum_{k=0}^{M} (-1)^k b_k x(n - k) \qquad (4.5.18)$$

Example 4.5.3

Convert the lowpass filter described by the difference equation

$$y(n) = 0.9y(n-1) + 0.1x(n)$$

into a highpass filter.

Solution   The difference equation for the highpass filter, according to (4.5.18), is

$$y(n) = -0.9y(n-1) + 0.1x(n)$$

and its frequency response is

$$H_{\text{hp}}(\omega) = \frac{0.1}{1 + 0.9e^{-j\omega}}$$

The reader may verify that $H_{\text{hp}}(\omega)$ is indeed highpass.

4.5.3 Digital Resonators

A *digital resonator* is a special two-pole bandpass filter with the pair of complex-conjugate poles located near the unit circle as shown in Fig. 4.48(a). The magnitude of the frequency response of the filter is shown in Fig. 4.48(b). The name resonator refers to the fact that the filter has a large magnitude response (i.e., it resonates) in the vicinity of the pole location. The angular position of the pole determines the resonant frequency of the filter. Digital resonators are useful in many applications, including simple bandpass filtering and speech generation.

In the design of a digital resonator with a resonant peak at or near $\omega = \omega_0$, we select the complex-conjugate poles at

$$p_{1,2} = re^{\pm j\omega_0} \qquad 0 < r < 1$$

In addition, we can select up to two zeros. Although there are many possible choices, two cases are of special interest. One choice is to locate the zeros at the origin. The other choice is to locate a zero at $z = 1$ and a zero at $z = -1$. This choice completely eliminates the response of the filter at frequencies $\omega = 0$ and $\omega = \pi$, and it is useful in many practical applications.

The system function of the digital resonator with zeros at the origin is

$$H(z) = \frac{b_0}{(1 - re^{j\omega_0}z^{-1})(1 - re^{-j\omega_0}z^{-1})} \tag{4.5.19}$$

$$H(z) = \frac{b_0}{1 - (2r\cos\omega_0)z^{-1} + r^2z^{-2}} \tag{4.5.20}$$

Since $|H(\omega)|$ has its peak at or near $\omega = \omega_0$, we select the gain b_0 so that $|H(\omega_0)| = 1$. From (4.5.19) we obtain

$$
\begin{aligned}
H(\omega_0) &= \frac{b_0}{(1 - re^{j\omega_0}e^{-j\omega_0})(1 - re^{-j\omega_0}e^{-j\omega_0})} \\
&= \frac{b_0}{(1 - r)(1 - re^{-j2\omega_0})}
\end{aligned}
\tag{4.5.21}
$$

and hence

$$|H(\omega_0)| = \frac{b_0}{(1-r)\sqrt{1 + r^2 - 2r\cos 2\omega_0}} = 1$$

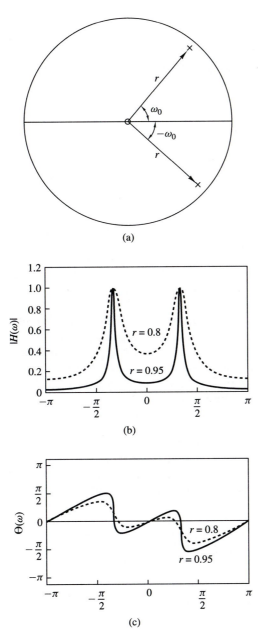

Figure 4.48 (a) Pole–zero pattern and (b) the corresponding magnitude and phase response of a digital resonator with (1) $r = 0.8$ and (2) $r = 0.95$.

Thus the desired normalization factor is

$$b_0 = (1 - r)\sqrt{1 + r^2 - 2r \cos 2\omega_0} \qquad (4.5.22)$$

The frequency response of the resonator in (4.5.19) can be expressed as

$$|H(\omega)| = \frac{b_0}{U_1(\omega)U_2(\omega)} \qquad (4.5.23)$$

$$\Theta(\omega) = 2\omega - \Phi_1(\omega) - \Phi_2(\omega)$$

where $U_1(\omega)$ and $U_2(\omega)$ are the magnitudes of the vectors from p_1 and p_2 to the point ω in the unit circle and $\Phi_1(\omega)$ and $\Phi_2(\omega)$ are the corresponding angles of these two vectors. The magnitudes $U_1(\omega)$ and $U_2(\omega)$ may be expressed as

$$U_1(\omega) = \sqrt{1 + r^2 - 2r \cos(\omega_0 - \omega)}$$

$$U_2(\omega) = \sqrt{1 + r^2 - 2r \cos(\omega_0 + \omega)} \qquad (4.5.24)$$

For any value of r, $U_1(\omega)$ takes its minimum value $(1 - r)$ at $\omega = \omega_0$. The product $U_1(\omega)U_2(\omega)$ reaches a minimum value at the frequency

$$\omega_r = \cos^{-1}\left(\frac{1 + r^2}{2r} \cos \omega_0\right) \qquad (4.5.25)$$

which defines precisely the resonant frequency of the filter. We observe that when r is very close to unity, $\omega_r \approx \omega_0$, which is the angular position of the pole. We also observe that as r approaches unity, the resonance peak becomes sharper because $U_1(\omega)$ changes more rapidly in relative size in the vicinity of ω_0. A quantitative measure of the sharpness of the resonance is provided by the 3-dB bandwidth $\Delta\omega$ of the filter. For values of r close to unity,

$$\Delta\omega \approx 2(1 - r) \qquad (4.5.26)$$

Figure 4.48 illustrates the magnitude and phase of digital resonators with $\omega_0 = \pi/3$, $r = 0.8$ and $\omega_0 = \pi/3$, $r = 0.95$. We note that the phase response undergoes its greatest rate of change near the resonant frequency.

If the zeros of the digital resonator are placed at $z = 1$ and $z = -1$, the resonator has the system function

$$H(z) = G\frac{(1 - z^{-1})(1 + z^{-1})}{(1 - re^{j\omega_0}z^{-1})(1 - re^{-j\omega_0}z^{-1})}$$

$$= G\frac{1 - z^{-2}}{1 - (2r \cos \omega_0)z^{-1} + r^2 z^{-2}} \qquad (4.5.27)$$

and a frequency response characteristic

$$H(\omega) = b_0 \frac{1 - e^{-j2\omega}}{[1 - re^{j(\omega_0 - \omega)}][1 - re^{-j(\omega_0 + \omega)}]} \qquad (4.5.28)$$

We observe that the zeros at $z = \pm 1$ affect both the magnitude and phase response of the resonator. For example, the magnitude response is

$$|H(\omega)| = b_0 \frac{N(\omega)}{U_1(\omega)U_2(\omega)} \qquad (4.5.29)$$

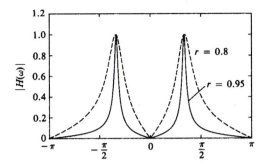

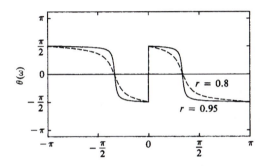

Figure 4.49 Magnitude and phase response of digital resonator with zeros at $\omega = 0$ and $\omega = \pi$ and (1) $r = 0.8$ and (2) $r = 0.95$.

where $N(\omega)$ is defined as

$$N(\omega) = \sqrt{2(1 - \cos 2\omega)}$$

Due to the presence of the zero factor, the resonant frequency is altered from that given by the expression in (4.5.25). The bandwidth of the filter is also altered. Although exact values for these two parameters are rather tedious to derive, we can easily compute the frequency response in (4.5.28) and compare the result with the previous case in which the zeros are located at the origin.

Figure 4.49 illustrates the magnitude and phase characteristics for $\omega_0 = \pi/3$, $r = 0.8$ and $\omega_0 = \pi/3$, $r = 0.95$. We observe that this filter has a slightly smaller bandwidth than the resonator, which has zeros at the origin. In addition, there appears to be a very small shift in the resonant frequency due to the presence of the zeros.

4.5.4 Notch Filters

A notch filter is a filter that contains one or more deep notches or, ideally, perfect nulls in its frequency response characteristic. Figure 4.50 illustrates the frequency response characteristic of a notch filter with nulls at frequencies ω_0 and ω_1. Notch filters are useful in many applications where specific frequency components must be eliminated. For example, instrumentation and recording systems require that the power-line frequency of 60 Hz and its harmonics be eliminated.

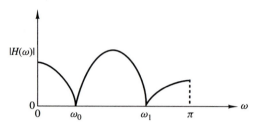

Figure 4.50 Frequency response characteristic of a notch filter.

To create a null in the frequency response of a filter at a frequency ω_0, we simply introduce a pair of complex-conjugate zeros on the unit circle at an angle ω_0. That is,

$$z_{1,2} = e^{\pm j\omega_0}$$

Thus the system function for an FIR notch filter is simply

$$
\begin{aligned}
H(z) &= b_0(1 - e^{j\omega_0}z^{-1})(1 - e^{-j\omega_0}z^{-1}) \\
&= b_0(1 - 2\cos\omega_0 z^{-1} + z^{-2})
\end{aligned}
\tag{4.5.30}
$$

As an illustration, Fig. 4.51 shows the magnitude response for a notch filter having a null at $\omega = \pi/4$.

The problem with the FIR notch filter is that the notch has a relatively large bandwidth, which means that other frequency components around the desired null are severely attenuated. To reduce the bandwidth of the null, we can resort to a more sophisticated, longer FIR filter designed according to criteria described in Chapter 8. Alternatively, we could, in an ad hoc manner, attempt to improve on the frequency response characteristics by introducing poles in the system function.

Suppose that we place a pair of complex-conjugate poles at

$$p_{1,2} = re^{\pm j\omega_0}$$

The effect of the poles is to introduce a resonance in the vicinity of the null and thus to reduce the bandwidth of the notch. The system function for the resulting filter is

$$H(z) = b_0 \frac{1 - 2\cos\omega_0 z^{-1} + z^{-2}}{1 - 2r\cos\omega_0 z^{-1} + r^2 z^{-2}} \tag{4.5.31}$$

The magnitude response $|H(\omega)|$ of the filter in (4.5.31) is plotted in Fig. 4.52 for $\omega_0 = \pi/4$, $r = 0.85$, and for $\omega_0 = \pi/4$, $r = 0.95$. When compared with the frequency response of the FIR filter in Fig. 4.51, we note that the effect of the poles is to reduce the bandwidth of the notch.

In addition to reducing the bandwidth of the notch, the introduction of a pole in the vicinity of the null may result in a small ripple in the passband of the filter due to the resonance created by the pole. The effect of the ripple can be reduced by introducing additional poles and/or zeros in the system function of the notch filter. The major problem with this approach is that it is basically an ad hoc, trial-and-error method.

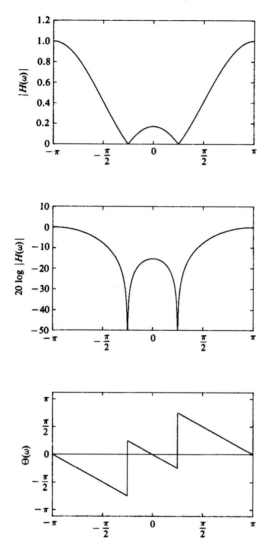

Figure 4.51 Frequency response characteristics of a notch filter with a notch at $\omega = \pi/4$ or $f = 1/8$; $H(z) = G[1 - 2\cos\omega_0 z^{-1} + z^{-2}]$.

4.5.5 Comb Filters

In its simplest form, a comb filter can be viewed as a notch filter in which the nulls occur periodically across the frequency band, hence the analogy to an ordinary comb that has periodically spaced teeth. Comb filters find applications in a wide range of practical systems such as in the rejection of power-line harmonics, in the separation of solar and lunar components from ionospheric measurements of electron concentration, and in the suppression of clutter from fixed objects in moving-target-indicator (MTI) radars.

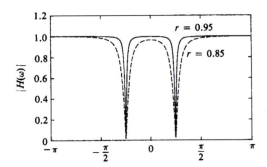

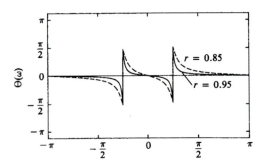

Figure 4.52 Frequency response characteristics of two notch filters with poles at (1) $r = 0.85$ and (2) $r = 0.95$; $H(z) = b_0[(1 - 2\cos\omega_0 z^{-1} + z^{-2})/(1 - 2r\cos\omega_0 z^{-1} + r^2 z^{-2})]$.

To illustrate a simple form of a comb filter, consider a moving average (FIR) filter described by the difference equation

$$y(n) = \frac{1}{M+1} \sum_{k=0}^{M} x(n-k) \tag{4.5.32}$$

The system function of this FIR filter is

$$
\begin{aligned}
H(z) &= \frac{1}{M+1} \sum_{k=0}^{M} z^{-k} \\
&= \frac{1}{M+1} \frac{[1 - z^{-(M+1)}]}{(1 - z^{-1})}
\end{aligned} \tag{4.5.33}
$$

and its frequency response is

$$H(\omega) = \frac{e^{-j\omega M/2}}{M+1} \frac{\sin\omega(\frac{M+1}{2})}{\sin(\omega/2)} \tag{4.5.34}$$

From (4.5.33) we observe that the filter has zeros on the unit circle at

$$z = e^{j2\pi k/(M+1)} \qquad k = 1, 2, 3, \ldots, M \tag{4.5.35}$$

Note that the pole at $z = 1$ is actually canceled by the zero at $z = 1$, so that in effect the FIR filter does not contain poles outside $z = 0$.

A plot of the magnitude characteristic of (4.5.34) clearly illustrates the existence of the periodically spaced zeros in frequency at $\omega_k = 2\pi k/(M+1)$ for $k = 1, 2, \ldots, M$. Figure 4.53 shows $|H(\omega)|$ for $M = 10$.

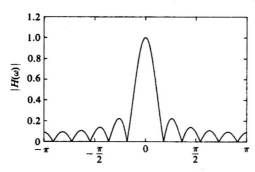

Figure 4.53 Magnitude response characteristic for the comb filter given by (5.4.32) with $M = 10$.

In more general terms, we can create a comb filter by taking an FIR filter with system function

$$H(z) = \sum_{k=0}^{M} h(k)z^{-k} \tag{4.5.36}$$

and replacing z by z^L, where L is a positive integer. Thus the new FIR filter has a system function

$$H_L(z) = \sum_{k=0}^{M} h(k)z^{-kL} \tag{4.5.37}$$

If the frequency response of the original FIR filter is $H(\omega)$, the frequency response of the FIR in (4.5.37) is

$$H_L(\omega) = \sum_{k=0}^{M} h(k)e^{-jkL\omega} \tag{4.5.38}$$
$$= H(L\omega)$$

Consequently, the frequency response characteristic $H_L(\omega)$ is simply an L-order repetition of $H(\omega)$ in the range $0 \le \omega \le 2\pi$. Figure 4.54 illustrates the relationship between $H_L(\omega)$ and $H(\omega)$ for $L = 5$.

Now, suppose that the original FIR filter with system function $H(z)$ has a spectral null (i.e., a zero), at some frequency ω_0. Then the filter with system function $H_L(z)$ has periodically spaced nulls at $\omega_k = \omega_0 + 2\pi k/L$, $k = 0, 1, 2, \ldots$, $L - 1$. As an illustration, Fig. 4.55 shows an FIR comb filter with $M = 3$ and $L = 3$. This FIR filter can be viewed as an FIR filter of length 10, but only four of the 10 filter coefficients are nonzero.

Let us now return to the moving average filter with system function given by (4.5.33). Suppose that we replace z by z^L. Then the resulting comb filter has the system function

$$H_L(z) = \frac{1}{M+1} \frac{1 - z^{-L(M+1)}}{1 - z^{-L}} \tag{4.5.39}$$

and a frequency response

$$H_L(\omega) = \frac{1}{M+1} \frac{\sin[\omega L(M+1)/2]}{\sin(\omega L/2)} e^{-j\omega LM/2} \tag{4.5.40}$$

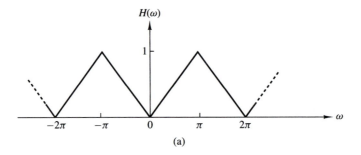

(a)

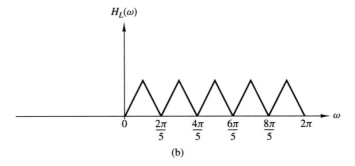

(b)

Figure 4.54 Comb filter with frequency response $H_L(\omega)$ obtained from $H(\omega)$.

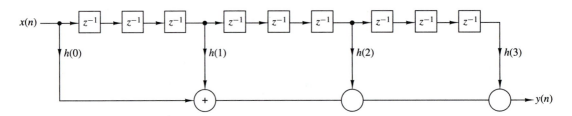

Figure 4.55 Realization of an FIR comb filter having $M = 3$ and $L = 3$.

This filter has zeros on the unit circle at

$$z_k = e^{j2\pi k/L(M+1)} \tag{4.5.41}$$

for all integer values of k except $k = 0, L, 2L, \ldots, ML$. Figure 4.56 illustrates $|H_L(\omega)|$ for $L = 5$ and $M = 10$.

The comb filter described by (4.5.39) finds application in the separation of solar and lunar spectral components in ionospheric measurements of electron concentration as described in the paper by Bernhardt et al. (1976). The solar period is $T_s = 24$ hours and results in a solar component of one cycle per day and its harmonics. The lunar period is $T_L = 24.84$ hours and provides spectral lines at 0.96618 cycle per day and its harmonics. Figure 4.57a shows a plot of the power density spectrum of the unfiltered ionospheric measurements of the electron con-

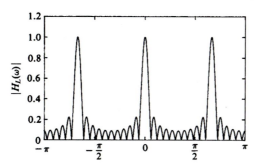

Figure 4.56   Magnitude response characteristic for a comb filter given by (4.5.40), with $L = 3$ and $M = 10$.

centration. Note that the weak lunar spectral components are almost hidden by the strong solar spectral components.

The two sets of spectral components can be separated by the use of comb filters. If we wish to obtain the solar components, we can use a comb filter with a narrow passband at multiples of one cycle per day. This can be achieved by selecting L such that $F_s/L = 1$ cycle per day, where F_s is the corresponding sampling frequency. The result is a filter that has peaks in its frequency response at multiples of one cycle per day. By selecting $M = 58$, the filter will have nulls at multiples of $(F_s/L)/(M + 1) = 1/59$ cycle per day. These nulls are very close to the lunar components and result in good rejection. Figure 4.57(b) illustrates

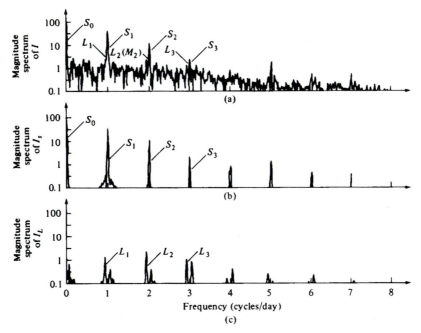

Figure 4.57   (a) Spectrum of unfiltered electron content data; (b) spectrum of output of solar filter; (c) spectrum of output of lunar filter. [From paper by Bernhardt et al. (1976). Reprinted with permission of the American Geophysical Union.]

the power spectral density of the output of the comb filter that isolates the solar components. A comb filter that rejects the solar components and passes the lunar components can be designed in a similar manner. Figure 4.57(c) illustrates the power spectral density at the output of such a lunar filter.

4.5.6 All-Pass Filters

An all-pass filter is defined as a system that has a constant magnitude response for all frequencies, that is,

$$|H(\omega)| = 1 \qquad 0 \le \omega \le \pi \qquad (4.5.42)$$

The simplest example of an all-pass filter is a pure delay system with system function

$$H(z) = z^{-k}$$

This system passes all signals without modification except for a delay of k samples. This is a trivial all-pass system that has a linear phase response characteristic.

A more interesting all-pass filter is described by the system function

$$H(z) = \frac{a_N + a_{N-1}z^{-1} + \cdots + a_1 z^{-N+1} + z^{-N}}{1 + a_1 z^{-1} + \cdots + a_N z^{-N}}$$

$$= \frac{\sum_{k=0}^{N} a_k z^{-N+k}}{\sum_{k=0}^{N} a_k z^{-k}} \qquad a_0 = 1 \qquad (4.5.43)$$

where all the filter coefficients $\{a_k\}$ are real. If we define the polynomial $A(z)$ as

$$A(z) = \sum_{k=0}^{N} a_k z^{-k} \qquad a_0 = 1$$

then (4.5.43) can be expressed as

$$H(z) = z^{-N} \frac{A(z^{-1})}{A(z)} \qquad (4.5.44)$$

Since

$$|H(\omega)|^2 = H(z)H(z^{-1})|_{z=e^{j\omega}} = 1$$

the system given by (4.5.44) is an all-pass system. Furthermore, if z_0 is a pole of $H(z)$, then $1/z_0$ is a zero of $H(z)$ (i.e., the poles and zeros are reciprocals of one another). Figure 4.58 illustrates typical pole–zero patterns for a single-pole, single-zero filter and a two-pole, two-zero filter. A plot of the phase characteristics of these filters is shown in Fig. 4.59 for $a = 0.6$ and $r = 0.9$, $\omega_0 = \pi/4$.

The most general form for the system function of an all-pass system with real coefficients, expressed in factored form in terms of poles and zeros, is

$$H_{\text{ap}}(z) = \prod_{k=1}^{N_R} \frac{z^{-1} - \alpha_k}{1 - \alpha_k z^{-1}} \prod_{k=1}^{N_c} \frac{(z^{-1} - \beta_k)(z^{-1} - \beta_k^*)}{(1 - \beta_k z^{-1})(1 - \beta_k^* z^{-1})} \qquad (4.5.45)$$

where there are N_R real poles and zeros and N_c complex-conjugate pairs of poles and zeros. For causal and stable systems we require that $-1 < \alpha_k < 1$ and $|\beta_k| < 1$.

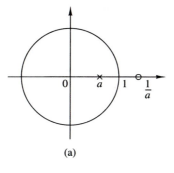

(a)

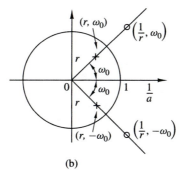

(b)

Figure 4.58 Pole–zero patterns of (a) a first-order and (b) a second-order all-pass filter.

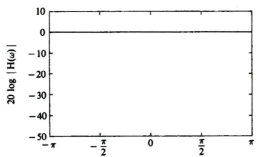

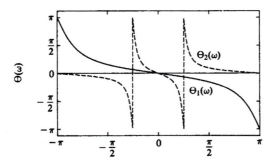

Figure 4.59 Frequency response characteristics of an all-pass filter with system functions
(1) $H(z) = (0.6 + z^{-1})/(1 + 0.6z^{-1})$,
(2) $H(z) = (r^2 - 2r \cos \omega_0 z^{-1} + z^{-2})/(1 - 2r \cos \omega_0 z^{-1} + r^2 z^{-2})$, $r = 0.9$, $\omega_0 = \pi/4$.

Expressions for the phase response and group delay of all-pass systems can easily be obtained using the method described in Section 4.4.6. For a single pole–single zero all-pass system we have

$$H_{\text{ap}}(\omega) = \frac{e^{j\omega} - re^{-j\theta}}{1 - re^{j\theta}e^{-j\omega}}$$

Hence

$$\Theta_{\text{ap}}(\omega) = -\omega - 2\tan^{-1}\frac{r\sin(\omega - \theta)}{1 - r\cos(\omega - \theta)}$$

and

$$\tau_z(\omega) = -\frac{d\Theta_{\text{ap}}(\omega)}{d\omega} = \frac{1 - r^2}{1 + r^2 - 2r\cos(\omega - \theta)} \tag{4.5.46}$$

We note that for a causal and stable system, $r < 1$ and hence $\tau_g(\omega) \geq 0$. Since the group delay of a higher-order pole–zero system consists of a sum of positive terms as in (4.5.46), the group delay will always be positive.

All-pass filters find application as phase equalizers. When placed in cascade with a system that has an undesired phase response, a phase equalizer is designed to compensate for the poor phase characteristics of the system and therefore to produce an overall linear-phase response.

4.5.7 Digital Sinusoidal Oscillators

A *digital sinusoidal oscillator* can be viewed as a limiting form of a two-pole resonator for which the complex-conjugate poles lie on the unit circle. From our previous discussion of second-order systems, we recall that a system with system function

$$H(z) = \frac{b_0}{1 + a_1 z^{-1} + a_2 z^{-2}} \tag{4.5.47}$$

and parameters

$$a_1 = -2r\cos\omega_0 \quad \text{and} \quad a_2 = r^2 \tag{4.5.48}$$

has complex-conjugate poles at $p = re^{\pm j\omega_0}$, and a unit sample response

$$h(n) = \frac{b_0 r^n}{\sin\omega_0}\sin(n + 1)\omega_0 u(n) \tag{4.5.49}$$

If the poles are placed on the unit circle ($r = 1$) and b_0 is set to $A\sin\omega_0$, then

$$h(n) = A\sin(n + 1)\omega_0 u(n) \tag{4.5.50}$$

Thus the impulse response of the second-order system with complex-conjugate poles on the unit circle is a sinusoid and the system is called a digital sinusoidal oscillator or a *digital sinusoidal generator*. A digital sinusoidal generator is a basic component of a digital frequency synthesizer.

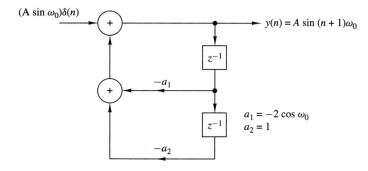

Figure 4.60   Digital sinusoidal generator.

The block diagram representation of the system function given by (4.5.47) is illustrated in Fig. 4.60. The corresponding difference equation for this system is

$$y(n) = -a_1 y(n-1) - y(n-2) + b_0 \delta(n) \tag{4.5.51}$$

where the parameters are $a_1 = -2 \cos \omega_0$ and $b_0 = A \sin \omega_0$, and the initial conditions are $y(-1) = y(-2) = 0$. By iterating the difference equation in (4.5.51), we obtain

$$y(0) = A \sin \omega_0$$

$$y(1) = 2 \cos \omega_0 y(0) = 2A \sin \omega_0 \cos \omega_0 = A \sin 2\omega_0$$

$$y(2) = 2 \cos \omega_0 y(1) - y(0)$$

$$= 2A \cos \omega_0 \sin 2\omega_0 - A \sin \omega_0$$

$$= A(4 \cos^2 \omega_0 - 1) \sin \omega_0$$

$$= 3A \sin \omega_0 - 4 \sin^3 \omega_0 = A \sin 3\omega_0$$

and so forth. We note that the application of the impulse at $n = 0$ serves the purpose of beginning the sinusoidal oscillation. Thereafter, the oscillation is self-sustaining because the system has no damping (i.e., $r = 1$).

It is interesting to note that the sinusoidal oscillation obtained from the system in (4.5.51) can also be obtained by setting the input to zero and setting the initial conditions to $y(-1) = 0$, $y(-2) = -A \sin \omega_0$. Thus the zero-input response to the second-order system described by the homogeneous difference equation

$$y(n) = -a_1 y(n-1) - y(n-2) \tag{4.5.52}$$

with initial conditions $y(-1) = 0$ and $y(-2) = -A \sin \omega_0$, is exactly the same as the response of (4.5.51) to an impulse excitation. In fact, the difference equation in (4.5.52) can be obtained directly from the trigonometric identity

$$\sin \alpha + \sin \beta = 2 \sin \frac{\alpha + \beta}{2} \cos \frac{\alpha - \beta}{2} \tag{4.5.53}$$

where, by definition, $\alpha = (n+1)\omega_0$, $\beta = (n-1)\omega_0$, and $y(n) = sin(n+1)\omega_0$.

In some practical applications involving modulation of two sinusoidal carrier signals in phase quadrature, there is a need to generate the sinusoids $A \sin \omega_0 n$ and $A \cos \omega_0 n$. These signals can be generated from the so-called *coupled-form oscillator*, which can be obtained from the trigonometric formulas

$$\cos(\alpha + \beta) = \cos\alpha \cos\beta - \sin\alpha \sin\beta$$

$$\sin(\alpha + \beta) = \sin\alpha \cos\beta + \cos\alpha \sin\beta$$

where, by definition, $\alpha = n\omega_0$, $\beta = \omega_0$, and

$$y_c(n) = \cos n\omega_0 u(n) \tag{4.5.54}$$

$$y_s(n) = \sin n\omega_0 u(n) \tag{4.5.55}$$

Thus we obtain the two coupled difference equations

$$y_c(n) = (\cos\omega_0)y_c(n-1) - (\sin\omega_0)y_s(n-1) \tag{4.5.56}$$

$$y_s(n) = (\sin\omega_0)y_c(n-1) + (\cos\omega_0)y_s(n-1) \tag{4.5.57}$$

which can also be expressed in matrix form as

$$\begin{bmatrix} y_c(n) \\ y_s(n) \end{bmatrix} = \begin{bmatrix} \cos\omega_0 & -\sin\omega_0 \\ \sin\omega_0 & \cos\omega_0 \end{bmatrix} \begin{bmatrix} y_c(n-1) \\ y_s(n-1) \end{bmatrix} \tag{4.5.58}$$

The structure for the realization of the coupled-form oscillator is illustrated in Fig. 4.61. We note that this is a two-output system which is not driven by any input, but which requires the initial conditions $y_c(-1) = A\cos\omega_0$ and $y_s(-1) = -A\sin\omega_0$ in order to begin its self-sustaining oscillations.

Finally, it is interesting to note that (4.5.58) corresponds to vector rotation in the two-dimensional coordinate system with coordinates $y_c(n)$ and $y_s(n)$. As a consequence, the coupled-form oscillator can also be implemented by use of the so-called CORDIC algorithm [see the book by Kung et al. (1985)].

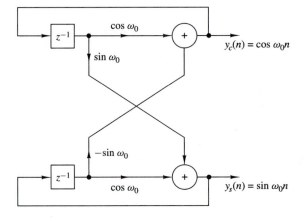

Figure 4.61 Realization of the coupled-form oscillator.

4.6 INVERSE SYSTEMS AND DECONVOLUTION

As we have seen, a linear time-invariant system takes an input signal $x(n)$ and produces an output signal $y(n)$, which is the convolution of $x(n)$ with the unit sample response $h(n)$ of the system. In many practical applications we are given an output signal from a system whose characteristics are unknown and we are asked to determine the input signal. For example, in the transmission of digital information at high data rates over telephone channels, it is well known that the channel distorts the signal and causes intersymbol interference among the data symbols. The intersymbol interference may cause errors when we attempt to recover the data. In such a case the problem is to design a corrective system which, when cascaded with the channel, produces an output that, in some sense, corrects for the distortion caused by the channel, and thus yields a replica of the desired transmitted signal. In digital communications such a corrective system is called an *equalizer*. In the general context of linear systems theory, however, we call the corrective system an *inverse system*, because the corrective system has a frequency response which is basically the reciprocal of the frequency response of the system that caused the distortion. Furthermore, since the distortive system yields an output $y(n)$ that is the convolution of the input $x(n)$ with the impulse response $h(n)$, the inverse system operation that takes $y(n)$ and produces $x(n)$ is called *deconvolution*.

If the characteristics of the distortive system are unknown, it is often necessary, when possible, to excite the system with a known signal, observe the output, compare it with the input, and in some manner, determine the characteristics of the system. For example, in the digital communication problem just described, where the frequency response of the channel is unknown, the measurement of the channel frequency response can be accomplished by transmitting a set of equal amplitude sinusoids, at different frequencies with a specified set of phases, within the frequency band of the channel. The channel will attenuate and phase shift each of the sinusoids. By comparing the received signal with the transmitted signal, the receiver obtains a measurement of the channel frequency response which can be used to design the inverse system. The process of determining the characteristics of the unknown system, either $h(n)$ or $H(\omega)$, by a set of measurements performed on the system is called *system identification*.

The term "deconvolution" is often used in seismic signal processing, and more generally, in geophysics to describe the operation of separating the input signal from the characteristics of the system which we wish to measure. The deconvolution operation is actually intended to identify the characteristics of the system, which in this case, is the earth, and can also be viewed as a system identification problem. The "inverse system," in this case, has a frequency response that is the reciprocal of the input signal spectrum that has been used to excite the system.

4.6.1 Invertibility of Linear Time-Invariant Systems

A system is said to be *invertible* if there is a one-to-one correspondence between its input and output signals. This definition implies that if we know the output sequence $y(n)$, $-\infty < n < \infty$, of an invertible system T, we can uniquely determine its input $x(n)$, $-\infty < n < \infty$. The *inverse system* with input $y(n)$ and output $x(n)$ is denoted by T^{-1}. Clearly, the cascade connection of a system and its inverse is equivalent to the identity system, since

$$w(n) = T^{-1}[y(n)] = T^{-1}\{T[x(n)]\} = x(n) \tag{4.6.1}$$

as illustrated in Fig. 4.62. For example, the systems defined by the input–output relations $y(n) = ax(n)$ and $y(n) = x(n-5)$ are invertible, whereas the input–output relations $y(n) = x^2(n)$ and $y(n) = 0$ represent noninvertible systems.

As indicated above, inverse systems are important in many practical applications, including geophysics and digital communications. Let us begin by considering the problem of determining the inverse of a given system. We limit our discussion to the class of linear time-invariant discrete-time systems.

Now, suppose that the linear time-invariant system T has an impulse response $h(n)$ and let $h_I(n)$ denote the impulse response of the inverse system T^{-1}. Then (4.6.1) is equivalent to the convolution equation

$$w(n) = h_I(n) * h(n) * x(n) = x(n) \tag{4.6.2}$$

But (4.6.2) implies that

$$h(n) * h_I(n) = \delta(n) \tag{4.6.3}$$

The convolution equation in (4.6.3) can be used to solve for $h_I(n)$ for a given $h(n)$. However, the solution of (4.6.3) in the time domain is usually difficult. A simpler approach is to transform (4.6.3) into the z-domain and solve for T^{-1}. Thus in the z-transform domain, (4.6.3) becomes

$$H(z)H_I(z) = 1$$

and therefore the system function for the inverse system is

$$H_I(z) = \frac{1}{H(z)} \tag{4.6.4}$$

If $H(z)$ has a rational system function

$$H(z) = \frac{B(z)}{A(z)} \tag{4.6.5}$$

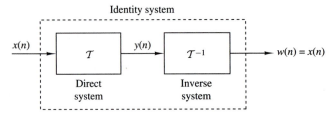

Figure 4.62 System T in cascade with its inverse T^{-1}.

then

$$H_I(z) = \frac{A(z)}{B(z)} \qquad (4.6.6)$$

Thus the zeros of $H(z)$ become the poles of the inverse system, and vice versa. Furthermore, if $H(z)$ is an FIR system, then $H_I(z)$ is an all-pole system, or if $H(z)$ is an all-pole system, then $H_I(z)$ is an FIR system.

Example 4.6.1

Determine the inverse of the system with impulse response

$$h(n) = (\tfrac{1}{2})^n u(n)$$

Solution   The system function corresponding to $h(n)$ is

$$H(z) = \frac{1}{1 - \tfrac{1}{2}z^{-1}} \qquad \text{ROC: } |z| > \tfrac{1}{2}$$

This system is both causal and stable. Since $H(z)$ is an all-pole system, its inverse is FIR and is given by the system function

$$H_I(z) = 1 - \tfrac{1}{2}z^{-1}$$

Hence its impulse response is

$$h_I(n) = \delta(n) - \tfrac{1}{2}\delta(n - 1)$$

Example 4.6.2

Determine the inverse of the system with impulse response

$$h(n) = \delta(n) - \tfrac{1}{2}\delta(n - 1)$$

Solution   This is an FIR system and its system function is

$$H(z) = 1 - \tfrac{1}{2}z^{-1} \qquad \text{ROC: } |z| > 0$$

The inverse system has the system function

$$H_I(z) = \frac{1}{H(z)} = \frac{1}{1 - \tfrac{1}{2}z^{-1}} = \frac{z}{z - \tfrac{1}{2}}$$

Thus $H_I(z)$ has a zero at the origin and a pole at $z = \tfrac{1}{2}$. In this case there are two possible regions of convergence and hence two possible inverse systems, as illustrated in Fig. 4.63. If we take the ROC of $H_I(z)$ as $|z| > \tfrac{1}{2}$, the inverse transform yields

$$h_I(n) = (\tfrac{1}{2})^n u(n)$$

which is the impulse response of a causal and stable system. On the other hand, if the ROC is assumed to be $|z| < \tfrac{1}{2}$, the inverse system has an impulse response

$$h_I(n) = -\left(\frac{1}{2}\right)^n u(-n - 1)$$

In this case the inverse system is anticausal and unstable.

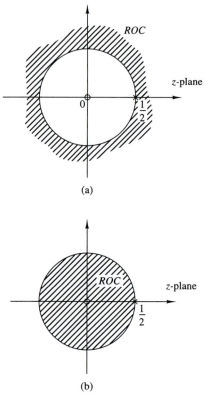

Figure 4.63 Two possible regions of convergence for $H(z) = z/(z - \frac{1}{2})$.

We observe that (4.6.3) cannot be solved uniquely by using (4.6.6) unless we specify the region of convergence for the system function of the inverse system.

In some practical applications the impulse response $h(n)$ does not possess a z-transform that can be expressed in closed form. As an alternative we may solve (4.6.3) directly using a digital computer. Since (4.6.3) does not, in general, possess a unique solution, we assume that the system and its inverse are causal. Then (4.6.3) simplifies to the equation

$$\sum_{k=0}^{n} h(k)h_I(n-k) = \delta(n) \tag{4.6.7}$$

By assumption, $h_I(n) = 0$ for $n < 0$. For $n = 0$ we obtain

$$h_I(0) = 1/h(0) \tag{4.6.8}$$

The values of $h_I(n)$ for $n \geq 1$ can be obtained recursively from the equation

$$h_I(n) = -\sum_{k=1}^{n} \frac{h(k)h_I(n-k)}{h(0)} \qquad n \geq 1 \tag{4.6.9}$$

This recursive relation can easily be programmed on a digital computer.

There are two problems associated with (4.6.9). First, the method does not work if $h(0) = 0$. However, this problem can easily be remedied by introducing an appropriate delay in the right-hand side of (4.6.7), that is, by replacing $\delta(n)$ by $\delta(n - m)$, where $m = 1$ if $h(0) = 0$ and $h(1) \neq 0$, and so on. Second, the recursion in (4.6.9) gives rise to round-off errors which grow with n and, as a result, the numerical accuracy of $h(n)$ deteriorates for large n.

Example 4.6.3

Determine the causal inverse of the FIR system with impulse response

$$h(n) = \delta(n) - \alpha\delta(n - 1)$$

Solution   Since $h(0) = 1$, $h(1) = -\alpha$, and $h(n) = 0$ for $n \geq \alpha$, we have

$$h_I(0) = 1/h(0) = 1$$

and

$$h_I(n) = \alpha h_I(n - 1) \qquad n \geq 1$$

Consequently,

$$h_I(1) = \alpha, \quad h_I(2) = \alpha^2, \quad \ldots, \quad h_I(n) = \alpha^n$$

which corresponds to a causal IIR system as expected.

4.6.2 Minimum-Phase, Maximum-Phase, and Mixed-Phase Systems

The invertibility of a linear time-invariant system is intimately related to the characteristics of the phase spectral function of the system. To illustrate this point, let us consider two FIR systems, characterized by the system functions

$$H_1(z) = 1 + \tfrac{1}{2}z^{-1} = z^{-1}(z + \tfrac{1}{2}) \tag{4.6.10}$$

$$H_2(z) = \tfrac{1}{2} + z^{-1} = z^{-1}(\tfrac{1}{2}z + 1) \tag{4.6.11}$$

The system in (4.6.10) has a zero at $z = -\tfrac{1}{2}$ and an impulse response $h(0) = 1$, $h(1) = 1/2$. The system in (4.6.11) has a zero at $z = -2$ and an impulse response $h(0) = 1/2$, $h(1) = 1$, which is the reverse of the system in (4.6.10). This is due to the reciprocal relationship between the zeros of $H_1(z)$ and $H_2(z)$.

In the frequency domain, the two systems are characterized by their frequency response functions, which can be expressed as

$$|H_1(\omega)| = |H_2(\omega)| = \sqrt{\tfrac{5}{4} + \cos \omega} \tag{4.6.12}$$

and

$$\Theta_1(\omega) = -\omega + \tan^{-1} \frac{\sin \omega}{\tfrac{1}{2} + \cos \omega} \tag{4.6.13}$$

$$\Theta_2(\omega) = -\omega + \tan^{-1} \frac{\sin \omega}{2 + \cos \omega} \tag{4.6.14}$$

The magnitude characteristics for the two systems are identical because the zeros of $H_1(z)$ and $H_2(z)$ are reciprocals.

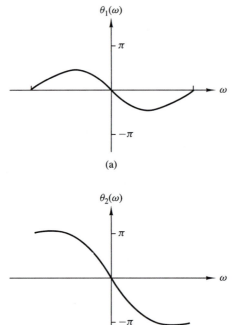

Figure 4.64 Phase response characteristics for the systems in (4.6.10) and (4.6.11).

The graphs of $\Theta_1(\omega)$ and $\Theta_2(\omega)$ are illustrated in Fig. 4.64. We observe that the phase characteristic $\Theta_1(\omega)$ for the first system begins at zero phase at the frequency $\omega = 0$ and terminates at zero phase at the frequency $\omega = \pi$. Hence the net phase change, $\Theta_1(\pi) - \Theta_1(0)$ is zero. On the other hand, the phase characteristic for the system with the zero outside the unit circle undergoes a net phase change $\Theta_2(\pi) - \Theta_2(0) = \pi$ radians. As a consequence of these different phase characteristics, we call the first system a *minimum-phase system* and the second system is called a *maximum-phase system*.

These definitions are easily extended to an FIR system of arbitrary length. To be specific, an FIR system of length $M + 1$ has M zeros. Its frequency response can be expressed as

$$H(\omega) = b_0(1 - z_1 e^{-j\omega})(1 - z_2 e^{-j\omega}) \cdots (1 - z_M e^{-j\omega}) \qquad (4.6.15)$$

where $\{z_i\}$ denote the zeros and b_0 is an arbitrary constant. When all the zeros are inside the unit circle, each term in the product of (4.6.15), corresponding to a real-valued zero, will undergo a net phase change of zero between $\omega = 0$ and $\omega = \pi$. Also, each pair of complex- conjugate factors in $H(\omega)$ will undergo a net phase change of zero. Therefore,

$$\measuredangle H(\pi) - \measuredangle H(0) = 0 \qquad (4.6.16)$$

and hence the system is called a minimum-phase system. On the other hand, when all the zeros are outside the unit circle, a real-valued zero will contribute a net phase change of π radians as the frequency varies from $\omega = 0$ to $\omega = \pi$, and each pair of complex-conjugate zeros will contribute a net phase change of 2π radians over the same range of ω. Therefore,

$$\measuredangle H(\pi) - \measuredangle H(0) = M\pi \qquad (4.6.17)$$

which is the largest possible phase change for an FIR system with M zeros. Hence the system is called maximum phase. It follows from the discussion above that

$$\measuredangle H_{\max}(\pi) \geq \measuredangle H_{\min}(\pi) \qquad (4.6.18)$$

If the FIR system with M zeros has some of its zeros inside the unit circle and the remaining zeros outside the unit circle, it is called a *mixed-phase system* or a *nonminimum-phase system*.

Since the derivative of the phase characteristic of the system is a measure of the time delay that signal frequency components undergo in passing through the system, a minimum-phase characteristic implies a minimum delay function, while a maximum-phase characteristic implies that the delay characteristic is also maximum.

Now suppose that we have an FIR system with real coefficients. Then the magnitude square value of its frequency response is

$$|H(\omega)|^2 = H(z)H(z^{-1})|_{z=e^{j\omega}} \qquad (4.6.19)$$

This relationship implies that if we replace a zero z_k of the system by its inverse $1/z_k$, the magnitude characteristic of the system does not change. Thus if we reflect a zero z_k that is inside the unit circle into a zero $1/z_k$ outside the unit circle, we see that the magnitude characteristic of the frequency response is invariant to such a change.

It is apparent from this discussion that if $|H(\omega)|^2$ is the magnitude square frequency response of an FIR system having M zeros, there are 2^M possible configurations for the M zeros, some of which are inside the unit circle and the remaining are outside the unit circle. Clearly, one configuration has all the zeros inside the unit circle, which corresponds to the minimum-phase system. A second configuration has all the zeros outside the unit circle, which corresponds to the maximum-phase system. The remaining $2^M - 2$ configurations correspond to mixed-phase systems. However, not all $2^M - 2$ mixed-phase configurations necessarily correspond to FIR systems with real-valued coefficients. Specifically, any pair of complex-conjugate zeros result in only two possible configurations, whereas a pair of real-valued zeros yield four possible configurations.

Example 4.6.4

Determine the zeros for the following FIR systems and indicate whether the system is minimum phase, maximum phase, or mixed phase.

$$H_1(z) = 6 + z^{-1} - z^{-2}$$
$$H_2(z) = 1 - z^{-1} - 6z^{-2}$$

$$H_3(z) = 1 - \tfrac{5}{2}z^{-1} - \tfrac{3}{2}z^{-2}$$

$$H_4(z) = 1 + \tfrac{5}{3}z^{-1} - \tfrac{2}{3}z^{-2}$$

Solution   By factoring the system functions we find the zeros for the four systems are

$$H_1(z) \longrightarrow z_{1,2} = -\tfrac{1}{2}, \tfrac{1}{3} \longrightarrow \text{ minimum phase}$$

$$H_2(z) \longrightarrow z_{1,2} = -2, 3 \longrightarrow \text{ maximum phase}$$

$$H_3(z) \longrightarrow z_{1,2} = -\tfrac{1}{2}, 3 \longrightarrow \text{ mixed phase}$$

$$H_4(z) \longrightarrow z_{1,2} = -2, \tfrac{1}{3} \longrightarrow \text{ mixed phase}$$

Since the zeros of the four systems are reciprocals of one another, it follows that all four systems have identical magnitude frequency response characteristics but different phase characteristics.

The minimum-phase property of FIR systems carries over to IIR systems that have rational system functions. Specifically, an IIR system with system function

$$H(z) = \frac{B(z)}{A(z)} \tag{4.6.20}$$

is called *minimum phase* if all its poles and zeros are inside the unit circle. For a stable and causal system [all roots of $A(z)$ fall inside the unit circle] the system is called *maximum phase* if all the zeros are outside the unit circle, and *mixed phase* if some, but not all, of the zeros are outside the unit circle.

This discussion brings us to an important point that should be emphasized. That is, a *stable* pole–zero system that is minimum phase has a stable inverse which is also minimum phase. The inverse system has the system function

$$H^{-1}(z) = \frac{A(z)}{B(z)} \tag{4.6.21}$$

Hence the minimum-phase property of $H(z)$ ensures the stability of the inverse system $H^{-1}(z)$ and the stability of $H(z)$ implies the minimum-phase property of $H^{-1}(z)$. Mixed-phase systems and maximum-phase systems result in unstable inverse systems.

Decomposition of nonminimum-phase pole–zero systems.   Any nonminimum-phase pole–zero system can be expressed as

$$H(z) = H_{\min}(z) H_{\text{ap}}(z) \tag{4.6.22}$$

where $H_{\min}(z)$ is a minimum-phase system and $H_{\text{ap}}(z)$ is an all-pass system. We demonstrate the validity of this assertion for the class of causal and stable systems with a rational system function $H(z) = B(z)/A(z)$. In general, if $B(z)$ has one or more roots outside the unit circle, we factor $B(z)$ into the product $B_1(z)B_2(z)$, where $B_1(z)$ has all its roots inside the unit circle and $B_2(z)$ has all its roots outside the unit circle. Then $B_2(z^{-1})$ has all its roots inside the unit circle. We define the minimum-phase system

$$H_{\min}(z) = \frac{B_1(z)B_2(z^{-1})}{A(z)}$$

and the all-pass system

$$H_{\mathrm{ap}}(z) = \frac{B_2(z)}{B_2(z^{-1})}$$

Thus $H(z) = H_{\min}(z)H_{\mathrm{ap}}(z)$. Note that $H_{\mathrm{ap}}(z)$ is a stable, all-pass, maximum-phase system.

Group delay of nonminimum-phase system.   Based on the decomposition of a nonminimum-phase system given by (4.6.22), we can express the group delay of $H(z)$ as

$$\tau_g(\omega) = \tau_g^{\min}(\omega) + \tau_g^{\mathrm{ap}}(\omega) \tag{4.6.23}$$

Since $\tau_g^{ap}(\omega) \geq 0$ for $0 \leq \omega \leq \pi$, it follows that $\tau_g(\omega) \geq \tau_g^{\min}(\omega)$, $0 \leq \omega \leq \pi$. From (4.6.23) we conclude that among all pole–zero systems having the same magnitude response, the minimum-phase system has the smallest group delay.

Partial energy of nonminimum-phase system.   The *partial energy* of a causal system with impulse response $h(n)$ is defined as

$$E(n) = \sum_{k=0}^{n} |h(k)|^2 \tag{4.6.24}$$

It can be shown that among all systems having the same magnitude response and the same total energy $E(\infty)$, the minimum-phase system has the largest partial energy [i.e., $E_{\min}(n) \geq E(n)$, where $E_{\min}(n)$ is the partial energy of the minimum-phase system].

4.6.3 System Identification and Deconvolution

Suppose that we excite an unknown linear time-invariant system with an input sequence $x(n)$ and we observe the output sequence $y(n)$. From the output sequence we wish to determine the impulse response of the unknown system. This is a problem in *system identification*, which can be solved by *deconvolution*. Thus we have

$$y(n) = h(n) * x(n)$$

$$= \sum_{k=-\infty}^{\infty} h(k)x(n-k) \tag{4.6.25}$$

An analytical solution of the deconvolution problem can be obtained by working with the z-transform of (4.6.25). In the z-transform domain we have

$$Y(z) = H(z)X(z)$$

and hence

$$H(z) = \frac{Y(z)}{X(z)} \tag{4.6.26}$$

$X(z)$ and $Y(z)$ are the z-transforms of the available input signal $x(n)$ and the observed output signal $y(n)$, respectively. This approach is appropriate only when there are closed-form expressions for $X(z)$ and $Y(z)$.

Example 4.6.5

A causal system produces the output sequence

$$y(n) = \begin{cases} 1, & n = 0 \\ \frac{7}{10}, & n = 1 \\ 0, & \text{otherwise} \end{cases}$$

when excited by the input sequence

$$x(n) = \begin{cases} 1, & n = 0 \\ -\frac{7}{10}, & n = 1 \\ \frac{1}{10}, & n = 2 \\ 0, & \text{otherwise} \end{cases}$$

Determine its impulse response and its input–output equation.

Solution   The system function is easily determined by taking the z-transforms of $x(n)$ and $y(n)$. Thus we have

$$H(z) = \frac{Y(z)}{X(z)} = \frac{1 + \frac{7}{10}z^{-1}}{1 - \frac{7}{10}z^{-1} + \frac{1}{10}z^{-2}}$$

$$= \frac{1 + \frac{7}{10}z^{-1}}{(1 - \frac{1}{2}z^{-1})(1 - \frac{1}{5}z^{-1})}$$

Since the system is causal, its ROC is $|z| > \frac{1}{2}$. The system is also stable since its poles lie inside the unit circle.

The input–output difference equation for the system is

$$y(n) = \frac{7}{10}y(n-1) - \frac{1}{10}y(n-2) + x(n) + \frac{7}{10}x(n-1)$$

Its impulse response is determined by performing a partial-fraction expansion of $H(z)$ and inverse transforming the result. This computation yields

$$h(n) = [4(\tfrac{1}{2})^n - 3(\tfrac{1}{5})^n]u(n)$$

We observe that (4.6.26) determines the unknown system uniquely if it is known that the system is causal. However, the example above is artificial, since the system response $\{y(n)\}$ is very likely to be infinite in duration. Consequently, this approach is usually impractical.

As an alternative, we can deal directly with the time-domain expression given by (4.6.25). If the system is causal, we have

$$y(n) = \sum_{k=0}^{n} h(k)x(n-k) \qquad n \geq 0$$

and hence

$$h(0) = \frac{y(0)}{x(0)}$$

$$h(n) = \frac{y(n) - \sum_{k=0}^{n-1} h(k)x(n-k)}{x(0)} \qquad n \geq 1 \tag{4.6.27}$$

This recursive solution requires that $x(0) \neq 0$. However, we note again that when $\{h(n)\}$ has infinite duration, this approach may not be practical unless we truncate the recursive solution at same stage [i.e., truncate $\{h(n)\}$].

Another method for identifying an unknown system is based on a crosscorrelation technique. Recall that the input–output crosscorrelation function derived in Section 2.6.5 is given as

$$r_{yx}(m) = \sum_{k=0}^{\infty} h(k)r_{xx}(m-k) = h(n) * r_{xx}(m) \tag{4.6.28}$$

where $r_{yx}(m)$ is the crosscorrelation sequence of the input $\{x(n)\}$ to the system with the output $\{y(n)\}$ of the system, and $r_{xx}(m)$ is the autocorrelation sequence of the input signal. In the frequency domain, the corresponding relationship is

$$S_{yx}(\omega) = H(\omega)S_{xx}(\omega) = H(\omega)|X(\omega)|^2$$

Hence

$$H(\omega) = \frac{S_{yx}(\omega)}{S_{xx}(\omega)} = \frac{S_{yx}(\omega)}{|X(\omega)|^2} \tag{4.6.29}$$

These relations suggest that the impulse response $\{h(n)\}$ or the frequency response of an unknown system can be determined (measured) by crosscorrelating the input sequence $\{x(n)\}$ with the output sequence $\{y(n)\}$, and then solving the deconvolution problem in (4.6.28) by means of the recursive equation in (4.6.27). Alternatively, we could simply compute the Fourier transform of (4.6.28) and determine the frequency response given by (4.6.29). Furthermore, if we select the input sequence $\{x(n)\}$ such that its autocorrelation sequence $\{r_{xx}(n)\}$, is a unit sample sequence, or equivalently, that its spectrum is flat (constant) over the passband of $H(\omega)$, the values of the impulse response $\{h(n)\}$ are simply equal to the values of the crosscorrelation sequence $\{r_{yx}(n)\}$.

In general, the crosscorrelation method described above is an effective and practical method for system identification. Another practical approach based on least-squares optimization is described in Chapter 8.

4.6.4 Homomorphic Deconvolution

The complex cepstrum, introduced in Section 4.2.7, is a useful tool for performing deconvolution in some applications such as seismic signal processing. To describe this method, let us suppose that $\{y(n)\}$ is the output sequence of a linear time-invariant system which is excited by the input sequence $\{x(n)\}$. Then

$$Y(z) = X(z)H(z) \tag{4.6.30}$$

where $H(z)$ is the system function. The logarithm of $Y(z)$ is

$$C_y(z) = \ln Y(z)$$
$$= \ln X(z) + \ln H(z) \qquad (4.6.31)$$
$$= C_x(z) + C_h(z)$$

Consequently, the complex cepstrum of the output sequence $\{y(n)\}$ is expressed as the sum of the cepstrum of $\{x(n)\}$ and $\{h(n)\}$, that is,

$$c_y(n) = c_x(n) + c_h(n) \qquad (4.6.32)$$

Thus we observe that convolution of the two sequences in the time domain corresponds to the summation of the cepstrum sequences in the cepstral domain. The system for performing these transformations is called a *homormorphic system* and is illustrated in Fig. 4.65.

In some applications, such as seismic signal processing and speech signal processing, the characteristics of the cepstral sequences $\{c_x(n)\}$ and $\{c_h(n)\}$ are sufficiently different so that they can be separated in the cepstral domain. Specifically, suppose that $\{c_h(n)\}$ has its main components (main energy) in the vicinity of small values of n, whereas $\{c_x(n)\}$ has its components concentrated at large values of n. We may say that $\{c_h(n)\}$ is "lowpass" and $\{c_x(n)\}$ is "highpass." We can then separate $\{c_h(n)\}$ from $\{c_x(n)\}$ using appropriate "lowpass" and "highpass" windows, as illustrated in Fig. 4.66. Thus

$$\hat{c}_h(n) = c_y(n)w_{\text{lp}}(n) \qquad (4.6.33)$$

and

$$\hat{c}_x(n) = c_y(n)w_{\text{hp}}(n) \qquad (4.6.34)$$

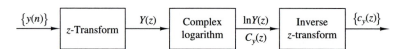

Figure 4.65 Homomorphic system for obtaining the cepstrum $\{c_y(n)\}$ of the sequence $\{y(n)\}$.

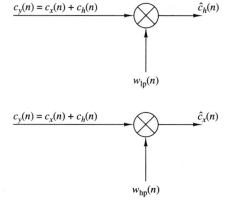

Figure 4.66 Separating the two cepstral components by "lowpass" and "highpass" windows.

where

$$w_{\mathrm{lp}}(n) = \begin{cases} 1, & |n| \leq N_1 \\ 0, & \text{otherwise} \end{cases} \qquad (4.6.35)$$

$$w_{\mathrm{hp}}(n) = \begin{cases} 0, & |n| \leq N_1 \\ 1, & |n| > N_1 \end{cases} \qquad (4.6.36)$$

Once we have separated the cepstrum sequences $\{\hat{c}_h(n)\}$ and $\{\hat{c}_x(n)\}$ by windowing, the sequences $\{\hat{x}(n)\}$ and $\{\hat{h}(n)\}$ are obtained by passing $\{\hat{c}_h(n)\}$ and $\{\hat{c}_x(n)\}$ through the inverse homomorphic system, shown in Fig. 4.67.

In practice, a digital computer would be used to compute the cepstrum of the sequence $\{y(n)\}$, to perform the windowing functions, and to implement the inverse homomorphic system shown in Fig. 4.67. In place of the z-transform and inverse z-transform, we would substitute a special form of the Fourier transform and its inverse. This special form, called the discrete Fourier transform, is described in Chapter 5.

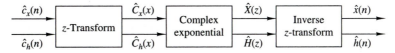

Figure 4.67 Inverse homomorphic system for recovering the sequences $\{x(n)\}$ and $\{h(n)\}$ from the corresponding cepstra.

4.7 SUMMARY AND REFERENCES

The Fourier series and the Fourier transform are the mathematical tools for analyzing the characteristics of signals in the frequency domain. The Fourier series is appropriate for representing a periodic signal as a weighted sum of harmonically related sinusoidal components, where the weighting coefficients represent the strengths of each of the harmonics, and the magnitude squared of each weighting coefficient represents the power of the corresponding harmonic. As we have indicated, the Fourier series is one of many possible orthogonal series expansions for a periodic signal. Its importance stems from the characteristic behavior of LTI systems, as we shall see in Chapter 5.

The Fourier transform is appropriate for representing the spectral characteristics of aperiodic signals with finite energy. The important properties of the Fourier transform were also presented in this chapter.

There are many excellent texts on Fourier series and Fourier transforms. For reference, we include the texts by Bracewell (1978), Davis (1963), Dym and McKean (1972), and Papoulis (1962).

In this chapter we also considered the frequency-domain characteristics of LTI systems. We showed that an LTI system is characterized in the frequency domain by its frequency response function $H(\omega)$, which is the Fourier transform of the impulse response of the system. We also observed that the frequency response function determines the effect of the system on any input signal. In fact, by transforming the input signal into the frequency domain, we observed that it is a simple matter to determine the effect of the system on the signal and to determine the system output. When viewed in the frequency domain, an LTI system performs spectral shaping or spectral filtering on the input signal.

The design of some simple IIR filters was also considered in this chapter from the viewpoint of pole–zero placement. By means of this method, we were able to design simple digital resonators, notch filters, comb filters, all-pass filters, and digital sinusoidal generators. The design of more complex IIR filters is treated in detail in Chapter 8, which also includes several references. Digital sinusoidal generators find use in frequency synthesis applications. A comprehensive treatment of frequency synthesis techniques is given in the text edited by Gorski-Popiel (1975).

Finally, we characterized LTI systems as either minimum-phase, maximum-phase, or mixed-phase, depending on the position of their poles and zeros in the frequency domain. Using these basic characteristics of LTI systems, we considered practical problems in inverse filtering, deconvolution, and system identification. We concluded with the description of a deconvolution method based on cepstral analysis of the output signal from a linear system.

A vast amount of technical literature exists on the topics of inverse filtering, deconvolution, and system identification. In the context of communications, system identification, and inverse filtering as they relate to channel equalization are treated in the book by Proakis (1995). Deconvolution techniques are widely used in seismic signal processing. For reference, we suggest the papers by Wood and Treitel (1975), Peacock and Treitel (1969), and the books by Robinson and Treitel (1978, 1980). Homomorphic deconvolution and its applications to speech processing is treated in the book by Oppenheim and Schafer (1989).

PROBLEMS

4.1 Consider the full-wave rectified sinusoid in Fig. P4.1.
   (a) Determine its spectrum $X_a(F)$.
   (b) Compute the power of the signal.

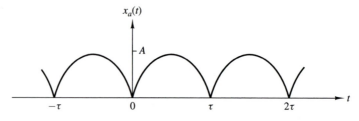

Figure P4.1

(c) Plot the power spectral density.

(d) Check the validity of Parseval's relation for this signal.

4.2 Compute and sketch the magnitude and phase spectra for the following signals ($a > 0$).

(a) $x_a(t) = \begin{cases} Ae^{-at}, & t \geq 0 \\ 0, & t < 0 \end{cases}$

(b) $x_a(t) = Ae^{-a|t|}$

4.3 Consider the signal

$$x(t) = \begin{cases} 1 - |t|/\tau, & |t| \leq \tau \\ 0, & \text{elsewhere} \end{cases}$$

(a) Determine and sketch its magnitude and phase spectra, $|X_a(F)|$ and $\angle X_a(F)$, respectively.

(b) Create a periodic signal $x_p(t)$ with fundamental period $T_p \geq 2\tau$, so that $x(t) = x_p(t)$ for $|t| < T_p/2$. What are the Fourier coefficients c_k for the signal $x_p(t)$?

(c) Using the results in parts (a) and (b), show that $c_k = (1/T_p)X_a(k/T_p)$.

4.4 Consider the following periodic signal:

$$x(n) = \{\ldots, 1, 0, 1, 2, 3, 2, 1, 0, 1, \ldots\}$$
$$\uparrow$$

(a) Sketch the signal $x(n)$ and its magnitude and phase spectra.

(b) Using the results in part (a), verify Parseval's relation by computing the power in the time and frequency domains.

4.5 Consider the signal

$$x(n) = 2 + 2\cos\frac{\pi n}{4} + \cos\frac{\pi n}{2} + \frac{1}{2}\cos\frac{3\pi n}{4}$$

(a) Determine and sketch its power density spectrum.

(b) Evaluate the power of the signal.

4.6 Determine and sketch the magnitude and phase spectra of the following periodic signals.

(a) $x(n) = 4\sin\dfrac{\pi(n-2)}{3}$

(b) $x(n) = \cos\dfrac{2\pi}{3}n + \sin\dfrac{2\pi}{5}n$

(c) $x(n) = \cos\dfrac{2\pi}{3}n\sin\dfrac{2\pi}{5}n$

(d) $x(n) = \{\ldots, -2, -1, 0, 1, 2, -2, -1, 0, 1, 2, \ldots\}$
$$\uparrow$$

(e) $x(n) = \{\ldots, -1, 2, 1, 2, -1, 0, -1, 2, 1, 2, \ldots\}$
$$\uparrow$$

(f) $x(n) = \{\ldots, 0, 0, 1, 1, 0, 0, 0, 1, 1, 0, 0, \ldots\}$
$$\uparrow$$

(g) $x(n) = 1, -\infty < n < \infty$

(h) $x(n) = (-1)^n, -\infty < n < \infty$

4.7 Determine the periodic signals $x(n)$, with fundamental period $N = 8$, if their Fourier coefficients are given by:

(a) $c_k = \cos\dfrac{k\pi}{4} + \sin\dfrac{3k\pi}{4}$

(b) $c_k = \begin{cases} \sin\dfrac{k\pi}{3}, & 0 \leq k \leq 6 \\ 0, & k = 7 \end{cases}$

(c) $\{c_k\} = \{\dots, 0, \frac{1}{4}, \frac{1}{2}, 1, 2, 1, \frac{1}{2}, \frac{1}{4}, 0 \dots\}$
$\uparrow$

4.8 Two DT signals, $s_k(n)$ and $s_l(n)$, are said to be orthogonal over an interval $[N_1, N_2]$ if

$$\sum_{n=N_1}^{N_2} s_k(n) s_l^*(n) = \begin{cases} A_k, & k = l \\ 0, & k \neq l \end{cases}$$

If $A_k = 1$, the signals are called orthonormal.

(a) Prove the relation

$$\sum_{n=0}^{N-1} e^{j2\pi kn/N} = \begin{cases} N, & k = 0, \pm N, \pm 2N, \dots \\ 0, & \text{otherwise} \end{cases}$$

(b) Illustrate the validity of the relation in part (a) by plotting for every value of $k = 1, 2, \dots, 6$, the signals $s_k(n) = e^{j(2\pi/6)kn}$, $n = 0, 1, \dots, 5$. [*Note:* For a given k, n the signal $s_k(n)$ can be represented as a vector in the complex plane.]

(c) Show that the harmonically related signals

$$s_k(n) = e^{j(2\pi/N)kn}$$

are orthogonal over any interval of length N.

4.9 Compute the Fourier transform of the following signals.

(a) $x(n) = u(n) - u(n - 6)$
(b) $x(n) = 2^n u(-n)$
(c) $x(n) = (\frac{1}{4})^n u(n + 4)$
(d) $x(n) = (\alpha^n \sin \omega_0 n) u(n) \qquad |\alpha| < 1$
(e) $x(n) = |\alpha|^n \sin \omega_0 n \qquad |\alpha| < 1$
(f) $x(n) = \begin{cases} 2 - (\frac{1}{2})n, & |n| \leq 4 \\ 0, & \text{elsewhere} \end{cases}$
(g) $x(n) = \{-2, -1, 0, 1, 2\}$
$\uparrow$
(h) $x(n) = \begin{cases} A(2M + 1 - |n|), & |n| \leq M \\ 0, & |n| > M \end{cases}$

Sketch the magnitude and phase spectra for parts (a), (f), and (g).

4.10 Determine the signals having the following Fourier transforms.

(a) $X(\omega) = \begin{cases} 0, & 0 \leq |\omega| \leq \omega_0 \\ 1, & \omega_0 < |\omega| \leq \pi \end{cases}$
(b) $X(\omega) = \cos^2 \omega$
(c) $X(\omega) = \begin{cases} 1, & \omega_0 - \delta\omega/2 \leq |\omega| \leq \omega_0 + \delta\omega/2 \\ 0, & \text{elsewhere} \end{cases}$
(d) The signal shown in Fig. P4.10.

4.11 Consider the signal

$$x(n) = \{1, 0, -1, 2, 3\}$$
$\uparrow$

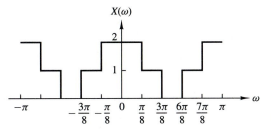

Figure P4.10

with Fourier transform $X(\omega) = X_R(\omega) + j(X_I(\omega))$. Determine and sketch the signal $y(n)$ with Fourier transform

$$Y(\omega) = X_I(\omega) + X_R(\omega)e^{j2\omega}$$

4.12 Determine the signal $x(n)$ if its Fourier transform is as given in Fig. P4.12.

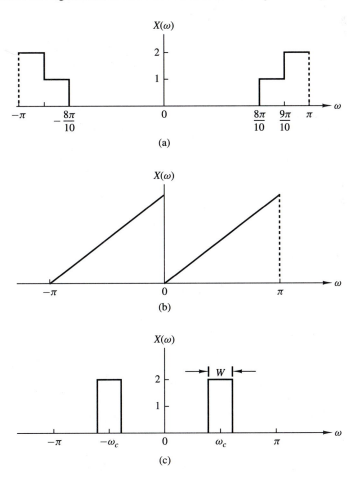

Figure P4.12

4.13 In Example 4.3.3, the Fourier transform of the signal

$$x(n) = \begin{cases} 1, & -M \leq n \leq M \\ 0, & \text{otherwise} \end{cases}$$

was shown to be

$$X(\omega) = 1 + 2 \sum_{n=1}^{M} \cos \omega n$$

Show that the Fourier transform of

$$x_1(n) = \begin{cases} 1, & 0 \leq n \leq M \\ 0, & \text{otherwise} \end{cases}$$

and

$$x_2(n) = \begin{cases} 1, & -M \leq n \leq -1 \\ 0, & \text{otherwise} \end{cases}$$

are, respectively,

$$X_1(\omega) = \frac{1 - e^{-j\omega(M+1)}}{1 - e^{-j\omega}}$$

$$X_2(\omega) = \frac{e^{j\omega} - e^{j\omega(M+1)}}{1 - e^{j\omega}}$$

Thus prove that

$$X(\omega) = X_1(\omega) + X_2(\omega)$$

$$= \frac{\sin(M + \frac{1}{2})\omega}{\sin(\omega/2)}$$

and therefore,

$$1 + 2 \sum_{n=1}^{M} \cos \omega n = \frac{\sin(M + \frac{1}{2})\omega}{\sin(\omega/2)}$$

4.14 Consider the signal

$$x(n) = \{-1, 2, -3, 2, -1\}$$
$$\uparrow$$

with Fourier transform $X(\omega)$. Compute the following quantities, without explicitly computing $X(\omega)$:

(a) $X(0)$ **(b)** $\angle X(\omega)$ **(c)** $\int_{-\pi}^{\pi} X(\omega) \, d\omega$ **(d)** $X(\pi)$
(e) $\int_{-\pi}^{\pi} |X(\omega)|^2 \, d\omega$

4.15 The center of gravity of a signal $x(n)$ is defined as

$$c = \frac{\displaystyle\sum_{n=-\infty}^{\infty} n x(n)}{\displaystyle\sum_{n=-\infty}^{\infty} x(n)}$$

and provides a measure of the "time delay" of the signal.

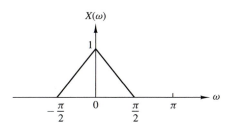

Figure P4.15

(a) Express c in terms of $X(\omega)$.

(b) Compute c for the signal $x(n)$ whose Fourier transform is shown in Fig. P4.15.

4.16 Consider the Fourier transform pair

$$a^n u(n) \overset{F}{\longleftrightarrow} \frac{1}{1 - ae^{-j\omega}} \qquad |a| < 1$$

Use the differentiation in frequency theorem and induction to show that

$$x(n) = \frac{(n + l - 1)!}{n!(l - 1)!} a^n u(n) \overset{F}{\longleftrightarrow} X(\omega) = \frac{1}{(1 - ae^{-j\omega})^l}$$

4.17 Let $x(n)$ be an arbitrary signal, not necessarily real-valued, with Fourier transform $X(\omega)$. Express the Fourier transforms of the following signals in terms of $X(\omega)$.

(a) $x^*(n)$

(b) $x^*(-n)$

(c) $y(n) = x(n) - x(n - 1)$

(d) $y(n) = \displaystyle\sum_{k=-\infty}^{n} x(k)$

(e) $y(n) = x(2n)$

(f) $y(n) = \begin{cases} x(n/2), & n \text{ even} \\ 0, & n \text{ odd} \end{cases}$

4.18 Determine and sketch the Fourier transforms $X_1(\omega)$, $X_2(\omega)$, and $X_3(\omega)$ of the following signals.

(a) $x_1(n) = \{1, 1, 1, 1, 1\}$
$\phantom{x_1(n) = \{}\uparrow$

(b) $x_2(n) = \{1, 0, 1, 0, 1, 0, 1, 0, 1\}$
$\phantom{x_2(n) = \{1, 0, }\uparrow$

(c) $x_3(n) = \{1, 0, 0, 1, 0, 0, 1, 0, 0, 1, 0, 0, 1\}$
$\phantom{x_3(n) = \{1, 0, 0, }\uparrow$

(d) Is there any relation between $X_1(\omega)$, $X_2(\omega)$, and $X_3(\omega)$? What is its physical meaning?

(e) Show that if

$$x_k(n) = \begin{cases} x\left(\dfrac{n}{k}\right), & \text{if } n/k \text{ integer} \\ 0, & \text{otherwise} \end{cases}$$

then

$$X_k(\omega) = X(k\omega)$$

4.19 Let $x(n)$ be a signal with Fourier transform as shown in Fig. P4.19. Determine and sketch the Fourier transforms of the following signals.

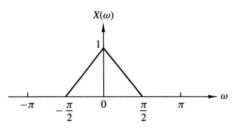

Figure P4.19

(a) $x_1(n) = x(n)\cos(\pi n/4)$ **(b)** $x_2(n) = x(n)\sin(\pi n/2)$
(c) $x_3(n) = x(n)\cos(\pi n/2)$ **(d)** $x_4(n) = x(n)\cos \pi n$
Note that these signal sequences are obtained by *amplitude modulation* of a carrier $\cos \omega_c n$ or $\sin \omega_c n$ by the sequence $x(n)$.

4.20 Consider an aperiodic signal $x(n)$ with Fourier transform $X(\omega)$. Show that the Fourier series coefficients C_k^y of the periodic signal

$$y(n) = \sum_{l=-\infty}^{\infty} x(n - lN)$$

are given by

$$C_k^y = \frac{1}{N} X\left(\frac{2\pi}{N} k\right) \qquad k = 0, 1, \ldots, N - 1$$

4.21 Prove that

$$X_N(\omega) = \sum_{n=-N}^{N} \frac{\sin \omega_c n}{\pi n} e^{-j\omega n}$$

may be expressed as

$$X_N(\omega) = \frac{1}{2\pi} \int_{-\omega_c}^{\omega_c} \frac{\sin[(2N + 1)(\omega - \theta/2)]}{\sin[(\omega - \theta)/2]} d\theta$$

4.22 A signal $x(n)$ has the following Fourier transform:

$$X(\omega) = \frac{1}{1 - ae^{-j\omega}}$$

Determine the Fourier transforms of the following signals:
(a) $x(2n + 1)$ **(b)** $e^{\pi n/2} x(n + 2)$
(b) $x(-2n)$ **(d)** $x(n)\cos(0.3\pi n)$
(c) $x(n) * x(n - 1)$ **(f)** $x(n) * x(-n)$

4.23 From a discrete-time signal $x(n)$ with Fourier transform $X(\omega)$, shown in Fig. P4.23, determine and sketch the Fourier transform of the following signals:
(a) $y_1(n) = \begin{cases} x(n), & n \text{ even} \\ 0, & n \text{ odd} \end{cases}$

(b) $y_2(n) = x(2n)$

(c) $y_3(n) = \begin{cases} x(n/2), & n \text{ even} \\ 0, & n \text{ odd} \end{cases}$

Note that $y_1(n) = x(n)s(n)$, where $s(n) = \{\ldots 0, 1, 0, 1, 0, 1, 0, 1, \ldots\}$
$\qquad\qquad\qquad\qquad\qquad\qquad\qquad\qquad\qquad\qquad\uparrow$

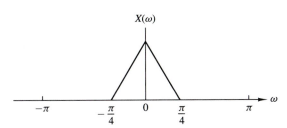

Figure P4.23

4.24 The following input–output pairs have been observed during the operation of various systems:

(a) $x(n) = (\frac{1}{2})^n \xrightarrow{T_1} y(n) = (\frac{1}{8})^n$

(b) $x(n) = (\frac{1}{2})^n u(n) \xrightarrow{T_2} y(n) = (\frac{1}{8})^n u(n)$

(c) $x(n) = e^{j\pi/5} \xrightarrow{T_3} y(n) = 3e^{j\pi/5}$

(d) $x(n) = e^{j\pi/5} u(n) \xrightarrow{T_4} y(n) = 3e^{j\pi/5u(n)}$

(e) $x(n) = x(n + N_1) \xrightarrow{T_5} y(n) = y(n + N_2)$ $N_1 \neq N_2$, N_1, N_2 prime

Determine their frequency response if each of the above systems is LTI.

4.25 (a) Determine and sketch the Fourier transform $W_R(\omega)$ of the rectangular sequence

$$w_R(n) = \begin{cases} 1, & 0 \le n \le M \\ 0, & \text{otherwise} \end{cases}$$

(b) Consider the triangular sequence

$$w_T(n) = \begin{cases} n, & 0 \le n \le M/2 \\ M - n, & M/2 < 2 \le M \\ 0, & \text{otherwise} \end{cases}$$

Determine and sketch the Fourier transform $W_T(\omega)$ of $w_T(n)$ by expressing it as the convolution of a rectangular sequence with itself.

(c) Consider the sequence

$$w_c(n) = \frac{1}{2}\left(1 + \cos\frac{2\pi n}{M}\right) w_R(n)$$

Determine and sketch $W_c(\omega)$ by using $W_R(\omega)$.

4.26 Consider an LTI system with impulse response $h(n) = (\frac{1}{2})^n u(n)$.

(a) Determine and sketch the magnitude and phase response $|H(\omega)|$ and $\angle H(\omega)$, respectively.

(b) Determine and sketch the magnitude and phase spectra for the input and output signals for the following inputs:

(1) $x(n) = \cos\dfrac{3\pi n}{10}, -\infty < n < \infty$

(2) $x(n) = \{\ldots, 1, 0, 0, 1, 1, 1, 0, 1, 1, 1, 0, 1, \ldots\}$
                       $\uparrow$

4.27 Determine and sketch the magnitude and phase response of the following systems:

(a) $y(n) = \frac{1}{2}[x(n) + x(n - 1)]$

(b) $y(n) = \frac{1}{2}[x(n) - x(n - 1)]$

(c) $y(n) = \frac{1}{2}[x(n + 1) - x(n - 1)]$

(d) $y(n) = \frac{1}{2}[x(n+1) + x(n-1)]$

(e) $y(n) = \frac{1}{2}[x(n) + x(n-2)]$

(f) $y(n) = \frac{1}{2}[x(n) - x(n-2)]$

(g) $y(n) = \frac{1}{3}[x(n) + x(n-1) + x(n-2)]$

(h) $y(n) = x(n) - x(n-8)$

(i) $y(n) = 2x(n-1) - x(n-2)$

(j) $y(n) = \frac{1}{4}[x(n) + x(n-1) + x(n-2) + x(n-3)]$

(k) $y(n) = \frac{1}{8}[x(n) + 3x(n-1) + 3x(n-2) + x(n-3)]$

(l) $y(n) = x(n-4)$

(m) $y(n) = x(n+4)$

(n) $y(n) = \frac{1}{4}[x(n) - 2x(n-1) + x(n-2)]$

4.28 An FIR filter is described by the difference equation

$$y(n) = x(n) + x(n-10)$$

(a) Compute and sketch its magnitude and phase response.

(b) Determine its response to the inputs

(1) $x(n) = \cos\frac{\pi}{10}n + 3\sin\left(\frac{\pi}{3}n + \frac{\pi}{10}\right)$ $-\infty < n < \infty$

(2) $x(n) = 10 + 5\cos\left(\frac{2\pi}{5}n + \frac{\pi}{2}\right)$ $-\infty < n < \infty$

4.29 Determine the transient and steady-state responses of the FIR *filter* shown in Fig. P4.29 to the input signal $x(n) = 10e^{j\pi n/2}u(n)$. Let $b = 2$ and $y(-1) = y(-2) = y(-3) = y(-4) = 0$.

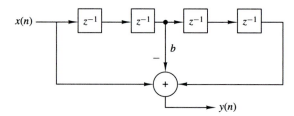

Figure P4.29

4.30 Consider the FIR filter

$$y(n) = x(n) + x(n-4)$$

(a) Compute and sketch its magnitude and phase response.

(b) Compute its response to the input

$$x(n) = \cos\frac{\pi}{2}n + \cos\frac{\pi}{4}n \qquad -\infty < n < \infty$$

(c) Explain the results obtained in part (b) in terms of the magnitude and phase responses obtained in part (a).

4.31 Determine the steady-state and transient responses of the system

$$y(n) = \frac{1}{2}[x(n) - x(n-2)]$$

to the input signal

$$x(n) = 5 + 3\cos\left(\frac{\pi}{2}n + 60°\right) \qquad -\infty < n < \infty$$

4.32 From our discussions it is apparent that an LTI system cannot produce frequencies at its output that are different from those applied in its input. Thus, if a system creates "new" frequencies, it must be nonlinear and/or time varying. Determine the frequency content of the outputs of the following systems to the input signal

$$x(n) = A\cos\frac{\pi}{4}n$$

(a) $y(n) = x(2n)$
(b) $y(n) = x^2(n)$
(c) $y(n) = (\cos \pi n)x(n)$

4.33 Determine and sketch the magnitude and phase response of the systems shown in Fig. P4.33(a) through (c).

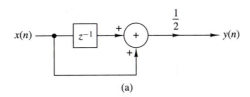

(a)

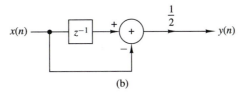

(b)

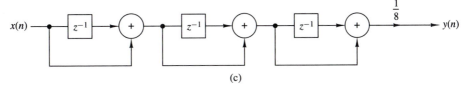

(c)

Figure P4.33

4.34 Determine the magnitude and phase response of the multipath channel

$$y(n) = x(n) + x(n - M)$$

At what frequencies does $H(\omega) = 0$?

4.35 Consider the filter

$$y(n) = 0.9y(n - 1) + bx(n)$$

(a) Determine b so that $|H(0)| = 1$.
(b) Determine the frequency at which $|H(\omega)| = 1/\sqrt{2}$.

(c) Is this filter lowpass, bandpass, or highpass?

(d) Repeat parts (b) and (c) for the filter $y(n) = -0.9y(n-1) + 0.1x(n)$.

4.36* *Harmonic distortion in digital sinusoidal generators* An ideal sinusoidal generator produces the signal

$$x(n) = \cos 2\pi f_0 n \qquad -\infty < n < \infty$$

which is periodic with fundamental period N if $f_0 = k_0/N$ and k_0, N are relatively prime numbers. The spectrum of such a "pure" sinusoid consist of two lines at $k = k_0$ and $k = N - k_0$ (we limit ourselves in the fundamental interval $0 \le k \le N - 1$). In practice, the approximations made in computing the samples of a sinusoid of relative frequency f_0 result in a certain amount of power falling into other frequencies. This spurious power results in distortion, which is referred to as *harmonic distortion*. Harmonic distortion is usually measured in terms of the *total harmonic distortion* (THD), which is defined as the ratio

$$\text{THD} = \frac{\text{spurious harmonic power}}{\text{total power}}$$

(a) Show that

$$\text{THD} = 1 - 2\frac{|c_{k_0}|^2}{P_x}$$

where

$$c_{k_0} = \frac{1}{N}\sum_{n=0}^{N-1} x(n)e^{-j(2\pi/N)k_0 n}$$

$$P_x = \frac{1}{N}\sum_{n=0}^{N-1} |x(n)|^2$$

(b) By using the Taylor approximation

$$\cos\phi = 1 - \frac{\phi^2}{2!} + \frac{\phi^4}{4!} - \frac{\phi^6}{6!} + \cdots$$

compute one period of $x(n)$ for $f_0 = 1/96$, $1/32$, $1/256$ by increasing the number of terms in the Taylor expansion from 2 to 8.

(c) Compute the THD and plot the power density spectrum for each sinusoid in part (b) as well as for the sinusoids obtained using the computer cosine function. Comment on the results.

4.37* *Measurement of the total harmonic distortion in quantized sinusoids* Let $x(n)$ be a periodic sinusoidal signal with frequency $f_0 = k/N$, that is,

$$x(n) = \sin 2\pi f_0 n$$

(a) Write a computer program that quantizes the signal $x(n)$ into b bits or equivalently into $L = 2^b$ levels by using rounding. The resulting signal is denoted by $x_q(n)$.

(b) For $f_0 = 1/50$ compute the THD of the quantized signals $x_q(n)$ obtained by using $b = 4, 6, 8$, and 16 bits.

(c) Repeat part (b) for $f_0 = 1/100$.

(d) Comment on the results obtained in parts (b) and (c).

4.38* Consider the discrete-time system

$$y(n) = ay(n-1) + (1-a)x(n) \qquad n \ge 0$$

where $a = 0.9$ and $y(-1) = 0$.

(a) Compute and sketch the output $y_i(n)$ of the system to the input signals

$$x_i(n) = \sin 2\pi f_i n \qquad 0 \le n \le 100$$

where $f_1 = \frac{1}{4}$, $f_2 = \frac{1}{5}$, $f_3 = \frac{1}{10}$, $f_4 = \frac{1}{20}$.

(b) Compute and sketch the magnitude and phase response of the system and use these results to explain the response of the system to the signals given in part (a).

4.39* Consider an LTI system with impulse response $h(n) = (\frac{1}{3})^{|n|}$

(a) Determine and sketch the magnitude and phase response $H(\omega)|$ and $\measuredangle H(\omega)$, respectively.

(b) Determine and sketch the magnitude and phase spectra for the input and output signals for the following inputs:

   (1) $x(n) = \cos \dfrac{3\pi n}{8}, -\infty < n < \infty$

   (2) $x(n) = \{\ldots, -1, 1, -1, 1, -1, 1, -1, 1, -1, 1, -1, 1, \ldots\}$
                          $\uparrow$

4.40* *Time-domain sampling*   Consider the continuous-time signal

$$x_a(t) = \begin{cases} e^{-j2\pi F_0 t}, & t \ge 0 \\ 0, & t < 0 \end{cases}$$

(a) Compute analytically the spectrum $X_a(F)$ of $x_a(t)$.

(b) Compute analytically the spectrum of the signal $x(n) = x_a(nT)$, $T = 1/F_s$.

(c) Plot the magnitude spectrum $|X_a(F)|$ for $F_0 = 10$ Hz.

(d) Plot the magnitude spectrum $|X(F)|$ for $F_s = 10, 20, 40$, and 100 Hz.

(e) Explain the results obtained in part (d) in terms of the aliasing effect.

4.41 Consider the digital filter shown in Fig. P4.41.

(a) Determine the input–output relation and the impulse response $h(n)$.

(b) Determine and sketch the magnitude $|H(\omega)|$ and the phase response $\measuredangle H(\omega)$ of the filter and find which frequencies are completely blocked by the filter.

(c) When $\omega_0 = \pi/2$, determine the output $y(n)$ to the input

$$x(n) = 3\cos\left(\frac{\pi}{3}n + 30°\right) \qquad -\infty < n < \infty$$

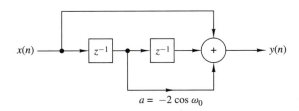

$a = -2\cos\omega_0$ **Figure P4.41**

4.42 Consider the FIR filter

$$y(n) = x(n) - x(n - 4)$$

(a) Compute and sketch its magnitude and phase response.

(b) Compute its response to the input

$$x(n) = \cos \frac{\pi}{2}n + \cos \frac{\pi}{4}n \qquad -\infty < n < \infty$$

(c) Explain the results obtained in part (b) in terms of the answer given in part (a).

4.43 Determine the steady-state response of the system

$$y(n) = \tfrac{1}{2}[x(n) - x(n-2)]$$

to the input signal

$$x(n) = 5 + 3\cos\left(\frac{\pi}{2}n + 60°\right) + 4\sin(\pi n + 45°) \qquad -\infty < n < \infty$$

4.44 Recall from Problem 4.32 that an LTI system cannot produce frequencies at its output that are different from those applied in its input. Thus if a system creates "new" frequencies, it must be nonlinear and/or time varying. Indicate whether the following systems are nonlinear and/or time varying and determine the output spectra when the input spectrum is

$$X(\omega) = \begin{cases} 1, & |\omega| \le \pi/4 \\ 0, & \pi/4 \le |\omega| \le \pi \end{cases}$$

(a) $y(n) = x(2n)$

(b) $y(n) = x^2(n)$

(c) $y(n) = (\cos \pi n)x(n)$

4.45 Consider an LTI system with impulse response

$$h(n) = \left[\left(\frac{1}{4}\right)^n \cos\left(\frac{\pi}{4}n\right)\right]u(n)$$

(a) Determine its system function $H(z)$.

(b) Is it possible to implement this system using a finite number of adders, multipliers, and unit delays? If yes, how?

(c) Provide a rough sketch of $|H(\omega)|$ using the pole–zero plot.

(d) Determine the response of the system to the input

$$x(n) = (\tfrac{1}{4})^n u(n)$$

4.46 An FIR filter is described by the difference equation

$$y(n) = x(n) - x(n-10)$$

(a) Compute and sketch its magnitude and phase response.

(b) Determine its response to the inputs

   (1) $x(n) = \cos \dfrac{\pi}{10}n + 3\sin\left(\dfrac{\pi}{3}n + \dfrac{\pi}{10}\right) \qquad -\infty < n < \infty$

   (2) $x(n) = 5 + 6\cos\left(\dfrac{2\pi}{5}n + \dfrac{\pi}{2}\right) \qquad -\infty < n < \infty$

4.47 The frequency response of an ideal bandpass filter is given by

$$H(\omega) = \begin{cases} 0, & |\omega| \le \dfrac{\pi}{8} \\ 1, & \dfrac{\pi}{8} < |\omega| < \dfrac{3\pi}{8} \\ 0, & \dfrac{3\pi}{8} \le |\omega| \le \pi \end{cases}$$

(a) Determine its impulse response
(b) Show that this impulse response can be expressed as the product of $\cos(n\pi/4)$ and the impulse response of a lowpass filter.

4.48 Consider the system described by the difference equation

$$y(n) = \tfrac{1}{2}y(n-1) + x(n) + \tfrac{1}{2}x(n-1)$$

(a) Determine its impulse response.
(b) Determine its frequency response:
   (1) From the impulse response
   (2) From the difference equation
(c) Determine its response to the input

$$x(n) = \cos\left(\frac{\pi}{2}n + \frac{\pi}{4}\right) \qquad -\infty < n < \infty$$

4.49 Sketch roughly the magnitude $|X(\omega)|$ of the Fourier transforms corresponding to the pole–zero patterns given in Fig. P4.49.

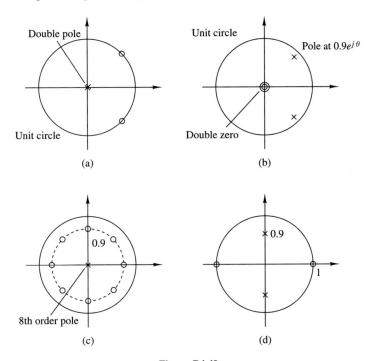

Figure P4.49

4.50 Design an FIR filter that completely blocks the frequency $\omega_0 = \pi/4$ and then compute its output if the input is

$$x(n) = \left(\sin\frac{\pi}{4}n\right)u(n)$$

for $n = 0, 1, 2, 3, 4$. Does the filter fulfill your expectations? Explain.

4.51 A digital filter is characterized by the following properties:
  (1) It is highpass and has one pole and one zero.
  (2) The pole is at a distance $r = 0.9$ from the origin of the z-plane.
  (3) Constant signals do not pass through the system.
  (a) Plot the pole–zero pattern of the filter and determine its system function $H(z)$.
  (b) Compute the magnitude response $|H(\omega)|$ and the phase response $\angle H(\omega)$ of the filter.
  (c) Normalize the frequency response $H(\omega)$ so that $|H(\pi)| = 1$.
  (d) Determine the input–output relation (difference equation) of the filter in the time domain.
  (e) Compute the output of the system if the input is

$$x(n) = 2\cos\left(\frac{\pi}{6}n + 45°\right) \qquad -\infty < n < \infty$$

  (You can use either algebraic or geometrical arguments.)

4.52 A causal first-order digital filter is described by the system function

$$H(z) = b_0 \frac{1 + bz^{-1}}{1 + az^{-1}}$$

  (a) Sketch the direct form I and direct form II realizations of this filter and find the corresponding difference equations.
  (b) For $a = 0.5$ and $b = -0.6$, sketch the pole–zero pattern. Is the system stable? Why?
  (c) For $a = -0.5$ and $b = 0.5$, determine b_0, so that the maximum value of $|H(\omega)|$ is equal to 1.
  (d) Sketch the magnitude response $|H(\omega)|$ and the phase response $\angle H(\omega)$ of the filter obtained in part (c).
  (e) In a specific application it is known that $a = 0.8$. Does the resulting filter amplify high frequencies or low frequencies in the input? Choose the value of b so as to improve the characteristics of this filter (i.e., make it a better lowpass or a better highpass filter).

4.53 Derive the expression for the resonant frequency of a two-pole filter with poles at $p_1 = re^{j\theta}$ and $p_2 = p_1^*$, given by (4.5.25).

4.54 Determine and sketch the magnitude and phase responses of the Hanning filter characterized by the (moving average) difference equation

$$y(n) = \tfrac{1}{4}x(n) + \tfrac{1}{2}x(n-1) + \tfrac{1}{4}x(n-2)$$

4.55 A causal LTI system excited by the input

$$x(n) = (\tfrac{1}{4})^n u(n) + u(-n-1)$$

produces an output $y(n)$ with z-transform

$$Y(z) = \frac{-\frac{3}{4}z^{-1}}{(1 - \frac{1}{4}z^{-1})(1 + z^{-1})}$$

  (a) Determine the system function $H(z)$ and its ROC.
  (b) Determine the output $y(n)$ of the system.
  (*Hint:* Pole cancellation increases the original ROC.)

4.56 Determine the coefficients of a linear-phase FIR filter

$$y(n) = b_0 x(n) + b_1 x(n-1) + b_2 x(n-2)$$

such that:
(a) It rejects completely a frequency component at $\omega_0 = 2\pi/3$.
(b) Its frequency response is normalized so that $H(0) = 1$.
(c) Compute and sketch the magnitude and phase response of the filter to check if it satisfies the requirements.

4.57 Determine the frequency response $H(\omega)$ of the following moving average filters.

(a) $y(n) = \dfrac{1}{2M+1} \displaystyle\sum_{k=-M}^{M} x(n-k)$

(b) $y(n) = \dfrac{1}{4M} x(n+M) + \dfrac{1}{2M} \displaystyle\sum_{k=-M+1}^{M-1} x(n-k) + \dfrac{1}{4M} x(n-M)$

Which filter provides better smoothing? Why?

4.58 The convolution $x(t)$ of two continuous-time signals $x_1(t)$ and $x_2(t)$, from which at least one is nonperiodic, is defined by

$$x(t) \overset{\Delta}{=} x_1(t) * x_2(t) \overset{\Delta}{=} \int_{-\infty}^{\infty} x_1(\lambda) x_2(t-\lambda) d\lambda$$

(a) Show that $X(F) = X_1(F) X_2(F)$, where $X_1(F)$ and $X_2(F)$ are the spectra of $x_1(t)$ and $x_2(t)$, respectively.
(b) Compute $x(t)$ if $x_1(t) = x_2(t) = \begin{cases} 1, & |t| < \tau/2 \\ 0, & \text{elsewhere} \end{cases}$.
(c) Determine the spectrum of $x(t)$ using the results in part (a).

4.59 Compute the magnitude and phase response of a filter with system function

$$H(z) = 1 + z^{-1} + z^{-2} + \cdots + z^{-8}$$

If the sampling frequency is $F_s = 1$ kHz, determine the frequencies of the analog sinusoids that cannot pass through the filter.

4.60 A second-order system has a double pole at $p_{1,2} = 0.5$ and two zeros at

$$z_{1,2} = e^{\pm j3\pi/4}$$

Using geometric arguments, choose the gain G of the filter so that $|H(0)| = 1$.

4.61 In this problem we consider the effect of a single zero on the frequency response of a system. Let $z = re^{j\theta}$ be a zero inside the unit circle ($r < 1$). Then

$$H_z(\omega) = 1 - re^{j\theta} e^{-j\omega}$$
$$= 1 - r\cos(\omega - \theta) + jr\sin(\omega - \theta)$$

(a) Show that the magnitude response is

$$|H_z(\omega)| = [1 - 2r\cos(\omega - \theta) + r^2]^{1/2}$$

or, equivalently,

$$20\log_{10} |H_z(\omega)| = 10\log_{10}[1 - 2r\cos(\omega - \theta) + r^2]$$

(b) Show that the phase response is given as

$$\Theta_z(\omega) = \tan^{-1}\frac{r\sin(\omega-\theta)}{1-r\cos(\omega-\theta)}$$

(c) Show that the group delay is given as

$$\tau_g^z(\omega) = \frac{r^2 - r\cos(\omega-\theta)}{1+r^2-2r\cos(\omega-\theta)}$$

(d) Plot the magnitude $|H(\omega)|_{dB}$, the phase $\Theta(\omega)$ and the group delay $\tau_g(\omega)$ for $r = 0.7$ and $\theta = 0$, $\pi/2$, and π.

4.62 In this problem we consider the effect of a single pole on the frequency response of a system. Hence, we let

$$H_p(\omega) = \frac{1}{1-re^{j\theta}e^{-j\omega}} \qquad r < 1$$

Show that

$$|H_p(\omega)|_{dB} = -|H_z(\omega)|_{dB}$$

$$\measuredangle H_p(\omega) = -\measuredangle H_z(\omega)$$

$$\tau_g^p(\omega) = -\tau_g^z(\omega)$$

where $H_z(\omega)$ and $\tau_g^z(\omega)$ are defined in Problem 4.61.

4.63 In this problem we consider the effect of complex-conjugate pair of poles and zeros on the frequency response of a system. Let

$$H_z(\omega) = (1-re^{j\theta}e^{-j\omega})(1-re^{-j\theta}e^{-j\omega})$$

(a) Show that the magnitude response in decibels is

$$|H_z(\omega)|_{dB} = 10\log_{10}[1+r^2-2r\cos(\omega-\theta)]$$
$$+ 10\log_{10}[1+r^2-2r\cos(\omega+\theta)]$$

(b) Show that the phase response is given as

$$\Theta_z(\omega) = \tan^{-1}\frac{r\sin(\omega-\theta)}{1-r\cos(\omega-\theta)} + \tan^{-1}\frac{r\sin(\omega+\theta)}{1-r\cos(\omega+\theta)}$$

(c) Show that the group delay is given as

$$\tau_g^z(\omega) = \frac{r^2-r\cos(\omega-\theta)}{1+r^2-2r\cos(\omega-\theta)} + \frac{r^2-r\cos(\omega+\theta)}{1+r^2-2r\cos(\omega+\theta)}$$

(d) If $H_p(\omega) = 1/H_z(\omega)$, show that

$$|H_p(\omega)|_{dB} = -|H_z(\omega)|_{dB}$$

$$\Theta_p(\omega) = -\Theta_z(\omega)$$

$$\tau_g^p(\omega) = -\tau_g^z(\omega)$$

(e) Plot $|H_p(\omega)|$, $\Theta_p(\omega)$ and $\tau_g^p(\omega)$ for $\tau = 0.9$, and $\theta = 0$, $\pi/2$.

4.64 Determine the 3-dB bandwidth of the filters $(0 < a < 1)$

$$H_1(z) = \frac{1-a}{1-az^{-1}}$$

$$H_2(z) = \frac{1-a}{2}\frac{1+z^{-1}}{1-az^{-1}}$$

Which is a better lowpass filter?

4.65 Design a digital oscillator with adjustable phase, that is, a digital filter which produces the signal

$$y(n) = \cos(\omega_0 n + \theta)u(n)$$

4.66 This problem provides another derivation of the structure for the coupled-form oscillator by considering the system

$$y(n) = ay(n-1) + x(n)$$

for $a = e^{j\omega_0}$.

Let $x(n)$ be real. Then $y(n)$ is complex. Thus

$$y(n) = y_R(n) + jy_I(n)$$

(a) Determine the equations describing a system with one input $x(n)$ and the two outputs $y_R(n)$ and $y_I(n)$.

(b) Determine a block diagram realization

(c) Show that if $x(n) = \delta(n)$, then

$$y_R(n) = (\cos\omega_0 n)u(n)$$

$$y_I(n) = (\sin\omega_0 n)u(n)$$

(d) Compute $y_R(n)$, $y_I(n)$, $n = 0, 1, \ldots, 9$ for $\omega_0 = \pi/6$. Compare these with the true values of the sine and cosine.

4.67 Consider a filter with system function

$$H(z) = b_0\frac{(1-e^{j\omega_0}z^{-1})(1-e^{-j\omega_0}z^{-1})}{(1-re^{j\omega_0}z^{-1})(1-re^{-j\omega_0}z^{-1})}$$

(a) Sketch the pole–zero pattern.

(b) Using geometric arguments, show that for $r \simeq 1$, the system is a notch filter and provide a rough sketch of its magnitude response if $\omega_0 = 60°$.

(c) For $\omega_0 = 60°$, choose b_0 so that the maximum value of $|H(\omega)|$ is 1.

(d) Draw a direct form II realization of the system

(e) Determine the approximate 3-dB bandwidth of the system.

4.68* Design an FIR digital filter that will reject a very strong 60-Hz sinusoidal interference contaminating a 200-Hz useful sinusoidal signal. Determine the gain of the filter so that the useful signal does not change amplitude. The filter works at a sampling frequency $F_s = 500$ samples/s. Compute the output of the filter if the input is a 60-Hz sinusoid or a 200-Hz sinusoid with unit amplitude. How does the performance of the filter compare with your requirements?

4.69 Determine the gain b_0 for the digital resonator described by (4.5.28) so that $|H(\omega_0)| = 1$.

4.70 Demonstrate that the difference equation given in (4.5.52) can be obtained by applying the trigonometric identity

$$\cos \alpha + \cos \beta = 2 \cos \frac{\alpha + \beta}{2} \cos \frac{\alpha - \beta}{2}$$

where $\alpha = (n+1)\omega_0$, $\beta = (n-1)\omega_0$, and $y(n) = \cos \omega_0 n$. Thus show that the sinusoidal signal $y(n) = A \cos \omega_0 n$ can be generated from (4.5.52) by use of the initial conditions $y(-1) = A \cos \omega_0$ and $y(-2) = A \cos 2\omega_0$.

4.71 Use the trigonometric identity in (4.5.53) with $\alpha = n\omega_0$ and $\beta = (n-2)\omega_0$ to derive the difference equation for generating the sinusoidal signal $y(n) = A \sin n\omega_0$. Determine the corresponding initial conditions.

4.72 Using the z-transform pairs 8 and 9 in Table 3.3, determine the difference equations for the digital oscillators that have impulse responses $h(n) = A \cos n\omega_0 u(n)$ and $h(n) = A \sin n\omega_0 u(n)$, respectively.

4.73 Determine the structure for the coupled-form oscillator by combining the structure for the digital oscillators obtained in Problem 4.72.

4.74 Convert the highpass filter with system function

$$H(z) = \frac{1 - z^{-1}}{1 - az^{-1}} \qquad a < 1$$

into a notch filter that rejects the frequency $\omega_0 = \pi/4$ and its harmonics.
(a) Determine the difference equation.
(b) Sketch the pole–zero pattern.
(c) Sketch the magnitude response for both filters.

4.75 Choose L and M for a lunar filter that must have narrow passbands at $(k \pm \Delta F)$ cycles/day, where $k = 1, 2, 3, \ldots$ and $\Delta F = 0.067726$.

4.76 (a) Show that the systems corresponding to the pole–zero patterns of Fig. 4.58 are all-pass.
(b) What is the number of delays and multipliers required for the efficient implementation of a second-order all-pass system?

4.77 A digital notch filter is required to remove an undesirable 60-Hz hum associated with a power supply in an ECG recording application. The sampling frequency used is $F_s = 500$ samples/s. **(a)** Design a second-order FIR notch filter and **(b)** a second-order pole–zero notch filter for this purpose. In both cases choose the gain b_0 so that $|H(\omega)| = 1$ for $\omega = 0$.

4.78 Determine the coefficients $\{h(n)\}$ of a highpass linear phase FIR filter of length $M = 4$ which has an antisymmetric unit sample response $h(n) = -h(M - 1 - n)$ and a frequency response that satisfies the condition

$$\left| H\left(\frac{\pi}{4}\right) \right| = \frac{1}{2} \qquad \left| H\left(\frac{3\pi}{4}\right) \right| = 1$$

4.79 In an attempt to design a four-pole bandpass digital filter with desired magnitude response

$$|H_d(\omega)| = \begin{cases} 1, & \dfrac{\pi}{6} \le \omega \le \dfrac{\pi}{2} \\ 0, & \text{elsewhere} \end{cases}$$

we select the four poles at

$$p_{1,2} = 0.8e^{\pm j2\pi/9}$$

$$p_{3,4} = 0.8e^{\pm j4\pi/9}$$

and four zeros at

$$z_1 = 1 \qquad z_2 = -1 \qquad z_{3,4} = e^{\pm 3\pi/4}$$

(a) Determine the value of the gain so that

$$\left| H\left(\frac{5\pi}{12}\right) \right| = 1$$

(b) Determine the system function $H(z)$.
(c) Determine the magnitude of the frequency response $H(\omega)$ for $0 \le \omega \le \pi$ and compare it with the desired response $|H_d(\omega)|$.

4.80 A discrete-time system with input $x(n)$ and output $y(n)$ is described in the frequency domain by the relation

$$Y(\omega) = e^{-j2\pi\omega}X(\omega) + \frac{dX(\omega)}{d\omega}$$

(a) Compute the response of the system to the input $x(n) = \delta(n)$.
(b) Check if the system is LTI and stable.

4.81 Consider an ideal lowpass filter with impulse response $h(n)$ and frequency response

$$H(\omega) = \begin{cases} 1, & |\omega| \le \omega_c \\ 0, & \omega_c < |\omega| < \pi \end{cases}$$

What is the frequency response of the filter defined by

$$g(n) = \begin{cases} h\left(\dfrac{n}{2}\right), & n \text{ even} \\ 0, & n \text{ odd} \end{cases}$$

4.82 Consider the system shown in Fig. P4.82. Determine its impulse response and its frequency response if the system $H(\omega)$ is:
(a) Lowpass with cutoff frequency ω_c.
(b) Highpass with cutoff frequency ω_c.

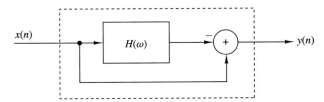

Figure P4.82

4.83 Frequency inverters have been used for many years for speech scrambling. Indeed, a voice signal $x(n)$ becomes unintelligible if we invert its spectrum as shown in Fig. P4.83.
(a) Determine how frequency inversion can be performed in the time domain.
(b) Design an unscrambler. (*Hint:* The required operations are very simple and can easily be done in real time.)

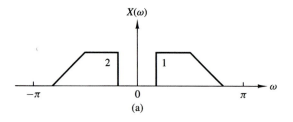

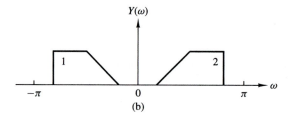

Figure P4.83 (a) Original spectrum; (b) frequency-inverted spectrum.

4.84 A lowpass filter is described by the difference equation

$$y(n) = 0.9y(n-1) + 0.1x(n)$$

(a) By performing a frequency translation of $\pi/2$, transform the filter into a bandpass filter.

(b) What is the impulse response of the bandpass filter?

(c) What is the major problem with the frequency translation method for transforming a prototype lowpass filter into a bandpass filter?

4.85 Consider a system with a real-valued impulse response $h(n)$ and frequency response

$$H(\omega) = |H(\omega)|e^{j\theta(\omega)}$$

The quantity

$$D = \sum_{n=-\infty}^{\infty} n^2 h^2(n)$$

provides a measure of the "effective duration" of $h(n)$.

(a) Express D in terms of $H(\omega)$.

(b) Show that D is minimized for $\theta(\omega) = 0$.

4.86 Consider the lowpass filter

$$y(n) = ay(n-1) + bx(n) \qquad 0 < a < 1$$

(a) Determine b so that $|H(0)| = 1$.

(b) Determine the 3-dB bandwidth ω_3 for the normalized filter in part (a).

(c) How does the choice of the parameter a affect ω_3?

(d) Repeat parts (a) through (c) for the highpass filter obtained by choosing $-1 < a < 0$.

4.87 Sketch the magnitude and phase response of the multipath channel

$$y(n) = x(n) + \alpha x(n-M) \qquad \alpha > 0$$

for $\alpha \ll 1$.

4.88 Determine the system functions and the pole-zero locations for the systems shown in Fig. P4.88(a) through (c), and indicate whether or not the systems are stable.

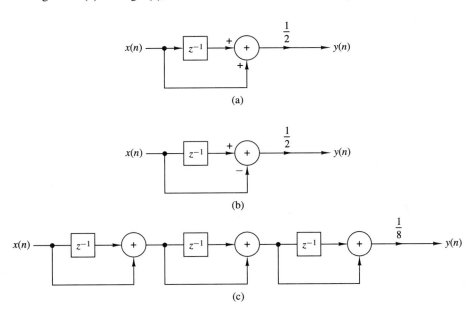

Figure P4.88

4.89 Determine and sketch the impulse response and the magnitude and phase responses of the FIR filter shown in Fig. P4.89 for $b = 1$ and $b = -1$.

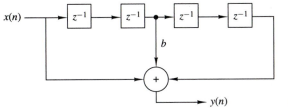

Figure P4.89

4.90 Consider the system

$$y(n) = x(n) - 0.95x(n - 6)$$

(a) Sketch its pole–zero pattern.
(b) Sketch its magnitude response using the pole–zero plot.
(c) Determine the system function of its causal inverse system.
(d) Sketch the magnitude response of the inverse system using the pole–zero plot.

4.91 Determine the impulse response and the difference equation for all possible systems specified by the system functions

(a) $H(z) = \dfrac{z^{-1}}{1 - z^{-1} - z^{-2}}$

(b) $H(z) = \dfrac{1}{1 - e^{-4a}z^{-4}}$ $0 < a < 1$

4.92 Determine the impulse response of a causal LTI system which produces the response

$$y(n) = \{1, -1, 3, -1, 6\}$$
$$\uparrow$$

when excited by the input signal

$$x(n) = \{1, 1, 2\}$$
$$\uparrow$$

4.93 The system

$$y(n) = \tfrac{1}{2}y(n - 1) + x(n)$$

is excited with the input

$$x(n) = (\tfrac{1}{4})^n u(n)$$

Determine the sequences $r_{xx}(l)$, $r_{hh}(l)$, $r_{xy}(l)$, and $r_{yy}(l)$.

4.94 Determine if the following FIR systems are minimum phase.
(a) $h(n) = \{10, 9, -7, -8, 0, 5, 3\}$
$\uparrow$

(b) $h(n) = \{5, 4, -3, -4, 0, 2, 1\}$
$\uparrow$

4.95 Can you determine the coefficients of the all-pole system

$$H(z) = \dfrac{1}{1 + \displaystyle\sum_{k=1}^{N} a_k z^{-k}}$$

if you know its order N and the values $h(0), h(1), \ldots, h(L-1)$ of its impulse response? How? What happens if you do not know N?

4.96 Consider a system with impulse response

$$h(n) = b_0 \delta(n) + b_1 \delta(n - D) + b_2 \delta(n - 2D)$$

(a) Explain why the system generates echoes spaced D samples apart.
(b) Determine the magnitude and phase response of the system.
(c) Show that for $|b_0 + b_2| << |b_1|$, the locations of maxima and minima of $|H(\omega)|^2$ are at

$$\omega = \pm\dfrac{k}{D}\pi \qquad k = 0, 1, 2, \ldots$$

(d) Plot $|H(\omega)|$ and $\measuredangle H(\omega)$ for $b_0 = 0.1$, $b_1 = 1$, and $b_2 = 0.05$ and discuss the results.

4.97 Consider the pole–zero system

$$H(z) = \dfrac{B(z)}{A(z)} = \dfrac{1 + bz^{-1}}{1 + az^{-1}} = \sum_{n=0}^{\infty} h(n)z^{-n}$$

(a) Determine $h(0)$, $h(1)$, $h(2)$, and $h(3)$ in terms of a and b.
(b) Let $r_{hh}(l)$ be the autocorrelation sequence of $h(n)$. Determine $r_{hh}(0)$, $r_{hh}(1)$, $r_{hh}(2)$, and $r_{hh}(3)$ in terms of a and b.

4.98 Let $x(n)$ be a real-valued minimum-phase sequence. Modify $x(n)$ to obtain another real-valued minimum-phase sequence $y(n)$ such that $y(0) = x(0)$ and $y(n) = |x(n)|$.

4.99 The frequency response of a stable LTI system is known to be real and even. Is the inverse system stable?

4.100 Let $h(n)$ be a real filter with nonzero linear or nonlinear phase response. Show that the following operations are equivalent to filtering the signal $x(n)$ with a zero-phase filter.

  (a) $g(n) = h(n) * x(n)$
    $f(n) = h(n) * g(-n)$
    $y(n) = f(-n)$
  (b) $g(n) = h(n) * x(n)$
    $f(n) = h(n) * x(-n)$
    $y(n) = g(n) + f(-n)$

(*Hint:* Determine the frequency response of the composite system $y(n) = H[x(n)]$.)

4.101 Check the validity of the following statements:

  (a) The convolution of two minimum-phase sequences is always minimum-phase sequence.
  (b) The sum of two minimum-phase sequences is always minimum phase.

4.102 Determine the minimum-phase system whose squared magnitude response is given by:

  (a) $|H(\omega)|^2 = \dfrac{\dfrac{5}{4} - \cos\omega}{\dfrac{10}{9} - \dfrac{2}{3}\cos\omega}$

  (b) $|H(\omega)|^2 = \dfrac{2(1 - a^2)}{(1 + a^2) - 2a\cos\omega} \qquad |a| < 1$

4.103 Consider an FIR system with the following system function:

$$H(z) = (1 - 0.8e^{j\pi/2}z^{-1})(1 - 0.8e^{-j\pi/2}z^{-1})(1 - 1.5e^{j\pi/4}z^{-1})(1 - 1.5e^{-j\pi/4}z^{-1})$$

  (a) Determine all systems that have the same magnitude response. Which is the minimum-phase system?
  (b) Determine the impulse response of all systems in part (a).
  (c) Plot the partial energy

$$E(n) = \sum_{k=0}^{n} h^2(n)$$

  for every system and use it to identify the minimum- and maximum-phase systems.

4.104 The causal system

$$H(z) = \frac{1}{1 + \displaystyle\sum_{k=1}^{N} a_k z^{-k}}$$

is known to be unstable.

  We modify this system by changing its impulse response $h(n)$ to

$$h'(n) = \lambda^n h(n)u(n)$$

(a) Show that by properly choosing λ we can obtain a new stable system.

(b) What is the difference equation describing the new system?

4.105* Given a signal $x(n)$, we can create echoes and reverberations by delaying and scaling the signal as follows

$$y(n) = \sum_{k=0}^{\infty} g_k x(n - kD)$$

where D is positive integer and $g_k > g_{k-1} > 0$.

(a) Explain why the comb filter

$$H(z) = \frac{1}{1 - az^{-D}}$$

can be used as a reverberator (i.e., as a device to produce artificial reverberations). (*Hint:* Determine and sketch its impulse response.)

(b) The all-pass comb filter

$$H(z) = \frac{z^{-D} - a}{1 - az^{-D}}$$

is used in practice to build digital reverberators by cascading three to five such filters and properly choosing the parameters a and D. Compute and plot the impulse response of two such reverberators each obtained by cascading three sections with the following parameters.

UNIT 1			UNIT 2		
Section	D	a	Section	D	a
1	50	0.7	1	50	0.7
2	40	0.665	2	17	0.77
3	32	0.63175	3	6	0.847

(c) The difference between echo and reverberation is that with pure echo there are clear repetitions of the signal, but with reverberations, there are not. How is this reflected in the shape of the impulse response of the reverberator? Which unit in part (b) is a better reverberator?

(d) If the delays D_1, D_2, D_3 in a certain unit are prime numbers, the impulse response of the unit is more "dense." Explain why.

(e) Plot the phase response of units 1 and 2 and comment on them.

(f) Plot $h(n)$ for D_1, D_2, and D_3 being nonprime. What do you notice?

More details about this application can be found in a paper by J. A. Moorer, "Signal Processing Aspects of Computer Music: A Survey," *Proc, IEEE,* vol. 65, No. 8, Aug. 1977, pp. 1108–1137.

4.106* By trial-and-error design a third-order lowpass filter with cutoff frequency at $\omega_c = \pi/9$ radians/sample interval. Start your search with

(a) $z_1 = z_2 = z_3 = 0$, $p_1 = r$, $p_{2,3} = re^{\pm j\omega_c}$, $r = 0.8$

(b) $r = 0.9$, $z_1 = z_2 = z_3 = -1$

4.107* A speech signal with bandwidth $B = 10$ kHz is sampled at $F_2 = 20$ kHz. Suppose that the signal is corrupted by four sinusoids with frequencies

$$F_1 = 10,000 \text{ Hz}, \qquad F_3 = 7778 \text{ Hz}$$

$$F_2 = 8889 \text{ Hz}, \qquad F_4 = 6667 \text{ Hz}$$

(a) Design a FIR filter that eliminates these frequency components.
(b) Choose the gain of the filter so that $|H(0)| = 1$ and then plot the log magnitude response and the phase response of the filter.
(c) Does the filter fulfill your objectives? Do you recommend the use of this filter in a practical application?

4.108* Compute and sketch the frequency response of a digital resonator with $\omega = \pi/6$ and $r = 0.6, 0.9, 0.99$. In each case, compute the bandwidth and the resonance frequency from the graph, and check if they are in agreement with the theoretical results.

4.109* The system function of a communication channel is given by

$$H(z) = (1 - 0.9e^{j0.4\pi}z^{-1})(1 - 0.9e^{-j0.4\pi}z^{-1})(1 - 1.5e^{j0.6\pi}z^{-1})(1 - 1.5e^{-j0.6\pi}z^{-1})$$

Determine the system function $H_c(z)$ of a causal and stable compensating system so that the cascade interconnection of the two systems has a flat magnitude response. Sketch the pole–zero plots and the magnitude and phase responses of all systems involved into the analysis process. [*Hint:* Use the decomposition $H(z) = H_{ap}(z)H_{min}(z)$.]

5

The Discrete Fourier Transform: Its Properties and Applications

Frequency analysis of discrete-time signals is usually and most conveniently performed on a digital signal processor, which may be a general-purpose digital computer or specially designed digital hardware. To perform frequency analysis on a discrete-time signal $\{x(n)\}$, we convert the time-domain sequence to an equivalent frequency-domain representation. We know that such a representation is given by the Fourier transform $X(\omega)$ of the sequence $\{x(n)\}$. However, $X(\omega)$ is a continuous function of frequency and therefore, it is not a computationally convenient representation of the sequence $\{x(n)\}$.

In this section we consider the representation of a sequence $\{x(n)\}$ by samples of its spectrum $X(\omega)$. Such a frequency-domain representation leads to the discrete Fourier transform (DFT), which is a powerful computational tool for performing frequency analysis of discrete-time signals.

5.1 FREQUENCY DOMAIN SAMPLING: THE DISCRETE FOURIER TRANSFORM

Before we introduce the DFT, we consider the sampling of the Fourier transform of an aperiodic discrete-time sequence. Thus, we establish the relationship between the sampled Fourier transform and the DFT.

5.1.1 Frequency-Domain Sampling and Reconstruction of Discrete-Time Signals

We recall that aperiodic finite-energy signals have continuous spectra. Let us consider such an aperiodic discrete-time signal $x(n)$ with Fourier transform

$$X(\omega) = \sum_{n=-\infty}^{\infty} x(n)e^{-j\omega n} \tag{5.1.1}$$

394

Suppose that we sample $X(\omega)$ periodically in frequency at a spacing of $\delta\omega$ radians between successive samples. Since $X(\omega)$ is periodic with period 2π, only samples in the fundamental frequency range are necessary. For convenience, we take N equidistant samples in the interval $0 \le \omega < 2\pi$ with spacing $\delta\omega = 2\pi/N$, as shown in Fig. 5.1. First, we consider the selection of N, the number of samples in the frequency domain.

If we evaluate (5.1.1) at $\omega = 2\pi k/N$, we obtain

$$X\left(\frac{2\pi}{N}k\right) = \sum_{n=-\infty}^{\infty} x(n)e^{-j2\pi kn/N} \qquad k = 0, 1, \ldots, N-1 \qquad (5.1.2)$$

The summation in (5.1.2) can be subdivided into an infinite number of summations, where each sum contains N terms. Thus

$$X\left(\frac{2\pi}{N}k\right) = \quad \cdots + \sum_{n=-N}^{-1} x(n)e^{-j2\pi kn/N} + \sum_{n=0}^{N-1} x(n)e^{-j2\pi kn/N}$$

$$+ \sum_{n=N}^{2N-1} x(n)e^{-j2\pi kn/N} + \cdots$$

$$= \sum_{l=-\infty}^{\infty} \sum_{n=lN}^{lN+N-1} x(n)e^{-j2\pi kn/N}$$

If we change the index in the inner summation from n to $n - lN$ and interchange the order of the summation, we obtain the result

$$X\left(\frac{2\pi}{N}k\right) = \sum_{n=0}^{N-1} \left[\sum_{l=-\infty}^{\infty} x(n - lN) \right] e^{-j2\pi kn/N} \qquad (5.1.3)$$

for $k = 0, 1, 2, \ldots, N-1$.

The signal

$$x_p(n) = \sum_{l=-\infty}^{\infty} x(n - lN) \qquad (5.1.4)$$

obtained by the periodic repetition of $x(n)$ every N samples, is clearly periodic with fundamental period N. Consequently, it can be expanded in a Fourier

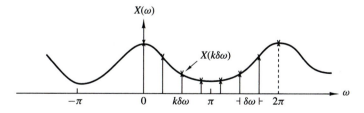

Figure 5.1   Frequency-domain sampling of the Fourier transform.

series as

$$x_p(n) = \sum_{k=0}^{N-1} c_k e^{j2\pi kn/N} \qquad n = 0, 1, \ldots, N-1 \tag{5.1.5}$$

with Fourier coefficients

$$c_k = \frac{1}{N} \sum_{n=0}^{N-1} x_p(n) e^{-j2\pi kn/N} \qquad k = 0, 1, \ldots, N-1 \tag{5.1.6}$$

Upon comparing (5.1.3) with (5.1.6), we conclude that

$$c_k = \frac{1}{N} X\left(\frac{2\pi}{N}k\right) \qquad k = 0, 1, \ldots, N-1 \tag{5.1.7}$$

Therefore,

$$x_p(n) = \frac{1}{N} \sum_{k=0}^{N-1} X\left(\frac{2\pi}{N}k\right) e^{j2\pi kn/N} \qquad n = 0, 1, \ldots, N-1 \tag{5.1.8}$$

The relationship in (5.1.8) provides the reconstruction of the periodic signal $x_p(n)$ from the samples of the spectrum $X(\omega)$. However, it does not imply that we can recover $X(\omega)$ or $x(n)$ from the samples. To accomplish this, we need to consider the relationship between $x_p(n)$ and $x(n)$.

Since $x_p(n)$ is the periodic extension of $x(n)$ as given by (5.1.4), it is clear that $x(n)$ can be recovered from $x_p(n)$ if there is no aliasing in the time domain, that is, if $x(n)$ is time-limited to less than the period N of $x_p(n)$. This situation is illustrated in Fig. 5.2, where without loss of generality, we consider a finite-duration

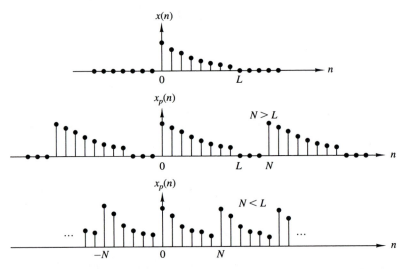

Figure 5.2 Aperiodic sequence $x(n)$ of length L and its periodic extension for $N \geq L$ (no aliasing) and $N < L$ (aliasing).

sequence $x(n)$, which is nonzero in the interval $0 \leq n \leq L - 1$. We observe that when $N \geq L$,

$$x(n) = x_p(n) \qquad 0 \leq n \leq N - 1$$

so that $x(n)$ can be recovered from $x_p(n)$ without ambiguity. On the other hand, if $N < L$, it is not possible to recover $x(n)$ from its periodic extension due to *time-domain aliasing*. Thus, we conclude that the spectrum of an aperiodic discrete-time signal with finite duration L, can be exactly recovered from its samples at frequencies $\omega_k = 2\pi k/N$, if $N \geq L$. The procedure is to compute $x_p(n)$, $n = 0, 1, \ldots, N-1$ from (5.1.8); then

$$x(n) = \begin{cases} x_p(n), & 0 \leq n \leq N - 1 \\ 0, & \text{elsewhere} \end{cases} \tag{5.1.9}$$

and finally, $X(\omega)$ can be computed from (5.1.1).

As in the case of continuous-time signals, it is possible to express the spectrum $X(\omega)$ directly in terms of its samples $X(2\pi k/N)$, $k = 0, 1, \ldots, N - 1$. To derive such an interpolation formula for $X(\omega)$, we assume that $N \geq L$ and begin with (5.1.8). Since $x(n) = x_p(n)$ for $0 \leq n \leq N - 1$,

$$x(n) = \frac{1}{N} \sum_{k=0}^{N-1} X\left(\frac{2\pi}{N}k\right) e^{j2\pi kn/N} \qquad 0 \leq n \leq N - 1 \tag{5.1.10}$$

If we use (5.1.1) and substitute for $x(n)$, we obtain

$$X(\omega) = \sum_{n=0}^{N-1} \left[\frac{1}{N} \sum_{k=0}^{N-1} X\left(\frac{2\pi}{N}k\right) e^{j2\pi kn/N} \right] e^{-j\omega n}$$

$$= \sum_{k=0}^{N-1} X\left(\frac{2\pi}{N}k\right) \left[\frac{1}{N} \sum_{n=0}^{N-1} e^{-j(\omega - 2\pi k/N)n} \right] \tag{5.1.11}$$

The inner summation term in the brackets of (5.1.11) represents the basic interpolation function shifted by $2\pi k/N$ in frequency. Indeed, if we define

$$P(\omega) = \frac{1}{N} \sum_{n=0}^{N-1} e^{-j\omega n} = \frac{1}{N} \frac{1 - e^{-j\omega N}}{1 - e^{-j\omega}}$$

$$= \frac{\sin(\omega N/2)}{N \sin(\omega/2)} e^{-j\omega(N-1)/2} \tag{5.1.12}$$

then (5.1.11) can be expressed as

$$X(\omega) = \sum_{k=0}^{N-1} X\left(\frac{2\pi}{N}k\right) P\left(\omega - \frac{2\pi}{N}k\right) \qquad N \geq L \tag{5.1.13}$$

The interpolation function $P(\omega)$ is not the familiar $(\sin\theta)/\theta$ but instead, it is a periodic counterpart of it, and it is due to the periodic nature of $X(\omega)$. The phase shift in (5.1.12) reflects the fact that the signal $x(n)$ is a causal, finite-duration sequence of length N. The function $\sin(\omega N/2)/(N\sin(\omega/2))$ is plotted in Fig. 5.3 for $N = 5$. We observe that the function $P(\omega)$ has the property

$$P\left(\frac{2\pi}{N}k\right) = \begin{cases} 1, & k = 0 \\ 0, & k = 1, 2, \ldots, N-1 \end{cases} \tag{5.1.14}$$

Consequently, the interpolation formula in (5.1.13) gives exactly the sample values $X(2\pi k/N)$ for $\omega = 2\pi k/N$. At all other frequencies, the formula provides a properly weighted linear combination of the original spectral samples.

The following example illustrates the frequency-domain sampling of a discrete-time signal and the time-domain aliasing that results.

Example 5.1.1

Consider the signal

$$x(n) = a^n u(n) \qquad 0 < a < 1$$

The spectrum of this signal is sampled at frequencies $\omega_k = 2\pi k/N$, $k = 0, 1, \ldots, N-1$. Determine the reconstructed spectra for $a = 0.8$ when $N = 5$ and $N = 50$.

Solution   The Fourier transform of the sequence $x(n)$ is

$$X(\omega) = \sum_{n=0}^{\infty} a^n e^{-j\omega n} = \frac{1}{1 - ae^{-j\omega}}$$

Suppose that we sample $X(\omega)$ at N equidistant frequencies $\omega_k = 2\pi k/N$, $k = 0, 1, \ldots, N-1$. Thus we obtain the spectral samples

$$X(\omega_k) \equiv X\left(\frac{2\pi k}{N}\right) = \frac{1}{1 - ae^{-j2\pi k/N}} \qquad k = 0, 1, \ldots, N-1$$

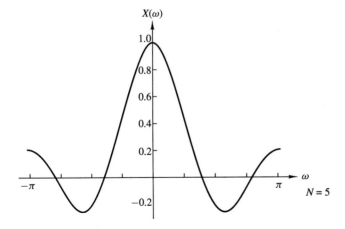

Figure 5.3   Plot of the function $[\sin(\omega N/2)]/[N\sin(\omega/2)]$.

The periodic sequence $x_p(n)$, corresponding to the frequency samples $X(2\pi k/N)$, $k = 0, 1, \ldots, N - 1$, can be obtained from either (5.1.4) or (5.1.8). Hence

$$x_p(n) = \sum_{l=-\infty}^{\infty} x(n - lN) = \sum_{l=-\infty}^{0} a^{n-lN}$$

$$= a^n \sum_{l=0}^{\infty} a^{lN} = \frac{a^n}{1 - a^N} \qquad 0 \le n \le N - 1$$

where the factor $1/(1 - a^N)$ represents the effect of aliasing. Since $0 < a < 1$, the aliasing error tends toward zero as $N \to \infty$.

For $a = 0.8$, the sequence $x(n)$ and its spectrum $X(\omega)$ are shown in Fig. 5.4a and b, respectively. The aliased sequences $x_p(n)$ for $N = 5$ and $N = 50$ and the corresponding spectral samples are shown in Fig. 5.4c and d, respectively. We note that the aliasing effects are negligible for $N = 50$.

If we define the aliased finite-duration sequence $x(n)$ as

$$\hat{x}(n) = \begin{cases} x_p(n), & 0 \le n \le N - 1 \\ 0, & \text{otherwise} \end{cases}$$

then its Fourier transform is

$$\hat{X}(\omega) = \sum_{n=0}^{N-1} \hat{x}(n)e^{-j\omega n} = \sum_{n=0}^{N-1} x_p(n)e^{-j\omega N}$$

$$= \frac{1}{1 - a^N} \cdot \frac{1 - a^N e^{-j\omega n}}{1 - ae^{-j\omega}}$$

Note that although $\hat{X}(\omega) \ne X(\omega)$, the sample values at $\omega_k = 2\pi k/N$ are identical. That is,

$$\hat{X}\left(\frac{2\pi}{N}k\right) = \frac{1}{1 - a^N} \cdot \frac{1 - a^N}{1 - ae^{-j2\pi kN}} = X\left(\frac{2\pi}{N}k\right)$$

5.1.2 The Discrete Fourier Transform (DFT)

The development in the preceding section is concerned with the frequency-domain sampling of an aperiodic finite-energy sequence $x(n)$. In general, the equally spaced frequency samples $X(2\pi k/N)$, $k = 0, 1, \ldots, N-1$, do not uniquely represent the original sequence $x(n)$ when $x(n)$ has infinite duration. Instead, the frequency samples $X(2\pi k/N)$, $k = 0, 1, \ldots, N - 1$, correspond to a periodic sequence $x_p(n)$ of period N, where $x_p(n)$ is an aliased version of $x(n)$, as indicated by the relation in (5.1.4), that is,

$$x_p(n) = \sum_{l=-\infty}^{\infty} x(n - lN) \qquad (5.1.15)$$

When the sequence $x(n)$ has a finite duration of length $L \le N$, then $x_p(n)$ is simply a periodic repetition of $x(n)$, where $x_p(n)$ over a single period is

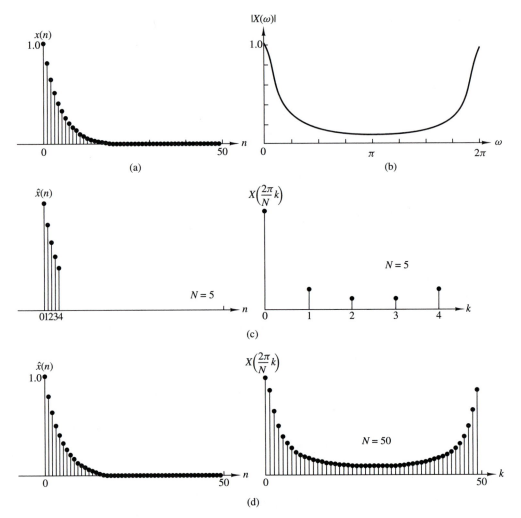

Figure 5.4 (a) Plot of sequence $x(n) = (0.8)^n u(n)$; (b) its Fourier transform (magnitude only); (c) effect of aliasing with $N = 5$; (d) reduced effect of aliasing with $N = 50$.

given as

$$x_p(n) = \begin{cases} x(n), & 0 \le n \le L - 1 \\ 0, & L \le n \le N - 1 \end{cases} \qquad (5.1.16)$$

Consequently, the frequency samples $X(2\pi k/N)$, $k = 0, 1, \ldots, N - 1$, uniquely represent the finite-duration sequence $x(n)$. Since $x(n) \equiv x_p(n)$ over a single period (padded by $N - L$ zeros), the original finite-duration sequence $x(n)$ can be obtained from the frequency samples $\{X(2\pi k/N)\}$ by means of the formula (5.1.8).

It is important to note that *zero padding* does not provide any additional information about the spectrum $X(\omega)$ of the sequence $\{x(n)\}$. The L equidistant samples of $X(\omega)$ are sufficient to reconstruct $X(\omega)$ using the reconstruction formula (5.1.13). However, padding the sequence $\{x(n)\}$ with $N - L$ zeros and computing an N-point DFT results in a "better display" of the Fourier transform $X(\omega)$.

In summary, a finite-duration sequence $x(n)$ of length L [i.e., $x(n) = 0$ for $n < 0$ and $n \geq L$] has a Fourier transform

$$X(\omega) = \sum_{n=0}^{L-1} x(n)e^{-j\omega n} \qquad 0 \leq \omega \leq 2\pi \qquad (5.1.17)$$

where the upper and lower indices in the summation reflect the fact that $x(n) = 0$ outside the range $0 \leq n \leq L - 1$. When we sample $X(\omega)$ at equally spaced frequencies $\omega_k = 2\pi k/N$, $k = 0, 1, 2, \ldots, N - 1$, where $N \geq L$, the resultant samples are

$$X(k) \equiv X\left(\frac{2\pi k}{N}\right) = \sum_{n=0}^{L-1} x(n)e^{-j2\pi kn/N}$$

$$X(k) = \sum_{n=0}^{N-1} x(n)e^{-j2\pi kn/N} \qquad k = 0, 1, 2, \ldots, N - 1 \qquad (5.1.18)$$

where for convenience, the upper index in the sum has been increased from $L - 1$ to $N - 1$ since $x(n) = 0$ for $n \geq L$.

The relation in (5.1.18) is a formula for transforming a sequence $\{x(n)\}$ of length $L \leq N$ into a sequence of frequency samples $\{X(k)\}$ of length N. Since the frequency samples are obtained by evaluating the Fourier transform $X(\omega)$ at a set of N (equally spaced) discrete frequencies, the relation in (5.1.18) is called the *discrete Fourier transform* (DFT) of $x(n)$. In turn, the relation given by (5.1.10), which allows us to recover the sequence $x(n)$ from the frequency samples

$$x(n) = \frac{1}{N} \sum_{k=0}^{N-1} X(k)e^{j2\pi kn/N} \qquad n = 0, 1, \ldots, N - 1 \qquad (5.1.19)$$

is called the *inverse DFT* (IDFT). Clearly, when $x(n)$ has length $L < N$, the N-point IDFT yields $x(n) = 0$ for $L \leq n \leq N - 1$. To summarize, the formulas for the DFT and IDFT are

DFT

$$X(k) = \sum_{n=0}^{N-1} x(n)e^{-j2\pi kn/N} \qquad k = 0, 1, 2, \ldots, N - 1 \qquad (5.1.18)$$

IDFT

$$x(n) = \frac{1}{N} \sum_{k=0}^{N-1} X(k)e^{j2\pi kn/N} \qquad n = 0, 1, 2, \ldots, N - 1 \qquad (5.1.19)$$

Example 5.1.2

A finite-duration sequence of length L is given as

$$x(n) = \begin{cases} 1, & 0 \le n \le L-1 \\ 0, & \text{otherwise} \end{cases}$$

Determine the N-point DFT of this sequence for $N \ge L$.

Solution The Fourier transform of this sequence is

$$X(\omega) = \sum_{n=0}^{L-1} x(n)e^{-j\omega n}$$

$$= \sum_{n=0}^{L-1} e^{-j\omega n} = \frac{1 - e^{-j\omega L}}{1 - e^{-j\omega}} = \frac{\sin(\omega L/2)}{\sin(\omega/2)} e^{-j\omega(L-1)/2}$$

The magnitude and phase of $X(\omega)$ are illustrated in Fig. 5.5 for $L = 10$. The N-point DFT of $x(n)$ is simply $X(\omega)$ evaluated at the set of N equally spaced frequencies $\omega_k = 2\pi k/N$, $k = 0, 1, \ldots, N - 1$. Hence

$$X(k) = \frac{1 - e^{-j2\pi kL/N}}{1 - e^{-j2\pi k/N}} \qquad k = 0, 1, \ldots, N - 1$$

$$= \frac{\sin(\pi kL/N)}{\sin(\pi k/N)} e^{-j\pi k(L-1)/N}$$

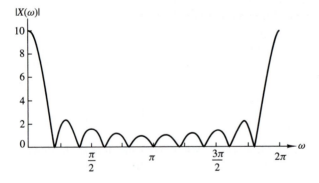

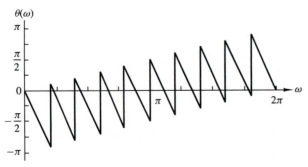

Figure 5.5 Magnitude and phase characteristics of the Fourier transform for signal in Example 5.1.2.

If N is selected such that $N = L$, then the DFT becomes

$$X(k) = \begin{cases} L, & k = 0 \\ 0, & k = 1, 2, \ldots, L-1 \end{cases}$$

Thus there is only one nonzero value in the DFT. This is apparent from observation of $X(\omega)$, since $X(\omega) = 0$ at the frequencies $\omega_k = 2\pi k/L$, $k \neq 0$. The reader should verify that $x(n)$ can be recovered from $X(k)$ by performing an L-point IDFT.

Although the L-point DFT is sufficient to uniquely represent the sequence $x(n)$ in the frequency domain, it is apparent that it does not provide sufficient detail to yield a good picture of the spectral characteristics of $x(n)$. If we wish to have better picture, we must evaluate (interpolate) $X(\omega)$ at more closely spaced frequencies, say $\omega_k = 2\pi k/N$, where $N > L$. In effect, we can view this computation as expanding the size of the sequence from L points to N points by appending $N - L$ zeros to the sequence $x(n)$, that is, zero padding. Then the N-point DFT provides finer interpolation than the L-point DFT.

Figure 5.6 provides a plot of the N-point DFT, in magnitude and phase, for $L = 10$, $N = 50$, and $N = 100$. Now the spectral characteristics of the sequence are more clearly evident, as one will conclude by comparing these spectra with the continuous spectrum $X(\omega)$.

5.1.3 The DFT as a Linear Transformation

The formulas for the DFT and IDFT given by (5.1.18) and (5.1.19) may be expressed as

$$X(k) = \sum_{n=0}^{N-1} x(n) W_N^{kn} \qquad k = 0, 1, \ldots, N-1 \qquad (5.1.20)$$

$$x(n) = \frac{1}{N} \sum_{k=0}^{N-1} X(k) W_N^{-kn} \qquad n = 0, 1, \ldots, N-1 \qquad (5.1.21)$$

where, by definition,

$$W_N = e^{-j2\pi/N} \qquad (5.1.22)$$

which is an Nth root of unity.

We note that the computation of each point of the DFT can be accomplished by N complex multiplications and $(N-1)$ complex additions. Hence the N-point DFT values can be computed in a total of N^2 complex multiplications and $N(N-1)$ complex additions.

It is instructive to view the DFT and IDFT as linear transformations on sequences $\{x(n)\}$ and $\{X(k)\}$, respectively. Let us define an N-point vector $\mathbf{x}_N$ of the signal sequence $x(n)$, $n = 0, 1, \ldots, N-1$, an N-point vector $\mathbf{X}_N$ of frequency

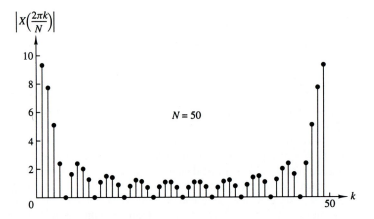

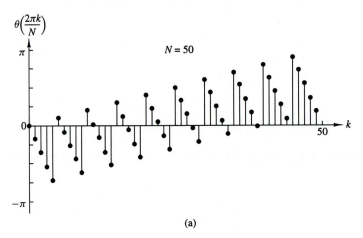

(a)

Figure 5.6   Magnitude and phase of an N-point DFT in Example 6.4.2; (a) $L = 10$, $N = 50$; (b) $L = 10$, $N = 100$.

samples, and an $N \times N$ matrix $\mathbf{W}_N$ as

$$\mathbf{x}_N = \begin{bmatrix} x(0) \\ x(1) \\ \vdots \\ x(N-1) \end{bmatrix}, \qquad \mathbf{X}_N = \begin{bmatrix} X(0) \\ X(1) \\ \vdots \\ X(N-1) \end{bmatrix}$$

$$\mathbf{W}_N = \begin{bmatrix} 1 & 1 & 1 \cdots & 1 \\ 1 & W_N & W_N^2 & \cdots & W_N^{N-1} \\ & W_N^2 & W_N^4 & \cdots & W_N^{2(N-1)} \\ \vdots & \vdots & \vdots & \vdots & \vdots \\ 1 & W_N^{N-1} & W_N^{2(N-1)} & \cdots & W_N^{(N-1)(N-1)} \end{bmatrix} \qquad (5.1.23)$$

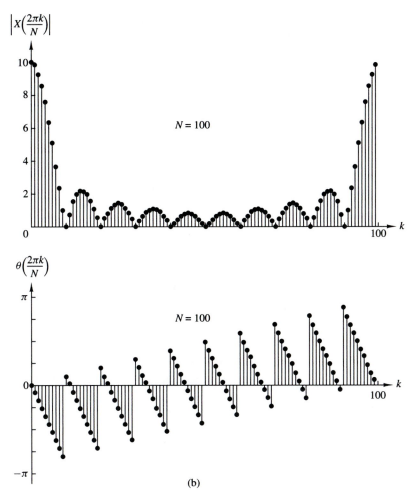

(b)

Figure 5.6 *continued*

With these definitions, the N-point DFT may be expressed in matrix form as

$$\mathbf{X}_N = \mathbf{W}_N \mathbf{x}_N \qquad (5.1.24)$$

where $\mathbf{W}_N$ is the matrix of the linear transformation. We observe that $\mathbf{W}_N$ is a symmetric matrix. If we assume that the inverse of $\mathbf{W}_N$ exists, then (5.1.24) can be inverted by premultiplying both sides by $\mathbf{W}_N^{-1}$. Thus we obtain

$$\mathbf{x}_N = \mathbf{W}_N^{-1} \mathbf{X}_N \qquad (5.1.25)$$

But this is just an expression for the IDFT.

In fact, the IDFT as given by (5.1.21), can be expressed in matrix form as

$$\mathbf{x}_N = \frac{1}{N} \mathbf{W}_N^* \mathbf{X}_N \qquad (5.1.26)$$

where $\mathbf{W}_N^*$ denotes the complex conjugate of the matrix $\mathbf{W}_N$. Comparison of (5.1.26) with (5.1.25) leads us to conclude that

$$\mathbf{W}_N^{-1} = \frac{1}{N}\mathbf{W}_N^* \tag{5.1.27}$$

which, in turn, implies that

$$\mathbf{W}_N\mathbf{W}_N^* = N\mathbf{I}_N \tag{5.1.28}$$

where $\mathbf{I}_N$ is an $N \times N$ identity matrix. Therefore, the matrix $\mathbf{W}_N$ in the transformation is an orthogonal (unitary) matrix. Furthermore, its inverse exists and is given as $\mathbf{W}_N^*/N$. Of course, the existence of the inverse of $\mathbf{W}_N$ was established previously from our derivation of the IDFT.

Example 5.1.3

Compute the DFT of the four-point sequence

$$x(n) = (0 \quad 1 \quad 2 \quad 3)$$

Solution   The first step is to determine the matrix $\mathbf{W}_4$. By exploiting the periodicity property of $\mathbf{W}_4$ and the symmetry property

$$W_N^{k+N/2} = -W_N^k$$

the matrix $\mathbf{W}_4$ may be expressed as

$$\mathbf{W}_4 = \begin{bmatrix} W_4^0 & W_4^0 & W_4^0 & W_4^0 \\ W_4^0 & W_4^1 & W_4^2 & W_4^3 \\ W_4^0 & W_4^2 & W_4^4 & W_4^6 \\ W_4^0 & W_4^3 & W_4^6 & W_4^9 \end{bmatrix} = \begin{bmatrix} 1 & 1 & 1 & 1 \\ 1 & W_4^1 & W_4^2 & W_4^3 \\ 1 & W_4^2 & W_4^0 & W_4^2 \\ 1 & W_4^3 & W_4^2 & W_4^1 \end{bmatrix}$$

$$= \begin{bmatrix} 1 & 1 & 1 & 1 \\ 1 & -j & -1 & j \\ 1 & -1 & 1 & -1 \\ 1 & j & -1 & -j \end{bmatrix}$$

Then

$$\mathbf{X}_4 = \mathbf{W}_4\mathbf{x}_4 = \begin{bmatrix} 6 \\ -2+2j \\ -2 \\ -2-2j \end{bmatrix}$$

The IDFT of $\mathbf{X}_4$ may be determined by conjugating the elements in $\mathbf{W}_4$ to obtain $\mathbf{W}_4^*$ and then applying the formula (5.1.26).

The DFT and IDFT are computational tools that play a very important role in many digital signal processing applications, such as frequency analysis (spectrum analysis) of signals, power spectrum estimation, and linear filtering. The importance of the DFT and IDFT in such practical applications is due to a large extent on the existence of computationally efficient algorithms, known collectively as fast

Fourier transform (FFT) algorithms, for computing the DFT and IDFT. This class of algorithms is described in Chapter 6.

5.1.4 Relationship of the DFT to Other Transforms

In this discussion we have indicated that the DFT is an important computational tool for performing frequency analysis of signals on digital signal processors. In view of the other frequency analysis tools and transforms that we have developed, it is important to establish the relationships between the DFT to these other transforms.

Relationship to the Fourier series coefficients of a periodic sequence. A periodic sequence $\{x_p(n)\}$ with fundamental period N can be represented in a Fourier series of the form

$$x_p(n) = \sum_{k=0}^{N-1} c_k e^{j2\pi nk/N} \qquad -\infty < n < \infty \tag{5.1.29}$$

where the Fourier series coefficients are given by the expression

$$c_k = \frac{1}{N} \sum_{n=0}^{N-1} x_p(n) e^{-j2\pi nk/N} \qquad k = 0, 1, \ldots, N-1 \tag{5.1.30}$$

If we compare (5.1.29) and (5.1.30) with (5.1.18) and (5.1.19), we observe that the formula for the Fourier series coefficients has the form of a DFT. In fact, if we define a sequence $x(n) = x_p(n), 0 \le n \le N-1$, the DFT of this sequence is simply

$$X(k) = N c_k \tag{5.1.31}$$

Furthermore, (5.1.29) has the form of an IDFT. Thus the N-point DFT provides the exact line spectrum of a periodic sequence with fundamental period N.

Relationship to the Fourier transform of an aperiodic sequence. We have already shown that if $x(n)$ is an aperiodic finite energy sequence with Fourier transform $X(\omega)$, which is sampled at N equally spaced frequencies $\omega_k = 2\pi k/N$, $k = 0, 1, \ldots, N-1$, the spectral components

$$X(k) = X(\omega)|_{\omega=2\pi k/N} = \sum_{n=-\infty}^{\infty} x(n) e^{-j2\pi nk/N} \qquad k = 0, 1, \ldots, N-1 \tag{5.1.32}$$

are the DFT coefficients of the periodic sequence of period N, given by

$$x_p(n) = \sum_{l=-\infty}^{\infty} x(n - lN) \tag{5.1.33}$$

Thus $x_p(n)$ is determined by aliasing $\{x(n)\}$ over the interval $0 \le n \le N-1$. The finite-duration sequence

$$\hat{x}(n) = \begin{cases} x_p(n), & 0 \le n \le N-1 \\ 0, & \text{otherwise} \end{cases} \tag{5.1.34}$$

bears no resemblance to the original sequence $\{x(n)\}$, unless $x(n)$ is of finite duration and length $L \leq N$, in which case

$$x(n) = \hat{x}(n) \qquad 0 \leq n \leq N - 1 \tag{5.1.35}$$

Only in this case will the IDFT of $\{X(k)\}$ yield the original sequence $\{x(n)\}$.

Relationship to the z-transform.   Let us consider a sequence $x(n)$ having the z-transform

$$X(z) = \sum_{n=-\infty}^{\infty} x(n)z^{-n} \tag{5.1.36}$$

with a ROC that includes the unit circle. If $X(z)$ is sampled at the N equally spaced points on the unit circle $z_k = e^{j2\pi k/N}$, $0, 1, 2, \ldots, N - 1$, we obtain

$$X(k) \equiv X(z)|_{z=e^{j2\pi nk/N}} \qquad k = 0, 1, \ldots, N - 1$$

$$= \sum_{n=-\infty}^{\infty} x(n)e^{-j2\pi nk/N} \tag{5.1.37}$$

The expression in (5.1.37) is identical to the Fourier transform $X(\omega)$ evaluated at the N equally spaced frequencies $\omega_k = 2\pi k/N$, $k = 0, 1, \ldots, N - 1$, which is the topic treated in Section 5.1.1.

If the sequence $x(n)$ has a finite duration of length N or less, the sequence can be recovered from its N-point DFT. Hence its z-transform is uniquely determined by its N-point DFT. Consequently, $X(z)$ can be expressed as a function of the DFT $\{X(k)\}$ as follows

$$X(z) = \sum_{n=0}^{N-1} x(n)z^{-n}$$

$$X(z) = \sum_{n=0}^{N-1} \left[\frac{1}{N} \sum_{k=0}^{N-1} X(k)e^{j2\pi kn/N} \right] z^{-n}$$

$$X(z) = \frac{1}{N} \sum_{k=0}^{N-1} X(k) \sum_{n=0}^{N-1} \left(e^{j2\pi k/N} z^{-1} \right)^n \tag{5.1.38}$$

$$X(z) = \frac{1 - z^{-N}}{N} \sum_{k=0}^{N-1} \frac{X(k)}{1 - e^{j2\pi k/N} z^{-1}}$$

When evaluated on the unit circle, (5.1.38) yields the Fourier transform of the finite-duration sequence in terms of its DFT, in the form

$$X(\omega) = \frac{1 - e^{-j\omega N}}{N} \sum_{k=0}^{N-1} \frac{X(k)}{1 - e^{-j(\omega - 2\pi k/N)}} \tag{5.1.39}$$

This expression for the Fourier transform is a polynomial (Lagrange) interpolation formula for $X(\omega)$ expressed in terms of the values $\{X(k)\}$ of the polynomial at a set of equally spaced discrete frequencies $\omega_k = 2\pi k/N$, $k = 0, 1, \ldots, N - 1$. With some algebraic manipulations, it is possible to reduce (5.1.39) to the interpolation formula given previously in (5.1.13).

Relationship to the Fourier series coefficients of a continuous-time signal.   Suppose that $x_a(t)$ is a continuous-time periodic signal with fundamental period $T_p = 1/F_0$. The signal can be expressed in a Fourier series

$$x_a(t) = \sum_{k=-\infty}^{\infty} c_k e^{j2\pi k F_0} \tag{5.1.40}$$

where $\{c_k\}$ are the Fourier coefficients. If we sample $x_a(t)$ at a uniform rate $F_s = N/T_p = 1/T$, we obtain the discrete-time sequence

$$x(n) \equiv x_a(nT) = \sum_{k=-\infty}^{\infty} c_k e^{j2\pi k F_0 nT} = \sum_{k=-\infty}^{\infty} c_k e^{j2\pi kn/N}$$

$$= \sum_{k=0}^{N-1} \left[\sum_{l=-\infty}^{\infty} c_{k-lN} \right] e^{j2\pi kn/N} \tag{5.1.41}$$

It is clear that (5.1.41) is in the form of an IDFT formula, where

$$X(k) = N \sum_{l=-\infty}^{\infty} c_{k-lN} \equiv N\tilde{c}_k \tag{5.1.42}$$

and

$$\tilde{c}_k = \sum_{l=-\infty}^{\infty} c_{k-lN} \tag{5.1.43}$$

Thus the $\{\tilde{c}_k\}$ sequence is an aliased version of the sequence $\{c_k\}$.

5.2 PROPERTIES OF THE DFT

In Section 5.1.2 we introduced the DFT as a set of N samples $\{X(k)\}$ of the Fourier transform $X(\omega)$ for a finite-duration sequence $\{x(n)\}$ of length $L \leq N$. The sampling of $X(\omega)$ occurs at the N equally spaced frequencies $\omega_k = 2\pi k/N$, $k = 0, 1, 2, \ldots, N-1$. We demonstrated that the N samples $\{X(k)\}$ uniquely represent the sequence $\{x(n)\}$ in the frequency domain. Recall that the DFT and inverse DFT (IDFT) for an N-point sequence $\{x(n)\}$ are given as

$$\text{DFT: } X(k) = \sum_{n=0}^{N-1} x(n) W_N^{kn} \qquad k = 0, 1, \ldots, N-1 \tag{5.2.1}$$

$$\text{IDFT: } x(n) = \frac{1}{N} \sum_{k=0}^{N-1} X(k) W_N^{-kn} \qquad n = 0, 1, \ldots, N-1 \tag{5.2.2}$$

where W_N is defined as

$$W_N = e^{-j2\pi/N} \tag{5.2.3}$$

In this section we present the important properties of the DFT. In view of the relationships established in Section 5.1.4 between the DFT and Fourier series, and Fourier transforms and z-transforms of discrete-time signals, we expect the properties of the DFT to resemble the properties of these other transforms and series. However, some important differences exist, one of which is the circular convolution property derived in the following section. A good understanding of these properties is extremely helpful in the application of the DFT to practical problems.

The notation used below to denote the N-point DFT pair $x(n)$ and $X(k)$ is

$$x(n) \underset{N}{\overset{\text{DFT}}{\longleftrightarrow}} X(k)$$

5.2.1 Periodicity, Linearity, and Symmetry Properties

Periodicity.   If $x(n)$ and X(k) are an N-point DFT pair, then

$$x(n + N) = x(n) \quad \text{for all } n \tag{5.2.4}$$

$$X(k + N) = X(k) \quad \text{for all } k \tag{5.2.5}$$

These periodicities in $x(n)$ and $X(k)$ follow immediately from formulas (5.2.1) and (5.2.2) for the DFT and IDFT, respectively.

We previously illustrated the periodicity property in the sequence $x(n)$ for a given DFT. However, we had not previously viewed the DFT $X(k)$ as a periodic sequence. In some applications it is advantageous to do this.

Linearity.   If

$$x_1(n) \underset{N}{\overset{\text{DFT}}{\longleftrightarrow}} X_1(k)$$

and

$$x_2(n) \underset{N}{\overset{\text{DFT}}{\longleftrightarrow}} X_2(k)$$

then for any real-valued or complex-valued constants a_1 and a_2,

$$a_1 x_1(n) + a_2 x_2(n) \underset{N}{\overset{\text{DFT}}{\longleftrightarrow}} a_1 X_1(k) + a_2 X_2(k) \tag{5.2.6}$$

This property follows immediately from the definition of the DFT given by (5.2.1).

Circular Symmetries of a Sequence.   As we have seen, the N-point DFT of a finite duration sequence, $x(n)$ of length $L \leq N$ is equivalent to the N-point DFT of a periodic sequence $x_p(n)$, of period N, which is obtained by periodically extending $x(n)$, that is,

$$x_p(n) = \sum_{l=-\infty}^{\infty} x(n - lN) \tag{5.2.7}$$

Now suppose that we shift the periodic sequence $x_p(n)$ by k units to the right. Thus we obtain another periodic sequence

$$x_p'(n) = x_p(n - k) = \sum_{l=-\infty}^{\infty} x(n - k - lN) \tag{5.2.8}$$

The finite-duration sequence

$$x'(n) = \begin{cases} x_p'(n), & 0 \le n \le N - 1 \\ 0, & \text{otherwise} \end{cases} \tag{5.2.9}$$

is related to the original sequence $x(n)$ by a circular shift. This relationship is illustrated in Fig. 5.7 for $N = 4$.

In general, the circular shift of the sequence can be represented as the index modulo N. Thus we can write

$$\begin{aligned} x'(n) &= x(n - k, \text{modulo } N) \\ &\equiv x((n - k))_N \end{aligned} \tag{5.2.10}$$

For example, if $k = 2$ and $N = 4$, we have

$$x'(n) = x((n - 2))_4$$

which implies that

$$x'(0) = x((-2))_4 = x(2)$$
$$x'(1) = x((-1))_4 = x(3)$$
$$x'(2) = x((0))_4 = x(0)$$
$$x'(3) = x((1))_4 = x(1)$$

Hence $x'(n)$ is simply $x(n)$ shifted circularly by two units in time, where the counterclockwise direction has been arbitrarily selected as the positive direction. Thus we conclude that a circular shift of an N-point sequence is equivalent to a linear shift of its periodic extension, and vice versa.

The inherent periodicity resulting from the arrangement of the N-point sequence on the circumference of a circle dictates a different definition of even and odd symmetry, and time reversal of a sequence.

An N-point sequence is called circularly *even* if it is symmetric about the point zero on the circle. This implies that

$$x(N - n) = x(n) \qquad 1 \le n \le N - 1 \tag{5.2.11}$$

An N-point sequence is called circularly *odd* if it is antisymmetric about the point zero on the circle. This implies that

$$x(N - n) = -x(n) \qquad 1 \le n \le N - 1 \tag{5.2.12}$$

The time reversal of an N-point sequence is attained by reversing its samples about the point zero on the circle. Thus the sequence $x((-n))_N$ is simply given as

$$x((-n))_N = x(N - n) \qquad 0 \le n \le N - 1 \tag{5.2.13}$$

This time reversal is equivalent to plotting $x(n)$ in a clockwise direction on a circle.

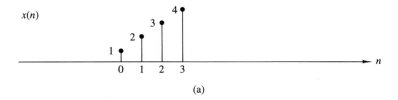

(a)

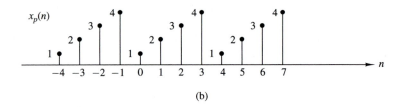

(b)

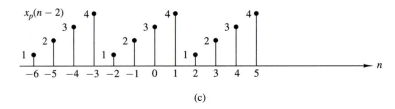

(c)

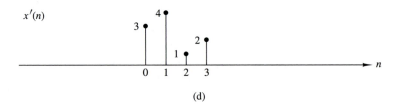

(d)

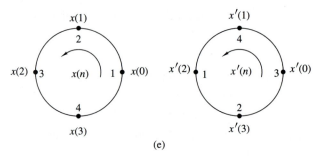

(e)

Figure 5.7 Circular shift of a sequence.

An equivalent definition of even and odd sequences for the associated periodic sequence $x_p(n)$ is given as follows

$$\text{even:}\quad x_p(n) = x_p(-n) = x_p(N - n)$$
$$\text{odd:}\quad x_p(n) = -x_p(-n) = -x_p(N - n) \tag{5.2.14}$$

If the periodic sequence is complex-valued, we have

$$\text{conjugate even:}\quad x_p(n) = x_p^*(N - n)$$
$$\text{conjugate odd:}\quad x_p(n) = -x_p^*(N - n) \tag{5.2.15}$$

These relationships suggest that we decompose the sequence $x_p(n)$ as

$$x_p(n) = x_{pe}(n) + x_{po}(n) \tag{5.2.16}$$

where

$$x_{pe}(n) = \tfrac{1}{2}[x_p(n) + x_p^*(N - n)]$$
$$x_{po}(n) = \tfrac{1}{2}[x_p(n) - x_p^*(N - n)] \tag{5.2.17}$$

Symmetry properties of the DFT.   The symmetry properties for the DFT can be obtained by applying the methodology previously used for the Fourier transform. Let us assume that the N-point sequence $x(n)$ and its DFT are both complex valued. Then the sequences can be expressed as

$$x(n) = x_R(n) + jx_I(n) \qquad 0 \le n \le N - 1 \tag{5.2.18}$$

$$X(k) = X_R(k) + jX_I(k) \qquad 0 \le k \le N - 1 \tag{5.2.19}$$

By substituting (5.2.18) into the expression for the DFT given by (5.2.1), we obtain

$$X_R(k) = \sum_{n=0}^{N-1} \left[x_R(n) \cos \frac{2\pi kn}{N} + x_I(n) \sin \frac{2\pi kn}{N} \right] \tag{5.2.20}$$

$$X_I(k) = -\sum_{n=0}^{N-1} \left[x_R(n) \sin \frac{2\pi kn}{N} - x_I(n) \cos \frac{2\pi kn}{N} \right] \tag{5.2.21}$$

Similarly, by substituting (5.2.19) into the expression for the IDFT given by (5.2.2), we obtain

$$x_R(n) = \frac{1}{N} \sum_{k=0}^{N-1} \left[X_R(k) \cos \frac{2\pi kn}{N} - X_I(k) \sin \frac{2\pi kn}{N} \right] \tag{5.2.22}$$

$$x_I(n) = \frac{1}{N} \sum_{k=0}^{N-1} \left[X_R(k) \sin \frac{2\pi kn}{N} + X_I(k) \cos \frac{2\pi kn}{N} \right] \tag{5.2.23}$$

Real-valued sequences.   If the sequence $x(n)$ is real, it follows directly from (5.2.1) that

$$X(N - k) = X^*(k) = X(-k) \tag{5.2.24}$$

Consequently, $|X(N - k)| = |X(k)|$ and $\angle X(N - k) = -\angle X(k)$. Furthermore, $x_I(n) = 0$ and therefore $x(n)$ can be determined from (5.2.22), which is another form for the IDFT.

Real and even sequences.   If $x(n)$ is real and even, that is,

$$x(n) = x(N - n) \qquad 0 \le n \le N - 1$$

then (5.2.21) yields $X_I(k) = 0$. Hence the DFT reduces to

$$X(k) = \sum_{n=0}^{N-1} x(n) \cos \frac{2\pi kn}{N} \qquad 0 \le k \le N - 1 \tag{5.2.25}$$

which is itself real-valued and even. Furthermore, since $X_I(k) = 0$, the IDFT reduces to

$$x(n) = \frac{1}{N} \sum_{k=0}^{N-1} X(k) \cos \frac{2\pi kn}{N} \qquad 0 \le n \le N - 1 \tag{5.2.26}$$

Real and odd sequences.   If $x(n)$ is real and odd, that is,

$$x(n) = -x(N - n) \qquad 0 \le n \le N - 1$$

then (5.2.20) yields $X_R(k) = 0$. Hence

$$X(k) = -j \sum_{n=0}^{N-1} x(n) \sin \frac{2\pi kn}{N} \qquad 0 \le k \le N - 1 \tag{5.2.27}$$

which is purely imaginary and odd. Since $X_R(k) = 0$, the IDFT reduces to

$$x(n) = j\frac{1}{N} \sum_{k=0}^{N-1} X(k) \sin \frac{2\pi kn}{N} \qquad 0 \le n \le N - 1 \tag{5.2.28}$$

Purely imaginary sequences.   In this case, $x(n) = jx_I(n)$. Consequently, (5.2.20) and (5.2.21) reduce to

$$X_R(k) = \sum_{n=0}^{N-1} x_I(n) \sin \frac{2\pi kn}{N} \tag{5.2.29}$$

$$X_I(k) = \sum_{n=0}^{N-1} x_I(n) \cos \frac{2\pi kn}{N} \tag{5.2.30}$$

We observe that $X_R(k)$ is odd and $X_I(k)$ is even.

If $x_I(n)$ is odd, then $X_I(k) = 0$ and hence $X(k)$ is purely real. On the other hand, if $x_I(n)$ is even, then $X_R(k) = 0$ and hence $X(k)$ is purely imaginary.

TABLE 5.1   SYMMETRY PROPERTIES OF THE DFT

N-Point Sequence $x(n)$, $0 \leq n \leq N - 1$	N-Point DFT
$x(n)$	$X(k)$
$x^*(n)$	$X^*(N - k)$
$x^*(N - n)$	$X^*(k)$
$x_R(n)$	$X_{ce}(k) = \frac{1}{2}[X(k) + X^*(N - k)]$
$jX_I(n)$	$X_{co}(k) = \frac{1}{2}[X(k) - X^*(N - k)]$
$x_{ce}(n) = \frac{1}{2}[x(n) + x^*(N - n)]$	$X_R(k)$
$x_{co}(n) = \frac{1}{2}[x(n) - x^*(N - n)]$	$jX_I(k)$

Real Signals	
Any real signal	$X(k) = X^*(N - k)$
$x(n)$	$X_R(k) = X_R(N - k)$
	$X_I(k) = -X_I(N - k)$
	$|X(k)| = |X(N - k)|$
	$\angle X(k) = -\angle X(N - k)$
$x_{ce}(n) = \frac{1}{2}[x(n) + x(N - n)]$	$X_R(k)$
$x_{co}(n) = \frac{1}{2}[x(n) - x(N - n)]$	$jX_I(k)$

The symmetry properties given above may be summarized as follows:

$$x(n) = x_R^e(n) + x_R^o(n) + jx_I^e(n) + jx_I^o(n)$$

$$X(k) = X_R^e(k) + X_R^o(k) + jX_I^e(k) + jX_I^o(k) \tag{5.2.31}$$

All the symmetry properties of the DFT can easily be deduced from (5.2.31). For example, the DFT of the sequence

$$x_{pe}(n) = \frac{1}{2}[x_p(n) + x_p^*(N - n)]$$

is

$$X_R(k) = X_R^e(k) + X_R^o(k)$$

The symmetry properties of the DFT are summarized in Table 5.1. Exploitation of these properties for the efficient computation of the DFT of special sequences is considered in some of the problems at the end of the chapter.

5.2.2 Multiplication of Two DFTs and Circular Convolution

Suppose that we have two finite-duration sequences of length N, $x_1(n)$ and $x_2(n)$. Their respective N-point DFTs are

$$X_1(k) = \sum_{n=0}^{N-1} x_1(n)e^{-j2\pi nk/N} \qquad k = 0, 1, \ldots, N - 1 \tag{5.2.32}$$

$$X_2(k) = \sum_{n=0}^{N-1} x_2(n)e^{-j2\pi nk/N} \qquad k = 0, 1, \ldots, N - 1 \tag{5.2.33}$$

If we multiply the two DFTs together, the result is a DFT, say $X_3(k)$, of a sequence $x_3(n)$ of length N. Let us determine the relationship between $x_3(n)$ and the sequences $x_1(n)$ and $x_2(n)$.

We have

$$X_3(k) = X_1(k)X_2(k) \qquad k = 0, 1, \ldots, N - 1 \qquad (5.2.34)$$

The IDFT of $\{X_3(k)\}$ is

$$x_3(m) = \frac{1}{N} \sum_{k=0}^{N-1} X_3(k) e^{j2\pi km/N}$$

$$\qquad (5.2.35)$$

$$= \frac{1}{N} \sum_{k=0}^{N-1} X_1(k)X_2(k) e^{j2\pi km/N}$$

Suppose that we substitute for $X_1(k)$ and $X_2(k)$ in (5.2.35) using the DFTs given in (5.2.32) and (5.2.33). Thus we obtain

$$x_3(m) = \frac{1}{N} \sum_{k=0}^{N-1} \left[\sum_{n=0}^{N-1} x_1(n) e^{-j2\pi kn/N} \right] \left[\sum_{l=0}^{N-1} x_2(l) e^{-j2\pi kl/N} \right] e^{j2\pi km/N}$$

$$\qquad (5.2.36)$$

$$= \frac{1}{N} \sum_{n=0}^{N-1} x_1(n) \sum_{l=0}^{N-1} x_2(l) \left[\sum_{k=0}^{N-1} e^{j2\pi k(m-n-l)/N} \right]$$

The inner sum in the brackets in (5.2.36) has the form

$$\sum_{k=0}^{N-1} a^k = \begin{cases} N, & a = 1 \\ \dfrac{1 - a^N}{1 - a}, & a \neq 1 \end{cases} \qquad (5.2.37)$$

where a is defined as

$$a = e^{j2\pi(m-n-l)/N}$$

We observe that $a = 1$ when $m - n - l$ is a multiple of N. On the other hand, $a^N = 1$ for any value of $a \neq 0$. Consequently, (5.2.37) reduces to

$$\sum_{k=0}^{N-1} a^k = \begin{cases} N, & l = m - n + pN = ((m - n))_N, \quad p \text{ an integer} \\ 0, & \text{otherwise} \end{cases} \qquad (5.2.38)$$

If we substitute the result in (5.2.38) into (5.2.36), we obtain the desired expression for $x_3(m)$ in the form

$$x_3(m) = \sum_{n=0}^{N-1} x_1(n)x_2((m - n))_N \qquad m = 0, 1, \ldots, N - 1 \qquad (5.2.39)$$

The expression in (5.2.39) has the form of a convolution sum. However, it is not the ordinary linear convolution that was introduced in Chapter 2, which relates the output sequence $y(n)$ of a linear system to the input sequence $x(n)$ and the impulse response $h(n)$. Instead, the convolution sum in (5.2.39) involves the index

$((m - n))_N$ and is called *circular convolution*. Thus we conclude that multiplication of the DFTs of two sequences is equivalent to the circular convolution of the two sequences in the time domain.

The following example illustrates the operations involved in circular convolution.

Example 5.2.1

Perform the circular convolution of the following two sequences:

$$x_1(n) = \{2, 1, 2, 1\}$$
$$\uparrow$$

$$x_2(n) = \{1, 2, 3, 4\}$$
$$\uparrow$$

Solution   Each sequence consists of four nonzero points. For the purposes of illustrating the operations involved in circular convolution, it is desirable to graph each sequence as points on a circle. Thus the sequences $x_1(n)$ and $x_2(n)$ are graphed as illustrated in Fig. 5.8(a). We note that the sequences are graphed in a counterclockwise direction on a circle. This establishes the reference direction in rotating one of the sequences relative to the other.

Now, $x_3(m)$ is obtained by circularly convolving $x_1(n)$ with $x_2(n)$ as specified by (5.2.39). Beginning with $m = 0$ we have

$$x_3(0) = \sum_{n=0}^{3} x_1(n) x_2((-n))_N$$

$x_2((-n))_4$ is simply the sequence $x_2(n)$ folded and graphed on a circle as illustrated in Fig. 5.8(b). In other words, the folded sequence is simply $x_2(n)$ graphed in a clockwise direction.

The product sequence is obtained by multiplying $x_1(n)$ with $x_2((-n))_4$, point by point. This sequence is also illustrated in Fig. 5.8(b). Finally, we sum the values in the product sequence to obtain

$$x_3(0) = 14$$

For $m = 1$ we have

$$x_3(1) = \sum_{n=0}^{3} x_1(n) x_2((1 - n))_4$$

It is easily verified that $x_2((1 - n))_4$ is simply the sequence $x_2((-n))_4$ rotated counterclockwise by one unit in time as illustrated in Fig. 5.8(c). This rotated sequence multiplies $x_1(n)$ to yield the product sequence, also illustrated in Fig. 5.8(c). Finally, we sum the values in the product sequence to obtain $x_3(1)$. Thus

$$x_3(1) = 16$$

For $m = 2$ we have

$$x_3(2) = \sum_{n=0}^{3} x_1(n) x_2((2 - n))_4$$

Now $x_2((2 - n))_4$ is the folded sequence in Fig. 5.8(b) rotated two units of time in the counterclockwise direction. The resultant sequence is illustrated in Fig. 5.8(d)

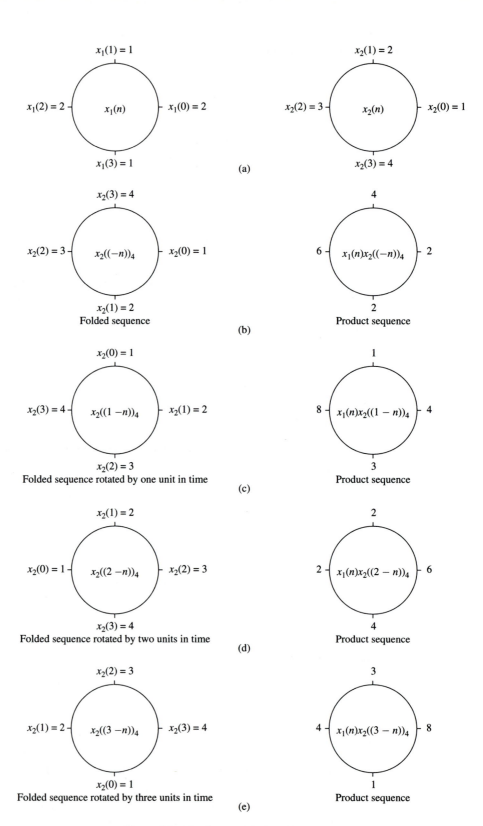

Figure 5.8 Circular convolution of two sequences.

along with the product sequence $x_1(n)x_2((2-n))_4$. By summing the four terms in the product sequence, we obtain

$$x_3(2) = 14$$

For $m = 3$ we have

$$x_3(3) = \sum_{n=0}^{3} x_1(n)x_2((3-n))_4$$

The folded sequence $x_2((-n))_4$ is now rotated by three units in time to yield $x_2((3-n))_4$ and the resultant sequence is multiplied by $x_1(n)$ to yield the product sequence as illustrated in Fig. 5.8(e). The sum of the values in the product sequence is

$$x_3(3) = 16$$

We observe that if the computation above is continued beyond $m = 3$, we simply repeat the sequence of four values obtained above. Therefore, the circular convolution of the two sequences $x_1(n)$ and $x_2(n)$ yields the sequence

$$x_3(n) = \{14, 16, 14, 16\}$$
$$\uparrow$$

From this example, we observe that circular convolution involves basically the same four steps as the ordinary *linear convolution* introduced in Chapter 2: *folding* (time reversing) one sequence, *shifting* the folded sequence, *multiplying* the two sequences to obtain a product sequence, and finally, *summing* the values of the product sequence. The basic difference between these two types of convolution is that, in circular convolution, the folding and shifting (rotating) operations are performed in a circular fashion by computing the index of one of the sequences modulo N. In linear convolution, there is no modulo N operation.

The reader can easily show from our previous development that either one of the two sequences may be folded and rotated without changing the result of the circular convolution. Thus

$$x_3(m) = \sum_{n=0}^{N-1} x_2(n)x_1((m-n))_N \qquad m = 0, 1, \ldots, N-1 \tag{5.2.40}$$

The following example serves to illustrate the computation of $x_3(n)$ by means of the DFT and IDFT.

Example 5.2.2

By means of the DFT and IDFT, determine the sequence $x_3(n)$ corresponding to the circular convolution of the sequences $x_1(n)$ and $x_2(n)$ given in Example 5.2.1.

Solution   First we compute the DFTs of $x_1(n)$ and $x_2(n)$. The four-point DFT of $x_1(n)$ is

$$X_1(k) = \sum_{n=0}^{3} x_1(n)e^{-j2\pi nk/4} \qquad k = 0, 1, 2, 3$$

$$= 2 + e^{-j\pi k/2} + 2e^{-j\pi k} + e^{-j3\pi k/2}$$

Thus

$$X_1(0) = 6 \qquad X_1(1) = 0 \qquad X_1(2) = 2 \qquad X_1(3) = 0$$

The DFT of $x_2(n)$ is

$$X_2(k) = \sum_{n=0}^{3} x_2(n)e^{-j2\pi nk/4} \qquad k = 0, 1, 2, 3$$

$$= 1 + 2e^{-j\pi k/2} + 3e^{-j\pi k} + 4e^{-j3\pi k/2}$$

Thus

$$X_2(0) = 10 \qquad X_2(1) = -2 + j2 \qquad X_2(2) = -2 \qquad X_2(3) = -2 - j2$$

When we multiply the two DFTs, we obtain the product

$$X_3(k) = X_1(k)X_2(k)$$

or, equivalently,

$$X_3(0) = 60 \qquad X_3(1) = 0 \qquad X_3(2) = -4 \qquad X_3(3) = 0$$

Now, the IDFT of $X_3(k)$ is

$$x_3(n) = \sum_{k=0}^{3} X_3(k)e^{j2\pi nk/4} \qquad n = 0, 1, 2, 3$$

$$= \tfrac{1}{4}(60 - 4e^{j\pi n})$$

Thus

$$x_3(0) = 14 \qquad x_3(1) = 16 \qquad x_3(2) = 14 \qquad x_3(3) = 16$$

which is the result obtained in Example 5.2.1 from circular convolution.

We conclude this section by formally stating this important property of the DFT.

Circular convolution.   If

$$x_1(n) \xleftrightarrow[N]{\text{DFT}} X_1(k)$$

and

$$x_2(n) \xleftrightarrow[N]{\text{DFT}} X_2(k)$$

then

$$x_1(n) \, \text{\textcircled{N}} \, x_2(n) \xleftrightarrow[N]{\text{DFT}} X_1(k)X_2(k) \qquad (5.2.41)$$

where $x_1(n) \, \text{\textcircled{N}} \, x_2(n)$ denotes the circular convolution of the sequence $x_1(n)$ and $x_2(n)$.

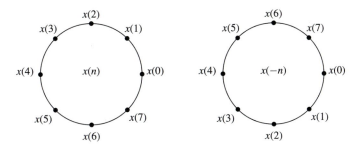

Figure 5.9   Time reversal of a sequence.

5.2.3 Additional DFT Properties

Time reversal of a sequence.   If

$$x(n) \overset{\text{DFT}}{\underset{N}{\longleftrightarrow}} X(k)$$

then

$$x((-n))_N = x(N-n) \overset{\text{DFT}}{\underset{N}{\longleftrightarrow}} X((-k))_N = X(N-k) \qquad (5.2.42)$$

Hence reversing the N-point sequence in time is equivalent to reversing the DFT values. Time reversal of a sequence $x(n)$ is illustrated in Fig. 5.9.

Proof. From the definition of the DFT in (5.2.1) we have

$$\text{DFT}\{x(N-n)\} = \sum_{n=0}^{N-1} x(N-n)e^{-j2\pi kn/N}$$

If we change the index from n to $m = N - n$, then

$$\text{DFT}\{x(N-n)\} = \sum_{m=0}^{N-1} x(m)e^{-j2\pi k(N-m)/N}$$

$$= \sum_{m=0}^{N-1} x(m)e^{j2\pi km/N}$$

$$= \sum_{m=0}^{N-1} x(m)e^{-j2\pi m(N-k)/N} = X(N-k)$$

We note that $X(N-k) = X((-k))_N$, $0 \le k \le N-1$.

Circular time shift of a sequence.   If

$$x(n) \overset{\text{DFT}}{\underset{N}{\longleftrightarrow}} X(k)$$

then

$$x((n-l))_N \xleftrightarrow[N]{\text{DFT}} X(k)e^{-j2\pi kl/N} \qquad (5.2.43)$$

Proof. From the definition of the DFT we have

$$\text{DFT}\{x((n-l))_N\} = \sum_{n=0}^{N-1} x((n-l))_N e^{-j2\pi kn/N}$$

$$= \sum_{n=0}^{l-1} x((n-l))_N e^{-j2\pi kn/N}$$

$$+ \sum_{n=l}^{N-1} x(n-l)e^{-j\pi kn/N}$$

But $x((n-l))_N = x(N-l+n)$. Consequently,

$$\sum_{n=0}^{l-1} x((n-l))_N e^{-j2\pi kn/N} = \sum_{n=0}^{l-1} x(N-l+n)e^{-j2\pi kn/N}$$

$$= \sum_{m=N-l}^{N-1} x(m)e^{-j2\pi k(m+l)/N}$$

Furthermore,

$$\sum_{n=l}^{N-1} x(n-l)e^{-j2\pi kn/N} = \sum_{m=0}^{N-1-l} x(m)e^{-j2\pi k(m+l)/N}$$

Therefore,

$$\text{DFT}\{x((n-l))\} = \sum_{m=0}^{N-1} x(m)e^{-j2\pi k(m+l)/N}$$

$$= X(k)e^{-j2\pi kl/N}$$

Circular frequency shift.   If

$$x(n) \xleftrightarrow[N]{\text{DFT}} X(k)$$

then

$$x(n)e^{j2\pi ln/N} \xleftrightarrow[N]{\text{DFT}} X((k-l))_N \qquad (5.2.44)$$

Hence, the multiplication of the sequence $x(n)$ with the complex exponential sequence $e^{j2\pi kn/N}$ is equivalent to the circular shift of the DFT by l units in frequency. This is the dual to the circular time-shifting property and its proof is similar to the latter.

Complex-conjugate properties.   If

$$x(n) \overset{\text{DFT}}{\underset{N}{\longleftrightarrow}} X(k)$$

then

$$x^*(n) \overset{\text{DFT}}{\underset{N}{\longleftrightarrow}} X^*((-k))_N = X^*(N - k) \qquad (5.2.45)$$

The proof of this property is left as an exercise for the reader. The IDFT of $X^*(k)$ is

$$\frac{1}{N} \sum_{k=0}^{N-1} X^*(k) e^{j2\pi kn/N} = \left[\frac{1}{N} \sum_{k=0}^{N-1} X(k) e^{j2\pi k(N-n)/N} \right]^*$$

Therefore,

$$x^*((-n))_N = x^*(N - n) \overset{\text{DFT}}{\underset{N}{\longleftrightarrow}} X^*(k) \qquad (5.2.46)$$

Circular correlation.   In general, for complex-valued sequences $x(n)$ and $y(n)$, if

$$x(n) \overset{\text{DFT}}{\underset{N}{\longleftrightarrow}} X(k)$$

and

$$y(n) \overset{\text{DFT}}{\underset{N}{\longleftrightarrow}} Y(k)$$

then

$$\tilde{r}_{xy}(l) \overset{\text{DFT}}{\underset{N}{\longleftrightarrow}} \tilde{R}_{xy}(k) = X(k) Y^*(k) \qquad (5.2.47)$$

where $\tilde{r}_{xy}(l)$ is the (unnormalized) circular crosscorrelation sequence, defined as

$$\tilde{r}_{xy}(l) = \sum_{n=0}^{N-1} x(n) y^*((n - l))_N$$

Proof. We can write $\tilde{r}_{xy}(l)$ as the circular convolution of $x(n)$ with $y^*(-n)$, that is,

$$\tilde{r}_{xy}(l) = x(l) \; \text{Ⓝ} \; y^*(-l)$$

Then, with the aid of the properties in (5.2.41) and (5.2.46), the N-point DFT of $\tilde{r}_{xy}(l)$ is

$$\tilde{R}_{xy}(k) = X(k) Y^*(k)$$

In the special case where $y(n) = x(n)$, we have the corresponding expression for the circular autocorrelation of $x(n)$,

$$\tilde{r}_{xx}(l) \overset{\text{DFT}}{\underset{N}{\longleftrightarrow}} \tilde{R}_{xx}(k) = |X(k)|^2 \qquad (5.2.48)$$

Multiplication of two sequences. If

$$x_1(n) \overset{\text{DFT}}{\underset{N}{\longleftrightarrow}} X_1(k)$$

and

$$x_2(n) \overset{\text{DFT}}{\underset{N}{\longleftrightarrow}} X_2(k)$$

then

$$x_1(n)x_2(n) \overset{\text{DFT}}{\underset{N}{\longleftrightarrow}} \frac{1}{N}X_1(k) \ \text{N}\ X_2(k) \qquad (5.2.49)$$

This property is the dual of (5.2.41). Its proof follows simply by interchanging the roles of time and frequency in the expression for the circular convolution of two sequences.

Parseval's theorem. For complex-valued sequences $x(n)$ and $y(n)$, in general, if

$$x(n) \overset{\text{DFT}}{\underset{N}{\longleftrightarrow}} X(k)$$

and

$$y(n) \overset{\text{DFT}}{\underset{N}{\longleftrightarrow}} Y(k)$$

then

$$\sum_{n=0}^{N-1} x(n)y^*(n) = \frac{1}{N} \sum_{k=0}^{N-1} X(k)Y^*(k) \qquad (5.2.50)$$

Proof. The property follows immediately from the circular correlation property in (5.2.47). We have

$$\sum_{n=0}^{N-1} x(n)y^*(n) = \tilde{r}_{xy}(0)$$

and

$$\tilde{r}_{xy}(l) = \frac{1}{N} \sum_{k=0}^{N-1} \tilde{R}_{xy}(k)e^{j2\pi kl/N}$$

$$= \frac{1}{N} \sum_{k=0}^{N-1} X(k)Y^*(k)e^{j2\pi kl/N}$$

Hence (5.2.50) follows by evaluating the IDFT at $l = 0$.

The expression in (5.2.50) is the general form of Parseval's theorem. In the special case where $y(n) = x(n)$, (5.2.50) reduces to

$$\sum_{n=0}^{N-1} |x(n)|^2 = \frac{1}{N} \sum_{k=0}^{N-1} |X(k)|^2 \qquad (5.2.51)$$

TABLE 5.2   PROPERTIES OF THE DFT

Property	Time Domain	Frequency Domain
Notation	$x(n), y(n)$	$X(k), Y(k)$
Periodicity	$x(n) = x(n + N)$	$X(k) = X(k + N)$
Linearity	$a_1 x_1(n) + a_2 x_2(n)$	$a_1 X_1(k) + a_2 X_2(k)$
Time reversal	$x(N - n)$	$X(N - k)$
Circular time shift	$x((n - l))_N$	$X(k)e^{-j2\pi kl/N}$
Circular frequency shift	$x(n)e^{j2\pi ln/N}$	$X((k - l))_N$
Complex conjugate	$x^*(n)$	$X^*(N - k)$
Circular convolution	$x_1(n) \,Ⓝ\, x_2(n)$	$X_1(k)X_2(k)$
Circular correlation	$x(n) \,Ⓝ\, y^*(-n)$	$X(k)Y^*(k)$
Multiplication of two sequences	$x_1(n)x_2(n)$	$\dfrac{1}{N}X_1(k) \,Ⓝ\, X_2(k)$
Parseval's theorem	$\displaystyle\sum_{n=0}^{N-1} x(n)y^*(n)$	$\dfrac{1}{N}\displaystyle\sum_{k=0}^{N-1} X(k)Y^*(k)$

which expresses the energy in the finite-duration sequence $x(n)$ in terms of the frequency components $\{X(k)\}$.

The properties of the DFT given above are summarized in Table 5.2.

5.3 LINEAR FILTERING METHODS BASED ON THE DFT

Since the DFT provides a discrete frequency representation of a finite-duration sequence in the frequency domain, it is interesting to explore its use as a computational tool for linear system analysis and, especially, for linear filtering. We have already established that a system with frequency response $H(\omega)$, when excited with an input signal that has a spectrum $X(\omega)$, possesses an output spectrum $Y(\omega) = X(\omega)H(\omega)$. The output sequence $y(n)$ is determined from its spectrum via the inverse Fourier transform. Computationally, the problem with this frequency-domain approach is that $X(\omega)$, $H(\omega)$, and $Y(\omega)$ are functions of the continuous variable ω. As a consequence, the computations cannot be done on a digital computer, since the computer can only store and perform computations on quantities at discrete frequencies.

On the other hand, the DFT does lend itself to computation on a digital computer. In the discussion that follows, we describe how the DFT can be used to perform linear filtering in the frequency domain. In particular, we present a computational procedure that serves as an alternative to time-domain convolution. In fact, the frequency-domain approach based on the DFT, is computationally more efficient than time-domain convolution due to the existence of efficient algorithms for computing the DFT. These algorithms, which are described in Chapter 6, are collectively called fast Fourier transform (FFT) algorithms.

5.3.1 Use of the DFT in Linear Filtering

In the preceding section it was demonstrated that the product of two DFTs is equivalent to the circular convolution of the corresponding time-domain sequences. Unfortunately, circular convolution is of no use to us if our objective is to determine the output of a linear filter to a given input sequence. In this case we seek a frequency-domain methodology equivalent to linear convolution.

Suppose that we have a finite-duration sequence $x(n)$ of length L which excites an FIR filter of length M. Without loss of generality, let

$$x(n) = 0, \qquad n < 0 \text{ and } n \geq L$$

$$h(n) = 0, \qquad n < 0 \text{ and } n \geq M$$

where $h(n)$ is the impulse response of the FIR filter.

The output sequence $y(n)$ of the FIR filter can be expressed in the time domain as the convolution of $x(n)$ and $h(n)$, that is

$$y(n) = \sum_{k=0}^{M-1} h(k)x(n-k) \tag{5.3.1}$$

Since $h(n)$ and $x(n)$ are finite-duration sequences, their convolution is also finite in duration. In fact, the duration of $y(n)$ is $L + M - 1$.

The frequency-domain equivalent to (5.3.1) is

$$Y(\omega) = X(\omega)H(\omega) \tag{5.3.2}$$

If the sequence $y(n)$ is to be represented uniquely in the frequency domain by samples of its spectrum $Y(\omega)$ at a set of discrete frequencies, the number of distinct samples must equal or exceed $L + M - 1$. Therefore, a DFT of size $N \geq L + M - 1$, is required to represent $\{y(n)\}$ in the frequency domain.

Now if

$$Y(k) \equiv Y(\omega)|_{\omega=2\pi k/N} \qquad k = 0, 1, \ldots, N-1$$

$$= X(\omega)H(\omega)|_{\omega=2\pi k/N} \qquad k = 0, 1, \ldots, N-1$$

then

$$Y(k) = X(k)H(k) \qquad k = 0, 1, \ldots, N-1 \tag{5.3.3}$$

where $\{X(k)\}$ and $\{H(k)\}$ are the N-point DFTs of the corresponding sequences $x(n)$ and $h(n)$, respectively. Since the sequences $x(n)$ and $h(n)$ have a duration less than N, we simply pad these sequences with zeros to increase their length to N. This increase in the size of the sequences does not alter their spectra $X(\omega)$ and $H(\omega)$, which are continuous spectra, since the sequences are aperiodic. However, by sampling their spectra at N equally spaced points in frequency (computing the N-point DFTs), we have increased the number of samples that represent these sequences in the frequency domain beyond the minimum number (L or M, respectively).

Since the $N = L + M - 1$-point DFT of the output sequence $y(n)$ is sufficient to represent $y(n)$ in the frequency domain, it follows that the multiplication of the N-point DFTs $X(k)$ and $H(k)$, according to (5.3.3), followed by the computation of the N-point IDFT, must yield the sequence $\{y(n)\}$. In turn, this implies that the N-point circular convolution of $x(n)$ with $h(n)$ must be equivalent to the linear convolution of $x(n)$ with $h(n)$. In other words, by increasing the length of the sequences $x(n)$ and $h(n)$ to N points (by appending zeros), and then circularly convolving the resulting sequences, we obtain the same result as would have been obtained with linear convolution. Thus with zero padding, the DFT can be used to perform linear filtering.

The following example illustrates the methodology in the use of the DFT in linear filtering.

Example 5.3.1

By means of the DFT and IDFT, determine the response of the FIR filter with impulse response

$$h(n) = \{1, 2, 3\}$$
$$\uparrow$$

to the input sequence

$$x(n) = \{1, 2, 2, 1\}$$
$$\uparrow$$

Solution   The input sequence has length $L = 4$ and the impulse response has length $M = 3$. Linear convolution of these two sequences produces a sequence of length $N = 6$. Consequently, the size of the DFTs must be at least six.

For simplicity we compute eight-point DFTs. We should also mention that the efficient computation of the DFT via the fast Fourier transform (FFT) algorithm is usually performed for a length N that is a power of 2. Hence the eight-point DFT of $x(n)$ is

$$X(k) = \sum_{n=0}^{7} x(n) e^{-j2\pi kn/8}$$

$$= 1 + 2e^{-j\pi k/4} + 2e^{-j\pi k/2} + e^{-j3\pi k/4} \qquad k = 0, 1, \ldots, 7$$

This computation yields

$$X(0) = 6 \qquad\qquad X(1) = \frac{2 + \sqrt{2}}{2} - j\left(\frac{4 + 3\sqrt{2}}{2}\right)$$

$$X(2) = -1 - j \qquad X(3) = \frac{2 - \sqrt{2}}{2} + j\left(\frac{4 - 3\sqrt{2}}{2}\right)$$

$$X(4) = 0 \qquad\qquad X(5) = \frac{2 - \sqrt{2}}{2} - j\frac{4 - 3\sqrt{2}}{2}$$

$$X(6) = -1 + j \qquad X(7) = \frac{2 + \sqrt{2}}{2} + j\left(\frac{4 + 3\sqrt{2}}{2}\right)$$

The eight-point DFT of $h(n)$ is

$$H(k) = \sum_{n=0}^{7} h(n)e^{-j2\pi kn/8}$$

$$= 1 + 2e^{-j\pi k/4} + 3e^{-j\pi k/2}$$

Hence

$$H(0) = 6, \qquad H(1) = 1 + \sqrt{2} - j\left(3 + \sqrt{2}\right), \qquad H(2) = -2 - j2$$

$$H(3) = 1 - \sqrt{2} + j\left(3 - \sqrt{2}\right), \qquad H(4) = 2$$

$$H(5) = 1 - \sqrt{2} - j\left(3 - \sqrt{2}\right), \qquad H(6) = -2 + j2$$

$$H(7) = 1 + \sqrt{2} + j\left(3 + \sqrt{2}\right)$$

The product of these two DFTs yields $Y(k)$, which is

$$Y(0) = 36, \qquad Y(1) = -14.07 - j17.48 \qquad Y(2) = j4 \qquad Y(3) = 0.07 + j0.515$$

$$Y(4) = 0, \qquad Y(5) = 0.07 - j0.515 \qquad Y(6) = -j4 \qquad Y(7) = -14.07 + j17.48$$

Finally, the eight-point IDFT is

$$y(n) = \sum_{k=0}^{7} Y(k)e^{j2\pi kn/8} \qquad n = 0, 1, \dots, 7$$

This computation yields the result

$$y(n) = \{1, 4, 9, 11, 8, 3, 0, 0\}$$
$$\uparrow$$

We observe that the first six values of $y(n)$ constitute the set of desired output values. The last two values are zero because we used an eight-point DFT and IDFT, when, in fact, the minimum number of points required is six.

Although the multiplication of two DFTs corresponds to circular convolution in the time domain, we have observed that padding the sequences $x(n)$ and $h(n)$ with a sufficient number of zeros forces the circular convolution to yield the same output sequence as linear convolution. In the case of the FIR filtering problem in Example 5.3.1, it is a simple matter to demonstrate that the six-point circular convolution of the sequences

$$h(n) = \{1, 2, 3, 0, 0, 0\} \qquad\qquad (5.3.4)$$
$$\uparrow$$

$$x(n) = \{1, 2, 2, 1, 0, 0\} \qquad\qquad (5.3.5)$$
$$\uparrow$$

results in the output sequence

$$y(n) = \{1, 4, 9, 11, 8, 3\} \qquad\qquad (5.3.6)$$
$$\uparrow$$

which is the same sequence obtained from linear convolution.

It is important for us to understand the aliasing that results in the time domain when the size of the DFTs is smaller than $L+M-1$. The following example focuses on the aliasing problem.

Example 5.3.2

Determine the sequence $y(n)$ that results from the use of four point DFTs in Example 5.3.1.

Solution   The four-point DFT of $h(n)$ is

$$H(k) = \sum_{n=0}^{3} h(n)e^{-j2\pi kn/4}$$

$$H(k) = 1 + 2e^{-j\pi k/2} + 3e^{-jk\pi} \qquad k = 0, 1, 2, 3$$

Hence

$$H(0) = 6, \qquad H(1) = -2 - j2, \qquad H(2) = 2, \qquad H(3) = -2 + j2$$

The four-point DFT of $x(n)$ is

$$X(k) = 1 + 2e^{-j\pi k/2} + 2e^{-j\pi k} + 3e^{-j3\pi k/2} \qquad k = 0, 1, 2, 3$$

Hence

$$X(0) = 6, \qquad X(1) = -1 - j, \qquad X(2) = 0, \qquad X(3) = -1 + j$$

The product of these two four-point DFTs is

$$\hat{Y}(0) = 36, \qquad \hat{Y}(1) = j4, \qquad \hat{Y}(2) = 0, \qquad \hat{Y}(3) = -j4$$

The four-point IDFT yields

$$\hat{y}(n) = \frac{1}{4} \sum_{k=0}^{3} \hat{Y}(k)e^{j2\pi kn/4} \qquad n = 0, 1, 2, 3$$

$$= \frac{1}{4}(36 + j4e^{j\pi n/2} - j4e^{j3\pi n/2})$$

Therefore,

$$\hat{y}(n) = \{9, 7, 9, 11\}$$
$$\uparrow$$

The reader can verify that the four-point circular convolution of $h(n)$ with $x(n)$ yields the same sequence $\hat{y}(n)$.

If we compare the result $\hat{y}(n)$, obtained from four-point DFTs with the sequence $y(n)$ obtained from the use of eight-point (or six-point) DFTs, the time-domain aliasing effects derived in Section 5.2.2 are clearly evident. In particular, $y(4)$ is aliased into $y(0)$ to yield

$$\hat{y}(0) = y(0) + y(4) = 9$$

Similarly, $y(5)$ is aliased into $y(1)$ to yield

$$\hat{y}(1) = y(1) + y(5) = 7$$

All other aliasing has no effect since $y(n) = 0$ for $n \geq 6$. Consequently, we have

$$\hat{y}(2) = y(2) = 9$$

$$\hat{y}(3) = y(3) = 11$$

Therefore, only the first two points of $\hat{y}(n)$ are corrupted by the effect of aliasing [i.e., $\hat{y}(0) \neq y(0)$ and $\hat{y}(1) \neq y(1)$]. This observation has important ramifications in the discussion of the following section, in which we treat the filtering of long sequences.

5.3.2 Filtering of Long Data Sequences

In practical applications involving linear filtering of signals, the input sequence $x(n)$ is often a very long sequence. This is especially true in some real-time signal processing applications concerned with signal monitoring and analysis.

Since linear filtering performed via the DFT involves operations on a block of data, which by necessity must be limited in size due to limited memory of a digital computer, a long input signal sequence must be segmented to fixed-size blocks prior to processing. Since the filtering is linear, successive blocks can be processed one at a time via the DFT and the output blocks are fitted together to form the overall output signal sequence.

We now describe two methods for linear FIR filtering a long sequence on a block-by-block basis using the DFT. The input sequence is segmented into blocks and each block is processed via the DFT and IDFT to produce a block of output data. The output blocks are fitted together to form an overall output sequence which is identical to the sequence obtained if the long block had been processed via time-domain convolution.

The two methods are called the *overlap-save method* and the *overlap-add method*. For both methods we assume that the FIR filter has duration M. The input data sequence is segmented into blocks of L points, where, by assumption, $L >> M$ without loss of generality.

Overlap-save method.   In this method the size of the input data blocks is $N = L + M - 1$ and the size of the DFTs and IDFT are of length N. Each data block consists of the last $M - 1$ data points of the previous data block followed by L new data points to form a data sequence of length $N = L + M - 1$. An N-point DFT is computed for each data block. The impulse response of the FIR filter is increased in length by appending $L - 1$ zeros and an N-point DFT of the sequence is computed once and stored. The multiplication of the two N-point DFTs $\{H(k)\}$ and $\{X_m(k)\}$ for the mth block of data yields

$$\hat{Y}_m(k) = H(k)X_m(k) \qquad k = 0, 1, \ldots, N - 1 \tag{5.3.7}$$

Then the N-point IDFT yields the result

$$\hat{Y}_m(n) = \{\hat{y}_m(0)\hat{y}_m(1) \cdots \hat{y}_m(M - 1)\hat{y}_m(M) \cdots \hat{y}_m(N - 1)\} \tag{5.3.8}$$

Since the data record is of length N, the first $M - 1$ points of $y_m(n)$ are corrupted by aliasing and must be discarded. The last L points of $y_m(n)$ are exactly the same as the result from linear convolution and, as a consequence,

$$\hat{y}_m(n) = y_m(n), n = M, M + 1, \ldots, N - 1 \qquad (5.3.9)$$

To avoid loss of data due to aliasing, the last $M - 1$ points of each data record are saved and these points become the first $M - 1$ data points of the subsequent record, as indicated above. To begin the processing, the first $M - 1$ points of the first record are set to zero. Thus the blocks of data sequences are

$$x_1(n) = \underbrace{\{0, 0, \ldots, 0,}_{M-1 \text{ points}} x(0), x(1), \ldots, x(L - 1)\} \qquad (5.3.10)$$

$$x_2(n) = \{\underbrace{x(L - M + 1), \ldots, x(L - 1),}_{\substack{M-1 \text{ data points} \\ \text{from } x_1(n)}} \underbrace{x(L), \ldots, x(2L - 1)}_{L \text{ new data points}}\} \qquad (5.3.11)$$

$$x_3(n) = \{\underbrace{x(2L - M + 1), \ldots, x(2L - 1),}_{\substack{M-1 \text{ data points} \\ \text{from } x_2(n)}} \underbrace{x(2L), \ldots, x(3L - 1)}_{L \text{ new data points}}\} \qquad (5.3.12)$$

and so forth. The resulting data sequences from the IDFT are given by (5.3.8), where the first $M - 1$ points are discarded due to aliasing and the remaining L points constitute the desired result from linear convolution. This segmentation of the input data and the fitting of the output data blocks together to form the output sequence are graphically illustrated in Fig. 5.10.

Overlap-add method.   In this method the size of the input data block is L points and the size of the DFTs and IDFT is $N = L + M - 1$. To each data block we append $M - 1$ zeros and compute the N-point DFT. Thus the data blocks may be represented as

$$x_1(n) = \{x(0), x(1), \ldots, x(L - 1), \underbrace{0, 0, \ldots, 0}_{M-1 \text{ zeros}}\} \qquad (5.3.13)$$

$$x_2(n) = \{x(L), x(L + 1), \ldots, x(2L - 1), \underbrace{0, 0, \ldots, 0}_{M-1 \text{ zeros}}\} \qquad (5.3.14)$$

$$x_3(n) = \{x(2L), \ldots, x(3L - 1), \underbrace{0, 0, \ldots, 0}_{M-1 \text{ zeros}}\} \qquad (5.3.15)$$

and so on. The two N-point DFTs are multiplied together to form

$$Y_m(k) = H(k)X_m(k) \qquad k = 0, 1, \ldots, N - 1 \qquad (5.3.16)$$

The IDFT yields data blocks of length N that are free of aliasing since the size of the DFTs and IDFT is $N = L + M - 1$ and the sequences are increased to N-points by appending zeros to each block.

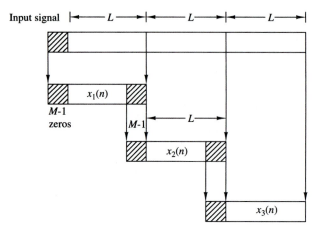

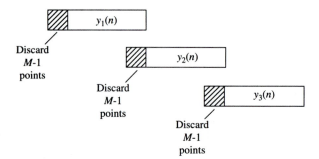

Figure 5.10   Linear FIR filtering by the overlap-save method.

Since each data block is terminated with $M - 1$ zeros, the last $M - 1$ points from each output block must be overlapped and added to the first $M - 1$ points of the succeeding block. Hence this method is called the overlap-add method. This overlapping and adding yields the output sequence

$$y(n) = \{y_1(0), y_1(1), \ldots, y_1(L - 1), y_1(L) + y_2(0), y_1(L + 1) +$$
$$y_2(1), \ldots, y_1(N - 1) + y_2(M - 1), y_2(M), \ldots\}$$

(5.3.17)

The segmentation of the input data into blocks and the fitting of the output data blocks to form the output sequence are graphically illustrated in Fig. 5.11.

At this point, it may appear to the reader that the use of the DFT in linear FIR filtering is not only an indirect method of computing the output of an FIR filter, but it may also be more expensive computationally since the input data must first be converted to the frequency domain via the DFT, multiplied by the DFT of the FIR filter, and finally, converted back to the time domain via the IDFT. On the contrary, however, by using the fast Fourier transform algorithm, as will be shown in Chapter 6, the DFTs and IDFT require fewer computations to compute the output sequence than the direct realization of the FIR filter in the time

Input data

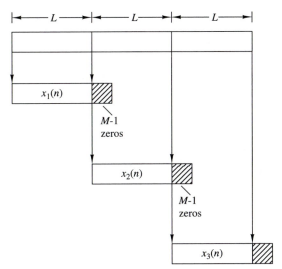

Output data

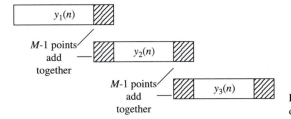

Figure 5.11 Linear FIR filtering by the overlap-add method.

domain. This computational efficiency is the basic advantage of using the DFT to compute the output of an FIR filter.

5.4 FREQUENCY ANALYSIS OF SIGNALS USING THE DFT

To compute the spectrum of either a continuous-time or discrete-time signal, the values of the signal for all time are required. However, in practice, we observe signals for only a finite duration. Consequently, the spectrum of a signal can only be approximated from a finite data record. In this section we examine the implications of a finite data record in frequency analysis using the DFT.

If the signal to be analyzed is an analog signal, we would first pass it through an antialiasing filter and then sample it at a rate $F_s \geq 2B$, where B is the bandwidth of the filtered signal. Thus the highest frequency that is contained in the sampled signal is $F_s/2$. Finally, for practical purposes, we limit the duration of the signal to the time interval $T_0 = LT$, where L is the number of samples and T

is the sample interval. As we shall observe in the following discussion, the finite observation interval for the signal places a limit on the frequency resolution; that is, it limits our ability to distinguish two frequency components that are separated by less than $1/T_0 = 1/LT$ in frequency.

Let $\{x(n)\}$ denote the sequence to be analyzed. Limiting the duration of the sequence to L samples, in the interval $0 \le n \le L - 1$, is equivalent to multiplying $\{x(n)\}$ by a rectangular window $w(n)$ of length L. That is,

$$\hat{x}(n) = x(n)w(n) \tag{5.4.1}$$

where

$$w(n) = \begin{cases} 1, & 0 \le n \le L - 1 \\ 0, & \text{otherwise} \end{cases} \tag{5.4.2}$$

Now suppose that the sequence $x(n)$ consists of a single sinusoid, that is,

$$x(n) = \cos \omega_0 n \tag{5.4.3}$$

Then the Fourier transform of the finite-duration sequence $x(n)$ can be expressed as

$$\hat{X}(\omega) = \tfrac{1}{2}[W(\omega - \omega_0) + W(\omega + \omega_0)] \tag{5.4.4}$$

where $W(\omega)$ is the Fourier transform of the window sequence, which is (for the rectangular window)

$$W(\omega) = \frac{\sin(\omega L/2)}{\sin(\omega/2)} e^{-j\omega(L-1)/2} \tag{5.4.5}$$

To compute $\hat{X}(\omega)$ we use the DFT. By padding the sequence $\hat{x}(n)$ with $N - L$ zeros, we can compute the N-point DFT of the truncated (L points) sequence $\{\hat{x}(n)\}$. The magnitude spectrum $|\hat{X}(k)| = |\hat{X}(\omega_k)|$ for $\omega_k = 2\pi k/N$, $k = 0, 1, \dots, N$, is illustrated in Fig. 5.12 for $L = 25$ and $N = 2048$. We note that the windowed spectrum $\hat{X}(\omega)$ is not localized to a single frequency, but instead it is spread out over the whole frequency range. Thus the power of the original signal sequence $\{x(n)\}$ that was concentrated at a single frequency has been spread by the window into the entire frequency range. We say that the power has "leaked out" into the entire frequency range. Consequently, this phenomenon, which is a characteristic of windowing the signal, is called *leakage*.

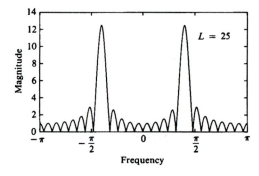

Figure 5.12 Magnitude spectrum for $L = 25$ and $n = 2048$, illustrating the occurrence of leakage.

Windowing not only distorts the spectral estimate due to the leakage effects, it also reduces spectral resolution. To illustrate this problem, let us consider a signal sequence consisting of two frequency components,

$$x(n) = \cos \omega_1 n + \cos \omega_2 n \tag{5.4.6}$$

When this sequence is truncated to L samples in the range $0 \le n \le L - 1$, the windowed spectrum is

$$\hat{X}(\omega) = \tfrac{1}{2}[W(\omega - \omega_1) + W(\omega - \omega_2) + W(\omega + \omega_1) + W(\omega + \omega_2)] \tag{5.4.7}$$

The spectrum $W(\omega)$ of the rectangular window sequence has its first zero crossing at $\omega = 2\pi/L$. Now if $|\omega_1 - \omega_2| < 2\pi/L$, the two window functions $W(\omega - \omega_1)$ and $W(\omega - \omega_2)$ overlap and, as a consequence, the two spectral lines in $x(n)$ are not distinguishable. Only if $(\omega_1 - \omega_2) \ge 2\pi/L$ will we see two separate lobes in the spectrum $\hat{X}(\omega)$. Thus our ability to resolve spectral lines of different frequencies is limited by the window main lobe width. Figure 5.13 illustrates the magnitude spectrum $|\hat{X}(\omega)|$, computed via the DFT, for the sequence

$$x(n) = \cos \omega_0 n + \cos \omega_1 n + \cos \omega_2 n \tag{5.4.8}$$

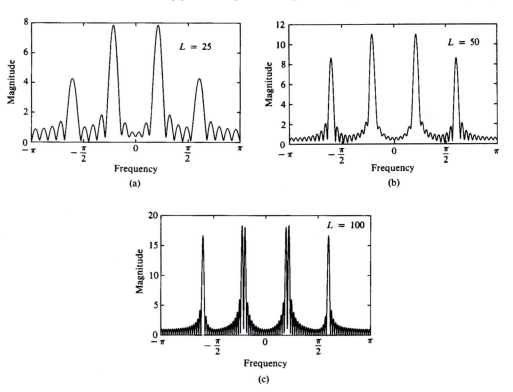

Figure 5.13 Magnitude spectrum for the signal given by (5.4.8), as observed through a rectangular window.

where $\omega_0 = 0.2\pi$, $\omega_1 = 0.22\pi$, and $\omega_2 = 0.6\pi$. The window lengths selected are $L = 25$, 50, and 100. Note that ω_0 and ω_1 are not resolvable for $L = 25$ and 50, but they are resolvable for $L = 100$.

To reduce leakage, we can select a data window $w(n)$ that has lower sidelobes in the frequency domain compared with the rectangular window. However, as we describe in more detail in Chapter 8, a reduction of the sidelobes in a window $W(\omega)$ is obtained at the expense of an increase in the width of the main lobe of $W(\omega)$ and hence a loss in resolution. To illustrate this point, let us consider the Hanning window, which is specified as

$$w(n) = \begin{cases} \frac{1}{2}(1 - \cos \frac{2\pi}{L-1}n), & 0 \le n \le L - 1 \\ 0, & \text{otherwise} \end{cases} \tag{5.4.9}$$

Figure 5.14 shows $|\hat{X}(\omega)|$ for the window of (5.4.9). Its sidelobes are significantly smaller than those of the rectangular window, but its main lobe is approximately twice as wide. Figure 5.15 shows the spectrum of the signal in (5.4.8), after it is windowed by the Hanning window, for $L = 50$, 75, and 100. The reduction of the sidelobes and the decrease in the resolution, compared with the rectangular window, is clearly evident.

For a general signal sequence $\{x(n)\}$, the frequency-domain relationship between the windowed sequence $\hat{x}(n)$ and the original sequence $x(n)$ is given by the convolution formula

$$\hat{X}(\omega) = \frac{1}{2\pi} \int_{-\pi}^{\pi} X(\theta)W(\omega - \theta)d\theta \tag{5.4.10}$$

The DFT of the windowed sequence $\hat{x}(n)$ is the sampled version of the spectrum $X(\omega)$. Thus we have

$$\hat{X}(k) \equiv \hat{X}(\omega)|_{\omega=2\pi k/N}$$
$$= \frac{1}{2\pi} \int_{-\pi}^{\pi} X(\theta)W\left(\frac{2\pi k}{N} - \theta\right)d\theta \qquad k = 0, 1, \ldots, N - 1 \tag{5.4.11}$$

Just as in the case of the sinusoidal sequence, if the spectrum of the window is relatively narrow in width compared to the spectrum $X(\omega)$ of the signal, the window function has only a small (smoothing) effect on the spectrum $X(\omega)$. On the other hand, if the window function has a wide spectrum compared to the width of

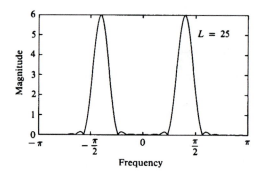

Figure 5.14 Magnitude spectrum of the Hanning window.

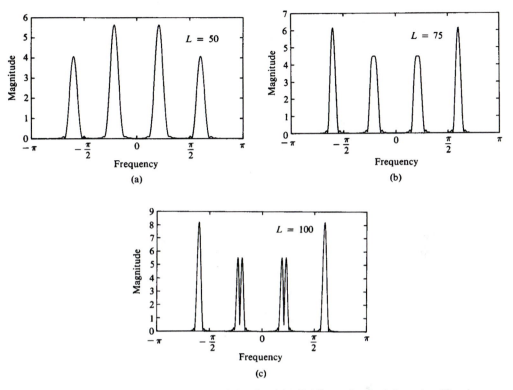

Figure 5.15   Magnitude spectrum of the signal in (5.4.8) as observed through a Hanning window.

$X(\omega)$, as would be the case when the number of samples L is small, the window spectrum masks the signal spectrum and, consequently, the DFT of the data reflects the spectral characteristics of the window function. Of course, this situation should be avoided.

Example 5.4.1

The exponential signal

$$x_a(t) = \begin{cases} e^{-t}, & t \geq 0 \\ 0, & t < 0 \end{cases}$$

is sampled at the rate $F_s = 20$ samples per second, and a block of 100 samples is used to estimate its spectrum. Determine the spectral characteristics of the signal $x_a(t)$ by computing the DFT of the finite-duration sequence. Compare the spectrum of the truncated discrete-time signal to the spectrum of the analog signal.

Solution   The spectrum of the analog signal is

$$X_a(F) = \frac{1}{1 + j2\pi F}$$

The exponential analog signal sampled at the rate of 20 samples per second yields the sequence

$$x(n) = e^{-nT} = e^{-n/20}, \qquad n \geq 0$$

$$= (e^{-1/20})^n = (0.95)^n, \qquad n \geq 0$$

Now, let

$$x(n) = \begin{cases} (0.95)^n, & 0 \leq n \leq 99 \\ 0, & \text{otherwise} \end{cases}$$

The N-point DFT of the $L = 100$ point sequence is

$$\hat{X}(k) = \sum_{k=0}^{99} \hat{x}(n)e^{-j2\pi k/N} \qquad k = 0, 1, \ldots, N-1$$

To obtain sufficient detail in the spectrum we choose $N = 200$. This is equivalent to padding the sequence $x(n)$ with 100 zeros.

The graph of the analog signal $x_a(t)$ and its magnitude spectrum $|X_a(F)|$ are illustrated in Fig. 5.16(a) and (b), respectively. The truncated sequence $x(n)$ and its $N = 200$ point DFT (magnitude) are illustrated in Fig. 5.16(c) and (d), respectively.

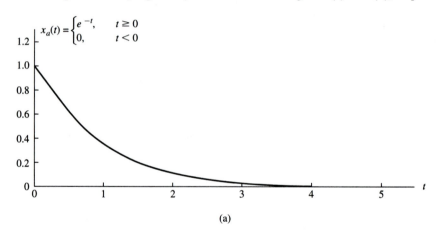

(a)

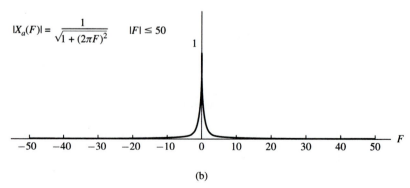

(b)

Figure 5.16   Effect of windowing (truncating) the sampled version of the analog signal in Example 5.4.1.

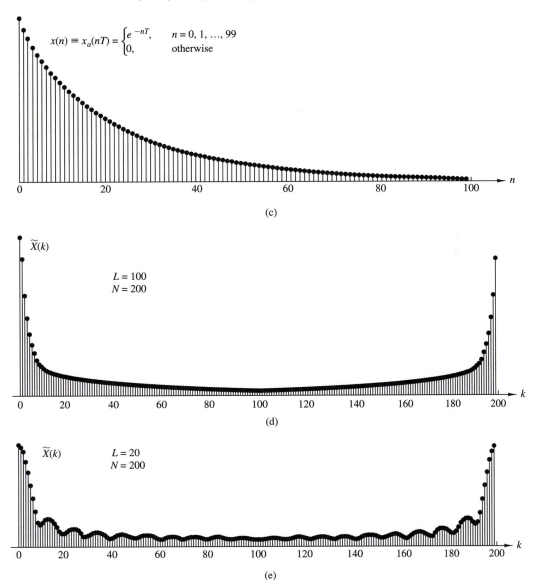

$$x(n) \equiv x_a(nT) = \begin{cases} e^{-nT}, & n = 0, 1, \ldots, 99 \\ 0, & \text{otherwise} \end{cases}$$

(c)

$\widetilde{X}(k)$

$L = 100$
$N = 200$

(d)

$\widetilde{X}(k)$ $L = 20$
$N = 200$

(e)

Figure 5.16 *Continued*

In this case the DFT $\{X(k)\}$ bears a close resemblance to the spectrum of the analog signal. The effect of the window function is relatively small.

On the other hand, suppose that a window function of length $L = 20$ is selected. Then the truncated sequence $x(n)$ is now given as

$$\hat{x}(n) = \begin{cases} (0.95)^n, & 0 \le n \le 19 \\ 0, & \text{otherwise} \end{cases}$$

Its $N = 200$ point DFT is illustrated in Fig. 5.16(e). Now the effect of the wider spectral window function is clearly evident. First, the main peak is very wide as a result of the wide spectral window. Second, the sinusoidal envelope variations in the spectrum away from the main peak are due to the large sidelobes of the rectangular window spectrum. Consequently, the DFT is no longer a good approximation of the analog signal spectrum.

5.5 SUMMARY AND REFERENCES

The major focus of this chapter was on the discrete Fourier transform, its properties and its applications. We developed the DFT by sampling the spectrum $X(\omega)$ of the sequence $x(n)$.

Frequency-domain sampling of the spectrum of a discrete-time signal is particularly important in the processing of digital signals. Of particular significance is the DFT, which was shown to uniquely represent a finite-duration sequence in the frequency domain. The existence of computationally efficient algorithms for the DFT, which are described in Chapter 6, make it possible to digitally process signals in the frequency domain much faster than in the time domain. The processing methods in which the DFT is especially suitable include linear filtering as described in this chapter and correlation, and spectrum analysis, which are treated in Chapters 6 and 12. A particularly lucid and concise treatment of the DFT and its application to frequency analysis is given in the book by Brigham (1988).

PROBLEMS

5.1 The first five points of the eight-point DFT of a real-valued sequence are {0.25, $0.125 - j0.3018, 0, 0.125 - j0.0518, 0$}. Determine the remaining three points.

5.2 Compute the eight-point circular convolution for the following sequences.

(a) $x_1(n) = \{1, 1, 1, 1, 0, 0, 0, 0\}$

$$x_2(n) = \sin \frac{3\pi}{8} n \qquad 0 \leq n \leq 7$$

(b) $x_1(n) = (\frac{1}{4})^n \qquad 0 \leq n \leq 7$

$$x_2(n) = \cos \frac{3\pi}{8} n \qquad 0 \leq n \leq 7$$

(c) Compute the DFT of the two circular convolution sequences using the DFTs of $x_1(n)$ and $x_2(n)$.

5.3 Let $X(k)$, $0 \leq k \leq N - 1$, be the N-point DFT of the sequence $x(n)$, $0 \leq n \leq N - 1$. We define

$$\hat{X}(k) = \begin{cases} X(k), & 0 \leq k \leq k_c, N - k_c \leq k \leq N - 1 \\ 0, & k_c < k < N - k_c \end{cases}$$

and we compute the inverse N-point DFT of $\hat{X}(k)$, $0 \leq k \leq N - 1$. What is the effect of this process on the sequence $x(n)$? Explain.

5.4 For the sequences

$$x_1(n) = \cos \frac{2\pi}{N} n \qquad x_2(n) = \sin \frac{2\pi}{N} n \qquad 0 \le n \le N - 1$$

determine the N-point:
- **(a)** Circular convolution $x_1(n) \, \textcircled{N} \, x_2(n)$
- **(b)** Circular correlation of $x_1(n)$ and $x_2(n)$
- **(c)** Circular autocorrelation of $x_1(n)$
- **(d)** Circular autocorrelation of $x_2(n)$

5.5 Compute the quantity

$$\sum_{n=0}^{N-1} x_1(n) x_2(n)$$

for the following pairs of sequences.
- **(a)** $x_1(n) = x_2(n) = \cos \dfrac{2\pi}{N} n \qquad 0 \le n \le N - 1$
- **(b)** $x_1(n) = \cos \dfrac{2\pi}{N} n \qquad x_2(n) = \sin \dfrac{2\pi}{N} n \qquad 0 \le n \le N - 1$
- **(c)** $x_1(n) = \delta(n) + \delta(n - 8) \qquad x_2(n) = u(n) - u(n - N)$

5.6 Determine the N-point DFT of the Blackman window

$$w(n) = 0.42 - 0.5 \cos \frac{2\pi n}{N - 1} + 0.08 \cos \frac{4\pi n}{N - 1} \qquad 0 \le n \le N - 1$$

5.7 If $X(k)$ is the DFT of the sequence $x(n)$, determine the N-point DFTs of the sequences

$$x_c(n) = x(n) \cos \frac{2\pi k n}{N} \qquad 0 \le n \le N - 1$$

and

$$x_s(n) = x(n) \sin \frac{2\pi k n}{N} \qquad 0 \le n \le N - 1$$

in terms of $X(k)$.

5.8 Determine the circular convolution of the sequences

$$x_1(n) = \{1, 2, 3, 1\}$$
$$\uparrow$$

$$x_2(n) = \{4, 3, 2, 2\}$$
$$\uparrow$$

using the time-domain formula in (5.2.39).

5.9 Use the four-point DFT and IDFT to determine the sequence

$$x_3(n) = x_1(n) \, \textcircled{N} \, x_2(n)$$

where $x_1(n)$ and $x_2(n)$ are the sequence given in Problem 5.8.

5.10 Compute the energy of the N-point sequence

$$x(n) = \cos \frac{2\pi k n}{N} \qquad 0 \le n \le N - 1$$

5.11 Given the eight-point DFT of the sequence

$$x(n) = \begin{cases} 1, & 0 \le n \le 3 \\ 0, & 4 \le n \le 7 \end{cases}$$

compute the DFT of the sequences:

(a) $x_1(n) = \begin{cases} 1, & n = 0 \\ 0, & 1 \le n \le 4 \\ 1, & 5 \le n \le 7 \end{cases}$

(b) $x_2(n) = \begin{cases} 0, & 0 \le n \le 1 \\ 1, & 2 \le n \le 5 \\ 0, & 6 \le n \le 7 \end{cases}$

5.12 Consider a finite-duration sequence

$$x(n) = \{0, 1, 2, 3, 4\}$$
$$\uparrow$$

(a) Sketch the sequence $s(n)$ with six-point DFT

$$S(k) = W_2^* X(k) \qquad k = 0, 1, \ldots, 6$$

(b) Determine the sequence $y(n)$ with six-point DFT $Y(k) = \text{Re}\,|X(k)|$.

(c) Determine the sequence $v(n)$ with six-point DFT $V(k) = \text{Im}\,|X(k)|$.

5.13 Let $x_p(n)$ be a periodic sequence with fundamental period N. Consider the following DFTs:

$$x_p(n) \overset{\text{DFT}}{\underset{N}{\longleftrightarrow}} X_1(k)$$

$$x_p(n) \overset{\text{DFT}}{\underset{3N}{\longleftrightarrow}} X_3(k)$$

(a) What is the relationship between $X_1(k)$ and $X_3(k)$?

(b) Verify the result in part (a) using the sequence

$$x_p(n) = \{\cdots 1, 2, 1, 2, 1, 2, 1, 2 \cdots\}$$
$$\uparrow$$

5.14 Consider the sequences

$$x_1(n) = \{0, 1, 2, 3, 4\} \qquad x_2(n) = \{0, 1, 0, 0, 0\} \qquad s(n) = \{1, 0, 0, 0, 0\}$$
$$\quad\uparrow \qquad\qquad\qquad\quad \uparrow \qquad\qquad\qquad\quad \uparrow$$

and their 5-point DFTs.

(a) Determine a sequence $y(n)$ so that $Y(k) = X_1(k)X_2(k)$.

(b) Is there a sequence $x_3(n)$ such that $S(k) = X_1(k)X_3(k)$?

5.15 Consider a causal LTI system with system function

$$H(z) = \frac{1}{1 - 0.5z^{-1}}$$

The output $y(n)$ of the system is known for $0 \le n \le 63$. Assuming that $H(z)$ is available, can you develop a 64-point DFT method to recover the sequence $x(n)$, $0 \le n \le 63$? Can you recover all values of $x(n)$ in this interval?

5.16* The impulse response of an LTI system is given by $h(n) = \delta(n) - \frac{1}{4}\delta(n - k_0)$. To determine the impulse response $g(n)$ of the inverse system, an engineer computes the N-point DFT $H(k)$, $N = 4k_0$, of $h(n)$ and then defines $g(n)$ as the inverse DFT of

$G(k) = 1/H(k), k = 0, 1, 2, \ldots, N-1$. Determine $g(n)$ and the convolution $h(n)*g(n)$, and comment on whether the system with impulse response $g(n)$ is the inverse of the system with impulse response $h(n)$.

5.17* Determine the eight-point DFT of the signal

$$x(n) = \{1, 1, 1, 1, 1, 1, 0, 0\}$$

and sketch its magnitude and phase.

5.18 A linear time-invariant system with frequency response $H(\omega)$ is excited with the periodic input

$$x(n) = \sum_{k=-\infty}^{\infty} \delta(n - kN)$$

Suppose that we compute the N-point DFT $Y(k)$ of the samples $y(n)$, $0 \leq n \leq N - 1$ of the output sequence. How is $Y(k)$ related to $H(\omega)$?

5.19 *DFT of real sequences with special symmetries*
  (a) Using the symmetry properties of Section 5.2 (especially the decomposition properties), explain how we can compute the DFT of two real symmetric (even) and two real antisymmetric (odd) sequences simultaneously using an N-point DFT only.
  (b) Suppose now that we are given four real sequences $x_i(n)$, $i = 1, 2, 3, 4$, that are all symmetric [i.e., $x_i(n) = x_i(N - n), 0 \leq n \leq N - 1$]. Show that the sequences

$$s_i(n) = x_i(n + 1) - x_i(n - 1)$$

  are antisymmetric [i.e., $s_i(n) = -s_i(N - n)$ and $s_i(0) = 0$].
  (c) Form a sequence $x(n)$ using $x_1(n)$, $x_2(n)$, $s_3(n)$, and $s_4(n)$ and show how to compute the DFT $X_i(k)$ of $x_i(n)$, $i = 1, 2, 3, 4$ from the N-point DFT $X(k)$ of $x(n)$.
  (d) Are there any frequency samples of $X_i(k)$ that cannot be recovered from $X(k)$? Explain.

5.20 *DFT of real sequences with odd harmonics only*   Let $x(n)$ be an N-point real sequence with N-point DFT $X(k)$ (N even). In addition, $x(n)$ satisfies the following symmetry property:

$$x\left(n + \frac{N}{2}\right) = -x(n) \qquad n = 0, 1, \ldots, \frac{N}{2} - 1$$

that is, the upper half of the sequence is the negative of the lower half.
  (a) Show that

$$X(k) = 0 \qquad k \text{ even}$$

  that is, the sequence has a spectrum with odd harmonics.
  (b) Show that the values of this odd-harmonic spectrum can be computed by evaluating the $N/2$-point DFT of a complex modulated version of the original sequence $x(n)$.

5.21 Let $x_a(t)$ be an analog signal with bandwidth $B = 3$ kHz. We wish to use a $N = 2^m$-point DFT to compute the spectrum of the signal with a resolution less than or equal to 50 Hz. Determine (a) the minimum sampling rate, (b) the minimum number of required samples, and (c) the minimum length of the analog signal record.

5.22 Consider the periodic sequence

$$x_p(n) = \cos \frac{2\pi}{10} n \qquad -\infty < n < \infty$$

with frequency $f_0 = \frac{1}{10}$ and fundamental period $N = 10$. Determine the 10-point DFT of the sequence $x(n) = x_p(n)$, $0 \le n \le N - 1$.

5.23 Compute the N-point DFTs of the signals
(a) $x(n) = \delta(n)$
(b) $x(n) = \delta(n - n_0) \quad 0 < n_0 < N$
(c) $x(n) = a^n \quad 0 \le n \le N - 1$
(d) $x(n) = \begin{cases} 1, & 0 \le n \le N/2 - 1 (N \text{ even}) \\ 0, & N/2 \le n \le N - 1 \end{cases}$
(e) $x(n) = e^{j(2\pi/N)k_0} \quad 0 \le n \le N - 1$
(f) $x(n) = \cos \dfrac{2\pi}{N} k_0 n \quad 0 \le n \le N - 1$
(g) $x(n) = \sin \dfrac{2\pi}{N} k_0 n \quad 0 \le n \le N - 1$
(h) $x(n) = \begin{cases} 1, & n \text{ even} \\ 0, & n \text{ odd} \quad 0 \le n \le N - 1 \end{cases}$

5.24 Consider the finite-duration signal

$$x(n) = \{1, 2, 3, 1\}$$

(a) Compute its four-point DFT by solving explicitly the 4-by-4 system of linear equations defined by the inverse DFT formula.
(b) Check the answer in part (a) by computing the four-point DFT, using its definition.

5.25 (a) Determine the Fourier transform $X(\omega)$ of the signal

$$x(n) = \{1, 2, 3, 2, 1, 0\}$$
$$\uparrow$$

(b) Compute the 6-point DFT $V(k)$ of the signal

$$v(n) = \{3, 2, 1, 0, 1, 2\}$$

(c) Is there any relation between $X(\omega)$ and $V(k)$? Explain.

5.26 Prove the identity

$$\sum_{l=-\infty}^{\infty} \delta(n + lN) = \frac{1}{N} \sum_{k=0}^{N-1} e^{j(2\pi/N)kn}$$

(*Hint:* Find the DFT of the periodic signal in the left-hand side.)

5.27 *Computation of the even and odd harmonics using the DFT* Let $x(n)$ be an N-point sequence with an N-point DFT $X(k)$ (N even)
(a) Consider the time-aliased sequence

$$y(n) = \begin{cases} \displaystyle\sum_{l=-\infty}^{\infty} x(n + lM), & 0 \le n \le M - 1 \\ 0, & \text{elsewhere} \end{cases}$$

What is the relationship between the M-point DFT $Y(k)$ of $y(n)$ and the Fourier transform $X(\omega)$ of $x(n)$?

(b) Let

$$y(n) = \begin{cases} x(n) + x\left(n + \dfrac{N}{2}\right), & 0 \le n \le N - 1 \\ 0, & \text{elsewhere} \end{cases}$$

and

$$y(n) \underset{N/2}{\overset{\text{DFT}}{\longleftrightarrow}} Y(k)$$

Show that $X(k) = Y(k/2)$, $k = 2, 4, \ldots, N - 2$.

(c) Use the results in parts (a) and (b) to develop a procedure that computes the odd harmonics of $X(k)$ using an $N/2$-point DFT.

5.28* *Frequency-domain sampling*   Consider the following discrete-time signal

$$x(n) = \begin{cases} a^{|n|}, & |n| \le L \\ 0, & |n| > L \end{cases}$$

where $a = 0.95$ and $L = 10$

(a) Compute and plot the signal $x(n)$.

(b) Show that

$$X(\omega) = \sum_{n=-\infty}^{\infty} x(n)e^{-j\omega n} = x(0) + 2\sum_{n-1}^{L} x(n)\cos\omega n$$

Plot $X(\omega)$ by computing it at $\omega = \pi k/100$, $k = 0, 1, \ldots, 100$.

(c) Compute

$$c_k = \frac{1}{N}X\left(\frac{2\pi}{N}K\right) \qquad k = 0, 1, \ldots, N - 1$$

for $N = 30$.

(d) Determine and plot the signal

$$\tilde{x}(n) = \sum_{k=0}^{N-1} c_k e^{j(2\pi/N)kn}$$

What is the relation between the signals $x(n)$ and $\tilde{x}(n)$? Explain.

(e) Compute and plot the signal $\tilde{x}_1(n) = \sum_{l=-\infty}^{\infty} x(n - lN)$, $-L \le n \le L$ for $N = 30$. Compare the signals $\tilde{x}(n)$ and $\tilde{x}_1(n)$.

(f) Repeat parts (c) to (e) for $N = 15$.

5.29* *Frequency-domain sampling*   The signal $x(n) = a^{|n|}$, $-1 < a < 1$ has a Fourier transform

$$X(\omega) = \frac{1 - a^2}{1 - 2a\cos\omega + a^2}$$

(a) Plot $X(\omega)$ for $0 \le \omega \le 2\pi$, $a = 0.8$.

Reconstruct and plot $X(\omega)$ from its samples $X(2\pi k/N)$, $0 \le k \le N - 1$ for:

(b) $N = 20$

(c) $N = 100$

(d) Compare the spectra obtained in parts (b) and (c) with the original spectrum $X(\omega)$ and explain the differences.

(e) Illustrate the time-domain aliasing when $N = 20$.

5.30* *Frequency analysis of amplitude-modulated discrete-time signal* The discrete-time signal

$$x(n) = \cos 2\pi f_1 n + \cos 2\pi f_2 n$$

where $f_1 = \frac{1}{18}$ and $f_2 = \frac{5}{128}$, modulates the amplitude of the carrier

$$x_c(n) = \cos 2\pi f_c n$$

where $f_c = \frac{50}{128}$. The resulting amplitude-modulated signal is

$$x_{am}(n) = x(n) \cos 2\pi f_c n$$

(a) Sketch the signals $x(n)$, $x_c(n)$, and $x_{am}(n)$, $0 \le n \le 255$.
(b) Compute and sketch the 128-point DFT of the signal $x_{am}(n)$, $0 \le n \le 127$.
(c) Compute and sketch the 128-point DFT of the signal $x_{am}(n)$, $0 \le n \le 99$.
(d) Compute and sketch the 256-point DFT of the signal $x_{am}(n)$, $0 \le n \le 179$.
(e) Explain the results obtained in parts (b) through (d), by deriving the spectrum of the amplitude-modulated signal and comparing it with the experimental results.

5.31* The sawtooth waveform in Fig. P5.31 can be expressed in the form of a Fourier series as

$$x(t) = \frac{2}{\pi} \left(\sin \pi t - \frac{1}{2} \sin 2\pi t + \frac{1}{3} \sin 3\pi t - \frac{1}{4} \sin 4\pi t \cdots \right)$$

(a) Determine the Fourier series coefficients c_k.
(b) Use an N-point subroutine to generate samples of this signal in the time domain using the first six terms of the expansion for $N = 64$ and $N = 128$. Plot the signal $x(t)$ and the samples generated, and comment on the results.

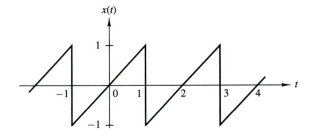

Figure P5.31

5.32 Recall that the Fourier transform of $x(t) = e^{j\Omega_0}$ is $X(j\Omega) = 2\pi\delta(\Omega - \Omega_0)$ and the Fourier transform of

$$p(t) = \begin{cases} 1, & 0 \le t \le T_0 \\ 0, & \text{otherwise} \end{cases}$$

is

$$P(j\Omega) = T_0 \frac{\sin \Omega T_0/2}{\Omega T_0/2} e^{-j\Omega T_0/2}$$

(a) Determine the Fourier transform $Y(j\Omega)$ of

$$y(t) = p(t)e^{j\Omega_0 t}$$

and roughly sketch $|Y(j\Omega)|$ versus Ω.

(b) Now consider the exponential sequence

$$x(n) = e^{j\omega_0 n}$$

where ω_0 is some arbitrary frequency in the range $0 < \omega_0 < \pi$ radians. Give the most general condition that ω_0 must satisfy in order for $x(n)$ to be periodic with period P (P is a positive integer).

(c) Let $y(n)$ be the finite-duration sequence

$$y(n) = x(n)w_N(n) = e^{j\omega_0 n}w_N(n)$$

where $w_N(n)$ is a finite-duration rectangular sequence of length N and where $x(n)$ is not necessarily periodic. Determine $Y(\omega)$ and roughly sketch $|Y(\omega)|$ for $0 \leq \omega \leq 2\pi$. What effect does N have in $|Y(\omega)|$? Briefly comment on the similarities and differences between $|Y(\omega)|$ and $|Y(j\Omega)|$.

(d) Suppose that

$$x(n) = e^{j(2\pi/P)n} \qquad P \text{ a positive integer}$$

and

$$y(n) = w_N(n)x(n)$$

where $N = lP$, l a positive integer. Determine and sketch the N-point DFT of $y(n)$. Relate your answer to the characteristics of $|Y(\omega)|$.

(e) Is the frequency sampling for the DFT in part (d) adequate for obtaining a rough approximation of $|Y(\omega)|$ directly from the magnitude of the DFT sequence $|Y(k)|$? If not, explain briefly how the sampling can be increased so that it will be possible to obtain a rough sketch of $|Y(\omega)|$ from an appropriate sequence $|Y(k)|$.

6

Efficient Computation of the DFT: Fast Fourier Transform Algorithms

As we have observed in the preceding chapter, the Discrete Fourier Transform (DFT) plays an important role in many applications of digital signal processing, including linear filtering, correlation analysis, and spectrum analysis. A major reason for its importance is the existence of efficient algorithms for computing the DFT.

The main topic of this chapter is the description of computationally efficient algorithms for evaluating the DFT. Two different approaches are described. One is a divide-and-conquer approach in which a DFT of size N, where N is a composite number, is reduced to the computation of smaller DFTs from which the larger DFT is computed. In particular, we present important computational algorithms, called fast Fourier transform (FFT) algorithms, for computing the DFT when the size N is a power of 2 and when it is a power of 4.

The second approach is based on the formulation of the DFT as a linear filtering operation on the data. This approach leads to two algorithms, the Goertzel algorithm and the chirp-z transform algorithm for computing the DFT via linear filtering of the data sequence.

6.1 EFFICIENT COMPUTATION OF THE DFT: FFT ALGORITHMS

In this section we present several methods for computing the DFT efficiently. In view of the importance of the DFT in various digital signal processing applications, such as linear filtering, correlation analysis, and spectrum analysis, its efficient computation is a topic that has received considerable attention by many mathematicians, engineers, and applied scientists.

Basically, the computational problem for the DFT is to compute the sequence $\{X(k)\}$ of N complex-valued numbers given another sequence of data $\{x(n)\}$ of length N, according to the formula

$$X(k) = \sum_{n=0}^{N-1} x(n) W_N^{kn} \qquad 0 \le k \le N - 1 \tag{6.1.1}$$

where

$$W_N = e^{-j2\pi/N} \tag{6.1.2}$$

In general, the data sequence $x(n)$ is also assumed to be complex valued.

Similarly, the IDFT becomes

$$x(n) = \frac{1}{N} \sum_{k=0}^{N-1} X(k) W_N^{-nk} \qquad 0 \le n \le N - 1 \tag{6.1.3}$$

Since the DFT and IDFT involve basically the same type of computations, our discussion of efficient computational algorithms for the DFT applies as well to the efficient computation of the IDFT.

We observe that for each value of k, direct computation of $X(k)$ involves N complex multiplications ($4N$ real multiplications) and $N - 1$ complex additions ($4N - 2$ real additions). Consequently, to compute all N values of the DFT requires N^2 complex multiplications and $N^2 - N$ complex additions.

Direct computation of the DFT is basically inefficient primarily because it does not exploit the symmetry and periodicity properties of the phase factor W_N. In particular, these two properties are:

$$\text{Symmetry property:} \quad W_N^{k+N/2} = -W_N^k \tag{6.1.4}$$

$$\text{Periodicity property:} \quad W_N^{k+N} = W_N^k \tag{6.1.5}$$

The computationally efficient algorithms described in this section, known collectively as fast Fourier transform (FFT) algorithms, exploit these two basic properties of the phase factor.

6.1.1 Direct Computation of the DFT

For a complex-valued sequence $x(n)$ of N points, the DFT may be expressed as

$$X_R(k) = \sum_{n=0}^{N-1} \left[x_R(n) \cos \frac{2\pi kn}{N} + x_I(n) \sin \frac{2\pi kn}{N} \right] \tag{6.1.6}$$

$$X_I(k) = -\sum_{n=0}^{N-1} \left[x_R(n) \sin \frac{2\pi kn}{N} - x_I(n) \cos \frac{2\pi kn}{N} \right] \tag{6.1.7}$$

The direct computation of (6.1.6) and (6.1.7) requires:

1. $2N^2$ evaluations of trigonometric functions.
2. $4N^2$ real multiplications.

3. $4N(N-1)$ real additions.

4. A number of indexing and addressing operations.

These operations are typical of DFT computational algorithms. The operations in items 2 and 3 result in the DFT values $X_R(k)$ and $X_I(k)$. The indexing and addressing operations are necessary to fetch the data $x(n)$, $0 \leq n \leq N-1$, and the phase factors and to store the results. The variety of DFT algorithms optimize each of these computational processes in a different way.

6.1.2 Divide-and-Conquer Approach to Computation of the DFT

The development of computationally efficient algorithms for the DFT is made possible if we adopt a divide-and-conquer approach. This approach is based on the decomposition of an N-point DFT into successively smaller DFTs. This basic approach leads to a family of computationally efficient algorithms known collectively as FFT algorithms.

To illustrate the basic notions, let us consider the computation of an N-point DFT, where N can be factored as a product of two integers, that is,

$$N = LM \tag{6.1.8}$$

The assumption that N is not a prime number is not restrictive, since we can pad any sequence with zeros to ensure a factorization of the form (6.1.8).

Now the sequence $x(n)$, $0 \leq n \leq N-1$, can be stored in either a one-dimensional array indexed by n or as a two-dimensional array indexed by l and m, where $0 \leq l \leq L-1$ and $0 \leq m \leq M-1$ as illustrated in Fig. 6.1. Note that l is the row index and m is the column index. Thus, the sequence $x(n)$ can be stored in a rectangular array in a variety of ways, each of which depends on the mapping of index n to the indexes (l, m).

For example, suppose that we select the mapping

$$n = Ml + m \tag{6.1.9}$$

This leads to an arrangement in which the first row consists of the first M elements of $x(n)$, the second row consists of the next M elements of $x(n)$, and so on, as illustrated in Fig. 6.2(a). On the other hand, the mapping

$$n = l + mL \tag{6.1.10}$$

stores the first L elements of $x(n)$ in the first column, the next L elements in the second column, and so on, as illustrated in Fig. 6.2(b).

A similar arrangement can be used to store the computed DFT values. In particular, the mapping is from the index k to a pair of indices (p, q), where $0 \leq p \leq L-1$ and $0 \leq q \leq M-1$. If we select the mapping

$$k = Mp + q \tag{6.1.11}$$

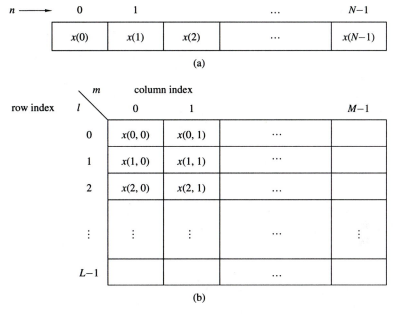

Figure 6.1   Two dimensional data array for storing the sequence $x(n)$, $0 \leq n \leq N - 1$.

the DFT is stored on a row-wise basis, where the first row contains the first M elements of the DFT $X(k)$, the second row contains the next set of M elements, and so on. On the other hand, the mapping

$$k = qL + p \tag{6.1.12}$$

results in a column-wise storage of $X(k)$, where the first L elements are stored in the first column, the second set of L elements are stored in the second column, and so on.

Now suppose that $x(n)$ is mapped into the rectangular array $x(l, m)$ and $X(k)$ is mapped into a corresponding rectangular array $X(p, q)$. Then the DFT can be expressed as a double sum over the elements of the rectangular array multiplied by the corresponding phase factors. To be specific, let us adopt a column-wise mapping for $x(n)$ given by (6.1.10) and the row-wise mapping for the DFT given by (6.1.11). Then

$$X(p, q) = \sum_{m=0}^{M-1} \sum_{l=0}^{L-1} x(l, m) W_N^{(Mp+q)(mL+l)} \tag{6.1.13}$$

But

$$W_N^{(Mp+q)(mL+l)} = W_N^{MLmp} W_N^{mLq} W_N^{Mpl} W_N^{lq} \tag{6.1.14}$$

However, $W_N^{Nmp} = 1$, $W_N^{mqL} = W_{N/L}^{mq} = W_M^{mq}$, and $W_N^{Mpl} = W_{N/M}^{pl} = W_L^{pl}$.

Row-wise $n = Ml + m$

l $\diagdown$ m	0	1	2	$\cdots$	$M - 1$
0	$x(0)$	$x(1)$	$x(2)$	$\cdots$	$x(M - 1)$
1	$x(M)$	$x(M + 1)$	$x(M + 2)$	$\cdots$	$x(2M - 1)$
2	$x(2M)$	$x(2M + 1)$	$x(2M + 2)$	$\cdots$	$x(3M - 1)$
	$\vdots$	$\vdots$	$\vdots$	$\cdots$	$\vdots$
$L - 1$	$x((L-1)M)$	$x((L-1)M+1)$	$x((L-1)M+2)$	$\cdots$	$x(LM - 1)$

(a)

Column-wise $n = l + mL$

l $\diagdown$ m	0	1	2	$\cdots$	$M - 1$
0	$x(0)$	$x(L)$	$x(2L)$	$\cdots$	$x((M-1)L)$
1	$x(1)$	$x(L + 1)$	$x(2L + 1)$	$\cdots$	$x((M-1)L+1)$
2	$x(2)$	$x(L + 2)$	$x(2L + 2)$	$\cdots$	$x((M-1)L+2)$
	$\vdots$	$\vdots$	$\vdots$	$\cdots$	$\vdots$
$L-1$	$x(L - 1)$	$x(2L - 1)$	$x(3L - 1)$	$\cdots$	$x(LM - 1)$

(b)

Figure 6.2   Two arrangements for the data arrays.

With these simplifications, (6.1.13) can be expressed as

$$X(p,q) = \sum_{l=0}^{L-1} \left\{ W_N^{lq} \left[\sum_{m=0}^{M-1} x(l,m) W_M^{mq} \right] \right\} W_L^{lp} \qquad (6.1.15)$$

The expression in (6.1.15) involves the computation of DFTs of length M and length L. To elaborate, let us subdivide the computation into three steps:

1. First, we compute the M-point DFTs

$$F(l,q) \equiv \sum_{m=0}^{M-1} x(l,m) W_M^{mq}, \qquad 0 \leq q \leq M - 1 \qquad (6.1.16)$$

for each of the rows $l = 0, 1, \ldots, L - 1$.

2. Second, we compute a new rectangular array $G(l, q)$ defined as

$$G(l, q) = W_N^{lq} F(l, q) \qquad \begin{matrix} 0 \le l \le L - 1 \\ 0 \le q \le M - 1 \end{matrix} \qquad (6.1.17)$$

3. Finally, we compute the L-point DFTs

$$X(p, q) = \sum_{l=0}^{L-1} G(l, q) W_L^{lp} \qquad (6.1.18)$$

for each column $q = 0, 1, \ldots, M - 1$, of the array $G(l, q)$.

On the surface it may appear that the computational procedure outlined above is more complex than the direct computation of the DFT. However, let us evaluate the computational complexity of (6.1.15). The first step involves the computation of L DFTs, each of M points. Hence this step requires LM^2 complex multiplications and $LM(M - 1)$ complex additions. The second step requires LM complex multiplications. Finally, the third step in the computation requires ML^2 complex multiplications and $ML(L - 1)$ complex additions. Therefore, the computational complexity is

$$\begin{aligned} &\text{Complex multiplications:} \quad N(M + L + 1) \\ &\text{Complex additions:} \quad\quad\ \ N(M + L - 2) \end{aligned} \qquad (6.1.19)$$

where $N = ML$. Thus the number of multiplications has been reduced from N^2 to $N(M + L + 1)$ and the number of additions has been reduced from $N(N - 1)$ to $N(M + L - 2)$.

For example, suppose that $N = 1000$ and we select $L = 2$ and $M = 500$. Then, instead of having to perform 10^6 complex multiplications via direct computation of the DFT, this approach leads to 503,000 complex multiplications. This represents a reduction by approximately a factor of 2. The number of additions is also reduced by about a factor of 2.

When N is a highly composite number, that is, N can be factored into a product of prime numbers of the form

$$N = r_1 r_2 \cdots r_\nu \qquad (6.1.20)$$

then the decomposition above can be repeated $(\nu - 1)$ more times. This procedure results in smaller DFTs, which, in turn, leads to a more efficient computational algorithm.

In effect, the first segmentation of the sequence $x(n)$ into a rectangular array of M columns with L elements in each column resulted in DFTs of sizes L and M. Further decomposition of the data in effect involves the segmentation of each row (or column) into smaller rectangular arrays which result in smaller DFTs. This procedure terminates when N is factored into its prime factors.

Example 6.1.1

To illustrate this computational procedure, let us consider the computation of an $N = 15$ point DFT. Since $N = 5 \times 3 = 15$, we select $L = 5$ and $M = 3$. In other words, we store the 15-point sequence $x(n)$ column-wise as follows:

Row 1: $x(0, 0) = x(0)$ $x(0, 1) = x(5)$ $x(0, 2) = x(10)$
Row 2: $x(1, 0) = x(1)$ $x(1, 1) = x(6)$ $x(1, 2) = x(11)$
Row 3: $x(2, 0) = x(2)$ $x(2, 1) = x(7)$ $x(2, 2) = x(12)$
Row 4: $x(3, 0) = x(3)$ $x(3, 1) = x(8)$ $x(3, 2) = x(13)$
Row 5: $x(4, 0) = x(4)$ $x(4, 1) = x(9)$ $x(4, 2) = x(14)$

Now, we compute the three-point DFTs for each of the five rows. This leads to the following 5×3 array:

$$
\begin{array}{ccc}
F(0,0) & F(0,1) & F(0,2) \\
F(1,0) & F(1,1) & F(1,2) \\
F(2,0) & F(2,1) & F(2,2) \\
F(3,0) & F(3,1) & F(3,2) \\
F(4,0) & F(4,1) & F(4,2)
\end{array}
$$

The next step is to multiply each of the terms $F(l, q)$ by the phase factors $W_N^{lq} = W_{15}^{lq}$, $0 \le l \le 4$ and $0 \le q \le 2$. This computation results in the 5×3 array:

Column 1	Column 2	Column 3
$G(0, 0)$	$G(0, 1)$	$G(0, 2)$
$G(1, 0)$	$G(1, 1)$	$G(1, 2)$
$G(2, 0)$	$G(2, 1)$	$G(2, 2)$
$G(3, 0)$	$G(3, 1)$	$G(3, 2)$
$G(4, 0)$	$G(4, 1)$	$G(4, 2)$

The final step is to compute the five-point DFTs for each of the three columns. This computation yields the desired values of the DFT in the form

$X(0, 0) = X(0)$ $X(0, 1) = X(1)$ $X(0, 2) = X(2)$
$X(1, 0) = X(3)$ $X(1, 1) = X(4)$ $X(1, 2) = X(5)$
$X(2, 0) = X(6)$ $X(2, 1) = X(7)$ $X(2, 2) = X(8)$
$X(3, 0) = X(9)$ $X(3, 1) = X(10)$ $X(3, 2) = X(11)$
$X(4, 0) = X(12)$ $X(4, 1) = X(13)$ $X(4, 2) = X(14)$

Figure 6.3 illustrates the steps in the computation.

It is interesting to view the segmented data sequence and the resulting DFT in terms of one-dimensional arrays. When the input sequence $x(n)$ and the output DFT $X(k)$ in the two-dimensional arrays are read across from row 1 through row 5, we obtain the following sequences:

INPUT ARRAY
$x(0)\ x(5)\ x(10)\ x(1)\ x(6)\ x(11)\ x(2)\ x(7)\ x(12)\ x(3)\ x(8)\ x(13)\ x(4)\ x(9)\ x(14)$

OUTPUT ARRAY
$X(0)\ X(1)\ X(2)\ X(3)\ X(4)\ X(5)\ X(6)\ X(7)\ X(8)\ X(9)\ X(10)\ X(11)\ X(12)\ X(13)\ X(14)$

We observe that the input data sequence is shuffled from the normal order in the computation of the DFT. On the other hand, the output sequence occurs in normal order. In this case the rearrangement of the input data array is due to the

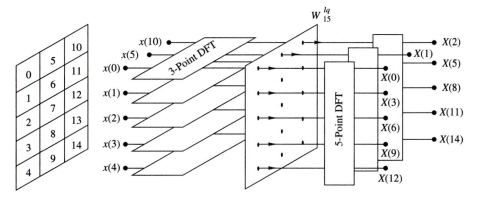

Figure 6.3   Computation of $N = 15$-point DFT by means of 3-point and 5-point DFTs.

segmentation of the one-dimensional array into a rectangular array and the order in which the DFTs are computed. This shuffling of either the input data sequence or the output DFT sequence is a characteristic of most FFT algorithms.

To summarize, the algorithm that we have introduced involves the following computations:

Algorithm 1

1. Store the signal column-wise.
2. Compute the M-point DFT of each row.
3. Multiply the resulting array by the phase factors W_N^{lq}.
4. Compute the L-point DFT of each column
5. Read the resulting array row-wise.

An additional algorithm with a similar computational structure can be obtained if the input signal is stored row-wise and the resulting transformation is column-wise. In this case we select as

$$n = Ml + m$$
$$k = qL + p \tag{6.1.21}$$

This choice of indices leads to the formula for the DFT in the form

$$X(p, q) = \sum_{m=0}^{M-1}\sum_{l=0}^{L-1} x(l, m) W_N^{pm} W_L^{pl} W_M^{qm}$$

$$= \sum_{m=0}^{M-1} W_M^{mq} \left[\sum_{l=0}^{L-1} x(l, m) W_L^{lp} \right] W_N^{mp} \tag{6.1.22}$$

Thus we obtain a second algorithm.

Algorithm 2

1. Store the signal row-wise.
2. Compute the L-point DFT at each column.
3. Multiply the resulting array by the factors W_N^{pm}.
4. Compute the M-point DFT of each row.
5. Read the resulting array column-wise.

The two algorithms given above have the same complexity. However, they differ in the arrangement of the computations. In the following sections we exploit the divide-and-conquer approach to derive fast algorithms when the size of the DFT is restricted to be a power of 2 or a power of 4.

6.1.3 Radix-2 FFT Algorithms

In the preceding section we described four algorithms for efficient computation of the DFT based on the divide-and-conquer approach. Such an approach is applicable when the number N of data points is not a prime. In particular, the approach is very efficient when N is highly composite, that is, when N can be factored as $N = r_1 r_2 r_3 \cdots r_v$, where the $\{r_j\}$ are prime.

Of particular importance as the case in which $r_1 = r_2 = \cdots = r_v \equiv r$, so that $N = r^v$. In such a case the DFTs are of size r, so that the computation of the N-point DFT has a regular pattern. The number r is called the radix of the FFT algorithm.

In this section we describe radix-2 algorithms, which are by far the most widely used FFT algorithms. Radix-4 algorithms are described in the following section.

Let us consider the computation of the $N = 2^v$ point DFT by the divide-and-conquer approach specified by (6.1.16) through (6.1.18). We select $M = N/2$ and $L = 2$. This selection results in a split of the N-point data sequence into two $N/2$-point data sequences $f_1(n)$ and $f_2(n)$, corresponding to the even-numbered and odd-numbered samples of $x(n)$, respectively, that is,

$$f_1(n) = x(2n)$$
$$f_2(n) = x(2n + 1), \quad n = 0, 1, \ldots, \frac{N}{2} - 1 \qquad (6.1.23)$$

Thus $f_1(n)$ and $f_2(n)$ are obtained by decimating $x(n)$ by a factor of 2, and hence the resulting FFT algorithm is called a decimation-in-time algorithm.

Now the N-point DFT can be expressed in terms of the DFTs of the decimated sequences as follows:

$$X(k) = \sum_{n=0}^{N-1} x(n) W_N^{kn} \qquad k = 0, 1, \ldots, N - 1$$

$$= \sum_{n \text{ even}} x(n) W_N^{kn} + \sum_{n \text{ odd}} x(n) W_N^{kn} \tag{6.1.24}$$

$$= \sum_{m=0}^{(N/2)-1} x(2m) W_N^{2mk} + \sum_{m=0}^{(N/2)-1} x(2m+1) W_N^{k(2m+1)}$$

But $W_N^2 = W_{N/2}$. With this substitution, (6.1.24) can be expressed as

$$
\begin{aligned}
X(k) &= \sum_{m=0}^{(N/2)-1} f_1(m) W_{N/2}^{km} + W_N^k \sum_{m=0}^{(N/2)-1} f_2(m) W_{N/2}^{km} \\
&= F_1(k) + W_N^k F_2(k) \qquad k = 0, 1, \ldots, N-1
\end{aligned} \tag{6.1.25}
$$

where $F_1(k)$ and $F_2(k)$ are the $N/2$-point DFTs of the sequences $f_1(m)$ and $f_2(m)$, respectively.

Since $F_1(k)$ and $F_2(k)$ are periodic, with period $N/2$, we have $F_1(k + N/2) = F_1(k)$ and $F_2(k + N/2) = F_2(k)$. In addition, the factor $W_N^{k+N/2} = -W_N^k$. Hence (6.1.25) can be expressed as

$$X(k) = F_1(k) + W_N^k F_2(k) \qquad k = 0, 1, \ldots, \frac{N}{2} - 1 \tag{6.1.26}$$

$$X\left(k + \frac{N}{2}\right) = F_1(k) - W_N^k F_2(k) \qquad k = 0, 1, \ldots, \frac{N}{2} - 1 \tag{6.1.27}$$

We observe that the direct computation of $F_1(k)$ requires $(N/2)^2$ complex multiplications. The same applies to the computation of $F_2(k)$. Furthermore, there are $N/2$ additional complex multiplications required to compute $W_N^k F_2(k)$. Hence the computation of $X(k)$ requires $2(N/2)^2 + N/2 = N^2/2 + N/2$ complex multiplications. This first step results in a reduction of the number of multiplications from N^2 to $N^2/2 + N/2$, which is about a factor of 2 for N large.

To be consistent with our previous notation, we may define

$$G_1(k) = F_1(k) \qquad k = 0, 1, \ldots, \frac{N}{2} - 1$$

$$G_2(k) = W_N^k F_2(k) \qquad k = 0, 1, \ldots, \frac{N}{2} - 1$$

Then the DFT $X(k)$ may be expressed as

$$
\begin{aligned}
X(k) &= G_1(k) + G_2(k) \qquad k = 0, 1, \ldots, \frac{N}{2} - 1 \\
X(k + \frac{N}{2}) &= G_1(k) - G_2(k) \qquad k = 0, 1, \ldots, \frac{N}{2} - 1
\end{aligned} \tag{6.1.28}
$$

This computation is illustrated in Fig. 6.4.

Having performed the decimation-in-time once, we can repeat the process for each of the sequences $f_1(n)$ and $f_2(n)$. Thus $f_1(n)$ would result in the two

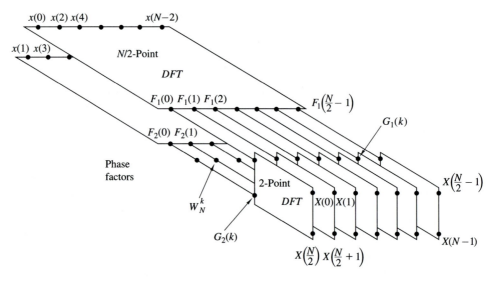

Figure 6.4   First step in the decimation-in-time algorithm.

$N/4$-point sequences

$$v_{11}(n) = f_1(2n) \qquad n = 0, 1, \ldots, \frac{N}{4} - 1$$

$$v_{12}(n) = f_1(2n + 1) \quad n = 0, 1, \ldots, \frac{N}{4} - 1$$

$$(6.1.29)$$

and $f_2(n)$ would yield

$$v_{21}(n) = f_2(2n) \qquad n = 0, 1, \ldots, \frac{N}{4} - 1$$

$$v_{22}(n) = f_2(2n + 1) \quad n = 0, 1, \ldots, \frac{N}{4} - 1$$

$$(6.1.30)$$

By computing $N/4$-point DFTs, we would obtain the $N/2$-point DFTs $F_1(k)$ and $F_2(k)$ from the relations

$$F_1(k) = V_{11}(k) + W_{N/2}^k V_{12}(k) \quad k = 0, 1, \ldots, \frac{N}{4} - 1$$

$$F_1\left(k + \frac{N}{4}\right) = V_{11}(k) - W_{N/2}^k V_{12}(k) \quad k = 0, 1, \ldots, \frac{N}{4} - 1$$

$$(6.1.31)$$

$$F_2(k) = V_{21}(k) + W_{N/2}^k V_{22}(k) \quad k = 0, 1, \ldots, \frac{N}{4} - 1$$

$$F_2\left(k + \frac{N}{4}\right) = V_{21}(k) - W_{N/2}^k V_{22}(k) \quad k = 0, \ldots, \frac{N}{4} - 1$$

$$(6.1.32)$$

where the $\{V_{ij}(k)\}$ are the $N/4$-point DFTs of the sequences $\{v_{ij}(n)\}$.

TABLE 6.1 COMPARISON OF COMPUTATIONAL COMPLEXITY FOR THE DIRECT COMPUTATION OF THE DFT VERSUS THE FFT ALGORITHM

Number of Points, N	Complex Multiplications in Direct Computation, N^2	Complex Multiplications in FFT Algorithm, $(N/2)\log_2 N$	Speed Improvement Factor
4	16	4	4.0
8	64	12	5.3
16	256	32	8.0
32	1,024	80	12.8
64	4,096	192	21.3
128	16,384	448	36.6
256	65,536	1,024	64.0
512	262,144	2,304	113.8
1,024	1,048,576	5,120	204.8

We observe that the computation of $\{V_{ij}(k)\}$ requires $4(N/4)^2$ multiplications and hence the computation of $F_1(k)$ and $F_2(k)$ can be accomplished with $N^2/4 + N/2$ complex multiplications. An additional $N/2$ complex multiplications are required to compute $X(k)$ from $F_1(k)$ and $F_2(k)$. Consequently, the total number of multiplications is reduced approximately by a factor of 2 again to $N^2/4 + N$.

The decimation of the data sequence can be repeated again and again until the resulting sequences are reduced to one-point sequences. For $N = 2^{\nu}$, this decimation can be performed $\nu = \log_2 N$ times. Thus the total number of complex multiplications is reduced to $(N/2)\log_2 N$. The number of complex additions is $N \log_2 N$. Table 6.1 presents a comparison of the number of complex multiplications in the FFT and in the direct computation of the DFT.

For illustrative purposes, Fig. 6.5 depicts the computation of an $N = 8$ point DFT. We observe that the computation is performed in three stages, beginning with the computations of four two-point DFTs, then two four-point DFTs, and

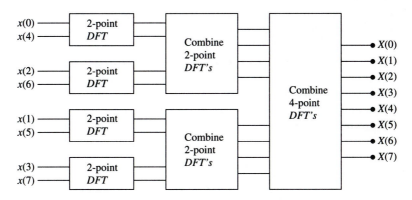

Figure 6.5 Three stages in the computation of an $N = 8$-point DFT.

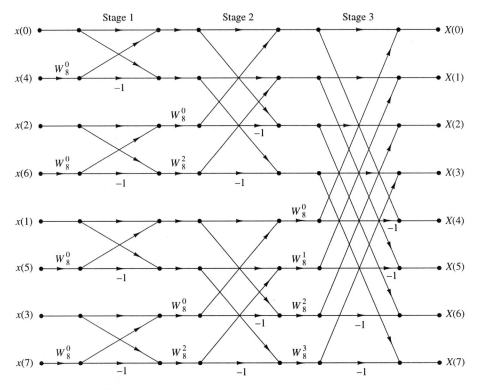

Figure 6.6 Eight-point decimation-in-time FFT algorithm.

finally, one eight-point DFT. The combination of the smaller DFTs to form the larger DFT is illustrated in Fig. 6.6 for $N = 8$.

Observe that the basic computation performed at every stage, as illustrated in Fig. 6.6, is to take two complex numbers, say the pair (a, b), multiply b by W_N^r, and then add and subtract the product from a to form two new complex numbers (A, B). This basic computation, which is shown in Fig. 6.7, is called a *butterfly* because the flow graph resembles a butterfly.

In general, each butterfly involves one complex multiplication and two complex additions. For $N = 2^\nu$, there are $N/2$ butterflies per stage of the computation process and $\log_2 N$ stages. Therefore, as previously indicated the total number of complex multiplications is $(N/2) \log_2 N$ and complex additions is $N \log_2 N$.

Once a butterfly operation is performed on a pair of complex numbers (a, b) to produce (A, B), there is no need to save the input pair (a, b). Hence we can

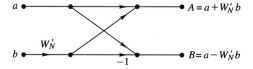

Figure 6.7 Basic butterfly computation in the decimation-in-time FFT algorithm.

store the result (A, B) in the same locations as (a, b). Consequently, we require a fixed amount of storage, namely, $2N$ storage registers, in order to store the results (N complex numbers) of the computations at each stage. Since the same $2N$ storage locations are used throughout the computation of the N-point DFT, we say that *the computations are done in place.*

A second important observation is concerned with the order of the input data sequence after it is decimated $(v - 1)$ times. For example, if we consider the case where $N = 8$, we know that the first decimation yields the sequence $x(0)$, $x(2)$, $x(4)$, $x(6)$, $x(1)$, $x(3)$, $x(5)$, $x(7)$, and the second decimation results in the sequence $x(0)$, $x(4)$, $x(2)$, $x(6)$, $x(1)$, $x(5)$, $x(3)$, $x(7)$. This *shuffling* of the input data sequence has a well-defined order as can be ascertained from observing Fig. 6.8, which illustrates the decimation of the eight-point sequence. By expressing the index n, in the sequence $x(n)$, in binary form, we note that the order of the decimated data sequence is easily obtained by reading the binary representation of the index n in reverse order. Thus the data point $x(3) \equiv x(011)$ is placed in position $m = 110$ or $m = 6$ in the decimated array. Thus we say that the data $x(n)$ after decimation is stored in bit-reversed order.

With the input data sequence stored in bit-reversed order and the butterfly computations performed in place, the resulting DFT sequence $X(k)$ is obtained in natural order (i.e., $k = 0, 1, \ldots, N - 1$). On the other hand, we should indicate that it is possible to arrange the FFT algorithm such that the input is left in natural order and the resulting output DFT will occur in bit-reversed order. Furthermore, we can impose the restriction that both the input data $x(n)$ and the output DFT $X(k)$ be in natural order, and derive an FFT algorithm in which the computations are not done in place. Hence such an algorithm requires additional storage.

Another important radix-2 FFT algorithm, called the decimation-in-frequency algorithm, is obtained by using the divide-and-conquer approach described in Section 6.1.2 with the choice of $M = 2$ and $L = N/2$. This choice of parameters implies a column-wise storage of the input data sequence. To derive the algorithm, we begin by splitting the DFT formula into two summations, one of which involves the sum over the first $N/2$ data points and the second sum involves the last $N/2$ data points. Thus we obtain

$$
\begin{aligned}
X(k) &= \sum_{n=0}^{(N/2)-1} x(n) W_N^{kn} + \sum_{n=N/2}^{N-1} x(n) W_N^{kn} \\
&= \sum_{n=0}^{(N/2)-1} x(n) W_N^{kn} + W_N^{Nk/2} \sum_{n=0}^{(N/2)-1} x\left(n + \frac{N}{2}\right) W_N^{kn}
\end{aligned}
\tag{6.1.33}
$$

Since $W_N^{kN/2} = (-1)^k$, the expression (6.1.33) can be rewritten as

$$
X(k) = \sum_{n=0}^{(N/2)-1} \left[x(n) + (-1)^k x\left(n + \frac{N}{2}\right) \right] W_N^{kn}
\tag{6.1.34}
$$

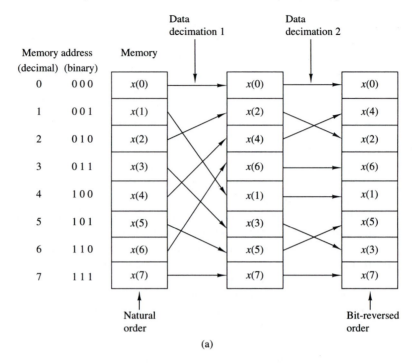

Figure 6.8   Shuffling of the data and bit reversal.

Now, let us split (decimate) $X(k)$ into the even- and odd-numbered samples. Thus we obtain

$$X(2k) = \sum_{n=0}^{(N/2)-1} \left[x(n) + x\left(n + \frac{N}{2} \right) \right] W_{N/2}^{kn} \qquad k = 0, 1, \ldots, \frac{N}{2} - 1 \qquad (6.1.35)$$

and

$$X(2k+1) = \sum_{n=0}^{(N/2)-1} \left\{ \left[x(n) - x\left(n + \frac{N}{2} \right) \right] W_N^n \right\} W_{N/2}^{kn} \qquad k = 0, 1, \ldots, \frac{N}{2} - 1$$

$$(6.1.36)$$

where we have used the fact that $W_N^2 = W_{N/2}$.

If we define the $N/2$-point sequences $g_1(n)$ and $g_2(n)$ as

$$g_1(n) = x(n) + x\left(n + \frac{N}{2}\right)$$

$$g_2(n) = \left[x(n) - x\left(n + \frac{N}{2}\right)\right] W_N^n \qquad n = 0, 1, 2, \ldots, \frac{N}{2} - 1$$

(6.1.37)

then

$$X(2k) = \sum_{n=0}^{(N/2)-1} g_1(n) W_{N/2}^{kn}$$

$$X(2k+1) = \sum_{n=0}^{(N/2)-1} g_2(n) W_{N/2}^{kn}$$

(6.1.38)

The computation of the sequences $g_1(n)$ and $g_2(n)$ according to (6.1.37) and the subsequent use of these sequences to compute the $N/2$-point DFTs are depicted in Fig. 6.9. We observe that the basic computation in this figure involves the butterfly operation illustrated in Fig. 6.10.

This computational procedure can be repeated through decimation of the $N/2$-point DFTs, $X(2k)$ and $X(2k + 1)$. The entire process involves $\nu = \log_2 N$

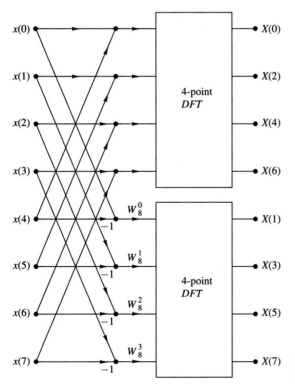

Figure 6.9 First stage of the decimation-in-frequency FFT algorithm.

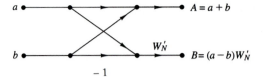

Figure 6.10 Basic butterfly computation in the decimation-in-frequency FFT algorithm.

stages of decimation, where each stage involves $N/2$ butterflies of the type shown in Fig. 6.10. Consequently, the computation of the N-point DFT via the decimation-in-frequency FFT algorithm, requires $(N/2) \log_2 N$ complex multiplications and $N \log_2 N$ complex additions, just as in the decimation-in-time algorithm. For illustrative purposes, the eight-point decimation-in-frequency algorithm is given in Fig. 6.11.

We observe from Fig. 6.11, that the input data $x(n)$ occurs in natural order, but the output DFT occurs in bit-reversed order. We also note that the computations are performed in place. However, it is possible to reconfigure the decimation-in-frequency algorithm so that the input sequence occurs in bit-reversed order while the output DFT occurs in normal order. Furthermore, if we abandon the requirement that the computations be done in place, it is also possible to have both the input data and the output DFT in normal order.

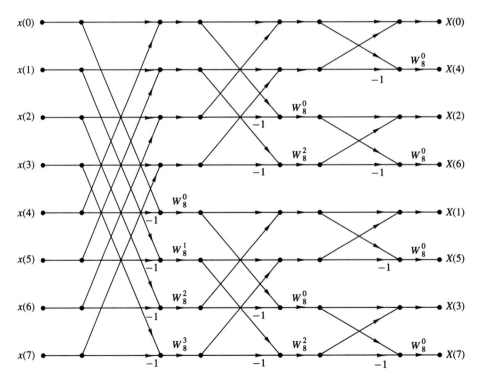

Figure 6.11 $N = 8$-point decimation-in-frequency FFT algorithmn.

6.1.4 Radix-4 FFT Algorithms

When the number of data points N in the DFT is a power of 4 (i.e., $N = 4^\nu$), we can, of course, always use a radix-2 algorithm for the computation. However, for this case, it is more efficient computationally to employ a radix-4 FFT algorithm.

Let us begin by describing a radix-4 decimation-in-time FFT algorithm, which is obtained by selecting $L = 4$ and $M = N/4$ in the divide-and-conquer approach described in Section 6.1.2. For this choice of L and M, we have $l, p = 0, 1, 2, 3; m, q = 0, 1, \ldots, N/4 - 1; n = 4m + l;$ and $k = (N/4)p + q$. Thus we split or decimate the N-point input sequence into four subsequences, $x(4n), x(4n + 1), x(4n + 2), x(4n + 3), n = 0, 1, \ldots, N/4 - 1$.

By applying (6.1.15) we obtain

$$X(p, q) = \sum_{l=0}^{3} \left[W_N^{lq} F(l, q) \right] W_4^{lp} \qquad p = 0, 1, 2, 3 \tag{6.1.39}$$

where $F(l, q)$ is given by (6.1.16), that is,

$$F(l, q) = \sum_{m=0}^{(N/4)-1} x(l, m) W_{N/4}^{mq} \qquad \begin{array}{l} l = 0, 1, 2, 3, \\ q = 0, 1, 2, \ldots, \dfrac{N}{4} - 1 \end{array} \tag{6.1.40}$$

and

$$x(l, m) = x(4m + l) \tag{6.1.41}$$

$$X(p, q) = X\left(\frac{N}{4}p + q\right) \tag{6.1.42}$$

Thus, the four $N/4$-point DFTs obtained from (6.1.40) are combined according to (6.1.39) to yield the N-point DFT. The expression in (6.1.39) for combining the $N/4$-point DFTs defines a radix-4 decimation-in-time butterfly, which can be expressed in matrix form as

$$\begin{bmatrix} X(0, q) \\ X(1, q) \\ X(2, q) \\ X(3, q) \end{bmatrix} = \begin{bmatrix} 1 & 1 & 1 & 1 \\ 1 & -j & -1 & j \\ 1 & -1 & 1 & -1 \\ 1 & j & -1 & -j \end{bmatrix} \begin{bmatrix} W_N^0 F(0, q) \\ W_N^q F(1, q) \\ W_N^{2q} F(2, q) \\ W_N^{3q} F(3, q) \end{bmatrix} \tag{6.1.43}$$

The radix-4 butterfly is depicted in Fig. 6.12(a) and in a more compact form in Fig. 6.12(b). Note that since $W_N^0 = 1$, each butterfly involves three complex multiplications, and 12 complex additions.

This decimation-in-time procedure can be repeated recursively ν times. Hence the resulting FFT algorithm consists of ν stages, where each stage contains $N/4$ butterflies. Consequently, the computational burden for the algorithm is $3\nu N/4 = (3N/8) \log_2 N$ complex multiplications and $(3N/2) \log_2 N$ complex additions. We note that the number of multiplications is reduced by 25%, but the number of additions has increased by 50% from $N \log_2 N$ to $(3N/2) \log_2 N$.

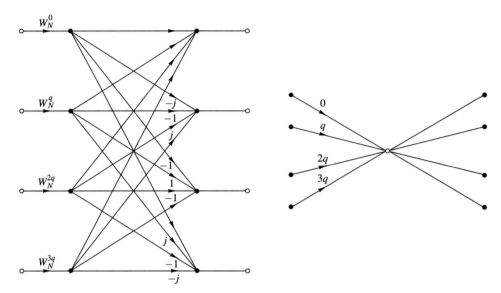

Figure 6.12 Basic butterfly computation in a radix-4 FFT algorithm.

It is interesting to note, however, that by performing the additions in two steps, it is possible to reduce the number of additions per butterfly from 12 to 8. This can be accomplished by expressing the matrix of the linear transformation in (6.1.43) as a product of two matrices as follows:

$$\begin{bmatrix} X(0,q) \\ X(1,q) \\ X(2,q) \\ X(3,q) \end{bmatrix} = \begin{bmatrix} 1 & 0 & 1 & 0 \\ 0 & 1 & 0 & -j \\ 1 & 0 & -1 & 0 \\ 0 & 1 & 0 & j \end{bmatrix} \begin{bmatrix} 1 & 0 & 1 & 0 \\ 1 & 0 & -1 & 0 \\ 0 & 1 & 0 & 1 \\ 0 & 1 & 0 & -1 \end{bmatrix} \begin{bmatrix} W_N^0 F(0,q) \\ W_N^q F(1,q) \\ W_N^{2q} F(2,q) \\ W_N^{3q} F(3,q) \end{bmatrix} \qquad (6.1.44)$$

Now each matrix multiplication involves four additions for a total of eight additions. Thus the total number of complex additions is reduced to $N \log_2 N$, which is identical to the radix-2 FFT algorithm. The computational savings results from the 25% reduction in the number of complex multiplications.

An illustration of a radix-4 decimation-in-time FFT algorithm is shown in Fig. 6.13 for $N = 16$. Note that in this algorithm, the input sequence is in normal order while the output DFT is shuffled. In the radix-4 FFT algorithm, where the decimation is by a factor of 4, the order of the decimated sequence can be determined by reversing the order of the number that represents the index n in a quaternary number system (i.e., the number system based on the digits 0, 1, 2, 3).

A radix-4 decimation-in-frequency FFT algorithm can be obtained by selecting $L = N/4$, $M = 4$; $l, p = 0, 1, \ldots, N/4 - 1$; $m, q = 0, 1, 2, 3$; $n = (N/4)m + l$; and $k = 4p + q$. With this choice of parameters, the general equation given by

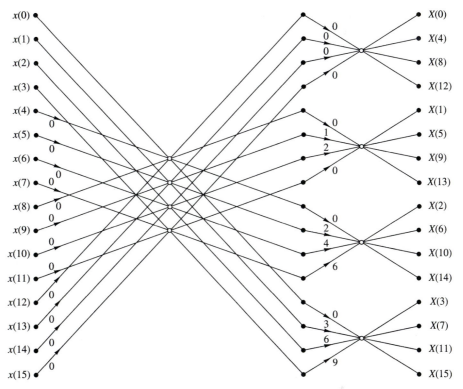

Figure 6.13   Sixteen-point radix-4 decimation-in-time algorithm with input in normal order and output in digit-reversed order.

(6.1.15) can be expressed as

$$X(p, q) = \sum_{l=0}^{(N/4)-1} G(l, q) W_{N/4}^{lp} \tag{6.1.45}$$

where

$$G(l, q) = W_N^{lq} F(l, q) \qquad \begin{array}{l} q = 0, 1, 2, 3 \\[4pt] l = 0, 1, \ldots, \dfrac{N}{4} - 1 \end{array} \tag{6.1.46}$$

and

$$F(l, q) = \sum_{m=0}^{3} x(l, m) W_4^{mq} \qquad \begin{array}{l} q = 0, 1, 2, 3 \\[4pt] l = 0, 1, 2, 3, \ldots, \dfrac{N}{4} - 1 \end{array} \tag{6.1.47}$$

We note that $X(p, q) = X(4p + q)$, $q = 0, 1, 2, 3$. Consequently, the N-point DFT is decimated into four $N/4$-point DFTs and hence we have a decimation-in-frequency FFT algorithm. The computations in (6.1.46) and (6.1.47) define

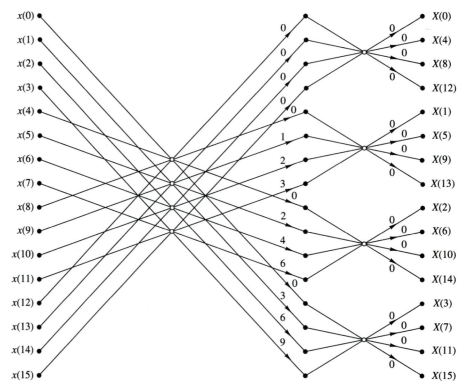

Figure 6.14 Sixteen-point, radix-4 decimation-in-frequency algorithm with input in normal order and output in digit-reversed order.

the basic radix-4 butterfly for the decimation-in-frequency algorithm. Note that the multiplications by the factors W_N^{lq} occur after the combination of the data points $x(l, m)$, just as in the case of the radix-2 decimation-in-frequency algorithm.

A 16-point radix-4 decimation-in-frequency FFT algorithm is shown in Fig. 6.14. Its input is in normal order and its output is in digit-reversed order. It has exactly the same computational complexity as the decimation-in-time radix-4 FFT algorithm.

For illustrative purposes, let us rederive the radix-4 decimation-in-frequency algorithm by breaking the N-point DFT formula into four smaller DFTs. We have

$$X(k) = \sum_{n=0}^{N-1} x(n) W_N^{kn}$$

$$= \sum_{n=0}^{N/4-1} x(n) W_N^{kn} + \sum_{n=N/4}^{N/2-1} x(n) W_N^{kn} + \sum_{n=N/2}^{3N/4-1} x(n) W_N^{kn} + \sum_{n=3N/4}^{N-1} x(n) W_N^{kn}$$

$$= \sum_{n=0}^{N/4-1} x(n) W_N^{kn} + W_N^{Nk/4} \sum_{n=0}^{N/4-1} x\left(n + \frac{N}{4}\right) W_N^{kn}$$

$$+ W_N^{kN/2} \sum_{n=0}^{N/4-1} x\left(n + \frac{N}{2}\right) W_N^{nk} + W_N^{3kN/4} \sum_{n=0}^{N/4-1} x\left(n + \frac{3N}{4}\right) W_N^{kn}$$

$$(6.1.48)$$

From the definition of the twiddle factors, we have

$$W_N^{kN/4} = (-j)^k \qquad W_N^{Nk/2} = (-1)^k \qquad W_N^{3Nk/4} = (j)^k \tag{6.1.49}$$

After substitution of (6.1.49) into (6.1.48), we obtain

$$X(k) = \sum_{n=0}^{N/4-1} \left[x(n) + (-j)^k x\left(n + \frac{N}{4}\right) \right. $$
$$\left. + (-1)^k x\left(n + \frac{N}{4}\right) + (j)^k x\left(n + \frac{3N}{4}\right) \right] W_N^{nk} \tag{6.1.50}$$

The relation in (6.1.50) is not an $N/4$-point DFT because the twiddle factor depends on N and not on $N/4$. To convert it into an $N/4$-point DFT, we subdivide the DFT sequence into four $N/4$-point subsequences, $X(4k)$, $X(4k+1)$, $X(4k+2)$, and $X(4k+3)$, $k = 0, 1, \ldots, N/4 - 1$. Thus we obtain the radix-4 decimation-in-frequency DFT as

$$X(4k) = \sum_{n=0}^{N/4-1} \left[x(n) + x\left(n + \frac{N}{4}\right) \right. \tag{6.1.51}$$
$$\left. + x\left(n + \frac{N}{2}\right) + x\left(n + \frac{3N}{4}\right) \right] W_N^0 W_{N/4}^{kn}$$

$$X(4k+1) = \sum_{n=0}^{N/4-1} \left[x(n) - jx\left(n + \frac{N}{4}\right) \right. \tag{6.1.52}$$
$$\left. - x\left(n + \frac{N}{2}\right) + jx\left(n + \frac{3N}{4}\right) \right] W_N^n W_{N/4}^{kn}$$

$$X(4k+2) = \sum_{n=0}^{N/4-1} \left[x(n) - x\left(n + \frac{N}{4}\right) \right. \tag{6.1.53}$$
$$\left. + x\left(n + \frac{N}{2}\right) - x\left(n + \frac{3N}{4}\right) \right] W_N^{2n} W_{N/4}^{kn}$$

$$X(4k+3) = \sum_{n=0}^{N/4-1} \left[x(n) + jx\left(n + \frac{N}{4}\right) \right. \tag{6.1.54}$$
$$\left. - x\left(n + \frac{N}{2}\right) - jx\left(n + \frac{3N}{4}\right) \right] W_N^{3n} W_{N/4}^{kn}$$

where we have used the property $W_N^{4kn} = W_{N/4}^{kn}$. Note that the input to each $N/4$-point DFT is a linear combination of four signal samples scaled by a twiddle factor. This procedure is repeated ν times, where $\nu = \log_4 N$.

6.1.5 Split-Radix FFT Algorithms

An inspection of the radix-2 decimation-in-frequency flowgraph shown in Fig. 6.11 indicates that the even-numbered points of the DFT can be computed independently of the odd-numbered points. This suggests the possibility of using different computational methods for independent parts of the algorithm with the objective of reducing the number of computations. The split-radix FFT (SRFFT) algorithms exploit this idea by using both a radix-2 and a radix-4 decomposition in the same FFT algorithm.

We illustrate this approach with a decimation-in-frequency SRFFT algorithm due to Duhamel (1986). First, we recall that in the radix-2 decimation-in-frequency FFT algorithm, the even-numbered samples of the N-point DFT are given as

$$X(2k) = \sum_{n=0}^{N/2-1} \left[x(n) + x\left(n + \frac{N}{2}\right) \right] W_{N/2}^{nk} \qquad k = 0, 1, \ldots, \frac{N}{2} - 1 \qquad (6.1.55)$$

Note that these DFT points can be obtained from an $N/2$-point DFT without any additional multiplications. Consequently, a radix-2 suffices for this computation.

The odd-numbered samples $\{X(2k+1)\}$ of the DFT require the premultiplication of the input sequence with the twiddle factors W_N^n. For these samples a radix-4 decomposition produces some computational efficiency because the four-point DFT has the largest multiplication-free butterfly. Indeed, it can be shown that using a radix greater than 4, does not result in a significant reduction in computational complexity.

If we use a radix-4 decimation-in-frequency FFT algorithm for the odd-numbered samples of the N-point DFT, we obtain the following $N/4$-point DFTs:

$$X(4k+1) = \sum_{n=0}^{N/4-1} \{[x(n) - x(n + N/2)] \qquad (6.1.56)$$

$$- j[x(n + N/4) - x(n + 3N/4)]\} W_N^n W_{N/4}^{kn}$$

$$X(4k+3) = \sum_{n=0}^{N/4-1} \{[x(n) - x(n + N/2)] \qquad (6.1.57)$$

$$+ j[x(n + N/4) - x(n + 3N/4)]\} W_N^{3n} W_{N/4}^{kn}$$

Thus the N-point DFT is decomposed into one $N/2$-point DFT without additional twiddle factors and two $N/4$-point DFTs with twiddle factors. The N-point DFT is obtained by successive use of these decompositions up to the last stage. Thus we obtain a decimation-in-frequency SRFFT algorithm.

Figure 6.15 shows the flow graph for an in-place 32-point decimation-in-frequency SRFFT algorithm. At stage A of the computation for $N = 32$, the

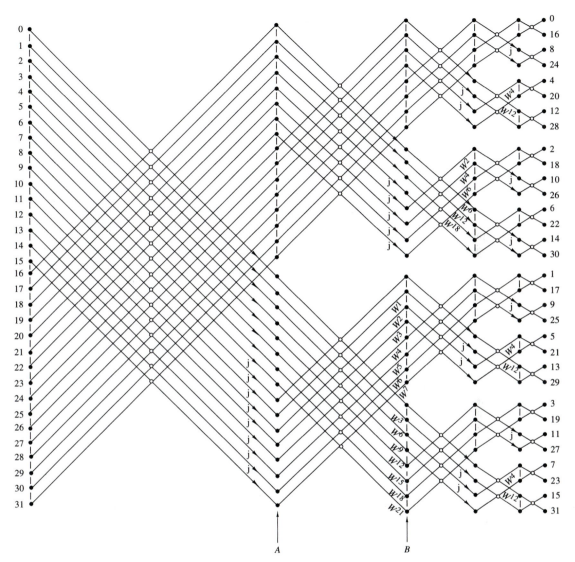

Figure 6.15 Length 32 split-radix FFT algorithms from paper by Duhamel (1986); reprinted with permission from the IEEE.

top 16 points constitute the sequence

$$g_0(n) = x(n) + x(n + N/2) \qquad 0 \le n \le 15 \qquad (6.1.58)$$

This is the sequence required for the computation of $X(2k)$. The next 8 points constitute the sequence

$$g_1(n) = x(n) - x(n + N/2) \qquad 0 \le n \le 7 \qquad (6.1.59)$$

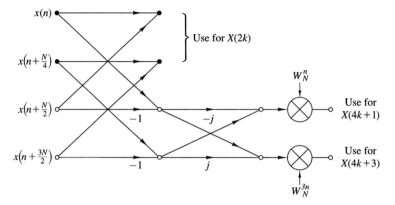

Figure 6.16 Butterfly for SRFFT algorithm.

The bottom eight points constitute the sequence $jg_2(n)$, where

$$g_2(n) = x(n + N/4) - x(n + 3N/4) \qquad 0 \le n \le 7 \qquad (6.1.60)$$

The sequences $g_1(n)$ and $g_2(n)$ are used in the computation of $X(4k + 1)$ and $X(4k+3)$. Thus, at stage A we have completed the first decimation for the radix-2 component of the algorithm. At stage B, the bottom eight points constitute the computation of $[g_1(n)+jg_2(n)]W_{32}^{3n}, 0 \le n \le 7$, which is used to compute $X(4k+3)$, $0 \le k \le 7$. The next eight points from the bottom constitute the computation of $[g_1(n) - jg_2(n)] \, W_{32}^n, 0 \le n \le 7$, which is used to compute $X(4k + 1), 0 \le k \le 7$. Thus at stage B, we have completed the first decimation for the radix-4 algorithm, which results in two 8-point sequences. Hence the basic butterfly computation for the SRFFT algorithm has the "L-shaped" form illustrated in Fig. 6.16.

Now we repeat the steps in the computation above. Beginning with the top 16 points at stage A, we repeat the decomposition for the 16-point DFT. In other words, we decompose the computation into an eight-point, radix-2 DFT and two four-point, radix-4 DFTs. Thus at stage B, the top eight points constitute the sequence (with $N = 16$)

$$g_0'(n) = g_0(n) + g_0(n + N/2) \qquad 0 \le n \le 7 \qquad (6.1.61)$$

and the next eight points constitute the two four-point sequences $g_1'(n)$ and $jg_2'(n)$, where

$$
\begin{aligned}
g_1'(n) &= g_0(n) - g_0(n + N/2) & 0 \le n \le 3 \\
g_2'(n) &= g_0(n + N/4) - g_0(n + 3N/4) & 0 \le n \le 3
\end{aligned}
\qquad (6.1.62)
$$

The bottom 16 points of stage B are in the form of two eight-point DFTs. Hence each eight-point DFT is decomposed into a four-point, radix-2 DFT and a four-point, radix-4 DFT. In the final stage, the computations involve the combination of two-point sequences.

Table 6.2 presents a comparison of the number of *nontrivial* real multiplications and additions required to perform an N-point DFT with complex-valued

TABLE 6.2 NUMBER OF NONTRIVIAL REAL MULTIPLICATIONS AND
ADDITIONS TO COMPUTE AN N-POINT COMPLEX DFT

	Real Multiplications				Real Additions			
N	Radix 2	Radix 4	Radix 8	Split Radix	Radix 2	Radix 4	Radix 8	Split Radix
16	24	20		20	152	148		148
32	88			68	408			388
64	264	208	204	196	1,032	976	972	964
128	712			516	2,504			2,308
256	1,800	1,392		1,284	5,896	5,488		5,380
512	4,360		3,204	3,076	13,566		12,420	12,292
1,024	10,248	7,856		7,172	30,728	28,336		27,652

Source: Extracted from Duhamel (1986).

data, using a radix-2, radix-4, radix-8, and a split-radix FFT. Note that the SRFFT algorithm requires the lowest number of multiplication and additions. For this reason, it is preferable in many practical applications.

Another type of SRFFT algorithm has been developed by Price (1990). Its relation to Duhamel's algorithm described previously can be seen by noting that the radix-4 DFT terms $X(4k+1)$ and $X(4k+3)$ involve the $N/4$-point DFTs of the sequences $[g_1(n) - jg_2(n)]W_N^n$ and $[g_1(n) + jg_2(n)]W_N^{3n}$, respectively. In effect, the sequences $g_1(n)$ and $g_2(n)$ are multiplied by the factor (vector) $(1, -j) = (1, W_{32}^8)$ and by W_N^n for the computation of $X(4k+1)$, while the computation of $X(4k+3)$ involves the factor $(1, j) = (1, W_{32}^{-8})$ and W_N^{3n}. Instead, one can rearrange the computation so that the factor for $X(4k+3)$ is $(-j, -1) = -(W_{32}^{-8}, 1)$. As a result of this phase rotation, the twiddle factors in the computation of $X(4k+3)$ become exactly the same as those for $X(4k+1)$, except that they occur in mirror image order. For example, at stage B of Fig. 6.15, the twiddle factors $W^{21}, W^{18}, \ldots, W^3$ are replaced by $W^1, W^2, \ldots, W^7$, respectively. This mirror-image symmetry occurs at every subsequent stage of the algorithm. As a consequence, the number of twiddle factors that must be computed and stored is reduced by a factor of 2 in comparison to Duhamel's algorithm. The resulting algorithm is called the "mirror" FFT (MFFT) algorithm.

An additional factor-of-2 savings in storage of twiddle factors can be obtained by introducing a 90° phase offset at the midpoint of each twiddle array, which can be removed if necessary at the output of the SRFFT computation. The incorporation of this improvement into the SRFFT (or the MFFT) results in another algorithm, also due to Price (1990), called the "phase" FFT (PFFT) algorithm.

6.1.6 Implementation of FFT Algorithms

Now that we have described the basic radix-2 and radix-4 FFT algorithms, let us consider some of the implementation issues. Our remarks apply directly to radix-2 algorithms, although similar comments may be made about radix-4 and higher-radix algorithms.

Basically, the radix-2 FFT algorithm consists of taking two data points at a time from memory, performing the butterfly computations and returning the resulting numbers to memory. This procedure is repeated many times ($(N \log_2 N)/2$ times) in the computation of an N-point DFT.

The butterfly computations require the twiddle factors $\{W_N^k\}$ at various stages in either natural or bit-reversed order. In an efficient implementation of the algorithm, the phase factors are computed once and stored in a table, either in normal order or in bit-reversed order, depending on the specific implementation of the algorithm.

Memory requirement is another factor that must be considered. If the computations are performed in place, the number of memory locations required is $2N$ since the numbers are complex. However, we can instead double the memory to 4N, thus simplifying the indexing and control operations in the FFT algorithms. In this case we simply alternate in the use of the two sets of memory locations from one stage of the FFT algorithm to the other. Doubling of the memory also allows us to have both the input sequence and the output sequence in normal order.

There are a number of other implementation issues regarding indexing, bit reversal, and the degree of parallelism in the computations. To a large extent, these issues are a function of the specific algorithm and the type of implementation, namely, a hardware or software implementation. In implementations based on a fixed-point arithmetic, or floating-point arithmetic on small machines, there is also the issue of round-off errors in the computation. This topic is considered in Section 6.4.

Although the FFT algorithms described previously were presented in the context of computing the DFT efficiently, they can also be used to compute the IDFT, which is

$$x(n) = \frac{1}{N} \sum_{k=0}^{N-1} X(k) W_N^{-nk} \tag{6.1.63}$$

The only difference between the two transforms is the normalization factor $1/N$ and the sign of the phase factor W_N. Consequently, an FFT algorithm for computing the DFT, can be converted to an FFT algorithm for computing the IDFT by changing the sign on all the phase factors and dividing the final output of the algorithm by N.

In fact, if we take the decimation-in-time algorithm that we described in Section 6.1.3, reverse the direction of the flow graph, change the sign on the phase factors, interchange the output and input, and finally, divide the output by N, we obtain a decimation-in-frequency FFT algorithm for computing the IDFT. On the other hand, if we begin with the decimation-in-frequency FFT algorithm described in Section 6.1.3 and repeat the changes described above, we obtain a decimation-in-time FFT algorithm for computing the IDFT. Thus it is a simple matter to devise FFT algorithms for computing the IDFT.

Finally, we note that the emphasis in our discussion of FFT algorithms was on radix-2, radix-4, and split-radix algorithms. These are by far the most widely used in practice. When the number of data points is not a power of 2 or 4, it is a simple matter to pad the sequence $x(n)$ with zeros such that $N = 2^v$ or $N = 4^v$.

The measure of complexity for FFT algorithms that we have emphasized is the required number of arithmetic operations (multiplications and additions). Although this is a very important benchmark for computational complexity, there are other issues to be considered in practical implementation of FFT algorithms. These include the architecture of the processor, the available instruction set, the data structures for storing twiddle factors, and other considerations.

For general-purpose computers, where the cost of the numerical operations dominate, radix-2, radix-4, and split-radix FFT algorithms are good candidates. However, in the case of special-purpose digital signal processors, featuring single-cycle multiply-and-accumulate operation, bit-reversed addressing, and a high degree of instruction parallelism, the structural regularity of the algorithm is equally important as arithmetic complexity. Hence for DSP processors, radix-2 or radix-4 decimation-in-frequency FFT algorithms are preferable in terms of speed and accuracy. The irregular structure of the SRFFT may render it less suitable for implementation on digital signal processors. Structural regularity is also important in the implementation of FFT algorithms on vector processors, multiprocessors, and in VLSI. Interprocessor communication is an important consideration in such implementations on parallel processors.

In conclusion, we have presented several important considerations in the implementation of FFT algorithms. Advances in digital signal processing technology, in hardware and software, will continue to influence the choice among FFT algorithms for various practical applications.

6.2 APPLICATIONS OF FFT ALGORITHMS

The FFT algorithms described in the preceding section find application in a variety of areas, including linear filtering, correlation, and spectrum analysis. Basically, the FFT algorithm is used as an efficient means to compute the DFT and the IDFT.

In this section we consider the use of the FFT algorithm in linear filtering and in the computation of the crosscorrelation of two sequences. The use of the FFT in spectrum analysis is considered in Chapter 12. In addition we illustrate how to enhance the efficiency of the FFT algorithm by forming complex-valued sequences from real-valued sequences prior to the computation of the DFT.

6.2.1 Efficient Computation of the DFT of Two Real Sequences

The FFT algorithm is designed to perform complex multiplications and additions, even though the input data may be real valued. The basic reason for this situation is that the phase factors are complex and hence, after the first stage of the algorithm, all variables are basically complex-valued.

In view of the fact that the algorithm can handle complex-valued input sequences, we can exploit this capability in the computation of the DFT of two real-valued sequences.

Suppose that $x_1(n)$ and $x_2(n)$ are two real-valued sequences of length N, and let $x(n)$ be a complex-valued sequence defined as

$$x(n) = x_1(n) + jx_2(n) \qquad 0 \le n \le N-1 \tag{6.2.1}$$

The DFT operation is linear and hence the DFT of $x(n)$ can be expressed as

$$X(k) = X_1(k) + jX_2(k) \tag{6.2.2}$$

The sequences $x_1(n)$ and $x_2(n)$ can be expressed in terms of $x(n)$ as follows:

$$x_1(n) = \frac{x(n) + x^*(n)}{2} \tag{6.2.3}$$

$$x_2(n) = \frac{x(n) - x^*(n)}{2j} \tag{6.2.4}$$

Hence the DFTs of $x_1(n)$ and $x_2(n)$ are

$$X_1(k) = \frac{1}{2}\{DFT[x(n)] + DFT[x^*(n)]\} \tag{6.2.5}$$

$$X_2(k) = \frac{1}{2j}\{DFT[x(n)] - DFT[x^*(n)]\} \tag{6.2.6}$$

Recall that the DFT of $x^*(n)$ is $X^*(N-k)$. Therefore,

$$X_1(k) = \frac{1}{2}[X(k) + X^*(N-k)] \tag{6.2.7}$$

$$X_2(k) = \frac{1}{j2}[X(k) - X^*(N-k)] \tag{6.2.8}$$

Thus, by performing a single DFT on the complex-valued sequence $x(n)$, we have obtained the DFT of the two real sequences with only a small amount of additional computation that is involved in computing $X_1(k)$ and $X_2(k)$ from $X(k)$ by use of (6.2.7) and (6.2.8).

6.2.2 Efficient Computation of the DFT of a 2N-Point Real Sequence

Suppose that $g(n)$ is a real-valued sequence of $2N$ points. We now demonstrate how to obtain the $2N$-point DFT of $g(n)$ from computation of one N-point DFT involving complex-valued data. First, we define

$$x_1(n) = g(2n)$$

$$x_2(n) = g(2n+1) \tag{6.2.9}$$

Thus we have subdivided the $2N$-point real sequence into two N-point real sequences. Now we can apply the method described in the preceding section.

Let $x(n)$ be the N-point complex-valued sequence

$$x(n) = x_1(n) + jx_2(n) \qquad (6.2.10)$$

From the results of the preceding section, we have

$$X_1(k) = \frac{1}{2}[X(k) + X^*(N - k)]$$

$$\qquad (6.2.11)$$

$$X_2(k) = \frac{1}{2j}[X(k) - X^*(N - k)]$$

Finally, we must express the $2N$-point DFT in terms of the two N-point DFTs, $X_1(k)$ and $X_2(k)$. To accomplish this, we proceed as in the decimation-in-time FFT algorithm, namely,

$$G(k) = \sum_{n=0}^{N-1} g(2n) W_{2N}^{2nk} + \sum_{n=0}^{N-1} g(2n + 1) W_{2N}^{(2n+1)k}$$

$$= \sum_{n=0}^{N-1} x_1(n) W_N^{nk} + W_{2N}^k \sum_{n=0}^{N-1} x_2(n) W_N^{nk}$$

Consequently,

$$G(k) = X_1(k) + W_2^k N X_2(k) \qquad k = 0, 1, \ldots, N - 1$$

$$G(k + N) = X_1(k) - W_2^k N X_2(k) \qquad k = 0, 1, \ldots, N - 1$$

$$\qquad (6.2.12)$$

Thus we have computed the DFT of a $2N$-point real sequence from one N-point DFT and some additional computation as indicated by (6.2.11) and (6.2.12).

6.2.3 Use of the FFT Algorithm in Linear Filtering and Correlation

An important application of the FFT algorithm is in FIR linear filtering of long data sequences. In Chapter 5 we described two methods, the overlap-add and the overlap-save methods for filtering a long data sequence with an FIR filter, based on the use of the DFT. In this section we consider the use of these two methods in conjunction with the FFT algorithm for computing the DFT and the IDFT.

Let $h(n)$, $0 \le n \le M - 1$, be the unit sample response of the FIR filter and let $x(n)$ denote the input data sequence. The block size of the FFT algorithm is N, where $N = L + M - 1$ and L is the number of new data samples being processed by the filter. We assume that for any given value of M, the number L of data samples is selected so that N is a power of 2. For purposes of this discussion, we consider only radix-2 FFT algorithms.

The N-point DFT of $h(n)$, which is padded by $L - 1$ zeros, is denoted as $H(k)$. This computation is performed once via the FFT and the resulting N complex numbers are stored. To be specific we assume that the decimation-in-frequency

FFT algorithm is used to compute $H(k)$. This yields $H(k)$ in bit-reversed order, which is the way it is stored in memory.

In the overlap-save method, the first $M - 1$ data points of each data block are the last $M - 1$ data points of the previous data block. Each data block contains L new data points, such that $N = L + M - 1$. The N-point DFT of each data block is performed by the FFT algorithm. If the decimation-in-frequency algorithm is employed, the input data block requires no shuffling and the values of the DFT occur in bit-reversed order. Since this is exactly the order of $H(k)$, we can multiply the DFT of the data, say $X_m(k)$, with $H(k)$ and thus the result

$$Y_m(k) = H(k)X_m(k)$$

is also in bit-reversed order.

The inverse DFT (IDFT) can be computed by use of an FFT algorithm that takes the input in bit-reversed order and produces an output in normal order. Thus there is no need to shuffle any block of data either in computing the DFT or the IDFT.

If the overlap-add method is used to perform the linear filtering, the computational method using the FFT algorithm is basically the same. The only difference is that the N-point data blocks consist of L new data points and $M - 1$ additional zeros. After the IDFT is computed for each data block, the N-point filtered blocks are overlapped as indicated in Section 5.3.2, and the $M - 1$ overlapping data points between successive output records are added together.

Let us assess the computational complexity of the FFT method for linear filtering. For this purpose, the one-time computation of $H(k)$ is insignificant and can be ignored. Each FFT requires $(N/2) \log_2 N$ complex multiplications and $N \log_2 N$ additions. Since the FFT is performed twice, once for the DFT and once for the IDFT, the computational burden is $N \log_2 N$ complex multiplications and $2N \log_2 N$ additions. There are also N complex multiplications and $N - 1$ additions required to compute $Y_m(k)$. Therefore, we have $(N \log_2 2N)/L$ complex multiplications per output data point and approximately $(2N \log_2 2N)/L$ additions per output data point. The overlap-add method requires an incremental increase of $(M - 1)/L$ in the number of additions.

By way of comparison, a direct form realization of the FIR filter involves M real multiplications per output point if the filter is not linear phase, and $M/2$ if it is linear phase (symmetric). Also, the number of additions is $M - 1$ per output point (see Sec. 8.2).

It is interesting to compare the efficiency of the FFT algorithm with the direct form realization of the FIR filter. Let us focus on the number of multiplications, which are more time consuming than additions. Suppose that $M = 128 = 2^7$ and $N = 2^\nu$. Then the number of complex multiplications per output point for an FFT size of $N = 2^\nu$ is

$$c(\nu) = \frac{N \log_2 2N}{L} = \frac{2^\nu(\nu + 1)}{N - M + 1}$$

$$\approx \frac{2^\nu(\nu + 1)}{2^\nu - 2^7}$$

TABLE 6.3 COMPUTATIONAL COMPLEXITY

Size of FFT $v = \log_2 N$	$c(v)$ Number of Complex Multiplications per Output Point
9	13.3
10	12.6
11	12.8
12	13.4
14	15.1

The values of $c(v)$ for different values of v are given in Table 6.3. We observe that there is an optimum value of v which minimizes $c(v)$. For the FIR filter of size $M = 128$, the optimum occurs at $v = 10$.

We should emphasize that $c(v)$ represents the number of complex multiplications for the FFT-based method. The number of real multiplications is four times this number. However, even if the FIR filter has linear phase (see Sec. 8.2), the number of computations per output point is still less with the FFT-based method. Furthermore, the efficiency of the FFT method can be improved by computing the DFT of two successive data blocks simultaneously, according to the method just described. Consequently, the FFT-based method is indeed superior from a computational point of view when the filter length is relatively large.

The computation of the cross correlation between two sequences by means of the FFT algorithm is similar to the linear FIR filtering problem just described. In practical applications involving crosscorrelation, at least one of the sequences has finite duration and is akin to the impulse response of the FIR filter. The second sequence may be a long sequence which contains the desired sequence corrupted by additive noise. Hence the second sequence is akin to the input to the FIR filter. By time reversing the first sequence and computing its DFT, we have reduced the cross correlation to an equivalent convolution problem (i.e., a linear FIR filtering problem). Therefore, the methodology we developed for linear FIR filtering by use of the FFT applies directly.

6.3 A LINEAR FILTERING APPROACH TO COMPUTATION OF THE DFT

The FFT algorithm takes N points of input data and produces an output sequence of N points corresponding to the DFT of the input data. As we have shown, the radix-2 FFT algorithm performs the computation of the DFT in $(N/2) \log_2 N$ multiplications and $N \log_2 N$ additions for an N-point sequence.

There are some applications where only a selected number of values of the DFT are desired, but the entire DFT is not required. In such a case, the FFT algorithm may no longer be more efficient than a direct computation of the desired values of the DFT. In fact, when the desired number of values of the DFT is less than $\log_2 N$, a direct computation of the desired values is more efficient.

The direct computation of the DFT can be formulated as a linear filtering operation on the input data sequence. As we will demonstrate, the linear filter takes the form of a parallel bank of resonators where each resonator selects one of the frequencies $\omega_k = 2\pi k/N$, $k = 0, 1, \ldots, N - 1$, corresponding to the N frequencies in the DFT.

There are other applications in which we require the evaluation of the z-transform of a finite-duration sequence at points other than the unit circle. If the set of desired points in the z-plane possesses some regularity, it is possible to also express the computation of the z-transform as a linear filtering operation. In this connection, we introduce another algorithm, called the chirp-z transform algorithm, which is suitable for evaluating the z-transform of a set of data on a variety of contours in the z-plane. This algorithm is also formulated as a linear filtering of a set of input data. As a consequence, the FFT algorithm can be used to compute the chirp-z transform and thus to evaluate the z-transform at various contours in the z-plane, including the unit circle.

6.3.1 The Goertzel Algorithm

The Goertzel algorithm exploits the periodicity of the phase factors $\{W_N^k\}$ and allows us to express the computation of the DFT as a linear filtering operation. Since $W_N^{-kN} = 1$, we can multiply the DFT by this factor. Thus

$$X(k) = W_N^{-kN} \sum_{m=0}^{N-1} x(m) W_N^{km} = \sum_{m=0}^{N-1} x(m) W_N^{-k(N-m)} \tag{6.3.1}$$

We note that (6.3.1) is in the form of a convolution. Indeed, if we define the sequence $y_k(n)$ as

$$y_k(n) = \sum_{m=0}^{N-1} x(m) W_N^{-k(n-m)} \tag{6.3.2}$$

then it is clear that $y_k(n)$ is the convolution of the finite-duration input sequence $x(n)$ of length N with a filter that has an impulse response

$$h_k(n) = W_N^{-kn} u(n) \tag{6.3.3}$$

The output of this filter at $n = N$ yields the value of the DFT at the frequency $\omega_k = 2\pi k/N$. That is,

$$X(k) = y_k(n)|_{n=N} \tag{6.3.4}$$

as can be verified by comparing (6.3.1) with (6.3.2).

The filter with impulse response $h_k(n)$ has the system function

$$H_k(z) = \frac{1}{1 - W_N^{-k} z^{-1}} \tag{6.3.5}$$

This filter has a pole on the unit circle at the frequency $\omega_k = 2\pi k/N$. Thus, the entire DFT can be computed by passing the block of input data into a parallel bank of N single-pole filters (resonators), where each filter has a pole at the corresponding frequency of the DFT.

Instead of performing the computation of the DFT as in (6.3.2), via convolution, we can use the difference equation corresponding to the filter given by (6.3.5) to compute $y_k(n)$ recursively. Thus we have

$$y_k(n) = W_N^{-k} y_k(n-1) + x(n) \qquad y_k(-1) = 0 \tag{6.3.6}$$

The desired output is $X(k) = y_k(N)$, for $k = 0, 1, \ldots, N-1$. To perform this computation, we can compute once and store the phase factors W_N^{-k}.

The complex multiplications and additions inherent in (6.3.6) can be avoided by combining the pairs of resonators possessing complex-conjugate poles. This leads to two-pole filters with system functions of the form

$$H_k(z) = \frac{1 - W_N^k z^{-1}}{1 - 2\cos(2\pi k/N)z^{-1} + z^{-2}} \tag{6.3.7}$$

The direct form II realization of the system illustrated in Fig. 6.17 is described by the difference equation

$$v_k(n) = 2\cos\frac{2\pi k}{N} v_k(n-1) - v_k(n-2) + x(n) \tag{6.3.8}$$

$$y_k(n) = v_k(n) - W_N^k v_k(n-1) \tag{6.3.9}$$

with initial conditions $v_k(-1) = v_k(-2) = 0$.

The recursive relation in (6.3.8) is iterated for $n = 0, 1, \ldots, N$, but the equation in (6.3.9) is computed only once at time $n = N$. Each iteration requires one real multiplication and two additions. Consequently, for a real input sequence $x(n)$, this algorithm requires $N+1$ real multiplications to yield not only $X(k)$ but also, due to symmetry, the value of $X(N-k)$.

The Goertzel algorithm is particularly attractive when the DFT is to be computed at a relatively small number M of values, where $M \leq \log_2 N$. Otherwise, the FFT algorithm is a more efficient method.

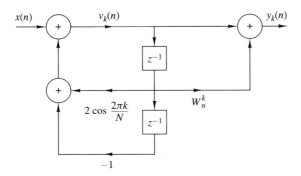

Figure 6.17 Direct form II realization of two-pole resonator for computing the DFT.

6.3.2 The Chirp-z Transform Algorithm

The DFT of an N-point data sequence x(n) has been viewed as the z-transform of $x(n)$ evaluated at N equally spaced points on the unit circle. It has also been viewed as N equally spaced samples of the Fourier transform of the data sequence $x(n)$. In this section we consider the evaluation of $X(z)$ on other contours in the z-plane, including the unit circle.

Suppose that we wish to compute the values of the z-transform of $x(n)$ at a set of points $\{z_k\}$. Then,

$$X(z_k) = \sum_{n=0}^{N-1} x(n) z_k^{-n} \qquad k = 0, 1, \ldots, L-1 \tag{6.3.10}$$

For example, if the contour is a circle of radius r and the z_k are N equally spaced points, then

$$z_k = r e^{j2\pi kn/N} \qquad k = 0, 1, 2, \ldots, N-1$$

$$X(z_k) = \sum_{n=0}^{N-1} [x(n) r^{-n}] e^{-j2\pi kn/N} \qquad k = 0, 1, 2, \ldots, N-1 \tag{6.3.11}$$

In this case the FFT algorithm can be applied on the modified sequence $x(n) r^{-n}$.

More generally, suppose that the points z_k in the z-plane fall on an arc which begins at some point

$$z_0 = r_0 e^{j\theta_0}$$

and spirals either in toward the origin or out away from the origin such that the points $\{z_k\}$ are defined as

$$z_k = r_0 e^{j\theta_0} (R_0 e^{j\phi_0})^k \qquad k = 0, 1, \ldots, L-1 \tag{6.3.12}$$

Note that if $R_0 < 1$, the points fall on a contour that spirals toward the origin and if $R_0 > 1$, the contour spirals away from the origin. If $R_0 = 1$, the contour is a circular arc of radius r_0. If $r_0 = 1$ and $R_0 = 1$, the contour is an arc of the unit circle. The latter contour would allow us to compute the frequency content of the sequence $x(n)$ at a dense set of L frequencies in the range covered by the arc without having to compute a large DFT, that is, a DFT of the sequence $x(n)$ padded with many zeros to obtain the desired resolution in frequency. Finally, if $r_0 = R_0 = 1$, $\theta_0 = 0$, $\phi_0 = 2\pi/N$, and $L = N$, the contour is the entire unit circle and the frequencies are those of the DFT. The various contours are illustrated in Fig. 6.18.

When points $\{z_k\}$ in (6.3.12) are substituted into the expression for the z-transform, we obtain

$$X(z_k) = \sum_{n=0}^{N-1} x(n) z_k^{-n}$$

$$= \sum_{n=0}^{N-1} x(n) (r_0 e^{j\theta_0})^{-n} V^{-nk} \tag{6.3.13}$$

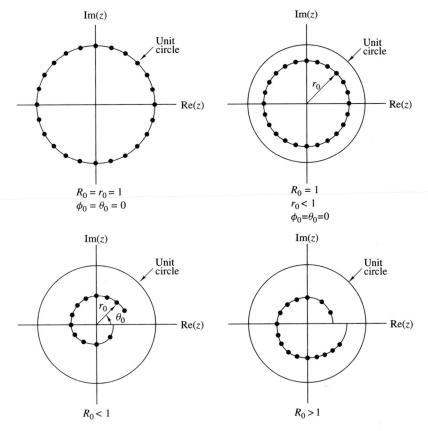

Figure 6.18 Some examples of contours on which we may evaluate the z-transform.

where, by definition,

$$V = R_0 e^{j\phi_0} \tag{6.3.14}$$

We can express (6.3.13) in the form of a convolution, by noting that

$$nk = \tfrac{1}{2}[n^2 + k^2 - (k-n)^2] \tag{6.3.15}$$

Substitution of (6.3.15) into (6.3.13) yields

$$X(z_k) = V^{-k^2/2} \sum_{n=0}^{N-1} [x(n)(r_0 e^{j\theta_0})^{-n} V^{-n^2/2}] V^{(k-n)^2/2} \tag{6.3.16}$$

Let us define a new sequence $g(n)$ as

$$g(n) = x(n)(r_0 e^{j\theta_0})^{-n} V^{-n^2/2} \tag{6.3.17}$$

Then (6.3.16) can be expressed as

$$X(z_k) = V^{-k^2/2} \sum_{n=0}^{N-1} g(n) V^{(k-n)^2/2} \tag{6.3.18}$$

The summation in (6.3.18) can be interpreted as the convolution of the sequence $g(n)$ with the impulse response $h(n)$ of a filter, where

$$h(n) = V^{n^2/2} \qquad (6.3.19)$$

Consequently, (6.3.18) may be expressed as

$$
\begin{aligned}
X(z_k) &= V^{-k^2/2} y(k) \\
&= \frac{y(k)}{h(k)} \qquad k = 0, 1, \ldots, L-1
\end{aligned}
\qquad (6.3.20)
$$

where $y(k)$ is the output of the filter

$$y(k) = \sum_{n=0}^{N-1} g(n)h(k-n) \qquad k = 0, 1, \ldots, L-1 \qquad (6.3.21)$$

We observe that both $h(n)$ and $g(n)$ are complex-valued sequences.

The sequence $h(n)$ with $R_0 = 1$ has the form of a complex exponential with argument $\omega n = n^2 \phi_0/2 = (n\phi_0/2)n$. The quantity $n\phi_0/2$ represents the frequency of the complex exponential signal, which increases linearly with time. Such signals are used in radar systems and are called *chirp signals*. Hence the z-transform evaluated as in (6.3.18) is called the *chirp-z transform*.

The linear convolution in (6.3.21) is most efficiently done by use of the FFT algorithm. The sequence $g(n)$ is of length N. However, $h(n)$ has infinite duration. Fortunately, only a portion $h(n)$ is required to compute the L values of $X(z)$.

Since we will compute the convolution in (6.3.1) via the FFT, let us consider the circular convolution of the N-point sequence $g(n)$ with an M-point section of $h(n)$, where $M > N$. In such a case, we know that the first $N-1$ points contain aliasing and that the remaining $M - N + 1$ points are identical to the result that would be obtained from a linear convolution of $h(n)$ with $g(n)$. In view of this, we should select a DFT of size

$$M = L + N - 1$$

which would yield L valid points and $N-1$ points corrupted by aliasing.

The section of $h(n)$ that is needed for this computation corresponds to the values of $h(n)$ for $-(N-1) \leq n \leq (L-1)$, which is of length $M = L + N - 1$, as observed from (6.3.21). Let us define the sequence $h_1(n)$ of length M as

$$h_1(n) = h(n - N + 1) \qquad n = 0, 1, \ldots, M-1 \qquad (6.3.22)$$

and compute its M-point DFT via the FFT algorithm to obtain $H_1(k)$. From $x(n)$ we compute $g(n)$ as specified by (6.3.17), pad $g(n)$ with $L-1$ zeros, and compute its M-point DFT to yield $G(k)$. The IDFT of the product $Y_1(k) = G(k)H_1(k)$ yields the M-point sequence $y_1(n)$, $n = 0, 1, \ldots, M-1$. The first $N-1$ points of $y_1(n)$ are corrupted by aliasing and are discarded. The desired values are $y_1(n)$ for $N - 1 \leq n \leq M - 1$, which correspond to the range $0 \leq n \leq L - 1$ in (6.3.21), that is,

$$y(n) = y_1(n + N - 1) \qquad n = 0, 1, \ldots, L - 1 \qquad (6.3.23)$$

Alternatively, we can define a sequence $h_2(n)$ as

$$h_2(n) = \begin{cases} h(n), & 0 \le n \le L - 1 \\ h(n - N - L + 1), & L \le n \le M - 1 \end{cases} \qquad (6.3.24)$$

The M-point DFT of $h_2(n)$ yields $H_2(k)$, which when multiplied by $G(k)$ yields $Y_2(k) = G(k)H_2(k)$. The IDFT of $Y_2(k)$ yields the sequence $y_2(n)$ for $0 \le n \le M-1$. Now the desired values of $y_2(n)$ are in the range $0 \le n \le L - 1$, that is,

$$y(n) = y_2(n) \qquad n = 0, 1, \ldots, L - 1 \qquad (6.3.25)$$

Finally, the complex values $X(z_k)$ are computed by dividing $y(k)$ by $h(k)$, $k = 0, 1, \ldots, L - 1$, as specified by (6.3.20).

In general, the computational complexity of the chirp-z transform algorithm described above is of the order of $M \log_2 M$ complex multiplications, where $M = N + L - 1$. This number should be compared with the product, $N \cdot L$, the number of computations required by direct evaluation of the z-transform. Clearly, if L is small, direct computation is more efficient. However, if L is large, then the chirp-z transform algorithm is more efficient.

The chirp-z transform method has been implemented in hardware to compute the DFT of signals. For the computation of the DFT, we select $r_0 = R_0 = 1$, $\theta_0 = 0$, $\phi_0 = 2\pi/N$, and $L = N$. In this case

$$V^{-n^2/2} = e^{-j\pi n^2/N}$$

$$= \cos \frac{\pi n^2}{N} - j \sin \frac{\pi n^2}{N} \qquad (6.3.26)$$

The chirp filter with impulse response

$$h(n) = V^{n^2/2}$$

$$= \cos \frac{\pi n^2}{N} + j \sin \frac{\pi n^2}{N} \qquad (6.3.27)$$

$$= h_r(n) + jh_i(n)$$

has been implemented as a pair of FIR filters with coefficients $h_r(n)$ and $h_i(n)$, respectively. Both *surface acoustic wave* (SAW) devices and *charge coupled devices* (CCD) have been used in practice for the FIR filters. The cosine and sine sequences given in (6.3.26) needed for the premultiplications and postmultiplications are usually stored in a read-only memory (ROM). Furthermore, we note that if only the magnitude of the DFT is desired, the postmultiplications are unnecessary. In this case,

$$|X(z_k)| = |y(k)| \qquad k = 0, 1, \ldots, N - 1 \qquad (6.3.28)$$

as illustrated in Fig. 6.19. Thus the linear FIR filtering approach using the chirp-z transform has been implemented for the computation of the DFT.

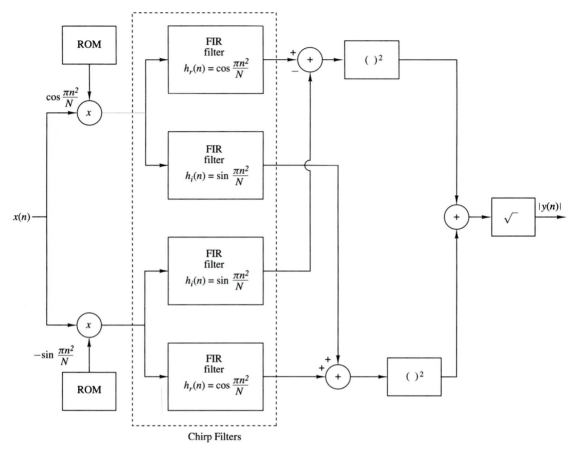

Figure 6.19   Block diagram illustrating the implementation of the chirp-z transform for computing the DFT (magnitude only).

6.4 QUANTIZATION EFFECTS IN THE COMPUTATION OF THE DFT*

As we have observed in our previous discussions, the DFT plays an important role in many digital signal processing applications, including FIR filtering, the computation of the correlation between signals, and spectral analysis. For this reason it is important for us to know the effect of quantization errors in its computation. In particular, we shall consider the effect of round-off errors due to the multiplications performed in the DFT with fixed-point arithmetic.

The model that we shall adopt for characterizing round-off errors in multiplication is the additive white noise model that we use in the statistical analysis of round-off errors in IIR and FIR filters (see Fig. 7.34). Although the statistical

---

*It is recommended that the reader review Section 7.5 prior to reading this section.

analysis is performed for rounding, the analysis can be easily modified to apply to truncation in two's-complement arithmetic (see Sec. 7.5.3).

Of particular interest is the analysis of round-off errors in the computation of the DFT via the FFT algorithm. However, we shall first establish a benchmark by determining the round-off errors in the direct computation of the DFT.

6.4.1 Quantization Errors in the Direct Computation of the DFT

Given a finite-duration sequence $\{x(n)\}$, $0 \leq n \leq N - 1$, the DFT of $\{x(n)\}$ is defined as

$$X(k) = \sum_{n=0}^{N-1} x(n) W_N^{kn} \qquad k = 0, 1, \ldots, N - 1 \qquad (6.4.1)$$

where $W_N = e^{-j2\pi/N}$. We assume that in general, $\{x(n)\}$ is a complex-valued sequence. We also assume that the real and imaginary components of $\{x(n)\}$ and $\{W_N^{kn}\}$ are represented by b bits. Consequently, the computation of the product $x(n) W_N^{kn}$ requires four real multiplications. Each real multiplication is rounded from $2b$ bits to b bits, and hence there are four quantization errors for each complex-valued multiplication.

In the direct computation of the DFT, there are N complex-valued multiplications for each point in the DFT. Therefore, the total number of real multiplications in the computation of a single point in the DFT is $4N$. Consequently, there are $4N$ quantization errors.

Let us evaluate the variance of the quantization errors in a fixed-point computation of the DFT. First, we make the following assumptions about the statistical properties of the quantization errors.

1. The quantization errors due to rounding are uniformly distributed random variables in the range $(-\Delta/2, \Delta/2)$ where $\Delta = 2^{-b}$.
2. The $4N$ quantization errors are mutually uncorrelated.
3. The $4N$ quantization errors are uncorrelated with the sequence $\{x(n)\}$.

Since each of the quantization errors has a variance

$$\sigma_e^2 = \frac{\Delta^2}{12} = \frac{2^{-2b}}{12} \qquad (6.4.2)$$

the variance of the quantization errors from the $4N$ multiplications is

$$\sigma_q^2 = 4N\sigma_e^2$$
$$= \frac{N}{3} \cdot 2^{-2b} \qquad (6.4.3)$$

Hence the variance of the quantization error is proportional to the size of DFT. Note that when N is a power of 2 (i.e., $N = 2^{\nu}$), the variance can be expressed as

$$\sigma_q^2 = \frac{2^{-2(b-v/2)}}{3} \tag{6.4.4}$$

This expression implies that every fourfold increase in the size N of the DFT requires an additional bit in computational precision to offset the additional quantization errors.

To prevent overflow, the input sequence to the DFT requires scaling. Clearly, an upper bound on $|X(k)|$ is

$$|X(k)| \le \sum_{n=0}^{N-1} |x(n)| \tag{6.4.5}$$

If the dynamic range in addition is $(-1, 1)$, then $|X(k)| < 1$ requires that

$$\sum_{n=0}^{N-1} |x(n)| < 1 \tag{6.4.6}$$

If $|x(n)|$ is initially scaled such that $|x(n)| < 1$ for all n, then each point in the sequence can be divided by N to ensure that (6.4.6) is satisfied.

The scaling implied by (6.4.6) is extremely severe. For example, suppose that the signal sequence $\{x(n)\}$ is white and, after scaling, each value $|x(n)|$ of the sequence is uniformly distributed in the range $(-1/N, 1/N)$. Then the variance of the signal sequence is

$$\sigma_x^2 = \frac{(2/N)^2}{12} = \frac{1}{3N^2} \tag{6.4.7}$$

and the variance of the output DFT coefficients $|X(k)|$ is

$$\sigma_X^2 = N\sigma_x^2$$
$$= \frac{1}{3N} \tag{6.4.8}$$

Thus the signal-to-noise power ratio is

$$\frac{\sigma_X^2}{\sigma_q^2} = \frac{2^{2b}}{N^2} \tag{6.4.9}$$

We observe that the scaling is responsible for reducing the SNR by N and the combination of scaling and quantization errors result in a total reduction that is proportional to N^2. Hence scaling the input sequence $\{x(n)\}$ to satisfy (6.4.6) imposes a severe penalty on the signal-to-noise ratio in the DFT.

Example 6.4.1

Use (6.4.9) to determine the number of bits required to compute the DFT of a 1024-point sequence with a SNR of 30 dB.

Solution   The size of the sequence is $N = 2^{10}$. Hence the SNR is

$$10 \log_{10} \frac{\sigma_X^2}{\sigma_q^2} = 10 \log_{10} 2^{2b-20}$$

For an SNR of 30 dB, we have

$$3(2b - 20) = 30$$

$$b = 15 \text{ bits}$$

Note that the 15 bits is the precision for both multiplication and addition.

Instead of scaling the input sequence $\{x(n)\}$, suppose we simply require that $|x(n)| < 1$. Then we must provide a sufficiently large dynamic range for addition such that $|X(k)| < N$. In such a case, the variance of the sequence $\{|x(n)|\}$ is $\sigma_x^2 = \frac{1}{3}$, and hence the variance of $|X(k)|$ is

$$\sigma_X^2 = N\sigma_x^2 = \frac{N}{3} \tag{6.4.10}$$

Consequently, the SNR is

$$\frac{\sigma_X^2}{\sigma_q^2} = 2^{2b} \tag{6.4.11}$$

If we repeat the computation in Example 6.4.1, we find that the number of bits required to achieve a SNR of 30 dB is $b = 5$ bits. However, we need an additional 10 bits for the accumulator (the adder) to accommodate the increase in the dynamic range for addition. Although we did not achieve any reduction in the dynamic range for addition, we have managed to reduce the precision in multiplication from 15 bits to 5 bits, which is highly significant.

6.4.2 Quantization Errors in FFT Algorithms

As we have shown, the FFT algorithms require significantly fewer multiplications than the direct computation of the DFT. In view of this we might conclude that the computation of the DFT via an FFT algorithm will result in smaller quantization errors. Unfortunately, that is not the case, as we will demonstrate.

Let us consider the use of fixed-point arithmetic in the computation of a radix-2 FFT algorithm. To be specific, we select the radix-2, decimation-in-time algorithm illustrated in Fig. 6.20 for the case $N = 8$. The results on quantization errors that we obtain for this radix-2 FFT algorithm are typical of the results obtained with other radix-2 and higher radix algorithms.

We observe that each butterfly computation involves one complex-valued multiplication or, equivalently, four real multiplications. We ignore the fact that some butterflies contain a trivial multiplication by ± 1. If we consider the butterflies that affect the computation of any one value of the DFT, we find that, in general, there are $N/2$ in the first stage of the FFT, $N/4$ in the second stage, $N/8$ in the third state, and so on, until the last stage, where there is only one. Consequently, the number of butterflies per output point is

$$2^{\nu-1} + 2^{\nu-2} + \cdots + 2 + 1 = 2^{\nu-1} \left[1 + \left(\tfrac{1}{2}\right) + \cdots + \left(\tfrac{1}{2}\right)^{\nu-1} \right]$$

$$= 2^{\nu}[1 - \left(\tfrac{1}{2}\right)^{\nu}] = N - 1 \tag{6.4.12}$$

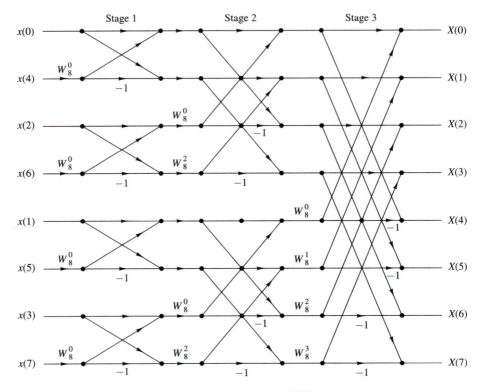

Figure 6.20 Decimation-in-time FFT algorithm.

For example, the butterflies that affect the computation of $X(3)$ in the eight-point FFT algorithm of Fig. 6.20 are illustrated in Fig. 6.21.

The quantization errors introduced in each butterfly propagate to the output. Note that the quantization errors introduced in the first stage propagate through $(v - 1)$ stages, those introduced in the second stage propagate through $(v - 2)$ stages, and so on. As these quantization errors propagate through a number of subsequent stages, they are phase shifted (phase rotated) by the phase factors W_N^{kn}. These phase rotations do not change the statistical properties of the quantization errors and, in particular, the variance of each quantization error remains invariant.

If we assume that the quantization errors in each butterfly are uncorrelated with the errors in other butterflies, then there are $4(N - 1)$ errors that affect the output of each point of the FFT. Consequently, the variance of the total quantization error at the output is

$$\sigma_q^2 = 4(N - 1)\frac{\Delta^2}{12} \approx \frac{N\Delta^2}{3} \tag{6.4.13}$$

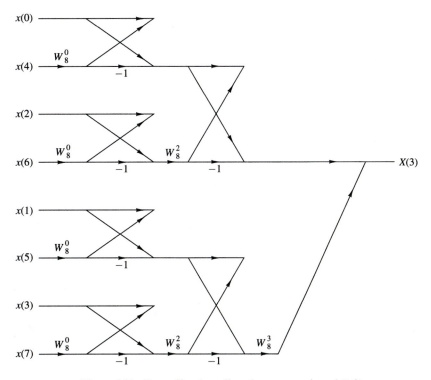

Figure 6.21 Butterflies that affect the computation of $X(3)$.

where $\Delta = 2^{-b}$. Hence

$$\sigma_q^2 = \frac{N}{3} \cdot 2^{-2b} \tag{6.4.14}$$

This is exactly the same result that we obtained for the direct computation of the DFT.

The result in (6.4.14) should not be surprising. In fact, the FFT algorithm does not reduce the number of multiplications required to compute a single point of the DFT. It does, however, exploit the periodicities in W_N^{kn} and thus reduces the number of multiplications in the computation of the entire block of N points in the DFT.

As in the case of the direct computation of the DFT, we must scale the input sequence to prevent overflow. Recall that if $|x(n)| < 1/N$, $0 \le n \le N - 1$, then $|X(k)| < 1$ for $0 \le k \le N - 1$. Thus overflow is avoided. With this scaling, the relations in (6.4.7), (6.4.8), and (6.4.9), obtained previously for the direct computation of the DFT, apply to the FFT algorithm as well. Consequently, the same SNR is obtained for the FFT.

Since the FFT algorithm consists of a sequence of stages, where each stage contains butterflies that involve pairs of points, it is possible to devise a differ-ent scaling strategy that is not as severe as dividing each input point by N. This alternative scaling strategy is motivated by the observation that the intermediate values $|X_n(k)|$ in the $n = 1, 2, \ldots, \nu$ stages of the FFT algorithm satisfy the conditions (see Problem 6.35)

$$\max[|X_{n+1}(k)|, |X_{n+1}(l)|] \geq \max[|X_n(k)|, |X_n(l)|]$$
$$\max[|X_{n+1}(k)|, |X_{n+1}(l)|] \leq 2\max[|X_n(k)|, |X_n(l)|]$$
(6.4.15)

In view of these relations, we can distribute the total scaling of $1/N$ into each of the stages of the FFT algorithm. In particular, if $|x(n)| < 1$, we apply a scale factor of $\frac{1}{2}$ in the first stage so that $|x(n)| < \frac{1}{2}$. Then the output of each subsequent stage in the FFT algorithm is scaled by $\frac{1}{2}$, so that after ν stages we have achieved an overall scale factor of $(\frac{1}{2})^\nu = 1/N$. Thus overflow in the computation of the DFT is avoided.

This scaling procedure does not affect the signal level at the output of the FFT algorithm, but it significantly reduces the variance of the quantization errors at the output. Specifically, each factor of $\frac{1}{2}$ reduces the variance of a quantization error term by a factor of $\frac{1}{4}$. Thus the $4(N/2)$ quantization errors introduced in the first stage are reduced in variance by $(\frac{1}{4})^{\nu-1}$, the $4(N/4)$ quantization errors introduced in the second stage are reduced in variance by $(\frac{1}{4})^{\nu-2}$, and so on. Consequently, the total variance of the quantization errors at the output of the FFT algorithm is

$$
\begin{aligned}
\sigma_q^2 &= \frac{\Delta^2}{12} \left\{ 4\left(\frac{N}{2}\right)\left(\frac{1}{4}\right)^{\nu-1} + 4\left(\frac{N}{4}\right)\left(\frac{1}{4}\right)^{\nu-2} + 4\left(\frac{N}{8}\right)\left(\frac{1}{4}\right)^{\nu-3} + \cdots + 4 \right\} \\
&= \frac{\Delta^2}{3} \left\{ \left(\frac{1}{2}\right)^{\nu-1} + \left(\frac{1}{2}\right)^{\nu-2} + \cdots + \frac{1}{2} + 1 \right\} \\
&= \frac{2\Delta^2}{3} \left[1 - \left(\frac{1}{2}\right)^\nu \right] \approx \frac{2}{3} \cdot 2^{-2b}
\end{aligned}
$$
(6.4.16)

where the factor $(\frac{1}{2})^\nu$ is negligible.

We now observe that (6.4.16) is no longer proportional to N. On the other hand, the signal has the variance $\sigma_X^2 = 1/3N$, as given in (6.4.8). Hence the SNR is

$$
\begin{aligned}
\frac{\sigma_X^2}{\sigma_q^2} &= \frac{1}{2N} \cdot 2^{2b} \\
&= 2^{2b-\nu-1}
\end{aligned}
$$
(6.4.17)

Thus, by distributing the scaling of $1/N$ uniformly throughout the FFT algorithm, we have achieved an SNR that is inversely proportional to N instead of N^2.

Example 6.4.2

Determine the number of bits required to compute an FFT of 1024 points with an SNR of 30 dB when the scaling is distributed as described above.

Solution   The size of the FFT is $N = 2^{10}$. Hence the SNR according to (6.4.17) is

$$10 \log_{10} 2^{2b - \nu - 1} = 30$$

$$3(2b - 11) = 30$$

$$b = \frac{21}{2} (11 \text{ bits})$$

This can be compared with the 15 bits required if all the scaling is performed in the first stage of the FFT algorithm.

6.5 SUMMARY AND REFERENCES

The focus of this chapter was on the efficient computation of the DFT. We demonstrated that by taking advantage of the symmetry and periodicity properties of the exponential factors W_N^{kn}, we can reduce the number of complex multiplications needed to compute the DFT from N^2 to $N \log_2 N$ when N is a power of 2. As we indicated, any sequence can be augmented with zeros, such that $N = 2^\nu$.

For decades, FFT-type algorithms were of interest to mathematicians who were concerned with computing values of Fourier series by hand. However, it was not until Cooley and Tukey (1965) published their well-known paper that the impact and significance of the efficient computation of the DFT was recognized. Since then the Cooley–Tukey FFT algorithm and its various forms, for example, the algorithms of Singleton (1967, 1969), have had a tremendous influence on the use of the DFT in convolution, correlation, and spectrum analysis. For a historical perspective on the FFT algorithm, the reader is referred to the paper by Cooley et al. (1967).

The split-radix FFT (SRFFT) algorithm described in Section 9.3.5 is due to Duhamel and Hollmann (1984, 1986). The "mirror" FFT (MFFT) and "phase" FFT (PFFT) algorithms were described to the authors by R. Price. The exploitation of symmetry properties in the data to reduce the computation time are described in a paper by Swarztrauber (1986).

Over the years, a number of tutorial papers have been published on FFT algorithms. We cite the early papers by Brigham and Morrow (1967), Cochran et al. (1967), Bergland (1969), and Cooley et al. (1967, 1969).

The recognition that the DFT can be arranged and computed as a linear convolution is also highly significant. Goertzel (1968) indicated that the DFT can be computed via linear filtering, although the computational savings of this approach is rather modest, as we have observed. More significant is the work of Bluestein (1970), who demonstrated that the computation of the DFT can be formulated as a chirp linear filtering operation. This work led to the development of the chirp-z transform algorithm by Rabiner et al. (1969).

In addition to the FFT algorithms described in this chapter, there are other efficient algorithms for computing the DFT, some of which further reduce the number of multiplications, but usually require more additions. Of particular importance is an algorithm due to Rader and Brenner (1976), the class of prime factor algorithms, such as the Good algorithm (1971), and the Winograd algorithm (1976, 1978). For a description of these and related algorithms, the reader may refer to the text by Blahut (1985).

PROBLEMS

6.1 Show that each of the numbers

$$e^{j(2\pi/N)k} \qquad 0 \le k \le N-1$$

corresponds to an Nth root of unity. Plot these numbers as phasors in the complex plane and illustrate, by means of this figure, the orthogonality property

$$\sum_{n=0}^{N-1} e^{j(2\pi/N)kn} e^{-j(2\pi/N)ln} = \begin{cases} N, & \text{if } k = l \\ 0, & \text{if } k \ne l \end{cases}$$

6.2 (a) Show that the phase factors can be computed recursively by

$$W_N^{ql} = W_N^q W_N^{q(l-1)}$$

(b) Perform this computation once using single-precision floating-point arithmetic and once using only four significant digits. Note the deterioration due to the accumulation of round-off errors in the later case.

(c) Show how the results in part (b) can be improved by resetting the result to the correct value $-j$, each time $ql = N/4$.

6.3 Let $x(n)$ be a real-valued N-point ($N = 2^\nu$) sequence. Develop a method to compute an N-point DFT $X'(k)$, which contains only the odd harmonics [i.e., $X'(k) = 0$ if k is even] by using only a real $N/2$-spoint DFT.

6.4 A designer has available a number of eight-point FFT chips. Show explicitly how he should interconnect three such chips in order to compute a 24-point DFT.

6.5 The z-transform of the sequence $x(n) = u(n) - u(n-7)$ is sampled at five points on the unit circle as follows

$$x(k) = X(z)|_z = e^{j2\pi k/5} \qquad k = 0, 1, 2, 3, 4$$

Determine the inverse DFT $x'(n)$ of $X(k)$. Compare it with $x(n)$ and explain the results.

6.6 Consider a finite-duration sequence $x(n)$, $0 \le n \le 7$, with z-transform $X(z)$. We wish to compute $X(z)$ at the following set of values:

$$z_k = 0.8 e^{j[(2\pi k/8)+(\pi/8)]} \qquad 0 \le k \le 7$$

(a) Sketch the points $\{z_k\}$ in the complex plane.

(b) Determine a sequence $s(n)$ such that its DFT provides the desired samples of $X(z)$.

6.7 Derive the radix-2 decimation-in-time FFT algorithm given by (6.1.26) and (6.1.27) as a special case of the more general algorithmic procedure given by (6.1.16) through (6.1.18).

6.8 Compute the eight-point DFT of the sequence

$$x(n) = \begin{cases} 1, & 0 \le n \le 7 \\ 0, & \text{otherwise} \end{cases}$$

by using the decimation-in-frequency FFT algorithm described in the text.

6.9 Derive the signal flow graph for the $N = 16$ point, radix-4 decimation-in-time FFT algorithm in which the input sequence is in normal order and the computations are done in place.

6.10 Derive the signal flow graph for the $N = 16$ point, radix-4 decimation-in-frequency FFT algorithm in which the input sequence is in digit-reversed order and the output DFT is in normal order.

6.11 Compute the eight-point DFT of the sequence

$$x(n) = \left\{ \frac{1}{2}, \frac{1}{2}, \frac{1}{2}, \frac{1}{2}, 0, 0, 0, 0 \right\}$$

using the in-place radix-2 decimation-in-time and radix-2 decimation-in-frequency algorithms. Follow exactly the corresponding signal flow graphs and keep track of all the intermediate quantities by putting them on the diagrams.

6.12 Compute the 16-point DFT of the sequence

$$x(n) = \cos \frac{\pi}{2} n \qquad 0 \le n \le 15$$

using the radix-4 decimation-in-time algorithm.

6.13 Consider the eight-point decimation-in-time (DIT) flow graph in Fig. 6.6.
   (a) What is the gain of the "signal path" that goes from $x(7)$ to $X(2)$?
   (b) How many paths lead from the input to a given output sample? Is this true for every output sample?
   (c) Compute $X(3)$ using the operations dictated by this flow graph.

6.14 Draw the flow graph for the decimation-in-frequency (DIF) SRFFT algorithm for $N = 16$. What is the number of nontrivial multiplications?

6.15 Derive the algorithm and draw the $N = 8$ flow graph for the DIT SRFFT algorithm. Compare your flow graph with the DIF radix-2 FFT flow graph shown in Fig. 6.11.

6.16 Show that the product of two complex numbers $(a+jb)$ and $(c+jd)$ can be performed with three real multiplications and five additions using the algorithm

$$x_R = (a - b)d + (c - d)a$$

$$x_I = (a - b)d + (c + d)b$$

where

$$x = x_R + jx_I = (a + jb)(c + jd)$$

6.17 Explain how the DFT can be used to compute N equispaced samples of the z-transform, of an N-point sequence, on a circle of radius r.

6.18 A real-valued N-point sequence $x(n)$ is called DFT bandlimited if its DFT $X(k) = 0$ for $k_0 \leq k \leq N - k_0$. We insert $(L - 1)N$ zeros in the middle of $X(k)$ to obtain the following LN-point DFT

$$X'(k) = \begin{cases} X(k), & 0 \leq k \leq k_0 - 1 \\ 0, & k_0 \leq k \leq LN - k_0 \\ X(k + N - LN), & LN - k_0 + 1 \leq k \leq LN - 1 \end{cases}$$

Show that

$$Lx'(Ln) = x(n) \qquad 0 \leq n \leq N - 1$$

where

$$x'(n) \overset{DFT}{\underset{LN}{\longleftrightarrow}} X'(k)$$

Explain the meaning of this type of processing by working out an example with $N = 4$, $L = 1$, and $X(k) = \{1, 0, 0, 1\}$.

6.19 Let $X(k)$ be the N-point DFT of the sequence $x(n)$, $0 \leq n \leq N - 1$. What is the N-point DFT of the sequence $s(n) = X(n)$, $0 \leq n \leq N - 1$?

6.20 Let $X(k)$ be the N-point DFT of the sequence $x(n)$, $0 \leq n \leq N - 1$. We define a $2N$-point sequence $y(n)$ as

$$y(n) = \begin{cases} x\left(\dfrac{n}{2}\right), & n \text{ even} \\ 0, & n \text{ odd} \end{cases}$$

Express the $2N$-point DFT of $y(n)$ in terms of $X(k)$.

6.21 **(a)** Determine the z-transform $W(z)$ of the Hanning window $w(n) = \left(1 - \cos\frac{2\pi n}{N-1}\right)/2$.
   (b) Determine a formula to compute the N-point DFT $X_w(k)$ of the signal $x_w(n) = w(n)x(n)$, $0 \leq n \leq N - 1$, from the N-point DFT $X(k)$ of the signal $x(n)$.

6.22 Create a DFT coefficient table that uses only $N/4$ memory locations to store the first quadrant of the sine sequence (assume N even).

6.23 Determine the computational burden of the algorithm given by (6.2.12) and compare it with the computational burden required in the $2N$-point DFT of $g(n)$. Assume that the FFT algorithm is a radix-2 algorithm.

6.24 Consider an IIR system described by the difference equation

$$y(n) = -\sum_{k=1}^{N} a_k y(n - k) + \sum_{k=0}^{M} b_k x(n - k)$$

Describe a procedure that computes the frequency response $H\left(\dfrac{2\pi}{N}k\right)$, $k = 0, 1, \ldots,$ $N - 1$ using the FFT algorithm ($N = 2^v$).

6.25 Develop a radix-3 decimation-in-time FFT algorithm for $N = 3^v$ and draw the corresponding flow graph for $N = 9$. What is the number of required complex multiplications? Can the operations be performed in place?

6.26 Repeat Problem 6.25 for the DIF case.

6.27 *FFT input and output pruning* In many applications we wish to compute only a few points M of the N-point DFT of a finite-duration sequence of length L (i.e., $M \ll N$ and $L \ll N$).

(a) Draw the flow graph of the radix-2 DIF FFT algorithm for $N = 16$ and eliminate [i.e., prune] all signal paths that originate from zero inputs assuming that only $x(0)$ and $x(1)$ are nonzero.

(b) Repeat part (a) for the radix-2 DIT algorithm.

(c) Which algorithm is better if we wish to compute all points of the DFT? What happens if we want to compute only the points $X(0)$, $X(1)$, $X(2)$, and $X(3)$? Establish a rule to choose between DIT and DIF pruning depending on the values of M and L.

(d) Give an estimate of saving in computations in terms of M, L, and N.

6.28 *Parallel computation of the DFT* Suppose that we wish to compute an $N = 2^p 2^v$ point DFT using 2^p digital signal processors (DSPs). For simplicity we assume that $p = v = 2$. In this case each DSP carries out all the computations that are necessary to compute 2^v DFT points.

(a) Using the radix-2 DIF flow graph, show that to avoid data shuffling, the entire sequence $x(n)$ should be loaded to the memory of each DSP.

(b) Identify and redraw the portion of the flow graph that is executed by the DSP that computes the DFT samples $X(2)$, $X(10)$, $X(6)$, and $X(14)$.

(c) Show that, if we use $M = 2^p$ DSPs, the computation speed-up S is given by

$$S = M \frac{\log_2 N}{\log_2 N - \log_2 M + 2(M - 1)}$$

6.29 Develop an inverse radix-2 DIT FFT algorithm starting with the definition. Draw the flow graph for computation and compare with the corresponding flow graph for the direct FFT. Can the IFFT flow graph be obtained from the one for the direct FFT?

6.30 Repeat Problem 6.29 for the DIF case.

6.31 Show that an FFT on data with Hermitian symmetry can be derived by reversing the flow graph of an FFT for real data.

6.32 Determine the system function $H(z)$ and the difference equation for the system that uses the Goertzel algorithm to compute the DFT value $X(N - k)$.

6.33 (a) Suppose that $x(n)$ is a finite-duration sequence of $N = 1024$ points. It is desired to evaluate the z-transform $X(z)$ of the sequence at the points

$$z_k = e^{j(2\pi/1024)k} \qquad k = 0, 100, 200, \ldots, 1000$$

by using the most efficient method or algorithm possible. Describe an algorithm for performing this computation efficiently. Explain how you arrived at your answer by giving the various options or algorithms that can be used.

(b) Repeat part (a) if $X(z)$ is to be evaluated at

$$z_k = 2(0.9)^k e^{j[(2\pi/5000)k + \pi/2]} \qquad k = 0, 1, 2, \ldots, 999$$

6.34 Repeat the analysis for the variance of the quantization error, carried out in Section 6.4.2, for the decimation-in-frequency radix-2 FFT algorithm.

6.35 The basic butterfly in the radix-2 decimation-in-time FFT algorithm is

$$X_{n+1}(k) = X_n(k) + W_N^m X_n(l)$$
$$X_{n+1}(l) = X_n(k) - W_N^m X_n(l)$$

(a) If we require that $|X_n(k)| < \frac{1}{2}$ and $|X_n(l)| < \frac{1}{2}$, show that

$$|\operatorname{Re}[X_{n+1}(k)]| < 1, \qquad |\operatorname{Re}[X_{n+1}(l)]| < 1$$

$$|\operatorname{Im}[X_{n+1}(k)]| < 1, \qquad |\operatorname{Im}[X_{n+1}(l)]| < 1$$

Thus overflow does not occur.

(b) Prove that

$$\max[|X_{n+1}(k)|, |X_{n+1}(l)|] \geq \max[|X_n(k)|, |X_n(l)|]$$

$$\max[|X_{n+1}(k)|, |X_{n+1}(l)|] \leq 2 \max[|X_n(k)|, |X_n(l)|]$$

6.36* *Computation of the DFT*   Use an FFT subroutine to compute the following DFTs and plot the magnitudes $|X(k)|$ of the DFTs.

(a) The 64-point DFT of the sequence

$$x(n) = \begin{cases} 1, & n = 0, 1, \ldots, 15 \quad (N_1 = 16) \\ 0, & \text{otherwise} \end{cases}$$

(b) The 64-point DFT of the sequence

$$x(n) = \begin{cases} 1, & n = 0, 1, \ldots, 7 \quad (N_1 = 8) \\ 0, & \text{otherwise} \end{cases}$$

(c) The 128-point DFT of the sequence in part (a).

(d) The 64-point DFT of the sequence

$$x(n) = \begin{cases} 10e^{j(\pi/8)n}, & n = 0, 1, \ldots, 63 \quad (N_1 = 64) \\ 0, & \text{otherwise} \end{cases}$$

Answer the following questions.

(1) What is the frequency interval between successive samples for the plots in parts (a), (b), (c), and (d)?

(2) What is the value of the spectrum at zero frequency (dc value) obtained from the plots in parts (a), (b), (c), (d)?
From the formula

$$X(k) = \sum_{n=0}^{N-1} x(n)e^{-j(2\pi/N)nk}$$

compute the theoretical values for the dc value and check these with the computer results.

(3) In plots (a), (b), and (c), what is the *frequency interval* between successive nulls in the spectrum? What is the relationship between N_1 of the sequence $x(n)$ and the frequency interval between successive nulls?

(4) Explain the difference between the plots obtained from parts (a) and (c).

6.37* *Identification of pole positions in a system*   Consider the system described by the difference equation

$$y(n) = -r^2 y(n-2) + x(n)$$

(a) Let $r = 0.9$ and $x(n) = \delta(n)$. Generate the output sequence $y(n)$ for $0 \leq n \leq 127$. Compute the $N = 128$ point DFT $\{Y(k)\}$ and plot $\{|Y(k)|\}$.

(b) Compute the $N = 128$ point DFT of the sequence

$$w(n) = (0.92)^{-n} y(n)$$

where $y(n)$ is the sequence generated in part (a). Plot the DFT values $|W(k)|$. What can you conclude from the plots in parts (a) and (b)?

(c) Let $r = 0.5$ and repeat part (a).

(d) Repeat part (b) for the sequence

$$w(n) = (0.55)^{-n} y(n)$$

where $y(n)$ is the sequence generated in part (c). What can you conclude from the plots in parts (c) and (d)?

(e) Now let the sequence generated in part (c) be corrupted by a sequence of "measurement" noise which is Gaussian with zero mean and variance $\sigma^2 = 0.1$. Repeat parts (c) and (d) for the noise-corrupted signal.

7
Implementation of
Discrete-Time Systems

The focus of this chapter is on the realization of linear time-invariant discrete-time systems in either software or hardware. As we noted in Chapter 2, there are various configurations or structures for the realization of any FIR and IIR discrete-time system. In Chapter 2 we described the simplest of these structures, namely, the direct-form realizations. However, there are other more practical structures that offer some distinct advantages, especially when quantization effects are taken into consideration.

Of particular importance are the cascade, parallel, and lattice structures, which exhibit robustness in finite-word-length implementations. Also described in this chapter is the frequency-sampling realization for an FIR system, which often has the advantage of being computationally efficient when compared with alternative FIR realizations. Other important filter structures are obtained by employing a state-space formulation for linear time-invariant systems. An analysis of systems characterized by the state-variable form is presented in both the time and frequency domains.

In addition to describing the various structures for the realization of discrete-time systems, we also treat problems associated with quantization effects in the implementation of digital filters using finite-precision arithmetic. This treatment includes the effects on the filter frequency response characteristics resulting from coefficient quantization and the round-off noise effects inherent in the digital implementation of discrete-time systems.

7.1 STRUCTURES FOR THE REALIZATION OF DISCRETE-TIME SYSTEMS

Let us consider the important class of linear time-invariant discrete-time systems characterized by the general linear constant-coefficient difference equation

$$y(n) = -\sum_{k=1}^{N} a_k y(n-k) + \sum_{k=0}^{M} b_k x(n-k) \qquad (7.1.1)$$

As we have shown by means of the z-transform, such a class of linear time-invariant discrete-time systems are also characterized by the rational system function

$$H(z) = \frac{\displaystyle\sum_{k=0}^{M} b_k z^{-k}}{1 + \displaystyle\sum_{k=1}^{N} a_k z^{-k}} \tag{7.1.2}$$

which is a ratio of two polynomials in z^{-1}. From the latter characterization, we obtain the zeros and poles of the system function, which depend on the choice of the system parameters $\{b_k\}$ and $\{a_k\}$ and which determine the frequency response characteristics of the system.

Our focus in this chapter is on the various methods of implementing (7.1.1) or (7.1.2) in either hardware, or in software on a programmable digital computer. We shall show that (7.1.1) or (7.1.2) can be implemented in a variety of ways depending on the form in which these two characterizations are arranged.

In general, we can view (7.1.1) as a computational procedure (an algorithm) for determining the output sequence $y(n)$ of the system from the input sequence $x(n)$. However, in various ways, the computations in (7.1.1) can be arranged into equivalent sets of difference equations. Each set of equations defines a computational procedure or an algorithm for implementing the system. From each set of equations we can construct a block diagram consisting of an interconnection of delay elements, multipliers, and adders. In Section 2.5 we referred to such a block diagram as a *realization* of the system or, equivalently, as a *structure* for realizing the system.

If the system is to be implemented in software, the block diagram or, equivalently, the set of equations that are obtained by rearranging (7.1.1), can be converted into a program that runs on a digital computer. Alternatively, the structure in block diagram form implies a hardware configuration for implementing the system.

Perhaps, the one issue that may not be clear to the reader at this point is why we are considering any rearrangements of (7.1.1) or (7.1.2). Why not just implement (7.1.1) or (7.1.2) directly without any rearrangement? If either (7.1.1) or (7.1.2) is rearranged in some manner, what are the benefits gained in the corresponding implementation?

These are the important questions which are answered in this chapter. At this point in our development, we simply state that the major factors that influence our choice of a specific realization are computational complexity, memory requirements, and finite-word-length effects in the computations.

Computational complexity refers to the number of arithmetic operations (multiplications, divisions, and additions) required to compute an output value $y(n)$ for the system. In the past, these were the only items used to measure computational complexity. However, with recent developments in the design and fabrication of rather sophisticated programmable digital signal processing chips, other factors, such as the number of times a fetch from memory is performed or the number of times a comparison between two numbers is performed per output sample, have become important in assessing the computational complexity of a given realization of a system.

Memory requirements refers to the number of memory locations required to store the system parameters, past inputs, past outputs, and any intermediate computed values.

Finite-word-length effects or finite-precision effects refer to the quantization effects that are inherent in any digital implementation of the system, either in hardware or in software. The parameters of the system must necessarily be represented with finite precision. The computations that are performed in the process of computing an output from the system must be rounded- off or truncated to fit within the limited precision constraints of the computer or the hardware used in the implementation. Whether the computations are performed in fixed-point or floating-point arithmetic is another consideration. All these problems are usually called finite-word-length effects and are extremely important in influencing our choice of a system realization. We shall see that different structures of a system, which are equivalent for infinite precision, exhibit different behavior when finite-precision arithmetic is used in the implementation. Therefore, it is very important in practice to select a realization that is not very sensitive to finite-word-length effects.

Although these three factors are the major ones in influencing our choice of the realization of a system of the type described by either (7.1.1) or (7.1.2), other factors, such as whether the structure or the realization lends itself to parallel processing, or whether the computations can be pipelined, may play a role in our selection of the specific implementation. These additional factors are usually important in the realization of more complex digital signal processing algorithms.

In our discussion of alternative realizations, we concentrate on the three major factors just outlined. Occasionally, we will include some additional factors that may be important in some implementations.

7.2 STRUCTURES FOR FIR SYSTEMS

In general, an FIR system is described by the difference equation

$$y(n) = \sum_{k=0}^{M-1} b_k x(n-k) \tag{7.2.1}$$

or, equivalently, by the system function

$$H(z) = \sum_{k=0}^{M-1} b_k z^{-k} \tag{7.2.2}$$

Furthermore, the unit sample response of the FIR system is identical to the coefficients $\{b_k\}$, that is,

$$h(n) = \begin{cases} b_n, & 0 \le n \le M - 1 \\ 0, & \text{otherwise} \end{cases} \tag{7.2.3}$$

The length of the FIR filter is selected as M to conform with the established notation in the technical literature.

We shall present several methods for implementing an FIR system, beginning with the simplest structure, called the direct form. A second structure is the cascade-form realization. The third structure that we shall describe is the frequency-sampling realization. Finally, we present a lattice realization of an FIR system. In this discussion we follow the convention often used in the technical literature, which is to use $\{h(n)\}$ for the parameters of an FIR system.

In addition to the four realizations indicated above, an FIR system can be realized by means of the DFT, as described in Section 6.2. From one point of view, the DFT can be considered as a computational procedure rather than a structure for an FIR system. However, when the computational procedure is implemented in hardware, there is a corresponding structure for the FIR system. In practice, hardware implementations of the DFT are based on the use of the fast Fourier transform (FFT) algorithms described in Chapter 6.

7.2.1 Direct-Form Structure

The direct-form realization follows immediately from the nonrecursive difference equation given by (7.2.1) or, equivalently, by the convolution summation

$$y(n) = \sum_{k=0}^{M-1} h(k)x(n-k) \tag{7.2.4}$$

The structure is illustrated in Fig. 7.1.

We observe that this structure requires $M - 1$ memory locations for storing the $M - 1$ previous inputs, and has a complexity of M multiplications and $M - 1$ additions per output point. Since the output consists of a weighted linear combination of $M - 1$ past values of the input and the weighted current value of the input, the structure in Fig. 7.1, resembles a tapped delay line or a transversal

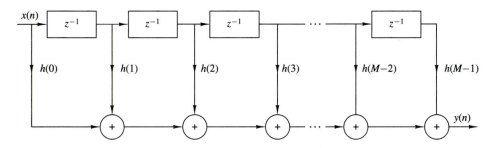

Figure 7.1   Direct-form realization of FIR system.

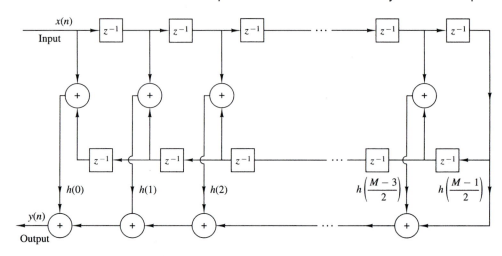

Figure 7.2   Direct-form realization of linear-phase FIR system (M odd).

system. Consequently, the direct-form realization is often called a transversal or tapped-delay-line filter.

When the FIR system has linear phase, as described in Section 8.2, the unit sample response of the system satisfies either the symmetry or asymmetry condition

$$h(n) = \pm h(M - 1 - n) \tag{7.2.5}$$

For such a system the number of multiplications is reduced from M to $M/2$ for M even and to $(M-1)/2$ for M odd. For example, the structure that takes advantage of this symmetry is illustrated in Fig. 7.2 for the case in which M is odd.

7.2.2 Cascade-Form Structures

The cascade realization follows naturally from the system function given by (7.2.2). It is a simple matter to factor $H(z)$ into second-order FIR systems so that

$$H(z) = \prod_{k=1}^{K} H_k(z) \tag{7.2.6}$$

where

$$H_k(z) = b_{k0} + b_{k1}z^{-1} + b_{k2}z^{-2} \qquad k = 1, 2, \ldots, K \tag{7.2.7}$$

and K is the integer part of $(M + 1)/2$. The filter parameter b_0 may be equally distributed among the K filter sections, such that $b_0 = b_{10}b_{20}\cdots b_{K0}$ or it may be assigned to a single filter section. The zeros of $H(z)$ are grouped in pairs to produce the second-order FIR systems of the form (7.2.7). It is always desirable to form pairs of complex-conjugate roots so that the coefficients $\{b_{ki}\}$ in (7.2.7) are real valued. On the other hand, real-valued roots can be paired in any arbitrary manner. The cascade-form realization along with the basic second-order section are shown in Fig. 7.3.

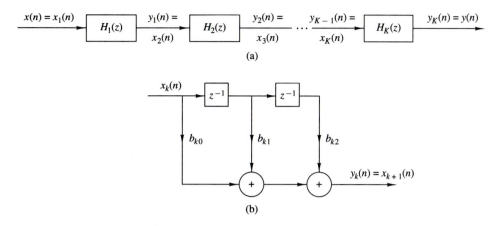

(a)

(b)

Figure 7.3 Cascade realization of an FIR system.

In the case of linear-phase FIR filters, the symmetry in $h(n)$ implies that the zeros of $H(z)$ also exhibit a form of symmetry. In particular, if z_k and z_k^* are a pair of complex-conjugate zeros then $1/z_k$ and $1/z_k^*$ are also a pair of complex-conjugate zeros (see Sec. 8.2). Consequently, we gain some simplification by forming fourth-order sections of the FIR system as follows

$$H_k(z) = c_{k0}(1 - z_k z^{-1})(1 - z_k^* z^{-1})(1 - z^{-1}/z_k)(1 - z^{-1}/z_k^*)$$
$$= c_{k0} + c_{k1} z^{-1} + c_{k2} z^{-2} + c_{k1} z^{-3} + z^{-4} \tag{7.2.8}$$

where the coefficients $\{c_{k1}\}$ and $\{c_{k2}\}$ are functions of z_k. Thus, by combining the two pairs of poles to form a fourth-order filter section, we have reduced the number of multiplications from six to three (i.e., by a factor of 50%). Figure 7.4 illustrates the basic fourth-order FIR filter structure.

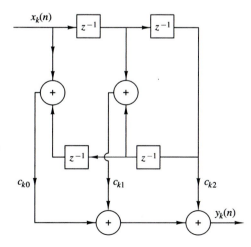

Figure 7.4 Fourth-order section in a cascade realization of an FIR system.

7.2.3 Frequency-Sampling Structures[†]

The frequency-sampling realization is an alternative structure for an FIR filter in which the parameters that characterize the filter are the values of the desired frequency response instead of the impulse response $h(n)$. To derive the frequency-sampling structure, we specify the desired frequency response at a set of equally spaced frequencies, namely

$$\omega_k = \frac{2\pi}{M}(k + \alpha) \qquad k = 0, 1, \ldots, \frac{M-1}{2} \quad M \text{ odd}$$

$$k = 0, 1, \ldots, \frac{M}{2} - 1 \quad M \text{ even}$$

$$\alpha = 0 \text{ or } \tfrac{1}{2}$$

and solve for the unit sample response $h(n)$ from these equally spaced frequency specifications. Thus we can write the frequency response as

$$H(\omega) = \sum_{n=0}^{M-1} h(n)e^{-j\omega n}$$

and the values of $H(\omega)$ at frequencies $\omega_k = (2\pi/M)(k + \alpha)$ are simply

$$H(k + \alpha) = H\left(\frac{2\pi}{M}(k + \alpha)\right)$$

$$= \sum_{n=0}^{M-1} h(n)e^{-j2\pi(k+\alpha)n/M} \qquad k = 0, 1, \ldots, M - 1 \tag{7.2.9}$$

The set of values $\{H(k+\alpha)\}$ are called the frequency samples of $H(\omega)$. In the case where $\alpha = 0$, $\{H(k)\}$ corresponds to the M-point DFT of $\{h(n)\}$.

It is a simple matter to invert (7.2.9) and express $h(n)$ in terms of the frequency samples. The result is

$$h(n) = \frac{1}{M} \sum_{k=0}^{M-1} H(k + \alpha)e^{j2\pi(k+\alpha)n/M} \qquad n = 0, 1, \ldots, M - 1 \tag{7.2.10}$$

When $\alpha = 0$, (7.2.10) is simply the IDFT of $\{H(k)\}$. Now if we use (7.2.10) to substitute for $h(n)$ in the z-transform $H(z)$, we have

$$H(z) = \sum_{n=0}^{M-1} h(n)z^{-n}$$

$$= \sum_{n=0}^{M-1} \left[\frac{1}{M} \sum_{k=0}^{M-1} H(k + \alpha)e^{j2\pi(k+\alpha)n/M} \right] z^{-n} \tag{7.2.11}$$

[†]The reader may also refer to Section 8.2.3 for additional discussion of frequency-sampling FIR filters.

By interchanging the order of the two summations in (7.2.11) and performing the summation over the index n we obtain

$$
\begin{aligned}
H(z) &= \sum_{k=0}^{M-1} H(k+\alpha) \left[\frac{1}{M} \sum_{n=0}^{M-1} (e^{j2\pi(k+\alpha)/M} z^{-1})^n \right] \\
&= \frac{1 - z^{-M} e^{j2\pi\alpha}}{M} \sum_{k=0}^{M-1} \frac{H(k+\alpha)}{1 - e^{j2\pi(k+\alpha)/M} z^{-1}}
\end{aligned}
\tag{7.2.12}
$$

Thus the system function $H(z)$ is characterized by the set of frequency samples $\{H(k+\alpha)\}$ instead of $\{h(n)\}$.

We view this FIR filter realization as a cascade of two filters [i.e., $H(z) = H_1(z)H_2(z)$]. One is an all-zero filter, or a comb filter, with system function

$$
H_1(z) = \frac{1}{M}(1 - z^{-M} e^{j2\pi\alpha})
\tag{7.2.13}
$$

Its zeros are located at equally spaced points on the unit circle at

$$
z_k = e^{j2\pi(k+\alpha)/M} \qquad k = 0, 1, \ldots, M-1
$$

The second filter with system function

$$
H_2(z) = \sum_{k=0}^{M-1} \frac{H(k+\alpha)}{1 - e^{j2\pi(k+\alpha)/M} z^{-1}}
\tag{7.2.14}
$$

consists of a parallel bank of single-pole filters with resonant frequencies

$$
p_k = e^{j2\pi(k+\alpha)/M} \qquad k = 0, 1, \ldots, M-1
$$

Note that the pole locations are identical to the zero locations and that both occur at $\omega_k = 2\pi(k+\alpha)/M$, which are the frequencies at which the desired frequency response is specified. The gains of the parallel bank of resonant filters are simply the complex-valued parameters $\{H(k+\alpha)\}$. This cascade realization is illustrated in Fig. 7.5.

When the desired frequency response characteristic of the FIR filter is narrowband, most of the gain parameters $\{H(k+\alpha)\}$ are zero. Consequently, the corresponding resonant filters can be eliminated and only the filters with nonzero gains need be retained. The net result is a filter that requires fewer computations (multiplications and additions) than the corresponding direct-form realization. Thus we obtain a more efficient realization.

The frequency-sampling filter structure can be simplified further by exploiting the symmetry in $H(k+\alpha)$, namely, $H(k) = H^*(M-k)$ for $\alpha = 0$ and

$$
H\left(k+\tfrac{1}{2}\right) = H^*\left(M-k-\tfrac{1}{2}\right) \qquad \text{for } \alpha = \tfrac{1}{2}
$$

These relations are easily deduced from (7.2.9). As a result of this symmetry, a pair of single-pole filters can be combined to form a single two-pole filter with

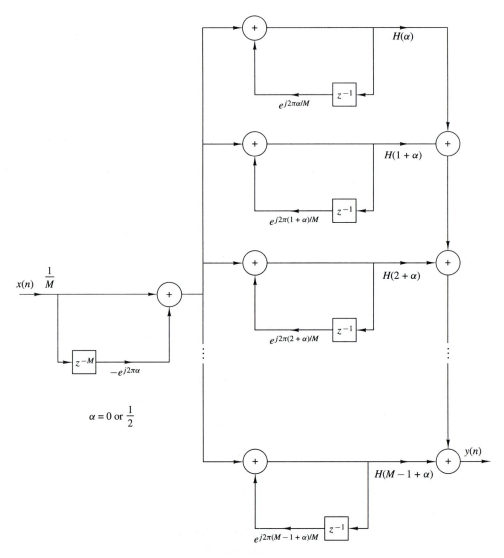

Figure 7.5 Frequency-sampling realization of FIR filter.

real-valued parameters. Thus for $\alpha = 0$ the system function $H_2(z)$ reduces to

$$H_2(z) = \frac{H(0)}{1 - z^{-1}} + \sum_{k=1}^{(M-1)/2} \frac{A(k) + B(k)z^{-1}}{1 - 2\cos(2\pi k/M)z^{-1} + z^{-2}} \qquad M \text{ odd}$$

$$H_2(z) = \frac{H(0)}{1 - z^{-1}} + \frac{H(M/2)}{1 + z^{-1}} + \sum_{k=1}^{(M/2)-1} \frac{A(k) + B(k)z^{-1}}{1 - 2\cos(2\pi k/M)z^{-1} + z^{-2}} \qquad M \text{ even}$$

$$(7.2.15)$$

where, by definition,

$$A(k) = H(k) + H(M - k)$$

$$B(k) = H(k)e^{-j2\pi k/M} + H(M - k)e^{j2\pi k/M} \qquad (7.2.16)$$

Similar expressions can be obtained for $\alpha = \frac{1}{2}$.

Example 7.2.1

Sketch the block diagram for the direct-form realization and the frequency-sampling realization of the $M = 32$, $\alpha = 0$, linear-phase (symmetric) FIR filter which has frequency samples

$$H\left(\frac{2\pi k}{32}\right) = \begin{cases} 1, & k = 0, 1, 2 \\ \dfrac{1}{2}, & k = 3 \\ 0, & k = 4, 5, \ldots, 15 \end{cases}$$

Compare the computational complexity of these two structures.

Solution   Since the filter is symmetric, we exploit this symmetry and thus reduce the number of multiplications per output point by a factor of 2, from 32 to 16 in the direct-form realization. The number of additions per output point is 31. The block diagram of the direct realization is illustrated in Fig. 7.6.

We use the form in (7.2.13) and (7.2.15) for the frequency-sampling realization and drop all terms that have zero-gain coefficients {H(k)}. The nonzero coefficients are $H(k)$ and the corresponding pairs are $H(M - k)$, for $k = 0, 1, 2, 3$. The block diagram of the resulting realization is shown in Fig. 7.7. Since $H(0) = 1$, the single-pole filter requires no multiplication. The three double-pole filter sections require three multiplications each for a total of nine multiplications. The total number of additions is 13. Therefore, the frequency-sampling realization of this FIR filter is computationally more efficient than the direct-form realization.

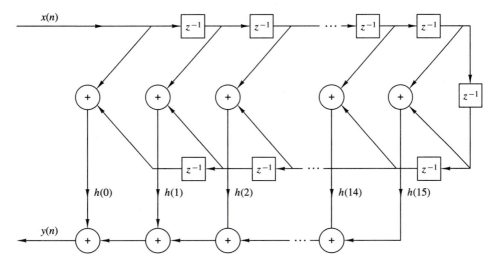

Figure 7.6   Direct-form realization of $M = 32$ FIR filter.

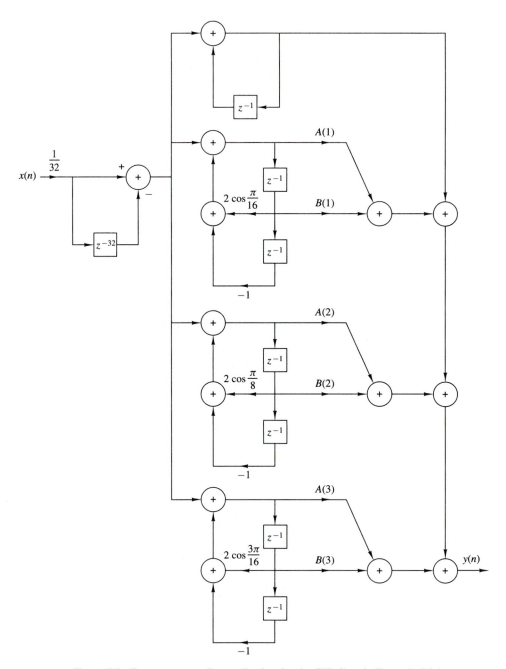

Figure 7.7 Frequency-sampling realization for the FIR filter in Example 7.2.1.

7.2.4 Lattice Structure

In this section we introduce another FIR filter structure, called the lattice filter or lattice realization. Lattice filters are used extensively in digital speech processing and in the implementation of adaptive filters.

Let us begin the development by considering a sequence of FIR filters with system functions

$$H_m(z) = A_m(z) \qquad m = 0, 1, 2, \ldots, M - 1 \tag{7.2.17}$$

where, by definition, $A_m(z)$ is the polynomial

$$A_m(z) = 1 + \sum_{k=1}^{m} \alpha_m(k) z^{-k} \qquad m \geq 1 \tag{7.2.18}$$

and $A_0(z) = 1$. The unit sample response of the mth filter is $h_m(0) = 1$ and $h_m(k) = \alpha_m(k)$, $k = 1, 2, \ldots, m$. The subscript m on the polynomial $A_m(z)$ denotes the degree of the polynomial. For mathematical convenience, we define $\alpha_m(0) = 1$.

If $\{x(n)\}$ is the input sequence to the filter $A_m(z)$ and $\{y(n)\}$ is the output sequence, we have

$$y(n) = x(n) + \sum_{k=1}^{m} \alpha_m(k) x(n - k) \tag{7.2.19}$$

Two direct-form structures of the FIR filter are illustrated in Fig. 7.8.

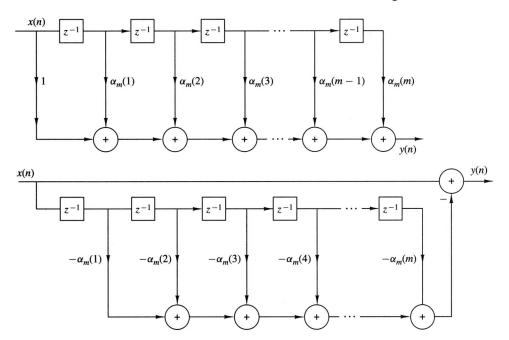

Figure 7.8 Direct-form realization of the FIR prediction filter.

In Chapter 11, we show that the FIR structures shown in Fig. 7.8 are intimately related with the topic of linear prediction, where

$$\hat{x}(n) = -\sum_{k=1}^{m} \alpha_m(k)x(n-k) \tag{7.2.20}$$

is the one-step forward predicted value of $x(n)$, based on m past inputs, and $y(n) = x(n) - \hat{x}(n)$, given by (7.2.19), represents the prediction error sequence. In this context, the top filter structure in Fig. 7.8 is called a *prediction error filter*.

Now suppose that we have a filter of order $m = 1$. The output of such a filter is

$$y(n) = x(n) + \alpha_1(1)x(n-1) \tag{7.2.21}$$

This output can also be obtained from a first-order or single-stage lattice filter, illustrated in Fig. 7.9, by exciting both of the inputs by $x(n)$ and selecting the output from the top branch. Thus the output is exactly (7.2.21), if we select $K_1 = \alpha_1(1)$. The parameter K_1 in the lattice is called a reflection coefficient and it is identical to the *reflection coefficient* introduced in the Schür–Cohn stability test described in Section 3.6.7.

Next, let us consider an FIR filter for which $m = 2$. In this case the output from a direct-form structure is

$$y(n) = x(n) + \alpha_2(1)x(n-1) + \alpha_2(2)x(n-2) \tag{7.2.22}$$

By cascading two lattice stages as shown in Fig. 7.10, it is possible to obtain the same output as (7.2.22). Indeed, the output from the first stage is

$$f_1(n) = x(n) + K_1 x(n-1)$$
$$g_1(n) = K_1 x(n) + x(n-1) \tag{7.2.23}$$

The output from the second stage is

$$f_2(n) = f_1(n) + K_2 g_1(n-1)$$
$$g_2(n) = K_2 f_1(n) + g_1(n-1) \tag{7.2.24}$$

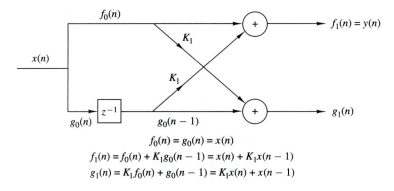

$$f_0(n) = g_0(n) = x(n)$$
$$f_1(n) = f_0(n) + K_1 g_0(n-1) = x(n) + K_1 x(n-1)$$
$$g_1(n) = K_1 f_0(n) + g_0(n-1) = K_1 x(n) + x(n-1)$$

Figure 7.9 Single-stage lattice filter.

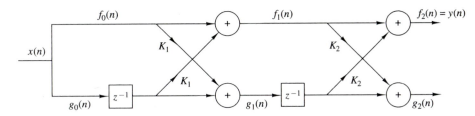

Figure 7.10 Two-stage lattice filter.

If we focus our attention on $f_2(n)$ and substitute for $f_1(n)$ and $g_1(n-1)$ from (7.2.23) into (7.2.24), we obtain

$$f_2(n) = x(n) + K_1 x(n-1) + K_2[K_1 x(n-1) + x(n-2)]$$
$$= x(n) + K_1(1 + K_2)x(n-1) + K_2 x(n-2) \tag{7.2.25}$$

Now (7.2.25) is identical to the output of the direct-form FIR filter as given by (7.2.22), if we equate the coefficients, that is,

$$\alpha_2(2) = K_2 \qquad \alpha_2(1) = K_1(1 + K_2) \tag{7.2.26}$$

or, equivalently,

$$K_2 = \alpha_2(2) \qquad K_1 = \frac{\alpha_2(1)}{1 + \alpha_2(2)} \tag{7.2.27}$$

Thus the reflection coefficients K_1 and K_2 of the lattice can be obtained from the coefficients $\{\alpha_m(k)\}$ of the direct-form realization.

By continuing this process, one can easily demonstrate by induction, the equivalence between an mth-order direct-form FIR filter and an m-order or m-stage lattice filter. The lattice filter is generally described by the following set of order-recursive equations:

$$f_0(n) = g_0(n) = x(n) \tag{7.2.28}$$

$$f_m(n) = f_{m-1}(n) + K_m g_{m-1}(n-1) \qquad m = 1, 2, \ldots, M-1 \tag{7.2.29}$$

$$g_m(n) = K_m f_{m-1}(n) + g_{m-1}(n-1) \qquad m = 1, 2, \ldots, M-1 \tag{7.2.30}$$

Then the output of the $(M-1)$-stage filter corresponds to the output of an $(M-1)$-order FIR filter, that is,

$$y(n) = f_{M-1}(n)$$

Figure 7.11 illustrates an $(M-1)$-stage lattice filter in block diagram form along with a typical stage that shows the computations specified by (7.2.29) and (7.2.30).

As a consequence of the equivalence between an FIR filter and a lattice filter, the output $f_m(n)$ of an m-stage lattice filter can be expressed as

$$f_m(n) = \sum_{k=0}^{m} \alpha_m(k)x(n-k) \qquad \alpha_m(0) = 1 \tag{7.2.31}$$

Since (7.2.31) is a convolution sum, it follows that the z-transform relationship is

$$F_m(z) = A_m(z)X(z)$$

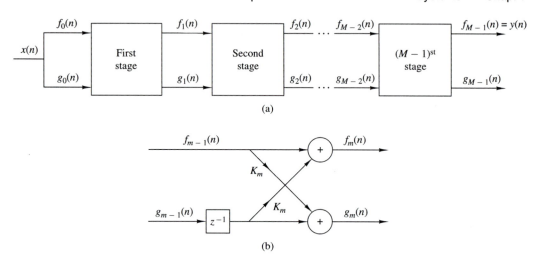

Figure 7.11 $(M-1)$-stage lattice filter.

or, equivalently,

$$A_m(z) = \frac{F_m(z)}{X(z)} = \frac{F_m(z)}{F_0(z)} \tag{7.2.32}$$

The other output component from the lattice, namely, $g_m(n)$, can also be expressed in the form of a convolution sum as in (7.2.31), by using another set of coefficients, say $\{\beta_m(k)\}$. That this in fact is the case, becomes apparent from observation of (7.2.23) and (7.2.24). From (7.2.23) we note that the filter coefficients for the lattice filter that produces $f_1(n)$ are $\{1, K_1\} = \{1, \alpha_1(1)\}$ while the coefficients for the filter with output $g_1(n)$ are $\{K_1, 1\} = \{\alpha_1(1), 1\}$. We note that these two sets of coefficients are in reverse order. If we consider the two-stage lattice filter, with the output given by (7.2.24), we find that $g_2(n)$ can be expressed in the form

$$
\begin{aligned}
g_2(n) &= K_2 f_1(n) + g_1(n-1) \\
&= K_2[x(n) + K_1 x(n-1)] + K_1 x(n-1) + x(n-2) \\
&= K_2 x(n) + K_1(1 + K_2)x(n-1) + x(n-2) \\
&= \alpha_2(2)x(n) + \alpha_2(1)x(n-1) + x(n-2)
\end{aligned}
$$

Consequently, the filter coefficients are $\{\alpha_2(2), \alpha_2(1), 1\}$, whereas the coefficients for the filter that produces the output $f_2(n)$ are $\{1, \alpha_2(1), \alpha_2(2)\}$. Here, again, the two sets of filter coefficients are in reverse order.

From this development it follows that the output $g_m(n)$ from an m-stage lattice filter can be expressed by the convolution sum of the form

$$g_m(n) = \sum_{k=0}^{m} \beta_m(k)x(n-k) \tag{7.2.33}$$

where the filter coefficients $\{\beta_m(k)\}$ are associated with a filter that produces $f_m(n) = y(n)$ but operates in reverse order. Consequently,

$$\beta_m(k) = \alpha_m(m - k) \qquad k = 0, 1, \ldots, m \qquad (7.2.34)$$

with $\beta_m(m) = 1$.

In the context of linear prediction, suppose that the data $x(n), x(n - 1), \ldots,$ $x(n - m + 1)$ is used to linearly predict the signal value $x(n - m)$ by use of a linear filter with coefficients $\{-\beta_m(k)\}$. Thus the predicted value is

$$\hat{x}(n - m) = -\sum_{k=0}^{m-1} \beta_m(k)x(n - k) \qquad (7.2.35)$$

Since the data are run in reverse order through the predictor, the prediction performed in (7.2.35) is called *backward prediction*. In contrast, the FIR filter with system function $A_m(z)$ is called a *forward predictor*.

In the z-transform domain, (7.2.33) becomes

$$G_m(z) = B_m(z)X(z) \qquad (7.2.36)$$

or, equivalently,

$$B_m(z) = \frac{G_m(z)}{X(z)} \qquad (7.2.37)$$

where $B_m(z)$ represents the system function of the FIR filter with coefficients $\{\beta_m(k)\}$, that is,

$$B_m(z) = \sum_{k=0}^{m} \beta_m(k)z^{-k} \qquad (7.2.38)$$

Since $\beta_m(k) = \alpha_m(m - k)$, (7.2.38) may be expressed as

$$
\begin{aligned}
B_m(z) &= \sum_{k=0}^{m} \alpha_m(m - k)z^{-k} \\
&= \sum_{l=0}^{m} \alpha_m(l)z^{l-m} \\
&= z^{-m} \sum_{l=0}^{m} \alpha_m(l)z^{l} \\
&= z^{-m} A_m(z^{-1})
\end{aligned}
\qquad (7.2.39)
$$

The relationship in (7.2.39) implies that the zeros of the FIR filter with system function $B_m(z)$ are simply the reciprocals of the zeros of $A_m(z)$. Hence $B_m(z)$ is called the reciprocal or *reverse* polynomial of $A_m(z)$.

Now that we have established these interesting relationships between the direct-form FIR filter and the lattice structure, let us return to the recursive lattice equations in (7.2.28) through (7.2.30) and transfer them to the z-domain. Thus we have

$$F_0(z) = G_0(z) = X(z) \tag{7.2.40}$$

$$F_m(z) = F_{m-1}(z) + K_m z^{-1} G_{m-1}(z) \qquad m = 1, 2, \ldots, M - 1 \tag{7.2.41}$$

$$G_m(z) = K_m F_{m-1}(z) + z^{-1} G_{m-1}(z) \qquad m = 1, 2, \ldots, M - 1 \tag{7.2.42}$$

If we divide each equation by $X(z)$, we obtain the desired results in the form

$$A_0(z) = B_0(z) = 1 \tag{7.2.43}$$

$$A_m(z) = A_{m-1}(z) + K_m z^{-1} B_{m-1}(z) \qquad m = 1, 2, \ldots, M - 1 \tag{7.2.44}$$

$$B_m(z) = K_m A_{m-1}(z) + z^{-1} B_{m-1}(z) \qquad m = 1, 2, \ldots, M - 1 \tag{7.2.45}$$

Thus a lattice stage is described in the z-domain by the matrix equation

$$\begin{bmatrix} A_m(z) \\ B_m(z) \end{bmatrix} = \begin{bmatrix} 1 & K_m \\ K_m & 1 \end{bmatrix} \begin{bmatrix} A_{m-1}(z) \\ z^{-1} B_{m-1}(z) \end{bmatrix} \tag{7.2.46}$$

Before concluding this discussion, it is desirable to develop the relationships for converting the lattice parameters $\{K_i\}$, that is, the reflection coefficients, to the direct-form filter coefficients $\{\alpha_m(k)\}$, and vice versa.

Conversion of lattice coefficients to direct-form filter coefficients.

The direct-form FIR filter coefficients $\{\alpha_m(k)\}$ can be obtained from the lattice coefficients $\{K_i\}$ by using the following relations:

$$A_0(z) = B_0(z) = 1 \tag{7.2.47}$$

$$A_m(z) = A_{m-1}(z) + K_m z^{-1} B_{m-1}(z) \qquad m = 1, 2, \ldots, M - 1 \tag{7.2.48}$$

$$B_m(z) = z^{-m} A_m(z^{-1}) \qquad m = 1, 2, \ldots, M - 1 \tag{7.2.49}$$

The solution is obtained recursively, beginning with $m = 1$. Thus we obtain a sequence of $(M - 1)$ FIR filters, one for each value of m. The procedure is best illustrated by means of an example.

Example 7.2.2

Given a three-stage lattice filter with coefficients $K_1 = \frac{1}{4}$, $K_2 = \frac{1}{2}$, $K_3 = \frac{1}{3}$, determine the FIR filter coefficients for the direct-form structure.

Solution We solve the problem recursively, beginning with (7.2.48) for $m = 1$. Thus we have

$$A_1(z) = A_0(z) + K_1 z^{-1} B_0(z)$$

$$= 1 + K_1 z^{-1} = 1 + \tfrac{1}{4} z^{-1}$$

Hence the coefficients of an FIR filter corresponding to the single-stage lattice are $\alpha_1(0) = 1$, $\alpha_1(1) = K_1 = \frac{1}{4}$. Since $B_m(z)$ is the reverse polynomial of $A_m(z)$, we have

$$B_1(z) = \tfrac{1}{4} + z^{-1}$$

Next we add the second stage to the lattice. For $m = 2$, (7.2.48) yields

$$A_2(z) = A_1(z) + K_2 z^{-1} B_1(z)$$

$$= 1 + \tfrac{3}{8} z^{-1} + \tfrac{1}{2} z^{-2}$$

Hence the FIR filter parameters corresponding to the two-stage lattice are $\alpha_2(0) = 1$, $\alpha_2(1) = \tfrac{3}{8}$, $\alpha_2(2) = \tfrac{1}{2}$. Also,

$$B_2(z) = \tfrac{1}{2} + \tfrac{3}{8} z^{-1} + z^{-2}$$

Finally, the addition of the third stage to the lattice results in the polynomial

$$A_3(z) = A_2(z) + K_3 z^{-1} B_2(z)$$

$$= 1 + \tfrac{13}{24} z^{-1} + \tfrac{5}{8} z^{-2} + \tfrac{1}{3} z^{-3}$$

Consequently, the desired direct-form FIR filter is characterized by the coefficients

$$\alpha_3(0) = 1 \qquad \alpha_3(1) = \tfrac{13}{24} \qquad \alpha_3(2) = \tfrac{5}{8} \qquad \alpha_3(3) = \tfrac{1}{3}$$

As this example illustrates, the lattice structure with parameters $K_1, K_2, \ldots, K_m$, corresponds to a class of m direct-form FIR filters with system functions $A_1(z)$, $A_2(z), \ldots, A_m(z)$. It is interesting to note that a characterization of this class of m FIR filters in direct form requires $m(m + 1)/2$ filter coefficients. In contrast, the lattice-form characterization requires only the m reflection coefficients $\{K_i\}$. The reason that the lattice provides a more compact representation for the class of m FIR filters is simply due to the fact that the addition of stages to the lattice does not alter the parameters of the previous stages. On the other hand, the addition of the mth stage to a lattice with $(m - 1)$ stages results in a FIR filter with system function $A_m(z)$ that has coefficients totally different from the coefficients of the lower-order FIR filter with system function $A_{m-1}(z)$.

A formula for determining the filter coefficients $\{\alpha_m(k)\}$ recursively can be easily derived from polynomial relationships in (7.2.47) through (7.2.49). From the relationship in (7.2.48) we have

$$A_m(z) = A_{m-1}(z) + K_m z^{-1} B_{m-1}(z)$$

$$\sum_{k=0}^{m} \alpha_m(k) z^{-k} = \sum_{k=0}^{m-1} \alpha_{m-1}(k) z^{-k} + K_m \sum_{k=0}^{m-1} \alpha_{m-1}(m - 1 - k) z^{-(k+1)} \qquad (7.2.50)$$

By equating the coefficients of equal powers of z^{-1} and recalling that $\alpha_m(0) = 1$ for $m = 1, 2, \ldots, M - 1$, we obtain the desired recursive equation for the FIR filter coefficients in the form

$$\alpha_m(0) = 1 \qquad (7.2.51)$$

$$\alpha_m(m) = K_m \qquad (7.2.52)$$

$$\alpha_m(k) = \alpha_{m-1}(k) + K_m \alpha_{m-1}(m - k)$$

$$= \alpha_{m-1}(k) + \alpha_m(m) \alpha_{m-1}(m - k) \qquad \begin{matrix} 1 \le k \le m - 1 \\ m = 1, 2, \ldots, M - 1 \end{matrix} \qquad (7.2.53)$$

We note that (7.2.51) through (7.2.53) are simply the Levinson–Durbin recursive equations given in Chapter 11.

Conversion of direct-form FIR filter coefficients to lattice coefficients.
Suppose that we are given the FIR coefficients for the direct-form realization or, equivalently, the polynomial $A_m(z)$, and we wish to determine the corresponding lattice filter parameters $\{K_i\}$. For the m-stage lattice we immediately obtain the parameter $K_m = \alpha_m(m)$. To obtain K_{m-1} we need the polynomials $A_{m-1}(z)$ since, in general, K_m is obtained from the polynomial $A_m(z)$ for $m = M-1, M-2, \ldots, 1$. Consequently, we need to compute the polynomials $A_m(z)$ starting from $m = M-1$ and "stepping down" successively to $m = 1$.

The desired recursive relation for the polynomials is easily determined from (7.2.44) and (7.2.45). We have

$$A_m(z) = A_{m-1}(z) + K_m z^{-1} B_{m-1}(z)$$
$$= A_{m-1}(z) + K_m[B_m(z) - K_m A_{m-1}(z)]$$

If we solve for $A_{m-1}(z)$, we obtain

$$A_{m-1}(z) = \frac{A_m(z) - K_m B_m(z)}{1 - K_m^2} \qquad m = M-1, M-2, \ldots, 1 \qquad (7.2.54)$$

which is just the step-down recursion used in the Schür–Cohn stability test described in Section 3.6.7. Thus we compute all lower-degree polynomials $A_m(z)$ beginning with $A_{M-1}(z)$ and obtain the desired lattice coefficients from the relation $K_m = \alpha_m(m)$. We observe that the procedure works as long as $|K_m| \neq 1$ for $m = 1, 2, \ldots, M-1$.

Example 7.2.3

Determine the lattice coefficients corresponding to the FIR filter with system function

$$H(z) = A_3(z) = 1 + \tfrac{13}{24}z^{-1} + \tfrac{5}{8}z^{-2} + \tfrac{1}{3}z^{-3}$$

Solution   First we note that $K_3 = \alpha_3(3) = \tfrac{1}{3}$. Furthermore,

$$B_3(z) = \tfrac{1}{3} + \tfrac{5}{8}z^{-1} + \tfrac{13}{24}z^{-2} + z^{-3}$$

The step-down relationship in (7.2.54) with $m = 3$ yields

$$A_2(z) = \frac{A_3(z) - K_3 B_3(z)}{1 - K_3^2}$$
$$= 1 + \tfrac{3}{8}z^{-1} + \tfrac{1}{2}z^{-2}$$

Hence $K_2 = \alpha_2(2) = \tfrac{1}{2}$ and $B_2(z) = \tfrac{1}{2} + \tfrac{3}{8}z^{-1} + z^{-1}$. By repeating the step-down recursion in (7.2.51), we obtain

$$A_1(z) = \frac{A_2(z) - K_2 B_2(z)}{1 - K_2^2}$$
$$= 1 + \tfrac{1}{4}z^{-1}$$

Hence $K_1 = \alpha_1(1) = \tfrac{1}{4}$.

From the step-down recursive equation in (7.2.54), it is relatively easy to obtain a formula for recursively computing K_m, beginning with $m = M - 1$ and stepping down to $m = 1$. For $m = M - 1, M - 2, \ldots, 1$ we have

$$K_m = \alpha_m(m) \quad \alpha_{m-1}(0) = 1 \tag{7.2.55}$$

$$\alpha_{m-1}(k) = \frac{\alpha_m(k) - K_m \beta_m(k)}{1 - K_m^2}$$

$$= \frac{\alpha_m(k) - \alpha_m(m)\alpha_m(m - k)}{1 - \alpha_m^2(m)} \quad 1 \le k \le m - 1 \tag{7.2.56}$$

which is again the recursion we introduced in the Schür–Cohn stability test.

As indicated above, the recursive equation in (7.2.56) breaks down if any lattice parameters $|K_m| = 1$. If this occurs, it is indicative of the fact that the polynomial $A_{m-1}(z)$ has a root on the unit circle. Such a root can be factored out from $A_{m-1}(z)$ and the iterative process in (7.2.56) is carried out for the reduced-order system.

7.3 STRUCTURES FOR IIR SYSTEMS

In this section we consider different IIR systems structures described by the difference equation in (7.1.1) or, equivalently, by the system function in (7.1.2). Just as in the case of FIR systems, there are several types of structures or realizations, including direct-form structures, cascade-form structures, lattice structures, and lattice-ladder structures. In addition, IIR systems lend themselves to a parallel-form realization. We begin by describing two direct-form realizations.

7.3.1 Direct-Form Structures

The rational system function as given by (7.1.2) that characterizes an IIR system can be viewed as two systems in cascade, that is,

$$H(z) = H_1(z)H_2(z) \tag{7.3.1}$$

where $H_1(z)$ consists of the zeros of $H(z)$, and $H_2(z)$ consists of the poles of $H(z)$,

$$H_1(z) = \sum_{k=0}^{M} b_k z^{-k} \tag{7.3.2}$$

and

$$H_2(z) = \frac{1}{1 + \sum_{k=1}^{N} a_k z^{-k}} \tag{7.3.3}$$

In Section 2.5.1 we describe two different direct-form realizations, characterized by whether $H_1(z)$ precedes $H_2(z)$, or vice versa. Since $H_1(z)$ is an FIR system, its direct-form realization was illustrated in Fig. 7.1. By attaching the all-pole

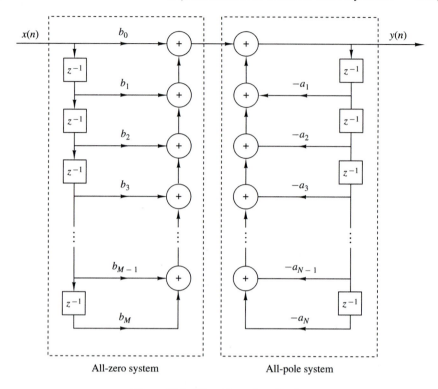

Figure 7.12   Direct form I realization.

system in cascade with $H_1(z)$, we obtain the direct form I realization depicted in Fig. 7.12. This realization requires $M + N + 1$ multiplications, $M + N$ additions, and $M + N + 1$ memory locations.

If the all-pole filter $H_2(z)$ is placed before the all-zero filter $H_1(z)$, a more compact structure is obtained as illustrated in Section 2.5.1. Recall that the difference equation for the all-pole filter is

$$w(n) = -\sum_{k=1}^{N} a_k w(n - k) + x(n) \tag{7.3.4}$$

Since $w(n)$ is the input to the all-zero system, its output is

$$y(n) = \sum_{k=0}^{M} b_k w(n - k) \tag{7.3.5}$$

We note that both (7.3.4) and (7.3.5) involve delayed versions of the sequence $\{w(n)\}$. Consequently, only a single delay line or a single set of memory locations is required for storing the past values of $\{w(n)\}$. The resulting structure that implements (7.3.4) and (7.3.5) is called a direct form II realization and is depicted in Fig. 7.13. This structure requires $M + N + 1$ multiplications, $M + N$ additions,

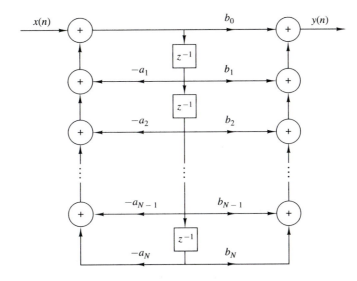

Figure 7.13   Direct form II realization ($N = M$).

and the maximum of $\{M, N\}$ memory locations. Since the direct form II realization minimizes the number of memory locations, it is said to be *canonic*. However, we should indicate that other IIR structures also possess this property, so that this terminology is perhaps unjustified.

The structures in Figs. 7.12 and 7.13 are both called "direct form" realizations because they are obtained directly from the system function $H(z)$ without any rearrangement of $H(z)$. Unfortunately, both structures are extremely sensitive to parameter quantization, in general, and are not recommended in practical applications. This topic is discussed in detail in Section 7.6, where we demonstrate that when N is large, a small change in a filter coefficient due to parameter quantization, results in a large change in the location of the poles and zeros of the system.

7.3.2 Signal Flow Graphs and Transposed Structures

A signal flow graph provides an alternative, but equivalent, graphical representation to a block diagram structure that we have been using to illustrate various system realizations. The basic elements of a flow graph are branches and nodes. A signal flow graph is basically a set of directed branches that connect at nodes. By definition, the signal out of a branch is equal to the branch gain (system function) times the signal into the branch. Furthermore, the signal at a node of a flow graph is equal to the sum of the signals from all branches connecting to the node.

To illustrate these basic notions, let us consider the two-pole and two-zero IIR system depicted in block diagram form in Fig. 7.14a. The system block

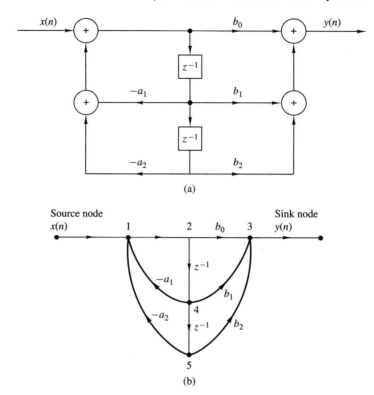

Figure 7.14 Second-order filter structure (a) and its signal flow graph (b).

diagram can be converted to the signal flow graph shown in Fig. 7.14b. We note that the flow graph contains five nodes labeled 1 through 5. Two of the nodes $(1, 3)$ are summing nodes (i.e., they contain adders), while the other three nodes represent branching points. Branch transmittances are indicated for the branches in the flow graph. Note that a delay is indicated by the branch transmittance z^{-1}. When the branch transmittance is unity, it is left unlabeled. The input to the system originates at a *source node* and the output signal is extracted at a *sink node*.

We observe that the signal flow graph contains the same basic information as the block diagram realization of the system. The only apparent difference is that *both* branch points and adders in the block diagram are represented by nodes in the signal flow graph.

The subject of linear signal flow graphs is an important one in the treatment of networks and many interesting results are available. One basic notion involves the transformation of one flow graph into another without changing the basic input–output relationship. Specifically, one technique that is useful in deriving new system structures for FIR and IIR systems stems from the *transposition* or *flow-graph reversal theorem*. This theorem simply states that if we reverse the directions of all branch transmittances and interchange the input and output in the flow graph, the system function remains unchanged. The resulting structure is called a *transposed structure* or a *transposed form*.

For example, the transposition of the signal flow graph in Fig. 7.14b is illustrated in Fig. 7.15a. The corresponding block diagram realization of the transposed form is depicted in Fig. 7.15b. It is interesting to note that the transposition of the original flow graph resulted in branching nodes becoming adder nodes, and vice versa. In Section 7.5 we provide a proof of the transposition theorem by using state-space techniques.

Let us apply the transposition theorem to the direct form II structure. First, we reverse all the signal flow directions in Fig. 7.13. Second, we change nodes into adders and adders into nodes, and finally, we interchange the input and the output. These operations result in the transposed direct form II structure shown in Fig. 7.16. This structure can be redrawn as in Fig. 7.17, which shows the input on the left and the output on the right.

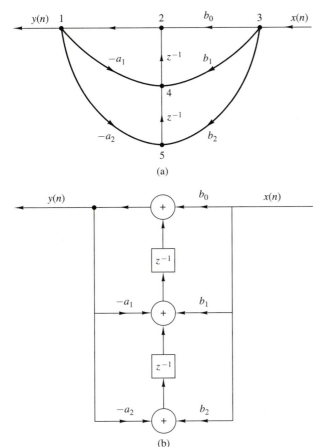

Figure 7.15 Signal flow graph of transposed structure (a) and its realization (b).

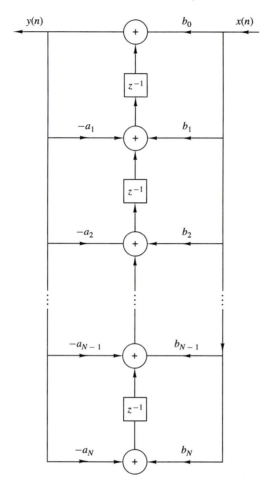

Figure 7.16 Transposed direct form II structure.

The transposed direct form II realization that we have obtained can be described by the set of difference equations

$$y(n) = w_1(n-1) + b_0 x(n) \tag{7.3.6}$$

$$w_k(n) = w_{k+1}(n-1) - a_k y(n) + b_k x(n) \qquad k = 1, 2, \ldots, N-1 \tag{7.3.7}$$

$$w_N(n) = b_N x(n) - a_N y(n) \tag{7.3.8}$$

Without loss of generality, we have assumed that $M = N$ in writing equations. It is also clear from observation of Fig. 7.17 that this set of difference equations is equivalent to the single difference equation

$$y(n) = -\sum_{k=1}^{N} a_k y(n-k) + \sum_{k=0}^{M} b_k x(n-k) \tag{7.3.9}$$

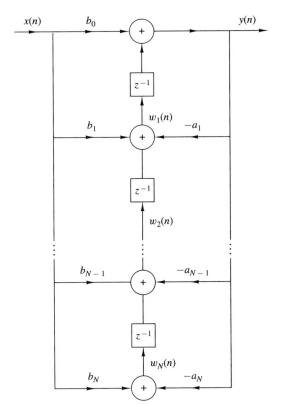

Figure 7.17 Transposed direct form II structure.

Finally, we observe that the transposed direct form II structure requires the same number of multiplications, additions, and memory locations as the original direct form II structure.

Although our discussion of transposed structures has been concerned with the general form of an IIR system, it is interesting to note that an FIR system, obtained from (7.3.9) by setting the $a_k = 0$, $k = 1, 2, \ldots, N$, also has a transposed direct form as illustrated in Fig. 7.18. This structure is simply obtained from Fig. 7.17 by setting $a_k = 0$, $k = 1, 2, \ldots, N$. This transposed form realization may

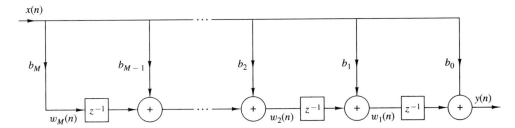

Figure 7.18 Transposed FIR structure.

be described by the set of difference equations

$$w_M(n) = b_M x(n) \tag{7.3.10}$$

$$w_k(n) = w_{k+1}(n-1) + b_k x(n) \qquad k = M-1, M-2, \dots, 1 \tag{7.3.11}$$

$$y(n) = w_1(n-1) + b_0 x(n) \tag{7.3.12}$$

In summary, Table 7.1 illustrates the direct-form structures and the corresponding difference equations for a basic two-pole and two-zero IIR system with system function

$$H(z) = \frac{b_0 + b_1 z^{-1} + b_2 z^{-2}}{1 + a_1 z^{-1} + a_2 z^{-2}} \tag{7.3.13}$$

This is the basic building block in the cascade realization of high-order IIR systems, as described in the following section. Of the three direct-form structures given in Table 7.1, the direct form II structures are preferable due to the smaller number of memory locations required in their implementation.

Finally, we note that in the z-domain, the set of difference equations describing a linear signal flow graph constitute a linear set of equations. Any rearrangement of such a set of equations is equivalent to a rearrangement of the signal flow graph to obtain a new structure, and vice versa.

7.3.3 Cascade-Form Structures

Let us consider a high-order IIR system with system function given by (7.1.2). Without loss of generality we assume that $N \geq M$. The system can be factored into a cascade of second-order subsystems, such that $H(z)$ can be expressed as

$$H(z) = \prod_{k=1}^{K} H_k(z) \tag{7.3.14}$$

where K is the integer part of $(N+1)/2$. $H_k(z)$ has the general form

$$H_k(z) = \frac{b_{k0} + b_{k1} z^{-1} + b_{k2} z^{-2}}{1 + a_{k1} z^{-1} + a_{k2} z^{-2}} \tag{7.3.15}$$

As in the case of FIR systems based on a cascade-form realization, the parameter b_0 can be distributed equally among the K filter sections so that $b_0 = b_{10} b_{20} \dots b_{K0}$.

The coefficients $\{a_{ki}\}$ and $\{b_{ki}\}$ in the second-order subsystems are real. This implies that in forming the second-order subsystems or quadratic factors in (7.3.15), we should group together a pair of complex-conjugate poles and we should group together a pair of complex-conjugate zeros. However, the pairing of two complex-conjugate poles with a pair of complex-conjugate zeros or real-valued zeros to form a subsystem of the type given by (7.3.15), can be done arbitrarily. Furthermore, any two real-valued zeros can be paired together to form a quadratic factor and, likewise, any two real-valued poles can be paired together to form a quadratic factor. Consequently, the quadratic factor in the numerator of (7.3.15) may consist

TABLE 7.1 SOME SECOND-ORDER MODULES FOR DISCRETE-TIME SYSTEMS

Structure	Implementation Equations	System Function

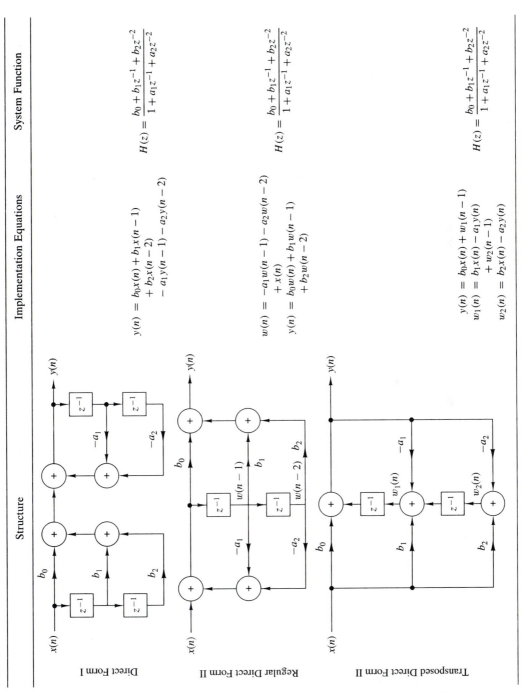

Direct Form I

$$y(n) = b_0 x(n) + b_1 x(n-1)$$
$$+ b_2 x(n-2)$$
$$- a_1 y(n-1) - a_2 y(n-2)$$

$$H(z) = \frac{b_0 + b_1 z^{-1} + b_2 z^{-2}}{1 + a_1 z^{-1} + a_2 z^{-2}}$$

Regular Direct Form II

$$w(n) = -a_1 w(n-1) - a_2 w(n-2)$$
$$+ x(n)$$
$$y(n) = b_0 w(n) + b_1 w(n-1)$$
$$+ b_2 w(n-2)$$

$$H(z) = \frac{b_0 + b_1 z^{-1} + b_2 z^{-2}}{1 + a_1 z^{-1} + a_2 z^{-2}}$$

Transposed Direct Form II

$$y(n) = b_0 x(n) + w_1(n-1)$$
$$w_1(n) = b_1 x(n) - a_1 y(n)$$
$$+ w_2(n-1)$$
$$w_2(n) = b_2 x(n) - a_2 y(n)$$

$$H(z) = \frac{b_0 + b_1 z^{-1} + b_2 z^{-2}}{1 + a_1 z^{-1} + a_2 z^{-2}}$$

of either a pair of real roots or a pair of complex-conjugate roots. The same statement applies to the denominator of (7.3.15).

If $N > M$, some of the second-order subsystems have numerator coefficients that are zero, that is, either $b_{k2} = 0$ or $b_{k1} = 0$ or both $b_{k2} = b_{k1} = 0$ for some k. Furthermore, if N is odd, one of the subsystems, say $H_k(z)$, must have $a_{k2} = 0$, so that the subsystem is of first order. To preserve the modularity in the implementation of $H(z)$, it is often preferable to use the basic second-order subsystems in the cascade structure and have some zero-valued coefficients in some of the subsystems.

Each of the second-order subsystems with system function of the form (7.3.15) can be realized in either direct form I, or direct form II, or transposed direct form II. Since there are many ways to pair the poles and zeros of H(z) into a cascade of second-order sections, and several ways to order the resulting subsystems, it is possible to obtain a variety of cascade realizations. Although all cascade realizations are equivalent for infinite precision arithmetic, the various realizations may differ significantly when implemented with finite-precision arithmetic.

The general form of the cascade structure is illustrated in Fig. 7.19. If we use the direct form II structure for each of the subsystems, the computational algorithm for realizing the IIR system with system function $H(z)$ is described by the following set of equations.

$$y_0(n) = x(n) \tag{7.3.16}$$

$$w_k(n) = -a_{k1}w_k(n-1) - a_{k2}w_k(n-2) + y_{k-1}(n) \qquad k = 1, 2, \ldots, K \tag{7.3.17}$$

$$y_k(n) = b_{k0}w_k(n) + b_{k1}w_k(n-1) + b_{k2}w_k(n-2) \qquad k = 1, 2, \ldots, K \tag{7.3.18}$$

$$y(n) = y_K(n) \tag{7.3.19}$$

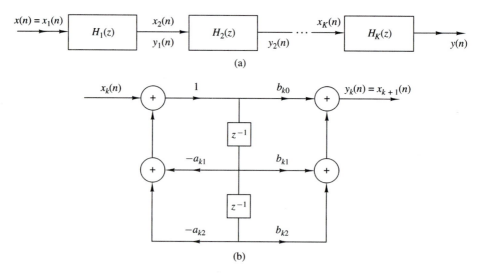

Figure 7.19 Cascade structure of second-order systems and a realization of each second-order section.

Thus this set of equations provides a complete description of the cascade structure based on direct form II sections.

7.3.4 Parallel-Form Structures

A parallel-form realization of an IIR system can be obtained by performing a partial-fraction expansion of $H(z)$. Without loss of generality, we again assume that $N \geq M$ and that the poles are distinct. Then, by performing a partial-fraction expansion of $H(z)$, we obtain the result

$$H(z) = C + \sum_{k=1}^{N} \frac{A_k}{1 - p_k z^{-1}} \tag{7.3.20}$$

where $\{p_k\}$ are the poles, $\{A_k\}$ *are* the coefficients (residues) in the partial-fraction expansion, and the constant C is defined as $C = b_N/a_N$. The structure implied by (7.3.20) is shown in Fig. 7.20. It consists of a parallel bank of single-pole filters.

In general, some of the poles of $H(z)$ may be complex valued. In such a case, the corresponding coefficients A_k are also complex valued. To avoid multiplications by complex numbers, we can combine pairs of complex-conjugate poles to form two-pole subsystems. In addition, we can combine, in an arbitrary manner,

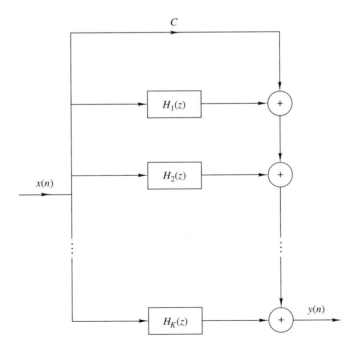

Figure 7.20   Parallel structure of IIR system.

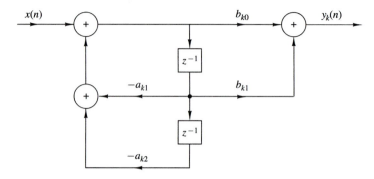

Figure 7.21 Structure of second-order section in a parallel IIR system realization.

pairs of real-valued poles to form two-pole subsystems. Each of these subsystems has the form

$$H_k(z) = \frac{b_{k0} + b_{k1}z^{-1}}{1 + a_{k1}z^{-1} + a_{k2}z^{-2}} \tag{7.3.21}$$

where the coefficients $\{b_{ki}\}$ and $\{a_{ki}\}$ are real-valued system parameters. The overall function can now be expressed as

$$H(z) = C + \sum_{k=1}^{K} H_k(z) \tag{7.3.22}$$

where K is the integer part of $(N+1)/2$. When N is odd, one of the $H_k(z)$ is really a single-pole system (i.e., $b_{k1} = a_{k2} = 0$).

The individual second-order sections which are the basic building blocks for $H(z)$ can be implemented in either of the direct forms or in a transposed direct form. The direct form II structure is illustrated in Fig. 7.21. With this structure as a basic building block, the parallel-form realization of the FIR system is described by the following set of equations

$$w_k(n) = -a_{k1}w_k(n-1) - a_{k2}w_k(n-2) + x(n) \qquad k = 1, 2, \dots, K \tag{7.3.23}$$

$$y_k(n) = b_{k0}w_k(n) + b_{k1}w_k(n-1) \qquad k = 1, 2, \dots, K \tag{7.3.24}$$

$$y(n) = Cx(n) + \sum_{k=1}^{K} y_k(n) \tag{7.3.25}$$

Example 7.3.1

Determine the cascade and parallel realizations for the system described by the system function

$$H(z) = \frac{10(1 - \frac{1}{2}z^{-1})(1 - \frac{2}{3}z^{-1})(1 + 2z^{-1})}{(1 - \frac{3}{4}z^{-1})(1 - \frac{1}{8}z^{-1})[1 - (\frac{1}{2} + j\frac{1}{2})z^{-1}][1 - (\frac{1}{2} - j\frac{1}{2})z^{-1}]}$$

Solution The cascade realization is easily obtained from this form. One possible pairing of poles and zeros is

$$H_1(z) = \frac{1 - \frac{2}{3}z^{-1}}{1 - \frac{7}{8}z^{-1} + \frac{3}{32}z^{-2}}$$

$$H_2(z) = \frac{1 + \frac{3}{2}z^{-1} - z^{-2}}{1 - z^{-1} + \frac{1}{2}z^{-2}}$$

and hence

$$H(z) = 10H_1(z)H_2(z)$$

The cascade realization is depicted in Fig. 7.22a.

To obtain the parallel-form realization, $H(z)$ must be expanded in partial fractions. Thus we have

$$H(z) = \frac{A_1}{1 - \frac{3}{4}z^{-1}} + \frac{A_2}{1 - \frac{1}{8}z^{-1}} + \frac{A_3}{1 - (\frac{1}{2} + j\frac{1}{2})z^{-1}} + \frac{A_3^*}{1 - (\frac{1}{2} - j\frac{1}{2})z^{-1}}$$

where A_1, A_2, A_3, and A_3^* are to be determined. After some arithmetic we find that

$$A_1 = 2.93, \quad A_2 = -17.68, \quad A_3 = 12.25 - j14.57, \quad A_3^* = 12.25 + j14.57$$

upon recombining pairs of poles, we obtain

$$H(z) = \frac{-14.75 - 12.90z^{-1}}{1 - \frac{7}{8}z^{-1} + \frac{3}{32}z^{-2}} + \frac{24.50 + 26.82z^{-1}}{1 - z^{-1} + \frac{1}{2}z^{-2}}$$

The parallel-form realization is illustrated in Fig. 7.22b.

7.3.5 Lattice and Lattice-Ladder Structures for IIR Systems

In Section 7.2.4 we developed a lattice filter structure that is equivalent to an FIR system. In this section we extend the development to IIR systems.

Let us begin with an all-pole system with system function

$$H(z) = \frac{1}{1 + \displaystyle\sum_{k=1}^{N} a_N(k)z^{-k}} = \frac{1}{A_N(z)} \tag{7.3.26}$$

The direct form realization of this system is illustrated in Fig. 7.23. The difference equation for this IIR system is

$$y(n) = -\sum_{k=1}^{N} a_N(k)y(n - k) + x(n) \tag{7.3.27}$$

It is interesting to note that if we interchange the roles of input and output [i.e., interchange $x(n)$ with $y(n)$ in (7.3.27)], we obtain

$$x(n) = -\sum_{k=1}^{N} a_N(k)x(n - k) + y(n)$$

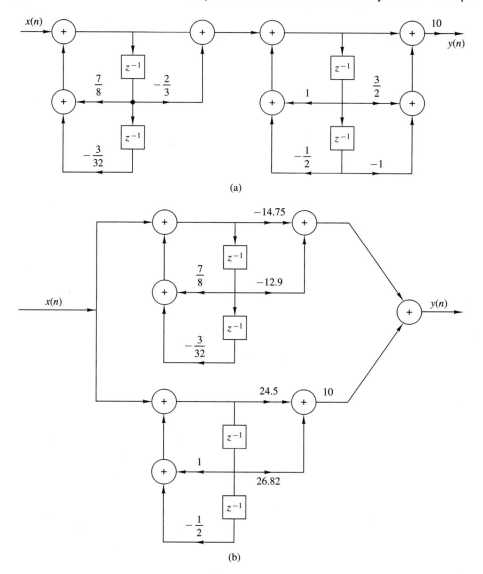

Figure 7.22   Cascade and parallel realizations for the system in Example 7.3.1.

or, equivalently,

$$y(n) = x(n) + \sum_{k=1}^{N} a_N(k)x(n-k) \qquad (7.3.28)$$

We note that the equation in (7.3.28) describes an FIR system having the system function $H(z) = A_N(z)$, while the system described by the difference equation in (7.3.27) represents an IIR system with system function $H(z) = 1/A_N(z)$.

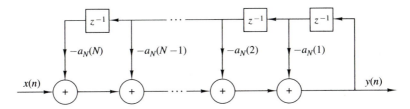

Figure 7.23 Direct-form realization of an all-pole system.

One system can be obtained from the other simply by interchanging the roles of the input and output.

Based on this observation, we shall use the all-zero (FIR) lattice described in Section 7.2.4 to obtain a lattice structure for an all-pole IIR system by interchanging the roles of the input and output. First, we take the all-zero lattice filter illustrated in Fig. 7.11 and then redefine the input as

$$x(n) = f_N(n) \tag{7.3.29}$$

and the output as

$$y(n) = f_0(n) \tag{7.3.30}$$

These are exactly the opposite of the definitions for the all-zero lattice filter. These definitions dictate that the quantities $\{f_m(n)\}$ be computed in descending order [i.e., $f_N(n)$, $f_{N-1}(n), \ldots$]. This computation can be accomplished by rearranging the recursive equation in (7.2.29) and thus solving for $f_{m-1}(n)$ in terms of $f_m(n)$, that is,

$$f_{m-1}(n) = f_m(n) - K_m g_{m-1}(n-1) \qquad m = N, N-1, \ldots, 1$$

The equation (7.2.30) for $g_m(n)$ remains unchanged.

The result of these changes is the set of equations

$$f_N(n) = x(n) \tag{7.3.31}$$

$$f_{m-1}(n) = f_m(n) - K_m g_{m-1}(n-1) \qquad m = N, N-1, \ldots, 1 \tag{7.3.32}$$

$$g_m(n) = K_m f_{m-1}(n) + g_{m-1}(n-1) \qquad m = N, N-1, \ldots, 1 \tag{7.3.33}$$

$$y(n) = f_0(n) = g_0(n) \tag{7.3.34}$$

which correspond to the structure shown in Fig. 7.24.

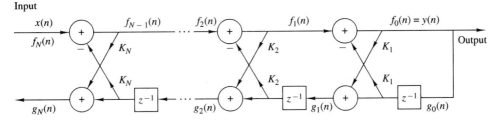

Figure 7.24 Lattice structure for an all-pole IIR system.

To demonstrate that the set of equations (7.3.31) through (7.3.34) represent an all-pole IIR system, let us consider the case where N = 1. The equations reduce to

$$x(n) = f_1(n)$$
$$f_0(n) = f_1(n) - K_1 g_0(n-1)$$
$$g_1(n) = K_1 f_0(n) + g_0(n-1) \qquad (7.3.35)$$
$$y(n) = f_0(n)$$
$$= x(n) - K_1 y(n-1)$$

Furthermore, the equation for $g_1(n)$ can be expressed as

$$g_1(n) = K_1 y(n) + y(n-1) \qquad (7.3.36)$$

We observe that (7.3.35) represents a first-order all-pole IIR system while (7.3.36) represents a first-order FIR system. The pole is a result of the feedback introduced by the solution of the $\{f_m(n)\}$ in descending order. This feedback is depicted in Fig. 7.25a.

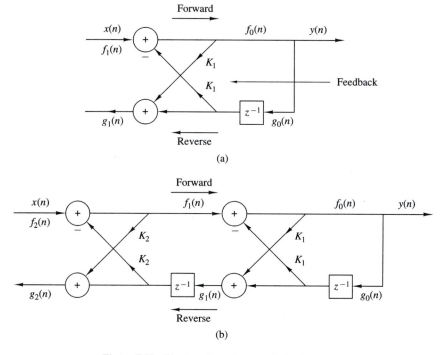

Figure 7.25 Single-pole and two-pole lattice system.

Next, let us consider the case $N = 2$, which corresponds to the structure in Fig. 7.25b. The equations corresponding to this structure are

$$f_2(n) = x(n)$$

$$f_1(n) = f_2(n) - K_2 g_1(n-1)$$

$$g_2(n) = K_2 f_1(n) + g_1(n-1)$$

$$f_0(n) = f_1(n) - K_1 g_0(n-1) \qquad (7.3.37)$$

$$g_1(n) = K_1 f_0(n) + g_0(n-1)$$

$$y(n) = f_0(n) = g_0(n)$$

After some simple substitutions and manipulations we obtain

$$y(n) = -K_1(1 + K_2)y(n-1) - K_2 y(n-2) + x(n) \qquad (7.3.38)$$

$$g_2(n) = K_2 y(n) + K_1(1 + K_2)y(n-1) + y(n-2) \qquad (7.3.39)$$

Clearly, the difference equation in (7.3.38) represents a two-pole IIR system, and the relation in (7.3.39) is the input–output equation for a two-zero FIR system. Note that the coefficients for the FIR system are identical to those in the IIR system except that they occur in reverse order.

In general, these conclusions hold for any N. Indeed, with the definition of $A_m(z)$ given in (7.2.32), the system function for the all-pole IIR system is

$$H_a(z) = \frac{Y(z)}{X(z)} = \frac{F_0(z)}{F_m(z)} = \frac{1}{A_m(z)} \qquad (7.3.40)$$

Similarly, the system function of the all-zero (FIR) system is

$$H_b(z) = \frac{G_m(z)}{Y(z)} = \frac{G_m(z)}{G_0(z)} = B_m(z) = z^{-m} A_m(z^{-1}) \qquad (7.3.41)$$

where we used the previously established relationships in (7.2.36) through (7.2.42). Thus the coefficients in the FIR system $H_b(z)$ are identical to the coefficients in $A_m(z)$, except that they occur in reverse order.

It is interesting to note that the all-pole lattice structure has an all-zero path with input $g_0(n)$ and output $g_N(n)$, which is identical to its counterpart all-zero path in the all-zero lattice structure. The polynomial $B_m(z)$, which represents the system function of the all-zero path common to both lattice structures, is usually called the *backward system function*, because it provides a backward path in the all-pole lattice structure.

From this discussion the reader should observe that the all-zero and all-pole lattice structures are characterized by the same set of lattice parameters, namely, $K_1, K_2, \ldots, K_N$. The two lattice structures differ only in the interconnections of their signal flow graphs. Consequently, the algorithms for converting between the system parameters $\{\alpha_m(k)\}$ in the direct form realization of an FIR system, and the parameters of its lattice counterpart apply as well to the all-pole structure.

We recall that the roots of the polynomial $A_N(z)$ lie inside the unit circle if and only if the lattice parameters $|K_m| < 1$ for all $m = 1, 2, \ldots, N$. Therefore, the all-pole lattice structure is a stable system if and only if its parameters $|K_m| < 1$ for all m.

In practical applications the all-pole lattice structure has been used to model the human vocal tract and a stratified earth. In such cases the lattice parameters, $\{K_m\}$ have the physical significance of being identical to reflection coefficients in the physical medium. This is the reason that the lattice parameters are often called *reflection coefficients*. In such applications, a stable model of the medium requires that the reflection coefficients, obtained by performing measurements on output signals from the medium, be less than unity.

The all-pole lattice provides the basic building block for lattice-type structures that implement IIR systems that contain both poles and zeros. To develop the appropriate structure, let us consider an IIR system with system function

$$H(z) = \frac{\displaystyle\sum_{k=0}^{M} c_M(k)z^{-k}}{1 + \displaystyle\sum_{k=1}^{N} a_N(k)z^{-k}} = \frac{C_M(z)}{A_N(z)} \tag{7.3.42}$$

where the notation for the numerator polynomial has been changed to avoid confusion with our previous development. Without loss of generality, we assume that $N \geq M$.

In the direct form II structure, the system in (7.3.42) is described by the difference equations

$$w(n) = -\sum_{k=1}^{N} a_N(k)w(n-k) + x(n) \tag{7.3.43}$$

$$y(n) = \sum_{k=0}^{M} c_M(k)w(n-k) \tag{7.3.44}$$

Note that (7.3.43) is the input–output of an all-pole IIR system and that (7.3.44) is the input–output of an all-zero system. Furthermore, we observe that the output of the all-zero system is simply a linear combination of delayed outputs from the all-pole system. This is easily seen by observing the direct form II structure redrawn as in Fig. 7.26.

Since zeros result from forming a linear combination of previous outputs we can carry over this observation to construct a pole–zero IIR system using the all-pole lattice structure as the basic building block. We have already observed that $g_m(n)$ is a linear combination of present and past outputs. In fact, the system

$$H_b(z) = \frac{G_m(z)}{Y(z)} = B_m(z)$$

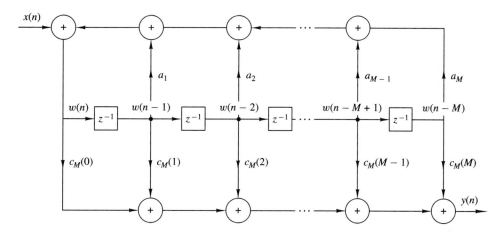

Figure 7.26   Direct form II realization of IIR system.

is an all-zero system. Therefore, any linear combination of $\{g_m(n)\}$ is also an all-zero system.

Thus we begin with an all-pole lattice structure with parameters K_m, $1 \leq m \leq N$, and we add a *ladder* part by taking as the output a weighted linear combination of $\{g_m(n)\}$. The result is a pole–zero IIR system which has the *lattice-ladder* structure shown in Fig. 7.27 for $M = N$. Its output is

$$y(n) = \sum_{m=0}^{M} v_m g_m(n) \tag{7.3.45}$$

where $\{v_m\}$ are the parameters that determine the zeros of the system. The system

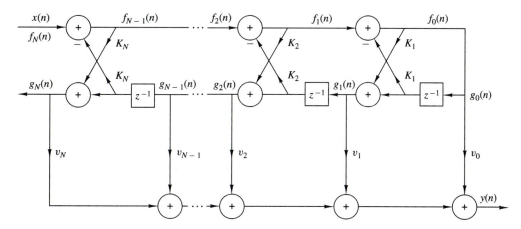

Figure 7.27   Lattice-ladder structure for the realization of a pole–zero system.

function corresponding to (7.3.45) is

$$
H(z) = \frac{Y(z)}{X(z)}
$$

$$
= \sum_{m=0}^{M} v_m \frac{G_m(z)}{X(z)}
$$

(7.3.46)

Since $X(z) = F_N(z)$ and $F_0(z) = G_0(z)$, (7.3.46) can be written as

$$
H(z) = \sum_{m=0}^{M} v_m \frac{G_m(z)}{G_0(z)} \frac{F_0(z)}{F_N(z)}
$$

$$
= \sum_{m=0}^{M} v_m \frac{B_m(z)}{A_N(z)}
$$

(7.3.47)

$$
= \frac{\sum_{m=0}^{M} v_m B_m(z)}{A_N(z)}
$$

If we compare (7.3.41) with (7.3.47), we conclude that

$$
C_M(z) = \sum_{m=0}^{M} v_m B_m(z)
$$

(7.3.48)

This is the desired relationship that can be used to determine the weighting coefficients $\{v_m\}$. Thus, we have demonstrated that the coefficients of the numerator polynomial $C_M(z)$ determine the ladder parameters $\{v_m\}$, whereas the coefficients in the denominator polynomial $A_N(z)$ determine the lattice parameters $\{K_m\}$.

Given the polynomials $C_M(z)$ and $A_N(z)$, where $N \geq M$, the parameters of the all-pole lattice are determined first, as described previously, by the conversion algorithm given in Section 7.2.4, which converts the direct form coefficients into lattice parameters. By means of the step-down recursive relations given by (7.2.54), we obtain the lattice parameters $\{K_m\}$ and the polynomials $B_m(z)$, $m = 1$, $2, \ldots, N$.

The ladder parameters are determined from (7.3.48), which can be expressed as

$$
C_m(z) = \sum_{k=0}^{m-1} v_k B_k(z) + v_m B_m(z)
$$

(7.3.49)

or, equivalently, as

$$
C_m(z) = C_{m-1}(z) + v_m B_m(z)
$$

(7.3.50)

Thus $C_m(z)$ can be computed recursively from the reverse polynomials $B_m(z)$, $m = 1, 2, \ldots, M$. Since $\beta_m(m) = 1$ for all m, the parameters v_m, $m = 0, 1, \ldots, M$ can be determined by first noting that

$$
v_m = c_m(m) \qquad m = 0, 1, \ldots, M
$$

(7.3.51)

Then, by rewriting (7.3.50) as

$$C_{m-1}(z) = C_m(z) - v_m B_m(z) \qquad\qquad (7.3.52)$$

and running this recursive relation backward in m (i.e., $m = M, M - 1, \ldots, 2$), we obtain $c_m(m)$ and therefore the ladder parameters according to (7.3.51).

The lattice-ladder filter structures that we have presented require the minimum amount of memory but not the minimum number of multiplications. Although lattice structures with only one multiplier per lattice stage exist, the two multiplier-per-stage lattice that we have described, is by far the most widely used in practical applications. In conclusion, the modularity, the built-in stability characteristics embodied in the coefficients $\{K_m\}$, and its robustness to finite-word-length effects make the lattice structure very attractive in many practical applications, including speech processing systems, adaptive filtering, and geophysical signal processing.

7.4 STATE-SPACE SYSTEM ANALYSIS AND STRUCTURES

Up to this point our treatment of linear time-invariant systems has been limited to an *input–output* or *external description* of the characteristics of the system. In other words, the system was characterized by mathematical equations that relate the input signal to the output signal. In this section we introduce the basic concepts in the state-space description of linear time-invariant causal systems. Although the *state-space* or *internal description* of the system still involves a relationship between the input and output signals, it also involves an additional set of variables, called *state variables*. Furthermore, the mathematical equations describing the system, its input, and its output are usually divided into two parts:

1. A set of mathematical equations relating the state variables to the input signal.
2. A second set of mathematical equations relating the state variables and the current input to the output signal.

The state variables provide information about all the internal signals in the system. As a result, the state-space description provides a more detailed description of the system than the input–output description. Although our treatment of state-space analysis is confined primarily to single input–single output linear time-invariant causal systems, the state-space techniques can also be applied to nonlinear systems, time-variant systems, and multiple input–multiple output systems. In fact, it is in the characterization and analysis of multiple input–multiple output systems that the power and importance of state-space methods are clearly evident.

Both input–output and state-variable descriptions of a system are useful in practice. The description we use depends on the problem, the available information, and the questions to be answered. In our presentation, the emphasis is on the use of state-space techniques in system analysis, and in the development of state-space structures for the realization of discrete-time systems.

7.4.1 State-Space Descriptions of Systems Characterized by Difference Equations

As we have already observed, the determination of the output of a system requires that we know the input signal and the set of initial conditions at the time the input is applied. If a system is not relaxed initially, say at time n_0, then knowledge of the input signal $x(n)$ for $n \geq n_0$ is not sufficient to uniquely determine the output $y(n)$ for $n \geq n_0$. The initial conditions of the system at $n = n_0$ must also be known and taken into account. This set of initial conditions is called the state of the system at $n = n_0$. Hence *we define the* state *of a system at time n_0 as the amount of information that must be provided at time n_0, which, together with the input signal $x(n)$ for $n \geq n_0$, uniquely determine the output of the system for all $n \geq n_0$.*

From this definition we infer that the concept of state leads to a decomposition of a system into two parts, a part that contains memory, and a memoryless component. The information stored in the memory component constitutes the set of initial conditions and is called the *state of the system*. The current output of the system then becomes a function of the current value of the input and the current state. Thus, to determine the output of the system at a given time, we need the current value of the state and the current input. Since the current value of the input is available, we only need to provide a mechanism for updating the state of the system recursively. Consequently, the state of the system at time $n_0 + 1$ should depend on the state of the system at time n_0 and the value of the input signal $x(n)$ at $n = n_0$.

The following example illustrates the approach in formulating a state-space description of a system. Let us consider a linear time-invariant causal system described by the difference equation

$$y(n) = - \sum_{k=1}^{3} a_k y(n-k) + \sum_{k=0}^{3} b_k x(n-k) \tag{7.4.1}$$

The direct form II realization for the system is shown in Fig. 7.28.

As state variables, we use the contents of the system memory registers, counting them from the bottom, as shown in Fig. 7.28. We recall that the output of a delay element represents the present value stored in the register and the input represents the next value to be stored in the memory. Consequently, with the aid of Fig. 7.28, we can write

$$v_1(n+1) = v_2(n)$$

$$v_2(n+1) = v_3(n) \tag{7.4.2}$$

$$v_3(n+1) = -a_3 v_1(n) - a_2 v_2(n) - a_1 v_3(n) + x(n)$$

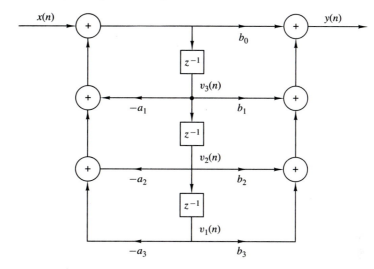

Figure 7.28 Direct form II realization of system described by the difference equation in (7.5.1).

It is interesting to note that the state-variable formulation for the third-order system of (7.4.1) involves three first-order difference equations given by (7.4.2). In general, an nth-order system can be described by n first-order difference equations.

The output equation, which expresses $y(n)$ in terms of the state variables and the present input value $x(n)$, can also be obtained by referring to Fig. 7.28. We have

$$y(n) = b_0 v_3(n+1) + b_3 v_1(n) + b_2 v_2(n) + b_1 v_3(n)$$

We can eliminate $v_3(n+1)$ by using the last equation in (7.4.2). Thus we obtain the desired output equation

$$y(n) = (b_3 - b_0 a_3)v_1(n) + (b_2 - b_0 a_2)v_2(n) + (b_1 - b_0 a_1)v_3(n) + b_0 x(n) \qquad (7.4.3)$$

If we put (7.4.2) and (7.4.3) into matrix form we have

$$\begin{bmatrix} v_1(n+1) \\ v_2(n+1) \\ v_3(n+1) \end{bmatrix} = \begin{bmatrix} 0 & 1 & 0 \\ 0 & 0 & 1 \\ -a_3 & -a_2 & -a_1 \end{bmatrix} \begin{bmatrix} v_1(n) \\ v_2(n) \\ v_3(n) \end{bmatrix} + \begin{bmatrix} 0 \\ 0 \\ 1 \end{bmatrix} x(n) \qquad (7.4.4)$$

and

$$y(n) = [(b_3 - b_0 a_3)\ (b_2 - b_0 a_2)\ (b_1 - b_0 a_1)] \begin{bmatrix} v_1(n) \\ v_2(n) \\ v_3(n) \end{bmatrix} + b_0 x(n) \qquad (7.4.5)$$

The equations (7.4.4) and (7.4.5) provide a complete description of the system. Furthermore, the variables $v_1(n)$, $v_2(n)$, and $v_3(n)$, which summarize all the necessary past information, are the *state variables* of the system. We also observe that as indicated previously, equations (7.4.4) and (7.4.5) split the system into two component parts, a dynamic (memory) subsystem and a static (memoryless) subsystem. We say that this set of equations provides a *state-space description* of the system.

By generalizing the previous example, it can easily be seen that the Nth-order system described by

$$y(n) = -\sum_{k=1}^{N} a_k y(n-k) + \sum_{k=0}^{N} b_k x(n-k) \tag{7.4.6}$$

can be expressed as a linear time-invariant state-space realization by the relations

State equation

$$\mathbf{v}(n+1) = \mathbf{F}\mathbf{v}(n) + \mathbf{q}x(n) \tag{7.4.7}$$

Output equation

$$y(n) = \mathbf{g}^t \mathbf{v}(n) + dx(n) \tag{7.4.8}$$

where the elements of $\mathbf{F}$, $\mathbf{q}$, $\mathbf{g}$, and d are constants (i.e., they do not change as a function of the time index n), given by

$$\mathbf{F} = \begin{bmatrix} 0 & 1 & 0 & \cdot & \cdot & & \cdot & 0 \\ 0 & 0 & 1 & 0 & \cdot & & \cdot & 0 \\ \vdots & \vdots & & & & & & \vdots \\ 0 & 0 & & \cdot & \cdot & 0 & & 1 \\ -a_N & -a_{n-1} & \cdot & & \cdot & \cdot & -a_2 & -a_1 \end{bmatrix} \qquad \mathbf{q} = \begin{bmatrix} 0 \\ 0 \\ \vdots \\ 0 \\ 1 \end{bmatrix} \tag{7.4.9}$$

$$\mathbf{g} = \begin{bmatrix} b_N - b_0 a_N \\ b_{N-1} - b_0 a_{N-1} \\ \vdots \\ b_1 - b_0 a_1 \end{bmatrix}$$

Any discrete-time system whose input $x(n)$, output $y(n)$, and state $\mathbf{v}(n)$, for all $n \geq n_0$, are related by the state-space equations above, where $\mathbf{F}$, $\mathbf{q}$, $\mathbf{g}$, and d are arbitrary but fixed quantities, will be called *linear* and *time invariant*. If at least one of the quantities in $\mathbf{F}$, $\mathbf{q}$, $\mathbf{g}$, or d depends on time, the system becomes *time variant*.

We will refer to (7.4.7) through (7.4.8) as the *linear time-invariant state-space model*, which can be represented by the simple vector-matrix block diagram in Fig. 7.29. In this figure the double lines represent vector quantities and the blocks represent the vector or matrix coefficients.

Example 7.4.1

Determine the state-space equations for the transposed direct form II structure shown in Fig. 7.30.

Solution   The validity of this structure can be seen if we rewrite (7.4.1) as

$$y(n) = \sum_{k=1}^{3} [b_k x(n-k) - a_k y(n-k)] + b_0 x(n)$$

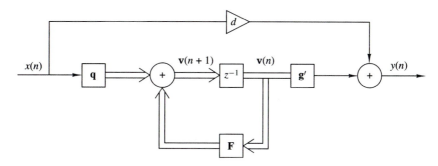

Figure 7.29   General state-space description of a linear time-invariant system.

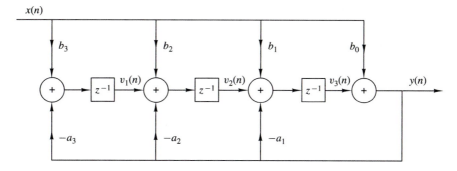

Figure 7.30   State-space realization for the system described by (7.4.1).

Due to the linearity and time invariance of the system, instead of first delaying the signals $x(n)$ and $y(n)$ and then computing the terms $b_k x(n - k) - a_k y(n - k)$ as in Fig. 7.28, we first compute the terms $b_k x(n) - a_k y(n)$ and then delay them.

If we use the state variables indicated in Fig. 7.30, we obtain

$$
\begin{bmatrix} v_1(n+1) \\ v_2(n+1) \\ v_3(n+1) \end{bmatrix} = \begin{bmatrix} 0 & 0 & -a_3 \\ 1 & 0 & -a_2 \\ 0 & 1 & -a_1 \end{bmatrix} \begin{bmatrix} v_1(n) \\ v_2(n) \\ v_3(n) \end{bmatrix} = \begin{bmatrix} b_3 - b_0 a_3 \\ b_2 - b_0 a_2 \\ b_1 - b_0 a_1 \end{bmatrix} x(n) \qquad (7.4.10)
$$

$$
y(n) = \begin{bmatrix} 0 & 0 & 1 \end{bmatrix} \begin{bmatrix} v_1(n) \\ v_2(n) \\ v_3(n) \end{bmatrix} + b_0 x(n) \qquad (7.4.11)
$$

The state-space description specified by (7.4.4) and (7.4.5) is known as a *type 1* state-space realization, whereas the one described by (7.4.10) and (7.4.11) is called a *type 2* state-space realization.

7.4.2 Solution of the State-Space Equations

There are several methods for solving the state-space equations. Here we discuss a recursive solution which makes use of the fact that the state-space equations are a set of linear first-order difference equations.

For the N-dimensional state-space model

$$\mathbf{v}(n+1) = \mathbf{F}\mathbf{v}(n) + \mathbf{q}x(n) \tag{7.4.12}$$

$$y(n) = \mathbf{g}^t\mathbf{v}(n) + dx(n) \tag{7.4.13}$$

and given the initial condition $\mathbf{v}(n_0)$, we have for $n > n_0$,

$$\mathbf{v}(n_0 + 1) = \mathbf{F}\mathbf{v}(n_0) + \mathbf{q}x(n)$$

$$\mathbf{v}(n_0 + 2) = \mathbf{F}\mathbf{v}(n_0 + 1) + \mathbf{q}x(n_0 + 1)$$

$$= \mathbf{F}^2\mathbf{v}(n_0) + \mathbf{F}\mathbf{q}x(n_0) + \mathbf{q}x(n_0 + 1)$$

where $\mathbf{F}^2$ represents the matrix product $\mathbf{FF}$ and $\mathbf{Fq}$ is the product of the matrix $\mathbf{F}$ and the vector $\mathbf{q}$. If we continue as in the one-dimensional case, we obtain, for $n > n_0$,

$$\mathbf{v}(n) = \mathbf{F}^{n-n_0}\mathbf{v}(n_0) + \sum_{k=n_0}^{n-1} \mathbf{F}^{n-1-k}\mathbf{q}x(k) \tag{7.4.14}$$

The matrix $\mathbf{F}^0$ is defined as the $N \times N$ identity matrix, having unity on the main diagonal and zeros elsewhere. The matrix $\mathbf{F}^{i-j}$ is often denoted as $\Phi(i - j)$, that is,

$$\Phi(i - j) = \mathbf{F}^{i-j} \tag{7.4.15}$$

for any positive integers $i \geq j$. This matrix is called the *state transition matrix* of the system.

The output of the system is obtained by substituting (7.4.14) into (7.4.13). The result of this substitution is

$$y(n) = \mathbf{g}^t\mathbf{F}^{n-n_0}\mathbf{v}(n_0) + \sum_{k=n_0}^{n-1} \mathbf{g}^t\mathbf{F}^{n-1-k}\mathbf{q}x(k) + dx(n)$$

$$= \mathbf{g}^t\Phi(n - n_0)\mathbf{v}(n_0) + \sum_{k=n_0}^{n-1} \mathbf{g}^t\Phi(n - 1 - k)\mathbf{q}x(k) + dx(n) \tag{7.4.16}$$

From this general result, we can determine the output for two special cases. First, the zero-input response of the system is

$$y_{zi}(n) = \mathbf{g}^t\mathbf{F}^{n-n_0}\mathbf{v}(n_0) = \mathbf{g}^t\Phi(n - n_0)\mathbf{v}(n_0) \tag{7.4.17}$$

On the other hand, the zero-state response is

$$y_{zs}(n) = \sum_{k=n_0}^{n-1} \mathbf{g}^t\Phi(n - 1 - k)\mathbf{q}x(k) + dx(n) \tag{7.4.18}$$

Clearly, the N-dimensional state-space system is zero-input linear, zero-state linear, and since $y(n) = y_{zi}(n) + y_{zs}(n)$, it is linear. Furthermore, since any system described by a linear constant-coefficient difference equation can be put in the state-space form, it is linear, in agreement with the results obtained in Section 2.4.

7.4.3 Relationships Between Input–Output and State-Space Descriptions

From our previous discussion we have seen that there is no unique choice for the state variables of a causal system. Furthermore, different choices for the state vector lead to different structures for the realization of the same system. Hence, in general, the input–output relationship does not uniquely describe the internal structure of the system.

To illustrate these assertions, let us consider an N-dimensional system with the state-space representation

$$\mathbf{v}(n+1) = \mathbf{F}\mathbf{v}(n) + \mathbf{q}x(n) \tag{7.4.19}$$

$$y(n) = \mathbf{g}'\mathbf{v}(n) + dx(n) \tag{7.4.20}$$

Let $\mathbf{P}$ be any $N \times N$ matrix whose inverse matrix $\mathbf{P}^{-1}$ exists. We define a new state vector $\hat{\mathbf{v}}(n)$ as

$$\hat{\mathbf{v}}(n) = \mathbf{P}\mathbf{v}(n) \tag{7.4.21}$$

Then

$$\mathbf{v}(n) = \mathbf{P}^{-1}\hat{\mathbf{v}}(n) \tag{7.4.22}$$

If (7.4.19) is premultiplied by $\mathbf{P}$, we obtain

$$\mathbf{P}\mathbf{v}(n+1) = \mathbf{P}\mathbf{F}\mathbf{v}(n) + \mathbf{P}\mathbf{q}x(n)$$

By using (7.4.22), the state equation above becomes

$$\hat{\mathbf{v}}(n+1) = (\mathbf{P}\mathbf{F}\mathbf{P}^{-1})\hat{\mathbf{v}}(n) + (\mathbf{P}\mathbf{q})x(n) \tag{7.4.23}$$

Similarly, with the aid of (7.4.22) the output equation (7.4.20) becomes

$$y(n) = (\mathbf{g}'\mathbf{P}^{-1})\hat{\mathbf{v}}(n) + dx(n) \tag{7.4.24}$$

Now, we define a new system parameter matrix $\hat{\mathbf{F}}$ and the vectors $\hat{\mathbf{q}}$ and $\hat{\mathbf{g}}$ as

$$\hat{\mathbf{F}} = \mathbf{P}\mathbf{F}\mathbf{P}^{-1}$$

$$\hat{\mathbf{q}} = \mathbf{P}\mathbf{q} \tag{7.4.25}$$

$$\hat{\mathbf{g}}' = \mathbf{g}'\mathbf{P}^{-1}$$

With these definitions, the state equations can be expressed in terms of the new system quantities as

$$\hat{\mathbf{v}}(n+1) = \hat{\mathbf{F}}\hat{\mathbf{v}}(n) + \hat{\mathbf{q}}x(n) \tag{7.4.26}$$

$$y(n) = \hat{\mathbf{g}}'\hat{\mathbf{v}}(n) + dx(n) \tag{7.4.27}$$

If we compare (7.4.19) and (7.4.20) with (7.4.26) and (7.4.27), we observe that by a simple linear transformation of the state variables, we have generated a new set of state equations and an output equation, in which the input $x(n)$ and the output $y(n)$ are unchanged. Since there is an infinite number of choices of the transformation matrix $\mathbf{P}$, there is also an infinite number of state-space equations and structures for a system. Some of these structures are different, while some others are very similar, differing only by scale factors.

Associated with any state-space realization of a system is the concept of a *minimal realization*. A state-space realization is said to be *minimal* if the dimension of the state space (the number of state variables) is the smallest of all possible realizations. Since each state variable represents a quantity that must be stored and updated at every time instant n, it follows that a minimal realization is one that requires the smallest number of delays (storage registers). We recall that the direct form II realization requires the smallest number of storages registers, and consequently, a state-space realization based on the contents of the delay elements results in a minimal realization. Similarly, an FIR system realized as a direct form structure leads to a minimal state-space realization if the values of the storage registers are defined as the state variables. On the other hand, the direct form I realization of an IIR system does not lead to a minimal realization.

Now, let us determine the impulse response of the system from the state-space realization. The impulse response provides one of the links between the input–output and state-space description of systems.

By definition the impulse response $h(n)$ of a system is the zero-state response of the system to the excitation $x(n) = \delta(n)$. Hence it can be obtained from equation (7.4.16) if we set $n_0 = 0$ (the time we apply the input), $\mathbf{v}(0) = 0$, and $x(n) = \delta(n)$. Thus the impulse response of the system described by (7.4.19) and (7.4.20) is given by

$$
\begin{aligned}
h(n) &= \mathbf{g}^t \mathbf{F}^{n-1} \mathbf{q} u(n-1) + d\delta(n) \\
&= \mathbf{g}^t \mathbf{\Phi}(n-1) \mathbf{q} u(n-1) + d\delta(n)
\end{aligned}
\tag{7.4.28}
$$

Given a state-space description, it is straightforward to determine the impulse response from (7.4.28). However, the inverse is not easy since there is an infinite number of state-space realizations for the same input–output description.

The transpose system. The transpose of a matrix $\mathbf{F}$ is obtained by interchanging its columns and rows, and it is denoted by $\mathbf{F}^t$. For example,

$$
\mathbf{F} = \begin{bmatrix} f_{11} & f_{12} & \cdots & f_{1N} \\ f_{21} & f_{22} & \cdots & f_{2N} \\ \vdots & \vdots & \cdots & \vdots \\ f_{N1} & f_{N2} & \cdots & f_{NN} \end{bmatrix}, \qquad
\mathbf{F}^t = \begin{bmatrix} f_{11} & f_{21} & \cdots & f_{N1} \\ f_{12} & f_{22} & \cdots & f_{N2} \\ \vdots & \vdots & \cdots & \vdots \\ f_{1N} & f_{2N} & \cdots & f_{NN} \end{bmatrix}
$$

Now define the *transpose system* (7.4.19)–(7.4.20) as

$$
\mathbf{v}'(n+1) = \mathbf{F}^t \mathbf{v}'(n) + \mathbf{g}x(n)
\tag{7.4.29}
$$

$$
y'(n) = \mathbf{q}^t \mathbf{v}'(n) + dx(n)
\tag{7.4.30}
$$

According to (7.4.28), the impulse response of this system is given as

$$
h'(n) = \mathbf{q}^t (\mathbf{F}^t)^{n-1} \mathbf{g} u(n-1) + d\delta(n)
\tag{7.4.31}
$$

From matrix algebra we know that $(\mathbf{F}^t)^{n-1} = (\mathbf{F}^{n-1})^t$. Hence

$$h'(n) = \mathbf{q}'(\mathbf{F}^{n-1})'\mathbf{g}u(n-1) + d\delta(n)$$

We claim that $h'(n) = h(n)$. Indeed, the term $\mathbf{q}'(\mathbf{F}^{n-1})'\mathbf{g}$ is a scalar. Hence it is equal to its transpose. Consequently,

$$[\mathbf{q}'(\mathbf{F}^{n-1})'\mathbf{g}]^t = \mathbf{g}'(\mathbf{F})^{n-1}\mathbf{q}$$

Since this is true, it follows that (7.4.31) is identical to (7.4.28) and, therefore, $h'(n) = h(n)$. Thus a *single input–single output system and its transpose have identical impulse responses and hence the same input–output relationship*. To support this claim further, we note that the type 1 and type 2 state-space realizations, described by (7.4.3), (7.4.4), (7.4.10), and (7.4.11) are transpose structures, which stem from the same input–output relationship (7.4.1).

We have introduced the transpose structure because it provides an easy method for generating a new structure. However, sometimes this new structure may either differ trivially or be identical to the original one.

The diagonal system.   A closed-form solution of the state-space equations is easily obtained when the system matrix $\mathbf{F}$ is diagonal. Hence, by finding a matrix $\mathbf{P}$ so that $\hat{\mathbf{F}} = \mathbf{P}\mathbf{F}\mathbf{P}^{-1}$ is diagonal, the solution of the state equations is simplified considerably. The diagonalization of the matrix $\mathbf{F}$ can be accomplished by first determining the eigenvalues and eigenvectors of the matrix.

A number λ is an *eigenvalue* of $\mathbf{F}$ and a nonzero vector $\mathbf{u}$ is the associated *eigenvector* if

$$\mathbf{F}\mathbf{u} = \lambda\mathbf{u} \tag{7.4.32}$$

To determine the eigenvalues of $\mathbf{F}$, we note that

$$(\mathbf{F} - \lambda\mathbf{I})\mathbf{u} = 0 \tag{7.4.33}$$

This equation has a (nontrivial) nonzero solution $\mathbf{u}$ if the matrix $\mathbf{F} - \lambda\mathbf{I}$ is singular [i.e., if $(\mathbf{F} - \lambda\mathbf{I})$ is noninvertible], which is the case if the determinant of $(\mathbf{F} - \lambda\mathbf{I})$ is zero, that is, if

$$\det (\mathbf{F} - \lambda\mathbf{I}) = 0 \tag{7.4.34}$$

This determinant in (7.4.34) yields the *characteristic polynomial* of the matrix $\mathbf{F}$. For an $N \times N$ matrix $\mathbf{F}$, the characteristic polynomial of $\mathbf{F}$ is degree N and hence it has N roots, say λ_i, $i = 1, 2, \ldots, N$. The roots may be distinct or some roots may be repeated. In any case, for each root λ_i, we can determine a vector $\mathbf{u}_i$, called the eigenvector corresponding to the eigenvalue λ_i, from the equation

$$\mathbf{F}\mathbf{u}_i = \lambda_i\mathbf{u}_i$$

These eigenvectors are orthogonal, that is, $\mathbf{u}_i'\mathbf{u}_j = 0$, for $i \neq j$.

If we form a matrix $\mathbf{U}$ whose columns consist of the eigenvectors $\{\mathbf{u}_i\}$, that is,

$$\mathbf{U} = \begin{bmatrix} \uparrow & \uparrow & & \uparrow \\ \mathbf{u}_1 & \mathbf{u}_2 & \cdots & \mathbf{u}_N \\ \downarrow & \downarrow & & \downarrow \end{bmatrix}$$

then the matrix $\hat{\mathbf{F}} = \mathbf{U}^{-1}\mathbf{F}\mathbf{U}$ is diagonal. Thus we have solved for the matrix that diagonalizes $\mathbf{F}$.

The following example illustrates the procedure of diagonalizing $\mathbf{F}$.

Example 7.4.2

The Fibonacci sequence, which is the sequence $\{1, 1, 2, 3, 5, 8, 13, \ldots\}$, can be generated as the impulse response of the system that satisfies the state-space equations

$$\mathbf{v}(n+1) = \begin{bmatrix} 0 & 1 \\ 1 & 1 \end{bmatrix} \mathbf{v}(n) + \begin{bmatrix} 0 \\ 1 \end{bmatrix} x(n)$$

$$y(n) = \begin{bmatrix} 1 & 1 \end{bmatrix} \mathbf{v}(n) + x(n)$$

Determine the impulse response $\{h(n)\}$ of the system.

Solution   Now we wish to determine an equivalent system

$$\hat{\mathbf{v}}(n+1) = \hat{\mathbf{F}}\hat{\mathbf{v}}(n) + \hat{\mathbf{q}}x(n)$$

$$y(n) = \hat{\mathbf{g}}'\hat{\mathbf{v}}(n) + dx(n)$$

such that the matrix $\hat{\mathbf{F}}$ is diagonal. From (7.4.25) we recall that the two systems are equivalent if

$$\hat{\mathbf{F}} = \mathbf{P}\mathbf{F}\mathbf{P}^{-1} \qquad \hat{\mathbf{q}} = \mathbf{P}\mathbf{q} \qquad \hat{\mathbf{g}}' = \mathbf{g}'\mathbf{P}^{-1}$$

Given $\mathbf{F}$, the problem is to determine a matrix $\mathbf{P}$ such that $\hat{\mathbf{F}} = \mathbf{P}\mathbf{F}\mathbf{P}^{-1}$ is a diagonal matrix.

First, we compute the determinant in (7.4.34). We have

$$\det(\mathbf{F} - \lambda\mathbf{I}) = \det \begin{bmatrix} -\lambda & 1 \\ 1 & 1-\lambda \end{bmatrix} = \lambda^2 - \lambda - 1 = 0$$

or

$$\lambda_1 = \frac{1+\sqrt{5}}{2} \qquad \lambda_2 = \frac{1-\sqrt{5}}{2}$$

To find the eigenvector $\mathbf{u}_1$ corresponding to λ_1, we have

$$\begin{bmatrix} 0 & 1 \\ 1 & 1 \end{bmatrix} \mathbf{u}_1 = \lambda_1 \mathbf{u}_1 \qquad \text{or} \qquad \mathbf{u}_1 = \begin{bmatrix} 1 \\ \lambda_1 \end{bmatrix}$$

Similarly, we obtain

$$\mathbf{u}_2 = \begin{bmatrix} 1 \\ \lambda_2 \end{bmatrix}$$

We observe that $\mathbf{u}_1'\mathbf{u}_2 = 1 + \lambda_1\lambda_2 = 0$ (i.e., the eigenvectors are orthogonal). Now matrix $\mathbf{U}$, whose columns are the eigenvectors of $\mathbf{F}$, is

$$\mathbf{U} = \begin{bmatrix} 1 & 1 \\ \lambda_1 & \lambda_2 \end{bmatrix}$$

Then the matrix $\mathbf{U}^{-1}\mathbf{F}\mathbf{U}$ is diagonal. Indeed, it easily follows that

$$\hat{\mathbf{F}} = \mathbf{U}^{-1}\mathbf{F}\mathbf{U} = \begin{bmatrix} \lambda_1 & 0 \\ 0 & \lambda_2 \end{bmatrix}$$

and since the transformation matrix is $\mathbf{P} = \mathbf{U}^{-1}$, we have

$$\mathbf{P} = \frac{1}{\lambda_2 - \lambda_1} \begin{bmatrix} \lambda_2 & -1 \\ -\lambda_1 & 1 \end{bmatrix}$$

Thus the diagonal matrix $\hat{\mathbf{F}}$ has the form

$$\hat{\mathbf{F}} = \begin{bmatrix} \lambda_1 & 0 \\ 0 & \lambda_2 \end{bmatrix}$$

where the diagonal elements are the eigenvalues of the characteristic polynomial.
Furthermore, we obtain

$$\hat{\mathbf{q}} = \mathbf{Pq} = \begin{bmatrix} \dfrac{1}{\sqrt{5}} \\ -\dfrac{1}{\sqrt{5}} \end{bmatrix}$$

and

$$\begin{aligned} \hat{\mathbf{g}}' &= \mathbf{g}'\mathbf{P}^{-1} = \mathbf{g}'\mathbf{U} \\ &= \begin{bmatrix} \dfrac{3+\sqrt{5}}{2} & \dfrac{3-\sqrt{5}}{2} \end{bmatrix} \end{aligned}$$

The impulse response of this equivalent diagonal system is

$$\begin{aligned} h(n) &= \hat{\mathbf{g}}'\hat{\mathbf{F}}\hat{\mathbf{q}}u(n-1) + d\delta(n) \\ &= \frac{1}{\sqrt{5}}\left[\left(\frac{3+\sqrt{5}}{2}\right)\left(\frac{1+\sqrt{5}}{2}\right)^{n-1} \right. \\ &\qquad \left. - \left(\frac{3-\sqrt{5}}{2}\right)\left(\frac{1-\sqrt{5}}{2}\right)^{n-1} \right] u(n-1) + \delta(n) \end{aligned}$$

which is the general formula for the Fibonacci sequence.

An alternative expression can be found by noting that the Fibonacci sequence can be considered as the zero-input response of the system described by the difference equation

$$y(n) = y(n-1) + y(n-2) + x(n)$$

with initial conditions $y(-1) = 1$, $y(-2) = -1$. From the type 1 state-space realization, we note that $v_1(0) = y(-2) = -1$ and $v_2(0) = y(-1) = 1$. Hence

$$\begin{bmatrix} \hat{v}_1(0) \\ \hat{v}_2(0) \end{bmatrix} = \mathbf{P}\begin{bmatrix} v_1(0) \\ v_2(0) \end{bmatrix} = \frac{-1}{5}\begin{bmatrix} \dfrac{-3+\sqrt{5}}{2} \\ \dfrac{3+\sqrt{5}}{2} \end{bmatrix}$$

and the zero-input response is

$$\begin{aligned} y_{zi}(n) &= \hat{\mathbf{g}}'\hat{\mathbf{F}}^n\hat{\mathbf{v}}(0) \\ &= \frac{1}{\sqrt{5}}\left[\left(\frac{1+\sqrt{5}}{2}\right)^n - \left(\frac{1-\sqrt{5}}{2}\right)^n \right] u(n) \end{aligned}$$

This is the more familiar form for the Fibonacci sequence, where the first term of the sequence is zero, that is, the sequences is $\{0, 1, 1, 2, 3, 5, 8, \ldots\}$.

This example illustrates the method for diagonalizing the matrix $\mathbf{F}$. The diagonal system yields a set of N decoupled, first-order linear difference equations that are easily solved to yield the state and the output of the system.

It is important to note that the eigenvalues of the matrix $\mathbf{F}$ are identical to the roots of the characteristic polynomial, which are obtained from the homogeneous difference equation that characterizes the system. For example, the system that generates the Fibonacci sequence is characterized by the homogeneous difference equation

$$y(n) - y(n - 1) - y(n - 2) = 0 \tag{7.4.35}$$

Recall that the solution is obtained by assuming that the homogeneous solution has the form

$$y_h(n) = \lambda^n$$

Substitution of this solution into (7.4.35) yields the characteristic polynomial

$$\lambda^2 - \lambda - 1 = 0$$

But this is exactly the same characteristic polynomial obtained from the determinant of $(\mathbf{F} - \lambda\mathbf{I})$.

Since the state-variable realization of the system is not unique, the matrix $\mathbf{F}$ is also not unique. However, the eigenvalues of the system are unique, that is, they are invariant to any nonsingular linear transformation of $\mathbf{F}$. Consequently, the characteristic polynomial of $\mathbf{F}$ can be determined either from evaluating the determinant of $(\mathbf{F} - \lambda\mathbf{I})$ or from the difference equation characterizing the system.

In conclusion, the state-space description provides an alternative characterization of the system that is equivalent to the input–output description. One advantage of the state-variable formulation is that it provides us with the additional information concerning the internal (state) variables of the system, information that is not easily obtained from the input–output description. Furthermore, the state-variable formulation of a linear time-invariant system allows us to represent the system by a set of (usually coupled) first-order difference equations. The decoupling of the equations can be achieved by means of a linear transformation that can be obtained by solving for the eigenvalues and eigenvectors of the system. The decoupled equations are then relatively simple to solve. More important, however, the state-space formulation provides a powerful, yet straightforward method for dealing with systems that have multiple inputs and multiple outputs (MIMO). Although we have not considered such systems in our study, it is in the treatment of MIMO systems where the true power and the beauty of the space-space formulation can be fully appreciated.

7.4.4 State-Space Analysis in the z-Domain

The state-space analysis in the previous sections has been performed in the time domain. However, as we have observed previously, the analysis of linear time-invariant discrete-time systems can also be carried out in the z-transform domain, often with greater ease. In this section we treat the state-space representation of linear time-invariant discrete-time systems in the z-transform domain.

Let us consider the state-space equation

$$\mathbf{v}(n+1) = \mathbf{F}\mathbf{v}(n) + \mathbf{q}x(n) \qquad (7.4.36)$$

If we define the vector $\mathbf{V}(z)$ as

$$\mathbf{V}(z) = \begin{bmatrix} V_1(z) \\ V_2(z) \\ \vdots \\ V_N(z) \end{bmatrix} \qquad (7.4.37)$$

then (7.4.36) can be expressed in matrix form as

$$z\mathbf{V}(z) = \mathbf{F}\mathbf{V}(z) + \mathbf{q}X(z) \qquad (7.4.38)$$

The two terms involving $\mathbf{V}(z)$ can be collected together and the resulting equation can be used to solve for $\mathbf{V}(z)$. Thus

$$(z\mathbf{I} - \mathbf{F})\mathbf{V}(z) = \mathbf{q}X(z)$$
$$\mathbf{V}(z) = (z\mathbf{I} - \mathbf{F})^{-1}\mathbf{q}X(z) \qquad (7.4.39)$$

The inverse z-transform of (7.4.39) yields the solution for the state equations.

Next, we turn our attention to the output equation, which is given as

$$y(n) = \mathbf{g}^t\mathbf{v}(n) + dx(n) \qquad (7.4.40)$$

The z-transform of (7.4.40) is

$$Y(z) = \mathbf{g}^t\mathbf{V}(z) + dX(z) \qquad (7.4.41)$$

By using the solution in (7.4.39) we can eliminate the state vector $\mathbf{V}(z)$ in (7.4.41). Thus we obtain

$$Y(z) = [\mathbf{g}^t(z\mathbf{I} - \mathbf{F})^{-1}\mathbf{q} + d]X(z) \qquad (7.4.42)$$

which is the z-transform of the zero-state response of the system. The system function is easily obtained from (7.4.42) as

$$H(z) = \frac{Y(z)}{X(z)} = \mathbf{g}^t(z\mathbf{I} - \mathbf{F})^{-1}\mathbf{q} + d \qquad (7.4.43)$$

The state equation given by (7.4.39), the output equation given by (7.4.42) and the system function given by (7.4.43) all have in common the factor $(z\mathbf{I} - \mathbf{F})^{-1}$. This is a fundamental quantity that is related to the z-transform of the state transition matrix of the system. The relationship is easily established by computing the

z-transform of the impulse response $h(n)$, which is given by (7.4.28). Thus we have

$$H(z) = \sum_{n=0}^{\infty} h(n)z^{-n}$$

$$= \sum_{n=0}^{\infty} [\mathbf{g}^t \mathbf{F}^{n-1}\mathbf{q}u(n-1) + d\delta(n)]z^{-n} \tag{7.4.44}$$

$$= \mathbf{g}^t \left(\sum_{n=1}^{\infty} \mathbf{F}^{n-1}z^{-n} \right) \mathbf{q} + d$$

The term in parentheses in (7.4.44) can be written as

$$\sum_{n=1}^{\infty} \mathbf{F}^{n-1}z^{-n} = z^{-1}(\mathbf{I} + \mathbf{F}z^{-1} + \mathbf{F}^2 z^{-2} + \cdots) \tag{7.4.45}$$

$$= z^{-1}(\mathbf{I} - \mathbf{F}z^{-1})^{-1} = (z\mathbf{I} - \mathbf{F})^{-1}$$

If we substitute the result in (7.4.45) into (7.4.44), we obtain the expression for $H(z)$ as given in (7.4.43).

Since the state transition matrix is given by

$$(n) = \mathbf{F}^n \tag{7.4.46}$$

the z-transform of (n) is

$$\sum_{n=0}^{\infty} \mathbf{F}^n z^{-n} = \mathbf{I} + \mathbf{F}z^{-1} + \mathbf{F}^2 z^{-2} + \mathbf{F}^3 z^{-3} + \cdots \tag{7.4.47}$$

$$= (\mathbf{I} - \mathbf{F}z^{-1})^{-1} = z(z\mathbf{I} - \mathbf{F})^{-1}$$

The relation in (7.4.47) provides a simple method for determining the state transition matrix by means of z-transforms. We recall that

$$(z\mathbf{I} - \mathbf{F})^{-1} = \frac{\text{adj}(z\mathbf{I} - \mathbf{F})}{\det(z\mathbf{I} - \mathbf{F})} \tag{7.4.48}$$

where adj($\mathbf{A}$) denotes the *adjoint matrix* of $\mathbf{A}$ and det ($\mathbf{A}$) denotes the determinant of the matrix $\mathbf{A}$. Substitution of (7.4.48) into (7.4.43) yields the result

$$H(z) = \mathbf{g}^t \frac{\text{adj}(z\mathbf{I} - \mathbf{F})}{\det(z\mathbf{I} - \mathbf{F})} \mathbf{q} + d \tag{7.4.49}$$

Consequently, the denominator $D(z)$ of the system function $H(z)$, which contains the poles of the system is simply

$$D(z) = \det(z\mathbf{I} - \mathbf{F}) \tag{7.4.50}$$

But the $\det(z\mathbf{I} - \mathbf{F})$ is just the characteristic polynomial of $\mathbf{F}$. Its roots, which are the poles of system, are the eigenvalues of the matrix $\mathbf{F}$.

Example 7.4.3

Determine the system function $H(z)$, the impulse response $h(n)$, and the state transition matrix $\Phi(n)$ of the system that generates the Fibonacci sequence. This system is described by the state-space equation

$$\mathbf{v}(n+1) = \begin{bmatrix} 0 & 1 \\ 1 & 1 \end{bmatrix} \mathbf{v}(n) + \begin{bmatrix} 0 \\ 1 \end{bmatrix} x(n)$$

$$y(n) = \begin{bmatrix} 1 & 1 \end{bmatrix} \mathbf{v}(n) + x(n)$$

Solution   First, we determine $H(z)$ and $h(n)$ by computing $(z\mathbf{I} - \mathbf{F})^{-1}$. We have

$$(z\mathbf{I} - \mathbf{F})^{-1} = \begin{bmatrix} z & -1 \\ -1 & z-1 \end{bmatrix}^{-1} = \frac{1}{z^2 - z - 1} \begin{bmatrix} z-1 & 1 \\ 1 & z \end{bmatrix}$$

Hence

$$H(z) = \frac{1}{z^2 - z - 1} \begin{bmatrix} 1 & 1 \end{bmatrix} \begin{bmatrix} z-1 & 1 \\ 1 & z \end{bmatrix} \begin{bmatrix} 0 \\ 1 \end{bmatrix} + 1$$

$$= \frac{z^2}{z^2 - z - 1} = \frac{1}{1 - z^{-1} - z^{-2}}$$

By inverting $H(z)$, we obtain $h(n)$ in the form

$$h(n) = \frac{1}{\sqrt{5}} \left[\left(\frac{1+\sqrt{5}}{2} \right)^{n+1} - \left(\frac{1-\sqrt{5}}{2} \right)^{n+1} \right] u(n)$$

We note that the poles of $H(z)$ are $p_1 = (1+\sqrt{5})/2$ and $p_2 = (1-\sqrt{5})/2$. Since $|p_1| > 1$, the system that generates the Fibonacci sequence is unstable.

The state transition matrix $\Phi(n)$ has the z-transform

$$z(z\mathbf{I} - \mathbf{F})^{-1} = \frac{1}{z^2 - z - 1} \begin{bmatrix} z^2 - z & z \\ z & z^2 \end{bmatrix}$$

The four elements of $\Phi(n)$ are obtained by computing the inverse transform of the four elements of $z(z\mathbf{I} - \mathbf{F})^{-1}$. Thus we obtain

$$\Phi(n) = \begin{bmatrix} \phi_{11}(n) & \phi_{12}(n) \\ \phi_{21}(n) & \phi_{22}(n) \end{bmatrix}$$

where

$$\phi_{11}(n) = \left[\frac{1+\sqrt{5}}{2\sqrt{5}} \left(\frac{1-\sqrt{5}}{2} \right)^n - \frac{1-\sqrt{5}}{2\sqrt{5}} \left(\frac{1+\sqrt{5}}{2} \right)^n \right] u(n)$$

$$\phi_{12}(n) = \phi_{21}(n) = \frac{1}{\sqrt{5}} \left[\left(\frac{1+\sqrt{5}}{2} \right)^n - \left(\frac{1-\sqrt{5}}{2} \right)^n \right] u(n)$$

$$\phi_{22}(n) = \frac{1}{\sqrt{5}} \left[\left(\frac{1+\sqrt{5}}{2} \right)^{n+1} - \left(\frac{1-\sqrt{5}}{2} \right)^{n+1} \right] u(n)$$

We note that the impulse response $h(n)$ can also be computed from (7.4.28) by using the state transition matrix.

This analysis method applies specifically to the computation of the zero-state response of the system. This is the consequence of the fact that we have used the two-sided z-transform.

If we wish to determine the total response of the system, beginning at a nonzero state, say $\mathbf{v}(n_0)$, we must use the one-sided z-transform. Thus, for a given initial state $\mathbf{v}(n_0)$ and a given input $x(n)$ for $n \geq n_0$, we can determine the state vector $\mathbf{v}(n)$ for $n \geq n_0$ and the output $y(n)$ for $n \geq n_0$, by means of the one-sided z-transform.

In this development we assume that $n_0 = 0$, without loss of generality. Then, given $x(n)$ for $n \geq 0$, and a causal system, described by the state equations in (7.4.36), the one-sided z-transform of the state equations is

$$z\mathbf{V}^+(z) - z\mathbf{v}(0) = \mathbf{F}\mathbf{V}^+(z) + \mathbf{q}X(z)$$

or, equivalently,

$$\mathbf{V}^+(z) = z(z\mathbf{I} - \mathbf{F})^{-1}\mathbf{v}(0) + (z\mathbf{I} - \mathbf{F})^{-1}\mathbf{q}X(z) \tag{7.4.51}$$

Note that $X^+(z) = X(z)$, since $x(n)$ is assumed to be causal.

Similarly, the z-transform of the output equation given by (7.4.40) is

$$Y^+(z) = \mathbf{g}^t\mathbf{V}^+(z) + dX(z) \tag{7.4.52}$$

If we substitute for $\mathbf{V}^+(z)$ from (7.4.51) into (7.4.52), we obtain the result

$$Y^+(z) = z\mathbf{g}^t(z\mathbf{I} - \mathbf{F})^{-1}\mathbf{v}(0) + [\mathbf{g}^t(z\mathbf{I} - \mathbf{F})^{-1}\mathbf{q} + d]X(z) \tag{7.4.53}$$

Of the terms on the right-hand side of (7.4.53), the first represents the zero-input response of the system due to the initial conditions, while the second represents the zero-state response of the system that we obtained previously. Consequently, (7.4.53) constitutes the total response of the system, which can be expressed in the time domain by inverting (7.4.53). The result of this inversion yields the form for $y(n)$ given previously by (7.4.16).

7.4.5 Additional State-Space Structures

In Section 7.4.2 we described how state-space equations can be obtained from a given structure and, conversely, how to obtain a realization of the system given the state equations. In this section we revisit the parallel-form and cascade-form realizations described previously and consider these structures in the context of a state-space formulation.

The parallel-form state-space structure is obtained by expanding the system function $H(z)$ into a partial-fraction expansion, developing the state-space formulation for each term in the expansion and the corresponding structure, and finally, connecting all the structures in parallel. We illustrate the procedure under the assumption that the poles are distinct and $N = M$.

The system function $H(z)$ can be expressed as

$$H(z) = C + \sum_{k=1}^{N} \frac{B_k}{z - p_k} \tag{7.4.54}$$

Note that this is a different expansion from that given in (7.3.20). The output of the system is

$$Y(z) = H(z)X(z) = CX(z) + \sum_{k=1}^{N} B_k Y_k(z) \qquad (7.4.55)$$

where, by definition,

$$Y_k(z) = \frac{X(z)}{z - p_k} \qquad k = 1, 2, \ldots, N \qquad (7.4.56)$$

In the time domain, the equations in (7.4.56) become

$$y_k(n + 1) = p_k y_k(n) + x(n) \qquad k = 1, 2, \ldots, N \qquad (7.4.57)$$

We define the state variables as

$$v_k(n) = y_k(n) \qquad k = 1, 2, \ldots, N \qquad (7.4.58)$$

Then the difference equations in (7.4.57) become

$$v_k(n + 1) = p_k v_k(n) + x(n) \qquad k = 1, 2, \ldots, N \qquad (7.4.59)$$

The state equations in (7.4.59) can be expressed in matrix form as

$$\mathbf{v}(n + 1) = \begin{bmatrix} p_1 & 0 & \cdots & 0 \\ 0 & p_2 & \cdots & 0 \\ & & \ddots & \\ 0 & & & p_N \end{bmatrix} \mathbf{v}(n) + \begin{bmatrix} 1 \\ 1 \\ \vdots \\ 1 \end{bmatrix} x(n) \qquad (7.4.60)$$

and the output equation is

$$y(n) = [\, B_1 \quad B_2 \quad \cdots \quad B_N \,]\mathbf{v}(n) + Cx(n) \qquad (7.4.61)$$

This parallel-form realization is called the *normal form* representation, because the matrix **F** is diagonal, and hence the state variables are uncoupled. An alternative structure is obtained by pairing complex-conjugate poles and any two real-valued poles to form second-order sections, which can be realized by using either type 1 or type 2 state-space structures.

The *cascade-form* state-space structure can be obtained by factoring $H(z)$ into a product of first-order and second-order sections, as described in Section 7.2.2, and then implementing each section by using either type 1 or type 2 state-space structures.

Let us consider the state-space representation of a single second-order section involving a pair of complex-conjugate poles. The system function is

$$\begin{aligned} H(z) &= \frac{b_0 + b_1 z^{-1} + b_2 z^{-2}}{1 + a_1 z^{-1} + a_2 z^{-2}} \\[2mm] &= \frac{b_0 z^2 + b_1 z + b_2}{z^2 + a_1 z + a_2} \\[2mm] &= b_0 + \frac{A}{z - p} + \frac{A^*}{z - p^*} \end{aligned} \qquad (7.4.62)$$

The output of this system can be expressed as

$$Y(z) = b_0 X(z) + \frac{AX(z)}{z - p} + \frac{A^* X(z)}{z - p^*} \qquad (7.4.63)$$

We define the quantity

$$S(z) = \frac{AX(z)}{z - p} \tag{7.4.64}$$

This relationship can be expressed in the time domain as

$$s(n + 1) = ps(n) + Ax(n) \tag{7.4.65}$$

Since $s(n)$, p, and A are complex valued, we define $s(n)$ as

$$s(n) = v_1(n) + jv_2(n)$$

$$p = \alpha_1 + j\alpha_2 \tag{7.4.66}$$

$$A = q_1 + jq_2$$

Upon substitution of these relations into (7.4.65) and separating its real and imaginary parts, we obtain

$$v_1(n + 1) = \alpha_1 v_1(n) - \alpha_2 v_2(n) + q_1 x(n)$$
$$v_2(n + 1) = \alpha_2 v_1(n) + \alpha_1 v_2(n) + q_2 x(n) \tag{7.4.67}$$

We choose $v_1(n)$ and $v_2(n)$ as the state variables and thus obtain the coupled pair of state equations which can be expressed in matrix form as

$$\mathbf{v}(n + 1) = \begin{bmatrix} \alpha_1 & -\alpha_2 \\ \alpha_2 & \alpha_1 \end{bmatrix} \mathbf{v}(n) + \begin{bmatrix} q_1 \\ q_2 \end{bmatrix} x(n) \tag{7.4.68}$$

The output equation can be expressed as

$$y(n) = b_0 x(n) + s(n) + s^*(n) \tag{7.4.69}$$

Upon substitution for $s(n)$ in (7.4.69), we obtain the desired result for the output in the form

$$y(n) = \begin{bmatrix} 2 & 0 \end{bmatrix} \mathbf{v}(n) + b_0 x(n) \tag{7.4.70}$$

A realization for the second-order section is shown in Fig. 7.31. It is simply called the *coupled-form* state-space realization. This structure, which is used as the building block in the implementation of cascade-form realizations for higher-order IIR systems, exhibits low sensitivity to finite-word-length effects.

7.5 REPRESENTATION OF NUMBERS

Up to this point we have considered the implementation of discrete-time systems without being concerned about the finite-word-length effects that are inherent in any digital realization, whether it be in hardware or in software. In fact, we have analyzed systems that are modeled as linear when, in fact, digital realizations of such systems are inherently nonlinear.

In this and the following two sections, we consider the various forms of quantization effects that arise in digital signal processing. Although we describe

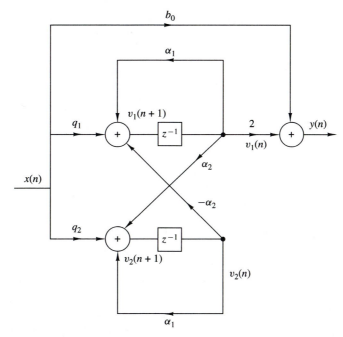

Figure 7.31 Coupled-form state-space realization of a two-pole, two-zero IIR system.

floating-point arithmetic operations briefly, our major concern is with fixed-point realizations of digital filters.

In this section we consider the representation of numbers for digital computations. The main characteristic of digital arithmetic is the limited (usually fixed) number of digits used to represent numbers. This constraint leads to finite numerical precision in computations, which leads to round-off errors and nonlinear effects in the performance of digital filters. We now provide a brief introduction to digital arithmetic.

7.5.1 Fixed-Point Representation of Numbers

The representation of numbers in a fixed-point format is a generalization of the familiar decimal representation of a number as a string of digits with a decimal point. In this notation, the digits to the left of the decimal point represent the integer part of the number, and the digits to the right of the decimal point represent the fractional part of the number. Thus a real number X can be represented as

$$X = (b_{-A}, \ldots, b_{-1}, b_0, b_1, \ldots, b_B)_r$$

$$= \sum_{i=-A}^{B} b_i r^{-i} \qquad 0 \le b_i \le (r-1) \tag{7.5.1}$$

where b_i represents the digit, r is the radix or base, A is the number of integer digits, and B is the number of fractional digits. As an example, the decimal number $(123.45)_{10}$ and the binary number $(101.01)_2$ represent the following sums:

$$(123.45)_{10} = 1 \times 10^2 + 2 \times 10^1 + 3 \times 10^0 + 4 \times 10^{-1} + 5 \times 10^{-2}$$

$$(101.01)_2 = 1 \times 2^2 + 0 \times 2^1 + 1 \times 2^0 + 0 \times 2^{-1} + 1 \times 2^{-2}$$

Let us focus our attention on the binary representation since it is the most important for digital signal processing. In this case $r = 2$ and the digits $\{b_i\}$ are called binary digits or bits and take the values $\{0, 1\}$. The binary digit b_{-A} is called the most significant bit (MSB) of the number, and the binary digit b_B is called the least significant bit (LSB). The "binary point" between the digits b_0 and b_1 does not exist physically in the computer. Simply, the logic circuits of the computer are designed so that the computations result in numbers that correspond to the assumed location of this point.

By using an n-bit integer format ($A = n - 1$, $B = 0$), we an represent unsigned integers with magnitude in the range 0 to $2^n - 1$. Usually, we use the fraction format ($A = 0$, $B = n - 1$), with a binary point between b_0 and b_1, that permits numbers in the range from 0 to $1 - 2^{-n}$. Note that any integer or mixed number can be represented in a fraction format by factoring out the term r^A in (7.5.1). In the sequel we focus our attention on the binary fraction format because mixed numbers are difficult to multiply and the number of bits representing an integer cannot be reduced by truncation or rounding.

There are three ways to represent negative numbers. This leads to three formats for the representation of signed binary fractions. The format for positive fractions is the same in all three representations, namely,

$$X = 0.b_1 b_2 \cdots b_B = \sum_{i=1}^{B} b_i \cdot 2^{-i}, \quad X \geq 0 \tag{7.5.2}$$

Note that the MSB b_0 is set to zero to represent the positive sign. Consider now the negative fraction

$$X = -0.b_1 b_2 \cdots b_B = -\sum_{i=1}^{B} b_i \cdot 2^{-i} \tag{7.5.3}$$

This number can be represented using one of the following three formats.

Sign-magnitude format. In this format, the MSB is set to 1 to represent the negative sign,

$$X_{\text{SM}} = 1.b_1 b_2 \cdots b_B \quad \text{for } X \leq 0 \tag{7.5.4}$$

One's-complement format. In this format the negative numbers are represented as

$$X_{1\text{C}} = 1.\bar{b}_1 \bar{b}_2 \cdots \bar{b}_B \quad X \leq 0 \tag{7.5.5}$$

where $\bar{b}_i = 1 - b_i$ is the one's complement of b_i. Thus if X is a positive number, the corresponding negative number is determined by complementing (changing 1's to 0's and 0's to 1's) all the bits. An alternative definition for X_{1C} can be obtained by noting that

$$X_{1C} = 1 \times 2^0 + \sum_{i=1}^{B}(1 - b_i) \cdot 2^{-i} = 2 - 2^{-B}|X| \qquad (7.5.6)$$

Two's-complement format.   In this format a negative number is represented by forming the two's complement of the corresponding positive number. In other words, the negative number is obtained by subtracting the positive number from 2.0. More simply, the two's complement is formed by complementing the positive number and adding one LSB. Thus

$$X_{2C} = 1.\bar{b}_1\bar{b}_2\cdots\bar{b}_B + 00\cdots01 \qquad X < 0 \qquad (7.5.7)$$

where + represents modulo-2 addition that ignores any carry generated from the sign bit. For example, the number $-\frac{3}{8}$ is simply obtained by complementing 0011 ($\frac{3}{8}$) to obtain 1100 and then adding 0001. This yields 1101, which represents $-\frac{3}{8}$ in two's complement.

From (7.5.6) and (7.5.7) is can easily be seen that

$$X_{2C} = X_{1C} + 2^{-B} = 2 - |X| \qquad (7.5.8)$$

To demonstrate that (7.5.7) truly represents a negative number, we use the identity

$$1 = \sum_{i=1}^{B} 2^{-i} + 2^{-B} \qquad (7.5.9)$$

The negative number X in (7.5.3) can be expressed as

$$X_{2C} = -\sum_{i=1}^{B} b_i \cdot 2^{-i} + 1 - 1$$

$$= -1 + \sum_{i=1}^{B}(1 - b_i)2^{-i} + 2^{-B}$$

$$= -1 + \sum_{i=1}^{B} \bar{b}_i \cdot 2^{-1} + 2^{-B}$$

which is exactly the two's-complement representation of (7.5.7).

In summary, the value of a binary string $b_0 b_1 \cdots b_B$ depends on the format used. For positive numbers, $b_0 = 0$, and the number is given by (7.5.2). For negative numbers, we use these corresponding formulas for the three formats.

Example 7.5.1

Express the fraction $\frac{7}{8}$ and $-\frac{7}{8}$ in sign-magnitude, two's-complement, and one's-complement format.

Solution $X = \frac{7}{8}$ is represented as $2^{-1} + 2^{-2} + 2^{-3}$, so that $X = 0.111$. In sign-magnitude format, $X = -\frac{7}{8}$ is represented as 1.111. In one's complement, we have

$$X_{1C} = 1.000$$

In two's complement, the result is

$$X_{2C} = 1.000 + 0.001 = 1.001$$

The basic arithmetic operations of addition and multiplication depend on the format used. For one's-complement and two's-complement formats, addition is carried out by adding the numbers bit by bit. The formats differ only in the way in which a carry bit affects the MSB. For example, $\frac{4}{8} - \frac{3}{8} = \frac{1}{8}$. In two's complement, we have

$$0100 \oplus 1101 = 0001$$

where $\oplus$ indicates modulo-2 addition. Note that the carry bit, if present in the MSB, is dropped. On the other hand, in one's- complement arithmetic, the carry in the MSB, if present, is carried around to the LSB. Thus the computation $\frac{4}{8} - \frac{3}{8} = \frac{1}{8}$ becomes

$$0100 \oplus 1100 = 0000 \oplus 0001 = 0001$$

Addition in the sign-magnitude format is more complex and can involve sign checks, complementing, and the generation of a carry. On the other hand, direct multiplication of two sign- magnitude numbers is relatively straightforward, whereas a special algorithm is usually employed for one's complement and two's complement multiplication.

Most fixed-point digital signal processors use two's-complement arithmetic. Hence, the range for $(B+1)-$bit numbers is from -1 to $1-2^{-B}$. These numbers can be viewed in a wheel format as shown in Fig. 7.32 for $B = 2$. Two's-complement arithmetic is basically arithmetic modulo-2^{B+1} [i.e., any number that falls outside

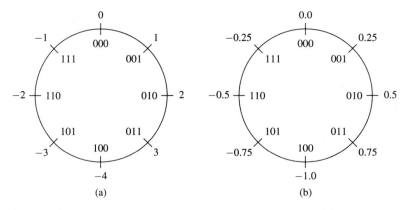

Figure 7.32   Counting wheel for 3-bit two's-complement numbers (a) integers and (b) functions.

the range (overflow or underflow) is reduced to this range by subtracting an appropriate multiple of 2^{B+1}]. This type of arithmetic can be viewed as counting using the wheel of Fig. 7.32. A very important property of two's- complement addition is that if the final sum of a string of numbers $X_1, X_2, \ldots, X_N$ is within the range, it will be computed correctly, even if individual partial sums result in overflows. This and other characteristics of two's-complement arithmetic are considered in Problem 7.44.

In general, the multiplication of two fixed-point numbers each of b bits in length results in a product of $2b$ bits in length. In fixed-point arithmetic, the product is either truncated or rounded back to b bits. As a result we have a truncation or round-off error in the b least significant bits. The characterization of such errors is treated below.

7.5.2 Binary Floating-Point Representation of Numbers

A fixed-point representation of numbers allows us to cover a range of numbers, say, $x_{\max} - x_{\min}$ with a resolution

$$\Delta = \frac{x_{\max} - x_{\min}}{m - 1}$$

where $m = 2^b$ is the number of levels and b is the number of bits. A basic characteristic of the fixed-point representation is that the resolution is fixed. Furthermore, Δ increases in direct proportion to an increase in the dynamic range.

A floating-point representation can be employed as a means for covering a larger dynamic range. The binary floating-point representation commonly used in practice, consists of a mantissa M, which is the fractional part of the number and falls in the range $\frac{1}{2} \le M < 1$, multiplied by the exponential factor 2^E, where the exponent E is either a positive or negative integer. Hence a number X is represented as

$$X = M \cdot 2^E$$

The mantissa requires a sign bit for representing positive and negative numbers, and the exponent requires an additional sign bit. Since the mantissa is a signed fraction, we can use any of the four fixed-point representations just described.

For example, the number $X_1 = 5$ is represented by the following mantissa and exponent:

$$M_1 = 0.101000$$

$$E_1 = 011$$

while the number $X_2 = \frac{3}{8}$ is represented by the following mantissa and exponent

$$M_2 = 0.110000$$

$$E_2 = 101$$

where the leftmost bit in the exponent represents the sign bit.

If the two numbers are to be multiplied, the mantissas are multiplied and the exponents are added. Thus the product of these two numbers is

$$X_1 X_2 = M_1 M_2 \cdot 2^{E_1 + E_2}$$
$$= (0.011110) \cdot 2^{010}$$
$$= (0.111100) \cdot 2^{001}$$

On the other hand, the addition of the two floating-point numbers requires that the exponents be equal. This can be accomplished by shifting the mantissa of the smaller number to the right and compensating by increasing the corresponding exponent. Thus the number X_2 can be expressed as

$$M_2 = 0.000011$$
$$E_2 = 011$$

With $E_2 = E_1$, we can add the two numbers X_1 and X_2. The result is

$$X_1 + X_2 = (0.101011) \cdot 2^{011}$$

It should be observed that the shifting operation required to equalize the exponent of X_2 with that for X_1 results in loss of precision, in general. In this example the six-bit mantissa was sufficiently long to accommodate a shift of four bits to the right for M_2 without dropping any of the ones. However, a shift of five bits would have caused the loss of a single bit and a shift of six bits to the right would have resulted in a mantissa of $M_2 = 0.000000$, unless we round upward after shifting so that $M_2 = 0.000001$.

Overflow occurs in the multiplication of two floating-point numbers when the sum of the exponents exceeds the dynamic range of the fixed-point representation of the exponent.

In comparing a fixed-point representation with a floating-point representation, each with the same number of total bits, it is apparent that the floating-point representation allows us to cover a larger dynamic range by varying the resolution across the range. The resolution decreases with an increase in the size of successive numbers. In other words, the distance between two successive floating-point numbers increases as the numbers increase in size. It is this variable resolution that results in a larger dynamic range. Alternatively, if we wish to cover the same dynamic range with both fixed-point and floating-point representations, the floating-point representation provides finer resolution for small numbers but coarser resolution for the larger numbers. In contrast, the fixed-point representation provides a uniform resolution throughout the range of numbers.

For example, if we have a computer with a word size of 32 bits, it is possible to represent 2^{32} numbers. If we wish to represent the positive integers beginning with zero, the largest possible integer that can be accommodated is

$$2^{32} - 1 = 4{,}294{,}967{,}295$$

The distance between successive numbers (the resolution) is 1. Alternatively, we can designate the leftmost bit as the sign bit and use the remaining 31 bits for the magnitude. In such a case a fixed-point representation allows us to cover the range

$$-(2^{31} - 1) = -2,147,483,647 \quad \text{to} \quad (2^{31} - 1) = 2,147,483,647$$

again with a resolution of 1.

On the other hand, suppose that we increase the resolution by allocating 10 bits for a fractional part, 21 bits for the integer part, and 1 bit for the sign. Then this representation allows us to cover the dynamic range

$$-(2^{31} - 1) \cdot 2^{-10} = -(2^{21} - 2^{-10}) \quad \text{to} \quad (2^{31} - 1) \cdot 2^{-10} = 2^{21} - 2^{-10}$$

or, equivalently,

$$-2,097,151.999 \quad \text{to} \quad 2,097,151.999$$

In this case, the resolution is 2^{-10}. Thus, the dynamic range has been decreased by a factor of approximately 1000 (actually 2^{10}), while the resolution has been increased by the same factor.

For comparison, suppose that the 32-bit word is used to represent floating-point numbers. In particular, let the mantissa be represented by 23 bits plus a sign bit and let the exponent be represented by 7 bits plus a sign bit. Now, the smallest number in magnitude will have the representation,

$$\begin{array}{cccc} \text{sign} & \text{23 bits} & \text{sign} & \text{7 bits} \\ 0. & 100 \cdots 0 & 1 & 1111111 \end{array} = \tfrac{1}{2} \times 2^{-127} \approx 0.3 \times 10^{-38}$$

At the other extreme, the largest number that can be represented with this floating-point representation is

$$\begin{array}{cccc} \text{sign} & \text{23 bits} & \text{sign} & \text{7 bits} \\ 0 & 111 \cdots 1 & 0 & 1111111 \end{array} = (1 - 2^{-23}) \times 2^{127} \approx 1.7 \times 10^{38}$$

Thus, we have achieved a dynamic range of approximately 10^{76}, but with varying resolution. In particular, we have fine resolution for small numbers and coarse resolution for larger numbers.

The representation of zero poses some special problems. In general, only the mantissa has to be zero, but not the exponent. The choice of M and E, the representation of zero, the handling of overflows, and other related issues have resulted in various floating-point representations on different digital computers. In an effort to define a common floating-point format, the Institute of Electrical and Electronic Engineers (IEEE) introduced the IEEE 754 standard, which is widely used in practice. For a 32-bit machine, the IEEE 754 standard single-precision, floating-point number is represented as $X = (-1)^s \cdot 2^{E-127}(M)$, where

0	1	8	9	31
S	E		M	

This number has the following interpretations:

> If $E = 255$ and $M \neq 0$, then X is not a number
> If $E = 255$ and $M = 0$, then $X = (-1)^S \cdot \infty$
> If $0 < E < 255$, then $X = (-1)^S \cdot 2^{E-127}(1.M)$
> If $E = 0$ and $M \neq 0$, then $X = (-1)^S \cdot 2^{-126}(0.M)$
> If $E = 0$ and $M = 0$, then $X = (-1)^S \cdot 0$

where $0.M$ is a fraction and $1.M$ is a mixed number with one integer bit and 23 fractional bits. For example, the number

0	1 0 0 0 0 0 1 0	1 0 1 0 $\cdots$ 0 0
S	E	M

has the value $X = -1^0 \times 2^{130-127} \times 1.1010\ldots0 = 2^3 \times \frac{13}{8} = 13$. The magnitude range of the 32-bit IEEE 754 floating-point numbers is from $2^{-126} \times 2^{-23}$ to $(2-2^{-23}) \times 2^{127}$ (i.e., from 1.18×10^{-38} to 3.40×10^{38}). Computations with numbers outside this range result in either underflow or overflow.

7.5.3 Errors Resulting from Rounding and Truncation

In performing computations such as multiplications with either fixed-point or floating-point arithmetic, we are usually faced with the problem of quantizing a number via truncation or rounding, from a given level of precision to a level of lower precision. The effect of rounding and truncation is to introduce an error whose value depends on the number of bits in the original number relative to the number of bits after quantization. The characteristics of the errors introduced through either truncation or rounding depend on the particular form of number representation.

To be specific, let us consider a fixed-point representation in which a number x is quantized from b_u bits to b bits. Thus the number

$$x = \overbrace{0.1011 \cdots 01}^{b_u}$$

consisting of b_u bits prior to quantization is represented as

$$x = \overbrace{0.101 \cdots 1}^{b}$$

after quantization, where $b < b_u$. For example, if x represents the sample of an analog signal, then b_u may be taken as infinite. In any case if the quantizer truncates the value of x, the truncation error is defined as

$$E_t = Q_t(x) - x \tag{7.5.10}$$

First, we consider the range of values of the error for sign-magnitude and two's-complement representation. In both of these representations, the positive numbers have identical representations. For positive numbers, truncation results in a number that is smaller than the unquantized number. Consequently, the truncation error resulting from a reduction of the number of significant bits from b_u to b is

$$- (2^{-b} - 2^{-b_u}) \le E_t \le 0 \qquad (7.5.11)$$

where the largest error arises from discarding $b_u - b$ bits, all of which are ones.

In the case of negative fixed-point numbers based on the sign-magnitude representation, the truncation error is positive, since truncation basically reduces the magnitude of the numbers. Consequently, for negative numbers, we have

$$0 \le E_t \le (2^{-b} - 2^{-b_u}) \qquad (7.5.12)$$

In the two's-complement representation, the negative of a number is obtained by subtracting the corresponding positive number from 2. As a consequence, the effect of truncation on a negative number is to increase the magnitude of the negative number. Consequently, $x > Q_t(x)$ and hence

$$- (2^{-b} - 2^{-b_u}) \le E_t \le 0 \qquad (7.5.13)$$

Hence we conclude that *the truncation error for the sign-magnitude representation is symmetric about zero and falls in the range*

$$- (2^{-b} - 2^{-b_u}) \le E_t \le (2^{-b} - 2^{-b_u}) \qquad (7.5.14)$$

On the other hand, *for two's-complement representation, the truncation error is always negative and falls in the range*

$$- (2^{-b} - 2^{-b_u}) \le E_t \le 0 \qquad (7.5.15)$$

Next, let us consider the quantization errors due to rounding of a number. A number x, represented by b_u bits before quantization and b bits after quantization, incurs a quantization error

$$E_r = Q_r(x) - x \qquad (7.5.16)$$

Basically, rounding involves only the magnitude of the number and, consequently, the round-off error is independent of the type of fixed-point representation. The maximum error that can be introduced through rounding is $(2^{-b} - 2^{-b_u})/2$ and this can be either positive or negative, depending on the value of x. Therefore, *the round-off error is symmetric about zero and falls in the range*

$$- \tfrac{1}{2}(2^{-b} - 2^{-b_u}) \le E_r \le \tfrac{1}{2}(2^{-b} - 2^{-b_u}) \qquad (7.5.17)$$

These relationships are summarized in Fig. 7.33 when x is a continuous signal amplitude ($b_u = \infty$).

In a floating-point representation, the mantissa is either rounded or truncated. Due to the nonuniform resolution, the corresponding error in a floating-point representation is proportional to the number being quantized. An appropriate

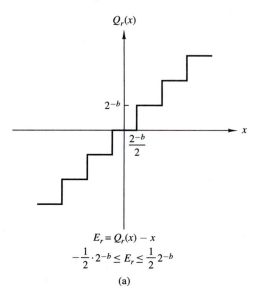

$$E_r = Q_r(x) - x$$

$$-\frac{1}{2} \cdot 2^{-b} \le E_r \le \frac{1}{2} 2^{-b}$$

(a)

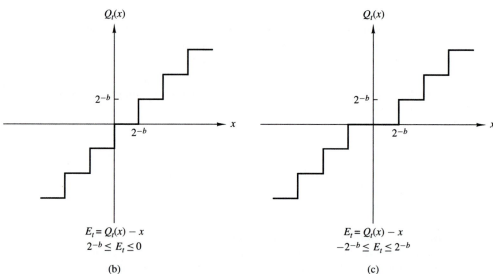

$$E_t = Q_t(x) - x$$
$$2^{-b} \le E_t \le 0$$

(b)

$$E_t = Q_t(x) - x$$
$$-2^{-b} \le E_t \le 2^{-b}$$

(c)

Figure 7.33 Quantization errors in rounding and truncation: (a) rounding; (b) truncation in two's complement; (c) truncation in sign-magnitude.

representation for the quantized value is

$$Q(x) = x + ex \tag{7.5.18}$$

where e is called the relative error. Now

$$Q(x) - x = ex \tag{7.5.19}$$

In the case of truncation based on two's-complement representation of the mantissa, we have

$$-2^E 2^{-b} < e_t x < 0 \tag{7.5.20}$$

for positive numbers. Since $2^{E-1} \leq x < 2^E$, it follows that

$$-2^{-b+1} < e_t \leq 0 \qquad x > 0 \tag{7.5.21}$$

On the other hand, for a negative number in two's-complement representation, the error is

$$0 \leq e_t x < 2^E 2^{-b}$$

and hence

$$0 \leq e_t < 2^{-b+1} \qquad x < 0 \tag{7.5.22}$$

In the case where the mantissa is rounded, the resulting error is symmetric relative to zero and has a maximum value of $\pm 2^{-b}/2$. Consequently, the round-off error becomes

$$-2^E \cdot 2^{-b}/2 < e_r x \leq 2^E \cdot 2^{-b}/2 \tag{7.5.23}$$

Again, since x falls in the range $2^{E-1} \leq x < 2^E$, we divide through by 2^{E-1} so that

$$-2^{-b} < e_r \leq 2^{-b} \tag{7.5.24}$$

In arithmetic computations involving quantization via truncation and rounding, it is convenient to adopt a statistical approach to the characterization of such errors. The quantizer can be modeled as introducing an additive noise to the unquantized value x. Thus we can write

$$Q(x) = x + \epsilon$$

where $\epsilon = E_r$ for rounding and $\epsilon = E_t$ for truncation. This model is illustrated in Fig. 7.34.

Since x can be any number that falls within any of the levels of the quantizer, the quantization error is usually modeled as a random variable that falls

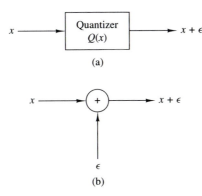

(a)

(b)

Figure 7.34 Additive noise model for the nonlinear quantization process: (a) actual system; (b) model for quantization.

within the limits specified. This random variable is assumed to be uniformly distributed within the ranges specified for the fixed-point representations. Furthermore, in practice, $b_u \gg b$, so that we can neglect the factor of 2^{-b_u} in the formulas given below. Under these conditions, the probability density functions for the round-off and truncation errors in the two fixed-point representations are illustrated in Fig. 7.35. We note that in the case of truncation of the two's-complement representation of the number, the average value of the error has a bias of $2^{-b}/2$, whereas in all other cases just illustrated, the error has an average value of zero.

We shall use this statistical characterization of the quantization errors in our treatment of such errors in digital filtering and in the computation of the DFT for fixed-point implementation.

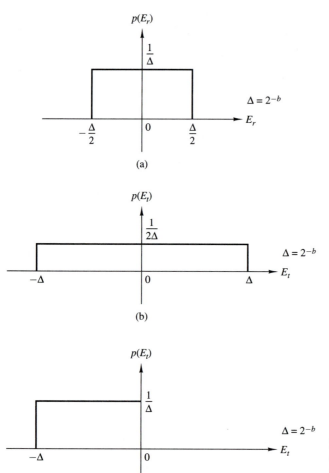

Figure 7.35 Statistical characterization of quantization errors: (a) round-off error; (b) truncation error for sign-magnitude; (c) truncation error for two's complement.

7.6 QUANTIZATION OF FILTER COEFFICIENTS

In the realization of FIR and IIR filters in hardware or in software on a general-purpose computer, the accuracy with which filter coefficients can be specified is limited by the word length of the computer or the length of the register provided to store the coefficients. Since the coefficients used in implementing a given filter are not exact, the poles and zeros of the system function will, in general, be different from the desired poles and zeros. Consequently, we obtain a filter having a frequency response that is different from the frequency response of the filter with unquantized coefficients.

In Section 7.6.1, we demonstrate that the sensitivity of the filter frequency response characteristics to quantization of the filter coefficients is minimized by realizing a filter having a large number of poles and zeros as an interconnection of second-order filter sections. This leads us to the parallel-form and cascade-form realizations in which the basic building blocks are second-order filter sections.

7.6.1 Analysis of Sensitivity to Quantization of Filter Coefficients

To illustrate the effect of quantization of the filter coefficients in a direct-form realization of an IIR filter, let us consider a general IIR filter with system function

$$H(z) = \frac{\sum_{k=0}^{M} b_k z^{-k}}{1 + \sum_{k=1}^{N} a_k z^{-k}} \tag{7.6.1}$$

The direct-form realization of the IIR filter with quantized coefficients has the system function

$$\overline{H}(z) = \frac{\sum_{k=0}^{M} \overline{b}_k z^{-k}}{1 + \sum_{k=1}^{N} \overline{a}_k z^{-k}} \tag{7.6.2}$$

where the quantized coefficients $\{\overline{b}_k\}$ and $\{\overline{a}_k\}$ can be related to the unquantized coefficients $\{b_k\}$ and $\{a_k\}$ by the relations

$$\overline{a}_k = a_k + \Delta a_k \quad k = 1, 2, \ldots, N$$

$$\overline{b}_k = b_k + \Delta b_k \quad k = 0, 1, \ldots, M \tag{7.6.3}$$

and $\{\Delta a_k\}$ and $\{\Delta b_k\}$ represent the quantization errors.

The denominator of $H(z)$ may be expressed in the form

$$D(z) = 1 + \sum_{k=0}^{N} a_k z^{-k} = \prod_{k=1}^{N} (1 - p_k z^{-1}) \tag{7.6.4}$$

where $\{p_k\}$ are the poles of $H(z)$. Similarly, we can express the denominator of $\overline{H}(z)$ as

$$\overline{D}(z) = \prod_{k=1}^{N} (1 - \overline{p}_k z^{-1}) \tag{7.6.5}$$

where $\overline{p}_k = p_k + \Delta p_k$, $k = 1, 2, \ldots, N$, and Δp_k is the error or perturbation resulting from the quantization of the filter coefficients.

We shall now relate the perturbation Δp_k to the quantization errors in the $\{a_k\}$.

The perturbation error Δp_i can be expressed as

$$\Delta p_i = \sum_{k=1}^{N} \frac{\partial p_i}{\partial a_k} \Delta a_k \tag{7.6.6}$$

where $\partial p_i / \partial a_k$, the partial derivative of p_i with respect to a_k, represents the incremental change in the pole p_i due to a change in the coefficient a_k. Thus the total error Δp_i is expressed as a sum of the incremental errors due to changes in each of the coefficients $\{a_k\}$.

The partial derivatives $\partial p_i / \partial a_k$, $k = 1, 2, \ldots, N$, can be obtained by differentiating $D(z)$ with respect to each of the $\{a_k\}$. First we have

$$\left(\frac{\partial D(z)}{\partial a_k} \right)_{z=p_i} = \left(\frac{\partial D(z)}{\partial z} \right)_{z=p_i} \left(\frac{\partial p_i}{\partial a_k} \right) \tag{7.6.7}$$

Then

$$\frac{\partial p_i}{\partial a_k} = \frac{(\partial D(z)/\partial a_k)_{z=p_i}}{(\partial D(z)/\partial z)_{z=p_i}} \tag{7.6.8}$$

The numerator of (7.6.8) is

$$\left(\frac{\partial D(z)}{\partial a_k} \right)_{z=p_i} = -z^{-k}|_{z=p_i} = -p_i^{-k} \tag{7.6.9}$$

The denominator of (7.6.8) is

$$\left(\frac{\partial D(z)}{\partial z} \right)_{z=p_i} = \left\{ \frac{\partial}{\partial z} \left[\prod_{l=1}^{N} (1 - p_l z^{-1}) \right] \right\}_{z=p_i}$$

$$= \left\{ \sum_{\substack{k=1}}^{N} \frac{p_k}{z^2} \prod_{\substack{l=1 \\ l \neq k}}^{N} (1 - p_l z^{-1}) \right\}_{z=p_i} \qquad (7.6.10)$$

$$= \frac{1}{p_i^N} \prod_{\substack{l=1 \\ l \neq i}}^{N} (p_i - p_l)$$

Therefore, (7.6.8) can be expressed as

$$\frac{\partial p_i}{\partial a_k} = \frac{-p_i^{N-k}}{\displaystyle\prod_{\substack{l=1 \\ l \neq i}}^{N} (p_i - p_l)} \qquad (7.6.11)$$

Substitution of the result in (7.6.11) into (7.6.6) yields the total perturbation error Δp_i in the form

$$\Delta p_i = - \sum_{k=1}^{N} \frac{p_i^{N-k}}{\displaystyle\prod_{\substack{l=1 \\ l \neq i}}^{N} (p_i - p_l)} \Delta a_k \qquad (7.6.12)$$

This expression provides a measure of the sensitivity of the ith pole to changes in the coefficients $\{a_k\}$. An analogous result can be obtained for the sensitivity of the zeros to errors in the parameters $\{b_k\}$.

The terms $(p_i - p_l)$ in the denominator of (7.6.12) represent vectors in the z-plane from the poles $\{p_l\}$ to the pole p_i. If the poles are tightly clustered as they are in a narrowband filter, as illustrated in Fig. 7.36, the lengths $|p_i - p_l|$ are small for the poles in the vicinity of p_i. These small lengths will contribute to large errors and hence a large perturbation error Δp_i results.

The error Δp_i can be minimized by maximizing the lengths $|p_i - p_l|$. This can be accomplished by realizing the high-order filter with either single-pole or double-pole filter sections. In general, however, single-pole (and single-zero) filter sections have complex-valued poles and require complex-valued arithmetic operations for their realization. This problem can be avoided by combining complex-valued poles (and zeros) to form second-order filter sections. Since the complex-valued poles are usually sufficiently far apart, the perturbation errors $\{\Delta p_i\}$ are minimized. As a consequence, the resulting filter with quantized coefficients more closely approximates the frequency response characteristics of the filter with unquantized coefficients.

It is interesting to note that even in the case of a two-pole filter section, the structure used to realize the filter section plays an important role in the errors

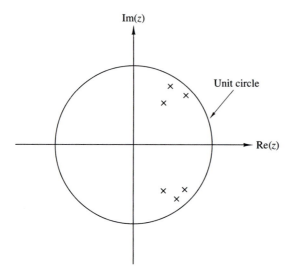

Figure 7.36 Pole positions for a bandpass IIR filter.

caused by coefficient quantization. To be specific, let us consider a two-pole filter with system function

$$H(z) = \frac{1}{1 - (2r\cos\theta)z^{-1} + r^2 z^{-2}} \tag{7.6.13}$$

This filter has poles at $z = re^{\pm j\theta}$. When realized as shown in Fig. 7.37, it has two coefficients, $a_1 = 2r\cos\theta$ and $a_2 = -r^2$. With infinite precision it is possible to achieve an infinite number of pole positions. Clearly, with finite precision (i.e., quantized coefficients a_1 and a_2), the possible pole positions are also finite. In fact, when b bits are used to represent the magnitudes of a_1 and a_2, there are at most $(2^b - 1)^2$ possible positions for the poles in each quandrant, excluding the case $a_1 = 0$ and $a_2 = 0$.

For example, suppose that $b = 4$. Then there are 15 possible nonzero values for a_1. There are also 15 possible values for r^2. We illustrate these possible values in Fig. 7.38 for the first quandrant of the z-plane only. There are 169 possible pole

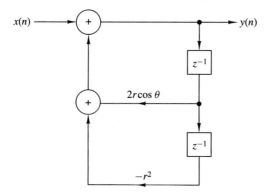

Figure 7.37 Realization of a two-pole IIR filter.

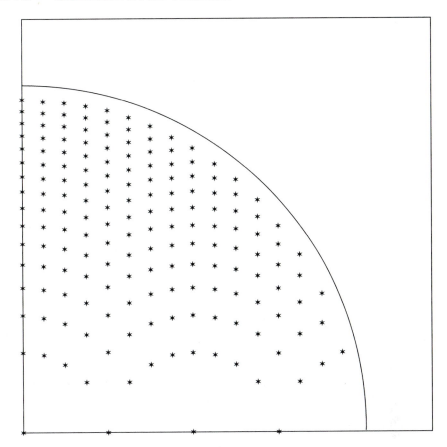

Figure 7.38 Possible pole positions for two-pole IIR filter realization in Fig. 7.37.

positions in this case. The nonuniformity in their positions is due to the fact that we are quantizing r^2, whereas the pole positions lie on a circular arc of radius r. Of particular significance is the sparse set of poles for values of θ near zero and, due to symmetry, near $\theta = \pi$. This situation would be highly unfavorable for lowpass filters and highpass filters which normally have poles clustered near $\theta = 0$ and $\theta = \pi$.

An alternative realization of the two-pole filter is the coupled-form realization illustrated in Fig. 7.39. The two coupled equations are

$$y_1(n) = x(n) + r\cos\theta \; y_1(n-1) - r\sin\theta \; y(n-1)$$

$$y(n) = r\sin\theta \; y_1(n-1) + r\cos\theta \; y(n-1)$$

(7.6.14)

By transforming these two equations into the z-domain, it is a simple matter to show that

$$\frac{Y(z)}{X(z)} = H(z) = \frac{(r\sin\theta)z^{-1}}{1 - (2r\cos\theta)z^{-1} + r^2 z^{-2}}$$

(7.6.15)

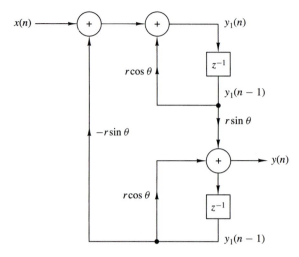

Figure 7.39 Coupled-form realization of a two-pole IIR filter.

In the coupled form we observe that there are also two coefficients, $\alpha_1 = r \sin \theta$ and $\alpha_2 = r \cos \theta$. Since they are both linear in r, the possible pole positions are now equally spaced points on a rectangular grid, as shown in Fig. 7.40. As a consequence, the pole positions are now uniformly distributed inside the unit circle, which is a more desirable situation than the previous realization, especially for lowpass filters. (There are 198 possible pole positions in this case.) However, the price that we pay for this uniform distribution of pole positions is an increase in computations. The coupled-form realization requires four multiplications per output point, whereas the realization in Fig. 7.37 requires only two multiplications per output point.

It is interesting to compare the coupled-form realization of Fig. 7.39 with the coupled (or normal) form state-space structure of Fig. 7.31. The poles of the state-space structure are directly related to its coefficients, since α_1 and $\pm \alpha_2$ are the real and imaginary parts of the roots. Since $\alpha_1 = r \cos \Theta$ and $\alpha_2 = r \sin \Theta$, it is clear that quantizing α_1 and α_2 results in a rectangular grid of possible pole positions, as shown in Fig. 7.40.

Since there are various ways in which one can realize a second-order filter section, there are obviously many possibilities for different pole locations with quantized coefficients. Ideally, we should select a structure that provides us with a dense set of points in the regions where the poles lie. Unfortunately, however, there is no simple and systematic method for determining the filter realization that yields this desired result.

Given that a higher-order IIR filter should be implemented as a combination of second-order sections, we still must decide whether to employ a parallel configuration or a cascade configuration. In other words, we must decide between the realization

$$H(z) = \prod_{k=1}^{K} \frac{b_{k0} + b_{k1}z^{-1} + b_{k2}z^{-2}}{1 + a_{k1}z^{-1} + a_{k2}z^{-2}} \tag{7.6.16}$$

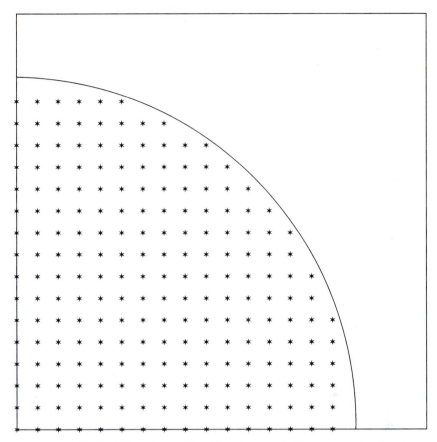

Figure 7.40 Possible pole positions for the coupled-form two-pole filter in Fig. 7.39.

and the realization

$$H(z) = \sum_{k=1}^{K} \frac{c_{k0} + c_{k1}z^{-1}}{1 + a_{k1}z^{-1} + a_{k2}z^{-2}} \qquad (7.6.17)$$

If the IIR filter has zeros on the unit circle, as is generally the case with elliptic and Chebyshev type II filters, each second-order section in the cascade configuration of (7.6.16) contains a pair of complex-conjugate zeros. The coefficients $\{b_k\}$ directly determine the location of these zeros. If the $\{b_k\}$ are quantized, the sensitivity of the system response to the quantization errors is easily and directly controlled by allocating a sufficiently large number of bits to the representation of the $\{b_{ki}\}$. In fact, we can easily evaluate the perturbation effect resulting from quantizing the coefficients $\{b_{ki}\}$ to some specified precision. Thus we have direct control of both the poles and the zeros that result from the quantization process.

On the other hand, the parallel realization of $H(z)$ provides direct control of the poles of the system only. The numerator coefficients $\{c_{k0}\}$ and $\{c_{k1}\}$ do not specify the location of the zeros directly. In fact, the $\{c_{k0}\}$ and $\{c_{k1}\}$ are obtained by performing a partial-fraction expansion of $H(z)$. Hence they do not directly influence the location of the zeros, but only indirectly through a combination of all the factors of $H(z)$. As a consequence, it is more difficult to determine the effect of quantization errors in the coefficients $\{c_{ki}\}$, on the location of the zeros of the system.

It is apparent that quantization of the parameters $\{c_{ki}\}$ is likely to produce a significant perturbation of the zero positions and usually, it is sufficiently large in fixed-point implementations to move the zeros off the unit circle. This is a highly undesirable situation, which can be easily remedied by use of a floating-point representation. In any case the cascade form is more robust in the presence of co-efficient quantization and should be the preferred choice in practical applications, especially where a fixed-point representation is employed.

Example 7.6.1

Determine the effect of parameter quantization on the frequency response of the 7-order elliptic filter given in Table 8.11 when it is realized as a cascade of second-order sections.

Solution   The coefficients for the elliptic filter given in Table 8.11 are specified for the cascade form to six significant digits. We quantized these coefficients to four and then three significant digits (by rounding) and plotted the magnitude (in decibels) and the phase of the frequency response. The results are shown in Fig. 7.41 along the frequency response of the filter with unquantized (six significant digits) coefficients. We observe that there is an insignificant degradation due to coefficient quantization for the cascade realization.

Example 7.6.2

Repeat the computation of the frequency response for the elliptic filter considered in Example 7.6.1 when it is realized in the parallel form with second-order sections.

Solution   The system function for the 7-order elliptic filter given in Table 8.11 is

$$H(z) = \frac{0.2781304 + 0.0054373108z^{-1}}{1 - 0.790103z^{-1}}$$

$$+ \frac{-0.3867805 + 0.3322229z^{-1}}{1 - 1.517223z^{-1} + 0.714088z^{-2}}$$

$$+ \frac{0.1277036 - 0.1558696z^{-1}}{1 - 1.421773z^{-1} + 0.861895z^{-2}}$$

$$+ \frac{-0.015824186 + 0.38377356z^{-1}}{1 - 1.387447z^{-1} + 0.962242z^{-2}}$$

The frequency response of this filter with coefficients quantized to four digits is shown in Fig. 7.42a. When this result is compared with the frequency response in Fig. 7.41, we observe that the zeros in the parallel realization have been perturbed sufficiently so that the nulls in the magnitude response are now at -80, -85, and -92 dB. The phase response has also been perturbed by a small amount.

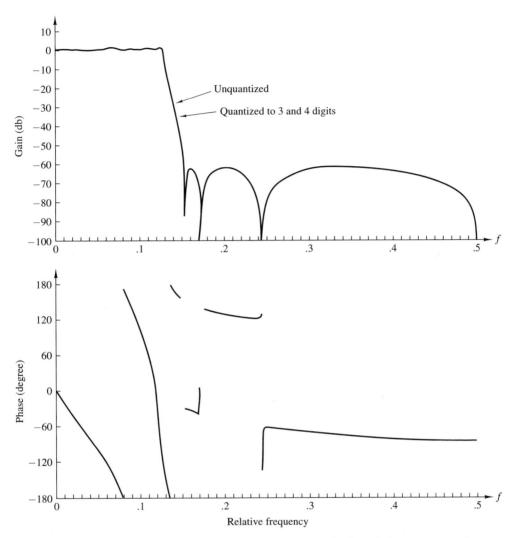

Figure 7.41 Effect of coefficient quantization of the magnitude and phase response of an $N = 7$ elliptic filter realized in cascade form.

When the coefficients are quantized to three significant digits, the frequency response characteristic deteriorates significantly, in both magnitude and phase, as illustrated in Fig. 7.42b. It is apparent from the magnitude response that the zeros are no longer on the unit circle as a result of the quantization of the coefficients. This result clearly illustrates the sensitivity of the zeros to quantization of the coefficients in the parallel form.

When compared with the results of Example 7.6.1, it is also apparent that the cascade form is definitely more robust to parameter quantization than the parallel form.

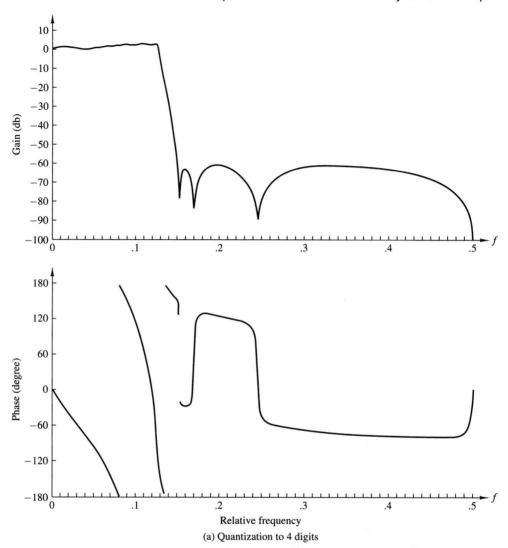

Figure 7.42 Effect of coefficient quantization of the magnitude and phase response of an $N = 7$ elliptic filter realized in cascade form: (a) quantization to four digits; (b) quantization to three digits.

7.6.2 Quantization of Coefficients in FIR Filters

As indicated in the preceding section, the sensitivity analysis performed on the poles of a system also applies directly to the zeros of the IIR filters. Consequently, an expression analogous to (7.6.12) can be obtained for the zeros of an FIR filter. In effect, we should generally realize FIR filters with a large number of zeros as

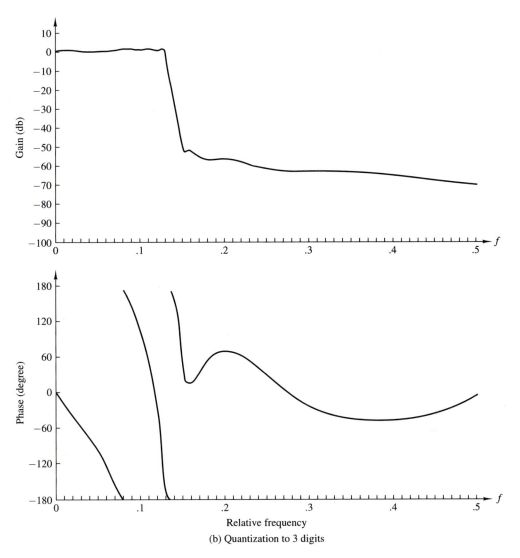

(b) Quantization to 3 digits

Figure 7.42   *Continued*

a cascade of second-order and first-order filter sections to minimize the sensitivity to coefficient quantization.

Of particular interest in practice is the realization of linear phase FIR filters. The direct-form realizations shown in Figs. 7.1 and 7.2 maintain the linear-phase property even when the coefficients are quantized. This follows easily from the observation that the system function of a linear-phase FIR filter satisfies the property

$$H(z) = \pm z^{-(M-1)} H(z^{-1})$$

independent of whether the coefficients are quantized or unquantized (see Section 8.2). Consequently, coefficient quantization does not affect the phase characteristic of the FIR filter, but affects only the magnitude. As a result, coefficient quantization effects are not as severe on a linear-phase FIR filter, since the only effect is in the magnitude.

Example 7.6.3

Determine the effect of parameter quantization on the frequency response of an $M = 32$ linear-phase FIR bandpass filter. The filter is realized in the direct form.

Solution   The frequency response of a linear-phase FIR bandpass filter with unquantized coefficients is illustrated in Fig. 7.43a. When the coefficients are quantized to four significant digits, the effect on the frequency response is insignificant. However, when the coefficients are quantized to three significant digits, the sidelobes increased by several decibels, as illustrated in Fig. 7.43b. This result indicates that we should use a minimum of 10 bits to represent the coefficients of this FIR filter and, preferably, 12 to 14 bits, if possible.

From this example we learn that a minimum of 10 bits is required to represent the coefficients in a direct-form realization of an FIR filter of moderate length. As the filter length increases, the number of bits per coefficient must be increased to maintain the same error in the frequency response characteristic of the filter.

For example, suppose that each filter coefficient is rounded to $(b + 1)$ bits. Then the maximum error in a coefficient value is bounded as

$$-2^{-(b+1)} < e_h(n) < 2^{-(b+1)}$$

Since the quantized values may be represented as $\overline{h}(n) = h(n) + e_h(n)$, the error in the frequency response is

$$E_M(\omega) = \sum_{n=0}^{M-1} e_h(n) e^{-j\omega n}$$

Since $e_h(n)$ is zero mean, it follows that $E_M(\omega)$ is also zero mean. Assuming that the coefficient error sequence $e_h(n)$, $0 \le n \le M - 1$, is uncorrelated, the variance of the error $E_M(\omega)$ in the frequency response is just the sum of the variances of the M terms. Thus we have

$$\sigma_E^2 = \frac{2^{-2(b+1)}}{12} M = \frac{2^{-2(b+2)}}{3} M$$

Here we note that the variance of the error in $H(\omega)$ increases linearly with M. Hence the standard deviation of the error in $H(\omega)$ is

$$\sigma_E = \frac{2^{-(b+2)}}{\sqrt{3}} \sqrt{M}$$

Consequently, for every factor of 4 increase in M, the precision in the filter coefficients must be increased by one additional bit to maintain the standard deviation fixed. This result, taken together with the results of Example 7.6.3, implies that

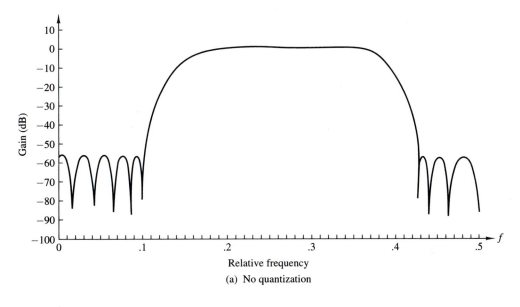

(a) No quantization

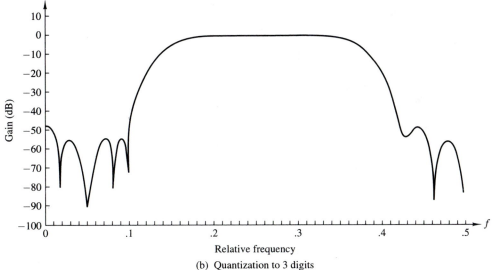

(b) Quantization to 3 digits

Figure 7.43   Effect of coefficient quantization of the magnitude of an $M = 32$ linear-phase FIR filter realized in direct form: (a) no quantization; (b) quantization to three digits.

the frequency error remains tolerable for filter lengths up to 256, provided that filter coefficients are represented by 12 to 13 bits. If the word length of the digital signal processor is less than 12 bits or if the filter length exceeds 256, the filter should be implemented as a cascade of smaller length filters to reduce the precision requirements.

In a cascade realization of the form

$$H(z) = G \prod_{k=1}^{K} H_k(z) \qquad (7.6.18)$$

where the second-order sections are given as

$$H_k(z) = 1 + b_{k1}z^{-1} + b_{k2}z^{-2} \qquad (7.6.19)$$

the coefficients of complex-valued zeros are expressed as $b_{k1} = -2r_k \cos\theta_k$ and $b_{k2} = r_k^2$. Quantization of b_{k1} and b_{k2} result in zero locations as shown in Fig. 7.38, except that the grid extends to points outside the unit circle.

A problem may arise, in this case, in maintaining the linear-phase property, because the quantized pair of zeros at $z = (1/r_k)e^{\pm j\theta_k}$ may not be the mirror image of the quantized zeros at $z = r_k e^{\pm j\theta_k}$. This problem can be avoided by rearranging the factors corresponding to the mirror-image zero. That is, we can write the mirror-image factor as

$$\left(1 - \frac{2}{r_k}\cos\theta_k z^{-1} + \frac{1}{r_k^2}z^{-2}\right) = \frac{1}{r_k^2}(r_k^2 - 2r_k \cos\theta_k z^{-1} + z^{-2}) \qquad (7.6.20)$$

The factors $\{1/r_k^2\}$ can be combined with the overall gain factor G, or they can be distributed in each of the second-order filters. The factor in (7.6.20) contains exactly the same parameters as the factor $(1 - 2r_k \cos\theta_k z^{-1} + r_k^2 z^{-2})$, and consequently, the zeros now occur in mirror-image pairs even when the parameters are quantized.

In this brief treatment we have given the reader an introduction to the problems of coefficient quantization in IIR and FIR filters. We have demonstrated that a high-order filter should be reduced to a cascade (for FIR or IIR filters) or a parallel (for IIR filters) realization to minimize the effects of quantization errors in the coefficients. This is especially important in fixed-point realizations in which the coefficients are represented by a relatively small number of bits.

7.7 ROUND-OFF EFFECTS IN DIGITAL FILTERS

In Section 7.5 we characterized the quantization errors that occur in arithmetic operations performed in a digital filter. The presence of one or more quantizers in the realization of a digital filter results in a nonlinear device with characteristics that may be significantly different from the ideal linear filter. For example, a recursive digital filter may exhibit undesirable oscillations in its output, as shown in the following section, even in the absence of an input signal.

As a result of the finite-precision arithmetic operations performed in the digital filter, some registers may overflow if the input signal level becomes large.

Overflow represents another form of undesirable nonlinear distortion on the desired signal at the output of the filter. Consequently, special care must be exercised to scale the input signal properly, either to prevent overflow completely or, at least, to minimize its rate of occurrence.

The nonlinear effects due to finite-precision arithmetic make it extremely difficult to precisely analyze the performance of a digital filter. To perform an analysis of quantization effects, we adopt a statistical characterization of quantization errors which, in effect, results in a linear model for the filter. Thus we are able to quantify the effects of quantization errors in the implementation of digital filters. Our treatment is limited to fixed-point realizations where quantization effects are very important.

7.7.1 Limit-Cycle Oscillations in Recursive Systems

In the realization of a digital filter, either in digital hardware or in software on a digital computer, the quantization inherent in the finite- precision arithmetic operations render the system nonlinear. In recursive systems, the nonlinearities due to the finite-precision arithmetic operations often cause periodic oscillations to occur in the output, even when the input sequence is zero or some nonzero constant value. Such oscillations in recursive systems are called *limit cycles* and are directly attributable to round-off errors in multiplication and overflow errors in addition.

To illustrate the characteristics of a limit-cycle oscillation, let us consider a single-pole system described by the linear difference equation

$$y(n) = ay(n-1) + x(n) \qquad (7.7.1)$$

where the pole is at $z = a$. The ideal system is realized as shown in Fig. 7.44. On the other hand, the actual system, which is described by the nonlinear difference equation

$$v(n) = Q[av(n-1)] + x(n) \qquad (7.7.2)$$

is realized as shown in Fig. 7.45.

Suppose that the actual system in Fig. 7.45 is implemented with fixed-point arithmetic based on four bits for the magnitude plus a sign bit. The quantization that takes place after multiplication is assumed to round the resulting product upward.

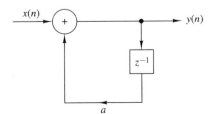

Figure 7.44 Ideal single-pole recursive system.

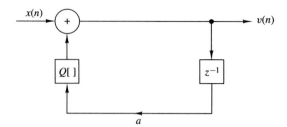

Figure 7.45　Actual nonlinear system.

In Table 7.2 we list the response of the actual system for four different locations of the pole $z = a$, and an input $x(n) = \beta\delta(n)$, where $\beta = 15/16$, which has the binary representation 0.1111. Ideally, the response of the system should decay toward zero exponentially [i.e., $y(n) = a^n \to 0$ as $n \to \infty$]. In the actual system, however, the response $v(n)$ reaches a steady-state periodic output sequence with a period that depends on the value of the pole. When the pole is positive, the oscillations occur with a period $N_p = 1$, so that the output reaches a constant value of $\frac{1}{16}$ for $a = \frac{1}{2}$ and $\frac{1}{8}$ for $a = \frac{3}{4}$. On the other hand, when the pole is negative, the output sequence oscillates between positive and negative values ($\pm\frac{1}{16}$ for $a = -\frac{1}{2}$ and $\pm\frac{1}{8}$ for $a = -\frac{3}{4}$). Hence the period is $N_p = 2$.

These limit cycles occur as a result of the quantization effects in multiplications. When the input sequence $x(n)$ to the filter becomes zero, the output of the filter then, after a number of iterations, enters into the limit cycle. The output remains in the limit cycle until another input of sufficient size is applied that drives the system out of the limit cycle. Similarly, zero-input limit cycles occur from nonzero initial conditions with the input $x(n) = 0$. The amplitudes of the output during a limit cycle are confined to a range of values that is called the *dead band* of the filter.

TABLE 7.2　LIMIT CYCLES FOR LOWPASS SINGLE-POLE FILTER

n	$a = 0.1000$ $= \frac{1}{2}$		$a = 1.1000$ $= -\frac{1}{2}$		$a = 0.1100$ $= \frac{3}{4}$		$a = 1.1100$ $= -\frac{3}{4}$	
0	0.1111	$\left(\frac{15}{16}\right)$	0.1111	$\left(\frac{15}{16}\right)$	0.1011	$\left(\frac{11}{16}\right)$	0.1011	$\left(\frac{11}{16}\right)$
1	0.1000	$\left(\frac{8}{16}\right)$	1.1000	$\left(-\frac{8}{16}\right)$	0.1000	$\left(\frac{8}{16}\right)$	1.1000	$\left(-\frac{8}{16}\right)$
2	0.0100	$\left(\frac{4}{16}\right)$	0.0100	$\left(\frac{4}{16}\right)$	0.0110	$\left(\frac{6}{16}\right)$	0.0110	$\left(\frac{6}{16}\right)$
3	0.0010	$\left(\frac{2}{16}\right)$	1.0010	$\left(-\frac{2}{16}\right)$	0.0101	$\left(\frac{5}{16}\right)$	1.0101	$\left(-\frac{5}{16}\right)$
4	0.0001	$\left(\frac{1}{16}\right)$	0.0001	$\left(\frac{1}{16}\right)$	0.0100	$\left(\frac{4}{16}\right)$	0.0100	$\left(\frac{4}{16}\right)$
5	0.0001	$\left(\frac{1}{16}\right)$	1.0001	$\left(-\frac{1}{16}\right)$	0.0011	$\left(\frac{3}{16}\right)$	1.0011	$\left(-\frac{3}{16}\right)$
6	0.0001	$\left(\frac{1}{16}\right)$	0.0001	$\left(\frac{1}{16}\right)$	0.0010	$\left(\frac{2}{16}\right)$	0.0010	$\left(\frac{2}{16}\right)$
7	0.0001	$\left(\frac{1}{16}\right)$	1.0001	$\left(-\frac{1}{16}\right)$	0.0010	$\left(\frac{2}{16}\right)$	1.0010	$\left(-\frac{2}{16}\right)$
8	0.0001	$\left(\frac{1}{16}\right)$	0.0001	$\left(\frac{1}{16}\right)$	0.0010	$\left(\frac{2}{16}\right)$	0.0010	$\left(\frac{2}{16}\right)$

It is interesting to note that when the response of the single-pole filter is in the limit cycle, the actual nonlinear system operates as an equivalent linear system with a pole at $z = 1$ when the pole is positive and $z = -1$ when the pole is negative. That is,

$$Q_r[av(n-1)] = \begin{cases} v(n-1), & a > 0 \\ -v(n-1), & a < 0 \end{cases} \qquad (7.7.3)$$

Since the quantized product $av(n-1)$ is obtained by rounding, it follows that the quantization error is bounded as

$$|Q_r[av(n-1)] - av(n-1)| \le \tfrac{1}{2} \cdot 2^{-b} \qquad (7.7.4)$$

where b is the number of bits (exclusive of sign) used in the representation of the pole a and $v(n)$. Consequently, (7.7.4) and (7.7.3) leads to

$$|v(n-1)| - |av(n-1)| \le \tfrac{1}{2} \cdot 2^{-b}$$

and hence

$$|v(n-1)| \le \frac{\tfrac{1}{2} \cdot 2^{-b}}{1 - |a|} \qquad (7.7.5)$$

The expression in (7.7.5) defines the dead band for a single-pole filter. For example, when $b = 4$ and $|a| = \tfrac{1}{2}$, we have a dead band with a range of amplitudes $(-\tfrac{1}{16}, \tfrac{1}{16})$. When $b = 4$ and $|a| = \tfrac{3}{4}$, the dead band increases to $(-\tfrac{1}{8}, \tfrac{1}{8})$.

The limit-cycle behavior in a two-pole filter is much more complex and a larger variety of oscillations can occur. In this case the ideal two-pole system is described by the linear difference equation,

$$y(n) = a_1 y(n-1) + a_2 y(n-2) + x(n) \qquad (7.7.6)$$

whereas the actual system is described by the nonlinear difference equation

$$v(n) = Q_r[a_1 v(n-1)] + Q_r[a_2 v(n-2)] + x(n) \qquad (7.7.7)$$

When the filter coefficients satisfy the condition $a_1^2 < -4a_2$, the poles of the system occur at

$$z = re^{\pm j\theta}$$

where $a_2 = -r^2$ and $a_1 = 2r \cos \theta$. As in the case of the single-pole filter, when the system is in a zero-input or zero-state limit cycle,

$$Q_r[a_2 v(n-2)] = -v(n-2) \qquad (7.7.8)$$

In other words, the system behaves as an oscillator with complex-conjugate poles on the unit circle (i.e., $a_2 = -r^2 = -1$). Rounding the product $av(n-2)$ implies that

$$|Q_r[a_2 v(n-2)] - a_2 v(n-2)| \le \tfrac{1}{2} \cdot 2^{-b} \qquad (7.7.9)$$

Upon substitution of (7.7.8) into (7.7.9), we obtain the result

$$|v(n-2)| - |a_2 v(n-2)| \le \tfrac{1}{2} \cdot 2^{-b}$$

or equivalently,

$$|v(n-2)| \leq \frac{\frac{1}{2} \cdot 2^{-b}}{1 - |a_2|} \tag{7.7.10}$$

The expression in (7.7.10) defines the dead band of the two-pole filter with complex-congugate poles. We observe that the dead-band limits depend only on $|a_2|$. The parameter $a_1 = 2r \cos\theta$ determines the frequency of oscillation.

Another possible limit-cycle mode with zero input, which occurs as a result of rounding the multiplications, corresponds to an equivalent second-order system with poles at $z = \pm 1$. In this case it was shown by Jackson (1969) that the two-pole filter exhibits oscillations with an amplitude that falls in the dead band bounded by $2^{-b}/(1 - |a_1| - a_2)$.

It is interesting to note that these limit cycles result from rounding the product of the filter coefficients with the previous outputs, $v(n - 1)$ and $v(n - 2)$. Instead of rounding, we may choose to truncate the products to b bits. With truncation, we can eliminate many, although not all, of the limit cycles as shown by Claasen et al. (1973). However, recall that truncation results in a biased error unless the sign-magnitude representation is used, in which case the truncation error is symmetric about zero. In general, this bias is undesirable in digital filter implementation.

In a parallel realization of a high-order IIR system, each second-order filter section exhibits its own limit-cycle behavior, with no interaction among the second-order filter sections. Consequently, the output is the sum of the zero-input limit cycles from the individual sections. In the case of a cascade realization for a high-order IIR system, the limit cycles are much more difficult to analyze. In particular, when the first filter section exhibits a zero-input limit cycle, the output limit cycle is filtered by the succeeding sections. If the frequency of the limit cycle falls near a resonance frequency in a succeeding filter section, the amplitude of the sequence is enhanced by the resonance characteristic. In general, we must be careful to avoid such situations.

In addition to limit cycles caused by rounding the result of multiplications, there are limit cycles caused by overflows in addition. An overflow in addition of two or more binary numbers occurs when the sum exceeds the word size available in the digital implementation of the system. For example, let us consider the second-order filter section illustrated in Fig. 7.46, in which the addition is performed in two's-complement arithmetic. Thus we can write the output $y(n)$ as

$$y(n) = g[a_1 y(n-1) + a_2 y(n-2) + x(n)] \tag{7.7.11}$$

where the function $g[\cdot]$ represents the two's-complement addition. It is easily verified that the function $g(v)$ versus v is described by the graph in Fig. 7.47.

Recall that the range of values of the parameters (a_1, a_2) for a stable filter is given by the stability triangle in Fig. 3.15. However, these conditions are no longer sufficient to prevent overflow oscillation with two's-complement arithmetic.

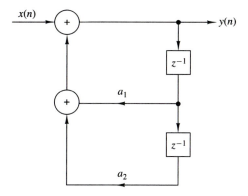

Figure 7.46 Two-pole filter realization.

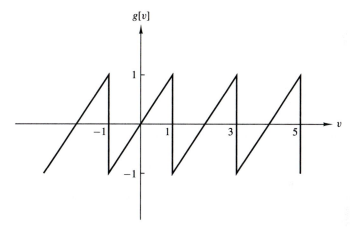

Figure 7.47 Characteristic functional relationship for two's complement addition of two or more numbers.

In fact, it can easily be shown that a necessary and sufficient condition for ensuring that no zero-input overflow limit cycles occur is

$$|a_1| + |a_2| < 1 \qquad (7.7.12)$$

which is extremely restrictive and hence an unreasonable constraint to impose on any second-order section.

An effective remedy for curing the problem of overflow oscillations is to modify the adder characteristic, as illustrated in Fig. 7.48, so that it performs saturation arithmetic. Thus when an overflow (or underflow) is sensed, the output of the adder will be the full-scale value of ± 1. The distortion caused by this nonlinearity in the adder is usually small provided that saturation occurs infrequently. The use of such a nonlinearity does not preclude the need for scaling of the signals and the system parameters, as described in the following section.

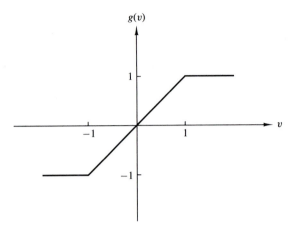

Figure 7.48 Characteristic functional relationship for addition with clipping at ±1.

7.7.2 Scaling to Prevent Overflow

Saturation arithmetic as just described eliminates limit cycles due to overflow, on the one hand, but on the other hand, it causes undesirable signal distortion due to the nonlinearity of the clipper. In order to limit the amount of nonlinear distortion, it is important to scale the input signal and the unit sample response, between the input and any internal summing node in the system, such that overflow becomes a rare event.

For fixed-point arithmetic, let us first consider the extreme condition that overflow is not permitted at any node of the system. Let $y_k(n)$ denote the response of the system at the kth node when the input sequence is $x(n)$ and the unit sample response between the node and the input is $h_k(n)$. Then

$$|y_k(n)| = \left| \sum_{m=-\infty}^{\infty} h_k(m)x(n-m) \right| \leq \sum_{m=-\infty}^{\infty} |h_k(m)||x(n-m)|$$

Suppose that $x(n)$ is upper bounded by A_x. Then

$$|y_k(n)| \leq A_x \sum_{m=-\infty}^{\infty} |h_k(m)| \qquad \text{for all } n \qquad (7.7.13)$$

Now, if the dynamic range of the computer is limited to $(-1, 1)$, the condition

$$|y_k(n)| < 1$$

can be satisfied by requiring that the input $x(n)$ be scaled such that

$$A_x < \frac{1}{\displaystyle\sum_{m=-\infty}^{\infty} |h_k(m)|} \qquad (7.7.14)$$

for all possible nodes in the system. The condition in (7.7.14) is both necessary and sufficient to prevent overflow.

The condition in (7.7.14) is overly conservative, however, to the point where the input signal may be scaled too much. In such a case, much of the precision used to represent $x(n)$ is lost. This is especially true for narrowband sequences, such as sinusoids, where the scaling implied by (7.7.14) is extremely severe. For narrowband signals we can use the frequency response characteristics of the system in determining the appropriate scaling. Since $|H(\omega)|$ represents the gain of the system at frequency ω, a less severe and reasonably adequate scaling is to require that

$$A_x < \frac{1}{\max_{0 \le \omega \le \pi} |H_k(\omega)|} \tag{7.7.15}$$

where $H_k(\omega)$ is the Fourier transform of $\{h_k(n)\}$.

In the case of an FIR filter, the condition in (7.7.14) reduces to

$$A_x < \frac{1}{\displaystyle\sum_{m=0}^{M-1} |h_k(m)|} \tag{7.7.16}$$

which is now a sum over the M nonzero terms of the filter unit sample response.

Another approach to scaling is to scale the input so that

$$\sum_{n=-\infty}^{\infty} |y_k(n)|^2 \le C^2 \sum_{n=-\infty}^{\infty} |x(n)|^2 = C^2 E_x \tag{7.7.17}$$

From Parseval's theorem we have

$$\sum_{n=-\infty}^{\infty} |y_k(n)|^2 = \frac{1}{2\pi} \int_{-\pi}^{\pi} |H(\omega)X(\omega)|^2 d\omega$$

$$\le E_x \frac{1}{2\pi} \int_{-\pi}^{\pi} |H(\omega)|^2 d\omega \tag{7.7.18}$$

By combining (7.7.17) with (7.7.18), we obtain

$$C^2 \le \frac{1}{\displaystyle\sum_{n=-\infty}^{\infty} |h_k(n)|^2} = \frac{1}{(1/2\pi) \displaystyle\int_{-\pi}^{\pi} |H(\omega)|^2 d\omega} \tag{7.7.19}$$

If we compare the different scaling factors given above, we find that

$$\left[\sum_{n=-\infty}^{\infty} |h_k(n)|^2 \right]^{1/2} \le \max_{\omega} |H_k(\omega)| \le \sum_{n=-\infty}^{\infty} |h_k(n)| \tag{7.7.20}$$

Clearly, (7.7.14) is the most pessimistic constraint.

In the following section we observe the ramifications of this scaling on the output signal-to-noise (power) ratio (SNR) from a first-order and a second-order filter section.

7.7.3 Statistical Characterization of Quantization Effects in Fixed-Point Realizations of Digital Filters

It is apparent from our treatment in the previous section that an analysis of quantization errors in digital filtering, based on deterministic models of quantization effects, is not a very fruitful approach. The basic problem is that the nonlinear effects in quantizing the products of two numbers and in clipping the sum of two numbers to prevent overflow are not easily modeled in large systems that contain many multipliers and many summing nodes.

To obtain more general results on the quantization effects in digital filters, we shall model the quantization errors in multiplication as an additive noise sequence $e(n)$, just as we did in characterizing the quantization errors in A/D conversion of an analog signal. For addition, we consider the effect of scaling the input signal to prevent overflow.

Let us begin our treatment with the characterization of the round-off noise in a single-pole filter which is implemented in fixed-point arithmetic and is described by the nonlinear difference equation

$$v(n) = Q_r[av(n-1)] + x(n) \tag{7.7.21}$$

The effect of rounding the product $av(n-1)$ is modeled as a noise sequence $e(n)$ added to the actual product $av(n-1)$, that is,

$$Q_r[av(n-1)] = av(n-1) + e(n) \tag{7.7.22}$$

With this model for the quantization error, the system under consideration is described by the *linear difference equation*

$$v(n) = av(n-1) + x(n) + e(n) \tag{7.7.23}$$

The corresponding system is illustrated in block diagram form in Fig. 7.49.

It is apparent from (7.7.23) that the output sequence $v(n)$ of the filter can be separated into two components. One is the response of the system to the input sequence $x(n)$. The second is the response of the system to the additive quantization noise $e(n)$. In fact, we can express the output sequence $v(n)$ as a sum of these two components, that is,

$$v(n) = y(n) + q(n) \tag{7.7.24}$$

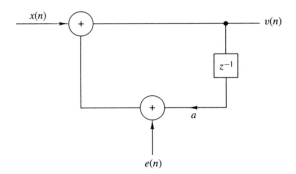

Figure 7.49 Additive noise model for the quantization error in a single-pole filter.

where $y(n)$ represents the response of the system to $x(n)$, and $q(n)$ represents the response of the system to the quantization error $e(n)$. Upon substitution from (7.7.24) for $v(n)$ into (7.7.23), we obtain

$$y(n) + q(n) = ay(n-1) + aq(n-1) + x(n) + e(n) \qquad (7.7.25)$$

To simplify the analysis, we make the following assumptions about the error sequence $e(n)$.

1. For any n, the error sequence $\{e(n)\}$ is uniformly distributed over the range $(-\frac{1}{2} \cdot 2^{-b}, \frac{1}{2} \cdot 2^{-b})$. This implies that the mean value of $e(n)$ is zero and its variance is

$$\sigma_e^2 = \frac{2^{-2b}}{12} \qquad (7.7.26)$$

2. The error $\{e(n)\}$ is a stationary white noise sequence. In other words, the error $e(n)$ and the error $e(m)$ are uncorrelated for $n \neq m$.
3. The error sequence $\{e(n)\}$ is uncorrelated with the signal sequence $\{x(n)\}$.

The last assumption allows us to separate the difference equation in (7.7.25) into two uncoupled difference equations, namely,

$$y(n) = ay(n-1) + x(n) \qquad (7.7.27)$$

$$q(n) = aq(n-1) + e(n) \qquad (7.7.28)$$

The difference equation in (7.7.27) represents the input–output relation for the desired system and the difference equation in (7.7.28) represents the relation for the quantization error at the output of the system.

To complete the analysis, we make use of two important relationships developed in Appendix A. The first is the relationship for the mean value of the output $q(n)$ of a linear shift-invariant filter with impulse response $h(n)$ when excited by a random sequence $e(n)$ having a mean value m_e. The result is

$$m_q = m_e \sum_{n=-\infty}^{\infty} h(n) \qquad (7.7.29)$$

or, equivalently,

$$m_q = m_e H(0) \qquad (7.7.30)$$

where $H(0)$ is the value of the frequency response $H(\omega)$ of the filter evaluated at $\omega = 0$.

The second important relationship is the expression for the autocorrelation sequence of the output $q(n)$ of the filter with impulse response $h(n)$ when the input random sequence $e(n)$ has an autocorrelation $\gamma_{ee}(n)$. This result is

$$\gamma_{qq}(n) = \sum_{k=-\infty}^{\infty} \sum_{l=-\infty}^{\infty} h(k)h(l)\gamma_{ee}(k-l+n) \qquad (7.7.31)$$

In the important special case where the random sequence is white (spectrally flat), the autocorrelation $\gamma_{ee}(n)$ is a unit sample sequence scaled by the variance σ_e^2, that is,

$$\gamma_{ee}(n) = \sigma_e^2 \delta(n) \qquad (7.7.32)$$

Upon substituting (7.7.32) into (7.7.31), we obtain the desired result for the auto-correlation sequence at the output of a filter excited by white noise, namely,

$$\gamma_{qq}(n) = \sigma_e^2 \sum_{k=-\infty}^{\infty} h(k)h(k+n) \qquad (7.7.33)$$

The variance σ_q^2 of the output noise is simply obtained by evaluating $\gamma_{qq}(n)$ at $n = 0$. Thus

$$\sigma_q^2 = \sigma_e^2 \sum_{k=-\infty}^{\infty} h^2(k) \qquad (7.7.34)$$

and with the aid of Parseval's theorem, we have the alternative expression

$$\sigma_q^2 = \frac{\sigma_e^2}{2\pi} \int_{-\pi}^{\pi} |H(\omega)|^2 d\omega \qquad (7.7.35)$$

In the case of the single-pole filter under consideration, the unit sample response is

$$h(n) = a^n u(n) \qquad (7.7.36)$$

Since the quantization error due to rounding has zero mean, the mean value of the error at the output of the filter is $m_q = 0$. The variance of the error at the output of the filter is

$$\sigma_q^2 = \sigma_e^2 \sum_{k=0}^{\infty} a^{2k}$$

$$= \frac{\sigma_e^2}{1 - a^2} \qquad (7.7.37)$$

We observe that the noise power σ_q^2 at the output of the filter is enhanced relative to the input noise power σ_e^2 by the factor $1/(1-a^2)$. This factor increases as the pole is moved closer to the unit circle.

To obtain a clearer picture of the effect of the quantization error, we should also consider the effect of scaling the input. Let us assume that the input sequence $\{x(n)\}$ is a white noise sequence (wideband signal), whose amplitude has been scaled according to (7.7.14) to prevent overflows in addition. Then

$$A_x < 1 - |a|$$

If we assume that $x(n)$ is uniformly distributed in the range $(-A_x, A_x)$, then, according to (7.7.31) and (7.7.34), the signal power at the output of the filter is

$$\sigma_y^2 = \sigma_x^2 \sum_{k=0}^{\infty} a^{2k}$$

$$= \frac{\sigma_x^2}{1 - a^2}$$

(7.7.38)

where $\sigma_x^2 = (1 - |a|)^2/3$ is the variance of the input signal. The ratio of the signal power σ_y^2 to the quantization error power σ_q^2, which is called the signal-to-noise ratio (SNR), is simply

$$\frac{\sigma_y^2}{\sigma_q^2} = \frac{\sigma_x^2}{\sigma_e^2}$$

$$= (1 - |a|)^2 \cdot 2^{2(b+1)}$$

(7.7.39)

This expression for the output SNR clearly illustrates the severe penalty paid as a consequence of the scaling of the input, expecially when the pole is near the unit circle. By comparison, if the input is not scaled and the adder has a sufficient number of bits to avoid overflow, then the signal amplitude may be confined to the range $(-1, 1)$. In this case, $\sigma_x^2 = \frac{1}{3}$, which is independent of the pole position. Then

$$\frac{\sigma_y^2}{\sigma_q^2} = 2^{2(b+1)}$$

(7.7.40)

The difference between the SNRs in (7.7.40) and (7.7.39) clearly demonstrates the need to use more bits in addition than in multiplication. The number of additional bits depends on the position of the pole and should be increased as the pole is moved closer to the unit circle.

Next, let us consider a two-pole filter with infinite precision which is described by the linear difference equation

$$y(n) = a_1 y(n - 1) + a_2 y(n - 2) + x(n)$$

(7.7.41)

where $a_1 = 2r\cos\theta$ and $a_2 = -r^2$. When the two products are rounded, we have a system which is described by the nonlinear difference equation

$$v(n) = Q_r[a_1 v(n - 1)] + Q_r[a_2 v(n - 2)] + x(n)$$

(7.7.42)

This system is illustrated in block diagram form in Fig. 7.50.

Now there are two multiplications, and hence two quantization errors are produced for each output. Consequently, we should introduce two noise sequences $e_1(n)$ and $e_2(n)$, which correspond to the quantizer outputs

$$Q_r[a_1 v(n - 1)] = a_1 v(n - 1) + e_1(n)$$

$$Q_r[a_2 v(n - 2)] = a_2 v(n - 2) + e_2(n)$$

(7.7.43)

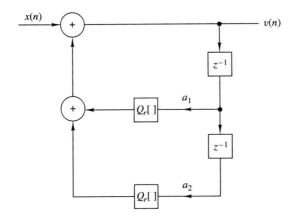

Figure 7.50 Two-pole digital filter with rounding quantizers.

A block diagram for the corresponding model is shown in Fig. 7.51. Note that the error sequences $e_1(n)$ and $e_2(n)$ can be moved directly to the input of the filter.

As in the case of the first-order filter, the output of the second-order filter can be separated into two components, the desired signal component and the quantization error component. The former is described by the difference equation

$$y(n) = a_1 y(n-1) + a_2 y(n-2) + x(n) \tag{7.7.44}$$

while the latter satisfies the difference equation

$$q(n) = a_1 q(n-1) + a_2 q(n-2) + e_1(n) + e_2(n) \tag{7.7.45}$$

It is reasonable to assume that the two sequences $e_1(n)$ and $e_2(n)$ are uncorrelated. Now the second-order filter has a unit sample response

$$h(n) = \frac{r^n}{\sin\theta}\sin(n+1)\theta u(n) \tag{7.7.46}$$

Hence

$$\sum_{n=0}^{\infty} h^2(n) = \frac{1+r^2}{1-r^2}\frac{1}{r^4+1-2r^2\cos 2\theta} \tag{7.7.47}$$

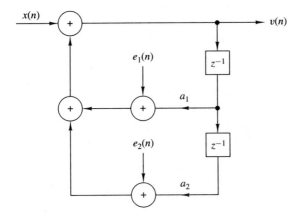

Figure 7.51 Additive noise model for the quantization errors in a two-pole filter realization.

By applying (7.7.34), we obtain the variance of the quantization errors at the output of the filter in the form

$$\sigma_q^2 = \sigma_e^2 \left(\frac{1+r^2}{1-r^2} \frac{1}{r^4+1-2r^2 \cos 2\theta} \right) \tag{7.7.48}$$

In the case of the signal component, if we scale the input as in (7.7.14) to avoid overflow, the power in the output signal is

$$\sigma_y^2 = \sigma_x^2 \sum_{n=0}^{\infty} h^2(n) \tag{7.7.49}$$

where the power in the input signal $x(n)$ is given by the variance

$$\sigma_x^2 = \frac{1}{3 \left[\displaystyle\sum_{n=0}^{\infty} |h(n)| \right]^2} \tag{7.7.50}$$

Consequently, the SNR at the output of the two-pole filter is

$$\frac{\sigma_y^2}{\sigma_q^2} = \frac{\sigma_x^2}{\sigma_e^2} = \frac{2^{2(b+1)}}{\left[\displaystyle\sum_{n=0}^{\infty} |h(n)| \right]^2} \tag{7.7.51}$$

Although it is difficult to determine the exact value of the denominator term in (7.7.51), it is easy to obtain an upper and a lower bound. In particular, $|h(n)|$ is upper bounded as

$$|h(n)| \le \frac{1}{\sin \theta} r^n \qquad n \ge 0 \tag{7.7.52}$$

so that

$$\sum_{n=0}^{\infty} |h(n)| \le \frac{1}{\sin \theta} \sum_{n=0}^{\infty} r^n = \frac{1}{(1-r) \sin \theta} \tag{7.7.53}$$

The lower bound may be obtained by noting that

$$|H(\omega)| = \left| \sum_{n=0}^{\infty} h(n) e^{-j\omega n} \right| \le \sum_{n=0}^{\infty} |h(n)|$$

But

$$H(\omega) = \frac{1}{(1 - re^{j\theta} e^{-j\omega})(1 - re^{-j\theta} e^{-j\omega})}$$

At $\omega = \theta$, which is the resonant frequency of the filter, we obtain the largest value of $|H(\omega)|$. Hence

$$\sum_{n=0}^{\infty} |h(n)| \ge |H(\theta)| = \frac{1}{(1-r)\sqrt{1+r^2 - 2r \cos 2\theta}} \tag{7.7.54}$$

Therefore, the SNR is bounded from above and below according to the relation

$$2^{2(b+1)}(1-r)^2 \sin^2 \theta \le \frac{\sigma_y^2}{\sigma_q^2} \le 2^{2(b+1)}(1-r)^2(1+r^2 - 2r\cos 2\theta) \qquad (7.7.55)$$

For example, when $\theta = \pi/2$, the expression in (7.7.55) reduces to

$$2^{2(b+1)}(1-r)^2 \le \frac{\sigma_y^2}{\sigma_q^2} \le 2^{2(b+1)}(1-r)^2(1+r)^2 \qquad (7.7.56)$$

The dominant term in this bound is $(1-r)^2$ which acts to reduce the SNR dramatically as the poles move toward the unit circle. Hence the effect of scaling in the second-order filter is more severe than in the single-pole filter. Note that if $d = 1 - r$ is the distance of the pole from the unit circle, the SNR in (7.7.56) is reduced by d^2, whereas in the single-pole filter the reduction is proportional to d. These results serve to reinforce the earlier statement regarding the use of more bits in addition than in multiplication as a mechanism for avoiding the severe penalty due to scaling.

The analysis of the quantization effects in a second-order filter can be applied directly to higher-order filters based on a parallel realization. In this case each second-order filter section is independent of all the other sections, and therefore the total quantization noise power at the output of the parallel bank is simply the linear sum of the quantization noise powers of each of the individual sections. On the other hand, the cascade realization is more difficult to analyze. For the cascade interconnection, the noise generated in any second-order filter section is filtered by the succeeding sections. As a consequence, there is the issue of how to pair together real-valued poles to form second-order sections and how to arrange the resulting second-order filters to minimize the total noise power at the output of the high-order filter. This general topic was investigated by Jackson (1970a, b), who showed that poles close to the unit circle should be paired with nearby zeros to reduce the gain of each second-order section. In ordering the second-order sections in cascade, a reasonable strategy is to place the sections in the order of decreasing maximum frequency gain. In this case the noise power generated in the early high-gain section is not boosted significantly by the latter sections.

The following example illustrates the point that proper ordering of sections in a cascade realization is important in controlling the round-off noise at the output of the overall filter.

Example 7.7.1

Determine the variance of the round-off noise at the output of the two cascade realizations of the filter with system function

$$H(z) = H_1(z)H_2(z)$$

where

$$H_1(z) = \frac{1}{1 - \frac{1}{2}z^{-1}}$$

$$H_2(z) = \frac{1}{1 - \frac{1}{4}z^{-1}}$$

Solution   Let $h(n)$, $h_1(n)$, and $h_2(n)$ represent the unit sample responses corresponding to the system functions $H(z)$, $H_1(z)$, and $H_2(z)$, respectively. It follows that

$$h_1(n) = (\tfrac{1}{2})^n u(n) \qquad h_2(n) = (\tfrac{1}{4})^n u(n)$$
$$h(n) = [2(\tfrac{1}{2})^n - (\tfrac{1}{4})^n]u(n)$$

The two cascade realizations are shown in Fig. 7.52.

In the first cascade realization, the variance of the output is

$$\sigma_{q1}^2 = \sigma_e^2 \left[\sum_{n=0}^{\infty} h^2(n) + \sum_{n=0}^{\infty} h_2^2(n) \right]$$

In the second cascade realization, the variance of the output noise is

$$\sigma_{q2}^2 = \sigma_e^2 \left[\sum_{n=0}^{\infty} h^2(n) + \sum_{n=0}^{\infty} h_1^2(n) \right]$$

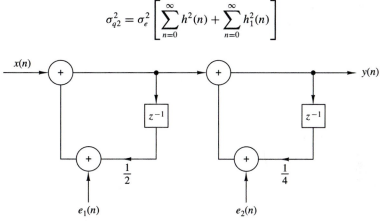

(a) Cascade realization I

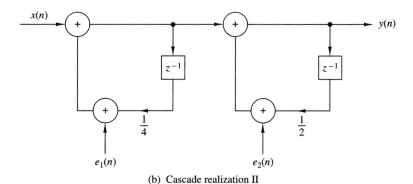

(b) Cascade realization II

Figure 7.52   Two cascade realizations in Example 7.8.1: (a) cascade realization I; (b) cascade realization II.

Now

$$\sum_{n=0}^{\infty} h_1^2(n) = \frac{1}{1 - \frac{1}{4}} = \frac{4}{3}$$

$$\sum_{n=0}^{\infty} h_2^2(n) = \frac{1}{1 - \frac{1}{16}} = \frac{16}{15}$$

$$\sum_{m=0}^{\infty} h^2(n) = \frac{4}{1 - \frac{1}{4}} - \frac{4}{1 - \frac{1}{8}} + \frac{1}{1 - \frac{1}{16}} = 1.83$$

Therefore,

$$\sigma_{q1}^2 = 2.90\sigma_e^2$$

$$\sigma_{q2}^2 = 3.16\sigma_e^2$$

and the ratio of noise variances is

$$\frac{\sigma_{q2}^2}{\sigma_{q1}^2} = 1.09$$

Consequently, the noise power in the second cascade realization is 9% larger than the first realization.

7.8 SUMMARY AND REFERENCES

From the treatment in this chapter we have seen that there are various realizations of discrete-time systems. FIR systems can be realized in a direct form, a cascade form, a frequency sampling form, and a lattice form. IIR systems can also be realized in a direct form, a cascade form, a lattice or a lattice-ladder form, and in a parallel form.

For any given system described by a linear constant-coefficient difference equation, these realizations are equivalent in that they represent the same system and produce the same output for any given input, provided that the internal computations are performed with infinite precision. However, the various structures are not equivalent when they are realized with finite- precision arithmetic.

The state-space formulation provides an internal description of a system and, as a consequence, we obtained additional system realizations, called *state-space realizations*. These realizations represent additional possible structures that provide good alternative candidate realizations for the system.

Three important factors are presented for choosing among the various FIR and IIR system realizations. These factors are computational complexity, memory requirements, and finite-word-length effects. Depending on either the time-domain or the frequency-domain characteristics of a system, some structures may require less computation and/or less memory than others. Hence our selection must consider these two important factors.

Much research has been done over the past two decades on state-space representation and realization of systems. For reference, we cite the books by Chen

(1970), DeRusso et al. (1965), Zadeh and Desoer (1963), and Gupta (1966). The use of state-space filter structures in the realization of IIR systems has been proposed by Mullis and Roberts (1976a,b), and further developed by Hwang (1977), Jackson et al. (1979), Jackson (1979), Mills et al. (1981), and Bomar (1985).

In deriving the transposed structures in Section 7.3, we introduced several concepts and operations on signal flow graphs. Signal flow graphs are treated in depth in the books by Mason and Zimmerman (1960) and Chow and Cassignol (1962).

Another important structure for IIR systems, a *wave digital filter*, has been investigated by Fettweis (1971) and further developed by Sedlmeyer and Fettweis (1973). A treatment of this filter structure can also be found in the book by Antoniou (1979).

Finite-word-length effects are an important factor in the implementation of digital signal processing systems. In this chapter we described the effects of a finite word length in digital filtering. In particular, we considered the following problems dealing with finite-word length effects:

1. Parameter quantization in digital filters
2. Round-off noise in multiplication
3. Overflow in addition
4. Limit cycles

These four effects are internal to the filter and influence the method by which the system will be implemented. In particular, we demonstrated that high-order systems, especially IIR systems, should be realized by using second-order sections as building blocks. We advocated the use of the direct form II realization, either the conventional or the transposed form.

Effects of round-off errors in fixed-point implementations of FIR and IIR filter structures have been investigated by many researchers. We cite the papers by Gold and Rader (1966), Rader and Gold (1967b), Jackson (1970a,b), Liu (1971), Chan and Rabiner (1973a,b,c), and Oppenheim and Weinstein (1972).

As an alternative to the use of direct form II second-order filters as building blocks for high-order filters, we can use second-order state-variable forms. Such state-variable forms can be optimized with respect to the state transition matrix to minimize round-off errors. The optimization leads to minimum-round-off-noise second-order state-variable filters that are highly robust for implementing both narrowband and wideband filters.

For a treatment of minimum-round-off-noise second-order state-space realizations, the reader can refer to the papers of Mullis and Roberts (1976a,b), Hwang (1977), Jackson et al. (1979), Mills et al. (1981), Bomar (1985), and the book by Roberts and Mullis (1987).

Limit-cycle oscillations occur in IIR filters as a result of quantization effects in fixed-point multiplication and rounding. Investigation of limit cycles in digital filtering and their characteristic behavior is treated in the papers by Parker and

Hess (1971), Brubaker and Gowdy (1972), Sandberg and Kaiser (1972), and Jackson (1969, 1979). The latter paper deals with limit cycles in state-space structures. Methods have also been devised to eliminate limit cycles caused by round-off errors. For example, the papers by Barnes and Fam (1977), Fam and Barnes (1979), Chang (1981), Butterweck et al. (1984), and Auer (1987) discuss this problem. Overflow oscillations have been treated in the paper by Ebert et al. (1969).

The effects of parameter quantization has been treated in a number of papers. We cite for reference the work of Rader and Gold (1967b), Knowles and Olcayto (1968), Avenhaus and Schuessler (1970), Herrmann and Schuessler (1970b), Chan and Rabiner (1973c), and Jackson (1976).

Finally, we mention that the lattice and lattice-ladder filter structures are known to be robust in fixed-point implementations. For a treatment of these types of filters, the reader is referred to the papers of Gray and Markel (1973), Makhoul (1978), and Morf et al. (1977) and to the book by Markel and Gray (1976).

PROBLEMS

7.1 Determine a direct form realization for the following linear phase filters.

    (a) $h(n) = \{1, 2, 3, 4, 3, 2, 1\}$
                $\uparrow$

    (b) $h(n) = \{1, 2, 3, 3, 2, 1\}$
                $\uparrow$

7.2 Consider an FIR filter with system function

$$H(z) = 1 + 2.88z^{-1} + 3.4048z^{-2} + 1.74z^{-3} + 0.4z^{-4}$$

Sketch the direct form and lattice realizations of the filter and determine in detail the corresponding input–output equations. Is the system minimum phase?

7.3 Determine the system function and the impulse response of the system shown in Fig. P7.3.

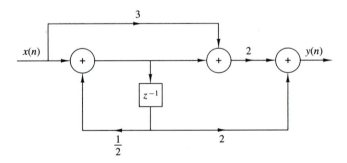

                                                                       Figure P7.3

7.4 Determine the system function and the impulse response of the system shown in Fig. P7.4.

7.5 Determine the transposed structure of the systems in Fig. P7.4 and verify that both the original and the transposed system have the same system function.

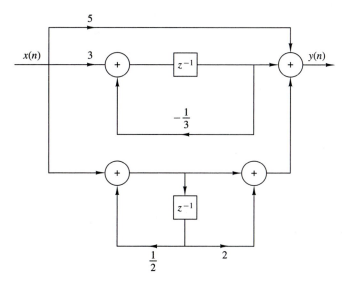

Figure P7.4

7.6 Determine a_1, a_2, and c_1 and c_0 in terms of b_1 and b_2 so that the two systems in Fig. P7.6 are equivalent.

7.7 Consider the filter shown in Fig. P7.7.
   (a) Determine its system function.
   (b) Sketch the pole–zero plot and check for stability if
       (1) $b_0 = b_2 = 1$, $b_1 = 2$, $a_1 = 1.5, a_2 = -0.9$
       (2) $b_0 = b_2 = 1$, $b_1 = 2$, $a_1 = 1, a_2 = -2$
   (c) Determine the response to $x(n) = \cos(\pi n/3)$ if $b_0 = 1$, $b_1 = b_2 = 0$, $a_1 = 1$, and $a_2 = -0.99$.

7.8 Consider an LTI system, initially at rest, described by the difference equation

$$y(n) = \tfrac{1}{4}y(n-2) + x(n)$$

   (a) Determine the impulse response, $h(n)$, of the system.
   (b) What is the response of the system to the input signal

$$x(n) = [(\tfrac{1}{2})^n + (-\tfrac{1}{2})^n]u(n)$$

   (c) Determine the direct form II, parallel-form, and cascade-form realizations for this system.
   (d) Sketch roughly the magnitude response $|H(\omega)|$ of this system.

7.9 Obtain the direct form I, direct form II, cascade, and parallel structures for the following systems.
   (a) $y(n) = \tfrac{3}{4}y(n-1) - \tfrac{1}{8}y(n-2) + x(n) + \tfrac{1}{3}x(n-1)$
   (b) $y(n) = -0.1y(n-1) + 0.72y(n-2) + 0.7x(n) - 0.252x(n-2)$
   (c) $y(n) = -0.1y(n-1) + 0.2y(n-2) + 3x(n) + 3.6x(n-1) + 0.6x(n-2)$
   (d) $H(z) = \dfrac{2(1 - z^{-1})(1 + \sqrt{2}z^{-1} + z^{-2})}{(1 + 0.5z^{-1})(1 - 0.9z^{-1} + 0.81z^{-2})}$

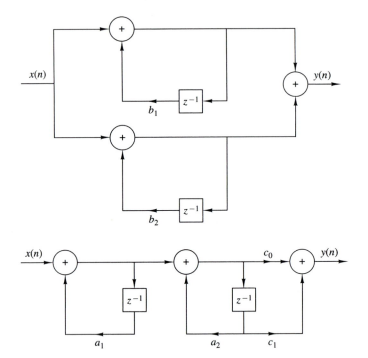

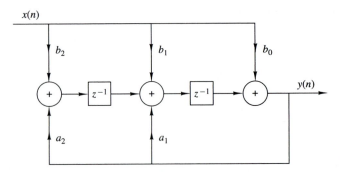

Figure P7.6

Figure P7.7

(e) $y(n) = \frac{1}{2}y(n-1) + \frac{1}{4}y(n-2) + x(n) + x(n-1)$

(f) $y(n) = y(n-1) - \frac{1}{2}y(n-2) + x(n) - x(n-1) + x(n-2)$

Which of the systems above are stable?

7.10 Show that the systems in Fig. P7.10 are equivalent.

7.11 Determine all the FIR filters which are specified by the lattice parameters $K_1 = \frac{1}{2}$, $K_2 = 0.6$, $K_3 = -0.7$, and $K_4 = \frac{1}{3}$.

7.12 Determine the set of difference equations for describing a realization of an IIR system based on the use of the transposed direct form II structure for the second-order subsystems.

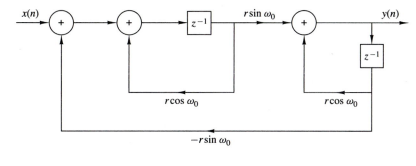

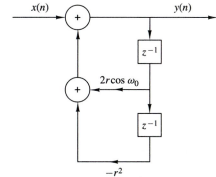

Figure P7.10

7.13* Write a program that implements a parallel-form realization based on transposed direct form II second-order modules.

7.14* Write a program that implements a cascade-form realization based on regular direct form II second-order modules.

7.15 Determine the parameters $\{K_m\}$ of the lattice filter corresponding to the FIR filter described by the system function

$$H(z) = A_2(z) = 1 + 2z^{-1} + z^{-2}$$

7.16 (a) Determine the zeros and sketch the zero pattern for the FIR lattice filter with parameters

$$K_1 = \tfrac{1}{2}, \qquad K_2 = -\tfrac{1}{3}, \qquad K_3 = 1$$

(b) The same as in part (a) but with $K_3 = -1$.

(c) You should have found that all the zeros lie exactly on the unit circle. Can this result be generalized? How?

(d) Sketch the phase response of the filters in parts (a) and (b). What did you notice? Can this result be generalized? How?

7.17 Consider an FIR lattice filter with coefficients $K_1 = 0.65$, $K_2 = -0.34$, and $K_3 = 0.8$.

(a) Find its impulse response by tracing a unit impulse input through the lattice structure.

(b) Draw the equivalent direct-form structure.

7.18 Consider a causal IIR system with system function

$$H(z) = \frac{1 + 2z^{-1} + 3z^{-2} + 2z^{-3}}{1 + 0.9z^{-1} - 0.8z^{-2} + 0.5z^{-3}}$$

  (a) Determine the equivalent lattice-ladder structure.
  (b) Check if the system is stable.

7.19 Determine the input–output relationship, the system function, and plot the pole–zero pattern for the discrete-time system shown in Fig. P7.19.

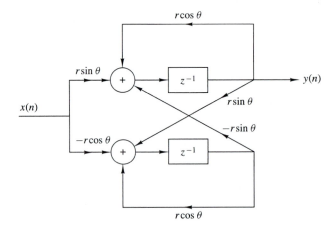

Figure P7.19

7.20 Determine the coupled-form state-space realization for the digital resonator

$$H(z) = \frac{1}{1 - (2r\cos\omega_0)z^{-1} + r^2 z^{-2}}$$

7.21 (a) Determine the impulse response of an FIR lattice filter with parameters $K_1 = 0.6$, $K_2 = 0.3$, $K_3 = 0.5$, and $K_4 = 0.9$.
  (b) Sketch the direct form and lattice all-zero and all-pole filters specified by the K-parameters given in part (a).

7.22 (a) Sketch the lattice realization for the resonator

$$H(z) = \frac{1}{1 - (2r\cos\omega_0)z^{-1} + r^2 z^{-2}}$$

  (b) What happens if $r = 1$?

7.23 Sketch the lattice-ladder structure for the system

$$H(z) = \frac{1 - 0.8z^{-1} + 0.15z^{-2}}{1 + 0.1z^{-1} - 0.72z^{-2}}$$

7.24 Determine a state-space model and the corresponding realization for the following FIR system:

$$y(n) = \sum_{k=0}^{M} b_k x(n-k)$$

7.25 Determine the state-space model for the system described by

$$y(n) = y(n-1) + 0.11y(n-2) + x(n)$$

and sketch the type 1 and type 2 state-space realizations.

7.26 Determine the type 1 and type 2 state-space realizations for the Fibonacci system and its diagonal form.

7.27 By means of the z-transform, determine the impulse response of the system described by the state-space parameters

$$\mathbf{F} = \begin{bmatrix} 0 & 0.11 \\ 1 & 1 \end{bmatrix}, \qquad \mathbf{q} = \begin{bmatrix} 0.11 \\ 1 \end{bmatrix}, \qquad \mathbf{g} = \begin{bmatrix} 0 \\ 1 \end{bmatrix}, \qquad d = 1$$

7.28 Determine the characteristic polynomial of the coupled-form state-space structure described by (7.4.68) and solve for the roots.

7.29 Determine the transpose structure for the coupled-form state-space structure shown in Fig. 7.31.

7.30 Consider a pole–zero system with system function

$$H(z) = \frac{(1 - 0.5e^{j\pi/4}z^{-1})(1 - 0.5e^{-j\pi/4}z^{-1})}{(1 - 0.8e^{j\pi/3}z^{-1})(1 - 0.8e^{-j\pi/3}z^{-1})}$$

(a) Sketch the regular and transpose direct form II realizations of the system.
(b) Determine and sketch the type 1 and type 2 state-space realizations.
(c) Determine the impulse response of the system by inverting $H(z)$ and by using state-space techniques.
(d) Determine the coupled-form state-space realization.
(e) Repeat parts (a) through (d) for the system obtained by changing the angle of the poles from $\pi/3$ to $\pi/4$.

7.31 (a) Determine a parallel and a cascade realization of the system

$$H(z) = \frac{1 + z^{-1}}{(1 - z^{-1})(1 - 0.8e^{j\pi/4}z^{-1})(1 - 0.8e^{-j\pi/4}z^{-1})}$$

(b) Determine the type 1 and type 2 state-space descriptions of the system in part (a).

7.32 Show how to use a lattice structure to implement the following all-pass filter

$$H(z) = \frac{0.5 + 0.2z^{-1} - 0.6z^{-2} + z^{-3}}{1 - 0.6z^{-1} + 0.2z^{-2} + 0.5z^{-3}}$$

Is the system stable?

7.33 Consider a system described by the following state-space equations:

$$\mathbf{v}(n+1) = \begin{bmatrix} 0 & 1 \\ -0.81 & 1 \end{bmatrix} \mathbf{v}(n) + \begin{bmatrix} 0 \\ 1 \end{bmatrix} x(n)$$

$$y(n) = \begin{bmatrix} -1.81 & 1 \end{bmatrix} \mathbf{v}(n) + x(n)$$

(a) Determine the characteristic polynomial and the eigenvalues of the system.
(b) Determine the state transition matrix $\mathbf{\Phi}(n)$ for $n \geq 0$.
(c) Determine the system function and the impulse response of the system.
(d) Compute the step response of the system if $\mathbf{v}(0) = \begin{bmatrix} 0 & 1 \end{bmatrix}^t$.
(e) Sketch a state-space realization for the system.

7.34 Repeat Problem 7.33 if the system is described by the state-space equations

$$\mathbf{v}(n+1) = \begin{bmatrix} 0 & 1 \\ -2 & -3 \end{bmatrix} \mathbf{v}(n) + \begin{bmatrix} 0 \\ 1 \end{bmatrix} x(n)$$

$$y(n) = \begin{bmatrix} 1 & 0 \end{bmatrix} \mathbf{v}(n)$$

7.35 Repeat Problem 7.33 for the system described by the state-space equations

$$\mathbf{v}(n+1) = \begin{bmatrix} -0.3 & 0.4 \\ 0.4 & -0.3 \end{bmatrix} \mathbf{v}(n) + \begin{bmatrix} 1 \\ 0 \end{bmatrix} x(n)$$

$$y(n) = \begin{bmatrix} 1 & 1 \end{bmatrix} \mathbf{v}(n) + x(n)$$

7.36 Consider the system

$$y(n) = 0.9y(n-1) - 0.08y(n-2) + x(n) + x(n-1)$$

(a) Determine the type 1 and type 2 state-space realizations of the system.
(b) Determine the parallel and cascade state-space realizations of the system.
(c) Determine the impulse response of the system by at least two different methods.

7.37 Consider the causal system

$$y(n) = \tfrac{3}{4}y(n-1) - \tfrac{1}{8}y(n-2) + x(n) + \tfrac{1}{3}x(n-1)$$

(a) Determine its system function.
(b) Determine the type 1 state-space model.
(c) Determine the state transition matrix $\mathbf{\Phi}(n) = \mathbf{F}^n$, for any n, using z-transform techniques.
(d) Determine the system function using the formula

$$H(z) = \mathbf{g}^t (z\mathbf{I} - \mathbf{F})^{-1}\mathbf{q} + d$$

Compare the answer with that in part (a).
(e) Compute the characteristic polynomial $\det(z\mathbf{I}-\mathbf{F})$ and check if the system is stable.

7.38 Determine the impulse response of the system

$$\mathbf{F} = \begin{bmatrix} 0 & 0.11 \\ 1 & 1 \end{bmatrix}, \qquad \mathbf{q} = \begin{bmatrix} 0.11 \\ 1 \end{bmatrix}, \qquad \mathbf{g} = \begin{bmatrix} 0 \\ 1 \end{bmatrix}, \qquad d = 1$$

using the z-transform approach.

7.39 A discrete-time system is described by the following state-space model:

$$\mathbf{v}(n+1) = \mathbf{F}\mathbf{v}(n) + \mathbf{q}x(n)$$

$$y(n) = \mathbf{g}^t\mathbf{v}(n) + dx(n)$$

where

$$\mathbf{F} = \begin{bmatrix} 0 & 1 \\ -\frac{5}{16} & -1 \end{bmatrix}, \qquad \mathbf{q} = \begin{bmatrix} 0 \\ 1 \end{bmatrix}, \qquad \mathbf{g} = \begin{bmatrix} \frac{11}{8} \\ 2 \end{bmatrix}, \qquad d = 2$$

(a) Sketch the corresponding state-space structure.
(b) Calculate the impulse response for $n = 0, 1, \dots, 5$ and for $n = 17$ by using the state-space approach
(c) Find the difference equation description of the system.
(d) Repeat part (b) by using the difference equation.
(e) Sketch the direct form II implementation of the system.

7.40 Determine the state-space parameters **F**, **q**, **g**, and d for:
   (a) the all-zero lattice structure
   (b) the all-pole lattice structure

7.41 The generic floating-point format for a DSP microprocessor is the following:

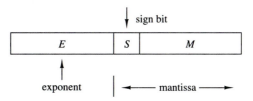

The value of the number X is given by

$$X = \begin{cases} 01.M \times 2^E & \text{if } S = 0 \\ 10.M \times 2^E & \text{if } S = 1 \\ 0 & \text{if } E \text{ is the most negative two's-complement value} \end{cases}$$

Determine the range of positive and negative numbers for the following two formats:

(a)

15	12	11	10	0
E		S	M		short format

(b)

31	24	23	22	0
E		S	M		single-precision format

7.42 Consider the IIR recursive filter shown in Fig. P7.42 and let $h_F(n)$, $h_R(n)$, and $h(n)$ denote the impulse responses of the FIR section, the recursive section, and the overall filter, respectively.
   (a) Find all the causal and stable recursive second-order sections with integer coefficients (a_1, a_2) and determine and sketch their impulse responses and frequency responses. These filters do not require complicated multiplications or quantization after multiplications.

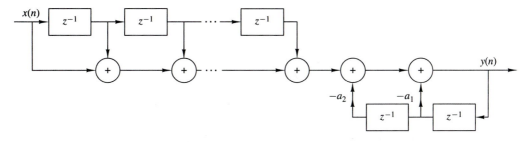

Figure P7.42

(b) Show that three of the sections obtained in part (a) can be obtained by interconnection of other sections.

(c) Find a difference equation that describes the impulse response $h(n)$ of the filter
and determine the conditions for the overall filter to be FIR.

(d) Rederive the results in parts (a) to (c) using z-domain considerations.

7.43 This problem illustrates the development of digital filter structures using Horner's
rule for polynomial evaluation. To this end consider the polynomial

$$p(x) = \alpha_p x^p + a_{p-1} x^{p-1} + \cdots + a_1 x + a_0$$

which computes $p(x)$ with the minimum cost of p multiplications and p additions.

(a) Draw the structures corresponding to the factorizations

$$H_1(z) = b_0(1 + b_1 z^{-1}(1 + b_2 z^{-1}(1 + b_3 z^{-1})))$$

$$H(z) = b_0(z^{-3} + (b_1 z^{-2} + (b_2 z^{-1} + b_3)))$$

and determine the system function, number of delay elements, and arithmetic
operations for each structure

(b) Draw the Horner structure for the following linear-phase system:

$$H(z) = z^{-1} \left[\alpha_0 + \sum_{k=1}^{3} (z^{-k} + z^k) \alpha_k \right]$$

7.44 Let x_1 and x_2 be $(b+1)$-bit binary numbers with magnitude less than 1. To compute
the sum of x_1 and x_2 using two's-complement representation we treat them as $(b+1)$-
bit unsigned numbers, we perform addition modulo-2 and ignore any carry after the
sign bit.

(a) Show that if the sum of two numbers with the same sign has the opposite sign,
this corresponds to overflow.

(b) Show that when we compute the sum of several numbers using two's-complement
representation, the result will be correct, even if there are overflows, if the correct
sum is less than 1 in magnitude. Illustrate this argument by constructing a simple
example with three numbers.

7.45 Consider the system described by the difference equation

$$y(n) = ay(n-1) - ax(n) + x(n-1)$$

(a) Show that it is all-pass.

(b) Obtain the direct form II realization of the system

(c) If you quantize the coefficients of the system in part (b), is it still all- pass?

(d) Obtain a realization by rewriting the difference equation as

$$y(n) = a[y(n-1) - x(n)] + x(n-1)$$

(e) If you quantize the coefficients of the system in part (d), is it still all-pass?

7.46 Consider the system

$$y(n) = \tfrac{1}{2} y(n-1) + x(n)$$

(a) Compute its response to the input $x(n) = (\tfrac{1}{4})^n u(n)$ assuming infinite-precision
arithmetic.

(b) Compute the response of the system $y(n)$, $0 \le n \le 5$ to the same input, assuming finite-precision sign-and-magnitude fractional arithmetic with five bits (i.e., the sign bit plus four fractional bits). The quantization is performed by truncation.

(c) Compare the results obtained in parts (a) and (b).

7.47 The input to the system

$$y(n) = 0.999y(n - 1) + x(n)$$

is quantized to $b = 8$ bits. What is the power produced by the quantization noise at the output of the filter?

7.48 Consider the system

$$y(n) = 0.875y(n - 1) - 0.125y(n - 2) + x(n)$$

(a) Compute its poles and design the cascade realization of the system.

(b) Quantize the coefficients of the system using truncation, maintaining a sign bit plus three other bits. Determine the poles of the resulting system.

(c) Repeat part (b) for the same precision using rounding.

(d) Compare the poles obtained in parts (b) and (c) with those in part (a). Which realization is better? Sketch the frequency responses of the systems in parts (a), (b), and (c).

7.49 Consider the system

$$H(z) = \frac{1 - \frac{1}{2}z^{-1}}{(1 - \frac{1}{4}z^{-1})(1 + \frac{1}{4}z^{-1})}$$

(a) Draw all possible realizations of the system.

(b) Suppose that we implement the filter with fixed-point sign-and-magnitude fractional arithmetic using $(b + 1)$ bits (one bit is used for the sign). Each resulting product is rounded into b bits. Determine the variance of the round-off noise created by the multipliers at the output of each one of the realizations in part (a).

7.50 The first-order filter shown in Fig. P7.50 is implemented in four-bit (including sign) fixed-point two's-complement fractional arithmetic. Products are rounded to four-bit representation. Using the input $x(n) = 0.10\delta(n)$, determine:

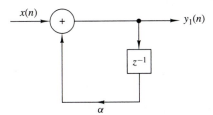

Figure P7.50

(a) The first five outputs if $\alpha = 0.5$. Does the filter go into a limit cycle?

(b) The first five outputs if $\alpha = 0.75$. Does the filter go into a limit cycle?

7.51 The digital system shown in Fig. P7.51 uses a six-bit (including sign) fixed-point two's-complement A/D converter with rounding, and the filter $H(z)$ is implemented using eight-bit (including sign) fixed-point two's-complement fractional arithmetic with rounding. The input $x(t)$ is a zero-mean uniformly distributed random process having autocorrelation $\gamma_{xx}(\tau) = 3\delta(\tau)$. Assume that the A/D converter can handle input values up to ± 1.0 without overflow.

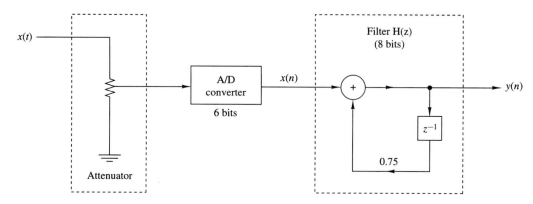

Figure P7.51

(a) What value of attenuation should be applied prior to the A/D converter to assure that it does not overflow?

(b) With the attenuation above, what is the signal-to-quantization noise ratio (SNR) at the A/D converter output?

(c) The six-bit A/D samples can be left-justified, right-justified, or centered in the eight-bit word used as the input to the digital filter. What is the correct strategy to use for maximum SNR at the filter output without overflow?

(d) What is the SNR at the output of the filter due to all quantization noise sources?

7.52 Shown in Fig. P7.52 is the coupled-form implementation of a two-pole filter with poles at $x = re^{\pm j\theta}$. There are four real multiplications per output point. Let $e_i(n)$, $i = 1, 2, 3, 4$ represent the round-off noise in a fixed-point implementation of the filter. Assume that the noise sources are zero-mean mutually uncorrelated stationary white noise sequences. For each n the probability density function $p(e)$ is uniform in the range $-\Delta/2 \le e \le \Delta/2$, where $\Delta = 2^{-b}$.

(a) Write the two coupled difference equations for $y(n)$ and $v(n)$, including the noise sources and the input sequence $x(n)$.

(b) From these two difference equations, show that the filter system functions $H_1(z)$ and $H_2(z)$ between the input noise terms $e_1(n) + e_2(n)$ and $e_3(n) + e_4(n)$ and the output $y(n)$ are:

$$H_1(z) = \frac{r \sin \theta z^{-1}}{1 - 2r \cos \theta z^{-1} + r^2 z^{-2}}$$

$$H_2(z) = \frac{1 - r \cos \theta z^{-1}}{1 - 2r \cos \theta z^{-1} + r^2 z^{-2}}$$

We know that

$$H(z) = \frac{1}{1 - 2r \cos \theta z^{-1} + r^2 z^{-2}} \Rightarrow h(n) = \frac{1}{\sin \theta} r^n \sin(n + 1)\theta u(n)$$

Determine $h_1(n)$ and $h_2(n)$.

(c) Determine a closed-form expression for the variance of the total noise from $e_i(n)$, $i = 1, 2, 3, 4$ at the output of the filter.

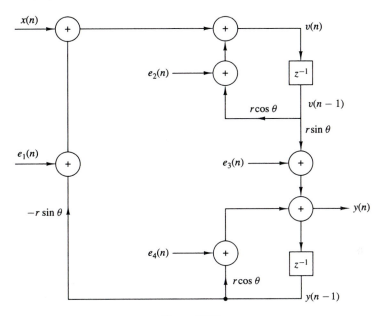

Figure P7.52

7.53 Determine the variance of the round-off noise at the output of the two cascade realizations of the filter shown in Fig. P7.53, with system function

$$H(z) = H_1(z)H_2(z)$$

where

$$H_1(z) = \frac{1}{1 - \frac{1}{2}z^{-1}}$$

$$H_2(z) = \frac{1}{1 - \frac{1}{4}z^{-1}}$$

7.54 *Quantization effects in direct-form FIR filters*   Consider a direct-form realization of an FIR filter of length M. Suppose that the multiplication of each coefficient with the corresponding signal sample is performed in fixed-point arithmetic with b bits and each product is rounded to b bits. Determine the variance of the quantization noise at the output of the filter by using a statistical characterization of the round-off noise as in Section 7.7.3.

7.55* Consider the system specified by the system function

$$H(z) = \frac{B(z)}{A(z)}$$

$$= \left[G_1 \frac{(1 - 0.8e^{j\pi/4}z^{-1})(1 - 0.8e^{-j\pi/4}z^{-1})}{(1 - \frac{1}{2}z^{-1})(1 + \frac{1}{3}z^{-1})} \right] \left[G_2 \frac{(1 + \frac{1}{4}z^{-1})(1 - \frac{5}{8}z^{-1})}{(1 - 0.8e^{j\pi/3}z^{-1})(1 - 0.8e^{-j\pi/3}z^{-1})} \right]$$

(a) Choose G_1 and G_2 so that the gain of each second-order section at $\omega = 0$ is equal to 1.

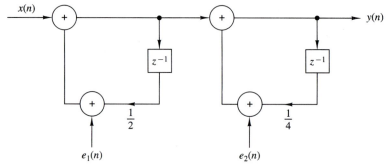

(a) Cascade realization I

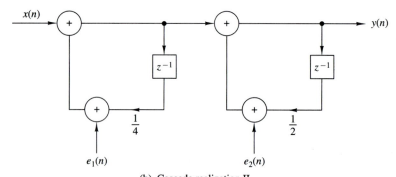

(b) Cascade realization II

Figure P7.53

   (b) Sketch the direct form 1, direct form 2, and cascade realizations of the system.
   (c) Write a program that implements the direct form 1 and direct form 2, and compute the first 100 samples of the impulse response and the step response of the system.
   (d) Plot the results in part (c) to illustrate the proper functioning of the programs.

7.56* Consider the system given in Problem 7.55 with $G_1 = G_2 = 1$.
   (a) Determine a lattice realization for the system

$$H(z) = B(z)$$

   (b) Determine a lattice realization for the system

$$H(z) = \frac{1}{A(z)}$$

   (c) Determine a lattice-ladder realization for the system $H(z) = B(z)/A(z)$.
   (d) Write a program for the implementation of the lattice-ladder structure in part (c).
   (e) Determine and sketch the first 100 samples of the impulse responses of the systems in parts (a) through (c) by working with the lattice structures.
   (f) Compute and sketch the first 100 samples of the convolution of impulse responses in parts (a) and (b). What did you find? Explain your results.

7.57* Consider the system given in Problem 7.55.

    (a) Determine the parallel-form structure and write a program for its implementation.

    (b) Sketch a parallel structure using second-order coupled-form state-space sections.

    (c) Write a program for the implementation of the structure in part (b).

    (d) Verify the programs in parts (a) and (c) by computing and sketching the impulse response of the system.

8

Design of Digital Filters

With the background that we have developed in the preceding chapters, we are now in a position to treat the subject of digital filter design. We shall describe several methods for designing FIR and IIR digital filters.

In the design of frequency-selective filters, the desired filter characteristics are specified in the frequency domain in terms of the desired magnitude and phase response of the filter. In the filter design process, we determine the coefficients of a causal FIR or IIR filter that closely approximates the desired frequency response specifications. The issue of which type of filter to design, FIR or IIR, depends on the nature of the problem and on the specifications of the desired frequency response.

In practice, FIR filters are employed in filtering problems where there is a requirement for a linear-phase characteristic within the passband of the filter. If there is no requirement for a linear-phase characteristic, either an IIR or an FIR filter may be employed. However, as a general rule, an IIR filter has lower sidelobes in the stopband than an FIR filter having the same number of parameters. For this reason, if some phase distortion is either tolerable or unimportant, an IIR filter is preferable, primarily because its implementation involves fewer parameters, requires less memory and has lower computational complexity.

In conjunction with our discussion of digital filter design, we describe frequency transformations in both the analog and digital domains for transforming a lowpass prototype filter into either another lowpass, bandpass, bandstop, or highpass filter.

Today, FIR and IIR digital filter design is greatly facilitated by the availability of numerous computer software programs. In describing the various digital filter design methods in this chapter, our primary objective is to give the reader the background necessary to select the filter that best matches the application and satisfies the design requirements.

8.1 GENERAL CONSIDERATIONS

In Section 4.5, we described the characteristics of ideal filters and demonstrated that such filters are not causal and therefore, are not physically realizable. In this section, the issue of causality and its implications is considered in more detail. Following this discussion, we present the frequency response characteristics of causal FIR and IIR digital filters.

8.1.1 Causality and Its Implications

Let us consider the issue of causality in more detail by examining the impulse response $h(n)$ of an ideal lowpass filter with frequency response characteristic

$$H(\omega) = \begin{cases} 1, & |\omega| \leq \omega_c \\ 0, & \omega_c < \omega \leq \pi \end{cases} \tag{8.1.1}$$

The impulse response of this filter is

$$h(n) = \begin{cases} \dfrac{\omega_c}{\pi}, & n = 0 \\ \dfrac{\omega_c}{\pi} \dfrac{\sin \omega_c n}{\omega_c n}, & n \neq 0 \end{cases} \tag{8.1.2}$$

A plot of $h(n)$ for $\omega_c = \pi/4$ is illustrated in Fig. 8.1. It is clear that the ideal lowpass filter is noncausal and hence it cannot be realized in practice.

One possible solution is to introduce a large delay n_0 in $h(n)$ and arbitrarily to set $h(n) = 0$ for $n < n_0$. However, the resulting system no longer has an ideal frequency response characteristic. Indeed, if we set $h(n) = 0$ for $n < n_0$, the Fourier series expansion of $H(\omega)$ results in the Gibbs phenomenon, as will be described in Section 8.2.

Although this discussion is limited to the realization of a lowpass filter, our conclusions hold, in general, for all the other ideal filter characteristics. In brief, none of the ideal filter characteristics previously illustrated in Fig. 4.43 are causal, hence all are physically unrealizable.

A question that naturally arises at this point is the following: What are the necessary and sufficient conditions that a frequency response characteristic $H(\omega)$

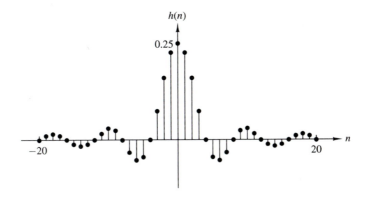

Figure 8.1 Unit sample response of an ideal lowpass filter.

must satisfy in order for the resulting filter to be causal? The answer to this question is given by the Paley–Wiener theorem, which can be stated as follows:

Paley–Wiener Theorem. If $h(n)$ has finite energy and $h(n) = 0$ for $n < 0$, then [for a reference, see Wiener and Paley (1934)]

$$\int_{-\pi}^{\pi} |\ln |H(\omega)|| d\omega < \infty \tag{8.1.3}$$

Conversely, if $|H(\omega)|$ is square integrable and if the integral in (8.1.3) is finite, then we can associate with $|H(\omega)|$ a phase response $\Theta(\omega)$, so that the resulting filter with frequency response

$$H(\omega) = |H(\omega)|e^{j\Theta(\omega)}$$

is causal.

One important conclusion that we draw from the Paley–Wiener theorem is that the magnitude function $|H(\omega)|$ can be zero at some frequencies, but it cannot be zero over any finite band of frequencies, since the integral then becomes infinite. Consequently, any ideal filter is noncausal.

Apparently, causality imposes some tight constraints on a linear time-invariant system. In addition to the Paley–Wiener condition, causality also implies a strong relationship between $H_R(\omega)$ and $H_I(\omega)$, the real and imaginary components of the frequency response $H(\omega)$. To illustrate this dependence, we decompose $h(n)$ into an even and an odd sequence, that is,

$$h(n) = h_e(n) + h_o(n) \tag{8.1.4}$$

where

$$h_e(n) = \tfrac{1}{2}[h(n) + h(-n)] \tag{8.1.5}$$

and

$$h_o(n) = \tfrac{1}{2}[h(n) - h(-n)] \tag{8.1.6}$$

Now, if $h(n)$ is causal, it is possible to recover $h(n)$ from its even part $h_e(n)$ for $0 \leq n \leq \infty$ or from its odd component $h_o(n)$ for $1 \leq n \leq \infty$.

Indeed, it can be easily seen that

$$h(n) = 2h_e(n)u(n) - h_e(0)\delta(n) \qquad n \geq 0 \tag{8.1.7}$$

and

$$h(n) = 2h_o(n)u(n) + h(0)\delta(n) \qquad n \geq 1 \tag{8.1.8}$$

Since $h_o(n) = 0$ for $n = 0$, we cannot recover $h(0)$ from $h_o(n)$ and hence we also must know $h(0)$. In any case, it is apparent that $h_o(n) = h_e(n)$ for $n \geq 1$, so there is a strong relationship between $h_o(n)$ and $h_e(n)$.

If $h(n)$ is absolutely summable (i.e., BIBO stable), the frequency response $H(\omega)$ exists, and

$$H(\omega) = H_R(\omega) + jH_I(\omega) \tag{8.1.9}$$

In addition, if $h(n)$ is real valued and causal, the symmetry properties of the Fourier transform imply that

$$h_e(n) \overset{F}{\longleftrightarrow} H_R(\omega)$$

$$h_o(n) \overset{F}{\longleftrightarrow} H_I(\omega)$$

$$\text{(8.1.10)}$$

Since $h(n)$ is completely specified by $h_e(n)$, it follows that $H(\omega)$ is completely determined if we know $H_R(\omega)$. Alternatively, $H(\omega)$ is completely determined from $H_I(\omega)$ and $h(0)$. In short, $H_R(\omega)$ and $H_I(\omega)$ are interdependent and cannot be specified independently if the system is causal. Equivalently, the magnitude and phase responses of a causal filter are interdependent and hence cannot be specified independently.

Given $H_R(\omega)$ for a corresponding real, even, and absolutely summable sequence $h_e(n)$, we can determine $H(\omega)$. The following example illustrates the procedure.

Example 8.1.1

Consider a stable LTI system with real and even impulse response $h(n)$. Determine $H(\omega)$ if

$$H_R(\omega) = \frac{1 - a\cos\omega}{1 - 2a\cos\omega + a^2}, \quad |a| < 1$$

Solution   The first step is to determine $h_e(n)$. This can be done by noting that

$$H_R(\omega) = H_R(z)|_{z=e^{j\omega}}$$

where

$$H_R(z) = \frac{1 - a(z + z^{-1})/2}{1 - a(z + z^{-1}) + a^2} = \frac{z - a(z^2 + 1)/2}{(z - a)(1 - az)}$$

The ROC has to be restricted by the poles at $p_1 = a$ and $p_2 = 1/a$ and should include the unit circle. Hence the ROC is $|a| < |z| < 1/|a|$. Consequently, $h_e(n)$ is a two-sided sequence, with the pole at $z = a$ contributing to the causal part and $p_2 = 1/a$ contributing to the anticausal part. By using a partial-fraction expansion, we obtain

$$h_e(n) = \tfrac{1}{2}a^{|n|} + \tfrac{1}{2}\delta(n) \qquad (8.1.11)$$

By substituting (8.1.11) into (8.1.7), we obtain $h(n)$ as

$$h(n) = a^n u(n)$$

Finally, we obtain the Fourier transform of $h(n)$ as

$$H(\omega) = \frac{1}{1 - ae^{-j\omega}}$$

The relationship between the real and imaginary components of the Fourier transform of an absolutely summable, causal, and real sequence can be easily established from (8.1.7). The Fourier transform relationship for (8.1.7) is

$$H(\omega) = H_R(\omega) + jH_I(\omega) = \frac{1}{\pi}\int_{-\pi}^{\pi} H_R(\lambda)U(\omega - \lambda)d\lambda - h_e(0) \qquad (8.1.12)$$

where $U(\omega)$ is the Fourier transform of the unit step sequence $u(n)$. Although the unit step sequence is not absolutely summable, it has a Fourier transform (see Section 4.2.8).

$$U(\omega) = \pi \delta(\omega) + \frac{1}{1 - e^{-j\omega}}$$

$$= \pi \delta(\omega) + \frac{1}{2} - j \frac{1}{2} \cot \frac{\omega}{2} \qquad -\pi \le \omega \le \pi \qquad (8.1.13)$$

By substituting (8.1.13) into (8.1.12) and carrying out the integration, we obtain the relation between $H_R(\omega)$ and $H_I(\omega)$ as

$$H_I(\omega) = -\frac{1}{2\pi} \int_{-\pi}^{\pi} H_R(\lambda) \cot \frac{\omega - \lambda}{2} d\lambda \qquad (8.1.14)$$

Thus $H_I(\omega)$ is uniquely determined from $H_R(\omega)$ through this integral relationship. The integral is called a *discrete Hilbert transform*. It is left as an exercise to the reader to establish the relationship for $H_R(\omega)$ in terms of the discrete Hilbert transform of $H_I(\omega)$.

To summarize, causality has very important implications in the design of frequency-selective filters. These are: (a) the frequency response $H(\omega)$ cannot be zero, except at a finite set of points in frequency; (b) the magnitude $|H(\omega)|$ cannot be constant in any finite range of frequencies and the transition from passband to stopband cannot be infinitely sharp [this is a consequence of the Gibbs phenomenon, which results from the truncation of $h(n)$ to achieve causality]; and (c) the real and imaginary parts of $H(\omega)$ are interdependent and are related by the discrete Hilbert transform. As a consequence, the magnitude $|H(\omega)|$ and phase $\Theta(\omega)$ of $H(\omega)$ cannot be chosen arbitrarily.

Now that we know the restrictions that causality imposes on the frequency response characteristic and the fact that ideal filters are not achievable in practice, we limit our attention to the class of linear time-invariant systems specified by the difference equation

$$y(n) = -\sum_{k=1}^{N} a_k y(n - k) + \sum_{k=0}^{M} b_k x(n - k)$$

which are causal and physically realizable. As we have demonstrated, such systems have a frequency response

$$H(\omega) = \frac{\displaystyle\sum_{k=0}^{M} b_k e^{-j\omega k}}{1 + \displaystyle\sum_{k=1}^{N} a_k e^{-j\omega k}} \qquad (8.1.15)$$

The basic digital filter design problem is to approximate any of the ideal frequency response characteristics with a system that has the frequency response (8.1.15), by properly selecting the coefficients $\{a_k\}$ and $\{b_k\}$. The approximation problem is treated in detail in Sections 8.2 and 8.3, where we discuss techniques for digital filter design.

8.1.2 Characteristics of Practical Frequency-Selective Filters

As we observed from our discussion of the preceding section, ideal filters are noncausal and hence physically unrealizable for real-time signal processing applications. Causality implies that the frequency response characteristic $H(\omega)$ of the filter cannot be zero, except at a finite set of points in the frequency range. In addition, $H(\omega)$ cannot have an infinitely sharp cutoff from passband to stopband, that is, $H(\omega)$ cannot drop from unity to zero abruptly.

Although the frequency response characteristics possessed by ideal filters may be desirable, they are not absolutely necessary in most practical applications. If we relax these conditions, it is possible to realize causal filters that approximate the ideal filters as closely as we desire. In particular, it is not necessary to insist that the magnitude $|H(\omega)|$ be constant in the entire passband of the filter. A small amount of ripple in the passband, as illustrated in Fig. 8.2, is usually tolerable. Similarly, it is not necessary for the filter response $|H(\omega)|$ to be zero in the stopband. A small, nonzero value or a small amount of ripple in the stopband is also tolerable.

The transition of the frequency response from passband to stopband defines the *transition band* or *transition region* of the filter, as illustrated in Fig. 8.2. The band-edge frequency ω_p defines the edge of the passband, while the frequency ω_s denotes the beginning of the stopband. Thus the width of the transition band is $\omega_s - \omega_p$. The width of the passband is usually called the *bandwidth* of the filter. For example, if the filter is lowpass with a passband edge frequency ω_p, its bandwidth is ω_p.

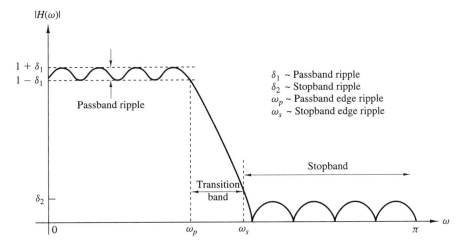

Figure 8.2 Magnitude characteristics of physically realizable filters.

If there is ripple in the passband of the filter, its value is denoted as δ_1, and the magnitude $|H(\omega)|$ varies between the limits $1 \pm \delta_1$. The ripple in the stopband of the filter is denoted as δ_2.

To accommodate a large dynamic range in the graph of the frequency response of any filter, it is common practice to use a logarithmic scale for the magnitude $|H(\omega)|$. Consequently, the ripple in the passband is $20 \log_{10} \delta_1$ decibels, and that in the stopband is $20 \log_{10} \delta_2$.

In any filter design problem we can specify (1) the maximum tolerable passband ripple, (2) the maximum tolerable stopband ripple, (3) the passband edge frequency ω_p, and (4) the stopband edge frequency ω_s. Based on these specifications, we can select the parameters $\{a_k\}$ and $\{b_k\}$ in the frequency response characteristic, given by (8.1.15), which best approximates the desired specification. The degree to which $H(\omega)$ approximates the specifications depends in part on the criterion used in the selection of the filter coefficients $\{a_k\}$ and $\{b_k\}$ as well as on the numbers (M, N) of coefficients.

In the following section we present a method for designing linear-phase FIR filters.

8.2 DESIGN OF FIR FILTERS

In this section we describe several methods for designing FIR filters. Our treatment is focused on the important class of linear-phase FIR filters.

8.2.1 Symmetric and Antisymmetric FIR Filters

An FIR filter of length M with input $x(n)$ and output $y(n)$ is described by the difference equation

$$y(n) = b_0 x(n) + b_1 x(n-1) + \cdots + b_{M-1} x(n-M+1)$$

$$= \sum_{k=0}^{M-1} b_k x(n-k) \tag{8.2.1}$$

where $\{b_k\}$ is the set of filter coefficients. Alternatively, we can express the output sequence as the convolution of the unit sample response $h(n)$ of the system with the input signal. Thus we have

$$y(n) = \sum_{k=0}^{M-1} h(k) x(n-k) \tag{8.2.2}$$

where the lower and upper limits on the convolution sum reflect the causality and finite-duration characteristics of the filter. Clearly, (8.2.1) and (8.2.2) are identical in form and hence it follows that $b_k = h(k)$, $k = 0, 1, \ldots, M-1$.

The filter can also be characterized by its system function

$$H(z) = \sum_{k=0}^{M-1} h(k) z^{-k} \tag{8.2.3}$$

which we view as a polynomial of degree $M - 1$ in the variable z^{-1}. The roots of this polynomial constitute the zeros of the filter.

An FIR filter has linear phase if its unit sample response satisfies the condition

$$h(n) = \pm h(M - 1 - n) \qquad n = 0, 1, \ldots, M - 1 \tag{8.2.4}$$

When the symmetry and antisymmetry conditions in (8.2.4) are incorporated into (8.2.3), we have

$$H(z) = h(0) + h(1)z^{-1} + h(2)z^{-2} + \cdots + h(M - 2)z^{-(M-2)} + h(M - 1)z^{-(M-1)}$$

$$= z^{-(M-1)/2} \left\{ h\left(\frac{M-1}{2}\right) + \sum_{n=0}^{(M-3)/2} h(n)\left[z^{(M-1-2k)/2} \pm z^{-(M-1-2k)/2}\right] \right\} \quad M \text{ odd}$$

$$= z^{-(M-1)/2} \sum_{n=0}^{(M/2)-1} h(n)[z^{(M-1-2k)/2} \pm z^{-(M-1-2k)/2}] \qquad M \text{ even} \tag{8.2.5}$$

Now, if we substitute z^{-1} for z in (8.2.3) and multiply both sides of the resulting equation by $z^{-(M-1)}$, we obtain

$$z^{-(M-1)} H(z^{-1}) = \pm H(z) \tag{8.2.6}$$

This result implies that the roots of the polynomial $H(z)$ are identical to the roots of the polynomial $H(z^{-1})$. Consequently, the roots of $H(z)$ must occur in reciprocal pairs. In other words, if z_1 is a root or a zero of $H(z)$, then $1/z_1$ is also a root. Furthermore, if the unit sample response $h(n)$ of the filter is real, complex-valued roots must occur in complex-conjugate pairs. Hence, if z_1 is a complex-valued root, z_1^* is also a root. As a consequence of (8.2.6), $H(z)$ also has a zero at $1/z_1^*$. Figure 8.3 illustrates the symmetry that exists in the location of the zeros of a linear-phase FIR filter.

The frequency response characteristics of linear-phase FIR filters are obtained by evaluating (8.2.5) on the unit circle. This substitution yields the expression for $H(\omega)$.

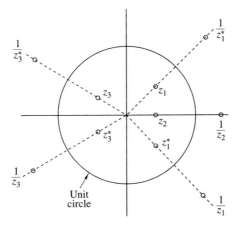

Figure 8.3 Symmetry of zero locations for a linear-phase FIR filter.

When $h(n) = h(M - 1 - n)$, $H(\omega)$ can be expressed as

$$H(\omega) = H_r(\omega)e^{-j\omega(M-1)/2} \tag{8.2.7}$$

where $H_r(\omega)$ is a real function of ω and can be expressed as

$$H_r(\omega) = h\left(\frac{M-1}{2}\right) + 2 \sum_{n=0}^{(M-3)/2} h(n) \cos \omega \left(\frac{M-1}{2} - n\right) \qquad M \text{ odd} \tag{8.2.8}$$

$$H_r(\omega) = 2 \sum_{n=0}^{(M/2)-1} h(n) \cos \omega \left(\frac{M-1}{2} - n\right) \qquad M \text{ even} \tag{8.2.9}$$

The phase characteristic of the filter for both M odd and M even is

$$\Theta(\omega) = \begin{cases} -\omega\left(\dfrac{M-1}{2}\right), & \text{if } H_r(\omega) > 0 \\[2mm] -\omega\left(\dfrac{M-1}{2}\right) + \pi, & \text{if } H_r(\omega) < 0 \end{cases} \tag{8.2.10}$$

When

$$h(n) = -h(M - 1 - n)$$

the unit sample response is *antisymmetric*. For M odd, the center point of the antisymmetric $h(n)$ is $n = (M - 1)/2$. Consequently,

$$h\left(\frac{M-1}{2}\right) = 0$$

However, if M is even, each term in $h(n)$ has a matching term of opposite sign.

It is straightforward to show that the frequency response of an FIR filter with an antisymmetric unit sample response can be expressed as

$$H(\omega) = H_r(\omega)e^{j[-\omega(M-1)/2+\pi/2]} \tag{8.2.11}$$

where

$$H_r(\omega) = 2 \sum_{n=0}^{(M-3)/2} h(n) \sin \omega \left(\frac{M-1}{2} - n\right) \qquad M \text{ odd} \tag{8.2.12}$$

$$H_r(\omega) = 2 \sum_{n=0}^{(M/2)-1} h(n) \sin \omega \left(\frac{M-1}{2} - n\right) \qquad M \text{ even} \tag{8.2.13}$$

The phase characteristic of the filter for both M odd and M even is

$$\Theta(\omega) = \begin{cases} \dfrac{\pi}{2} - \omega\left(\dfrac{M-1}{2}\right), & \text{if } H_r(\omega) > 0 \\[2mm] \dfrac{3\pi}{2} - \omega\left(\dfrac{M-1}{2}\right), & \text{if } H_r(\omega) < 0 \end{cases} \tag{8.2.14}$$

These general frequency response formulas can be used to design linear-phase FIR filters with symmetric and antisymmetric unit sample responses. We note that, for a symmetric $h(n)$, the number of filter coefficients that specify the frequency response is $(M+1)/2$ when M is odd or $M/2$ when M is even. On the other hand, if the unit sample response is antisymmetric,

$$h\left(\frac{M-1}{2}\right) = 0$$

so that there are $(M-1)/2$ filter coefficients when M is odd and $M/2$ coefficients when M is even to be specified.

The choice of a symmetric or antisymmetric unit sample response depends on the application. As we shall see later, a symmetric unit sample response is suitable for some applications, while an antisymmetric unit sample response is more suitable for other applications. For example, if $h(n) = -h(M-1-n)$ and M is odd, (8.2.12) implies that $H_r(0) = 0$ and $H_r(\pi) = 0$. Consequently, (8.2.12) is not suitable as either a lowpass filter or a highpass filter. Similarly, the antisymmetric unit sample response with M even also results in $H_r(0) = 0$, as can be easily verified from (8.2.13). Consequently, we would not use the antisymmetric condition in the design of a lowpass linear-phase FIR filter. On the other hand, the symmetry condition $h(n) = h(M-1-n)$ yields a linear-phase FIR filter with a nonzero response at $\omega = 0$, if desired, that is,

$$H_r(0) = h\left(\frac{M-1}{2}\right) + 2\sum_{n=0}^{(M-3)/2} h(n), \qquad M \text{ odd} \tag{8.2.15}$$

$$H_r(0) = 2\sum_{n=0}^{(M/2)-1} h(n), \qquad M \text{ even} \tag{8.2.16}$$

In summary, the problem of FIR filter design is simply to determine the M coefficients $h(n)$, $n = 0, 1, \ldots, M-1$, from a specification of the desired frequency response $H_d(\omega)$ of the FIR filter. The important parameters in the specification of $H_d(\omega)$ are given in Fig. 8.2.

In the following subsections we describe design methods based on specification of $H_d(\omega)$.

8.2.2 Design of Linear-Phase FIR Filters Using Windows

In this method we begin with the desired frequency response specification $H_d(\omega)$ and determine the corresponding unit sample response $h_d(n)$. Indeed, $h_d(n)$ is related to $H_d(\omega)$ by the Fourier transform relation

$$H_d(\omega) = \sum_{n=0}^{\infty} h_d(n)e^{-j\omega n} \tag{8.2.17}$$

where

$$h_d(n) = \frac{1}{2\pi}\int_{-\pi}^{\pi} H_d(\omega)e^{j\omega n}\,d\omega \tag{8.2.18}$$

Thus, given $H_d(\omega)$, we can determine the unit sample response $h_d(n)$ by evaluating the integral in (8.2.18).

In general, the unit sample response $h_d(n)$ obtained from (8.2.17) is infinite in duration and must be truncated at some point, say at $n = M - 1$, to yield an FIR filter of length M. Truncation of $h_d(n)$ to a length $M - 1$ is equivalent to multiplying $h_d(n)$ by a "rectangular window," defined as

$$w(n) = \begin{cases} 1, & n = 0, 1, \ldots, M - 1 \\ 0, & \text{otherwise} \end{cases} \tag{8.2.19}$$

Thus the unit sample response of the FIR filter becomes

$$\begin{aligned} h(n) &= h_d(n)w(n) \\ &= \begin{cases} h_d(n), & n = 0, 1, \ldots, M - 1 \\ 0, & \text{otherwise} \end{cases} \end{aligned} \tag{8.2.20}$$

It is instructive to consider the effect of the window function on the desired frequency response $H_d(\omega)$. Recall that multiplication of the window function $w(n)$ with $h_d(n)$ is equivalent to convolution of $H_d(\omega)$ with $W(\omega)$, where $W(\omega)$ is the frequency-domain representation (Fourier transform) of the window function, that is,

$$W(\omega) = \sum_{n=0}^{M-1} w(n)e^{-j\omega n} \tag{8.2.21}$$

Thus the convolution of $H_d(\omega)$ with $W(\omega)$ yields the frequency response of the (truncated) FIR filter. That is,

$$H(\omega) = \frac{1}{2\pi} \int_{-\pi}^{\pi} H_d(v)W(\omega - v)dv \tag{8.2.22}$$

The Fourier transform of the rectangular window is

$$\begin{aligned} W(\omega) &= \sum_{n=0}^{M-1} e^{-j\omega n} \\ &= \frac{1 - e^{-j\omega M}}{1 - e^{-j\omega}} = e^{-j\omega(M-1)/2} \frac{\sin(\omega M/2)}{\sin(\omega/2)} \end{aligned} \tag{8.2.23}$$

This window function has a magnitude response

$$|W(\omega)| = \frac{|\sin(\omega M/2)|}{|\sin(\omega/2)|} \qquad -\pi \leq \omega \leq \pi \tag{8.2.24}$$

and a piecewise linear phase

$$\Theta(\omega) = \begin{cases} -\omega\left(\dfrac{M-1}{2}\right), & \text{when } \sin(\omega M/2) \geq 0 \\ -\omega\left(\dfrac{M-1}{2}\right) + \pi, & \text{when } \sin(\omega M/2) < 0 \end{cases} \tag{8.2.25}$$

The magnitude response of the window function is illustrated in Fig. 8.4 for $M = 31$ and 61. The width of the main lobe [width is measured to the first zero of $W(\omega)$]

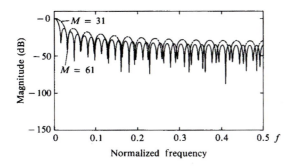

Figure 8.4 Frequency response for rectangular window of lengths (a) $M = 31$, (b) $M = 61$.

is $4\pi/M$. Hence, as M increases, the main lobe becomes narrower. However, the sidelobes of $|W(\omega)|$ are relatively high and remain unaffected by an increase in M. In fact, even though the width of each sidelobe decreases with an increase in M, the height of each sidelobe increases with an increase in M in such a manner that the area under each sidelobe remains invariant to changes in M. This characteristic behavior is not evident from observation of Fig. 8.4 because $W(\omega)$ has been normalized by M such that the normalized peak values of the sidelobes remain invariant to an increase in M.

The characteristics of the rectangular window play a significant role in determining the resulting frequency response of the FIR filter obtained by truncating $h_d(n)$ to length M. Specifically, the convolution of $H_d(\omega)$ with $W(\omega)$ has the effect of smoothing $H_d(\omega)$. As M is increased, $W(\omega)$ becomes narrower, and the smoothing provided by $W(\omega)$ is reduced. On the other hand, the large sidelobes of $W(\omega)$ result in some undesirable ringing effects in the FIR filter frequency response $H(\omega)$, and also in relatively larger sidelobes in $H(\omega)$. These undesirable effects are best alleviated by the use of windows that do not contain abrupt discontinuities in their time-domain characteristics, and have correspondingly low sidelobes in their frequency-domain characteristics.

Table 8.1 lists several window functions that possess desirable frequency response characteristics. Figure 8.5 illustrates the time-domain characteristics of the windows. The frequency response characteristics of the Hanning, Hamming, and Blackman windows are illustrated in Figs. 8.6 through 8.8. All of these window functions have significantly lower sidelobes compared with the rectangular window. However, for the same value of M, the width of the main lobe is also wider for these windows compared to the rectangular window. Consequently, these window functions provide more smoothing through the convolution operation in the frequency domain, and as a result, the transition region in the FIR filter response is wider. To reduce the width of this transition region, we can simply increase the length of the window which results in a larger filter. Table 8.2 summarizes these important frequency-domain features of the various window functions.

The window technique is best described in terms of a specific example. Suppose that we want to design a symmetric lowpass linear-phase FIR filter having a

TABLE 8.1   WINDOW FUNCTIONS FOR FIR FILTER DESIGN

Name of window	Time-domain sequence, $h(n), 0 \leq n \leq M - 1$
Bartlett (triangular)	$1 - \dfrac{2\left\|n - \dfrac{M-1}{2}\right\|}{M-1}$
Blackman	$0.42 - 0.5\cos\dfrac{2\pi n}{M-1} + 0.08\cos\dfrac{4\pi n}{M-1}$
Hamming	$0.54 - 0.46\cos\dfrac{2\pi n}{M-1}$
Hanning	$\dfrac{1}{2}\left(1 - \cos\dfrac{2\pi n}{M-1}\right)$
Kaiser	$\dfrac{I_0\left[\alpha\sqrt{\left(\dfrac{M-1}{2}\right)^2 - \left(n - \dfrac{M-1}{2}\right)^2}\right]}{I_0\left[\alpha\left(\dfrac{M-1}{2}\right)\right]}$
Lanczos	$\left\{\dfrac{\sin\left[2\pi\left(n - \dfrac{M-1}{2}\right)\middle/(M-1)\right]}{2\pi\left(n - \dfrac{M-1}{2}\right)\middle/\left(\dfrac{M-1}{2}\right)}\right\}^L \qquad L > 0$
Tukey	$1, \left\|n - \dfrac{M-1}{2}\right\| \leq \alpha\dfrac{M-1}{2} \qquad 0 < \alpha < 1$ $\dfrac{1}{2}\left[1 + \cos\left(\dfrac{n - (1+a)(M-1)/2}{(1-\alpha)(M-1)/2}\pi\right)\right]$ $\alpha(M-1)/2 \leq \left\|n - \dfrac{M-1}{2}\right\| \leq \dfrac{M-1}{2}$

desired frequency response

$$H_d(\omega) = \begin{cases} 1e^{-j\omega(M-1)/2}, & 0 \leq |\omega| \leq \omega_c \\ 0, & \text{otherwise} \end{cases} \qquad (8.2.26)$$

A delay of $(M - 1)/2$ units is incorporated into $H_d(\omega)$ in anticipation of forcing the filter to be of length M. The corresponding unit sample response, obtained by evaluating the integral in (8.2.18), is

$$h_d(n) = \frac{1}{2\pi}\int_{-\omega_c}^{\omega_c} e^{j\omega\left(n - \frac{M-1}{2}\right)}d\omega$$

$$= \frac{\sin\omega_c\left(n - \dfrac{M-1}{2}\right)}{\pi\left(n - \dfrac{M-1}{2}\right)} \qquad n \neq \frac{M-1}{2} \qquad (8.2.27)$$

Clearly, $h_d(n)$ is noncausal and infinite in duration.

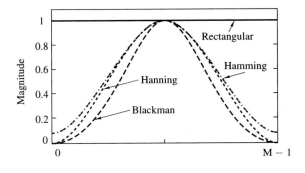

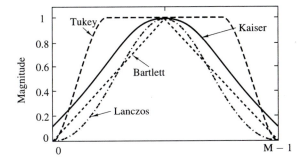

Figure 8.5 Shapes of several window functions.

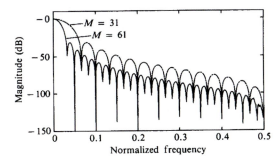

Figure 8.6 Frequency responses of Hanning window for (a) $M = 31$ and (b) $M = 61$.

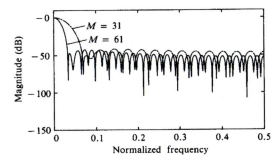

Figure 8.7 Frequency responses for Hamming window for (a) $M = 31$ and (b) $M = 61$.

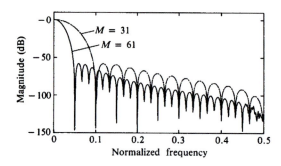

Figure 8.8 Frequency responses for Blackman window for (a) $M = 31$ and (b) $M = 61$.

TABLE 8.2 IMPORTANT FREQUENCY-DOMAIN CHARACTERISTICS OF SOME WINDOW FUNCTIONS

Type of window	Approximate transition width of main lobe	Peak sidelobe (dB)
Rectangular	$4\pi/M$	-13
Bartlett	$8\pi/M$	-27
Hanning	$8\pi/M$	-32
Hamming	$8\pi/M$	-43
Blackman	$12\pi/M$	-58

If we multiply $h_d(n)$ by the rectangular window sequence in (8.2.19), we obtain an FIR filter of length M having the unit sample response

$$h(n) = \frac{\sin \omega_c \left(n - \dfrac{M-1}{2} \right)}{\pi \left(n - \dfrac{M-1}{2} \right)} \qquad 0 \le n \le M-1 \quad n \ne \frac{M-1}{2} \qquad (8.2.28)$$

If M is selected to be odd, the value of $h(n)$ at $n = (M-1)/2$ is

$$h\left(\frac{M-1}{2} \right) = \frac{\omega_c}{\pi} \qquad (8.2.29)$$

The magnitude of the frequency response $H(\omega)$ of this filter is illustrated in Fig. 8.9 for $M = 61$ and $M = 101$. We observe that relatively large oscillations or ripples occur near the band edge of the filter. The oscillations increase in frequency as M increases, but they do not diminish in amplitude. As indicated previously, these large oscillations are the direct result of the large sidelobes existing in the frequency characteristic $W(\omega)$ of the rectangular window. As this window function is convolved with the desired frequency response characteristic $H_d(\omega)$, the oscillations occur as the large constant area sidelobes of $W(\omega)$ move across the discontinuity that exists in $H_d(\omega)$. Since (8.2.17) is basically a Fourier series representation of $H_d(\omega)$, the multiplication of $h_d(n)$ with a rectangular window is identical to truncating the Fourier series representation of the desired filter char-

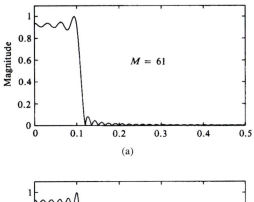

(a)

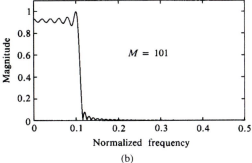

Normalized frequency

(b)

Figure 8.9 Lowpass filter designed with a rectangular window (a) $M = 61$ and (b) $M = 101$.

acteristic $H_d(\omega)$. The truncation of the Fourier series is known to introduce ripples in the frequency response characteristic $H(\omega)$ due to the nonuniform convergence of the Fourier series at a discontinuity. The oscillatory behavior near the band edge of the filter is called the *Gibbs phenomenon.*

To alleviate the presence of large oscillations in both the passband and the stopband, we should use a window function that contains a taper and decays toward zero gradually, instead of abruptly, as it occurs in a rectangular window. Figures 8.10 through 8.13, illustrate the frequency response of the resulting filter when some of the window functions listed in Table 8.1 are used to taper $h_d(n)$. As illustrated in Figs. 8.10 through 8.13, the window functions do indeed eliminate the ringing effects at the band edge and do result in lower sidelobes at the expense of an increase in the width of the transition band of the filter.

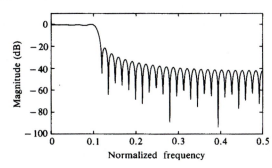

Normalized frequency

Figure 8.10 Lowpass FIR filter designed with rectangular window ($M = 61$).

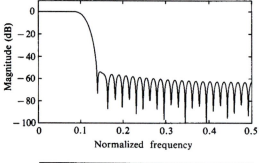

Figure 8.11 Lowpass FIR filter designed with Hamming window ($M = 61$).

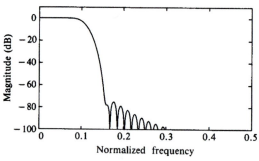

Figure 8.12 Lowpass FIR filter designed with Blackman window ($M = 61$).

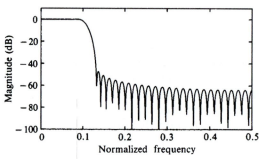

Figure 8.13 Lowpass FIR filter designed with $\alpha = 4$ Kaiser window ($M = 61$).

8.2.3 Design of Linear-Phase FIR Filters by the Frequency-Sampling Method

In the frequency sampling method for FIR filter design, we specify the desired frequency response $H_d(\omega)$ at a set of equally spaced frequencies, namely

$$\omega_k = \tfrac{2\pi}{M}(k + \alpha) \qquad k = 0, 1, \ldots, \frac{M-1}{2} \quad M \text{ odd}$$

$$k = 0, 1, \ldots, \frac{M}{2} - 1 \quad M \text{ even} \qquad (8.2.30)$$

$$\alpha = 0 \quad \text{or} \quad \tfrac{1}{2}$$

and solve for the unit sample response $h(n)$ of the FIR filter from these equally spaced frequency specifications. To reduce sidelobes, it is desirable to optimize the frequency specification in the transition band of the filter. This optimization can be accomplished numerically on a digital computer by means of linear programming techniques as shown by Rabiner et al. (1970).

In this section we exploit a basic symmetry property of the sampled frequency response function to simplify the computations. Let us begin with the desired frequency response of the FIR filter, which is [for simplicity, we drop the subscript in $H_d(\omega)$],

$$H(\omega) = \sum_{n=0}^{M-1} h(n)e^{-j\omega n} \tag{8.2.31}$$

Suppose that we specify the frequency response of the filter at the frequencies given by (8.2.30). Then from (8.2.31) we obtain

$$H(k + \alpha) \equiv H\left(\frac{2\pi}{M}(k + \alpha)\right)$$

$$H(k + \alpha) \equiv \sum_{n=0}^{M-1} h(n)e^{-j2\pi(k+\alpha)n/M} \qquad k = 0, 1, \ldots, M - 1 \tag{8.2.32}$$

It is a simple matter to invert (8.2.32) and express $h(n)$ in terms of $H(k+\alpha)$. If we multiply both sides of (8.2.32) by the exponential, $\exp(j2\pi km/M)$, $m = 0$, $1, \ldots, M - 1$, and sum over $k = 0, 1, \ldots, M - 1$, the right-hand side of (8.2.32) reduces to $Mh(m)\exp(-j2\pi\alpha m/M)$. Thus we obtain

$$h(n) = \frac{1}{M}\sum_{k=0}^{M-1} H(k + \alpha)e^{j2\pi(k+\alpha)n/M} \qquad n = 0, 1, \ldots, M - 1 \tag{8.2.33}$$

The relationship in (8.2.33) allows us to compute the values of the unit sample response $h(n)$ from the specification of the frequency samples $H(k + \alpha)$, $k = 0$, $1, \ldots, M - 1$. Note that when $\alpha = 0$, (8.2.32) reduces to the discrete Fourier transform (DFT) of the sequence $\{h(n)\}$ and (8.2.33) reduces to the inverse DFT (IDFT).

Since $\{h(n)\}$ is real, we can easily show that the frequency samples $\{H(k+\alpha)\}$ satisfy the symmetry condition

$$H(k + \alpha) = H^*(M - k - \alpha) \tag{8.2.34}$$

This symmetry condition, along with the symmetry conditions for $\{h(n)\}$, can be used to reduce the frequency specifications from M points to $(M+1)/2$ points for M odd and $M/2$ points for M even. Thus the linear equations for determining $\{h(n)\}$ from $\{H(k + \alpha)\}$ are considerably simplified.

In particular, if (8.2.11) is sampled at the frequencies $\omega_k = 2\pi(k + \alpha)/M$, $k = 0, 1, \ldots, M - 1$, we obtain

$$H(k + \alpha) = H_r\left(\frac{2\pi}{M}(k + \alpha)\right) e^{j[\beta\pi/2 - 2\pi(k+\alpha)(M-1)/2M]} \tag{8.2.35}$$

where $\beta = 0$ when $\{h(n)\}$ is symmetric and $\beta = 1$ when $\{h(n)\}$ is antisymmetric. A simplication occurs by defining a set of real frequency samples $\{G(k+m)\}$

$$G(k+\alpha) = (-1)^k H_r\left(\frac{2\pi}{M}(k+\alpha)\right) \qquad k = 0, 1, \ldots, M-1 \qquad (8.2.36)$$

We use (8.2.36) in (8.2.35) to eliminate $H_r(\omega_k)$. Thus we obtain

$$H(k+\alpha) = G(k+\alpha)e^{j\pi k}e^{j[\beta\pi/2 - 2\pi(k+\alpha)(M-1)/2M]} \qquad (8.2.37)$$

Now the symmetry condition for $H(k+\alpha)$ given in (8.2.34) translates into a corresponding symmetry condition for $G(k+\alpha)$, which can be exploited by substituting into (8.2.33), to simplify the expressions for the FIR filter impulse response $\{h(n)\}$ for the four cases $\alpha = 0$, $\alpha = \frac{1}{2}$, $\beta = 0$, and $\beta = 1$. The results are summarized in Table 8.3. The detailed derivations are left as exercises for the reader.

Although the frequency sampling method provides us with another means for designing linear-phase FIR filters, its major advantage lies in the efficient frequency sampling structure, which is obtained when most of the frequency samples are zero, as demonstrated in Section 7.2.3.

The following examples illustrate the design of linear-phase FIR filters based on the frequency sampling method. The optimum values for the samples in the transition band are obtained from the tables in Appendix C which are taken from the paper by Rabiner et al. (1970).

Example 8.2.1

Determine the coefficients of a linear-phase FIR filter of length $M = 15$ which has a symmetric unit sample response and a frequency response that satisfies the conditions

$$H_r\left(\frac{2\pi k}{15}\right) = \begin{cases} 1, & k = 0, 1, 2, 3 \\ 0.4, & k = 4 \\ 0, & k = 5, 6, 7 \end{cases}$$

Solution   Since $h(n)$ is symmetric and the frequencies are selected to correspond to the case $\alpha = 0$, we use the corresponding formula in Table 8.3 to evaluate $h(n)$. In this case

$$G(k) = (-1)^k H_r\left(\frac{2\pi k}{15}\right) \qquad k = 0, 1, \ldots, 7$$

The result of this computation is

$$
\begin{aligned}
h(0) &= h(14) = -0.014112893 \\
h(1) &= h(13) = -0.001945309 \\
h(2) &= h(12) = 0.04000004 \\
h(3) &= h(11) = 0.01223454 \\
h(4) &= h(10) = -0.09138802 \\
h(5) &= h(9) = -0.01808986 \\
h(6) &= h(8) = 0.3133176 \\
h(7) & = 0.52
\end{aligned}
$$

TABLE 8.3   UNIT SAMPLE RESPONSE: $h(n) = \pm h(M - 1 - n)$

Symmetric

$$H(k) = G(k)e^{j\pi k/M} \qquad k = 0, 1, \ldots, M - 1$$

$$G(k) = (-1)^k H_r\left(\frac{2\pi k}{M}\right) \qquad G(k) = -G(M - k)$$

$\alpha = 0$

$$h(n) = \frac{1}{M}\left\{G(0) + 2\sum_{k=1}^{U} G(k)\cos\frac{2\pi k}{M}\left(n + \tfrac{1}{2}\right)\right\}$$

$$U = \begin{cases} \dfrac{M - 1}{2}, & M \text{ odd} \\[2mm] \dfrac{M}{2} - 1, & M \text{ even} \end{cases}$$

$$H\left(k + \tfrac{1}{2}\right) = G\left(k + \tfrac{1}{2}\right)e^{-j\pi/2}e^{j\pi(2k+1)/2M}$$

$\alpha = \tfrac{1}{2}$

$$G\left(k + \tfrac{1}{2}\right) = (-1)^k H_r\left[\frac{2\pi}{M}\left(k + \tfrac{1}{2}\right)\right]$$

$$G\left(k + \tfrac{1}{2}\right) = G\left(M - k - \tfrac{1}{2}\right)$$

$$h(n) = \frac{2}{M}\sum_{k=0}^{U} G\left(k + \tfrac{1}{2}\right)\sin\frac{2\pi}{M}\left(k + \tfrac{1}{2}\right)\left(n + \tfrac{1}{2}\right)$$

Antisymmetric

$$H(k) = G(k)e^{j\pi/2}e^{j\pi k/M} \qquad k = 0, 1, \ldots, M - 1$$

$$G(k) = (-1)^k H_r\left(\frac{2\pi k}{M}\right) \qquad G(k) = G(M - k)$$

$\alpha = 0$

$$h(n) = -\frac{2}{M}\sum_{k=1}^{(M-1)/2} G(k)\sin\frac{2\pi k}{M}\left(n + \tfrac{1}{2}\right) \qquad M \text{ odd}$$

$$h(n) = \frac{1}{M}\left\{(-1)^{n+1}G(M/2) - 2\sum_{k=1}^{(M/2)-1} G(k)\sin\frac{2\pi}{M}k\left(n + \tfrac{1}{2}\right)\right. \qquad M \text{ even}$$

$$H\left(k + \tfrac{1}{2}\right) = G\left(k + \tfrac{1}{2}\right)e^{j\pi(2k+1)/2M}$$

$\alpha = \tfrac{1}{2}$

$$G\left(k + \tfrac{1}{2}\right) = (-1)^k H_r\left[\frac{2\pi}{M}\left(k + \tfrac{1}{2}\right)\right]$$

$$G\left(k + \tfrac{1}{2}\right) = -G\left(M - k - \tfrac{1}{2}\right); G(M/2) = 0 \text{ for } M \text{ odd}$$

$$h(n) = \frac{2}{M}\sum_{k=0}^{V} G\left(k + \tfrac{1}{2}\right)\cos\frac{2\pi}{M}\left(k + \tfrac{1}{2}\right)\left(n + \tfrac{1}{2}\right)$$

$$V = \begin{cases} \dfrac{M - 3}{2}, & M \text{ odd} \\[2mm] \dfrac{M}{2} - 1, & M \text{ even} \end{cases}$$

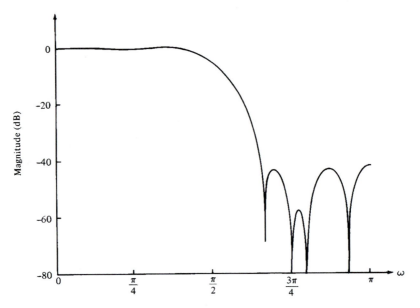

Figure 8.14   Frequency response of linear-phase FIR filter in Example 8.2.1.

The frequency response characteristic of this filter is shown in Fig. 8.14. We should emphasize that $H_r(\omega)$ is exactly equal to the values given by the specifications above at $\omega_k = 2\pi k/15$.

Example 8.2.2

Determine the coefficients of a linear-phase FIR filter of length $M = 32$ which has a symmetric unit sample response and a frequency response that satisfies the condition

$$H_r\left(\frac{2\pi(k+\alpha)}{32}\right) = \begin{cases} 1, & k = 0, 1, 2, 3, 4, 5 \\ T_1, & k = 6 \\ 0, & k = 7, 8, \ldots, 15 \end{cases}$$

where $T_1 = 0.3789795$ for $\alpha = 0$, and $T_1 = 0.3570496$ for $\alpha = \frac{1}{2}$. These values of T_1 were obtained from the tables of optimum transition parameters given in Appendix C.

Solution   The appropriate equations for this computation are given in Table 8.3 for $\alpha = 0$ and $\alpha = \frac{1}{2}$. These computations yield the unit sample responses shown in Table 8.4. The corresponding frequency response characteristics are illustrated in Figs. 8.15 and 8.16, respectively. Note that the bandwidth of the filter for $\alpha = \frac{1}{2}$ is wider than that for $\alpha = 0$.

The optimization of the frequency samples in the transition region of the frequency response can be explained by evaluating the system function $H(z)$, given by (7.2.12), on the unit circle and using the relationship in (8.2.37) to express $H(\omega)$

TABLE 8.4

M = 32	M = 32
ALPHA = 0.	ALPHA = 0.5
T1 = 0.3789795E+00	T1 = 0.3570496E+00

h(0) = -0.7141978E-02	h(0) = -0.4089120E-02
h(1) = -0.3070801E-02	h(1) = -0.9973779E-02
h(2) = 0.5891327E-02	h(2) = -0.7379891E-02
h(3) = 0.1349923E-01	h(3) = 0.5949799E-02
h(4) = 0.8087033E-02	h(4) = 0.1727056E-01
h(5) = -0.1107258E-01	h(5) = 0.7878412E-02
h(6) = -0.2420687E-01	h(6) = -0.1798590E-01
h(7) = -0.9446550E-02	h(7) = -0.2670584E-01
h(8) = 0.2544464E-01	h(8) = 0.3778549E-02
h(9) = 0.3985050E-01	h(9) = 0.4191022E-01
h(10) = 0.2753036E-02	h(10) = 0.2839344E-01
h(11) = -0.5913959E-01	h(11) = -0.4163144E-01
h(12) = -0.6841660E-01	h(12) = -0.8254962E-01
h(13) = 0.3175741E-01	h(13) = 0.2802212E-02
h(14) = 0.2080981E+00	h(14) = 0.2013655E+00
h(15) = 0.3471138E+00	h(15) = 0.3717532E+00

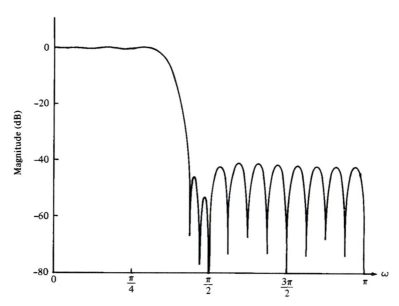

Figure 8.15   Frequency response of linear-phase FIR filter in Example 8.2.2 ($M = 32$ and $\alpha = 0$).

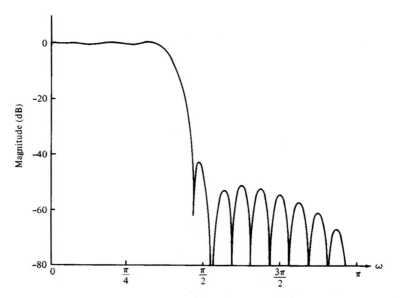

Figure 8.16 Frequency response of linear-phase FIR filter in Example 8.2.2 ($M = 32$ and $\alpha = \frac{1}{2}$).

in terms of $G(k + \alpha)$. Thus for the symmetric filter we obtain

$$H(\omega) = \left\{ \frac{\sin\left(\dfrac{\omega M}{2} - \pi\alpha\right)}{M} \sum_{k=0}^{M-1} \frac{G(k+\alpha)}{\sin\left[\dfrac{\omega}{2} - \dfrac{\pi}{M}(k+\alpha)\right]} \right\} e^{-j\omega(M-1)/2} \qquad (8.2.38)$$

where

$$G(k+\alpha) = \begin{cases} -G(M-k), & \alpha = 0 \\ G(M-k-\frac{1}{2}), & \alpha = \frac{1}{2} \end{cases} \qquad (8.2.39)$$

Similarly, for the antisymmetric linear-phase FIR filter we obtain

$$H(\omega) = \left\{ \frac{\sin\left(\dfrac{\omega M}{2} - \pi\alpha\right)}{M} \sum_{k=0}^{M-1} \frac{G(k+\alpha)}{\sin\left[\dfrac{\omega}{2} - \dfrac{\pi}{M}(k+\alpha)\right]} \right\} e^{-j\omega(M-1)/2} e^{j\pi/2} \qquad (8.2.40)$$

where

$$G(k+\alpha) = \begin{cases} G(M-k), & \alpha = 0 \\ -G\left(M-k-\frac{1}{2}\right), & \alpha = \frac{1}{2} \end{cases} \qquad (8.2.41)$$

With these expressions for the frequency response $H(\omega)$ given in terms of the desired frequency samples $\{G(k+\alpha)\}$, we can easily explain the method for selecting the parameters $\{G(k + \alpha)\}$ in the transition band which result in minimizing the peak sidelobe in the stopband. In brief, the values of $G(k+\alpha)$ in the passband are set to $(-1)^k$ and those in the stopband are set to zero. For any choice of $G(k + \alpha)$ in the transition band, the value of $H(\omega)$ is computed at a dense set of frequencies (e.g., at $\omega_n = 2\pi n/K$, $n = 0, 1, \ldots, K - 1$, where, for example, $K = 10M$). The value of the maximum sidelobe is determined, and the values of the parameters $\{G(k + \alpha)\}$ in the transition band are changed in a direction of steepest descent, which, in effect, reduces the maximum sidelobe. The computation of $H(\omega)$ is now repeated with the new choice of $\{G(k + \alpha)\}$. The maximum sidelobe of $H(\omega)$ is again determined and the values of the parameters $\{G(k + \alpha)\}$ in the transition band are adjusted in a direction of steepest descent that, in turn, reduces the sidelobe. This interactive process is performed until it converges to the optimum choice of the parameters $\{G(k + \alpha)\}$ in the transition band.

There is a potential problem in the frequency-sampling realization of the FIR linear-phase filter. The frequency sampling realization of the FIR filter introduces poles and zeros at equally spaced points on the unit circle. In the ideal situation, the zeros cancel the poles and, consequently, the actual zeros of $H(z)$ are determined by the selection of the frequency samples $\{H(k + \alpha)\}$. In a practical implementation of the frequency-sampling realization, however, quantization effects preclude a perfect cancellation of the poles and zeros. In fact, the location of poles on the unit circle provide no damping of the round-off noise that is introduced in the computations. As a result, such noise tends to increase with time and, ultimately, may destroy the normal operation of the filter.

To mitigate this problem, we can move both the poles and zeros from the unit circle to a circle just inside the unit circle, say at radius $r = 1 - \epsilon$, where ϵ is a very small number. Thus the system function of the linear-phase FIR filter becomes

$$H(z) = \frac{1 - r^M z^{-M} e^{j2\pi\alpha}}{M} \sum_{k=0}^{M-1} \frac{H(k + \alpha)}{1 - re^{j2\omega\pi(k+\alpha)/M} z^{-1}} \qquad (8.2.42)$$

The corresponding two-pole filter realization given in Section 7.2.3 can be modified accordingly. The damping provided by selecting $r < 1$ ensures that roundoff noise will be bounded and thus instability is avoided.

8.2.4 Design of Optimum Equiripple Linear-Phase FIR Filters

The window method and the frequency-sampling method are relatively simple techniques for designing linear-phase FIR filters. However, they also possess some minor disadvantages, described in Section 8.2.6, which may render them undesirable for some applications. A major problem is the lack of precise control of the critical frequencies such as ω_p and ω_s.

The filter design method described in this section is formulated as a Chebyshev approximation problem. It is viewed as an optimum design criterion in the sense that the weighted approximation error between the desired frequency response and the actual frequency response is spread evenly across the passband

and evenly across the stopband of the filter minimizing the maximum error. The resulting filter designs have ripples in both the passband and the stopband.

To describe the design procedure, let us consider the design of a lowpass filter with passband edge frequency ω_p and stopband edge frequency ω_s. From the general specifications given in Fig. 8.2, in the passband, the filter frequency response satisfies the condition

$$1 - \delta_1 \le H_r(\omega) \le 1 + \delta_1 \qquad |\omega| \le \omega_p \tag{8.2.43}$$

Similarly, in the stopband, the filter frequency response is specified to fall between the limits $\pm\delta_2$, that is,

$$-\delta_2 \le H_r(\omega) \le \delta_2 \qquad |\omega| > \omega_s \tag{8.2.44}$$

Thus δ_1 represents the ripple in the passband and δ_2 represents the attenuation or ripple in the stopband. The remaining filter parameter is M, the filter length or the number of filter coefficients.

Let us focus on the four different cases that result in a linear-phase FIR filter. These cases were treated in Section 8.2.2 and are summarized below.

Case 1: Symmetric unit sample response $h(n) = h(M-1-n)$ and M Odd.
In this case, the real-valued frequency response characteristic $H_r(\omega)$ is

$$H_r(\omega) = h\left(\frac{M-1}{2}\right) + 2\sum_{n=0}^{(M-3)/2} h(n) \cos\omega\left(\frac{M-1}{2} - n\right) \tag{8.2.45}$$

If we let $k = (M-1)/2 - n$ and define a new set of filter parameters $\{a(k)\}$ as

$$a(k) = \begin{cases} h\left(\dfrac{M-1}{2}\right), & k = 0 \\ 2h\left(\dfrac{M-1}{2} - k\right), & k = 1, 2, \ldots, \dfrac{M-1}{2} \end{cases} \tag{8.2.46}$$

then (8.2.45) reduces to the compact form

$$H_r(\omega) = \sum_{k=0}^{(M-1)/2} a(k) \cos\omega k \tag{8.2.47}$$

Case 2: Symmetric unit sample response $h(n) = h(M-1-n)$ and M Even.
In this case, $H_r(\omega)$ is expressed as

$$H_r(\omega) = 2\sum_{n=0}^{(M/2)-1} h(n) \cos\omega\left(\frac{M-1}{2} - n\right) \tag{8.2.48}$$

Again, we change the summation index from n to $k = M/2 - n$ and define a new set of filter parameters $\{b(k)\}$ as

$$b(k) = 2h\left(\frac{M}{2} - k\right), k = 1, 2, \ldots, M/2 \tag{8.2.49}$$

With these substitutions (8.2.48) becomes

$$H_r(\omega) = \sum_{k=1}^{M/2} b(k) \cos \omega \left(k - \frac{1}{2} \right) \tag{8.2.50}$$

In carrying out the optimization, it is convenient to rearrange (8.2.50) further into the form

$$H_r(\omega) = \cos \frac{\omega}{2} \sum_{k=0}^{(M/2)-1} \tilde{b}(k) \cos \omega k \tag{8.2.51}$$

where the coefficients $\{\tilde{b}(k)\}$ are linearly related to the coefficients $\{b(k)\}$. In fact, it can be shown that the relationship is

$$\tilde{b}(0) = \tfrac{1}{2} b(1)$$

$$\tilde{b}(k) = 2b(k) - \tilde{b}(k-1) \qquad k = 1, 2, 3, \dots, \frac{M}{2} - 2 \tag{8.2.52}$$

$$\tilde{b}\left(\frac{M}{2} - 1 \right) = 2b \left(\frac{M}{2} \right)$$

Case 3: Antisymmetric unit sample response $h(n) = -h(M-1-n)$ and M Odd.   The real-valued frequency response characteristic $H_r(\omega)$ for this case is

$$H_r(\omega) = 2 \sum_{n=0}^{(M-3)/2} h(n) \sin \omega \left(\frac{M-1}{2} - n \right) \tag{8.2.53}$$

If we change the summation in (8.2.53) from n to $k = (M-1)/2 - n$ and define a new set of filter parameters $\{c(k)\}$ as

$$c(k) = 2h \left(\frac{M-1}{2} - k \right) \qquad k = 1, 2, \dots, (M-1)/2 \tag{8.2.54}$$

then (8.2.53) becomes

$$H_r(\omega) = \sum_{k=1}^{(M-1)/2} c(k) \sin \omega k \tag{8.2.55}$$

As in the previous case, it is convenient to rearrange (8.2.55) into the form

$$H_r(\omega) = \sin \omega \sum_{k=0}^{(M-3)/2} \tilde{c}(k) \cos \omega k \tag{8.2.56}$$

where the coefficients $\{\tilde{c}(k)\}$ are linearly related to the parameters $\{c(k)\}$. This desired relationship can be derived from (8.2.55) and (8.2.56) and is simply given as

$$\tilde{c}\left(\frac{M-3}{2}\right) = c\left(\frac{M-1}{2}\right)$$

$$\tilde{c}\left(\frac{M-5}{2}\right) = 2c\left(\frac{M-3}{2}\right)$$

$$\vdots \qquad\qquad \vdots \qquad\qquad\qquad (8.2.57)$$

$$\tilde{c}(k-1) - \tilde{c}(k+1) = 2c(k) \qquad 2 \le k \le \frac{M-5}{2}$$

$$\tilde{c}(0) + \tfrac{1}{2}\tilde{c}(2) = c(1)$$

Case 4: Antisymmetric unit sample response $h(n) = -h(M-1-n)$ and M Even. In this case, the real-valued frequency response characteristic $H_r(\omega)$ is

$$H_r(\omega) = 2 \sum_{n=0}^{(M/2)-1} h(n) \sin \omega \left(\frac{M-1}{2} - n\right) \qquad (8.2.58)$$

A change in the summation index from n to $k = M/2 - n$ combined with a definition of a new set of filter coefficients $\{d(k)\}$, related to $\{h(n)\}$ according to

$$d(k) = 2h\left(\frac{M}{2} - k\right) \qquad k = 1, 2, \ldots, \frac{M}{2} \qquad (8.2.59)$$

results in the expression

$$H_r(\omega) = \sum_{k=1}^{M/2} d(k) \sin \omega \left(k - \frac{1}{2}\right) \qquad (8.2.60)$$

As in the previous two cases, we find it convenient to rearrange (8.2.60) into the form

$$H_r(\omega) = \sin \frac{\omega}{2} \sum_{k=0}^{(M/2)-1} \tilde{d}(k) \cos \omega k \qquad (8.2.61)$$

where the new filter parameters $\{\tilde{d}(k)\}$ are related to $\{d(k)\}$ as follows:

$$\tilde{d}\left(\frac{M}{2} - 1\right) = 2d\left(\frac{M}{2}\right)$$

$$\tilde{d}(k-1) - \tilde{d}(k) = 2d(k) \qquad 2 \le k \le \frac{M}{2} - 1 \qquad (8.2.62)$$

$$\tilde{d}(0) - \tfrac{1}{2}\tilde{d}(1) = d(1)$$

The expressions for $H_r(\omega)$ in these four cases are summarized in Table 8.5. We note that the rearrangements that we made in cases 2, 3, and 4 have allowed

TABLE 8.5 REAL-VALUED FREQUENCY RESPONSE
FUNCTIONS FOR LINEAR-PHASE FIR FILTERS

Filter type	$Q(\omega)$	$P(\omega)$
$h(n) = h(M - 1 - n)$ M odd (case 1)	1	$\displaystyle\sum_{k=0}^{(M-1)/2} a(k)\cos\omega k$
$h(n) = h(M - 1 - n)$ M even (case 2)	$\cos\dfrac{\omega}{2}$	$\displaystyle\sum_{k=0}^{(M/2)-1} \tilde{b}(k)\cos\omega k$
$h(n) = -h(M - 1 - n)$ M odd (case 3)	$\sin\omega$	$\displaystyle\sum_{k=0}^{(M-3)/2} \tilde{c}(k)\cos\omega k$
$h(n) = -h(M - 1 - n)$ M even (case 4)	$\sin\dfrac{\omega}{2}$	$\displaystyle\sum_{k=0}^{(M/2)-1} \tilde{d}(k)\cos\omega k$

us to express $H_r(\omega)$ as

$$H_r(\omega) = Q(\omega)P(\omega) \tag{8.2.63}$$

where

$$Q(\omega) = \begin{cases} 1 & \text{case 1} \\ \cos\dfrac{\omega}{2} & \text{case 2} \\ \sin\omega & \text{case 3} \\ \sin\dfrac{\omega}{2} & \text{case 4} \end{cases} \tag{8.2.64}$$

and $P(\omega)$ has the common form

$$P(\omega) = \sum_{k=0}^{L} \alpha(k)\cos\omega k \tag{8.2.65}$$

with $\{\alpha(k)\}$ representing the parameters of the filter, which are linearly related to the unit sample response $h(n)$ of the FIR filter. The upper limit L in the sum is $L = (M - 1)/2$ for Case 1, $L = (M - 3)/2$ for Case 3, and $L = M/2 - 1$ for Case 2 and Case 4.

In addition to the common framework given above for the representation of $H_r(\omega)$, we also define the real-valued desired frequency response $H_{dr}(\omega)$ and the weighting function $W(\omega)$ on the approximation error. The real-valued desired frequency response $H_{dr}(\omega)$ is simply defined to be unity in the passband and zero in the stopband. For example, Fig. 8.17 illustrates several different types of characteristics for $H_{dr}(\omega)$. The weighting function on the approximation error allows us to choose the relative size of the errors in the different frequency bands (i.e., in the passband and in the stopband). In particular, it is convenient to normalize $W(\omega)$ to unity in the stopband and set $W(\omega) = \delta_2/\delta_1$ in the passband,

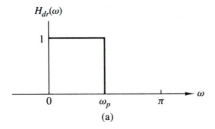

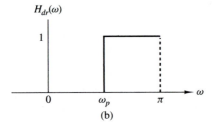

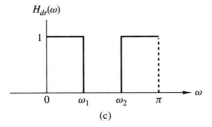

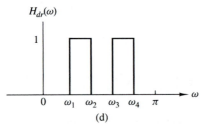

Figure 8.17 Desired frequency response characteristics for different types of filters.

that is,

$$W(\omega) = \begin{cases} \delta_2/\delta_1, & \omega \text{ in the passband} \\ 1, & \omega \text{ in the stopband} \end{cases} \tag{8.2.66}$$

Then we simply select $W(\omega)$ in the passband to reflect our emphasis on the relative size of the ripple in the stopband to the ripple in the passband.

With the specification of $H_{dr}(\omega)$ and $W(\omega)$, we can now define the weighted approximation error as

$$E(\omega) = W(\omega)[H_{dr}(\omega) - H_r(\omega)]$$
$$= W(\omega)[H_{dr}(\omega) - Q(\omega)P(\omega)]$$

$$= W(\omega) Q(\omega) \left[\frac{H_{dr}(\omega)}{Q(\omega)} - P(\omega) \right] \qquad (8.2.67)$$

For mathematical convenience, we define a modified weighting function $\hat{W}(\omega)$ and a modified desired frequency response $\hat{H}_{dr}(\omega)$ as

$$\hat{W}(\omega) = W(\omega) Q(\omega)$$

$$\hat{H}_{dr}(\omega) = \frac{H_{dr}(\omega)}{Q(\omega)} \qquad (8.2.68)$$

Then the weighted approximation error may be expressed as

$$E(\omega) = \hat{W}(\omega)[\hat{H}_{dr}(\omega) - P(\omega)] \qquad (8.2.69)$$

for all four different types of linear-phase FIR filters.

Given the error function $E(\omega)$, the Chebyshev approximation problem is basically to determine the filter parameters $\{\alpha(k)\}$ that minimize the maximum absolute value of $E(\omega)$ over the frequency bands in which the approximation is to be performed. In mathematical terms, we seek the solution to the problem

$$\min_{\text{over } \{\alpha(k)\}} \left[\max_{\omega \in S} |E(\omega)| \right] = \min_{\text{over } \{a(k)\}} \left[\max_{\omega \in S} |\hat{W}(\omega)[\hat{H}_{dr}(\omega) - \sum_{k=0}^{L} \alpha(k) \cos \omega k]| \right]$$

$$(8.2.70)$$

where S represents the set (disjoint union) of frequency bands over which the optimization is to be performed. Basically, the set S consists of the passbands and stopbands of the desired filter.

The solution to this problem is due to Parks and McClellan (1972a), who applied a theorem in the theory of Chebyshev approximation. It is called the *alternation theorem*, which we state without proof.

Alternation Theorem: Let S be a compact subset of the interval $[0, \pi)$. A necessary and sufficient condition for

$$P(\omega) = \sum_{k=0}^{L} \alpha(k) \cos \omega k$$

to be the unique, best weighted Chebyshev approximation to $\hat{H}_{dr}(\omega)$ in S, is that the error function $E(\omega)$ exhibit at least $L + 2$ extremal frequencies in S. That is, there must exist at least $L+2$ frequencies $\{\omega_i\}$ in S such that $\omega_1 < \omega_2 < \cdots < \omega_{L+2}$, $E(\omega_i) = -E(\omega_{i+1})$, and

$$|E(\omega_i)| = \max_{\omega \in S} |E(\omega)| \qquad i = 1, 2, \ldots, L + 2$$

We note that the error function $E(\omega)$ alternates in sign between two successive extremal frequencies. Hence the theorem is called the alternation theorem.

To elaborate on the alternation theorem, let us consider the design of a lowpass filter with passband $0 \le \omega \le \omega_p$ and stopband $\omega_s \le \omega \le \pi$. Since the desired frequency response $H_{dr}(\omega)$ and the weighting function $W(\omega)$ are piecewise constant, we have

$$
\frac{dE(\omega)}{d\omega} = \frac{d}{d\omega}\{W(\omega)[H_{dr}(\omega) - H_r(\omega)]\}
$$

$$
= -\frac{dH_r(\omega)}{d\omega} = 0
$$

Consequently, the frequencies $\{\omega_i\}$ corresponding to the peaks of $E(\omega)$ also correspond to peaks at which $H_r(\omega)$ meets the error tolerance. Since $H_r(\omega)$ is a trigonometric polynomial of degree L, for Case 1, for example,

$$
H_r(\omega) = \sum_{k=0}^{L} \alpha(k)\cos\omega k
$$

$$
= \sum_{k=0}^{L} \alpha(k)\left[\sum_{n=0}^{k} \beta_{nk}(\cos\omega)^n\right]
$$

$$
= \sum_{k=0}^{L} \alpha'(k)(\cos\omega)^k \tag{8.2.71}
$$

it follows that $H_r(\omega)$ can have at most $L-1$ local maxima and minima in the open interval $0 < \omega < \pi$. In addition, $\omega = 0$ and $\omega = \pi$ are usually extrema of $H_r(\omega)$ and, also, of $E(\omega)$. Therefore, $H_r(\omega)$ has at most $L+1$ extremal frequencies. Furthermore, the band-edge frequencies ω_p and ω_s are also extrema of $E(\omega)$, since $|E(\omega)|$ is maximum at $\omega = \omega_p$ and $\omega = \omega_s$. As a consequence, there are at most $L+3$ extremal frequencies in $E(\omega)$ for the unique, best approximation of the ideal lowpass filter. On the other hand, the alternation theorem states that there are at least $L+2$ extremal frequencies in $E(\omega)$. Thus the error function for the lowpass filter design has either $L+3$ or $L+2$ extrema. In general, filter designs that contain more than $L+2$ alternations or ripples are called *extra ripple filters*. When the filter design contains the maximum number of alternations, it is called a *maximal ripple filter*.

The alternation theorem guarantees a unique solution for the Chebyshev optimization problem in (8.2.70). At the desired extremal frequencies $\{\omega_n\}$, we have the set of equations

$$
\hat{W}(\omega_n)[\hat{H}_{dr}(\omega_n) - P(\omega_n)] = (-1)^n\delta \qquad n = 0, 1, \ldots, L+1 \tag{8.2.72}
$$

where δ represents the maximum value of the error function $E(\omega)$. In fact, if we select $W(\omega)$ as indicated by (8.2.66), it follows that $\delta = \delta_2$.

The set of linear equations in (8.2.72) can be rearranged as

$$
P(\omega_n) + \frac{(-1)^n\delta}{\hat{W}(\omega_n)} = \hat{H}_{dr}(\omega_n) \qquad n = 0, 1, \ldots, L+1
$$

or, equivalently, in the form

$$
\sum_{k=0}^{L} \alpha(k)\cos\omega_n k + \frac{(-1)^n\delta}{\hat{W}(\omega_n)} = \hat{H}_{dr}(\omega_n) \qquad n = 0, 1, \ldots, L+1 \tag{8.2.73}
$$

If we treat the $\{a(k)\}$ and δ as the parameters to be determined, (8.2.73) can be expressed in matrix form as

$$
\begin{bmatrix}
1 & \cos \omega_0 & \cos 2\omega_0 & \cdots & \cos L\omega_0 & \dfrac{1}{\hat{W}(\omega_0)} \\
1 & \cos \omega_1 & \cos 2\omega_1 & \cdots & \cos L\omega_1 & \dfrac{-1}{\hat{W}(\omega_1)} \\
& \vdots & & & & \\
1 & \cos \omega_{L+1} & \cos 2\omega_{L+1} & \cdots & \cos L\omega_{L+1} & \dfrac{(-1)^{L+1}}{\hat{W}(\omega_{L+1})}
\end{bmatrix}
\begin{bmatrix}
\alpha(0) \\
\alpha(1) \\
\vdots \\
\alpha(L) \\
\delta
\end{bmatrix}
=
\begin{bmatrix}
\hat{H}_{dr}(\omega_0) \\
\hat{H}_{dr}(\omega_1) \\
\vdots \\
\hat{H}_{dr}(\omega_{L+1})
\end{bmatrix}
$$

$$(8.2.74)$$

Initially, we know neither the set of extremal frequencies $\{\omega_n\}$ nor the parameters $\{\alpha(k)\}$ and δ. To solve for the parameters, we use an iterative algorithm, called the *Remez exchange algorithm* [see Rabiner et al. (1975)], in which we begin by guessing at the set of extremal frequencies, determine $P(\omega)$ and δ, and then compute the error function $E(\omega)$. From $E(\omega)$ we determine another set of $L+2$ extremal frequencies and repeat the process iteratively until it converges to the optimal set of extremal frequencies. Although the matrix equation in (8.2.74) can be used in the iterative procedure, matrix inversion is time consuming and inefficient.

A more efficient procedure, suggested in the paper by Rabiner et al. (1975), is to compute δ analytically, according to the formula

$$
\delta = \frac{\gamma_0 \hat{H}_{dr}(\omega_0) + \gamma_1 \hat{H}_{dr}(\omega_1) + \cdots + \gamma_{L+1} \hat{H}_{dr}(\omega_{L+1})}{\dfrac{\gamma_0}{\hat{W}(\omega_0)} - \dfrac{\gamma_1}{\hat{W}(\omega_1)} + \cdots + \dfrac{(-1)^{L+1}\gamma_{L+1}}{\hat{W}(\omega_{L+1})}}
\tag{8.2.75}
$$

where

$$
\gamma_k = \prod_{\substack{n=0 \\ n \neq k}}^{L+1} \frac{1}{\cos \omega_k - \cos \omega_n}
\tag{8.2.76}
$$

The expression for δ in (8.2.75) follows immediately from the matrix equation in (8.2.74). Thus with an initial guess at the $L+2$ extremal frequencies, we compute δ.

Now since $P(\omega)$ is a trigometric polynomial of the form

$$
P(\omega) = \sum_{k=0}^{L} \alpha(k) x^k \qquad x = \cos \omega
$$

and since we know that the polynomial at the points $x_n \equiv \cos \omega_n, n = 0, 1, \dots, L+1$, has the corresponding values

$$
P(\omega_n) = \hat{H}_{dr}(\omega_n) - \frac{(-1)^n \delta}{\hat{W}(\omega_n)} \qquad n = 0, 1, \dots, L+1
\tag{8.2.77}
$$

we can use the Lagrange interpolation formula for $P(\omega)$. Thus $P(\omega)$ can be expressed as [see Hamming (1962)]

$$P(\omega) = \frac{\displaystyle\sum_{k=0}^{L} P(\omega_k)[\beta_k/(x - x_k)]}{\displaystyle\sum_{k=0}^{L} [\beta_k/(x - x_k)]} \tag{8.2.78}$$

where $P(\omega_n)$ is given by (8.2.77), $x = \cos\omega$, $x_k = \cos\omega_k$, and

$$\beta_k = \prod_{\substack{n=0 \\ n \neq k}}^{L} \frac{1}{x_k - x_n} \tag{8.2.79}$$

Having the solution for $P(\omega)$, we can now compute the error function $E(\omega)$ from

$$E(\omega) = \hat{W}(\omega)[\hat{H}_{dr}(\omega) - P(\omega)] \tag{8.2.80}$$

on a dense set of frequency points. Usually, a number of points equal to $16M$, where M is the length of the filter, suffices. If $|E(\omega)| \geq \delta$ for some frequencies on the dense set, then a new set of frequencies corresponding to the $L+2$ largest peaks of $|E(\omega)|$ are selected and the computational procedure beginning with (8.2.75) is repeated. Since the new set of $L + 2$ extremal frequencies are selected to correspond to the peaks of the error function $|E(\omega)|$, the algorithm forces δ to increase in each iteration until it converges to the upper bound and hence to the optimum solution for the Chebyshev approximation problem. In other words, when $|E(\omega)| \leq \delta$ for all frequencies on the dense set, the optimal solution has been found in terms of the polynomial $H(\omega)$.

A flowchart of the algorithm is shown in Fig. 8.18 and is due to Remez (1957).

Once the optimal solution has been obtained in terms of $P(\omega)$, the unit sample response $h(n)$ can be computed directly, without having to compute the parameters $\{\alpha(k)\}$. In effect, we have determined

$$H_r(\omega) = Q(\omega)P(\omega)$$

which can be evaluated at $\omega = 2\pi k/M$, $k = 0, 1, \dots, (M - 1)/2$, for M odd, or $M/2$ for M even. Then, depending on the type of filter being designed, $h(n)$ can be determined from the formulas given in Table 8.3.

A computer program written by Parks and McClellan (1972b) is available for designing linear phase FIR filters based on the Chebyshev approximation criterion and implemented with the Remez exchange algorithm. This program can be used to design lowpass, highpass or bandpass filters, differentiators, and Hilbert transformers. The latter two types of filters are described in the following sections. A number of software packages for designing equiripple linear-phase FIR filters are now available.

The Parks–McClellan program requires a number of input parameters which determine the filter characteristics. In particular, the following parameters must

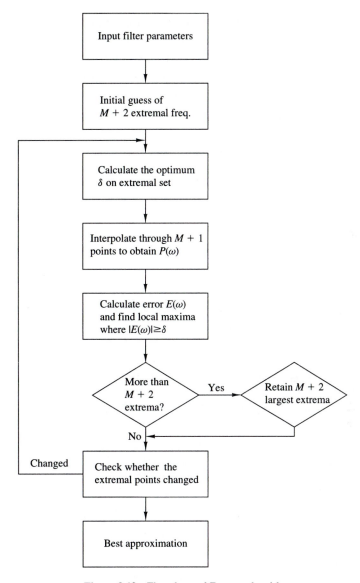

Figure 8.18 Flowchart of Remez algorithm.

be specified:

LINE 1

NFILT: The filter length, denoted above as M.

JTYPE: Type of filter:
             JTYPE $= 1$ results in a multiple passband/stopband filter.

JTYPE $= 2$ results in a differentiator.

JTYPE $= 3$ results in a Hilbert transformer.

NBANDS: The number of frequency bands from 2 (for a lowpass filter) to a maximum of 10 (for a multiple-band filter).

LGRID: The grid density for interpolating the error function $E(\omega)$. The default value is 16 if left unspecified.

LINE 2

EDGE: The frequency bands specified by lower and upper cutoff frequencies, up to a maximum of 10 bands (an array of size 20, maximum). The frequencies are given in terms of the variable $f = \omega/2\pi$, where $f = 0.5$ corresponds to the folding frequency.

LINE 3

FX: An array of maximum size 10 that specifies the desired frequency response $H_{dr}(\omega)$ in each band.

LINE 4

WTX: An array of maximum size 10 that specifies the weight function in each band.

The following examples demonstrate the use of this program to design a lowpass and a bandpass filter.

Example 8.2.3

Design a lowpass filter of length $M = 61$ with a passband edge frequency $f_p = 0.1$ and a stopband edge frequency $f_s = 0.15$.

Solution The lowpass filter is a two-band filter with passband edge frequencies $(0, 0.1)$ and stopband edge frequencies $(0.15, 0.5)$. The desired response is $(1, 0)$ and the weight function is arbitrarily selected as $(1, 1)$.

$$61, 1, 2$$
$$0.0, 0.1, 0.15, 0.5$$
$$1.0, 0.0$$
$$1.0, 1.0$$

The result of this design is illustrated in Table 8.6, which gives the filter coefficients. The frequency response is shown in Fig. 8.19. The resulting filter has a stopband attenuation of -56 dB and a passband ripple of 0.0135 dB.

If we increase the length of the filter to $M = 101$ while maintaining all the other parameters given above the same, the resulting filter has the frequency response characteristic shown in Fig. 8.20. Now, the stopband attenuation is -85 dB and the passband ripple is reduced to 0.00046 dB.

We should indicate that it is possible to increase the attenuation in the stopband by keeping the filter length fixed, say at $M = 61$, and decreasing the weighting function $W(\omega) = \delta_2/\delta_1$ in the passband. With $M = 61$ and a weighting function

TABLE 8.6 PARAMETERS FOR LOWPASS FILTER DESIGN IN EXAMPLE 8.2.3

```
              FINITE IMPULSE RESPONSE (FIR)
              LINEAR PHASE DIGITAL FILTER DESIGN
                 REMEZ EXCHANGE ALGORITHM
                    FILTER LENGTH = 61
                ***** IMPULSE RESPONSE *****
          H( 1) = -0.12109351E-02 = H( 61)
          H( 2) = -0.67270687E-03 = H( 60)
          H( 3) =  0.98090240E-04 = H( 59)
          H( 4) =  0.13536664E-02 = H( 58)
          H( 5) =  0.22969784E-02 = H( 57)
          H( 6) =  0.19963495E-02 = H( 56)
          H( 7) =  0.97026095E-04 = H( 55)
          H( 8) = -0.26466695E-02 = H( 54)
          H( 9) = -0.45133103E-02 = H( 53)
          H(10) = -0.37704944E-02 = H( 52)
          H(11) =  0.13079655E-04 = H( 51)
          H(12) =  0.51791356E-02 = H( 50)
          H(13) =  0.84883478E-02 = H( 49)
          H(14) =  0.69532110E-02 = H( 48)
          H(15) =  0.71037059E-04 = H( 47)
          H(16) = -0.90407897E-02 = H( 46)
          H(17) = -0.14723047E-01 = H( 45)
          H(18) = -0.11958945E-01 = H( 44)
          H(19) = -0.29799214E-04 = H( 43)
          H(20) =  0.15713422E-01 = H( 42)
          H(21) =  0.25657151E-01 = H( 41)
          H(22) =  0.21057373E-01 = H( 40)
          H(23) =  0.68637768E-04 = H( 39)
          H(24) = -0.28902054E-01 = H( 38)
          H(25) = -0.49118541E-01 = H( 37)
          H(26) = -0.42713970E-01 = H( 36)
          H(27) = -0.50114304E-04 = H( 35)
          H(28) =  0.73574215E-01 = H( 34)
          H(29) =  0.15782040E+00 = H( 33)
          H(30) =  0.22465512E+00 = H( 32)
          H(31) =  0.25007001E+00 = H( 31)
```

	BAND 1	BAND 2
LOWER BAND EDGE	0.0000000	0.1500000
UPPER BAND EDGE	0.1000000	0.5000000
DESIRED VALUE	1.0000000	0.0000000
WEIGHTING	1.0000000	1.0000000
DEVIATION	0.0015537	0.0015537
DEVIATION IN DB	0.0134854	-56.1724014

```
EXTREMAL FREQUENCIES--MAXIMA OF THE ERROR CURVE
0.0000000   0.0252016   0.0423387   0.0584677   0.0735887
0.0866935   0.0957661   0.1000000   0.1500000   0.1540323
0.1631048   0.1762097   0.1903225   0.2054435   0.2215725
0.2377015   0.2538306   0.2699596   0.2860886   0.3022176
0.3183466   0.3354837   0.3516127   0.3677417   0.3848788
0.4010078   0.4171368   0.4342739   0.4504029   0.4665320
0.4836690   0.5000000
```

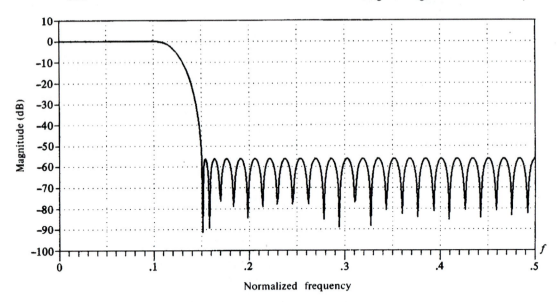

Figure 8.19   Frequency response of $M = 61$ FIR filter in Example 8.2.3.

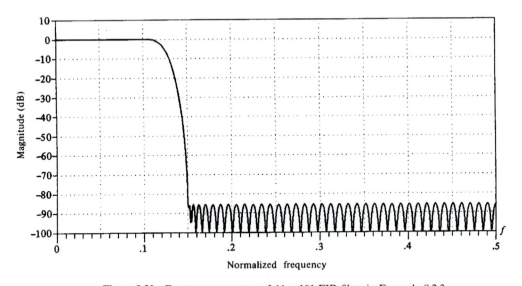

Figure 8.20   Frequency response of $M = 101$ FIR filter in Example 8.2.3.

$(0.1, 1)$, we obtain a filter that has a stopband attenuation of -65 dB and a pass-band ripple of 0.049 dB.

Example 8.2.4

Design a bandpass filter of length $M = 32$ with passband edge frequencies $f_{p1} = 0.2$ and $f_{p2} = 0.35$ and stopband edge frequencies of $f_{s1} = 0.1$ and $f_{s2} = 0.425$.

Solution   This passband filter is a three-band filter with a stopband range of $(0, 0.1)$, a passband range of $(0.2, 0.35)$, and a second stopband range of $(0.425, 0.5)$. The weighting function is selected as $(10.0, 1.0, 10.0)$, or as $(1.0, 0.1, 1.0)$, and the desired response in the three bands is $(0.0, 1.0, 0.0)$. Thus the input parameters to the program are

$$32, 1, 3$$
$$0.0, 0.1, 0.2, 0.35, 0.425, 0.5$$
$$0.0, 1.0, 0.0$$
$$10.0, 1.0, 10.0$$

The results of this design are shown in Table 8.7, which gives the filter coefficients. We note that the ripple in the stopbands δ_2 is 10 times smaller than the ripple in

TABLE 8.7   PARAMETERS FOR BANDPASS FILTER IN EXAMPLE 8.2.4

```
                    FINITE IMPULSE RESPONSE (FIR)
                  LINEAR PHASE DIGITAL FILTER DESIGN
                     REMEZ EXCHANGE ALGORITHM
                        BANDPASS FILTER
                      FILTER LENGTH = 32
                 ***** IMPULSE RESPONSE *****
                 H( 1) = -0.57534026E-02 = H( 32)
                 H( 2) =  0.99026691E-03 = H( 31)
                 H( 3) =  0.75733471E-02 = H( 30)
                 H( 4) = -0.65141204E-02 = H( 29)
                 H( 5) =  0.13960509E-01 = H( 28)
                 H( 6) =  0.22951644E-02 = H( 27)
                 H( 7) = -0.19994041E-01 = H( 26)
                 H( 8) =  0.71369656E-02 = H( 25)
                 H( 9) = -0.39657373E-01 = H( 24)
                 H(10) =  0.11260066E-01 = H( 23)
                 H(11) =  0.66233635E-01 = H( 22)
                 H(12) = -0.10497202E-01 = H( 21)
                 H(13) =  0.85136160E-01 = H( 20)
                 H(14) = -0.12024988E+00 = H( 19)
                 H(15) = -0.29678580E+00 = H( 18)
                 H(16) =  0.30410913E+00 = H( 17)
```

	BAND 1	BAND 2	BAND 3
LOWER BAND EDGE	0.0000000	0.2000000	0.4250000
UPPER BAND EDGE	0.1000000	0.3500000	0.5000000
DESIRED VALUE	0.0000000	1.0000000	0.0000000
WEIGHTING	10.0000000	1.0000000	10.0000000
DEVIATION	0.0015131	0.0151312	0.0015131
DEVIATION IN DB	-56.4025536	0.1304428	-56.4025536

```
EXTREMAL FREQUENCIES--MAXIMA OF THE ERROR CURVE
0.0000000    0.0273438    0.0527344    0.0761719    0.0937500
0.1000000    0.2000000    0.2195313    0.2527344    0.2839844
0.3132813    0.3386719    0.3500000    0.4250000    0.4328125
0.4503906    0.4796875
```

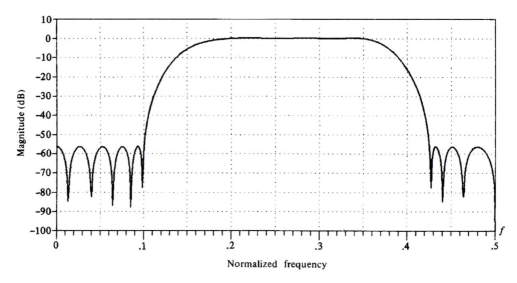

Figure 8.21   Frequency response of $M = 32$ FIR filter in Example 8.2.4.

the passband due to the fact that errors in the stopband were given a weight of 10 compared to the passband weight of unity. The frequency response of the bandpass filter is illustrated in Fig. 8.21.

These examples serve to illustrate the relative ease with which optimal low-pass, highpass, bandstop, bandpass, and more general multiband linear-phase FIR filters can be designed based on the Chebyshev approximation criterion implemented by means of the Remez exchange algorithm. In the next two sections we consider the design of differentiators and Hilbert transformers.

8.2.5 Design of FIR Differentiators

Differentiators are used in many analog and digital systems to take the derivative of a signal. An ideal differentiator has a frequency response that is linearly proportional to frequency. Similarly, an ideal digital differentiator is defined as one that has the frequency response

$$H_d(\omega) = j\omega \qquad -\pi \leq \omega \leq \pi \qquad (8.2.81)$$

The unit sample response corresponding to $H_d(\omega)$ is

$$h_d(n) = \frac{1}{2\pi} \int_{-\pi}^{\pi} H_d(\omega) e^{j\omega n} d\omega$$

$$= \frac{1}{2\pi} \int_{-\pi}^{\pi} j\omega e^{j\omega n} d\omega$$

$$= \frac{\cos \pi n}{n} \qquad -\infty < n < \infty, n \neq 0 \qquad (8.2.82)$$

We observe that the ideal differentiator has an antisymmetric unit sample response [i.e., $h_d(n) = -h_d(-n)$]. Hence, $h_d(0) = 0$.

In this section we consider the design of linear-phase FIR differentiators based on the Chebyshev approximation criterion. In view of the fact that the ideal differentiator has an antisymmetric unit sample response, we shall confine our attention to FIR designs in which $h(n) = -h(M - 1 - n)$. Hence we consider the filter types classified in the preceding section, as Case 3 and Case 4.

We recall that in Case 3, where M is odd, the real-valued frequency response of the FIR filter $H_r(\omega)$ has the characteristic that $H_r(0) = 0$. A zero response at zero frequency is just the condition that the differentiator should satisfy, and we see from Table 8.5 that both filter types satisfy this condition. However, if a full-band differentiator is desired, this is impossible to achieve with an FIR filter having an odd number of coefficients, since $H_r(\pi) = 0$ for M odd. In practice, however, full-band differentiators are rarely required.

In most cases of practical interest, the desired frequency response characteristic need only be linear over the limited frequency range $0 \leq \omega \leq 2\pi f_p$, where f_p is called the bandwidth of the differentiator. In the frequency range $2\pi f_p < \omega \leq \pi$, the desired response may be either left unconstrained or constrained to be zero.

In the design of FIR differentiators based on the Chebyshev approximation criterion, the weighting function $W(\omega)$ is specified in the program as

$$W(\omega) = \frac{1}{\omega} \qquad 0 \leq \omega \leq 2\pi f_p \qquad (8.2.83)$$

in order that the relative ripple in the passband be a constant. Thus the absolute error between the desired response ω and the approximation $H_r(\omega)$ increases as ω varies from 0 to $2\pi f_p$. However, the weighting function in (8.2.83) ensures that the relative error

$$\delta = \max_{0 \leq \omega \leq 2\pi f_p} \{W(\omega)[\omega - H_r(\omega)]\}$$

$$= \max_{0 \leq \omega \leq 2\pi f_p} \left[1 - \frac{H_r(\omega)}{\omega}\right] \qquad (8.2.84)$$

is fixed within the passband of the differentiator.

Example 8.2.5

Use the Remez algorithm to design a linear-phase FIR differentiator of length $M = 60$. The passband edge frequency is 0.1 and the stopband edge frequency is 0.15.

Solution   The input parameters to the program are

$$60, \quad 2, \quad 2$$
$$0.0, \quad 0.1, \quad 0.15, \quad 0.5$$
$$1.0, \quad 0.0$$
$$1.0, \quad 1.0$$

The results of this design including the filter coefficients are shown in Table 8.8. The frequency response characteristic is illustrated in Fig. 8.22. Also shown in the same figure is the approximation error over the passband $0 \leq f \leq 0.1$ of the filter.

TABLE 8.8   PARAMETERS FOR FIR DIFFERENTIATOR IN EXAMPLE 8.2.5

```
                    FINITE IMPULSE RESPONSE (FIR)
                   LINEAR-PHASE DIGITAL FILTER DESIGN
                       REMEZ EXCHANGE ALGORITHM
                            DIFFERENTIATOR
                         FILTER LENGTH = 60
                    ***** IMPULSE RESPONSE *****
             H( 1) = -0.12478075E-02 = -H( 60)
             H( 2) = -0.15713560E-02 = -H( 59)
             H( 3) =  0.36846737E-02 = -H( 58)
             H( 4) =  0.19298020E-02 = -H( 57)
             H( 5) =  0.14264141E-02 = -H( 56)
             H( 6) = -0.17615277E-02 = -H( 55)
             H( 7) = -0.43110573E-02 = -H( 54)
             H( 8) = -0.46953405E-02 = -H( 53)
             H( 9) = -0.14105244E-02 = -H( 52)
             H(10) =  0.41694222E-02 = -H( 51)
             H(11) =  0.85736215E-02 = -H( 50)
             H(12) =  0.79813031E-02 = -H( 49)
             H(13) =  0.11833385E-02 = -H( 48)
             H(14) = -0.87396065E-02 = -H( 47)
             H(15) = -0.15401847E-01 = -H( 46)
             H(16) = -0.12878445E-01 = -H( 45)
             H(17) = -0.18826872E-03 = -H( 44)
             H(18) =  0.16620506E-01 = -H( 43)
             H(19) =  0.26741523E-01 = -H( 42)
             H(20) =  0.20892018E-01 = -H( 41)
             H(21) = -0.18584095E-02 = -H( 40)
             H(22) = -0.31109909E-01 = -H( 39)
             H(23) = -0.48822176E-01 = -H( 38)
             H(24) = -0.38673453E-01 = -H( 37)
             H(25) =  0.36760122E-02 = -H( 36)
             H(26) =  0.65462478E-01 = -H( 35)
             H(27) =  0.12066317E+00 = -H( 34)
             H(28) =  0.14182134E+00 = -H( 33)
             H(29) =  0.11403757E+00 = -H( 32)
             H(30) =  0.43620080E-01 = -H( 31)
```

	BAND 1	BAND 2
LOWER BAND EDGE	0.0000000	0.1500000
UPPER BAND EDGE	0.1000000	0.5000000
DESIRED SLOPE	10.0000000	0.0000000
WEIGHTING	1.0000000	1.0000000
DEVIATION	0.0073580	0.0073580

```
EXTREMAL FREQUENCIES---MAXIMA OF THE ERROR CURVE
0.0010417    0.0156250    0.0312500    0.0468750    0.0614583
0.0750000    0.0875000    0.0968750    0.1000000    0.1500000
0.1552083    0.1666667    0.1822916    0.1979166    0.2156249
0.2322916    0.2489582    0.2666668    0.2843754    0.3020839
0.3187508    0.3364594    0.3541680    0.3718765    0.3906268
0.4083354    0.4260439    0.4447942    0.4625027    0.4812530
0.5000000
```

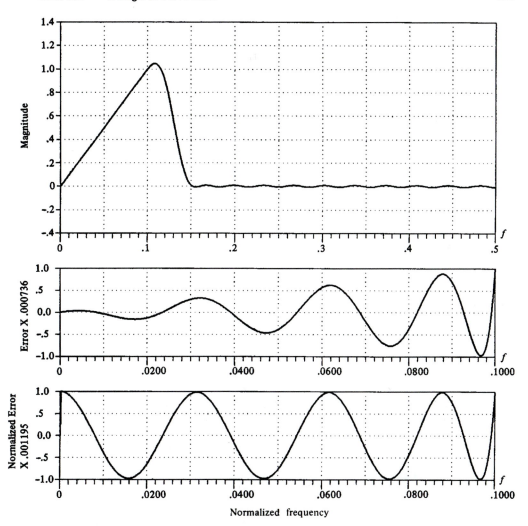

Figure 8.22 Frequency response and approximation error for $M = 60$ FIR differentiator of Example 8.2.5.

The important parameters in a differentiator are its length M, its bandwidth {band-edge frequency} f_p, and the peak relative error δ of the approximation. The interrelationship among these three parameters can be easily displayed parametrically. In particular, the value of $20 \log_{10} \delta$ versus f_p with M as a parameter is shown in Fig. 8.23 for M even and in Fig. 8.24 for M odd. These results, due to Rabiner and Schafter (1974a), are useful in the selection of the filter length, given specifications on the inband ripple and the cutoff frequency f_p.

A comparison of the graphs in Figs. 8.23 and 8.24 reveals that even-length differentiators result in a significantly smaller approximation error δ than comparable odd-length differentiators. Designs based on M odd are particularly poor if

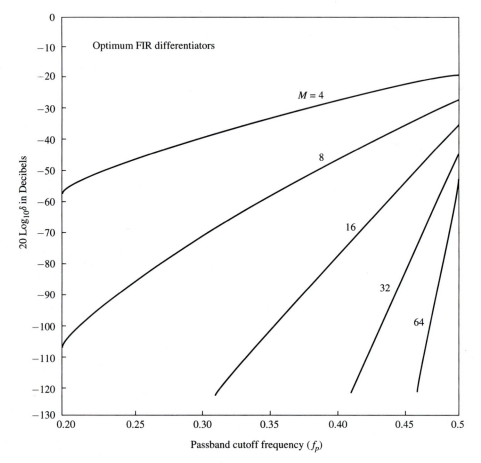

Figure 8.23   Curves of $20 \log_{10} \delta$ versus f_p for $M = 4$, 8, 16, 32, and 64. [From paper by Rabiner and Schafer (1974a). Reprinted with permission of AT&T.]

the bandwidth exceeds $f_p = 0.45$. The problem is basically the zero in the frequency response at $\omega = \pi (f = 1/2)$. When $f_p < 0.45$, good designs are obtained for M odd, but comparable-length differentiators with M even are always better in the sense that the approximation error is smaller.

In view of the obvious advantage of even-length over odd-length differentiators, a conclusion might be that even-length differentiators are always preferable in practical systems. This is certainly true for many applications. However, we should note that the signal delay introduced by any linear-phase FIR filter is $(M - 1)/2$, which is not an integer when M is even. In many practical applications, this is unimportant. In some applications where it is desirable to have an integer-valued delay in the signal at the output of the differentiator, we must select M to be odd.

These numerical results are based on designs resulting from the Chebyshev approximation criterion. We wish to indicate it is also possible and relatively

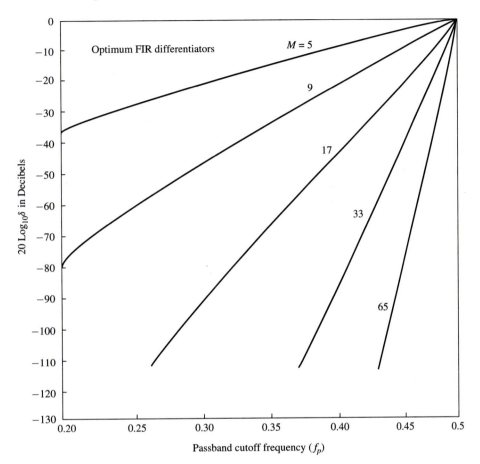

Figure 8.24   Curves of $20 \log_{10} \delta$ versus F_p for $M = 5, 9, 17, 33$ and 65. [From paper by Rabiner and Schafer (1974a). Reprinted with permission of AT&T.]

easy to design linear phase FIR differentiators based on the frequency sampling method. For example, Fig. 8.25 illustrates the frequency response characteristics of a wideband ($f_p = 0.5$) differentiator, of length $M = 30$. The graph of the absolute value of the approximation error as a function of frequency is also shown in this figure.

8.2.6 Design of Hilbert Transformers

An ideal Hilbert transformer is an all-pass filter that imparts a 90° phase shift on the signal at its input. Hence the frequency response of the ideal Hilbert transformer is specified as

$$H_d(\omega) = \begin{cases} -j, & 0 < \omega \le \pi \\ j, & -\pi < \omega < 0 \end{cases} \qquad (8.2.85)$$

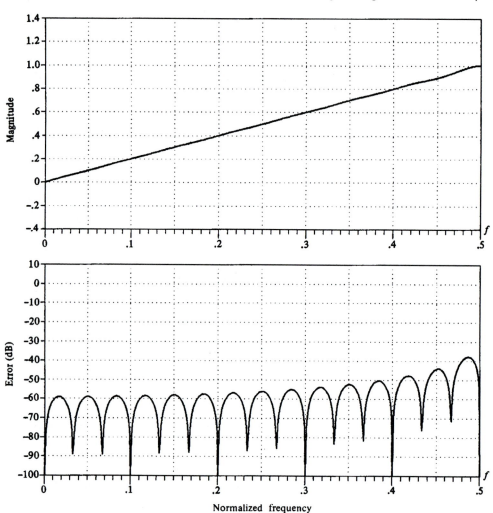

Figure 8.25 Frequency response and approximation error for $M = 30$ FIR differentiator designed by frequency sampling method.

Hilbert transformers are frequently used in communication systems and signal processing, as, for example, in the generation of single-sideband modulated signals, radar signal processing, and speech signal processing.

The unit sample response of an ideal Hilbert transformer is

$$h_d(n) = \frac{1}{2\pi} \int_{-\pi}^{\pi} H_d(\omega) e^{j\omega n} d\omega$$

$$= \frac{1}{2\pi} \left(\int_{-\pi}^{0} j e^{j\omega n} d\omega - \int_{0}^{\pi} j e^{j\omega n} d\omega \right)$$

$$= \begin{cases} \dfrac{2}{\pi} \dfrac{\sin^2(\pi n/2)}{n}, & n \neq 0 \\ 0, & n = 0 \end{cases} \tag{8.2.86}$$

As expected, $h_d(n)$ is infinite in duration and noncausal. We note that $h_d(n)$ is antisymmetric [i.e., $h_d(n) = -h_d(-n)$]. In view of this characteristic, we focus our attention on the design of linear-phase FIR Hilbert transformers with antisymmetric unit sample response [i.e., $h(n) = -h(M - 1 - n)$]. We also observe that our choice of an antisymmetric unit sample response is consistent with having a purely imaginary frequency response characteristic $H_d(\omega)$.

We recall once again that when $h(n)$ is antisymmetric, the real-valued frequency response characteristic $H_r(\omega)$ is zero at $\omega = 0$ for both M odd and even and at $\omega = \pi$ when M is odd. Clearly, then, it is impossible to design an all-pass digital Hilbert transformer. Fortunately, in practical signal processing applications, an all-pass Hilbert transformer is unnecessary. Its bandwidth need only cover the bandwidth of the signal to be phase shifted. Consequently, we specify the desired real-valued frequency response of a Hilbert transform filter as

$$H_{dr}(\omega) = 1 \qquad 2\pi f_l \leq \omega \leq 2\pi f_u \tag{8.2.87}$$

where f_l and f_u are the lower and upper cutoff frequencies, respectively.

It is interesting to note that the ideal Hilbert transformer with unit sample response $h_d(n)$ as given in (8.2.86) is zero for n even. This property is retained by the FIR Hilbert transformer under some symmetry conditions. In particular, let us consider the Case 3 filter type for which

$$H_r(\omega) = \sum_{k=1}^{(M-1)/2} c(k) \sin \omega k \tag{8.2.88}$$

and suppose that $f_l = 0.5 - f_u$. This ensures a symmetric passband about the midpoint frequency $f = 0.25$. If we have this symmetry in the frequency response, $H_r(\omega) = H_r(\pi - \omega)$ and hence (8.2.88) yields

$$\sum_{k=1}^{(M-1)/2} c(k) \sin \omega k = \sum_{k=1}^{(M-1)/2} c(k) \sin k(\pi - \omega)$$

$$= \sum_{k=1}^{(M-1)/2} c(k) \sin \omega k \cos \pi k$$

$$= \sum_{k=1}^{(M-1)/2} c(k)(-1)^{k+1} \sin \omega k$$

or equivalently,

$$\sum_{k=1}^{(M-1)/2} [1 - (-1)^{k+1}] c(k) \sin \omega k = 0 \tag{8.2.89}$$

Clearly, $c(k)$ must be equal to zero for $k = 0, 2, 4, \ldots$.

Now, the relationship between $\{c(k)\}$ and the unit sample response $\{h(n)\}$ is, from (8.2.54),

$$c(k) = 2h\left(\frac{M-1}{2} - k\right)$$

or, equivalently,

$$h\left(\frac{M-1}{2} - k\right) = \frac{1}{2}c(k) \tag{8.2.90}$$

If $c(k)$ is zero for $k = 0, 2, 4, \ldots$, then (8.2.90) yields

$$h(k) = \begin{cases} 0, & k = 0, 2, 4, \ldots, \text{ for } \dfrac{M-1}{2} \text{ even} \\[2mm] 0, & k = 1, 3, 5, \ldots, \text{ for } \dfrac{M-1}{2} \text{ odd} \end{cases} \tag{8.2.91}$$

Unfortunately, (8.2.91) holds only for M odd. It does not hold for M even. This means that for comparable values of M, the case M odd is preferable since the computational complexity (number of multiplications and additions per output point) is roughly one half of that for M even.

When the design of the Hilbert transformer is performed by the Chebyshev approximation criterion using the Remez algorithm, we select the filter coefficients to minimize the peak approximation error

$$\begin{aligned} \delta &= \max_{2\pi f_l \le \omega \le 2\pi f_u} [H_{dr}(\omega) - H_r(\omega)] \\ &= \max_{2\pi f_l \le \omega \le 2\pi f_u} [1 - H_r(\omega)] \end{aligned} \tag{8.2.92}$$

Thus the weighting function is set to unity and the optimization is performed over the single frequency band (i.e., the passband of the filter).

Example 8.2.6

Design a Hilbert transformer with parameters $M = 31$, $f_l = 0.05$, and $f_u = 0.45$.

Solution   We observe that the frequency response is symmetric, since $f_u = 0.5 - f_l$. The parameters for executing the Remez algorithm are

$$31, \quad 3, \quad 1$$
$$0.05, \quad 0.45$$
$$1.0$$
$$1.0$$

The result of this design is the unit sample response coefficients and the peak approximation error $\delta = 0.0026803$ or -51.4 dB given in Table 8.9. We observe that, indeed, every other value of $h(n)$ is essentially zero (these values are of the order of 10^{-7}). The frequency response of the Hilbert transformer is shown in Fig. 8.26.

TABLE 8.9 PARAMETERS FOR FIR HILBERT TRANSFORM FILTER IN
EXAMPLE 8.2.6

```
                    FINITE IMPULSE RESPONSE (FIR)
                 LINEAR PHASE DIGITAL FILTER DESIGN
                    REMEZ EXCHANGE ALGORITHM
                       HILBERT TRANSFORMER
                      FILTER LENGTH = 31
                ***** IMPULSE RESPONSE *****
            H( 1) =  0.41957516E-02 = -H( 31)
            H( 2) =  0.64310257E-07 = -H( 30)
            H( 3) =  0.92822444E-02 = -H( 29)
            H( 4) =  0.52693927E-07 = -H( 28)
            H( 5) =  0.18835988E-01 = -H( 27)
            H( 6) =  0.82308283E-07 = -H( 26)
            H( 7) =  0.34401190E-01 = -H( 25)
            H( 8) =  0.93328794E-07 = -H( 24)
            H( 9) =  0.59551738E-01 = -H( 23)
            H(10) =  0.50821171E-07 = -H( 22)
            H(11) =  0.10303782E+00 = -H( 21)
            H(12) =  0.17612138E-07 = -H( 20)
            H(13) =  0.19683167E+00 = -H( 19)
            H(14) = -0.23977606E-07 = -H( 18)
            H(15) =  0.63135374E+00 = -H( 17)
            H(16) =  0.0
                                      BAND 1
             LOWER BAND EDGE        0.0500000
             UPPER BAND EDGE        0.4500000
             DESIRED VALUE          1.0000000
             WEIGHTING              1.0000000
             DEVIATION              0.0026803
       EXTREMAL FREQUENCIES---MAXIMA OF THE ERROR CURVE
     0.0500000   0.0562500   0.0750000   0.1000000   0.1291666
     0.1583333   0.1874999   0.2187499   0.2499998   0.2812498
     0.3124998   0.3416664   0.3708331   0.3999997   0.4249997
     0.4437497
```

Rabiner and Schafer (1974b) have investigated the characteristics of Hilbert transformer designs for both M odd and M even. If the filter design is restricted to a symmetric frequency response, then there are basically three parameters of interest, M, δ, and f_l. Figure 8.27 is a plot of $20 \log_{10} \delta$ versus f_l with M as a parameter. We observe that for comparable values of M, there is no performance advantage of using M odd over M even, and vice versa. However, the computational complexity in implementing a filter for M odd is less by a factor of 2 over M even as previously indicated. Therefore, M odd is preferable in practice.

For design purposes, the graphs in Fig. 8.27 suggest that, as a rule of thumb,

$$M f_l \approx -0.61 \log_{10} \delta \qquad (8.2.93)$$

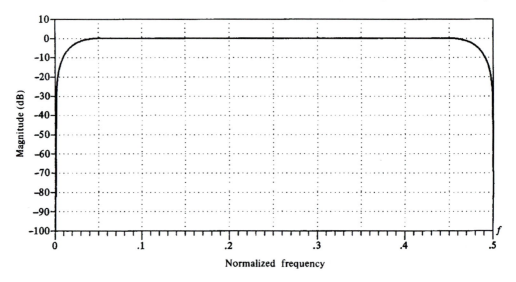

Figure 8.26 Frequency response of FIR Hilbert transform filter in Example 8.6.6.

Hence this formula can be used to estimate the size of one of the three basic filter parameters when the other two parameters are specified.

In concluding this section, we wish to show that Hilbert transformers can also be designed by the window method and the frequency sampling method. For example, Fig. 8.28 illustrates the frequency response of an $M = 31$ Hilbert transformer designed using the frequency sampling method. The corresponding values of the unit sample response are given in Table 8.10. A comparison of these filter parameters with those given in Table 8.9 indicates some small differences. In particular, it appears that the Chebyshev approximation criterion gives significantly smaller values for the filter coefficients that should be zero. In general, the Chebyshev approximation criterion results in better filter designs.

8.2.7 Comparison of Design Methods for Linear-Phase FIR Filters

Historically, the design method based on the use of windows to truncate the impulse response $h_d(n)$ and obtaining the desired spectral shaping, was the first method proposed for designing linear-phase FIR filters. The frequency-sampling method and the Chebyshev approximation method were developed in the 1970s and have since become very popular in the design of practical linear-phase FIR filters.

The major disadvantage of the window design method is the lack of precise control of the critical frequencies, such as ω_p and ω_s, in the design of a lowpass FIR filter. The values of ω_p and ω_s, in general, depend on the type of window and the filter length M.

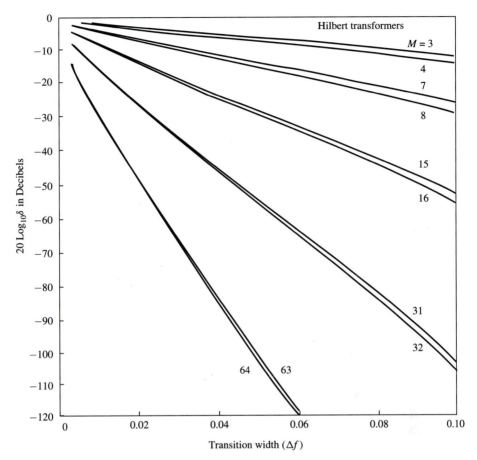

Figure 8.27   Curves of $20 \log_{10} \delta$ versus Δf for $M = 3, 4, 7, 8, 15, 16, 31, 32, 63,$ 64. [From paper by Rabiner and Schafer (1974b). Reprinted with permission of AT&T.]

The frequency sampling method provides an improvement over the window design method, since $H_r(\omega)$ is specified at the frequencies $\omega_k = 2\pi k/M$ or $\omega_k = \pi(2k + 1)/M$ and the transition band is a multiple of $2\pi/M$. This filter design method is particularly attractive when the FIR filter is realized either in the frequency domain by means of the DFT or in any of the frequency sampling realizations. The attractive feature of these realizations is that $H_r(\omega_k)$ is either zero or unity at all frequencies, except in the transition band.

The Chebyshev approximation method provides total control of the filter specifications, and, as a consequence, it is usually preferable over the other two methods. For a lowpass filter, the specifications are given in terms of the parameters ω_p, ω_s, δ_1, δ_2, and M. We can specify the parameters ω_p, ω_s, M and δ, and optimize the filters relative to δ_2. By spreading the approximation error over the

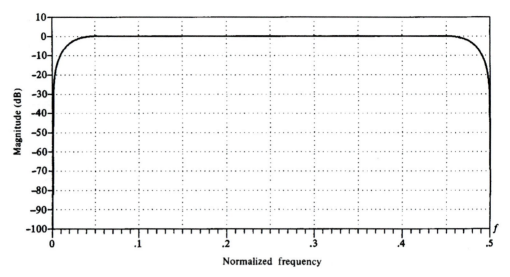

Figure 8.28 Frequency response of $M = 31$ FIR Hilbert transform filter designed by the frequency sampling method.

passband and the stopband of the filter, this method results in an optimal filter design, in the sense that for a given set of specifications just described, the maximum sidelobe level is minimized.

The Chebyshev design procedure based on the Remez exchange algorithm requires that we specify the length of the filter, the critical frequencies ω_p and ω_s, and the ratio δ_2/δ_1. However, it is more natural in filter design to specify ω_p, ω_s, δ_1, and δ_2 and to determine the filter length that satisfies the specifications. Although there is no simple formula to determine the filter length from these specifications, a number of approximations have been proposed for estimating M from ω_p, ω_s, δ_1, and δ_2. A particularly simple formula attributed to Kaiser for approximating M is

$$\hat{M} = \frac{-20 \log_{10} \left(\sqrt{\delta_1 \delta_2} \right) - 13}{14.6 \, \Delta f} + 1 \qquad (8.2.94)$$

where Δf is the transition band, defined as $\Delta f = (\omega_s - \omega_p)/2\pi$. This formula has been given in the paper by Rabiner et al. (1975). A more accurate formula proposed by Herrmann et al. (1973) is

$$\hat{M} = \frac{D_\infty(\delta_1, \delta_2) - f(\delta_1, \delta_2)(\Delta f)^2}{\Delta f} + 1 \qquad (8.2.95)$$

where, by definition,

$$D_\infty(\delta_1, \delta_2) = [0.005309(\log_{10} \delta_1)^2 + 0.07114(\log_{10} \delta_1) - 0.4761](\log_{10} \delta_2)$$
$$- [0.00266(\log_{10} \delta_1)^2 + 0.5941 \log_{10} \delta_1 + 0.4278] \qquad (8.2.96)$$

$$f(\delta_1, \delta_2) = 11.012 + 0.51244(\log_{10} \delta_1 - \log_{10} \delta_2) \qquad (8.2.97)$$

TABLE 8.10 PARAMETERS A FOR *M* = 31 HILBERT
TRANSFORM FILTER DESIGNED BY THE FREQUENCY-
SAMPLING METHOD

```
LINEAR PHASE FIR HILBERT TRANSFORM
FREQUENCY SAMPLING METHOD

FILTER LENGTH =31
LOWER CUTOFF FREQUENCY (RELATIVE) = 0.5000000E-01
UPPER CUTOFF FREQUENCY (RELATIVE) = 0.4500000E+00

IMPULSE RESPONSE:
                        H( 0) = -0.1342662E-03
                        H( 1) =  0.2133148E-02
                        H( 2) =  0.4848863E-02
                        H( 3) =  0.2286159E-02
                        H( 4) =  0.1423532E-01
                        H( 5) =  0.1517075E-02
                        H( 6) =  0.3001805E-01
                        H( 7) =  0.5263533E-03
                        H( 8) =  0.5574721E-01
                        H( 9) = -0.2281570E-03
                        H(10) =  0.1001032E+00
                        H(11) = -0.5338326E-03
                        H(12) =  0.1949848E+00
                        H(13) = -0.3994641E-03
                        H(14) =  0.6307253E+00
                        H(15) = -0.9335956E-06
                        H(16) = -0.6307245E+00
                        H(17) =  0.3996222E-03
                        H(18) = -0.1949853E+00
                        H(19) =  0.5341307E-03
                        H(20) = -0.1001035E+00
                        H(21) =  0.2285338E-03
                        H(22) = -0.5574735E-01
                        H(23) = -0.5263340E-03
                        H(24) = -0.3001794E-01
                        H(25) = -0.1517240E-02
                        H(26) = -0.1423557E-01
                        H(27) = -0.2285915E-02
                        H(28) = -0.4848215E-02
                        H(29) = -0.2133800E-02
                        H(30) =  0.1344162E-03
```

These formulas are extremely useful in obtaining a good estimate of the filter length required to achieve the given specifications Δf, δ_1, and δ_2. The estimate is used to carry out the design and if the resulting δ exceeds the specified δ_2, the length can be increased until we obtain a sidelobe level that meets the specifications.

8.3 DESIGN OF IIR FILTERS FROM ANALOG FILTERS

Just as in the design of FIR filters, there are several methods that can be used to design digital filters having an infinite-duration unit sample response. The techniques described in this section are all based on converting an analog filter into a digital filter. Analog filter design is a mature and well developed field, so it is not surprising that we begin the design of a digital filter in the analog domain and then convert the design into the digital domain.

An analog filter can be described by its system function.

$$H_a(s) = \frac{B(s)}{A(s)} = \frac{\displaystyle\sum_{k=0}^{M} \beta_k s^k}{\displaystyle\sum_{k=0}^{N} \alpha_k s^k} \tag{8.3.1}$$

where $\{\alpha_k\}$ and $\{\beta_k\}$ are the filter coefficients, or by its impulse response, which is related to $H_a(s)$ by the Laplace transform

$$H_a(s) = \int_{-\infty}^{\infty} h(t)e^{-st}dt \tag{8.3.2}$$

Alternatively, the analog filter having the rational system function $H(s)$ given in (8.3.1), can be described by the linear constant-coefficient differential equation

$$\sum_{k=0}^{N} \alpha_k \frac{d^k y(t)}{dt^k} = \sum_{k=0}^{M} \beta_k \frac{d^k x(t)}{dt^k} \tag{8.3.3}$$

where $x(t)$ denotes the input signal and $y(t)$ denotes the output of the filter.

Each of these three equivalent characterizations of an analog filter leads to alternative methods for converting the filter into the digital domain, as will be described in Sections 8.3.1 through 8.3.4. We recall that an analog linear time-invariant system with system function $H(s)$ is stable if all its poles lie in the left half of the s-plane. Consequently, if the conversion technique is to be effective, it should possess the following desirable properties:

1. The $j\Omega$ axis in the s-plane should map into the unit circle in the z-plane. Thus there will be a direct relationship between the two frequency variables in the two domains.

2. The left-half plane (LHP) of the s-plane should map into the inside of the unit circle in the z-plane. Thus a stable analog filter will be converted to a stable digital filter.

We mentioned in the preceding section that physically realizable and stable IIR filters cannot have linear phase. Recall that a linear-phase filter must have a system function that satisfies the condition

$$H(z) = \pm z^{-N} H(z^{-1}) \qquad (8.3.4)$$

where z^{-N} represents a delay of N units of time. But if this were the case, the filter would have a mirror-image pole outside the unit circle for every pole inside the unit circle. Hence the filter would be unstable. Consequently, a causal and stable IIR filter cannot have linear phase.

If the restriction on physical realizability is removed, it is possible to obtain a linear-phase IIR filter, at least in principle. This approach involves performing a time reversal of the input signal $x(n)$, passing $x(-n)$ through a digital filter $H(z)$, time-reversing the output of $H(z)$, and finally, passing the result through $H(z)$ again. This signal processing is computationally cumbersome and appears to offer no advantages over linear-phase FIR filters. Consequently, when an application requires a linear-phase filter, it should be an FIR filter.

In the design of IIR filters, we shall specify the desired filter characteristics for the magnitude response only. This does not mean that we consider the phase response unimportant. Since the magnitude and phase characteristics are related, as indicated in Section 8.1, we specify the desired magnitude characteristics and accept the phase response that is obtained from the design methodology.

8.3.1 IIR Filter Design by Approximation of Derivatives

One of the simplest methods for converting an analog filter into a digital filter is to approximate the differential equation in (8.3.3) by an equivalent difference equation. This approach is often used to solve a linear constant-coefficient differential equation numerically on a digital computer.

For the derivative $dy(t)/dt$ at time $t = nT$, we substitute the *backward difference* $[y(nT) - y(nT - 1)]/T$. Thus

$$\left. \frac{dy(t)}{dt} \right|_{t=nT} = \frac{y(nT) - y(nT - T)}{T}$$

$$= \frac{y(n) - y(n - 1)}{T} \qquad (8.3.5)$$

where T represents the sampling interval and $y(n) \equiv y(nT)$. The analog differentiator with output $dy(t)/dt$ has the system function $H(s) = s$, while the digital system that produces the output $[y(n) - y(n - 1)]/T$ has the system function $H(z) = (1 - z^{-1})/T$. Consequently, as shown in Fig. 8.29, the frequency-domain

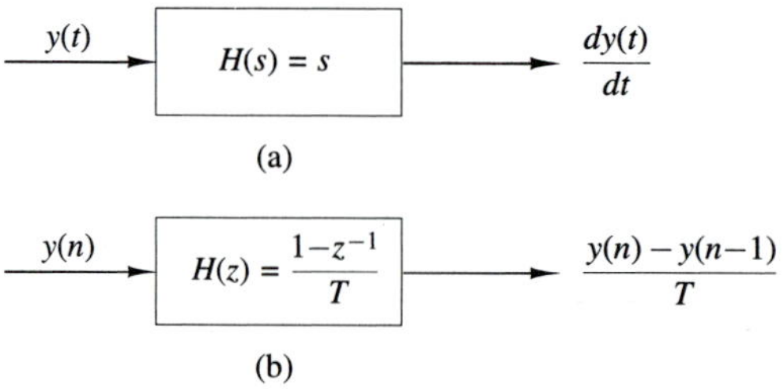

(a)

(b)

Figure 8.29 Substitution of the backward difference for the derivative implies the mapping $s = (1 - z^{-1})/T$.

equivalent for the relationship in (8.3.5) is

$$s = \frac{1 - z^{-1}}{T} \tag{8.3.6}$$

The second derivative $d^2 y(t)/dt^2$ is replaced by the second difference, which is derived as follows:

$$
\begin{aligned}
\frac{d^2 y(t)}{dt^2}\bigg|_{t=nT} &= \frac{d}{dt}\left[\frac{dy(t)}{dt}\right]_{t=nT} \\
&= \frac{[y(nT) - y(nT - T)]/T - [y(nT - T) - y(nT - 2T)]/T}{T} \\
&= \frac{y(n) - 2y(n-1) + y(n-2)}{T^2}
\end{aligned} \tag{8.3.7}
$$

In the frequency domain, (8.3.7) is equivalent to

$$s^2 = \frac{1 - 2z^{-1} + z^{-2}}{T^2} = \left(\frac{1 - z^{-1}}{T}\right)^2 \tag{8.3.8}$$

It easily follows from the discussion that the substitution for the kth derivative of $y(t)$ results in the equivalent frequency-domain relationship

$$s^k = \left(\frac{1 - z^{-1}}{T}\right)^k \tag{8.3.9}$$

Consequently, the system function for the digital IIR filter obtained as a result of the approximation of the derivatives by finite differences is

$$H(z) = H_a(s)|_{s=(1-z^{-1})/T} \tag{8.3.10}$$

where $H_a(s)$ is the system function of the analog filter characterized by the differential equation given in (8.3.3).

Let us investigate the implications of the mapping from the s-plane to the z-plane as given by (8.3.6) or, equivalently,

$$z = \frac{1}{1 - sT} \tag{8.3.11}$$

If we substitute $s = j\Omega$ in (8.2.11), we find that

$$z = \frac{1}{1 - j\Omega T}$$

$$= \frac{1}{1 + \Omega^2 T^2} + j \frac{\Omega T}{1 + \Omega^2 T^2} \tag{8.3.12}$$

As Ω varies from $-\infty$ to ∞, the corresponding locus of points in the z-plane is a circle of radius $\frac{1}{2}$ and with center at $z = \frac{1}{2}$, as illustrated in Fig. 8.30.

It is easily demonstrated that the mapping in (8.3.11) takes points in the LHP of the s-plane into corresponding points inside this circle in the z-plane and points in the RHP of the s-plane are mapped into points outside this circle. Consequently, this mapping has the desirable property that a stable analog filter is transformed into a stable digital filter. However, the possible location of the poles of the digital filter are confined to relatively small frequencies and, as a consequence, the mapping is restricted to the design of lowpass filters and bandpass filters having relatively small resonant frequencies. It is not possible, for example, to transform a highpass analog filter into a corresponding highpass digital filter.

In an attempt to overcome the limitations in the mapping given above, more complex substitutions for the derivatives have been proposed. In particular, an Lth-order difference of the form

$$\frac{dy(t)}{dt}\bigg|_{t=nT} = \frac{1}{T} \sum_{k=1}^{L} \alpha_k \frac{y(nT + kT) - y(nT - kT)}{T} \tag{8.3.13}$$

has been proposed, where $\{\alpha_k\}$ are a set of parameters that can be selected to optimize the approximation. The resulting mapping between the s-plane and the z-plane is now

$$s = \frac{1}{T} \sum_{k=1}^{L} \alpha_k (z^k - z^{-k}) \tag{8.3.14}$$

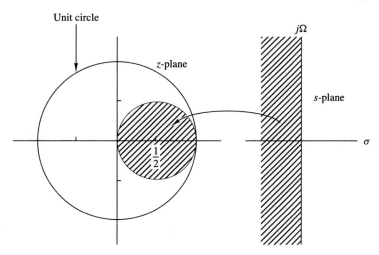

Figure 8.30 The mapping $s = (1 - z^{-1})/T$ takes LHP in the s-plane into points inside the circle of radius $\frac{1}{2}$ and center $z = \frac{1}{2}$ in the z-plane.

When $z = e^{j\omega}$, we have

$$s = j\frac{2}{T}\sum_{k=1}^{L}\alpha_k \sin \omega k \qquad (8.3.15)$$

which is purely imaginary. Thus

$$\Omega = \frac{2}{T}\sum_{k=1}^{L}\alpha_k \sin \omega k \qquad (8.3.16)$$

is the resulting mapping between the two frequency variables. By proper choice of the coefficients $\{\alpha_k\}$ it is possible to map the $j\Omega$-axis into the unit circle. Furthermore, points in the LHP in s can be mapped into points inside the unit circle in z.

Despite achieving the two desirable characteristics with the mapping of (8.3.16), the problem of selecting the set of coefficients $\{\alpha_k\}$ remains. In general, this is a difficult problem. Since simpler techniques exist for converting analog filters into IIR digital filters, we shall not emphasize the use of the Lth-order difference as a substitute for the derivative.

Example 8.3.1

Convert the analog bandpass filter with system function

$$H_a(s) = \frac{1}{(s + 0.1)^2 + 9}$$

into a digital IIR filter by use of the backward difference for the derivative.

Solution   Substitution for s from (8.3.6) into $H(s)$ yields

$$H(z) = \frac{1}{\left(\dfrac{1 - z^{-1}}{T} + 0.1\right)^2 + 9}$$

$$= \frac{T^2/(1 + 0.2T + 9.01T^2)}{1 - \dfrac{2(1 + 0.1T)}{1 + 0.2T + 9.01T^2}z^{-1} + \dfrac{1}{1 + 0.2T + 9.01T^2}z^{-2}}$$

The system function $H(z)$ has the form of a resonator provided that T is selected small enough (e.g., $T \leq 0.1$), in order for the poles to be near the unit circle. Note that the condition $a_1^2 < 4a_2$ is satisfied, so that the poles are complex valued.

For example, if $T = 0.1$, the poles are located at

$$p_{1,2} = 0.91 \pm j0.27$$

$$= 0.949e^{\pm j16.5°}$$

We note that the range of resonant frequencies is limited to low frequencies, due to the characteristics of the mapping. The reader is encouraged to plot the frequency response $H(\omega)$ of the digital filter for different values of T and compare the results with the frequency response of the analog filter.

Example 8.3.2

Convert the analog bandpass filter in Example 8.3.1 into a digital IIR filter by use of the mapping

$$s = \frac{1}{T}(z - z^{-1})$$

Solution   By substituting for s in $H(s)$, we obtain

$$H(z) = \frac{1}{\left(\dfrac{z - z^{-1}}{T} + 0.1\right)^2 + 9}$$

$$= \frac{z^2 T^2}{z^4 + 0.2Tz^3 + (2 + 9.01T^2)z^2 - 0.2Tz + 1}$$

We observe that this mapping has introduced two additional poles in the conversion from $H_a(s)$ to $H(z)$. As a consequence, the digital filter is significantly more complex than the analog filter. This is a major drawback to the mapping given above.

8.3.2 IIR Filter Design by Impulse Invariance

In the impulse invariance method, our objective is to design an IIR filter having a unit sample response $h(n)$ that is the sampled version of the impulse response of the analog filter. That is,

$$h(n) \equiv h(nT) \qquad n = 0, 1, 2, \ldots \qquad (8.3.17)$$

where T is the sampling interval.

To examine the implications of (8.3.17), we refer back to Section 4.2.9. Recall that when a continuous time signal $x_a(t)$ with spectrum $X_a(F)$ is sampled at a rate $F_s = 1/T$ samples per second, the spectrum of the sampled signal is the periodic repetition of the scaled spectrum $F_s X_a(F)$ with period F_s. Specifically, the relationship is

$$X(f) = F_s \sum_{k=-\infty}^{\infty} X_a[(f - k)F_s] \qquad (8.3.18)$$

where $f = F/F_s$ is the normalized frequency. Aliasing occurs if the sampling rate F_s is less than twice the highest frequency contained in $X_a(F)$.

Expressed in the context of sampling the impulse response of an analog filter with frequency response $H_a(F)$, the digital filter with unit sample response $h(n) \equiv h_a(nT)$ has the frequency response

$$H(f) = F_s \sum_{k=-\infty}^{\infty} H_a[(f - k)F_s] \qquad (8.3.19)$$

or, equivalently,

$$H(\omega) = F_s \sum_{k=-\infty}^{\infty} H_a[(\omega - 2\pi k)F_s] \qquad (8.3.20)$$

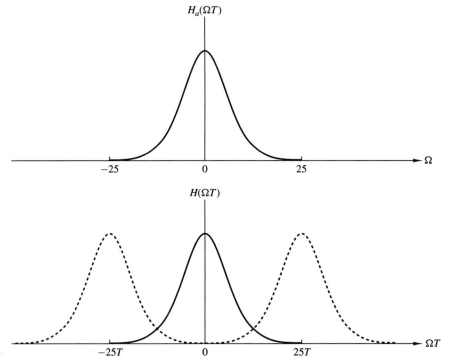

Figure 8.31 Frequency response $H_a(\Omega)$ of the analog filter and frequency response of the corresponding digital filter with aliasing.

or

$$H(\Omega T) = \frac{1}{T} \sum_{k=-\infty}^{\infty} H_a\left(\Omega - \frac{2\pi k}{T}\right) \qquad (8.3.21)$$

Figure 8.31 depicts the frequency response of a lowpass analog filter and the frequency response of the corresponding digital filter.

It is clear that the digital filter with frequency response $H(\omega)$ has the frequency response characteristics of the corresponding analog filter if the sampling interval T is selected sufficiently small to completely avoid or at least minimize the effects of aliasing. It is also clear that the impulse invariance method is inappropriate for designing highpass filters due the to spectrum aliasing that results from the sampling process.

To investigate the mapping of points between the z-plane and the s-plane implied by the sampling process, we rely on a generalization of (8.3.21) which relates the z-transform of $h(n)$ to the Laplace transform of $h_a(t)$. This relationship is

$$H(z)|_{z=e^{sT}} = \frac{1}{T} \sum_{k=-\infty}^{\infty} H_a\left(s - j\frac{2\pi k}{T}\right) \qquad (8.3.22)$$

where

$$H(z) = \sum_{n=0}^{\infty} h(n)z^{-n}$$

$$H(z)|_{z=e^{sT}} = \sum_{n=0}^{\infty} h(n)e^{-sTn} \qquad (8.3.23)$$

Note that when $s = j\Omega$, (8.3.22) reduces to (8.3.21), where the factor of j in $H_a(\Omega)$ is suppressed in our notation.

Let us consider the mapping of points from the s-plane to the z-plane implied by the relation

$$z = e^{sT} \qquad (8.3.24)$$

If we substitute $s = \sigma + j\Omega$ and express the complex variable z in polar form as $z = re^{j\omega}$, (8.3.24) becomes

$$re^{j\omega} = e^{\sigma T}e^{j\Omega T}$$

Clearly, we must have

$$r = e^{\sigma T}$$

$$\omega = \Omega T \qquad (8.3.25)$$

Consequently, $\sigma < 0$ implies that $0 < r < 1$ and $\sigma > 0$ implies that $r > 1$. When $\sigma = 0$, we have $r = 1$. Therefore, the LHP in s is mapped inside the unit circle in z and the RHP in s is mapped outside the unit circle in z.

Also, the $j\Omega$-axis is mapped into the unit circle in z as indicated above. However, the mapping of the $j\Omega$-axis into the unit circle is not one-to-one. Since ω is unique over the range $(-\pi, \pi)$, the mapping $\omega = \Omega T$ implies that the interval $-\pi/T \leq \Omega \leq \pi/T$ maps into the corresponding values of $-\pi \leq \omega \leq \pi$. Furthermore, the frequency interval $\pi/T \leq \Omega \leq 3\pi/T$ also maps into the interval $-\pi \leq \omega \leq \pi$ and, in general, so does the interval $(2k-1)\pi/T \leq \Omega \leq (2k+1)\pi/T$, when k is an integer. Thus the mapping from the analog frequency Ω to the frequency variable ω in the digital domain is many-to-one, which simply reflects the effects of aliasing due to sampling. Figure 8.32 illustrates the mapping from the s-plane to the z-plane for the relation in (8.3.24).

To explore further the effect of the impulse invariance design method on the characteristics of the resulting filter, let us express the system function of the analog filter in partial-fraction form. On the assumption that the poles of the analog filter are distinct, we can write

$$H_a(s) = \sum_{k=1}^{N} \frac{c_k}{s - p_k} \qquad (8.3.26)$$

where $\{p_k\}$ are the poles of the analog filter and $\{c_k\}$ are the coefficients in the partial-fraction expansion. Consequently,

$$h_a(t) = \sum_{k=1}^{N} c_k e^{p_k t} \qquad t \geq 0 \qquad (8.3.27)$$

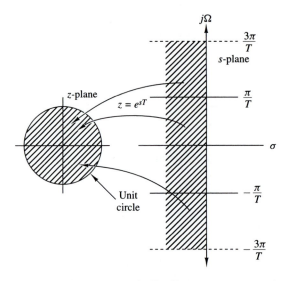

Figure 8.32 The mapping of $z = e^{sT}$ maps strips of width $2\pi/T$ (for $\sigma < 0$) in the s-plane into points in the unit circle in the z-plane.

If we sample $h_a(t)$ periodically at $t = nT$, we have

$$h(n) = h_a(nT)$$

$$= \sum_{k=1}^{N} c_k e^{p_k T n} \tag{8.3.28}$$

Now, with the substitution of (8.3.28), the system function of the resulting digital IIR filter becomes

$$H(z) = \sum_{n=0}^{\infty} h(n) z^{-n}$$

$$= \sum_{n=0}^{\infty} \left(\sum_{k=1}^{N} c_k e^{p_k T n} \right) z^{-n}$$

$$= \sum_{k=1}^{N} c_k \sum_{n=0}^{\infty} (e^{p_k T} z^{-1})^n \tag{8.3.29}$$

The inner sum in (8.3.29) converges because $p_k < 0$ and yields

$$\sum_{n=0}^{\infty} (e^{p_k T} z^{-1})^n = \frac{1}{1 - e^{p_k T} z^{-1}} \tag{8.3.30}$$

Therefore, the system function of the digital filter is

$$H(z) = \sum_{k=1}^{N} \frac{c_k}{1 - e^{p_k T} z^{-1}} \tag{8.3.31}$$

We observe that the digital filter has poles at

$$z_k = e^{p_k T} \qquad k = 1, 2, \ldots, N \tag{8.3.32}$$

Although the poles are mapped from the s-plane to the z-plane by the relationship in (8.3.32), we should emphasize that the zeros in the two domains do not satisfy the same relationship. Therefore, the impulse invariance method does not correspond to the simple mapping of points given by (8.3.24).

The development that resulted in $H(z)$ given by (8.3.31) was based on a filter having distinct poles. It can be generalized to include multiple-order poles. For brevity, however, we shall not attempt to generalize (8.3.31).

Example 8.3.3

Convert the analog filter with system function

$$H_a(s) = \frac{s + 0.1}{(s + 0.1)^2 + 9}$$

into a digital IIR filter by means of the impulse invariance method.

Solution   We note that the analog filter has a zero at $s = -0.1$ and a pair of complex-conjugate poles at

$$p_k = -0.1 \pm j3$$

as illustrated in Fig. 8.33.

We do not have to determine the impulse response $h_a(t)$ in order to design the digital IIR filter based on the method of impulse invariance. Instead, we directly determine $H(z)$, as given by (8.2.31), from the partial-fraction expansion of $H_a(s)$. Thus we have

$$H(s) = \frac{\frac{1}{2}}{s + 0.1 - j3} + \frac{\frac{1}{2}}{s + 0.1 + j3}$$

Then

$$H(z) = \frac{\frac{1}{2}}{1 - e^{-0.1T}e^{j3T}z^{-1}} + \frac{\frac{1}{2}}{1 - e^{-0.1T}e^{-j3T}z^{-1}}$$

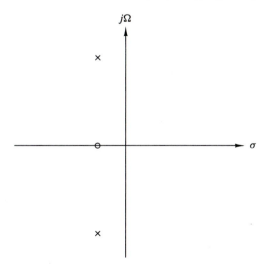

Figure 8.33 Pole–zero locations for analog filter in Example 8.3.3.

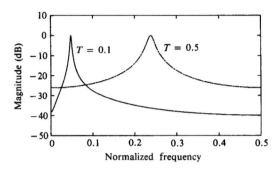

Figure 8.34 Frequency response of digital filter in Example 8.3.3.

Since the two poles are complex conjugates, we can combine them to form a single two-pole filter with system function

$$H(z) = \frac{1 - (e^{-0.1T} \cos 3T)z^{-1}}{1 - (2e^{-0.1T} \cos 3T)z^{-1} + e^{-0.2T}z^{-1}}$$

The magnitude of the frequency response characteristic of this filter is plotted in Fig. 8.34 for $T = 0.1$ and $T = 0.5$. For purpose of comparison, we have also plotted the magnitude of the frequency response of the analog filter in Fig. 8.35, We note that aliasing is significantly more prevalent when $T = 0.5$ than when $T = 0.1$. Also, note the shift of the resonant frequency as T changes.

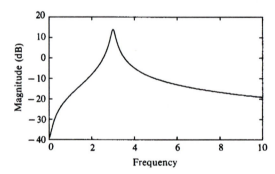

Figure 8.35 Frequency response of analog filter in Example 8.3.3.

The preceding example illustrates the importance of selecting a small value for T to minimize the effect of aliasing. Due to the presence of aliasing, the impulse invariance method is appropriate for the design of lowpass and bandpass filters only.

8.3.3 IIR Filter Design by the Bilinear Transformation

The IIR filter design techniques described in the preceding two sections have a severe limitation in that they are appropriate only for lowpass filters and a limited class of bandpass filters.

In this section we describe a mapping from the s-plane to the z-plane, called the bilinear transformation, that overcomes the limitation of the other two design methods described previously. The bilinear transformation is a conformal mapping that transforms the $j\Omega$-axis into the unit circle in the z-plane only once, thus avoiding aliasing of frequency components. Furthermore, all points in the LHP of s are mapped inside the unit circle in the z-plane and all points in the RHP of s are mapped into corresponding points outside the unit circle in the z-plane.

The bilinear transformation can be linked to the trapezoidal formula for numerical integration. For example, let us consider an analog linear filter with system function

$$H(s) = \frac{b}{s+a} \tag{8.3.33}$$

This system is also characterized by the differential equation

$$\frac{dy(t)}{dt} + ay(t) = bx(t) \tag{8.3.34}$$

Instead of substituting a finite difference for the derivative, suppose that we integrate the derivative and approximate the integral by the trapezoidal formula. Thus

$$y(t) = \int_{t_0}^{t} y'(\tau)d\tau + y(t_0) \tag{8.3.35}$$

where $y'(t)$ denotes the derivative of $y(t)$. The approximation of the integral in (8.3.35) by the trapezoidal formula at $t = nT$ and $t_0 = nT - T$ yields

$$y(nT) = \frac{T}{2}[y'(nT) + y'(nT - T)] + y(nT - T) \tag{8.3.36}$$

Now the differential equation in (8.3.34) evaluated at $t = nT$ yields

$$y'(nT) = -ay(nT) + bx(nT) \tag{8.3.37}$$

We use (8.3.37) to substitute for the derivative in (8.3.36) and thus obtain a difference equation for the equivalent discrete-time system. With $y(n) \equiv y(nT)$ and $x(n) \equiv x(nT)$, we obtain the result

$$\left(1 + \frac{aT}{2}\right) y(n) - \left(1 - \frac{aT}{2}\right) y(n-1) = \frac{bT}{2}[x(n) + x(n-1)] \tag{8.3.38}$$

The z-transform of this difference equation is

$$\left(1 + \frac{aT}{2}\right) Y(z) - \left(1 - \frac{aT}{2}\right) z^{-1} Y(z) = \frac{bT}{2}(1 + z^{-1}) X(z)$$

Consequently, the system function of the equivalent digital filter is

$$H(z) = \frac{Y(z)}{X(z)} = \frac{(bT/2)(1 + z^{-1})}{1 + aT/2 - (1 - aT/2)z^{-1}}$$

or, equivalently,

$$H(z) = \frac{b}{\frac{2}{T}\left(\frac{1 - z^{-1}}{1 + z^{-1}}\right) + a} \tag{8.3.39}$$

Clearly, the mapping from the s-plane to the z-plane is

$$s = \frac{2}{T} \left(\frac{1 - z^{-1}}{1 + z^{-1}} \right) \qquad (8.3.40)$$

This is called the *bilinear transformation*.

Although our derivation of the bilinear transformation was performed for a first-order differential equation, it holds, in general, for an Nth-order differential equation.

To investigate the characteristics of the bilinear transformation, let

$$z = re^{j\omega}$$

$$s = \sigma + j\Omega$$

Then (8.3.40) can be expressed as

$$s = \frac{2}{T} \frac{z - 1}{z + 1}$$

$$= \frac{2}{T} \frac{re^{j\omega} - 1}{re^{j\omega} + 1}$$

$$= \frac{2}{T} \left(\frac{r^2 - 1}{1 + r^2 + 2r\cos\omega} + j\frac{2r\sin\omega}{1 + r^2 + 2r\cos\omega} \right)$$

Consequently,

$$\sigma = \frac{2}{T} \frac{r^2 - 1}{1 + r^2 + 2r\cos\omega} \qquad (8.3.41)$$

$$\Omega = \frac{2}{T} \frac{2r\sin\omega}{1 + r^2 + 2r\cos\omega} \qquad (8.3.42)$$

First, we note that if $r < 1$, then $\sigma < 0$, and if $r > 1$, then $\sigma > 0$. Consequently, the LHP in s maps into the inside of the unit circle in the z-plane and the RHP in s maps into the outside of the unit circle. When $r = 1$, then $\sigma = 0$ and

$$\Omega = \frac{2}{T} \frac{\sin\omega}{1 + \cos\omega}$$

$$= \frac{2}{T} \tan\frac{\omega}{2} \qquad (8.3.43)$$

or, equivalently,

$$\omega = 2\tan^{-1}\frac{\Omega T}{2} \qquad (8.3.44)$$

The relationship in (8.3.44) between the frequency variables in the two domains is illustrated in Fig. 8.36. We observe that the entire range in Ω is mapped only once into the range $-\pi \le \omega \le \pi$. However, the mapping is highly nonlinear. We observe a frequency compression or *frequency warping*, as it is usually called, due to the nonlinearity of the arctangent function.

It is also interesting to note that the bilinear transformation maps the point $s = \infty$ into the point $z = -1$. Consequently, the single-pole lowpass filter in

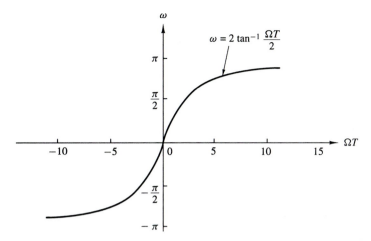

Figure 8.36 Mapping between the frequency variables ω and Ω resulting from the bilinear transformation.

(8.3.33), which has a zero at $s = \infty$, results in a digital filter that has a zero at $z = -1$.

Example 8.3.4

Convert the analog filter with system function

$$H_a(s) = \frac{s + 0.1}{(s + 0.1)^2 + 16}$$

into a digital IIR filter by means of the bilinear transformation. The digital filter is to have a resonant frequency of $\omega_r = \pi/2$.

Solution   First, we note that the analog filter has a resonant frequency $\Omega_r = 4$. This frequency is to be mapped into $\omega_r = \pi/2$ by selecting the value of the parameter T. From the relationship in (8.3.43), we must select $T = \frac{1}{2}$ in order to have $\omega_r = \pi/2$. Thus the desired mapping is

$$s = 4\left(\frac{1 - z^{-1}}{1 + z^{-1}}\right)$$

The resulting digital filter has the system function

$$H(z) = \frac{0.128 + 0.006z^{-1} - 0.122z^{-2}}{1 + 0.0006z^{-1} + 0.975z^{-2}}$$

We note that the coefficient of the z^{-1} term in the denominator of $H(z)$ is extremely small and can be approximated by zero. Thus we have the system function

$$H(z) = \frac{0.128 + 0.006z^{-1} - 0.122z^{-2}}{1 + 0.975z^{-2}}$$

This filter has poles at

$$p_{1,2} = 0.987e^{\pm j\pi/2}$$

and zeros at

$$z_{1,2} = -1, 0.95$$

Therefore, we have succeeded in designing a two-pole filter that resonates near $\omega = \pi/2$.

In this example the parameter T was selected to map the resonant frequency of the analog filter into the desired resonant frequency of the digital filter. Usually, the design of the digital filter begins with specifications in the digital domain, which involve the frequency variable ω. These specifications in frequency are converted to the analog domain by means of the relation in (8.3.43). The analog filter is then designed that meets these specifications and converted to a digital filter by means of the bilinear transformation in (8.3.40). In this procedure, the parameter T is transparent and may be set to any arbitrary value (e.g., $T = 1$). The following example illustrates this point.

Example 8.3.5

Design a single-pole lowpass digital filter with a 3-dB bandwidth of 0.2π, using the bilinear transformation applied to the analog filter

$$H(s) = \frac{\Omega_c}{s + \Omega_c}$$

where Ω_c is the 3-dB bandwidth of the analog filter.

Solution The digital filter is specified to have its -3-dB gain at $\omega_c = 0.2\pi$. In the frequency domain of the analog filter $\omega_c = 0.2\pi$ corresponds to

$$\Omega_c = \frac{2}{T} \tan 0.1\pi$$

$$= \frac{0.65}{T}$$

Thus the analog filter has the system function

$$H(s) = \frac{0.65/T}{s + 0.65/T}$$

This represents our filter design in the analog domain.

Now, we apply the bilinear transformation given by (8.3.40) to convert the analog filter into the desired digital filter. Thus we obtain

$$H(z) = \frac{0.245(1 + z^{-1})}{1 - 0.509z^{-1}}$$

where the parameter T has been divided out.

The frequency response of the digital filter is

$$H(\omega) = \frac{0.245(1 + e^{-j\omega})}{1 - 0.509e^{-j\omega}}$$

At $\omega = 0$, $H(0) = 1$, and at $\omega = 0.2\pi$, we have $|H(0.2\pi)| = 0.707$, which is the desired response.

8.3.4 The Matched-*z* Transformation

Another method for converting an analog filter into an equivalent digital filter is to map the poles and zeros of $H(s)$ directly into poles and zeros in the z-plane. Suppose that the system function of the analog filter is expressed in the factored form

$$H(s) = \frac{\displaystyle\prod_{k=1}^{M}(s - z_k)}{\displaystyle\prod_{k=1}^{N}(s - p_k)} \tag{8.3.45}$$

where $\{z_k\}$ are the zeros and $\{p_k\}$ are the poles of the filter. Then the system function for the digital filter is

$$H(z) = \frac{\displaystyle\prod_{k=1}^{M}(1 - e^{z_k T}z^{-1})}{\displaystyle\prod_{k=1}^{N}(1 - e^{p_k T}z^{-1})} \tag{8.3.46}$$

where T is the sampling interval. Thus each factor of the form $(s - a)$ in $H(s)$ is mapped into the factor $(1 - e^{aT}z^{-1})$. This mapping is called the *matched-z transformation*.

   We observe that the poles obtained from the matched-z transformation are identical to the poles obtained with the impulse invariance method. However, the two techniques result in different zero positions.

   To preserve the frequency response characteristic of the analog filter, the sampling interval in the matched-z transformation must be properly selected to yield the pole and zero locations at the equivalent position in the z-plane. Thus aliasing must be avoided by selecting T sufficiently small.

8.3.5 Characteristics of Commonly Used Analog Filters

As we have seen from our discussion above, IIR digital filters can easily be obtained by beginning with an analog filter and then using a mapping to transform the s-plane into the z-plane. Thus the design of a digital filter is reduced to designing an appropriate analog filter and then performing the conversion from $H(s)$ to $H(z)$, in such a way so as to preserve as much as possible, the desired characteristics of the analog filter.

   Analog filter design is a well-developed field and many books have been written on the subject. In this section we briefly describe the important characteristics of commonly used analog filters and introduce the relevant filter parameters. Our discussion is limited to lowpass filters. Subsequently, we describe several frequency transformations that convert a lowpass prototype filter into either a bandpass, highpass, or band-elimination filter.

Butterworth filters. Lowpass Butterworth filters are all-pole filters characterized by the magnitude-squared frequency response

$$|H(\Omega)|^2 = \frac{1}{1 + (\Omega/\Omega_c)^{2N}} = \frac{1}{1 + \epsilon^2(\Omega/\Omega_p)^{2N}} \tag{8.3.47}$$

where N is the order of the filter, Ω_c is its -3-dB frequency (usually called the cutoff frequency), Ω_p is the passband edge frequency, and $1/(1 + \epsilon^2)$ is the band-edge value of $|H(\Omega)|^2$. Since $H(s)H(-s)$ evaluated at $s = j\Omega$ is simply equal to $|H(\Omega)|^2$, it follows that

$$H(s)H(-s) = \frac{1}{1 + (-s^2/\Omega_c^2)^N} \tag{8.3.48}$$

The poles of $H(s)H(-s)$ occur on a circle of radius Ω_c at equally spaced points. From (8.3.48) we find that

$$\frac{-s^2}{\Omega_c^2} = (-1)^{1/N} = e^{j(2k+1)\pi/N} \qquad k = 0, 1, \ldots, N - 1$$

and hence

$$s_k = \Omega_c e^{j\pi/2} e^{j(2k+1)\pi/2N} \qquad k = 0, 1, \ldots, N - 1 \tag{8.3.49}$$

For example, Fig. 8.37 illustrates the pole positions for an $N = 4$ and $N = 5$ Butterworth filters.

The frequency response characteristics of the class of Butterworth filters are shown in Fig. 8.38 for several values of N. We note that $|H(\Omega)|^2$ is monotonic in both the passband and stopband. The order of the filter required to meet an attenuation δ_2 at a specified frequency Ω_s is easily determined from (8.3.47). Thus at $\Omega = \Omega_s$ we have

$$\frac{1}{1 + \epsilon^2(\Omega_s/\Omega_p)^{2N}} = \delta_2^2$$

and hence

$$N = \frac{\log[(1/\delta_2^2) - 1]}{2\log(\Omega_s/\Omega_c)} = \frac{\log(\delta/\epsilon)}{\log(\Omega_s/\Omega_p)} \tag{8.3.50}$$

where, by definition, $\delta_2 = 1/\sqrt{1 + \delta^2}$. Thus the Butterworth filter is completely characterized by the parameters N, δ_2, ϵ, and the ratio Ω_s/Ω_p.

Example 8.3.6

Determine the order and the poles of a lowpass Butterworth filter that has a -3-dB bandwidth of 500 Hz and an attenuation of 40 dB at 1000 Hz.

Solution The critical frequencies are the -3-dB frequency Ω_c and the stopband frequency Ω_s, which are

$$\Omega_c = 1000\pi$$

$$\Omega_s = 2000\pi$$

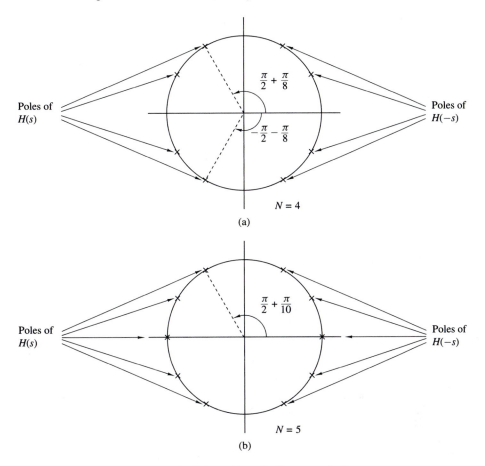

$$\frac{\pi}{2} + \frac{\pi}{8}$$

$$-\frac{\pi}{2} - \frac{\pi}{8}$$

Poles of
$H(s)$

Poles of
$H(-s)$

$N = 4$

(a)

$$\frac{\pi}{2} + \frac{\pi}{10}$$

Poles of
$H(s)$

Poles of
$H(-s)$

$N = 5$

(b)

Figure 8.37   Pole positions for Butterworth filters.

For an attenuation of 40 dB, $\delta_2 = 0.01$. Hence from (8.3.50) we obtain

$$N = \frac{\log_{10}(10^4 - 1)}{2 \log_{10} 2}$$

$$= 6.64$$

To meet the desired specifications, we select $N = 7$. The pole positions are

$$s_k = 1000\pi \, e^{j[\pi/2 + (2k+1)\pi/14]} \qquad k = 0, 1, 2, \ldots, 6$$

Chebyshev filters.   There are two types of Chebyshev filters. Type I Chebyshev filters are all-pole filters that exhibit equiripple behavior in the passband and a monotonic characteristic in the stopband. On the other hand, the family of type II Chebyshev filters contains both poles and zeros and exhibits a

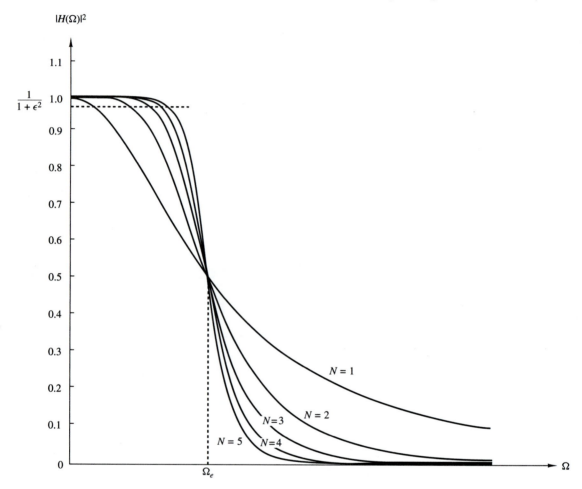

Figure 8.38   Frequency response of Butterworth filters.

monotonic behavior in the passband and an equiripple behavior in the stopband. The zeros of this class of filters lie on the imaginary axis in the s-plane.

The magnitude squared of the frequency response characteristic of a type I Chebyshev filter is given as

$$|H(\Omega)|^2 = \frac{1}{1 + \epsilon^2 T_N^2(\Omega/\Omega_p)} \tag{8.3.51}$$

where ϵ is a parameter of the filter related to the ripple in the passband and $T_N(x)$ is the Nth-order Chebyshev polynomial defined as

$$T_N(x) = \begin{cases} \cos(N\cos^{-1}x), & |x| \le 1 \\ \cosh(N\cosh^{-1}x), & |x| > 1 \end{cases} \tag{8.3.52}$$

The Chebyshev polynomials can be generated by the recursive equation

$$T_{N+1}(x) = 2x T_N(x) - T_{N-1}(x) \qquad N = 1, 2, \ldots \tag{8.3.53}$$

where $T_0(x) = 1$ and $T_1(x) = x$. From (8.3.53) we obtain $T_2(x) = 2x^2 - 1$, $T_3(x) = 4x^3 - 3x$, and so on.

Some of the properties of these polynomials are as follows:

1. $|T_N(x)| \leq 1$ for all $|x| \leq 1$.
2. $T_N(1) = 1$ for all N.
3. All the roots of the polynomial $T_N(x)$ occur in the interval $-1 \leq x \leq 1$.

The filter parameter ϵ is related to the ripple in the passband, as illustrated in Fig. 8.39, for N odd and N even. For N odd, $T_N(0) = 0$ and hence $|H(0)|^2 = 1$. On the other hand, for N even, $T_N(0) = 1$ and hence $|H(0)|^2 = 1/(1 + \epsilon^2)$. At the band edge frequency $\Omega = \Omega_p$, we have $T_N(1) = 1$, so that

$$\frac{1}{\sqrt{1 + \epsilon^2}} = 1 - \delta_1$$

or, equivalently,

$$\epsilon^2 = \frac{1}{(1 - \delta_1)^2} - 1 \tag{8.3.54}$$

where δ_1 is the value of the passband ripple.

The poles of a type I Chebyshev filter lie on an ellipse in the s-plane with major axis

$$r_1 = \Omega_p \frac{\beta^2 + 1}{2\beta} \tag{8.3.55}$$

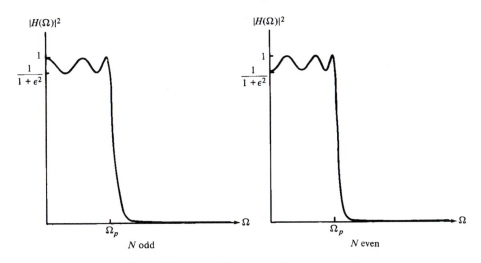

Figure 8.39   Type I Chebyshev filter characteristic.

and minor axis

$$r_2 = \Omega_p \frac{\beta^2 - 1}{2\beta} \tag{8.3.56}$$

where β is related to ϵ according to the equation

$$\beta = \left[\frac{\sqrt{1 + \epsilon^2} + 1}{\epsilon} \right]^{1/N} \tag{8.3.57}$$

The pole locations are most easily determined for a filter of order N by first locating the poles for an equivalent Nth-order Butterworth filter that lie on circles of radius r_1 or radius r_2, as illustrated in Fig. 8.40. If we denote the angular positions of the poles of the Butterworth filter as

$$\phi_k = \frac{\pi}{2} + \frac{(2k + 1)\pi}{2N} \qquad k = 0, 1, 2, \ldots, N - 1 \tag{8.3.58}$$

then the positions of the poles for the Chebyshev filter lie on the ellipse at the coordinates (x_k, y_k), $k = 0, 1, \ldots, N - 1$, where

$$x_k = r_2 \cos \phi_k, \qquad k = 0, 1, \ldots, N - 1$$
$$y_k = r_1 \sin \phi_k, \qquad k = 0, 1, \ldots, N - 1 \tag{8.3.59}$$

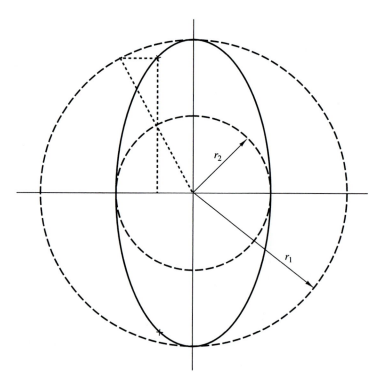

Figure 8.40 Determination of the pole locations for a Chebyshev filter.

A type II Chebyshev filter contains zeros as well as poles. The magnitude squared of its frequency response is given as

$$|H(\Omega)|^2 = \frac{1}{1 + \epsilon^2[T_N^2(\Omega_s/\Omega_p)/T_N^2(\Omega_s/\Omega)]} \tag{8.3.60}$$

where $T_N(x)$ is, again, the Nth-order Chebyshev polynomial and Ω_s is the stopband frequency as illustrated in Fig. 8.41. The zeros are located on the imaginary axis at the points

$$s_k = j\frac{\Omega_s}{\sin\phi_k} \qquad k = 0, 1, \ldots, N-1 \tag{8.3.61}$$

The poles are located at the points (v_k, w_k), where

$$v_k = \frac{\Omega_s x_k}{\sqrt{x_k^2 + y_k^2}} \qquad k = 0, 1, \ldots, N-1 \tag{8.3.62}$$

$$w_k = \frac{\Omega_s y_k}{\sqrt{x_k^2 + y_k^2}} \qquad k = 0, 1, \ldots, N-1 \tag{8.3.63}$$

where $\{x_k\}$ and $\{y_k\}$ are defined in (8.3.59) with β now related to the ripple in the stopband through the equation

$$\beta = \left[\frac{1 + \sqrt{1 - \delta_2^2}}{\delta_2}\right]^{1/N} \tag{8.3.64}$$

From this description, we observe that the Chebyshev filters are characterized by the parameters N, ϵ, δ_2, and the ratio Ω_s/Ω_p. For a given set of specifications

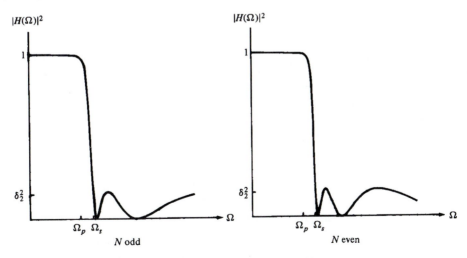

Figure 8.41 Type II Chebyshev filters.

on ϵ, δ_2, and Ω_s/Ω_p, we can determine the order of the filter from the equation

$$N = \frac{\log\left[\left(\sqrt{1-\delta_2^2} + \sqrt{1-\delta_2^2(1+\epsilon^2)}\right)/\epsilon\delta_2\right]}{\log\left[(\Omega_s/\Omega_p) + \sqrt{(\Omega_s/\Omega_p)^2 - 1}\right]} \tag{8.3.65}$$

$$= \frac{\cosh^{-1}(\delta/\epsilon)}{\cosh^{-1}(\Omega_s/\Omega_p)}$$

where, by definition, $\delta_2 = 1/\sqrt{1+\delta^2}$.

Example 8.3.7

Determine the order and the poles of a type I lowpass Chebyshev filter that has a 1-dB ripple in the passband, a cutoff frequency $\Omega_p = 1000\pi$, a stopband frequency of 2000π, and an attenuation of 40 dB or more for $\Omega \geq \Omega_s$.

Solution First, we determine the order of the filter. We have

$$10\log_{10}(1+\epsilon^2) = 1$$

$$1+\epsilon^2 = 1.259$$

$$\epsilon^2 = 0.259$$

$$\epsilon = 0.5088$$

Also,

$$20\log_{10}\delta_2 = -40$$

$$\delta_2 = 0.01$$

Hence from (8.3.65) we obtain

$$N = \frac{\log_{10} 196.54}{\log_{10}(2+\sqrt{3})}$$

$$= 4.0$$

Thus a type I Chebyshev filter having four poles meets the specifications.

The pole positions are determined from the relations in (8.3.55) through (8.3.59). First, we compute β, r_1, and r_2. Hence

$$\beta = 1.429$$

$$r_1 = 1.06\Omega_p$$

$$r_2 = 0.365\Omega_p$$

The angles $\{\phi_k\}$ are

$$\phi_k = \frac{\pi}{2} + \frac{(2k+1)\pi}{8} \qquad k = 0, 1, 2, 3$$

Therefore, the poles are located at

$$x_1 + jy_1 = -0.1397\Omega_p \pm j0.979\Omega_p$$

$$x_2 + jy_2 = -0.337\Omega_p \pm j0.4056\Omega_p$$

The filter specifications in Example 8.3.7 are very similar to the specifications given in Example 8.3.6, which involved the design of a Butterworth filter. In that case the number of poles required to meet the specifications was seven. On the other hand, the Chebyshev filter required only four. This result is typical of such comparisons. In general, the Chebyshev filter meets the specifications with a fewer number of poles than the corresponding Butterworth filter. Alternatively, if we compare a Butterworth filter to a Chebyshev filter having the same number of poles and the same passband and stopband specifications, the Chebyshev filter will have a smaller transition bandwidth. For a tabulation of the characteristics of Chebyshev filters and their pole–zero locations, the interested reader is referred to the handbook of Zverev (1967).

Elliptic filters.   Elliptic (or Cauer) filters exhibit equiripple behavior in both the passband and the stopband, as illustrated in Fig. 8.42 for N odd and N even. This class of filters contains both poles and zeros and is characterized by the magnitude-squared frequency response

$$|H(\Omega)|^2 = \frac{1}{1 + \epsilon^2 U_N(\Omega/\Omega_p)} \tag{8.3.66}$$

where $U_N(x)$ is the Jacobian elliptic function of order N, which has been tabulated by Zverev (1967), and ϵ is a parameter related to the passband ripple. The zeros lie on the $j\Omega$-axis.

We recall from our discussion of FIR filters that the most efficient designs occur when we spread the approximation error equally over the passband and the stopband. Elliptic filters accomplish this objective and, as a consequence, are the most efficient from the viewpoint of yielding the smallest-order filter for a given

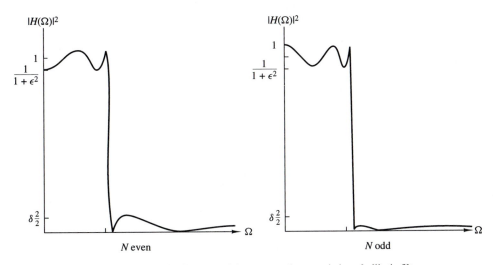

Figure 8.42   Magnitude-squared frequency characteristics of elliptic filters.

set of specifications. Equivalently, we can say that for a given order and a given set of specifications, an elliptic filter has the smallest transition bandwidth.

The filter order required to achieve a given set of specifications in passband ripple δ_1, stopband ripple δ_2, and transition ratio Ω_p/Ω_s is given as

$$N = \frac{K(\Omega_p/\Omega_s)K\left(\sqrt{1-(\epsilon^2/\delta^2)}\right)}{K(\epsilon/\delta)K\left(\sqrt{1-(\Omega_p/\Omega_s)^2}\right)} \tag{8.3.67}$$

where $K(x)$ is the complete elliptic integral of the first kind, defined as

$$K(x) = \int_0^{\pi/2} \frac{d\theta}{\sqrt{1-x^2\sin^2\theta}} \tag{8.3.68}$$

and $\delta_2 = 1/\sqrt{1+\delta^2}$. Values of this integral have been tabulated in a number of texts [e.g., the books by Jahnke and Emde (1945) and Dwight (1957)]. The passband ripple is $10\log_{10}(1+\epsilon^2)$.

We shall not attempt to describe elliptic functions in any detail because such a discussion would take us too far afield. Suffice to say that computer programs are available for designing elliptic filters from the frequency specifications indicated above.

In view of the optimality of elliptic filters, the reader may question the reason for considering the class of Butterworth or the class of Chebyshev filters in practical applications. One important reason that these other types of filters might be preferable in some applications is that they possess better phase response characteristics. The phase response of elliptic filters is more nonlinear in the passband than a comparable Butterworth filter or a Chebyshev filter, especially near the band edge.

Bessel filters. Bessel filters are a class of all-pole filters that are characterized by the system function

$$H(s) = \frac{1}{B_N(s)} \tag{8.3.69}$$

where $B_N(s)$ is the Nth-order Bessel polynomial. These polynomials can be expressed in the form

$$B_N(s) = \sum_{k=0}^{N} a_k s^k \tag{8.3.70}$$

where the coefficients $\{a_k\}$ are given as

$$a_k = \frac{(2N-k)!}{2^{N-k}k!(N-k)!} \qquad k = 0, 1, \ldots, N \tag{8.3.71}$$

Alternatively, the Bessel polynomials may be generated recursively from the relation

$$B_N(s) = (2N-1)B_{N-1}(s) + s^2 B_{N-2}(s) \tag{8.3.72}$$

with $B_0(s) = 1$ and $B_1(s) = s + 1$ as initial conditions.

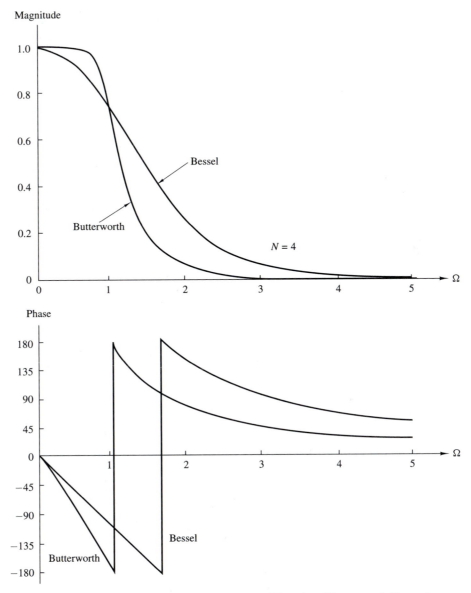

Figure 8.43 Magnitude and phase responses of Bessel and Butterworth filters of order $N = 4$.

An important characteristic of Bessel filters is the linear-phase response over the passband of the filter. For example, Fig. 8.43 shows a comparison of the magnitude and phase responses of a Bessel filter and Butterworth filter of order $N = 4$. We note that the Bessel filter has a larger transition bandwidth, but its phase is linear within the passband. However, we should emphasize that the linear-phase chacteristics of the analog filter are destroyed in the process of converting the filter into the digital domain by means of the transformations decribed previously.

8.3.6 Some Examples of Digital Filter Designs Based on the Bilinear Transformation

In this section we present several examples of digital filter designs obtained from analog filters by applying the bilinear transformation to convert $H(s)$ to $H(z)$. These filters designs are performed with the aid of one of several software packages now available for use on a personal computer.

A lowpass filter is designed to meet specifications of a maximum ripple of $\frac{1}{2}$ dB in the passband, 60-dB attenuation in the stopband, a passband edge frequency of $\omega_p = 0.25\pi$, and a stopband edge frequency of $\omega_s = 0.30\pi$.

A Butterworth filter of order $N = 37$ is required to satisfy the specifications. Its frequency response characteristics are illustrated in Fig. 8.44. If a Chebyshev filter is used, a filter of order $N = 13$ satisfies the specifications. The frequency response characteristics for a type I and type II Chebyshev filters are shown in Figs. 8.45 and 8.46, respectively. The type I filter has a passband ripple of 0.31 dB. Finally, an elliptic filter of order $N = 7$ is designed which also satisfied the specifications. For illustrative purposes, we show in Table 8.11, the numerical values for the filter parameters and the resulting frequency specifications are shown in Fig. 8.47. The following notation is used for the parameters in the function $H(z)$:

$$H(z) = \prod_{i=1}^{K} \frac{b(i, 0) + b(i, 1)z^{-1} + b(i, 2)z^{-2}}{1 + a(i, 1)z^{-1} + a(i, 2)z^{-2}} \tag{8.3.73}$$

Although we have described only lowpass analog filters in the preceding section, it is a simple matter to convert a lowpass analog filter into a bandpass, bandstop, or highpass analog filter by a frequency transformation, as is described in Section 8.4. The bilinear transformation is then applied to convert the analog filter into an equivalent digital filter. As in the case of the lowpass filters described above, the entire design can be carried out on a computer.

8.4 FREQUENCY TRANSFORMATIONS

The treatment in the preceding section is focused primarily on the design of lowpass IIR filters. If we wish to design a highpass or a bandpass or a bandstop filter, it is a simple matter to take a lowpass prototype filter (Butterworth, Chebyshev, elliptic, Bessel) and perform a frequency transformation.

One possibility is to perform the frequency transformation in the analog domain and then to convert the analog filter into a corresponding digital filter by a mapping of the s-plane into the z-plane. An alternative approach is first to

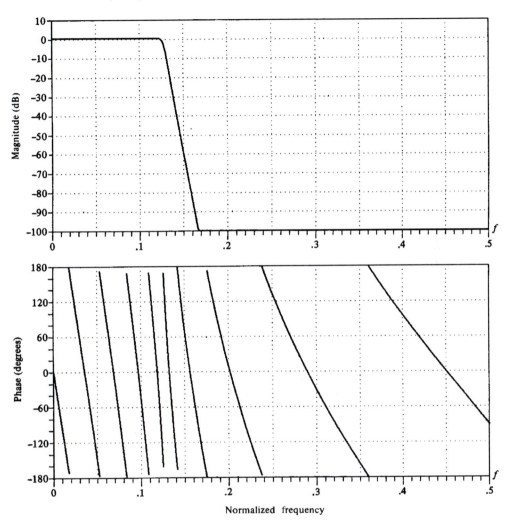

Figure 8.44   Frequency response characteristics of a 37-order Butterworth filter.

convert the analog lowpass filter into a lowpass digital filter and then to transform the lowpass digital filter into the desired digital filter by a digital transformation. In general, these two approaches yield different results, except for the bilinear transformation, in which case the resulting filter designs are identical. These two approaches are described below.

8.4.1 Frequency Transformations in the Analog Domain

First, we consider frequency transformations in the analog domain. Suppose that we have a lowpass filter with passband edge frequency Ω_p and we wish to convert

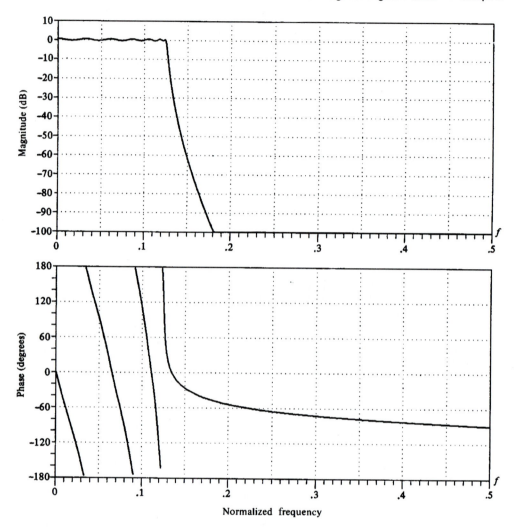

Figure 8.45   Frequency response characteristics of a 13-order type I Chebyshev filter.

it to another lowpass filter with passband edge frequency Ω_p'. The transformation that accomplishes this is

$$s \longrightarrow \frac{\Omega_p}{\Omega_p'} s \qquad \text{(lowpass to lowpass)} \qquad (8.4.1)$$

Thus we obtain a lowpass filter with system function $H_l(s) = H_p[(\Omega_p/\Omega_p')s]$, where $H_p(s)$ is the system function of the prototype filter with passband edge frequency Ω_p.

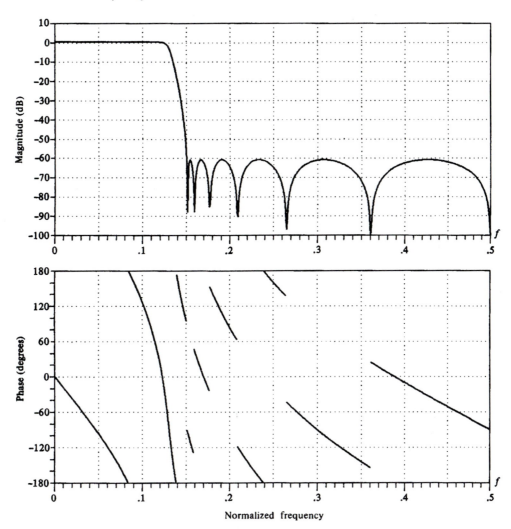

Figure 8.46 Frequency response characteristics of a 13-order type II Chebyshev filter.

If we wish to convert a lowpass filter into a highpass filter with passband edge frequency Ω'_p, the desired transformation is

$$s \longrightarrow \frac{\Omega_p \Omega'_p}{s} \qquad \text{(lowpass to highpass)} \qquad (8.4.2)$$

The system function of the highpass filter is $H_h(s) = H_p(\Omega_p \Omega'_p / s)$.

The transformation for converting a lowpass analog filter with passband edge frequency Ω_p into a band filter, having a lower band edge frequency Ω_l and an upper band edge frequency Ω_u, can be accomplished by first converting the lowpass

TABLE 8.11   FILTER COEFFICIENTS FOR A 7-ORDER ELLIPTIC FILTER

```
              INFINITE IMPULSE RESPONSE (IIR)
                 ELLIPTIC LOWPASS FILTER
                 UNQUANTIZED COEFFICIENTS
         FILTER ORDER = 7
         SAMPLING FREQUENCY = 2.000 KILOHERTZ


         I.   A(I, 1)     A(I, 2)    B(I, 0)    B(I, 1)     B(I, 2)

         1    -.790103    .000000    .104948    .104948    .000000
         2   -1.517223    .714088    .102450   -.007817    .102232
         3   -1.421773    .861895    .420100   -.399842    .419864
         4   -1.387447    .962252    .714929   -.826743    .714841
              *** CHARACTERISTICS OF DESIGNED FILTER ***
                           BAND 1        BAND 2
         LOWER BAND EDGE    .00000        .30000
         UPPER BAND EDGE    .25000       1.00000
         NOMINAL GAIN      1.00000        .00000
         NOMINAL RIPPLE     .05600        .00100
         MAXIMUM RIPPLE     .04910        .00071
         RIPPLE IN DB       .41634     -63.00399
```

filter into another lowpass filter having a band edge frequency $\Omega'_p = 1$ and then performing the transformation

$$s \longrightarrow \frac{s^2 + \Omega_l \Omega_u}{s(\Omega_u - \Omega_l)} \qquad \text{(lowpass to bandpass)} \qquad (8.4.3)$$

Equivalently, we can accomplish the same result in a single step by means of the transformation

$$s \longrightarrow \Omega_p \frac{s^2 + \Omega_l \Omega_u}{s(\Omega_u - \Omega_l)} \qquad \text{(lowpass to bandpass)} \qquad (8.4.4)$$

where

$$\Omega_l = \text{lower band edge frequency}$$

$$\Omega_u = \text{upper band edge frequency}$$

Thus we obtain

$$H_b(s) = H_p \left(\Omega_p \frac{s^2 + \Omega_l \Omega_u}{s(\Omega_u - \Omega_l)} \right)$$

Finally, if we wish to convert a lowpass analog filter with band edge frequency Ω_p into a bandstop filter, the transformation is simply the inverse of (8.4.3) with the additional factor Ω_p serving to normalize for the band edge frequency of the lowpass filter. Thus the transformation is

$$s \longrightarrow \Omega_p \frac{s(\Omega_u - \Omega_l)}{s^2 + \Omega_u \Omega_l} \qquad \text{(lowpass to bandstop)} \qquad (8.4.5)$$

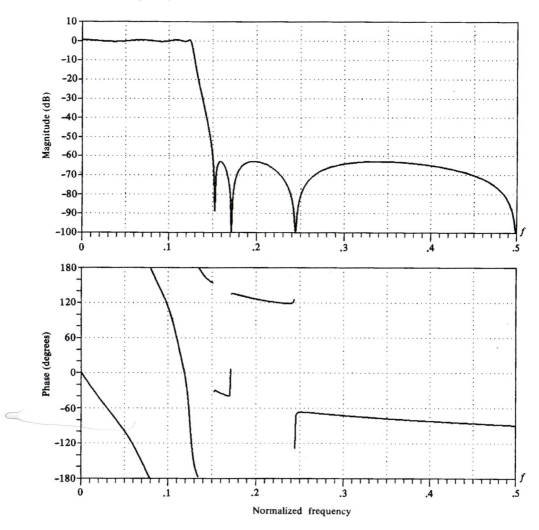

Figure 8.47 Frequency response characteristics of a 7-order elliptic filter.

which leads to

$$H_{bs}(s) = H_p \left(\Omega_p \frac{s(\Omega_u - \Omega_l)}{s^2 + \Omega_u \Omega_l} \right)$$

The mappings in (8.4.1), (8.4.2), (8.4.3), and (8.4.5) are summarized in Table 8.12. The mappings in (8.4.4) and (8.4.5) are nonlinear and may appear to distort the frequency response characteristics of the lowpass filter. However, the effects of the nonlinearity on the frequency response are minor, primarily affecting the frequency scale but preserving the amplitude response characteristics of the filter. Thus an equiripple lowpass filter is transformed into an equiripple bandpass or bandstop or highpass filter.

TABLE 8.12 FREQUENCY TRANSFORMATIONS FOR
ANALOG FILTERS (PROTOTYPE LOWPASS FILTER
HAS BAND EDGE FREQUENCY Ω_P)

Type of transformation	Transformation	Band edge frequencies of new filter
Lowpass	$s \longrightarrow \dfrac{\Omega_p}{\Omega'_p} s$	Ω'_p
Highpass	$s \longrightarrow \dfrac{\Omega_p \Omega'_p}{s}$	Ω'_p
Bandpass	$s \longrightarrow \Omega_p \dfrac{s^2 + \Omega_l \Omega_u}{s(\Omega_u - \Omega_l)}$	Ω_l, Ω_u
Bandstop	$s \longrightarrow \Omega_p \dfrac{s(\Omega_u - \Omega_c)}{s^2 + \Omega_u \Omega_l}$	Ω_l, Ω_u

Example 8.4.1

Transform the single-pole lowpass Butterworth filter with system function

$$H(s) = \frac{\Omega_p}{s + \Omega_p}$$

into a bandpass filter with upper and lower band edge frequencies Ω_u and Ω_l, respectively.

Solution   The desired transformation is given by (8.3.4). Thus we have

$$H(s) = \frac{1}{\dfrac{s^2 + \Omega_l \Omega_u}{s(\Omega_u - \Omega_l)} + 1}$$

$$= \frac{(\Omega_u - \Omega_l)s}{s^2 + (\Omega_u - \Omega_l)s + \Omega_l \Omega_u}$$

The resulting filter has a zero at s = 0 and poles at

$$s = \frac{-(\Omega_u - \Omega_l) \pm \sqrt{\Omega_u^2 + \Omega_l^2 - 6\Omega_u \Omega_l}}{2}$$

8.4.2 Frequency Transformations in the Digital Domain

As in the analog domain, frequency transformations can be performed on a digital lowpass filter to convert it to either a bandpass, bandstop, or highpass filter. The transformation involves replacing the variable z^{-1} by a rational function $g(z^{-1})$, which must satisfy the following properties.

1. The mapping $z^{-1} \longrightarrow g(z^{-1})$ must map points inside the unit circle in the z-plane into itself.
2. The unit circle must also be mapped into itself.

Condition (2) implies that for $r = 1$,

$$e^{-j\omega} = g(e^{-j\omega}) \equiv g(\omega)$$
$$= |g(\omega)|e^{j \arg[g(\omega)]}$$

It is clear that we must have $|g(\omega)| = 1$ for all ω. That is, the mapping must be all-pass. Hence it is of the form

$$g(z^{-1}) = \pm \prod_{k=1}^{n} \frac{z^{-1} - a_k}{1 - a_k z^{-1}} \tag{8.4.6}$$

where $|a_k| < 1$ to ensure that a stable filter is transformed into another stable filter (i.e., to satisfy condition 1).

From the general form in (8.4.6), we obtain the desired set of digital transformations for converting a prototype digital lowpass filter into either a bandpass, a bandstop, a highpass, or another lowpass digital filter. These transformations are tabulated in Table 8.13.

TABLE 8.13 FREQUENCY TRANSFORMATION FOR DIGITAL FILTERS (PROTOTYPE LOWPASS FILTER HAS BAND EDGE FREQUENCY ω_p)

Type of transformation	Transformation	Parameters
Lowpass	$z^{-1} \longrightarrow \dfrac{z^{-1} - a}{1 - az^{-1}}$	$\omega_p' = $ band edge frequency of new filter $a = \dfrac{\sin[(\omega_p - \omega_p')/2]}{sin[(\omega_p + \omega_p')/2]}$
Highpass	$z^{-1} \longrightarrow -\dfrac{z^{-1} + a}{1 + az^{-1}}$	$\omega_p' = $ band edge frequency new filter $a = -\dfrac{\cos[(\omega_p + \omega_p')/2]}{\cos[(\omega_p - \omega_p')/2]}$
Bandpass	$z^{-1} \longrightarrow -\dfrac{z^{-2} - a_1 z^{-1} + a_2}{a_2 z^{-2} - a_1 z^{-1} + 1}$	$\omega_l = $ lower band edge frequency $\omega_u = $ upper band edge frequency $a_1 = -2\alpha K/(K + 1)$ $a_2 = (K - 1)/(K + 1)$ $\alpha = \dfrac{\cos[(\omega_u + \omega_l)/2]}{\cos[(\omega_u - \omega_l)/2]}$ $K = \cot \dfrac{\omega_u - \omega_l}{2} \tan \dfrac{\omega_p}{2}$
Bandstop	$z^{-1} \longrightarrow -\dfrac{z^{-2} - a_1 z^{-1} + a_2}{a_2 z^{-2} - a_1 z^{-1} + 1}$	$\omega_l = $ lower band edge frequency $\omega_u = $ upper band edge frequency $a_1 = -2\alpha/(K + 1)$ $a_2 = (1 - K)/(1 + K)$ $\alpha = \dfrac{\cos[(\omega_u + \omega_l)/2]}{\cos[(\omega_u - \omega_l)/2]}$ $K = \tan \dfrac{\omega_u - \omega_l}{2} \tan \dfrac{\omega_p}{2}$

Example 8.4.2

Convert the single-pole lowpass Butterworth filter with system function

$$H(z) = \frac{0.245(1 + z^{-1})}{1 - 0.509z^{-1}}$$

into a bandpass filter with upper and lower cutoff frequencies ω_u and ω_l, respectively. The lowpass filter has 3-dB bandwidth $\omega_p = 0.2\pi$ (see Example 8.3.5).

Solution The desired transformation is

$$z^{-1} \longrightarrow -\frac{z^{-2} - a_1 z^{-1} + a_2}{a_2 z^{-2} - a_1 z^{-1} + 1}$$

where a_1 and a_2 are defined in Table 8.13. Substitution into $H(z)$ yields

$$H(z) = \frac{0.245 \left[1 - \dfrac{z^{-2} - a_1 z^{-1} + a_2}{a_2 z^{-2} - a_1 z^{-1} + 1} \right]}{1 + 0.509 \left(\dfrac{z^{-2} - a_1 z^{-1} + a_2}{a_2 z^{-2} - a_1 z^{-1} + 1} \right)}$$

$$= \frac{0.245(1 - a_2)(1 - z^{-2})}{(1 + 0.509a_2) - 1.509a_1 z^{-1} + (a_2 + 0.509)z^{-2}}$$

Note that the resulting filter has zeros at $z = \pm 1$ and a pair of poles that depend on the choice of ω_u and ω_l.

For example, suppose that $\omega_u = 3\pi/5$ and $\omega_l = 2\pi/5$. Since $\omega_p = 0.2\pi$, we find that $K = 1$, $a_2 = 0$, and $a_1 = 0$. Then

$$H(z) = \frac{0.245(1 - z^{-2})}{1 + 0.509z^{-2}}$$

This filter has poles at $z = \pm j0.713$ and hence resonates at $\omega = \pi/2$.

Since a frequency transformation can be performed either in the analog domain or in the digital domain, the filter designer has a choice as to which approach to take. However, some caution must be exercised depending on the types of filters being designed. In particular, we know that the impulse invariance method and the mapping of derivatives are inappropriate to use in designing highpass and many bandpass filters, due to the aliasing problem. Consequently, one would not employ an analog frequency transformation followed by conversion of the result into the digital domain by use of these two mappings. Instead, it is much better to perform the mapping from an analog lowpass filter into a digital lowpass filter by either of these mappings, and then to perform the frequency transformation in the digital domain. Thus the problem of aliasing is avoided.

In the case of the bilinear transformation, where aliasing is not a problem, it does not matter whether the frequency transformation is performed in the analog domain or in the digital domain. In fact, in this case only, the two approaches result in identical digital filters.

8.5 DESIGN OF DIGITAL FILTERS BASED ON LEAST-SQUARES METHOD

Except for the impulse invariance method, the design techniques for IIR filters described in Section 8.3 involved the conversion of an analog filter into a digital filter by some mapping from the s-plane to the z-plane. As an alternative, one can design digital IIR filters directly in the z-domain without reference to the analog domain.

We now describe several methods for designing digital filters directly. In the first three techniques, the Padé approximation method and least-squares design methods, the specifications are given in the time domain and the design is carried out in the time domain. The final section describes a least-squares technique in which the design is carried out in the frequency domain.

8.5.1 Padé Approximation Method

Suppose that the desired impulse response $h_d(n)$ is specified for $n \geq 0$. The filter to be designed has the system function

$$
H(z) = \frac{\displaystyle\sum_{k=0}^{M} b_k z^{-k}}{1 + \displaystyle\sum_{k=1}^{N} a_k z^{-k}}
$$

$$
= \sum_{k=0}^{\infty} h(k) z^{-k} \tag{8.5.1}
$$

where $h(k)$ is its unit sample response. The filter has $L = M + N + 1$ parameters, namely, the coefficients $\{a_k\}$ and $\{b_k\}$, which can be selected to minimize some error criterion.

The least-squares error criterion is often used in optimization problems of this type. Suppose that we minimize the sum of the squared errors

$$
\mathcal{E} = \sum_{n=0}^{U} [h_d(n) - h(n)]^2 \tag{8.5.2}
$$

with respect to the filter parameters $\{a_k\}$ and $\{b_k\}$, where U is some preselected upper limit in the summation.

In general, $h(n)$ is a nonlinear function of the filter parameters and hence the minimization of $\mathcal{E}$ involves the solution of a set of nonlinear equations. However, if we select the upper limit as $U = L - 1$, it is possible to match $h(n)$ perfectly to the desired response $h_d(n)$ for $0 \leq n \leq M + N$. This can be achieved in the following manner.

The difference equation for the desired filter is

$$
y(n) = -a_1 y(n-1) - a_2 y(n-2) - \cdots - a_N y(n-N)
$$
$$
+ b_0 x(n) + b_1 x(n-1) + \cdots + b_M x(n-M) \tag{8.5.3}
$$

Suppose that the input to the filter is a unit sample [i.e., $x(n) = \delta(n)$]. Then the response of the filter is $y(n) = h(n)$ and hence (8.5.3) becomes

$$h(n) = -a_1 h(n-1) - a_2 h(n-2) - \cdots - a_N h(n-N)$$
$$+ b_0 \delta(n) + b_1 \delta(n-1) + \cdots + b_M \delta(n-M) \tag{8.5.4}$$

Since $\delta(n-k) = 0$ except for $n = k$, (8.5.4) reduces to

$$h(n) = -a_1 h(n-1) - a_2 h(n-2) - \cdots - a_N h(n-N) + b_n \qquad 0 \le n \le M \tag{8.5.5}$$

For $n > M$, (8.5.4) becomes

$$h(n) = -a_1 h(n-1) - a_2 h(n-2) - \cdots - a_N h(n-N) \tag{8.5.6}$$

The set of linear equations in (8.5.5) and (8.5.6) can be used to solve for the filter parameters $\{a_k\}$ and $\{b_k\}$. We set $h(n) = h_d(n)$ for $0 \le n \le M + N$, and use the linear equations in (8.5.6) to solve for the filter parameters $\{a_k\}$. Then we use values for the $\{a_k\}$ in (8.5.5) and solve for the parameters $\{b_k\}$. Thus we obtain a perfect match between $h(n)$ and the desired response $h_d(n)$ for the first L values of the impulse response. This design technique is usually called the *Padé approximation procedure*.

The degree to which this design technique produces acceptable filter designs depends in part on the number of filter coefficients selected. Since the design method matches $h_d(n)$ only up to the number of filter parameters, the more complex the filter, the better the approximation to $h_d(n)$ for $0 \le n \le M + N$. However, this is also the major limitation with the Padé approximation method, namely, the resulting filter must contain a large number of poles and zeros. For this reason, the Padé approximation method has found limited use in filter designs for practical applications.

Example 8.5.1

Suppose that the desired unit sample response is

$$h_d(n) = 2(\tfrac{1}{2})^n u(n)$$

Determine the parameters of the filter with system function

$$H(z) = \frac{b_0 + b_1 z^{-1}}{1 + a_1 z^{-1}}$$

using the Padé approximation technique.

Solution In this simple example, $H(z)$ can provide a perfect match to $H_d(z)$, by selecting $b_0 = 2$, $b_1 = 0$, and $a_1 = -\tfrac{1}{2}$. Let us apply the Padé approximation to see if we indeed obtain the same result.

With $\delta(n)$ as the input to $H(z)$, we obtain the output

$$h(n) = -a_1 h(n-1) + b_0 \delta(n) + b_1 \delta(n-1)$$

For $n > 1$, we have

$$h(n) = -a_1 h(n-1)$$

or, equivalently,

$$h_d(n) = -a_1 h_d(n-1)$$

With the substitution for $h_d(n)$, we obtain $a_1 = -\frac{1}{2}$. To solve for b_0 and b_1, we use the form (8.5.5) with $h(n) = h_d(n)$. Thus

$$h_d(n) = \tfrac{1}{2}h_d(n-1) + b_0\delta(n) + b_1\delta(n-1)$$

For $n = 0$ this equation yields $b_0 = 2$. For $n = 1$ we obtain the result $b_1 = 0$. Thus $H(z) = H_d(z)$.

This example illustrates that the Padé approximation results in a perfect match to $H_d(z)$ when the desired system function is rational and we have prior knowledge of the number of poles and zeros in the system. In general, however, this is not the case in practice, since $h_d(n)$ is determined from some desired frequency response specifications $H_d(\omega)$. In such a case the Padé approximation may not result in a good filter design. To illustrate a potential problem and suggest a solution, let us consider the following examples.

Example 8.5.2

A fourth-order Butterworth filter has the system function

$$H_d(z) = \frac{4.8334 \times 10^{-3}(z+1)^4}{(z^2 - 1.3205z + 0.6326)(z^2 - 1.0482z + 0.2959)}$$

The unit sample response corresponding to $H_d(z)$ is illustrated in Fig. 8.48. Use the Padé approximation method to approximate $H_d(z)$.

Solution   We observe that the desired filter has $M = 4$ zeros and $N = 4$ poles. It is instructive to determine the coefficients in the Padé approximation when the number of zeros and/or poles are not identical to the desired number of filter parameters.

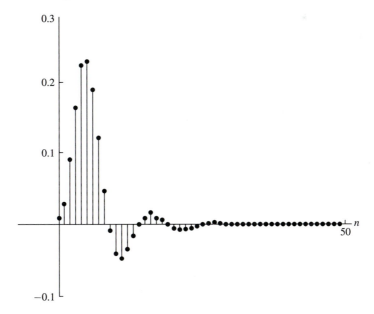

Figure 8.48   Impulse response $h_d(n)$ of digital Butterworth filter in Example 8.5.2.

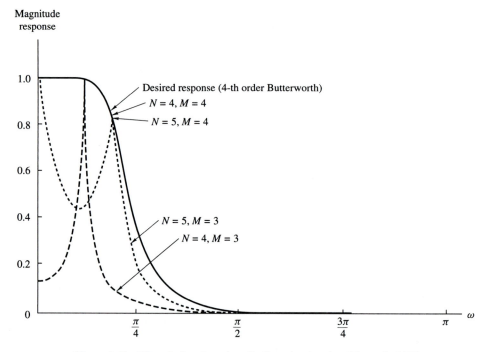

Figure 8.49 Filter designs based on Padè approximation (Example 8.5.2).

In Fig. 8.49 we plot the frequency response of the filter obtained by the Padé approximation method. We have considered four cases: $M = 3, N = 5$; $M = 3, N = 4$; $M = 4, N = 4$; $M = 4, N = 5$. We observe that when $M = 3$, the resulting frequency response is a relatively poor approximation to the desired response. However, an increase in the number of poles from $N = 4$ to $N = 5$ appears to compensate in part for the lack of the one zero. When M is increased from three to four, we obtain a perfect match with the desired Butterworth filter not only for $N = 4$ but for $N = 5$, and, in fact, for larger values of N.

Example 8.5.3

A three-pole and three-zero type II lowpass Chebyshev digital filter has the system function

$$H_d(z) = \frac{0.3060(1 + z^{-1})(0.2652 - 0.09z^{-1} + 0.2652z^{-2})}{(1 - 0.3880z^{-1})(1 - 1.1318z^{-1} + 0.5387z^{-2})}$$

Its unit sample response is illustrated in Fig. 8.50. Use the Padé approximation method to approximate $H_d(z)$.

Solution By following the same procedure as in Example 8.5.2, we determined the Padé approximation of $H_d(z)$ based on the selection of $M = 2, N = 3$; $M = 2, N = 4$; $M = 3, N = 3$; $M = 3, N = 4$. The frequency responses of the resulting designs are illustrated in Fig. 8.51.

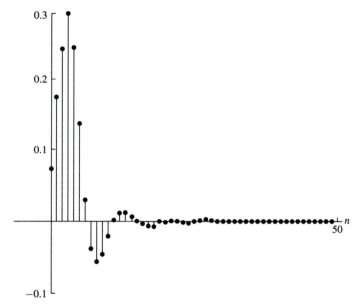

Figure 8.50 Impulse response $h_d(n)$ of type II Chebyshev digital filter given in Example 8.5.3.

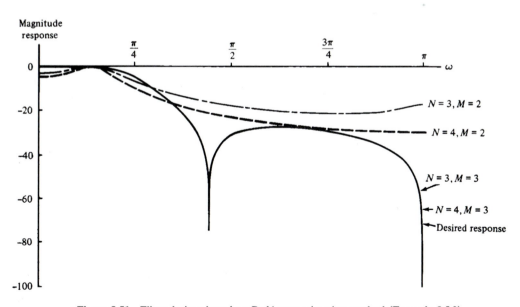

Figure 8.51 Filter designs based on Padé approximation method (Example 8.5.3).

As in Example 8.5.2, we note that when we underestimate the number of zeros we obtain a relatively poor design, as evidenced by the two cases in which $M = 2$. On the other hand, if $M = 3$, we obtain a perfect match for $N = 3$ and $N = 4$.

These two examples suggest that an effective approach in using the Padé approximation is to try different values of M and N until the frequency responses of the resulting filters converge to the desired frequency response within some small, acceptable approximation error. However, in practice, this approach appears to be cumbersome.

8.5.2 Least-Squares Design Methods

Again, let us assume that $h_d(n)$ is specified for $n \geq 0$. We begin with the simple case in which the digital filter to be designed contains only poles, that is,

$$H(z) = \frac{b_0}{1 + \sum_{k=1}^{N} a_k z^{-k}} \qquad (8.5.7)$$

Now, consider the cascade connection of the desired filter $H_d(z)$ with the reciprocal, all-zero filter $1/H(z)$, as illustrated in Fig. 8.52. Now suppose that the cascade configuration in Fig. 8.52 is excited by the unit sample sequence $\delta(n)$. Thus the input to the inverse system $1/H(z)$ is $h_d(n)$ and the output is $y(n)$. Ideally, $y_d(n) = \delta(n)$. The actual output is

$$y(n) = \frac{1}{b_0} \left[h_d(n) + \sum_{k=1}^{N} a_k h_d(n - k) \right] \qquad (8.5.8)$$

The condition that $y_d(0) \equiv y(0) = 1$ is satisfied by selecting $b_0 = h_d(0)$. For $n > 0$, $y(n)$ represents the error between the desired output $y_d(n) = 0$ and the actual output. Hence the parameters $\{a_k\}$ are selected to minimize the sum of

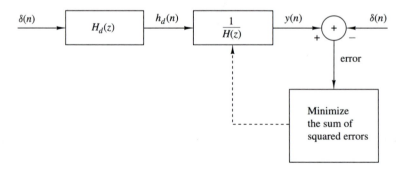

Figure 8.52   Least-squares inverse filter design method.

squares of the error sequence,

$$\mathcal{E} = \sum_{n=1}^{\infty} y^2(n)$$

$$= \frac{\sum_{n=1}^{\infty} i \left[h_d(n) + \sum_{k=1}^{N} a_k h_d(n-k) \right]^2}{h_d^2(0)} \tag{8.5.9}$$

By differentiating with respect to the parameters $\{a_k\}$, it is easily established that we obtain the set of linear equations of the form

$$\sum_{k=1}^{N} a_k r_{hh}(k, l) = -r_{hh}(l, 0) \qquad l = 1, 2, \ldots, N \tag{8.5.10}$$

where, by definition,

$$r_{hh}(k, l) = \sum_{n=1}^{\infty} h_d(n-k) h_d(n-l)$$

$$= \sum_{n=0}^{\infty} h_d(n) h_d(n+k-l) = r_{hh}(k-l) \tag{8.5.11}$$

The solution of (8.5.10) yields the desired parameters for the inverse system $1/H(z)$. Thus we obtain the coefficients of the all-pole filter.

In a practical design problem, the desired impulse response $h_d(n)$ is specified for a finite set of points, say $0 \le n \le L$, where $L \gg N$. In such a case, the correlation sequence $r_{dd}(k)$ can be computed from the finite sequence $h_d(n)$ as

$$\hat{r}_{hh}(k-l) = \sum_{n=0}^{L-|k-l|} h_d(n) h_d(n+k-l) \qquad 0 \le k-l \le N \tag{8.5.12}$$

and these values can be used to solve the set of linear equations in (8.5.10).

The least-squares method can also be used in a pole–zero approximation for $H_d(z)$. If the filter $H(z)$ that approximates $H_d(z)$ has both poles and zeros, its response to the unit impulse $\delta(n)$ is

$$h(n) = -\sum_{k=1}^{N} a_k h(n-k) + \sum_{k=0}^{M} b_k \delta(n-k) \qquad n \ge 0 \tag{8.5.13}$$

or, equivalently,

$$h(n) = -\sum_{k=1}^{N} a_k h(n-k) + b_n \qquad 0 \le n \le M \tag{8.5.14}$$

For $n > M$, (8.5.13) reduces to

$$h(n) = -\sum_{k=1}^{N} a_k h(n-k) \qquad n > M \tag{8.5.15}$$

Clearly, if $H_d(z)$ is a pole–zero filter, its response to $\delta(n)$ would satisfy the same equations (8.5.13) through (8.5.15). In general, however, it does not. Nevertheless, we can use the desired response $h_d(n)$ for $n > M$ to construct an estimate of $h_d(n)$, according to (8.5.15). That is,

$$\hat{h}_d(n) = -\sum_{k=1}^{N} a_k h_d(n-k) \tag{8.5.16}$$

Then we can select the filter parameters $\{a_k\}$ to minimize the sum of squared errors between the desired response $h_d(n)$ and the estimate $\hat{h}_d(n)$ for $n > M$. Thus we have

$$\mathcal{E}_1 = \sum_{n=M+1}^{\infty} [h_d(n) - \hat{h}_d(n)]^2$$

$$= \sum_{n=M+1}^{\infty} \left[h_d(n) + \sum_{k=1}^{N} a_k h_d(n-k) \right]^2 \tag{8.5.17}$$

The minimization of $\mathcal{E}_1$, with respect to the pole parameters $\{a_k\}$, leads to the set of linear equations

$$\sum_{l=1}^{N} a_l r_{hh}(k, l) = -r_{hh}(k, 0) \qquad k = 1, 2, \ldots, N \tag{8.5.18}$$

where $r_{hh}(k, l)$ is now defined as

$$r_{hh}(k, l) = \sum_{n=M+1}^{\infty} h_d(n-k) h_d(n-l) \tag{8.5.19}$$

Thus these linear equations yield the filter parameters $\{a_k\}$. Note that these equations reduce to the all-pole filter approximation when M is set to zero.

The parameters $\{b_k\}$ that determine the zeros of the filter can be obtained simply from (8.5.14), where $h(n) = h_d(n)$, by substitution of the values $\{\hat{a}_k\}$ obtained by solving (8.5.18). Thus

$$b_n = h_d(n) + \sum_{k=1}^{N} \hat{a}_k h_d(n-k) \qquad 0 \le n \le M \tag{8.5.20}$$

Therefore, the parameters $\{\hat{a}_k\}$ that determine the poles are obtained by the method of least squares while the parameters $\{b_k\}$ that determine the zeros are obtained by the the Padé approximation method. The foregoing approach for determining the poles and zeros of $H(z)$ is sometimes called *Prony's method*.

The least-squares method provides good estimates for the pole parameters $\{a_k\}$. However, Prony's method may not be as effective in estimating the parameters $\{b_k\}$, primarily because the computation in (8.5.20) is not based on the least-squares method.

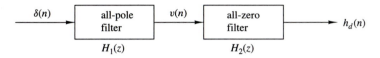

Figure 8.53   Least-squares method for determining the poles and zeros of a filter.

An alternative method in which both sets of parameters $\{a_k\}$ and $\{b_k\}$ are determined by application of the least-squares method has been proposed by Shanks (1967). In Shanks' method, the parameters $\{a_k\}$ are computed on the basis of the least-squares criterion, according to (8.5.18), as indicated above. This yields the estimates $\{\hat{a}_k\}$, which allow us to synthesize the all-pole filter.

$$H_1(z) = \frac{1}{1 + \sum_{k=1}^{N} \hat{a}_k z^{-k}} \tag{8.5.21}$$

The response of this filter to the impulse $\delta(n)$ is

$$v(n) = -\sum_{k=1}^{N} \hat{a}_k v(n-k) + \delta(n) \qquad n \geq 0 \tag{8.5.22}$$

If the sequence $\{v(n)\}$ is used to excite an all-zero filter with system function

$$H_2(z) = \sum_{k=0}^{M} b_k z^{-k} \tag{8.5.23}$$

as illustrated in Fig. 8.53, its response is

$$\hat{h}_d(n) = \sum_{k=0}^{M} b_k v(n-k) \tag{8.5.24}$$

Now we can define an error sequence $e(n)$ as

$$e(n) = h_d(n) - \hat{h}_d(n)$$

$$= h_d(n) - \sum_{k=0}^{M} b_k v(n-k) \tag{8.5.25}$$

and, consequently, the parameters $\{b_k\}$ can also be determined by means of the least-squares criterion, namely, from the minimization of

$$\mathcal{E}_2 = \sum_{n=0}^{\infty} \left[h_d(n) - \sum_{k=0}^{M} b_k v(n-k) \right]^2 \tag{8.5.26}$$

Thus we obtain a set of linear equations for the parameters $\{b_k\}$, in the form

$$\sum_{k=0}^{M} b_k r_{vv}(k, l) = r_{hv}(l) \qquad l = 0, 1, \ldots, M \tag{8.5.27}$$

where, by definition,

$$r_{vv}(k, l) = \sum_{n=0}^{\infty} v(n-k)v(n-l) \tag{8.5.28}$$

$$r_{hv}(k) = \sum_{n=0}^{\infty} h_d(n)v(n-k) \tag{8.5.29}$$

Example 8.5.4

Approximate the fourth-order Butterworth filter given in Example 8.5.2 by means of an all-pole filter using the least-squares inverse design method.

Solution   From the desired impulse response $h_d(n)$, which is illustrated in Fig. 8.48, we computed the autocorrelation sequence $r_{hh}(k, l) = r_{hh}(k-l)$ and solved to set of linear equations in (8.5.10) to obtain the filter coefficients. The results of this computation are given in Table 8.14 for $N = 3, 4, 5, 10$, and 15. In Table 8.15 we list the poles of the filter designs for $N = 3, 4$, and 5 along with the actual poles of the fourth-order Butterworth filter. We note that the poles obtained from the designs are far from the actual poles of the desired filter.

TABLE 8.14   ESTIMATES OF FILTER COEFFICIENTS
$\{a_k\}$ IN LEAST-SQUARES INVERSE FILTER DESIGN
METHOD

$N =$	3		$N =$	15
$a_1 =$	$0.254295E + 01$		$a_1 =$	2.993620
$a_2 =$	$-0.241800E + 01$		$a_2 =$	-1.143053
$a_3 =$	$0.853829E + 00$		$a_3 =$	-12.132861
$N =$	4		$a_4 =$	39.663433
$a_1 =$	$0.319047E + 01$		$a_5 =$	-75.749001
$a_2 =$	$-0.425176E + 01$		$a_6 =$	109.247757
$a_3 =$	$0.278234E + 01$		$a_7 =$	-129.513794
$a_4 =$	$0.758375E + 00$		$a_8 =$	131.026794
$N =$	5		$a_9 =$	-114.905266
$a_1 =$	$0.368733E + 01$		$a_{10} =$	87.449211
$a_2 =$	$-0.607422E + 01$		$a_{11} =$	-57.031906
$a_3 =$	$0.556726E + 01$		$a_{12} =$	30.915134
$a_4 =$	$-0.284813E + 01$		$a_{13} =$	-13.124536
$a_5 =$	$0.654996E + 00$		$a_{14} =$	3.879295
			$a_{15} =$	-0.597313
$N =$	10			
$a_1 =$	5.008451			
$a_2 =$	-12.660761			
$a_3 =$	21.557365			
$a_4 =$	-27.804110			
$a_5 =$	28.683949			
$a_6 =$	-24.058558			
$a_7 =$	16.156847			
$a_8 =$	-8.247148			
$a_9 =$	2.854789			
$a_{10} =$	-0.502956			

TABLE 8.15 ESTIMATES OF POLE
POSITIONS IN LEAST-SQUARES
INVERSE FILTER DESIGN METHOD
(EXAMPLE 8.5.4)

Number of poles	Pole positions
$N = 3$	0.9305
	$0.8062 \pm j0.5172$
$N = 4$	$0.8918 \pm j0.2601$
	$0.7037 \pm j0.6194$
$N = 5$	0.914
	$0.8321 \pm j0.4307$
	$0.5544 \pm j0.7134$
$N = 4$	$0.6603 \pm j0.4435$
Butterworth	$0.5241 \pm j0.1457$
filter	

The frequency responses of the filter designs are plotted in Fig. 8.54. We note that when N is small, the approximation to the desired filter is poor. As N is increased to $N = 10$ and $N = 15$, the approximation improves significally. However, even for $N = 15$, there are large ripples in the passband of the filter response. It is apparent that this method, which is based on an all-pole approximation, does not provide good approximations to filters that contain zeros.

Example 8.5.5

Approximate the type II Chebyshev lowpass filter given in Example 8.5.3 by means of the three least-squares methods described above.

Solution   The results of the filter designs obtained by means of the least-squares inverse method, Prony's method and Shanks' method, are illustrated in Fig. 8.55. The filter parameters obtained from these design methods are listed in Table 8.16.

The frequency response characteristics in Fig. 8.55 illustrate that the least-squares inverse (all-pole) design method yields poor designs when the filter contains zeros. On the other hand, both Prony's method and Shanks' method yield very good designs when the number of poles and zeros equals or exceeds the number of poles and zeros in the actual filter. Thus the inclusion of zeros in the approximation has a significant effect in the resulting filter design.

8.5.3 FIR Least-Squares Inverse (Wiener) Filters

In the preceding section we described the use of the least-squares error criterion in the design of pole–zero filters. In this section we use a similar approach to determine a least-squares FIR inverse filter to a desired filter.

The inverse to a linear time-invariant system with impulse response $h(n)$ and system function $H(z)$ is defined as the system whose impulse response $h_I(n)$ and

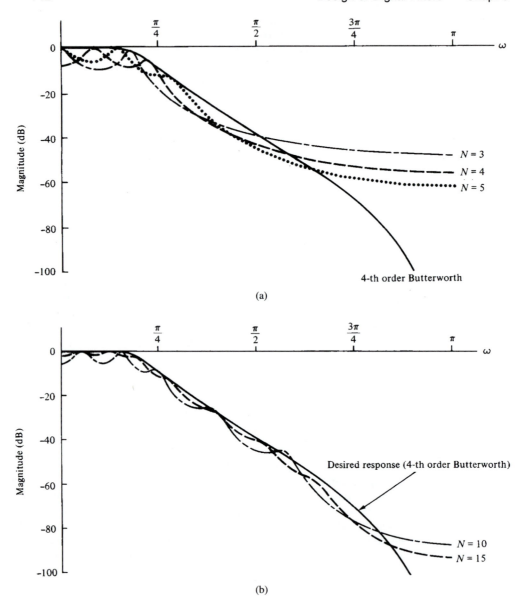

Figure 8.54 Magnitude responses for filter designs based on the least-squares inverse filter method.

system function $H_I(z)$, satisfy the respective equations.

$$h(n) * h_I(n) = \delta(n) \tag{8.5.30}$$

$$H(z)H_I(z) = 1 \tag{8.5.31}$$

In general, $H_I(z)$ is IIR, unless $H(z)$ is an all-pole system, in which case $H_I(z)$ is FIR.

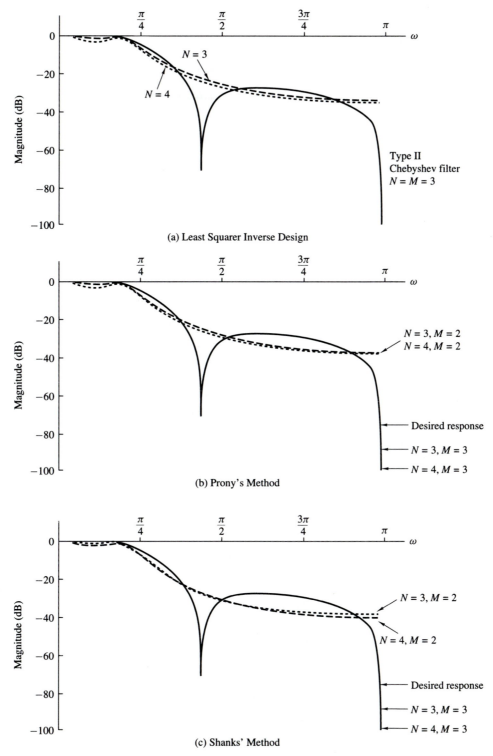

Figure 8.55 Filter designs based on least-squares methods (Example 8.5.5): (a) least-squares design; (b) Prony's method; (c) Shank's method.

TABLE 8.16 POLE–ZERO LOCATIONS FOR FILTER DESIGNS IN EXAMPLE 8.5.5

Chebyshev Filter:
Zeros: $-1, 0.1738311 \pm j0.9847755$
Poles: $0.3880, 0.5659 \pm j0.467394$

Filter Order	Poles in Least-Squares Inverse
$N = 3$	0.8522
	$0.6544 \pm j0.6224$
$N = 4$	$0.7959 \pm j0.3248$
	$0.4726 \pm j0.7142$

Filter Order	Prony's Method Poles	Prony's Method Zeros	Shanks' Method Poles	Shanks' Method Zeros
$N = 3$	0.5332		0.5348	
$M = 2$	$0.6659 \pm j0.4322$	$-0.1497 \pm j0.4925$	$0.6646 \pm j0.4306$	$-0.2437 \pm j0.5918$
$N = 4$	0.7092		0.7116	
	-0.2919		-0.2921	
$M = 2$	$0.6793 \pm j0.4863$	$-0.1982 \pm j0.37$	$0.6783 \pm j0.4855$	$0.306 \pm j0.4482$
$N = 3$	0.3881	-1	0.3881	-1
$M = 3$	$0.5659 \pm j0.4671$	$0.1736 \pm j0.9847$	$0.5659 \pm j0.4671$	$0.1738 \pm j0.9848$
$N = 4$	-0.00014	-1	-0.00014	-1
	0.388		0.388	
$M = 3$	$0.5661 \pm j0.4672$	$0.1738 \pm j0.9848$	$0.566 \pm j0.4671$	$0.1738 \pm j0.9848$

In many practical applications, it is desirable to restrict the inverse filter to be FIR. Obviously, one simple method is to truncate $h_I(n)$. In so doing, we incur a total squared approximation error equal to

$$\mathcal{E}_t = \sum_{n=M+1}^{\infty} h_I^2(n) \tag{8.5.32}$$

where $M + 1$ is the length of the truncated filter and $\mathcal{E}_t$ represents the energy in the tail of the impulse response $h_I(n)$.

Alternatively, we can use the least-squares error criterion to optimize the $M + 1$ coefficients of the FIR filter. First, let $d(n)$ denote the *desired output sequence* of the FIR filter of length $M + 1$ and let $h(n)$ be the input sequence. Then, if $y(n)$ is the output sequence of the filter, as illustrated in Fig. 8.56, the error sequence between the desired output and the actual output is

$$e(n) = d(n) - \sum_{k=0}^{M} b_k h(n - k) \tag{8.5.33}$$

where the $\{b_k\}$ are the FIR filter coefficients.

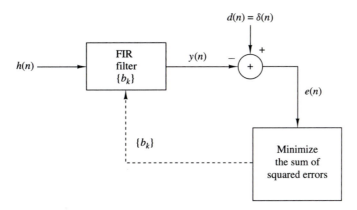

Figure 8.56   Least-squares FIR inverse filter.

The sum of squares of the error sequence is

$$\mathcal{E} = \sum_{n=0}^{\infty} \left[d(n) - \sum_{k=0}^{M} b_k h(n-k) \right]^2 \tag{8.5.34}$$

When $\mathcal{E}$ is minimized with respect to the filter coefficients, we obtain the set of linear equations

$$\sum_{k=0}^{M} b_k r_{hh}(k-l) = r_{dh}(l) \qquad l = 0, 1, \ldots, M \tag{8.5.35}$$

where $r_{hh}(l)$ is the autocorrelation of $h(n)$, defined as

$$r_{hh}(l) = \sum_{n=0}^{\infty} h(n) h(n-l) \tag{8.5.36}$$

and $r_{dh}(n)$ is the crosscorrelation between the desired output $d(n)$ and the input sequence $h(n)$, defined as

$$r_{dh}(l) = \sum_{n=0}^{\infty} d(n) h(n-l) \tag{8.5.37}$$

The optimum, in the least-squares sense, FIR filter that satisfies the linear equations in (8.5.35) is called the *Wiener filter*, after the famous mathematician Norbert Wiener, who introduced optimum least-squares filtering methods in engineering [see book by Wiener (1949)].

If the optimum least-squares FIR filter is to be an approximate inverse filter, the desired response is

$$d(n) = \delta(n) \tag{8.5.38}$$

The crosscorrelation between $d(n)$ and $h(n)$ reduces to

$$r_{dh}(l) = \begin{cases} h(0) & l = 0 \\ 0 & \text{otherwise} \end{cases} \tag{8.5.39}$$

Therefore, the coefficients of the least-squares FIR filter are obtained from the solution of the linear equations in (8.5.35), which can be expressed in matrix form as

$$
\begin{bmatrix}
r_{hh}(0) & r_{hh}(1) & r_{hh}(2) & \cdots & r_{hh}(M) \\
r_{hh}(1) & r_{hh}(0) & r_{hh}(1) & \cdots & r_{hh}(M-1) \\
\vdots & \vdots & \vdots & & \vdots \\
r_{hh}(M) & r_{hh}(M-1) & \cdots & \cdots & r_{hh}(0)
\end{bmatrix}
\begin{bmatrix}
b_0 \\ b_1 \\ \vdots \\ b_M
\end{bmatrix}
=
\begin{bmatrix}
h(0) \\ 0 \\ \vdots \\ 0
\end{bmatrix}
\tag{8.5.40}
$$

We observe that the matrix is not only symmetric but it also has the special property that all the elements along any diagonal are equal. Such a matrix is called a Toeplitz matrix and lends itself to efficient inversion by means of an algorithm due to Levinson (1947) and Durbin (1959), which requires a number of computations proportional to M^2 instead of the usual M^3. The Levinson–Durbin algorithm is described in Chapter 11.

The minimum value of the least-squares error obtained with the optimum FIR filter is

$$
\mathcal{E}_{min} = \sum_{n=0}^{\infty} \left[d(n) - \sum_{k=0}^{M} b_k h(n-k) \right] d(n)
$$

$$
= \sum_{n=0}^{\infty} d^2(n) - \sum_{k=0}^{M} b_k r_{dh}(k)
\tag{8.5.41}
$$

In the case where the FIR filter is the least-squares inverse filter, $d(n) = \delta(n)$ and $r_{dh}(n) = h(0)\delta(n)$. Therefore,

$$
\mathcal{E}_{min} = 1 - h(0)b_0
\tag{8.5.42}
$$

Example 8.5.6

Determine the least-squares FIR inverse filter of length 2 to the system with impulse response

$$
h(n) = \begin{cases}
1, & n = 0 \\
-\alpha, & n = 1 \\
0, & \text{otherwise}
\end{cases}
$$

where $|\alpha| < 1$. Compare the least-squares solution with the approximate inverse obtained by truncating $h_I(n)$.

Solution   Since the system has a system function $H(z) = 1 - \alpha z^{-1}$, the exact inverse is IIR and is given by

$$
H_I(z) = \frac{1}{1 - \alpha z^{-1}}
$$

or, equivalently,

$$
h_I(n) = \alpha^n u(n)
$$

If this is truncated after n terms, the residual energy in the tail is

$$\mathcal{E}_t = \sum_{k=n}^{\infty} \alpha^{2k}$$

$$\mathcal{E}_t = \alpha^{2n}(1 + \alpha^2 + \alpha^4 + \cdots)$$

$$\mathcal{E}_t = \frac{\alpha^{2n}}{1 - \alpha^2}$$

From (8.5.40) the least-squares FIR filter of length 2 satisfies the equations

$$\begin{bmatrix} 1+\alpha^2 & -\alpha \\ -\alpha & 1+\alpha^2 \end{bmatrix} \begin{bmatrix} b_0 \\ b_1 \end{bmatrix} = \begin{bmatrix} 1 \\ 0 \end{bmatrix}$$

which have the solution

$$b_0 = \frac{1+\alpha^2}{1+\alpha^2+\alpha^4}$$

$$b_1 = \frac{\alpha}{1+\alpha^2+\alpha^4}$$

For purposes of comparison, the truncated inverse filter of length 2 has the coefficients $b_0 = 1$, $b_1 = \alpha$.

The least-squares error is

$$\mathcal{E}_{min} = \frac{\alpha^4}{1+\alpha^2+\alpha^4}$$

which compares with

$$\mathcal{E}_t = \frac{\alpha^4}{1-\alpha^2}$$

for the truncated approximate inverse. Clearly, $\mathcal{E}_t > \mathcal{E}_{min}$, so that the least-squares FIR inverse filter is superior.

In this example, the impulse response $h(n)$ of the system was minimum phase. In such a case we selected the desired response to be $d(0) = 1$ and $d(n) = 0$, $n \geq 1$. On the other hand, if the system is nonminimum phase, a delay should be inserted in the desired response in order to obtain a good filter design. The value of the appropriate delay depends on the characteristics of $h(n)$. In any case we can compute the least-squares error filter for different delays and select the filter that produces the smallest error. The following example illustrates the effect of the delay.

Example 8.5.7

Determine the least-squares FIR inverse of length 2 to the system with impulse response

$$h(n) = \begin{cases} -\alpha, & n = 0 \\ 1, & n = 1 \\ 0, & \text{otherwise} \end{cases}$$

where $|\alpha| < 1$.

Solution   This is a maximum-phase system. If we select $d(n) = [1 \quad 0]$ we obtain the same solution as in Example 8.5.6, with a minimum least-squares error

$$\mathcal{E}_{\min} = 1 - h(0)b_0$$

$$= 1 + \alpha \frac{1 + \alpha^2}{1 + \alpha^2 + \alpha^4}$$

If $0 < \alpha < 1$, then $\mathcal{E}_{\min} > 1$, which represents a poor inverse filter. If $-1 < \alpha < 0$, then $\mathcal{E}_{\min} < 1$. In particular, for $\alpha = \frac{1}{2}$, we obtain $\mathcal{E}_{\min} = 1.57$. For $\alpha = -\frac{1}{2}$, $\mathcal{E}_{\min} = 0.81$, which is still a very large value for the squared error.

Now suppose that the desired response is specified as $d(n) = \delta(n - 1)$. Then the set of equations for the filter coefficients, obtained from (8.5.35), are the solution to the equations

$$\begin{bmatrix} 1 + \alpha^2 & -\alpha \\ -\alpha & 1 + \alpha^2 \end{bmatrix} \begin{bmatrix} b_0 \\ b_1 \end{bmatrix} = \begin{bmatrix} h(1) \\ h(0) \end{bmatrix} = \begin{bmatrix} 1 \\ -\alpha \end{bmatrix}$$

The solution of these equations is

$$b_0 = \frac{1}{1 + \alpha^2 + \alpha^4}$$

$$b_1 = \frac{-\alpha^3}{1 + \alpha^2 + \alpha^4}$$

The least-squares error, given by (8.5.41), is

$$\mathcal{E}_{\min} = 1 - b_0 r_{dh}(0) - b_1 r_{dh}(1)$$

$$= 1 - b_0 h(1) - b_1 h(0)$$

$$\mathcal{E}_{\min} = 1 - \frac{1}{1 + \alpha^2 + \alpha^4} + \frac{\alpha^4}{1 + \alpha^2 + \alpha^4}$$

$$\mathcal{E}_{\min} = 1 - \frac{1 - \alpha^4}{1 + \alpha^2 + \alpha^4}$$

In particular, suppose that $\alpha = \pm\frac{1}{2}$. Then $\mathcal{E}_{\min} = 0.29$. Consequently, the desired response $d(n) = \delta(n - 1)$ results in a significantly better inverse filter. Further improvement is possible by increasing the length of the inverse filter.

In general, when the desired response is specified to contain a delay D, then the crosscorrelation $r_{dh}(l)$, defined in (8.5.37), becomes

$$r_{dh}(l) = h(D - l) \qquad l = 0, 1, \ldots, M$$

The set of linear equations for the coefficients of the least-squares FIR inverse filter given by (8.5.35) reduce to

$$\sum_{k=0}^{M} b_k r_{hh}(k - l) = h(D - l) \qquad l = 0, 1, \ldots, M \tag{8.5.43}$$

Then the expression for the corresponding least-squares error, given in general by

(8.5.41), becomes

$$\mathcal{E}_{\min} = 1 - \sum_{k=0}^{M} b_k h(D - k) \tag{8.5.44}$$

Least-squares FIR inverse filters are often used in many practical applications for deconvolution, including communications and seismic signal processing.

8.5.4 Design of IIR Filters in the Frequency Domain

The IIR filter design methods described in Sections 8.5.1 through 8.5.3 are carried out in the time domain. There are also direct design techniques for IIR filters that can be performed in the frequency domain. In this section we describe a filter parameter optimization technique carried out in the frequency domain that is representative of frequency-domain design methods.

The design is most easily carried out with the system function for the IIR filter expressed in the cascade form as

$$H(z) = G \prod_{k=1}^{K} \frac{1 + \beta_{k1} z^{-1} + \beta_{k2} z^{-2}}{1 + \alpha_{k1} z^{-1} + \alpha_{k2} z^{-2}} \tag{8.5.45}$$

where the filter gain G and the filter coefficients $\{\alpha_{k1}\}$, $\{\alpha_{k2}\}$, $\{\beta_{k1}\}$, $\{\beta_{k2}\}$ are to be determined. The frequency response of the filter can be expressed as

$$H(\omega) = GA(\omega)e^{j\Theta(\omega)} \tag{8.5.46}$$

where

$$A(\omega) = \prod_{k=1}^{K} \left| \frac{1 + \beta_{k1} z^{-1} + \beta_{k2} z^{-2}}{1 + \alpha_{k1} z^{-1} + \alpha_{k2} z^{-2}} \right|_{z=e^{j\omega}} \tag{8.5.47}$$

and $\Theta(\omega)$ is the phase response.

Instead of dealing with the phase of the filter, it is more convenient to deal with the envelope delay as a function of frequency, which is

$$\tau_g(\omega) = -\frac{d\Theta(\omega)}{d\omega} \tag{8.5.48}$$

or, equivalently,

$$\tau_g(\omega) = \tau_g(z)|_{z=e^{j\omega}}$$
$$= -\left[\frac{d\Theta(z)}{dz} \right]_{z=e^{j\omega}} \frac{dz}{d\omega} \tag{8.5.49}$$

It can be shown that $\tau_g(z)$ can be expressed as

$$\tau_g(z) = \mathrm{Re} \left\{ \sum_{k=1}^{K} \left[\frac{\beta_{k1} z + 2\beta_{k2}}{z^2 + \beta_{k1} z + \beta_{k2}} - \frac{\alpha_{k1} z + 2\alpha_{k2}}{z^2 + \alpha_{k1} z + \alpha_{k2}} \right] \right\} \tag{8.5.50}$$

where $Re(u)$ denotes the real part of the complex-valued quantity u.

Now suppose that the desired magnitude and delay characteristics $A(\omega)$ and $\tau_g(\omega)$ are specified at arbitrarily chosen discrete frequencies $\omega_1, \omega_2, \ldots, \omega_L$ in the range $0 \leq |\omega| \leq \pi$. Then the error in magnitude at the frequency ω_k is $GA(\omega_k) - A_d(\omega_k)$ where $A_d(\omega_k)$ is the desired magnitude response at ω_k. Similarly, the error in delay at ω_k can be defined as $\tau_g(\omega_k) - \tau_d(\omega_k)$, where $\tau_d(\omega_k)$ is the desired delay response. However, the choice of $\tau_d(\omega_k)$ is complicated by the difficulty in assigning a nominal delay to the filter. Hence, we are led to define the error in delay as $\tau_g(\omega_k) - \tau_g(\omega_0) - \tau_d(\omega_k)$, where $\tau_g(\omega_0)$ is the filter delay at some nominal center frequency in the passband of the filter and $\tau_d(\omega_k)$ is the desired delay response of the filter relative to $\tau_g(\omega_0)$. By defining the error in delay in this manner, we are willing to accept a filter having whatever nominal delay $\tau_g(\omega_0)$ results from the optimization procedure.

As a performance index for determining the filter parameters, one can choose any arbitrary function of the errors in magnitude and delay. To be specific, let us select the total weighted least-squares error over all frequencies $\omega_1, \omega_2, \ldots, \omega_L$, that is,

$$\mathcal{E}(\mathbf{p}, G) = (1 - \lambda) \sum_{n=1}^{L} w_n [GA(\omega_n) - A_d(\omega_n)]^2$$

$$+ \lambda \sum_{n=1}^{L} v_n [\tau_g(\omega_n) - \tau_g(\omega_0) - \tau_d(\omega_n)]^2 \qquad (8.5.51)$$

where $\mathbf{p}$ denotes the $4K$-dimensional vector of filter coefficients $\{\alpha_{k1}\}$, $\{\alpha_{k2}\}$, $\{\beta_{k1}\}$, and $\{\beta_{k2}\}$, and λ, $\{w_n\}$, and $\{v_n\}$ are weighting factors selected by the designer. Thus the emphasis on the errors affecting the design may be placed entirely on the magnitude ($\lambda = 0$), or on the delay ($\lambda = 1$) or, perhaps, equally weighted between magnitude and delay ($\lambda = 1/2$). Similarly, the weighting factors in frequency $\{w_n\}$ and $\{v_n\}$ determine the relative emphasis on the errors as a function of frequency.

The squared-error function $E(\mathbf{p}, G)$ is a nonlinear function of $(4K + 1)$ parameters. The gain G that minimizes $\mathcal{E}$ is easily determined and given by the relation

$$\hat{G} = \frac{\displaystyle\sum_{n=1}^{L} w_n A(\omega_n) A_d(\omega_n)}{\displaystyle\sum_{n=1}^{L} w_n A^2(\omega_n)} \qquad (8.5.52)$$

The optimum gain G can be substituted in (8.5.51) to yield

$$\mathcal{E}(\mathbf{p}, \hat{G}) = (1 - \lambda) \sum_{n=1}^{L} w_n [\hat{G} A(\omega_n) - A_d(\omega_n)]^2$$

$$+ \lambda \sum_{n=1}^{L} v_n [\tau_g(\omega_n) - \tau_g(\omega_0) - \tau_d(\omega_n)]^2 \qquad (8.5.53)$$

Due to the nonlinear nature of $\mathcal{E}(\mathbf{p}, \hat{G})$, its minimization over the remaining $4K$ parameters is performed by an iterative numerical optimization method such as the Fletcher and Powell method (1963). One begins the iterative process by assuming an initial set of parameter values, say $\mathbf{p}^{(0)}$. With the initial values substituted in (8.5.51), we obtain the least-squares error $\mathcal{E}(\mathbf{p}^{(0)}, \hat{G})$. If we also evaluate the partial derivatives $\partial \mathcal{E}/\partial \alpha_{k1}$, $\partial \mathcal{E}/\partial \alpha_{k2}$, $\partial \mathcal{E}/\partial \beta_{k1}$, and $\partial \mathcal{E}/\partial \beta_{k2}$ at the initial value $\mathbf{p}^{(0)}$, we can use this first derivative information to change the initial values of the parameters in a direction that leads toward the minimum of the function $\mathcal{E}(\mathbf{p}, \hat{G})$ and thus to a new set of parameters $\mathbf{p}^{(1)}$.

Repetition of the above steps results in an iterative algorithm which is described mathematically by the recursive equation

$$\mathbf{p}^{(m+1)} = \mathbf{p}^{(m)} - \Delta^{(m)} \mathbf{Q}^{(m)} \mathbf{g}^{(m)} \qquad m = 0, 1, 2, \ldots$$

where $\Delta^{(m)}$ is a scalar representing the step size of the iteration, $\mathbf{Q}^{(m)}$ is a $(4K \times 4K)$ matrix, which is an estimate of the Hessian, and $\mathbf{g}^{(m)}$ is a $(4K \times 1)$ vector consisting of the four K-dimensional vectors of gradient components of $\mathcal{E}$ (i.e., $\partial \mathcal{E}/\partial \alpha_{k1}$, $\partial \mathcal{E}/\partial \alpha_{k2}$, $\partial \mathcal{E}/\partial \beta_{k1}$, $\partial \mathcal{E}/\partial \beta_{k2}$), evaluated at $\alpha_{k1} = \alpha_{k1}^{(m)}$, $\alpha_{k2} = \alpha_{k2}^{(m)}$, $\beta_{k1} = \beta_{k1}^{(m)}$, and $\beta_{k2} = \beta_{k2}^{(m)}$. This iterative process is terminated when the gradient components are nearly zero and the value of the function $\mathcal{E}(\mathbf{p}, \hat{G})$ does not change appreciably from one iteration to another.

The stability constraint is easily incorporated into the computer program through the parameter vector $\mathbf{p}$. When $|\alpha_{k2}| > 1$ for any $k = 1, \ldots, K$, the parameter α_{k2} is forced back inside the unit circle and the iterative process continued. A similar process can be used to force zeros inside the unit circle if a minimum-phase filter is desired.

The major difficulty with any iterative procedure that searches for the parameter values that minimize a nonlinear function is that the process may converge to a local minimum instead of a global minimum. Our only recourse around this problem is to start the iterative process with different values for the parameters and observe the end result.

Example 8.5.8

Let us design a lowpass filter using the Fletcher–Powell optimization procedure just described. The filter is to have a bandwidth of 0.3π and a rejection band commencing at 0.45π. The delay distortion can be ignored by selecting the weighting factor $\lambda = 0$.

Solution We have selected a two-stage ($K = 2$) or four-pole and four-zero filter which we believe is adequate to meet the transition band and rejection requirements. The magnitude response is specified at 19 equally spaced frequencies, which is considered a sufficiently dense set of points to realize a good design. Finally, a set of uniform weights is selected.

This filter has the response shown in Fig. 8.57. It has a remarkable resemblance to the response of the elliptic lowpass filter shown in Fig. 8.58, which was designed to have the same passband ripple and transition region as the computer-generated filter. A small but noticeable difference between the elliptic filter and the computer-generated filter is the somewhat flatter delay response of the latter relative to the former.

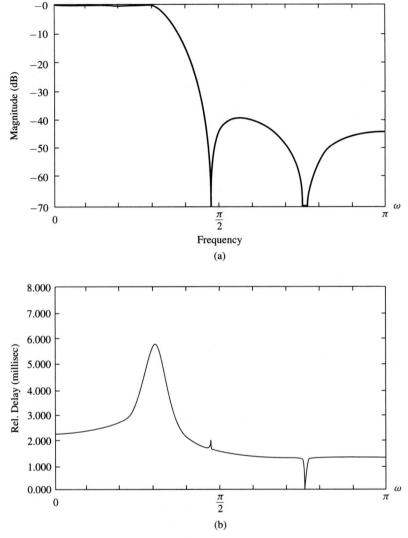

Figure 8.57 Filter designed by Fletcher–Powell optimization method (Example 8.5.8).

Example 8.5.9

Design an IIR filter with magnitude characteristics

$$A_d(\omega) = \begin{cases} \sin\omega, & 0 \le |\omega| \le \dfrac{\pi}{2} \\ 0, & \dfrac{\pi}{2} < |\omega| < \pi \end{cases}$$

and a constant envelope delay in the passband.

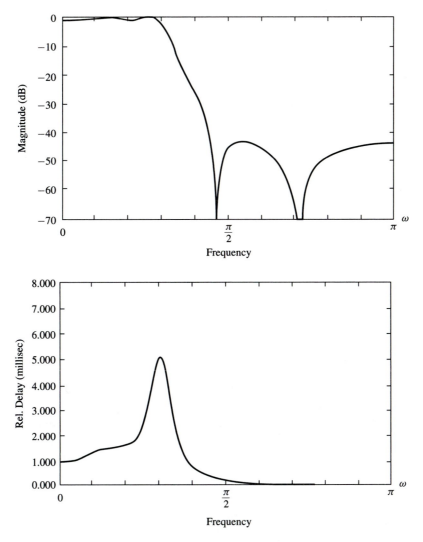

Figure 8.58 Amplitude and delay response for elliptic filter.

Solution The desired filter is called a modified duobinary filter and finds application in high-speed digital communications modems. The frequency response was specified at the frequencies illustrated in Fig. 8.59. The envelope delay was left unspecified in the stopband and selected to be flat in the passband. Equal weighting coefficients $\{w_n\}$ and $\{v_n\}$ were selected. A weighting factor of $\lambda = 1/2$ was selected.

A two-stage (four-pole, four-zero) filter is designed to meet the foregoing specifications. The result of the design is illustrated in Fig. 8.60. We note that the magnitude characteristic is reasonably well matched to $\sin \omega$ in the passband, but the stopband attenuation peaks at about -25 dB, which is rather large. The envelope delay characteristic is relatively flat in the passband.

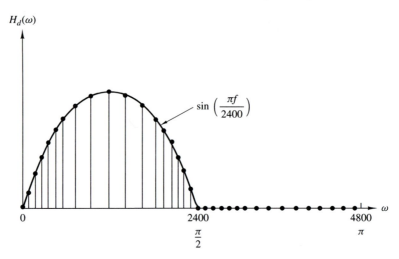

Figure 8.59 Frequency response of an ideal modified duobinary filter.

A four-stage (eight-pole, eight-zero) filter having the same frequency response specifications was also designed. This design produced better results, especially in the stopband where the attenuation peaked at -36 dB. The envelope delay was also considerably flatter.

8.6 SUMMARY AND REFERENCES

We have described in some detail the most important techniques for designing FIR and IIR digital filters based on either frequency-domain specifications expressed in terms of a desired frequency response $H_d(\omega)$, or in terms of the desired impulse response $h_d(n)$.

As a general rule, FIR filters are used in applications where there is a need for a linear-phase filter. This requirement occurs in many applications, especially in telecommunications, where there is a requirement to separate (demultiplex) signals such as data that have been frequency-division multiplexed, without distorting these signals in the process of demultiplexing. Of the several methods described for designing FIR filters, the frequency sampling design method and the optimum Chebyshev approximation method yield the best designs.

IIR filters are generally used in applications where some phase distortion is tolerable. Of the class of IIR filters, elliptic filters are the most efficient to implement in the sense that for a given set of specifications, an elliptic filter has a lower order or fewer coefficients than any other IIR filter type. When compared with FIR filters, elliptic filters are also considerably more efficient. In view of this, one might consider the use of an elliptic filter to obtain the desired frequency selectivity, followed then by an all-pass phase equalizer that compensates for the phase distortion in the elliptic filter. However, attempts to accomplish this have resulted in filters with a number of coefficients in the cascade combination that

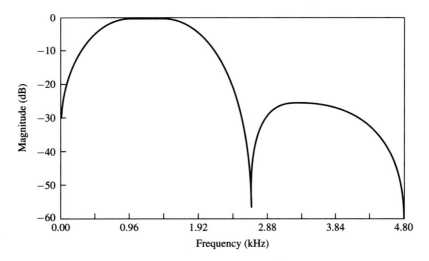

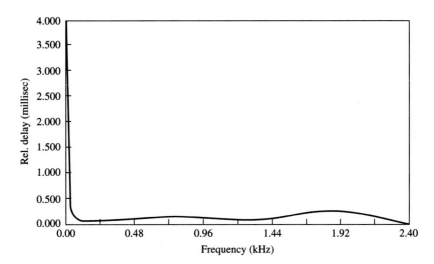

Figure 8.60 Frequency response of filter in Example 8.5.9. Designed by the Fletcher–Powell optimization method.

equaled or exceeded the number of coefficients in an equivalent linear-phase FIR filter. Consequently, no reduction in complexity is achievable in using phase-equalized elliptic filters.

In addition to the filter design methods based on the transformation of analog filters into the digital domain, we also presented several methods in which the design is done directly in the discrete-time domain. The least-squares method is particularly appropriate for designing IIR filters. The least-squares method is also used for the design of FIR Wiener filters.

Such a rich literature now exists on the design of digital filters that it is not possible to cite all the important references. We shall cite only a few. Some of the early work on digital filter design was done by Kaiser (1963, 1966), Steiglitz (1965), Golden and Kaiser (1964), Rader and Gold (1967a), Shanks (1967), Helms (1968), Gibbs (1969, 1970), and Gold and Rader (1969).

The design of analog filters is treated in the classic books by Storer (1957), Guillemin (1957), Weinberg (1962), and Daniels (1974).

The frequency sampling method for filter design was first proposed by Gold and Jordan (1968, 1969), and optimized by Rabiner et al. (1970). Additional results were published by Herrmann (1970), Herrmann and Schuessler (1970a), and Hofstetter et al. (1971). The Chebyshev (minimax) approximation method for designing linear-phase FIR filters was proposed by Parks and McClellan (1972a,b) and discussed further by Rabiner et al. (1975). The design of elliptic digital filters is treated in the book by Gold and Rader (1969) and in the paper by Gray and Markel (1976). The latter includes a computer program for designing digital elliptic filters.

The use of frequency transformations in the digital domain was proposed by Constantinides (1967, 1968, 1970). These transformations are appropriate only for IIR filters. The reader should note that when these transformations are applied to a lowpass FIR filter, the resulting filter is IIR.

Direct design techniques for digital filters have been considered in a number of papers, including Shanks (1967), Burrus and Parks (1970), Steiglitz (1970), Deczky (1972), Brophy and Salazar (1973), and Bandler and Bardakjian (1973).

PROBLEMS

8.1 Design an FIR linear phase, digital filter approximating the ideal frequency response

$$H_d(\omega) = \begin{cases} 1, & \text{for } |\omega| \leq \dfrac{\pi}{6} \\ 0, & \text{for } \dfrac{\pi}{6} < |\omega| \leq \pi \end{cases}$$

    (a) Determine the coefficients of a 25-tap filter based on the window method with a rectangular window.

    (b) Determine and plot the magnitude and phase response of the filter.

    (c) Repeat parts (a) and (b) using the Hamming window.

    (d) Repeat parts (a) and (b) using a Bartlett window.

8.2 Repeat Problem 8.1 for a bandstop filter having the ideal response

$$H_d(\omega) = \begin{cases} 1, & \text{for } |\omega| \leq \dfrac{\pi}{6} \\ 0, & \text{for } \dfrac{\pi}{6} < |\omega| < \dfrac{\pi}{3} \\ 1, & \text{for } \dfrac{\pi}{3} \leq |\omega| \leq \pi \end{cases}$$

8.3 Redesign the filter of Problem 8.1 using the Hanning and Blackman windows.

8.4 Redesign the filter of Problem 8.2 using the Hanning and Blackman windows.

8.5 Determine the unit sample response $\{h(n)\}$ of a linear-phase FIR filter of length $M = 4$ for which the frequency response at $\omega = 0$ and $\omega = \pi/2$ is specified as

$$H_r(0) = 1 \qquad H_r\left(\frac{\pi}{2}\right) = \frac{1}{2}$$

8.6 Determine the coefficients $\{h(n)\}$ of a linear-phase FIR filter of length $M = 15$ which has a symmetric unit sample response and a frequency response that satisfies the condition

$$H_r\left(\frac{2\pi k}{15}\right) = \begin{cases} 1, & k = 0, 1, 2, 3 \\ 0, & k = 4, 5, 6, 7 \end{cases}$$

8.7 Repeat the filter design problem in Problem 8.6 with the frequency response specifications

$$H_r\left(\frac{2\pi k}{15}\right) = \begin{cases} 1 & k = 0, 1, 2, 3 \\ 0.4 & k = 4 \\ 0 & k = 5, 6, 7 \end{cases}$$

8.8 The ideal analog differentiator is described by

$$y_a(t) = \frac{dx_a(t)}{dt}$$

where $x_a(t)$ is the input and $y_a(t)$ the output signal.

(a) Determine its frequency response by exciting the system with the input $x_a(t) = e^{j2\pi Ft}$.

(b) Sketch the magnitude and phase response of an ideal analog differentiator band-limited to B hertz.

(c) The ideal digital differentiator is defined as

$$H(\omega) = j\omega \qquad |\omega| \le \pi$$

Justify this definition by comparing the frequency response $|H(\omega)|$, $\measuredangle H(\omega)$ with that in part (b).

(d) By computing the frequency response $H(\omega)$, show that the discrete-time system

$$y(n) = x(n) - x(n-1)$$

is a good approximation of a differentiator at low frequencies.

(e) Compute the response of the system to the input

$$x(n) = A\cos(\omega_0 n + \theta)$$

8.9 Use the window method with a Hamming window to design a 21-tap differentiator as shown in Fig. P8.9. Compute and plot the magnitude and phase response of the resulting filter.

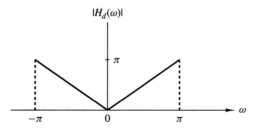

Figure P8.9

8.10 Use the matched-z transformation to convert the analog filter with system function

$$H(s) = \frac{s + 0.1}{(s + 0.1)^2 + 9}$$

into a digital IIR filter. Select $T = 0.1$ and compare the location of the zeros in $H(z)$ with the locations of the zeros obtained by applying the impulse invariance method in the conversion of $H(s)$.

8.11 Convert the analog bandpass filter designed in Example 8.4.1 into a digital filter by means of the bilinear transformation. Thereby derive the digital filter characteristic obtained in Example 8.4.2 by the alternative approach and verify that the bilinear transformation applied to the analog filter results in the same digital bandpass filter.

8.12 An ideal analog integrator is described by the system function $H_a(s) = 1/s$. A digital integrator with system function $H(z)$ can obtained by use of the bilinear transformation. That is,

$$H(z) = \frac{T}{2} \frac{1 + z^{-1}}{1 - z^{-1}} \equiv H_a(s)|_{s = (2/T)(1 - z^{-1})/(1 + z^{-1})}$$

(a) Write the difference equation for the digital integrator relating the input $x(n)$ to the output $y(n)$.

(b) Roughly sketch the magnitude $|H_a(j\Omega)|$ and phase $\Theta(\Omega)$ of the analog integrator.

(c) It is easily verified that the frequency response of the digital integrator is

$$H(\omega) = -j \frac{T}{2} \frac{\cos(\omega/2)}{\sin(\omega/2)} = -j \frac{T}{2} \cot \frac{\omega}{2}$$

Roughly sketch $|H(\omega)|$ and $\theta(\omega)$.

(d) Compare the magnitude and phase characteristics obtained in parts (b) and (c). How well does the digital integrator match the magnitude and phase characteristics of the analog integrator?

(e) The digital integrator has a pole at $z = 1$. If you implement this filter on a digital computer, what restrictions might you place on the input signal sequence $x(n)$ to avoid computational difficulties?

8.13 A z-plane pole–zero plot for a certain digital filter is shown in Fig. P8.13. The filter has unity gain at dc.

(a) Determine the system function in the form

$$H(z) = A \left[\frac{(1 + a_1 z^{-1})(1 + b_1 z^{-1} + b_2 z^{-2})}{(1 + c_1 z^{-1})(1 + d_1 z^{-1} + d_2 z^{-2})} \right]$$

giving numerical values for the parameters A, a_1, b_1, b_2, c_1, d_1, and d_2.

(b) Draw block diagrams showing numerical values for path gains in the following forms:

(1) Direct form II (canonic form)

(2) Cascade form (make each section canonic, with real coefficients)

8.14 Consider the pole–zero plot shown in Fig. P8.14.

(a) Does it represent an FIR filter?

(b) Is it a linear-phase system?

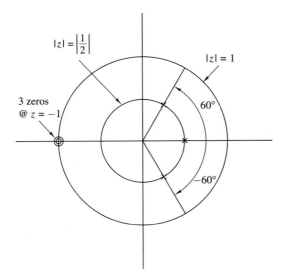

Figure P8.13

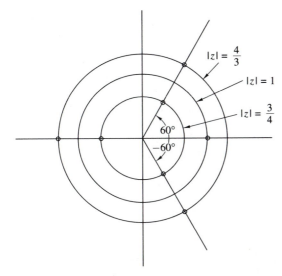

Figure P8.14

(c) Give a direct form realization that exploits all symmetries to minimize the number of multiplications. Show all path gains.

8.15* A digital low-pass filter is required to meet the following specifications:

Passband ripple: ≤ 1 dB

Passband edge: 4 kHz

Stopband attenuation: ≥ 40 dB

Stopband edge: 6 kHz

Sample rate: 24 kHz

The filter is to be designed by performing a bilinear transformation on an analog system function. Determine what order Butterworth, Chebyshev, and elliptic analog designs must be used to meet the specifications in the digital implementation.

8.16* An IIR digital low-pass filter is required to meet the following specifications:
Passband ripple (or peak-to-peak ripple): ≤ 0.5 dB
Passband edge: 1.2 kHz
Stopband attenuation: ≥ 40 dB
Stopband edge: 2.0 kHz
Sample rate: 8.0 kHz
Use the design formulas in the book to determine the required filter order for
(a) A digital Butterworth filter
(b) A digital Chebyshev filter
(c) A digital elliptic filter

8.17* Determine the system function $H(z)$ of the lowest-order Chebyshev digital filter that meets the following specifications:
(a) 1-dB ripple in the passband $0 \leq |\omega| \leq 0.3\pi$.
(b) At least 60 dB attentuation in the stopband $0.35\pi \leq |\omega| \leq \pi$. Use the bilinear transformation.

8.18* Determine the system function $H(z)$ of the lowest-order Chebyshev digital filter that meets the following specifications:
(a) $\frac{1}{2}$-dB ripple in the passband $0 \leq |\omega| \leq 0.24\pi$.
(b) At least 50-dB attenuation in the stopband $0.35\pi \leq |\omega| \leq \pi$. Use the bilinear transformation.

8.19* An analog signal $x(t)$ consists of the sum of two components $x_1(t)$ and $x_2(t)$. The spectral characteristics of $x(t)$ are shown in the sketch in Fig. P8.19. The signal $x(t)$ is bandlimited to 40 kHz and it is sampled at a rate of 100 kHz to yield the sequence $x(n)$.

It is desired to suppress the signal $x_2(t)$ by passing the sequence $x(n)$ through a digital lowpass filter. The allowable amplitude distortion on $|X_1(f)|$ is $\pm 2\%$ ($\delta_1 = 0.02$) over the range $0 \leq |F| \leq 15$ kHz. Above 20 kHz, the filter must have an attenuation of at least 40 dB ($\delta_2 = 0.01$).
(a) Use the Remez exchange algorithm to design the *minimum*-order linear-phase FIR filter that meets the specifications above. From the plot of the magnitude characteristic of the filter frequency response, give the actual specifications achieved by the filter.
(b) Compare the order M obtained in part (a) with the approximate formulas given in equations (8.2.94) and (8.2.95).
(c) For the order M obtained in part (a), design an FIR digital lowpass filter using the window technique and the Hamming window. Compare the frequency response characteristics of this design with those obtained in part (a).

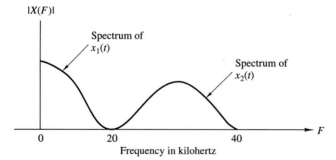

Figure P8.14

(d) Design the *minimum*-order elliptic filter that meets the given amplitude specifications. Compare the frequency response of the elliptic filter with that of the FIR filter in part (a).

(e) Compare the complexity of implementing the FIR filter in part (a) versus the elliptic filter obtained in part (d). Assume that the FIR filter is implemented in the direct form and the elliptic filter is implemented as a cascade of two-pole filters. Use storage requirements and the number of multiplications per output point in the comparison of complexity.

8.20 The impulse response of an analog filter is shown in Fig. P8.20.

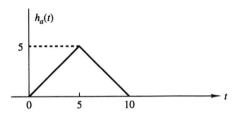

Figure P8.20

(a) Let $h(n) = h_a(nT)$, where $T = 1$, be the impulse response of a discrete-time filter. Determine the system function $H(z)$ and the frequency response $H(\omega)$ for this FIR filter.

(b) Sketch (roughly) $|H(\omega)|$ and compare this frequency response characteristic with $|H_a(j\Omega)|$.

(c) The FIR filter with unit sample response $h(n)$ given above is to be approximated by a second-order IIR filter of the form

$$G(z) = \frac{b_0 z^{-1}}{1 - a_1 z^{-1} - a_2 z^{-2}}$$

Use the least-squares inverse design procedure to determine the values of the coefficients b_0, a_1, and a_2.

8.21 In this problem you will be comparing some of the characteristics of analog and digital implementations of the single-pole low-pass analog system

$$H_a(s) = \frac{1}{s + \alpha} \Leftrightarrow h_a(t) = e^{-\alpha t}$$

(a) What is the gain at dc? At what radian frequency is the analog frequency response 3 dB down from its dc value? At what frequency is the analog frequency reponse zero? At what time has the analog impulse response decayed to $1/e$ of its initial value?

(b) Give the digital system function $H(z)$ for the impulse-invariant design for this filter. What is the gain at dc? Give an expression for the 3-dB radian frequency. At what (real-valued) frequency is the response zero? How many samples are there in the unit sample time-domain response before it has decayed to $1/e$ of its initial value?

(c) "Prewarp" the parameter α and perform the bilinear transformation to obtain the digital system function $H(z)$ from the analog design. What is the gain at dc? At what (real-valued) frequency is the response zero? Give an expression for the 3-dB radian frequency. How many samples in the unit sample time-domain response before it has decayed to $1/e$ of its initial value?

8.22 We wish to design a FIR bandpass filter having a duration $M = 201$. $H_d(\omega)$ represents the ideal characteristic of the noncausal bandpass filter as shown in Fig. P8.22.

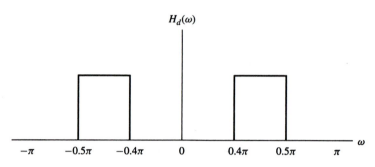

Figure P8.22

(a) Determine the unit sample (impulse) response $h_d(n)$ corresponding to $H_d(\omega)$.
(b) Explain how you would use the Hamming window

$$w(n) = 0.54 + 0.46 \cos\left(\frac{2\pi}{M-1}\right) \qquad -\frac{M-1}{2} \leq n \leq \frac{M-1}{2}$$

to design a FIR bandpass filter having an impulse response $h(n)$ for $0 \leq n \leq 200$.
(c) Suppose that you were to design the FIR filter with $M = 201$ by using the frequency sampling technique in which the DFT coefficients $H(k)$ are specified instead of $h(n)$. Give the values of $H(k)$ for $0 \leq k \leq 200$ corresponding to $H_d(e^{j\omega})$ and indicate how the frequency response of the actual filter will differ from the ideal. Would the actual filter represent a good design? Explain your answer.

8.23 We wish to design a digital bandpass filter from a second-order analog lowpass Butterworth filter prototype using the bilinear transformation. The specifications on the digital filter are shown in Fig. P8.23(a). The cutoff frequencies (measured at the half power points) for the digital filter should lie at $\omega = 5\pi/12$ and $\omega = 7\pi/12$.
The analog protoype is given by

$$H(s) = \frac{1}{s^2 + \sqrt{2}s + 1}$$

with the half-power point at $\Omega = 1$.
(a) Determine the system function for the digital bandpass filter.
(b) Using the same specs on the digital filter as in part (a), determine which of the analog bandpass prototype filters shown in Fig. P.8.23(b) could be transformed directly using the bilinear transformation to give the proper digital filter. Only the plot of the magnitude squared of the frequency is given.

8.24 Figure P8.24 shows a digital filter designed using the frequency sampling method.
(a) Sketch a z-plane pole–zero plot for this filter.
(b) Is the filter lowpass, highpass, or bandpass?
(c) Determine the magnitude response $|H(\omega)|$ at the frequencies $\omega_k = \pi k/6$ for $k = 0, 1, 2, 3, 4, 5, 6$.

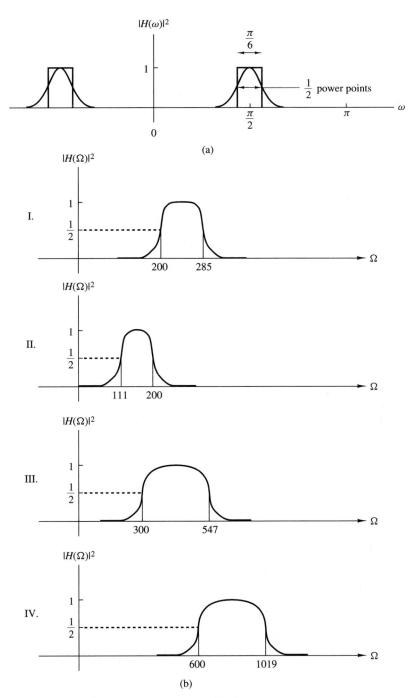

Figure P8.23

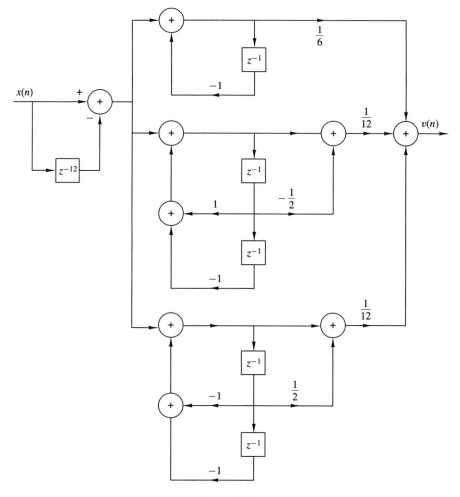

Figure P8.24

(d) Use the results of part (c) to sketch the magnitude response for $0 \le \omega \le \pi$ and confirm your answer to part (b).

8.25 An analog signal of the form $x_a(t) = a(t) \cos 2000\pi t$ is bandlimited to the range $900 \le F \le 1100$ Hz. It is used as an input to the system shown in Fig. P8.25.

(a) Determine and sketch the spectra for the signals $x(n)$ and $w(n)$.

(b) Use a Hamming window of length $M = 31$ to design a lowpass linear phase FIR filter $H(\omega)$ that passes $\{a(n)\}$.

(c) Determine the sampling rate of the A/D converter that would allow us to eliminate the frequency conversion in Fig. P8.25.

8.26 *System identification*   Consider an unknown LTI system and an FIR system model as shown in Fig. P8.26. Both systems are excited by the same input sequence $\{x(n)\}$. The problem is to determine the coefficients $\{h(n), 0 \le n \le M - 1\}$ of the FIR model

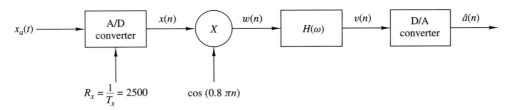

Figure P8.25

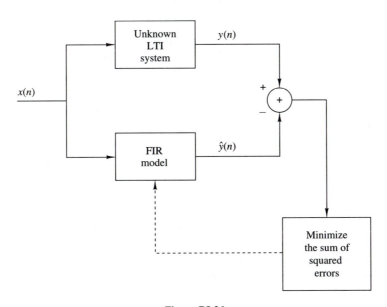

Figure P8.26

of the system to minimize the average squared error between the outputs of the two systems.

(a) Use the least-squares criterion to determine the equations for the optimum FIR filter coefficients.

(b) Repeat part (a) if the output of the unknown system is corrupted by an additive white noise $\{w(n)\}$ sequence with variance σ_w^2.

8.27 Determine the least-squares FIR inverse of length 3 to the system with impulse response

$$h(n) = \begin{cases} 2, & n = 0 \\ 1, & n = 1 \\ 0, & \text{otherwise} \end{cases}$$

Also, determine the minimum squared error $\mathcal{E}_{\min}$.

8.28 Determine the least-squares FIR inverse filter of length 3 for the system with impulse response $h(n)$ given in Example 8.5.6, when $\alpha = \frac{1}{2}$ and the desired response is specified as $d(n) = \delta(n - 2)$. Also compute the minimum least-squares error.

8.29* A linear time-invariant system has an input sequence $x(n)$ and an output sequence $y(n)$. The user has access only to the system output $y(n)$. In addition, the following information is available.

(a) The input signal is periodic with a given fundamental period N and has a flat spectral envelope, that is,

$$x(n) = \sum_{k=0}^{N-1} c_k^x e^{j(2\pi/N)kn} \qquad \text{all } n$$

where $c_k^x = 1$ for all k.

(b) The system $H(z)$ is all-pole, that is,

$$H(z) = \frac{1}{1 + \sum_{k=1}^{P} a_k z^{-k}}$$

but the order p and the coefficients $(a_k, 1 \le k \le p)$ are unknown. Is it possible to determine the order p and the numerical values of the coefficients $\{a_k, 1 \le k \le p\}$ by taking measurements on the output $y(n)$? If yes, explain how. Is this possible for every value of p?

(c) Repeat Problem 8.31 for a system with system function

$$H(z) = \frac{\sum_{k=0}^{M} b_k z^{-k}}{1 + \sum_{k=1}^{P} a_k z^{-k}}$$

(d) *FIR system modeling* Consider an "unknown" FIR system with impulse response $h(n)$, $0 \le n \le 11$, given by

$$h(0) = h(11) = 0.309828 \times 10^{-1}$$
$$h(1) = h(10) = 0.416901 \times 10^{-1}$$
$$h(2) = h(9) = -0.577081 \times 10^{-1}$$
$$h(3) = h(8) = -0.852502 \times 10^{-1}$$
$$h(4) = h(7) = 0.147157 \times 10^{0}$$
$$h(5) = h(6) = 0.449188 \times 10^{0}$$

A potential user has access to the input and output of the system but does not have any information about its impulse response other than that it is FIR. In an effort to determine the impulse response of the system, the user excites it with a zero mean, random sequence $x(n)$ uniformly distributed in the range $[-0.5, 0.5]$, and records the signal $x(n)$ and the corresponding output $y(n)$ for $0 \le n \le 199$.

(1) By using the available information that the unknown system is FIR, the user employs the method of least-squares to obtain an FIR model $h(n)$, $0 \le n \le M - 1$. Set up the system of linear equations, specifying the parameters $h(0), h(1), \ldots, h(M-1)$. Specify formulas we should use to determine the necessary autocorrelation and crosscorrelation values.

(2) Since the order of the system is unknown, the user decides to try models of different orders and check the corresponding total squared error. Clearly, this error will be zero (or very close to it if the order of the model becomes equal to the order of the system). Compute the FIR models $h_M(n)$, $0 \leq n \leq M - 1$ for $M = 8, 9, 10, 11, 12, 13, 14$ as well as the corresponding total squared errors E_M, $M = 8, 9, \ldots, 14$. What do you observe?

(3) Determine and plot the frequency response of the system and the models for $M = 11, 12, 13$. Comment on the results.

(4) Suppose now that the output of the system is corrupted by additive noise, so instead of the signal $y(n)$, $0 \leq n \leq 199$, we have available the signal

$$v(n) = y(n) + 0.01w(n)$$

where $w(n)$ is a Gaussian random sequence with zero mean and variance $\sigma^2 = 1$.

Repeat part (b) by using $v(n)$ instead of $y(n)$ and comment on the results. The quality of the model can be also determined by the quantity

$$Q = \frac{\sum_{n=0}^{\infty}[h(n) - \hat{h}(n)]^2}{\sum_{n=0}^{\infty} h^2(n)}$$

9

Sampling and Reconstruction of Signals

In Chapters 1 and 4 we treated the sampling of continuous-time signals and demonstrated that if the signals are bandlimited, it is possible to reconstruct the original signal from the samples, provided that the sampling rate is at least twice the highest frequency contained in the signal. We also briefly described the subsequent operations of quantization and coding that are necessary to convert an analog signal to a digital signal appropriate for digital processing.

In this chapter we consider time-domain sampling, analog-to-digital (A/D) conversion (quantization and coding), and digital-to-analog (D/A) conversion (signal reconstruction) in greater depth. First, we consider the sampling of the special class of signals that are characterized as bandpass signals. Then we treat analog-to-digital converters and their characteristics. Of particular interest is the use of oversampling and sigma-delta modulation in the design of high precision A/D converters. The final topic of the chapter is digital-to-analog conversion or, simply, the reconstruction of the continuous-time signal from its sampled values.

9.1 SAMPLING OF BANDPASS SIGNALS

Our main focus in this section is the sampling of bandpass signals. We begin by describing the time and frequency domain representations of bandpass signals.

9.1.1 Representation of Bandpass Signals

Suppose that a real-valued signal $x(t)$ has a frequency content concentrated in a narrow band of frequencies in the vicinity of a frequency F_c, as shown in Fig. 9.1. Our objective is to develop a mathematical representation of such signals. First, we construct a signal that contains only the positive frequencies in $x(t)$. Such a signal can be expressed as

$$X_+(F) = 2V(F)X(F) \tag{9.1.1}$$

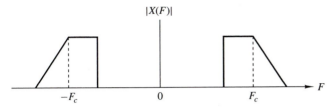

Figure 9.1 Spectrum of a bandpass signal.

where $X(F)$ is the Fourier transform of $x(t)$ and $V(F)$ is the unit step function. The equivalent time-domain expression for (9.1.1) is

$$x_+(t) = \int_{-\infty}^{\infty} X_+(F)e^{j2\pi Ft}dF$$

$$= F^{-1}[2V(F)] \star F^{-1}[X(F)] \tag{9.1.2}$$

The signal $x_+(t)$ is called the *analytic signal* or the *pre-envelope* of $x(t)$. We note that $F^{-1}[X(F)] = x(t)$ and

$$F^{-1}[2V(f)] = \delta(t) + \frac{j}{\pi t} \tag{9.1.3}$$

Hence,

$$x_+(t) = \left[\delta(t) + \frac{j}{\pi t}\right] \star x(t)$$

$$= x(t) + j\frac{1}{\pi t} \star x(t) \tag{9.1.4}$$

We define $\hat{x}(t)$ as

$$\hat{x}(t) = \frac{1}{\pi t} \star x(t)$$

$$= \frac{1}{\pi} \int_{-\infty}^{\infty} \frac{x(\tau)}{t - \tau} d\tau \tag{9.1.5}$$

The signal $\hat{x}(t)$ can be viewed as the output of the filter with impulse response

$$h(t) = \frac{1}{\pi t}, \qquad -\infty < t < \infty \tag{9.1.6}$$

when excited by the input signal $x(t)$. Such a filter is called a *Hilbert transformer*. The frequency response of this filter is simply

$$H(F) = \int_{-\infty}^{\infty} h(t)e^{-j2\pi Ft}dt$$

$$= \frac{1}{\pi} \int_{-\infty}^{\infty} \frac{1}{t}e^{-2\pi Ft}dt$$

$$= \begin{cases} -j & (F > 0) \\ 0 & (F = 0) \\ j & (F < 0) \end{cases} \tag{9.1.7}$$

We observe that $|H(F)| = 1$ and that the phase response $\Theta(F) = -\frac{1}{2}\pi$ for $F > 0$

and $\Theta(F) = \frac{1}{2}\pi$ for $F < 0$. Therefore, this filter is basically a 90° phase shifter for all frequencies in the input signal, and it is akin to the discrete-time Hilbert transform filter described in Section 8.2.6.

The analytic signal $x_+(t)$ is a bandpass signal. We can obtain an equivalent lowpass representation by performing a frequency translation of $X_+(F)$. Thus, we define $X_l(F)$ as

$$X_l(F) = X_+(F + F_c) \tag{9.1.8}$$

The equivalent time-domain relation is

$$x_l(t) = x_+(t)e^{-j2\pi F_c t}$$
$$= [x(t) + j\hat{x}(t)]e^{-j2\pi F_c t} \tag{9.1.9}$$

or, equivalently,

$$x(t) + j\hat{x}(t) = x_l(t)e^{j2\pi F_c t} \tag{9.1.10}$$

In general, the signal $x_l(t)$ is complex-valued (see Problem 9.3), and can be expressed as

$$x_l(t) = u_c(t) + ju_s(t) \tag{9.1.11}$$

If we substitute for $x_l(t)$ in (9.1.10) and equate real and imaginary parts on each side, we obtain the relations

$$x(t) = u_c(t)\cos 2\pi F_c t - u_s(t)\sin 2\pi F_c t \tag{9.1.12}$$

$$\hat{x}(t) = u_c(t)\sin 2\pi F_c t + u_s(t)\cos 2\pi F_c t \tag{9.1.13}$$

The expression (9.1.12) is the desired form for the representation of a bandpass signal. The low-frequency signal components $u_c(t)$ and $u_s(t)$ can be viewed as amplitude modulations impressed on the carrier components $\cos 2\pi F_c t$ and $\sin 2\pi F_c t$, respectively. Since these carrier components are in phase quadrature, $u_c(t)$ and $u_s(t)$ are called the *quadrature components* of the bandpass signal $x(t)$.

Another representation of the signal in (9.1.12) is

$$x(t) = \text{Re}\{[u_c(t) + ju_s(t)]e^{j2\pi F_c t}\}$$
$$= \text{Re}[x_l(t)e^{j2\pi F_c t}] \tag{9.1.14}$$

where Re denotes the real part of the complex-valued quantity in the brackets following. The lowpass signal $x_l(t)$ is usually called the *complex envelope* of the real signal $x(t)$, and is basically the *equivalent lowpass signal*.

Finally, a third possible representation of a bandpass signal is obtained by expressing $x_l(t)$ as

$$x_l(t) = a(t)e^{j\theta(t)} \tag{9.1.15}$$

where

$$a(t) = \sqrt{u_c^2(t) + u_s^2(t)} \tag{9.1.16}$$

$$\theta(t) = \tan^{-1}\frac{u_s(t)}{u_c(t)} \tag{9.1.17}$$

Then

$$x(t) = \text{Re}[x_l(t)e^{j2\pi F_c t}]$$

$$= \text{Re}[a(t)e^{j[2\pi F_c t + \theta(t)]}]$$

$$= a(t)\cos[2\pi F_c t + \theta(t)] \qquad (9.1.18)$$

The signal $a(t)$ is called the *envelope* of $x(t)$, and $\theta(t)$ is called the *phase* of $x(t)$. Therefore, (9.1.12), (9.1.14), and (9.1.18) are equivalent representations of bandpass signals.

The Fourier transform of $x(t)$ is

$$X(F) = \int_{-\infty}^{\infty} x(t)e^{-j2\pi F t}\,dt$$

$$= \int_{-\infty}^{\infty} \{\text{Re}[x_l(t)e^{j2\pi F_c t}]\}e^{-j2\pi F t}\,dt \qquad (9.1.19)$$

Use of the identity

$$\text{Re}(\xi) = \tfrac{1}{2}(\xi + \xi^*) \qquad (9.1.20)$$

in (9.1.19) yields the result

$$X(F) = \frac{1}{2}\int_{-\infty}^{\infty}[x_l(t)e^{j2\pi F_c t} + x_l^*(t)e^{-j2\pi F_c t}]e^{-j2\pi F t}\,dt$$

$$= \tfrac{1}{2}[X_l(F - F_c) + X_l^*(-F - F_c)] \qquad (9.1.21)$$

where $X_l(F)$ is the Fourier transform of $x_l(t)$. This is the basic relationship between the spectrum of the real bandpass signal $x(t)$ and the spectrum of the equivalent lowpass signal $x_l(t)$.

It is apparent from (9.1.21) that the spectrum of the bandpass signal $x(t)$ can be obtained from the spectrum of the complex signal $x_l(t)$ by a frequency translation. To be more precise, suppose that the spectrum of the signal $x_l(t)$ is as shown in Fig. 9.2(a). Then the spectrum $X(F)$ for positive frequencies is simply $X_l(F)$ translated in frequency to the right by F_c and scaled in amplitude by $\tfrac{1}{2}$. The spectrum $X(F)$ for negative frequencies is obtained by first folding $X_l(F)$ about $F = 0$ to obtain $X_l(-F)$, conjugating $X_l(-F)$ to obtain $X_l^*(-F)$, translating $X_l^*(-F)$ in frequency to the left by F_c, and scaling the result by $\tfrac{1}{2}$. The folding and conjugation of $X_l(F)$ for the negative-frequency component of the spectrum result in a magnitude spectrum $|X(F)|$ that is even and a phase spectrum $\angle X(F)$ that is odd as shown in Fig. 9.2(b). These symmetry properties must hold since the signal $x(t)$ is real valued. However, they do not apply to the spectrum of the equivalent complex signal $X_l(t)$.

The development above implies that *any bandpass signal $x(t)$ can be represented by an equivalent lowpass signal $x_l(t)$*. In general, the equivalent lowpass signal $x_l(t)$ is complex valued, whereas the bandpass signal $x(t)$ is real. The latter can be obtained from the former through the time-domain relation in (9.1.14) or through the frequency-domain relation in (9.1.21).

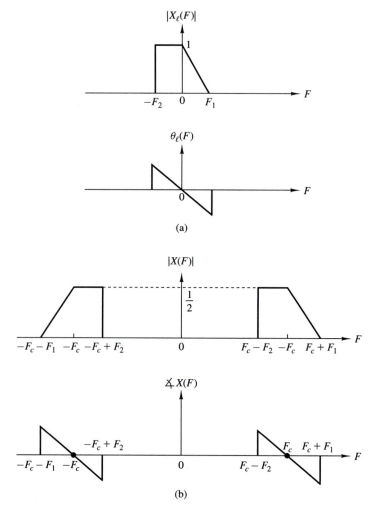

Figure 9.2   (a) Spectrum of the lowpass signal and (b) the corresponding spectrum for the bandpass signal.

9.1.2 Sampling of Bandpass Signals

We have already demonstrated that a continuous-time signal with highest frequency B can be uniquely represented by samples taken at the minimum rate (Nyquist rate) of $2B$ samples per second. However, if the signal is a bandpass signal with frequency components in the band $B_1 \leq F \leq B_2$, as shown in Fig. 9.3, a blind application of the sampling theorem would have us sampling the signal at a rate of $2B_2$ samples per second.

    If that were the case and B_2 was an extremely high frequency, it would certainly be advantageous to perform a frequency shift of the bandpass signal by

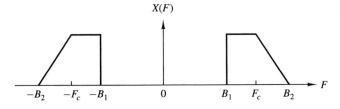

Figure 9.3 Bandpass signal with frequency components in the range $B_1 \leq F \leq B_2$.

an amount

$$F_c = \frac{B_1 + B_2}{2} \tag{9.1.22}$$

and sampling the equivalent lowpass signal. Such a frequency shift can be achieved by multiplying the bandpass signal as given in (9.1.12) by the quadrature carriers $\cos 2\pi F_c t$ and $\sin 2\pi F_c t$ and lowpass filtering the products to eliminate the signal components at $2F_c$. Clearly, the multiplication and the subsequent filtering are first performed in the analog domain and then the outputs of the filters are sampled. The resulting equivalent lowpass signal has a bandwidth $B/2$, where $B = B_2 - B_1$. Therefore, it can be represented uniquely by samples taken at the rate of B samples per second for each of the quadrature components. Thus the sampling can be performed on each of the lowpass filter outputs at the rate of B samples per second, as indicated in Fig. 9.4. Therefore, the resulting rate is $2B$ samples per second.

In view of the fact that frequency conversion to lowpass allows us to reduce the sampling rate to $2B$ samples per second, it should be possible to sample the bandpass signal at a comparable rate. In fact, it is.

Suppose that the upper frequency $F_c + B/2$ is a multiple of the bandwidth B (i.e., $F_c + B/2 = kB$), where k is a positive integer. If we sample $x(t)$ at the rate

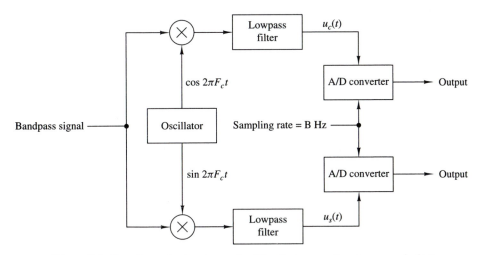

Figure 9.4 Sampling of a bandpass signal by first converting to an equivalent low-pass signal.

$2B = 1/T$ samples per second, we have

$$x(nT) = u_c(nT) \cos 2\pi F_c nT - u_s(nT) \sin 2\pi F_c nT$$

$$= u_c(nT) \cos \frac{\pi n(2k-1)}{2} - u_s(nT) \sin \frac{\pi n(2k-1)}{2} \qquad (9.1.23)$$

where the last step is obtained by substituting $F_c = kB - B/2$ and $T = 1/2B$.

For n even, say $n = 2m$, (9.1.23) reduces to

$$x(2mT) \equiv x(mT_1) = u_c(mT_1) \cos \pi m(2k-1) = (-1)^m u_c(mT_1) \qquad (9.1.24)$$

where $T_1 = 2T = 1/B$. For n odd, say $n = 2m - 1$, (9.1.23) reduces to

$$x(2mT - T) \equiv x\left(mT_1 - \frac{T_1}{2}\right) = u_s\left(mT_1 - \frac{T_1}{2}\right)(-1)^{m+k+1} \qquad (9.1.25)$$

Therefore, the even-numbered samples of $x(t)$, which occur at the rate of B samples per second, produce samples of the lowpass signal component $u_c(t)$. The odd-numbered samples of $x(t)$, which also occur at the rate of B samples per second, produce samples of the lowpass signal component $u_s(t)$.

Now, the samples $\{u_c(mT_1)\}$ and the samples $\{u_s(mT_1 - T_1/2)\}$ can be used to reconstruct the equivalent lowpass signals. Thus, according to the sampling theorem for lowpass signals with $T_1 = 1/B$,

$$u_c(t) = \sum_{m=-\infty}^{\infty} u_c(mT_1) \frac{\sin(\pi/T_1)(t - mT_1)}{(\pi/T_1)(t - mT_1)} \qquad (9.1.26)$$

$$u_s(t) = \sum_{m=-\infty}^{\infty} u_s\left(mT_1 - \frac{T_1}{2}\right) \frac{\sin(\pi/T_1)(t - mT_1 + T_1/2)}{(\pi T_1)(t - mT_1 + T_1/2)} \qquad (9.1.27)$$

Furthermore, the relations in (9.1.24) and (9.1.25) allow us to express $u_c(t)$ and $u_s(t)$ directly in terms of samples of $x(t)$. Now, since $x(t)$ is expressed as

$$x(t) = u_c(t) \cos 2\pi F_c t - u_s(t) \sin 2\pi F_c t \qquad (9.1.28)$$

substitution from (9.1.27), (9.1.26), (9.1.25), and (9.1.24) into (9.1.28) yields

$$x(t) = \sum_{m=-\infty}^{\infty} \left\{ (-1)^m x(2mT) \frac{\sin(\pi/2T)(t - 2mT)}{(\pi/2T)(t - 2mT)} \cos 2\pi F_c t \right.$$

$$\left. + (-1)^{m+k} x((2m-1)T) \frac{\sin(\pi/2T)(t - 2mT + T)}{(\pi/2T)(t - 2mT + T)} \sin 2\pi F_c t \right\} \qquad (9.1.29)$$

But

$$(-1)^m \cos 2\pi F_c t = \cos 2\pi F_c(t - 2mT)$$

and

$$(-1)^{m+k} \sin 2\pi F_c t = \cos 2\pi F_c(t - 2mT + T)$$

With these substitutions, (9.1.29) reduces to

$$x(t) = \sum_{m=-\infty}^{\infty} x(mT) \frac{\sin(\pi/2T)(t - mT)}{(\pi/2T)(t - mT)} \cos 2\pi F_c(t - mT) \tag{9.1.30}$$

where $T = 1/2B$. This is the desired reconstruction formula for the bandpass signal $x(t)$, with samples taken at the rate of $2B$ samples per second, for the special case in which the upper band frequency $F_c + B/2$ is a multiple of the signal bandwidth B.

In the general case, where only the condition $F_c \geq B/2$ is assumed to hold, let us define the integer part of the ratio $F_c + B/2$ to B as

$$r = \left\lfloor \frac{F_c + B/2}{B} \right\rfloor \tag{9.1.31}$$

While holding the upper cutoff frequency $F_c + B/2$ constant, we increase the bandwidth from B to B' such that

$$\frac{F_c + B/2}{B'} = r \tag{9.1.32}$$

Furthermore, it is convenient to define a new center frequency for the increased bandwidth signal as

$$F_c' = F_c + \frac{B}{2} - \frac{B'}{2} \tag{9.1.33}$$

Clearly, the increased signal bandwidth B' includes the original signal spectrum of bandwidth B.

Now the upper cutoff frequency $F_c + B/2$ is a multiple of B'. Consequently, the signal reconstruction formula in (9.1.30) holds with F_c replaced by F_c' and T replaced by T', where $T' = 1/2B'$, that is,

$$x(t) = \sum_{n=-\infty}^{\infty} x(nT') \frac{\sin(\pi/2T')(t - mT')}{(\pi/2T')(t - mT')} \cos 2\pi F_c'(t - mT') \tag{9.1.34}$$

This proves that $x(t)$ can be represented by samples taken at the uniform rate $1/T' = 2Br'/r$, where r' is the ratio

$$r' = \frac{F_c + B/2}{B} = \frac{F_c}{B} + \frac{1}{2} \tag{9.1.35}$$

and $r = \lfloor r' \rfloor$.

We observe that when the upper cutoff frequency $F_c + B/2$ is not an integer multiple of the bandwidth B, the sampling rate for the bandpass signal must be increased by the factor r'/r. However, note that as F_c/B increases, the ratio r'/r tends toward unity. Consequently, the percent increase in sampling rate tends to zero.

The derivation given above also illustrates the fact that the lowpass signal components $u_c(t)$ and $u_s(t)$ can be expressed in terms of samples of the bandpass signal. Indeed, from (9.1.24), (9.1.25), (9.1.26), and (9.1.27), we obtain the result

$$u_c(t) = \sum_{n=-\infty}^{\infty} (-1)^n x(2nT') \frac{\sin(\pi/2T')(t - 2nT')}{(\pi/2T')(t - 2nT')} \tag{9.1.36}$$

and

$$u_s(t) = \sum_{n=-\infty}^{\infty} (-1)^{n+r+1} x(2nT' - T') \frac{\sin(\pi/2T')(t - 2nT' + T')}{(\pi/2T')(t - 2nT' + T')} \tag{9.1.37}$$

where $r = \lfloor r' \rfloor$.

In conclusion, we have demonstrated that a bandpass signal can be represented uniquely by samples taken at a rate

$$2B \leq F_s < 4B$$

where B is the bandwidth of the signal. The lower limit applies when the upper frequency $F_c + B/2$ is a multiple of B. The upper limit on F_s is obtained under worst-case conditions when $r = 1$ and $r' \approx 2$.

9.1.3 Discrete-Time Processing of Continuous-Time Signals

As indicated in our introductory remarks in Chapter 1, there are numerous applications where it is advantageous to process continuous-time (analog) signals on a digital signal processor. Figure 9.5 illustrates the general configuration of the system for digital processing of an analog signal. In designing the processing to be performed, we must first select the bandwidth of the signal to be processed since the signal bandwidth determines the minimum sampling rate. For example, a speech signal, which is to be transmitted digitally, can contain frequency components above 3000 Hz, but for purposes of speech intelligibility and speaker identification, the preservation of frequency components below 3000 Hz is sufficient. Consequently, it would be inefficient from a processing viewpoint to preserve the higher-frequency components and wasteful of channel bandwidth to transmit the extra bits needed to represent these higher-frequency components in the speech signal. Once the desired frequency band is selected we can specify the sampling rate and the characteristics of the prefilter, which is also called an antialiasing filter.

Antialiasing filter. The antialiasing filter is an analog filter which has a twofold purpose. First, it ensures that the bandwidth of the signal to be sampled is limited to the desired frequency range. Thus any frequency components of the signal above the folding frequency $F_s/2$ are sufficiently attenuated so that the

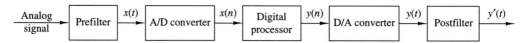

Figure 9.5 Configuration of system for digital processing of an analog signal.

amount of signal distortion due to aliasing is negligible. For example, the speech signal to be transmitted digitally over a telephone channel would be filtered by a lowpass filter having a passband extending to 3000 Hz, a transition band of approximately 400 to 500 Hz, and a stopband above 3400 to 3500 Hz. The speech signal may be sampled at 8000 Hz and hence the folding frequency would be 4000 Hz. Thus aliasing would be negligible.

Another reason for using an antialiasing filter is to limit the additive noise spectrum and other interference, which often corrupts the desired signal. Usually, additive noise is wideband and exceeds the bandwidth of the desired signal. By prefiltering we reduce the additive noise power to that which falls within the bandwidth of the desired signal and we reject the out-of-band noise.

Ideally, we would like to employ a filter with steep cutoff frequency response characteristics and with no delay distortion within the passband. Practically, however, we are constrained to employ filters that have a finite-width transition region, are relatively simple to implement, and introduce some tolerable amount of delay distortion. Very stringent filter specifications, such as a narrow transition region, result in very complex filters. In practice, we may choose to sample the signal well above the Nyquist rate and thus relax the design specifications on the antialiasing filter.

Once we have specified the prefilter requirements and have selected the desired sampling rate, we can proceed with the design of the digital signal processing operations to be performed on the discrete-time signal. The selection of the sampling rate $F_s = 1/T$, where T is the sampling interval, not only determines the highest frequency ($F_s/2$) that is preserved in the analog signal, but also serves as a scale factor that influences the design specifications for digital filters and any other discrete-time systems through which the signal is processed.

For example, suppose that we have an analog signal to be differentiated that has a bandwidth of 3000 Hz. Although differentiation can be performed directly on the analog signal, we choose to do it digitally in discrete time. Hence we sample the signal at the range $F_s = 8000$ Hz and design a digital differentiator as described in Sec. 8.2.4. In this case, the sampling rate $F_s = 8000$ Hz establishes the folding frequency of 4000 Hz, which corresponds to the frequency $\omega = \pi$ in the discrete-time signal. Hence the signal bandwidth of 3000 Hz corresponds to the frequency $\omega_c = 0.75\pi$. Consequently, the discrete-time differentiator for processing the signal would be designed to have a passband of $0 \leq |\omega| \leq 0.75\pi$.

As another example of digital processing, the speech signal that is bandlimited to 3000 Hz and sampled at 8000 Hz may be separated into two or more frequency subbands by digital filtering, and each subband of speech is digitally encoded with different precision, as is done in subband coding (see Section 10.9.5 for more details). The frequency response characteristics of the digital filters for separating the 0- to 3000-Hz signal into subbands are specified relative to the folding frequency of 4000 Hz, which corresponds to the frequency $\omega = \pi$ for the discrete-time signal. Thus we may process any continuous-time signal in the discrete-time domain by performing equivalent operations in discrete time.

The one implicit assumption that we have made in this discussion on the equivalence of continuous-time and discrete-time signal processing is that the quantization error in analog-to-digital conversion and round-off errors in digital signal processing are negligible. These issues are further discussed in this chapter. However, we should emphasize that analog signal processing operations cannot be done very precisely either, since electronic components in analog systems have tolerances and they introduce noise during their operation. In general, a digital system designer has better control of tolerances in a digital signal processing system than an analog system designer who is designing an equivalent analog system.

9.2 ANALOG-TO-DIGITAL CONVERSION

The discussion in Section 9.1 focused on the conversion of continuous-time signals to discrete-time signals using an ideal sampler and ideal interpolation. In this section we deal with the devices for performing these conversions from analog to digital.

Recall that the process of converting a continuous-time (analog) signal to a digital sequence that can be processed by a digital system requires that we quantize the sampled values to a finite number of levels and represent each level by a number of bits. The electronic device that performs this conversion from an analog signal to a digital sequence is called an *analog-to-digital (A/D) converter* (ADC). On the other hand, a *digital-to-analog (D/A) converter* (DAC) takes a digital sequence and produces at its output a voltage or current proportional to the size of the digital word applied to its input. D/A conversion is treated in Section 9.3.

Figure 9.6(a) shows a block diagram of the basic elements of an A/D converter. In this section we consider the performance requirements for these elements. Although we focus mainly on ideal system characteristics, we shall also mention some key imperfections encountered in practical devices and indicate how they affect the performance of the converter. We concentrate on those aspects that are more relevant to signal processing applications. The practical aspects of A/D converters and related circuitry can be found in the manufacturers' specifications and data sheets.

9.2.1 Sample-and-Hold

In practice, the sampling of an analog signal is performed by a sample-and-hold (S/H) circuit. The sampled signal is then quantized and converted to digital form. Usually, the S/H is integrated into the A/D converter.

The S/H is a digitally controlled analog circuit that tracks the analog input signal during the sample mode, and then holds it fixed during the hold mode to the instantaneous value of the signal at the time the system is switched from the

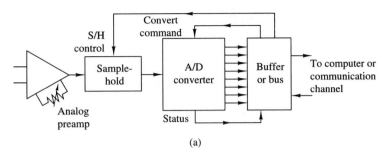

(a)

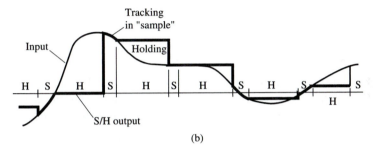

(b)

Figure 9.6 (a) Block diagram of basic elements of an A/D converter; (b) time-domain response of an ideal S/H circuit.

sample mode to the hold mode. Figure 9.6(b) shows the time-domain response of an ideal S/H circuit (i.e., a S/H that responds instantaneously and accurately).

The goal of the S/H is to continuously sample the input signal and then to hold that value constant as long as it takes for the A/D converter to obtain its digital representation. The use of an S/H allows the A/D converter to operate more slowly compared to the time actually used to acquire the sample. In the absence of a S/H, the input signal must not change by more than one-half of the quantization step during the conversion, which may be an impractical constraint. Consequently, the S/H is crucial in high-resolution (12 bits per sample or higher) digital conversion of signals that have large bandwidths (i.e., they change very rapidly).

An ideal S/H introduces no distortion in the conversion process and is accurately modeled as an ideal sampler. However, time-related degradations such as errors in the periodicity of the sampling process ("jitter"), nonlinear variations in the duration of the sampling aperture, and changes in the voltage held during conversion ("droop") do occur in practical devices.

The A/D converter begins the conversion after it receives a convert command. The time required to complete the conversion should be less than the duration of the hold mode of the S/H. Furthermore, the sampling period T should be larger than the duration of the sample mode and the hold mode.

In the following sections we assume that the S/H introduces negligible errors and we focus on the digital conversion of the analog samples.

9.2.2 Quantization and Coding

The basic task of the A/D converter is to convert a continuous range of input amplitudes into a discrete set of digital code words. This conversion involves the processes of *quantization* and *coding*. Quantization is a nonlinear and noninvertible process that maps a given amplitude $x(n) \equiv x(nT)$ at time $t = nT$ into an amplitude x_k, taken from a finite set of values. The procedure is illustrated in Fig. 9.7(a), where the signal amplitude range is divided into L intervals

$$I_k = \{x_k < x(n) \leq x_{k+1}\} \qquad k = 1, 2, \ldots, L \qquad (9.2.1)$$

by the $L + 1$ *decision levels* $x_1, x_L, \ldots, x_{L+1}$. The possible outputs of the quantizer (i.e., the quantization levels) are denoted as $\hat{x}_1, \hat{x}_2, \ldots, \hat{x}_L$. The operation of the quantizer is defined by the relation

$$x_q(n) \equiv Q[x(n)] = \hat{x}_k \qquad \text{if } x(n) \in I_k \qquad (9.2.2)$$

In most digital signal processing operations the mapping in (9.2.2) is independent of n (i.e., the quantization is memoryless and is simply denoted as $x_q = Q[x]$). Furthermore, in signal processing we often use *uniform* or *linear quantizers* defined by

$$\hat{x}_{k+1} - \hat{x}_k = \Delta \qquad k = 1, 2, \ldots, L - 1$$

$$\hspace{5cm} (9.2.3)$$

$$x_{k+1} - x_k = \Delta \qquad \text{for finite } x_k, x_{k+1}$$

where Δ is the *quantizer step size*. Uniform quantization is usually a requirement if the resulting digital signal is to be processed by a digital system. However, in transmission and storage applications of signals such as speech, nonlinear and time-variant quantizers are frequently used.

If a zero is assigned a quantization level, the quantizer is of the *midtread* type. If zero is assigned a decision level, the quantizer is called a *midrise* type.

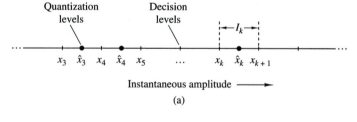

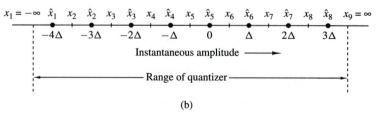

Figure 9.7   Quantization process and an example of a midtread quantizer.

Figure 9.7(b) illustrates a midtread quantizer with $L = 8$ levels. In theory, the extreme decision levels are taken as $x_1 = -\infty$ and $x_{L+1} = \infty$, to cover the total *dynamic range* of the input signal. However, practical A/D converters can handle only a finite range. Hence we define the *range R* of the quantizer by assuming that $I_1 = I_L = \Delta$. For example, the range of the quantizer shown in Fig. 9.7(b) is equal to 8Δ. In practice, the term *full-scale range* (FSR) is used to describe the range of an A/D converter for bipolar signals (i.e., signals with both positive and negative amplitudes). The term *full scale* (FS) is used for unipolar signals.

It can be easily seen that the quantization error $e_q(n)$ is always in the range $-\Delta/2$ to $\Delta/2$:

$$-\frac{\Delta}{2} < e_q(n) \le \frac{\Delta}{2} \tag{9.2.4}$$

In other words, the instantaneous quantization error cannot exceed half of the quantization step. If the dynamic range of the signal, defined as $x_{\max} - x_{\min}$, is larger than the range of the quantizer, the samples that exceed the quantizer range are clipped, resulting in a large (greater than $\Delta/2$) quantization error.

The operation of the quantizer is better described by the quantization characteristic function, illustrated in Fig. 9.8 for a midtread quantizer with eight

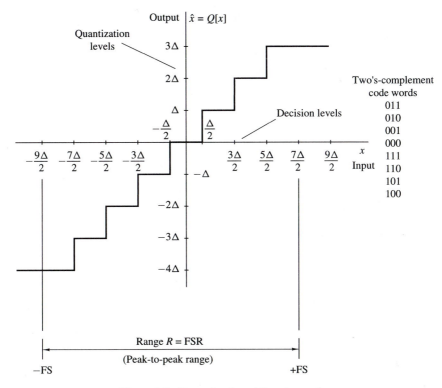

Figure 9.8 Example of a midtread quantizer.

quantization levels. This characteristic is preferred in practice over the midriser because it provides an output that is insensitive to infinitesimal changes of the input signal about zero. Note that the input amplitudes of a midtread quantizer are rounded to the nearest quantization levels.

The *coding* process in an A/D converter assigns a unique binary number to each quantization level. If we have L levels, we need at least L different binary numbers. With a word length of $b + 1$ bits we can represent 2^{b+1} distinct binary numbers. Hence we should have $2^{b+1} \geq L$ or, equivalently, $b + 1 \geq \log_2 L$. Then the step size or the *resolution* of the A/D converter is given by

$$\Delta = \frac{R}{2^{b+1}} \tag{9.2.5}$$

where R is the range of the quantizer.

There are various binary coding schemes, each with its advantages and disadvantages. Table 9.1 illustrates some existing schemes for 3-bit binary coding. These number representation schemes were described in detail in Section 7.5.

The two's-complement representation is used in most digital signal processors. Thus it is convenient to use the same system to represent digital signals because we can operate on them directly without any extra format conver-

TABLE 9.1 COMMONLY USED BIPOLAR CODES

Number	Decimal Fraction Positive Reference	Decimal Fraction Negative Reference	Sign + Magnitude	Two's Complement	Offset Binary	One's Complement
+7	$+\frac{7}{8}$	$-\frac{7}{8}$	0 1 1 1	0 1 1 1	1 1 1 1	0 1 1 1
+6	$+\frac{6}{8}$	$-\frac{6}{8}$	0 1 1 0	0 1 1 0	1 1 1 0	0 1 1 0
+5	$+\frac{5}{8}$	$-\frac{5}{8}$	0 1 0 1	0 1 0 1	1 1 0 1	0 1 0 1
+4	$+\frac{4}{8}$	$-\frac{4}{8}$	0 1 0 0	0 1 0 0	1 1 0 0	0 1 0 0
+3	$+\frac{3}{8}$	$-\frac{3}{8}$	0 0 1 1	0 0 1 1	1 0 1 1	0 0 1 1
+2	$+\frac{2}{8}$	$-\frac{2}{8}$	0 0 1 0	0 0 1 0	1 0 1 0	0 0 1 0
+1	$+\frac{1}{8}$	$-\frac{1}{8}$	0 0 0 1	0 0 0 1	1 0 0 1	0 0 0 1
0	0+	0−	0 0 0 0	0 0 0 0	1 0 0 0	0 0 0 0
0	0−	0+	1 0 0 0	(0 0 0 0)	(1 0 0 0)	1 1 1 1
−1	$-\frac{1}{8}$	$+\frac{1}{8}$	1 0 0 1	1 1 1 1	0 1 1 1	1 1 1 0
−2	$-\frac{2}{8}$	$+\frac{2}{8}$	1 0 1 0	1 1 1 0	0 1 1 0	1 1 0 1
−3	$-\frac{3}{8}$	$+\frac{3}{8}$	1 0 1 1	1 1 0 1	0 1 0 1	1 1 0 0
−4	$-\frac{4}{8}$	$+\frac{4}{8}$	1 1 0 0	1 1 0 0	0 1 0 0	1 0 1 1
−5	$-\frac{5}{8}$	$+\frac{5}{8}$	1 1 0 1	1 0 1 1	0 0 1 1	1 0 1 0
−6	$-\frac{6}{8}$	$+\frac{6}{8}$	1 1 1 0	1 0 1 0	0 0 1 0	1 0 0 1
−7	$-\frac{7}{8}$	$+\frac{7}{8}$	1 1 1 1	1 0 0 1	0 0 0 1	1 0 0 0
−8	$-\frac{8}{8}$	$+\frac{8}{8}$		(1 0 0 0)	(0 0 0 0)	

sion. In general, a $(b + 1)$-bit binary fraction of the form $\beta_0 \beta_1 \beta_2 \cdots \beta_B$ has the value

$$-\beta_0 \cdot 2^0 + \beta_1 \cdot 2^{-1} + \beta_2 \cdot 2^{-2} + \cdots + \beta_b \cdot 2^{-b}$$

if we use the two's-complement representation. Note that β_0 is the most significant bit (MSB) and β_b is the least significant bit (LSB). Although the binary code used to represent the quantization levels is important for the design of the A/D converter and the subsequent numerical computations, it does not have any effect in the performance of the quantization process. Thus in our subsequent discussions we ignore the process of coding when we analyze the performance of A/D converters.

Figure 9.9(a) shows the characteristic of an ideal 3-bit A/D converter. The only degradation introduced by an ideal converter is the quantization error, which can be reduced by increasing the number of bits. This error, that dominates the performance of practical A/D converters, is analyzed in the next section.

Practical A/D converters differ from ideal converters in several ways. Various degradations are usually encountered in practice. A number of these performance degradations are illustrated in Fig. 9.9(b)–(e). We note that practical A/D converters may have *offset* error (the first transition may not occur at exactly $+\frac{1}{2}$LSB), *scale-factor* (or gain) error (the difference between the values at which the first transition and the last transition occur is not equal to FS − 2LSB), and *linearity* error (the differences between transition values are not all equal or uniformly changing). If the *differential linearity* error is large enough, it is possible for one or more code words to be missed. Performance data on commercially available A/D converters are specified in the manufacturers' data sheets.

9.2.3 Analysis of Quantization Errors

To determine the effects of quantization on the performance of an A/D converter, we adopt a statistical approach. The dependence of the quantization error on the characteristics of the input signal and the nonlinear nature of the quantizer make a deterministic analysis intractable, except in very simple cases.

In the statistical approach, we assume that the quantization error is random in nature. We model this error as noise that is added to the original (unquantized) signal. If the input analog signal is within the range of the quantizer, the quantization error $e_q(n)$ is bounded in magnitude [i.e., $|e_q(n)| < \Delta/2$], and the resulting error is called *granular noise*. When the input falls outside the range of the quantizer (clipping), $e_q(n)$ becomes unbounded and results in *overload noise*. This type of noise can result in severe signal distortion when it occurs. Our only remedy is to scale the input signal so that its dynamic range falls within the range of the quantizer. The following analysis is based on the assumption that there is no overload noise.

The mathematical model for the quantization error $e_q(n)$ is shown in Fig. 9.10. To carry out the analysis, we make the following assumptions about the statistical

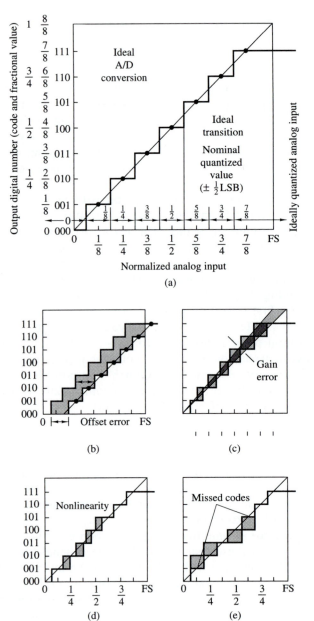

Figure 9.9 Characteristics of ideal and practical A/D converters.

properties of $e_q(n)$:

1. The error $e_q(n)$ is uniformly distributed over the range $-\Delta/2 < e_q(n) < \Delta/2$.

2. The error sequence $\{e_q(n)\}$ is a stationary white noise sequence. In other words, the error $e_q(n)$ and the error $e_q(m)$ for $m \neq n$ are uncorrelated.

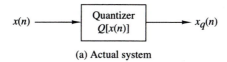

(a) Actual system

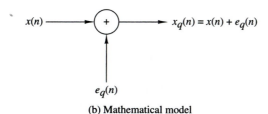

(b) Mathematical model

Figure 9.10 Mathematical model of quantization noise.

3. The error sequence $\{e_q(n)\}$ is uncorrelated with the signal sequence $x(n)$.
4. The signal sequence $x(n)$ is zero mean and stationary.

These assumptions do not hold, in general. However, they do hold when the quantization step size is small and the signal sequence $x(n)$ traverses several quantization levels between two successive samples.

Under these assumptions, the effect of the additive noise $e_q(n)$ on the desired signal can be quantified by evaluating the signal-to-quantization noise (power) ratio (SQNR), which can be expressed on a logarithmic scale (in decibels or dB) as

$$\text{SQNR} = 10\log_{10}\frac{P_x}{P_n} \tag{9.2.6}$$

where $P_x = \sigma_x^2 = E[x^2(n)]$ is the signal power and $P_n = \sigma_e^2 = E[e_q^2(n)]$ is the power of the quantization noise.

If the quantization error is uniformly distributed in the range $(-\Delta/2, \Delta/2)$ as shown in Fig. 9.11, the mean value of the error is zero and the variance (the quantization noise power) is

$$P_n = \sigma_e^2 = \int_{-\Delta/2}^{\Delta/2} e^2 p(e)\,de = \frac{1}{\Delta}\int_{-\Delta/2}^{\Delta/2} e^2\,de = \frac{\Delta^2}{12} \tag{9.2.7}$$

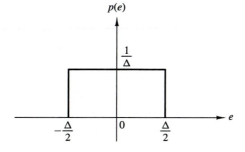

Figure 9.11 Probability density function for the quantization error.

By combining (9.2.5) with (9.2.7) and substituting the result into (9.2.6), the expression for the SQNR becomes

$$\text{SQNR} = 10 \log \frac{P_x}{P_n} = 20 \log \frac{\sigma_x}{\sigma_e}$$

$$= 6.02b + 16.81 - 20 \log \frac{R}{\sigma_x} \quad \text{dB} \tag{9.2.8}$$

The last term in (9.2.8) depends on the range R of the A/D converter and the statistics of the input signal. For example, if we assume that $x(n)$ is Gaussian distributed and the range of the quantizer extends from $-3\sigma_x$ to $3\sigma_x$ (i.e., $R = 6\sigma_x$), then less than 3 out of every 1000 input signal amplitudes would result in an overload on the average. For $R = 6\sigma_x$, (9.2.8) becomes

$$\text{SQNR} = 6.02b + 1.25 \ \text{dB}$$

The formula in (9.2.8) is frequently used to specify the precision needed in an A/D converter. It simply means that each additional bit in the quantizer increases the signal-to-quantization noise ratio by 6 dB. (It is interesting to note that the same result was derived in Section 1.4 for a sinusoidal signal using a deterministic approach.) However, we should bear in mind the conditions under which this result has been derived.

Due to limitations in the fabrication of A/D converters, their performance falls short of the theoretical value given by (9.2.8). As a result, the effective number of bits may be somewhat less than the number of bits in the A/D converter. For instance, a 16-bit converter may have only an effective 14 bits of accuracy.

9.2.4 Oversampling A/D Converters

The basic idea in oversampling A/D converters is to increase the sampling rate of the signal to the point where a low-resolution quantizer suffices. By oversampling, we can reduce the dynamic range of the signal values between successive samples and thus reduce the resolution requirements on the quantizer. As we have observed in the preceding section, the variance of the quantization error in A/D conversion is $\sigma_e^2 = \Delta^2/12$, where $\Delta = R/2^{b+1}$. Since the dynamic range of the signal, which is proportional to its standard deviation σ_x, should match the range R of the quantizer, it follows that Δ is proportional to σ_x. Hence for a given number of bits, the power of the quantization noise is proportional to the variance of the signal to be quantized. Consequently, for a given fixed SQNR, a reduction in the variance of the signal to be quantized allows us to reduce the number of bits in the quantizer.

The basic idea for reducing the dynamic range leads us to consider *differential quantization*. To illustrate this point, let us evaluate the variance of the difference between two successive signal samples. Thus we have

$$d(n) = x(n) - x(n-1) \tag{9.2.9}$$

The variance of $d(n)$ is

$$
\begin{aligned}
\sigma_d^2 &= E[d^2(n)] = E\{[x(n) - x(n-1)]^2\} \\
&= E[x^2(n)] - 2E[x(n)x(n-1)] + E[x^2(n-1)] \\
&= 2\sigma_x^2[1 - \gamma_{xx}(1)]
\end{aligned}
\tag{9.2.10}
$$

where $\gamma_{xx}(1)$ is the value of the autocorrelation sequence $\gamma_{xx}(m)$ of $x(n)$ evaluated at $m = 1$. If $\gamma_{xx}(1) > 0.5$, we observe that $\sigma_d^2 < \sigma_x^2$. Under this condition, it is better to quantize the difference $d(n)$ and to recover $x(n)$ from the quantized values $\{d_q(n)\}$. To obtain a high correlation between successive samples of the signal, we require that the sampling rate be significantly higher than the Nyquist rate.

An even better approach is to quantize the difference

$$
d(n) = x(n) - ax(n-1)
\tag{9.2.11}
$$

where a is a parameter selected to minimize the variance in $d(n)$. This leads to the result (see Problem 9.7) that the optimum choice of a is

$$
a = \frac{\gamma_{xx}(1)}{\gamma_{xx}(0)} = \frac{\gamma_{xx}(1)}{\sigma_x^2}
$$

and

$$
\sigma_d^2 = \sigma_x^2[1 - a^2]
\tag{9.2.12}
$$

In this case, $\sigma_d^2 < \sigma_x^2$, since $0 \le a \le 1$. The quantity $ax(n-1)$ is called a first-order predictor of $x(n)$.

Figure 9.12 shows a more general *differential predictive* signal quantizer system. This system is used in speech encoding and transmission over telephone channels and is known as differential pulse code modulation (DPCM). The goal of the predictor is to provide an estimate $\hat{x}(n)$ of $x(n)$ from a linear combination of past values of $x(n)$, so as to reduce the dynamic range of the difference signal $d(n) = x(n) - \hat{x}(n)$. Thus a predictor of order p has the form

$$
\hat{x}(n) = \sum_{k=1}^{p} a_k x(n-k)
\tag{9.2.13}
$$

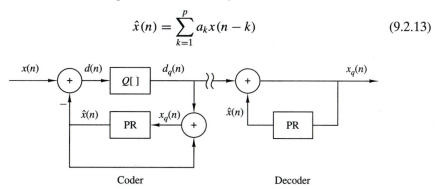

Coder   Decoder

Figure 9.12 Encoder and decoder for differential predictive signal quantizer system.

The use of the feedback loop around the quantizer as shown in Fig. 9.12 is necessary to avoid the accumulation of quantization errors at the decoder. In this configuration, the error $e(n) = d(n) - d_q(n)$ is

$$e(n) = d(n) - d_q(n) = x(n) - \hat{x}(n) - d_q(n) = x(n) - x_q(n)$$

Thus the error in the reconstructed quantized signal $x_q(n)$ is equal to the quantization error for the sample $d(n)$. The decoder for DPCM that reconstructs the signal from the quantized values is also shown in Fig. 9.12.

The simplest form of differential predictive quantization is called *delta modulation* (DM). In DM, the quantizer is a simple 1-bit (two-level) quantizer and the predictor is a first-order predictor, as shown in Fig. 9.13(a). Basically, DM provides a staircase approximation of the input signal. At every sampling instant, the sign of the difference between the input sample $x(n)$ and its most recent staircase approximation $\hat{x}(n) = ax_q(n - 1)$ is determined, and then the staircase signal is updated by a step Δ in the direction of the difference.

From Fig. 9.13(a) we observe that

$$x_q(n) = ax_q(n - 1) + d_q(n) \qquad (9.2.14)$$

which is the discrete-time equivalent of an analog integrator. If $a = 1$, we have an ideal accumulator (integrator) whereas the choice $a < 1$ results in a "leaky integrator." Figure 9.13(c) shows an analog model that illustrates the basic principle for the practical implementation of a DM system. The analog lowpass filter is necessary for the rejection of out-of-band components in the frequency range between B and $F_s/2$, since $F_s >> B$ due to oversampling.

The crosshatched areas in Fig. 9.13(b) illustrate two types of quantization error in DM, slope-overload distortion and granular noise. Since the maximum slope Δ/T in $x(n)$ is limited by the step size, slope-overload distortion can be avoided if $\max |dx(t)/dt| \le \Delta/T$. The granular noise occurs when the DM tracks a relatively flat (slowly changing) input signal. We note that increasing Δ reduces overload distortion but increases the granular noise, and vice versa.

One way to reduce these two types of distortion is to use an integrator in front of the DM, as shown in Fig. 9.14(a). This has two effects. First, it emphasizes the low frequencies of $x(t)$ and increases the correlation of the signal into the DM input. Second, it simplifies the DM decoder because the differentiator (inverse system) required at the decoder is canceled by the DM integrator. Hence the decoder is simply a lowpass filter, as shown in Fig. 9.14(a). Furthermore, the two integrators at the encoder can be replaced by a single integrator placed before the comparator, as shown in Fig. 9.14(b). This system is known as *sigma-delta modulation* (SDM).

SDM is an ideal candidate for A/D conversion. Such a converter takes advantage of the high sampling rate and spreads the quantization noise across the band up to $F_s/2$. Since $F_s >> B$, the noise in the signal-free band $B \le F \le F_s/2$ can be removed by appropriate digital filtering. To illustrate this principle, let

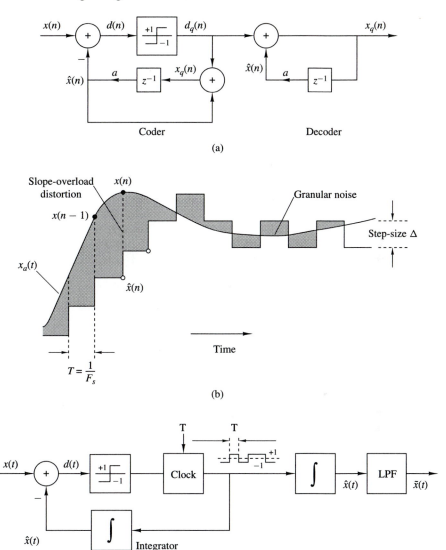

Figure 9.13   Delta modulation system and two types of quantization errors.

us consider the discrete-time model of SDM, shown in Fig. 9.15, where we have assumed that the comparator (1-bit quantizer) is modeled by an additive white noise source with variance $\sigma_e^2 = \Delta^2/12$. The integrator is modeled by the discrete-time system with system function

$$H(z) = \frac{z^{-1}}{1 - z^{-1}} \tag{9.2.15}$$

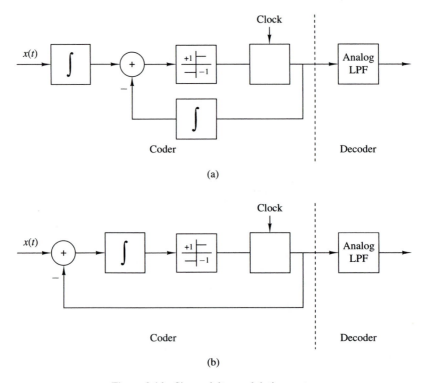

Figure 9.14   Sigma-delta modulation system.

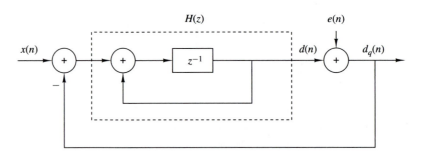

Figure 9.15   Discrete-time model of sigma-delta modulation.

The z-transform of the sequence $\{d_q(n)\}$ is

$$
D_q(z) = \frac{H(z)}{1 + H(z)} X(z) + \frac{1}{1 + H(z)} E(z) \tag{9.2.16}
$$

$$
= H_s(z) X(z) + H_n(z) E(z)
$$

where $H_s(z)$ and $H_n(z)$ are the signal and noise system functions, respectively. A good SDM system has a flat frequency response $H_s(\omega)$ in the signal frequency band $0 \le F \le B$. On the other hand, $H_n(z)$ should have high attenuation in the frequency band $0 \le F \le B$ and low attenuation in the band $B \le F \le F_s/2$.

For the first-order SDM system with the integrator specified by (9.2.15), we have

$$H_s(z) = z^{-1} \qquad H_n(z) = 1 - z^{-1} \tag{9.2.17}$$

Thus $H_s(z)$ does not distort the signal. The performance of the SDM system is therefore determined by the noise system function $H_n(z)$, which has a magnitude frequency response

$$|H_n(F)| = 2 \left| \sin \frac{\pi F}{F_s} \right| \tag{9.2.18}$$

as shown in Fig. 9.16. The in-band quantization noise variance is given as

$$\sigma_n^2 = \int_{-B}^{B} |H_n(F)|^2 S_e(F) dF \tag{9.2.19}$$

where $S_e(F) = \sigma_e^2/F_s$ is the power spectral density of the quantization noise. From this relationship we note that doubling F_s (increasing the sampling rate by a factor of 2), while keeping B fixed, reduces the power of the quantization noise by 3 dB. This result is true for any quantizer. However, additional reduction may be possible by properly choosing the filter $H(z)$.

For the first-order SDM, it can be shown (see Problem 9.10) that for $F_s \gg 2B$, the in-band quantization noise power is

$$\sigma_n^2 \approx \frac{1}{3} \pi^2 \sigma_e^2 \left(\frac{2B}{F_s} \right)^3 \tag{9.2.20}$$

Note that doubling the sampling frequency reduces the noise power by 9 dB of which 3 dB is due to the reduction in $S_e(F)$ and 6 dB is due to the filter characteristic $H_n(F)$. An additional 6-dB reduction can be achieved by using a double integrator (see Problem 9.11).

In summary, the noise power σ_n^2 can be reduced by increasing the sampling rate to spread the quantization noise power over a larger frequency band $(-F_s/2, F_s/2)$, and then shaping the noise power spectral density by means of an

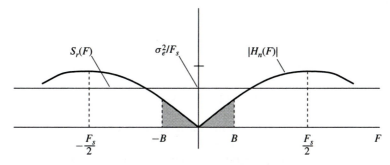

Figure 9.16 Frequency (magnitude) response of noise system function.

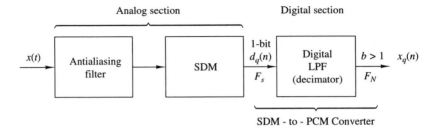

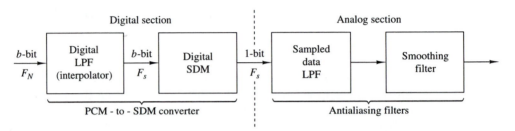

Figure 9.17 Basic elements of an oversampling A/D converter.

appropriate filter. Thus, SDM provides a 1-bit quantized signal at a sampling frequency $F_s = 2IB$, where the oversampling (interpolation) factor I determines the SNR of the SDM quantizer.

Next, we explain how to convert this signal into a b-bit quantized signal at the Nyquist rate. First, we recall that the SDM decoder is an analog lowpass filter with a cutoff frequency B. The output of this filter is an approximation to the input signal $x(t)$. Given the 1-bit signal $d_q(n)$ at sampling frequency F_s, we can obtain a signal $x_q(n)$ at a lower sampling frequency, say the Nyquist rate of $2B$ or somewhat faster, by resampling the output of the lowpass filter at the $2B$ rate. To avoid aliasing, we first filter out the out-of-band $(B, F_s/2)$ noise by processing the wideband signal. The signal is then passed through the lowpass filter and resampled (downsampled) at the lower rate. The downsampling process is called *decimation* and is treated in great detail in Chapter 10.

For example, if the interpolation factor is $I = 256$, the A/D converter output can be obtained by averaging successive non-overlapping blocks of 128 bits. This averaging would result in a digital signal with a range of values from zero to $256(b \approx 8$ bits) at the Nyquist rate. The averaging process also provides the required antialiasing filtering.

Figure 9.17 illustrates the basic elements of an oversampling A/D converter. Oversampling A/D converters for voice-band (3-kHz) signals are currently fabricated as integrated circuits. Typically, they operate at a 2-MHz sampling rate, downsample to 8 kHz, and provide 16-bit accuracy.

9.3 DIGITAL-TO-ANALOG CONVERSION

In Section 4.2.9 we demonstrated that a bandlimited lowpass analog signal, which has been sampled at the Nyquist rate (or faster), can be reconstructed from its samples without distortion. The ideal reconstruction formula or ideal interpolation formula derived in Section 4.2.9 is

$$x(t) = \sum_{n=-\infty}^{\infty} x(nT)\frac{\sin(\pi/T)(t - nT)}{(\pi/T)(t - nT)} \tag{9.3.1}$$

where the sampling interval $T = 1/F_s = 1/2B$, F_s is the sampling frequency and B is the bandwidth of the analog signal.

We have viewed the reconstruction of the signal $x(t)$ from its samples as an interpolation problem and have described the function

$$g(t) = \frac{\sin(\pi t/T)}{\pi t/T} \tag{9.3.2}$$

as the ideal interpolation function. The interpolation formula for $x(t)$, given by (9.3.1), is basically a linear superposition of time-shifted versions of $g(t)$, with each $g(t - nT)$ weighted by the corresponding signal sample $x(nT)$.

Alternatively, we can view the reconstruction of the signal from its samples as a linear filtering process in which a discrete-time sequence of short pulses (ideally impulses) with amplitudes equal to the signal samples, excites an analog filter, as illustrated in Fig. 9.18. The analog filter corresponding to the ideal interpolator has a frequency response

$$H(F) = \begin{cases} T, & |F| \le \dfrac{1}{2T} = \dfrac{F_s}{2} \\[2mm] 0, & |F| > \dfrac{1}{2T} \end{cases} \tag{9.3.3}$$

$H(F)$ is simply the Fourier transform of the interpolation function $g(t)$. In other words, $H(F)$ is the frequency response of an analog reconstruction filter whose

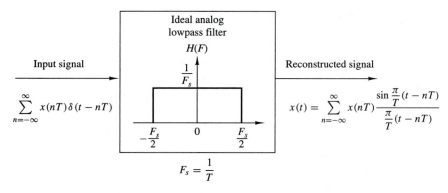

Figure 9.18 Signal reconstruction viewed as a filtering process.

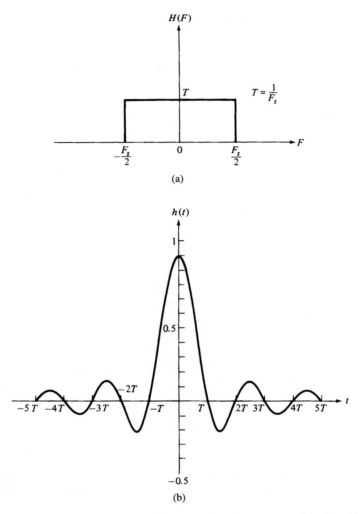

Figure 9.19 Frequency response (a) and the impulse response (b) of an ideal low-pass filter.

impulse response is $h(t) = g(t)$. As shown in Fig. 9.19, the ideal reconstruction filter is an ideal lowpass filter and its impulse response extends for all time. Hence the filter is noncausal and physically nonrealizable. Although the interpolation filter with impulse response given by (9.3.1) can be approximated closely with some delay, the resulting function is still impractical for most applications where D/A conversion is required.

In this section we present some practical, albeit nonideal, interpolation techniques and interpret them as linear filters. Although many sophisticated polynomial interpolation techniques can be devised and analyzed, our discussion is limited to constant and linear interpolation. Quadratic and higher polynomial interpolation is often used in numerical analysis, but is it less likely to be used in digital signal processing.

9.3.1 Sample and Hold

In practice, D/A conversion is usually performed by combining a D/A converter with a sample-and-hold (S/H) and followed by a lowpass (smoothing) filter, as shown in Fig. 9.20. The D/A converter accepts at its input, electrical signals that correspond to a binary word, and produces an output voltage or current that is proportional to the value of the binary word. Ideally, its input–output characteristic is as shown in Fig. 9.21(a) for a 3-bit bipolar signal. The line connecting the dots is a straight line through the origin. In practical D/A converters, the line connecting the dots may deviate from the ideal. Some of the typical deviations from ideal are offset errors, gain errors, and nonlinearities in the input–output characteristic. These types of errors are illustrated in Fig. 9.21(b).

 An important parameter of a D/A converter is its *settling time*, which is defined as the time required for the output of the D/A converter to reach and remain within a given fraction (usually, $\pm\frac{1}{2}$LSB) of the final value, after application of the input code word. Often, the application of the input code word results in a high-amplitude transient, called a "glitch." This is especially the case when two consecutive code words to the A/D differ by several bits. The usual way to remedy this problem is to use a S/H circuit designed to serve as a "deglitcher." Hence the basic task of the S/H is to hold the output of the D/A converter constant at the previous output value until the new sample at the output of the D/A reaches steady state, then it samples and holds the new value in the next sampling interval. Thus the S/H approximates the analog signal by a series of rectangular pulses whose height is equal to the corresponding value of the signal pulse. Figure 9.22(a) illustrates the approximation of the analog signal $x(t)$ by a S/H. As shown, the approximation, denoted as $\hat{x}(t)$, is basically a staircase function which takes the signal sample from the D/A converter and holds it for T seconds. When the next sample arrives, it jumps to the next value and holds it for T seconds, and so on.

 When viewed as a linear filter, as shown in Fig. 9.22(b), the S/H has an impulse response

$$h(t) = \begin{cases} 1, & 0 \le t \le T \\ 0, & \text{otherwise} \end{cases} \tag{9.3.4}$$

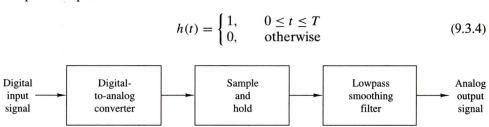

Figure 9.20   Basic operations in converting a digital signal into an analog signal.

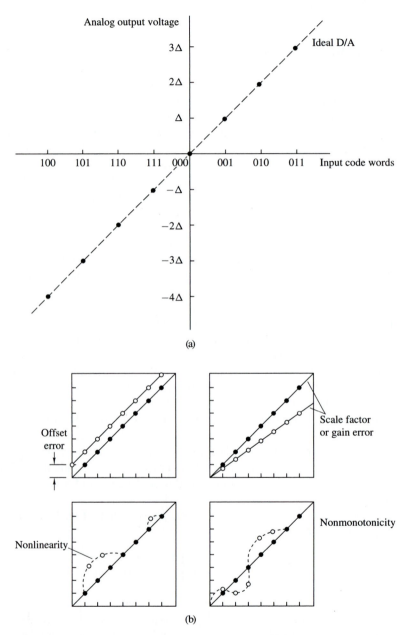

Figure 9.21 (a) Ideal D/A converter characteristic and (b) typical deviations from ideal performance in practical D/A converters.

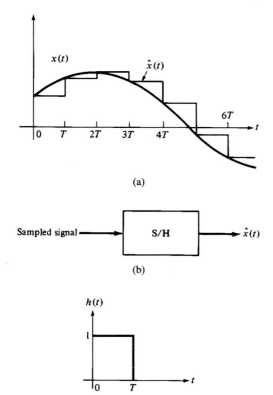

(a)

(b)

(c)

Figure 9.22 (a) Approximation of an analog signal by a staircase; (b) linear filtering interpretation; (c) impulse response of the S/H.

This is illustrated in Fig. 9.22(c). The corresponding frequency response is

$$H(F) = \int_{-\infty}^{\infty} h(t)e^{-j2\pi Ft}dt$$

$$= \int_{0}^{T} e^{-j2\pi Ft}dt \tag{9.3.5}$$

$$= T\left(\frac{\sin \pi FT}{\pi FT}\right)e^{-j\pi FT}$$

The magnitude and phase of $H(F)$ are plotted in Figs. 9.23. For comparison, the frequency response of the ideal interpolator is superimposed on the magnitude characteristics.

It is apparent that the S/H does not possess a sharp cutoff frequency response characteristic. This is due to a large extent to the sharp transitions of its impulse response $h(t)$. As a consequence, the S/H passes undesirable aliased frequency components (frequencies above $F_s/2$) to its output. To remedy this problem, it is common practice to filter $\hat{x}(t)$ by passing it through a lowpass filter

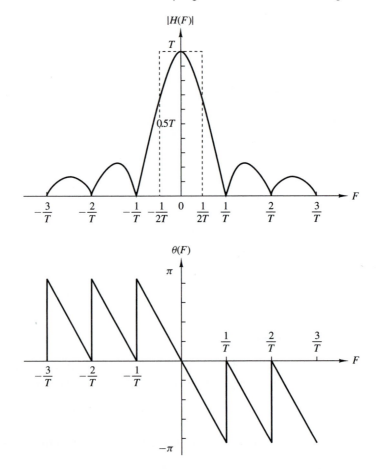

Figure 9.23   Frequency response charactersitics of the S/H.

which highly attenuates frequency components above $F_s/2$. In effect, the lowpass filter following the S/H smooths the signal $\hat{x}(t)$ by removing the sharp discontinuities.

9.3.2 First-Order Hold

A first-order hold approximates $x(t)$ by straight-line segments which have a slope that is determined by the current sample $x(nT)$ and the previous sample $x(nT-T)$. An illustration of this signal reconstruction techniques is given in Fig. 9.24.

The mathematical relationship between the input samples and the output waveform is

$$\hat{x}(t) = x(nT) + \frac{x(nT) - x(nT-T)}{T}(t - nT) \qquad nT \le t < (n+1)T \qquad (9.3.6)$$

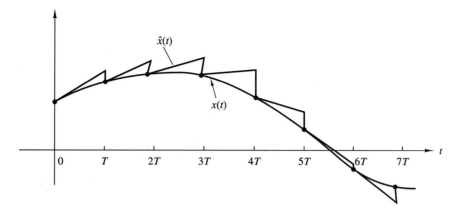

Figure 9.24   Signal reconstruction with a first-order hold.

When viewed as a linear filter, the impulse response of the first-order hold is

$$
h(t) = \begin{cases} 1 + \dfrac{t}{T}, & 0 \le t \le T \\[2mm] 1 - \dfrac{t}{T}, & T \le t < 2T \\[2mm] 0, & \text{otherwise} \end{cases} \tag{9.3.7}
$$

This impulse response is depicted in Fig. 9.25(a). The Fourier transform of $h(t)$ yields the frequency response, which can be expressed in the form

$$
H(F) = T(1 + 4\pi F^2 T^2)^{1/2} \left(\frac{\sin \pi FT}{\pi FT} \right)^2 e^{j\Theta(F)} \tag{9.3.8}
$$

where the phase $\Theta(F)$ is

$$
\Theta(F) = -\pi FT + \tan^{-1} 2\pi FT \tag{9.3.9}
$$

These frequency response characteristics are graphically illustrated in Fig. 9.25(b) and (c).

Since this reconstruction technique also suffers from distortion due to passage of frequency components above $F_s/2$, as can be observed from Fig. 9.25(b), it is followed by a lowpass filter that significantly attenuates frequencies above the folding frequency $F_s/2$.

The peaks in $H(F)$ within the band $|F| \le F_s/2$ may be undesirable in some applications. In such a case it is possible to modify the impulse response by reducing the slope by some factor $\beta < 1$. This results in the impulse response $h(t)$ illustrated in Fig. 9.26(a). The corresponding frequency response is given by

$$
H(F) = T \left[1 - \beta + \beta(1 + j2\pi FT) \frac{\sin \pi FT}{\pi FT} e^{-j\pi FT} \right] \frac{\sin \pi FT}{\pi FT} \tag{9.3.10}
$$

The magnitude $|H(F)|$ is illustrated in Fig. 9.26(b) for $\beta = 0.5$, $\beta = 0.3$, and $\beta = 0.1$. We note that the peak in $H(F)$ is relatively small for $\beta = 0.3$ and

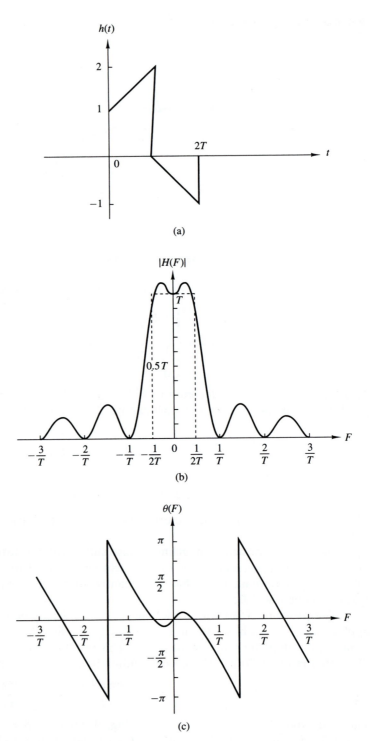

Figure 9.25 Impulse response (a) and frequency response characteristics (b) and (c) for a first-order hold.

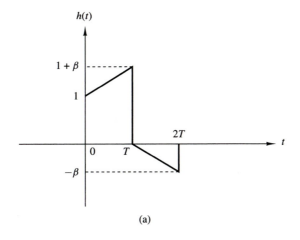

(a)

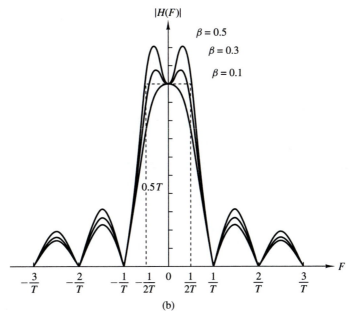

(b)

Figure 9.26 Impulse response (a) and frequency (magnitude) response (b) for a modified first-order hold.

does not exist when $\beta = 0.1$. Thus this modified first-order hold exhibits better frequency response characteristics in the frequency range $|F| \leq F_s/2$.

9.3.3 Linear Interpolation with Delay

The first-order hold performs signal reconstruction by computing the slope of the straight line based on the current sample $x(nT)$ and the past sample $x(nT - T)$ of the signal. In effect, this technique *linearly extrapolates* or attempts to *linearly predict* the next sample of the signal based on the samples $x(nT)$ and $x(nT - T)$. As a consequence, the estimated signal waveform $\hat{x}(t)$ contains jumps at the sample points.

The jumps in $\hat{x}(t)$ can be avoided by providing a one-sample delay in the reconstruction process. Then successive sample points can be connected by straight-line segments. Thus the resulting interpolated signal $\hat{x}(t)$ can be expressed as

$$\hat{x}(t) = x(nT - T) + \frac{x(nT) - x(nT - T)}{T}(t - nT) \qquad nT \le t < (n+1)T \qquad (9.3.11)$$

We observe that at $t = nT$, $\hat{x}(nT) = x(nT - T)$ and at $t = nT + T$, $\hat{x}(nT + T) = x(nT)$. Therefore, $x(t)$ has an inherent delay of T seconds in interpolating the actual signal $x(t)$. Figure 9.27 illustrates this linear interpolation technique.

Viewed as a linear filter, the linear interpolator with a T-second delay has an impulse response

$$h(t) = \begin{cases} t/T, & 0 \le t < T \\ 2 - t/T, & T \le t < 2T \\ 0, & \text{otherwise} \end{cases} \qquad (9.3.12)$$

The corresponding frequency response is

$$H(F) = \int_0^T \frac{t}{T} e^{-j2\pi Ft} dt + \int_T^{2T} \left(2 - \frac{t}{T}\right) e^{-j2\pi Ft} dt$$

$$= T \left(\frac{\sin \pi FT}{\pi FT}\right)^2 e^{-j2\pi Ft} \qquad (9.3.13)$$

The impulse response and frequency response characteristics of this interpolation filter are illustrated in Fig. 9.28. We observe that the magnitude characteristic falls off rapidly and contains small sidelobes beyond the sampling frequency F_s. Furthermore, its phase characteristic is linear due to the delay T. By following this interpolator with a lowpass filter that has a sharp cutoff beyond the frequency $F_s/2$, the high-frequency components in $\hat{x}(t)$ can be further reduced.

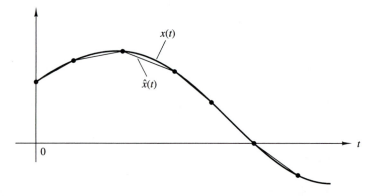

Figure 9.27 Linear interpolation of $x(t)$ with a T-second delay.

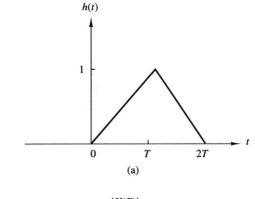

(a)

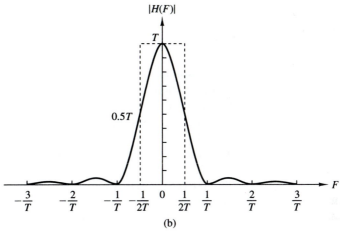

(b)

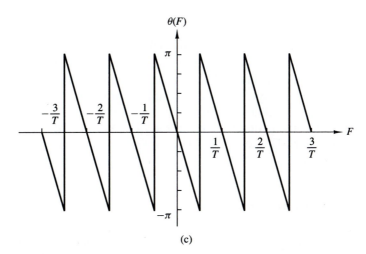

(c)

Figure 9.28 Impulse response (a) and frequency response characteristics (b) and (c) for the linear interpolator with delay.

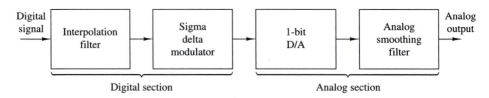

Figure 9.29 Elements of an oversampling D/A converter.

This concludes our discussion of signal reconstruction based on simple interpolation techniques. The techniques that we have described are easily incorporated into the design of practical D/A converters for the reconstruction of analog signals from digital signals. We shall consider interpolation again in Chapter 10 in the context of changing the sampling rate in a digital signal processing system.

9.3.4 Oversampling D/A Converters

The elements of an oversampling D/A converter are shown in Fig. 9.29. As we observe, it is subdivided into a digital front end followed by an analog section. The digital section consists of an interpolator whose function is to increase the sampling rate by some factor I, and then is followed by a SDM. The interpolator simply increases the digital sampling rate by inserting $I - 1$ zeros between successive low rate samples. The resulting signal is then processed by a digital filter with cutoff frequency $F_c = B/F_s$ in order to reject the images (replicas) of the input signal spectrum. This higher rate signal is fed to the SDM, which creates a noise-shaped 1-bit sample. Each 1-bit sample is fed to the 1-bit D/A, which provides the analog interface to the antialiasing and smoothing filters. The output analog filters have a passband of $0 \leq F \leq B$ hertz and serve to smooth the signal and to remove the quantization noise in the frequency band $B \leq F \leq F_s/2$. In effect, the oversampling D/A converter uses SDM with the roles of the analog and digital sections reversed compared to the A/D converter.

In practice, oversampling D/A (and A/D) converters have many advantages over the more conventional D/A (and A/D) converters. First, the high sampling rate and the subsequent digital filtering minimize or remove the need for complex and expensive analog antialiasing filters. Furthermore, any analog noise introduced during the conversion phase is filtered out. Also, there is no need for S/H circuits. Oversampling SDM A/D and D/A converters are very robust with respect to variations in the analog-circuit parameters, are inherently linear, and have low cost.

9.4 SUMMARY AND REFERENCES

The major focus of this chapter was on the sampling and reconstruction of signals. In particular, we treated the sampling of continuous-time signals and the subsequent operation of A/D conversion. These are necessary operations in the digital processing of analog signals, either on a general-purpose computer or on a custom-designed digital signal processor. The related issue of D/A conversion was also treated. In addition to the conventional A/D and D/A conversion techniques, we also described another type of A/D and D/A conversion, based on the principle of oversampling and a type of waveform encoding called sigma-delta modulation. Sigma-delta conversion technology is especially suitable for audio band signals due to their relatively small bandwidth (less than 20 kHz) and in some applications, the requirements for high fidelity.

The sampling theorem was introduced by Nyquist (1928) and later popularized in the classic paper by Shannon (1949). D/A and A/D conversion techniques are treated in a book by Sheingold (1986). Oversampling A/D and D/A conversion has been treated in the technical literature. Specifically, we cite the work of Candy (1986), Candy et al. (1981) and Gray (1990).

PROBLEMS

9.1 Consider the sampling of the bandpass signal whose spectrum is illustrated in Fig. P9.1. Determine the minimum sampling rate F_s to avoid aliasing.

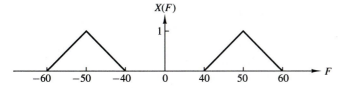

Figure P9.1

9.2 Consider the sampling of the bandpass signal whose spectrum is illustrated in Fig. P9.2. Determine the minimum sampling rate F_s to avoid aliasing.

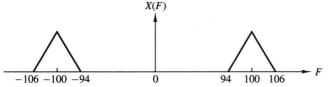

Figure P9.2

9.3 Prove that $x_l(t)$ is generally a complex-valued signal and give the condition under which it is real. Assume that $x(t)$ is a real-valued bandpass signal.

9.4 Consider the two systems shown in Fig. P9.4.
  (a) Sketch the spectra of the various signals if $x_a(t)$ has the Fourier transform shown in Fig. 9.4(b) and $F_s = 2B$. How are $y_1(t)$ and $y_2(t)$ related to $x_a(t)$?
  (b) Determine $y_1(t)$ and $y_2(t)$ if $x_a(t) = \cos 2\pi F_0 t$, $F_0 = 20$ Hz, and $F_s = 50$ Hz or $F_s = 30$ Hz.

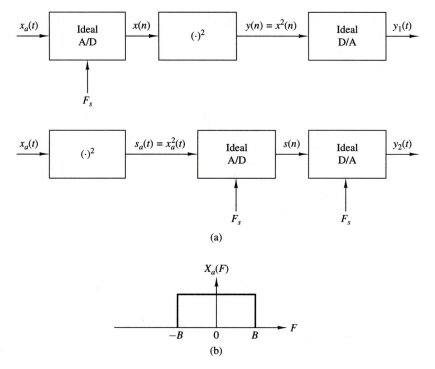

(a)

(b)

Figure P9.4

9.5 A continuous-time signal $x_a(t)$ with bandwidth B and its echo $x_a(t - \tau)$ arrive simultaneously at a TV receiver. The received analog signal

$$s_a(t) = x_a(t) + \alpha x_a(t - \tau) \qquad |\alpha| < 1$$

is processed by the system shown in Fig. P9.5. Is it possible to specify F_s and $H(z)$ so that $y_a(t) = x_a(t)$ [i.e., remove the "ghost" $x_a(t - \tau)$ from the received signal]?

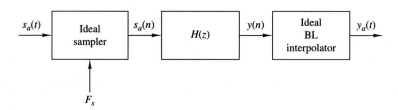

Figure P9.5

9.6 A bandlimited continuous-time signal $x_a(t)$ is sampled at a sampling frequency $F_s \geq 2B$. Determine the energy E_d of the resulting discrete-time signal $x(n)$ as a function of the energy of the analog signal, E_a, and the sampling period $T = 1/F_s$.

9.7 Let $x(n)$ be a zero-mean stationary process with variance σ_x^2 and autocorrelation $\gamma_x(l)$.

(a) Show that the variance σ_d^2 of the first-order prediction error

$$d(n) = x(n) - ax(n-1)$$

is given by

$$\sigma_d^2 = \sigma_x^2[1 + a^2 - 2a\rho_x(1)]$$

where $\rho_x(1) = \gamma_x(1)/\gamma_x(0)$ is the normalized autocorrelation sequence.

(b) Show that σ_d^2 attains its minimum value

$$\sigma_d^2 = \sigma_x^2[1 - \rho_x^2(1)]$$

for $a = \gamma_x(1)/\gamma_x(0) = \rho_x(1)$.

(c) Under what conditions is $\sigma_d^2 < \sigma_x^2$?

(d) Repeat steps (a) to (c) for the second-order prediction error

$$d(n) = x(n) - a_1 x(n-1) - a_2 x(n-2)$$

9.8 Consider a DM coder with input $x(n) = A\cos(2\pi n F/F_s)$. What is the condition for avoiding slope overload? Illustrate this condition graphically.

9.9 Let $x_a(t)$ be a bandlimited signal with fixed bandwidth B and variance σ_x^2.

(a) Show that the signal-to-quantization noise ratio, SQNR $= 10\log_{10}(\sigma_x^2/\sigma_e^2)$, increases by 3 dB each time we double the sampling frequency F_s. Assume that the quantization noise model discussed in Section 9.2.3 is valid.

(b) If we wish to increase the SQNR of a quantizer by doubling its sampling frequency, what is the most efficient way to do it? Should we choose a linear multibit A/D converter or an oversampling one?

9.10 Consider the first-order SDM model shown in Fig. 9.15.

(a) Show that the quantization noise power in the signal band $(-B, B)$ is given by

$$\sigma_n^2 = \frac{2\sigma_e^2}{\pi}\left[\frac{2\pi B}{F_s} - \sin\left(2\pi\frac{B}{F_s}\right)\right]$$

(b) Using a two-term Taylor series expansion of the sine function and assuming that $F_s \gg B$, show that

$$\sigma_n^2 \approx \frac{1}{3}\pi^2\sigma_e^2\left(\frac{2B}{F_s}\right)^3$$

9.11 Consider the second-order SDM model shown in Fig. P9.11.

(a) Determine the signal and noise system functions $H_s(z)$ and $H_n(z)$, respectively.

(b) Plot the magnitude response for the noise system function and compare it with the one for the first-order SDM. Can you explain the 6-dB difference from these curves?

(c) Show that the in-band quantization noise power σ_n^2 is given approximately by

$$\sigma_n^2 \approx \frac{\pi\sigma_e^2}{5}\left(\frac{2B}{F_s}\right)^5$$

which implies a 15-dB increase for every doubling of the sampling frequency.

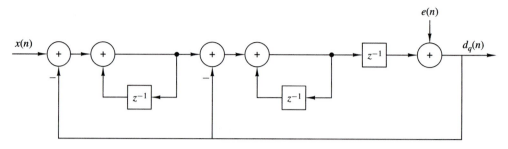

Figure P9.11

9.12 Figure P9.12 illustrates the basic idea for a lookup table based sinusoidal signal generator. The samples of one period of the signal

$$x(n) = \cos\left(\frac{2\pi}{N}n\right) \qquad n = 0, 1, \ldots, N - 1$$

are stored in memory. A digital sinusoidal signal is generated by stepping through the table and wrapping around at the end when the angle exceeds 2π. This can be done by using modulo-N addressing (i.e., using a "circular" buffer). Samples of $x(n)$ are feeding the ideal D/A converter every T seconds.

(a) Show that by changing F_s we can adjust the frequency F_0 of the resulting analog sinusoid.

(b) Suppose now that $F_s = 1/T$ is fixed. How many distinct analog sinusoids can be generated using the given lookup table? Explain.

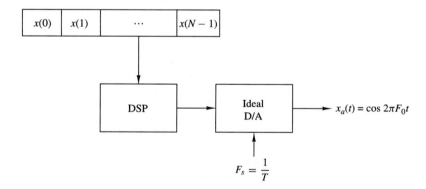

Figure P9.12

9.13 Suppose that we represent an analog bandpass filter by the frequency response

$$H(F) = C(F - F_c) + C^*(-F - F_c)$$

where $C(f)$ is the frequency response of an equivalent lowpass filter, as shown in Fig. P9.13.

(a) Show that the impulse response $c(t)$ of the equivalent lowpass filter is related to the impulse response $h(t)$ of the bandpass filter as follows:

$$h(t) = 2 \operatorname{Re}[c(t)e^{j2\pi F_c t}]$$

(b) Suppose that the bandpass system with frequency response $H(F)$ is excited by a bandpass signal of the form

$$x(t) = \operatorname{Re}[u(t)e^{j2\pi F_c t}]$$

where $u(t)$ is the equivalent lowpass signal. Show that the filter output may be expressed as

$$y(t) = \operatorname{Re}[v(t)e^{j2\pi F_c t}]$$

where

$$v(t) = \int^{\omega} c(\tau)u(t-\tau)d\tau$$

(*Hint:* Use the frequency domain to prove this result.)

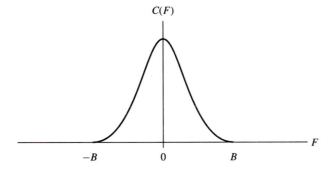

Figure P9.13

9.14* Consider the sinusoidal signal generator in Fig. P9.14, where both the stored sinusoidal data

$$x(n) = \cos\left(\frac{2A}{N}n\right) \qquad 0 \le n \le N-1$$

and the sampling frequency $F_s = 1/T$ are fixed. An engineer wishing to produce a sinusoid with period $2N$ suggests that we use either zero-order or first-order (linear) interpolation to double the number of samples per period in the original sinusoid as illustrated in Fig. P9.14(a).

(a) Determine the signal sequences $y(n)$ generated using zero-order interpolation and linear interpolation and then compute the total harmonic distortion (THD) in each case for $N = 32, 64, 128$.

(b) Repeat part (a) assuming that all sample values are quantized to 8 bits.

(c) Show that the interpolated signal sequences $y(n)$ can be obtained by the system shown in Fig. P9.14(b). The first module inserts one zero sample between successive samples of $x(n)$. Determine the system $H(z)$ and sketch its magnitude response for the zero-order interpolation and for the linear interpolation cases. Can you explain the difference in performance in terms of the frequency response functions?

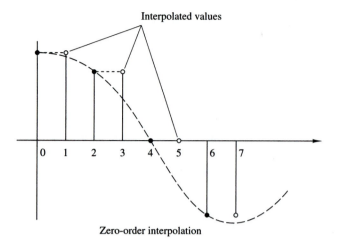

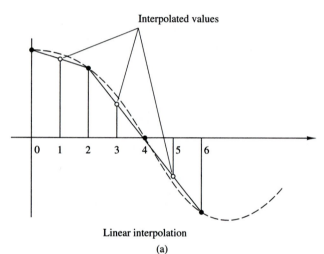

(a)

Figure P9.13   (a)

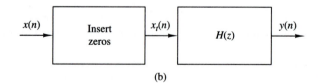

(b)

Figure P9.13   (b)

(d) Determine and sketch the spectra of the resulting sinusoids in each case both analytically [using the results in part (c)] and evaluating the DFT of the resulting signals.

(e) Sketch the spectra of $x_i(n)$ and $y(n)$, if $x(n)$ has the spectrum shown in Fig. P9.14(c) for both zero-order and linear interpolation. Can you suggest a better choice for $H(z)$?

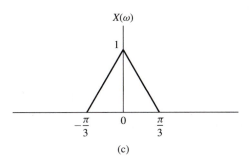

(c)   **Figure P9.13   (c)**

9.15 Let $x_a(t)$ be a time-limited signal; that is, $x_a(t) = 0$ for $|t| > \tau$, with Fourier transform $X_a(F)$. The function $X_a(F)$ is sampled with sampling interval $\delta F = 1/T_s$.
(a) Show that the function

$$x_p(t) = \sum_{n=-\infty}^{\infty} x_a(t - nT_s)$$

can be expressed as a Fourier series with coefficients

$$c_k = \frac{1}{T_s} X_a(k\delta F)$$

(b) Show that $X_a(F)$ can be recovered from the samples $X_a(k\delta F)$, $-\infty < k < \infty$ if $T_s \geq 2\tau$.
(c) Show that if $T_s < 2\tau$, there is "time-domain aliasing" that prevents exact reconstruction of $X_a(F)$.
(d) Show that if $T_s \geq 2\tau$, perfect reconstruction of $X_a(F)$ from the samples $X(k\delta F)$ is possible using the interpolation formula

$$X_a(F) = \sum_{k=-\infty}^{\infty} X_a(k\delta F) \frac{\sin[(\pi/\delta F)(F - k\delta F)]}{(\pi/\delta F)(F - k\delta F)}$$

10

Multirate Digital Signal Processing

In many practical applications of digital signal processing, one is faced with the problem of changing the sampling rate of a signal, either increasing it or decreasing it by some amount. For example, in telecommunication systems that transmit and receive different types of signals (e.g., teletype, facsimile, speech, video, etc.), there is a requirement to process the various signals at different rates commensurate with the corresponding bandwidths of the signals. The process of converting a signal from a given rate to a different rate is called *sampling rate conversion*. In turn, systems that employ multiple sampling rates in the processing of digital signals are called *multirate digital signal processing systems*.

Sampling rate conversion of a digital signal can be accomplished in one of two general methods. One method is to pass the digital signal through a D/A converter, filter it if necessary, and then to resample the resulting analog signal at the desired rate (i.e., to pass the analog signal through an A/D converter). The second method is to perform the sampling rate conversion entirely in the digital domain.

One apparent advantage of the first method is that the new sampling rate can be arbitrarily selected and need not have any special relationship to the old sampling rate. A major disadvantage, however, is the signal distortion, introduced by the D/A converter in the signal reconstruction, and by the quantization effects in the A/D conversion. Sampling rate conversion performed in the digital domain avoids this major disadvantage.

In this chapter we describe sampling rate conversion and multirate signal processing in the digital domain. First we describe sampling rate conversion by a rational factor and present several methods for implementing the rate converter, including single-stage and multistage implementations. Then, we describe a method for sampling rate conversion by an arbitrary factor and discuss its implementation. Finally, we present several applications of sampling rate conversion in multirate signal processing systems, which include the implementation of narrowband filters, digital filter banks, and quadrature mirror filters. We also discuss the use of quadrature mirror filters in subband coding, transmultiplexers, and finally over-sampling A/D and D/A converters.

10.1 INTRODUCTION

The process of sampling rate conversion in the digital domain can be viewed as a linear filtering operation, as illustrated in Fig. 10.1(a). The input signal $x(n)$ is characterized by the sampling rate $F_x = 1/T_x$ and the output signal $y(m)$ is characterized by the sampling rate $F_y = 1/T_y$, where T_x and T_y are the corresponding sampling intervals. In the main part of our treatment, the ratio F_y/F_x is constrained to be rational,

$$\frac{F_y}{F_x} = \frac{I}{D}$$

where D and I are relatively prime integers. We shall show that the linear filter is characterized by a time-variant impulse response, denoted as $h(n, m)$. Hence the input $x(n)$ and the output $y(m)$ are related by the convolution summation for time-variant systems.

The sampling rate conversion process can also be understood from the point of view of digital resampling of the same analog signal. Let $x(t)$ be the analog signal that is sampled at the first rate F_x to generate $x(n)$. The goal of rate conversion is to obtain another sequence $y(m)$ directly from $x(n)$, which is equal to the sampled values of $x(t)$ at a second rate F_y. As is depicted in Fig. 10.1(b), $y(m)$ is a time-shifted version of $x(n)$. Such a time shift can be

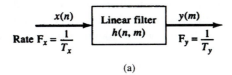

(a)

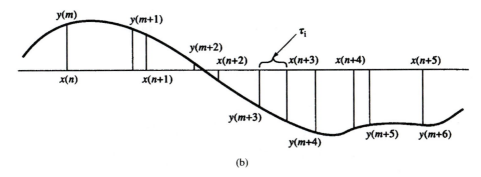

(b)

Figure 10.1   Sampling rate conversion viewed as a linear filtering process.

realized by using a linear filter that has a flat magnitude response and a linear phase response (i.e., it has a frequency response of $e^{-j\omega\tau_i}$, where τ_i is the time delay generated by the filter). If the two sampling rates are not equal, the required amount of time shifting will vary from sample to sample, as shown in Fig. 10.1(b). Thus the rate converter can be implemented using a set of linear filters that have the same flat magnitude response but generate different time delays.

Before considering the general case of sampling rate conversion, we shall consider two special cases. One is the case of sampling rate reduction by an integer factor D, and the second is the case of a sampling rate increase by an integer factor I. The process of reducing the sampling rate by a factor D (downsampling by D) is called *decimation*. The process of increasing the sampling rate by an integer factor I (upsampling by I) is called *interpolation*.

10.2 DECIMATION BY A FACTOR *D*

Let us assume that the signal $x(n)$ with spectrum $X(\omega)$ is to be downsampled by an integer factor D. The spectrum $X(\omega)$ is assumed to be nonzero in the frequency interval $0 \leq |\omega| \leq \pi$ or, equivalently, $|F| \leq F_x/2$. We know that if we reduce the sampling rate simply by selecting every Dth value of $x(n)$, the resulting signal will be an aliased version of $x(n)$, with a folding frequency of $F_x/2D$. To avoid aliasing, we must first reduce the bandwidth of $x(n)$ to $F_{\max} = F_x/2D$ or, equivalently, to $\omega_{\max} = \pi/D$. Then we may downsample by D and thus avoid aliasing.

The decimation process is illustrated in Fig. 10.2. The input sequence $x(n)$ is passed through a lowpass filter, characterized by the impulse response $h(n)$ and a frequency response $H_D(\omega)$, which ideally satisfies the condition

$$H_D(\omega) = \begin{cases} 1, & |\omega| \leq \pi/D \\ 0, & \text{otherwise} \end{cases} \qquad (10.2.1)$$

Thus the filter eliminates the spectrum of $X(\omega)$ in the range $\pi/D < \omega < \pi$. Of course, the implication is that only the frequency components of $x(n)$ in the range $|\omega| \leq \pi/D$ are of interest in further processing of the signal.

The output of the filter is a sequence $v(n)$ given as

$$v(n) = \sum_{k=0}^{\infty} h(k)x(n-k) \qquad (10.2.2)$$

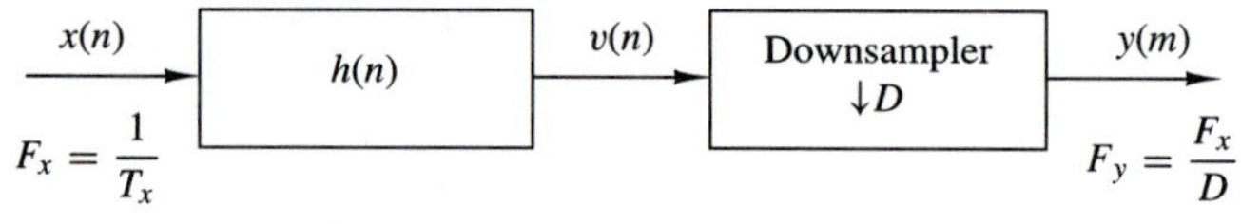

Figure 10.2   Decimation by a factor D.

which is then downsampled by the factor D to produce $y(m)$. Thus

$$y(m) = v(mD)$$

$$= \sum_{k=0}^{\infty} h(k)x(mD - k) \qquad (10.2.3)$$

Although the filtering operation on $x(n)$ is linear and time invariant, the downsampling operation in combination with the filtering results in a time-variant system. This is easily verified. Given the fact that $x(n)$ produces $y(m)$, we note that $x(n - n_0)$ does not imply $y(n - n_0)$ unless n_0 is a multiple of D. Consequently, the overall linear operation (linear filtering followed by downsampling) on $x(n)$ is not time invariant.

The frequency-domain characteristics of the output sequence $y(m)$ can be obtained by relating the spectrum of $y(m)$ to the spectrum of the input sequence $x(n)$. First, it is convenient to define a sequence $\tilde{v}(n)$ as

$$\tilde{v}(n) = \begin{cases} v(n), & n = 0, \pm D, \pm 2D, \ldots \\ 0, & \text{otherwise} \end{cases} \qquad (10.2.4)$$

Clearly, $\tilde{v}(n)$ can be viewed as a sequence obtained by multiplying $v(n)$ with a periodic train of impulses $p(n)$, with period D, as illustrated in Fig. 10.3. The discrete Fourier series representation of $p(n)$ is

$$p(n) = \frac{1}{D} \sum_{k=0}^{D-1} e^{j2\pi kn/D} \qquad (10.2.5)$$

Hence

$$\tilde{v}(n) = v(n)p(n) \qquad (10.2.6)$$

and

$$y(m) = \tilde{v}(mD) = v(mD)p(mD) = v(mD) \qquad (10.2.7)$$

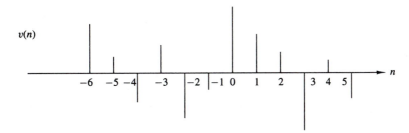

Figure 10.3 Multiplication of $v(n)$ with a periodic impulse train $p(n)$ with period $D = 3$.

Now the z-transform of the output sequence $y(m)$ is

$$Y(z) = \sum_{m=-\infty}^{\infty} y(m)z^{-m}$$

$$= \sum_{m=-\infty}^{\infty} \tilde{v}(mD)z^{-m} \qquad (10.2.8)$$

$$Y(z) = \sum_{m=-\infty}^{\infty} \tilde{v}(m)z^{-m/D}$$

where the last step follows from the fact that $\tilde{v}(m) = 0$, except at multiples of D. By making use of the relations in (10.2.5) and (10.2.6) in (10.2.8), we obtain

$$Y(z) = \sum_{m=-\infty}^{\infty} v(m) \left[\frac{1}{D} \sum_{k=0}^{D-1} e^{j2\pi mk/D} \right] z^{-m/D}$$

$$= \frac{1}{D} \sum_{k=0}^{D-1} \sum_{m=-\infty}^{\infty} v(m)(e^{-j2\pi k/D} z^{1/D})^{-m}$$

$$= \frac{1}{D} \sum_{k=0}^{D-1} V(e^{-j2\pi k/D} z^{1/D}) \qquad (10.2.9)$$

$$= \frac{1}{D} \sum_{k=0}^{D-1} H_D(e^{-j2\pi k/D} z^{1/D}) X(e^{-j2\pi k/D} z^{1/D})$$

where the last step follows from the fact that $V(z) = H_D(z)X(z)$.

By evaluating $Y(z)$ in the unit circle, we obtain the spectrum of the output signal $y(m)$. Since the rate of $y(m)$ is $F_y = 1/T_y$, the frequency variable, which we denote as ω_y, is in radians and is relative to the sampling rate F_y,

$$\omega_y = \frac{2\pi F}{F_y} = 2\pi F T_y \qquad (10.2.10)$$

Since the sampling rates are related by the expression

$$F_y = \frac{F_x}{D} \qquad (10.2.11)$$

it follows that the frequency variables ω_y and

$$\omega_x = \frac{2\pi F}{F_x} = 2\pi F T_x \qquad (10.2.12)$$

are related by

$$\omega_y = D\omega_x \qquad (10.2.13)$$

Thus, as expected, the frequency range $0 \leq |\omega_x| \leq \pi/D$ is stretched into the corresponding frequency range $0 \leq |\omega_y| \leq \pi$ by the downsampling process.

We conclude that the spectrum $Y(\omega_y)$, which is obtained by evaluating (10.2.9) on the unit circle, can be expressed as

$$Y(\omega_y) = \frac{1}{D} \sum_{k=0}^{D-1} H_D\left(\frac{\omega_y - 2\pi k}{D}\right) X\left(\frac{\omega_y - 2\pi k}{D}\right) \tag{10.2.14}$$

With a properly designed filter $H_D(\omega)$, the aliasing is eliminated and, consequently, all but the first term in (10.2.14) vanish. Hence

$$\begin{aligned} Y(\omega_y) &= \frac{1}{D} H_D\left(\frac{\omega_y}{D}\right) X\left(\frac{\omega_y}{D}\right) \\ &= \frac{1}{D} X\left(\frac{\omega_y}{D}\right) \end{aligned} \tag{10.2.15}$$

for $0 \le |\omega_y| \le \pi$. The spectra for the sequences $x(n)$, $v(n)$, and $y(m)$ are illustrated in Fig. 10.4.

10.3 INTERPOLATION BY A FACTOR I

An increase in the sampling rate by an integer factor of I can be accomplished by interpolating $I - 1$ new samples between successive values of the signal. The interpolation process can be accomplished in a variety of ways. We shall describe a process that preserves the spectral shape of the signal sequence $x(n)$.

Let $v(m)$ denote a sequence with a rate $F_y = I F_x$, which is obtained from $x(n)$ by adding $I - 1$ zeros between successive values of $x(n)$. Thus

$$v(m) = \begin{cases} x(m/I), & m = 0, \pm I, \pm 2I, \ldots \\ 0, & \text{otherwise} \end{cases} \tag{10.3.1}$$

and its sampling rate is identical to the rate of $y(m)$. This sequence has a z-transform

$$\begin{aligned} V(z) &= \sum_{m=-\infty}^{\infty} v(m) z^{-m} \\ &= \sum_{m=-\infty}^{\infty} x(m) z^{-mI} \tag{10.3.2} \\ &= X(z^I) \end{aligned}$$

The corresponding spectrum of $v(m)$ is obtained by evaluating (10.3.2) on the unit circle. Thus

$$V(\omega_y) = X(\omega_y I) \tag{10.3.3}$$

where ω_y denotes the frequency variable relative to the new sampling rate F_y (i.e., $\omega_y = 2\pi F/F_y$). Now the relationship between sampling rates is $F_y = I F_x$ and hence, the frequency variables ω_x and ω_y are related according to the formula

$$\omega_y = \frac{\omega_x}{I} \tag{10.3.4}$$

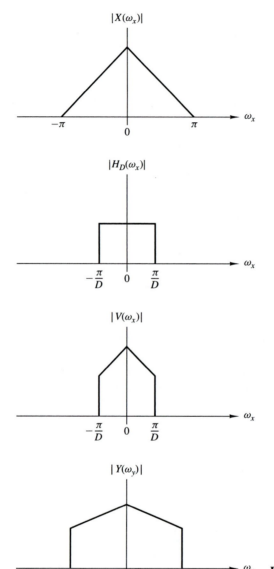

Figure 10.4 Spectra of signals in the decimation of $x(n)$ by a factor D.

The spectra $X(\omega_x)$ and $V(\omega_y)$ are illustrated in Fig. 10.5. We observe that the sampling rate increase, obtained by the addition of $I-1$ zero samples between successive values of $x(n)$, results in a signal whose spectrum $V(\omega_y)$ is an I-fold periodic repetition of the input signal spectrum $X(\omega_x)$.

Since only the frequency components of $x(n)$ in the range $0 \le \omega_y \le \pi/I$ are unique, the images of $X(\omega)$ above $\omega_y = \pi/I$ should be rejected by passing the sequence $v(m)$ through a lowpass filter with frequency response $H_I(\omega_y)$ that

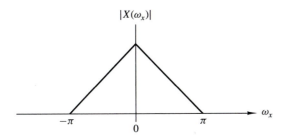

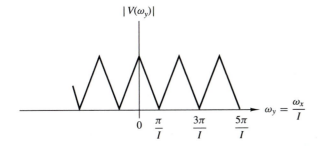

Figure 10.5 Spectra of $x(n)$ and $v(n)$ where $V(\omega_y) = X(\omega_y I)$.

ideally has the characteristic

$$H_I(\omega_y) = \begin{cases} C, & 0 \le |\omega_y| \le \pi/I \\ 0, & \text{otherwise} \end{cases} \tag{10.3.5}$$

where C is a scale factor required to properly normalize the output sequence $y(m)$. Consequently, the output spectrum is

$$Y(\omega_y) = \begin{cases} CX(\omega_y I), & 0 \le |\omega_y| \le \pi/I \\ 0, & \text{otherwise} \end{cases} \tag{10.3.6}$$

The scale factor C is selected so that the output $y(m) = x(m/I)$ for $m = 0$, $\pm I, +2I, \ldots$. For mathematical convenience, we select the point $m = 0$. Thus

$$y(0) = \frac{1}{2\pi} \int_{-\pi}^{\pi} Y(\omega_y) d\omega_y$$
$$= \frac{C}{2\pi} \int_{-\pi/I}^{\pi/I} X(\omega_y I) d\omega_y \tag{10.3.7}$$

Since $\omega_y = \omega_x/I$, (10.3.7) can be expressed as

$$y(0) = \frac{C}{I} \frac{1}{2\pi} \int_{-\pi}^{\pi} X(\omega_x) d\omega_x$$
$$= \frac{C}{I} x(0) \tag{10.3.8}$$

Therefore, $C = I$ is the desired normalization factor.

Finally, we indicate that the output sequence $y(m)$ can be expressed as a convolution of the sequence $v(n)$ with the unit sample response $h(n)$ of the lowpass filter. Thus

$$y(m) = \sum_{k=-\infty}^{\infty} h(m-k)v(k) \qquad (10.3.9)$$

Since $v(k) = 0$ except at multiples of I, where $v(kI) = x(k)$, (10.3.9) becomes

$$y(m) = \sum_{k=-\infty}^{\infty} h(m-kI)x(k) \qquad (10.3.10)$$

10.4 SAMPLING RATE CONVERSION BY A RATIONAL FACTOR *I/D*

Having discussed the special cases of decimation (downsampling by a factor D) and interpolation (upsampling by a factor I), we now consider the general case of sampling rate conversion by a rational factor I/D. Basically, we can achieve this sampling rate conversion by first performing interpolation by the factor I and then decimating the output of the interpolator by the factor D. In other words, a sampling rate conversion by the rational factor I/D is accomplished by cascading an interpolator with a decimator, as illustrated in Fig. 10.6.

We emphasize that the importance of performing the interpolation first and the decimation second, is to preserve the desired spectral characteristics of $x(n)$. Furthermore, with the cascade configuration illustrated in Fig. 10.6, the two filters with impulse response $\{h_u(l)\}$ and $\{h_d(l)\}$ are operated at the same rate, namely IF_x and hence can be combined into a single lowpass filter with impulse response $h(l)$ as illustrated in Fig. 10.7. The frequency response $H(\omega_v)$ of the combined filter must incorporate the filtering operations for both interpolation and decimation, and hence it should ideally possess the frequency response characteristic

$$H(\omega_v) = \begin{cases} I, & 0 \le |\omega_v| \le \min(\pi/D, \pi/I) \\ 0, & \text{otherwise} \end{cases} \qquad (10.4.1)$$

where $\omega_v = 2\pi F/F_v = 2\pi F/IF_x = \omega_x/I$.

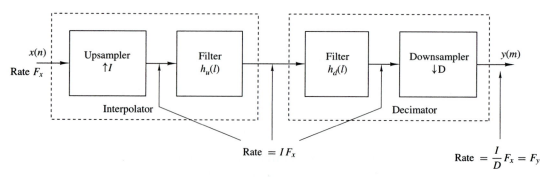

Figure 10.6 Method for sampling rate conversion by a factor I/D.

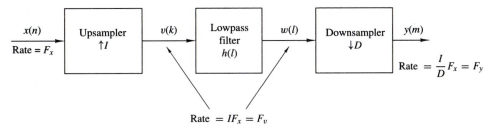

Figure 10.7 Method for sampling rate conversion by a factor I/D.

In the time domain, the output of the upsampler is the sequence

$$v(l) = \begin{cases} x(l/I), & l = 0, \pm I, \pm 2I, \ldots \\ 0, & \text{otherwise} \end{cases} \tag{10.4.2}$$

and the output of the linear time-invariant filter is

$$w(l) = \sum_{k=-\infty}^{\infty} h(l-k)v(k)$$

$$= \sum_{k=-\infty}^{\infty} h(l-kI)x(k) \tag{10.4.3}$$

Finally, the output of the sampling rate converter is the sequence $\{y(m)\}$, which is obtained by downsampling the sequence $\{w(l)\}$ by a factor of D. Thus

$$y(m) = w(mD)$$

$$= \sum_{k=-\infty}^{\infty} h(mD - kI)x(k) \tag{10.4.4}$$

It is illuminating to express (10.4.4) in a different form by making a change in variable. Let

$$k = \left\lfloor \frac{mD}{I} \right\rfloor - n \tag{10.4.5}$$

where the notation $\lfloor r \rfloor$ denotes the largest integer contained in r. With this change in variable, (10.4.4) becomes

$$y(m) = \sum_{n=-\infty}^{\infty} h\left(mD - \left\lfloor \frac{mD}{I} \right\rfloor I + nI\right) x\left(\left\lfloor \frac{mD}{I} \right\rfloor - n\right) \tag{10.4.6}$$

We note that

$$mD - \left\lfloor \frac{mD}{I} \right\rfloor I = mD \qquad \text{modulo } I$$

$$= (mD)_I$$

Consequently, (10.4.6) can be expressed as

$$y(m) = \sum_{n=-\infty}^{\infty} h(nI + (mD)_I)x\left(\left\lfloor \frac{mD}{I} \right\rfloor - n\right) \tag{10.4.7}$$

It is apparent from this form that the output $y(m)$ is obtained by passing the input sequence $x(n)$ through a time-variant filter with impulse response

$$g(n, m) = h(nI + (mD)_I) \qquad -\infty < m, n < \infty \qquad (10.4.8)$$

where $h(k)$ is the impulse response of the time-invariant lowpass filter operating at the sampling rate IF_x. We further observe, that for any integer k,

$$
\begin{aligned}
g(n, m + kI) &= h(nI + (mD + kDI)_I) \\
&= h(nI + (mD)_I) \qquad\qquad (10.4.9) \\
&= g(n, m)
\end{aligned}
$$

Hence $g(n, m)$ is periodic in the variable m with period I.

The frequency-domain relationships can be obtained by combining the results of the interpolation and decimation processes. Thus the spectrum at the output of the linear filter with impulse response $h(l)$ is

$$
\begin{aligned}
V(\omega_v) &= H(\omega_v)X(\omega_v I) \\
&= \begin{cases} IX(\omega_v I), & 0 \le |\omega_v| \le \min(\pi/D, \pi/I) \\ 0, & \text{otherwise} \end{cases} \qquad (10.4.10)
\end{aligned}
$$

The spectrum of the output sequence $y(m)$, obtained by decimating the sequence $v(n)$ by a factor of D, is

$$Y(\omega_y) = \frac{1}{D} \sum_{k=0}^{D-1} V\left(\frac{\omega_y - 2\pi k}{D}\right) \qquad (10.4.11)$$

where $\omega_y = D\omega_v$. Since the linear filter prevents aliasing as implied by (10.4.10), the spectrum of the output sequence given by (10.4.11) reduces to

$$Y(\omega_y) = \begin{cases} \dfrac{I}{D}X\left(\dfrac{\omega_y}{D}\right), & 0 \le |\omega_y| \le \min\left(\pi, \dfrac{\pi D}{I}\right) \\ 0, & \text{otherwise} \end{cases} \qquad (10.4.12)$$

10.5 FILTER DESIGN AND IMPLEMENTATION FOR SAMPLING-RATE CONVERSION

As indicated in the discussion above, sampling rate conversion by a factor I/D can be achieved by first increasing the sampling rate by I, accomplished by inserting $I - 1$ zeros between successive values of the input signal $x(n)$, followed by linear filtering of the resulting sequence to eliminate the unwanted images of $X(\omega)$, and finally, by downsampling the filtered signal by the factor D. In this section we consider the design and implementation of the linear filter.

10.5.1 Direct-Form FIR Filter Structures

In principle, the simplest realization of the filter is the direct-form FIR structure with system function

$$H(z) = \sum_{k=0}^{M-1} h(k) z^{-k} \tag{10.5.1}$$

where $\{h(k)\}$ is the unit sample response of the FIR filter. The lowpass filter can be designed to have linear phase, a specified passband ripple and stopband attenuation. Any of the standard, well known FIR filter design techniques (e.g., window method, frequency sampling method) can be used to carry out this design. Thus we will have the filter parameters $\{h(k)\}$, which allow us to implement the FIR filter directly as shown in Fig. 10.8.

Although the direct-form FIR filter realization illustrated in Fig. 10.8 is simple, it is also very inefficient. The inefficiency results from the fact that the upsampling process introduces $I - 1$ zeros between successive points of the input signal. If I is large, most of the signal components in the FIR filter are zero. Consequently, most of the multiplications and additions result in zeros. Furthermore, the downsampling process at the output of the filter implies that only one out of

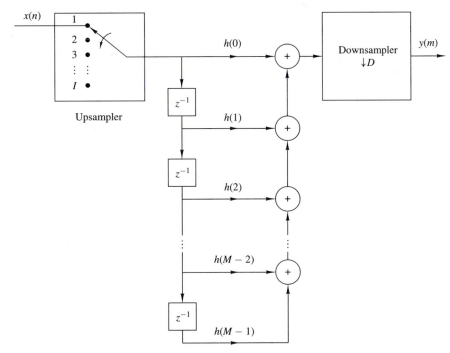

Figure 10.8 Direct-form realization of FIR filter in sampling rate conversion by factor I/D.

every D output samples is required at the output of the filter. Consequently, only one out of every D possible values at the output of the filter should be computed.

To develop a more efficient filter structure, let us begin with a decimator that reduces the sampling rate by an integer factor D. From our previous discussion, the decimator is obtained by passing the input sequence $x(n)$ through an FIR filter and then downsampling the filter output by a factor D, as illustrated in Fig. 10.9a. In this configuration, the filter is operating at the high sampling rate F_x, while only one out of every D output samples is actually needed. The logical solution to this inefficiency problem is to embed the downsampling operation within the filter, as illustrated in the filter realization given in Fig. 10.9b. In this filter structure, all the multiplications and additions are performed at the lower sampling rate F_x/D. Thus we have achieved the desired efficiency. Additional reduction in computation can be achieved by exploiting the symmetry characteristics of $\{h(k)\}$. Figure 10.10 illustrates an efficient realization of the decimator in which the FIR filter has linear phase, and hence $\{h(k)\}$ is symmetric.

Next, let us consider the efficient implementation of an interpolator, which is realized by first inserting $I - 1$ zeros between samples of $x(n)$ and then filtering the resulting sequence. The direct-form realization is illustrated in Fig. 10.11. The major problem with this structure is that the filter computations are performed at the high sampling rate IF_x. The desired simplification is achieved by first using the transposed form of the FIR filter, as illustrated in Fig. 10.12a, and then embedding the upsampler within the filter, as shown in Fig. 10.12b. Thus, all the filter multiplications are performed at the low rate F_x, while the upsampling process introduces $I - 1$ zeros in each of the filter branches of the structure shown in Fig. 10.12b. The reader can easily verify that the two filter structures in Fig. 10.12 are equivalent.

It is interesting to note that the structure of the interpolator, shown in Fig. 10.12b, can be obtained by transposing the structure of the decimator shown in Fig. 10.9. We observe that the transpose of a decimator is an interpolator, and vice versa. These relationships are illustrated in Fig. 10.13, where (b) is obtained by transposing (a) and (d) is obtained by transposing (c). Consequently, a decimator is the dual of an interpolator, and vice versa. From these relationships, it follows that there is an interpolator whose structure is the dual of the decimator shown in Fig. 10.10, which exploits the symmetry in $h(n)$.

10.5.2 Polyphase Filter Structures

The computational efficiency of the filter structure shown in Fig. 10.12 can also be achieved by reducing the large FIR filter of length M into a set of smaller filters of length $K = M/I$, where M is selected to be a multiple of I. To demonstrate this point, let us consider the interpolator given in Fig. 10.11. Since the upsampling process inserts $I - 1$ zeros between successive values of $x(n)$, only K out of the M input values stored in the FIR filter at any one time are nonzero. At one time instant, these nonzero values coincide and are multiplied by the filter coefficients $h(0), h(I), h(2I), \ldots, h(M - I)$. In the following time instant, the

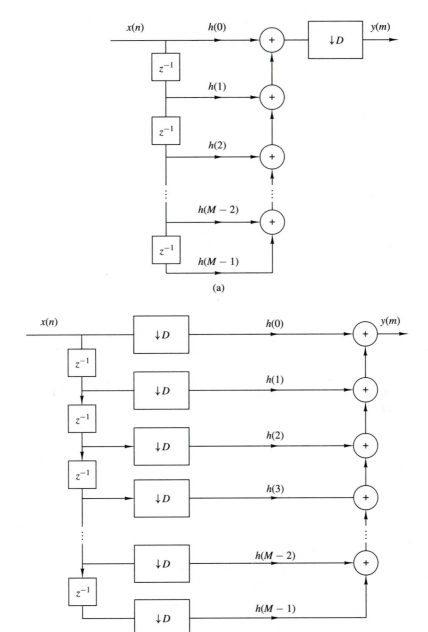

Figure 10.9 Decimation by a factor D.

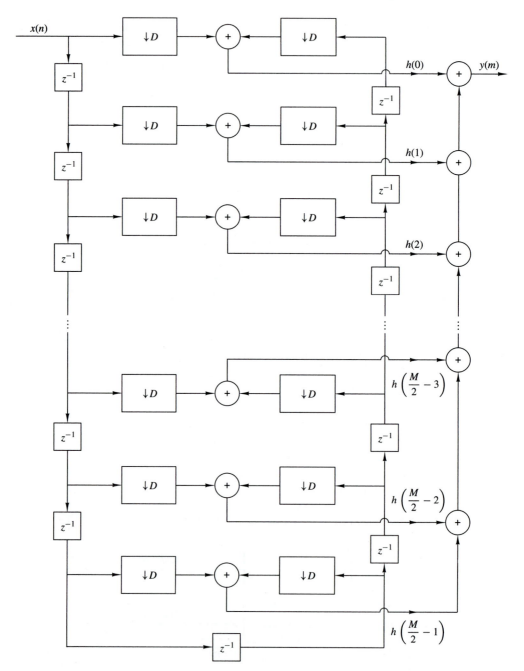

Figure 10.10 Efficient realization of a decimator that exploits the symmetry in the FIR filter.

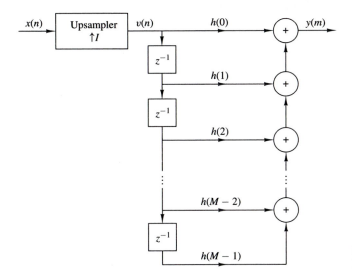

Figure 10.11 Direct-form realization of FIR filter in interpolation by a factor I.

nonzero values of the input sequence coincide and are multiplied by the filter coefficients $h(1)$, $h(I + 1)$, $h(2I + 1)$, ... $h(M - I + 1)$, and so on. This observation leads us to define a set of smaller filters, called polyphase filters, with unit sample responses

$$p_k(n) = h(k + nI) \qquad \begin{aligned} & k = 0, 1, \ldots, I - 1 \\ & n = 0, 1, \ldots, K - 1 \end{aligned} \qquad (10.5.2)$$

where $K = M/I$ is an integer.

From this discussion it follows that the set of I polyphase filters can be arranged as a parallel realization, and the output of each filter can be selected by a commutator as illustrated in Fig. 10.14. The rotation of the commutator is in the counterclockwise direction beginning with the point at $m = 0$. Thus, the polyphase filters perform the computations at the low sampling rate F_x, and the rate conversion results from the fact that I output samples are generated, one from each of the filters, for each input sample.

The decomposition of $\{h(k)\}$ into the set of I subfilters with impulse response $p_k(n)$, $k = 0, 1, \ldots, I - 1$, is consistent with our previous observation that the input signal was being filtered by a periodically time-variant linear filter with impulse response

$$g(n, m) = h(nI + (mD)_I) \qquad (10.5.3)$$

where $D = 1$ in the case of the interpolator. We noted previously that $g(n, m)$ varies periodically with period I. Consequently, a different set of coefficients are used to generate the set of I output samples $y(m)$, $m = 0, 1, \ldots, I - 1$.

Additional insight can be gained about the characteristics of the set of polyphase subfilters by noting that $p_k(n)$ is obtained from $h(n)$ by decimation with a factor I. Consequently, if the original filter frequency response $H(\omega)$ is flat over the range $0 \le |\omega| \le \omega/I$, each of the polyphase subfilters possesses a relatively flat

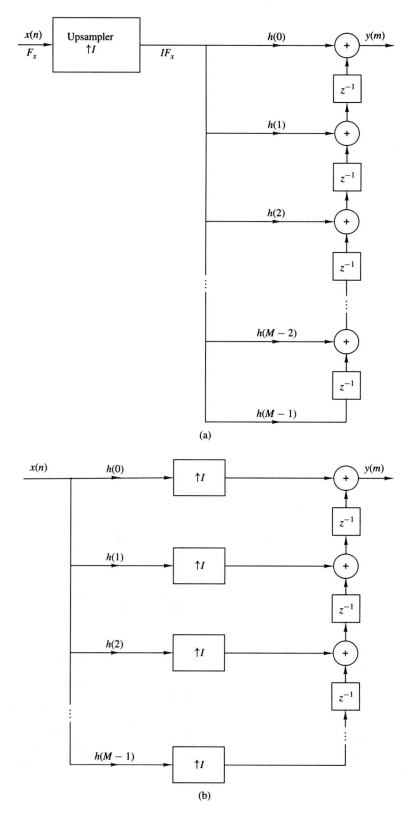

Figure 10.12 Efficient realization of an interpolator.

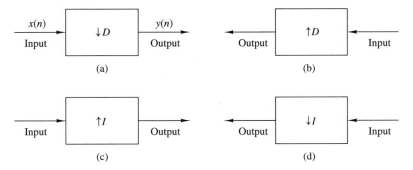

Figure 10.13   Duality relationships obtained through transposition.

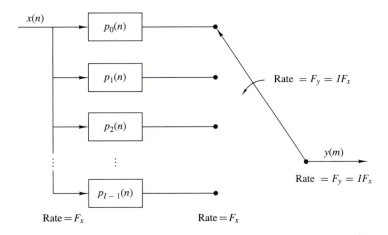

Figure 10.14   Interpolation by use of polyphase filters.

response over the range $0 \leq |\omega| \leq \pi$ (i.e., the polyphase subfilters are basically all-pass filters and differ primarily in their phase characteristics). This explains the reason for the term "polyphase" in describing these filters.

The polyphase filter can also be viewed as a set of I subfilters connected to a common delay line. Ideally, the kth subfilter will generate a forward time shift of $(k/I)T_x$, for $k = 0, 1, 2, \ldots, I - 1$, relative to the zeroth subfilter. Therefore, if the zeroth filter generates zero delay, the frequency response of the kth subfilter is

$$p_k(\omega) = e^{j\omega k/I}$$

A time shift of an integer number of input sampling intervals (e.g., lT_x) can be generated by shifting the input data in the delay line by l samples and using the same subfilters. By combining these two methods, we can generate an output that is shifted forward by an amount $(l + i/I)T_x$ relative to the previous output.

By transposing the interpolator structure in Fig. 10.14, we obtain a commutator structure for a decimator based on the parallel bank of polyphase filters, as illustrated in Fig. 10.15. The unit sample responses of the polyphase filters are

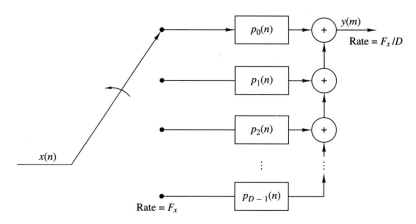

Figure 10.15 Decimation by use of polyphase filters.

now defined as

$$p_k(n) = h(k + nD) \qquad k = 0, 1, \ldots, D - 1$$

$$n = 0, 1, \ldots, K - 1 \qquad (10.5.4)$$

where $K = M/D$ is an integer when M is selected to be a multiple of D. The commutator rotates in a counterclockwise direction starting with the filter $p_0(n)$ at $m = 0$.

Although the two commutator structures for the interpolator and the decimator just described rotate in a counterclockwise direction, it is also possible to derive an equivalent pair of commutator structures having a clockwise rotation. In this alternative formulation, the sets of polyphase filters are defined to have impulse responses

$$p_k(n) = h(nI - k) \qquad k = 0, 1, \ldots, I - 1 \qquad (10.5.5)$$

$$p_k(n) = h(nD - k) \qquad k = 0, 1, \ldots, D - 1 \qquad (10.5.6)$$

for the interpolator and decimator, respectively.

10.5.3 Time-Variant Filter Structures

Having described the filter implementation for a decimator and an interpolator, let us now consider the general problem of sampling rate conversion by the factor I/D. In the general case of sampling rate conversion by a factor I/D, the filtering can be accomplished by means of the linear time-variant filter described by the response function

$$g(n, m) = h(nI - (mD)_I) \qquad (10.5.7)$$

where $h(n)$ is the impulse response of the low-pass FIR filter, which ideally, has the frequency response specified by (10.4.1). For convenience we select the length of the FIR filter $\{h(n)\}$ to a multiple of I (i.e., $M = KI$). As a consequence, the set of coefficients $\{g(n, m)\}$ for each $m = 0, 1, 2, \ldots, I - 1$, contains K elements.

Since $g(n, m)$ is also periodic with period I, as demonstrated in (10.4.9), it follows that the output $y(m)$ can be expressed as

$$y(m) = \sum_{n=0}^{K-1} g\left(n, m - \left\lfloor \frac{m}{I} \right\rfloor I\right) x\left(\left\lfloor \frac{mD}{I} \right\rfloor - n\right) \tag{10.5.8}$$

Conceptually, we can think of performing the computations specified by (10.5.8) by processing blocks of data of length K by a set of K filter coefficients $g(n, m - \lfloor m/I \rfloor I)$, $n = 0, 1, \ldots, K - 1$. There are I such sets of coefficients, one set for each block of I output points of $y(m)$. For each block of I output points, there is a corresponding block of D input points of $x(n)$ that enter in the computation.

The block processing algorithm for computing (10.5.8) can be visualized as illustrated in Fig. 10.16. A block of D input samples is buffered and shifted into a second buffer of length K, one sample at a time. The shifting from the input buffer to the second buffer occurs at a rate of one sample each time the quantity $\lfloor mD/I \rfloor$ increases by one. For each output sample $y(l)$, the samples from the second buffer are multiplied by the corresponding set of filter coefficients $g(n, l)$ for $n = 0, 1, \ldots, K - 1$, and the K products are accumulated to give $y(l)$, for $l = 0$,

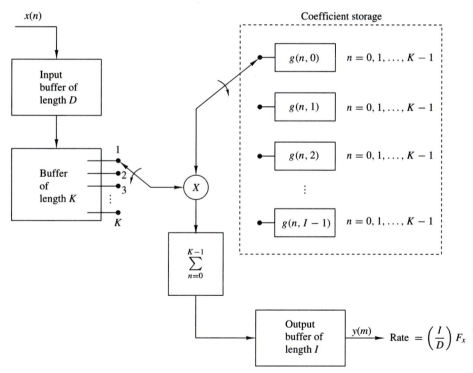

Figure 10.16 Efficient implementation of sampling-rate conversion by block processing.

$1, \ldots, I - 1$. Thus this computation produces I outputs. It is then repeated for a new set of D input samples, and so on.

An alternative method for computing the output of the sample rate converter, specified by (10.5.8), is by means of an FIR filter structure with periodically varying filter coefficients. Such a structure is illustrated in Fig. 10.17. The input samples $x(n)$ are passed into a shift register that operates at the sampling rate F_x and is of length $K = M/I$, where M is the length of the time-invariant FIR filter specified by the frequency response given by (10.4.1). Each stage of the register is connected to a hold-and-sample device that serves to couple the input sample rate F_x to the output sample rate $F_y = (I/D)F_x$. The sample at the input to each hold-and-sample device is held until the next input sample arrives and then is discarded. The output samples of the hold-and-sample device are taken at times mD/I, $m = 0, 1, 2, \ldots$. When both the input and output sampling times coincide (i.e., when mD/I is an integer), the input to the hold-and-sample is changed first and then the output samples the new input. The K outputs from the K hold-and-sample devices are multiplied by the periodically time-varying coefficients $g(n, m - \lfloor m/I \rfloor I)$, for $n = 0$, $1, \ldots, K - 1$, and the resulting products are summed to yield $y(m)$. The computations at the output of the hold-and-sample devices are repeated at the output sampling rate of $F_y = (I/D)F_x$.

Finally, rate conversion by a rational factor I/D can also be performed by use of a polyphase filter having I subfilters. If we assume that the mth sample

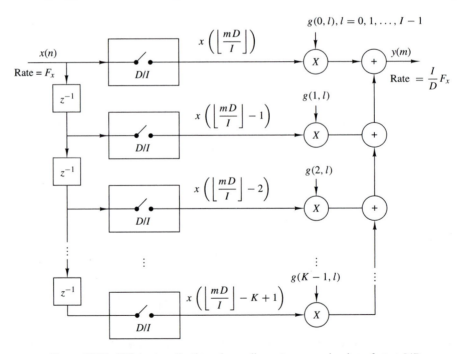

Figure 10.17 Efficient realization of sampling-rate conversion by a factor I/D.

$y(m)$ is computed by taking the output of the i_mth subfilter with input data $x(n)$, $x(n-1), \ldots, x(n-K+1)$, in the delay line, the next sample $y(m+1)$ is taken from the (i_{m+1})st subfilter after shifting l_{m+1} new samples in the delay lines where $i_{m+1} = (i_m + D)_{\text{mod } I}$ and l_{m+1} is the integer part of $(i_m + D)/I$. The integer i_{m+1} should be saved to be used in determining the subfilter from which the next sample is taken.

Let us now demonstrate the filter design procedure, first in the design of a decimator, second in the design of an interpolator, and finally, in the design of a rational sample-rate converter.

Example 10.5.1

Design a decimator that downsamples an input signal $x(n)$ by a factor $D = 2$. Use the Remez algorithm to determine the coefficients of the FIR filter that has a 0.1-dB ripple in the passband and is down by at least 30 dB in the stopband. Also determine the polyphase filter structure in a decimator realization that employs polyphase filters.

Solution   A filter of length $M = 30$ achieves the design specifications given above. The impulse response of the FIR filter is given in Table 10.1 and the frequency response is illustrated in Fig. 10.18. Note that the cutoff frequency is $\omega_c = \pi/2$.

The polyphase filters obtained from $h(n)$ have impulse responses

$$p_k(n) = h(2n + k) \qquad k = 0, 1; \qquad n = 0, 1, \ldots, 14$$

Note that $p_0(n) = h(2n)$ and $p_1(n) = h(2n+1)$. Hence one filter consists of the even-numbered samples of $h(n)$ and the other filter consists of the odd-numbered samples of $h(n)$.

Example 10.5.2

Design an interpolator that increases the input sampling rate by a factor of $I = 5$. Use the Remez algorithm to determine the coefficients of the FIR filter that has 0.1-dB ripple in the passband and is down by at least 30 dB in the stopband. Also, determine the polyphase filter structure in an interpolator realization based on polyphase filters.

Solution   A filter of length $M = 30$ achieves the design specifications given above. The frequency response of the FIR filter is illustrated in Fig. 10.19 and its coefficients are given in Table 10.2. The cutoff frequency is $\omega_c = \pi/5$.

The polyphase filters obtained from $h(n)$ have impulse responses

$$p_k(n) = h(5n + k) \qquad k = 0, 1, 2, 3, 4$$

Consequently, each filter has length 6.

Example 10.5.3

Design a sample-rate converter that increases the sampling rate by a factor 2.5. Use the Remez algorithm to determine the coefficients of the FIR filter that has 0.1-dB ripple in the passband and is down by at least 30 dB in the stopband. Specify the sets of time-varying coefficients $g(n, m)$ used in the realization of the sampling-rate converter according to the structure in Fig. 10.17.

Solution   The FIR filter that meets the specifications of this problem is exactly the same as the filter designed in Example 10.5.2. Its bandwidth is $\pi/5$.

TABLE 10.1 COEFFICIENTS OF LINEAR-PHASE FIR FILTER IN EXAMPLE 10.5.1

```
                  FINITE IMPULSE RESPONSE (FIR)
                LINEAR-PHASE DIGITAL FILTER DESIGN
                    REMEZ EXCHANGE ALGORITHM


                     FILTER LENGTH = 30

               *****  IMPULSE RESPONSE  *****
          H( 1)  =   0.60256165E-02  =  H( 30)
          H( 2)  =  -0.12817143E-01  =  H( 29)
          H( 3)  =  -0.28582066E-02  =  H( 28)
          H( 4)  =   0.13663346E-01  =  H( 27)
          H( 5)  =  -0.46688961E-02  =  H( 26)
          H( 6)  =  -0.19704415E-01  =  H( 25)
          H( 7)  =   0.15984623E-01  =  H( 24)
          H( 8)  =   0.21384886E-01  =  H( 23)
          H( 9)  =  -0.34979440E-01  =  H( 22)
          H(10)  =  -0.15615522E-01  =  H( 21)
          H(11)  =   0.64006113E-01  =  H( 20)
          H(12)  =  -0.73451772E-02  =  H( 19)
          H(13)  =  -0.11873185E+00  =  H( 18)
          H(14)  =   0.98047845E-01  =  H( 17)
          H(15)  =   0.49225068E+00  =  H( 16)
```

	BAND 1	BAND 2
LOWER BAND EDGE	0.0000000	0.3100000
UPPER BAND EDGE	0.2500000	0.5000000
DESIRED VALUE	1.0000000	0.0000000
WEIGHTING	2.0000000	1.0000000
DEVIATION	0.0107151	0.0214302
DEVIATION IN DB	0.0925753	-33.3794746

```
EXTREMAL FREQUENCIES-MAXIMA OF THE ERROR CURVE
0.0000000    0.0416667    0.0791667    0.1166666    0.1520833
0.1854166    0.2145832    0.2395832    0.2500000    0.3100000
0.3225000    0.3495833    0.3808333    0.4141666    0.4474999
0.4829165
```

The coefficients of the filter are given by (10.4.8) as

$$g(n, m) = h(nI + (mD)_I)$$

$$= h\left(nI + mD - \left\lfloor \frac{mD}{I} \right\rfloor I \right)$$

By substituting $I = 5$ and $D = 2$, we obtain

$$g(n, m) = h\left(5n + 2m - 5 \left\lfloor \frac{2m}{5} \right\rfloor \right)$$

By evaluating $g(n, m)$ for $n = 0, 1, \ldots, 5$ and $m = 0, 1, \ldots, 4$ we obtain the following

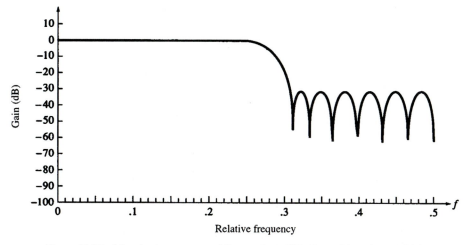

Figure 10.18 Magnitude response of linear-phase FIR filter of length $M = 30$ in Example 10.5.1.

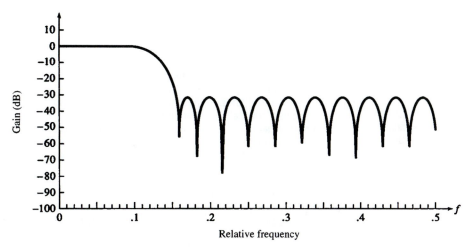

Figure 10.19 Magnitude response of linear-phase FIR filter of length $M = 30$ in Example 10.5.2.

coefficients for the time-variant filter:

$$g(0, m) = \{h(0) \quad h(2) \quad h(4) \quad h(1) \quad h(3)\}$$
$$g(1, m) = \{h(5) \quad h(7) \quad h(9) \quad h(6) \quad h(8)\}$$
$$g(2, m) = \{h(10) \quad h(12) \quad h(14) \quad h(11) \quad h(13)\}$$
$$g(3, m) = \{h(15) \quad h(17) \quad h(19) \quad h(16) \quad h(18)\}$$
$$g(4, m) = \{h(20) \quad h(22) \quad h(24) \quad h(21) \quad h(23)\}$$
$$g(5, m) = \{h(25) \quad h(27) \quad h(29) \quad h(26) \quad h(28)\}$$

A polyphase filter implementation would employ five subfilters, each of length six. To decimate the output of the polyphase filters by a factor of $D = 2$ simply means

TABLE 10.2 COEFFICIENTS OF LINEAR-PHASE FIR FILTER IN
EXAMPLE 10.5.2

```
            FINITE IMPULSE RESPONSE (FIR)
          LINEAR-PHASE DIGITAL FILTER DESIGN
               REMEZ EXCHANGE ALGORITHM

                 FILTER LENGTH = 30

            ***** IMPULSE RESPONSE *****
          H( 1) =   0.63987216E-02 = H( 30)
          H( 2) =  -0.14761304E-01 = H( 29)
          H( 3) =  -0.10886577E-02 = H( 28)
          H( 4) =  -0.28714957E-02 = H( 27)
          H( 5) =   0.10486430E-01 = H( 26)
          H( 6) =   0.21477142E-01 = H( 25)
          H( 7) =   0.19479362E-01 = H( 24)
          H( 8) =  -0.31067431E-03 = H( 23)
          H( 9) =  -0.30053033E-01 = H( 22)
          H(10) =  -0.49877029E-01 = H( 21)
          H(11) =  -0.37371285E-01 = H( 20)
          H(12) =   0.18482896E-01 = H( 19)
          H(13) =   0.10747141E+00 = H( 18)
          H(14) =   0.19951098E+00 = H( 17)
          H(15) =   0.25794828E+00 = H( 16)
```

	BAND 1	BAND 2
LOWER BAND EDGE	0.0000000	0.1600000
UPPER BAND EDGE	0.1000000	0.5000000
DESIRED VALUE	1.0000000	0.0000000
WEIGHTING	3.0000000	1.0000000
DEVIATION	0.0097524	0.0292572
DEVIATION IN DB	0.0842978	-30.6753349

```
EXTREMAL FREQUENCIES-MAXIMA OF THE ERROR CURVE
0.0000000   0.0333333   0.0645834   0.0895833   0.1000000
0.1600000   0.1745833   0.2016666   0.2370833   0.2704166
0.3058332   0.3412498   0.3766665   0.4120831   0.4474997
0.4829164
```

that we take every other output from the polyphase filters. Thus the first output $y(0)$ is taken from $p_0(n)$, the second output $y(1)$ is taken from $p_2(n)$, the third from $p_4(n)$, the fourth from $p_1(n)$, the fifth from $p_3(n)$, and so on.

10.6 MULTISTAGE IMPLEMENTATION OF SAMPLING-RATE CONVERSION

In practical applications of sampling-rate conversion we often encounter decimation factors and interpolation factors that are much larger than unity. For example, suppose that we are given the task of altering the sampling rate by the factor

$I/D = 130/63$. Although, in theory, this rate alteration can be achieved exactly, the implementation would require a bank of 130 polyphase filters and may be computationally inefficient. In this section we consider methods for performing sampling-rate conversion for either $D \gg 1$ and/or $I \gg 1$ in multiple stages.

First, let us consider interpolation by a factor $I \gg 1$ and let us assume that I can be factored into a product of positive integers as

$$I = \prod_{i=1}^{L} I_i \tag{10.6.1}$$

Then, interpolation by a factor I can be accomplished by cascading L stages of interpolation and filtering, as shown in Fig. 10.20. Note that the filter in each of the interpolators eliminates the images introduced by the upsampling process in the corresponding interpolator.

In a similar manner, decimation by a factor D, where D may be factored into a product of positive integers as

$$D = \prod_{i=1}^{J} D_i \tag{10.6.2}$$

can be implemented as a cascade of J stages of filtering and decimation as illustrated in Fig. 10.21. Thus the sampling rate at the output of the ith stage is

$$F_i = \frac{F_{i-1}}{D_i} \qquad i = 1, 2, \ldots, J \tag{10.6.3}$$

where the input rate for the sequence $\{x(n)\}$ is $F_0 = F_x$.

To ensure that no aliasing occurs in the overall decimation process, we can design each filter stage to avoid aliasing within the frequency band of interest. To

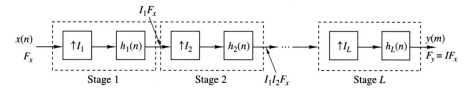

Figure 10.20 Multistage implementation of interpolation by a factor I.

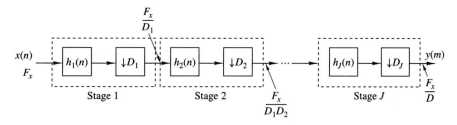

Figure 10.21 Multistage implementation of decimation by a factor D.

elaborate, let us define the desired passband and the transition band in the overall decimator as

$$\text{Passband: } 0 \le F \le F_{\text{pc}}$$

$$\text{Transition band: } F_{\text{pc}} \le F \le F_{\text{sc}}$$

(10.6.4)

where $F_{\text{sc}} \le F_x/2D$. Then, aliasing in the band $0 \le F \le F_{\text{sc}}$ is avoided by selecting the frequency bands of each filter stage as follows:

$$\text{Passband: } 0 \le F \le F_{\text{pc}}$$

$$\text{Transition band: } F_{\text{pc}} \le F \le F_i - F_{\text{sc}}$$

$$\text{Stopband: } F_i - F_{\text{sc}} \le F \le \frac{F_{i-1}}{2}$$

(10.6.5)

For example, in the first filter stage we have $F_1 = F_x/D_1$, and the filter is designed to have the following frequency bands:

$$\text{Passband: } 0 \le F \le F_{\text{pc}}$$

$$\text{Transition band: } F_{\text{pc}} \le F \le F_1 - F_{\text{sc}}$$

$$\text{Stopband: } F_1 - F_{\text{sc}} \le F \le \frac{F_0}{2}$$

(10.6.6)

After decimation by D_1, there is aliasing from the signal components that fall in the filter transition band, but the aliasing occurs at frequencies above F_{sc}. Thus there is no aliasing in the frequency band $0 \le F \le F_{\text{sc}}$. By designing the filters in the subsequent stages to satisfy the specifications given in (10.6.5), we ensure that no aliasing occurs in the primary frequency band $0 \le F \le F_{\text{sc}}$.

Example 10.6.1

Consider an audio-band signal with a nominal bandwidth of 4 kHz that has been sampled at a rate of 8 kHz. Suppose that we wish to isolate the frequency components below 80 Hz with a filter that has a passband $0 \le F \le 75$ and a transition band $75 \le F \le 80$. Hence $F_{\text{pc}} = 75$ Hz and $F_{\text{sc}} = 80$. The signal in the band $0 \le F \le 80$ may be decimated by the factor $D = F_x/2F_{\text{sc}} = 50$. We also specify that the filter have a passband ripple $\delta_1 = 10^{-2}$ and a stopband ripple of $\delta_2 = 10^{-4}$.

The length of the linear phase FIR filter required to satisfy these specifications can be estimated from one of the well known formulas given in the literature. Recall that a particularly simple formula for approximating the length M, attributed to Kaiser, is

$$\hat{M} = \frac{-10\log_{10}\delta_1\delta_2 - 13}{14.6\Delta f} + 1$$

(10.6.7)

where Δf is the normalized (by the sampling rate) width of the transition region [i.e., $\Delta f = (F_{\text{sc}} - F_{\text{pc}})/F_s$]. A more accurate formula proposed by Herrmann et al. (1973) is

$$\hat{M} = \frac{D_\infty(\delta_1, \delta_2) - f(\delta_1, \delta_2)(\Delta f)^2}{\Delta f} + 1$$

(10.6.8)

where $D_\infty(\delta_1, \delta_2)$ and $f(\delta_1, \delta_2)$ are defined as

$$D_\infty(\delta_1, \delta_2) = [0.005309(\log_{10}\delta_1)^2 + 0.07114(\log_{10}\delta_1)$$
$$-0.4761]\log_{10}\delta_2$$
$$-[0.00266(\log_{10}\delta_1)^2 + 0.5941\log_{10}\delta_1 + 0.4278] \qquad (10.6.9)$$
$$f(\delta_1, \delta_2) = 11.012 + 0.51244[\log_{10}\delta_1 - \log_{10}\delta_2] \qquad (10.6.10)$$

Now a single FIR filter followed by a decimator would require (using the Kaiser formula) a filter of (approximate) length

$$\hat{M} = \frac{-10\log_{10}10^{-6} - 13}{14.6(5/8000)} + 1 \approx 5152$$

As an alternative, let us consider a two-stage decimation process with $D_1 = 25$ and $D_2 = 2$. In the first stage we have the specifications $F_1 = 320$ Hz and

> Passband: $0 \leq F \leq 75$
> Transition band: $75 < F \leq 240$
> $$\Delta f = \frac{165}{8000}$$
> $$\delta_{11} = \frac{\delta_1}{2} \qquad d_{21} = \delta_2$$

Note that we have reduced the passband ripple δ_1 by a factor of 2, so that the total passband ripple in the cascade of the two filters does not exceed δ_1. On the other hand, the stopband ripple is maintained at δ_2 in both stages. Now the Kaiser formula yields an estimate of M_1 as

$$\hat{M}_1 = \frac{-10\log_{10}\delta_{11}\delta_{21} - 13}{14.6\Delta f} + 1 \approx 167$$

For the second stage, we have $F_2 = F_1/2 = 160$ and the specifications

> Passband: $0 \leq F \leq 75$
> Transition band: $75 < F \leq 80$
> $$\Delta f = \frac{5}{320}$$
> $$\delta_{12} = \frac{\delta_1}{2} \qquad \delta_{22} = \delta_2$$

Hence the estimate of the length M_2 of the second filter is

$$\hat{M}_2 \approx 220$$

Therefore, the total length of the two FIR filters is approximately $\hat{M}_1 + \hat{M}_2 = 387$. This represents a reduction in the filter length by a factor of more than 13.

The reader is encouraged to repeat the computation above with $D_1 = 10$ and $D_2 = 5$.

It is apparent from the computations in Example 10.6.1 that the reduction in the filter length results from increasing the factor Δf, which appears in the denominator in (10.6.7) and (10.6.8). By decimating in multiple stages, we are able to increase the width of the transition region through a reduction in the sampling rate.

In the case of a multistage interpolator, the sampling rate at the output of the ith stage is

$$F_{i-1} = I_i F_i \qquad i = J, J - 1, \ldots, 1$$

and the output rate is $F_0 = I F_J$ when the input sampling rate is F_J. The corresponding frequency band specifications are

$$\text{Passband: } 0 \leq F \leq F_p$$

$$\text{Transition band: } F_p < F \leq F_i - F_{sc}$$

The following example illustrates the advantages of multistage interpolation.

Example 10.6.2

Let us reverse the filtering problem described in Example 10.6.1 by beginning with a signal having a passband $0 \leq F \leq 75$ and a transition band of $75 \leq F \leq 80$. We wish to interpolate by a factor of 50. By selecting $I_1 = 2$ and $I_2 = 25$, we have basically a transposed form of the decimation problem considered in Example 10.6.1. Thus we can simply transpose the two-stage decimator to achieve the two-stage interpolator with $I_1 = 2$, $I_2 = 25$, $\hat{M}_1 \approx 220$, and $\hat{M}_2 \approx 167$.

10.7 SAMPLING-RATE CONVERSION OF BANDPASS SIGNALS

A bandpass signal is a signal with frequency content concentrated in a narrow band of frequencies above zero frequency. The center frequency F_c of the signal is generally much larger than the bandwidth B (i.e., $F_c \gg B$). Bandpass signals arise frequently in practice, most notably in communications, where information bearing signals such as speech and video are translated in frequency and then transmitted over such channels as wire lines, microwave radio, and satellites.

In this section we consider the decimation and interpolation of bandpass signals. We begin by noting that any bandpass signal has an equivalent lowpass representation, obtained by a simple frequency translation of the bandpass signal. For example, the bandpass signal with spectrum $X(F)$ shown in Fig. 10.22a can be translated to lowpass by means of a frequency translation of F_c, where F_c is an appropriate choice of frequency (usually, the center frequency) within the bandwidth occupied by the bandpass signal. Thus we obtain the equivalent lowpass signal as illustrated in Fig. 10.22b.

From Section 9.1 we recall that an analog bandpass signal can be represented as

$$
\begin{aligned}
x(t) &= A(t) \cos[2\pi F_c t + \theta(t)] \\
&= A(t) \cos \theta(t) \cos 2\pi F_c t - A(t) \sin \theta(t) \sin 2\pi F_c t \\
&= u_c(t) \cos 2\pi F_c t - u_s(t) \sin 2\pi F_c t \\
&= \text{Re}[x_l(t) e^{j2\pi F_c t}]
\end{aligned}
\tag{10.7.1}
$$

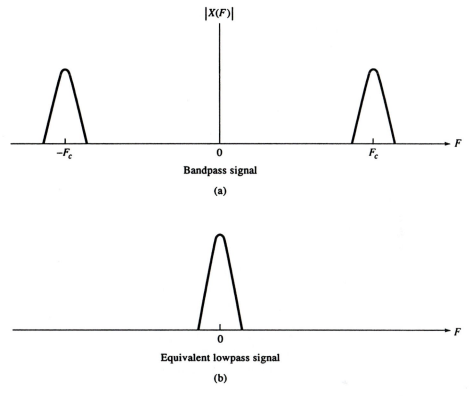

Figure 10.22   Bandpass signal and its equivalent lowpass representation.

where, by definition,

$$u_c(t) = A(t) \cos \theta(t) \tag{10.7.2}$$

$$u_s(t) = A(t) \sin \theta(t) \tag{10.7.3}$$

$$x_l(t) = u_c(t) + ju_s(t) \tag{10.7.4}$$

$A(t)$ is called the *amplitude* or *envelope* of the signal, $\theta(t)$ is the phase, and $u_c(t)$ and $u_s(t)$ are called the *quadrature components* of the signal.

Physically, the translation of $x(t)$ to lowpass involves multiplying (mixing) $x(t)$ by the quadrature carriers $\cos 2\pi F_c t$ and $\sin 2\pi F_c t$ and then lowpass filtering the two products to eliminate the frequency components generated around the frequency $2F_c$ (the double frequency terms). Thus all the information content contained in the bandpass signal is preserved in the lowpass signal, and hence the latter is equivalent to the former. This fact is obvious from the spectral representation of the bandpass signal, which can be written as

$$X(F) = \tfrac{1}{2}[X_l(F - F_c) + X_l^*(-F - F_c)] \tag{10.7.5}$$

where $X_l(f)$ is the Fourier transform of the equivalent lowpass signal $x_l(t)$ and $X(F)$ is the Fourier transform of $x(t)$.

It was shown in Section 9.1 that a bandpass signal of bandwidth B can be uniquely represented by samples taken at a rate of $2B$ samples per second, provided that the upper band (highest) frequency is a multiple of the signal bandwidth B. On the other hand, if the upper band frequency is not a multiple of B, the sampling rate must be increased by a small amount to avoid aliasing. In any case, the sampling rate for the bandpass signal is bounded from above and below as

$$2B \leq F_s \leq 4B \tag{10.7.6}$$

The representation of discrete-time bandpass signals is basically the same as that for analog signals given by (10.7.1) with the substitution of $t = nT$, where T is the sampling interval.

10.7.1 Decimation and Interpolation by Frequency Conversion

The mathematical equivalence between the bandpass signal $x(t)$ and its equivalent lowpass representation $x_l(t)$ provides one method for altering the sampling rate of the signal. Specifically, we can take the bandpass signal which has been sampled at rate F_x, convert it to lowpass through the frequency conversion process illustrated in Fig. 10.23, and perform the sampling-rate conversion on the lowpass signal using the methods described previously. The lowpass filters for obtaining the two quadrature components can be designed to have linear phase within the

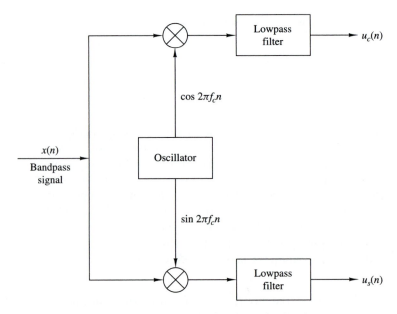

Figure 10.23 Conversion of a bandpass signal to lowpass.

bandwidth of the signal and to approximate the ideal frequency response characteristic

$$H(\omega) = \begin{cases} 1, & |\omega| \le \omega_B/2 \\ 0, & \text{otherwise} \end{cases} \tag{10.7.7}$$

where ω_B is the bandwidth of the discrete-time bandpass signal ($\omega_B \le \pi$).

If decimation is to be performed by an integer factor D, the antialiasing filter preceding the decimator can be combined with the lowpass filter used for frequency conversion into a single filter that approximates the ideal frequency response

$$H_D(\omega) = \begin{cases} 1, & |\omega| \le \omega_D/D \\ 0, & \text{otherwise} \end{cases} \tag{10.7.8}$$

where ω_D is any desired frequency in the range $0 \le \omega_D \le \pi$. For example, we may select $\omega_D = \omega_B/2$ if we are interested only in the frequency range $0 \le \omega \le \omega_B/2D$ of the original signal.

If interpolation is to be performed by an integer factor I on the frequency-translated signal, the filter used to reject the images in the spectrum should be designed to approximate the lowpass filter characteristic

$$H_I(\omega) = \begin{cases} I, & |\omega| \le \omega_B/2I \\ 0, & \text{otherwise} \end{cases} \tag{10.7.9}$$

We note that in the case of interpolation, the lowpass filter normally used to reject the double-frequency components is redundant and may be omitted. Its function is essentially served by the image rejection filter $H_I(\omega)$.

Finally, we indicate that sampling-rate conversion by any rational factor I/D can be accomplished on the bandpass signal as illustrated in Fig. 10.24. Again, the lowpass filter for rejecting the double-frequency components generated in the frequency-conversion process can be omitted. Its function is simply served by the image-rejection/antialiasing filter following the interpolator, which is designed to approximate the ideal frequency response characteristic:

$$H(\omega) = \begin{cases} I, & 0 \le |\omega| \le \min(\omega_B/2D, \omega_B/2I) \\ 0, & \text{otherwise} \end{cases} \tag{10.7.10}$$

Once the sampling rate of the quadrature signal components has been altered by either decimation or interpolation or both, a bandpass signal can be regenerated by amplitude modulating the quadrature carriers $\cos \omega_c n$ and $\sin \omega_c n$ by the corresponding signal components and then adding the two signals. The center

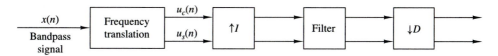

Figure 10.24 Sampling rate conversion of a bandpass signal.

frequency ω_c is any desirable frequency in the range

$$\min(\omega_B/2D, \omega_B/2I) \leq \omega_c \leq \pi \qquad (10.7.11)$$

10.7.2 Modulation-Free Method for Decimation and Interpolation

By restricting the frequency range for the signal whose frequency is to be altered, it is possible to avoid the carrier modulation process and to achieve frequency translation directly. In this case we exploit the frequency translation property inherent in the process of decimation and interpolation.

To be specific, let us consider the decimation of the sampled bandpass signal whose spectrum is shown in Fig. 10.25. Note that the signal spectrum is confined to the frequency range

$$\frac{m\pi}{D} \leq \omega \leq \frac{(m+1)\pi}{D} \qquad (10.7.12)$$

where m is a positive integer. A bandpass filter would normally be used to eliminate signal frequency components outside the desired frequency range. Then direct decimation of the bandpass signal by the factor D results in the spectrum shown in Fig. 10.26a, for m odd, and Fig. 10.26b for m even. In the case where m is odd, there is an inversion of the spectrum of the signal. This inversion can be undone by multiplying each sample of the decimated signal by $(-1)^n$, $n = 0$, $1, \ldots$. Note that violation of the bandwidth constraint given by (10.7.12) results in signal aliasing.

Modulation-free interpolation of a bandpass signal by an integer factor I can be accomplished in a similar manner. The process of upsampling by inserting zeros between samples of $x(n)$ produces I images in the band $0 \leq \omega \leq \pi$. The desired image can be selected by bandpass filtering. Note that the process of interpolation also provides us with the opportunity to achieve frequency translation of the spectrum.

Finally, modulation-free sampling rate conversion for a bandpass signal by a rational factor I/D can be accomplished by cascading a decimator with an interpolator in a manner that depends on the choice of the parameters D and I. A bandpass filter preceding the sampling converter is usually required to isolate the signal frequency band of interest. Note that this approach provides us with a modulation-free method for achieving frequency translation of a signal by selecting $D = I$.

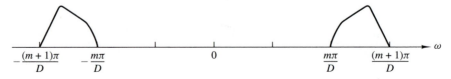

Figure 10.25   Spectrum of a bandpass signal.

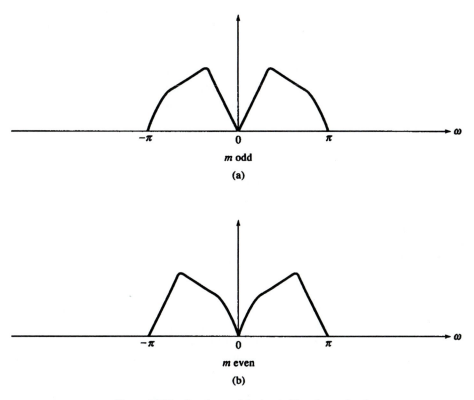

Figure 10.26   Spectrum of decimated bandpass signal.

10.8 SAMPLING-RATE CONVERSION BY AN ARBITRARY FACTOR

In the previous sections of this chapter, we have shown how to perform sampling rate conversion exactly by a rational number I/D. In some applications, it is either inefficient or, sometimes impossible to use such an exact rate conversion scheme. We first consider the following two cases.

Case 1.   We need to perform rate conversion by the rational number I/D, where I is a large integer (e.g., $I/D = 1023/511$). Although we can achieve exact rate conversion by this number, we would need a polyphase filter with 1023 subfilters. Such an exact implementation is obviously inefficient in memory usage because we need to store a large number of filter coefficients.

Case 2.   In some applications, the exact conversion rate is not known when we design the rate converter, or the rate is continuously changing during the conversion process. For example, we may encounter the situation where the input and output samples are controlled by two independent clocks. Even though it is still possible to define a nominal conversion rate that is a rational number, the actual rate would be slightly different, depending on the frequency difference between the two clocks. Obviously, it is not possible to design an exact rate converter in this case.

To implement sampling rate conversion for applications similar to these cases, we resort to nonexact rate conversion schemes. Unavoidably, a nonexact scheme will introduce some distortion in the converted output signal. (It should be noted that distortion exists even in an exact rational rate converter because the polyphase filter is never ideal.) Such a converter will be adequate, as long as the total distortion does not exceed the specification required in the application.

Depending on the application requirements and implementation constraints, we can use first-order, second-order, or higher-order approximations. We shall describe first-order and second-order approximation methods and provide an analysis of the resulting timing errors.

10.8.1 First-Order Approximation

Let us denote the arbitrary conversion rate by r and suppose that the input to the rate converter is the sequence $\{x(n)\}$. We need to generate a sequence of output samples separated in time by T_x/r, where T_x is the sample interval for $\{x(n)\}$. By constructing a polyphase filter with a large number of subfilters as just described, we can approximate such a sequence with a nonuniformly spaced sequence. Without loss of generality, we can express $1/r$ as

$$\frac{1}{r} = \frac{k}{I} + \beta$$

where k and I are positive integers and β is a number in the range

$$0 < \beta < \frac{1}{I}$$

Consequently, $1/r$ is bounded from above and below as

$$\frac{k}{I} < \frac{1}{r} < \frac{k+1}{I}$$

I corresponds to the interpolation factor, which will be determined to satisfy the specification on the amount of tolerable distortion introduced by rate conversion. I is also equal to the number of polyphase filters.

For example, suppose that $r = 2.2$ and that we have determined, as we will demonstrate, that $I = 6$ polyphase filters are required to meet the distortion specification. Then

$$\frac{k}{I} \equiv \frac{2}{6} < \frac{1}{r} < \frac{3}{6} \equiv \frac{k+1}{I}$$

so that $k = 2$. The time spacing between samples of the interpolated sequence is T_x/I. However, the desired conversion rate $r = 2.2$ for $I = 6$ corresponds to a decimation factor of 2.727, which falls between $k = 2$ and $k = 3$. In the first-order approximation, we achieve the desired decimation rate by selecting the output

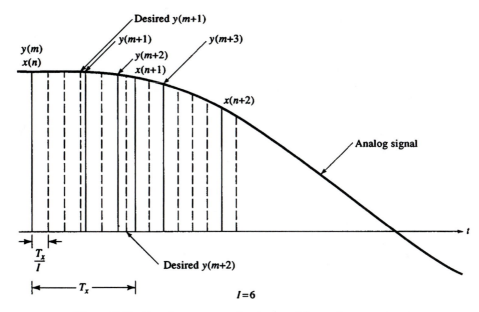

Figure 10.27   Sample rate conversion by use of first-order approximation.

sample from the polyphase filter closest in time to the desired sampling time. This is illustrated in Fig. 10.27 for $I = 6$.

In general, to perform rate conversion by a factor r, we employ a polyphase filter to perform interpolation and therefore to increase the frequency of the original sequence of a factor of I. The time spacing between the samples of the interpolated sequence is equal to T_x/I. If the ideal sampling time of the mth sample, $y(m)$, of the desired output sequence is between the sampling times of two samples of the interpolated sequence, we select the sample closer to $y(m)$ as its approximation.

Let us assume that the mth selected sample is generated by the (i_m)th subfilter using the input samples $x(n)$, $x(n-1)$, ..., $x(n-K+1)$ in the delay line. The normalized sampling time error (i.e., the time difference between the selected sampling time and the desired sampling time normalized by T_x) is denoted by t_m. The sign of t_m is positive if the desired sampling time leads the selected sampling time, and negative otherwise. It is easy to show that $|t_m| \leq 0.5/I$. The normalized time advance from the mth output $y(m)$ to the $(m+1)$st output $y(m+1)$ is equal to $(1/r) + t_m$.

To compute the next output, we first determine a number closest to $i_m/I + 1/r + t_m + k_m/I$ that is of the form $l_{m-1} + i_{m+1}/I$, where both l_{m+1} and i_{m+1} are integers and $i_{m+1} < I$. Then, the $(m+1)$st output $y(m+1)$ is computed using the (i_{m+1})th subfilter after shifting the signal in the delay line by l_{m+1} input samples. The normalized timing error for the $(m+1)$th sample is $t_{m+1} = (i_m/I + 1/r + t_m) - (l_{m+1} + i_{m+1}/I)$. It is saved for the computation of the next output sample.

By increasing the number of subfilters used, we can arbitrarily increase the conversion accuracy. However, we also require more memory to store the large number of filter coefficients. Hence it is desirable to use as few subfilters as possible while keeping the distortion in the converted signal below the specification. The distortion introduced due to the sampling-time approximation is most conveniently evaluated in the frequency domain.

Suppose that the input data sequence $\{x(n)\}$ has a flat spectrum from $-\omega_x$ to ω_x, where $\omega_x < \pi$, with a magnitude A. Its total power can be computed using Parseval's theorem, namely,

$$P_s = \frac{1}{2\pi} \int_{-\omega_x}^{\omega_x} |X(\omega)|^2 d\omega = \frac{A^2 \omega_x}{\pi} \tag{10.8.1}$$

From this discussion given, we know that for each output $y(m)$, the time difference between the desired filter and the filter actually used is t_m, where $|t_m| \leq 0.5/I$. Hence the frequency response of these filters can be written as $e^{j\omega\tau}$ and $e^{j\omega(\tau-t_m)}$, respectively. When I is large, ωt_m is small. By ignoring high-order errors, we can write the difference between the frequency responses as

$$e^{j\omega\tau} - e^{j\omega(\tau-t_m)} = e^{j\omega\tau}(1 - e^{-j\omega t_m})$$
$$= e^{j\omega\tau}(1 - \cos \omega t_m + j \sin \omega t_m) \approx j e^{j\omega\tau} \omega t_m \tag{10.8.2}$$

By using the bound $|t_m| \leq 0.5/I$, we obtain an upper bound for the total error power as

$$P_e = \frac{1}{2\pi} \int_{-\omega_x}^{\omega_x} |X(\omega)e^{j\omega\tau} - X(\omega)e^{j\omega(\tau-t_m)}|^2 d\omega \approx \frac{1}{2\pi} \int_{-\omega_x}^{\omega_x} |X(\omega) j e^{j\omega\tau} \omega t_m|^2 d\omega$$

$$\leq \frac{1}{2\pi} \int_{-\omega_x}^{\omega_x} A^2 \left(\frac{0.5}{I}\right)^2 \omega^2 d\omega = \frac{A^2 \omega_x^3}{12\pi I^2} \tag{10.8.3}$$

This bound shows that the error power is inversely proportional to the square of the number of subfilters I. Therefore, the error magnitude is inversely proportional to I. Hence we call the approximation of the rate conversion method described above a first-order approximation. By using (10.8.3) and (10.8.1), the ratio of the signal-to-distortion due to a sampling-time error for the first-order approximation, denoted as SD$_t$R1, is lower bounded as

$$\text{SD}_t\text{R1} = \frac{P_s}{P_e} \geq \frac{12 I^2}{\omega_x^2} \tag{10.8.4}$$

It can be seen from (10.8.4) that the signal-to-distortion ratio is proportional to the square of the number of subfilters.

Example 10.8.1

Suppose that the input signal has a flat spectrum between -0.8π and 0.8π. Determine the number of subfilters to achieve a signal-to-distortion ratio of 50 dB.

Solution   To achieve an $SD_tR > 10^5$, we set $SD_1R1 = 12I^2/\omega_x^2$ equal to 10^5. Thus we find that

$$I \approx \omega_x \sqrt{\frac{10^5}{12}} \approx 230 \text{ subfilters}$$

10.8.2 Second-Order Approximation (Linear Interpolation)

The disadvantage of the first-order approximation method is the large number of subfilters needed to achieve a specified distortion requirement. In the following discussion, we describe a method that uses linear interpolation to achieve the same performance with a reduced number of subfilters.

The implementation of the linear interpolation method is very similar to the first-order approximation discussed above. Instead of using the sample from the interpolating filter closest to the desired conversion output as the approximation, we compute two adjacent samples with the desired sampling time falling between their sampling times, as is illustrated in Fig. 10.28. The normalized time spacing

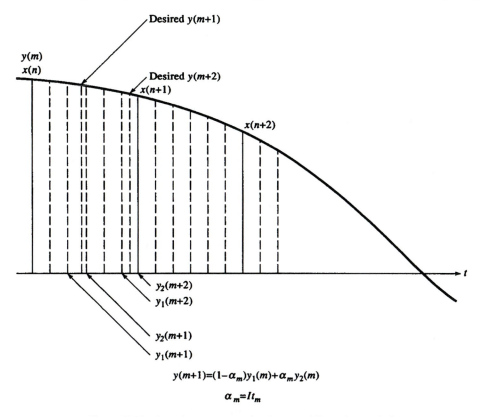

Figure 10.28 Sample rate conversion by use of linear interpolation.

between these two samples is $1/I$. Assuming that the sampling time of the first sample lags the desired sampling time by t_m, the sampling time of the second sample is then leading the desired sampling time by $(1/I) - t_m$. If we denote these two samples by $y_1(m)$ and $y_2(m)$ and use linear interpolation, we can compute the approximation to the desired output as

$$y(m) = (1 - \alpha_m)y_1(m) + \alpha_m y_2(m) \tag{10.8.5}$$

where $\alpha_m = It_m$. Note that $0 \leq \alpha_m \leq 1$.

The implementation of linear interpolation is similar to that for the first-order approximation. Normally, both $y_1(m)$ and $y_2(m)$ are computed using the ith and $(i+1)$th subfilters, respectively, with the same set of input data samples in the delay line. The only exception is in the boundary case, where $i = I - 1$. In this case we use the $(I-1)$th subfilter to compute $y_1(m)$, but the second sample $y_2(m)$ is computed using the zeroth subfilter after new input data are shifted into the delay line.

To analyze the error introduced by the second-order approximation, we first write the frequency responses of the desired filter and the two subfilters used to compute $y_1(m)$ and $y_2(m)$, as $e^{j\omega\tau}$, $e^{j\omega(\tau-t_m)}$, and $e^{j\omega(t-t_m+1/I)}$, respectively. Because linear interpolation is a linear operation, we can also use linear interpolation to compute the frequency response of the filter that generates $y(m)$ as

$$(1 - It_m)e^{j\omega(\tau-t_m)} + It_m e^{j\omega(\tau-t_m+1/I)}$$

$$= e^{j\omega\tau}[(1 - \alpha_m)e^{-j\omega t_m} + a_m e^{j\omega(-t_m+1/I)}]$$

$$= e^{j\omega\tau}(1 - \alpha_m)(\cos\omega t_m - j\sin\omega t_m) \tag{10.8.6}$$

$$+ e^{j\omega\tau}\alpha_m[\cos\omega(1/I - t_m) + j\sin\omega(1/I - t_m)]$$

By ignoring high-order errors, we can write the difference between the desired frequency responses and the one given by (10.8.6) as

$$e^{j\omega\tau} - (1 - \alpha_m)e^{j\omega(\tau-t_m)} - \alpha_m e^{j\omega(t-t_m+1/I)}$$

$$= e^{j\omega\tau}\{[1 - (1 - \alpha_m)\cos\omega t_m - a_m\cos\omega(1/I - t_m)]$$

$$+ j[(1 - \alpha_m)\sin\omega t_m - \alpha_m\sin\omega(1/I - t_m)]\} \tag{10.8.7}$$

$$\approx e^{j\omega\tau}\left[\omega^2(1 - \alpha_m)\frac{\alpha_m}{I^2}\right]$$

Using $(1 - \alpha_m)\alpha_m \leq \frac{1}{4}$, we obtain an upper bound for the total error power as

$$P_e = \frac{1}{2\pi}\int_{-\omega_x}^{\omega_x} |X(\omega)[e^{j\omega\tau} - (1 - \alpha_m)e^{j\omega(\tau-t_m)} - \alpha_m e^{j\omega(\tau-t_m+1/I)}]|^2 d\omega$$

$$\approx \frac{1}{2\pi}\int_{-\omega_x}^{\omega_x} \left|X(\omega)e^{j\omega\tau}\left[\omega^2(1 - \alpha_m)\frac{\alpha_m}{I^2}\right]\right|^2 d\omega \tag{10.8.8}$$

$$\leq \frac{1}{2\pi}\int_{-\omega_x}^{\omega_x} A^2\left(\frac{0.25}{I^2}\right)^2 \omega^4 d\omega = \frac{A^2\omega_x^5}{80\pi I^4}$$

This result indicates that the error magnitude is inversely proportional to I^2. Hence we call the approximation using linear interpolation a *second-order approximation*. Using (10.8.8) and (10.8.1), the ratio of signal-to-distortion due to a sampling time error for the second-order approximation, denoted by $SD_t R2$, is bounded from below as

$$SD_1 R2 = \frac{P_s}{P_e} \geq \frac{80 I^4}{\omega_x^4} \tag{10.8.9}$$

Therefore, the signal-to-distortion ratio is proportional to the fourth power of the number of subfilters.

Example 10.8.2

Determine the number of subfilters required to meet the specifications given in Example 10.8.1 when linear interpolation is employed.

Solution   To achieve $SD_1 R > 10^5$, we set $SD_1 R2 = 80 I^4 / \omega_x^4$ equal to 10^5. Thus we obtain

$$I \approx \omega_x \sqrt[4]{\frac{10^5}{80}} \approx 15 \text{ subfilters.}$$

From this example we see that the required number of subfilters for the second-order approximation is reduced by a factor of about 15 compared to the first-order approximation. However, we now need to compute two interpolated samples in this case, instead of one for the first-order approximation. Hence we have doubled the computational complexity.

Linear interpolation is the simplest case of the class of approximation methods based on Lagrange polynomials. It is also possible to use higher-order Lagrange polynomial approximations (interpolation) to further reduce the number of subfilters required to meet specifications. However, the second-order approximation seems sufficient for most practical applications. The interested reader is referred to the paper by Ramstad (1984) for higher-order Lagrange interpolation methods.

10.9 APPLICATIONS OF MULTIRATE SIGNAL PROCESSING

There are numerous practical applications of multirate signal processing. In this section we describe a few of these applications.

10.9.1 Design of Phase Shifters

Suppose that we wish to design a network that delays the signal $x(n)$ by a fraction of a sample. Let us assume that the delay is a rational fraction of a sampling

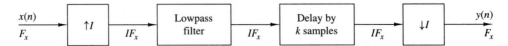

Figure 10.29   Method for generating a delay in a discrete-time signal.

interval T_x [i.e., $d = (k/I)T_x$, where k and I are relatively prime positive integers]. In the frequency domain, the delay corresponds to a linear phase shift of the form

$$\Theta(\omega) = -\frac{k\omega}{I} \qquad (10.9.1)$$

The design of an all-pass linear-phase filter is relatively difficult. However, we can use the methods of sample-rate conversion to achieve a delay of $(k/I)T_x$, exactly, without introducing any significant distortion in the signal. To be specific, let us consider the system shown in Fig. 10.29. The sampling rate is increased by a factor I using a standard interpolator. The lowpass filter eliminates the images in the spectrum of the interpolated signal, and its output is delayed by k samples at the sampling rate IF_x. The delayed signal is decimated by a factor $D = I$. Thus we have achieved the desired delay of $(k/I)T_x$.

An efficient implementation of the interpolator is the polyphase filter illustrated in Fig. 10.30. The delay of k samples is achieved by placing the initial position of the commutator at the output of the kth subfilter. Since decimation by

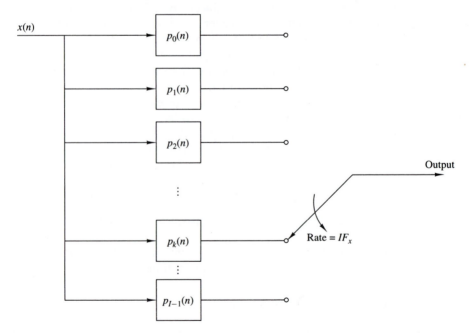

Figure 10.30   Polyphase filter structure for implementing the system shown in Fig. 10.29.

$D = I$ means that we take one out of every I samples from the polyphase filter, the commutator position can be fixed to the output of the kth subfilter. Thus a delay in k/I can be achieved by using only the kth subfilter of the polyphase filter. We note that the polyphase filter introduces an additional delay of $(M - 1)/2$ samples, where M is the length of its impulse response.

Finally, we mention that if the desired delay is a nonrational factor of the sample interval T_x, either the first-order or second-order approximation method described in Section 10.8 can be used to obtain the delay.

10.9.2 Interfacing of Digital Systems with Different Sampling Rates

In practice we frequently encounter the problem of interfacing two digital systems that are controlled by independently operating clocks. An analog solution to this problem is to convert the signal from the first system to analog form and then resample it at the input to the second system using the clock in this system. However, a simpler approach is one where the interfacing is done by a digital method using the basic sample-rate conversion methods described in this chapter.

To be specific, let us consider interfacing the two systems with independent clocks as shown in Fig. 10.31. The output of system A at rate F_x is fed to an interpolator which increases the sampling rate by I. The output of the interpolator is fed at the rate $I F_x$ to a digital sample-and-hold which serves as the interface to system B at the high sampling rate $I F_x$. Signals from the digital sample-and-hold are read out into system B at the clock rate $D F_y$ of system B. Thus the output rate from the sample-and-hold is not synchronized with the input rate.

In the special case where $D = I$ and the two clock rates are comparable but not identical, some samples at the output of the sample-and-hold may be repeated or dropped at times. The amount of signal distortion resulting from this method can be kept small if the interpolator/decimator factor is large. By using linear interpolation in place of the digital sample-and-hold, as we described in Section 10.8, we can further reduce the distortion and thus reduce the size of the interpolator factor.

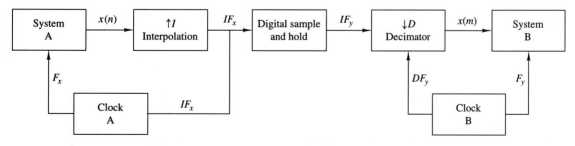

Figure 10.31   Interfacing of two digital systems with different sampling rates.

10.9.3 Implementation of Narrowband Lowpass Filters

In Section 10.6 we demonstrated that a multistage implementation of sampling-rate conversion often provides for a more efficient realization, especially when the filter specifications are very tight (e.g., a narrow passband and a narrow transition band). Under similar conditions, a lowpass, linear-phase FIR filter may be more efficiently implemented in a multistage decimator-interpolator configuration. To be more specific, we can employ a multistage implementation of a decimator of size D, followed by a multistage implementation of an interpolator of size I, where $I = D$.

We demonstrate the procedure by means of an example for the design of a lowpass filter which has the same specifications as the filter that is given in Example 10.6.1.

Example 10.9.1

Design a linear-phase FIR filter that satisfies the following specifications:

Sampling frequency:	8000 Hz
Passband:	$0 \leq F \leq 75$
Transition band:	$75 \leq F \leq 80$
Stopband	$80 \leq F \leq 4000$
Passband ripple:	$\delta_1 = 10^{-2}$
Stopband ripple:	$\delta_2 = 10^{-4}$

Solution If this filter were designed as a single-rate linear-phase FIR filter, the length of the filter required to meet the specifications is (from Kaiser's formula)

$$\hat{M} \approx 5152$$

Now, suppose that we employ a multirate implementation of the lowpass filter based on a decimation and interpolation factor of $D = I = 100$. A single-stage implementation of the decimator-interpolator requires an FIR filter of length

$$\hat{M}_1 = \frac{-10\log_{10}(\delta_1\delta_2/2) - 13}{14.6\Delta f} + 1 \approx 5480$$

However, there is a significant savings in computational complexity by implementing the decimator and interpolator filters using their corresponding polyphase filters. If we employ linear-phase (symmetric) decimation and interpolation filters, the use of polyphase filters reduces the multiplication rate by a factor of 100.

A significantly more efficient implementation is obtained by using two stages of decimation followed by two stages of interpolation. For example, suppose that we select $D_1 = 50$, $D_2 = 2$, $I_1 = 2$, and $I_2 = 50$. Then the required filter lengths are

$$\hat{M}_1 = \frac{-10\log(\delta_1\delta_2/4) - 13}{14.6\Delta f} + 1 \approx 177$$

$$\hat{M}_2 = \frac{-10\log_{10}(\delta_1\delta_2/4) - 13}{14.6\Delta f} + 1 \approx 233$$

Thus we obtain a reduction in the overall filter length of $2(5480)/2(177+233) \approx 13.36$. In addition, we obtain further reduction in the multiplication rate by using polyphase filters. For the first stage of decimation, the reduction in multiplication rate is 50, while for the second stage the reduction in multiplication rate is 100. Further reductions can be obtained by increasing the number of stages of decimation and interpolation.

10.9.4 Implementation of Digital Filter Banks

Filter banks are generally categorized as two types, *analysis filter banks* and *synthesis filter banks*. An analysis filter bank consists of a set of filters, with system functions $\{H_z(k)\}$, arranged in a parallel bank as illustrated in Fig. 10.32a. The frequency response characteristics of this filter bank splits the signal into a corresponding number of subbands. On the other hand, a synthesis filter bank consists of a set of filters with system functions $\{G_k(z)\}$, arranged as shown in Fig. 10.32b, with corresponding inputs $\{y_k(n)\}$. The outputs of the filters are summed to form the synthesized signal $\{x(n)\}$.

Filter banks are often used for performing spectrum analysis and signal synthesis. When a filter bank is employed in the computation of the discrete Fourier

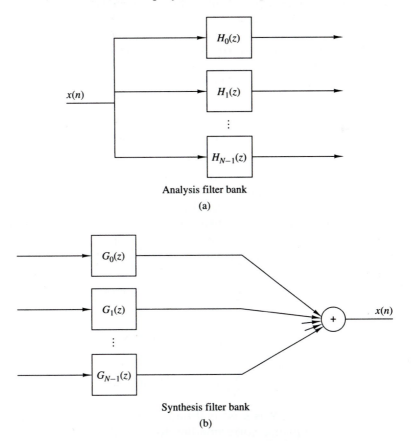

Analysis filter bank

(a)

Synthesis filter bank

(b)

Figure 10.32   A digital filter bank.

transform (DFT) of a sequence $\{x(n)\}$, the filter bank is called a DFT filter bank. An analysis filter bank consisting of N filters $\{H_k(z), k = 0, 1, \ldots, N-1\}$ is called a uniform DFT filter bank if $H_k(z), k = 1, 2, \ldots, N-1$, are derived from a prototype filter $H_0(z)$, where

$$H_k(\omega) = H_0\left(\omega - \frac{2\pi k}{N}\right) \qquad k = 1, 2, \ldots, N-1 \tag{10.9.2}$$

Hence the frequency response characteristics of the filters $\{H_k(z), k = 0, 1, \ldots, N-1\}$ are simply obtained by uniformly shifting the frequency response of the prototype filter by multiples of $2\pi/N$. In the time domain the filters are characterized by their impulse responses, which can be expressed as

$$h_k(n) = h_0(n)e^{j2\pi nk/N} \qquad k = 0, 1, \ldots, N-1 \tag{10.9.3}$$

where $\{h_0(n)\}$ is the impulse response of the prototype filter.

The uniform DFT analysis filter bank can be realized as shown in Fig. 10.33a, where the frequency components in the sequence $\{x(n)\}$ are translated in frequency to lowpass by multiplying $x(n)$ with the complex exponentials $\exp(-j2\pi nk/N), k = 1, \ldots, N-1$, and the resulting product signals are passed through a lowpass filter with impulse response $\{h_0(n)\}$. Since the output of the lowpass filter is relatively narrow in bandwidth, the signal can be decimated by a factor $D \leq N$. The resulting decimated output signal can be expressed as

$$X_k(m) = \sum_n h_0(mD - n)x(n)e^{-j2\pi nk/N} \qquad \begin{aligned} &k = 0, 1, \ldots, N-1 \\ &m = 0, 1, \ldots \end{aligned} \tag{10.9.4}$$

where $\{X_k(m)\}$ are samples of the DFT at frequencies $\omega_k = 2\pi k/N$.

The corresponding synthesis filter for each element in the filter bank can be viewed as shown in Fig. 10.33b, where the input signal sequences $\{Y_k(m), k = 0, 1, \ldots, N-1\}$ are upsampled by a factor of $I = D$, filtered to remove the images, and translated in frequency by multiplication by the complex exponentials $\{\exp(j2\pi nk/N), k = 0, 1, \ldots, N-1\}$. The resulting frequency-translated signals from the N filters are then summed. Thus we obtain the sequence

$$\begin{aligned} v(n) &= \frac{1}{N}\sum_{k=0}^{N-1} e^{j2\pi nk/N}\left[\sum_m Y_k(m)g_0(n - mI)\right] \\ &= \sum_m g_0(n - mI)\left[\frac{1}{N}\sum_{k=0}^{N-1} Y_k(m)e^{j2\pi nk/N}\right] \\ &= \sum_m g_0(n - mI)y_n(m) \end{aligned} \tag{10.9.5}$$

where the factor $1/N$ is a normalization factor, $\{y_n(m)\}$ represent samples of the inverse DFT sequence corresponding to $\{Y_k(m)\}$, $\{g_0(n)\}$ is the impulse response of the interpolation filter, and $I = D$.

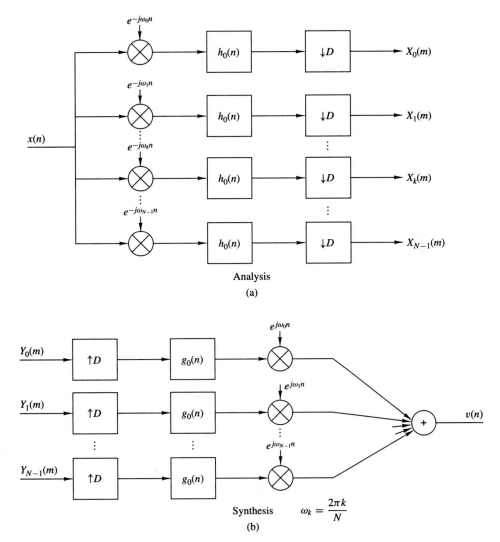

Figure 10.33 A uniform DFT filter bank.

The relationship between the output $\{X_k(n)\}$ of the analysis filter bank and the input $\{Y_k(m)\}$ to the synthesis filter bank depends on the application. Usually, $\{Y_k(m)\}$ is a modified version of $\{X_k(m)\}$, where the specific modification is determined by the application.

An alternative realization of the analysis and synthesis filter banks is illustrated in Fig. 10.34. The filters are realized as bandpass filters with impulse responses

$$h_k(n) = h_0(n)e^{j2\pi nk/N} \qquad k = 0, 1, \ldots, N-1 \tag{10.9.6}$$

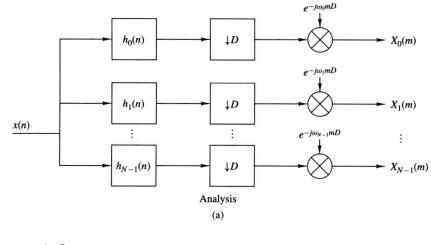

Analysis

(a)

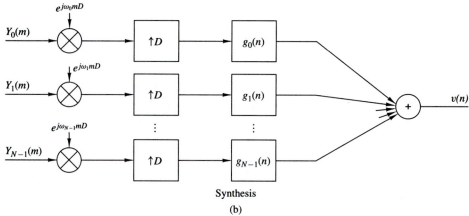

Synthesis

(b)

Figure 10.34   Alternative realization of a uniform DFT filter bank.

The output of each bandpass filter is decimated by a factor D and multiplied by $\exp(-j2\pi mk/N)$ to produce the DFT sequence $\{X_k(m)\}$. The modulation by the complex exponential allows us to shift the spectrum of the signal from $\omega_k = 2\pi k/N$ to $\omega_0 = 0$. Hence this realization is equivalent to the realization given in Fig. 10.33. The filter bank output can be written as

$$X_k(m) = \left[\sum_n x(n)h_0(mD - n)e^{j2\pi k(mD-n)/N} \right] e^{-j2\pi mkD/N} \qquad (10.9.7)$$

The corresponding filter bank synthesizer can be realized as shown in Fig. 10.34b, where the input sequences are first multiplied by the exponential factors $[\exp(j2\pi kmD/N)]$, upsampled by the factor $I = D$, and the resulting sequences are filtered by the bandpass interpolation filters with impulse responses

$$g_k(n) = g_0(n)e^{j2\pi nk/N} \tag{10.9.8}$$

where $\{g_0(n)\}$ is the impulse response of the prototype filter. The outputs of these filters are then summed to yield

$$v(n) = \frac{1}{N} \sum_{k=0}^{N-1} \left\{ \sum_m [Y_k(m)e^{j2\pi kmI/N}]g_k(n - mI) \right\} \tag{10.9.9}$$

where $I = D$.

In the implementation of digital filters banks, computational efficiency can be achieved by use of polyphase filters for decimation and interpolation. Of particular interest is the case where the decimation factor D is selected to be equal to the number N of frequency bands. When $D = N$, we say that the filter bank is *critically sampled*.

For the analysis filter bank, let us define a set of $N = D$ polyphase filters with impulse responses

$$p_k(n) = h_0(nN - k) \qquad k = 0, 1, \ldots, N - 1 \tag{10.9.10}$$

and the corresponding set of decimated input sequences

$$x_k(n) = x(nN + k) \qquad k = 0, 1, \ldots, N - 1 \tag{10.9.11}$$

Note that this definition of $\{p_k(n)\}$ implies that the commutator for the decimator rotates clockwise.

The structure of the analysis filter bank based on the use of polyphase filters can be obtained by substituting (10.9.10) and (10.9.11) into (10.9.7) and rearranging the summation into the form

$$X_k(m) = \sum_{n=0}^{N-1} \left[\sum_l p_n(l)x_n(m - l) \right] e^{-j2\pi nk/N} \qquad k = 0, 1, \ldots, D - 1 \tag{10.9.12}$$

where $N = D$. Note that the inner summation represents the convolution of $\{p_n(l)\}$ with $\{x_n(l)\}$. The outer summation represents the N-point DFT of the filter outputs. The filter structure corresponding to this computation is illustrated in Fig. 10.35. Each sweep of the commutator results in N outputs, denoted as $\{r_n(m), n = 0, 1, \ldots, N - 1\}$ from the N polyphase filters. The N-point DFT of this sequence yields the spectral samples $\{X_k(m)\}$. For large values of N, the FFT algorithm provides an efficient means for computing the DFT.

Now suppose that the spectral samples $\{X_k(m)\}$ are modified in some manner, prescribed by the application, to produce $\{Y_k(m)\}$. A filter bank synthesis filter based on a polyphase filter structure can be realized in a similar manner. First, we define the impulse response of the N $(D = I = N)$ polyphase filters for the interpolation filter as

$$q_k(n) = g_0(nN + k) \qquad k = 0, 1, \ldots, N - 1 \tag{10.9.13}$$

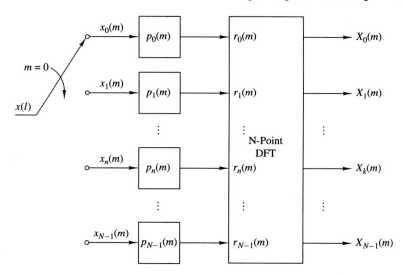

Figure 10.35 Digital filter bank structure for the computation of (10.9.12).

and the corresponding set of output signals as

$$v_k(n) = v(nN + k) \qquad k = 0, 1, \ldots, N-1 \qquad (10.9.14)$$

Note that this definition of $\{q_k(n)\}$ implies that the commutator for the interpolator rotates counterclockwise.

By substituting (10.9.13) into (10.9.5), we can express the output $v_l(n)$ of the lth polyphase filter as

$$v_l(n) = \sum_m q_l(n-m)\left[\frac{1}{N}\sum_{k=0}^{N-1} Y_k(m)e^{j2\pi kl/N}\right] \qquad l = 0, 1, \ldots, N-1 \qquad (10.9.15)$$

The term in brackets is the N-point inverse DFT of $\{Y_k(m)\}$, which we denote as $\{y_l(m)\}$. Hence

$$v_l(n) = \sum_m q_l(n-m)y_l(m) \qquad l = 0, 1, \ldots, N-1 \qquad (10.9.16)$$

The synthesis structure corresponding to (10.9.16) is shown in Fig. 10.36. It is interesting to note that by defining the polyphase interpolation filter as in (10.9.13), the structure in Fig. 10.36 is the transpose of the polyphase analysis filter shown in Fig. 10.35.

In our treatment of digital filter banks we considered the important case of critically sampled DFT filter banks, where $D = N$. Other choices of D and N can be employed in practice, but the implementation of the filters becomes more complex. Of particular importance is the oversampled DFT filter bank, where $N = KD$, D denotes the decimation factor and K is an integer that specifies the

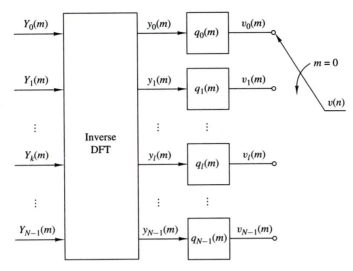

Figure 10.36   Digital filter bank structure for the computation of (10.9.16).

oversampling factor. In this case it can be shown that the polyphase filter bank structures for the analysis and synthesis filters can be implemented by use of N subfilters and N-point DFTs and inverse DFTs.

10.9.5 Subband Coding of Speech Signals

A variety of techniques have been developed to efficiently represent speech signals in digital form for either transmission or storage. Since most of the speech energy is contained in the lower frequencies, we would like to encode the lower-frequency band with more bits than the high-frequency band. Subband coding is a method, where the speech signal is subdivided into several frequency bands and each band is digitally encoded separately.

   An example of a frequency subdivision is shown in Fig. 10.37a. Let us assume that the speech signal is sampled at a rate F_s samples per second. The first frequency subdivision splits the signal spectrum into two equal-width segments, a lowpass signal ($0 \leq F \leq F_x/4$) and a highpass signal ($F_s/4 \leq F \leq F_s/2$). The second frequency subdivision splits the lowpass signal from the first stage into two equal bands, a lowpass signal ($0 < F \leq F_s/8$) and a highpass signal ($F_s/8 \leq F \leq F_s/4$). Finally, the third frequency subdivision splits the lowpass signal from the second stage into two equal bandwidth signals. Thus the signal is subdivided into four frequency bands, covering three octaves, as shown in Fig. 10.37b.

   Decimation by a factor of 2 is performed after frequency subdivision. By allocating a different number of bits per sample to the signal in the four subbands, we can achieve a reduction in the bit rate of the digitalized speech signal.

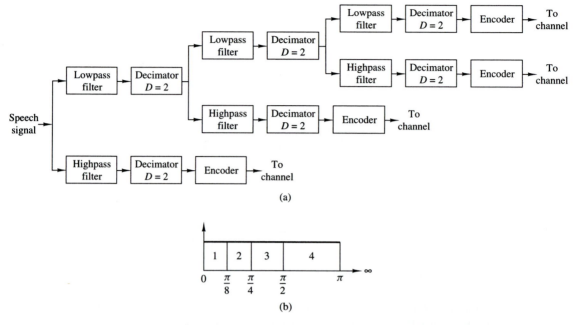

Figure 10.37 Block diagram of a subband speech coder.

Filter design is particularly important in achieving good performance in subband coding. Aliasing resulting from decimation of the subband signals must be negligible. It is clear that we cannot use brickwall filter characteristics as shown in Fig. 10.38a, since such filters are physically unrealizable. A particularly practical solution to the aliasing problem is to use *quadrature mirror filters* (QMF), which have the frequency response characteristics shown in Fig. 10.38b. These filters are described in the following section.

The synthesis method for the subband encoded speech signal is basically the reverse of the encoding process. The signals in adjacent lowpass and highpass frequency bands are interpolated, filtered, and combined as shown in Fig. 10.39. A pair of QMF is used in the signal synthesis for each octave of the signal.

Subband coding is also an effective method to achieve data compression in image signal processing. By combining subband coding with vector quantization for each subband signal, Safranek et al. (1988) have obtained coded images with approximately $\frac{1}{2}$ bit per pixel, compared with 8 bits per pixel for the uncoded image.

In general, subband coding of signals is an effective method for achieving bandwidth compression in a digital representation of the signal, when the signal energy is concentrated in a particular region of the frequency band. Multirate signal processing notions provide efficient implementations of the subband encoder.

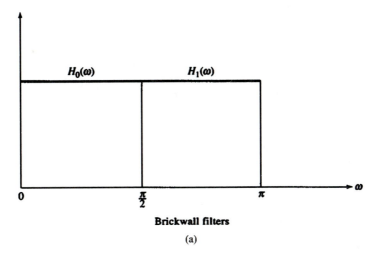

Brickwall filters

(a)

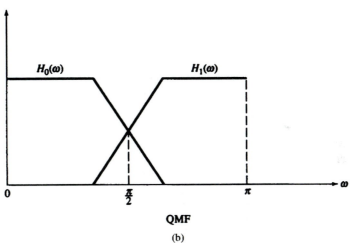

QMF

(b)

Figure 10.38 Filter characteristics for subband coding.

10.9.6 Quadrature Mirror Filters

The basic building block in applications of quadrature mirror filters (QMF) is the two-channel QMF bank shown in Fig. 10.40. This is a multirate digital filter structure that employs two decimators in the "signal analysis" section and two interpolators in the "signal synthesis" section. The lowpass and highpass filters in the analysis section have impulse responses $h_0(n)$ and $h_1(n)$, respectively. Similarly, the lowpass and highpass filters contained in the synthesis section have impulse responses $g_0(n)$ and $g_1(n)$, respectively.

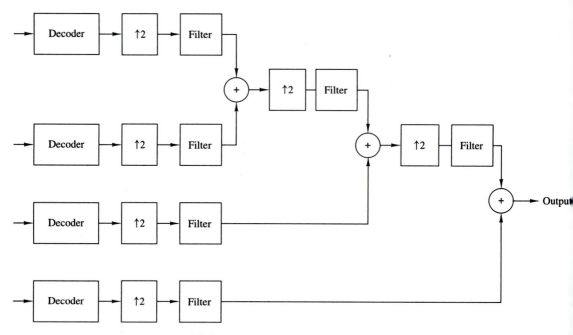

Figure 10.39 Synthesis of subband-encoded signals.

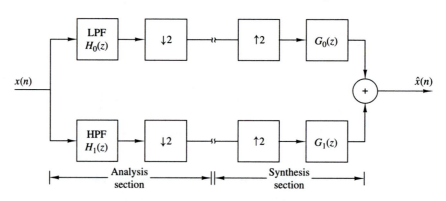

Figure 10.40 Two-channel QMF bank.

The Fourier transforms of the signals at the outputs of the two decimators are

$$X_{a0}(\omega) = \frac{1}{2}\left[X\left(\frac{\omega}{2}\right)H_0\left(\frac{\omega}{2}\right) + X\left(\frac{\omega - 2\pi}{2}\right)H_0\left(\frac{\omega - 2\pi}{2}\right)\right]$$

$$X_{a1}(\omega) = \frac{1}{2}\left[X\left(\frac{\omega}{2}\right)H_1\left(\frac{\omega}{2}\right) + X\left(\frac{\omega - 2\pi}{2}\right)H_1\left(\frac{\omega - 2\pi}{2}\right)\right]$$

(10.9.17)

If $X_{s0}(\omega)$ and $X_{s1}(\omega)$ represent the two inputs to the synthesis section, the output is simply

$$\hat{X}(\omega) = X_{s0}(2\omega)G_0(\omega) + X_{s1}(2\omega)G_1(\omega) \qquad (10.9.18)$$

Now, suppose that we connect the analysis filter to the corresponding synthesis filter, so that $X_{a0}(\omega) = X_{s0}(\omega)$ and $X_{a1}(\omega) = X_{s1}(\omega)$. Then, by substituting from (10.9.17) into (10.9.18), we obtain

$$\hat{X}(\omega) = \tfrac{1}{2}[H_0(\omega)G_0(\omega) + H_1(\omega)G_1(\omega)]X(\omega)$$
$$+ \tfrac{1}{2}[H_0(\omega - \pi)G_0(\omega) + H_1(\omega - \pi)G_1(\omega)]X(\omega - \pi) \qquad (10.9.19)$$

The first term in (10.9.19) is the desired signal output from the QMF bank. The second term represents the effect of aliasing, which we would like to eliminate. Hence we require that

$$H_0(\omega - \pi)G_0(\omega) + H_1(\omega - \pi)G_1(\omega) = 0 \qquad (10.9.20)$$

This condition can be simply satisfied by selecting $G_0(\omega)$ and $G_1(\omega)$ as

$$G_0(\omega) = H_1(\omega - \pi) \qquad G_1(\omega) = -H_0(\omega - \pi) \qquad (10.9.21)$$

Thus the second term in (10.9.19) vanishes.

To elaborate, let us assume that $H_0(\omega)$ is a lowpass filter and $H_1(\omega)$ is a mirror-image highpass filter. Then we can express $H_0(\omega)$ and $H_1(\omega)$ as

$$H_0(\omega) = H(\omega)$$
$$H_1(\omega) = H(\omega - \pi) \qquad (10.9.22)$$

where $H(\omega)$ is the frequency response of a lowpass filter. In the time domain, the corresponding relations are

$$h_0(n) = h(n)$$
$$h_1(n) = (-1)^n h(n) \qquad (10.9.23)$$

As a consequence, $H_0(\omega)$ and $H_1(\omega)$ have mirror-image symmetry about the frequency $\omega = \pi/2$, as shown in Fig. 10.38b. To be consistent with the constraint in (10.9.21), we select the lowpass filter $G_0(\omega)$ as

$$G_0(\omega) = 2H(\omega) \qquad (10.9.24)$$

and the highpass filter $G_1(\omega)$ as

$$G_1(\omega) = -2H(\omega - \pi) \qquad (10.9.25)$$

In the time domain, these relations become

$$g_0(n) = 2h(n)$$
$$g_1(n) = -2(-1)^n h(n) \qquad (10.9.26)$$

The scale factor of 2 in $g_0(n)$ and $g_1(n)$ corresponds to the interpolation factor used to normalize the overall frequency response of the QMF. With this choice of the filter characteristics, the component due to aliasing vanishes. Thus the aliasing resulting from decimation in the analysis section of the QMF bank is perfectly canceled by the image signal spectrum that arises due to interpolation. As a result, the two-channel QMF behaves as a linear, time-invariant system.

If we substitute for $H_0(\omega)$, $H_1(\omega)$, $G_0(\omega)$, and $G_1(\omega)$ into the first term of (10.9.19), we obtain

$$\hat{X}(\omega) = [H^2(\omega) - H^2(\omega - \pi)]X(\omega) \qquad (10.9.27)$$

Ideally, the two-channel QMF bank should have unity gain,

$$|H^2(\omega) - H^2(\omega - \pi)| = 1 \qquad \text{for all } \omega \qquad (10.9.28)$$

where $H(\omega)$ is the frequency response of a lowpass filter. Furthermore, it is also desirable for the QMF to have linear phase.

Now, let us consider the use of a linear phase filter $H(\omega)$. Hence $H(\omega)$ may be expressed in the form

$$H(\omega) = H_r(\omega)e^{-j\omega(N-1)/2} \qquad (10.9.29)$$

where N is the filter length. Then

$$\begin{aligned} H^2(\omega) &= H_r^2(\omega)e^{-j\omega(N-1)} \\ &= |H(\omega)|^2 e^{-j\omega(N-1)} \end{aligned} \qquad (10.9.30)$$

and

$$\begin{aligned} H^2(\omega - \pi) &= H_r^2(\omega - \pi)e^{-j(\omega-\pi)(N-1)} \\ &= (-1)^{N-1}|H(\omega - \pi)|^2 e^{-j\omega(N-1)} \end{aligned} \qquad (10.9.31)$$

Therefore, the overall transfer function of the two-channel QMF which employs linear-phase FIR filters is

$$\frac{\hat{X}(\omega)}{X(\omega)} = [|H(\omega)|^2 - (-1)^{N-1}|H(\omega - \pi)|^2]e^{-j\omega(N-1)} \qquad (10.9.32)$$

Note that the overall filter has a delay of $N - 1$ samples and a magnitude characteristic

$$A(\omega) = |H(\omega)|^2 - (-1)^{N-1}|H(\omega - \pi)|^2 \qquad (10.9.33)$$

We also note that when N is odd, $A(\pi/2) = 0$, because $|H(\pi/2)| = |H(3\pi/2)|$. This is an undesirable property for a QMF design. On the other hand, when N is even,

$$A(\omega) = |H(\omega)|^2 + |H(\omega - \pi)|^2 \qquad (10.9.34)$$

which avoids the problem of a zero at $\omega = \pi/2$. For N even, the ideal two-channel QMF should satisfy the condition

$$A(\omega) = |H(\omega)|^2 + |H(\omega - \pi)|^2 = 1 \qquad \text{for all } \omega \qquad (10.9.35)$$

which follows from (10.9.33). Unfortunately, the only filter frequency response function that satisfies (10.9.35) is the trivial function $|H(\omega)|^2 = \cos^2 a\omega$. Consequently, any nontrivial linear-phase FIR filter $H(\omega)$ introduces some amplitude distortion.

The amount of amplitude distortion introduced by a nontrivial linear phase FIR filter in the QMF can be minimized by optimizing the FIR filter coefficients. A particularly effective method is to select the filter coefficients of $H(\omega)$ such that $A(\omega)$ is made as flat as possible while simultaneously minimizing (or constraining) the stopband energy of $H(\omega)$. This approach leads to the minimization of the integral squared error

$$J = w \int_{\omega_s}^{\pi} |H(\omega)|^2 d\omega + (1 - w) \int_{0}^{\pi} [A(\omega) - 1]^2 d\omega \qquad (10.9.36)$$

where w is a weighting factor in the range $0 < w < 1$. In performing the optimization, the filter impulse response is constrained to be symmetric (linear phase). This optimization is easily done numerically on a digital computer. This approach has been used by Johnston (1980), and Jain and Crochiere (1984) to design two-channel QMFs. Tables of optimum filter coefficients have been tabulated by Johnston (1980).

As an alternative to linear-phase FIR filters, we can design an IIR filter that satisfies the all-pass constraint given by (10.9.28). For this purpose, elliptic filters provide especially efficient designs. Since the QMF would introduce some phase distortion, the signal at the output of the QMF can be passed through an all-pass phase equalizer designed to minimize phase distortion.

In addition to these two methods for QMF design, one can also design the two-channel QMFs to eliminate completely both amplitude and phase distortion as well as canceling aliasing distortion. Smith and Barnwell (1984) have shown that such *perfect reconstruction QMF* can be designed by relaxing the linear-phase condition of the FIR lowpass filter $H(\omega)$. To achieve perfect reconstruction, we begin by designing a linear-phase FIR halfband filter of length $2N - 1$.

A half-band filter is defined as a zero-phase FIR filter whose impulse response $\{b(n)\}$ satisfies the condition

$$b(2n) = \begin{cases} \text{constant}, & n = 0 \\ 0, & n \neq 0 \end{cases} \qquad (10.9.37)$$

Hence all the even-numbered samples are zero except at $n = 0$. The zero-phase requirement implies that $b(n) = b(-n)$. The frequency response of such a filter is

$$B(\omega) \sum_{n=-K}^{K} b(n)e^{-j\omega n} \qquad (10.9.38)$$

where K is odd. Furthermore, $B(\omega)$ satisfies the condition $B(\omega) + B(\pi - \omega)$ is equal to a constant for all frequencies. The typical frequency response characteristic of a half-band filter is shown in Fig. 10.41. We note that the filter response is symmetric with respect to $\pi/2$, the band edges frequencies ω_p and ω_s are symmetric about

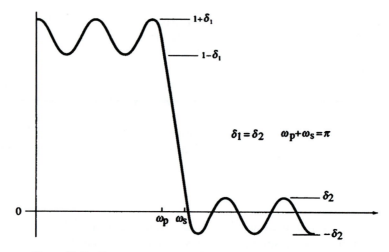

Figure 10.41 Frequency response characteristic of FIR half-band filter.

$\omega = \pi/2$, and the peak passband and stopband errors are equal. We also note that the filter can be made causal by introducing a delay of K samples.

Now, suppose that we design an FIR half-band filter of length $2N - 1$, where N is even, with frequency response as shown in Fig. 10.42(a). From $B(\omega)$ we construct another half-band filter with frequency response

$$B_+(\omega) = B(\omega) + \delta e^{-j\omega(N-1)} \tag{10.9.39}$$

as shown in Fig. 10.42(b). Note that $B_+(\omega)$ is nonnegative and hence it has the spectral factorization

$$B_+(z) = H(z)H(z^{-1})z^{-(N-1)} \tag{10.9.40}$$

or, equivalently,

$$B_+(\omega) = |H(\omega)|^2 e^{-j\omega(N-1)} \tag{10.9.41}$$

where $H(\omega)$ is the frequency response of an FIR filter of length N with real coefficients. Due to the symmetry of $B_+(\omega)$ with respect to $\omega = \pi/2$, we also have

$$B_+(z) + (-1)^{N-1}B_+(-z) = \alpha z^{-(N-1)} \tag{10.9.42}$$

or, equivalently,

$$B_+(\omega) + (-1)^{N-1}B_+(\omega - \pi) = \alpha e^{-j\omega(N-1)} \tag{10.9.43}$$

where α is a constant. Thus, by substituting (10.9.40) into (10.9.42), we obtain

$$H(z)H(z^{-1}) + H(-z)H(-z^{-1}) = \alpha \tag{10.9.44}$$

Since $H(z)$ satisfies (10.9.44) and since aliasing is eliminated when we have $G_0(z) = H_1(-z)$ and $G_1(z) = -H_0(-z)$, it follows that these conditions are satisfied by

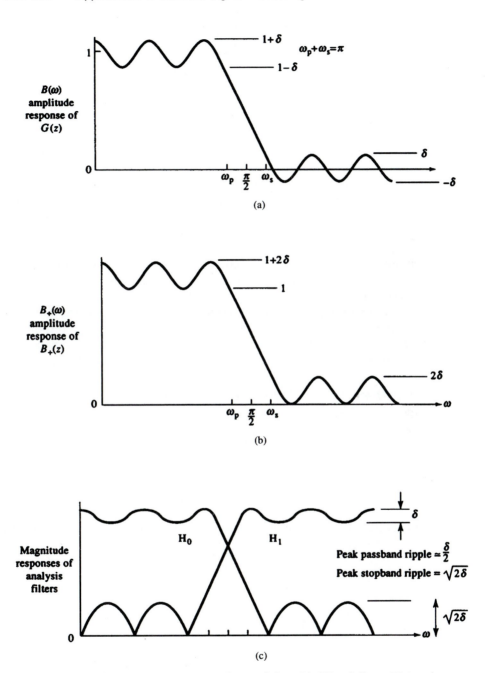

Figure 10.42 Frequency response characteristics of half-band filters $B(\omega)$ and $B_+(\omega)$. (From Vaidyanathan (1987))

choosing $H_1(z)$, $G_0(z)$, and $G_1(z)$ as

$$H_0(z) = H(z)$$

$$H_1(z) = -z^{-(N-1)}H_0(-z^{-1})$$

$$G_0(z) = z^{-(N-1)}H_0(z^{-1})$$

$$G_1(z) = z^{-(N-1)}H_1(z^{-1}) = -H_0(-z)$$

(10.9.45)

Thus aliasing distortion is eliminated and since $\hat{X}(\omega)/X(\omega)$ is a constant, the QMF performs perfect reconstruction so that $x(n) = \alpha x(n - N + 1)$. However, we note that $H(z)$ is not a linear-phase filter.

The FIR filters $H_0(z)$, $H_1(z)$, $G_0(z)$, and $G_1(z)$ in the two-channel QMF bank are efficiently realized as polyphase filters. Since $I = D = 2$, two polyphase filters are implemented for each decimator and two for each interpolator. However, if we employ linear-phase FIR filters, the symmetry properties of the analysis filters and synthesis filters allow us to simplify the structure and reduce the number of polyphase filters in the analysis section to two filters and to another two filters in the synthesis section.

To demonstrate this construction, let us assume that the filters are linear-phase FIR filters of length N (N even), which have impulse responses given by (10.9.23). Then the outputs of the analysis filter pair, after decimation by a factor of 2, can be expressed as

$$X_{ak}(m) = \sum_{n=-\infty}^{\infty} (-1)^{kn} h(n) x(2m - n) \qquad k = 0, 1$$

$$= \sum_{i=0}^{1} \sum_{l=-\infty}^{\infty} (-1)^{k(2l+1)} h(2l + 1) x(2m - 2l - i)$$

$$= \sum_{l=0}^{N-1} h(2l) x(2m - 2l) + (-1)^k \sum_{l=0}^{N-1} h(2l + 1) x(2m - 2l - 1) \qquad (10.9.46)$$

Now let us define the impulse response of two polyphase filters of length $N/2$ as

$$p_i(m) = h(2m + i) \qquad i = 0, 1 \qquad (10.9.47)$$

Then (10.9.46) can be expressed as

$$X_{ak}(m) = \sum_{l=0}^{N/2-1} p_0(m) x(2(m - l))$$

$$+ (-1)^k \sum_{l=0}^{N/2-1} p_1(m) x(2m - 2l - 1) \qquad k = 0, 1$$

(10.9.48)

This expression corresponds to the polyphase filter structure for the analysis section shown in Fig. 10.43. Note that the commutator rotates counterclockwise

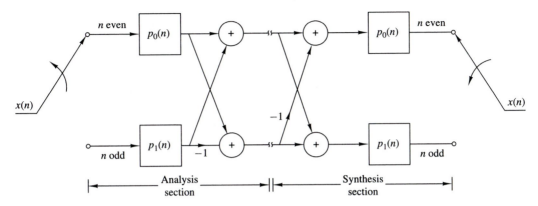

Figure 10.43 Polyphase filter structure for the QMF bank.

and that the filter with impulse response $\{p_0(m)\}$ processes the even-numbered samples of the input sequence and the filter with impulse response $\{p_1(m)\}$ processes the odd-numbered samples of the input signal.

In a similar manner, by using (10.9.26), we can obtain the structure for the polyphase synthesis section, which is also shown in Fig. 10.43. This derivation is left as an exercise for the reader (Problem 10.16). Note that the commutator also rotates counterclockwise.

Finally, we observe that the polyphase filter structure shown in Fig. 10.43 is approximately four times more efficient than the direct-form FIR filter realization.

10.9.7 Transmultiplexers

Another application of multirate signal processing is in the design and implementation of digital transmultiplexers which are devices for converting between time-division-multiplexed (TDM) signals and frequency-division-multiplexed (FDM) signals.

In a transmultiplexer for TDM-to-FDM conversion, the input signal $\{x(n)\}$ is a time-division multiplexed signal consisting of L signals, which are separated by a commutator switch. Each of these L signals are then modulated on different carrier frequencies to obtain an FDM signal for transmission. In a transmultiplexer for FDM-to-TDM conversion, the composite signal is separated by filtering into the L signal components which are then time-division multiplexed.

In telephony, single-sideband transmission is used with channels spaced at a nominal 4-kHz bandwidth. Twelve channels are usually stacked in frequency to form a basic group channel, with a bandwidth of 48 kHz. Larger bandwidth FDM signals are formed by frequency translation of multiple groups into adjacent frequency bands. We shall confine our discussion to digital transmultiplexers for 12-channel FDM and TDM signals.

Let us first consider FDM-to-TDM conversion. The analog FDM signal is passed through an A/D converter as shown in Fig. 10.44a. The digital signal is then

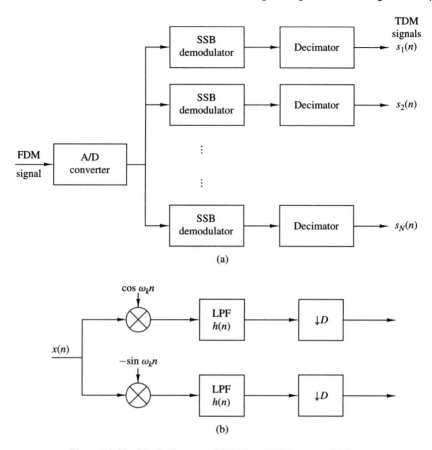

Figure 10.44   Block diagram of FDM-to-TDM transmultiplexer.

demodulated to baseband by means of single-sideband demodulators. The output of each demodulator is decimated and fed to commutator of the TDM system.

To be specific, let us assume that the 12-channel FDM signal is sampled at the Nyquist rate of 96 kHz and passed through a filter-bank demodulator. The basic building block in the FDM demodulator consists of a frequency converter, a lowpass filter, and a decimator, as illustrated in Fig. 10.44b. Frequency conversion can be efficiently implemented by the DFT filter bank described previously. The lowpass filter and decimator are efficiently implemented by use of the polyphase filter structure. Thus the basic structure for the FDM-to-TDM converter has the form of a DFT filter bank analyzer. Since the signal in each channel occupies a 4-kHz bandwidth, its Nyquist rate is 8 kHz, and hence the polyphase filter output can be decimated by a factor of 12. Consequently, the TDM commutator is operating at a rate of 12×8 kHz or 96 kHz.

In TDM-to-FDM conversion, the 12-channel TDM signal is demultiplexed into the 12 individual signals, where each signal has a rate of 8 kHz. The signal

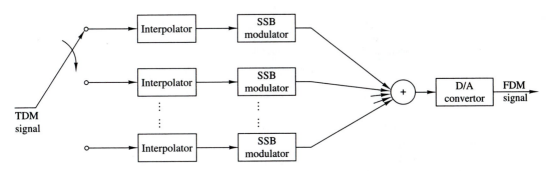

Figure 10.45   Block diagram of TDM-to-FDM transmultiplexer.

in each channel is interpolated by a factor of 12 and frequency converted by a single-sideband modulator, as shown in Fig. 10.45. The signal outputs from the 12 single-sideband modulators are summed and fed to the D/A converter. Thus we obtain the analog FDM signal for transmission. As in the case of FDM-to-TDM conversion, the interpolator and the modulator filter are combined and efficiently implemented by use of a polyphase filter. The frequency translation can be accomplished by the DFT. Consequently, the TDM-to-FDM converter encompasses the basic principles introduced previously in our discussion of DFT filter bank synthesis.

10.9.8 Oversampling A/D and D/A Conversion

Our treatment of oversampling A/D and D/A converters in Chapter 9 provides another example of multirate signal processing. Recall that an oversampling A/D converter is implemented by a cascade of an analog sigma-delta modulator (SDM) followed by a digital antialiasing decimation filter and a digital highpass filter as shown in Fig. 10.46. The analog SDM produces a 1-bit per sample output at a very high sampling rate. This 1-bit per sample output is passed through a digital lowpass filter, which provides a high-precision (multiple-bit) output that is decimated to a lower sampling rate. This output is then passed to a digital highpass filter that serves to attenuate the quantization noise at the lower frequencies.

The reverse operations take place in an oversampling D/A converter, as shown in Fig. 10.47. As illustrated in this figure, the digital signal is passed through a highpass filter whose output is fed to a digital interpolator (upsampler and anti-imaging filter). This high-sampling-rate signal is the input to the digital SDM that provides a high-sampling-rate 1-bit per sample output. The 1-bit per sample output

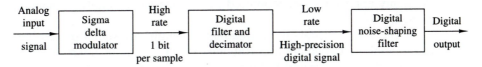

Figure 10.46   Diagram of oversampling A/D converter

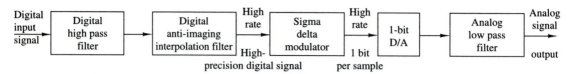

Figure 10.47   Diagram of oversampling D/A converter

is then converted to an analog signal by lowpass filtering and further smoothing with analog filters.

Figure 10.48 illustrates the block diagram of a commercial (Analog Devices ADSP-28 msp02) codec (encoder and decoder) for voice-band signals based on sigma-delta A/D and D/A converters and analog front-end circuits needed as an interface to the analog voice-band signals. The nominal sampling rate (after decimation) is 8 kHz and the sampling rate of the SDM is 1 MHz. The codec has a 65-dB SNR and harmonic distortion performance.

10.10 SUMMARY AND REFERENCES

The need for sampling rate conversion arises frequently in digital signal processing applications. In this chapter we first treated sampling rate reduction (decimation) and sampling rate increase (interpolation) by integer factors and then demonstrated how the two processes can be combined to obtain sampling rate conversion by any rational factor. Later, in Section 10.8, we described a method to achieve sampling rate conversion by an arbitrary factor.

In general, the implementation of sampling rate conversion requires the use of a linear time-variant filter. We described methods for implementing such filters, including the class of polyphase filter structures, which are especially simple to implement. We also described the use of multistage implementations of multirate conversion as a means of simplifying the complexity of the filter required to meet the specifications.

In the special case where the signal to be resampled is a bandpass signal, we described two methods for performing the sampling rate conversion, one of which involves frequency conversion, while the second is a direct conversion method that does not employ modulation.

Finally, we described a number of applications that employ multirate signal processing, including the implementation of narrowband filters, phase shifters, filter banks, subband speech coders, and transmultiplexers. These are just a few of the many applications encountered in practice where multirate signal processing is used.

The first comprehensive treatment of multirate signal processing was given in the book by Crochiere and Rabiner (1983). In the technical literature, we cite the papers by Schafer and Rabiner (1973), and Crochiere and Rabiner (1975, 1976, 1981, 1983). The use of interpolation methods to achieve sampling rate conversion

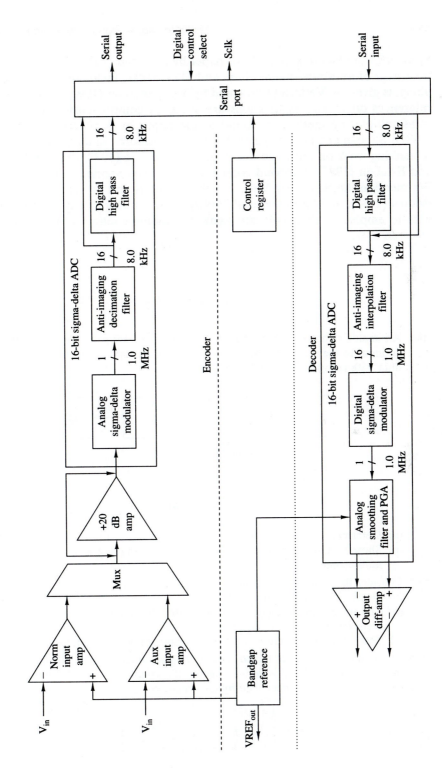

Figure 10.48 Diagram of Analog Devices ADSP-28 codec

845

by an arbitrary factor is treated in a paper by Ramstad (1984). A thorough tutorial treatment of multirate digital filters and filter banks, including quadrature mirror filters, is given by Vetterli (1987), and by Vaidyanathan (1990, 1993), where many references on various applications are cited. A comprehensive survey of digital transmultiplexing methods is found in the paper by Scheuermann and Gockler (1981). Subband coding of speech has been considered in many publications. The pioneering work on this topic was done by Crochiere (1977, 1981) and by Garland and Esteban (1980). Subband coding has also been applied to coding of images. We mention the papers by Vetterli (1984), Woods and O'Neil (1986), Smith and Eddins (1988), and Safranek et al. (1988) as just a few examples. In closing, we wish to emphasize that multirate signal processing continues to be a very active research area.

PROBLEMS

10.1 An analog signal $x_a(t)$ is bandlimited to the range $900 \leq F \leq 1100$ Hz. It is used as an input to the system shown in Fig. P10.1. In this system, $H(\omega)$ is an ideal lowpass filter with cutoff frequency $F_c = 125 Hz$.

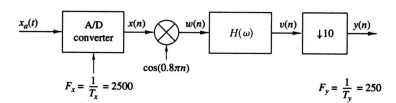

Figure P10.1

(a) Determine and sketch the spectra for the signals $x(n)$, $w(n)$, $v(n)$, and $y(n)$.

(b) Show that it is possible to obtain $y(n)$ by sampling $x_a(t)$ with period $T = 4$ milliseconds.

10.2 Consider the signal $x(n) = a^n u(n)$, $|a| < 1$.

(a) Determine the spectrum $X(\omega)$.

(b) The signal $x(n)$ is applied to a decimator that reduces the rate by a factor of 2. Determine the output spectrum.

(c) Show that the spectrum in part (b) is simply the Fourier transform of $x(2n)$.

10.3 The sequence $x(n)$ is obtained by sampling an analog signal with period T. From this signal a new signal is derived having the sampling period $T/2$ by use of a linear interpolation method described by the equation

$$y(n) = \begin{cases} x(n/2), & n \text{ even} \\ \frac{1}{2}\left[x\left(\frac{n-1}{2}\right) + x\left(\frac{n+1}{2}\right)\right], & n \text{ odd} \end{cases}$$

(a) Show that this linear interpolation scheme can be realized by basic digital signal processing elements.

(b) Determine the spectrum of $y(n)$ when the spectrum of $x(n)$ is

$$X(\omega) = \begin{cases} 1, & 0 \le |\omega| \le 0.2\pi \\ 0, & \text{otherwise} \end{cases}$$

(c) Determine the spectrum of $y(n)$ when the spectrum of $x(n)$ is

$$X(\omega) = \begin{cases} 1, & 0.7\pi \le |\omega| \le 0.9\pi \\ 0, & \text{otherwise} \end{cases}$$

10.4 Consider a signal $x(n)$ with Fourier transform

$$X(\omega) = 0 \text{ for } \quad \begin{aligned} \omega_m &< |\omega| \le \pi \\ f_m &< |f| \le \tfrac{1}{2} \end{aligned}$$

(a) Show that the signal $x(n)$ can be recovered from its samples $x(mD)$ if the sampling frequency $\omega_s = 2\pi/D \ge 2\omega_m$ ($f_s = 1/D \ge 2f_m$).

(b) Show that $x(n)$ can be reconstructed using the formula

$$x(n) = \sum_{k=-\infty}^{\infty} x(kD) h_r(n - kD)$$

where

$$h_r(n) = \frac{\sin(2\pi f_c n)}{2\pi n} \quad \begin{aligned} f_m &< f_c < f_s - f_m \\ \omega_m &< \omega_c < \omega_s - \omega_m \end{aligned}$$

(c) Show that the bandlimited interpolation in part (b) can be thought as a two-step process: First, increasing the sampling rate by a factor of D by inserting $(D-1)$ zero samples between successive samples of the decimated signal $x_a(n) = x(nD)$ and second, filtering the resulting signal using an ideal lowpass filter with cutoff frequency ω_c.

10.5 In this problem we illustrate the concepts of sampling and decimation for discrete-time signals. To this end consider a signal $x(n)$ with Fourier transform $X(\omega)$ as in Fig. P10.5.

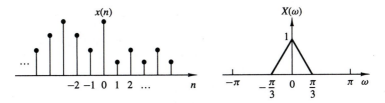

Figure P10.5

(a) Sampling $x(n)$ with a sampling period $D = 2$ results to the signal

$$x_s(n) = \begin{cases} x(n), & n = 0, \pm 2, \pm 4, \ldots \\ 0, & n = \pm 1, \pm 3, \pm 5, \ldots \end{cases}$$

Compute and sketch the signal $x_s(n)$ and its Fourier transform $X_s(\omega)$. Can we reconstruct $x(n)$ from $x_s(n)$? How?

(b) Decimating $x(n)$ by a factor of $D = 2$ produces the signal

$$x_d(n) = x(2n) \qquad \text{all} \quad n$$

Show that $X_d(\omega) = X_s(\omega/2)$. Plot the signal $x_d(n)$ and its transform $X_d(\omega)$. Do we lose any information when we decimate the sampled signal $x_s(n)$?

10.6* Design a decimator that downsamples an input signal $x(n)$ by a factor $D = 5$. Use the Remez algorithm to determine the coefficients of the FIR filter that has 0.1-dB ripple in the passband ($0 \le \omega \le \pi/5$) and is down by at least 30 dB in the stopband. Also determine the corresponding polyphase filter structure for implementing the decimator.

10.7* Design an interpolator that increases the input sampling rate by a factor of $I = 2$. Use the Remez algorithm to determine the coefficients of the FIR filter that has a 0.1-dB ripple in the passband ($0 \le \omega \le \pi/2$) and is down by at least 30 dB in the stopband. Also, determine the corresponding polyphase filter structure for implementing the interpolator.

10.8* Design a sample-rate converter that reduces the sampling rate by a factor $\frac{2}{3}$. Use the Remez algorithm to determine the coefficients of the FIR filter that has a 0.1-dB ripple in the passband and is down by at least 30 dB in the stopband. Specify the sets of time-variant coefficients $g(n, m)$ and the corresponding coefficients in the polyphase filter realization of the sample-rate converter.

10.9 Consider the two different ways of cascading a decimator with an interpolator shown in Fig. P10.9.

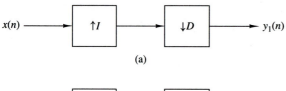

(a)

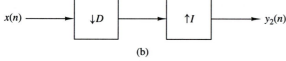

(b)   **Figure P10.9**

(a) If $D = I$, show that the outputs of the two configurations are different. Hence, in general, the two systems are not identical.

(b) Show that the two systems are identical if and only if D and I are relatively prime.

10.10 Prove the equivalence of the two decimator and interpolator configurations shown in Fig. P10.10. These equivalent relations are called the "noble identities" (see Vaidyanathan, 1990).

10.11 Consider an arbitrary digital filter with transfer function

$$H(z) = \sum_{n=-\infty}^{\infty} h(n)z^{-n}$$

(a) Perform a two-component polyphase decomposition of $H(z)$ by grouping the even-numbered samples $h_0(n) = h(2n)$ and the odd-numbered samples $h_1(n) =$

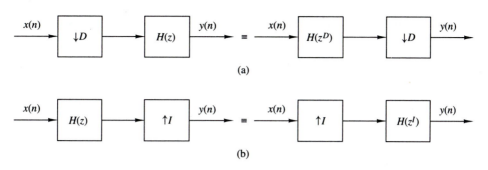

Figure P10.10

$h(2n + 1)$. Thus show that $H(z)$ can be expressed as

$$H(z) = H_0(z^2) + z^{-1}H_1(z^2)$$

and determine $H_0(z)$ and $H_1(z)$.

(b) Generalize the result in part (a) by showing that $H(z)$ can be decomposed into an D-component polyphase filter structure with transfer function

$$H(z) = \sum_{k=0}^{D-1} z^{-k} H_k(z^D)$$

Determine $H_k(z)$.

(c) For the IIR filter with transfer function

$$H(z) = \frac{1}{1 - az^{-1}}$$

determine $H_0(z)$ and $H_1(z)$ for the two-component decomposition.

10.12 Design a two-stage decimator for the following specifications

	$D = 100$
Passband:	$0 \le F \le 50$
Transition band:	$50 \le F \le 55$
Input sampling rate:	10,000 Hz
Ripple:	$\delta_1 = 10^{-1}, \delta_2 = 10^{-3}$

10.13 Design a linear phase FIR filter that satisfies the following specifications based on a single-stage and a two-stage multirate structure.

Sampling rate:	10,000 Hz
Passband:	$0 \le F \le 60$
Transition band:	$60 \le F \le 65$
Ripple:	$\delta_1 = 10^{-1}, \delta_2 = 10^{-3}$

10.14 Prove that the half-band filter that satisfies (10.9.43) is always odd and the even coefficients are zero.

10.15 Design one-stage and two-stage interpolators to meet the following specification:

$$I = 20$$

Input sampling rate:	10,000 Hz
Passband:	$0 \le F \le 90$
Transition band:	$90 \le F \le 100$
Ripple:	$\delta_1 = 10^{-2}, \delta_2 = 10^{-3}$

10.16 By using (10.9.26) derive the equations corresponding to the structure for the poly-phase synthesis section shown in Fig. 10.43.

10.17 Show that the transpose of an L-stage interpolator for increasing the sampling rate by an integer factor I is equivalent to an L-stage decimator that decreases the sampling rate by a factor $D = I$.

10.18 Sketch the polyphase filter structure for achieving a time advance of $(k/I)T_x$ in a sequence $x(n)$.

10.19 Prove the following expressions for an interpolator of order I.

(a) The impulse response $h(n)$ can be expressed as

$$h(n) = \sum_{k=0}^{I-1} p_k(n - k)$$

where

$$p_k(n) = \begin{cases} p_k(n/I), & n = 0, \pm I, \pm 2I, \ldots \\ 0, & \text{otherwise} \end{cases}$$

(b) $H(z)$ may be expressed as

$$H(z) = \sum_{k=0}^{I-1} z^{-k} p_k(z)$$

(c)
$$P_k(z) = \frac{1}{I} \sum_{n=-\infty}^{\infty} \sum_{l=0}^{I-1} h(n) e^{j2\pi l(n-k)/I} z^{-(n-k)/I}$$

$$P_k(\omega) = \frac{1}{I} \sum_{l=0}^{I-1} H\left(\omega - \frac{2\pi l}{I}\right) e^{j(\omega - 2\pi l)k/I}$$

10.20* *Zoom-frequency analysis*　Consider the system in Fig. P10.20a.

(a) Sketch the spectrum of the signal $y(n) = y_R(n) + jy_I(n)$ if the input signal $x(n)$ has the spectrum shown in Fig. P10.20b.

(b) Suppose that we are interested in the analysis of the frequencies in the band $f_0 \le f \le f_0 + \Delta f$, where $f_0 = \pi/6$ and $\Delta f = \pi/3$. Determine the cutoff of a lowpass filter and the decimation factor D required to retain the information contained in this band of frequencies.

(c) Assume that

$$x(n) = \sum_{k=0}^{p-1} \left(1 - \frac{k}{2p}\right) \cos 2\pi f_k n$$

where $p = 40$ and $f_k = k/p$, $k = 0, 1, \ldots, p-1$. Compute and plot the 1024-point DFT of $x(n)$.

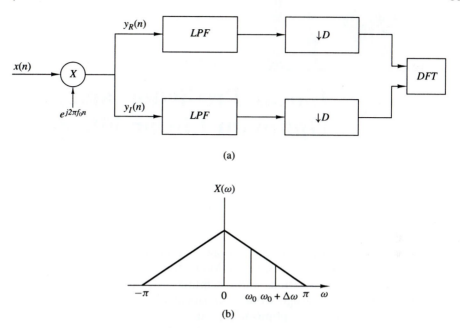

(a)

(b)

Figure P10.19

(d) Repeat part (b) for the signal $x(n)$ given in part (c) by using an appropriately designed lowpass linear phase FIR filter to determine the decimated signal $s(n) = s_R(n) + js_I(n)$.

(e) Compute the 1024-point DFT of $s(n)$ and investigate to see if you have obtained the expected results.

11

Linear Prediction and Optimum Linear Filters

The design of filters to perform signal estimation is a problem that frequently arises in the design of communication systems, control systems, in geophysics, and in many other applications and disciplines. In this chapter we treat the problem of optimum filter design from a statistical viewpoint. The filters are constrained to be linear and the optimization criterion is based on the minimization of the mean-square error. As a consequence, only the second-order statistics (autocorrelation and crosscorrelation functions) of a stationary process are required in the determination of the optimum filters.

Included in this treatment is the design of optimum filters for linear prediction. Linear prediction is a particularly important topic in digital signal processing, with applications in a variety of areas, such as speech signal processing, image processing, and noise suppression in communication systems. As we shall observe, determination of the optimum linear filter for prediction requires the solution of a set of linear equations that have some special symmetry. To solve these linear equations, we describe two algorithms, the Levinson–Durbin algorithm and the Schur algorithm, which provide the solution to the equations through computationally efficient procedures that exploit the symmetry properties.

The last section of the chapter treats an important class of optimum filters called Wiener filters. Wiener filters are widely used in many applications involving the estimation of signals corrupted with additive noise.

11.1 INNOVATIONS REPRESENTATION OF A STATIONARY RANDOM PROCESS

In this section we demonstrate that a wide-sense stationary random process can be represented as the output of a causal and causally invertible linear system excited by a white noise process. The condition that the system is causally invertible also allows us to represent the wide-sense stationary random process by the output of the inverse system, which is a white noise process.

852

Let us consider a wide-sense stationary process $\{x(n)\}$ with autocorrelation sequence $\{\gamma_{xx}(m)\}$ and power spectral density $\Gamma_{xx}(f)$, $|f| \leq \frac{1}{2}$. We assume that $\Gamma_{xx}(f)$ is real and continuous for all $|f| \leq \frac{1}{2}$. The z-transform of the autocorrelation sequence $\{\gamma_{xx}(m)\}$ is

$$\Gamma_{xx}(z) = \sum_{m=-\infty}^{\infty} \gamma_{xx}(m)z^{-m} \tag{11.1.1}$$

from which we obtain the power spectral density by evaluating $\Gamma_{xx}(z)$ on the unit circle [i.e. by substituting $z = \exp(j2\pi f)$].

Now, let us assume that $\log \Gamma_{xx}(z)$ is analytic (possesses derivatives of all orders) in an annular region in the z-plane that includes the unit circle (i.e., $r_1 < |z| < r_2$ where $r_1 < 1$ and $r_2 > 1$). Then, $\log \Gamma_{xx}(z)$ can be expanded in a Laurent series of the form

$$\log \Gamma_{xx}(z) = \sum_{m=-\infty}^{\infty} v(m)z^{-m} \tag{11.1.2}$$

where the $\{v(m)\}$ are the coefficients in the series expansion. We can view $\{v(m)\}$ as the sequence with z-transform $V(z) = \log \Gamma_{xx}(z)$. Equivalently, we can evaluate $\log \Gamma_{xx}(z)$ on the unit circle,

$$\log \Gamma_{xx}(f) = \sum_{m=-\infty}^{\infty} v(m)e^{-j2\pi fm} \tag{11.1.3}$$

so that the $\{v(m)\}$ are the Fourier coefficients in the Fourier series expansion of the periodic function $\log \Gamma_{xx}(f)$. Hence

$$v(m) = \int_{-\frac{1}{2}}^{\frac{1}{2}} [\log \Gamma_{xx}(f)]e^{j2\pi fm}df \qquad m = 0, \pm 1, \ldots \tag{11.1.4}$$

We observe that $v(m) = v(-m)$, since $\Gamma_{xx}(f)$ is a real and even function of f.

From (11.1.2) it follows that

$$\Gamma_{xx}(z) = \exp\left[\sum_{m=-\infty}^{\infty} v(m)z^{-m} \right] \tag{11.1.5}$$

$$= \sigma_w^2 H(z)H(z^{-1})$$

where, by definition, $\sigma_w^2 = \exp[v(0)]$ and

$$H(z) = \exp\left[\sum_{m=1}^{\infty} v(m)z^{-m} \right], \qquad |z| > r_1 \tag{11.1.6}$$

If (11.1.5) is evaluated on the unit circle, we have the equivalent representation of the power spectral density as

$$\Gamma_{xx}(f) = \sigma_w^2 |H(f)|^2 \tag{11.1.7}$$

We note that

$$\log \Gamma_{xx}(f) = \log \sigma_w^2 + \log H(f) + \log H^*(f)$$

$$= \sum_{m=-\infty}^{\infty} v(m)e^{-j2\pi fm}$$

From the definition of $H(z)$ given by (11.1.6), it is clear that the causal part of the Fourier series in (11.1.3) is associated with $H(z)$ and the anticausal part is associated with $H(z^{-1})$. The Fourier series coefficients $\{v(m)\}$ are the *cepstral coefficients* and the sequence $\{v(m)\}$ is called the cepstrum of the sequence $\{\gamma_{xx}(m)\}$, as defined in Section 4.2.7.

The filter with system function $H(z)$ given by (11.1.6) is analytic in the region $|z| > r_1 < 1$. Hence, in this region, it has a Taylor series expansion as a causal system of the form

$$H(z) = \sum_{m=0}^{\infty} h(n)z^{-n} \tag{11.1.8}$$

The output of this filter in response to a white noise input sequence $w(n)$ with power spectral density σ_w^2 is a stationary random process $\{x(n)\}$ with power spectral density $\Gamma_{xx}(f) = \sigma_w^2 |H(f)|^2$. Conversely, the stationary random process $\{x(n)\}$ with power spectral density $\Gamma_{xx}(f)$ can be transformed into a white noise process by passing $\{x(n)\}$ through a linear filter with system function $1/H(z)$. We call this filter a noise *whitening filter*. Its output, denoted as $\{w(n)\}$ is called the *innovations process* associated with the stationary random process $\{x(n)\}$. These two relationships are illustrated in Fig. 11.1.

The representation of the stationary stochastic process $\{x(n)\}$ as the output of an IIR filter with system function $H(z)$ given by (11.1.8) and excited by a white noise sequence $\{w(n)\}$ is called the *Wold representation*.

11.1.1 Rational Power Spectra

Let us now restrict our attention to the case where the power spectral density of the stationary random process $\{x(n)\}$ is a rational function, expressed as

$$\Gamma_{xx}(z) = \sigma_w^2 \frac{B(z)B(z^{-1})}{A(z)A(z^{-1})} \qquad r_1 < |z| < r_2 \tag{11.1.9}$$

where the polynomials $B(z)$ and $A(z)$ have roots that fall inside the unit circle in the z-plane. Then the linear filter $H(z)$ for generating the random process $\{x(n)\}$

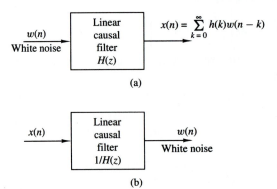

(a)

(b)

Figure 11.1 Filters for generating (a) the random process $x(n)$ from white noise and (b) the inverse filter.

from the white noise sequence $\{w(n)\}$ is also rational and is expressed as

$$
H(z) = \frac{B(z)}{A(z)} = \frac{\displaystyle\sum_{k=0}^{q} b_k z^{-k}}{1 + \displaystyle\sum_{k=1}^{p} a_k z^{-k}} \qquad |z| > r_1 \tag{11.1.10}
$$

where $\{b_k\}$ and $\{a_k\}$ are the filter coefficients that determine the location of the zeros and poles of $H(z)$, respectively. Thus $H(z)$ is causal, stable, and minimum phase. Its reciprocal $1/H(z)$ is also a causal, stable, and minimum-phase linear system. Therefore, the random process $\{x(n)\}$ uniquely represents the statistical properties of the innovations process $\{w(n)\}$, and vice versa.

For the linear system with the rational system function $H(z)$ given by (11.1.10), the output $x(n)$ is related to the input $w(n)$ by the difference equation

$$
x(n) + \sum_{k=1}^{p} a_k x(n-k) = \sum_{k=0}^{q} b_k w(n-k) \tag{11.1.11}
$$

We will distinguish among three specific cases.

Autoregressive (AR) process. $b_0 = 1$, $b_k = 0$, $k > 0$. In this case, the linear filter $H(z) = 1/A(z)$ is an all-pole filter and the difference equation for the input–output relationship is

$$
x(n) + \sum_{k=1}^{p} a_k x(n-k) = w(n) \tag{11.1.12}
$$

In turn, the noise-whitening filter for generating the innovations process is an all-zero filter.

Moving average (MA) process. $a_k = 0$, $k \geq 1$. In this case, the linear filter $H(z) = B(z)$ is an all-zero filter and the difference equation for the input–output relationship is

$$
x(n) = \sum_{k=0}^{q} b_k w(n-k) \tag{11.1.13}
$$

The noise-whitening filter for the MA process is an all-pole filter.

Autoregressive, moving average (ARMA) process.   In this case, the linear filter $H(z) = B(z)/A(z)$ has both finite poles and zeros in the z-plane and the corresponding difference equation is given by (11.1.11). The inverse system for generating the innovation process from $x(n)$ is also a pole–zero system of the form $1/H(z) = A(z)/B(z)$.

11.1.2 Relationships Between the Filter Parameters and the Autocorrelation Sequence

When the power spectral density of the stationary random process is a rational function, there is a basic relationship between the autocorrelation sequence

$\{\gamma_{xx}(m)\}$ and the parameters $\{a_k\}$ and $\{b_k\}$ of the linear filter $H(z)$ that generates the process by filtering the white noise sequence $w(n)$. This relationship can be obtained by multiplying the difference equation in (11.1.11) by $x^*(n - m)$ and taking the expected value of both sides of the resulting equation. Thus we have

$$E[x(n)x^*(n - m)] = -\sum_{k=1}^{p} a_k E[x(n - k)x^*(n - m)]$$

$$+ \sum_{k=0}^{q} b_k E[w(n - k)x^*(n - m)] \tag{11.1.14}$$

Hence

$$\gamma_{xx}(m) = -\sum_{k=1}^{p} a_k \gamma_{xx}(m - k) + \sum_{k=0}^{q} b_k \gamma_{wx}(m - k) \tag{11.1.15}$$

where $\gamma_{wx}(m)$ is the cross-correlation sequence between $w(n)$ and $x(n)$.

The crosscorrelation $\gamma_{wx}(m)$ is related to the filter impulse response. That is,

$$\gamma_{wx}(m) = E[x^*(n)w(n + m)]$$

$$= E\left[\sum_{k=0}^{\infty} h(k)w^*(n - k)w(n + m)\right] \tag{11.1.16}$$

$$= \sigma_w^2 h(-m)$$

where, in the last step, we have used the fact that the sequence $w(n)$ is white. Hence

$$\gamma_{wx}(m) = \begin{cases} 0, & m > 0 \\ \sigma_w^2 h(-m), & m \leq 0 \end{cases} \tag{11.1.17}$$

By combining (11.1.17) with (11.1.15), we obtain the desired relationship

$$\gamma_{xx}(m) = \begin{cases} -\displaystyle\sum_{k=1}^{p} a_k \gamma_{xx}(m - k), & m > q \\ -\displaystyle\sum_{k=1}^{p} a_k \gamma_{xx}(m - k) + \sigma_w^2 \displaystyle\sum_{k=0}^{q-m} h(k)b_{k+m}, & 0 \leq m \leq q \\ \gamma_{xx}^*(-m), & m < 0 \end{cases} \tag{11.1.18}$$

This represents a nonlinear relationship between $\gamma_{xx}(m)$ and the parameters $\{a_k\}$, $\{b_k\}$.

The relationship in (11.1.18) applies, in general, to the ARMA process. For an AR process, (11.1.18) simplifies to

$$\gamma_{xx}(m) = \begin{cases} -\displaystyle\sum_{k=1}^{p} a_k \gamma_{xx}(m - k), & m > 0 \\ -\displaystyle\sum_{k=1}^{p} a_k \gamma_{xx}(m - k) + \sigma_w^2, & m = 0 \\ \gamma_{xx}^*(-m), & m < 0 \end{cases} \tag{11.1.19}$$

Thus we have a linear relationship between $\gamma_{xx}(m)$ and the $\{a_k\}$ parameters. These equations, called the *Yule–Walker* equations, can be expressed in the matrix form

$$
\begin{bmatrix}
\gamma_{xx}(0) & \gamma_{xx}(-1) & \gamma_{xx}(-2) & \cdots & \gamma_{xx}(-p) \\
\gamma_{xx}(1) & \gamma_{xx}(0) & \gamma_{xx}(-1) & \cdots & \gamma_{xx}(-p+1) \\
\vdots & \vdots & \vdots & & \vdots \\
\gamma_{xx}(p) & \gamma_{xx}(p-1) & \gamma_{xx}(p-2) & \cdots & \gamma_{xx}(0)
\end{bmatrix}
\begin{bmatrix}
1 \\ a_1 \\ \vdots \\ a_p
\end{bmatrix}
=
\begin{bmatrix}
\sigma_w^2 \\ 0 \\ \vdots \\ 0
\end{bmatrix}
\qquad (11.1.20)
$$

This correlation matrix is Toeplitz, and hence it can be efficiently inverted by use of the algorithms described in Section 11.3. *→ diagonal elements are the same !*

Finally, by setting $a_k = 0$, $1 \le k \le p$, and $h(k) = b_k$, $0 \le k \le q$, in (11.1.18), we obtain the relationship for the autocorrelation sequence in the case of a MA *the relationship between* process, namely,

the problem: the coeff. b_i and γ_{xx} are non-linear

$$
\gamma_{xx}(m) =
\begin{cases}
\sigma_w^2 \displaystyle\sum_{k=0}^{q} b_k b_{k+m}, & 0 \le m \le q \\
0, & m > q \\
\gamma_{xx}^*(-m), & m < 0
\end{cases}
\qquad (11.1.21)
$$

11.2 FORWARD AND BACKWARD LINEAR PREDICTION

Linear prediction is an important topic in digital signal processing with many practical applications. In this section we consider the problem of linearly predicting the value of a stationary random process either forward in time or backward in time. This formulation leads to lattice filter structures and to some interesting connections to parametric signal models.

11.2.1 Forward Linear Prediction

Let us begin with the problem of predicting a future value of a stationary random process from observation of past values of the process. In particular, we consider the *one-step forward linear predictor*, which forms the prediction of the value $x(n)$ by a weighted linear combination of the past values $x(n-1), x(n-2), \ldots, x(n-p)$. Hence the linearly predicted value of $x(n)$ is

$$
\hat{x}(n) = -\sum_{k=1}^{p} a_p(k) x(n-k)
\qquad (11.2.1)
$$

where the $\{-a_p(k)\}$ represent the weights in the linear combination. These weights are called the *prediction coefficients* of the one-step forward linear predictor of *order p*. The negative sign in the definition of $x(n)$ is for mathematical convenience and conforms with current practice in the technical literature.

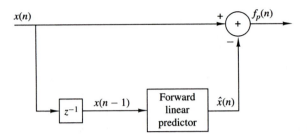

Figure 11.2   Forward linear prediction

The difference between the value $x(n)$ and the predicted value $x(n)$ is called the *forward prediction error*, denoted as $f_p(n)$:

$$f_p(n) = x(n) - \hat{x}(n)$$

$$= x(n) + \sum_{k=1}^{p} a_p(k)x(n-k) \qquad (11.2.2)$$

We view linear prediction as equivalent to linear filtering where the predictor is embedded in the linear filter, as shown in Fig. 11.2. This is called a *prediction-error filter* with input sequence $\{x(n)\}$ and output sequence $\{f_p(n)\}$. An equivalent realization for the prediction-error filter is shown in Fig. 11.3. This realization is a direct-form FIR filter with system function

$$A_p(z) = \sum_{k=0}^{p} a_p(k)z^{-k} \qquad (11.2.3)$$

where, by definition, $a_p(0) = 1$.

As shown in Section 7.2.4, the direct-form FIR filter is equivalent to an all-zero lattice filter. The lattice filter is generally described by the following set of

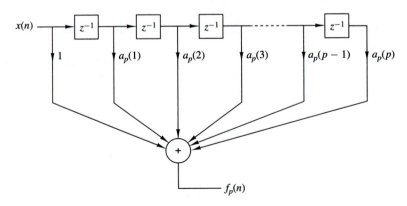

Figure 11.3   Prediction-error filter

order-recursive equation:

$$f_0(n) = g_0(n) = x(n)$$

$$f_m(n) = f_{m-1}(n) + K_m g_{m-1}(n-1) \qquad m = 1, 2, \ldots, p \qquad (11.2.4)$$

$$g_m(n) = K_m^* f_{m-1}(n) + g_{m-1}(n-1) \qquad m = 1, 2, \ldots, p$$

where $\{K_m\}$ are the reflection coefficients and $g_m(n)$ is the backward prediction error defined in the following section. Note that for complex-valued data, the conjugate of K_m is used in the equation for $g_m(n)$. Figure 11.4 illustrates a p-stage lattice filter in block diagram form along with a typical stage showing the computations given by (11.2.4).

As a consequence of the equivalence between the direct-form prediction-error FIR filter and the FIR lattice filter, the output of the p-stage lattice filter is expressed as

$$f_p(n) = \sum_{k=0}^{p} a_p(k)x(n-k) \qquad a_p(0) = 1 \qquad (11.2.5)$$

Since (11.2.5) is a convolution sum, the z-transform relationship is

$$F_p(z) = A_p(z)X(z) \qquad (11.2.6)$$

or, equivalently,

$$A_p(z) = \frac{F_p(z)}{X(z)} = \frac{F_p(z)}{F_0(z)} \qquad (11.2.7)$$

The mean-square value of the forward linear prediction error $f_p(n)$ is

$$\mathcal{E}_p^f = E[|f_p(n)|^2]$$

$$= \gamma_{xx}(0) + 2\,\mathrm{Re}\left[\sum_{k=1}^{p} a_p^*(k)\gamma_{xx}(k)\right] + \sum_{k=1}^{p}\sum_{l=1}^{p} a_p^*(l)a_p(k)\gamma_{xx}(l-k) \qquad (11.2.8)$$

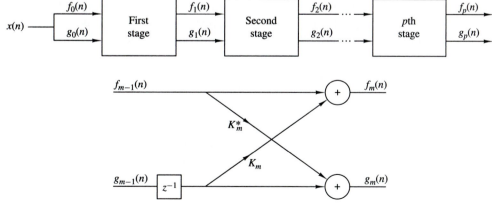

Figure 11.4   *p*-stage lattice filter

$\mathcal{E}_p^f$ is a quadratic function of the predictor coefficients and its minimization leads to the set of linear equations

$$\gamma_{xx}(l) = -\sum_{k=1}^{p} a_p(k)\gamma_{xx}(l-k) \qquad l = 1, 2, \ldots, p \tag{11.2.9}$$

These are called the *normal equations* for the coefficients of the linear predictor. The minimum mean-square prediction error is simply

$$\min[\mathcal{E}_p^f] \equiv E_p^f = \gamma_{xx}(0) + \sum_{k=1}^{p} a_p(k)\gamma_{xx}(-k) \tag{11.2.10}$$

In the following section we extend the development above to the problem of predicting the value of a time series in the opposite direction, namely, backward in time.

11.2.2 Backward Linear Prediction

Let us assume that we have the data sequence $x(n), x(n-1), \ldots, x(n-p+1)$ from a stationary random process and we wish to predict the value $x(n-p)$ of the process. In this case we employ a *one-step backward linear predictor* of order p. Hence

$$\hat{x}(n-p) = -\sum_{k=0}^{p-1} b_p(k)x(n-k) \tag{11.2.11}$$

The difference between the value $x(n-p)$ and the estimate $\hat{x}(n-p)$ is called the *backward prediction error*, denoted as $g_p(n)$:

$$g_p(n) = x(n-p) + \sum_{k=0}^{p-1} b_p(k)x(n-k)$$

$$\tag{11.2.12}$$

$$= \sum_{k=0}^{p} b_p(k)x(n-k) \qquad b_p(p) = 1$$

The backward linear predictor can be realized either by a direct-form FIR filter structure similar to the structure shown in Fig. 11.2 or as a lattice structure. The lattice structure shown in Fig. 11.4 provides the backward linear predictor as well as the forward linear predictor.

The weighting coefficients in the backward linear predictor are the complex conjugates of the coefficients for the forward linear predictor, but they occur in reverse order. Thus we have

$$b_p(k) = a_p^*(p-k) \qquad k = 0, 1, \ldots, p \tag{11.2.13}$$

In the z-domain, the convolution sum in (11.2.12) becomes

$$G_p(z) = B_p(z)X(z) \tag{11.2.14}$$

or, equivalently,

$$B_p(z) = \frac{G_p(z)}{X(z)} = \frac{G_p(z)}{G_0(z)} \tag{11.2.15}$$

where $B_p(z)$ represents the system function of the FIR filter with coefficients $b_p(k)$.

Since $b_p(k) = a^*(p - k)$, $G_p(z)$ is related to $A_p(z)$

$$
\begin{aligned}
B_p(z) &= \sum_{k=0}^{p} b_p(k) z^{-k} \\
&= \sum_{k=0}^{p} a_p^*(p - k) z^{-k} \\
&= z^{-p} \sum_{k=0}^{p} a_p^*(k) z^{k} \\
&= z^{-p} A_p^*(z^{-1})
\end{aligned}
\tag{11.2.16}
$$

The relationship in (11.2.16) implies that the zeros of the FIR filter with system function $B_p(z)$ are simply the (conjugate) reciprocals of the zeros of $A_p(z)$. Hence $B_p(z)$ is called the reciprocal or *reverse polynomial* of $A_p(z)$.

Now that we have established these interesting relationships between the forward and backward one-step predictors, let us return to the recursive lattice equations in (11.2.4) and transform them to the z-domain. Thus we have

$$
F_0(z) = G_0(z) = X(z)
$$

$$
F_m(z) = F_{m-1}(z) + K_m z^{-1} G_{m-1}(z) \qquad m = 1, 2, \ldots, p \tag{11.2.17}
$$

$$
G_m(z) = K_m^* F_{m-1}(z) + z^{-1} G_{m-1}(z) \qquad m = 1, 2, \ldots, p
$$

If we divide each equation by $X(z)$, we obtain the desired results in the form

$$
A_0(z) = B_0(z) = 1
$$

$$
A_m(z) = A_{m-1}(z) + K_m z^{-1} B_{m-1}(z) \qquad m = 1, 2, \ldots, p \tag{11.2.18}
$$

$$
B_m(z) = K_m^* A_{m-1}(z) + z^{-1} B_{m-1}(z) \qquad m = 1, 2, \ldots, p
$$

Thus a lattice filter is described in the z-domain by the matrix equation

$$
\begin{bmatrix} A_m(z) \\ B_m(z) \end{bmatrix} = \begin{bmatrix} 1 & K_m z^{-1} \\ K_m^* & z^{-1} \end{bmatrix} \begin{bmatrix} A_{m-1}(z) \\ B_{m-1}(z) \end{bmatrix}
\tag{11.2.19}
$$

The relations in (11.2.17) for $A_m(z)$ and $B_m(z)$ allow us to obtain the direct-form FIR filter coefficients $\{a_m(k)\}$ from the reflection coefficients $\{K_m\}$, and vice versa. These relationships were given in Section 7.2.4 by (7.2.51) through (7.2.53).

The lattice structure with parameters $K_1, K_2, \ldots, K_p$ corresponds to a class of p direct-form FIR filters with system functions $A_1(z), A_2(z), \ldots, A_p(z)$. It is interesting to note that a characterization of this class of p FIR filters in direct form requires $p(p+1)/2$ filter coefficients. In contrast, the lattice-form characterization requires only the p reflection coefficients $\{K_i\}$. The reason the lattice provides a more compact representation for the class of p FIR filters is because appending stages to the lattice does not alter the parameters of the previous stages. On the other hand, appending the pth stage to a lattice with $(p-1)$ stages is equivalent to increasing the length of an FIR filter by one coefficient. The resulting FIR filter with system function $A_p(z)$ has coefficients totally different from the coefficients of the lower-order FIR filter with system function $A_{p-1}(z)$.

The formula for determining the filter coefficients $\{a_p(k)\}$ recursively is easily derived from polynomial relationships (11.2.18). We have

$$A_m(z) = A_{m-1}(z) + K_m z^{-1} B_{m-1}(z)$$

$$\sum_{k=0}^{m} a_m(k)z^{-k} = \sum_{k=0}^{m-1} a_{m-1}(k)z^{-k} + K_m \sum_{k=0}^{m-1} a_{m-1}^*(m-1-k)z^{-(k+1)} \qquad (11.2.20)$$

By equating the coefficients of equal powers of z^{-1} and recalling that $a_m(0) = 1$ for $m = 1, 2, \ldots, p$, we obtain the desired recursive equation for the FIR filter coefficients in the form

$$a_m(0) = 1$$

$$a_m(m) = K_m$$

$$a_m(k) = a_{m-1}(k) + K_m a_{m-1}^*(m-k) \qquad (11.2.21)$$

$$= a_{m-1}(k) + a_m(m)a_{m-1}^*(m-k) \qquad 1 \le k \le m-1$$

$$m = 1, 2, \ldots, p$$

The conversion formula from the direct-form FIR filter coefficients $\{a_p(k)\}$ to the lattice reflection coefficients $\{K_i\}$ is also very simple. For the p-stage lattice we immediately obtain the reflection coefficient $K_p = a_p(p)$. To obtain $K_{p-1}, \ldots, K_1$, we need the polynomials $A_m(z)$ for $m = p-1, \ldots, 1$. From (11.2.19) we obtain

$$A_{m-1}(z) = \frac{A_m(z) - K_m B_m(z)}{1 - |K_m|^2} \qquad m = p, \ldots, 1 \qquad (11.2.22)$$

which is just a step-down recursion. Thus we compute all lower-degree polynomials $A_m(z)$ beginning with $A_{p-1}(z)$ and obtain the desired lattice reflection coefficients from the relation $K_m = a_m(m)$. We observe that the procedure works as long as $|K_m| \ne 1$ for $m = 1, 2, \ldots, p-1$. From this step-down recursion for the polynomials, it is relatively easy to obtain a formula for recursively and directly computing K_m, $m = p-1, \ldots, 1$. For $m = p-1, \ldots, 1$, we have

$$K_m = a_m(m)$$

$$a_{m-1}(k) = \frac{a_m(k) - K_m b_m(k)}{1 - |K_m|^2} \qquad (11.2.23)$$

$$= \frac{a_m(k) - a_m(m)a_m^*(m-k)}{1 - |a_m(m)|^2}$$

which is just the recursion in the Schür–Cohn stability test for the polynomial $A_m(z)$.

As just indicated, the recursive equation in (11.2.23) breaks down if any of the lattice parameters $|K_m| = 1$. If this occurs, it is indicative that the polynomial $A_{m-1}(z)$ has a root located on the unit circle. Such a root can be factored out from $A_{m-1}(z)$ and the iterative process in (11.2.23) carried out for the reduced-order system.

Finally, let us consider the minimization of the mean-square error in a backward linear predictor. The backward prediction error is

$$g_p(n) = x(n - p) + \sum_{k=0}^{p-1} b_p(k)x(n - k)$$

$$= x(n - p) + \sum_{k=1}^{p} a_p^*(k)x(n - p + k) \tag{11.2.24}$$

and its mean-square value is

$$\mathcal{E}_p^b = E[|g_p(n)|^2] \tag{11.2.25}$$

The minimization of $\mathcal{E}_p^b$ with respect to the prediction coefficients yields the same set of linear equations as in (11.2.9). Hence the minimum mean-square error is

$$\min[\mathcal{E}_p^b] \equiv E_p^b = E_p^f \tag{11.2.26}$$

which is given by (11.2.10).

11.2.3 The Optimum Reflection Coefficients for the Lattice Forward and Backward Predictors

In Sections 11.2.1 and 11.2.2 we derived the set of linear equations which provide the predictor coefficients that minimize the mean-square value of the prediction error. In this section we consider the problem of optimizing the reflection coefficients in the lattice predictor and expressing the reflection coefficients in terms of the forward and backward prediction errors.

The forward prediction error in the lattice filter is expressed as

$$f_m(n) = f_{m-1}(n) + K_m g_{m-1}(n - 1) \tag{11.2.27}$$

The minimization of $E[|f_m(n)|^2]$ with respect to the reflection coefficient K_m yields the result

$$K_m = \frac{-E[f_{m-1}(n)g_{m-1}^*(n - 1)]}{E[|g_{m-1}(n - 1)|^2]} \tag{11.2.28}$$

or, equivalently,

$$K_m = \frac{-E[f_{m-1}(n)g_{m-1}^*(n - 1)]}{\sqrt{E_{m-1}^f E_{m-1}^b}} \tag{11.2.29}$$

where $E_{m-1}^f = E_{m-1}^b = E[|g_{m-1}(n - 1)|^2] = E[|f_{m-1}(n)|^2]$.

We observe that the optimum choice of the reflection coefficients in the lattice predictor is the negative of the (normalized) crosscorrelation coefficients between the forward and backward errors in the lattice.* Since it is apparent from (11.2.28) that $|K_m| \leq 1$, it follows that the minimum mean-square value of the prediction error, which can be expressed recursively as

$$E_m^f = (1 - |K_m|^2) E_{m-1}^f \tag{11.2.30}$$

is a monotonically decreasing sequence.

11.2.4 Relationship of an AR Process to Linear Prediction

The parameters of an $AR(p)$ process are intimately related to a predictor of order p for the same process. To see the relationship, we recall that in an $AR(p)$ process, the autocorrelation sequence $\{\gamma_{xx}(m)\}$ is related to the parameters $\{a_k\}$ by the Yule–Walker equations given in (11.1.19) or (11.1.20). The corresponding equations for the predictor of order p are given by (11.2.9) and (11.2.10).

A direct comparison of these two sets of relations reveals that there is a one-to-one correspondence between the parameters $\{a_k\}$ of the $AR(p)$ process and the predictor coefficients $\{a_p(k)\}$ of the pth-order predictor. In fact, if the underlying process $\{x(n)\}$ is $AR(p)$, the prediction coefficients of the pth-order predictor are identical to $\{a_k\}$. Furthermore, the minimum MSE in the pth-order predictor E_p^f is identical to σ_w^2, the variance of the white noise process. In this case, the prediction-error filter is a noise-whitening filter which produces the innovations sequence $\{w(n)\}$.

11.3 SOLUTION OF THE NORMAL EQUATIONS

In the preceding section we observed that the minimization of the mean-square value of the forward prediction error resulted in a set of linear equations for the coefficients of the predictor given by (11.2.9). These equations, called the normal equations, may be expressed in the compact form

$$\sum_{k=0}^{p} a_p(k) \gamma_{xx}(l - k) = 0 \qquad \begin{array}{l} l = 1, 2, \ldots, p \\ a_p(0) = 1 \end{array} \tag{11.3.1}$$

The resulting minimum MSE (MMSE) is given by (11.2.10). If we augment (11.2.10) to the normal equations given by (11.3.1) we obtain the set of *augmented normal equations*, which may be expressed as

$$\sum_{k=0}^{p} a_p(k) \gamma_{xx}(l - k) = \begin{cases} E_p^f, & l = 0 \\ 0, & l = 1, 2, \ldots, p \end{cases} \tag{11.3.2}$$

We also noted that if the random process is an $AR(p)$ process, the MMSE $E_p^f = \sigma_w^2$.

---

*The normalized crosscorrelation coefficients between the forward and backward error in the lattice (i.e., $\{-K_m\}$) are often called the partial correlation (PARCOR) coefficients.

In this section we describe two computationally efficient algorithms for solving the normal equations. One algorithm, originally due to Levinson (1947) and modified by Durbin (1959), is called the Levinson–Durbin algorithm. This algorithm is suitable for serial processing and has a computation complexity of $O(p^2)$. The second algorithm, due to Schür (1917), also computes the reflection coefficients in $O(p^2)$ operations but with parallel processors, the computations can be performed in $O(p)$ time. Both algorithms exploit the Toeplitz symmetry property inherent in the autocorrelation matrix.

We begin by describing the Levinson–Durbin algorithm.

11.3.1 The Levinson–Durbin Algorithm

The Levinson-Durbin algorithm is a computationally efficient algorithm for solving the normal equations in (11.3.1) for the prediction coefficients. This algorithm exploits the special symmetry in the autocorrelation matrix

$$
p = \begin{bmatrix}
\gamma_{xx}(0) & \gamma_{xx}^*(1) & \cdots & \gamma_{xx}^*(p-1) \\
\gamma_{xx}(1) & \gamma_{xx}(0) & \cdots & \gamma_{xx}^*(p-2) \\
\vdots & \vdots & & \\
\gamma_{xx}(p-1) & \gamma_{xx}(p-2) & \cdots & \gamma_{xx}(0)
\end{bmatrix} \tag{11.3.3}
$$

Note that $\Gamma_p(i, j) = \Gamma_p(i - j)$, so that the autocorrelation matrix is a *Toeplitz matrix*. Since $\Gamma_p(i, j) = \Gamma_p^*(j, i)$, the matrix is also Hermitian.

The key to the Levinson–Durbin method of solution, that exploits the Toeplitz property of the matrix, is to proceed recursively, beginning with a predictor of order $m = 1$ (one coefficient) and then to increase the order recursively, using the lower-order solutions to obtain the solution to the next-higher order. Thus the solution to the first-order predictor obtained by solving (11.3.1) is

$$
a_1(1) = -\frac{\gamma_{xx}(1)}{\gamma_{xx}(0)} \tag{11.3.4}
$$

and the resulting MMSE is

$$
\begin{aligned}
E_1^f &= \gamma_{xx}(0) + a_1(1)\gamma_{xx}(-1) \\
&= \gamma_{xx}(0)[1 - |a_1(1)|^2]
\end{aligned} \tag{11.3.5}
$$

Recall that $a_1(1) = K_1$, the first reflection coefficient in the lattice filter.

The next step is to solve for the coefficients $\{a_2(1), a_2(2)\}$ of the second-order predictor and express the solution in terms of $a_1(1)$. The two equations obtained from (11.3.1) are

$$
\begin{aligned}
a_2(1)\gamma_{xx}(0) + a_2(2)\gamma_{xx}^*(1) &= -\gamma_{xx}(1) \\
a_2(1)\gamma_{xx}(1) + a_2(2)\gamma_{xx}(0) &= -\gamma_{xx}(2)
\end{aligned} \tag{11.3.6}
$$

By using the solution in (11.3.4) to eliminate $\gamma_{xx}(1)$, we obtain the solution

$$a_2(2) = -\frac{\gamma_{xx}(2) + a_1(1)\gamma_{xx}(1)}{\gamma_{xx}(0)[1 - |a_1(1)|^2]}$$

$$= -\frac{\gamma_{xx}(2) + a_1(1)\gamma_{xx}(1)}{E_1^f} \tag{11.3.7}$$

$$a_2(1) = a_1(1) + a_2(2)a_1^*(1)$$

Thus we have obtained the coefficients of the second-order predictor. Again, we note that $a_2(2) = K_2$, the second reflection coefficient in the lattice filter.

Proceeding in this manner, we can express the coefficients of the mth-order predictor in terms of the coefficients of the $(m-1)$st-order predictor. Thus we can write the coefficient vector $\mathbf{a}_m$ as the sum of two vectors, namely,

$$\mathbf{a}_m = \begin{bmatrix} a_m(1) \\ a_m(2) \\ \vdots \\ a_m(m) \end{bmatrix} = \begin{bmatrix} \mathbf{a}_{m-1} \\ \cdots \\ 0 \end{bmatrix} + \begin{bmatrix} \mathbf{d}_{m-1} \\ \cdots \\ K_m \end{bmatrix} \tag{11.3.8}$$

where $\mathbf{a}_{m-1}$ is the predictor coefficient vector of the $(m-1)$st-order predictor and the vector $\mathbf{d}_{m-1}$ and the scalar K_m are to be determined. Let us also partition the $m \times m$ autocorrelation matrix Γ_{xx} as

$$\Gamma_m = \begin{bmatrix} \Gamma_{m-1} & \gamma_{m-1}^{b*} \\ \gamma_{m-1}^{bt} & \gamma_{xx}(0) \end{bmatrix} \tag{11.3.9}$$

where $\gamma_{m-1}^{bt} = [\gamma_{xx}(m-1) \quad \gamma_{xx}(m-2) \quad \cdots \quad \gamma_{xx}(1)] = (\gamma_{m-1}^b)^t$, the asterisk (*) denotes the complex conjugate, and γ_m^t denotes the transpose of γ_m. The superscript b on γ_{m-1} denotes the vector $\gamma_{m-1}^t = [\gamma_{xx}(1) \quad \gamma_{xx}(2) \quad \cdots \quad \gamma_{xx}(m-1)]$ with elements taken in reverse order.

The solution to the equation $\Gamma_m \mathbf{a}_m = -\gamma_m$ can be expressed as

$$\begin{bmatrix} \Gamma_{m-1} & \gamma_{m-1}^{b*} \\ \gamma_{m-1}^{bt} & \gamma_{xx}(0) \end{bmatrix} \left\{ \begin{bmatrix} \mathbf{a}_{m-1} \\ 0 \end{bmatrix} + \begin{bmatrix} \mathbf{d}_{m-1} \\ K_m \end{bmatrix} \right\} = -\begin{bmatrix} \gamma_{m-1} \\ \gamma_{xx}(m) \end{bmatrix} \tag{11.3.10}$$

This is the key step in the Levinson–Durbin algorithm. From (11.3.10) we obtain two equations, namely,

$$\Gamma_{m-1}\mathbf{a}_{m-1} + \Gamma_{m-1}\mathbf{d}_{m-1} + K_m\gamma_{m-1}^{b*} = -\gamma_{m-1} \tag{11.3.11}$$

$$\gamma_{m-1}^{bt}\mathbf{a}_{m-1} + \gamma_{m-1}^{bt}\mathbf{d}_{m-1} + K_m\gamma_{xx}(0) = -\gamma_{xx}(m) \tag{11.3.12}$$

Since $\Gamma_{m-1}\mathbf{a}_{m-1} = -\gamma_{m-1}$, (11.3.11) yields the solution

$$\mathbf{d}_{m-1} = -K_m\Gamma_{m-1}^{-1}\gamma_{m-1}^{b*} \tag{11.3.13}$$

But γ_{m-1}^{b*} is just γ_{m-1} with elements taken in reverse order and conjugated. Therefore, the solution in (11.3.13) is simply

$$\mathbf{d}_{m-1} = K_m \mathbf{a}_{m-1}^{b*} = K_m \begin{bmatrix} a_{m-1}^*(m-1) \\ a_{m-1}^*(m-2) \\ . \\ a_{m-1}^*(1) \end{bmatrix} \tag{11.3.14}$$

The scalar equation (11.3.12) can now be used to solve for K_m. If we eliminate $\mathbf{d}_{m-1}$ in (11.3.12) by using (11.3.14), we obtain

$$K_m[\gamma_{xx}(0) + \boldsymbol{\gamma}_{m-1}^{bt} \mathbf{a}_{m-1}^{b*}] + \boldsymbol{\gamma}_{m-1}^{bt} \mathbf{a}_{m-1} = -\gamma_{xx}(m)$$

Hence

$$K_m = -\frac{\gamma_{xx}(m) + \boldsymbol{\gamma}_{m-1}^{bt} \mathbf{a}_{m-1}}{\gamma_{xx}(0) + \boldsymbol{\gamma}_{m-1}^{bt} \mathbf{a}_{m-1}^{b*}} \tag{11.3.15}$$

Therefore, by substituting the solutions in (11.3.14) and (11.3.15) into (11.3.8), we obtain the desired recursion for the predictor coefficients in the Levinson–Durbin algorithm as

$$a_m(m) = K_m = -\frac{\gamma_{xx}(m) + \boldsymbol{\gamma}_{m-1}^{bt} \mathbf{a}_{m-1}}{\gamma_{xx}(0) + \boldsymbol{\gamma}_{m-1}^{bt} \mathbf{a}_{m-1}^{b*}} = -\frac{\gamma_{xx}(m) + \boldsymbol{\gamma}_{m-1}^{bt} \mathbf{a}_{m-1}}{E_m^f} \tag{11.3.16}$$

$$a_m(k) = a_{m-1}(k) + K_m a_{m-1}^*(m-k)$$

$$= a_{m-1}(k) + a_m(m) a_{m-1}^*(m-k) \qquad \begin{matrix} k = 1, 2, \ldots, m-1 \\ m = 1, 2, \ldots, p \end{matrix} \tag{11.3.17}$$

The reader should note that the recursive relation in (11.3.17) is identical to the recursive relation in (11.2.21) for the predictor coefficients, obtained from the polynomials $A_m(z)$ and $B_m(z)$. Furthermore, K_m is the reflection coefficient in the mth stage of the lattice predictor. This development clearly illustrates that the Levinson–Durbin algorithm produces the reflection coefficients for the optimum lattice prediction filter as well as the coefficients of the optimum direct-form FIR predictor.

Finally, let us determine the expression for the MMSE. For the mth-order predictor, we have

$$E_m^f = \gamma_{xx}(0) + \sum_{k=1}^m a_m(k) \gamma_{xx}(-k)$$

$$= \gamma_{xx}(0) + \sum_{k=1}^m [a_{m-1}(k) + a_m(m) a_{m-1}^*(m-k)] \gamma_{xx}(-k) \tag{11.3.18}$$

$$= E_{m-1}^f[1 - |a_m(m)|^2] = E_{m-1}^f(1 - |K_m|^2) \qquad m = 1, 2, \ldots, p$$

where $E_0^f = \gamma_{xx}(0)$. Since the reflection coefficients satisfy the property that $|K_m| \leq 1$, the MMSE for the sequence of predictors satisfies the condition

$$E_0^f \geq E_1^f \geq E_2^f \geq \cdots \geq E_p^f \tag{11.3.19}$$

This concludes the derivation of the Levinson–Durbin algorithm for solving the linear equations $\Gamma_m \mathbf{a}_m = -\gamma_m$, for $m = 0, 1, \ldots, p$. We observe that the linear equations have the special property that the vector on the right-hand side also appears as a vector in Γ_m. In the more general case, where the vector on the right-hand side is some other vector, say $\mathbf{c}_m$, the set of linear equations can be solved recursively by introducing a second recursive equation to solve the more general linear equations $\Gamma_m \mathbf{b}_m = \mathbf{c}_m$. The result is a *generalized Levinson–Durbin* algorithm (see Problem 11.12).

The Levinson–Durbin recursion given by (11.3.17) requires $O(m)$ multiplications and additions (operations) to go from stage m to stage $m + 1$. Therefore, for p stages it takes on the order of $1 + 2 + 3 + \cdots + p(p+1)/2$, or $O(p^2)$, operations to solve for the prediction filter coefficients, or the reflection coefficients, compared with $O(p^3)$ operations if we did not exploit the Toeplitz property of the correlation matrix.

If the Levinson–Durbin algorithm is implemented on a serial computer or signal processor, the required computation time is on the order of $O(p^2)$ time units. On the other hand, if the processing is performed on a parallel processing machine utilizing as many processors as necessary to exploit the full parallelism in the algorithm, the multiplications as well as the additions required to compute (11.3.17) can be carried out simultaneously. Therefore, this computation can be performed in $O(p)$ time units. However, the computation in (11.3.16) for the reflection coefficients takes additional time. Certainly, the inner products involving the vectors $\mathbf{a}_{m-1}$ and γ_{m-1}^b can be computed simultaneously by employing parallel processors. However, the addition of these products cannot be done simultaneously, but instead, require $O(\log p)$ time units. Hence, the computations in the Levinson–Durbin algorithm, when performed by p parallel processors, can be accomplished in $O(p \log p)$ time.

In the following section we describe another algorithm, due to Schür (1917), that avoids the computation of inner products, and therefore is more suitable for parallel computation of the reflection coefficients.

11.3.2 The Schür Algorithm

The Schür algorithm is intimately related to a recursive test for determining the positive definiteness of a correlation matrix. To be specific, let us consider the autocorrelation matrix Γ_{p+1} associated with the augmented normal equations given by (11.3.2). From the elements of this matrix we form the function

$$R_0(z) = \frac{\gamma_{xx}(1)z^{-1} + \gamma_{xx}(2)z^{-2} + \cdots + \gamma_{xx}(p)z^{-p}}{\gamma_{xx}(0) + \gamma_{xx}(1)z^{-1} + \cdots + \gamma_{xx}(p)z^{-p}} \qquad (11.3.20)$$

and the sequence of functions $R_m(z)$ defined recursively as

$$R_m(z) = \frac{R_{m-1}(z) - R_{m-1}(\infty)}{z^{-1}[1 - R_{m-1}^*(\infty)R_{m-1}(z)]} \qquad m = 1, 2, \ldots \qquad (11.3.21)$$

Schür's theorem states that a necessary and sufficient condition for the correlation matrix to be positive definite is that $|R_m(\infty)| < 1$ for $m = 1, 2, \ldots, p$.

Let us demonstrate that the condition for positive definiteness of the autocorrelation matrix Γ_{p+1} is equivalent to the condition that the reflection coefficients in the equivalent lattice filter satisfy the condition $|K_m| < 1$, $m = 1, 2, \ldots, p$.

First, we note that $R_0(\infty) = 0$. Then, from (11.3.21) we have

$$R_1(z) = \frac{\gamma_{xx}(1) + \gamma_{xx}(2)z^{-1} + \cdots + \gamma_{xx}(p)z^{-p+1}}{\gamma_{xx}(0) + \gamma_{xx}(1)z^{-1} + \cdots + \gamma_{xx}(p)z^{-p}} \qquad (11.3.22)$$

Hence $R_1(\infty) = \gamma_{xx}(1)/\gamma_{xx}(0)$. We observe that $R_1(\infty) = -K_1$.

Second, we compute $R_2(z)$ according to (11.3.21) and evaluate the result at $z = \infty$. Thus we obtain

$$R_2(\infty) = \frac{\gamma_{xx}(2) + K_1\gamma_{xx}(1)}{\gamma_{xx}(0)(1 - |K_1|^2)}$$

Again, we observe that $R_2(\infty) = -K_2$. By continuing this procedure, we find that $R_m(\infty) = -K_m$, for $m = 1, 2, \ldots, p$. Hence the condition $|R_m(\infty)| < 1$ for $m = 1, 2, \ldots, p$, is identical to the condition $|K_m| < 1$ for $m = 1, 2, \ldots, p$, and ensures the positive definiteness of the autocorrelation matrix Γ_{p+1}.

Since the reflection coefficients can be obtained from the sequence of functions $R_m(z)$, $m = 1, 2, \ldots, p$, we have another method for solving the normal equations. We call this method the *Schür algorithm*.

Schür algorithm.   Let us first rewrite $R_m(z)$ as

$$R_m(z) = \frac{P_m(z)}{Q_m(z)} \qquad m = 0, 1, \ldots, p \qquad (11.3.23)$$

where

$$P_0(z) = \gamma_{xx}(1)z^{-1} + \gamma_{xx}(2)z^{-2} + \cdots + \gamma_{xx}(p)z^{-p}$$
$$Q_0(z) = \gamma_{xx}(0) + \gamma_{xx}(1)z^{-1} + \cdots + \gamma_{xx}(p)z^{-p} \qquad (11.3.24)$$

Since $K_0 = 0$ and $K_m = -R_m(\infty)$ for $m = 1, 2, \ldots, p$, the recursive equation (11.3.21) implies the following recursive equations for the polynomials $P_m(z)$ and $Q_m(z)$:

$$\begin{bmatrix} P_m(z) \\ Q_m(z) \end{bmatrix} = \begin{bmatrix} 1 & K_{m-1} \\ K_{m-1}^* z^{-1} & z^{-1} \end{bmatrix} \begin{bmatrix} P_{m-1}(z) \\ Q_{m-1}(z) \end{bmatrix} \qquad m = 1, 2, \ldots, p \qquad (11.3.25)$$

Thus we have

$$P_1(z) = P_0(z) = \gamma_{xx}(1)z^{-1} + \gamma_{xx}(2)z^{-2} + \cdots + \gamma_{xx}(p)z^{-p}$$
$$Q_1(z) = z^{-1}Q_0(z) = \gamma_{xx}(0)z^{-1} + \gamma_{xx}(1)z^{-2} + \cdots + \gamma_{xx}(p)z^{-p-1} \qquad (11.3.26)$$

and

$$K_1 = -\left.\frac{P_1(z)}{Q_1(z)}\right|_{z=\infty} = -\frac{\gamma_{xx}(1)}{\gamma_{xx}(0)} \qquad (11.3.27)$$

Next the reflection coefficient K_2 is obtained by determining $P_2(z)$ and $Q_2(z)$ from (11.3.25), dividing $P_2(z)$ by $Q_2(z)$ and evaluating the result at $z = \infty$. Thus we find that

$$
\begin{aligned}
P_2(z) &= P_1(z) + K_1 Q_1(z) \\
&= [\gamma_{xx}(2) + K_1 \gamma_{xx}(1)]z^{-2} + \cdots \\
&\quad + [\gamma_{xx}(p) + K_1 \gamma_{xx}(p-1)]z^{-p} \\
Q_2(z) &= z^{-1}[Q_1(z) + K_1^* P_1(z)] \\
&= [\gamma_{xx}(0) + K_1^* \gamma_{xx}(1)]z^{-2} + \cdots \\
&\quad + [\gamma_{xx}(p-2) + K_1^* \gamma_{xx}(p-1)]z^{-p}
\end{aligned}
\tag{11.3.28}
$$

where the terms involving z^{-p-1} have been dropped. Thus we observe that the recursive equation in (11.3.25) is equivalent to (11.3.21).

Based on these relationships, the Schür algorithm is described by the following recursive procedure.

Initialization. Form the $2x(p+1)$ generator matrix

$$
\mathbf{G}_0 = \begin{bmatrix} 0 & \gamma_{xx}(1) & \gamma_{xx}(2) & \cdots & \gamma_{xx}(p) \\ \gamma_{xx}(0) & \gamma_{xx}(1) & \gamma_{xx}(2) & \cdots & \gamma_{xx}(p) \end{bmatrix}
\tag{11.3.29}
$$

where the elements of the first row are the coefficients of $P_0(z)$ and the elements of the second row are the coefficients of $Q_0(z)$.

Step 1. Shift the second row of the generator matrix to the right by one place and discard the last element of this row. A zero is placed in the vacant position. Thus we obtain a new generator matrix,

$$
\mathbf{G}_1 = \begin{bmatrix} 0 & \gamma_{xx}(1) & \gamma_{xx}(1) & \gamma_{xx}(2) & \cdots & \gamma_{xx}(p) \\ 0 & \gamma_{xx}(0) & \gamma_{xx}(1) & \gamma_{xx}(1) & \cdots & \gamma_{xx}(p-1) \end{bmatrix}
\tag{11.3.30}
$$

The (negative) ratio of the elements in the second column yield the reflection coefficient $K_1 = -\gamma_{xx}(1)/\gamma_{xx}(0)$.

Step 2. Multiply the generator matrix by the 2×2 matrix

$$
\mathbf{V}_1 = \begin{bmatrix} 1 & K_1 \\ K_1^* & 1 \end{bmatrix}
\tag{11.3.31}
$$

Thus we obtain

$$
\mathbf{V}_1 \mathbf{G}_1 = \begin{bmatrix} 0 & 0 & \gamma_{xx}(2) + K_1 \gamma_{xx}(1) & \cdots & \gamma_{xx}(p) + K_1 \gamma_{xx}(p-1) \\ 0 & \gamma_{xx}(0) + K_1^* \gamma_{xx}(1) & \cdots & & \gamma_{xx}(p-1) + K_1^* \gamma_{xx}(p) \end{bmatrix}
\tag{11.3.32}
$$

Step 3. Shift the second row of $\mathbf{V}_1 \mathbf{G}_1$ by one place to the right and thus form the new generator matrix

$$
\mathbf{G}_2 = \begin{bmatrix} 0 & 0 & \gamma_{xx}(2) + K_1 \gamma_{xx}(1) & \cdots & \gamma_{xx}(p) + K_1 \gamma_{xx}(p-1) \\ 0 & 0 & \gamma_{xx}(0) + K_1^* \gamma_{xx}(1) & \cdots & \gamma_{xx}(p-2) + K_1^* \gamma_{xx}(p-1) \end{bmatrix}
\tag{11.3.33}
$$

The negative ratio of the elements in the third column of G_2 yields K_2.

Steps 2 and 3 are repeated until we have solved for all p reflection coefficients. In general, the 2×2 matrix in step 2 is

$$\mathbf{V}_m = \begin{bmatrix} 1 & K_m \\ K_m^* & 1 \end{bmatrix} \tag{11.3.34}$$

and multiplication of $\mathbf{V}_m$ by $\mathbf{G}_m$ yields $\mathbf{V}_m \mathbf{G}_m$. In step 3 we shift the second row of $\mathbf{V}_m \mathbf{G}_m$ one place to the right and obtain the new generator matrix $\mathbf{G}_{m+1}$.

We observe that the shifting operation of the second row in each iteration is equivalent to multiplication by the delay operator z^{-1} in the second recursive equation in (11.3.25). We also note that the division of the polynomial $P_m(z)$ by the polynomial $Q_m(z)$ and the evaluation of the quotient at $z = \infty$ is equivalent to dividing the elements in the $(m + 1)$st column of $\mathbf{G}_m$. The computation of the p reflection coefficients can be accomplished by use of parallel processors in $O(p)$ time units. Below we describe a pipelined architecture for performing these computations.

Another way of demonstrating the relationship of the Schur algorithm to the Levinson–Durbin algorithm and the corresponding lattice predictor is to determine the output of the lattice filter obtained when the input sequence is the correlation sequence $\{\gamma_{xx}(m), m = 0, 1, \ldots\}$. Thus, the first input to the lattice filter is $\gamma_{xx}(0)$, the second input is $\gamma_{xx}(1)$, and so on [i.e., $f_0(n) = \gamma_{xx}(n)$]. After the delay in the first stage, we have $g_0(n - 1) = \gamma_{xx}(n - 1)$. Hence, for $n = 1$, the ratio $f_0(1)/g_0(0) = \gamma_{xx}(1)/\gamma_{xx}(0)$, which is the negative of the reflection coefficient K_1. Alternatively, we can express this relationship as

$$f_0(1) + K_1 g_0(0) = \gamma_{xx}(1) + K_1 \gamma_{xx}(0) = 0$$

Furthermore, $g_0(0) = \gamma_{xx}(0) = E_0^f$. At time $n = 2$, the input to the second stage is, according to (11.2.4),

$$f_1(2) = f_0(2) + K_1 g_0(1) = \gamma_{xx}(2) + K_1 \gamma_{xx}(1)$$

and after the unit of delay in the second stage, we have

$$g_1(1) = K_1^* f_0(1) + g_0(0) = K_1^* \gamma_{xx}(1) + \gamma_{xx}(0)$$

Now the ratio $f_1(2)/g_1(1)$ is

$$\frac{f_1(2)}{g_1(1)} = \frac{\gamma_{xx}(2) + K_1 \gamma_{xx}(1)}{\gamma_{xx}(0) + K_1^* \gamma_{xx}(1)} = \frac{\gamma_{xx}(2) + K_1 \gamma_{xx}(1)}{E_1^f} = -K_2$$

Hence

$$f_1(2) + K_2 g_1(1) = 0$$

$$g_1(1) = E_1^f$$

By continuing in this manner, we can show that at the input to the mth lattice stage, the ratio $f_{m-1}(m)/g_{m-1}(m - 1) = -K_m$ and $g_{m-1}(m - 1) = E_{m-1}^f$. Consequently, the lattice filter coefficients obtained from the Levinson algorithm are identical to the coefficients obtained in the Schür algorithm. Furthermore, the lattice filter structure provides a method for computing the reflection coefficients in the lattice predictor.

A pipelined architecture for implementing the Schür algorithm.   Kung and Hu (1983) developed a pipelined lattice-type processor for implementing the Schür algorithm. The processor consists of a cascade of p lattice-type stages, where each stage consists of two processing elements (PEs), which we designate as upper PEs denoted as $A_1, A_2, \ldots, A_p$, and lower PEs denoted as $B_1, B_2, \ldots, B_p$, as shown in Fig. 11.5. The PE designated as A_1 is assigned the task of performing divisions. The remaining PEs perform one multiplication and one addition per iteration (one clock cycle).

Initially, the upper PEs are loaded with the elements of the first row of the generator matrix $\mathbf{G}_0$, as illustrated in Fig. 11.5. The lower PEs are loaded with the elements of the second row of the generator matrix $\mathbf{G}_0$. The computational process begins with the division PE, A_1, which computes the first reflection coefficient as $K_1 = -\gamma_{xx}(1)/\gamma_{xx}(0)$. The value of K_1 is sent simultaneously to all the PEs in the upper branch and lower branch.

The second step in the computation is to update the contents of all processing elements simultaneously. The contents of the upper and lower PEs are updated as follows:

$$\text{PE } A_m: A_m \leftarrow A_m + K_1 B_m \qquad m = 2, 3, \ldots, p$$
$$\text{PE } B_m: B_m \leftarrow B_m + K_1^* A_m \qquad m = 1, 2, \ldots, p$$

The third step involves the shifting of the contents of the upper PEs one place to the left. Thus we have

$$\text{PE } A_m: A_{m-1} \leftarrow A_m \qquad m = 2, 3, \ldots, p$$

At this point, PE A_1 contains $\gamma_{xx}(2) + K_1\gamma_{xx}(1)$ while PE B_1 contains $\gamma_{xx}(0) + K_1^*\gamma_{xx}(1)$. Hence the processor A_1 is ready to begin the second cycle by computing

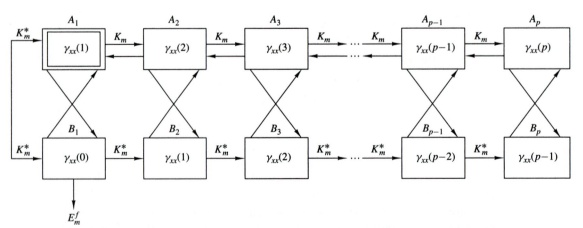

Figure 11.5 Pipelined parallel processor for computing the reflection coefficients

the second reflection coefficient $K_2 = -A_1/B_1$. The three computational steps beginning with the division A_1/B_1 are repeated until all p reflection coefficients are computed. Note that PE B_1 provides the minimum mean-square error E_m^f for each iteration.

If τ_d denotes the time for PE A_1 to perform a (complex) division and τ_{ma} is the time required to perform a (complex) multiplication and an addition, the time required to compute all p reflection coefficients is $p(\tau_d + \tau_{ma})$ for the Schür algorithm.

11.4 PROPERTIES OF THE LINEAR PREDICTION-ERROR FILTERS

Linear prediction filters possess several important properties, which we now describe. We begin by demonstrating that the forward prediction-error filter is minimum phase.

Minimum-phase property of the forward prediction-error filter. We have already demonstrated that the reflection coefficients $\{K_i\}$ are correlation coefficients, and consequently, $|K_i| \le 1$ for all i. This condition and the relation $E_m^f = (1 - |K_m|^2)E_{m-1}^f$ can be used to show that the zeros of the prediction-error filter are either all inside the unit circle or they are all on the unit circle.

First, we show that if $E_p^f > 0$, the zeros $|z_i| < 1$ for every i. The proof is by induction. Clearly, for $p = 1$ the system function for the prediction-error filter is

$$A_1(z) = 1 + K_1 z^{-1} \qquad (11.4.1)$$

Hence $z_1 = -K_1$ and $E_1^f = (1 - |K_1|^2)E_0^f > 0$. Now, suppose that the hypothesis is true for $p - 1$. Then, if z_i is a root of $A_p(z)$, we have from (11.2.16) and (11.2.18),

$$A_p(z_i) = A_{p-1}(z_i) + K_p z_i^{-1} B_{p-1}(z_i)$$
$$= A_{p-1}(z_i) + K_p z_i^{-p} A_{p-1}^*\left(\frac{1}{z_i}\right) = 0 \qquad (11.4.2)$$

Hence

$$\frac{1}{K_p} = -\frac{z_i^{-p} A_{p-1}^*(1/z_i)}{A_{p-1}(z_i)} \equiv Q(z_i) \qquad (11.4.3)$$

We note that the function $Q(z)$ is all pass. In general, an all-pass function of the form

$$P(z) = \prod_{k=1}^{N} \frac{z z_k^* + 1}{z + z_k} \qquad |z_k| < 1 \qquad (11.4.4)$$

satisfies the property that $|P(z)| > 1$ for $|z| < 1$, $|P(z)| = 1$ for $|z| = 1$, and $|P(z)| < 1$ for $|z| > 1$. Since $Q(z) = -P(z)/z$, it follows that $|z_i| < 1$ if $|Q(z)| > 1$. Clearly, this is the case since $Q(z_i) = 1/K_p$ and $E_p^f > 0$.

On the other hand, suppose that $E_{p-1}^f > 0$ and $E_p^f = 0$. In this case $|K_p| = 1$ and $|Q(z_i)| = 1$. Since the MMSE is zero, the random process $x(n)$ is called *predictable* or *deterministic*. Specifically, a purely sinusoidal random process of the form

$$x(n) = \sum_{k=1}^{M} \alpha_k e^{j(n\omega_k + \theta_k)} \tag{11.4.5}$$

where the phases $\{\theta_k\}$ are statistically independent and uniformly distributed over $(0, 2\pi)$, has the autocorrelation

$$\gamma_{xx}(m) = \sum_{k=1}^{M} \alpha_k^2 e^{jm\omega_k} \tag{11.4.6}$$

and the power density spectrum

$$\Gamma_{xx}(f) = \sum_{k=1}^{M} \alpha_k^2 \delta(f - f_k) \qquad f_k = \frac{\omega_k}{2\pi} \tag{11.4.7}$$

This process is predictable with a predictor of order $p \geq M$.

To demonstrate the validity of the statement, consider passing this process through a prediction error filter of order $p \geq M$. The MSE at the output of this filter is

$$\mathcal{E}_p^f = \int_{-1/2}^{1/2} \Gamma_{xx}(f)|A_p(f)|^2 df$$

$$= \int_{-1/2}^{1/2} \left[\sum_{k=1}^{M} \alpha_k^2 \delta(f - f_k) \right] |A_p(f)|^2 df \tag{11.4.8}$$

$$= \sum_{k=1}^{M} \alpha_k^2 |A_p(f_k)|^2$$

By choosing M of the p zeros of the prediction-error filter to coincide with the frequencies $\{f_k\}$, the MSE $\mathcal{E}_p^f$ can be forced to zero. The remaining $p - M$ zeros can be selected arbitrarily to be anywhere inside the unit circle.

Finally, the reader can prove that if a random process consists of a mixture of a continuous power spectral density and a discrete spectrum, the prediction-error filter must have all its roots inside the unit circle.

Maximum-phase property of the backward prediction-error filter.   The system function for the backward prediction error filter of order p is

$$B_p(z) = z^{-p} A_p^*(z^{-1}) \tag{11.4.9}$$

Consequently, the roots of $B_p(z)$ are the reciprocals of the roots of the forward prediction-error filter with system function $A_p(z)$. Hence if $A_p(z)$ is minimum phase, then $B_p(z)$ is maximum phase. However, if the process $x(n)$ is predictable, all the roots of $B_p(z)$ lie on the unit circle.

Whitening property.   Suppose that the random process $x(n)$ is an $AR(p)$ stationary random process that is generated by passing white noise with variance σ_w^2 through an all-pole filter with system function

$$H(z) = \frac{1}{1 + \sum\limits_{k=1}^{p} a_k z^{-k}} \tag{11.4.10}$$

Then the prediction-error filter of order p has the system function

$$A_p(z) = 1 + \sum_{k=1}^{p} a_p(k) z^{-k} \tag{11.4.11}$$

where the predictor coefficients $a_p(k) = a_k$. The response of the prediction-error filter is a white noise sequence $\{w(n)\}$. In this case the prediction-error filter whitens the input random process $x(n)$ and is called a whitening filter, as indicated in Section 11.2.

More generally, even if the input process $x(n)$ is not an AR process, the prediction-error filter attempts to remove the correlation among the signal samples of the input process. As the order of the predictor is increased, the predictor output $\hat{x}(n)$ becomes a closer approximation to $x(n)$ and hence the difference $f(n) = \hat{x}(n) - x(n)$ approaches a white noise sequence.

Orthogonality of the backward prediction errors.   The backward prediction errors $\{g_m(k)\}$ from different stages in the FIR lattice filter are orthogonal. That is,

$$E[g_m(n)g_l^*(n)] = \begin{cases} 0, & 0 \le l \le m-1 \\ E_m^b, & l = m \end{cases} \tag{11.4.12}$$

This property is easily proved by substituting for $g_m(n)$ and $g_l^*(n)$ into (11.4.12) and carrying out the expectation. Thus

$$E[g_m(n)g_l^*(n)] = \sum_{k=0}^{m} b_m(k) \sum_{j=0}^{l} b_l^*(j) E[x(n-k)x^*(n-j)] \tag{11.4.13}$$

$$= \sum_{j=0}^{l} b_l^*(j) \sum_{k=0}^{m} b_m(k) \gamma_{xx}(j-k)$$

But the normal equations for the backward linear predictor require that

$$\sum_{k=0}^{m} b_m(k) \gamma_{xx}(j-k) = \begin{cases} 0, & j = 1, 2, \ldots, m-1 \\ E_m^b, & j = m \end{cases} \tag{11.4.14}$$

Therefore,

$$E[g_m(n)g_l^*(n)] = \begin{cases} E_m^b = E_m^f, & m = l \\ 0, & 0 \le l \le m-1 \end{cases} \tag{11.4.15}$$

Additional properties.   There are a number of other interesting properties regarding the forward and backward prediction errors in the FIR lattice filter. These are given here for real-valued data. Their proof is left as an exercise for the reader.

(a) $E[f_m(n)x(n-i)] = 0,$ $1 \le i \le m$

(b) $E[g_m(n)x(n-i)] = 0,$ $0 \le i \le m-1$

(c) $E[f_m(n)x(n)] = E[g_m(n)x(n-m)] = E_m$

(d) $E[f_i(n)f_j(n)] = E_{\max}(i, j)$

(e) $E[f_i(n)f_j(n-t)] = 0,$ for $\begin{cases} 1 \le t \le i-j, & i > j \\ -1 \ge t \ge i-j, & i < j \end{cases}$

(f) $E[g_i(n)g_j(n-t)] = 0,$ for $\begin{cases} 0 \le t \le i-j, & i > j \\ 0 \ge t \ge i-j+1, & i < j \end{cases}$

(g) $E[f_i(n+i)f_j(n+j)] = \begin{cases} E_i, & i = j \\ 0, & i \ne j \end{cases}$

(h) $E[g_i(n+i)g_j(n+j)] = E_{\max\,(i,j)}$

(i) $E[f_i(n)g_j(n)] = \begin{cases} K_j E_i, & i \ge j, \\ 0, & i < j \end{cases}$ $i, j \ge 0,$ $K_0 = 1$

(j) $E[f_i(n)g_i(n-1)] = -K_{i+1}E_i$

(k) $E[g_i(n-1)x(n)] = E[f_i(n+1)x(n-1)] = -K_{i+1}E_i$

(l) $E[f_i(n)g_j(n-1)] = \begin{cases} 0, & i > j \\ -K_{j+1}E_i, & i \le j \end{cases}$

11.5 AR LATTICE AND ARMA LATTICE-LADDER FILTERS

In Section 11.2 we showed the relationship of the all-zero FIR lattice to linear prediction. The linear predictor with transfer function,

$$A_p(z) = 1 + \sum_{k=1}^{p} a_p(k)z^{-k} \tag{11.5.1}$$

when excited by an input random process $\{x(n)\}$, produces an output that approaches a white noise sequence as $p \to \infty$. On the other hand, if the input process is an AR(p), the output of $A_p(z)$ is white. Since $A_p(z)$ generates a MA(p) process when excited with a white noise sequence, the all-zero lattice is sometimes called a MA lattice.

In the following section, we develop the lattice structure for the inverse filter $1/A_p(z)$, called the AR lattice, and the lattice-ladder structure for an ARMA process.

11.5.1 AR Lattice Structure

Let us consider an all-pole system with system function

$$H(z) = \frac{1}{1 + \displaystyle\sum_{k=1}^{p} a_p(k) z^{-k}} \qquad (11.5.2)$$

The difference equation for this IIR system is

$$y(n) = -\sum_{k=1}^{p} a_p(k) y(n-k) + x(n) \qquad (11.5.3)$$

Now suppose that we interchange the roles of the input and output [i.e., interchange $x(n)$ with $y(n)$ in (11.5.3)] obtaining the difference equation

$$x(n) = -\sum_{k=1}^{p} a_p(k) x(n-k) + y(n)$$

or, equivalently,

$$y(n) = x(n) + \sum_{k=1}^{p} a_p(k) x(n-k) \qquad (11.5.4)$$

We observe that (11.5.4) is a difference equation for an FIR system with system function $A_p(z)$. Thus an all-pole IIR system can be converted to an all-zero system by interchanging the roles of the input and output.

Based on this observation, we can obtain the structure of an AR(p) lattice from a MA(p) lattice by interchanging the input with the output. Since the MA(p) lattice has $y(n) = f_p(n)$ as its output and $x(n) = f_0(n)$ is the input, we let

$$\begin{aligned} x(n) &= f_p(n) \\ y(n) &= f_0(n) \end{aligned} \qquad (11.5.5)$$

These definitions dictate that the quantities $\{f_m(n)\}$ be computed in descending order. This computation can be accomplished by rearranging the recursive equation for $\{f_m(n)\}$ in (11.2.4) and solving for $f_{m-1}(n)$ in terms of $f_m(n)$. Thus we obtain

$$f_{m-1}(n) = f_m(n) - K_m g_{m-1}(n-1) \qquad m = p, p-1, \ldots, 1$$

The equation for $g_m(n)$ remains unchanged. The result of these changes is the set of equations

$$\begin{aligned} x(n) &= f_p(n) \\ f_{m-1}(n) &= f_m(n) - K_m g_{m-1}(n-1) \\ g_m(n) &= K_m^* f_{m-1}(n) + g_{m-1}(n-1) \\ y(n) &= f_0(n) = g_0(n) \end{aligned} \qquad (11.5.6)$$

The corresponding structure for the AR(p) lattice is shown in Fig. 11.6. Note that the all-pole lattice structure has an all-zero path with input $g_0(n)$ and output $g_p(n)$,

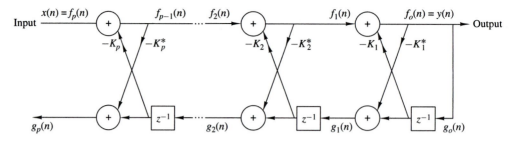

Figure 11.6 Lattice structure for an all-pole system

which is identical to the all-zero path in the Ma(p) lattice structure. This is not surprising, since the equation for $g_m(n)$ is identical in the two lattice structures.

We also observe that the AR(p) and MA(p) lattice structures are characterized by the same parameters, namely, the reflection coefficients $\{K_i\}$. Consequently, the equations given in (11.2.21) and (11.2.23) for converting between the system parameters $\{a_p(k)\}$ in the direct-form realizations of the all-zero system $A_p(z)$ and the lattice parameters $\{K_i\}$ of the MA(p) lattice structure, apply as well to the all-pole structures.

11.5.2 ARMA Processes and Lattice-Ladder Filters

The all-pole lattice provides the basic building block for lattice-type structures that implement IIR systems that contain both poles and zeros. To construct the appropriate structure, let us consider an IIR system with system function

$$H(z) = \frac{\displaystyle\sum_{k=0}^{q} c_q(k)z^{-k}}{1 + \displaystyle\sum_{k=1}^{p} a_p(k)z^{-k}} = \frac{C_q(z)}{A_p(z)} \tag{11.5.7}$$

Without loss of generality, we assume that $p \geq q$.

This system is described by the difference equations

$$v(n) = -\sum_{k=1}^{p} a_p(k)v(n-k) + x(n)$$

$$y(n) = \sum_{k=0}^{q} c_q(k)v(n-k) \tag{11.5.8}$$

obtained by viewing the system as a cascade of an all-pole system followed by an all-zero system. From (11.5.8) we observe that the output $y(n)$ is simply a linear combination of delayed outputs from the all-pole system.

Since zeros result from forming a linear combination of previous outputs, we can carry over this observation to construct a pole–zero system using the all-pole lattice structure as the basic building block. We have clearly observed that $g_m(n)$ in the all-pole lattice can be expressed as a linear combination of present and past outputs. In fact, the system

$$H_b(z) \equiv \frac{G_m(z)}{Y(z)} = B_m(z) \tag{11.5.9}$$

is an all-zero system. Therefore, any linear combination of $\{g_m(n)\}$ is also an all-zero filter.

Let us begin with an all-pole lattice filter with coefficients K_m, $1 \le m \le p$, and add a *ladder* part by taking as the output, a weighted linear combination of $\{g_m(n)\}$. The result is a pole–zero filter that has the *lattice-ladder* structure shown in Fig. 11.7. Its output is

$$y(n) = \sum_{k=0}^{q} \beta_k g_k(n) \tag{11.5.10}$$

where $\{\beta_k\}$ are the parameters that determine the zeros of the system. The system function corresponding to (11.5.10) is

$$H(z) = \frac{Y(z)}{X(z)} = \sum_{k=0}^{q} \beta_k \frac{G_k(z)}{X(z)} \tag{11.5.11}$$

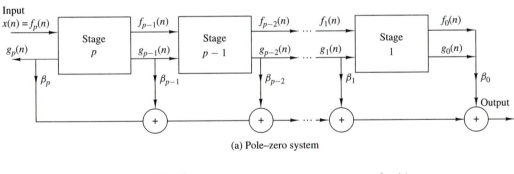

(a) Pole–zero system

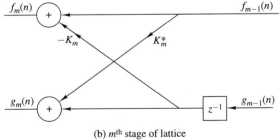

(b) m^{th} stage of lattice

Figure 11.7   Lattice-ladder structure for pole-zero system

Since $X(z) = F_p(z)$ and $F_0(z) = G_0(z)$, (11.5.11), can be expressed as

$$
\begin{aligned}
H(z) &= \sum_{k=0}^{q} \beta_k \frac{G_k(z)}{G_0(z)} \frac{F_0(z)}{F_p(z)} \\
&= \frac{1}{A_p(z)} \sum_{k=0}^{q} \beta_k B_k(z)
\end{aligned}
\tag{11.5.12}
$$

Therefore,

$$
C_q(z) = \sum_{k=0}^{q} \beta_k B_k(z)
\tag{11.5.13}
$$

This is the desired relationship that can be used to determine the weighting coefficients $\{\beta_k\}$ as previously shown in Section 7.3.5.

Given the polynomials $C_q(z)$ and $A_p(z)$, where $p \geq q$, the reflection coefficients $\{K_i\}$ are determined first from the coefficients $\{a_p(k)\}$. By means of the step-down recursive relation given by (11.2.22), we also obtain the polynomials $B_k(z)$, $k = 1, 2, \ldots, p$. Then the ladder parameters can be obtained from (11.5.13), which can be expressed as

$$
\begin{aligned}
C_m(z) &= \sum_{k=0}^{m-1} \beta_k B_k(z) + \beta_m B_m(z) \\
&= C_{m-1}(z) + \beta_m B_m(z)
\end{aligned}
\tag{11.5.14}
$$

or, equivalently,

$$
C_{m-1}(z) = C_m(z) - \beta_m B_m(z) \qquad m = p, p - 1, \ldots, 1
\tag{11.5.15}
$$

By running this recursive relation backward, we can generate all the lower-degree polynomials, $C_m(z)$, $m = p - 1, \ldots, 1$. Since $b_m(m) = 1$, the parameters β_m are determined from (11.5.15) by setting

$$
\beta_m = c_m(m) \qquad m = p, p - 1, \ldots, 1, 0
$$

When excited by a white noise sequence, this lattice-ladder filter structure generates an ARMA(p, q) process that has a power density spectrum

$$
\Gamma_{xx}(f) = \sigma_w^2 \frac{|C_q(f)|^2}{|A_p(f)|^2}
\tag{11.5.16}
$$

and an autocorrelation function that satisfies (11.1.18), where σ_w^2 in the variance of the input white noise sequence.

11.6 WIENER FILTERS FOR FILTERING AND PREDICTION

In many practical applications we are given an input signal $\{x(n)\}$, consisting of the sum of a desired signal $\{s(n)\}$ and an undesired noise or interference $\{w(n)\}$, and we are asked to design a filter that suppresses the undesired interference

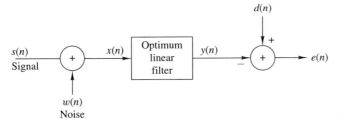

Figure 11.8 Model for linear estimation problem

component. In such a case, the objective is to design a system that filters out the additive interference while preserving the characteristics of the desired signal $\{s(n)\}$.

In this section we treat the problem of signal estimation in the presence of an additive noise disturbance. The estimator is constrained to be a linear filter with impulse response $\{h(n)\}$, designed so that its output approximates some specified desired signal sequence $\{d(n)\}$. Figure 11.8 illustrates the linear estimation problem.

The input sequence to the filter is $x(n) = s(n) + w(n)$, and its output sequence is $y(n)$. The difference between the desired signal and the filter output is the error sequence $e(n) = d(n) - y(n)$.

We distinguish three special cases:

1. If $d(n) = s(n)$, the linear estimation problem is referred to as *filtering*.
2. If $d(n) = s(n + D)$, where $D > 0$, the linear estimation problem is referred to as signal *prediction*. Note that this problem is different than the prediction considered earlier in this chapter, where $d(n) = x(n + D)$, $D \geq 0$.
3. If $d(n) = s(n - D)$, where $D > 0$, the linear estimation problem is referred to as signal *smoothing*.

Our treatment will concentrate on filtering and prediction.

The criterion selected for optimizing the filter impulse response $\{h(n)\}$ is the minimization of the mean-square error. This criterion has the advantages of simplicity and mathematical tractability.

The basic assumptions are that the sequences $\{s(n)\}$, $\{w(n)\}$, and $\{d(n)\}$ are zero mean and wide-sense stationary. The linear filter will be assumed to be either FIR or IIR. If it is IIR, we assume that the input data $\{x(n)\}$ are available over the infinite past. We begin with the design of the optimum FIR filter. The optimum linear filter, in the sense of minimum mean-square error (MMSE), is called a *Wiener filter*.

11.6.1 FIR Wiener Filter

Suppose that the filter is constrained to be of length M with coefficients $\{h_k, 0 \leq k \leq M - 1\}$. Hence its output $y(n)$ depends on the finite data record $x(n)$,

$x(n-1), \ldots, x(n-M+1)$,

$$y(n) = \sum_{k=0}^{M-1} h(k)x(n-k) \tag{11.6.1}$$

The mean-square value of the error between the desired output $d(n)$ and $y(n)$ is

$$\mathcal{E}_M = E|e(n)|^2$$

$$= E\left|d(n) - \sum_{k=0}^{M-1} h(k)x(n-k)\right| \tag{11.6.2}$$

Since this is a quadratic function of the filter coefficients, the minimization of $\mathcal{E}_M$ yields the set of linear equations

$$\sum_{k=0}^{M-1} h(k)\gamma_{xx}(l-k) = \gamma_{dx}(l) \qquad l = 0, 1, \ldots, M-1 \tag{11.6.3}$$

where $\gamma_{xx}(k)$ is the autocorrelation of the input sequence $\{x(n)\}$ and $\gamma_{dx}(k) = E[d(n)x^*(n-k)]$ is the crosscorrelation between the desired sequence $\{d(n)\}$ and the input sequence $\{x(n), 0 \leq n \leq M-1\}$. The set of linear equations that specify the optimum filter is called the *Wiener–Hopf equation*. These equations are also called the normal equations, encountered earlier in the chapter in the context of linear one-step prediction.

In general, the equations in (11.6.3) can be expressed in matrix form as

$$\mathbf{\Gamma}_M \mathbf{h}_M = \mathbf{\gamma}_d \tag{11.6.4}$$

where $\mathbf{\Gamma}_M$ is an $M \times M$ (Hermitian) Toeplitz matrix with elements $\Gamma_{lk} = \gamma_{xx}(l-k)$ and $\mathbf{\gamma}_d$ is the $M \times 1$ crosscorrelation vector with elements $\gamma_{dx}(l), l = 0, 1, \ldots, M-1$. The solution for the optimum filter coefficients is

$$\mathbf{h}_{\text{opt}} = \mathbf{\Gamma}_M^{-1} \mathbf{\gamma}_d \tag{11.6.5}$$

and the resulting minimum MSE achieved by the Wiener filter is

$$\text{MMSE}_M = \min_{\mathbf{h}_M} \mathcal{E}_M = \sigma_d^2 - \sum_{k=0}^{M-1} h_{\text{opt}}(k)\gamma_{dx}^*(k) \tag{11.6.6}$$

or, equivalently,

$$\text{MMSE}_M = \sigma_d^2 - \mathbf{\gamma}_d^{*t}\mathbf{\Gamma}_M^{-1}\mathbf{\gamma}_d \tag{11.6.7}$$

where $\sigma_d^2 = E|d(n)|^2$.

Let us consider some special cases of (11.6.3). If we are dealing with filtering, the $d(n) = s(n)$. Furthermore, if $s(n)$ and $w(n)$ are uncorrelated random sequences, as is usually the case in practice, then

$$\gamma_{xx}(k) = \gamma_{ss}(k) + \gamma_{ww}(k)$$

$$\gamma_{dx}(k) = \gamma_{ss}(k) \tag{11.6.8}$$

and the normal equations in (11.6.3) become

$$\sum_{k=0}^{M-1} h(k)[\gamma_{ss}(l-k) + \gamma_{ww}(l-k)] = \gamma_{ss}(l) \qquad l = 0, 1, \ldots, M-1 \qquad (11.6.9)$$

If we are dealing with prediction, then $d(n) = s(n + D)$ where $D > 0$. Assuming that $s(n)$ and $w(n)$ are uncorrelated random sequences, we have

$$\gamma_{dx}(k) = \gamma_{ss}(l + D) \qquad (11.6.10)$$

Hence the equations for the Wiener prediction filter become

$$\sum_{k=0}^{M-1} h(k)[\gamma_{ss}(l-k) + \gamma_{ww}(l-k)] = \gamma_{ss}(l + D) \qquad l = 0, 1, \ldots, M-1 \qquad (11.6.11)$$

In all these cases, the correlation matrix to be inverted is Toeplitz. Hence the (generalized) Levinson–Durbin algorithm may be used to solve for the optimum filter coefficients.

Example 11.6.1

Let us consider a signal $x(n) = s(n) + w(n)$, where $s(n)$ is an AR(1) process that satisfies the difference equation

$$s(n) = 0.6s(n-1) + v(n)$$

where $\{v(n)\}$ is a white noise sequence with variance $\sigma_v^2 = 0.64$, and $\{w(n)\}$ is a white noise sequence with variance $\sigma_w^2 = 1$. We will design a Wiener filter of length $M = 2$ to estimate $\{s(n)\}$.

Solution   Since $\{s(n)\}$ is obtained by exciting a single-pole filter by white noise, the power spectral density of $s(n)$ is

$$\Gamma_{ss}(f) = \sigma_v^2 |H(f)|^2$$

$$= \frac{0.64}{|1 - 0.6e^{-j2\pi f}|^2}$$

$$= \frac{0.64}{1.36 - 1.2 \cos 2\pi f}$$

The corresponding autocorrelation sequence $\{\gamma_{ss}(m)\}$ is

$$\gamma_{ss}(m) = (0.6)^{|m|}$$

The equations for the filter coefficients are

$$2h(0) + 0.6h(1) = 1$$
$$0.6h(0) + 2h(1) = 0.6$$

normal eq.

Solution of these equations yields the result

$$h(0) = 0.451 \qquad h(1) = 0.165$$

The corresponding minimum MSE is

$$\text{MMSE}_2 = 1 - h(0)\gamma_{ss}(0) - h(1)\gamma_{ss}(1)$$

$$= 1 - 0.451 - (0.165)(0.6)$$

$$= 0.45$$

This error can be reduced further by increasing the length of the Wiener filter (see Problem 11.27).

11.6.2 Orthogonality Principle in Linear Mean-Square Estimation

The normal equations for the optimum filter coefficients given by (11.6.3) can be obtained directly by applying the orthogonality principle in linear mean-square estimation. Simply stated, the mean-square error $\mathcal{E}_M$ in (11.6.2) is a minimum if the filter coefficients $\{h(k)\}$ are selected such that the error is orthogonal to each of the data points in the estimate,

$$E[e(n)x^*(n-l)] = 0 \qquad l = 0, 1, \ldots, M-1 \tag{11.6.12}$$

where

$$e(n) = d(n) - \sum_{k=0}^{M-1} h(k)x(n-k) \tag{11.6.13}$$

Conversely, if the filter coefficients satisfy (11.6.12), the resulting MSE is a minimum.

When viewed geometrically, the output of the filter, which is the estimate

$$\hat{d}(n) = \sum_{k=0}^{M-1} h(k)x(n-k) \tag{11.6.14}$$

is a vector in the subspace spanned by the data $\{x(k), 0 \leq k \leq M-1\}$. The error $e(n)$ is a vector from $d(n)$ to $\hat{d}(n)$ [i.e., $d(n) = e(n) + \hat{d}(n)$], as shown in Fig. 11.9. The orthogonality principle states that the length $\mathcal{E}_M = E|e(n)|^2$ is a minimum when $e(n)$ is perpendicular to the data subspace [i.e., $e(n)$ is orthogonal to each data point $x(k), 0 \leq k \leq M-1$].

We note that the solution obtained from the normal equations in (11.6.3) is unique if the data $\{x(n)\}$ in the estimate $\hat{d}(n)$ are *linearly independent*. In this case, the correlation matrix $\boldsymbol{\Gamma}_M$ is nonsingular. On the other hand, if the data are linearly dependent, the rank of $\boldsymbol{\Gamma}_M$ is less than M and therefore the solution is not unique. In this case, the estimate $\hat{d}(n)$ can be expressed as a linear combination of a reduced set of linearly independent data points equal to the rank of $\boldsymbol{\Gamma}_M$.

Since the MSE is minimized by selecting the filter coefficients to satisfy the orthogonality principle, the residual minimum MSE is simply

$$\text{MMSE}_M = E[e(n)d^*(n)] \tag{11.6.15}$$

which yields the result given in (11.6.6).

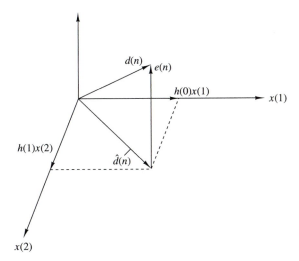

Figure 11.9 Geometric interpretation of linear MSE problem

11.6.3 IIR Wiener Filter

In the preceding section we constrained the filter to be FIR and obtained a set of M linear equations for the optimum filter coefficients. In this section we allow the filter to be infinite in duration (IIR) and the data sequence to be infinite as well. Hence the filter output is

$$y(n) = \sum_{k=0}^{\infty} h(k)x(n-k) \tag{11.6.16}$$

The filter coefficients are selected to minimize the mean-square error between the desired output $d(n)$ and $y(n)$, that is,

$$\mathcal{E}_{\infty} = E|e(n)|^2$$

$$= E\left|d(n) - \sum_{k=0}^{\infty} h(k)x(n-k)\right|^2 \tag{11.6.17}$$

Application of the orthogonality principle leads to the Wiener–Hopf equation,

$$\sum_{k=0}^{\infty} h(k)\gamma_{xx}(l-k) = \gamma_{dx}(l) \qquad l \geq 0 \tag{11.6.18}$$

The residual MMSE is simply obtained by application of the condition given by (11.6.15). Thus we obtain

$$\text{MMSE}_{\infty} = \min_{\mathbf{h}} \mathcal{E}_{\infty} = \sigma_d^2 - \sum_{k=0}^{\infty} h_{\text{opt}}(k)\gamma_{dx}^*(k) \tag{11.6.19}$$

The Wiener–Hopf equation given by (11.6.18) cannot be solved directly with z-transform techniques, because the equation holds only for $l \geq 0$. We shall solve for the optimum IIR Wiener filter based on the innovations representation of the stationary random process $\{x(n)\}$.

Recall that a stationary random process $\{x(n)\}$ with autocorrelation $\gamma_{xx}(k)$ and power spectral density $\Gamma_{xx}(f)$ can be represented by an equivalent innovations process, $\{i(n)\}$ by passing $\{x(n)\}$ through a noise-whitening filter with system function $1/G(z)$, where $G(z)$ is the minimum-phase part obtained from the spectral factorization of $\Gamma_{xx}(z)$:

$$\Gamma_{xx}(z) = \sigma_i^2 G(z)G(z^{-1}) \tag{11.6.20}$$

Hence $G(z)$ is analytic in the region $|z| > r_1$, where $r_1 < 1$.

Now, the optimum Wiener filter can be viewed as the cascade of the whitening filter $1/G(z)$ with a second filter, say $Q(z)$, whose output $y(n)$ is identical to the output of the optimum Wiener filter. Since

$$y(n) = \sum_{k=0}^{\infty} q(k)i(n-k) \tag{11.6.21}$$

and $e(n) = d(n) - y(n)$, application of the orthogonality principle yields the new Wiener–Hopf equation as

$$\sum_{k=0}^{\infty} q(k)\gamma_{ii}(l-k) = \gamma_{di}(l) \qquad l \geq 0 \tag{11.6.22}$$

But since $\{i(n)\}$ is white, it follows that $\gamma_{ii}(l-k) = 0$ unless $l = k$. Thus we obtain the solution as

$$q(l) = \frac{\gamma_{di}(l)}{\gamma_{ii}(0)} = \frac{\gamma_{di}(l)}{\sigma_i^2} \qquad l \geq 0 \tag{11.6.23}$$

The z-transform of the sequence $\{q(l)\}$ is

$$Q(z) = \sum_{k=0}^{\infty} q(k)z^{-k}$$
$$= \frac{1}{\sigma_i^2} \sum_{k=0}^{\infty} \gamma_{di}(k)z^{-k} \tag{11.6.24}$$

If we denote the z-transform of the two-sided crosscorrelation sequence $\gamma_{di}(k)$ by $\Gamma_{di}(z)$:

$$\Gamma_{di}(z) = \sum_{k=-\infty}^{\infty} \gamma_{di}(k)z^{-k} \tag{11.6.25}$$

and define $[\Gamma_{di}(z)]_+$ as

$$[\Gamma_{di}(z)]_+ = \sum_{k=0}^{\infty} \gamma_{di}(k)z^{-k} \tag{11.6.26}$$

then

$$Q(z) = \frac{1}{\sigma_i^2}[\Gamma_{di}(z)]_+ \tag{11.6.27}$$

To determine $[\Gamma_{di}(z)]_+$, we begin with the output of the noise-whitening filter, which can be expressed as

$$i(n) = \sum_{k=0}^{\infty} v(k)x(n-k) \tag{11.6.28}$$

where $\{v(k), k \geq 0\}$ is the impulse response of the noise-whitening filter,

$$\frac{1}{G(z)} \equiv V(z) = \sum_{k=0}^{\infty} v(k)z^{-k} \tag{11.6.29}$$

Then

$$\gamma_{di}(k) = E[d(n)i^*(n-k)]$$

$$= \sum_{m=0}^{\infty} v(m)E[d(n)x^*(n-m-k)] \tag{11.6.30}$$

$$= \sum_{m=0}^{\infty} v(m)\gamma_{dx}(k+m)$$

The z-transform of the crosscorrelation $\gamma_{di}(k)$ is

$$\Gamma_{di}(z) = \sum_{k=-\infty}^{\infty} \left[\sum_{m=0}^{\infty} v(m)\gamma_{dx}(k+m) \right] z^{-k}$$

$$= \sum_{m=0}^{\infty} v(m) \sum_{k=-\infty}^{\infty} \gamma_{dx}(k+m)z^{-k}$$

$$\tag{11.6.31}$$

$$= \sum_{m=0}^{\infty} v(m)z^m \sum_{k=-\infty}^{\infty} \gamma_{dx}(k)z^{-k}$$

$$= V(z^{-1})\Gamma_{dx}(z) = \frac{\Gamma_{dx}(z)}{G(z^{-1})}$$

Therefore,

$$Q(z) = \frac{1}{\sigma_i^2} \left[\frac{\Gamma_{dx}(z)}{G(z^{-1})} \right]_+ \tag{11.6.32}$$

Finally, the optimum IIR Wiener filter has the system function

$$H_{\text{opt}}(z) = \frac{Q(z)}{G(z)}$$

$$\tag{11.6.33}$$

$$= \frac{1}{\sigma_i^2 G(z)} \left[\frac{\Gamma_{dx}(z)}{G(z^{-1})} \right]_+$$

In summary, the solution for the optimum IIR Wiener filter requires that we perform a spectral factorization of $\Gamma_{xx}(z)$ to obtain $G(z)$, the minimum-phase component, and then we solve for the causal part of $\Gamma_{dx}(z)/G(z^{-1})$. The following example illustrates the procedure.

Example 11.6.2

Let us determine the optimum IIR Wiener filter for the signal given in Example 11.6.1.

Solution   For this signal we have

$$\Gamma_{xx}(z) = \Gamma_{ss}(z) + 1 = \frac{1.8(1 - \frac{1}{3}z^{-1})(1 - \frac{1}{3}z)}{(1 - 0.6z^{-1})(1 - 0.6z)}$$

where $\sigma_i^2 = 1.8$ and

$$G(z) = \frac{1 - \frac{1}{3}z^{-1}}{1 - 0.6z^{-1}}$$

The z-transform of the crosscorrelation $\gamma_{dx}(m)$ is

$$\Gamma_{dx}(z) = \Gamma_{ss}(z) = \frac{0.64}{(1 - 0.6z^{-1})(1 - 0.6z)}$$

Hence

$$\left[\frac{\Gamma_{dx}(z)}{G(z^{-1})} \right]_{+} = \left[\frac{0.64}{(1 - \frac{1}{3}z)(1 - 0.6z^{-1})} \right]_{+}$$

$$= \left[\frac{0.8}{1 - 0.6z^{-1}} + \frac{0.266z}{1 - \frac{1}{3}z} \right]_{+}$$

$$= \frac{0.8}{1 - 0.6z^{-1}}$$

The optimum IIR filter has the system function

$$H_{\text{opt}}(z) = \frac{1}{1.8} \left(\frac{1 - 0.6z^{-1}}{1 - \frac{1}{3}z^{-1}} \right) \left(\frac{0.8}{1 - 0.6z^{-1}} \right)$$

$$= \frac{\frac{4}{9}}{1 - \frac{1}{3}z^{-1}}$$

and an impulse response

$$h_{\text{opt}}(n) = \frac{4}{9} \left(\frac{1}{3} \right)^n \qquad n \geq 0$$

We conclude this section by expressing the minimum MSE given by (11.6.19) in terms of the frequency-domain characteristics of the filter. First, we note that $\sigma_d^2 \equiv E|d(n)|^2$ is simply the value of the autocorrelation sequence $\{\gamma_{dd}(k)\}$ evaluated at $k = 0$. Since

$$\gamma_{dd}(k) = \frac{1}{2\pi j} \oint_C \Gamma_{dd}(z) z^{k-1} dz \tag{11.6.34}$$

it follows that

$$\sigma_d^2 = \gamma_{dd}(0) = \frac{1}{2\pi j} \oint_C \frac{\Gamma_{dd}(z)}{z} dz \tag{11.6.35}$$

where the contour integral is evaluated along a closed path encircling the origin in the region of convergence of $\Gamma_{dd}(z)$.

The second term in (11.6.19) is also easily transformed to the frequency domain by application of Parseval's theorem. Since $h_{opt}(k) = 0$ for $k < 0$, we have

$$\sum_{k=-\infty}^{\infty} h_{opt}(k)\gamma_{dx}^*(k) = \frac{1}{2\pi j}\oint_C H_{opt}(z)\Gamma_{dx}(z^{-1})z^{-1}dz \qquad (11.6.36)$$

where C is a closed contour encircling the origin that lies within the common region of convergence of $H_{opt}(z)$ and $\Gamma_{dx}(z^{-1})$.

By combining (11.6.35) with (11.6.36), we obtain the desired expression for the MMSE$_\infty$ in the form

$$\text{MMSE}_\infty = \frac{1}{2\pi j}\oint_C[\Gamma_{dd}(z) - H_{opt}(z)\Gamma_{dx}(z^{-1})]z^{-1}dz \qquad (11.6.37)$$

Example 11.6.3

For the optimum Wiener filter derived in Example 11.6.2, the minimum MSE is

$$\text{MMSE}_\infty = \frac{1}{2\pi j}\oint_C\left[\frac{0.3555}{(z - \frac{1}{3})(1 - 0.6z)}\right]dz$$

There is a single pole inside the unit circle at $z = \frac{1}{3}$. By evaluating the residue at the pole, we obtain

$$\text{MMSE}_\infty = 0.444$$

We observe that this MMSE is only slightly smaller than that for the optimum two-tap Wiener filter in Example 11.6.1.

11.6.4 Noncausal Wiener Filter

In the preceding section we constrained the optimum Wiener filter to be causal [i.e., $h_{opt}(n) = 0$ for $n < 0$]. In this section we drop this condition and allow the filter to include both the infinite past and the infinite future of the sequence $\{x(n)\}$ in forming the output $y(n)$, that is,

$$y(n) = \sum_{k=-\infty}^{\infty} h(k)x(n - k) \qquad (11.6.38)$$

The resulting filter is physically unrealizable. It can also be viewed as a *smoothing filter* in which the infinite future signal values are used to smooth the estimate $\hat{d}(n) = y(n)$ of the desired signal $d(n)$.

Application of the orthogonality principle yields the Wiener–Hopf equation for the noncausal filter in the form

$$\sum_{k=-\infty}^{\infty} h(k)\gamma_{xx}(l - k) = \gamma_{dx}(l) \qquad -\infty < l < \infty \qquad (11.6.39)$$

and the resulting MMSE$_{nc}$ as

$$\text{MMSE}_{nc} = \sigma_d^2 - \sum_{k=-\infty}^{\infty} h(k)\gamma_{dx}^*(k) \qquad (11.6.40)$$

Since (11.6.39) holds for $-\infty < l < -\infty$, this equation can be transformed directly to yield the optimum noncausal Wiener filter as

$$H_{nc}(z) = \frac{\Gamma_{dx}(z)}{\Gamma_{xx}(z)} \tag{11.6.41}$$

The MMSE$_{nc}$ can also be simply expressed in the z-domain as

$$\text{MMSE}_{nc} = \frac{1}{2\pi j} \oint_C [\Gamma_{dd}(z) - H_{nc}(z)\Gamma_{dx}(z^{-1})]z^{-1}dz \tag{11.6.42}$$

In the following example we compare the form of the optimal noncausal filter with the optimal causal filter obtained in the previous section.

Example 11.6.4

The optimum noncausal Wiener filter for the signal characteristics given in Example 11.6.1 is given by (11.6.41), where

$$\Gamma_{dx}(z) = \Gamma_{ss}(z) = \frac{0.64}{(1 - 0.6z^{-1})(1 - 0.6z)}$$

and

$$\Gamma_{xx}(z) = \Gamma_{ss}(z) + 1$$
$$= \frac{2(1 - 0.3z^{-1} - 0.3z)}{(1 - 0.6z^{-1})(1 - 0.6z)}$$

Then,

$$H_{nc}(z) = \frac{0.3555}{(1 - \frac{1}{3}z^{-1})(1 - \frac{1}{3}z)}$$

This filter is clearly noncausal.

The minimum MSE achieved by this filter is determined from evaluating (11.6.42). The integrand is

$$\frac{1}{z}\Gamma_{ss}(z)[1 - H_{nc}(z)] = \frac{0.3555}{(z - \frac{1}{3})(1 - \frac{1}{3}z)}$$

The only pole inside the unit circle is $z = \frac{1}{3}$. Hence the residue is

$$\left.\frac{0.3555}{1 - \frac{1}{3}z}\right|_{z=\frac{1}{3}} = \frac{0.3555}{8/9} = 0.40$$

Hence the minimum achievable MSE obtained with the optimum noncausal Wiener filter is

$$\text{MMSE}_{nc} = 0.40$$

Note that this is lower than the MMSE for the causal filter, as expected.

11.7 SUMMARY AND REFERENCES

The major focal point in this chapter is the design of optimum linear systems for linear prediction and filtering. The criterion for optimality is the minimization of the mean-square error between a specified desired filter output and the actual filter output.

In the development of linear prediction, we demonstrated that the equations for the forward and backward prediction errors specified a lattice filter whose parameters, the reflection coefficients $\{K_m\}$, were simply related to the filter coefficients $\{a_m(k)\}$ of the direct form FIR linear predictor and the associated prediction error filter. The optimum filter coefficients $\{K_m\}$ and $\{a_m(k)\}$ are easily obtained from the solution of the normal equations.

We described two computationally efficient algorithms for solving the normal equations, the Levinson–Durbin algorithm and the Schür algorithm. Both algorithms are suitable for solving a Toeplitz system of linear equations and have a computational complexity of $O(p^2)$ when executed on a single processor. However, with full parallel processing, the Schür algorithm solves the normal equations in $O(p)$ time, whereas the Levinson–Durbin algorithm requires $O(p \log p)$ time.

In addition to the all-zero lattice filter resulting from linear prediction, we also derived the AR lattice (all-pole) filter structure and the ARMA lattice-ladder (pole–zero) filter structure. Finally, we described the design of the class of optimum linear filters, called Wiener filters.

Linear estimation theory has had a long and rich history of development over the past four decades. Kailath (1974) presents a historical account of the first three decades. The pioneering work of Wiener (1949) on optimum linear filtering for statistically stationary signals is especially significant. The generalization of the Wiener filter theory to dynamical systems with random inputs was developed by Kalman (1960) and Kalman and Bucy (1961). Kalman filters are treated in the books by Meditch (1969), Brown (1983), and Chui and Chen (1987). The monograph by Kailath (1981) treats both Wiener and Kalman filters.

There are numerous references on linear prediction and lattice filters. Tutorial treatments on these subjects have been published in the journal papers by Makhoul (1975, 1978) and Friedlander (1982a, b). The books by Haykin (1991), Markel and Gray 1976), and Tretter (1976) provide comprehensive treatments of these subjects. Applications of linear prediction to spectral analysis are found in the books by Kay (1988) and Marple (1987), to geophysics in the book Robinson and Treitel (1980), and to adaptive filtering in the book by Haykin (1991).

The Levinson–Durbin algorithm for solving the normal equations recursively was given by Levinson (1947) and later modified by Durbin (1959). Variations of this classical algorithm, called *split Levinson algorithms*, have been developed by Delsarte and Genin (1986) and by Krishna (1988). These algorithms exploit additional symmetries in the Toeplitz correlation matrix and save about a factor of 2 in the number of multiplications.

The Schür algorithm was originally described by Schür (1917) in a paper published in German. An English translation of this paper appears in the book edited by Gohberg (1986). The Schür algorithm is intimately related to the polynomials $\{A_m(z)\}$, which can be interpreted as orthogonal polynomials. A treatment of orthogonal polynomials is given in the books by Szegö (1967), Grenander and Szegö (1958), and Geronimus (1958). The thesis of Vieira (1977) and the papers by Kailath et al. (1978), Delsarte et al. (1978), and Youla and Kazanjian (1978) provide additional results on orthogonal polynomials. Kailath (1985, 1986) provides tutorial treatments of the Schür algorithm and its relationship to orthogonal polynomials and the Levinson–Durbin algorithm. The pipelined parallel processing structure for computing the reflection coefficients based on the Schür algorithm and the related problem of solving Toeplitz systems of linear equations is described in the paper by Kung and Hu (1983). Finally, we should mention that some additional computational efficiency can be achieved in the Schür algorithm, by further exploiting symmetry properties of Toeplitz matrices, as described by Krishna (1988). This leads to the so-called split-Schür algorithm, which is analogous to the split-Levinson algorithm.

PROBLEMS

11.1 The power density spectrum of an AR process $\{x(n)\}$ is given as

$$\Gamma_{xx}(\omega) = \frac{\sigma_w^2}{|A(\omega)|^2}$$

$$= \frac{25}{|1 - e^{-j\omega} + \frac{1}{2}e^{-j2\omega}|^2}$$

where σ_w^2 is the variance of the input sequence.
  (a) Determine the difference equation for generating the AR process when the excitation is white noise.
  (b) Determine the system function for the whitening filter.

11.2 An ARMA process has an autocorrelation $\{\gamma_{xx}(m)\}$ whose z-transform is given as

$$\Gamma_{xx}(z) = 9\frac{(z - \frac{1}{3})(z - 3)}{(z - \frac{1}{2})(z - 2)} \quad \frac{1}{2} < |z| < 2$$

  (a) Determine the filter $H(z)$ for generating $\{x(n)\}$ from a white noise input sequence. Is $H(z)$ unique? Explain.
  (b) Determine a stable linear whitening filter for the sequence $\{x(n)\}$.

11.3 Consider the ARMA process generated by the difference equation

$$x(n) = 1.6x(n - 1) - 0.63x(n - 2) + w(n) + 0.9w(n - 1)$$

  (a) Determine the system function of the whitening filter and its poles and zeros.
  (b) Determine the power density spectrum of $\{x(n)\}$.

11.4 Determine the lattice coefficients corresponding to the FIR filter with system function

$$H(z) = A_3(z) = 1 + \tfrac{13}{24}z^{-1} + \tfrac{5}{8}z^{-2} + \tfrac{1}{3}z^{-3}$$

11.5 Determine the reflection coefficients $\{K_m\}$ of the lattice filter corresponding to the FIR filter described by the system function

$$H(z) = A_2(z) = 1 + 2z^{-1} + \tfrac{1}{3}z^{-2}$$

11.6 (a) Determine the zeros and sketch the zero pattern for the FIR lattice filter with reflection coefficients

$$K_1 = \tfrac{1}{2} \qquad K_2 = -\tfrac{1}{3} \qquad K_3 = 1$$

(b) Repeat part (a) but with $K_3 = -1$.

(c) You should have found that the zeros lie on the unit circle. Can this result be generalized? How?

11.7 Determine the impulse response of the FIR filter that is described by the lattice coefficients $K_1 = 0.6$, $K_2 = 0.3$, $K_3 = 0.5$, and $K_4 = 0.9$.

11.8 In Section 11.2.4 we indicated that the noise-whitening filter $A_p(z)$ for a causal AR(p) process is a forward linear prediction-error filter of order p. Show that the backward linear prediction-error filter of order p is the noise-whitening filter of the corresponding anticausal AR(p) process.

11.9 Use the orthogonality principle to determine the normal equations and the resulting minimum MSE for a forward predictor of order p that predicts m samples ($m > 1$) into the future (m-step forward predictor). Sketch the prediction error filter.

11.10 Repeat Problem 11.9 for an m-step backward predictor.

11.11 Determine a Levinson–Durbin recursive algorithm for solving for the coefficients of a backward prediction-error filter. Use the result to show that coefficients of the forward and backward predictors can be expressed recursively as

$$\mathbf{a}_m = \begin{bmatrix} \mathbf{a}_{m-1} \\ 0 \end{bmatrix} + K_m \begin{bmatrix} \mathbf{b}_{m-1} \\ 1 \end{bmatrix}$$

$$\mathbf{b}_m = \begin{bmatrix} \mathbf{b}_{m-1} \\ 0 \end{bmatrix} + K_m^* \begin{bmatrix} \mathbf{a}_{m-1} \\ 1 \end{bmatrix}$$

11.12 The Levinson–Durbin algorithm described in Section 11.3.1 solved the linear equations

$$\mathbf{\Gamma}_m \mathbf{a}_m = -\boldsymbol{\gamma}_m$$

where the right-hand side of this equation has elements of the autocorrelation sequence that are also elements of the matrix $\mathbf{\Gamma}$. Let us consider the more general problem of solving the linear equations

$$\mathbf{\Gamma}_m \mathbf{b}_m = \mathbf{c}_m$$

where $\mathbf{c}_m$ is an arbitrary vector. (The vector $\mathbf{b}_m$ is not related to the coefficients of the backward predictor.) Show that the solution to $\mathbf{\Gamma}_m \mathbf{b}_m = \mathbf{c}_m$ can be obtained from a *generalized Levinson–Durbin* algorithm which is given recursively as

$$b_m(m) = \frac{c(m) - \boldsymbol{\gamma}_{m-1}^{bt} \mathbf{b}_{m-1}}{E_{m-1}^f}$$

$$b_m(k) = b_{m-1}(k) - b_m(m)a_{m-1}^*(m-k) \qquad \begin{matrix} k = 1, 2, \ldots, m-1 \\ m = 1, 2, \ldots, p \end{matrix}$$

where $b_1(1) = c(1)/\gamma_{xx}(0) = c(1)/E_0^f$ and $a_m(k)$ is given by (11.3.17). Thus a second recursion is required to solve the equation $\mathbf{\Gamma}_m \mathbf{b}_m = \mathbf{c}_m$.

11.13 Use the generalized Levinson–Durbin algorithm to solve the normal equations recursively for the m-step forward and backward predictors.

11.14 Show that the transformation

$$\mathbf{V}_m = \begin{bmatrix} 1 & K_m \\ K_m^* & 1 \end{bmatrix}$$

in the Schur algorithm satisfies the special property

$$\mathbf{V}_m \mathbf{J} \mathbf{V}_m^t = (1 - |K_m|^2)\mathbf{J}$$

where

$$\mathbf{J} = \begin{bmatrix} 1 & 0 \\ 0 & -1 \end{bmatrix}$$

Thus $\mathbf{V}_m$ is called a J-rotation matrix. Its role is to rotate or hyperbolate the row of $\mathbf{G}_m$ to lie along the first coordinate direction (Kailath, 1985).

11.15 Prove the additional properties (a) through (l) of the prediction- error filters given in Section 11.4.

11.16 Extend the additional properties (a) through (l) of the prediction error filters given in Section 11.4 to complex-valued signals.

11.17 Determine the reflection coefficient K_3 in terms of the autocorrelations $\{\gamma_{xx}(m)\}$ from the Schür algorithm and compare your result with the expression for K_3 obtained from the Levinson–Durbin algorithm.

11.18 Consider a infinite-length ($p = \infty$) one-step forward predictor for a stationary random process $\{x(n)\}$ with a power density spectrum of $\Gamma_{xx}(f)$. Show that the mean-square error of the prediction-error filter can be expressed as

$$E_\infty^f = 2\pi \exp\{ \int_{-1/2}^{1/2} \ln \Gamma_{xx}(f) \, df \}$$

11.19 Determine the output of an infinite-length ($p = \infty$) m-step forward predictor and the resulting mean-square error when the input signal is a first-order autoregressive process of the form

$$x(n) = ax(n-1) + w(n)$$

11.20 An AR(3) process $\{x(n)\}$ is characterized by the autocorrelation sequence $\gamma_{xx}(0) = 1$, $\gamma_{xx}(1) = \frac{1}{2}$, $\gamma_{xx}(2) = \frac{1}{8}$, and $\gamma_{xx}(3) = \frac{1}{64}$.
(a) Use the Schür algorithm to determine the three reflection coefficients K_1, K_2, and K_3.
(b) Sketch the lattice filter for synthesizing $\{x(n)\}$ from a white noise excitation.

11.21 The purpose of this problem is to show that the polynomials $\{A_m(z)\}$, which are the system functions of the forward prediction-error filters of order m, $m = 0, 1, \ldots, p$, can be interpreted as orthogonal on the unit circle. Toward this end, suppose that $\Gamma_{xx}(f)$ is the power spectral density of a zero-mean random process $\{x(n)\}$ and let $\{A_m(z)\}$, $m = 0, 1, \ldots, p\}$, be the system functions of the corresponding prediction-error filters. Show that the polynomials $\{A_m(z)\}$ satisfy the orthogonality property

$$\int_{-1/2}^{1/2} \Gamma_{xx}(f) A_m(f) A_n^*(f) df = E_m^f \delta_{mn} \qquad m, n = 0, 1, \ldots, p$$

11.22 Determine the system function of the all-pole filter described by the lattice coefficients $K_1 = 0.6$, $K_2 = 0.3$, $K_3 = 0.5$, and $K_4 = 0.9$.

11.23 Determine the parameters and sketch the lattice-ladder filter structure for the system with system function

$$H(z) = \frac{1 - 0.8z^{-1} + 0.15z^{-2}}{1 + 0.1z^{-1} - 0.72z^{-2}}$$

11.24 Consider a signal $x(n) = s(n) + w(n)$, where $s(n)$ is an AR(1) process that satisfies the difference equation

$$s(n) = 0.8s(n-1) + v(n)$$

where $\{v(n)\}$ is a white noise sequence with variance $\sigma_v^2 = 0.49$ and $\{w(n)\}$ is a white noise sequence with variance $\sigma_w^2 = 1$. The processes $\{v(n)\}$ and $\{w(n)\}$ are uncorrelated.

(a) Determine the autocorrelation sequences $\{\gamma_{ss}(m)\}$ and $\{\gamma_{xx}(m)\}$.
(b) Design a Wiener filter of length $M = 2$ to estimate $\{s(n)\}$.
(c) Determine the MMSE for $M = 2$.

11.25 Determine the optimum causal IIR Wiener filter for the signal given in Problem 11.24 and the corresponding MMSE_∞.

11.26 Determine the system function for the noncausal IIR Wiener filter for the signal given in Problem 11.24 and the corresponding MMSE_{nc}.

11.27 Determine the optimum FIR Wiener filter of length $M = 3$ for the signal in Example 11.6.1 and the corresponding MMSE_3. Compare MMSE_3 with MMSE_2 and comment on the difference.

11.28 An AR(2) process is defined by the difference equation

$$x(n) = x(n-1) - 0.6x(n-2) + w(n)$$

where $\{w(n)\}$ is a white noise process with variance σ_w^2. Use the Yule–Walker equations to solve for the values of the autocorrelation $\gamma_{xx}(0)$, $\gamma_{xx}(1)$, and $\gamma_{xx}(2)$.

11.29 An observed random process $\{x(n)\}$ consists of the sum of an AR(p) process of the form

$$s(n) = -\sum_{k=1}^{p} a_p(k)s(n-k) + v(n)$$

and a white noise process $\{w(n)\}$ with variance σ_w^2. The random process $\{v(n)\}$ is also white with variance σ_v^2. The sequences $\{v(n)\}$ and $\{w(n)\}$ are uncorrelated.

Show that the observed process $\{x(n) = s(n) + w(n)\}$ is ARMA(p, p) and determine the coefficients of the numerator polynomial (MA component) in the corresponding system function.

12

Power Spectrum Estimation

In this chapter we are concerned with the estimation of the spectral characteristics of signals characterized as random processes. Many of the phenomena that occur in nature are best characterized statistically in terms of averages. For example, meteorological phenomena such as the fluctuations in air temperature and pressure are best characterized statistically as random processes. Thermal noise voltages generated in resistors and electronic devices are additional examples of physical signals that are well modeled as random processes.

Due to the random fluctuations in such signals, we must adopt a statistical viewpoint, which deals with the average characteristics of random signals. In particular, the autocorrelation function of a random process is the appropriate statistical average that we will use for characterizing random signals in the time domain, and the Fourier transform of the autocorrelation function, which yields the power density spectrum, provides the transformation from the time domain to the frequency domain.

Power spectrum estimation methods have a relatively long history. For a historical perspective, the reader is referred to the paper by Robinson (1982) and the book by Marple (1987). Our treatment of this subject covers the classical power spectrum estimation methods based on the periodogram, originally introduced by Schuster (1898), and by Yule (1927), who originated the modern model-based or parametric methods. These methods were subsequently developed and applied by Walker (1931), Bartlett (1948), Parzen (1957), Blackman and Tukey (1958), Burg (1967), and others. We also describe the method of Capon (1969) and methods based on eigenanalysis of the data correlation matrix.

12.1 ESTIMATION OF SPECTRA FROM FINITE-DURATION OBSERVATIONS OF SIGNALS

The basic problem that we consider in this chapter is the estimation of the power density spectrum of a signal from the observation of the signal over a finite time interval. As we will see, the finite record length of the data sequence is a major limitation on the quality of the power spectrum estimate. When dealing with signals that are statistically stationary, the longer the data record, the better the estimate that can be extracted from the data. On the other hand, if the signal statistics are nonstationary, we cannot select an arbitrarily long data record to estimate the spectrum. In such a case, the length of the data record that we select is determined by the rapidity of the time variations in the signal statistics. Ultimately, our goal is to select as short a data record as possible that still allows us to resolve the spectral characteristics of different signal components in the data record that have closely spaced spectra.

One of the problems that we encounter with classical power spectrum estimation methods based on a finite-length data record is the distortion of the spectrum that we are attempting to estimate. This problem occurs in both the computation of the spectrum for a deterministic signal and the estimation of the power spectrum of a random signal. Since it is easier to observe the effect of the finite length of the data record on a deterministic signal, we treat this case first. Thereafter, we consider only random signals and the estimation of their power spectra.

12.1.1 Computation of the Energy Density Spectrum

Let us consider the computation of the spectrum of a deterministic signal from a finite sequence of data. The sequence $x(n)$ is usually the result of sampling a continuous-time signal $x_a(t)$ at some uniform sampling rate F_s. Our objective is to obtain an estimate of the true spectrum from a finite-duration sequence $x(n)$.

Recall that if $x(t)$ is a finite-energy signal, that is,

$$E = \int_{-\infty}^{\infty} |x_a(t)|^2 dt < \infty$$

then its Fourier transform exists and is given as

$$X_a(F) = \int_{-\infty}^{\infty} x_a(t) e^{-j2\pi Ft} dt$$

From Parseval's theorem we have

$$E = \int_{-\infty}^{\infty} |x_a(t)|^2 dt = \int_{-\infty}^{\infty} |X_a(F)|^2 dF \tag{12.1.1}$$

The quantity $|X_a(F)|^2$ represents the distribution of signal energy as a function of frequency, and hence it is called the energy density spectrum of the signal, that is,

$$S_{xx}(F) = |X_a(F)|^2 \tag{12.1.2}$$

as described in Chapter 4. Thus the total energy in the signal is simply the integral of $S_{xx}(F)$ over all F [i.e., the total area under $S_{xx}(F)$].

It is also interesting to note that $S_{xx}(F)$ can be viewed as the Fourier transform of another function, $R_{xx}(\tau)$, called the *autocorrelation function* of the finite-energy signal $x_a(t)$, defined as

$$R_{xx}(\tau) = \int_{-\infty}^{\infty} x_a^*(t)x_a(t+\tau)dt \tag{12.1.3}$$

Indeed, it easily follows that

$$\int_{-\infty}^{\infty} R_{xx}(\tau)e^{-j2\pi F\tau}d\tau = S_{xx}(F) = |X_a(F)|^2 \tag{12.1.4}$$

so that $R_{xx}(\tau)$ and $S_{xx}(F)$ are a Fourier transform pair.

Now suppose that we compute the energy density spectrum of the signal $x_a(t)$ from its samples taken at the rate F_s samples per second. To ensure that there is no spectral aliasing resulting from the sampling process, the signal is assumed to be prefiltered, so that, for practical purposes, its bandwidth is limited to B hertz. Then the sampling frequency F_s is selected such that $F_s > 2B$.

The sampled version of $x_a(t)$ is a sequence $x(n)$, $-\infty < n < \infty$, which has a Fourier transform (voltage spectrum)

$$X(\omega) = \sum_{n=-\infty}^{\infty} x(n)e^{-j\omega n}$$

or, equivalently,

$$X(f) = \sum_{n=-\infty}^{\infty} x(n)e^{-j2\pi fn} \tag{12.1.5}$$

Recall that $X(f)$ can be expressed in terms of the voltage spectrum of the analog signal $x_a(t)$ as

$$X\left(\frac{F}{F_s}\right) = F_s \sum_{k=-\infty}^{\infty} X_a(F-kF_s) \tag{12.1.6}$$

where $f = F/F_s$ is the normalized frequency variable.

In the absence of aliasing, within the fundamental range $|F| \leq F_s/2$, we have

$$X\left(\frac{F}{F_s}\right) = F_s X_a(F), \quad |F| \leq F_s/2 \tag{12.1.7}$$

Hence the voltage spectrum of the sampled signal is identical to the voltage spectrum of the analog signal. As a consequence, the energy density spectrum of the sampled signal is

$$S_{xx}\left(\frac{F}{F_s}\right) = \left|X\left(\frac{F}{F_s}\right)\right|^2 = F_s^2|X_a(F)|^2 \tag{12.1.8}$$

We can proceed further by noting that the autocorrelation of the sampled signal, which is defined as

$$r_{xx}(k) = \sum_{n=-\infty}^{\infty} x^*(n)x(n+k) \tag{12.1.9}$$

has the Fourier transform (Wiener–Khintchine theorem)

$$S_{xx}(f) = \sum_{k=-\infty}^{\infty} r_{xx}(k)e^{-j2\pi kf} \tag{12.1.10}$$

Hence the energy density spectrum can be obtained by the Fourier transform of the autocorrelation of the sequence $\{x(n)\}$.

The relations above lead us to distinguish between two distinct methods for computing the energy density spectrum of a signal $x_a(t)$ from its samples $x(n)$. One is the *direct method*, which involves computing the Fourier transform of $\{x(n)\}$, and then

$$S_{xx}(f) = |X(f)|^2$$

$$= \left| \sum_{n=-\infty}^{\infty} x(n)e^{-j2\pi fn} \right|^2 \tag{12.1.11}$$

The second approach is called the *indirect method* because it requires two steps. First, the autocorrelation $r_{xx}(k)$ is computed from $x(n)$ and then the Fourier transform of the autocorrelation is computed as in (12.1.10) to obtain the energy density spectrum.

In practice, however, only the finite-duration sequence $x(n)$, $0 \le n \le N-1$, is available for computing the spectrum of the signal. In effect, limiting the duration of the sequence $x(n)$ to N points is equivalent to multiplying $x(n)$ by a rectangular window. Thus we have

$$\tilde{x}(n) = x(n)w(n) = \begin{cases} x(n), & 0 \le n \le N-1 \\ 0, & \text{otherwise} \end{cases} \tag{12.1.12}$$

From our discussion of FIR filter design based on the use of windows to limit the duration of the impulse response, we recall that multiplication of two sequences is equivalent to convolution of their voltage spectra. Consequently, the frequency-domain relation corresponding to (12.1.12) is

$$\tilde{X}(f) = X(f) * W(f)$$

$$= \int_{-1/2}^{1/2} X(\alpha)W(f-\alpha)d\alpha \tag{12.1.13}$$

Recall from our discussion in Section 8.2.1 that convolution of the window function $W(f)$ with $X(f)$ smooths the spectrum $X(f)$, provided that the spectrum $W(f)$ is relatively narrow compared to $X(f)$. But this condition implies that the window $w(n)$ be sufficiently long (i.e., N must be sufficiently large) such that $W(f)$ is narrow compared to $X(f)$. Even if $W(f)$ is narrow compared to $X(f)$, the convolution of $X(f)$ with the sidelobes of $W(f)$ results in sidelobe energy in $\tilde{X}(f)$, in frequency bands where the true signal spectrum $X(f) = 0$. This sidelobe energy is called *leakage*. The following example illustrates the leakage problem.

Example 12.1.1

A signal with (voltage) spectrum

$$X(f) = \begin{cases} 1, & |f| \le 0.1 \\ 0, & \text{otherwise} \end{cases}$$

is convolved with the rectangular window of length $N = 61$. Determine the spectrum of $\tilde{X}(f)$ given by (12.1.13).

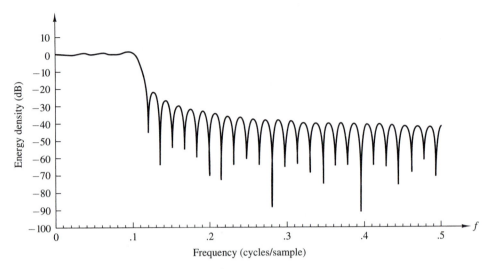

Figure 12.1 Spectrum obtained by convolving an $M = 61$ rectangular window with the ideal lowpass spectrum in Example 12.1.1.

Solution The spectral characteristic $W(f)$ for the length $N = 61$ rectangular window is illustrated in Fig. 8.2(b). Note that the width of the main lobe of the window function is $\Delta\omega = 4\pi/61$ or $\Delta f = 2/61$, which is narrow compared to $X(f)$.

The convolution of $X(f)$ with $W(f)$ is illustrated in Fig. 12.1. We note that energy has leaked into the frequency band $0.1 < |f| \le 0.5$, where $X(f) = 0$. A part of this is due to the width of the main lobe in $W(f)$, which causes a broadening or smearing of $X(f)$ outside the range $|f| \le 0.1$. However, the sidelobe energy in $\tilde{X}(f)$ is due to the presence of the sidelobes of $W(f)$, which are convolved with $X(f)$. The smearing of $X(f)$ for $|f| > 0.1$ and the sidelobes in the range $0.1 \le |f| \le 0.5$ constitute the leakage.

Just as in the case of FIR filter design, we can reduce sidelobe leakage by selecting windows that have low sidelobes. This implies that the windows have a smooth time-domain cutoff instead of the abrupt cutoff in the rectangular window. Although such window functions reduce sidelobe leakage, they result in an increase in smoothing or broadening of the spectral characteristic $X(f)$. For example, the use of a Blackman window of length $N = 61$ in Example 12.1.1 results in the spectral characteristic $\tilde{X}(f)$ shown in Fig. 12.2. The sidelobe leakage has certainly been reduced, but the spectral width has been increased by about 50%.

The broadening of the spectrum being estimated due to windowing is particularly a problem when we wish to resolve signals with closely spaced frequency components. For example, the signal with spectral characteristic $X(f) = X_1(f) + X_2(f)$, as shown in Fig. 12.3, cannot be resolved as two separate signals unless the width of the window function is significantly narrower than the frequency separation Δf. Thus we observe that using smooth time-domain windows reduces leakage at the expense of a decrease in frequency resolution.

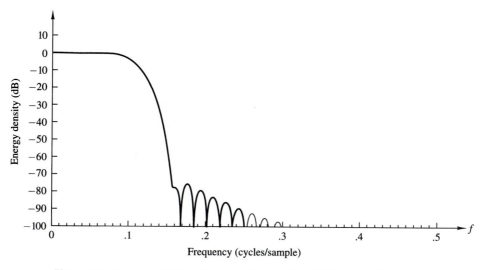

Figure 12.2   Spectrum obtained by convolving an $M = 61$ Blackman window with the ideal lowpass spectrum in Example 12.1.1.

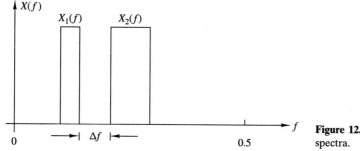

Figure 12.3   Two narrowband signal spectra.

It is clear from this discussion that the energy density spectrum of the windowed sequence $\{x(n)\}$ is an approximation of the desired spectrum of the sequence $\{x(n)\}$. The spectral density obtained from $\{\tilde{x}(n)\}$ is

$$S_{\tilde{x}\tilde{x}}(f) = |\tilde{X}(f)|^2 = \left| \sum_{n=0}^{N-1} \tilde{x}(n)e^{-j2\pi fn} \right|^2 \tag{12.1.14}$$

The spectrum given by (12.1.14) can be computed numerically at a set of N frequency points by means of the DFT. Thus

$$\tilde{X}(k) = \sum_{n=0}^{N-1} \tilde{x}(n)e^{-j2\pi kn/N} \tag{12.1.15}$$

Then

$$|\tilde{X}(k)|^2 = S_{\tilde{x}\tilde{x}}(f)|_{f=k/N} = S_{\tilde{x}\tilde{x}}\left(\frac{k}{N}\right) \tag{12.1.16}$$

and hence

$$S_{\tilde{x}\tilde{x}}\left(\frac{k}{N}\right) = \left|\sum_{n=0}^{N-1} \tilde{x}(n)e^{-j2\pi kn/N}\right|^2 \tag{12.1.17}$$

which is a distorted version of the true spectrum $S_{xx}(k/N)$.

12.1.2 Estimation of the Autocorrelation and Power Spectrum of Random Signals: The Periodogram

The finite-energy signals considered in the preceding section possess a Fourier transform and are characterized in the spectral domain by their energy density spectrum. On the other hand, the important class of signals characterized as stationary random processes do not have finite energy and hence do not possess a Fourier transform. Such signals have finite average power and hence are characterized by a *power density spectrum*. If $x(t)$ is a stationary random process, its autocorrelation function is

$$\gamma_{xx}(\tau) = E[x^*(t)x(t+\tau)] \tag{12.1.18}$$

where $E[\cdot]$ denotes the statistical average. Then, via the Wiener–Khintchine theorem, the power density spectrum of the stationary random process is the Fourier transform of the autocorrelation function, that is,

$$\Gamma_{xx}(F) = \int_{-\infty}^{\infty} \gamma_{xx}(\tau)e^{-j2\pi F\tau}dt \tag{12.1.19}$$

In practice, we deal with a single realization of the random process from which we estimate the power spectrum of the process. We do not know the true autocorrelation function $\gamma_{xx}(\tau)$ and as a consequence, we cannot compute the Fourier transform in (12.1.19) to obtain $\Gamma_{xx}(F)$. On the other hand, from a single realization of the random process we can compute the time-average autocorrelation function

$$R_{xx}(\tau) = \frac{1}{2T_0}\int_{-T_0}^{T_0} x^*(t)x(t+\tau)dt \tag{12.1.20}$$

where $2T_0$ is the observation interval. If the stationary random process is *ergodic* in the first and second moments (mean and autocorrelation function), then

$$\gamma_{xx}(\tau) = \lim_{T_0 \to \infty} R_{xx}(\tau)$$

$$= \lim_{T_0 \to \infty} \frac{1}{2T_0}\int_{-T_0}^{T_0} x^*(t)x(t+\tau)dt \tag{12.1.21}$$

This relation justifies the use of the time-average autocorrelation $R_{xx}(\tau)$ as an estimate of the statistical autocorrelation function $\gamma_{xx}(\tau)$. Furthermore, the Fourier transform of $R_{xx}(\tau)$ provides an estimate $P_{xx}(F)$ of the power density spectrum, that is,

$$P_{xx}(F) = \int_{-T_0}^{T_0} R_{xx}(\tau)e^{-j2\pi F\tau}d\tau$$

$$= \frac{1}{2T_0}\int_{-T_0}^{T_0}\left[\int_{-T_0}^{T_0} x^*(t)x(t+\tau)dt\right]e^{-j2\pi F\tau}d\tau \qquad (12.1.22)$$

$$= \frac{1}{2T_0}\left|\int_{-T_0}^{T_0} x(t)e^{-j2\pi Ft}dt\right|^2$$

The actual power density spectrum is the expected value of $P_{xx}(F)$ in the limit as $T_0 \to \infty$,

$$\Gamma_{xx}(F) = \lim_{T_0 \to \infty} E[P_{xx}(F)]$$

$$= \lim_{T_0 \to \infty} E\left[\frac{1}{2T_0}\left|\int_{-T_0}^{T_0} x(t)e^{-j2\pi Ft}dt\right|^2\right] \qquad (12.1.23)$$

From (12.1.20) and (12.1.22) we again note the two possible approaches to computing $P_{xx}(F)$, the direct method as given by (12.1.22) or the indirect method, in which we obtain $R_{xx}(\tau)$ first and then compute the Fourier transform.

We shall consider the estimation of the power density spectrum from samples of a single realization of the random process. In particular, we assume that $x_a(t)$ is sampled at a rate $F_s > 2B$, where B is the highest frequency contained in the power density spectrum of the random process. Thus we obtain a finite-duration sequence $x(n)$, $0 \le n \le N - 1$, by sampling $x_a(t)$. From these samples we can compute the time-average autocorrelation sequence

$$r'_{xx}(m) = \frac{1}{N-m}\sum_{n=0}^{N-m-1} x^*(n)x(n+m) \qquad m = 0, 1, \ldots, N-1$$

$$\qquad (12.1.24)$$

$$r'_{xx}(m) = \frac{1}{N-|m|}\sum_{n=|m|}^{N-1} x^*(n)x(n+m) \qquad m = -1, -2, \ldots, 1-N$$

and then compute the Fourier transform

$$P'_{xx}(f) = \sum_{m=-N+1}^{N-1} r'_{xx}(m)e^{-j2\pi fm} \qquad (12.1.25)$$

The normalization factor $N - |m|$ in (12.1.24) results in an estimate with mean value

$$E[r'_{xx}(m)] = \frac{1}{N-|m|}\sum_{n=0}^{N-m-1} E[x^*(n)x(n+m)]$$

$$\qquad (12.1.26)$$

$$= \gamma_{xx}(m)$$

where $\gamma_{xx}(m)$ is the true (statistical) autocorrelation sequence of $x(n)$. Hence $r'_{xx}(m)$ is an unbiased estimate of the autocorrelation function $\gamma_{xx}(m)$. The variance of the estimate $r'_{xx}(m)$ is approximately

$$\text{var}[r'_{xx}(m)] \approx \frac{N}{[N-|m|]^2} \sum_{n=-\infty}^{\infty} \left[|\gamma_{xx}(n)|^2 + \gamma_{xx}^*(n-m)\gamma_{xx}(n+m) \right] \qquad (12.1.27)$$

which is a result given by Jenkins and Watts (1968). Clearly,

$$\lim_{N\to\infty} \text{var}[r'_{xx}(m)] = 0 \qquad (12.1.28)$$

provided that

$$\sum_{n=-\infty}^{\infty} |\gamma_{xx}(n)|^2 < \infty$$

Since $E[r'_{xx}(m)] = \gamma_{xx}(m)$ and the variance of the estimate converges to zero as $N \to \infty$, the estimate $r'_{xx}(m)$ is said to be *consistent*.

For large values of the lag parameter m, the estimate $r'_{xx}(m)$ given by (12.1.24) has a large variance, especially as m approaches N. This is due to the fact that fewer data points enter into the estimate for large lags. As an alternative to (12.1.24) we can use the estimate

$$r_{xx}(m) = \frac{1}{N} \sum_{n=0}^{N-m-1} x^*(n)x(n+m) \qquad 0 \le m \le N-1$$

$$\qquad (12.1.29)$$

$$r_{xx}(m) = \frac{1}{N} \sum_{n=|m|}^{N-1} x^*(n)x(n+m) \qquad m = -1, -2, \ldots, 1-N$$

which has a bias of $|m|\gamma_{xx}(m)/N$, since its mean value is

$$E[r_{xx}(m)] = \frac{1}{N} \sum_{n=0}^{N-m-1} E[x^*(n)x(n+m)]$$

$$\qquad (12.1.30)$$

$$= \frac{N-|m|}{N}\gamma_{xx}(m) = \left(1 - \frac{|m|}{N}\right)\gamma_{xx}(m)$$

However, this estimate has a smaller variance, given approximately as

$$\text{var}[r_{xx}(m)] \approx \frac{1}{N} \sum_{n=-\infty}^{\infty} \left[|\gamma_{xx}(n)|^2 + \gamma_{xx}^*(n-m)\gamma_{xx}(n+m) \right] \qquad (12.1.31)$$

We observe that $r_{xx}(m)$ is *asymptotically unbiased*, that is,

$$\lim_{N\to\infty} E[r_{xx}(m)] = \gamma_{xx}(m) \qquad (12.1.32)$$

and its variance converges to zero as $N \to \infty$. Therefore, the estimate $r_{xx}(m)$ is also a *consistent estimate* of $\gamma_{xx}(m)$.

We shall use the estimate $r_{xx}(m)$ given by (12.1.29) in our treatment of power spectrum estimation. The corresponding estimate of the power density spectrum is

$$P_{xx}(f) = \sum_{m=-(N-1)}^{N-1} r_{xx}(m)e^{-j2\pi fm} \qquad (12.1.33)$$

If we substitute for $r_{xx}(m)$ from (12.1.29) into (12.1.33), the estimate $P_{xx}(f)$ can also be expressed as

$$P_{xx}(f) = \frac{1}{N} \left| \sum_{n=0}^{N-1} x(n)e^{-j2\pi fn} \right|^2 = \frac{1}{N}|X(f)|^2 \qquad (12.1.34)$$

where $X(f)$ is the Fourier transform of the sample sequence $x(n)$. This well known form of the power density spectrum estimate is called the *periodogram*. It was originally introduced by Schuster (1898) to detect and measure "hidden periodicities" in data.

From (12.1.33), the average value of the periodogram estimate $P_{xx}(f)$ is

$$E[P_{xx}(f)] = E\left[\sum_{m=-(N-1)}^{N-1} r_{xx}(m)e^{-j2\pi fm} \right] = \sum_{m=-(N-1)}^{N-1} E[r_{xx}(m)]e^{-j2\pi fm}$$

$$E[P_{xx}(f)] = \sum_{m=-(N-1)}^{N-1} \left(1 - \frac{|m|}{N}\right) \gamma_x(m)e^{-j2\pi fm} \qquad (12.1.35)$$

The interpretation that we give to (12.1.35) is that the mean of the estimated spectrum is the Fourier transform of the windowed autocorrelation function

$$\tilde{\gamma}_{xx}(m) = \left(1 - \frac{|m|}{N}\right) \gamma_{xx}(m) \qquad (12.1.36)$$

where the window function is the (triangular) Bartlett window. Hence the mean of the estimated spectrum is

$$E[P_{xx}(f)] = \sum_{m=-\infty}^{\infty} \tilde{\gamma}_{xx}(m)e^{-j2\pi fm}$$

$$= \int_{-1/2}^{1/2} \Gamma_{xx}(\alpha)W_B(f - \alpha)d\alpha \qquad (12.1.37)$$

where $W_B(f)$ is the spectral characteristic of the Bartlett window. The relation (12.1.37) illustrates that the mean of the estimated spectrum is the convolution of the true power density spectrum $\Gamma_{xx}(f)$ with the Fourier transform $W_B(f)$ of the Bartlett window. Consequently, the mean of the estimated spectrum is a smoothed version of the true spectrum and suffers from the same spectral leakage problems which are due to the finite number of data points.

We observe that the estimated spectrum is asymptotically unbiased, that is,

$$\lim_{N\to\infty} E\left[\sum_{m=-(N-1)}^{N-1} r_{xx}(m)e^{-j2\pi fm} \right] = \sum_{m=-\infty}^{\infty} \gamma_{xx}(m)e^{-j2\pi fm} = \Gamma_{xx}(f)$$

However, in general, the variance of the estimate $P_{xx}(f)$ does not decay to zero as $N \to \infty$. For example, when the data sequence is a Gaussian random process, the variance is easily shown to be (see Problem 12.4)

$$\text{var}[P_{xx}(f)] = \Gamma_{xx}^2(f)\left[1 + \left(\frac{\sin 2\pi f N}{N\sin 2\pi f}\right)^2\right] \qquad (12.1.38)$$

which, in the limit as $N \to \infty$, becomes

$$\lim_{N\to\infty}\text{var}[P_{xx}(f)] = \Gamma_{xx}^2(f) \qquad (12.1.39)$$

Hence we conclude that the *periodogram is not a consistent estimate of the true power density spectrum* (i.e., it does not converge to the true power density spectrum).

In summary, the estimated autocorrelation $r_{xx}(m)$ is a consistent estimate of the true autocorrelation function $\gamma_{xx}(m)$. However, its Fourier transform $P_{xx}(f)$, the periodogram, is not a consistent estimate of the true power density spectrum. We observed that $P_{xx}(f)$ is an asymptotically unbiased estimate of $\Gamma_{xx}(f)$, but for a finite-duration sequence, the mean value of $P_{xx}(f)$ contains a bias, which from (12.1.37) is evident as a distortion of the true power density spectrum. Thus the estimated spectrum suffers from the smoothing effects and the leakage embodied in the Bartlett window. The smoothing and leakage ultimately limit our ability to resolve closely spaced spectra.

The problems of leakage and frequency resolution that we have just described as well as the problem that the periodogram is not a consistent estimate of the power spectrum, provide the motivation for the power spectrum estimation methods described in Sections 12.2, 12.3, and 12.4. The methods described in Section 12.2 are classical nonparametric methods, which make no assumptions about the data sequence. The emphasis of the classical methods is on obtaining a consistent estimate of the power spectrum through some averaging or smoothing operations performed directly on the periodogram or on the autocorrelation. As we will see, the effect of these operations is to reduce the frequency resolution further, while the variance of the estimate is decreased.

The spectrum estimation methods described in Section 12.3 are based on some model of how the data were generated. In general, the model-based methods that have been developed over the past two decades provide significantly higher resolution than do the classical methods.

Additional methods are described in Sections 12.4 and 12.5. One of these methods, due to Capon (1969), is based on minimizing the variance in the spectral estimate. The methods described in Section 12.5 are based on an eigenvalue/eigenvector decomposition of the data correlation matrix.

12.1.3 The Use of the DFT in Power Spectrum Estimation

As given by (12.1.14) and (12.1.34), the estimated energy density spectrum $S_{xx}(f)$ and the periodogram $P_{xx}(f)$, respectively, can be computed by use of the DFT, which in turn is efficiently computed by a FFT algorithm. If we have N data points, we compute as a minimum the N-point DFT. For example, the computation yields samples of the periodogram

$$P_{xx}\left(\frac{k}{N}\right) = \frac{1}{N}\left|\sum_{n=0}^{N-1} x(n)e^{-j2\pi nk/N}\right|^2 \qquad k = 0, 1, \ldots, N - 1 \qquad (12.1.40)$$

at the frequencies $f_k = k/N$.

In practice, however, such a sparse sampling of the spectrum does not provide a very good representation or a good picture of the continuous spectrum estimate $P_{xx}(f)$. This is easily remedied by evaluating $P_{xx}(f)$ at additional frequencies. Equivalently, we can effectively increase the length of the sequence by means of zero padding and then evaluate $P_{xx}(f)$ at a more dense set of frequencies. Thus if we increase the data sequence length to L points by means of zero padding and evaluate the L-point DFT, we have

$$P_{xx}\left(\frac{k}{L}\right) = \frac{1}{N}\left|\sum_{n=0}^{N-1} x(n)e^{-j2\pi nk/L}\right|^2 \qquad k = 0, 1, \ldots, L - 1 \qquad (12.1.41)$$

We emphasize that zero padding and evaluating the DFT at $L > N$ points does not improve the frequency resolution in the spectral estimate. It simply provides us with a method for interpolating the values of the measured spectrum at more frequencies. The frequency resolution in the spectral estimate $P_{xx}(f)$ is determined by the length N of the data record.

Example 12.1.2

A sequence of $N = 16$ samples is obtained by sampling an analog signal consisting of two frequency components. The resulting discrete-time sequence is

$$x(n) = \sin 2\pi(0.135)n + \cos 2\pi(0.135 + \Delta f)n \qquad n = 0, 1, \ldots, 15$$

where Δf is the frequency separation. Evaluate the power spectrum $P(f) = (1/N)|X(f)|^2$ at the frequencies $f_k = k/L$, $k = 0, 1, \ldots, L - 1$, for $L = 8, 16, 32$, and 128 for values of $\Delta f = 0.06$ and $\Delta f = 0.01$.

Solution   By zero padding, we increase the data sequence to obtain the power spectrum estimate $P_{xx}(k/L)$. The results for $\Delta f = 0.06$ are plotted in Fig. 12.4. Note that zero padding does not change the resolution, but it does have the effect of interpolating the spectrum $P_{xx}(f)$. In this case the frequency separation Δf is sufficiently large so that the two frequency components are resolvable.

The spectral estimates for $\Delta f = 0.01$ are shown in Fig. 12.5. In this case the two spectral components are not resolvable. Again, the effect of zero padding is to provide more interpolation, thus giving us a better picture of the estimated spectrum. It does not improve the frequency resolution.

When only a few points of the periodogram are needed, the Goertzel algorithm described in Chapter 6 may provide a more efficient computation. Since the Goertzel algorithm has been interpreted as a linear filtering approach to computing the DFT, it is clear that the periodogram estimate can be obtained by passing

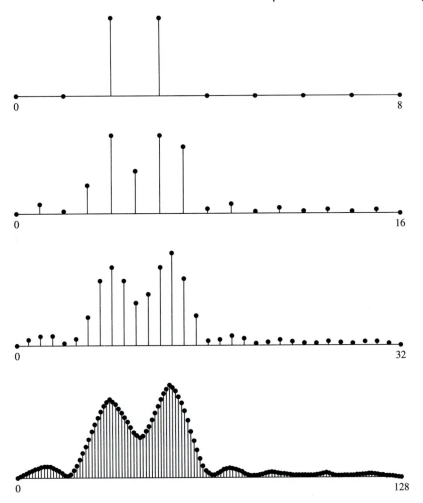

Figure 12.4   Spectra of two sinusoids with frequency separation $\Delta f = 0.06$.

the signal through a bank of parallel tuned filters and squaring their outputs (see Problem 12.5).

12.2 NONPARAMETRIC METHODS FOR POWER SPECTRUM ESTIMATION

The power spectrum estimation methods described in this section are the classical methods developed by Bartlett (1948), Blackman and Tukey (1958), and Welch (1967). These methods make no assumption about how the data were generated and hence are called *nonparametric*.

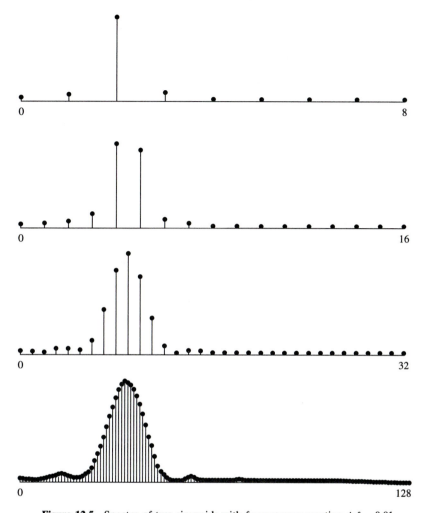

Figure 12.5   Spectra of two sinusoids with frequency separation $\Delta f = 0.01$.

Since the estimates are based entirely on a finite record of data, the frequency resolution of these methods is, at best, equal to the spectral width of the rectangular window of length N, which is approximately $1/N$ at the -3-dB points. We shall be more precise in specifying the frequency resolution of the specific methods. All the estimation techniques described in this section decrease the frequency resolution in order to reduce the variance in the spectral estimate.

First, we describe the estimates and derive the mean and variance of each. A comparison of the three methods is given in Section 12.2.4. Although the spectral estimates are expressed as a function of the continuous frequency variable f, in practice, the estimates are computed at discrete frequencies via the FFT algorithm. The FFT-based computational requirements are considered in Section 12.2.5.

12.2.1 The Bartlett Method: Averaging Periodograms

Bartlett's method for reducing the variance in the periodogram involves three steps. First, the N-point sequence is subdivided into K nonoverlapping segments, where each segment has length M. This results in the K data segments

$$x_i(n) = x(n + iM) \qquad i = 0, 1, \ldots, K - 1 \tag{12.2.1}$$

$$n = 0, 1, \ldots, M - 1$$

For each segment, we compute the periodogram

$$P_{xx}^{(i)}(f) = \frac{1}{M} \left| \sum_{n=0}^{M-1} x_i(n) e^{-j2\pi fn} \right|^2 \qquad i = 0, 1, \ldots, K - 1 \tag{12.2.2}$$

Finally, we average the periodograms for the K segments to obtain the Bartlett power spectrum estimate [Bartlett (1948)]

$$P_{xx}^B(f) = \frac{1}{K} \sum_{i=0}^{K-1} P_{xx}^{(i)}(f) \tag{12.2.3}$$

The statistical properties of this estimate are easily obtained. First, the mean value is

$$E[P_{xx}^B(f)] = \frac{1}{K} \sum_{i=0}^{K-1} E[P_{xx}^{(i)}(f)] \tag{12.2.4}$$

$$= E[P_{xx}^{(i)}(f)]$$

From (12.1.35) and (12.1.37) we have the expected value for the single periodogram as

$$E[P_{xx}^{(i)}(f)] = \sum_{m=-(M-1)}^{M-1} \left(1 - \frac{|m|}{M}\right) \gamma_{xx}(m) e^{-j2\pi fm}$$

$$= \frac{1}{M} \int_{-1/2}^{1/2} \Gamma_{xx}(\alpha) \left(\frac{\sin \pi (f - \alpha) M}{\sin \pi (f - \alpha)} \right)^2 d\alpha \tag{12.2.5}$$

where

$$W_B(f) = \frac{1}{M} \left(\frac{\sin \pi f M}{\sin \pi f} \right)^2 \tag{12.2.6}$$

is the frequency characteristics of the Bartlett window

$$w_B(n) = \begin{cases} 1 - \dfrac{|m|}{M}, & |m| \le M - 1 \\ 0, & \text{otherwise} \end{cases} \tag{12.2.7}$$

From (12.2.5) we observe that the true spectrum is now convolved with the frequency characteristic $W_B(f)$ of the Bartlett window. The effect of reducing the length of the data from N points to $M = N/K$ results in a window whose spectral width has been increased by a factor of K. Consequently, the frequency resolution has been reduced by a factor K.

In return for this reduction in resolution, we have reduced the variance. The variance of the Bartlett estimate is

$$\text{var}[P_{xx}^B(f)] = \frac{1}{K^2} \sum_{i=0}^{K-1} \text{var}[P_{xx}^{(i)}(f)] \tag{12.2.8}$$

$$= \frac{1}{K} \text{var}[P_{xx}^{(i)}(f)]$$

If we make use of (12.1.38) in (12.2.8), we obtain

$$\text{var}[P_{xx}^B(f)] = \frac{1}{K} \Gamma_{xx}^2(f) \left[1 + \left(\frac{\sin 2\pi f M}{M \sin 2\pi f} \right)^2 \right] \tag{12.2.9}$$

Therefore, the variance of the Bartlett power spectrum estimate has been reduced by the factor K.

12.2.2 The Welch Method: Averaging Modified Periodograms

Welch (1967) made two basic modifications to the Bartlett method. First, he allowed the data segments to overlap. Thus the data segments can be represented as

$$x_i(n) = x(n + iD) \qquad n = 0, 1, \ldots, M - 1 \tag{12.2.10}$$

$$i = 0, 1, \ldots, L - 1$$

where iD is the starting point for the ith sequence. Observe that if $D = M$, the segments do not overlap and the number L of data segments is identical to the number K in the Bartlett method. However, if $D = M/2$, there is 50% overlap between successive data segments and $L = 2K$ segments are obtained. Alternatively, we can form K data segments each of length $2M$.

The second modification made by Welch to the Bartlett method is to window the data segments prior to computing the periodogram. The result is a "modified" periodogram

$$\tilde{P}_{xx}^{(i)}(f) = \frac{1}{MU} \left| \sum_{n=0}^{M-1} x_i(n) w(n) e^{-j2\pi f n} \right|^2 \qquad i = 0, 1, \ldots, L - 1 \tag{12.2.11}$$

where U is a normalization factor for the power in the window function and is selected as

$$U = \frac{1}{M} \sum_{n=0}^{M-1} w^2(n) \tag{12.2.12}$$

The Welch power spectrum estimate is the average of these modified periodograms, that is,

$$P_{xx}^W(f) = \frac{1}{L} \sum_{i=0}^{L-1} \tilde{P}_{xx}^{(i)}(f) \tag{12.2.13}$$

The mean value of the Welch estimate is

$$E[P_{xx}^W(f)] = \frac{1}{L} \sum_{i=0}^{L-1} E[\tilde{P}_{xx}^{(i)}(f)]$$

(12.2.14)

$$= E[\tilde{P}_{xx}^{(i)}(f)]$$

But the expected value of the modified periodogram is

$$E[\tilde{P}_{xx}^{(i)}(f)] = \frac{1}{MU} \sum_{n=0}^{M-1} \sum_{m=0}^{M-1} w(n)w(m) E[x_i(n)x_i^*(m)] e^{-j2\pi f(n-m)}$$

(12.2.15)

$$= \frac{1}{MU} \sum_{n=0}^{M-1} \sum_{m=0}^{M-1} w(n)w(m) \gamma_{xx}(n-m) e^{-j2\pi f(n-m)}$$

Since

$$\gamma_{xx}(n) = \int_{-1/2}^{1/2} \Gamma_{xx}(\alpha) e^{j2\pi \alpha n} d\alpha$$

(12.2.16)

substitution for $\gamma_{xx}(n)$ from (12.2.16) into (12.2.15) yields

$$E[\tilde{P}_{xx}^{(i)}(f)] = \frac{1}{MU} \int_{-1/2}^{1/2} \Gamma_{xx}(\alpha) \left[\sum_{n=0}^{M-1} \sum_{m=0}^{M-1} w(n)w(m) e^{-j2\pi(n-m)(f-\alpha)} \right] d\alpha$$

(12.2.17)

$$= \int_{-1/2}^{1/2} \Gamma_{xx}(\alpha) W(f-\alpha) d\alpha$$

where, by definition,

$$W(f) = \frac{1}{MU} \left| \sum_{n=0}^{M-1} w(n) e^{-j2\pi fn} \right|^2$$

(12.2.18)

The normalization factor U ensures that

$$\int_{-1/2}^{1/2} W(f) df = 1$$

(12.2.19)

The variance of the Welch estimate is

$$\text{var}[P_{xx}^W(f)] = \frac{1}{L^2} \sum_{i=0}^{L-1} \sum_{j=0}^{L-1} E[\tilde{P}_{xx}^{(i)}(f)\tilde{P}_{xx}^{(j)}(f)] - \{E[P_{xx}^W(f)]\}^2$$

(12.2.20)

In the case of no overlap between successive data segments $(L = K)$, Welch has shown that

$$\text{var}[P_{xx}^W(f)] = \frac{1}{L} \text{var}[\tilde{P}_{xx}^{(i)}(f)]$$

(12.2.21)

$$\approx \frac{1}{L} \Gamma_{xx}^2(f)$$

In the case of 50% overlap between successive data segments ($L = 2K$), the variance of the Welch power spectrum estimate with the Bartlett (triangular) window, also derived in the paper by Welch, is

$$\text{var}[P_{xx}^{W}(f)] \approx \frac{9}{8L}\Gamma_{xx}^{2}(f) \tag{12.2.22}$$

Although we considered only the triangular window in the computation of the variance, other window functions may be used. In general, they will yield a different variance. In addition, one may vary the data segment overlapping by either more or less than the 50% considered in this section in an attempt to improve the relevant characteristics of the estimate.

12.2.3 The Blackman and Tukey Method: Smoothing the Periodogram

Blackman and Tukey (1958) proposed and analyzed the method in which the sample autocorrelation sequence is windowed first and then Fourier transformed to yield the estimate of the power spectrum. The rationale for windowing the estimated autocorrelation sequence $r_{xx}(m)$ is that, for large lags, the estimates are less reliable because a smaller number ($N - m$) of data points enter into the estimate. For values of m approaching N, the variance of these estimates is very high, and hence these estimates should be given a smaller weight in the formation of the estimated power spectrum. Thus the Blackman–Tukey estimate is

$$P_{xx}^{BT}(f) = \sum_{m=-(M-1)}^{M-1} r_{xx}(m)w(m)e^{-j2\pi fm} \tag{12.2.23}$$

where the window function $w(n)$ has length $2M - 1$ and is zero for $|m| \geq M$. With this definition for $w(n)$, the limits on the sum in (12.2.23) can be extended to $(-\infty, \infty)$. Hence the frequency-domain equivalent expression for (12.2.23) is the convolution integral

$$P_{xx}^{BT}(f) = \int_{-1/2}^{1/2} P_{xx}(\alpha)W(f - \alpha)d\alpha \tag{12.2.24}$$

where $P_{xx}(f)$ is the periodogram. It is clear from (12.2.24) that the effect of windowing the autocorrelation is to smooth the periodogram estimate, thus decreasing the variance in the estimate at the expense of reducing the resolution.

The window sequence $w(n)$ should be symmetric (even) about $m = 0$ to ensure that the estimate of the power spectrum is real. Furthermore, it is desirable to select the window spectrum to be nonnegative, that is,

$$W(f) \geq 0 \qquad |f| \leq 1/2 \tag{12.2.25}$$

This condition ensures that $P_{xx}^{BT}(f) \geq 0$ for $|f| \leq 1/2$, which is a desirable property for any power spectrum estimate. We should indicate, however, that some of the window functions we have introduced do not satisfy this condition. For example, in spite of their low sidelobe levels, the Hamming and Hann (or Hanning) windows do not satisfy the property in (12.2.25) and, consequently, may result in negative spectrum estimates in some parts of the frequency range.

The expected value of the Blackman–Tukey power spectrum estimate is

$$E[P_{xx}^{BT}(f)] = \int_{-1/2}^{1/2} E[P_{xx}(\alpha)]W(f-\alpha)d\alpha \qquad (12.2.26)$$

where from (12.1.37) we have

$$E[P_{xx}(\alpha)] = \int_{-1/2}^{1/2} \Gamma_{xx}(\theta)W_B(\alpha-\theta)d\theta \qquad (12.2.27)$$

and $W_B(f)$ is the Fourier transform of the Bartlett window. Substitution of (12.2.27) into (12.2.26) yields the double convolution integral

$$E[P_{xx}^{BT}(f)] = \int_{-1/2}^{1/2}\int_{-1/2}^{1/2} \Gamma_{xx}(\theta)W_B(\alpha-\theta)W(f-\alpha)d\alpha d\theta \qquad (12.2.28)$$

Equivalently, by working in the time domain, the expected value of the Blackman–Tukey power spectrum estimate is

$$E[P_{xx}^{BT}(f)] = \sum_{m=-(M-1)}^{M-1} E[r_{xx}(m)]w(m)e^{-j2\pi fm}$$

$$\qquad (12.2.29)$$

$$= \sum_{m=-(M-1)}^{M-1} \gamma_{xx}(m)w_B(m)w(m)e^{-j2\pi fm}$$

where the Bartlett window is

$$w_B(m) = \begin{cases} 1 - \dfrac{|m|}{N}, & |m| < N \\ 0, & \text{otherwise} \end{cases} \qquad (12.2.30)$$

Clearly, we should select the window length for $w(n)$ such that $M \ll N$, that is, $w(n)$ should be narrower than $w_B(m)$ to provide additional smoothing of the periodogram. Under this condition, (12.2.28) becomes

$$E[P_{xx}^{BT}(f)] \approx \int_{-1/2}^{1/2} \Gamma_{xx}(\theta)W(f-\theta)d\theta \qquad (12.2.31)$$

since

$$\int_{-1/2}^{1/2} W_B(\alpha-\theta)W(f-\alpha)d\alpha = \int_{-1/2}^{1/2} W_B(\alpha)W(f-\theta-\alpha)d\alpha$$

$$\qquad (12.2.32)$$

$$\approx W(f-\theta)$$

The variance of the Blackman–Tukey power spectrum estimate is

$$\text{var}[P_{xx}^{BT}(f)] = E\{[P_{xx}^{BT}(f)^2]\} - \{E[P_{xx}^{BT}(f)]\}^2 \qquad (12.2.33)$$

where the mean can be approximated as in (12.2.31). The second moment in (11.2.33) is

$$E\{[P_{xx}^{BT}(f)]^2\} = \int_{-1/2}^{1/2} \int_{-1/2}^{1/2} E[P_{xx}(\alpha)P_{xx}(\theta)]W(f-\alpha)W(f-\theta)d\alpha d\theta \quad (12.2.34)$$

On the assumption that the random process is Gaussian (see Problem 12.5), we find that

$$E[P_{xx}(\alpha)P_{xx}(\theta)] = \Gamma_{xx}(\alpha)\Gamma_{xx}(\theta)\left\{1 + \left[\frac{\sin\pi(\theta+\alpha)N}{N\sin\pi(\theta+\alpha)}\right]^2 + \left[\frac{\sin\pi(\theta-\alpha)N}{N\sin\pi(\theta-\alpha)}\right]^2\right\}$$

$$(12.2.35)$$

Substitution of (12.2.35) into (12.2.34) yields

$$E\{[P_{xx}^{BT}(f)]^2\} = \left[\int_{-1/2}^{1/2}\Gamma_{xx}(\theta)W(f-\theta)d\theta\right]^2$$

$$+ \int_{-1/2}^{1/2}\int_{-1/2}^{1/2}\Gamma_{xx}(\alpha)\Gamma_{xx}(\theta)W(f-\alpha)W(f-\theta)$$

$$\times \left\{\left[\frac{\sin\pi(\theta+\alpha)N}{N\sin\pi(\theta+\alpha)}\right]^2 + \left[\frac{\sin\pi(\theta-\alpha)N}{N\sin\pi(\theta-\alpha)}\right]^2\right\} d\alpha\,d\theta \quad (12.2.36)$$

The first term in (12.2.36) is simply the square of the mean of $P_{xx}^{BT}(f)$, which is to be subtracted out according to (12.2.33). This leaves the second term in (12.2.36), which constitutes the variance. For the case in which $N \gg M$, the functions $\sin\pi(\theta+\alpha)N/N\sin\pi(\theta+\alpha)$ and $\sin\pi(\theta-\alpha)N/N\sin\pi(\theta-\alpha)$ are relatively narrow compared to $W(f)$ in the vicinity of $\theta = -\alpha$ and $\theta = \alpha$, respectively. Therefore,

$$\int_{-1/2}^{1/2}\Gamma_{xx}(\theta)W(f-\theta)\left\{\left[\frac{\sin\pi(\theta+\alpha)N}{N\sin\pi(\theta+\alpha)}\right]^2 + \left[\frac{\sin\pi(\theta-\alpha)N}{N\sin\pi(\theta-\alpha)}\right]^2\right\}d\theta$$

$$(12.2.37)$$

$$\approx \frac{\Gamma_{xx}(-\alpha)W(f+\alpha) + \Gamma_{xx}(\alpha)W(f-\alpha)}{N}$$

With this approximation, the variance of $P_{xx}^{BT}(f)$ becomes

$$\text{var}[P_{xx}^{BT}(f)] \approx \frac{1}{N}\int_{-1/2}^{1/2}\Gamma_{xx}(\alpha)W(f-\alpha)[\Gamma_{xx}(-\alpha)W(f+\alpha) + \Gamma_{xx}(\alpha)W(f-\alpha)]d\alpha$$

$$\approx \frac{1}{N}\int_{-1/2}^{1/2}\Gamma_{xx}^2(\alpha)W^2(f-\alpha)d\alpha \quad (12.2.38)$$

where in the last step we made the approximation

$$\int_{-1/2}^{1/2}\Gamma_{xx}(\alpha)\Gamma_{xx}(-\alpha)W(f-\alpha)W(f+\alpha)d\alpha \approx 0 \quad (12.2.39)$$

We shall make one additional approximation in (12.2.38). When $W(f)$ is narrow compared to the true power spectrum $\Gamma_{xx}(f)$, (12.2.38) is further approximated as

$$\text{var}[P_{xx}^{BT}(f)] \approx \Gamma_{xx}^2(f)\left[\frac{1}{N}\int_{-1/2}^{1/2}W^2(\theta)d\theta\right]$$

$$\approx \Gamma_{xx}^2(f)\left[\frac{1}{N}\sum_{m=-(M-1)}^{M-1}w^2(m)\right] \tag{12.2.40}$$

12.2.4 Performance Characteristics of Nonparametric Power Spectrum Estimators

In this section we compare the quality of the Bartlett, Welch, and Blackman and Tukey power spectrum estimates. As a measure of quality, we use the ratio of its variance to the square of the mean of the power spectrum estimate that is,

$$Q_A = \frac{\{E[P_{xx}^A(f)]\}^2}{\text{var}[P_{xx}^A(f)]} \tag{12.2.41}$$

where $A = B$, W, or BT for the three power spectrum estimates. The reciprocal of this quantity, called the *variability*, can also be used as a measure of performance.

For reference, the periodogram has a mean and variance

$$E[P_{xx}(f)] = \int_{-1/2}^{1/2}\Gamma_{xx}(\theta)W_B(f-\theta)d\theta \tag{12.2.42}$$

$$\text{var}[P_{xx}(f)] = \Gamma_{xx}^2(f)\left[1+\left(\frac{\sin 2\pi f N}{N\sin 2\pi f}\right)^2\right] \tag{12.2.43}$$

where

$$W_B(f) = \frac{1}{N}\left(\frac{\sin \pi f N}{\sin \pi f}\right)^2 \tag{12.2.44}$$

For large N (i.e., $N \to \infty$),

$$E[P_{xx}(f)] \to \Gamma_{xx}(f)\int_{-1/2}^{1/2}W_B(\theta)d\theta = w_B(0)\Gamma_{xx}(f) = \Gamma_{xx}(f)$$

$$\text{var}[P_{xx}(f)] \to \Gamma_{xx}^2(f) \tag{12.2.45}$$

Hence, as indicated previously, the periodogram is an asymptotically unbiased estimate of the power spectrum, but it is not consistent because its variance does not approach zero as N increases toward infinity.

Asymptotically, the periodogram is characterized by the quality factor

$$Q_P = \frac{\Gamma_{xx}^2(f)}{\Gamma_{xx}^2(f)} = 1 \tag{12.2.46}$$

The fact that Q_P is fixed and independent of the data length N is another indication of the poor quality of this estimate.

Bartlett power spectrum estimate.   The mean and variance of the Bartlett power spectrum estimate are

$$E[P_{xx}^B(f)] = \int_{-1/2}^{1/2} \Gamma_{xx}(\theta) W_B(f - \theta) d\theta \tag{12.2.47}$$

$$\text{var}[P_{xx}^B(f)] = \frac{1}{K} \Gamma_{xx}^2(f) \left[1 + \left(\frac{\sin 2\pi f M}{M \sin 2\pi f} \right)^2 \right] \tag{12.2.48}$$

and

$$W_B(f) = \frac{1}{M} \left(\frac{\sin \pi f M}{\sin \pi f} \right)^2 \tag{12.2.49}$$

As $N \to \infty$ and $M \to \infty$, while $K = N/M$ remains fixed, we find that

$$E[P_{xx}^B(f)] \to \Gamma_{xx}(f) \int_{-1/2}^{1/2} W_B(f) df = \Gamma_{xx}(f) w_B(0) = \Gamma_{xx}(f)$$

$$\tag{12.2.50}$$

$$\text{var}[P_{xx}^B(f)] \to \frac{1}{K} \Gamma_{xx}^2(f)$$

We observe that the Bartlett power spectrum estimate is asymptotically unbiased and if K is allowed to increase with an increase in N, the estimate is also consistent. Hence, asymptotically, this estimate is characterized by the quality factor

$$Q_B = K = \frac{N}{M} \tag{12.2.51}$$

The frequency resolution of the Bartlett estimate, measured by taking the 3-dB width of the main lobe of the rectangular window, is

$$\Delta f = \frac{0.9}{M} \tag{12.2.52}$$

Hence, $M = 0.9/\Delta f$ and the quality factor becomes

$$Q_B = \frac{N}{0.9/\Delta f} = 1.1 N \Delta f \tag{12.2.53}$$

Welch power spectrum estimate.   The mean and variance of the Welch power spectrum estimate are

$$E[P_{xx}^W(f)] = \int_{-1/2}^{1/2} \Gamma_{xx}(\theta) W(f - \theta) d\theta \tag{12.2.54}$$

where

$$W(f) = \frac{1}{MU} \left| \sum_{n=0}^{M-1} w(n) e^{-j2\pi fn} \right|^2 \tag{12.2.55}$$

and

$$\text{var}[P_{xx}^W(f)] = \begin{cases} \dfrac{1}{L} \Gamma_{xx}^2(f), & \text{for no overlap} \\[2ex] \dfrac{9}{8L} \Gamma_{xx}^2(f), & \begin{array}{l} \text{for 50\% overlap and} \\ \text{triangular window} \end{array} \end{cases} \tag{12.2.56}$$

As $N \to \infty$ and $M \to \infty$, the mean converges to

$$E[P_{xx}^W(f)] \to \Gamma_{xx}(f) \tag{12.2.57}$$

and the variance converges to zero, so that the estimate is consistent.

Under the two conditions given by (12.2.56) the quality factor is

$$Q_W = \begin{cases} L = \dfrac{N}{M}, & \text{for no overlap} \\ \dfrac{8L}{9} = \dfrac{16N}{9M}, & \text{for 50\% overlap and} \\ & \text{triangular window} \end{cases} \tag{12.2.58}$$

On the other hand, the spectral width of the triangular window at the 3-dB points is

$$\Delta f = \frac{1.28}{M} \tag{12.2.59}$$

Consequently, the quality factor expressed in terms of Δf and N is

$$Q_W = \begin{cases} 0.78 N \Delta f, & \text{for no overlap} \\ 1.39 N \Delta f, & \text{for 50\% overlap and} \\ & \text{triangular window} \end{cases} \tag{12.2.60}$$

Blackman–Tukey power spectrum estimate.   The mean and variance of this estimate are approximated as

$$E[P_{xx}^{BT}(f)] \approx \int_{-1/2}^{1/2} \Gamma_{xx}(\theta) W(f - \theta) d\theta$$

$$\text{var}[P_{xx}^{BT}(f)] \approx \Gamma_{xx}^2(f) \left[\frac{1}{N} \sum_{m=-(M-1)}^{M-1} w^2(m) \right] \tag{12.2.61}$$

where $w(m)$ is the window sequence used to taper the estimated autocorrelation sequence. For the rectangular and Bartlett (triangular) windows we have

$$\frac{1}{N} \sum_{n=-(M-1)}^{M-1} w^2(n) = \begin{cases} 2M/N, & \text{rectangular window} \\ 2M/3N, & \text{triangular window} \end{cases} \tag{12.2.62}$$

It is clear from (12.2.61) that the mean value of the estimate is asymptotically unbiased. Its quality factor for the triangular window is

$$Q_{BT} = 1.5 \frac{N}{M} \tag{12.2.63}$$

Since the window length is $2M - 1$, the frequency resolution measured at the 3-dB points is

$$\Delta f = \frac{1.28}{2M} = \frac{0.64}{M} \tag{12.2.64}$$

and hence

$$Q_{BT} = \frac{1.5}{0.64} N \Delta f = 2.34 N \Delta f \tag{12.2.65}$$

TABLE 12.1 QUALITY OF POWER SPECTRUM ESTIMATES

Estimate	Quality Factor
Bartlett	$1.11 N \Delta f$
Welch (50% overlap)	$1.39 N \Delta f$
Blackman–Tukey	$2.34 N \Delta f$

These results are summarized in Table 12.1. It is apparent from the results we have obtained that the Welch and Blackman–Tukey power spectrum estimates are somewhat better than the Bartlett estimate. However, the differences in performance are relatively small. The main point is that the quality factor increases with an increase in the length N of the data. This characteristic behavior is not shared by the periodogram estimate. Furthermore, the quality factor depends on the product of the data length N and the frequency resolution Δf. For a desired level of quality, Δf can be decreased (frequency resolution increased) by increasing the length N of the data, and vice versa.

12.2.5 Computational Requirements of Nonparametric Power Spectrum Estimates

The other important aspect of the nonparametric power spectrum estimates is their computational requirements. For this comparison we assume the estimates are based on a fixed amount of data N and a specified resolution Δf. The radix-2 FFT algorithm is assumed in all the computations. We shall count only the number of complex multiplications required to compute the power spectrum estimate.

Bartlett power spectrum estimate

$$\text{FFT length} = M = 0.9/\Delta f$$

$$\text{Number of FFTs} = \frac{N}{M} = 1.11 N \Delta f$$

$$\text{Number of computations} = \frac{N}{M}\left(\frac{M}{2}\log_2 M\right) = \frac{N}{2}\log_2 \frac{0.9}{\Delta f}$$

Welch power spectrum estimate (50% overlap)

$$\text{FFT length} = M = 1.28/\Delta f$$

$$\text{Number of FFTs} = \frac{2N}{M} = 1.56 N \Delta f$$

$$\text{Number of computations} = \frac{2N}{M}\left(\frac{M}{2}\log_2 M\right) = N\log_2 \frac{1.28}{\Delta f}$$

In addition to the $2N/M$ FFTs, there are additional multiplications required for windowing the data. Each data record requires M multiplications. Therefore, the total number of computations is

$$\text{Total computations} = 2N + N \log_2 \frac{1.28}{\Delta f} = N \log_2 \frac{5.12}{\Delta f}$$

Blackman–Tukey power spectrum estimate. In the Blackman–Tukey method, the autocorrelation $r_{xx}(m)$ can be computed efficiently via the FFT algorithm. However, if the number of data points is large, it may not be possible to compute one N-point DFT. For example, we may have $N = 10^5$ data points but only the capacity to perform 1024-point DFTs. Since the autocorrelation sequence is windowed to $2M - 1$ points where $M \ll N$, it is possible to compute the desired $2M - 1$ points of $r_{xx}(m)$ by segmenting the data into $K = N/2M$ records and then computing $2M$-point DFTs and one $2M$-point IDFT via the FFT algorithm. Rader (1970) has described a method for performing this computation (see Problem 12.7).

If we base the computational complexity of the Blackman–Tukey method on this approach, we obtain the following computational requirements.

$$\text{FFT length} = 2M = 1.28/\Delta f$$

$$\text{Number of FFTs} = 2K + 1 = 2\left(\frac{N}{2M}\right) + 1 \approx \frac{N}{M}$$

$$\text{Number of computations} = \frac{N}{M}(M \log_2 2M) = N \log_2 \frac{1.28}{\Delta f}$$

We can neglect the additional M multiplications required to window the autocorrelation sequence $r_{xx}(m)$, since it is a relatively small number. Finally, there is the additional computation required to perform the Fourier transform of the windowed autocorrelation sequence. The FFT algorithm can be used for this computation with some zero padding for purposes of interpolating the spectral estimate. As a result of these additional computations, the number of computations is increased by a small amount.

From these results we conclude that the Welch method requires a little more computational power than do the other two methods. The Bartlett method apparently requires the smallest number of computations. However, the differences in the computational requirements of the three methods are relatively small.

12.3 PARAMETRIC METHODS FOR POWER SPECTRUM ESTIMATION

The nonparametric power spectrum estimation methods described in the preceding section are relatively simple, well understood, and easy to compute using the FFT algorithm. However, these methods require the availability of long data records in order to obtain the necessary frequency resolution required in many applications. Furthermore, these methods suffer from spectral leakage effects, due to windowing, that are inherent in finite-length data records. Often, the spectral leakage masks weak signals that are present in the data.

From one point of view, the basic limitation of the nonparametric methods is the inherent assumption that the autocorrelation estimate $r_{xx}(m)$ is zero for $m \geq N$, as implied by (12.1.33). This assumption severely limits the frequency resolution and the quality of the power spectrum estimate that is achieved. From another viewpoint, the inherent assumption in the periodogram estimate is that the data are periodic with period N. Neither one of these assumptions is realistic.

In this section we describe power spectrum estimation methods that do not require such assumptions. In fact, these methods *extrapolate* the values of the autocorrelation for lags $m \geq N$. Extrapolation is possible if we have some *a priori* information on how the data were generated. In such a case a model for the signal generation can be constructed with a number of parameters that can be estimated from the observed data. From the model and the estimated parameters, we can compute the power density spectrum implied by the model.

In effect, the modeling approach eliminates the need for window functions and the assumption that the autocorrelation sequence is zero for $|m| \geq N$. As a consequence, *parametric* (model-based) power spectrum estimation methods avoid the problem of leakage and provide better frequency resolution than do the FFT-based, nonparametric methods described in the preceding section. This is especially true in applications where short data records are available due to time-variant or transient phenomena.

The parametric methods considered in this section are based on modeling the data sequence $x(n)$ as the output of a linear system characterized by a rational system function of the form

$$H(z) = \frac{B(z)}{A(z)} = \frac{\displaystyle\sum_{k=0}^{q} b_k z^{-k}}{1 + \displaystyle\sum_{k=1}^{p} a_k z^{-k}} \tag{12.3.1}$$

The corresponding difference equation is

$$x(n) = -\sum_{k=1}^{p} a_k x(n-k) + \sum_{k=0}^{q} b_k w(n-k) \tag{12.3.2}$$

where $w(n)$ is the input sequence to the system and the observed data, $x(n)$, represents the output sequence.

In power spectrum estimation, the input sequence is not observable. However, if the observed data are characterized as a stationary random process, then the input sequence is also assumed to be a stationary random process. In such a case the power density spectrum of the data is

$$\Gamma_{xx}(f) = |H(f)|^2 \Gamma_{ww}(f)$$

where $\Gamma_{ww}(f)$ is the power density spectrum of the input sequence and $H(f)$ is the frequency response of the model.

Since our objective is to estimate the power density spectrum $\Gamma_{xx}(f)$, it is convenient to assume that the input sequence $w(n)$ is a zero-mean white noise sequence with autocorrelation

$$\gamma_{ww}(m) = \sigma_w^2 \delta(m)$$

where σ_w^2 is the variance (i.e., $\sigma_w^2 = E[|w(n)|^2]$). Then the power density spectrum of the observed data is simply

$$\Gamma_{xx}(f) = \sigma_w^2 |H(f)|^2 = \sigma_w^2 \frac{|B(f)|^2}{|A(f)|^2} \tag{12.3.3}$$

In Section 11.1 we described the representation of a stationary random process as given by (12.3.3).

In the model-based approach, the spectrum estimation procedure consists of two steps. Given the data sequence $x(n)$, $0 \le n \le N - 1$, we estimate the parameters $\{a_k\}$ and $\{b_k\}$ of the model. Then from these estimates, we compute the power spectrum estimate according to (12.3.3).

Recall that the random process $x(n)$ generated by the pole–zero model in (12.3.1) or (12.3.2) is called an *autoregressive–moving average* (ARMA) *process* of order (p, q) and it is usually denoted as ARMA (p, q). If $q = 0$ and $b_0 = 1$, the resulting system model has a system function $H(z) = 1/A(z)$ and its output $x(n)$ is called an *autoregressive* (AR) *process* of order p. This is denoted as AR(p). The third possible model is obtained by setting $A(z) = 1$, so that $H(z) = B(z)$. Its output $x(n)$ is called a *moving average* (MA) *process* of order q and denoted as MA(q).

Of these three linear models the AR model is by far the most widely used. The reasons are twofold. First, the AR model is suitable for representing spectra with narrow peaks (resonances). Second, the AR model results in very simple linear equations for the AR parameters. On the other hand, the MA model, as a general rule, requires many more coefficients to represent a narrow spectrum. Consequently, it is rarely used by itself as a model for spectrum estimation. By combining poles and zeros, the ARMA model provides a more efficient representation, from the viewpoint of the number of model parameters, of the spectrum of a random process.

The decomposition theorem due to Wold (1938) asserts that any ARMA or MA process can be represented uniquely by an AR model of possibly infinite order, and any ARMA or AR process can be represented by a MA model of possibly infinite order. In view of this theorem, the issue of model selection reduces to selecting the model that requires the smallest number of parameters that are also easy to compute. Usually, the choice in practice is the AR model. The ARMA model is used to a lesser extent.

Before describing methods for estimating the parameters in an AR(p), MA(q), and ARMA(p, q) models, it is useful to establish the basic relationships between the model parameters and the autocorrelation sequence $\gamma_{xx}(m)$. In addition, we relate the AR model parameters to the coefficients in a linear predictor for the process $x(n)$.

12.3.1 Relationships Between the Autocorrelation and the Model Parameters

In Section 11.1.2 we established the basic relationships between the autocorrelation $\{\gamma_{xx}(m)\}$ and the model parameters $\{a_k\}$ and $\{b_k\}$. For the ARMA(p, q) process, the relationship given by (11.1.18) is

$$
\gamma_{xx}(m) = \begin{cases} -\displaystyle\sum_{k=1}^{p} a_k \gamma_{xx}(m - k), & m > q \\[2mm] -\displaystyle\sum_{k=1}^{p} a_k \gamma_{xx}(m - k) + \sigma_w^2 \sum_{k=0}^{q-m} h(k) b_{k+m}, & 0 \le m \le q \\[2mm] \gamma_{xx}^*(-m), & m < 0 \end{cases} \tag{12.3.4}
$$

The relationships in (12.3.4) provide a formula for determining the model parameters $\{a_k\}$ by restricting our attention to the case $m > q$. Thus the set of linear equations

$$
\begin{bmatrix} \gamma_{xx}(q) & \gamma_{xx}(q - 1) & \cdots & \gamma_{xx}(q - p + 1) \\ \gamma_{xx}(q + 1) & \gamma_{xx}(q) & \cdots & \gamma_{xx}(q + p + 2) \\ \vdots & \vdots & & \\ \gamma_{xx}(q + p - 1) & \gamma_{xx}(q + p - 2) & \cdots & \gamma_{xx}(q) \end{bmatrix} \begin{bmatrix} a_1 \\ a_2 \\ \vdots \\ a_p \end{bmatrix}
$$

$$
= - \begin{bmatrix} \gamma_{xx}(q + 1) \\ \gamma_{xx}(q + 2) \\ \vdots \\ \gamma_{xx}(q + p) \end{bmatrix} \tag{12.3.5}
$$

can be used to solve for the model parameters $\{a_k\}$ by using estimates of the autocorrelation sequence in place of $\gamma_{xx}(m)$ for $m \ge q$. This problem is discussed in Section 12.3.8.

Another interpretation of the relationship in (12.3.5) is that the values of the autocorrelation $\gamma_{xx}(m)$ for $m > q$ are uniquely determined from the pole parameters $\{a_k\}$ and the values of $\gamma_{xx}(m)$ for $0 \le m \le p$. Consequently, the linear system model automatically extends the values of the autocorrelation sequence $\gamma_{xx}(m)$ for $m > p$.

If the pole parameters $\{a_k\}$ are obtained from (12.3.5), the result does not help us in determining the MA parameters $\{b_k\}$, because the equation

$$
\sigma_w^2 \sum_{k=0}^{q-m} h(k) b_{k+m} = \gamma_{xx}(m) + \sum_{k=1}^{p} a_k \gamma_{xx}(m - k) \qquad 0 \le m \le q
$$

depends on the impulse response $h(n)$. Although the impulse response can be expressed in terms of the parameters $\{b_k\}$ by long division of $B(z)$ with the known $A(z)$, this approach results in a set of nonlinear equations for the MA parameters.

If we adopt an $AR(p)$ model for the observed data, the relationship between the AR parameters and the autocorrelation sequence is obtained by setting $q = 0$ in (12.3.4). Thus we obtain

$$
\gamma_{xx}(m) = \begin{cases} -\displaystyle\sum_{k=1}^{p} a_k \gamma_{xx}(m-k), & m > 0 \\ -\displaystyle\sum_{k=1}^{p} a_k \gamma_{xx}(m-k) + \sigma_w^2, & m = 0 \\ \gamma_{xx}^*(-m), & m < 0 \end{cases} \tag{12.3.6}
$$

In this case, the AR parameters $\{a_k\}$ are obtained from the solution of the Yule–Walker or normal equations

$$
\begin{bmatrix} \gamma_{xx}(0) & \gamma_{xx}(-1) & \vdots & \gamma_{xx}(-p+1) \\ \gamma_{xx}(1) & \gamma_{xx}(0) & \vdots & \gamma_{xx}(-p+2) \\ \cdots & \cdots & & \vdots \\ \gamma_{xx}(p-1) & \gamma_{xx}(p-2) & \vdots & \gamma_{xx}(0) \end{bmatrix} \begin{bmatrix} a_1 \\ a_2 \\ \vdots \\ a_p \end{bmatrix} = - \begin{bmatrix} \gamma_{xx}(1) \\ \gamma_{xx}(2) \\ \vdots \\ \gamma_{xx}(p) \end{bmatrix} \tag{12.3.7}
$$

and the variance σ_w^2 can be obtained from the equation

$$
\sigma_w^2 = \gamma_{xx}(0) + \sum_{k=1}^{p} a_k \gamma_{xx}(-k) \tag{12.3.8}
$$

The equations in (12.3.7) and (12.3.8) are usually combined into a single matrix equation of the form

$$
\begin{bmatrix} \gamma_{xx}(0) & \gamma_{xx}(-1) & \cdots & \gamma_{xx}(-p) \\ \gamma_{xx}(1) & \gamma_{xx}(0) & \cdots & \gamma_{xx}(-p+1) \\ \vdots & \vdots & & \vdots \\ \gamma_{xx}(p) & \gamma_{xx}(p-1) & \cdots & \gamma_{xx}(0) \end{bmatrix} \begin{bmatrix} 1 \\ a_1 \\ \vdots \\ a_p \end{bmatrix} = \begin{bmatrix} \sigma_w^2 \\ 0 \\ \vdots \\ 0 \end{bmatrix} \tag{12.3.9}
$$

Since the correlation matrix in (12.3.7), or in (12.3.9), is Toeplitz, it can be efficiently inverted by use of the Levinson–Durbin algorithm.

Thus all the system parameters in the $AR(p)$ model are easily determined from knowledge of the autocorrelation sequence $\gamma_{xx}(m)$ for $0 \le m \le p$. Furthermore, (12.3.6) can be used to extend the autocorrelation sequence for $m > p$, once the $\{a_k\}$ are determined.

Finally, for completeness, we indicate that in a $MA(q)$ model for the observed data, the autocorrelation sequence $\gamma_{xx}(m)$ is related to the MA parameters $\{b_k\}$ by the equation

$$
\gamma_{xx}(m) = \begin{cases} \sigma_w^2 \displaystyle\sum_{k=0}^{q} b_k b_{k+m}, & 0 \le m \le q \\ 0, & m > q \\ \gamma_{xx}^*(-m), & m < 0 \end{cases} \tag{12.3.10}
$$

which was established in Section 11.1.

With this background established, we now describe the power spectrum estimation methods for the AR(p), ARMA(p, q), and MA(q) models.

12.3.2 The Yule–Walker Method for the AR Model Parameters

In the Yule–Walker method we simply estimate the autocorrelation from the data and use the estimates in (12.3.7) to solve for the AR model parameters. In this method it is desirable to use the biased form of the autocorrelation estimate,

$$r_{xx}(m) = \frac{1}{N} \sum_{n=0}^{N-m-1} x^*(n)x(n+m) \qquad m \geq 0 \tag{12.3.11}$$

to ensure that the autocorrelation matrix is positive semidefinite. The result is a stable AR model. Although stability is not a critical issue in power spectrum estimation, it is conjectured that a stable AR model best represents the data.

The Levinson–Durbin algorithm described in Chapter 11 with $r_{xx}(m)$ substituted for $\gamma_{xx}(m)$ yields the AR parameters. The corresponding power spectrum estimate is

$$P_{xx}^{YW}(f) = \frac{\hat{\sigma}_{wp}^2}{|1 + \sum_{k=1}^{p} \hat{a}_p(k)e^{-j2\pi fk}|^2} \tag{12.3.12}$$

where $\hat{a}_p(k)$ are estimates of the AR parameters obtained from the Levinson–Durbin recursions and

$$\hat{\sigma}_{wp}^2 = \hat{E}_p^f = r_{xx}(0) \prod_{k=1}^{p}[1 - |\hat{a}_k(k)|^2] \tag{12.3.13}$$

is the estimated minimum mean-square value for the pth-order predictor. An example illustrating the frequency resolution capabilities of this estimator is given in Section 12.3.9.

In estimating the power spectrum of sinusoidal signals via AR models, Lacoss (1971) showed that spectral peaks in an AR spectrum estimate are proportional to the square of the power of the sinusoidal signal. On the other hand, the area under the peak in the power density spectrum is linearly proportional to the power of the sinusoid. This characteristic behavior holds for all AR model-based estimation methods.

12.3.3 The Burg Method for the AR Model Parameters

The method devised by Burg (1968) for estimating the AR parameters can be viewed as an order-recursive least-squares lattice method, based on the minimization of the forward and backward errors in linear predictors, with the constraint that the AR parameters satisfy the Levinson–Durbin recursion.

To derive the estimator, suppose that we are given the data $x(n)$, $n = 0$, $1, \ldots, N - 1$, and let us consider the forward and backward linear prediction estimates of order m, given as

$$\hat{x}(n) = -\sum_{k=1}^{m} a_m(k)x(n-k)$$

$$\hat{x}(n-m) = -\sum_{k=1}^{m} a_m^*(k)x(n+k-m)$$

(12.3.14)

and the corresponding forward and backward errors $f_m(n)$ and $g_m(n)$ defined as $f_m(n) = x(n) - \hat{x}(n)$ and $g_m(n) = x(n-m) - \hat{x}(n-m)$ where $a_m(k)$, $0 \leq k \leq m - 1$, $m = 1, 2, \ldots, p$, are the prediction coefficients. The least-squares error is

$$\mathcal{E}_m = \sum_{n=m}^{N-1} [|f_m(n)|^2 + |g_m(n)|^2]$$

(12.3.15)

This error is to be minimized by selecting the prediction coefficients, subject to the constraint that they satisfy the Levinson–Durbin recursion given by

$$a_m(k) = a_{m-1}(k) + K_m a_{m-1}^*(m-k) \qquad 1 \leq k \leq m - 1$$

(12.3.16)

$$1 \leq m \leq p$$

where $K_m = a_m(m)$ is the mth reflection coefficient in the lattice filter realization of the predictor. When (12.3.16) is substituted into the expressions for $f_m(n)$ and $g_m(n)$, the result is the pair of order-recursive equations for the forward and backward prediction errors given by (11.2.4).

Now, if we substitute from (11.2.4) into (12.3.16) and perform the minimization of $\mathcal{E}_m$ with respect to the complex-valued reflection coefficient K_m, we obtain the result

$$\hat{K}_m = \frac{-\sum_{n=m}^{N-1} f_{m-1}(n)g_{m-1}^*(n-1)}{\dfrac{1}{2}\sum_{n=m}^{N-1} [|f_{m-1}(n)|^2 + |g_{m-1}(n-1)|^2]} \qquad m = 1, 2, \ldots, p$$

(12.3.17)

The term in the numerator of (12.3.17) is an estimate of the crosscorrelation between the forward and backward prediction errors. With the normalization factors in the denominator of (12.3.17), it is apparent that $|K_m| < 1$, so that the all-pole model obtained from the data is stable. The reader should note the similarity of (12.3.17) to its statistical counterparts given by (11.2.29).

We note that the denominator in (12.3.17) is simply the least-squares estimate of the forward and backward errors, E_{m-1}^f and E_{m-1}^b, respectively. Hence (12.3.17) can be expressed as

$$
\hat{K}_m = \frac{-\displaystyle\sum_{n=m}^{N-1} f_{m-1}(n)g_{m-1}^*(n-1)}{\frac{1}{2}[\hat{E}_{m-1}^f + \hat{E}_{m-1}^b]} \qquad m = 1, 2, \ldots, p \tag{12.3.18}
$$

where $\hat{E}_{m-1}^f + \hat{E}_{m-1}^b$ is an estimate of the total squared error E_m. We leave it as an exercise for the reader to verify that the denominator term in (12.3.18) can be computed in an order-recursive fashion according to the relation

$$
\hat{E}_m = (1 - |\hat{K}_m|^2)\hat{E}_{m-1} - |f_{m-1}(m-1)|^2 - |g_{m-1}(m-2)|^2 \tag{12.3.19}
$$

where $\hat{E}_m \equiv \hat{E}_m^f + \hat{E}_m^b$ is the total least-squares error. This result is due to Andersen (1978).

To summarize, the Burg algorithm computes the reflection coefficients in the equivalent lattice structure as specified by (12.3.18) and (12.3.19), and the Levinson–Durbin algorithm is used to obtain the AR model parameters. From the estimates of the AR parameters, we form the power spectrum estimate

$$
P_{xx}^{BU}(f) = \frac{\hat{E}_p}{\left|1 + \displaystyle\sum_{k=1}^{p} \hat{a}_p(k)e^{-j2\pi fk}\right|^2} \tag{12.3.20}
$$

The major advantages of the Burg method for estimating the parameters of the AR model are (1) it results in high frequency resolution, (2) it yields a stable AR model, and (3) it is computationally efficient.

The Burg method is known to have several disadvantages, however. First, it exhibits spectral line splitting at high signal-to-noise ratios. [see the paper by Fougere et al. (1976)]. By line splitting, we mean that the spectrum of $x(n)$ may have a single sharp peak, but the Burg method may result in two or more closely spaced peaks. For high-order models, the method also introduces spurious peaks. Furthermore, for sinusoidal signals in noise, the Burg method exhibits a sensitivity to the initial phase of a sinusoid, especially in short data records. This sensitivity is manifest as a frequency shift from the true frequency, resulting in a phase dependent frequency bias. For more details on some of these limitations the reader is referred to the papers of Chen and Stegen (1974), Ulrych and Clayton (1976), Fougere et al. (1976), Kay and Marple (1979), Swingler (1979a, 1980), Herring (1980), and Thorvaldsen (1981).

Several modifications have been proposed to overcome some of the more important limitations of the Burg method: namely, the line splitting, spurious peaks, and frequency bias. Basically, the modifications involve the introduction of a weighting (window) sequence on the squared forward and backward errors. That is, the least-squares optimization is performed on the weighted squared errors

$$\mathcal{E}_m^{WB} = \sum_{n=m}^{N-1} w_m(n)[|f_m(n)|^2 + |g_m(n)|^2] \qquad (12.3.21)$$

which, when minimized, results in the reflection coefficient estimates

$$\hat{K}_m = \frac{\displaystyle\sum_{n=m}^{N-1} w_{m-1}(n) f_{m-1}(n) g_{m-1}^*(n-1)}{\dfrac{1}{2}\displaystyle\sum_{n=m}^{N-1} w_{m-1}(n)[|f_{m-1}(n)|^2 + |g_{m-1}(n-1)|^2]} \qquad (12.3.22)$$

In particular, we mention the use of a Hamming window used by Swingler (1979b), a quadratic or parabolic window used by Kaveh and Lippert (1983), the energy weighting method used by Nikias and Scott (1982), and the data-adaptive energy weighting used by Helme and Nikias (1985).

These windowing and energy weighting methods have proved effective in reducing the occurrence of line splitting and spurious peaks, and are also effective in reducing frequency bias.

The Burg method for power spectrum estimation is usually associated with *maximum entropy spectrum estimation*, a criterion used by Burg (1967, 1975) as a basis for AR modeling in parametric spectrum estimation. The problem considered by Burg was how best to extrapolate from the given values of the autocorrelation sequence $\gamma_{xx}(m)$, $0 \le m \le p$, the values for $m > p$, such that the entire autocorrelation sequence is positive semidefinite. Since an infinite number of extrapolations are possible, Burg postulated that the extrapolations be made on the basis of maximizing uncertainty (entropy) or randomness, in the sense that the spectrum $\Gamma_{xx}(f)$ of the process is the flattest of all spectra which have the given autocorrelation values $\gamma_{xx}(m)$, $0 \le m \le p$. In particular the entropy per sample is proportional to the integral [see Burg (1975)]

$$\int_{-1/2}^{1/2} \ln \Gamma_{xx}(f) df \qquad (12.3.23)$$

Burg found that the maximum of this integral subject to the $(p+1)$ constraints

$$\int_{-1/2}^{1/2} \Gamma_{xx}(f) e^{j2\pi fm} df = \gamma_{xx}(m) \qquad 0 \le m \le p \qquad (12.3.24)$$

is the AR(p) process for which the given autocorrelation sequence $\gamma_{xx}(m)$, $0 \le m \le p$ is related to the AR parameters by the equation (12.3.6). This solution provides an additional justification for the use of the AR model in power spectrum estimation.

In view of Burg's basic work in maximum entropy spectral estimation, the Burg power spectrum estimation procedure is often called the *maximum entropy method* (MEM). We should emphasize, however, that the maximum entropy spectrum is identical to the AR-model spectrum only when the exact autocorrelation

$\gamma_{xx}(m)$ is known. When only an estimate of $\gamma_{xx}(m)$ is available for $0 \le m \le p$, the AR-model estimates of Yule–Walker and Burg are not maximum entropy spectral estimates. The general formulation for the maximum entropy spectrum based on estimates of the autocorrelation sequence results in a set of nonlinear equations. Solutions for the maximum entropy spectrum with measurement errors in the correlation sequence have been obtained by Newman (1981) and Schott and McClellan (1984).

12.3.4 Unconstrained Least-Squares Method for the AR Model Parameters

As described in the preceding section, the Burg method for determining the parameters of the AR model is basically a least-squares lattice algorithm with the added constraint that the predictor coefficients satisfy the Levinson recursion. As a result of this constraint, an increase in the order of the AR model requires only a single parameter optimization at each stage. In contrast to this approach, we may use an unconstrained least-squares algorithm to determine the AR parameters.

To elaborate, we form the forward and backward linear prediction estimates and their corresponding forward and backward errors as indicated in (12.3.14) and (12.3.15). Then we minimize the sum of squares of both errors, that is,

$$\mathcal{E}_p = \sum_{n=p}^{N-1} [|f_p(n)|^2 + |g_p(n)|^2]$$

$$= \sum_{n=p}^{N-1} \left[\left| x(n) + \sum_{k=1}^{p} a_p(k)x(n-k) \right|^2 + \left| x(n-p) + \sum_{k=1}^{p} a_p^*(k)x(n+k-p) \right|^2 \right]$$

(12.3.25)

which is the same performance index as in the Burg method. However, we do not impose the Levinson–Durbin recursion in (12.3.25) for the AR parameters. The unconstrained minimization of $\mathcal{E}_p$ with respect to the prediction coefficients yields the set of linear equations

$$\sum_{k=1}^{p} a_p(k)r_{xx}(l,k) = -r_{xx}(l,0) \qquad l = 1, 2, \ldots, p$$

(12.3.26)

where, by definition, the autocorrelation $r_{xx}(l,k)$ is

$$r_{xx}(l,k) = \sum_{n=p}^{N-1} [x(n-k)x^*(n-l) + x(n-p+l)x^*(n-p+k)]$$

(12.3.27)

The resulting residual least-squares error is

$$\mathcal{E}_p^{LS} = r_{xx}(0,0) + \sum_{k=1}^{p} \hat{a}_p(k)r_{xx}(0,k)$$

(12.3.28)

Hence the unconstrained least-squares power spectrum estimate is

$$P_{xx}^{LS}(f) = \frac{\mathcal{E}_p^{LS}}{\left| 1 + \sum_{k=1}^{p} \hat{a}_p(k) e^{-j2\pi fk} \right|^2} \tag{12.3.29}$$

The correlation matrix in (12.3.27), with elements $r_{xx}(l, k)$, is not Toeplitz, so that the Levinson–Durbin algorithm cannot be applied. However, the correlation matrix has sufficient structure to make it possible to devise computationally efficient algorithms with computational complexity proportional to p^2. Marple (1980) devised such an algorithm, which has a lattice structure and employs Levinson–Durbin-type order recursions and additional time recursions.

This form of the unconstrained least-squares method described has also been called the *unwindowed data* least-squares method. It has been proposed for spectrum estimation in several papers, including the papers by Burg (1967), Nuttall (1976), and Ulrych and Clayton (1976). Its performance characteristics have been found to be superior to the Burg method, in the sense that the unconstrained least-squares method does not exhibit the same sensitivity to such problems as line splitting, frequency bias, and spurious peaks. In view of the computational efficiency of Marple's algorithm, which is comparable to the efficiency of the Levinson–Durbin algorithm, the unconstrained least-squares method is very attractive. With this method there is no guarantee that the estimated AR parameters yield a stable AR model. However, in spectrum estimation, this is not considered to be a problem.

12.3.5 Sequential Estimation Methods for the AR Model Parameters

The three power spectrum estimation methods described in the preceding sections for the AR model can be classified as block processing methods. These methods obtain estimates of the AR parameters from a block of data, say $x(n)$, $n = 0$, $1, \ldots, N - 1$. The AR parameters, based on the block of N data points, are then used to obtain the power spectrum estimate.

In situations where data are available on a continuous basis, we can still segment the data into blocks of N points and perform spectrum estimation on a block-by-block basis. This is often done in practice, for both real-time and non-real-time applications. However, in such applications, there is an alternative approach based on sequential (in time) estimation of the AR model parameters as each new data point becomes available. By introducing a weighting function into past data samples, it is possible to deemphasize the effect of older data samples as new data are received.

Sequential lattice methods based on recursive least squares can be employed to optimally estimate the prediction and reflection coefficients in the lattice realization of the forward and backward linear predictors. The recursive equations for the prediction coefficients relate directly to the AR model parameters. In addition to the order-recursive nature of these equations, as implied by the lattice structure, we can also obtain time-recursive equations for the reflection coefficients in the lattice and for the forward and backward prediction coefficients.

Sequential recursive least-squares algorithms are equivalent to the unconstrained least-squares, block processing method described in the preceding section. Hence the power spectrum estimates obtained by the sequential recursive least-squares method retain the desirable properties of the block processing algorithm described in Section 12.3.4. Since the AR parameters are being continuously estimated in a sequential estimation algorithm, power spectrum estimates can be obtained as often as desired, from once per sample to once every N samples. By properly weighting past data samples, the sequential estimation methods are particularly suitable for estimating and tracking time-variant power spectra resulting from nonstationary signal statistics.

The computational complexity of sequential estimation methods is generally proportional to p, the order of the AR process. As a consequence, sequential estimation algorithms are computationally efficient and, from this viewpoint, may offer some advantage over the block processing methods.

There are numerous references on sequential estimation methods. The papers by Griffiths (1975), Friedlander (1982a, b), and Kalouptsidis and Theodoridis (1987) are particularly relevant to the spectrum estimation problem.

12.3.6 Selection of AR Model Order

One of the most important aspects of the use of the AR model is the selection of the order p. As a general rule, if we select a model with too low an order, we obtain a highly smoothed spectrum. On the other hand, if p is selected too high, we run the risk of introducing spurious low-level peaks in the spectrum. We mentioned previously that one indication of the performance of the AR model is the mean-square value of the residual error, which, in general, is different for each of the estimators described above. The characteristic of this residual error is that it decreases as the order of the AR model is increased. We can monitor the rate of decrease and decide to terminate the process when the rate of decrease becomes relatively slow. It is apparent, however, that this approach may be imprecise and ill-defined, and other methods should be investigated.

Much work has been done by various researchers on this problem and many experimental results have been given in the literature [e.g., the papers by Gersch and Sharpe (1973), Ulrych and Bishop (1975), Tong (1975, 1977), Jones (1976), Nuttall (1976), Berryman (1978), Kaveh and Bruzzone (1979), and Kashyap (1980)].

Two of the better known criteria for selecting the model order have been proposed by Akaike (1969, 1974). With the first, called the *final prediction error*

(FPE) criterion, the order is selected to minimize the performance index

$$\text{FPE}(p) = \hat{\sigma}_{wp}^2 \left(\frac{N + p + 1}{N - p - 1} \right) \qquad (12.3.30)$$

where $\hat{\sigma}_{wp}^2$ is the estimated variance of the linear prediction error. This performance index is based on minimizing the mean-square error for a one-step predictor.

The second criterion proposed by Akaike (1974), called the *Akaike information criterion* (AIC), is based on selecting the order that minimizes

$$\text{AIC}(p) = \ln \hat{\sigma}_{wp}^2 + 2p/N \qquad (12.3.31)$$

Note that the term $\hat{\sigma}_{wp}^2$ decreases and therefore $\ln \hat{\sigma}_{wp}^2$ also decreases as the order of the AR model is increased. However, $2p/N$ increases with an increase in p. Hence a minimum value is obtained for some p.

An alternative information criterion, proposed by Rissanen (1983), is based on selecting the order that *minimizes the description length* (MDL), where MDL is defined as

$$\text{MDL}(p) = N \ln \hat{\sigma}_{wp}^2 + p \ln N \qquad (12.3.32)$$

A fourth criterion has been proposed by Parzen (1974). This is called the *criterion autoregressive transfer* (CAT) function and is defined as

$$\text{CAT}(p) = \left(\frac{1}{N} \sum_{k=1}^{p} \frac{1}{\bar{\sigma}_{wk}^2} \right) - \frac{1}{\hat{\sigma}_{wp}^2} \qquad (12.3.33)$$

where

$$\bar{\sigma}_{wk}^2 = \frac{N}{N - k} \hat{\sigma}_{wk}^2 \qquad (12.3.34)$$

The order p is selected to minimize $\text{CAT}(p)$.

In applying this criteria, the mean should be removed from the data. Since $\hat{\sigma}_{wk}^2$ depends on the type of spectrum estimate we obtain, the model order is also a function of the criterion.

The experimental results given in the references just cited indicate that the model-order selection criteria do not yield definitive results. For example, Ulrych and Bishop (1975), Jones (1976), and Berryman (1978), found that the FPE(p) criterion tends to underestimate the model order. Kashyap (1980) showed that the AIC criterion is statistically inconsistent as $N \to \infty$. On the other hand, the MDL information criterion proposed by Rissanen is statistically consistent. Other experimental results indicate that for small data lengths, the order of the AR model should be selected to be in the range $N/3$ to $N/2$ for good results. It is apparent that in the absence of any prior information regarding the physical process that resulted in the data, one should try different model orders and different criteria and, finally, consider the different results.

12.3.7 MA Model for Power Spectrum Estimation

As shown in Section 12.3.1, the parameters in a MA(q) model are related to the statistical autocorrelation $\gamma_{xx}(m)$ by (12.3.10). However,

$$B(z)B(z^{-1}) = D(z) = \sum_{m=-q}^{q} d_m z^{-m} \tag{12.3.35}$$

where the coefficients $\{d_m\}$ are related to the MA parameters by the expression

$$d_m = \sum_{k=0}^{q-|m|} b_k b_{k+m} \qquad |m| \le q \tag{12.3.36}$$

Clearly, then,

$$\gamma_{xx}(m) = \begin{cases} \sigma_w^2 d_m, & |m| \le q \\ 0, & |m| > q \end{cases} \tag{12.3.37}$$

and the power spectrum for the MA(q) process is

$$\Gamma_{xx}^{MA}(f) = \sum_{m=-q}^{q} \gamma_{xx}(m) e^{-j2\pi f m} \tag{12.3.38}$$

It is apparent from these expressions that we do not have to solve for the MA parameters $\{b_k\}$ to estimate the power spectrum. The estimates of the autocorrelation $\gamma_{xx}(m)$ for $|m| \le q$ suffice. From such estimates we compute the estimated MA power spectrum, given as

$$P_{xx}^{MA}(f) = \sum_{m=-q}^{q} r_{xx}(m) e^{-j2\pi f m} \tag{12.3.39}$$

which is identical to the classical (nonparametric) power spectrum estimate described in Section 12.1.

There is an alternative method for determining $\{b_k\}$ based on a high-order AR approximation to the MA process. To be specific, let the MA(q) process be modeled by an AR(p) model, where $p >> q$. Then $B(z) = 1/A(z)$, or equivalently, $B(z)A(z) = 1$. Thus the parameters $\{b_k\}$ and $\{a_k\}$ are related by a convolution sum, which can be expressed as

$$\hat{a}_n + \sum_{k=1}^{q} b_k \hat{a}_{n-k} = \begin{cases} 1, & n = 0 \\ 0, & n \neq 0 \end{cases} \tag{12.3.40}$$

where $\{\hat{a}_n\}$ are the parameters obtained by fitting the data to an AR(p) model.

Although this set of equations can be easily solved for the $\{b_k\}$, a better fit is obtained by using a least-squares error criterion. That is, we form the squared error

$$\mathcal{E} = \sum_{n=0}^{p} \left[\hat{a}_n + \sum_{k=1}^{q} b_k \hat{a}_{n-k} \right]^2 - 1 \qquad \hat{a}_0 = 1, \quad \hat{a}_k = 0, \quad k < 0 \tag{12.3.41}$$

which is minimized by selecting the MA(q) parameters $\{b_k\}$. The result of this minimization is

$$\hat{\mathbf{b}} = -\mathbf{R}_{aa}^{-1}\mathbf{r}_{aa} \tag{12.3.42}$$

where the elements of $\mathbf{R}_{aa}$ and $\mathbf{r}_{aa}$ are given as

$$R_{aa}(|i - j|) = \sum_{n=0}^{p-|i-j|} \hat{a}_n \hat{a}_{n+|i-j|} \qquad i, j = 1, 2, \ldots, q$$

$$r_{aa}(i) = \sum_{n=0}^{p-i} \hat{a}_n \hat{a}_n + i \qquad i = 1, 2, \ldots, q \tag{12.3.43}$$

This least squares method for determining the parameters of the MA(q) model is attributed to Durbin (1959). It has been shown by Kay (1988) that this estimation method is approximately the maximum likelihood under the assumption that the observed process is Gaussian.

The order q of the MA model may be determined empirically by several methods. For example, the AIC for MA models has the same form as for AR models,

$$\text{AIC}(q) = \ln \sigma_{wq}^2 + \frac{2q}{N} \tag{12.3.44}$$

where σ_{wq}^2 is an estimate of the variance of the white noise. Another approach, proposed by Chow (1972b), is to filter the data with the inverse MA(q) filter and test the filtered output for whiteness.

12.3.8 ARMA Model for Power Spectrum Estimation

The Burg algorithm, its variations, and the least-squares method described in the previous sections provide reliable high-resolution spectrum estimates based on the AR model. An ARMA model provides us with an opportunity to improve on the AR spectrum estimate, perhaps, by using fewer model parameters.

The ARMA model is particularly appropriate when the signal has been corrupted by noise. For example, suppose that the data $x(n)$ are generated by an AR system, where the system output is corrupted by additive white noise. The z-transform of the autocorrelation of the resultant signal can be expressed as

$$\Gamma_{xx}(z) = \frac{\sigma_w^2}{A(z)A(z^{-1})} + \sigma_n^2$$

$$= \frac{\sigma_w^2 + \sigma_n^2 A(z)A(z^{-1})}{A(z)A(z^{-1})} \tag{12.3.45}$$

where σ_n^2 is the variance of the additive noise. Therefore, the process $x(n)$ is ARMA(p, p), where p is the order of the autocorrelation process. This relationship provides some motivation for investigating ARMA models for power spectrum estimation.

As we have demonstrated in Section 12.3.1, the parameters of the ARMA model are related to the autocorrelation by the equation in (12.3.4). For lags

$|m| > q$, the equation involves only the AR parameters $\{a_k\}$. With estimates substituted in place of $\gamma_{xx}(m)$, we can solve the p equations in (12.3.5) to obtain $\hat{a}_k$. For high-order models, however, this approach is likely to yield poor estimates of the AR parameters due to the poor estimates of the autocorrelation for large lags. Consequently, this approach is not recommended.

A more reliable method is to construct an overdetermined set of linear equations for $m > q$, and to use the method of least squares on the set of overdetermined equations, as proposed by Cadzow (1979). To elaborate, suppose that the autocorrelation sequence can be accurately estimated up to lag M, where $M > p + q$. Then we can write the following set of linear equations:

$$\begin{bmatrix} r_{xx}(q) & r_{xx}(q-1) & \cdots & r_{xx}(q-p+1) \\ r_{xx}(q+1) & r_{xx}(q) & \cdots & r_{xx}(q-p+2) \\ \vdots & \vdots & & \\ r_{xx}(M-1) & r_{xx}(M-2) & & r_{xx}(M-p) \end{bmatrix} \begin{bmatrix} a_1 \\ a_2 \\ \vdots \\ a_p \end{bmatrix} = - \begin{bmatrix} r_{xx}(q+1) \\ r_{xx}(q+2) \\ \vdots \\ r_{xx}(M) \end{bmatrix}$$

(12.3.46)

or equivalently,

$$\mathbf{R}_{xx}\mathbf{a} = -\mathbf{r}_{xx} \qquad (12.3.47)$$

Since $\mathbf{R}_{xx}$ is of dimension $(M-q) \times p$, and $M-q > p$ we can use the least-squares criterion to solve for the parameter vector $\mathbf{a}$. The result of this minimization is

$$\hat{\mathbf{a}} = -(\mathbf{R}_{xx}^t \mathbf{R}_{xx})^{-1} \mathbf{R}_{xx}^t \mathbf{r}_{xx} \qquad (12.3.48)$$

This procedure is called the *least-squares modified Yule–Walker method*. A weighting factor can also be applied to the autocorrelation sequence to deemphasize the less reliable estimates for large lags.

Once the parameters for the AR part of the model have been estimated as indicated above, we have the system

$$\hat{A}(z) = 1 + \sum_{k=1}^{p} \hat{a}_k z^{-k} \qquad (12.3.49)$$

The sequence $x(n)$ can now be filtered by the FIR filter $\hat{A}(z)$ to yield the sequence

$$v(n) = x(n) + \sum_{k=1}^{p} \hat{a}_k x(n-k) \qquad n = 0, 1, \ldots, N-1 \qquad (12.3.50)$$

The cascade of the ARMA(p, q) model with $\hat{A}(z)$ is approximately the MA(q) process generated by the model $B(z)$. Hence we can apply the MA estimate given in the preceding section to obtain the MA spectrum. To be specific, the filtered sequence $v(n)$ for $p \le n \le N-1$ is used to form the estimated correlation sequences $r_{vv}(m)$, from which we obtain the MA spectrum

$$P_{vv}^{\text{MA}}(f) = \sum_{m=-q}^{q} r_{vv}(m) e^{-j2\pi fm} \qquad (12.3.51)$$

First, we observe that the parameters $\{b_k\}$ are not required to determine the power spectrum. Second, we observe that $r_{vv}(m)$ is an estimate of the autocorrelation for the MA model given by (12.3.10). In forming the estimate $r_{vv}(m)$, weighting (e.g., with the Bartlett window) may be used to deemphasize correlation estimates for large lags. In addition, the data may be filtered by a backward filter, thus creating another sequence, say $v^b(n)$, so that both $v(n)$ and $v^b(n)$ can be used in forming the estimate of the autocorrelation $r_{vv}(m)$, as proposed by Kay (1980). Finally, the estimated ARMA power spectrum is

$$\hat{P}_{xx}^{\text{ARMA}}(f) = \frac{P_{vv}^{\text{MA}}(f)}{\left|1 + \displaystyle\sum_{k=1}^{p} \hat{a}_k e^{-j2\pi fk}\right|^2} \tag{12.3.52}$$

The problem of order selection for the ARMA(p, q) model has been investigated by Chow (1972a, b) and Bruzzone and Kaveh (1980). For this purpose the minimum of the AIC index

$$\text{AIC}(p, q) = \ln \hat{\sigma}_{wpq}^2 + \frac{2(p+q)}{N} \tag{12.3.53}$$

can be used, where $\hat{\sigma}_{wpq}^2$ is an estimate of the variance of the input error. An additional test on the adequacy of a particular ARMA(p, q) model is to filter the data through the model and test for whiteness of the output data. This requires that the parameters of the MA model be computed from the estimated autocorrelation, using spectral factorization to determine $B(z)$ from $D(z) = B(z)B(z^{-1})$.

For additional reading on ARMA power spectrum estimation, the reader is referred to the papers by Graupe et al. (1975), Cadzow (1981, 1982), Kay (1980), and Friedlander (1982b).

12.3.9 Some Experimental Results

In this section we present some experimental results on the performance of AR and ARMA power spectrum estimates obtained by using artificially generated data. Our objective is to compare the spectral estimation methods on the basis of their frequency resolution, bias, and their robustness in the presence of additive noise.

The data consist of either one or two sinusoids and additive Gaussian noise. The two sinusoids are spaced Δf apart. Clearly, the underlying process is ARMA$(4, 4)$. The results that are shown employ an AR(p) model for these data. For high signal-to-noise ratios (SNRs) we expect the AR(4) to be adequate. However, for low SNRs, a higher-order AR model is needed to approximate the ARMA$(4, 4)$ process. The results given below are consistent with this statement. The SNR is defined as $10 \log_{10} A^2/2\sigma^2$, where σ^2 is variance of the additive noise and A is the amplitude of the sinusoid.

In Fig. 12.6 we illustrate the results for $N = 20$ data points based on an AR(4) model with a SNR $= 20$ dB and $\Delta f = 0.13$. Note that the Yule–Walker method gives an extremely smooth (broad) spectral estimate with small peaks. If Δf is

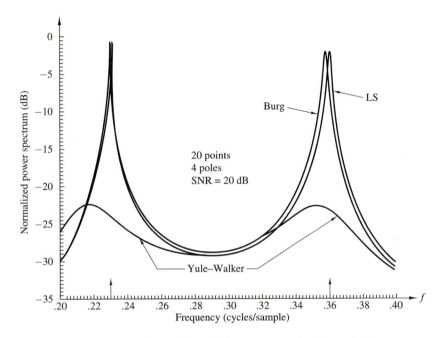

Figure 12.6   Comparison of AR spectrum estimation methods.

decreased to $\Delta f = 0.07$, the Yule–Walker method no longer resolves the peaks as illustrated in Fig. 12.7. Some bias is also evident in the Burg method. Of course, by increasing the number of data points the Yule–Walker method eventually is able to resolve the peaks. However, the Burg and least-squares methods are clearly superior for short data records.

The effect of additive noise on the estimate is illustrated in Fig. 12.8 for the least-squares method. The effect of filter order on the Burg and least-squares methods is illustrated in Figs. 12.9 and 12.10, respectively. Both methods exhibit spurious peaks as the order is increased to $p = 12$.

The effect of initial phase is illustrated in Figs. 12.11 and 12.12 for the Burg and least-squares methods. It is clear that the least-squares method exhibits less sensitivity to initial phase than the Burg algorithm.

An example of line splitting for the Burg method is shown in Fig. 12.13 with $p = 12$. It does not occur for the AR(8) model. The least-squares method did not exhibit line splitting under the same conditions. On the other hand, the line splitting on the Burg method disappeared with an increase in the number of data points N.

Figures 12.14 and 12.15 illustrate the resolution properties of the Burg and least-squares methods for $\Delta f = 0.07$ and $N = 20$ points at low SNR (3 dB). Since the additive noise process is ARMA, a higher-order AR model is required to provide a good approximation at low SNR. Hence the frequency resolution improves as the order is increased.

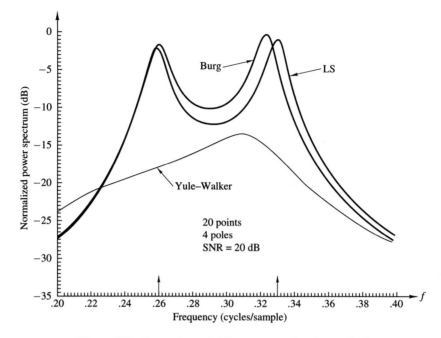

Figure 12.7 Comparison of AR spectrum estimation methods.

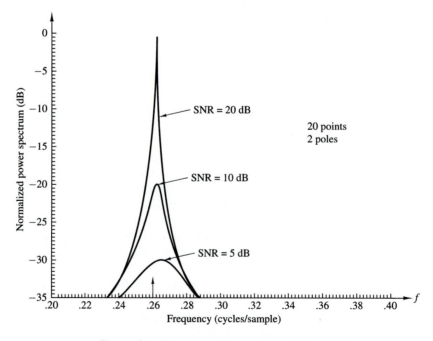

Figure 12.8 Effect of additive noise on LS method.

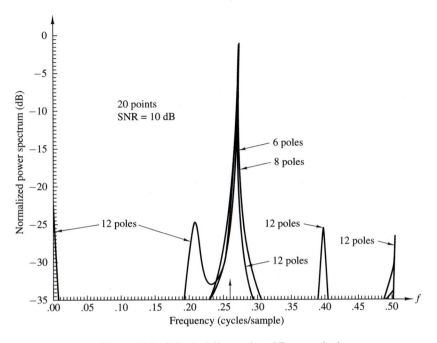

Figure 12.9   Effect of filter order of Burg method.

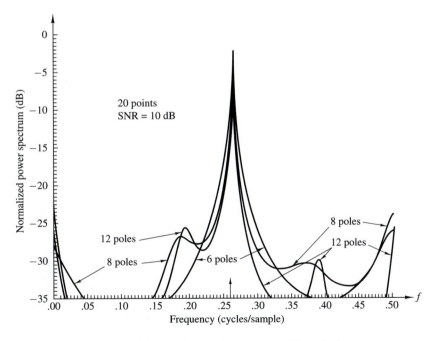

Figure 12.10   Effect of filter order on LS method.

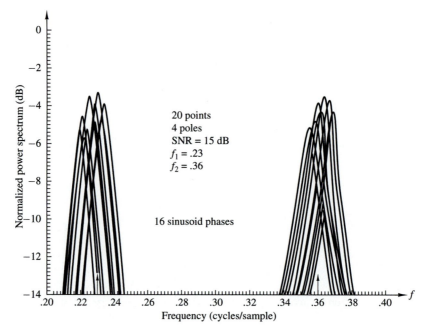

Figure 12.11   Effect of initial phase on Burg method.

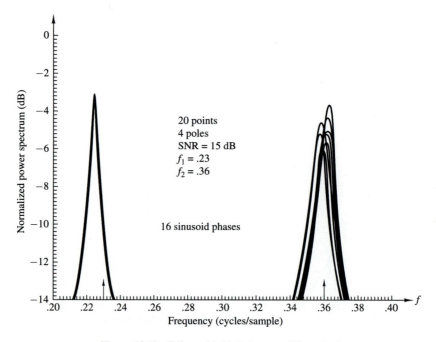

Figure 12.12   Effect of initial phase on LS method.

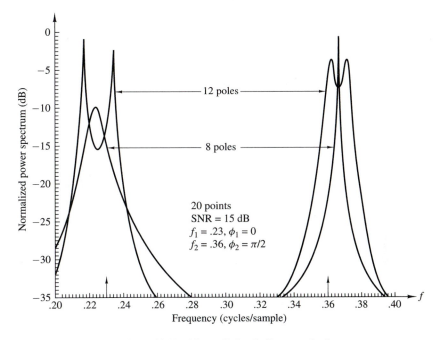

Figure 12.13   Line splitting in Burg method.

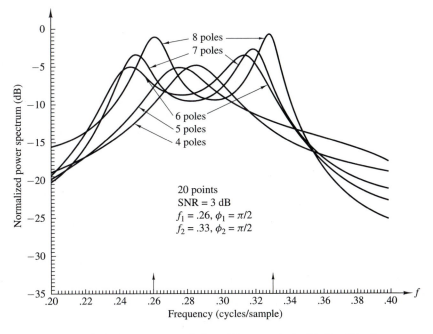

Figure 12.14   Frequency resolution of Burg method with $N = 20$ points.

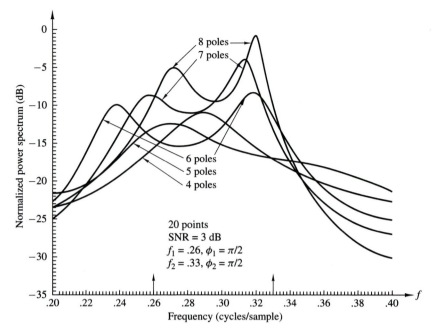

Figure 12.15 Frequency resolution of LS method with $N = 20$ points.

The FPE for the Burg method is illustrated in Fig. 12.16 for an SNR = 3 dB. For this SNR the optimum value is $p = 12$ according to the FPE criterion.

The Burg and least-squares methods were also tested with data from a narrowband process, obtained by exciting a four-pole (two pairs of complex-conjugate poles) narrowband filter and selecting a portion of the output sequence for the data record. Figure 12.17 illustrates the superposition of 20 data records of 20 points each. We observe a relatively small variability. In contrast, the Burg method exhibited a much larger variability, approximately a factor of 2 compared to the least-squares method. The results shown in Figs. 12.6 through 12.17 are taken from Poole (1981).

Finally, we show in Fig. 12.18 the ARMA(10, 10) spectral estimates obtained by Kay (1980) for two sinusoids in noise using the least-squares ARMA method described in Section 12.3.8, as an illustration of the quality of power spectrum estimation obtained with the ARMA model.

12.4 MINIMUM VARIANCE SPECTRAL ESTIMATION

The spectral estimator proposed by Capon (1969) was intended for use in large seismic arrays for frequency–wave number estimation. It was later adapted to single-time-series spectrum estimation by Lacoss (1971), who demonstrated that

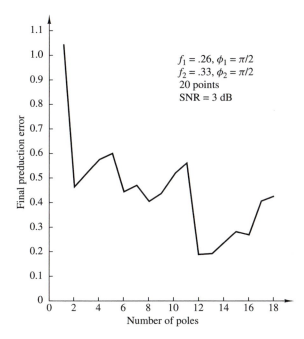

$f_1 = .26, \phi_1 = \pi/2$
$f_2 = .33, \phi_2 = \pi/2$
20 points
SNR = 3 dB

Figure 12.16 Final prediction error for Burg estimate.

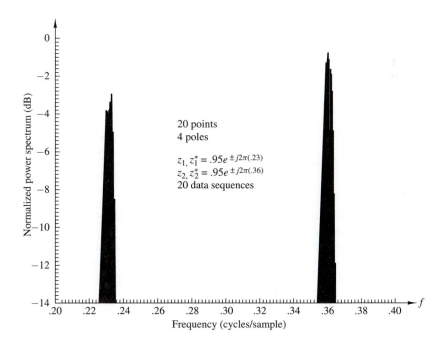

20 points
4 poles

$z_1, z_1^* = .95e^{\pm j2\pi(.23)}$
$z_2, z_2^* = .95e^{\pm j2\pi(.36)}$
20 data sequences

Figure 12.17 Effect of starting point in sequence on LS method.

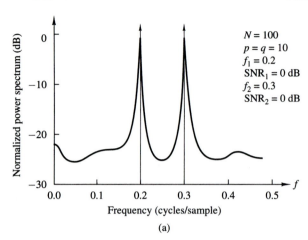

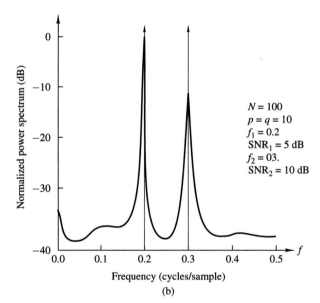

Figure 12.18 ARMA (10, 10) power spectrum estimates from paper by Kay (1980). Reprinted with permission from the IEEE.

the method provides a minimum variance unbiased estimate of the spectral components in the signal.

Following the development of Lacoss, let us consider an FIR filter with coefficients a_k, $0 \le k \le p$, to be determined. Unlike the linear prediction problem, we do not constrain a_0 to be unity. Then, if the observed data $x(n)$, $0 \le n \le N - 1$, are passed through the filter, the response is

$$y(n) = \sum_{k=0}^{p} a_k x(n - k) \equiv \mathbf{X}^t(n)\mathbf{a} \qquad (12.4.1)$$

$\mathbf{X}^t(n) = [\, x(n) \quad x(n-1) \quad \cdots \quad x(n-p) \,]$ is the data vector and $\mathbf{a}$ is the filter coefficient vector. If we assume that $E[x(n)] = 0$, the variance of the output sequence is

$$\sigma_y^2 = E[|y(n)|^2] = E[\mathbf{a}^{*t}\mathbf{X}^*(n)\mathbf{X}^t(n)\mathbf{a}]$$
$$= \mathbf{a}^{*t}\mathbf{\Gamma}_{xx}\mathbf{a} \qquad (12.4.2)$$

where $\mathbf{\Gamma}_{xx}$ is the autocorrelation matrix of the sequence $x(n)$, with elements $\gamma_{xx}(m)$.

The filter coefficients are selected so that at the frequency f_l, the frequency response of the FIR filter is normalized to unity, that is,

$$\sum_{k=0}^{p} a_k e^{-j2\pi k f_l} = 1$$

This constraint can also be written in matrix form as

$$\mathbf{E}^{*t}(f_l)\mathbf{a} = 1 \qquad (12.4.3)$$

where

$$\mathbf{E}^t(f_l) = [\, 1 \quad e^{j2\pi f_l} \quad \cdots \quad e^{j2\pi p f_l} \,]$$

By minimizing the variance σ_y^2 subject to the constraint (12.4.3), we obtain an FIR filter that passes the frequency component f_l undistorted, while components distant from f_l are severely attenuated. The result of this minimization is shown by Lacoss to lead to the coefficient vector

$$\hat{\mathbf{a}} = \mathbf{\Gamma}_{xx}^{-1}\mathbf{E}^*(f_l)/\mathbf{E}^t(f_l)\mathbf{\Gamma}_{xx}^{-1}\mathbf{E}^*(f_l) \qquad (12.4.4)$$

If $\hat{\mathbf{a}}$ is substituted into (12.4.2), we obtain the minimum variance

$$\sigma_{\min}^2 = \frac{1}{\mathbf{E}^t(f_l)\mathbf{\Gamma}_{xx}^{-1}\mathbf{E}^*(f_l)} \qquad (12.4.5)$$

The expression in (12.4.5) is the minimum variance power spectrum estimate at the frequency f_l. By changing f_l over the range $0 \le f_l \le 0.5$, we can obtain the power spectrum estimate. It should be noted that although $\mathbf{E}(f)$ changes with the choice of frequency, $\mathbf{\Gamma}_{xx}^{-1}$ is computed only once. As demonstrated by Lacoss (1971), the computation of the quadratic form $\mathbf{E}^t(f)\mathbf{\Gamma}_{xx}^{-1}\mathbf{E}^*(f)$ can be done with a single DFT.

With an estimate $\mathbf{R}_{xx}$ of the autocorrelation matrix substituted in place of $\mathbf{\Gamma}_{xx}$, we obtain the minimum variance power spectrum estimate of Capon as

$$P_{xx}^{\mathrm{MV}}(f) = \frac{1}{\mathbf{E}^t(f)\mathbf{R}_{xx}^{-1}\mathbf{E}^*(f)} \qquad (12.4.6)$$

It has been shown by Lacoss (1971) that this power spectrum estimator yields estimates of the spectral peaks proportional to the power at that frequency. In constrast, the AR methods described in Section 12.3 result in estimates of the spectral peaks proportional to the square of the power at that frequency.

This minimum variance method is basically a filter bank implementation for the spectrum estimator. It differs basically from the filter bank interpretation of the periodogram in that the filter coefficients in the Capon method are optimized.

Experiments on the performance of this method compared with the performance of the Burg method have been done by Lacoss (1971) and others. In general, the minimum variance estimate in (12.4.6) outperforms the nonparametric spectral estimators in frequency resolution, but it does not provide the high frequency resolution obtained from the AR methods of Burg and the unconstrained least squares. Extensive comparisons between the Burg method and the minimum variance method have been made in the paper by Lacoss. Furthermore, Burg (1972) demonstrated that for a known correlation sequence, the minimum variance spectrum is related to the AR model spectrum through the equation

$$\frac{1}{\Gamma_{xx}^{MV}(f)} = \frac{1}{p} \sum_{k=0}^{p} \frac{1}{\Gamma_{xx}^{AR}(f, k)} \qquad (12.4.7)$$

where $\Gamma_{xx}^{AR}(f, k)$ is the AR power spectrum obtained with an AR(k) model. Thus the reciprocal of the minimum variance estimate is equal to the average of the reciprocals of all spectra obtained with AR(k) models for $1 \le k \le p$. Since low-order AR models, in general, do not provide good resolution, the averaging operation in (12.4.7) reduces the frequency resolution in the spectral estimate. Hence we conclude that the AR power spectrum estimate of order p is superior to the minimum variance estimate of order $p + 1$.

The relationship given by (12.4.7) represents a frequency-domain relationship between the Capon minimum variance estimate and the Burg AR estimate. A time-domain relationship between these two estimates also can be established as shown by Musicus (1985). This has led to a computationally efficient algorithm for the minimum variance estimate.

Additional references to the method of Capon and comparisons with other estimators can be found in the literature. We cite the papers of Capon and Goodman (1971), Marzetta (1983), Marzetta and Lang (1983, 1984), Capon (1983), and McDonough (1983).

12.5 EIGENANALYSIS ALGORITHMS FOR SPECTRUM ESTIMATION

In Section 12.3.8 we demonstrated that an AR(p) process corrupted by additive (white) noise is equivalent to an ARMA(p, p) process. In this section we consider the special case in which the signal components are sinusoids corrupted by additive white noise. The algorithms are based on an eigen-decomposition of the correlation matrix of the noise-corrupted signal.

From our previous discussion on the generation of sinusoids in Chapter 4, we recall that a real sinusoidal signal can be generated via the difference equation,

$$x(n) = -a_1 x(n - 1) - a_2 x(n - 2) \qquad (12.5.1)$$

where $a_1 = 2 \cos 2\pi f_k$, $a_2 = 1$, and initially, $x(-1) = -1$, $x(-2) = 0$. This system has a pair of complex-conjugate poles (at $f = f_k$ and $f = -f_k$) and therefore generates the sinusoid $x(n) = \cos 2\pi f_k n$, for $n \ge 0$.

In general, a signal consisting of p sinusoidal components satisfies the difference equation

$$x(n) = -\sum_{m=1}^{2p} a_m x(n-m) \qquad (12.5.2)$$

and corresponds to the system with system function

$$H(z) = \frac{1}{1 + \displaystyle\sum_{m=1}^{2p} a_m z^{-m}} \qquad (12.5.3)$$

The polynomial

$$A(z) = 1 + \sum_{m=1}^{2p} a_m z^{-m} \qquad (12.5.4)$$

has $2p$ roots on the unit circle which correspond to the frequencies of the sinusoids.

Now, suppose that the sinusoids are corrupted by a white noise sequence $w(n)$ with $E[|w(n)|^2] = \sigma_w^2$. Then we observe that

$$y(n) = x(n) + w(n) \qquad (12.5.5)$$

If we substitute $x(n) = y(n) - w(n)$ in (12.5.2), we obtain

$$y(n) - w(n) = -\sum_{m=1}^{2p} [y(n-m) - w(n-m)] a_m$$

or, equivalently,

$$\sum_{m=0}^{2p} a_m y(n-m) = \sum_{m=0}^{2p} a_m w(n-m) \qquad (12.5.6)$$

where, by definition, $a_0 = 1$.

We observe that (12.5.6) is the difference equation for an ARMA(p, p) process in which both the AR and MA parameters are identical. This symmetry is a characteristic of the sinusoidal signals in white noise. The difference equation in (12.5.6) may be expressed in matrix form as

$$\mathbf{Y}^t \mathbf{a} = \mathbf{W}^t \mathbf{a} \qquad (12.5.7)$$

where $\mathbf{Y}^t = [\, y(n) \quad y(n-1) \quad \cdots \quad y(n-2p)\,]$ is the observed data vector of dimension $(2p+1)$, $\mathbf{W}^t = [\, w(n) \quad w(n-1) \quad \cdots \quad w(n-2p)\,]$ is the noise vector, and $\mathbf{a} = [1 \quad a_1 \quad \cdots \quad a_{2p}\,]$ is the coefficient vector.

If we premultiply (12.5.7) by $\mathbf{Y}$ and take the expected value, we obtain

$$E(\mathbf{YY}^t)\mathbf{a} = E(\mathbf{YW}^t)\mathbf{a} = E[(\mathbf{X}+\mathbf{W})\mathbf{W}^t]\mathbf{a}$$

$$\mathbf{\Gamma}_{yy}\mathbf{a} = \sigma_w^2 \mathbf{a} \qquad (12.5.8)$$

where we have used the assumption that the sequence $w(n)$ is zero mean and white, and $\mathbf{X}$ is a deterministic signal.

The equation in (12.5.8) is in the form of an eigenequation, that is,

$$(\mathbf{\Gamma}_{yy} - \sigma_w^2 \mathbf{I})\mathbf{a} = \mathbf{O} \tag{12.5.9}$$

where σ_w^2 is an eigenvalue of the autocorrelation matrix $\mathbf{\Gamma}_{yy}$. Then the parameter vector $\mathbf{a}$ is an eigenvector associated with the eigenvalue σ_w^2. The eigenequation in (12.5.9) forms the basis for the Pisarenko harmonic decomposition method.

12.5.1 Pisarenko Harmonic Decomposition Method

For p randomly-phased sinusoids in additive white noise, the autocorrelation values are

$$\gamma_{yy}(0) = \sigma_w^2 + \sum_{i=1}^{p} P_i$$

$$\gamma_{yy}(k) = \sum_{i=1}^{p} P_i \cos 2\pi f_i k \qquad k \neq 0 \tag{12.5.10}$$

where $P_i = A_i^2/2$ is the average power in the ith sinusoid and A_i is the corresponding amplitude. Hence we may write

$$\begin{bmatrix} \cos 2\pi f_1 & \cos 2\pi f_2 & \cdots & \cos 2\pi f_p \\ \cos 4\pi f_1 & \cos 4\pi f_2 & \cdots & \cos 4\pi f_p \\ \vdots & \vdots & & \vdots \\ \cos 2\pi p f_1 & \cos 2\pi p f_2 & \cdots & \cos 2\pi p f_p \end{bmatrix} \begin{bmatrix} P_1 \\ P_2 \\ \vdots \\ P_p \end{bmatrix} = \begin{bmatrix} \gamma_{yy}(1) \\ \gamma_{yy}(2) \\ \vdots \\ \gamma_{yy}(p) \end{bmatrix} \tag{12.5.11}$$

If we know the frequencies f_i, $1 \leq i \leq p$, we can use this equation to determine the powers of the sinusoids. In place of $\gamma_{xx}(m)$, we use the estimates $r_{xx}(m)$. Once the powers are known, the noise variance can be obtained from (12.5.10) as

$$\sigma_w^2 = r_{yy}(0) - \sum_{i=1}^{p} P_i \tag{12.5.12}$$

The problem that remains is to determine the p frequencies f_i, $1 \leq i \leq p$, which, in turn, require knowledge of the eigenvector $\mathbf{a}$ corresponding to the eigenvalue σ_w^2. Pisarenko (1973) observed [see also Papoulis (1984) and Grenander and Szegö (1958)] that for an ARMA process consisting of p sinusoids in additive white noise, the variance σ_w^2 corresponds to the minimum eigenvalue of $\mathbf{\Gamma}_{yy}$ when the dimension of the autocorrelation matrix equals or exceeds $(2p+1) \times (2p+1)$. The desired ARMA coefficient vector corresponds to the eigenvector associated with the minimum eigenvalue. Therefore, the frequencies f_i, $1 \leq i \leq p$ are obtained from the roots of the polynomial in (12.5.4), where the coefficients are the elements of the eigenvector $\mathbf{a}$ corresponding to the minimum eigenvalue σ_w^2.

In summary, the Pisarenko harmonic decomposition method proceeds as follows. First we estimate $\mathbf{\Gamma}_{yy}$ from the data (i.e., we form the autocorrelation matrix $\mathbf{R}_{yy}$). Then we find the minimum eigenvalue and the corresponding minimum eigenvector. The minimum eigenvector yields the parameters of the

ARMA$(2p, 2p)$ model. From (12.5.4.) we can compute the roots that constitute the frequencies $\{f_i\}$. By using these frequencies, we can solve (12.5.11) for the signal powers $\{P_i\}$ by substituting the estimates $r_{yy}(m)$ for $\gamma_{yy}(m)$.

As will be seen in the following example, the Pisarenko method is based on the use of a noise subspace eigenvector to estimate the frequencies of the sinusoids.

Example 12.5.1

Suppose that we are given the autocorrelation values $\gamma_{yy}(0) = 3$, $\gamma_{yy}(1) = 1$, and $\gamma_{yy}(2) = 0$ for a process consisting of a single sinusoid in additive white noise. Determine the frequency, its power, and the variance of the additive noise.

Solution   The correlation matrix is

$$\Gamma_{yy} = \begin{bmatrix} 3 & 1 & 0 \\ 1 & 3 & 1 \\ 0 & 1 & 3 \end{bmatrix}$$

The minimum eigenvalue is the smallest root of the characteristic polynomial

$$g(\lambda) = \begin{bmatrix} 3-\lambda & 1 & 0 \\ 1 & 3-\lambda & 1 \\ 0 & 1 & 3-\lambda \end{bmatrix} = (3-\lambda)(\lambda^2 - 6\lambda + 7) = 0$$

Therefore, the eigenvalues are $\lambda_1 = 3$, $\lambda_2 = 3 + \sqrt{2}$, $\lambda_3 = 3 - \sqrt{2}$.

The variance of the noise is

$$\sigma_w^2 = \lambda_{\min} = 3 - \sqrt{2}$$

The corresponding eigenvalue is the vector that satisfies (12.5.9), that is,

$$\begin{bmatrix} \sqrt{2} & 1 & 0 \\ 1 & \sqrt{2} & 1 \\ 0 & 1 & \sqrt{2} \end{bmatrix} \begin{bmatrix} 1 \\ a_1 \\ a_2 \end{bmatrix} = \begin{bmatrix} 0 \\ 0 \\ 0 \end{bmatrix}$$

The solution is $a_1 = -\sqrt{2}$ and $a_2 = 1$.

The next step is to use the value a_1 and a_2 to determine the roots of the polynomial in (12.5.4). We have

$$z^2 - \sqrt{2}z + 1 = 0$$

Thus

$$z_1, z_2 = \frac{1}{\sqrt{2}} \pm j \frac{1}{\sqrt{2}}$$

Note that $|z_1| = |z_2| = 1$, so that the roots are on the unit circle. The corresponding frequency is obtained from

$$z_i = e^{j2\pi f_1} = \frac{1}{\sqrt{2}} + j \frac{1}{\sqrt{2}}$$

which yields $f_1 = \frac{1}{8}$. Finally, the power of the sinusoid is

$$P_1 \cos 2\pi f_1 = \gamma_{yy}(1) = 1$$

$$P_1 = \sqrt{2}$$

and its amplitude is $A = \sqrt{2P_1} = \sqrt{2\sqrt{2}}$.

As a check on our computations, we have

$$\sigma_w^2 = \gamma_{yy}(0) - P_1$$

$$= 3 - \sqrt{2}$$

which agrees with $\lambda_{\min}$.

12.5.2 Eigen-decomposition of the Autocorrelation Matrix for Sinusoids in White Noise

In the previous discussion we assumed that the sinusoidal signal consists of p real sinusoids. For mathematical convenience we shall now assume that the signal consists of p complex sinusoids of the form

$$x(n) = \sum_{i=1}^{p} A_i e^{j(2\pi f_i n + \phi_i)} \tag{12.5.13}$$

where the amplitudes $\{A_i\}$ and the frequencies $\{f_i\}$ are unknown and the phases $\{\phi_i\}$ are statistically independent random variables uniformly distributed on $(0, 2\pi)$. Then the random process $x(n)$ is wide-sense stationary with autocorrelation function

$$\gamma_{xx}(m) = \sum_{i=1}^{p} P_i e^{j2\pi f_i m} \tag{12.5.14}$$

where, for complex sinusoids, $P_i = A_i^2$ is the power of the ith sinusoid.

Since the sequence observed is $y(n) = x(m) + w(n)$, where $w(n)$ is a white noise sequence with spectral density σ_w^2, the autocorrelation function for $y(n)$ is

$$\gamma_{yy}(m) = \gamma_{xx}(m) + \sigma_w^2 \delta(m) \qquad m = 0, \pm 1, \ldots, \pm(M-1) \tag{12.5.15}$$

Hence the $M \times M$ autocorrelation matrix for $y(n)$ can be expressed as

$$\Gamma_{yy} = \Gamma_{xx} + \sigma_w^2 \mathbf{I} \tag{12.5.16}$$

where Γ_{xx} is the autocorrelation matrix for the signal $x(n)$ and $\sigma_w^2 \mathbf{I}$ is the autocorrelation matrix for the noise. Note that if select $M > p$, Γ_{xx} which is of dimension $M \times M$ is not of full rank, because its rank is p. However, Γ_{yy} is full rank because $\sigma_w^2 \mathbf{I}$ is of rank M.

In fact, the signal matrix Γ_{xx} can be represented as

$$\Gamma_{xx} = \sum_{i=1}^{p} P_i \mathbf{s}_i \mathbf{s}_i^H \tag{12.5.17}$$

where H denotes the conjugate transpose and $\mathbf{s}_i$ is a signal vector of dimension M defined as

$$\mathbf{s}_i = [1, e^{j2\pi f_i}, e^{j4\pi f_i}, \ldots, e^{j2\pi(M-1)f_i}] \tag{12.5.18}$$

Since each vector (outer product) $\mathbf{s}_i \mathbf{s}_i^H$ is a matrix of rank 1 and since there are p vector products, the matrix Γ_{xx} is of rank p. Note that if the sinusoids were real, the correlation matrix Γ_{xx} has rank $2p$.

Now, let us perform an eigen-decomposition of the matrix $\boldsymbol{\Gamma}_{yy}$. Let the eigenvalues $\{\lambda_i\}$ be ordered in decreasing value with $\lambda_1 \geq \lambda_2 \geq \lambda_3 \geq \cdots \geq \lambda_M$ and let the corresponding eigenvectors be denoted as $\{\mathbf{v}_i, i = 1, \ldots, M\}$. We assume that the eigenvectors are normalized so that $\mathbf{v}_i^H \cdot \mathbf{v}_j = \delta_{ij}$. In the absence of noise the eigenvalues $\lambda_i, i = 1, 2, \ldots, p$, are nonzero while $\lambda_{p+1} = \lambda_{p+2} = \cdots = \lambda_M = 0$. Furthermore, it follows that the signal correlation matrix can be expressed as

$$\boldsymbol{\Gamma}_{xx} = \sum_{i=1}^{p} \lambda_i \mathbf{v}_i \mathbf{v}_i^H \qquad (12.5.19)$$

Thus, the eigenvectors $\mathbf{v}_i, i = 1, 2, \ldots, p$ span the signal subspace as do the signal vectors $\mathbf{s}_i, i = 1, 2, \ldots, p$. These p eigenvectors for the signal subspace are called the *principal eigenvectors* and the corresponding eigenvalues are called the *principal eigenvalues*.

In the presence of noise, the noise autocorrelation matrix in (12.5.16) can be represented as

$$\sigma_w^2 \mathbf{I} = \sigma_w^2 \sum_{i=1}^{M} \mathbf{v}_i \mathbf{v}_i^H \qquad (12.5.20)$$

By substituting (12.5.19) and (12.5.20) into (12.5.16), we obtain

$$\begin{aligned} \boldsymbol{\Gamma}_{yy} &= \sum_{i=1}^{p} \lambda_i \mathbf{v}_i \mathbf{v}_i^H + \sum_{i=1}^{M} \sigma_w^2 \mathbf{v}_i \mathbf{v}_i^H \\ &= \sum_{i=1}^{p} (\lambda_i + \sigma_w^2) \mathbf{v}_i \mathbf{v}_i^H + \sum_{i=p+1}^{M} \sigma_w^2 \mathbf{v}_i \mathbf{v}_i^H \end{aligned} \qquad (12.5.21)$$

This eigen-decomposition separates the eigenvectors into two sets. The set $\{\mathbf{v}_i, i = 1, 2, \ldots, p\}$, which are the principal eigenvectors, span the signal subspace, while the set $\{\mathbf{v}_i, i = p + 1, \ldots, M\}$, which are orthogonal to the principal eigenvectors, are said to belong to the noise subspace. Since the signal vectors $\{\mathbf{s}_i, i = 1, 2, \ldots, p\}$ are in the signal subspace, it follows that the $\{\mathbf{s}_i\}$ are simply linear combinations of the principal eigenvectors and are also orthogonal to the vectors in the noise subspace.

In this context we see that the Pisarenko method is based on an estimation of the frequencies by using the orthogonality property between the signal vectors and the vectors in the noise subspace. For complex sinusoids, if we select $M = p + 1$ (for real sinusoids we select $M = 2p + 1$), there is only a single eigenvector in the noise subspace (corresponding to the minimum eigenvalue) which must be orthogonal to the signal vectors. Thus we have

$$\mathbf{s}_i^H \mathbf{v}_{p+1} = \sum_{k=0}^{p} v_{p+1}(k+1) e^{-j2\pi f_i k} = 0 \qquad i = 1, 2, \ldots, p \qquad (12.5.22)$$

But (12.5.22) implies that the frequencies $\{f_i\}$ can be determined by solving for the zeros of the polynomial

$$V(z) = \sum_{n=0}^{p} v_{p+1}(k+1)z^{-k} \tag{12.5.23}$$

all of which lie on the unit circle. The angles of these roots are $2\pi f_i$, $i = 1$, $2, \ldots, p$.

When the number of sinusoids is unknown, the determination of p may prove to be difficult, especially if the signal level is not much higher than the noise level. In theory, if $M > p + 1$, there is a multiplicity $(M - p)$ of the minimum eigenvalue. However, in practice the $(M - p)$ small eigenvalues of $\mathbf{R}_{yy}$ will probably be different. By computing all the eigenvalues it may be possible to determine p by grouping the $M - p$ small (noise) eigenvalues into a set and averaging them to obtain an estimate of σ_w^2. Then, the average value can be used in (12.5.9) along with $\mathbf{R}_{yy}$ to determine the corresponding eigenvector.

12.5.3 MUSIC Algorithm

The multiple signal classification (MUSIC) method is also a noise subspace frequency estimator. To develop the method, let us first consider the "weighted" spectral estimate

$$P(f) = \sum_{k=p+1}^{M} w_k |\mathbf{s}^H(f)\mathbf{v}_k|^2 \tag{12.5.24}$$

where $\{\mathbf{v}_k, k = p + 1, \ldots, M\}$ are the eigenvectors in the noise subspace, $\{w_k\}$ are a set of positive weights, and $\mathbf{s}(f)$ is the complex sinusoidal vector

$$\mathbf{s}(f) = [1, e^{j2\pi f}, e^{j4\pi f}, \ldots, e^{j2\pi(M-1)f}] \tag{12.5.25}$$

Note that at $f = f_i$, $\mathbf{s}(f_i) \equiv \mathbf{s}_i$, so that at any one of the p sinusoidal frequency components of the signal, we have

$$P(f_i) = 0 \qquad i = 1, 2, \ldots, p \tag{12.5.26}$$

Hence, the reciprocal of $P(f)$ is a sharply peaked function of frequency and provides a method for estimating the frequencies of the sinusoidal components. Thus

$$\frac{1}{P(f)} = \frac{1}{\displaystyle\sum_{k=p+1}^{M} w_k \left|\mathbf{s}^H(f)\mathbf{v}_k\right|^2} \tag{12.5.27}$$

Although theoretically $1/P(f)$ is infinite at $f = f_i$, in practice the estimation errors result in finite values for $1/P(f)$ at all frequencies.

The MUSIC sinusoidal frequency estimator proposed by Schmidt (1981, 1986) is a special case of (12.5.27) in which the weights $w_k = 1$ for all k. Hence

$$P_{\text{MUSIC}}(f) = \frac{1}{\displaystyle\sum_{k=p+1}^{M} \left|\mathbf{s}^H(f)v_k\right|^2} \tag{12.5.28}$$

The estimate of the sinusoidal frequencies are the peaks of $P_{\text{MUSIC}}(f)$. Once the sinusoidal frequencies are estimated, the power of each of the sinusoids can be obtained by solving (12.5.11).

12.5.4 ESPRIT Algorithm

ESPRIT (estimation of signal parameters via rotational invariance techniques) is yet another method for estimating frequencies of a sum of sinusoids by use of an eigen-decomposition approach. As we observe from the development that follows, which is due to Roy et al. (1986), ESPRIT exploits an underlying rotational invariance of signal subspaces spanned by two temporally displaced data vectors.

We again consider the estimation of p complex-valued sinusoids in additive white noise. The received sequence is given by the vector

$$\mathbf{y}(n) = [y(n), y(n+1), \ldots, y(n+M-1)]^t$$
$$= \mathbf{x}(n) + \mathbf{w}(n) \tag{12.5.29}$$

where $\mathbf{x}(n)$ is the signal vector and $\mathbf{w}(n)$ is the noise vector. To exploit the deterministic character of the sinusoids, we define the time-displaced vector $\mathbf{z}(n) = \mathbf{y}(n+1)$. Thus

$$\mathbf{z}(n) = [z(n), z(n+1), \ldots, z(n+M-1)]^t$$
$$= [y(n+1), y(n+2), \ldots, y(n+M)]^t \tag{12.5.30}$$

With these definitions we can express the vectors $\mathbf{y}(n)$ and $\mathbf{z}(n)$ as

$$\mathbf{y}(n) = \mathbf{Sa} + \mathbf{w}(n)$$
$$\mathbf{z}(n) = \mathbf{S\Phi a} + \mathbf{w}(n) \tag{12.5.31}$$

where $\mathbf{a} = [a_1, a_2, \ldots, a_p]^t$, $a_i = A_i e^{j\phi i}$, and $\mathbf{\Phi}$ is a diagonal $p \times p$ matrix consisting of the relative phase between adjacent time samples of each of the complex sinusoids,

$$\mathbf{\Phi} = \text{diag}[e^{j2\pi f_1}, e^{j2\pi f_2}, \ldots, e^{j2\pi f_p}] \tag{12.5.32}$$

Note that the matrix $\mathbf{\Phi}$ relates the time-displaced vectors $\mathbf{y}(n)$ and $\mathbf{z}(n)$ and can be called a rotation operator. We also note that $\mathbf{\Phi}$ is unitary. The matrix $\mathbf{S}$ is the $M \times p$ Vandermonde matrix specified by the column vectors

$$\mathbf{s}_i = [1, e^{j2\pi f_i}, e^{j4\pi f_i}, \ldots, e^{j2\pi(M-1)f_i}] \qquad i = 1, 2, \ldots, p \tag{12.5.33}$$

Now the autocovariance matrix for the data vector $\mathbf{y}(n)$ is

$$\mathbf{\Gamma}_{yy} = E[\mathbf{y}(n)\mathbf{y}^H(n)]$$
$$= \mathbf{SPS}^H + \sigma_w^2 \mathbf{I} \tag{12.5.34}$$

where $\mathbf{P}$ is the $p \times p$ diagonal matrix consisting of the powers of the complex sinusoids,

$$\mathbf{P} = \text{diag}[|a_1|^2, |a_2|^2, \ldots, |a_p|^2]$$
$$= \text{diag}[P_1, P_2, \ldots, P_p] \tag{12.5.35}$$

We observe that $\mathbf{P}$ is a diagonal matrix since complex sinusoids of different frequencies are orthogonal over the infinite interval. However, we should emphasize that the ESPRIT algorithm does not require $\mathbf{P}$ to be diagonal. Hence the algorithm is applicable to the case in which the covariance matrix is estimated from finite data records.

The crosscovariance matrix of the signal vectors $\mathbf{y}(n)$ and $\mathbf{z}(n)$ is

$$\boldsymbol{\Gamma}_{yz} = E[\mathbf{y}(n)\mathbf{z}^H(n)] = \mathbf{S}\mathbf{P}\boldsymbol{\Phi}^H\mathbf{S}^H + \boldsymbol{\Gamma}_w \tag{12.5.36}$$

where

$$\boldsymbol{\Gamma}_w = E[\mathbf{w}(n)\mathbf{w}^H(n+1)]$$

$$= \sigma_w^2 \begin{bmatrix} 0 & 0 & 0 & \cdots & 0 & 0 \\ 1 & 0 & 0 & \cdots & 0 & 0 \\ 0 & 1 & 0 & \cdots & 0 & 0 \\ \vdots & \vdots & \vdots & \vdots & & \vdots \\ 0 & 0 & 0 & \cdots & 1 & 0 \end{bmatrix} \equiv \sigma_w^2 \mathbf{Q} \tag{12.5.37}$$

The auto and crosscovariance matrices $\boldsymbol{\Gamma}_{yy}$ and $\boldsymbol{\Gamma}_{yz}$ are given as

$$\boldsymbol{\Gamma}_{yy} = \begin{bmatrix} \gamma_{yy}(0) & \gamma_{yy}(1) & \cdots & \gamma_{yy}(M-1) \\ \gamma_{yy}^*(1) & \gamma_{yy}(0) & \cdots & \gamma_{yy}(M-2) \\ \vdots & \vdots & & \vdots \\ \gamma_{yy}^*(M-1) & \gamma_{yy}(M-2) & \cdots & \gamma_{yy}(0) \end{bmatrix} \tag{12.5.38}$$

$$\boldsymbol{\Gamma}_{yz} = \begin{bmatrix} \gamma_{yy}(1) & \gamma_{yy}(2) & \cdots & \gamma_{yy}(M) \\ \gamma_{yy}(0) & \gamma_{yy}(1) & \cdots & \gamma_{yy}(M-1) \\ \vdots & \vdots & & \vdots \\ \gamma_{yy}^*(M-2) & \gamma_{yy}^*(M-3) & \cdots & \gamma_{yy}(1) \end{bmatrix} \tag{12.5.39}$$

where $\gamma_{yy}(m) = E[y^*(n)y(n+m)]$. Note that both $\boldsymbol{\Gamma}_{yy}$ and $\boldsymbol{\Gamma}_{yz}$ are Toeplitz matrices.

Based on this formulation, the problem is to determine the frequencies $\{f_i\}$ and their powers $\{P_i\}$ from the autocorrelation sequence $\{\gamma_{yy}(m)\}$.

From the underlying model, it is clear that the matrix $\mathbf{S}\mathbf{P}\mathbf{S}^H$ has rank p. Consequently, $\boldsymbol{\Gamma}_{yy}$ given by (12.5.34) has $(M-p)$ identical eigenvalues equal to σ_w^2. Hence

$$\boldsymbol{\Gamma}_{yy} - \sigma_w^2 \mathbf{I} = \mathbf{S}\mathbf{P}\mathbf{S}^H \equiv \mathbf{C}_{yy} \tag{12.5.40}$$

From (12.5.36) we also have

$$\boldsymbol{\Gamma}_{yz} - \sigma_w^2 \boldsymbol{\Gamma}_w = \mathbf{S}\mathbf{P}\boldsymbol{\Phi}^H\mathbf{S}^H \equiv \mathbf{C}_{yz} \tag{12.5.41}$$

Now, let us consider the matrix $\mathbf{C}_{yy} - \lambda\mathbf{C}_{yz}$, which can be written as

$$\mathbf{C}_{yy} - \lambda\mathbf{C}_{yz} = \mathbf{S}\mathbf{P}(\mathbf{I} - \lambda\boldsymbol{\Phi}^H)\mathbf{S}^H \tag{12.5.42}$$

Clearly, the column space of $\mathbf{SPS}^H$ is identical to the column space of $\mathbf{SP\Phi}^H\mathbf{S}^H$. Consequently, the rank of $\mathbf{C}_{yy} - \lambda\mathbf{C}_{yz}$ is equal to p. However, we note that if $\lambda = \exp(j2\pi f_i)$, the ith row of $(\mathbf{I} - \lambda\mathbf{\Phi}^H)$ is zero and, hence the rank of $[\mathbf{I} - \mathbf{\Phi}^H\exp(j2\pi f_i)]$ is $p - 1$. But $\lambda_i = \exp(j2\pi f_i)$, $i = 1, 2, \ldots, p$, are the generalized eigenvalues of the matrix pair $(\mathbf{C}_{yy}, \mathbf{C}_{yz})$. Thus the p generalized eigenvalues $\{\lambda_i\}$ that lie on the unit circle correspond to the elements of the rotation operator $\mathbf{\Phi}$. The remaining $M - p$ generalized eigenvalues of the pair $\{\mathbf{C}_{yy}, \mathbf{C}_{yz}\}$ which correspond to the common null space of these matrices, are zero [i.e., the $(M - p)$ eigenvalues are at the origin in the complex plane].

Based on these mathematical relationships we can formulate an algorithm (ESPRIT) for estimating the frequencies $\{f_i\}$. The procedure is as follows:

1. From the data, compute the autocorrelation values $r_{yy}(m)$, $m = 1, 2, \ldots, M$, and form the matrices $\mathbf{R}_{yy}$ and $\mathbf{R}_{yz}$ corresponding to estimates of $\mathbf{\Gamma}_{yy}$ and $\mathbf{\Gamma}_{yz}$.

2. Compute the eigenvalues of $\mathbf{R}_{yy}$. For $M > p$, the minimum eigenvalue is an estimate of σ_w^2.

3. Compute $\hat{\mathbf{C}}_{yy} = \mathbf{R}_{yy} - \hat{\sigma}_w^2\mathbf{I}$ and $\hat{\mathbf{C}}_{yz} = \mathbf{R}_{yz} - \hat{\sigma}_w^2\mathbf{Q}$, where $\mathbf{Q}$ is defined in (12.5.37).

4. Compute the generalized eigenvalues of the matrix pair $\{\hat{\mathbf{C}}_{yy}, \hat{\mathbf{C}}_{yz}\}$. The p generalized eigenvalues of these matrices that lie on (or near) the unit circle determine the (estimate) elements of $\mathbf{\Phi}$ and hence the sinusoidal frequencies. The remaining $M - p$ eigenvalues will lie at (or near) the origin.

One method for determining the power in the sinusoidal components is to solve the equation in (12.5.11) with $r_{yy}(m)$ substituted for $\gamma_{yy}(m)$.

Another method is based on the computation of the generalized eigenvectors $\{\mathbf{v}_i\}$ corresponding to the generalized eigenvalues $\{\lambda_i\}$. We have

$$(\mathbf{C}_{yy} - \lambda_i\mathbf{C}_{yz})\mathbf{v}_i = \mathbf{SP}(\mathbf{I} - \lambda_i\mathbf{\Phi}^H)\mathbf{S}^H\mathbf{v}_i = 0 \qquad (12.5.43)$$

Since the column space of $(\mathbf{C}_{yy} - \lambda_i \cdot \mathbf{C}_{yz})$ is identical to the column space spanned by the vectors $\{\mathbf{s}_j, j \neq i\}$ given by (12.5.33), it follows that the generalized eigenvector $\mathbf{v}_i$ is orthogonal to $\mathbf{s}_j$, $j \neq i$. Since $\mathbf{P}$ is diagonal, it follows from (12.5.43) that the signal powers are

$$P_i = \frac{\mathbf{v}_i^H\mathbf{C}_{yy}\mathbf{v}_i}{|\mathbf{v}_i^H\mathbf{s}_i|^2} \qquad i = 1, 2, \ldots, p \qquad (12.5.44)$$

12.5.5 Order Selection Criteria

The eigenanalysis methods described in this section for estimating the frequencies and the powers of the sinusoids, also provide information about the number of sinusoidal components. If there are p sinusoids, the eigenvalues associated with the signal subspace are $\{\lambda_i + \sigma_w^2, i = 1, 2 \ldots, p\}$ while the remaining $(M - p)$ eigenvalues are all equal to σ_w^2. Based on this eigenvalue decomposition, a test can be designed that compares the eigenvalues with a specified threshold. An alternative method also uses the eigenvector decomposition of the estimated autocorrelation matrix of the observed signal and is based on matrix perturbation analysis. This method is decribed in the paper by Fuchs (1988).

Another approach based on an extension and modification of the AIC criterion to the eigen-decomposition method, has been proposed by Wax and Kailath (1985). If the eigenvalues of the sample autocorrelation matrix are ranked so that $\lambda_1 \geq \lambda_2 \geq \cdots \geq \lambda_M$, where $M > p$, the number of sinusoids in the signal subspace is estimated by selecting the minimum value of MDL(p), given as

$$\text{MDL}(p) = -\log\left[\frac{G(p)}{A(p)}\right]^N + E(p) \qquad (12.5.45)$$

where

$$G(p) = \prod_{i=p+1}^{M} \lambda_i \qquad p = 0, 1, \ldots, M - 1$$

$$A(p) = \left[\frac{1}{M - p} \sum_{i=p+1}^{M} \lambda_i\right]^{M-p} \qquad (12.5.46)$$

$$E(p) = \frac{1}{2}p(2M - p)\log N$$

N: number of samples used to estimate the M
    autocorrelation lags

Some results on the quality of this order selection criterion are given in the paper by Wax and Kailath (1985). The MDL criterion is guaranteed to be consistent.

12.5.6 Experimental Results

In this section we illustrate with an example, the resolution characteristics of the eigenanalysis-based spectral estimation algorithms and compare their performance with the model-based methods and nonparametric methods. The signal sequence is

$$x(n) = \sum_{i=1}^{4} A_i e^{j(2\pi f_i n + \phi_i)} + w(n)$$

where $A_i = 1$, $i = 1, 2, 3, 4$, $\{\phi_i\}$ are statistically independent random variables uniformly distributed on $(0, 2\pi)$, $\{w(n)\}$ is a zero-mean, white noise sequence with variance σ_w^2, and the frequencies are $f_1 = -0.222$, $f_2 = -0.166$, $f_3 = 0.10$, and $f_4 = 0.122$. The sequence $\{x(n), 0 \leq n \leq 1023\}$ is used to estimate the number of frequency components and the corresponding values of their frequencies for $\sigma_w^2 = 0.1, 0.5, 1.0$, and $M = 12$ (length of the estimated autocorrelation).

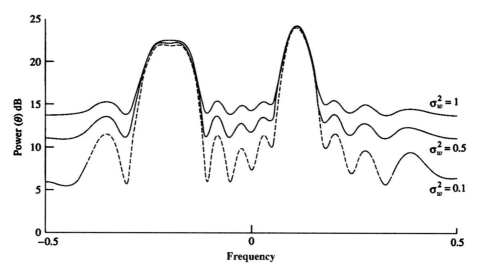

Figure 12.19   Power spectrum estimates from Blackman–Tukey method.

Figures 12.19, 12.20, 12.21, and 12.22 illustrate the estimated power spectra of the signal using the Blackman–Tukey method, the minimum variance method of Capon, the AR Yule–Walker method, and the MUSIC algorithm, respectively. The results from the ESPRIT algorithm are given in Table 12.2. From these results it is apparent that (1) the Blackman–Tukey method does not provide sufficient

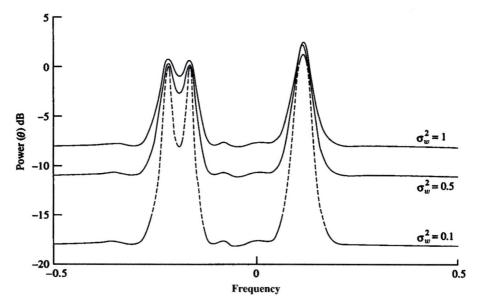

Figure 12.20   Power spectrum estimates from minimum variance method.

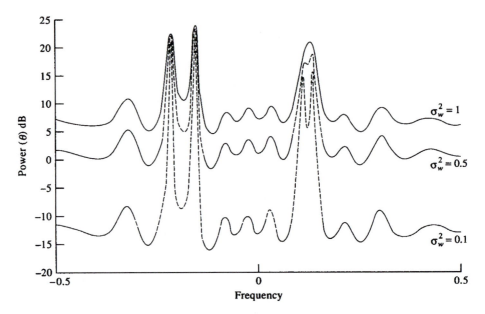

Figure 12.21   Power spectrum estimates from Yule–Walker AR method.

resolution to estimate the sinusoids from the data; (2) the minimum variance method of Capon resolves only the frequencies f_1, f_2 but not f_3 and f_4; (3) the AR methods resolve all frequencies for $\sigma_w^2 = 0.1$ and $\sigma_w^2 = 0.5$; and (4) the MUSIC and ESPRIT algorithms not only recover all four sinusoids, but their performance for different values of σ_w^2 is essentially indistinguishable. We further observe that

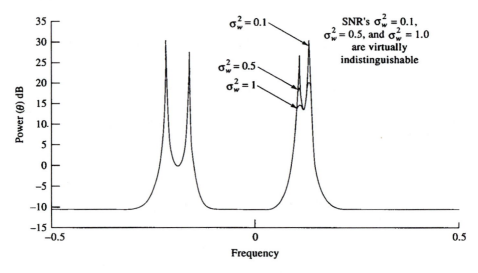

Figure 12.22   Power spectrum estimates from MUSIC algorithm.

TABLE 12.2   ESPRIT ALGORITHM

σ_w^2	$\hat{f}_1$	$\hat{f}_2$	$\hat{f}_3$	$\hat{f}_4$
0.1	−0.2227	−0.1668	−0.1224	−0.10071
0.5	−0.2219	−0.167	−0.121	0.0988
1.0	−0.222	−0.167	0.1199	0.1013
True values	−0.222	−0.166	0.122	0.100

the resolution properties of the minimum variance method and the AR method is a function of the noise variance. These results clearly demonstrate the power of the eigenanalysis-based algorithms in resolving sinusoids in additive noise.

In conclusion, we should emphasize that the high-resolution, eigenanalysis-based spectral estimation methods described in this section, namely MUSIC and ESPRIT, are not only applicable to sinusoidal signals, but apply more generally to the estimation of narrowband signals.

12.6 SUMMARY AND REFERENCES

Power spectrum estimation is one of the most important areas of research and applications in digital signal processing. In this chapter we have described the most important power spectrum estimation techniques and algorithms that have been developed over the past century, beginning with the nonparametric or classical methods based on the periodogram and concluding with the more modern parametric methods based on AR, MA, and ARMA linear models. Our treatment is limited in scope to single-time-series spectrum estimation methods, based on second moments (autocorrelation) of the statistical data.

The parametric and nonparametric methods that we described have been extended to multichannel and multidimensional spectrum estimation. The tutorial paper by McClellan (1982) treats the multidimensional spectrum estimation problem, while the paper by Johnson (1982) treats the multichannel spectrum estimation problem. Additional spectrum estimation methods have been developed for use with higher-order cumulants that involve the bispectrum and the trispectrum. A tutorial paper on these topics has been published by Nikias and Raghuveer (1987).

As evidenced from our previous discussion, power spectrum estimation is an area that has attracted many researchers and, as a result, thousands of papers have been published in the technical literature on this subject. Much of this work has been concerned with new algorithms and techniques, and modifications of existing techniques. Other work has been concerned with obtaining an understanding of the capabilities and limitations of the various power spectrum methods. In this context the statistical properties and limitations of the classical nonparametric methods have been thoroughly analyzed and are well understood. The parametric methods have also been investigated by many researchers, but the analysis of their performance is difficult and, consequently, fewer results are available. Some of the papers that have addressed the problem of performance characteristics of parametric methods are those of Kromer (1969), Lacoss (1971), Berk (1974), Baggeroer (1976), Sakai (1979), Swingler (1980), Lang and McClellan (1980), and Tufts and Kumaresan (1982).

In addition to the references already given in this chapter on the various methods for spectrum estimation and their performance, we should include for reference some of the tutorial and survey papers. In particular, we cite the tutorial paper by Kay and Marple (1981), which includes about 280 references, the paper by Brillinger (1974), and the Special Issue on Spectral Estimation of the *IEEE Proceedings*, September 1982. Another indication of the widespread interest in the subject of spectrum estimation and analysis is the recent publication of texts by Gardner (1987), Kay (1988), and Marple (1987), and the IEEE books edited by Childers (1978) and Kesler (1986).

Many computer programs as well as software packages that implement the various spectrum estimation methods described in this chapter are available. One software package is available through the IEEE (*Programs for Digital Signal Processing*, IEEE Press, 1979); others are available commercially.

PROBLEMS

12.1 (a) By expanding (12.1.23), taking the expected value, and finally taking the limit as $T_0 \to \infty$, show that the right-hand side converges to $\Gamma_{xx}(F)$.

(b) Prove that

$$\sum_{m=-N}^{N} r_{xx}(m)e^{-j2\pi fm} = \frac{1}{N}\left|\sum_{n=0}^{N-1} x(n)e^{-j2\pi fn}\right|^2$$

12.2 For zero mean, jointly Gaussian random variables, X_1, X_2, X_3, X_4, it is well known [see Papoulis (1984)] that

$$E(X_1X_2X_3X_4) = E(X_1X_2)E(X_3X_4) + E(X_1X_3)E(X_2X_4) + E(X_1X_4)E(X_2X_3)$$

Use this result to derive the mean-square value of $r'_{xx}(m)$, given by (12.1.27) and the variance, which is

$$\text{var}[r'_{xx}(m)] = E[|r'_{xx}(m)|^2] - |E[r'_{xx}(m)]|^2$$

12.3 By use of the expression for the fourth joint moment for Gaussian random variables, show that

(a) $E[P_{xx}(f_1)P_{xx}(f_2)] = \sigma_x^4\left\{1 + \left[\frac{\sin\pi(f_1+f_2)N}{N\sin\pi(f_1+f_2)}\right]^2 + \left[\frac{\sin\pi(f_1-f_2)N}{N\sin\pi(f_1-f_2)}\right]^2\right\}$

(b) $\text{cov}[P_{xx}(f_1)P_{xx}(f_2)] = \sigma_x^4 \left\{ \left[\dfrac{\sin \pi (f_1 + f_2)N}{N \sin \pi (f_1 + f_2)} \right]^2 \right.$

$$\left. + \left[\dfrac{\sin \pi (f_1 - f_2)N}{N \sin \pi (f_1 - f_2)} \right]^2 \right\}$$

(c) $\text{var}[P_{xx}(f)] = \sigma_x^4 \left\{ 1 + \left(\dfrac{\sin 2\pi f N}{N \sin 2\pi f} \right)^2 \right\}$ under the condition that the sequence $x(n)$

is a zero-mean white Gaussian noise sequence with variance σ_x^2.

12.4 Generalize the results in Problem 12.3 to a zero-mean Gaussian noise process with power density spectrum $\Gamma_{xx}(f)$. Then derive the variance of the periodogram $P_{xx}(f)$, as given by (12.1.38). (*Hint:* Assume that the colored Gaussian noise process is the output of a linear system excited by white Gaussian noise. Then use the appropriate relations given in Appendix A.)

12.5 Show that the periodogram values at frequencies $f_k = k/L$, $k = 0, 1, \ldots, L - 1$, given by (12.1.41) can be computed by passing the sequence through a bank of N IIR filters, where each filter has an impulse response

$$h_k(n) = e^{-j2\pi nk/N} u(n)$$

and then compute the magnitude-squared value of the filter outputs at $n = N$. Note that each filter has a pole on the unit circle at the frequency f_k.

12.6 Prove that the normalization factor given by (12.2.12) ensures that (12.2.19) is satisfied.

12.7 Let us consider the use of the DFT (computed via the FFT algorithm) to compute the autocorrelation of the complex-valued sequence $x(n)$, that is,

$$r_{xx}(m) = \frac{1}{N} \sum_{n=0}^{N-m-1} x^*(n)x(n + m), \ m \geq 0$$

Suppose the size M of the FFT is much smaller than that of the data length N. Specifically, assume that $N = KM$.

(a) Determine the steps needed to section $x(n)$ and compute $r_{xx}(m)$ for $-(M/2) + 1 \leq m \leq (M/2) - 1$, by using $4K$ M-point DFTs and one M-point IDFT

(b) Now consider the following three sequences $x_1(n)$, $x_2(n)$, and $x_3(n)$, each of duration M. Let the sequences $x_1(n)$ and $x_2(n)$ have arbitrary values in the range $0 \leq n \leq (M/2) - 1$, but are zero for $(M/2) \leq n \leq M - 1$. The sequence $x_3(n)$ is defined as

$$x_3(n) = \begin{cases} x_1(n), & 0 \leq \dfrac{M}{2} - 1 \\ x_2\left(n - \dfrac{M}{2}\right), & \dfrac{M}{2} \leq n \leq M - 1 \end{cases}$$

Determine a simple relationship among the M-point DFTs $X_1(k)$, $X_2(k)$, and $X_3(k)$.

(c) By using the result in part (b), show how the computation of the DFTs in part (a) can be reduced in number from $4K$ to $2K$.

12.8 The Bartlett method is used to estimate the power spectrum of a signal $x(n)$. We know that the power spectrum consists of a single peak with a 3-dB bandwidth of 0.01 cycle per sample, but we do not know the location of the peak.

(a) Assuming that N is large, determine the value of $M = N/K$ so that the spectral window is narrower than the peak.

(b) Explain why it is not advantageous to increase M beyond the value obtained in part (a).

12.9 Suppose we have $N = 1000$ samples from a sample sequence of a random process.

(a) Determine the frequency resolution of the Bartlett, Welch (50% overlap), and Blackman–Tukey methods for a quality factor $Q = 10$.

(b) Determine the record lengths (M) for the Bartlett, Welch (50% overlap), and Blackman–Tukey methods.

12.10 Consider the problem of continuously estimating the power spectrum from a sequence $x(n)$ based on averaging periodograms with exponential weighting into the past. Thus with $P_{xx}^{(0)}(f) = 0$, we have

$$P_{xx}^{(m)}(f) = w P_{xx}^{(m-1)}(f) + \frac{1-w}{M} \left| \sum_{n=0}^{M-1} x_m(n) e^{-j2\pi f n} \right|^2$$

where successive periodograms are assumed to be uncorrelated and w is the (exponential) weighting factor.

(a) Determine the mean and variance of $P_{xx}^{(m)}(f)$ for a Gaussian random process.

(b) Repeat the analysis of part (a) for the case in which the modified periodogram defined by Welch is used in the averaging with no overlap.

12.11 The periodogram in the Bartlett method can be expressed as

$$P_{xx}^{(i)}(f) = \sum_{m=-(M-1)}^{M-1} \left(1 - \frac{|m|}{M} \right) r_{xx}^{(i)}(m) e^{-j2\pi f m}$$

where $r_{xx}^{(i)}(m)$ is the estimated autocorrelation sequence obtained from the ith block of data. Show that $P_{xx}^{(i)}(f)$ can be expressed as

$$P_{xx}^{(i)}(f) = E^{*t}(f) R_{xx}^{(i)} E(f)$$

where

$$E(f) = [1 \quad e^{j2\pi f} \quad e^{j4\pi f} \quad \cdots \quad e^{j2\pi (M-1)f}]^t$$

and therefore,

$$P_{xx}^{B}(f) = \frac{1}{K} \sum_{k=1}^{K} E^{*t}(f) R_{xx}^{(k)} E(f)$$

12.12 Derive the recursive order-update equation given in (12.3.19).

12.13 Determine the mean and the autocorrelation of the sequence $x(n)$, which is the output of a ARMA $(1, 1)$ process described by the difference equation

$$x(n) = \tfrac{1}{2}x(n-1) + w(n) - w(n-1)$$

where $w(n)$ is a white noise process with variance σ_w^2.

12.14 Determine the mean and the autocorrelation of the sequence $x(n)$ generated by the MA(2) process described by the difference equation

$$x(n) = w(n) - 2w(n-1) + w(n-2)$$

where $w(n)$ is a white noise process with variance σ_w^2.

12.15 An MA(2) process has the autocorrelation sequence

$$
\gamma_{xx}(m) = \begin{cases}
6\sigma_w^2, & m = 0 \\
-4\sigma_w^2, & m = \pm 1 \\
-2\sigma_w^2, & m = \pm 2 \\
0, & \text{otherwise}
\end{cases}
$$

(a) Determine the coefficients of the MA(2) process that have the foregoing autocorrelation.

(b) Is the solution unique? If not, give all the possible solutions.

12.16 An MA(2) process has the autocorrelation sequence

$$
\gamma_{xx}(m) = \begin{cases}
\sigma_w^2, & m = 0 \\
-\dfrac{35}{62}\sigma_w^2, & m = \pm 1 \\
\dfrac{6}{62}\sigma_w^2, & m = \pm 2
\end{cases}
$$

(a) Determine the coefficients of the minimum-phase system for the MA(2) process.

(b) Determine the coefficients of the maximum-phase system for the MA(2) process.

(c) Determine the coefficients of the mixed-phase system for the MA(2) process.

12.17 Consider the linear system described by the difference equation

$$
y(n) = 0.8y(n-1) + x(n) + x(n-1)
$$

where $x(n)$ is a wide-sense stationary random process with zero mean and autocorrelation

$$
\gamma_{xx}(m) = \left(\tfrac{1}{2}\right)^{|m|}
$$

(a) Determine the power density spectrum of the output $y(n)$.

(b) Determine the autocorrelation $\gamma_{yy}(m)$ of the output.

(c) Determine the variance σ_y^2 of the output.

12.18 From (12.3.6) and (12.3.9) we note that an AR(p) stationary random process satisfies the equation

$$
\gamma_{xx}(m) + \sum_{k=1}^{p} a_p(k)\gamma_{xx}(m-k) = \begin{cases}
\sigma_w^2, & m = 0, \\
0, & 1 \le m \le p,
\end{cases}
$$

where $a_p(k)$ are the prediction coefficients of the linear predictor of order p and σ_w^2 is the minimum mean-square prediction error. If the $(p+1) \times (p+1)$ autocorrelation matrix Γ_{xx} in (12.3.9) is positive definite, prove that:

(a) The reflection coefficients $|K_m| < 1$ for $1 \le m \le p$.

(b) The polynomial

$$
A_p(z) = 1 + \sum_{k=1}^{p} a_p(k)z^{-k}
$$

has all its roots inside the unit circle (i.e., it is minimum phase).

12.19 Consider the AR(3) process generated by the equation

$$
x(n) = \tfrac{14}{24}x(n-1) + \tfrac{9}{24}x(n-2) - \tfrac{1}{24}x(n-3) + w(n)
$$

where $w(n)$ is a stationary white noise process with variance σ_w^2.

(a) Determine the coefficients of the optimum $p = 3$ linear predictor.
(b) Determine the autocorrelation sequence $\gamma_{xx}(m)$, $0 \le m \le 5$.
(c) Determine the reflection coefficients corresponding to the $p = 3$ linear predictor.

12.20* An AR(2) process is described by the difference equation

$$x(n) = 0.81x(n-2) + w(n)$$

where $w(n)$ is a white noise process with variance σ_w^2.
(a) Determine the parameters of the MA(2), MA(4), and MA(8) models which provide a minimum mean-square error fit to the data $x(n)$.
(b) Plot the true spectrum and those of the MA(q), $q = 2, 4, 8$, spectra and compare the results. Comment on how well the MA(q) models approximate the AR(2) process.

12.21 An MA(2) process is described by the difference equation

$$x(n) = w(n) + 0.81w(n-2)$$

where $w(n)$ is a white noise process with variance σ_w^2.
(a) Determine the parameters of the AR(2), AR(4), and AR(8) models that provide a minimum mean-square error fit to the data $x(n)$.
(b) Plot the true spectra m and those of the AR(p), $p = 2, 4, 8$, and compare the results. Comment on how well the AR(p) models approximate the MA(2) process.

12.22 The z-transform of the autocorrelation $\gamma_{xx}(m)$ of an ARMA(1, 1) process is

$$\Gamma_{xx}(z) = \sigma_w^2 H(z)H(z^{-1})$$

$$\Gamma_{xx}(z) = \frac{4\sigma_w^2}{9} \frac{5 - 2z - 2z^{-1}}{10 - 3z^{-1} - 3z}$$

(a) Determine the minimum-phase system function $H(z)$.
(b) Determine the system function $H(z)$ for a mixed-phase stable system.

12.23 Consider a FIR filter with coefficient vector

$$\begin{bmatrix} 1 & -2r\cos\theta & r^2 \end{bmatrix}$$

(a) Determine the reflection coefficients for the corresponding FIR lattice filter.
(b) Determine the values of the reflection coefficients in the limit as $r \to 1$.

12.24 An AR(3) process is characterized by the prediction coefficients

$$a_3(1) = -1.25, \qquad a_3(2) = 1.25, \qquad a_3(3) = -1$$

(a) Determine the reflection coefficients.
(b) Determine $\gamma_{xx}(m)$ for $0 \le m \le 3$.
(c) Determine the mean-square prediction error.

12.25 The autocorrelation sequence for a random process is

$$\gamma_{xx}(m) = \begin{cases} 1, & m = 0 \\ -0.5, & m = \pm 1 \\ 0.625, & m = \pm 2 \\ -0.6875, & m = \pm 3 \\ 0 & \text{otherwise} \end{cases}$$

Determine the system functions $A_m(z)$ for the prediction-error filters for $m = 1, 2, 3$, the reflection coefficients $\{K_m\}$, and the corresponding mean-square prediction errors.

12.26 **(a)** Determine the power spectra for the random processes generated by the following difference equations.

    (1) $x(n) = -0.81x(n-2) + w(n) - w(n-1)$

    (2) $x(n) = w(n) - w(n-2)$

    (3) $x(n) = -0.81x(n-2) + w(n)$

    where $w(n)$ is a white noise process with variance σ_w^2.

    (b) Sketch the spectra for the processes given in part (a).

    (c) Determine the autocorrelation $\gamma_{xx}(m)$ for the processes in (2) and (3).

12.27 The autocorrelation sequence for an AR process $x(n)$ is

$$\gamma_{xx}(m) = (\tfrac{1}{4})^{|m|}$$

    (a) Determine the difference equation for $x(n)$.

    (b) Is your answer unique? If not, give any other possible solutions.

12.28 Repeat Problem 12.27 for an AR process with autocorrelation

$$\gamma_{xx}(m) = a^{|m|}\cos\frac{\pi m}{2}$$

where $0 < a < 1$.

12.29 The Bartlett method is used to estimate the power spectrum of a signal from a sequence $x(n)$ consisting of $N = 2400$ samples.

    (a) Determine the smallest length M of each segment in the Bartlett method that yields a frequency resolution of $\Delta f = 0.01$.

    (b) Repeat part (a) for $\Delta f = 0.02$.

    (c) Determine the quality factors Q_B for parts (a) and (b).

12.30 Prove that a FIR filter with system function

$$A_p(z) = 1 + \sum_{k=1}^{p} a_p(k)z^{-k}$$

and reflection coefficients $|K_k| < 1$ for $1 \le k \le p-1$ and $|K_p| > 1$ is maximum phase [all the roots of $A_p(z)$ lie outside the unit circle].

12.31 A random process $x(n)$ is characterized by the power density spectrum

$$\Gamma_{xx}(f) = \sigma_w^2 \frac{|e^{j2\pi f} - 0.9|^2}{|e^{j2\pi f} - j0.9|^2|e^{j2\pi f} + j0.9|^2}$$

where σ_w^2 is a constant (scale factor).

    (a) If we view $\Gamma_{xx}(f)$ as the power spectrum at the output of a linear pole–zero system $H(z)$ driven by white noise, determine $H(z)$.

    (b) Determine the system function of a stable system (noise-whitening filter) that produces a white noise output when excited by $x(n)$.

12.32 The N-point DFT of a random sequence $x(n)$ is

$$X(k) = \sum_{n=0}^{N-1} x(n)e^{-j2\pi nk/N}$$

Assume that $E[x(n)] = 0$ and $E[x(n)x(n+m)] = \sigma_x^2\delta(m)$ [i.e., $x(n)$ is a white noise process].

(a) Determine the variance of $X(k)$.

(b) Determine the autocorrelation of $X(k)$.

12.33 Suppose that we represent an ARMA(p, q) process as a cascade of a MA(q) followed by an AR(p) model. The input–output equation for the MA(q) model is

$$v(n) = \sum_{k=0}^{q} b_k w(n - k)$$

where $w(n)$ is a white noise process. The input–output equation for the AR(p) model is

$$x(n) + \sum_{k=1}^{p} a_k x(n - k) = v(n)$$

(a) By computing the autocorrelation of $v(n)$, show that

$$\gamma_{vv}(m) = \sigma_w^2 \sum_{k=0}^{q-m} b_k^* b_{k+m}$$

(b) Show that

$$\gamma_{vv}(m) = \sum_{k=0}^{p} a_k \gamma_{vx}(m + k) \qquad a_o = 1$$

where $\gamma_{vx}(m) = E[v(n + m)x^*(n)]$:

12.34 Determine the autocorrelation $\gamma_{xx}(m)$ of the random sequence

$$x(n) = A \cos(\omega_1 n + \phi)$$

where the amplitude A and the frequency ω_1 are (known) constants and ϕ is a uniformly distributed random phase over the interval $(0, 2\pi)$.

12.35 Suppose that the AR(2) process in Problem 12.20 is corrupted by an additive white noise process $v(n)$ with variance σ_v^2. Thus we have

$$y(n) = x(n) + v(n)$$

(a) Determine the difference equation for $y(n)$ and thus demonstrate that $y(n)$ is an ARMA(2, 2) process. Determine the coefficients of the ARMA process.

(b) Generalize the result in part (a) to an AR(p) process

$$x(n) = - \sum_{k=1}^{p} a_k(xn - k) + w(n)$$

and

$$y(n) = x(n) + v(n)$$

12.36 (a) Determine the autocorrelation of the random sequence

$$x(n) = \sum_{k=1}^{K} A_k \cos(\omega_k n + \phi_k) + w(n)$$

where $\{A_k\}$ are constant amplitudes, $\{\omega_k\}$ are constant frequencies, and $\{\phi_k\}$ are mutually statistically independent and uniformly distributed random phases. The noise sequence $w(n)$ is white with variance σ_w^2.

(b) Determine the power density spectrum of $x(n)$.

12.37 The harmonic decomposition problem considered by Pisarenko can be expressed as
the solution to the equation

$$\mathbf{a}^{*t}\boldsymbol{\Gamma}_{yy}\mathbf{a} = \sigma_w^2 \mathbf{a}^{*t}\mathbf{a}$$

The solution for $\mathbf{a}$ can be obtained by minimizing the quadratic form $\mathbf{a}^{*t}\boldsymbol{\Gamma}_{yy}\mathbf{a}$ subject to
the constraint that $\mathbf{a}^{*t}\mathbf{a} = 1$. The constraint can be incorporated into the performance
index by means of a Lagrange multiplier. Thus the performance index becomes

$$\mathcal{E} = \mathbf{a}^{*t}\boldsymbol{\Gamma}_{yy}\mathbf{a} + \lambda(1 - \mathbf{a}^{*t}\mathbf{a})$$

By minimizing $\mathcal{E}$ with respect to $\mathbf{a}$, show that this formulation is equivalent to the
Pisarenko eigenvalue problem given in (12.5.9) with the Lagrange multiplier play-
ing the role of the eigenvalue. Thus show that the minimum of $\mathcal{E}$ is the minimum
eigenvalue σ_w^2.

12.38 The autocorrelation of a sequence consisting of a sinusoid with random phase in noise
is

$$\gamma_{xx}(m) = P \cos 2\pi f_1 m + \sigma_w^2 \delta(m)$$

where f_1 is the frequency of the sinusoidal, P is its power, and σ_w^2 is the variance of
the noise. Suppose that we attempt to fit an AR(2) model to the data.
   (a) Determine the optimum coefficients of the AR(2) model as a function of σ_w^2
   and f_1.
   (b) Determine the reflection coefficients K_1 and K_2 corresponding to the AR(2)
   model parameters.
   (c) Determine the limiting values of the AR(2) parameters and (K_1, K_2) as $\sigma_w^2 \to 0$.

12.39 This problem involves the use of crosscorrelation to detect a signal in noise and estimate the time delay in the signal. A signal $x(n)$ consists of a pulsed sinusoid corrupted
by a stationary zero-mean white noise sequence. That is,

$$x(n) = y(n - n_0) + w(n) \qquad 0 \le n \le N - 1$$

where $w(n)$ is the noise with variance σ_w^2 and the signal is

$$\begin{aligned} y(n) &= A \cos \omega_0 n, \qquad &0 \le n \le M - 1 \\ &= 0, \qquad &\text{otherwise} \end{aligned}$$

The frequency ω_0 is known but the delay n_0, which is a positive integer, is unknown,
and is to be determined by crosscorrelating $x(n)$ with $y(n)$. Assume that $N > M + n_0$.
Let

$$r_{xy}(m) = \sum_{n=0}^{N-1} y(n - m)x(n)$$

denote the crosscorrelation sequence between $x(n)$ and $y(n)$. In the absence of noise
this function exhibits a peak at delay $m = n_0$. Thus n_0 is determined with no error.
The presence of noise can lead to errors in determining the unknown delay.
   (a) For $m = n_0$, determine $E[r_{xy}(n_0)]$. Also, determine the variance, $\text{var}[r_{xy}(n_0)]$,
   due to the presence of the noise. In both calculations, assume that the double
   frequency term averages to zero. That is, $M \gg 2\pi/\omega_0$.
   (b) Determine the signal-to-noise ratio, defined as

$$\text{SNR} = \frac{\{E[r_{xy}(n_0)]\}^2}{\text{var}[r_{xy}(n_0)]}$$

(c) What is the effect of the pulse duration M on the SNR?

12.40* Generate 100 samples of a zero-mean white noise sequence $w(n)$ with variance $\sigma_w^2 = \frac{1}{12}$, by using a uniform random number generator.

(a) Compute the autocorrelation of $w(n)$ for $0 \le m \le 15$.

(b) Compute the periodogram estimate $P_{xx}(f)$ and plot it.

(c) Generate 10 different realizations of $w(n)$ and compute the corresponding sample autocorrelation sequences $r_k(m)$, $1 \le k \le 10$ and $0 \le m \le 15$.

(d) Compute and plot the average periodogram for part (c):

$$r_{av}(m) = \tfrac{1}{10} \sum_{k=1}^{10} r_k(m)$$

(e) Comment on the results in parts (a) through (d).

12.41* A random signal is generated by passing zero-mean white Gaussian noise with unit variance through a filter with system function

$$H(z) = \frac{1}{(1 + az^{-1} + 0.99z^{-2})(1 - az^{-1} + 0.98z^{-2})}$$

(a) Sketch a typical plot of the theoretical power spectrum $\Gamma_{xx}(f)$ for a small value of the parameter a (i.e., $0 < a < 0.1$). Pay careful attention to the value of the two spectral peaks and the value of $P_{xx}(\omega)$ for $\omega = \pi/2$.

(b) Let $a = 0.1$. Determine the section length M required to resolve the spectral peaks of $\Gamma_{xx}(f)$ when using Bartlett's method.

(c) Consider the Blackman–Tukey method of smoothing the periodogram. How many lags of the correlation estimate must be used to obtain resolution comparable to that of the Bartlett estimate considered in part (b)? How many data must be used if the variance of the estimate is to be comparable to that of a four-section Bartlett estimate?

(d) For $a = 0.05$, fit an AR(4) model to 100 samples of the data based on the Yule–Walker method and plot the power spectrum. Avoid transient effects by discarding the first 200 samples of the data.

(e) Repeat part (d) with the Burg method.

(f) Repeat parts (d) and (e) for 50 data samples and comment on similarities and differences in the results.

Appendix A

Random Signals, Correlation Functions, and Power Spectra

In this appendix we provide a brief review of the characterization of random signals in terms of statistical averages expressed in both the time domain and the frequency domain. The reader is assumed to have a background in probability theory and random processes, at the level given in the books of Helstrom (1990) and Peebles (1987).

Random Processes

Many physical phenomena encountered in nature are best characterized in statistical terms. For example, meteorological phenomena such as air temperature and air pressure fluctuate randomly as a function of time. Thermal noise voltages generated in the resistors of electronic devices, such as a radio or television receiver, are also randomly fluctuating phenomena. These are just a few examples of random signals. Such signals are usually modeled as infinite-duration infinite-energy signals.

Suppose that we take the set of waveforms corresponding to the air temperature in different cities around the world. For each city there is a corresponding waveform that is a function of time, as illustrated in Fig. A.1. The set of all possible waveforms is called an *ensemble* of time functions or, equivalently, a *random process*. The waveform for the temperature in any particular city is a *single realization* or a *sample function* of the random process.

Similarly, the thermal noise voltage generated in a resistor is a single realization or a sample function of the random process consisting of all noise voltage waveforms generated by the set of all resistors.

The set (ensemble) of all possible noise waveforms of a random process is denoted as $X(t, S)$, where t represents the time index and S represents the set (sample space) of all possible sample functions. A single waveform in the ensemble is denoted by $x(t, s)$. Usually, we drop the variable s (or S) for notational convenience, so that the random process is denoted as $X(t)$ and a single realization is denoted as $x(t)$.

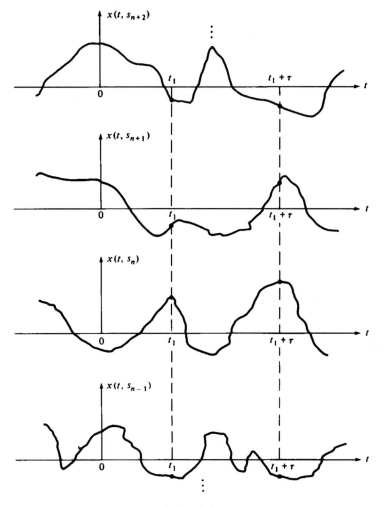

Figure A.1

Having defined a random process $X(t)$ as an ensemble of sample functions, let us consider the values of the process for any set of time instants $t_1 > t_2 > \cdots > t_n$, where n is any positive integer. In general, the samples $X_{t_i} \equiv x(t_i)$, $i = 1, 2, \ldots, n$ are n random variables characterized statistically by their joint probability density function (PDF) denoted as $p(x_{t_1}, x_{t_2}, \ldots, x_{t_n})$ for any n.

Stationary Random Processes

Suppose that we have n samples of the random process $X(t)$ at $t = t_i$, $i = 1$, $2, \ldots, n$, and another set of n samples displaced in time from the first set by an amount τ. Thus the second set of samples are $X_{t_i+\tau} \equiv X(t_i + \tau)$, $i = 1, 2, \ldots, n$, as shown in Fig. A.1. This second set of n random variables is characterized by the joint probability density function $p(x_{t_i+\tau}, \ldots, x_{t_n+\tau})$. The joint PDFs of the two sets of random variables may or may not be identical. When they are identical, then

$$p(x_{t_1}, x_{t_2}, \ldots, x_{t_n}) = p(x_{t_1+\tau}, x_{t_2+\tau}, \ldots, x_{t_n+\tau}) \tag{A.1}$$

for all τ and all n, then the random process is said to be *stationary in the strict sense*. In other words, the statistical properties of a stationary random process are invariant to a translation of the time axis. On the other hand, when the joint PDFs are different, the random process is nonstationary.

Statistical (Ensemble) Averages

Let us consider a random process $X(t)$ sampled at time instant $t = t_i$. Thus $X(t_i)$ is a random variable with PDF $p(x_{t_i})$. The *l*th *moment* of the random variable is defined as the *expected value* of $X^l(t_i)$, that is,

$$E(x_{t_i}^l) = \int_{-\infty}^{\infty} x_{t_i}^l p(x_{t_i}) dx_{t_i} \tag{A.2}$$

In general, the value of the *l*th moment depends on the time instant t_i, if the PDF of X_{t_i} depends on t_i. When the process is stationary, however, $p(x_{t_i+\tau}) = p(x_{t_i})$ for all τ. Hence the PDF is independent of time and, consequently, the *l*th moment is independent of time (a constant).

Next, let us consider the two random variables $X_{t_i} = X(t_i)$, $i = 1, 2$, corresponding to samples of $X(t)$ taken at $t = t_1$ and $t = t_2$. The statistical (ensemble) correlation between X_{t_1} and X_{t_2} is measured by the joint moment

$$E(X_{t_1} X_{t_2}) = \int_{-\infty}^{\infty} \int_{-\infty}^{\infty} x_{t_1} x_{t_2} p(x_{t_1} x_{t_2}) dx_1 dx_2 \tag{A.3}$$

Since the joint moment depends on the time instants t_1 and t_2, it is denoted as $\gamma_{xx}(t_1, t_2)$ and is called the *autocorrelation function* of the random process. When the process $X(t)$ is stationary, the joint PDF of the pair (X_{t_1}, X_{t_2}) is identical to the joint PDF of the pair $(X_{t_1+\tau}, X_{t_2+\tau})$ for any arbitrary τ. This implies that the autocorrelation function of $X(t)$ depends on the time difference $t_1 - t_2 = \tau$. Hence for a stationary real-valued random process the autocorrelation function is

$$\gamma_{xx}(\tau) = E[X_{t_1+\tau} X_{t_1}] \tag{A.4}$$

On the other hand,

$$\gamma_{xx}(-\tau) = E(X_{t_1-\tau} X_{t_1}) = E(X_{t_1'} X_{t_1'+\tau}) = \gamma_{xx}(\tau) \tag{A.5}$$

Therefore, $\gamma_{xx}(\tau)$ is an even function. We also note that $\gamma_{xx}(0) = E(X_{t_1}^2)$ is the *average power* of the random process.

There exist nonstationary processes with the property that the mean value of the process is a constant and the autocorrelation function satisfies the property $\gamma_{xx}(t_1, t_2) = \gamma_{xx}(t_1 - t_2)$. Such a process is called *wide-sense stationary*. Clearly, wide-sense stationarity is a less stringent condition than strict-sense stationarity. In our treatment we shall require only that the processes be wide-sense stationary.

Related to the autocorrelation function is the autocovariance function, which is defined as

$$c_{xx}(t_1, t_2) = E\{[X_{t_1} - m(t_1)][X_{t_2} - m(t_2)]\}$$
$$= \gamma_{xx}(t_1, t_2) - m(t_1)m(t_2) \tag{A.6}$$

where $m(t_1) = E(X_{t_1})$ and $m(t_2) = E(X_{t_2})$ are the mean values of X_{t_1} and X_{t_2}, respectively. When the process is stationary,

$$c_{xx}(t_1, t_2) = c_{xx}(t_1 - t_2) = c_{xx}(\tau) = \gamma_{xx}(\tau) - m_x^2 \tag{A.7}$$

where $\tau = t_1 - t_2$. Furthermore, the variance of the process is $\sigma_x^2 = c_{xx}(0) = \gamma_{xx}(0) - m_x^2$.

Statistical Averages for Joint Random Processes

Let $X(t)$ and $Y(t)$ be two random processes and let $X_{t_i} \equiv X(t_i)$, $i = 1, 2, \ldots, n$, and $Y_{t_j'} \equiv Y(t_j')$, $j = 1, 2, \ldots, m$, represent the random variables at times $t_1 > t_2 > \cdots > t_n$ and $t_1' > t_2' > \cdots > t_m'$, respectively. The two sets of random variables are characterized statistically by the joint PDF

$$p(x_{t_1}, x_{t_2}, \ldots, x_{t_n}, y_{t_1'}, y_{t_2'}, \ldots, y_{t_m'})$$

for any set of time instants $\{t_i\}$ and $\{t_j'\}$ and for any positive integer values of m and n.

The *crosscorrelation function* of $X(t)$ and $Y(t)$, denoted as $\gamma_{xy}(t_1, t_2)$, is defined by the joint moment

$$\gamma_{xy}(t_1, t_2) \equiv E(X_{t_1} Y_{t_2}) = \int_{-\infty}^{\infty} \int_{-\infty}^{\infty} x_{t_1} y_{t_2} p(x_{t_1}, y_{t_2}) dx_{t_1} dy_{t_2} \tag{A.8}$$

and the crosscovariance is

$$c_{xy}(t_1, t_2) = \gamma_{xy}(t_1, t_2) - m_x(t_1)m_y(t_2) \tag{A.9}$$

When the random processes are jointly and individually stationary, we have $\gamma_{xy}(t_1, t_2) = \gamma_{xy}(t_1 - t_2)$ and $c_{xy}(t_1, t_2) = c_{xy}(t_1 - t_2)$. In this case

$$\gamma_{xy}(-\tau) = E(X_{t_1} Y_{t_1 + \tau}) = E(X_{t_1' - \tau} Y_{t_1'}) = \gamma_{yx}(\tau) \tag{A.10}$$

The random processes $X(t)$ and $Y(t)$ are said to be statistically independent if and only if

$$p(x_{t_1}, x_{t_2}, \ldots, x_{t_n}, y_{t_1'}, y_{t_2'}, \ldots, y_{t_m'}) = p(x_{t_1}, \ldots, x_{t_n}) p(y_{t_1'}, \ldots, y_{t_m'})$$

for all choices of t_i, t_i' and for all positive integers n and m. The processes are said to be uncorrelated if

$$\gamma_{xy}(t_1, t_2) = E(X_{t_1})E(Y_{t_2}) \tag{A.11}$$

so that $c_{xy}(t_1, t_2) = 0$.

A complex-valued random process $Z(t)$ is defined as

$$Z(t) = X(t) + jY(t) \tag{A.12}$$

where $X(t)$ and $Y(t)$ are random processes. The joint PDF of the complex-valued random variables $Z_{t_i} \equiv Z(t_i)$, $i = 1, 2, \ldots$, is given by the joint PDF of the components (X_{t_i}, Y_{t_i}), $i = 1, 2, \ldots, n$. Thus the PDF that characterizes Z_{t_i}, $i = 1, 2, \ldots, n$ is

$$p(x_{t_1}, x_{t_2}, \ldots, x_{t_n}, y_{t_1}, y_{t_2}, \ldots, y_{t_n})$$

A complex-valued random process $Z(t)$ is encountered in the representation of the in-phase and quadrature components of the lowpass equivalent of a narrowband random signal or noise. An important characteristic of such a process is its autocorrelation function, which is defined as

$$\begin{aligned}
\gamma_{zz}(t_1, t_2) &= E(Z_{t_1} Z_{t_2}^*) \\
&= E[(X_{t_1} + jY_{t_1})(X_{t_2} - jY_{t_2})] \\
&= \gamma_{xx}(t_1, t_2) + \gamma_{yy}(t_1, t_2) + j[\gamma_{yx}(t_1, t_2) - \gamma_{xy}(t_1, t_2)]
\end{aligned} \tag{A.13}$$

When the random processes $X(t)$ and $Y(t)$ are jointly and individually stationary, the autocorrelation function of $Z(t)$ becomes

$$\gamma_{zz}(t_1, t_2) = \gamma_{zz}(t_1 - t_2) = \gamma_{zz}(\tau)$$

where $\tau = t_1 - t_2$. The complex conjugate of (A.13) is

$$\gamma_{zz}^*(\tau) = E(Z_{t_1}^* Z_{t_1 - \tau}) = \gamma_{zz}(-\tau) \tag{A.14}$$

Now, suppose that $Z(t) = X(t) + jY(t)$ and $W(t) = U(t) + jV(t)$ are two complex-valued random processes. Their crosscorrelation function is defined as

$$\begin{aligned}
\gamma_{zw}(t_1, t_2) &= E(Z_{t_1} W_{t_2}^*) \\
&= E[(X_{t_1} + jY_{t_1})(U_{t_2} - jV_{t_2})] \\
&= \gamma_{xu}(t_1, t_2) + \gamma_{yv}(t_1, t_2) + j[\gamma_{yu}(t_1, t_2) - \gamma_{xv}(t_1, t_2)]
\end{aligned} \tag{A.15}$$

When $X(t)$, $Y(t)$, $U(t)$, and $V(t)$ are pairwise stationary, the crosscorrelation functions in (A.15) become functions of the time difference $\tau = t_1 - t_2$. In addition, we have

$$\gamma_{zw}^*(\tau) = E(Z_{t_1}^* W_{t_1 - \tau}) = E(Z_{t_1 + \tau}^* W_{t_1}) = \gamma_{wz}(-\tau) \tag{A.16}$$

Power Density Spectrum

A stationary random process is an infinite-energy signal and hence its Fourier transform does not exist. The spectral characteristic of a random process is obtained according to the Wiener–Khinchine theorem, by computing the Fourier transform of the autocorrelation function. That is, the distribution of power with frequency is given by the function

$$\Gamma_{xx}(F) = \int_{-\infty}^{\infty} \gamma_{xx}(\tau)e^{-j2\pi F\tau}\,d\tau \qquad (A.17)$$

The inverse Fourier transform is given as

$$\gamma_{xx}(\tau) = \int_{-\infty}^{\infty} \Gamma_{xx}(F)e^{j2\pi F\tau}\,dF \qquad (A.18)$$

We observe that

$$\begin{aligned} \gamma_{xx}(0) &= \int_{-\infty}^{\infty} \Gamma_{xx}(F)\,dF \\ &= E(X_t^2) \geq 0 \end{aligned} \qquad (A.19)$$

Since $E(X_t^2) = \gamma_{xx}(0)$ represents the average power of the random process, which is the area under $\Gamma_{xx}(F)$, it follows that $\Gamma_{xx}(F)$ is the distribution of power as a function of frequency. For this reason, $\Gamma_{xx}(F)$ is called the *power density spectrum* of the random process.

   If the random process is real, $\gamma_{xx}(\tau)$ is real and even and hence $\Gamma_{xx}(F)$ is real and even. If the random process is complex valued, $\gamma_{xx}(\tau) = \gamma_{xx}^*(-\tau)$ and, hence

$$\begin{aligned} \Gamma_{xx}^*(F) &= \int_{-\infty}^{\infty} \gamma_{xx}^*(\tau)e^{j2\pi F\tau}\,d\tau = \int_{-\infty}^{\infty} \gamma_{xx}^*(-\tau)e^{-j2\pi F\tau}\,d\tau \\ &= \int_{-\infty}^{\infty} \gamma_{xx}(\tau)e^{-j2\pi F\tau}\,d\tau = \Gamma_{xx}(F) \end{aligned}$$

Therefore, $\Gamma_{xx}(F)$ is always real.

   The definition of the power density spectrum can be extended to two jointly stationary random processes $X(t)$ and $Y(t)$, which have a crosscorrelation function $\gamma_{xy}(\tau)$. The Fourier transform of $\gamma_{xy}(\tau)$ is

$$\Gamma_{xy}(F) = \int_{-\infty}^{\infty} \gamma_{xy}(\tau)e^{-j2\pi F\tau}\,d\tau \qquad (A.20)$$

which is called the *cross-power density spectrum*. It is easily shown that $\Gamma_{xy}^*(F) = \Gamma_{yx}(-F)$. For real random processes, the condition is $\Gamma_{yx}(F) = \Gamma_{xy}(-F)$.

Discrete-Time Random Signals

This characterization of continuous-time random signals can be easily carried over to discrete-time signals. Such signals are usually obtained by uniformly sampling a continuous-time random process.

   A discrete-time random process $X(n)$ consists of an ensemble of sample sequences $x(n)$. The statistical properties of $X(n)$ are similar to the characterization of $X(t)$, with the restriction that n is now an integer (time) variable. To be specific, we state the form for the important moments that we use in this text.

The lth moment of $X(n)$ is defined as

$$E(X_n^l) = \int_{-\infty}^{\infty} x_n^l p(x_n) dx_n \qquad (A.21)$$

and the autocorrelation sequence is

$$\gamma_{xx}(n, k) = E(X_n X_k) = \int_{-\infty}^{\infty} \int_{-\infty}^{\infty} x_n x_k p(x_n, x_k) dx_n dx_k \qquad (A.22)$$

Similarly, the autocovariance is

$$c_{xx}(n, k) = \gamma_{xx}(n, k) - E(X_n)E(X_k) \qquad (A.23)$$

For a stationary process, we have the special forms $(m = n - k)$

$$\gamma_{xx}(n - k) = \gamma_{xx}(m)$$
$$c_{xx}(n - k) = c_{xx}(m) = \gamma_{xx}(m) - m_x^2 \qquad (A.24)$$

where $m_x = E(X_n)$ is the mean of the random process. The variance is defined as $\sigma^2 = c_{xx}(0) = \gamma_{xx}(0) - m_x^2$.

For a complex-valued stationary process $Z(n) = X(n) + jY(n)$, we have

$$\gamma_{zz}(m) = \gamma_{xx}(m) + \gamma_{yy}(m) + j[\gamma_{yx}(m) - \gamma_{xy}(m)] \qquad (A.25)$$

and the crosscorrelation sequence of two complex-valued stationary sequences is

$$\gamma_{zw}(m) = \gamma_{xu}(m) + \gamma_{yv}(m) + j[\gamma_{yu}(m) - \gamma_{xv}(m)] \qquad (A.26)$$

As in the case of a continuous-time random process, a discrete-time random process has infinite energy but a finite average power and is given as

$$E(X_n^2) = \gamma_{xx}(0) \qquad (A.27)$$

By use of the Wiener–Khinchine theorem, we obtain the power density spectrum of the discrete-time random process by computing the Fourier transform of the autocorrelation sequence $\gamma_{xx}(m)$, that is,

$$\Gamma_{xx}(f) = \sum_{m=-\infty}^{\infty} \gamma_{xx}(m) e^{-j2\pi fm} \qquad (A.28)$$

The inverse transform relationship is

$$\gamma_{xx}(m) = \int_{-1/2}^{1/2} \Gamma_{xx}(f) e^{j2\pi fm} df \qquad (A.29)$$

We observe that the average power is

$$\gamma_{xx}(0) = \int_{-1/2}^{1/2} \Gamma_{xx}(f) df \qquad (A.30)$$

so that $\Gamma_{xx}(f)$ is the distribution of power as a function of frequency, that is, $\Gamma_{xx}(f)$ is the power density spectrum of the random process $X(n)$. The properties we have stated for $\Gamma_{xx}(F)$ also hold for $\Gamma_{xx}(f)$.

Time Averages for a Discrete-Time Random Process

Although we have characterized a random process in terms of statistical averages, such as the mean and the autocorrelation sequence, in practice, we usually have available a single realization of the random process. Let us consider the problem of obtaining the averages of the random process from a single realization. To accomplish this, the random process must be *ergodic*.

By definition, a random process $X(n)$ is ergodic if, with probability 1, all the statistical averages can be determined from a single sample function of the process. In effect, the random process is ergodic if time averages obtained from a single realization are equal to the statistical (ensemble) averages. Under this condition we can attempt to estimate the ensemble averages using time averages from a single realization.

To illustrate this point, let us consider the estimation of the mean and the autocorrelation of the random process from a single realization $x(n)$. Since we are interested only in these two moments, we define engodicity with respect to these parameters. For additional details on the requirements for mean ergodicity and autocorrelation ergodicity which are given below, the reader is referred to the book of Papoulis (1984).

Mean-Ergodic Process

Given a stationary random process $X(n)$ with mean

$$m_x = E(X_n)$$

let us form the *time average*

$$\hat{m}_x = \frac{1}{2N+1} \sum_{n=-N}^{N} x(n) \tag{A.31}$$

In general, we view $\hat{m}_x$ in (A.31) as an estimate of the statistical mean whose value will vary with the different realizations of the random process. Hence $\hat{m}_x$ is a random variable with a PDF $p(\hat{m}_x)$. Let us compute the expected value of $\hat{m}_x$ over all possible realizations of $X(n)$. Since the summation and the expectation are linear operations we can interchange them, so that

$$E(\hat{m}_x) = \frac{1}{2N+1} \sum_{n=-N}^{N} E[x(n)] = \frac{1}{2N+1} \sum_{n=-N}^{N} m_x = m_x \tag{A.32}$$

Since the mean value of the estimate is equal to the statistical mean, we say that the estimate $\hat{m}_x$ is unbiased.

Next, we compute the variance of $\hat{m}_x$. We have

$$\text{var}(\hat{m}_x) = E(|\hat{m}_x|^2) - |m_x|^2$$

But

$$E(|\hat{m}_x|^2) = \frac{1}{(2N+1)^2} \sum_{n=-N}^{N} \sum_{k=-N}^{N} E[x^*(n)x(k)]$$

$$= \frac{1}{(2N+1)^2} \sum_{n=-N}^{N} \sum_{k=-N}^{N} \gamma_{xx}(k-n)$$

$$= \frac{1}{2N+1} \sum_{m=-2N}^{2N} \left(1 - \frac{|m|}{2N+1}\right) \gamma_{xx}(m)$$

Therefore,

$$\text{var}(\hat{m}_x) = \frac{1}{2N+1} \sum_{m=-2N}^{2N} \left(1 - \frac{|m|}{2N+1}\right) \gamma_{xx} - |m_x|^2$$

$$= \frac{1}{2N+1} \sum_{m=-2N}^{2N} \left(1 - \frac{|m|}{2N+1}\right) c_{xx}(m) \tag{A.33}$$

If $\text{var}(m_x) \to 0$ as $N \to \infty$, the estimate converges with probability 1 to the statistical mean m_x. Therefore, the process $X(n)$ is mean ergodic if

$$\lim_{N\to\infty} \frac{1}{2N+1} \sum_{m=-2N}^{2N} \left(1 - \frac{|m|}{2N+1}\right) c_{xx}(m) = 0 \tag{A.34}$$

Under this condition, the estimate $\hat{m}_x$ in the limit as $N \to \infty$ becomes equal to the statistical mean, that is,

$$m_x = \lim_{N\to\infty} \frac{1}{2N+1} \sum_{n=-N}^{N} x(n) \tag{A.35}$$

Thus the time-average mean, in the limit as $N \to \infty$, is equal to the ensemble mean.

A sufficient condition for (A.34) to hold is if

$$\sum_{m=-\infty}^{\infty} |c_{xx}(m)| < \infty \tag{A.36}$$

which implies that $c_{xx}(m) \to 0$ as $m \to \infty$. This condition holds for most zero-mean processes encountered in the physical world.

Correlation-Ergodic Processes

Now, let us consider the estimate of the autocorrelation $\gamma_{xx}(m)$ from a single realization of the process. Following our previous notation, we denote the estimate (for a complex-valued signal, in general) as

$$r_{xx}(m) = \frac{1}{2N+1} \sum_{n=-N}^{N} x^*(n)x(n+m) \tag{A.37}$$

Again, we regard $r_{xx}(m)$ as a random variable for any given lag m, since it is a function of the particular realization. The expected value (mean value over all realizations) is

$$E[r_{xx}(m)] = \frac{1}{2N+1} \sum_{n=-N}^{N} E[x^*(n)x(n+m)]$$

$$= \frac{1}{2N+1} \sum_{n=-N}^{N} \gamma_{xx}(m) = \gamma_{xx}(m)$$

(A.38)

Therefore, the expected value of the time-average autocorrelation is equal to the statistical average. Hence we have an unbiased estimate of $\gamma_{xx}(m)$.

To determine the variance of the estimate $r_{xx}(m)$, we compute the expected value of $|r_{xx}(m)|^2$ and subtract the square of the mean value. Thus

$$\text{var}[r_{xx}(m)] = E[|r_{xx}(m)|^2] - |\gamma_{xx}(m)|^2$$

(A.39)

But

$$E[|r_{xx}(m)|^2] = \frac{1}{(2N+1)^2} \sum_{n=-N}^{N} \sum_{k=-N}^{N} E[x^*(n)x(n+m)x(k)x^*(k+m)]$$

(A.40)

The expected value of the term $x^*(n)x(n+m)x(k)x^*(k+m)$ is just the autocorrelation sequence of a random process defined as

$$v_m(n) = x^*(n)x(n+m)$$

Hence (A.40) may be expressed as

$$E[|r_{xx}(m)|^2] = \frac{1}{(2N+1)^2} \sum_{n=-N}^{N} \sum_{k=-N}^{N} \gamma_{vv}^{(m)}(n-k)$$

$$= \frac{1}{2N+1} \sum_{n=-2N}^{2N} \left(1 - \frac{|n|}{2N+1}\right) \gamma_{vv}^{(m)}(n)$$

(A.41)

and the variance is

$$\text{var}[r_{xx}(m)] = \frac{1}{2N+1} \sum_{n=-2N}^{2N} \left(1 - \frac{|n|}{2N+1}\right) \gamma_{vv}^{(m)}(n) - |\gamma_{xx}(m)|^2$$

(A.42)

If $\text{var}[r_{xx}(m)] \to 0$ as $N \to \infty$, the estimate $r_{xx}(m)$ converges with probability 1 to the statistical autocorrelation $\gamma_{xx}(m)$. Under these conditions, the process is correlation ergodic and the time-average correlation is identical to the statistical average, that is,

$$\lim_{N \to \infty} \frac{1}{2N+1} \sum_{n=-N}^{N} x^*(n)x(n+m) = \gamma_{xx}(m)$$

(A.43)

In our treatment of random signals, we assume that the random processes are mean ergodic and correlation ergodic, so that we can deal with time averages of the mean and the autocorrelation obtained from a single realization of the process.

Appendix B
Random Number Generators

In some of the examples given in the text, random numbers are generated to simulate the effect of noise on signals and to illustrate how the method of correlation can be used to detect the presence of a signal buried in noise. In the case of periodic signals, the correlation technique also allowed us to estimate the period of the signal.

In practice, random number generators are often used to simulate the effect of noiselike signals and other random phenomena encountered in the physical world. Such noise is present in electronic devices and systems and usually limits our ability to communicate over large distances and to be able to detect relatively weak signals. By generating such noise on a computer, we are able to study its effects through simulation of communication systems, radar detection systems, and the like and to assess the performance of such systems in the presence of noise.

Most computer software libraries include a uniform random number generator. Such a random number generator generates a number between zero and 1 with equal probability. We call the output of the random number generator a random variable. If A denotes such a random variable, its range is the interval $0 \leq A \leq 1$.

We know that the numerical output of a digital computer has limited precision, and as a consequence, it is impossible to represent the continuum of numbers in the interval $0 \leq A \leq 1$. However, we can assume that our computer represents each output by a large number of bits in either fixed point or floating point. Consequently, for all practical purposes, the number of outputs in the interval $0 \leq A \leq 1$ is sufficiently large, so that we are justified in assuming that any value in the interval is a possible output from the generator.

The uniform probability density function for the random variable A, denoted as $p(A)$, is illustrated in Fig. B.1a. We note that the average value or mean value of A, denoted as m_A, is $m_A = \frac{1}{2}$. The integral of the probability density function, which represents the area under $p(A)$, is called the probability distribution function

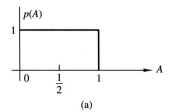

(a)

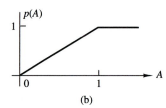

(b)   **Figure B.1**

of the random variable A and is defined as

$$F(A) = \int_{-\infty}^{A} p(x)dx \tag{B.1}$$

For any random variable, this area must always be unity, which is the maximum value that can be achieved by a distribution function. Hence

$$F(1) = \int_{-\infty}^{1} p(x)dx = 1 \tag{B.2}$$

and the range of $F(A)$ is $0 \le F(A) \le 1$ for $0 \le A \le 1$.

If we wish to generate uniformly distributed noise in an interval $(b, b + 1)$ it can simply be accomplished by using the output A of the random number generator and shifting it by an amount b. Thus a new random variable B can be defined as

$$B = A + b \tag{B.3}$$

which now has a mean value $m_B = b + \frac{1}{2}$. For example, if $b = -\frac{1}{2}$, the random variable B is uniformly distributed in the interval $(-\frac{1}{2}, \frac{1}{2})$, as shown in Fig. B.2a. Its probability distribution function $F(B)$ is shown in Fig. B.2b.

A uniformly distributed random variable in the range $(0, 1)$ can be used to generate random variables with other probability distribution functions. For example, suppose that we wish to generate a random variable C with probability distribution function $F(C)$, as illustrated in Fig. B.3. Since the range of $F(C)$ is the interval $(0, 1)$, we begin by generating a uniformly distributed random variable A in the range $(0, 1)$. If we set

$$F(C) = A \tag{B.4}$$

then

$$C = F^{-1}(A) \tag{B.5}$$

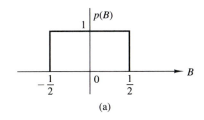

(a)

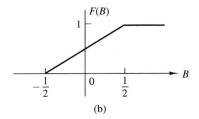

(b)   **Figure B.2**

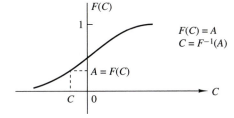

Figure B.3

Thus we solve (B.4) for C and the solution in (B.5) provides the value of C for which $F(C) = A$. By this means we obtain a new random variable C with probability distribution $F(C)$. This inverse mapping from A to C is illustrated in Fig. B.3.

Example B.1

Generate a random variable C that has the linear probability density function shown in Fig. B.4a, that is,

$$p(C) = \begin{cases} \dfrac{C}{2}, & 0 \le C \le 2 \\ 0, & \text{otherwise} \end{cases}$$

Solution   This random variable has a probability distribution function

$$F(C) = \begin{cases} 0, & C < 0 \\ \tfrac{1}{4}C^2, & 0 \le C \le 2 \\ 1, & C > 2 \end{cases}$$

which is illustrated in Fig. B.4b. We generate a uniformly distributed random variable A and set $F(C) = A$. Hence

$$F(C) = \tfrac{1}{4}C^2 = A$$

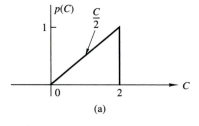

(a)

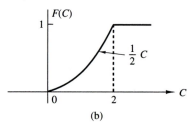

(b)   **Figure B.4**

Upon solving for C, we obtain

$$C = 2\sqrt{A}$$

Thus we generate a random variable C with probability function $F(C)$, as shown in Fig. B.4b.

In Example B.1 the inverse mapping $C = F^{-1}(A)$ was simple. In some cases it is not. This problem arises in trying to generate random numbers that have a normal distribution function.

Noise encountered in physical systems is often characterized by the normal or Gaussian probability distribution, which is illustrated in Fig. B.5. The probability

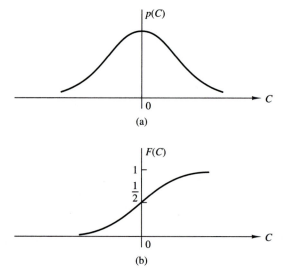

(a)

(b)   **Figure B.5**

density function is given by

$$p(C) = \frac{1}{\sqrt{2\pi}\sigma}e^{-C^2/2\sigma^2} \qquad -\infty < C < \infty \qquad (\text{B.6})$$

where σ^2 is the variance of C, which is a measure of the spread of the probability density function $p(C)$. The probability distribution function $F(C)$ is the area under $p(C)$ over the range $(-\infty, C)$. Thus

$$F(C) = \int_{-\infty}^{C} p(x)dx \qquad (\text{B.7})$$

Unfortunately, the integral in (B.7) cannot be expressed in terms of simple functions. Consequently, the inverse mapping is difficult to achieve.

A way has been found to circumvent this problem. From probability theory it is known that a (Rayleigh distributed) random variable R, with probability distribution function

$$F(R) = \begin{cases} 0, & R < 0 \\ 1 - e^{-R^2/2\sigma^2}, & R \geq 0 \end{cases} \qquad (\text{B.8})$$

is related to a pair of Gaussian random variables C and D, through the transformation

$$C = R\cos\Theta \qquad (\text{B.9})$$

$$D = R\sin\Theta \qquad (\text{B.10})$$

where Θ is a uniformly distributed variable in the interval $(0, 2\pi)$. The parameter σ^2 is the variance of C and D. Since (B.8) is easily inverted, we have

$$F(R) = 1 - e^{-R^2/2\sigma^2} = A$$

and hence

$$R = \sqrt{2\sigma^2 \ln[1/(1 - A)]} \qquad (\text{B.11})$$

where A is a uniformly distributed random variable in the interval $(0, 1)$. Now if we generate a second uniformly distributed random variable B and define

$$\Theta = 2\pi B \qquad (\text{B.12})$$

then from (B.9) and (B.10), we obtain two statistically independent Gaussian distributed random variables C and D.

The method described above is often used in practice to generate Gaussian distributed random variables. As shown in Fig. B.5, these random variables have a mean value of zero and a variance σ^2. If a nonzero mean Gaussian random variable is desired, then C and D can be translated by the addition of the mean value.

A subroutine implementing this method for generating Gaussian distributed random variables is given in Fig. B.6.

```
C        SUBROUTINE GAUSS CONVERTS A UNIFORM RANDOM
C        SEQUENCE XIN IN [0,1] TO A GAUSSIAN RANDOM
C        SEQUENCE WITH G(0,SIGMA**2)
C        PARAMETERS   :
C             XIN     :UNIFORM IN [0,1] RANDOM NUMBER
C               B     :UNIFORM IN [0,1] RANDOM NUMBER
C           SIGMA     :STANDARD DEVIATION OF THE GAUSSIAN
C            YOUT     :OUTPUT FROM THE GENERATOR
C
         SUBROUTINE GAUSS 9XIN,B,SIGMA,YOUT)
         PI=4.0*ATAN (1.0)
         B=2.0*PI*B
         R=SQRT (2.0*(SIGMA**2)*ALOG(1.0/(1.0-XIN)))
         YOUT=R*COS(B)
         RETURN
         END
C        NOTE: TO USE THE ABOVE SUBROUTINE FOR A
C              GAUSSIAN RANDOM NUMBER GENERATOR
C              YOU MUST PROVIDE AS INPUT TWO UNIFORM RANDOM NUMBERS
C              XIN AND B
C              XIN AND B MUST BE STATISTICALLY INDEPENDENT
C
```

Figure B.6 Subroutine for generating Gaussian random variables

Appendix C

Tables of Transition Coefficients for the Design of Linear-Phase FIR Filters

In Section 8.2.3 we described a design method for linear-phase FIR filters that involved the specification of $H_r(\omega)$ at a set of equally spaced frequencies $\omega_k = 2\pi(k+\alpha)/M$, where $\alpha = 0$ or $\alpha = \frac{1}{2}$, $k = 0, 1, \ldots, (M-1)/2$ for M odd and $k = 0, 1, 2, \ldots, (M/2) - 1$ for M even, where M is the length of the filter. Within the passband of the filter, we select $H_r(\omega_k) = 1$, and in the stopband, $H_r(\omega_k) = 0$. For frequencies in the transition band, the values of $H_r(\omega_k)$ are optimized to minimize the maximum sidelobe in the stopband. This is called a *minimax optimization criterion*.

   The optimization of the values of $H_r(\omega)$ in the transition band has been performed by Rabiner et al. (1970) and tables of transition values have been provided in the published paper. A selected number of the tables for lowpass FIR filters are included in this appendix.

   Four tables are given. Table C.1 lists the transition coefficients for the case $\alpha = 0$ and one coefficient in the transition band for both M odd and M even. Table C.2 lists the transition coefficients for the case $\alpha = 0$, and two coefficients in the transition band for M odd and M even. Table C.3 lists the transition coefficients for the case $\alpha = \frac{1}{2}$, M even and one coefficient in the transition band. Finally, Table C.4 lists the transition coefficients for the case $\alpha = \frac{1}{2}$, M even, and two coefficients in the transition band. The tables also include the level of the maximum sidelobe and a bandwidth parameter, denoted as BW.

   To use the tables, we begin with a set of specifications, including (1) the bandwidth of the filter, which can be defined as $(2\pi/M)(\mathrm{BW} + \alpha)$, where BW is the number of consecutive frequencies at which $H(\omega_k) = 1$, (2) the width of the transition region, which is roughly $2\pi/M$ times the number of transition coefficients, and (3) the maximum tolerable sidelobe in the stopband. The length of the filter can be selected from the tables to satisfy the specifications.

TABLE C.1 TRANSITION COEFFICIENTS FOR $\alpha = 0$

	M Odd			M Even	
BW	Minimax	T_1	BW	Minimax	T_1
	$M = 15$			$M = 16$	
1	−42.30932283	0.43378296	1	−39.75363827	0.42631836
2	−41.26299286	0.41793823	2	−37.61346340	0.40397949
3	−41.25333786	0.41047636	3	−36.57721567	0.39454346
4	−41.94907713	0.40405884	4	−35.87249756	0.38916626
5	−44.37124538	0.39268189	5	−35.31695461	0.38840332
6	−56.01416588	0.35766525	6	−35.51951933	0.40155639
	$M = 33$			$M = 32$	
1	−43.03163004	0.42994995	1	−42.24728918	0.42856445
2	−42.42527962	0.41042481	2	−41.29370594	0.40773926
3	−42.40898275	0.40141601	3	−41.03810358	0.39662476
4	−42.45948601	0.39641724	4	−40.93496323	0.38925171
6	−42.52403450	0.39161377	5	−40.85183477	0.37897949
8	−42.44085121	0.39039917	8	−40.75032616	0.36990356
10	−42.11079407	0.39192505	10	−40.54562140	0.35928955
12	−41.92705250	0.39420166	12	−39.93450451	0.34487915
14	−44.69430351	0.38552246	14	−38.91993237	0.34407349
15	−56.18293285	0.35360718			
	$M = 65$			$M = 64$	
1	−43.16935968	0.42919312	1	−42.96059322	0.42882080
2	−42.61945581	0.40903320	2	−42.30815172	0.40830689
3	−42.70906305	0.39920654	3	−42.32423735	0.39807129
4	−42.86997318	0.39335937	4	−42.43565893	0.39177246
5	−43.01999664	0.38950806	5	−42.55461407	0.38742065
6	−43.14578819	0.38679809	6	−42.66526604	0.38416748
10	−43.44808340	0.38129272	10	−43.01104736	0.37609863
14	−43.54684496	0.37946167	14	−43.28309965	0.37089233
18	−43.48173618	0.37955322	18	−43.56508827	0.36605225
22	−43.19538212	0.38162842	22	−43.96245098	0.35977783
26	−42.44725609	0.38746948	26	−44.60516977	0.34813232
30	−44.76228619	0.38417358	30	−43.81448936	0.29973144
31	−59.21673775	0.35282745			
	$M = 125$			$M = 128$	
1	−43.20501566	0.42899170	1	−43.15302420	0.42889404
2	−42.66971111	0.40867310	2	−42.59092569	0.40847778
3	−42.77438974	0.39868774	3	−42.67634487	0.39838257
4	−42.95051050	0.39268189	4	−42.84038544	0.39226685
6	−43.25854683	0.38579101	5	−42.99805641	0.38812256
8	−43.47917461	0.38195801	7	−43.25537014	0.38281250
10	−43.63750410	0.37954102	10	−43.52547789	0.3782638
18	−43.95589399	0.37518311	18	−43.93180990	0.37251587
26	−44.05913115	0.37384033	26	−44.18097305	0.36941528
34	−44.05672455	0.37371826	34	−44.40153408	0.36686401
42	−43.94708776	0.37470093	42	−44.67161417	0.36394653
50	−43.58473492	0.37797851	50	−45.17186594	0.35902100
58	−42.14925432	0.39086304	58	−46.92415667	0.34273681
59	−42.60623264	0.39063110	62	−49.46298973	0.28751221
60	−44.78062010	0.38383713			
61	−56.22547865	0.35263062			

Source: Rabiner et al. (1970); © 1970 IEEE; reprinted with permission.

TABLE C.2   TRANSITION COEFFICIENTS FOR $\alpha = 0$

	M Odd				M Even		
BW	Minimax	T_1	T_2	BW	Minimax	T_1	T_2
	$M = 15$				$M = 16$		
1	−70.60540585	0.09500122	0.58995418	1	−65.27693653	0.10703125	0.60559357
2	−69.26168156	0.10319824	0.59357118	2	−62.85937929	0.12384644	0.62201631
3	−69.91973495	0.10083618	0.58594327	3	−62.96594906	0.12827148	0.62855407
4	−75.51172256	0.08407953	0.55715312	4	−66.03942485	0.12130127	0.61952704
5	−103.45078300	0.05180206	0.49917424	5	−71.73997498	0.11066284	0.60979204
	$M = 33$				$M = 32$		
1	−70.60967541	0.09497070	0.58985167	1	−67.37020397	0.09610596	0.59045212
2	−68.16726971	0.10585937	0.59743846	2	−63.93104696	0.11263428	0.60560235
3	−67.13149548	0.10937500	0.59911696	3	−62.49787903	0.11931763	0.61192546
5	−66.53917217	0.10965576	0.59674101	5	−61.28204536	0.12541504	0.61824023
7	−67.23387909	0.10902100	0.59417456	7	−60.82049131	0.12907715	0.62307031
9	−67.85412312	0.10502930	0.58771575	9	−59.74928167	0.12068481	0.60685586
11	−69.08597469	0.10219727	0.58216391	11	−62.48683357	0.13004150	0.62821502
13	−75.86953640	0.08137207	0.54712777	13	−70.64571857	0.11017914	0.60670943
14	−104.04059029	0.05029373	0.49149549				
	$M = 65$				$M = 64$		
1	−70.66014957	0.09472656	0.58945943	1	−70.26372528	0.09376831	0.58789222
2	−68.89622307	0.10404663	059.476127	2	−67.20729542	0.10411987	0.59421778
3	−67.90234470	0.10720215	0.59577449	3	−65.80684280	0.10850220	0.59666158
4	−67.24003792	0.10726929	0.59415763	4	−64.95227051	0.11038818	0.59730067
5	−66.86065960	0.10689087	0.59253047	5	−64.42742348	0.11113281	0.59698496
9	−66.27561188	0.10548706	0.58845983	9	−63.41714096	0.10936890	0.59088884
13	−65.96417046	0.10466309	0.58660485	13	−62.72142410	0.10828857	0.58738641
17	−66.16404629	0.10649414	0.58862042	17	−62.37051868	0.11031494	0.58968142
21	−66.76456833	0.10701904	0.58894575	21	−62.04848146	0.11254273	0.59249461
25	−68.13407993	0.10327148	0.58320831	25	−61.88074064	0.11994629	0.60564501
29	−75.98313046	0.08069458	0.54500379	29	−70.05681992	0.10717773	0.59842159
30	−104.92083740	0.04978485	0.48965181				
	$M = 125$				$M = 128$		
1	−70.68010235	0.09464722	0.58933268	1	−70.58992958	0.09445190	0.58900996
2	−68.94157696	0.10390015	0.59450024	2	−68.62421608	0.10349731	0.59379058
3	−68.19352627	0.10682373	0.59508549	3	−67.66701698	0.10701294	0.59506081
5	−67.34261131	0.10668945	0.59187505	4	−66.95196629	0.10685425	0.59298926
7	−67.09767151	0.10587158	0.59821869	6	−66.32718945	0.10596924	0.58953845
9	−67.05801296	0.10523682	0.58738706	9	−66.01315498	0.10471191	0.58593906
17	−67.17504501	0.10372925	0.58358265	17	−65.89422417	0.10288086	0.58097354
25	−67.22918987	0.10316772	0.58224835	25	−65.92644215	0.10182495	0.57812308
33	−67.11609936	0.10303955	0.58198956	33	−65.95577812	0.10096436	0.57576437
41	−66.71271324	0.10313721	0.58245499	41	−65.97698021	0.10094604	0.57451694
49	−66.62364197	0.10561523	0.58629534	49	−65.67919827	0.09865112	0.56927420
57	−69.28378487	0.10061646	0.57812192	57	−64.61514568	0.09845581	0.56604486
58	−70.35782337	0.09663696	0.57121235	61	−71.76589394	0.10496826	0.59452277
59	−75.94707718	0.08054886	0.54451285				
60	−104.09012318	0.04991760	0.48963264				

Source: Rabiner et al. (1970); © 1970 IEEE; reprinted with permission.

TABLE C.3 TRANSITION
COEFFICIENTS FOR $\alpha = \frac{1}{2}$

BW	Minimax	T_1
	M = 16	
1	−51.60668707	0.26674805
2	−47.48000240	0.32149048
3	−45.19746828	0.34810181
4	−44.32862616	0.36308594
5	−45.68347692	0.36661987
6	−56.63700199	0.34327393
	M = 32	
1	−52.64991188	0.26073609
2	−49.39390278	0.30878296
3	−47.72596645	0.32984619
4	−46.68811989	0.34217529
6	−45.33436489	0.35704956
8	−44.30730963	0.36750488
10	−43.11168003	0.37810669
12	−42.97900438	0.38465576
14	−56.32780266	0.35030518
	M = 64	
1	−52.90375662	0.25923462
2	−49.74046421	0.30603638
3	−48.38088989	0.32510986
4	−47.47863007	0.33595581
5	−46.88655186	0.34287720
6	−46.46230555	0.34774170
10	−45.46141434	0.35859375
14	−44.85988188	0.36470337
18	−44.34302616	0.36983643
22	−43.69835377	0.37586059
26	−42.45641375	0.38624268
30	−56.25024033	0.35200195
	M = 128	
1	−52.96778202	0.25885620
2	−49.82771969	0.30534668
3	−48.51341629	0.32404785
4	−47.67455149	0.33443604
5	−47.11462021	0.34100952
7	−46.43420267	0.34880371
10	−45.88529110	0.35493774
18	−45.21660566	0.36182251
26	−44.87959814	0.36521607
34	−44.61497784	0.36784058
42	−44.32706451	0.37066040
50	−43.87646437	0.37500000
58	−42.30969715	0.38807373
62	−56.23294735	0.35241699

TABLE C.4   TRANSITION COEFFICIENTS FOR $\alpha = \frac{1}{2}$

BW	Minimax	T_1	T_2
	$M = 16$		
1	−77.26126766	0.05309448	0.41784180
2	−73.81026745	0.07175293	0.49369211
3	−73.02352142	0.07862549	0.51966134
4	−77.95156193	0.07042847	0.51158076
5	−105.23953247	0.04587402	0.46967784
	$M = 32$		
1	−80.49464130	0.04725342	0.40357383
2	−73.92513466	0.07094727	0.49129255
3	−72.40863037	0.08012695	0.52153983
5	−70.95047379	0.08935547	0.54805908
7	−70.22383976	0.09403687	0.56031410
9	−69.94402790	0.09628906	0.56637987
11	−70.82423878	0.09323731	0.56226952
13	−104.85642624	0.04882812	0.48479068
	$M = 64$		
1	−80.80974960	0.04658203	0.40168723
2	−75.11772251	0.06759644	0.48390015
3	−72.66662025	0.07886963	0.51850058
4	−71.85610867	0.08393555	0.53379876
5	−71.34401417	0.08721924	0.54311474
9	−70.32861614	0.09371948	0.56020256
13	−69.34809303	0.09761963	0.56903714
17	−68.06440258	0.10051880	0.57543691
21	−67.99149132	0.10289307	0.58007699
25	−69.32065105	0.10068359	0.57729656
29	−105.72862339	0.04923706	0.48767025
	$M = 128$		
1	−80.89347839	0.04639893	0.40117195
2	−77.22580583	0.06295776	0.47399521
3	−73.43786240	0.07648926	0.51361278
4	−71.93675232	0.08345947	0.53266251
6	−71.10850430	0.08880615	0.54769675
9	−70.53600121	0.09255371	0.55752959
17	−69.95890045	0.09628906	0.56676912
25	−69.29977322	0.09834595	0.57137301
33	−68.75139713	0.10077515	0.57594641
41	−67.89687920	0.10183716	0.57863142
49	−66.76120186	0.10264282	0.58123560
57	−69.21525860	0.10157471	0.57946395
61	−104.57432938	0.04970703	0.48900685

Source: Rabiner et al.   (1970); © 1970 IEEE; reprinted with permission.

As an illustration, the filter design for which $M = 15$ and

$$H_r\left(\frac{2\pi k}{M}\right) = \begin{cases} 1, & k = 0, 1, 2, 3 \\ T_1, & k = 4 \\ 0, & k = 5, 6, 7 \end{cases}$$

corresponds to $\alpha = 0$, BW $= 4$, since $H_r(\omega_k) = 1$ at the four consecutive frequencies $\omega_k = 2\pi k/15$, $k = 0, 1, 2, 3$, and the transition coefficient is T_1 at the frequency $\omega_k = 8\pi/15$. The value given in Table C.1 for $M = 15$ and BW $= 4$ is $T_1 = 0.40405884$. The maximum sidelobe is at -41.9 dB, according to Table C.1.

Appendix D
List of MATLAB Functions

In this Appendix, we list several MATLAB functions that the student can use to solve some of the problems numerically. The list includes the most relevant MATLAB functions for each of the chapters, but it is not exhaustive. However, this list is cumulative in the sense that once a function is listed in any chapter, it is not repeated in subsequent chapters. These MATLAB functions are obtained from two sources: (1) the student version of MATLAB and (2) the book entitled *Digital Signal Processing Using MATLAB*. (PWS Kent 1996), by V.K. Ingle and J.G. Proakis.

Our primary objective in listing these MATLAB functions is to inform the student who is not familiar with MATLAB of the existence of these functions and to encourage the student to use them in the solution of some of the homework problems.

CHAPTER 1

sin(x), cos(x), tan(x)	trigonometric functions
abs(x)	absolute values of a vector x with real or complex components.
real(x)	takes the real part of each components of the vector x.
imag(x)	takes the imaginary part of each component of the vector x.
conj(x)	complex-conjugate of each component of x.
exp(z)	$e^x(\cos y + j\sin y)$, where $z = x + jy$.
sum(x)	sum of the (real or complex) components of the vector x.
prod(x)	product of the (real or complex) components of the vector x.
angle(x)	computes the phase angles of each component of the vector x.
log(x)	computes the natural logarithm of each of the elements of x.

log10(x)   computes the logarithm to the base 10 of the elements of x.

sqrt(x)   computes the square root of the elements of x.

CHAPTER 2

conv(x, h)   convolution of the two (vector) sequences x and h.

fliplr(x)   folds the (vector) sequence x.

filter(b, a, x)   solves the difference equation with coefficients

$$a = [a_0, a_1, \ldots, a_n]$$

$$b = [b_0, b_1, \ldots, b_m]$$

$$x = \text{input sequence}$$

filter(b, 1, x)   implements an FIR filter with input x and coefficients b.

rand(1, N)   generates a length N random sequence that is uniform in the interval $(0, 1)$.

randn(1, N)   generates a length N sequence of Guassian random variables with zero mean and unit variance.

xcorr(x, y)   computes the crosscorrelation of the two sequences x and y.

xcorr(x)   computes the autocorrelation of the sequence x.

CHAPTER 3

roots(a)   computes the roots of the polynomial with coefficients

$$a = [a_0, a_1, \ldots, a_N]$$

residuez(b, a)   computes the residues in a partial fraction expansion, where

$$b = \text{coefficients of numerator polynomial}$$
$$[b_0, b_1, \ldots b_M]$$

$$a = \text{coefficients of denominator polynomial}$$
$$[a_0, a_1, \ldots a_N]$$

deconv(b, a)   computes the result of dividing b by a in a polynomial part p and a remainder r.

poly(r)   computes the coefficients of the polynomial p with roots r.

pzplotz(b, a)   plots the poles and zeros in the z-plane given the coefficient vectors b and a.

filter(b, a, x, xic)   implements the filter given by a difference equation with coefficient vector b and a, input x and initial conditions x_{ic}.

CHAPTER 4

freqz(b, a, N)   computes an N-point complex frequency response vector and an N-point frequency vector ω, uniform over the interval $0 < \omega < \pi$, for filter with coefficient vector b and a.

freqz(b, a, N, 'whole')   same computation as freqz(b, a, N), except that the frequence range is $0 < \omega < 2\pi$.

freqz(b, a, ω)   computes the frequency response of the system at the frequencies specified by the vector ω.

grpdelay(b, a, N)   computes the group delay of the filter with numerator polynomial having coefficients b and denominator polynomial with coefficients a, at N points over the interval $(0, \pi)$.

grpdelay(b, a, N, 'whole')   same as above, except that the frequency range is $0 < \omega < 2\pi$.

CHAPTER 5

dfs(x, N)   computes the discrete fourier series (DFS) coefficient array for the periodic signal sequence x with period N.

idfs(y, N)   computes the signal sequence from the DFS coefficient array y.

rem(n, N)   determines the remainder after dividing n by N.

mod(n, N)   computes n mod N.

dft(x, N)   computes the N-point DFT of the data sequence x.

idft(X, N)   computes the N-point inverse DFT of X.

ovrlpsav(x, h, N)   implements the overlap-save method to perform block convolution where N is the block length.

CHAPTER 6

fft(x, N)   implements a radix-2, N-point FFT algorithm.

ifft(X, N)   implements a radix-2, N-point inverse FFT algorithm.

fftshift(x)   rearranges the outputs of fft so that the zero frequency component is the center of the spectrum.

CHAPTER 7

dir2cas(b, a)   converts a direct form structure to the cascade form.

cas2dir(b_0, B, A)   converts a cascade structure to the direct form struc-
                                   ture.

dir2par(b, a)   converts direct form to the parallel form structure.

par2dir(C0B, A)   converts a parallel form to the direct form structure.

dir2latc(b)   converts a FIR direct form structure to an all-zero
                                   latice structure.

latc2dir(K)   converts an all-zero lattice structure to the direct form
                                   structure.

dir2ladr(b, a)   converts direct form IIR structure to pole-zero lattice-
                                   ladder structure.

ladr2dir(K, C)   converts a lattice-ladder structure to the direct form
                                   IIR structure.

casfiltr($b0, B, A, x$)   implements the cascade form IIR and FIR realization
                                   of a filter with input sequence x.

parfiltr(C, B, A, x)   implements the parallel form IIR realization of a dil-
                                   ter with input sequence x.

latcfilt(K, x)   implements the FIR lattice filter realization with input
                                   sequence x.

ladrfilt(K, C, x)   implements the lattice-ladder realization of a filter
                                   with input sequence x.

round(x)   rounds the components of the vector x to the nearest
                                   integer

fix(x)   rounds (truncates) the component of the vector x to
                                   the nearest integer toward zero.

sign(x)   each component of x is set to $+1$ if it is positive and
                                   -1 if it is negative.

ss2tf(A, B, C, D, iu)   computes the transfer function $H(x)$ of a system given
                                   the state-space description of the form

$$x = Ax + Bu$$

$$y = Cx + Du$$

from the iuth input.

ss2zp(A, D, C, D, iu)   computes the transfer function $H(s)$ and expresses it
                                   in factored form, thus, giving the poles and zeros of
                                   $H(s)$.

CHAPTER 8

boxcar(M)   generates an M-point rectangular window.

bartlett(M)   generates an M-point Bartlett window.

hanning(M)   generates an M-point Hanning window.

hamming(M**)**	generates an M-point Hamming window.
blackman(M**)**	generates an M-point Blackman window.
kaiser(M**)**	generates an M-point kaiser window.
buttap(N**)**	provides the coefficients of an analog lowpass Butterworth filter of order N, with normalized frequency, in cascade form.
chebiap(N, R_p**)**	provides the coefficients of an analog lowpass Chebyshev filter of order N, with normalized frequency and passband ripple R_p, in cascade form.
ellipap(N, R_p, A_s**)**	provides the coefficients of an analog lowpass elliptic filter of order N, passband ripple R_p, stopband attenuation A_2, with normalized frequency, in cascade form.
freqs(b, a, ω**)**	computes the frequency response of an analog filter, with ω in rad/sec.
butter($N, \omega n$**)**	designs a digital lowpass Butterworth filter of order N and cutoff frequency ωn.
cheby1($N, R_p, \omega n$**)**	designs a digital lowpass Chebyshev filter of order N passband ripple R_p, and cutoff frequency ωn.
cheby2($N, A_s, \omega n$**)**	designs a Type 2 lowpass Chebyshev filter or order N, stopband ripple A_s and cutoff frequency ωn.
ellip($N, R_p, A_s, \omega n$**)**	designs a digital lowpass elliptic filter of order N, passband ripple R_p, stopband ripple A_s, and cutoff frequency ωn.
bilinear($z, p, k,$ **fs)**	uses the bilinear transformation to convert an analog filter with zeros z, poles p, and gain k, into a digital filter, with fs being the sample frequency in Hz.
bilinear(num, den, fs)	uses the bilinear transformation to convert an analog filter with numerator polynomial coefficients num, denominator polynomial coefficients den, and sample frequency fs, into a digital filter.
remez(N, f, m**)**	uses the Remez algorithm to determine the coefficients of an optimum equiripple, linear phase FIR filter of length $N + 1$, from frequency specifications f and gains m for each band.
remez($N, f, m,$ **'ftype')**	same description as above with 'ftype' used to specify a Hilbert transform or differentiator.
butter($N, \omega n,$ **'high')**	designs a highpass Butterworth filter of order N and 3-dB cutoff frequency ωn.
butter($N, \omega n,$ **'bandpass')**	designs a $2N$-order bandpass Butterworth filter, with 3-dB passband $\omega 1 < \omega < \omega 2$, where $\omega n = [\omega 1, \omega 2]$.
cheby1($N, R_p, \omega n,$ **'high')**	designs a highpass Chebyshev filter of order N, passband ripple R_p, and cutoff frequency ωn.

ellilp$(N, R_p, A_s, \omega n)$ designs an elliptic bandpass filter of order N, pass-band ripple R_p, stopband attenuation A_s, and cutoff frequencies $\omega n = [\omega 1, \omega 2]$.

lp2bp(num, den, $\omega o, B\omega$)   transforms an analog lowpass filter to an analog band-pass filter.

lp2bs(num, den, $\omega o, B\omega$)   transforms an analog lowpass filter to an analog band-stop filter.

lp2hp(num, den, ωo)   transforms an analog lowpass filter to an analog high-pass filter.

lp2lp(num, den, ωo)   transforms an analog lowpass filter to an analog low-pass filter with cutoff frequency ωo.

polyfit(x, y, n)   finds a polynomial p such that $p(x)$ fits the data in a vector y in a least squares sense.

CHAPTER 9

spline(x, y, xi)   cubic spline data interpolation.

spline(nt_s, x, t)   uses cubic spline interpolation, where x and nt_s are arrays containing samples $x(n)$ at nt_s, and t is an array that contains a fine grid at which the function is evaluated.

CHAPTER 10

dnsample(x, M)   downsamples the sequence x by the factor M.

References and Bibliography

AKAIKE, H. 1969. "Power Spectrum Estimation Through Autoregression Model Fitting," *Ann. Inst. Stat. Math.*, Vol. 21, pp. 407–149.

AKAIKE, H. 1974. "A New Look at the Statistical Model Identification," *IEEE Trans. Automatic Control,* Vol. AC-19, pp. 716–723, December.

ANDERSEN, N. O. 1978. "Comments on the performance of Maximum Entropy Algorithm," *Proc. IEEE,* Vol. 66, pp. 1581–1582, November.

ANTONIOU, A. 1979. *Digital Filters: Analysis and Design*, McGraw-Hill, New York.

AUER, E. 1987. "A Digital Filter Structure Free of Limit Cycles," *Proc. 1987 ICASSP*, pp. 21.11.1–21.11.4, Dallas, Tex., April.

AVENHAUS, E., and SCHUESSLER, H. W. 1970. "On the Approximation Problem in the Design of Digital Filters with Limited Wordlength," *Arch. Elek. Ubertragung*, Vol. 24, pp. 571–572.

BAGGEROER, A. B. 1976. "Confidence Intervals for Regression (MEM) Spectral Estimates," *IEEE Trans. Information Theory*, Vol. IT-22, pp. 534–545, September.

BANDLER, J. W., and BARDAKJIAN, B. J. 1973. "Least pth Optimization of Recursive Digital Filters," *IEEE Trans. Audio and Electroacoustics*, Vol. AU-21, pp. 460–470. October.

BARNES, C. W., and FAM, A. T. 1977. "Minimum Norm Recursive Digital Filters That Are Free of Overflow Limit Cycles," *IEEE Trans. Circuits and Systems*, Vol. CAS-24, pp. 569–574, October.

BARTLETT, M. S. 1948. "Smoothing Periodograms from Time Series with Continuous Spectra," *Nature* (London), Vol. 161, pp. 686–687, May.

BARTLETT, M. S. 1961. *Stochastic processes*, Cambridge University Press, Cambridge.

BERGLAND, G. D. 1969. "A Guided Tour of the Fast Fourier Transform," *IEEE Spectrum*, Vol. 6, pp. 41–52, July.

BERK, K. N. 1974. "Consistent Autoregressive Spectral Estimates," *Ann. Stat.*, Vol. 2, pp. 489–502.

BERNHARDT, P. A., ANTONIADIS, D. A., and DA ROSA, A. V. 1976. "Lunar Perturbations in Columnar Electron Content and Their Interpretation in Terms of Dynamo Electrostatic Fields," *J. Geophys. Res.*, Vol. 81, pp. 5957–5963, December.

BERRYMAN, J. G. 1978. "Choice of Operator Length for Maximum Entropy Spectral Analysis," *Geophysics*, Vol. 43, pp. 1384–1391, December.

BLACKMAN, R. B., and TUKEY, J. W. 1958. *The Measurement of Power Spectra*, Dover, New York.

BLAHUT, R. E. 1985. *Fast Algorithms for Digital Signal Processing*, Addison-Wesley, Reading, Mass.

BLUESTEIN, L. I. 1970. "A Linear Filtering Approach to the Computation of the Discrete Fourier Transform," *IEEE Trans. Audio and Electroacoustics*, Vol. AU-18, pp. 451–455, December.

BOLT, B. A. 1988. *Earthquakes*, W. H. Freeman and Co., New York.

BOMAR, B. W. 1985. "New Second-Order State-Space Structures for Realizing Low Roundoff Noise Digital Filters," *IEEE Trans. Acoustics, Speech, and Signal Processing*, Vol. ASSP-33, pp. 106–110, February.

BRACEWELL, R. N. 1978. *The Fourier Transform and Its Applications*, 2nd ed., McGraw-Hill, New York.

BRIGHAM, E. O. 1988. *The Fast Fourier Transform and Its Applications*. Prentice Hall. Englewood Cliffs, N.J.

BRIGHAM, E. O., and MORROW, R. E. 1967. "The Fast Fourier Transform," *IEEE Spectrum*, Vol. 4, pp. 63–70, December.

BRILLINGER, D. R. 1974. "Fourier Analysis of Stationary Processes," *Proc. IEEE*, Vol. 62, pp. 1628–1643, December.

BROPHY, F., and SALAZAR, A. C. 1973, "Considerations of the Padé Approximant Technique in the Synthesis of Recursive Digital Filters," *IEEE Trans. Audio and Electroacoustics*, Vol. AU-21, pp. 500–505, December.

BROWN, J. L., JR., 1980. "First-Order Sampling of Bandpass Signals–A New Approach," *IEEE Trans. Information Theory*, Vol. IT-26, pp. 613–615, September.

BROWN, R. C. 1983. *Introduction to Random Signal Analysis and Kalman Filtering*, Wiley, New York.

BRUBAKER, T. A., and GOWDY, J. N. 1972. "Limit Cycles in Digital Filters," *IEEE Trans. Automatic Control*, Vol. AC-17, pp. 675–677, October.

BRUZZONE, S. P., and KAVEH, M. 1980. "On Some Suboptimum ARMA Spectral Estimators," *IEEE Trans. Acoustics, Speech, and Signal Processing*, Vol. ASSP-28, pp. 753–755, December.

BURG, J. P. 1967. "Maximum Entropy Spectral Analysis," *Proc. 37th Meeting of the Society of Exploration Geophysicists*, Oklahoma City, Okla., October. Reprinted in *Modern Spectrum Analysis*, D. G. Childres, ed. IEEE press, New York.

BURG, J. P. 1968. "A New Analysis Technique for Time Series Data." NATO Advanced Study Institute on Signal Processing with Emphasis on Underwater Acoustics, August 12–23. Reprinted in *Modern Spectrum Analysis*, D. G. Childers, ed., IEEE Press, New York.

BURG, J. P. 1972. "The Relationship Between Maximum Entropy and Maximum Likelihood Spectra," *Geophysics*, Vol. 37, pp. 375–376, April.

BURG, J. P. 1975. "Maximum Entropy Spectral Analysis," Ph.D. dissertation, Department of Geophysics, Stanford University, Stanford, Calif., May.

BURRUS, C. S., and PARKS, T. W. 1970. "Time-Domain Design of Recursive Digital Filters," *IEEE Trans. Audio and Electroacoustics*, Vol. 18, pp. 137–141, June.

BURRUS, C. S. and PARKS, T. W. 1985. *DFT/FFT and Convolution Algorithms*, Wiley, New York.

BUTTERWECK, H. J., VAN MEER, A. C. P., and VERKROOST, G. 1984. "New Second-Order Digital Filter Sections Without Limit Cycles," *IEEE Trans. Circuits and Systems*, Vol. CAS-31, pp. 141–146, February.

CADZOW, J. A. 1979. "ARMA Spectral Estimation: An Efficient Closed-Form Procedure," *Proc. RADC Spectrum Estimation Workshop*, pp. 81–97, Rome, N.Y., October.

CADZOW, J. A. 1981. "Autoregressive–Moving Average Spectral Estimation: A Model Equation Error Procedure," *IEEE Trans. Geoscience Remote Sensing*, Vol. GE-19, pp. 24–28, January.

CADZOW, J. A. 1982. "Spectral Estimation: An Overdetermined Rational Model Equation Approach," *Proc. IEEE*, Vol. 70, pp. 907–938, September.

CANDY, J. C. 1986. "Decimation for Sigma Delta Modulation," *IEEE Trans. Communications*, Vol. COM-34, pp. 72–76, January.

CANDY, J. C., WOOLEY, B. A., and BENJAMIN, D. J. 1981. "A Voiceband Codec with Digital Filtering," *IEEE Trans. Communications*, Vol. COM-29, pp. 815–830, June.

CAPON, J. 1969. "High-Resolution Frequency–Wavenumber Spectrum Analysis," *Proc. IEEE*, Vol. 57, pp. 1408–1418, August.

CAPON, J. 1983. "Maximum-Likelihood Spectral Estimation," in *Nonlinear Methods of Spectral Analysis*, 2nd ed., S. Haykin, ed., Springer-Verlag, New York.

CAPON, J., and GOODMAN, N. R. 1971. "Probability Distribution for Estimators of the Frequency-Wavenumber Spectrum, *Proc. IEEE*, Vol. 58, pp. 1785–1786, October.

CHAN, D. S. K., and RABINER, L. R. 1973a. "Theory of Roundoff Noise in Cascade Realizations of Finite Impulse Response Digital Filters," *Bell Syst. Tech. J.*, Vol. 52, pp. 329–345, March.

CHAN, D. S. K., and RABINER, L. R. 1973b. "An Algorithm for Minimizing Roundoff Noise in Cascade Realizations of Finite Impulse Response Digital Filters," *Bell Sys. Tech. J.*, Vol. 52, pp. 347–385, March.

CHAN, D. S. K., and RABINER, L. R. 1973c. "Analysis of Quantization Errors in the Direct Form for Finite Impulse Response Digital Filters," *IEEE Trans. Audio and Electroacoustics*, Vol. AU-21, pp. 354–366, August.

CHANG, T. 1981. "Suppression of Limit Cycles in Digital Filters Designed with One Magnitude-Truncation Quantizer," *IÉEE Trans. Circuits and Systems*, Vol. CAS-28, pp. 107–111, February.

CHEN, C. T. 1970. *Introduction to Linear System Theory*, Holt, Rinehart and Winston, New York.

CHEN, W. Y., and STEGEN, G. R. 1974. "Experiments with Maximum Entropy Power Spectra of Sinusoids," *J. Geophys. Res.*, Vol. 79, pp. 3019–3022, July.

CHILDERS, D. G., ed. 1978. *Modern Spectrum Analysis*, IEEE Press, New York.

CHOW, J. C. 1972a. "On the Estimation of the Order of a Moving-Average Process," *IEEE Trans. Automatic Control*, Vol. AC-17, pp. 386–387, June.

CHOW, J. C. 1972b. "On Estimating the Orders of an Autoregressive-Moving Average Process with Uncertain Observations," *IEEE Trans. Automatic Control*, Vol. AC-17, pp. 707–709, October.

CHOW, Y., and CASSIGNOL, E. 1962. *Linear Signal Flow Graphs and Applications*, Wiley, New York.

CHUI, C. K., and CHEN, G. 1987. *Kalman Filtering*, Springer-Verlag, New York.

CLAASEN, T. A. C. M., MECKLENBRAUKER, W. F. G., and PEEK, J. B. H. 1973. "Second-Order Digital Filter with Only One Magnitude-Truncation Quantizer and Having Practically No Limit Cycles," *Electron. Lett.*, Vol. 9, November.

COCHRAN, W. T., COOLEY, J. W., FAVIN, D. L., HELMS, H. D., KAENEL, R. A., LANG, W. W., MALING, G. C., NELSON, D. E., RADER, C. E., and WELCH, P. D. 1967. "What Is the Fast Fourier Transform," *IEEE Trans. Audio and Electroacoustics*, Vol. AU-15, pp. 45–55, June.

CONSTANTINIDES, A. G. 1967. "Frequency Transformations for Digital Filters," *Electron. Lett.*, Vol. 3, pp. 487–489, November.

CONSTANTINIDES, A. G. 1968. "Frequency Transformations for Digital Filters," *Electron. Lett.*, Vol. 4, pp. 115–116, April.

CONSTANTINIDES, A. G. 1970. "Spectral Transformations for Digital Filters," *Proc. IEE*, Vol. 117, pp. 1585–1590. August.

COOLEY, J. W., and TUKEY, J. W. 1965. "An Algorithm for the Machine Computation of Complex Fourier Series," *Math. Comp.*, Vol. 19, pp. 297–301, April.

COOLEY, J. W., LEWIS, P., and WELCH, P. D. 1967. "Historical Notes on the Fast Fourier Transform," *IEEE Trans. Audio and Electroacoustics*, Vol. AU-15, pp. 76–79, June.

COOLEY, J. W., LEWIS, P., and WELCH, P. D. 1969. "The Fast Fourier Transform and Its Applications," *IEEE Trans. Education*, Vol. E-12 pp. 27–34, March.

CROCHIERE, R. E. 1977. "On the Design of Sub-band Coders for Low Bit Rate Speech Communication," BELL SYST. TECH. J., Vol. 56, pp. 747–711, May–June.

CROCHIERE, R. E. 1981. "Sub-band Coding," *Bell Syst. Tech. J.*, Vol. 60, pp. 1633–1654, September.

CROCHIERE, R. E., and RABINER, L. R. 1975. "Optimum FIR Digital Filter Implementations for Decimation, Interpolation, and Narrowband Filtering," *IEEE Trans. Acoustics, Speech and Signal Processing*, Vol. ASSP-23, pp. 444–456, October.

CROCHIERE, R. E., and RABINER, L. R. 1976. "Further Considerations in the Design of Decimators and Interpolators," *IEEE Trans. Acoustics, Speech, and Signal Processing*, Vol. ASSP-24, pp. 296–311, August.

CROCHIERE, R. E., and RABINER, L. R. 1981. "Interpolation and Decimation of Digital Signals– A Tutorial Review," *Proc. IEEE*, Vol. 69, pp. 300–331, March.

CROCHIERE, R. E., and RABINER, L. R. 1983. *Multirate Digital Signal Processing*, Prentice Hall, Engelwood Cliffs, N.J.

DANIELS, R. W. 1974. *Approximation Methods for the Design of Passive, Active and Digital Filters*, McGraw-Hill, New York.

DAVENPORT, W. B., JR. 1970. *Probability and Random Processes: An Introduction for Applied Scientists and Engineers.* McGraw-Hill, New York.

DAVIS, H. F. 1963. *Fourier Series and Orthogonal Functions.* Allyn and Bacon, Boston.

DECZKY, A. G. 1972. "Synthesis of Recursive Digital Filters Using the Minimum p-Error Criterion," *IEEE Trans. Audio and Electroacoustics*, Vol. AU-20, pp. 257–263, October.

DELSARTE, P., and GENIN, Y. 1986. "The Split Levinson Algorithm," *IEEE Trans. Acoustics, Speech, and Signal Processing*, Vol. ASSP-34, pp. 470–478, June.

DELSARTE, P., GENIN, Y., and KAMP, Y. 1978. "Orthogonal Polynomial Matrices on the Unit Circle," *IEEE Trans. Circuits and Systems*, Vol. CAS-25, pp. 149–160. January.

DERUSSO, P. M., ROY, R. J., and CLOSE, C. M. 1965. *State Variables for Engineers*, Wiley, New York.

DUHAMEL, P. 1986. "Implementation of Split-Radix FFT Algorithms for Complex, Real, and Real-Symmetric Data," *IEEE Trans. Acoustics, Speech, and Signal Processing*, Vol. ASSP-34, pp. 285–295, April.

DUHAMEL, P., and HOLLMANN, H. 1984. "Split-Radix FFT Algorithm," *Electron. Lett.*, Vol. 20, pp. 14–16, January.

DURBIN, J. 1959. "Efficient Estimation of Parameters in Moving-Average Models," *Biometrika*, Vol. 46, pp. 306–316.

DWIGHT, H. B. 1957. *Tables of Integrals and Other Mathematical Data*, 3rd ed., Macmillan, New York.

DYM, H., and MCKEAN, H. P. 1972. *Fourier Series and Integrals*, Academic, New York.

EBERT, P. M., MAZO, J. E., and TAYLOR, M. G. 1969. "Overflow Oscillations in Digital Filters," *Bell Syst. Tech. J.*, Vol. 48, pp. 2999–3020, November.

FAM, A. T., and BARNES, C. W. 1979. "Non-minimal Realizations of Fixed-Point Digital Filters That Are Free of All Finite Wordlength Limit Cycles," *IEEE Trans. Acoustics, Speech, and Signal Processing*, Vol. ASSP-27, pp. 149–153, April.

FETTWEIS, A. 1971. "Some Principles of Designing Digital Filters Imitating Classical Filter Structures," *IEEE Trans. Circuit Theory*, Vol. CT-18, pp. 314–316, March.

FLETCHER, R., and POWELL, M. J. D. 1963. "A Rapidly Convergent Descent Method for Minimization," *Comput. J.*, Vol. 6, pp. 163–168.

FOUGERE, P. F., ZAWALICK, E. J., and RADOSKI, H. R. 1976. "Spontaneous Line Splitting in Maximum Entropy Power Spectrum Analysis," *Phys. Earth Planet. Inter.*, Vol. 12, 201–207, August.

FRIEDLANDER, B. 1982a. "Lattice Filters for Adaptive Processing," *Proc. IEEE*, Vol. 70, pp. 829–867, August.

FRIEDLANDER, B. 1982b. "Lattice Methods for Spectral Estimation," *Proc. IEEE*, Vol. 70, pp. 990–1017, September.

FUCHS, J. J. 1988. "Estimating the Number of Sinusoids in Additive White Noise," *IEEE Trans. Acoustics, Speech, and Signal Processing*, Vol. ASSP-36, pp. 1846–1853, December.

GANTMACHER, F. R. 1960. *The Theory of Matrices*, Vol. I., Chelsea, New York.

GARDNER, W. A. 1987. *Statistical Spectral Analysis: A Nonprobabilistic Theory*, Prentice Hall, Englewood Cliffs, N.J.

GARLAN, C., and ESTEBAN, D. 1980. "16 Kbps Real-Time QMF Sub-band Coding Implementation," *Proc. 1980 International Conference on Acoustics, Speech, and Signal Processing*, pp. 332–335, April.

GERONIMUS, L. Y. 1958. *Orthogonal Polynomials* (in Russian) (English translation by Consultant's Bureau, New York, 1961).

GERSCH, W., and SHARPE, D. R. 1973. "Estimation of Power Spectra with Finite-Order Autoregressive Models," *IEEE Trans. Automatic Control*, Vol. AC-18, pp. 367–369, August.

GIBBS, A. J. 1969. "An Introduction to Digital Filters," *Aust. Tellcommun. Res.*, Vol. 3, pp. 3–14, November.

GIBBS, A. J. 1970. "The Design of Digital Filters," *Aust. Telecommun. Res.*, Vol. 4, pp. 29–34, May.

GOERTZEL, G. 1968. "An Algorithm for the Evaluation of Finite Trigonometric Series," *Am. Math. Monthly*, Vol. 65, pp. 34–35, January.

GOHBERG, I., ed. 1986. *I. Schür Methods in Operator Theory and Signal Processing*, Birkhauser Verlag, Stuttgart, Germany.

GOLD, B., and JORDAN, K. L., JR. 1986. "A Note on Digital Filter Synthesis," *Proc. IEEE*, Vol. 56, pp. 1717–1718, October.

GOLD, B., and JORDAN, K. L., JR. 1969. "A Direct Search Procedure for Designing Finite Duration Impulse Response Filters," *IEEE Trans. Audio and Electroacoustics*, Vol. AU-17, pp. 33–36, March.

GOLD, B., and RADER, C. M. 1966. "Effects of Quantization Noise in Digital Filters." *Proc. AFIPS 1966 Spring Joint Computer Conference*, Vol. 28, pp. 213–219.

GOLD, B., and RADER, C. M. 1969. *Digital Processing of Signals*, McGraw-Hill, New York.

GOLDEN, R. M., and KAISER, J. F. 1964. "Design of Wideband Sampled Data Filters," *Bell Syst. Tech. J.*, Vol. 43, pp. 1533–1546, July.

GOOD, I. J. 1971. "The Relationship Between Two Fast Fourier Transforms," *IEEE Trans. Computers*, Vol. C-20, pp. 310–317.

GORSKI-POPIEL, J., ed. 1975. *Frequency Synthesis: Techniques and Applications*, IEEE Press, New York.

GRAUPE, D., KRAUSE, D. J., and MOORE, J. B. 1975. "Identification of Autoregressive-Moving Average Parameters of Time Series," *IEEE Trans. Automatic Control*, Vol. AC-20, pp. 104–107, February.

GRAY, A. H., and MARKEL, J. D. 1973. "Digital Lattice and Ladder Filter Synthesis," *IEEE Trans. Acoustics, Speech, and Signal Processing*, Vol. ASSP-21, pp. 491–500, December.

GRAY, A. H., and MARKEL, J. D. 1976. "A Computer Program for Designing Digital Elliptic Filters," *IEEE Trans. Acoustics, Speech, and Signal Processing*, Vol. ASSP-24, pp. 529–538, December.

GRAY, R. M. 1990. *Source Coding Theory*, Kluwer, Boston, MA.

GRENANDER, O., and SZEGÖ, G. 1958. *Teoplitz Forms and Their Applications*, University of California Press, Berkeley, Calif.

GRIFFITHS, L. J. 1975. "Rapid Measurements of Digital Instantaneous Frequency," *IEEE Trans. Acoustics, Speech, and Signal Processing*, Vol. ASSP-23, pp. 207–222, April.

GUILLEMIN, E. A. 1957. *Synthesis of Passive Networks*, Wiley, New York.

GUPTA, S. C. 1966. *Transform and State Variable Methods in Linear Systems*, Wiley, New York.

HAMMING, R. W. 1962. *Numerical Methods for Scientists and Engineers*, McGraw-Hill, New York.

HAYKIN, S. 1991. *Adaptive Filter Theory*, 2nd ed., Prentice Hall, Englewood Cliffs, N.J.

HELME, B., and NIKIAS, C. S. 1985. "Improved Spectrum Performance via a Data-Adaptive Weighted Burg Technique," *IEEE Trans. Acoustics, Speech, and Signal Processing*, Vol. ASSP-33, pp. 903–910, August,

HELMS, H. D. 1967. "Fast Fourier Transforms Method of Computing Difference Equations and Simulating Filters," *IEEE Trans. Audio and Electroacoustics*, Vol. AU-15, pp. 85–90, June.

HELMS, H. D. 1968. "Nonrecursive Digital Filters: Design Methods for Achieving Specifications on Frequency Response," *IEEE Trans. Audio and Electroacoustics*, Vol. AU-16, pp. 336–342, September.

HELSTROM, C. W. 1990. *Probability and Stochastic Processes for Engineers*, 2nd ed., Macmillan, New York.

HERRING, R. W. 1980. "The Cause of Line Splitting in Burg Maximum-Entropy Spectral Analysis," *IEEE Trans. Acoustics, Speech, and Signal Processing*, Vol. ASSP-28, pp. 692–701, December.

HERMANN, O. 1970. "Design of Nonrecursive Digital Filters with Linear Phase," *Electron. Lett.*, Vol. 6, pp. 328–329, November.

HERMANN, O., and SCHUESSLER, H. W. 1970a. "Design of Nonrecursive Digital Filters with Minimum Phase," *Electron. Lett.*, Vol. 6, pp. 329–330, November.

HERRMANN, O., and SCHUESSLER, H. W. 1970b. "On the Accuracy Problem in the Design of Nonrecursive Digital Filters," *Arch. Elek. Ubertragung*, Vol. 24, pp. 525–526.

HERRMANN, O., RABINER, L. R., and CHAN, D. S. K. (1973). "Practical Design Rules for Optimum Finite Impulse Response Lowpass Digital Filters," *Bell Syst. Tech. J.*, Vol. 52, pp. 769–799. July–August.

HILDEBRAND, F. B. 1952. *Methods of Applied Mathematics*, Prentice Hall, Englewood Cliffs, N.J.

HOFSTETTER, E., OPPENHEIM, A. V., and SIEGEL, J. 1971. "A New Technique for the Design of Nonrecursive Digital Filters," *Proc. 5th Annual Princeton Conference on Information Sciences and Systems*, pp. 64–72.

HOUSEHOLDER, A. S. 1964. *The Theory of Matrices in Numerical Analysis*, Blaisdell, Waltham, Mass.

HWANG, S. Y. 1977. "Minimum Uncorrelated Unit Noise in State Space Digital Filtering." *IEEE Trans. Acoustics, Speech, and Signal Processing*, Vol. ASSP-25, pp. 273–281. August.

JACKSON, L. B. 1969. "An Analysis of Limit Cycles Due to Multiplication Rounding in Recursive Digital (Sub) Filters," *Proc. 7th Annual Allerton Conference on Circuit and System Theory*, pp. 69–78.

JACKSON, L. B. 1970a. "On the Interaction of Roundoff Noise and Dynamic Range in Digital Filters," *Bell Syst. Tech. J.*, Vol. 49, pp. 159–184, February.

JACKSON, L. B. 1970b. "Roundoff Noise Analysis for Fixed-Point Digital Filters Realized in Cascade or Parallel Form," *IEEE Trans. Audio and Electroacoustics*, Vol. AU-18, pp. 107–122, June.

JACKSON, L. B. 1976. "Roundoff Noise Bounds Derived from Coefficients Sensitivities in Digital Filters," *IEEE Trans. Circuits and Systems*, Vol. CAS-23, pp. 481–485, August.

JACKSON, L. B. 1979. "Limit Cycles on State-Space Structures for Digital Filters," *IEEE Trans. Circuits and Systems*, Vol. CAS-26, pp. 67–68, January.

JACKSON, L. B., LINDGREN, A. G., and KIM, Y. 1979. "Optimal Synthesis of Second-Order State-Space Structures for Digital Filters," *IEEE Trans. Circuits and Systems*, Vol. CAS-26, pp. 149–153, March.

JAHNKE, E., and EMDE, F. 1945. *Tables of Functions*, 4th ed., Dover, New York.

JAIN, V. K., and CROCHIERE, R. E. 1984. "Quadrature Mirror Filter Design in the Time Domain," *IEEE Trans. Acoustics, Speech, and Signal Processing*, Vol. ASSP-32, pp. 353–361, April.

JENKINS, G. M., and WATTS, D. G. 1968. *Spectral Analysis and Its Applications*, Holden-Day, San Francisco.

JOHNSTON, J. D. 1980. "A Filter Family Designed for Use in Quadrature Mirror Filter Banks," *IEEE International Conference on Acoustics, Speech, and Signal Processing*, pp. 291–294, April.

JOHNSON, D. H. 1982. "The Application of Spectral Estimation Methods to Bearing Estimation Problems," *Proc. IEEE*, Vol. 70, pp. 1018–1028, September.

JONES, R. H. 1976. "Autoregression Order Selection," *Geophysics*, Vol. 41, pp. 771–773, August.

JURY, E. I. 1964. *Theory and Applications of the z-Transform Method*, Wiley, New York.

KAILATH, T. 1974. "A View of Three Decades of Linear Filter Theory," *IEEE Trans. Information Theory*, Vol. IT-20, pp. 146–181, March.

KAILATH, T. 1981. *Lectures on Wiener and Kalman Filtering*, 2nd Printing, Springer-Verlag, New York.

KAILATH, T. 1985. "Linear Estimation for Stationary and Near-Stationary Processes," in *Modern Signal Processing*, T. Kailath, ed., Hemisphere Publishing Corp., Washington, D.C.

KAILATH, T. 1986. "A Theorem of I. Schür and Its Impact on Modern Signal Processing," in Gohberg (1986).

KAILATH, T., VIEIRA, A. C. G, and MORF, M. 1978. "Inverses of Toeplitz Operators, Innovations, and Orthogonal Polynomials." *SIAM Rev.*, Vol. 20, pp. 1006–1019.

KAISER, J. F. 1963. "Design Methods for Sampled Data Filters." *Proc. First Allerton Conference on Circuit System Theory*, pp. 221–236, November.

KAISER, J. F. 1966. "Digital Filters," in *System Analysis by Digital Computer*, F. F. Kuo and J. F. Kaiser, Eds., Wiley, New York.

KALOUPTSIDIS, N., and THEODORIDIS, S. 1987. "Fast Adaptive Least-Squares Algorithms for Power Spectral Estimation," *IEEE Trans. Acoustics, Speech, and Signal Processing*, Vol. ASSP-35, pp. 661–670, May.

KALMAN, R. E. 1960. "A New Approach to Linear Filtering and Prediction Problems." *Trans. ASME, J. Basic Eng.*, Vol. 82D, pp. 35–45, March.

KALMAN, R. E., and BUCY, R. S. 1961. "New Results in Linear Filtering Theory," *Trans. ASME, J. Basic Eng.*, Vol. 83, pp. 95–108.

KASHYAP, R. L. 1980. "Inconsistency of the AIC Rule for Estimating the Order of Autoregressive Models," *IEEE Trans. Automatic Control*, Vol. AC-25, pp. 996–998, October

KAVEH, J., and BRUZZONE, S. P. 1979. "Order Determination for Autoregressive Spectral Estimation," *Record of the 1979 RADC Spectral Estimation Workshop*, pp. 139–145, Griffin Air Force Base, Rome, N.Y.

KAVEH, M., and LIPPERT, G. A. 1983. "An Optimum Tapered Burg Algorithm for Linear Prediction and Spectral Analysis," *IEEE Trans. Acoustics, Speech, and Signal Processing*, Vol. ASSP-31, pp. 438–444, April.

KAY, S. M. 1980. "A New ARMA Spectral Estimator," *IEEE Trans. Acoustics, Speech, and Signal Processing*, Vol. ASSP-28, pp. 585–588, October.

KAY, S. M. 1988. *Modern Spectral Estimation*, Prentice Hall, Englewood Cliffs, N. J.

KAY, S. M., and MARPLE, S. L., JR. 1979. "Sources of and Remedies for Spectral Line Splitting in Autoregressive Spectrum Analysis," *Proc. 1979 ICASSP*, pp. 151–154.

KAY, S. M., and MARPLE, S. L., JR. 1981. "Spectrum Analysis: A Modern Perspective," *Proc. IEEE*, Vol. 69, pp. 1380–1419, November.

KESLER, S. B., ed. 1986. *Modern Spectrum Analysis II*, IEEE Press, New York.

KNOWLES, J. B., and OLCAYTO, E. M. 1968. "Coefficient Accuracy and Digital Filter Response," *IEEE Trans. Circuit Theory*, Vol. CT-15, pp. 31–41, March.

KRISHNA, H. 1988. "New Split Levinson, Schür, and Lattice Algorithms for Digital Signal Processing," *Proc. 1988 International Conference on Acoustics, Speech, and Signal Processing*, pp. 1640–1642, New York, April.

KROMER, R. E. 1969. "Asymptotic Properties of the Autoregressive Spectral Estimator," Ph.D. dissertation, Department of Statistics, Stanford University, Stanford, Calif.

KUNG, S. Y., and HU, Y. H. 1983. "A Highly Concurrent Algorithm and Pipelined Architecture for Solving Toeplitz Systems," *IEEE Trans. Acoustics, Speech, and Signal Processing*, Vol. ASSP-31, pp. 66–76, January.

KUNG, S. Y., WHITEHOUSE, H. J., and KAILATH, T., eds. 1985. *VLSI and Modern Signal Processing*, Prentice Hall, Englewood Cliffs, N.J.

LACOSS, R. T. 1971. "Data Adaptive Spectral Analysis Methods," *Geophysics*, Vol. 36, pp. 661–675, August.

LANG, S. W., and McCLELLAN, J. H. 1980. "Frequency Estimation with Maximum Entropy Spectral Estimators," *IEEE Trans. Acoustics, Speech, and Signal Processing*, Vol. ASSP-28, pp. 716–724, December.

LEVINSON, N. 1947. "The Wiener RMS Error Criterion in Filter Design and Prediction," *J. Math. Phys.*, Vol. 25, pp. 261–278.

LEVY, H., and LESSMAN, F. 1961. *Finite Difference Equations*, Macmillan, New York.

LIU, B. 1971. "Effect of Finite Word Length on the Accuracy of Digital Filters—A Review," *IEEE Trans. Circuit Theory*, Vol. CT-18, pp. 670–677, November.

MAKHOUL, J. 1975. "Linear Prediction: A Tutorial Review," *Proc. IEEE*, Vol. 63, pp. 561–580, April.

MAKHOUL, J. 1978. "A Class of All-Zero Lattice Digital Filters: Properties and Applications," *IEEE Trans. Acoustics, Speech, and Signal Processing*, Vol. ASSP-26, pp. 304–314, August.

MARKEL, J. D., and GRAY, A. H., JR. 1976. *Linear Prediction of Speech*, Springer-Verlag, New York.

MARPLE, S. L., JR. 1980. "A New Autoregressive Spectrum Analysis Algorithm." *IEEE Trans. Acoustics, Speech, and Signal Processing*, Vol. ASSP-28, pp. 441–454, August.

MARPLE, S. L., JR. 1987. *Digital Spectral Analysis with Applications*, Prentice Hall, Englewood Cliffs, N.J.

MARZETTA, T. L. 1983. "A New Interpretation for Capon's Maximum Likelihood Method of Frequency–Wavenumber Spectral Estimation," *IEEE Trans. Acoustics, Speech, and Signal Processing*, Vol. ASSP-31, pp. 445–449, April.

MARZETTA, T. L., and LANG, S. W. 1983. "New Interpretations for the MLM and DASE Spectral Estimators," *Proc. 1983 ICASSP*, pp. 844–846, Boston, April.

MARZETTA, T. L., and LANG, S. W. 1984. "Power Spectral Density Bounds." *IEEE Trans. Information Theory*, Vol. IT-30, pp. 117–122, January.

MASON, S. J., and ZIMMERMAN, H. J. 1960. *Electronic Circuits, Signals and Systems*, Wiley, New York.

McCLELLAN, J. H. 1982. "Multidimensional Spectral Estimation," *Proc. IEEE*, Vol. 70, pp. 1029–1039, September.

McDONOUGH, R. N. 1983. "Application of the Maximum-Likelihood Method and the Maximum Entropy Method to Array Processing," in *Nonlinear Methods of Spectral Analysis*, 2nd ed., S. Haykin, ed., Springer-Verlag, New York.

McGILLEM, C. D., and COOPER, G. R. 1984. *Continuous and Discrete Signal and System Analysis*, 2nd ed., Holt Rinehart and Winston, New York.

MEDITCH, J. E. 1969. *Stochastic Optimal Linear Estimation and Control*, McGraw-Hill, New York.

MILLS, W. L., MULLIS, C. T., and ROBERTS, R. A. 1981. "Low Roundoff Noise and Normal Realizations of Fixed-Point IIR Digital Filters," *IEEE Trans. Acoustics, Speech, and Signal processing*, Vol. ASSP-29, pp. 893–903, August.

MORF, M., VIEIRA, A., and LEE, D. T. 1977. "Ladder Forms for Identification and Speech Processing," *Proc. 1977 IEEE Conference Decision and Control*, pp. 1074–1078, New Orleans, La., December.

MULLIS, C. T., and ROBERTS, R. A. 1976a. "Synthesis of Minimum Roundoff Noise Fixed-Point Digital Filters," *IEEE Trans. Circuits and Systems*, Vol. CAS-23, pp. 551–561, September.

MULLIS, C. T., and ROBERTS, R. A. 1976b. "Roundoff Noise in Digital Filters: Frequency Transformations and Invariants," *IEEE Trans. Acoustics, Speech and Signal Processing*, Vol. ASSP-24, pp. 538–549, December

MUSICUS, B. 1985. "Fast MLM Power Spectrum Estimation from Uniformly Spaced Correlations," *IEEE Trans. Acoustics, Speech, and Signal Proc.*, Vol. ASSP-33, pp. 1333–1335, October.

NEWMAN, W. I. 1981. "Extension to the Maximum Entropy Method III," *Proc. 1st ASSP Workshop on Spectral Estimation*, pp. 1.7.1–1.7.6, Hamilton, Ontario, Canada, August.

NIKIAS, C. L., and RAGHUVEER, M. R. 1987. "Bispectrum Estimation: A Digital Signal Processing Framework," *Proc. IEEE*, Vol. 75, pp. 869–891, July.

NIKIAS, C. L., and SCOTT, P. D. 1982. "Energy-Weighted Linear Predictive Spectral Estimation: A New Method Combining Robustness and High Resolution," *IEEE Trans. Acoustics, Speech, and Signal Processing*, Vol. ASSP-30, pp. 287–292, April.

NUTTALL, A. H. 1976. "Spectral Analysis of a Univariate Process with Bad Data Points, via Maximum Entropy and Linear Predictive Techniques," *NUSC Technical Report TR-5303*, New London, Conn., March.

NYQUIST, H. 1928. "Certain Topics in Telegraph Transmission Theory," *Trans. AIEE*, Vol. 47, pp. 617–644, April.

OPPENHEIM, A. V. 1978. *Applications of Digital Signal Processing*, Prentice Hall, Englewood Cliffs, N.J.

OPPENHEIM, A. V., and SCHAFER, R. W. 1989. *Discrete-Time Signal Processing*, Prentice Hall, Englewood Cliffs, N.J.

OPPENHEIM, A. V., and WEINSTEIN, C. W. 1972. "Effects of Finite Register Length in Digital Filters and the Fast Fourier Transform," *Proc. IEEE*, Vol. 60, pp. 957–976. August.

OPPENHEIM, A. V., and WILLSKY, A. S. 1983. *Signals and Systems*, Prentice Hall, Englewood Cliffs, N.J.

PAPOULIS, A. 1962 *The Fourier Integral and Its Applications*, McGraw-Hall, New York.

PAPOULIS, A. 1984. *Probability, Random Variables, and Stochastic Processes*, 2nd ed., McGraw-Hill, New York.

PARKER, S. R., and HESS, S. F. 1971. "Limit-Cycle Oscillations in Digital Filters," *IEEE Trans. Circuit Theory*, Vol. CT-18, pp. 687–696, November.

PARKS, T. W., and McCLELLAN, J. H. 1972a. "Chebyshev-Approximation for Nonrecursive Digital Filters with Linear Phase," *IEEE Trans. Circuit Theory*, Vol. CT-19, pp. 189–194, March.

PARKS, T. W., and McCLELLAN, J. H. 1972b. "A Program for the Design of Linear Phase Finite Impulse Response Digital Filters," *IEEE Trans. Audio and Electroacoustics*, Vol. AU-20, pp. 195–199, August.

PARZEN, E. 1957. "On Consistent Estimates of the Spectrum of a Stationary Time Series," *Am. Math. Stat.*, Vol. 28, pp. 329–348.

PARZEN, E. 1974. "Some Recent Advances in Time Series Modeling," *IEEE Trans. Automatic Control*, Vol. AC-19, pp. 723–730, December.

PEACOCK, K. L, and TREITEL, S. 1969. "Predictive Deconvolution–Theory and Practice," *Geophysics*, Vol. 34, pp. 155–169.

PEEBLES, P. Z., JR. 1987. *Probability, Random Variables, and Random Signal Principles*, 2nd ed., McGraw-Hill, New York.

PISARENKO, V. F. 1973. "The Retrieval of Harmonics from a Convariance Function," *Geophys. J. R. Astron. Soc.*, Vol. 33, pp. 347–366.

POOLE, M. A. 1981. *Autoregressive Methods of Spectral Analysis*, E.E. degree thesis, Department of Electrical and Computer Engineering, Northeastern University, Boston, May.

PRICE, R. 1990. "Mirror FFT and Phase FFT Algorithms," unpublished work, Raytheon Research Division, May.

PROAKIS, J. G. 1995. *Digital Communications*, 3rd ed., McGraw-Hill, New York.

RABINER, L. R., and SCHAEFER, R. W. 1974a. "On the Behavior of Minimax Relative Error FIR Digital Differentiators," *Bell Syst. Tech. J.*, Vol. 53, pp. 333–362, February.

RABINER, L. R., and SCHAEFER, R. W. 1974b. "On the Behavior of Minimax FIR Digital Hilbert Transformers," *Bell Syst. Tech. J.*, Vol. 53, pp. 363–394, February.

RABINER, L. R., and SCHAEFER, R. W. 1978. *Digital Processing of Speech Signals*, Prentice Hall, Englewood Cliffs, N.J.

RABINER, L. R., SCHAFER, R. W., and RADER, C. M. 1969. "The Chirp z-Transform Algorithm and Its Applications," *Bell Syst. Tech. J.*, Vol. 48, pp. 1249–1292, May–June.

RABINER, L. R., GOLD, B., and McGONEGAL, C. A. 1970. "An Approach to the Approximation Problem for Nonrecursive Digital Filters," *IEEE Trans. Audio and Electroacoustics*, Vol. AU-18, pp. 83–106, June.

RABINER, L. R., MCCLELLAN, J. H., and PARKS, T. W. 1975. "FIR Digital Filter Design Techniques Using Weighted Chebyshev Approximation," *Proc. IEEE*, Vol. 63, pp. 595–610, April.

RADER, C. M. 1970. An Improved Algorithm for High-Speed Auto-correlation with Applications to Spectral Estimation," *IEEE Trans. Audio and Electroacoustics*, Vol. AU-18, pp. 439–441, December.

RADER, C. M., and BRENNER, N. M. 1976. "A New Principle for Fast Fourier Transformation," *IEEE Trans. Acoustics, Speech and Signal Processing*, Vol. ASSP-24, pp. 264–266, June.

RADER, C. M., and GOLD, B. 1967a. "Digital Filter Design Techniques in the Frequency Domain," *Proc. IEEE*, Vol. 55, pp. 149–171, February.

RADER, C. M., and GOLD, B. 1967b. "Effects of Parameter Quantization on the Poles of a Digital Filter," *Proc. IEEE*, Vol. 55, pp. 688–689, May.

RAMSTAD, T. A. 1984. "Digital Methods for Conversion Between Arbitrary Sampling Frequencies," *IEEE Trans. Acoustics, Speech, and Signal Processing*, Vol. ASSP-32, pp. 577–591, June.

REMEZ, E. YA. 1957. *General Computational Methods of Chebyshev Approximation*, Atomic Energy Translation 4491, Kiev, USSR.

RISSANEN, J. 1983. "A Universal Prior for the Integers and Estimation by Minimum Description Length," *Ann. Stat.*, Vol. 11, pp. 417–431.

ROBERTS, R. A., and MULLIS, C. T. 1987. *Digital Signal Processing*, Addison-Wesley, Reading, Mass.

ROBINSON, E. A. 1962. *Random Wavelets and Cybernetic Systems*, Charles Griffin, London.

ROBINSON, E. A. 1982. "A Historical Perspective of Spectrum Estimation," *Proc. IEEE*, Vol. 70, pp. 885–907, September.

ROBINSON, E. A., and TREITEL, S. 1978. "Digital Signal Processing in Geophysics," in *Applications of Digital Signal Processing*, A. V. Oppenheim, ed., Prentice Hall, Englewood Cliffs, N.J.

ROBINSON, E. A., and TREITEL, S. 1980. *Geophysical Signal Analysis*, Prentice Hall, Englewood Cliffs, N.J.

ROY, R., PAULRAJ, A., and KAILATH, T. 1986. "ESPRIT: A Subspace Rotation Approach to Estimation of Parameters of Cisoids in Noise," *IEEE Trans. Acoustics, Speech, and Signal Processing*, Vol. ASSP-34, pp. 1340–1342, October.

SAFRANEK, R. J., MACKAY, K. JAYANT, N. W., and KIM, T. 1988. "Image Coding Based on Selective Quantization of the Reconstruction Noise in the Dominant Sub-band," *Proc. 1988 IEEE International Conference on Acoustics, Speech, and Signal Processing*, pp. 765–768, April.

SAKAI, H. 1979. "Statistical Properties of AR Spectral Analysis," *IEEE Trans. Acoustics, Speech, and Signal Processing*, Vol. ASSP-27, pp. 402–409, August.

SANDBERG, I. W., and KAISER, J. F. 1972. "A Bound on Limit Cycles in Fixed-Point Implementations of Digital Filters," *IEEE Trans. Audio and Electroacoustics*, Vol. AU-20, pp. 110–112, June.

SATORIUS, E. H., and ALEXANDER J. T. 1978. "High Resolution Spectral Analysis of Sinusoids in Correlated Noise," *Proc. 1978 ICASSP*, pp. 349–351, Tulsa, Okla., April 10–12.

SCHAFER, R. W., and RABINER, L. R. 1973. "A Digital Signal Processing Approach to Interpolation," *Proc. IEEE*, Vol. 61, pp. 692–702, June.

SCHEUERMANN, H., and GOCKLER, H. 1981. "A Comprehensive Survey of Digital Transmultiplexing Methods," *Proc. IEEE*, Vol. 69, pp. 1419–1450.

SCHMIDT, R. D. 1981. "A Signal Subspace Approach to Multiple Emitter Location and Spectral Estimation," Ph.D. dissertation, Department of Electrical Engineering, Stanford University, Stanford, Calif., November.

SCHMIDT, R. D. 1986. "Multiple Emitter Location and Signal Parameter Estimation," *IEEE Trans. Antennas and Propagation*, Vol. AP 34, pp. 276–280, March.

SCHOTT, J. P., and MCCLELLAN, J. H. 1984. "Maximum Entropy Power Spectrum Estimation with Uncertainty in Correlation Measurements," *IEEE Trans. Acoustics, Speech, and Signal Processing*, Vol. ASSP-32, pp. 410–418, April.

SCHÜR, I. 1917. "On Power Series Which Are Bounded in the Interior of the Unit Circle," *J. Reine Angew. Math.*, Vol. 147, pp. 205–232, Berlin. For an English translation of the paper, see Gohberg 1986.

SCHUSTER, SIR ARTHUR. 1898. "On the Investigation of Hidden Periodicities with Application to a Supposed Twenty-Six-Day Period of Meteorological Phenomena," *Terr. Mag.*, Vol. 3, pp. 13–41, March.

SEDLMEYER, A., and FETTWEIS, A. 1973. "Digital Filters with True Ladder Configuration," *Int. J. Circuit Theory Appl.*, Vol. 1, pp. 5–10, March.

SHANKS, J. L. 1967. "Recursion Filters for Digital Processing," *Geophysics*, Vol. 32, pp. 33–51, February.

SHANNON, C. E. 1949. "Communication in the Presence of Noise," *Proc. IRE*, pp. 10–21, January.

SHEINGOLD, D. H., ed. 1986. *Analog-Digital Conversion Handbook*, Prentice Hall, Englewood Cliffs, N.J.

SIEBERT, W. M. 1986. *Circuits, Signals and Systems*, McGraw-Hill, New York.

SINGLETON, R. C. 1967. "A Method for Computing the Fast Fourier Transform with Auxiliary Memory and Limit High speed Storage." *IEEE Trans. Audio and Electroacoustics*, Vol. AU-15, pp. 91–98, June.

SINGLETON, R. C. 1969. "An Algorithm for Computing the Mixed Radix Fast Fourier Transform," *IEEE Trans. Audio and Electroacoustics*, Vol. AU-17, pp. 93–103, June.

SMITH, M. J. T., and BARWELL, T. P. 1984. "A Procedure for Designing Exact Reconstruction Filter Banks for Tree Structured Subband Coders," *Proc. 1984 IEEE International Conference on Acoustics, Speech, and Signal Processing*, pp. 27.1.1–27.1.4, San Diego, March.

SMITH, M. J. T., and EDDINS, S. L. 1988. "Subband Coding of Images with Octave Band Tree Structures," *Proc. 1987 IEEE International Conference on Acoustics, Speech, and Signal Processing*, pp. 1382–1385, Dallas, April.

STEIGLITZ, K. 1965. "The Equivalence of Digital and Analog Signal Processing," *Inf. Control*, Vol. 8, pp. 455–467, October.

STEIGLITZ, K. 1970. "Computer-Aided Design of Recursive Digital Filters," *IEEE Trans. Audio and Electroacoustics*, Vol. AU-18, pp. 123–129, June.

STOCKHAM, T. G. 1966. "High Speed Convolution and Correlation," *1966 Spring Joint Computer Conference, AFIPS Proc.*, Vol. 28, pp. 229–233.

STORER, J. E. 1957. *Passive Network Synthesis*, McGraw-Hill, New York.

SWARZTRAUBER, P. 1986. "Symmetric FFT's," *Mathematics of Computation*, Vol. 47, pp. 323–346, July.

SWINGLER, D. N. 1979a. "A Comparison Between Burg's Maximum Entropy Method and a Nonrecursive Technique for the Spectral Analysis of Deterministic Signals," *J. Geophys. Res.*, Vol. 84, pp. 679–685, February.

SWINGLER, D. N. 1979b. "A Modified Burg Algorithm for Maximum Entropy Spectral Analysis," *Proc. IEEE*, Vol. 67, pp. 1368–1369, September.

SWINGLER, D. N. 1980. "Frequency Errors in MEM Processing," *IEEE Trans. Acoustics, Speech, and Signal Processing*, Vol. ASSP-28, pp. 257–259, April.

SZEGÖ, G. 1967. *Orthogonal Polynomials*, 3rd ed., Colloquium Publishers, no. 23, American Mathematical Society, Providence, R.I.

THORVALDSEN, T. 1981. "A Comparison of the Least-Squares Method and the Burg Method for Autoregressive Spectral Analysis," *IEEE Trans. Antennas and Propagation*, Vol. AP-29, pp. 675–679, July.

TONG, H. 1975. "Autoregressive Model Fitting with Noisy Data by Akaike's Information Criterion," *IEEE Trans. Information Theory*, Vol. IT-21, pp. 476–480, July.

TONG, H. 1977. "More on Autoregressive Model Fitting with Noise Data by Akaike's Information Criterion," *IEEE Trans. Information Theory*, Vol. IT-23, pp. 409–410, May.

TRETTER, S. A. 1976. *Introduction to Discrete-Time Signal Processing*, Wiley, New York.

TUFTS, D. W., and KUMARESAN, R. 1982. "Estimation of Frequencies of Multiple Sinusoids: Making Linear Prediction Perform Like Maximum Likelihood," *Proc. IEEE*, Vol. 70, pp. 975–989, September.

ULRYCH, T. J., and BISHOP, T. N. 1975. "Maximum Entropy Spectral Analysis and Autoregressive Decomposition," *Rev. Geophys. Space Phys.*, Vol. 13, pp. 183–200, February.

ULRYCH, T. J., and CLAYTON, R. W. 1976. "Time Series Modeling and Maximum Entropy," *Phys. Earth Planet. Inter.*, Vol. 12, pp. 188–200, August.

VAIDYANATHAN, P. P. 1990. "Multirate Digital Filters, Filter Banks, Polyphase Networks, and Applications: A Tutorial," *Proc. IEEE*, Vol. 78, pp. 56–93, January.

VAIDYANATHAN, P. P. 1993. *Multirate Systems and Filter Banks*, Prentice Hall, Englewood Cliffs, N.J.

VETTERLI, J. 1984. "Multi-dimensional Sub-band Coding: Some Theory and Algorithms," *Signal Processing*, Vol. 6, pp. 97–112, April.

VETTERLI, J. 1987. "A Theory of Multirate Filter Banks," *IEEE Trans. Acoustics, Speech, and Signal Processing*, Vol. ASSP-35, pp. 356–372, March.

VIEIRA, A. C. G. 1977. "Matrix Orthogonal Polynomials with Applications to Autoregressive Modeling and Ladder Forms," Ph.D. dissertation, Department of Electrical Engineering, Stanford University, Stanford, Calif., December.

WALKER, G. 1931. "On Periodicity in Series of Related Terms," *Proc. R. Soc.*, Ser. A. Vol. 313, pp. 518–532.

WAX, M., and KAILATH, T. 1985. "Detection of Signals by Information Theoretic Criteria." *IEEE Trans. Acoustics, Speech, and Signal Processing*, Vol. ASSP-32, pp. 387–392, April.

WEINBERG, L. 1962. *Network Analysis and Synthesis*, McGraw-Hill, New York.

WELCH, P. D. 1967. "The Use of Fast Fourier Transform for the Estimation of Power Spectra: A Method Based on Time Averaging over Short Modified Periodograms," *IEEE Trans. Audio and Electroacoustics*, Vol. AU-15, pp. 70–73, June.

WIENER, N. 1949. *Extrapolation, Interpolation and Smoothing of Stationary Time Series with Engineering Applications*, Wiley, New York.

WIENER, N., and PALEY, R. E. A. C. 1934. *Fourier Transforms in the Complex Domain*. American Mathematical Society, Providence, R.I.

WINOGRAD, S. 1976. "On Computing the Direte Fourier Transform," *Proc. Natl. Acad. Sci.*, Vol. 73, pp. 105–106.

WINOGRAD, S. 1978. "On Computing the Discrete Fourier Transform," *Math. Comp.*, Vol. 32, pp. 177–199.

WOLD, H. 1938. *A Study in the Analysis of Stationary Time Series*, reprinted by Almqvist & Wiksell, Stockholm, 1954.

WOOD, L. C., and TREITEL, S. 1975. "Seismic Signal Processing," *Proc. IEEE*, Vol. 63, pp. 649–661, April.

WOODS, J. W., and O'NEIL, S. D. 1986. "Subband Coding of Images," *IEEE Trans. Acoustics, Speech, and Signal Processing*, Vol. ASSP-34, pp. 1278–1288, October.

YOULA, D., and KAZANJIAN, N. 1978. "Bauer-Type Factorization of Positive Matrices and the Theory of Matrices Orthogonal on the Unit Circle," *IEEE Trans. Circuits and Systems*, Vol. CAS-25, pp. 57–69, January.

YULE, G. U. 1927. "On a Method of Investigating Periodicities in Disturbed Series with Special References to Wolfer's Sunspot Numbers," *Philos. Trans. R. Soc. London*, Ser. A. Vol. 226, pp. 267–298, July.

ZADEH, L. A., and DESOER, C. A. 1963. *Linear System Theory: The State-Space Approach*, McGraw-Hill, New York.

ZVEREV, A. I. 1967. *Handbook of Filter Synthesis*, Wiley, New York.

Index